FOURTH EDITION

ENVIRONMENTAL CHANGE & CHALLENGE
A CANADIAN PERSPECTIVE

Dearden • Mitchell

OXFORD
UNIVERSITY PRESS

Oxford University Press is a department of the University of Oxford.
It furthers the University's objective of excellence in research, scholarship, and education by publishing worldwide.
Oxford is a registered trade mark of Oxford University Press in the UK and in certain other countries.

Published in Canada by
Oxford University Press
8 Sampson Mews, Suite 204,
Don Mills, Ontario M3C 0H5 Canada

www.oupcanada.com

Copyright © Oxford University Press Canada 2012

The moral rights of the authors have been asserted

Database right Oxford University Press (maker)

First Edition published in 1998

Second Edition published in 2005

Third Edition published in 2009

All rights reserved. No part of this publication may be reproduced, stored in a retrieval system, or transmitted, in any form or by any means, without the prior permission in writing of Oxford University Press, or as expressly permitted by law, by licence, or under terms agreed with the appropriate reprographics rights organization. Enquiries concerning reproduction outside the scope of the above should be sent to the Permissions Department at the address above or through the following url: www.oupcanada.com/permission/permission_request.php

Every effort has been made to determine and contact copyright holders.
In the case of any omissions, the publisher will be pleased to make suitable acknowledgement in future editions.

Library and Archives Canada Cataloguing in Publication
Dearden, Philip
Environmental change and challenge : a Canadian perspective / Philip Dearden, Bruce Mitchell. — 4th ed.

Includes bibliographical references and index.
ISBN 978-0-19-544625-8

1. Environmental management—Canada—Textbooks. 2. Human ecology—Canada—Textbooks.
3. Nature—Effect of human beings on— Canada—Textbooks. 4. Global environmental change—Textbooks.
I. Mitchell, Bruce, 1944– II. Title.

GF511.D42 2012   333.70971   C2012-900956-3

Cover image: John E Marriott/All Canada Photos

Part- and chapter-opening photo credits: Page 1: © Nepal Images/Alamy; page 3: © Nikada/iStockphoto; page 44: © antony spencer/iStockphoto; page 49: © Sandra vom Stein/iStockphoto; page 83: © Ron and Patty Thomas Photography/iStockphoto; page 114: © Kipp Schoen/iStockphoto; page 152: © Lawrence Wiles/Alamy; page 156: © George Clerk/iStockphoto; page 172: © Ketian Chen/iStockphoto; page 298: © Andrew Penner/iStockphoto; page 2.1: © National Geographic Image Collection/Alamy; page 238: © brytta/iStockphoto; page 282: © Scott Cramer/iStockphoto; page 332: © kevin miller/iStockphoto; page 374: © Douglas Allen/iStockphoto; page 419: © Gunter Marx/Alamy; page 459: © Arpad Benedek/iStockphoto; page 491: © Jim Bradenburg/National Geographic Stock; page 538: © Bartosz Hadyniak/iStockphoto; page 540: © Caro/Alamy

Printed and bound in Canada

3 4 5 — 16 15 14

Contents Overview

Preface ix
Features xi
Acknowledgements xvi
About the Authors xvii

Part A Introduction 1

Chapter 1 Environment, Resources, and Society 3

Guest Statement: Perspective on the Tar Ponds (Tim Babcock) 12
International Guest Statement: The Resource Dilemma: Environment, Livelihood, and Poverty Nexus in Developing Countries (Peter O. Adeniyi) 23

Part B The Ecosphere 45

Chapter 2 Energy Flows and Ecosystems 49

Guest Statement: Landscape Ecology (Chris Malcolm) 72

Chapter 3 Ecosystems Are Dynamic 83

Guest Statement: How Will Forests Respond to Rising Atmospheric Carbon Dioxide? (Ze'ev Gedalof and Aaron Berg) 88
International Guest Statement: Life at the Crossroads: How Climate Change Threatens the Existence of the Maasai (Philip Osano) 109

Chapter 4 Ecosystems and Matter Cycling 114

International Guest Statement: Action-oriented Research on Community Recycling in São Paulo, Brazil (Jutta Gutberlet) 116

Part C Planning and Management: Philosophy, Process, and Product 153

Chapter 5 Planning and Management: Philosophy 156

International Guest Statement: Water Governance for the Twenty-First Century (Ali Memon) 159
Guest Statement: Visioning: Wanted from Individuals and Beyond (Dan Shrubsole) 161

Chapter 6   Planning and Management: Process, Method, and Product   172

Guest Statement: Cumulative Environmental Effects: Thinking beyond the Project in Environmental Assessment (Bram Noble)   181
International Guest Statement: Spatial Planning and Disaster Risk Reduction: Mediating Long-Term and Short-Term Interests: Lessons from Indonesia (Bakti Setiawan)   191

Part D   Resource and Environmental Management in Canada   199

Chapter 7   Climate Change   201

Guest Statement: Global Policy Challenges (Barry Smit)   231

Chapter 8   Oceans and Fisheries   238

Guest Statement: Public and Political Will Needed to Protect Our Oceans (Sabine Jessen)   274
International Guest Statement: The Rise and Fall of Industrial Fisheries (Daniel Pauly)   247

Chapter 9   Forests   232

Guest Statement: Forest Ownership, Forest Stewardship, Community Sustainability (Kevin Hanna)   294
International Guest Statement: Can the Private Sector Help Slow Deforestation? (Gijsbert Nollen)   325

Chapter 10   Agriculture   332

Guest Statement: Canada Feeding the World? (Peter Schroeder)   363

Chapter 11   Water   374

Guest Statement: How Becoming a Heritage River Can Influence Water Management (Barbara Veale)   405
International Guest Statement: The Grand Canal of China (Shuheng Li)   408

Chapter 12   Minerals and Energy   419

Guest Statement: Accountability in Resource Management: Independent Oversight of Proponents and Government (Patricia Fitzpatrick)   430
Guest Statement: Corporate Social Investment (Michael Hitch)   432

Chapter 13   Urban Environmental Management   459

Guest Statement: Urban Waste Management (Virginia W. Maclaren)   464
International Guest Statement: Natural Landscapes in Urban Environments and the Role of *Feng Shui* (Lawal Marafa)   481

Chapter 14   Endangered Species and Protected Areas   491

International Guest Statement: Tiger Conservation in Thailand (Anak Pattanavibool)   511
Guest Statement: Canada's Great Bear Rainforest (Ian McAllister)   533

Part E   Environmental Change and Challenge Revisited   539

Chapter 15   Making It Happen   540

Guest Statement: A Generation of Possibility (Skye Augustine)   552
Guest Statement: Sustainability in Higher Education: Learning It, Teaching It, and Doing It (Darren Bardati)   554

Appendix: Conservation Organizations   567
Glossary   572
References   583
Index   598

Detailed Contents

Preface ix
Features xi
Acknowledgements xvi
About the Authors xvii

Part A | Introduction 1

Chapter 1 Environment, Resources, and Society 3

Introduction: Change and Challenge 3
Defining Environment and Resources 5
Alternative Approaches to Understanding Complex Natural and Socio-economic Systems 5
Science-Based Management of Resources and Environment 7
The Sydney Tar Ponds, Cape Breton Regional Municipality, Nova Scotia 8
Sustainable Development, Sustainable Livelihoods, and Resilience 13
The Global Picture 16
Jurisdictional Arrangements for Environmental Management in Canada 28
Measuring Progress 30
Implications 37
Summary 39
Key Terms 41
Questions for Review and Critical Thinking 41
Related Websites 42
Further Readings 42

Part B | The Ecosphere 45

Chapter 2 Energy Flows and Ecosystems 49

Introduction 49
Energy 50
Energy Flows in Ecological Systems 53
Ecosystem Structure 65
Abiotic Components 66
Biodiversity 74
Implications 79
Summary 80
Key Terms 81
Questions for Review and Critical Thinking 82
Related Websites 82
Further Readings 82

Chapter 3 Ecosystems Are Dynamic 83

Introduction 83
Ecological Succession 84
Changing Ecosystems 92
Population Growth 100
Evolution, Speciation, and Extinction 104
Implications 110
Summary 111
Key Terms 112
Questions for Review and Critical Thinking 112
Related Websites 113
Further Readings 113

Chapter 4 Ecosystems and Matter Cycling 114

Introduction 114
Matter 114
Biogeochemical Cycles 115
The Hydrological Cycle 129
Biogeochemical Cycles and Human Activity 135
Implications 148
Summary 148
Key Terms 149
Questions for Review and Critical Thinking 150
Related Websites 150
Further Readings 151

Part C | Planning and Management: Philosophy, Process, and Product 153

Chapter 5 Planning and Management: Philosophy 156

Introduction   156
Planning and Management Components   157
Implications   168
Summary   169
Key Terms   169
Questions for Review and Critical Thinking   170
Related Websites   170
Further Readings   171

Chapter 6 Planning and Management: Process, Method, and Product 172

Introduction   172
Collaboration and Co-ordination   172
Stakeholders and Participatory Approaches   173
Communication   175
Adaptive Management   176
Impact and Risk Assessment   178
Dispute Resolution   184
Regional and Land-Use Planning   190
Implementation Barriers   193
Implications   193
Summary   194
Key Terms   195
Questions for Review and Critical Thinking   195
Related Websites   196
Further Readings   197

Part D | Resource and Environmental Management in Canada 199

Chapter 7 Climate Change 201

Introduction   201
Nature of Climate Change   202
Scientific Evidence Related to Climate Change   203
Modelling Climate Change   207
Scientific Explanations   209
Implications of Climate Change   209
Communicating Global Change   218
Kyoto Protocol   221
Policy and Action Options   229
Summary   234
Key Terms   235
Questions for Review and Critical Thinking   235
Related Websites   236
Further Readings   237

Chapter 8 Oceans and Fisheries 238

Introduction   238
Oceanic Ecosystems   239
Ocean Management Challenges   245
Global Responses   257
Canada's Oceans   258
Fisheries   259
Aboriginal Use of Marine Resources   266
Pollution   270
Some Canadian Responses   271
Aquaculture   275
Implications   278
Summary   279
Key Terms   280
Questions for Review and Critical Thinking   280
Related Websites   281
Further Readings   281

Chapter 9 Forests 282

Canada's Boreal Forest 282
An Overview of Canada's Forests 285
Forest Management Practices 294
Environmental and Social Impacts of Forest Management Practices 302
New Forestry 315
Canada's National Forest Strategies 317
Global Forest Strategies 321
Implications 324
Summary 327
Key Terms 329
Questions for Review and Critical Thinking 329
Related Websites 330
Further Readings 331

Chapter 10 Agriculture 332

Introduction 332
Agriculture as an Ecological Process 338
Modern Farming Systems in the Industrialized World 339
Trends in Canadian Agriculture 347
Environmental Challenges for Canadian Agriculture 349
Sustainable Food Production Systems 365
Organic Farming 367
Local Agriculture 368
Implications 369
Summary 370
Key Terms 372
Questions for Review and Critical Thinking 372
Related Websites 373
Further Readings 373

Chapter 11 Water 374

Introduction 374
Human Interventions in the Hydrological Cycle: Water Diversions 376
The James Bay Hydroelectric Project 377
Water Quality 383
Water Security: Protecting Quantity and Quality 387
Water as Hazard 397
Heritage Rivers 403
Hydrosolidarity 407
Water Ethics 411
Implications 413
Summary 414
Key Terms 416
Questions for Review and Critical Thinking 416
Related Websites 417
Further Readings 418

Chapter 12 Minerals and Energy 419

Introduction 419
Framing Issues and Questions 420
Non-Renewable Resources in Canada: Basic Information 421
Potash in Saskatchewan 422
Developing a Diamond Mine: Ekati, NWT 423
Energy Resources 434
Implications 454
Summary 454
Key Terms 456
Questions for Review and Critical Thinking 456
Related Websites 457
Further Readings 458

Chapter 13   Urban Environmental Management   459

Introduction   459
Sustainable Urban Development   460
Environmental Issues in Cities   466
Vulnerability of Urban Areas to Natural and Human-Induced Hazards   470
Urban Sustainability   473
Best Practice for Urban Environmental Management   479
Implications   487
Summary   488
Key Terms   489
Questions for Review and Critical Thinking   489
Related Websites   489
Further Readings   490

Chapter 14   Endangered Species and Protected Areas   491

Introduction   491
Valuing Biodiversity   493
Main Pressures Causing Extinction   496
Vulnerability to Extinction   508
Responses to the Loss of Biodiversity   512
Protected Areas   518
Implications   534
Summary   534
Key Terms   535
Questions for Review and Critical Thinking   536
Related Websites   536
Further Readings   537

Part E | Environmental Change and Challenge Revisited   539

Chapter 15   Making It Happen   540

Introduction   540
Global Perspectives   541
National Perspectives   547
Personal Perspectives   550
The Law of Everybody   559
Implications   563
Summary   564
Key Terms   565
Questions for Review and Critical Thinking   565
Related Websites   566
Further Readings   566

Appendix: Conservation Organizations   567
Glossary   572
References   583
Index   598

Preface

When we wrote the first edition of *Environmental Change and Challenge* more than a decade and a half ago, it was already becoming very obvious that the two themes of 'change' and 'challenge' were going to be major defining characteristics of the twenty-first century. However, the speed and magnitude with which change has occurred were often unanticipated. And with that rapid change have come massive challenges.

Scientists in the mid-1990s were well aware of global climate change, but the speed of change was expected to be a concern for the next rather than this generation. The Arctic Ocean was predicted to be ice-free in 50 to 100 years. Yet, following the colossal ice losses over the last several years as positive feedback loops kicked in, that prediction has been revised to within the next few years.

The challenges created by these and other changes will be profound and global. Sea levels will rise, communities will be flooded, ocean currents will change, rainfall patterns will alter, crops will fail, and billions of lives will be affected. As we revise this fourth edition millions of people are suffering from drought in Somalia while millions of others are coping with a giant flood that has inundated one of the world's major cities, Bangkok.

The reality of these changes is difficult for many people to believe, since it counters many of our most deep-seated beliefs. Treaties were signed with First Nations for 'as long as the rains fall, as long as the rivers flow, as long as the winds blow' because these were the immutable constructs of nature that were reliable. The Earth was also conceived as being so large that the impact of humans was trifling in comparison. Photos of our lonely planet floating through space taken from spacecraft helped to dispel this myth.

A fundamental change has taken place over the past couple of decades in the relationship between humans and our fragile planet. No longer is the planet a vast and wild place where change occurs on a geological time scale driven by natural forces; it has, in fact, become the 'greenhouse' of the greenhouse gas analogy in which wild nature is replaced by human constructs, and even the vast atmosphere and oceans reflect human desires as they become increasingly choked by the industrial wastes of a consumer society.

There has never been a more critical time when humans should know how the planet works and especially about the processes that drive our life support system. But environmental management is not only about managing natural systems; it is also about managing humans and our impacts on these systems. This book was written with these twin goals in mind: that students should gain a basic appreciation of how the planet works and also understand the impacts of humanity on these systems, the challenges created, and potential solutions.

The book is also focused primarily on Canada. Canada is a huge and beautiful country, one of the most magnificent places on Earth. Our geography, people, history, and political culture are different from those of the US and Europe. Canadians can also play a major role in what happens globally in terms of the environment. We are the world's second largest country in terms of area. We are also a rich country. In general, our citizens have a high quality of life and value the environment, but we also create some of the highest per capita impacts in the world in terms of carbon dioxide emissions, water use, and waste production. Changes need to take place. And those changes need to take place far more quickly than is currently the case. Our 'leaders' have often been willing to make those changes only if they perceive support for them. That support hinges on having a well-informed and active populace.

We believe it is critical that university students leave our universities when they graduate with a greater understanding of the planetary ecosystems that support life and of their impacts on ecosystems, as well as an awareness of what society and individuals can do to help improve the situation. If all university graduates came out thus informed and acted on this knowledge to create change in their own lifestyles and society, the prognosis for the future would be a little more optimistic.

This book was written for students taking a first course in environment to impart an understanding of the biosphere's function and to link basic environmental management principles to environmental and resource problems in a Canadian context. The book provides both a basic background for those who will go on to specialize in fields other than environment and a broad platform upon which more detailed courses on environment can build later.

Part A (Chapter 1) provides an overall introduction to environment, resources, and society and the role of science, both social and natural, in helping us to understand the relationship among them. This relationship is illustrated in more detail by a well-known Canadian case study, the Sydney Tar Ponds. We also provide a global and national context for environmental management and describe some approaches for assessing current progress in dealing with environmental challenges. If we do not know how we are doing, we can hardly judge with any degree of accuracy the severity of the problem or map out suitable strategies to address the problem.

Part B (Chapters 2–4) provides a basic primer on the environmental processes that constitute the Earth's life support system. Primary emphasis is on energy flows, biogeochemical cycles, and biotic responses, with reference to Canadian examples wherever possible. A strong emphasis is placed both here and in subsequent sections on making explicit links between these principles and examples illustrating the principles in action.

Part C (Chapters 5 and 6) reviews different philosophies, processes, and products that should characterize high-quality resource and environmental planning and management. Some refer to such attributes as elements of 'best practice'. Our hope is that by the end of these two chapters, you will be able to develop a mental checklist of the attributes you would expect to see used in planning and management and that you would advocate either as a team member addressing resource and environmental issues or as a member of civil society.

Part D (Chapters 7–14) takes the basic science of Part B and the management approaches of Part C and puts them together by focusing on environmental and resource management themes: climate change, oceans and fisheries, forests, agriculture, water, minerals and energy, urban environmental management, and endangered species and protected areas. In each chapter, we provide an overview of the current situation in Canada and the main management challenges. Selected international examples also are provided. Text boxes highlight particular case studies of interest and also illustrate the connectivity among the different themes.

The final section (Chapter 15) concludes the book with views from three perspectives—global, national, and personal. Here we emphasize solutions and the actions that individuals can take in moving towards a more sustainable society and introduce the 'law of everybody', suggesting that if everyone took a few conservation actions, they would add up to a massive contribution to the overall changes required. We question the ways in which values are taken into account in much environmental decision-making and also how development progress is measured. The Happy Planet Index, for example, is one international indicator that has been suggested as an alternative to measure progress that takes into account not only level of human well-being but also the costs of achieving that well-being.

Change and challenge are main themes of this book, and fundamental changes are required in the way by which society manages itself to meet the challenges that lie ahead. We hope that this book will help in some small way to contribute to producing the more sustainable future that must evolve over the next few years and encourage you, the reader, to become part of making this future a reality.

Philip Dearden
Bruce Mitchell

Features

New!

Current Changes and Challenges

Coverage of current events—the *Deepwater Horizon* oil spill, flooding in Pakistan, an earthquake and tsunami in Japan, bluefin tuna on the brink of extinction, drought and famine in East Africa, the birth of the Earth's seven billionth human—illustrates environmental changes and challenges happening every day all over the world.

xii | Features

Fresh National and Global Perspectives

'Guest Statement' features, including 18 entirely new and revised statements, highlight the work of engaged scientists making a difference in Canada and internationally in areas such as water governance, endangered species conservation, global policy change, industrial fishing, and disaster risk reduction.

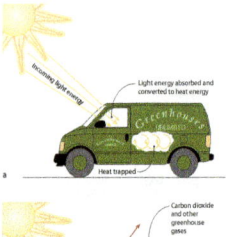
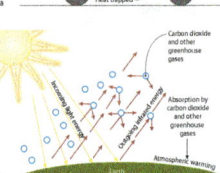

Revised Art Program

From carefully chosen images that present the issues in living colour to figures and tables that clearly present the most up-to-date data, the updated art program complements and augments the discussion in every chapter.

Focus and Perspectives on the Environment

Special-topic boxes demonstrate how concepts, approaches, and theories manifest in the real world, while quotations from multiple perspectives offer analysis and insight into key environmental issues.

Emphasis on Solutions

With a focus on global, national, and personal solutions, the text suggests actions that can be taken to move towards more sustainable ways of living.

xiv   Features

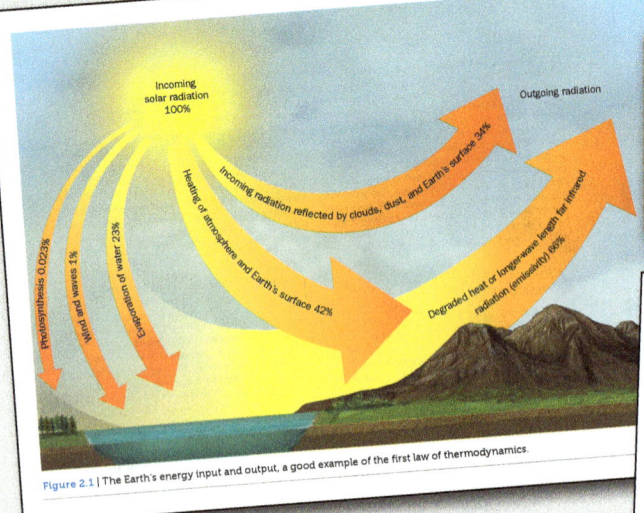

Primer on Scientific Processes
Thorough coverage of the environmental processes that form the Earth's life support systems allows readers to make the links between these scientific concepts and examples that illustrate these concepts in action.

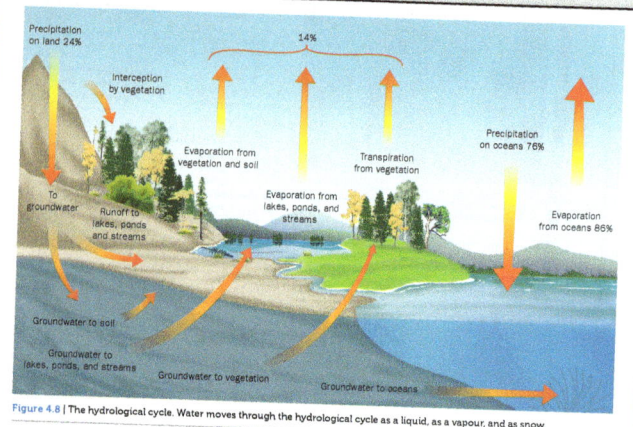

Coverage of Environmental Planning and Management
In-depth coverage of the philosophies, processes, and products that should characterize high-quality natural resources and environmental planning and management remains a unique feature in this new fourth edition.

Robust Online Ancillary Suite

The fourth edition of *Environmental Change and Challenge* is supported by a wide range of supplementary items for students and instructors alike, all designed to enrich and complete the learning and teaching experience.

For the Student

Available at www.oupcanada.com/DeardenMitchell4e.

- **Student Study Guide** of review material designed to aid student learning.

For Instructors

The following instructors' resources are available to qualifying adopters. Please contact your OUP Canada sales representative for more information.

- An exciting **Media Supplement** features video and audio material chosen specifically to complement content in each chapter and enhance students' experience of critical concepts and issues.
- An **Instructor's Manual**, **PowerPoint Slides**, and **Image Bank** make classroom presentation of material more engaging and relevant for students.
- A **Test Generator** allows instructors a wide array of options for sorting, editing, importing, and distributing questions.

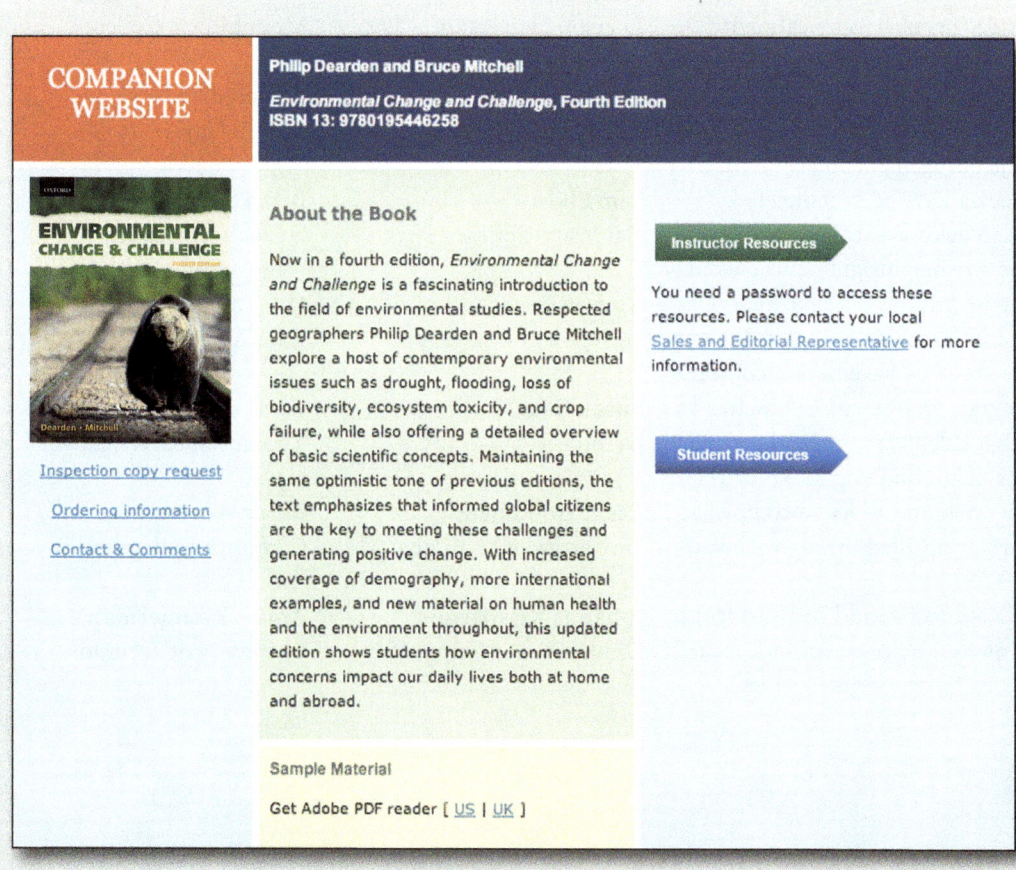

 www.oupcanada.com/DeardenMitchell4e

Acknowledgements

Bruce Mitchell gratefully acknowledges the research assistance provided by Joslyn Spurgeon for the fourth edition. Joslyn provided support in a variety of ways, all of which contributed to enhancing the final product.

Roland Hall, Department of Biology, University of Waterloo, and Mike English, Department of Geography and Environmental Studies, Wilfrid Laurier University, provided photographs related to the Athabasca oil sands development.

Tanya Collier MacDonald of the Cape Breton Tar Ponds Agency provided information about the ongoing remediation work as well as photographs related to the tar ponds.

Paul Parker in the Department of Geography and Environmental Management, University of Waterloo, provided a photograph of an 'energy-efficient' residential home.

Graeme Hayward, Manitoba Floodway Authority, arranged for the use of images of the construction on the Red River Floodway.

Frances Hannigan worked diligently to prepare the final versions of chapters before they were submitted to Oxford University Press. She also provided ongoing technical advice.

Joan Mitchell in so many ways provided support, allowing Bruce Mitchell the time needed to work on the fourth edition. This partnership has continued since 1965 and has involved many projects. Without her encouragement and constructive comments, this work would never have been completed.

Philip Dearden would like to acknowledge the invaluable assistance of Skye Augustine for her insights and encouragement and for her tenacity in finding relevant materials, and Dennis Jelinski for his comments on Chapter 3. Over the years, my many first-year students have been a constant source of inspiration and energy for the subject matter in the book, and it is to them that I dedicate the book. Special thanks go to family members, including dog Max, for their patience and understanding on home tasks uncompleted, walks not taken, and promises unfulfilled through the various editions of this book.

We are both indebted to Kate Skene and Jodi Lewchuk of Oxford University Press, as well as copy editor Richard Tallman, whose comments significantly improved the text. We also express our appreciation to Amy Gordon, who systematically verified sources, compiled entries for the secondary glossary, and identified options related to supplementary video material. We appreciate the time taken by our guest contributors to provide statements that help diversify and enliven the text.

Thanks also to the many reviewers whose comments have proved invaluable over the years. In addition to those who provided anonymous feedback on this fourth edition, the authors and publisher thank the following reviewers, whose thoughtful comments and suggestions have helped to shape all the editions of this text:

Darren Bardati
Bishop's University

Michael Bardecki
Ryerson University

Stephen Doyle
Okanagan College

Tim Elkin
Camosun College

Andrea Freeman
University of Calgary

Susan Gass
Dalhousie University

Leslie Goodman
University of Manitoba

Johanne Kristjanson
University of Manitoba

Robert McLeman
University of Ottawa

Barbara Jean McNicol
Mount Royal College

T. Meredith
McGill University

Brian S. Osborne
Queen's University

Hilary Sandford
Camosun College

Tom Waldichuk
Thompson Rivers University

Barry Weaver
Camosun College

Ann P. Zimmerman
University of Toronto

About the Authors

Philip Dearden

I grew up in Britain. Even though home was in one of the wilder parts of Britain, I was always struck with the biological impoverishment of my homeland and dreamed of living in a country where wild nature still existed. My dream was realized when I first came to Newfoundland as a graduate student in the early 1970s. Since that time I have travelled all over Canada, and most of the rest of the world, and have a strong appreciation of the beauty and grandeur of the Canadian landscape.

My main interest is in conservation, and I have taught courses and undertaken research on this topic, based at the University of Victoria, for 30 years. Throughout this period, I have taught large introductory classes in society and environment and loved every minute of it. I have a strong belief that the power of individual actions can help to make a better environmental future and that we need to support NGOs working in this area. I have held many positions in the Canadian Parks and Wilderness Society, including chair of the British Columbia chapter, and am currently a trustee emeritus.

My main field of research is conservation and protected areas, and I maintain active research programs in Canada, Asia, and Africa on this topic. I am a member of the IUCN's World Commission on Protected Areas and have advised many international bodies on protected area management. Author of more than 220 articles and nine books and monographs, including (with Rick Rollins) *Parks and Protected Areas in Canada: Planning and Management* (third edition, Oxford, 2009), I have also been recognized for excellence in teaching with an Alumni Outstanding Teacher Award and Maclean's Popular Professor recognition at the University of Victoria. I am Leader of the Marine Protected Area Research Group at the University of Victoria and like nothing better than to be out on the water sailing with my wife, Jittiya, and children, Kate and Theresa.

Bruce Mitchell

I was born and raised in Prince Rupert, a small community on the northwestern coast of British Columbia whose economy was strongly based on natural resources, especially forests and fish. Thus, from an early age, I recognized the importance of 'resources' and the 'environment'. As a graduate student, I focused on water resources, having become more and more aware of how critical this resource is for natural systems and humankind. Now, in the second decade of the twenty-first century, it is apparent that water is a key resource at a global scale.

As a high school student and an undergraduate, I worked summers as a shoreworker in a fish-processing plant and then as a deckhand on a troller, fishing for salmon. That provided first-hand experience with a resource-harvesting industry. But more importantly, it made me aware of how knowledgeable people who had not finished their formal education in the school system could be. This experience showed me that 'traditional ecological knowledge' is a remarkable source of understanding and insight. This became a lifelong lesson: experiential learning deserves respect, and those pursuing science should continuously look to such knowledge to enhance what they think they know based solely on science.

Studying geography provided me with a foundation to understand natural systems, as well as the way in which humans interact with or use them. As time passed, I became more and more convinced that many 'resource problems' were often 'people problems'. As a result, I oriented my work towards planning and management, while always remaining mindful of the need to draw on scientific research.

I have published about 150 articles and 27 books, have served as president of the Canadian Water Resources Association, have been a visiting professor at 11 universities in various countries, and received the Award for Scholarly Distinction from the Canadian Association of Geographers as well as a Distinguished Teacher Award from the University of Waterloo. I am a Fellow of the Royal Society of Canada and a Fellow of the International Water Resources Association, have been awarded the Massey Medal from the Royal Canadian Geographical Society, and am an honorary professor at four Chinese universities.

As a faculty member, I have had the opportunity to conduct research and consult both in Canada and in developing countries. The work in other countries has made me aware of how fortunate most Canadians are and of how our system of governance and civil society usually provides us with opportunities to become engaged in societal and environmental issues. Barriers and challenges exist, but Canadians have considerable latitude to participate in and shape our shared future.

On Spaceship Earth, there are no passengers;
we are all members of the crew.

—Marshall McLuhan

Part A
Introduction

The relationship among environment, resources, and society is one of the most important challenges, if not the most important challenge, facing humans on Earth. For many of Earth's human inhabitants, this relationship is an ongoing reality as they try to meet their everyday needs for food, water, and shelter. They do not need to be reminded of how important it is. To ignore the relationship is to perish.

For others, usually urban dwellers in developed countries, this reality seems distant. Food comes from the supermarket, water is piped into homes, even work and home environments have controlled temperature through central heating and air conditioning. Not until disruptions occur in these delivery systems—caused by floods, tsunamis, droughts, ice storms, earthquakes, hurricanes, insect infestations, or similar forces of nature—do many people realize that they, too, depend on the environment for survival, as has been true since before the dawn of human civilization.

This first section introduces some basic concepts regarding the relationship among environment, resources, and society and the ways by which we try to understand complex natural and socio-economic systems. There are many ways of knowing about environment. Here we concentrate mainly on the contribution of the natural and social sciences.

Science, especially environmental science, is becoming an increasingly collaborative undertaking. This collaboration involves not only workers from one discipline but those from many disciplines coming together to contribute their understanding of a particular phenomenon. An understanding of acid precipitation, for example, necessarily requires the input of chemists, biochemists, climatologists, geologists, hydrologists, geographers, biologists, health specialists, economists, and political and legal experts, to name a few. Each discipline has its own expertise and methods of approach, and they can be combined in different ways to yield more effective answers to environmental problems.

The need for a collaborative approach to environment is illustrated by the Sydney Tar Ponds, located in the Cape Breton Regional Municipality in Nova Scotia. Many of the problems resulting from steelmaking and the subsequent pollution of the Muggah Creek estuary came about through failure to effectively use scientific information known at the time. Even with the use of science, there can be high levels of uncertainty. How we can try to deal with uncertainty and change is one of the main themes of this book.

The Sydney Tar Ponds case also helps to illustrate some of the complexity entailed in trying to address just one environmental problem in one small place. The tar ponds should also make us pause and consider what desirable future we should be aiming for and striving to create. In that context, Chapter 1 considers three concepts relating to a vision for the future: sustainable development, sustainable livelihoods, and resilience. Sustainable development, popularized in 1987 with the publication of *Our Common Future*, the report of the World Commission on Environment and Development, has provoked much debate and disagreement, since different groups interpret it in ways that favour their values and interests. Despite conflicting views about what sustainable development means, it frequently appears in policies related to the environment and natural resources. Thus, it is important to have a critical appreciation of its strengths and limitations. More recently, attention is turning towards the concept of sustainable livelihoods as an alternative possible direction. Finally, resilience is also being proposed, and offers a quite different perspective from either of the other two concepts. Our intent here is to ensure that you understand what these concepts mean, what they offer, and what their weaknesses are.

Needless to say, the global situation is infinitely more complex than sustainability alone. The next section of Chapter 1 provides an overview of the global situation with regard to environment and society. What are some of the main trends pointing to future directions? Although there are disagreements about the rate and severity of environmental change, few claim that overall conditions are improving. One indicator, the Living Planet Index, suggests that the overall ecological health of the Earth has declined by 50 per cent since 1970.

However, there is good news. Predictions of global population, for example, are slightly lower than previously—9.15 billion people by the year 2050. One important dimension that has only shown growth, however, is resource use, fuelled mainly by the demands of consumers in developed countries—we are reminded of the old comic strip 'Pogo' in which the title character, a possum living in a swamp, famously proclaimed that 'We have found the enemy, and it is us.' If there is one fundamental message that we would like to convey, it is that the power of individuals to make decisions on a daily basis can reduce these pressures. Canadians have much to contribute in this regard, since we are among the most profligate consumers of energy and water in the world and are also among the most prolific producers of waste. Our society has developed into one of the most wasteful on the planet. Only we can turn that around.

The jurisdictional and governance arrangements for environmental management in Canada constitute one critical factor influencing our relationship with the environment and resources. Such arrangements are rarely taken into account by scientists and environmentalists, but they can be the most important factor when considering how and when a particular problem is going to be addressed. Canada is a large country, and the various levels of government are complex and often work poorly together.

Whether the context is global, national, or regional, we are interested in measuring our progress in addressing environmental change. As noted earlier, however, the situation is very complex, with far more variables and changes than we can possibly measure. The following section of the first chapter provides some background on how we try measuring progress through the use of indicators and outlines the various kinds of indicators and their strengths and weaknesses.

The chapter ends with the presentation of a simple framework that summarizes the process of environmental management. Throughout the book, we return to this framework to illustrate deficiencies in understanding or lack of connection between different elements of the framework.

Part A provides an overall introduction to environmental change and challenge with reference to the global, national, and regional levels. Most of the remainder of the book concentrates on Canada, although we consider global aspects throughout the book and return to a global perspective in the final chapter. Part B provides an overview of the main environmental processes we need to be familiar with to understand many environmental problems. Part C discusses some of the philosophical dimensions and best practices of various aspects of resource management. This is followed by Part D, in which we discuss various thematic aspects of resource management, such as fisheries, water, and climate change. The final section, Part E, draws together some of these themes, returns to global and national summaries of current trends, and points out some of the things that individuals can do to effect change for the better in the environment of tomorrow.

Chapter 1
Environment, Resources, and Society

Learning Objectives

- To appreciate different perspectives related to environment and resources.
- To understand different approaches to analyzing complex environmental and socio-economic systems.
- To understand the implications for change, complexity, uncertainty, and conflict relative to environmental issues and problems.
- To learn about various aspects that must be addressed to bring 'science' to bear on environmental and resource problems.
- To understand the significance of sustainable development, sustainable livelihoods, and resilience.
- To understand the nature of human population growth.
- To appreciate the impacts of overconsumption on global ecosystems.
- To understand relevant jurisdictional and governance arrangements in Canada.
- To recognize that Canada's natural environment and society are part of a global system.
- To describe different ways of tracking progress among nations on environmental matters over time.

Introduction: Change and Challenge

The year 2010 was the hottest year in Canada on record, with overall average temperatures three degrees warmer than average. Some areas were much warmer than the average. Canada is warming. Thirteen of the last 20 years are among the warmest on record.

Natural systems change. They have always changed and will always change. There is strong evidence, discussed later, that human activities have become a main driving force behind environmental change. Whatever the reason, it seems that changes are happening more abruptly and with greater magnitude than previously. They threaten societal well-being, and society must respond—and respond quickly.

Changes also occur as a result of shifts in human values, expectations, perceptions, and attitudes, which may have implications for future interactions between societies and natural systems. The value of the world economy has increased more than sixfold in the past 30 years. This increase was not merely the result of population growth; the chief cause has been increased consumption. Expectations have changed. Things seen as luxuries 50 or 60 years ago, such as TVs and

Perspectives on the Environment

On Change

The only constant is change.

—Heraclitus, ancient Greek philosopher who lived from 535 BC to 475 BC, some 100 years before Plato

There is nothing wrong with 'change', if it is the right direction.

—Winston Churchill, British Prime Minister (1940–5, 1951–5)

There is nothing more certain and unchanging than uncertainty and change.

—John F. Kennedy, President of the United States (1961–3)

automobiles, can now be found in some of the most remote societies on Earth.

Changes in natural and human systems generate challenges. If we wish to protect the integrity of biophysical systems yet also ensure that human needs are satisfied, questions arise about how to determine ecosystem integrity and how to define basic human needs. Such questions force us to think about conditions both *today* and in the *future*. Such questions also remind us that an understanding of environmental and resource systems requires both natural and social sciences. Neither alone provides sufficient understanding and insight to guide decisions. Finally, such questions pose fundamental challenges as to whether we can realistically expect to manage or control natural systems or whether we should focus on trying to manage human interactions with natural systems.

In this chapter, we begin by explaining what we mean by 'environment', 'resources', and 'society' and then consider alternative ways to understand systems, issues, and problems. A case study of the Sydney Tar Ponds in the Cape Breton Regional Municipality of Nova Scotia illustrates opportunities and challenges regarding resource and environmental systems as well as the importance of using both science and social science to inform public decisions and policy-making. The case study vividly demonstrates that decisions are often made in the context of changing conditions, incomplete knowledge and understanding, conflicting interests and values, trade-offs, and uncertainty.

These conditions apply not only to the Sydney Tar Ponds or, indeed, to Canada generally but also to the global stage. In this context, we provide an overview of some major environmental trends and the main issues that arise. There is no doubt that human population growth is a stress on this planet, but so are the consumption patterns of the more affluent sectors of society. These factors are leading to unprecedented changes in global systems. Of particular concern are the challenges posed by **global climate change**. It is important that we appreciate the role that Canada plays in global environmental change—both as the second largest country on Earth and as a source of major carbon resources, such as our forests and oil resources. The decisions made by Canada regarding these resources have global implications. Therefore, you should understand the governance aspects of environmental management in Canada to appreciate how decisions are made and how stakeholders such as you can become more involved.

The changes taking place are very complex. It is important that we grasp the essence of major changes and act accordingly. One way of doing this is to use the various **indicators** that measure environmental change and response, and we discuss the ways in which indicators, such as ecological footprints, are used in this regard. We conclude the chapter by identifying some key considerations regarding how scientific understanding and insight can be used to inform resource and environmental management and decisions.

A man and a woman walk down Laurier Boulevard next to a green Memorial Park in Brockville, Ontario, on 20 December 2006.

A cyclist uses the foot-bridge to cross the waterless Twenty Mile Creek at Balls Falls Park west of St Catharines, Ontario, 1 August 2007.

Defining Environment and Resources

The **environment** includes the atmosphere, hydrosphere, cryosphere, lithosphere, and **biosphere** in which humans, other living species, and non-animate phenomena exist. As an analogy, the environment is the habitat or home on which humans and others depend to survive. In contrast, **resources** are more specific and are normally thought of as such things as forests, wildlife, oceans, rivers and lakes, and minerals and petroleum.

Some consider resources to be only those components of the environment with utility for humans. From this perspective, coal and copper were part of the environment but were not resources until humans had the understanding to recognize their existence, the insight to appreciate how they could be used, and the skills or technology to access and apply them. In other words, in this perspective, elements of the environment do not become resources until they have value for humans. This is considered an **anthropocentric view** in the sense that value is defined relative to human interests, wants, and needs.

In contrast, another perspective sees resources as existing independently of human wants and needs. On that basis, components of the environment, such as temperate rain forests and grizzly bears, have value regardless of their immediate value for people. This perspective is labelled as **ecocentric** or **biocentric** because it values aspects of the environment simply because they exist and accepts that they have the right to exist.

In this book, we are interested in resources both as they have the potential to meet human needs and with regard to their own intrinsic value. Whichever category is emphasized, we often encounter change, complexity, uncertainty, and conflict. For example, different attitudes at different times may lead to an area being logged or used for mining or designated as a protected area. People and other animals drink water to live, and at the same time urban areas may compete with farmers for access to water. An area of significant value for its biodiversity or ecological integrity may be designated as a national or provincial park—but then might change into an ecosystem of less intrinsic value to humans as a result of ecological processes. For instance, if a fire sweeps through an old-growth forest, the question arises as to whether the fire should be allowed to burn because it is a natural part of ecosystem processes or whether humans should intervene to put it out.

Thus, recognizing anthropocentric and biocentric perspectives does not automatically resolve all the problems that scientists and managers face when dealing with the environment and resources. However, being aware of such viewpoints helps us to understand the positions that individuals or groups take with regard to what is appropriate action. You may wish to consider whether your perspective is more anthropocentric or biocentric. Does understanding your fundamental perspective help you to appreciate why your view about the right thing to do regarding a resource situation sometimes conflicts with what others would like to see happen?

The person who sees a manta ray underwater will never forget such a sight or dispute the right of such species to survive. However, a purely anthropocentric resource view will tend to look at their value more in terms of making money as material for bags, wallets, shoes, and jewellery, like the bracelet pictured here.

Alternative Approaches to Understanding Complex Natural and Socio-economic Systems

In this book, we examine many complex systems, and our understanding of these systems is often based on the knowledge derived from more than one discipline. As an individual, you should be aware of what insight you can contribute from a disciplinary, interdisciplinary, and/or transdisciplinary base and what knowledge you offer to others, from various disciplines, as part of a team. Systems have environmental, economic, and social components. The environmental component alone can be subdivided into aspects requiring expertise in disciplines such as biology, zoology, chemistry, geology, and geography. However, while humans have organized knowledge into disciplines for convenience and manageability, the 'real world' is not organized in that way, nor does it recognize disciplinary boundaries. As an individual, you can approach research from a disciplinary or cross-disciplinary perspective, so it is important to understand alternative ways of creating and applying knowledge. These alternative ways include at least the following.

1. *Disciplinary.* Disciplinary understanding is organized around the concepts, theories, assumptions, and methods associated with an academic discipline. Disciplines reflect a belief that specialization will result in more in-depth understanding, and this is correct. However, since systems of interest to environmental scientists and managers have many components, the danger of a disciplinary approach is that important connections with parts of the system not considered by a disciplinary specialist will not be taken into account. Some disciplines are broader than others and more open to interaction with other disciplines. Geography is one such example where the discipline specializes in synthesizing knowledge from many disciplines to understand differences among places.
2. *Multidisciplinary.* To obtain the in-depth insight of the disciplinary specialist but also gain the benefits of a broader view by drawing on specialists from various disciplines, the multidisciplinary approach emerged. In this approach, different specialists examine an issue, such as biodiversity, from their disciplinary perspectives, such as biology, economics, and law. The specialists work in isolation, or only with others from the same discipline or profession, and provide separate reports, which are submitted to one person or group, who then synthesizes the findings and insights. In this manner, both depth and breadth are achieved through synthesis of the findings of different specialists *after* they have completed their analyses.
3. *Cross-disciplinary.* While specialists in a multidisciplinary team work in isolation from one another, in cross-disciplinary research a disciplinary specialist 'crosses' the boundaries of other disciplines and borrows concepts, theories, methods, and empirical findings to enhance his or her disciplinary perspective. However, while in this approach the specialist deliberately crosses disciplinary boundaries to borrow from other disciplines, he or she does not actively engage with specialists from the other disciplines but simply draws on their ideas, approaches, and findings. This approach allows the investigator to make connections throughout an investigation that would not occur in a disciplinary or multidisciplinary approach, and this can be very positive. At the same time, it can also involve misunderstanding of the borrowed material, using theories, concepts, and methods out of context, and overlooking contradictory evidence, tests, or explanations in the discipline from which the borrowing is done.
4. *Interdisciplinary.* To overcome the limitations of the previous three approaches, interdisciplinary investigations involve disciplinary specialists crossing other disciplinary boundaries and engaging with other specialists *from the very beginning* of a research project. The objective is to achieve the benefits of both depth and breadth *from the outset*, as well as synthesis or integration, rather than at the end of the process, as occurs in the multidisciplinary approach. This approach requires more time than the second and third approaches, because a team of disciplinary specialists must meet at the start and then regularly throughout a study. In addition, the approach requires respect, trust, and mutual understanding among the disciplinary specialists, since it is common for one disciplinary specialist to question basic beliefs or assumptions that another specialist takes for granted. The approach also requires patience, because disciplinary specialists have to be prepared to learn the 'jargon' of other specialists so that clear communication can occur. Finally, an interdisciplinary approach requires team members to have considerable self-confidence and a willingness to acknowledge the weaknesses of their disciplines, since their disciplinary views will inevitably be challenged by others.
5. *Transdisciplinary.* A transdisciplinary approach extends the interdisciplinary perspective by seeking a holistic understanding that crosses or transcends boundaries of many disciplines. Furthermore, the problem or issue is usually not viewed as in the domain of any one discipline or profession. An example is 'health informatics', which brings together concepts and methods from medical and information sciences.

The same weight is given to the perspectives from each discipline or profession. This approach, as with interdisciplinarity, can be challenging given the reality of specialized vocabularies associated with different expertise, and the potential to be overwhelmed by huge volumes of data and insights, some of which may be contradictory. Brown, Harris, and Russell (2010) provide further insight about this approach.

Science-Based Management of Resources and Environment

In this book, we consider how understanding and insight from science can be used to inform management and decision-making. The nature of science is discussed in more detail in the introduction to Part B. Mills et al. (2001) provide five guidelines for contributions by scientists to effective management of resources and the environment.

1. *Focus the science on key issues, and communicate it in a policy-relevant form.* If science is to have value for managers, it must address pertinent management issues, and research must be conceived in a manner relevant to such issues. This stipulation does not preclude scientific research from addressing basic or fundamental questions. However, to be perceived as relevant to the needs of managers, scientific work must be focused on and be timely to the needs of managers. In that regard, while scientists can provide important input into establishing management goals, this task is properly in the domain of the value-laden process of decision-making and is not part of scientific research per se.
2. *Use scientific information to clarify issues, identify potential management options, and estimate consequences of decisions.* A basic challenge for managers is to determine whether a problem has been defined in an appropriate manner. Sometimes, because of complexity and uncertainty, managers may be unaware of questions that should be asked. In that regard, science, by helping to clarify relations and trends in systems, can clarify known issues and identify issues previously overlooked or unknown. Science can also help to calculate the implications of different options related to an issue or problem.
3. *Clearly and simply communicate key scientific findings to all participants.* While it is important for scientists to publish their results in peer-reviewed journals, if their work is to be relevant for managers then scientists must also share their findings in forums and formats accessible and understandable to non-specialists.
4. *Evaluate whether or not the final decision is consistent with scientific information.* Making relevant scientific information available and accessible is necessary but not sufficient. It must be considered and incorporated into decision-making. One way of ascertaining whether that is happening, and of putting pressure on decision-makers to do so, is to conduct systematic and formal evaluations of decisions to determine to what extent they have relied on science.
5. *Avoid advocacy of any particular solution.* There is much debate regarding this guideline, since some scientists believe that they should be advocates for solutions when their knowledge leads them to a preferred conclusion. Others maintain that if scientists and their evidence and interpretations are to be credible, they should not be seen to favour any particular solution. For example, if a scientist is a known advocate of the use of herbicides and pesticides to enhance agricultural production, would that person's evidence supporting the use of herbicides and pesticides be viewed as credible? Even if scientists can separate their basic values from their scientific understanding, there is a danger that they will be perceived to favour a particular viewpoint, leaving doubt in some people's minds as to whether data, interpretations, conclusions, and recommendations from such a scientist have been 'contaminated' by those values.

Perspectives on the Environment
Professional Judgement

Since there seldom is time to conduct new research in the middle of a major policy debate, there always will be holes, sometimes big ones, in the science information. The scientists will be asked to at least hypothesize relations that might fill those holes and that will require significant personal judgement. Often, tight time frames will not permit the sort of multiple rounds of peer review that are desirable and typical in the science arena. In these circumstances, faith in the objectivity and independence of the scientists is particularly important.

—Mills et al. (2001: 14)

The dilemma, of course, is that nobody is value-free or value-neutral, so to suggest that scientists can or should be value-free or 'objective' is difficult to sustain. However, there is a difference between being perceived to be open-minded in defining a problem or identifying alternative solutions and being known to uphold a particular view or position and consistently producing findings that support only that one particular view or position. A further complication occurs when there is insufficient evidence to support a conclusion and scientists are asked to provide a professional opinion. In such situations, the scientist will be viewed as more credible if he or she has no record of advocating a particular perspective.

The following case study of the Sydney Tar Ponds illustrates the different kinds of knowledge required to understand resource and environmental problems. It also shows that the ideal of having 'science' inform policy decisions is often fraught with challenges when both the natural environment and human perspectives are changing.

The Sydney Tar Ponds, Cape Breton Regional Municipality, Nova Scotia

Background

Sydney, Nova Scotia, merged in 1995 with other communities to form the Cape Breton Regional Municipality (CBRM), is located on the northern part of Cape Breton Island (Figure 1.1). Extensive deposits of coal and iron ore in the CBRM area, still referred today as 'Industrial Cape Breton', resulted in a long history of coal mining and steel production. These resources, in addition to a coastal fishery, forestry, and a striking natural landscape, have been the economic base for the community.

In the late nineteenth century the Industrial Revolution was being powered by coal. Geologists were aware that Cape Breton Island and mainland Nova Scotia had substantial coal deposits. Shafts were sunk, and men and boys began the hard work of digging out the coal. Production steadily grew, and shortly after 1891, when a rail line connected Sydney to Halifax, annual coal production had grown to 1.5 million tonnes. By 1893, the numerous small coal mines joined together as the Dominion Coal Company (DOMCO), and output continued to grow—to more than 6 million tonnes annually by 1913.

> ### Perspectives on the Environment
> #### Steel Production in Sydney
>
> By 1912 Cape Breton was the source of nearly half of all the steel produced in Canada, and Sydney's Dominion Iron and Steel Company had the largest share of the pie. For half a century business boomed.
>
> —Lahey (1998: 38)

The steady expansion of rail lines in North America led to a high demand for steel, prompting the American owner of DOMCO to form a partnership to create a new corporation, the Dominion Iron and Steel Company (DISCO). In 1899 construction started on the new steel plant located along Muggah Creek. A supply of coal was readily available, but more iron ore was needed, and it was procured from the iron ore mines at Wabana, Bell Island, Newfoundland.

If coal deposits contain too much sulphur, inferior coke is the result. Low-grade iron ore leads to the need for larger quantities of limestone to remove the impurities. To assess the quality of the basic inputs to the steelmaking process, science is essential. However, such science was not drawn upon when arrangements were being made for the basic raw materials. This was the beginning of what would become a pattern. As Barlow and May (2000, 11–12) observed:

In a rush to begin full operation, they failed to run the most basic tests on their coals and ores. DISCO tapped its first steel on New Year's Eve 1901, and right from the start there were problems.

The coal from Cape Breton seams was very high in sulphur, so far more coal had to be baked to produce usable coke. The iron ore from Wabana was full of impurities, such as silica and rock, so far more limestone was required to pull out the impurities as slag. The unusually large amounts of limestone required in the blast furnaces caused the furnace linings to deteriorate rapidly. . . .

The poor quality of the basic ingredients led to higher costs, less marketable and inferior products, and far more waste. In what had been Muggah Creek, the slag would eventually create a mountain range of waste, stretching hundreds of feet high and reducing the mouth of the estuary by nearly a mile.

The coal and steel operations continued for almost a century under various owners. The coke ovens, in which coal was baked at high temperatures to produce a higher-quality fuel source for the steel mill's open-hearth blast furnaces, were closed in 1988 (see photo, next page). The steel plant stopped operating in 2001. The closure of these operations was a serious blow to the economy of Sydney. However, even after the coke ovens closed, the legacy of millions of tonnes of toxic sludge remained a threat to adjacent residents and the environment. For decades, air pollution originating from the coke ovens and open-hearth furnaces had been clearly evident. Nevertheless, the implications of the air pollution for land, waterways, and the Muggah Creek watershed were in some cases unknown and in other cases only poorly understood.

The area labelled as the 'tar ponds' was created from the deposit of chemical by-products from the coking process, runoff of water used for cooling in the coke ovens and the steel mill, leaching from contaminated soil at the coke and steelmaking plants, a garbage dump site in the upper part of the watershed, and discharge of raw sewage from residential and commercial areas of Sydney. Other contaminants were produced from a cement factory, a gas and oil company, and a brick factory. The outcome was a chemical and bacteria-laden river system, including the **estuary** full of contaminated sediments, the latter referred to as the tar ponds, 'a two-kilometre-long stretch of contaminated water and sludge which federal officials refer to as the largest chemical waste site in Canada' (Lahey, 1998: 37). Various government surveys of this ecosystem identified polycyclic aromatic hydrocarbons (PAHs), polychlorinated biphenyls (PCBs), and other chemicals and metals among the pollutants. Muggah Creek and the estuary empty into Sydney harbour, and discovery of PAHs in those waters led to the closure of the lobster fishery. The studies determined that the soil on the coke ovens

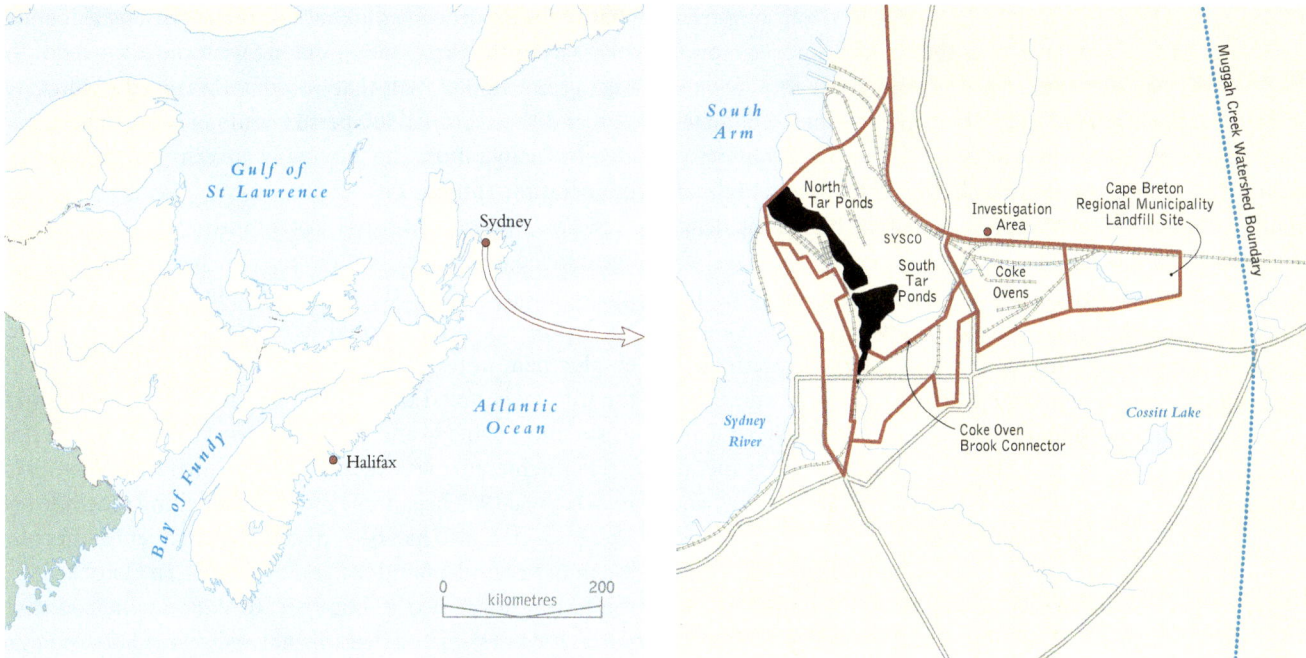

Figure 1.1 | On the detailed map (right), the highway that curves around Cossitt Lake and meets the Glace Bay Highway, which bisects the figure horizontally, now extends onward and links up with a road, SPAR (an acronym for Sydney Port Authority Road), on the other side of the CBRM landfill site. It then continues past the coke ovens site and into the main cleanup area. This link was created to facilitate easier access to the cleaned-up area for industrial and commercial purposes. *Source: Rainham (2002: 27), from Joint Action Group, Sydney, Cape Breton (1999).*

site was highly contaminated. The site also contained storage tanks holding chemical waste.

The tar ponds are in the lower part of the Muggah Creek watershed, and since Dominion Iron and Steel built its plant in 1899, what was once an active, navigable waterway and a habitat for fish and birds became a highly contaminated narrow tidal outlet. A causeway and bridge divide what used to be an estuary into a north pond and a south pond. Surveys by governments and consultants revealed that the two ponds contain about 700,000 tonnes of sediment contaminated with PAHs, including some 45,000 tonnes also containing PCBs at concentrations above 50 parts per million (ppm).

Challenges for Epidemiological Studies

Various studies began revealing that citizens of Sydney were encountering serious health problems. Unreleased government studies pointed to significantly higher cancer rates in Sydney compared to the rest of Canada. Health and Welfare Canada expressed concerns about the possible link to the toxic wastes from the coke ovens. One study found that the life expectancy for male and female residents in Sydney was as much as five years less than for the Canadian population as a whole. The primary contributions to reduced life expectancy were significantly higher levels of cancer and cardiovascular disease. Another study showed increased incidence of cancer in the Sydney population over a 45-year period. Yet another study indicated that rates of major birth anomalies are significantly higher among Sydney residents relative to the rest of Nova Scotia.

In the mid-1980s, Health and Welfare Canada alerted the Environment Canada Atlantic regional office about health concerns, and the regional official alerted its provincial counterpart. In contrast, the epidemiologist for Nova Scotia

Sydney, Nova Scotia, coke ovens, 1987. Aerial view of the coke ovens, showing the quenching plant with steam at left, coal pocket and batteries between exhaust stacks, centre, and conveyor leading from pocket to blending plant, right. By-product building is large brick structure, foreground. Smaller buildings are pump house, carpenter shop, and oil house (Coke Ovens, SYSCO, 1987, 90–221–19653, Beaton Institute, University College of Cape Breton, Sydney, NS).

took the position that the hazard depended on long-term exposure and that balanced against social and economic benefits of the coke ovens, it was reasonable to allow the coke operations to continue. The province then conducted its own investigation and concluded that unhealthy lifestyles (smoking, alcohol, poor diet such as fatty foods and high salt intake) were more likely to be causes of the higher incidence of poor health.

Other factors can confound our understanding. For example, Goodarzi and Mukhopadhyay (2000: 369–70) reviewed the possible source and level of concentrations of PAHs and metals from a sample of bore holes and lakes providing water to the Cape Breton Regional Municipality. They concluded that 'the origin of the PAHs and the priority metals in the drinking water of the Sydney basin is not known', and identified at least three possible origins for the PAHs and metals, highlighting that it is not uncommon during scientific investigations for alternative explanations to emerge.

Problems When Science Is Not Used to Inform Decisions

Serious thinking about remediation followed a 1980 federal survey of lobsters in Sydney harbour. The survey revealed that the lobsters were contaminated with cancer-causing PAH chemicals, as well as with mercury, cadmium, and lead. This finding led to closure of the lobster fishery in 1982 in the south arm of the harbour. Testing by Environment Canada indicated the obvious source—the steelmaking operations of SYSCO, or the Sydney Steel Corporation, an agency of the Nova Scotia government that had taken control of the failing private-sector operations.

In 1984, the consulting firm Acres International received a contract to determine the scope of the pollution and recommend options. Initial testing indicated that the tar ponds contained the equivalent of 540,000 tonnes (dry weight) of toxic waste, including 4.4 to 8.8 million pounds of PAHs. The sludge on the bottom of the estuary was judged to be between one and four metres deep. Acres focused on the challenges represented by the PAHs. Although PCBs had been identified in the earlier study of lobsters, they were not considered a problem because they were not a by-product of making coke or steel. The source of the PCBs was waste transformer fluid used in the steel mill. Another source of PCBs was the railway yard operated by Canadian National on the opposite side of Muggah Creek.

Random sampling based on drilling bore holes in the estuary revealed only small quantities of PCBs. From those results and its analysis, Acres identified three options: (1) leave the polluted sludge in place and cover it; (2) remove the sludge and store it somewhere else; and (3) remove the sludge and incinerate it. Acres estimated that incineration would destroy 99.99 per cent of the PAHs. However, PCBs are virtually indestructible at extreme temperatures and would have been changed into airborne dioxins and other poisons if incinerated. Given the three options, along with the estimate of the high proportion of PAHs that would be destroyed by incineration and the almost 1,500 person-years of work to be generated by incineration, the provincial government selected the incineration option.

In 1987, federal and Nova Scotia ministers of environment announced a $34.3 million package to cover excavation and incineration of the toxic waste in the tar ponds. The workers in the coke ovens would have the first opportunity for employment in the cleanup project. At the press conference, the ministers stated that the tar ponds were the worst toxic waste site in Canada and the second worst in North America.

The incinerator was supposed to be operational by 1990. However, as 1992 began, the project was behind schedule and over budget. It also had been decided to conduct further testing to better understand the contaminants. In October 1992, the testing identified a 'hot spot' of PCBs in the south tar pond. It was estimated that this hot spot had 4,000 tonnes of sludge contaminated with PCBs. Canadian law requires PCBs over 50 ppm to be incinerated at a minimum temperature of 1,200°C, and the sample from the south tar pond revealed concentrations up to 633 ppm. The incinerator, designed to destroy only PAHs, had a capacity up to 900°C.

By the fall of 1994, the problem created by the PCBs was not resolved, although the incinerator and dredging equipment were working. In late 1994, the province decided that the incinerator option was not viable. It called for tenders for new approaches, and all the bids were above $100 million. The province rejected all the proposals as too expensive. It subsequently invited one Nova Scotia consulting firm, Jacques Whitford, to determine what could be done for $20 million or less. In January 1996, the Nova Scotia Minister of Supply and Services announced that he had accepted the Jacques Whitford plan to use the slag piled up next to the tar ponds to fill in the ponds. Once that work had been done, grass and trees would be planted to create a park. This proposal was greeted with surprise and anger by the people of Sydney, none of whom had participated in the development of this 'solution'.

Jacques Whitford started the first phase of its work, a sampling of the tar ponds to determine the extent of the PCBs. By this time, 10 years had passed since the federal and provincial governments had announced the cleanup, and under federal law, PCBs cannot be buried. The intent was to identify the PCB-contaminated sludge and remove it to a disposal site in Quebec. Throughout the spring, the sampling continued, and by midsummer the estimate was that 45,000 tonnes of PCB-contaminated sludge existed, leaving Jacques Whitford to express reservations about its proposal. The outcome was that the 'encapsulation' option was rejected. To this point, $60 million had been spent, and a viable solution had not emerged.

The federal and provincial governments announced that they would pursue a more open and participatory approach and established a community–government committee to develop a cleanup plan. The committee was named the Joint Action Group (JAG). In 1998, an agreement was reached to clarify the relationship of JAG with the three levels of government, and $62 million were committed to complete studies, designs, and other preparations for the cleanup. In 2000, another consulting firm, Conestoga Rovers and Associates, was hired to manage the agreement intended to lead to the cleanup of the tar ponds as well as the coke ovens site. Over the next few years, a sewer system was built to divert tonnes of raw sewage flowing daily into the tar ponds; the derelict structures on the coke oven site were demolished and removed; and the old Sydney landfill was closed and capped.

Next Steps

In 2007, a $400 million cleanup of the tar ponds was announced by the federal and provincial governments. This was the fourth major initiative to deal with the legacy of pollutants in the tar ponds. The press release from the governments of Canada and Nova Scotia stated that the solution would involve solidification, containment, and capping of contaminated soils, to be followed by site development and long-term and ongoing monitoring and maintenance. Incineration would not be used. The recommendation not to use incineration was not based on advice from scientists or consultants, but because of strong rejection by the community. In announcing the cleanup initiative, the governments emphasized that the remediation proposal had been reviewed through an environmental assessment process that started during 2005. The independent Environmental Assessment Panel strongly recommended that remediation should focus on containment and capping of all materials in the tar ponds rather than incinerating some materials and containment and capping of the remainder.

Key milestones of the remediation work are:

- Late spring 2007: Building began for more than two kilometres of channels through the tar ponds to allow clean water from the Coke Oven Brook and Wash Brook to flow through the ponds without becoming contaminated.
- Spring and summer 2008: Sediments from the old channel of the Coke Oven Brook were removed and moved elsewhere on the site to be stabilized and solidified. High-density polyethylene liner was then placed on the bed of the brook and covered with gravel to allow water to drain off the site without becoming contaminated.
- Summer 2008 to early 2012: Sediments in the tar ponds were stabilized, solidified, and capped. The finished surface is to be planted with grass and other vegetation.
- Spring 2011: Construction began on the layered cap over the treated tar ponds sediment, with completion scheduled for summer 2013.
- Post-2012: The provincial government will begin a long-term monitoring program regarding the water quality of the harbour, and governments will consult with the community regarding a plan for future uses on the remediated sites.

An aerial view of the Sydney Tar Ponds site (top) shows the scope of the cleanup project, and an artist's rendering (middle) and close-up of tennis courts (bottom) give an on-the-ground view of some of the remediation work.

Guest Statement
Perspective on the Tar Ponds
Tim Babcock

I moved to the outskirts of Sydney (far away on the *other* side of town from the tar ponds) in the fall of 2003, after two decades of working in Indonesia on issues related to resource management. What contrasts between the two places, and yet what similarities!

One of the first things that visitors to Sydney would likely notice at that time, if they arrived by car along the Trans-Canada Highway from the mainland, was a sign on the outskirts (since removed) proclaiming Sydney to be the 'Steel Center of Eastern Canada'. But this is only a small example of the disjunctions one soon came across in the area. There was no sign, perhaps on the other side of the highway, proclaiming the existence of the largest polluted industrial site in Canada.

I soon learned that the tar ponds cleanup (for which a whole government agency had been set up) continued to be a major topic of discussion—and a source of frustration and anger—among the residents of the area. It was embedded in a much larger issue—the dismantling of the steel plant itself, the appropriate disposition and use of the land on which it stood, and, far more seriously, the devastating effects of the plant's closure on the economy, and on the very fabric of society, of Cape Breton Island. Cape Breton, the recently amalgamated Regional Municipality, was in a 'state of crisis'. Much uncertainty, at least in the public mind, still existed concerning the choice of cleanup measures, the amount of money available, and even when the program would get into high gear. Argument, not all of high quality, raged on concerning the appropriate means to clean up the site among the Sierra Club, the still extant though no longer funded Joint Action Group, members of the public, and the cleanup agency. We are no further ahead than we were in 1996, opined some informed individuals; others, with intimate knowledge, claimed that progress was 'rapid'.

At this point, many tens of millions of dollars had been spent to obtain what one presumes were the best scientific analyses and expert solutions available, and many more dollars had been spent on various forms of 'citizen participation' to discuss and propose a range of solutions. Yet the studies continued to be debated, and confusion, and much unhappiness, reigned. Some technical solutions recommended by the experts (e.g., incineration) were roundly rejected by the community once a further opportunity for their voices to be heard was provided through the Panel Review in 2006.

In mid-2011, calm appears to have settled, finally. 'Seeing is believing' appears to be the operative principle in the community. Remediation of the south pond is now complete

Sydney: Steel Center of Eastern Canada, March 2004.

and the overall cleanup has neared the halfway mark. In the larger area that includes the old steel mill site, much greenery now can be seen—large grassed-over areas; new trees planted; recreated brooks carrying clean water—ducks and foxes are sighted, community recreation facilities are appearing, and several commercial entities have established themselves. Aboriginal subcontractors are involved, and support is provided for women workers in 'non-traditional' jobs. The loudest complaint in recent times has concerned unpleasant odours, a nuisance more than anything else, resulting from digging up the contaminated soil for processing. The community has been reassured, and bright pink odour-suppressing foam has been sprayed over mountains of soil awaiting final treatment. Information is proactively supplied to the public, information tours of the site are held on an almost weekly basis, and in June 2011 the US cable program 'Today in America' filmed the 'good news' cleanup story for distribution across North America.

But the question remains: were the planning and consultation processes flawed, or were they the 'state of the art', the best that available Canadian money could buy?

Over the past several decades, Canada has spent hundreds of millions of dollars providing what is, one hopes, state-of-the-art advice to developing countries on how to better manage their environments and natural resources. But are we following 'best practice' here in Canada? Would we wish such a lengthy and complex process as the tar ponds case demonstrates on a poor country with equally serious pollution problems, probably affecting far more people, but with far fewer scientific, educational, and financial resources to deal with them? And we have so far left politics out of our analysis: this adds a further, almost unlimited layer of complexity and uncertainty to the situation, where individuals, groups, organizations, and governments will inevitably, and selectively, use the results of 'science' to further their own sincerely, or cynically, held positions. There are clearly no simple answers.

Tim Babcock has a Ph.D. in anthropology, and has worked for more than 30 years on various development programs in Indonesia and elsewhere.

The cost is shared by the federal government ($280 million) and the provincial government ($120 million) over the eight years scheduled for the cleanup. During the construction seasons, up to 150 people have been employed. Full opportunity for participation in the remediation has been provided to local companies and to Aboriginal businesses. Thirty per cent of the contracts for the coke ovens site were a part of the Aboriginal set-aside program. An environmental training program was also initiated in 2009 to continue to develop environmental remediation skills for Aboriginal workers. The solidification and stabilization process started in 2010 with the south pond. Sediments in the pond were solidified, and a multi-layered cap was begun in 2011 to be completed during the summer of 2012. The project reached its halfway point in the summer of 2011, and the cleanup was expected to be completed by 2014.

What has been learned from this experience so far? At least three lessons stand out. First, when basic science is not used from the outset to inform policy decisions related to environmental issues, the probability is high that funds will not be allocated to effective solutions. Second, even when science is used, understanding can be incomplete, and decisions will have to be taken in the face of considerable uncertainty. Third, when local stakeholders are not included in the process, challenges can be expected to proposed solutions. In this book, our goal is to help you enhance your understanding of scientific aspects of ecosystems and how that can be combined with 'best practices' related to management of resources and the environment.

Perspectives on the Environment
Lessons

It would appear that the residents of Sydney . . . have suffered and fought in isolation and perhaps in vain. We appear not to have learned one thing from their ordeal. We continue to talk about the 'trade-off' between jobs and the environment; jobs and health. The tar ponds saga should have taught us that such trade-offs are wrong economically, environmentally, and morally.

—Barlow and May (2000: 200)

The Sydney Tar Ponds situation highlights the challenges we can face from industrial legacies of pollution, but it is not unique. Other communities with similar challenges exist: the Lubicon First Nation in Alberta, surrounded by sour gas wells; the White Dog Reserve downstream from Dryden, Ontario, experiencing mercury contamination from a paper mill in Dryden; Deline, NWT, and its abandoned uranium mine; Yellowknife, NWT, with an abandoned gold mine; Port Hope, Ontario, with waste from a refinery for radium and uranium; Newcastle, NB, and Transcona, Manitoba, with abandoned wood preservative plants; and Squamish, BC, with heavy-metal pollution from an old chloralkali plant and the nearby abandoned Britannia mine site on Howe Sound. All of these communities face formidable challenges. These community-based problems also illustrate that resource and environmental problems can be located within or adjacent to settlements and indeed in highly urbanized landscapes. In Chapters 7, 12, and 13, other examples of environmental challenges in more urbanized areas will be presented.

Sustainable Development, Sustainable Livelihoods, and Resilience

The previous sections have highlighted that we have choices but that deciding what is the right thing to do is often difficult because of changing conditions, uncertainty, complexity, and conflicting views. Many believe that a key element required in making choices is a clear sense of where we want to go. What is the desirable future that we want to create? Without a clear sense of direction, it is challenging to know what matters deserve priority and how to resolve trade-offs when conflicts occur. The concept of 'vision' often comes up in regard to establishing a clear direction, and is discussed in Chapter 5.

In this section, however, we introduce three concepts—sustainable development, sustainable livelihoods, and resilience—often pointed to as representing the kind of future to which we should aspire.

Part A | Introduction

...nable Development

Perspectives on the Environment

Sustainable Development

Sustainable development is development that meets the needs of the present without compromising the ability of future generations to meet their own needs. It contains within it two key concepts:

- the concept of 'needs', in particular the essential needs of the world's poor, to which overriding priority should be given; and
- the idea of limitations imposed by the state of technology and social organization on the environment's ability to meet present and future needs.

—WCED, World Commission on Environment and Development (1987: 43)

Sustainable development as an environmental and economic objective, and as a philosophical starting point, emerged in the late 1980s through the work of the World Commission on Environment and Development. In directing us to pursue development that meets the needs of the present without compromising the ability of future generations to meet their own needs, sustainable development stipulates that we consider both intra- and intergenerational equity. However, sustainable development offers a major challenge, since it requires a re-examination of and shift in current values, policies, processes, and practices.

Sustainable development entails three strategic aspects. At one level, it presents a vision or direction regarding the nature of future societies. In sustainable societies, attention is given to meeting basic human needs, achieving equity and justice for present and future generations, realizing self-empowerment, protecting the integrity of biophysical systems, integrating environmental and economic considerations, and keeping future options open. The comments by Mary Ellen Turpel and Ovide Mercredi in Box 1.1 highlight that many of the ideals associated with sustainable development, such as a relation of stewardship between humans and the resources of nature, were recognized well before the concept was popularized in the 1980s.

At a second level, sustainable development emphasizes a system of governance and management characterized by openness, transparency, decentralization, and accessibility. It accepts the legitimacy of local or indigenous knowledge and seeks to incorporate such understanding with science-based knowledge when developing strategies and plans. It also recognizes that conditions change and much uncertainty exists. Thus, it is necessary to be flexible and adaptable, thereby allowing for policies and practices to be modified as experience accumulates. At a third level, and related to specific places or resource sectors, sustainable development seeks to ensure that economic, environmental, and social aspects are considered together and that trade-offs are visible and transparent to those affected.

The concept of sustainable development has generated both enthusiasm and frustration. The enthusiasm comes from those who believe that it provides a compelling vision for the twenty-first century, one with more attention to longer-term implications of development and to balancing economic, social, and environmental considerations. The phrase 'think globally and act locally' reminds us that while ultimately the planet is a single system in which actions in one part often have implications for other parts, resolution of problems also requires significant action at the local level, thereby stimulating self-empowerment, partnerships, and co-operative approaches to management and development (see Chapter 6).

The frustration has come from those who believe that 'sustainable development' is so vague that it can be defined in ways to suit different and often conflicting interests. Thus, developers like the concept because they can argue that growth must continue if basic human needs are to be met and if standards of living are to continue to rise. In contrast, environmentalists support the concept because they can use it to argue that environmental integrity must be given priority if there is to be long-term and equitable development.

How do you think such incompatibilities and tensions can be reconciled?

Sustainable Livelihoods

Sustainable development is often greeted with skepticism today. As a result, in some countries and regions, interest has shifted to the concept of **sustainable livelihoods**, viewed as more realistic and focused. The idea of sustainable livelihoods emphasizes the conditions necessary to ensure that basic human needs (e.g., food, shelter) are satisfied. However, the concept has been criticized by those who view it as too anthropocentric. Critics argue that other living creatures or inanimate components of ecosystems may be sacrificed or degraded to meet human needs.

Nonetheless, the concept of sustainable livelihoods does respond to the concerns of the World Commission on Environment and Development (WCED, 1987: 43) that 'Poverty is a major cause and effect of global environmental problems', that 'many present development trends leave increasing numbers of people poor and vulnerable', and that sustainable development contains within it two key concepts, one of which is 'the concept of "needs", in particular the essential needs of the world's poor, to which overriding priority should be given'.

For sustainable livelihoods, attention is given to ways for local people to meet basic needs (food, housing) as well as other needs related to security and dignity through

Environment in Focus

Box 1.1 Aboriginal Perspectives on Environment and Resources

First Nations peoples use the expression Turtle Island to refer to North America, which is thought of as a shell of a turtle surrounded by oceans. The images of the protective shell jutting out and the living creature within make a powerful metaphor for the connection with and respect for the land that all First Nations cultures share.

—Mary Ellen Turpel

We have always been here on this land we call Turtle Island, on our homelands given to us by the Creator, and we have a responsibility to care for and live in harmony with all of her creations. . . . There is harmony in the universe, among our relatives, among the animals, and among all creatures. . . we do not believe in competition, in the survival of the fittest. We believe all should be cared for in our Nations, that caring and sharing, not self-interest, must be our overriding aims. . . .

There are many resources in this country, and Canadians respond to this with massive resource development. We believe that people have forgotten the importance of protecting the environment for future generations. The peoples that I represent, those with an indigenous philosophy, have a world view that is different from that of the corporate mainstream. We have a view of the environment that does not stop all forms of development but allows it to proceed in a way that respects the environment and ensures that it is protected for future generations. . . . But to respect Mother Earth as a living entity is not easy, particularly when the preoccupation of economic development may well be to exploit natural resources rather than to preserve or sustain them. That endangers our common survival and the survival of future generations who are relying on us to preserve the planet for them.

—Ovide Mercredi

Source: Mercredi and Turpel (1993: 14, 16, 44, 45, 155–6).

meaningful work, at the same time minimizing environmental degradation, rehabilitating damaged environments, and addressing concerns about social justice. Strategies for sustainable livelihoods usually aim to create diverse opportunities, efficiency, and sufficiency relative to basic needs while also achieving social equity and sensitivity regarding environmental integrity.

Resilience

Walker and Salt (2006: 1) define **resilience** as 'the ability of a system to absorb disturbance and still retain its basic function and structure'. However, they observe that resource management 'best practice' normally focuses on optimizing particular goods or services from a natural resource system. Such optimization is usually achieved by taking specific components from the system through controlling other components. An example would be to increase crop production by using herbicides or pesticides to control organisms detrimental to such increased production, as discussed in Chapter 10. The ultimate goal is to move a system into some ideal state and sustain it in that state.

The dilemma, they note, is that reaching and sustaining an ideal state assumes that future changes in the system will be minor, incremental, and linear. In contrast, reality is often the opposite. Systems are frequently altered through 'lurching and non-linear' changes. And, in striving to use resource systems efficiently, one outcome can be that desirable redundancies are eliminated, since the goal is to retain only features with immediate value. The ultimate outcome is a drastic reduction in resilience. This perspective about resilience fundamentally challenges the goal of aiming for 'sustainable development', as defined earlier. In the view of Walter and Salt (ibid., 9):

> the more you optimize elements of a complex system of humans and nature for some specific goal, the more you diminish that system's resilience. A drive for an efficient optimal state outcome has the effect of making the total system more vulnerable to shocks and disturbances.

> The bottom line for sustainability is that any proposal for sustainable development that does not explicitly acknowledge a system's resilience is simply not going to keep delivering goods (or services). The key to sustainability lies in enhancing the resilience of social-ecological systems, not in optimizing isolated components of the system.

Do you think 'sustainable livelihoods' or 'resilience' provides appropriate alternatives to 'sustainable development'? If you conclude that any or all of these concepts contain major flaws, what alternative 'vision' would you propose to help guide us towards a desirable future condition?

The question of sustainable development and Canada's progress will be reviewed in a later section. The next section provides a quick snapshot of the global context for environmental management in the face of increasing human pressures.

The Global Picture

Our home, planet Earth, is different from all the other planets we know. As it hurtles through space at 107,200 kilometres per hour, an apparently infinite supply of energy from the sun fuels a life support system that should provide perpetual sustenance for Earth's passengers. Unfortunately, this seems not to be the case. Organisms are becoming extinct at rates unsurpassed for at least 65 million years. These extinctions cover all life forms and probably represent the largest orgy of extinction ever in the 4.5-billion-year history of the planet. Our seas are no longer the infinite sources of fish we thought they were. Our forests are dwindling at unprecedented rates. Even the atmosphere is changing in composition and making the spectre of significant climatic change a reality. Every raindrop that falls on this planet bears the indelible stamp of the one organism bringing about these changes—you and us.

Concern over this situation led to the request by the UN Secretary-General to undertake an assessment of the relationship between planetary ecosystems and the demands placed on them by human activity. Between 2001 and 2005, the **Millennium Ecosystem Assessment** was carried out to assess the consequences of ecosystem change for human well-being and to establish the scientific basis for actions needed to enhance the conservation and sustainable use of ecosystems and their contributions to human well-being. Some 1,360 experts from 95 countries were involved in the assessment (Box 1.2). The experts concluded that good evidence indicates many of the changes are non-linear, and once they start, the processes of degradation will increase rapidly. These positive feedback loops are discussed more extensively in Chapter 4.

Our home, planet Earth.

Population

One main variable that affects our impact on the planetary life support system is the number of passengers being supported. Although countless billions of passengers—from insects to the great blue whale—are on board planet Earth, we are mainly concerned with those who seem to be having the greatest impact on the system—humans, or *Homo sapiens*. This species, along with a few others such as rats and cockroaches, has experienced a staggering increase in population numbers over the past century.

The steep curve of population increase, shown in Box 1.3, coincides with the time that humans learned how to exploit the vast energy supplies of past *photosynthetic* activity lain down as coal and oil in the Earth's crust. Until then, energy supplies had been limited by daily inputs from the sun. The discovery of this new treasure house of energy allowed humans to increase food supplies dramatically and improve and greatly speed up the processing and transportation of materials (see Chapters 2 and 12 for more discussion on this). More than 7 billion humans now draw upon the planetary life support system for sustenance; before the Industrial Revolution, there were fewer than a billion. Another result of increased energy consumption is the pollution that now chokes this life support system and is causing unprecedented human-induced changes in global climate.

An estimated 4.3 people are born every second around the world. By the end of October 2011 the Earth supported 7 billion people. The United Nations forecasts an increase to 9.3 billion people by 2050 and 10.1 billion by 2100 (UN Population Division, 2011), representing more than 80 million additional people per year to feed. This scenario assumes that replacement-level fertility rates are maintained. A high-variant scenario, which assumes slightly higher fertility rates, places global population at 10.6 billion by mid-century and 15.8 billion by the end of the twenty-first century.

Much of the projected increase will occur in less developed countries, where populations, according to the UN's medium scenario, are predicted to grow by 33 per cent between 2005 and 2050, compared to only 2.4 per cent in developed countries (Figure 1.3). By 2050, the United Nations has forecast that the populations of the world's 50 least developed countries will increase by 56 per cent. China's massive population would continue to grow until 2030, when economic growth would trigger reductions in fertility, and level out at around 1.47 billion. India is predicted to overtake China as the most populous country on Earth by 2030 and continue to grow until 2060, when it would peak at 1.7 billion people. Nigeria would also experience rapid growth, with the population tripling to 288 million, while that of Bangladesh would double to 254 million.

The different trajectories of the developed and developing countries are epitomized by Nigeria and Japan. In Japan,

Symbolic 7 billionth human born on 31 October 2011
At the Community Health Centre in Mall, about 45 kilometres from Lucknow in India's most populous state of Uttar Pradesh, father and mother look on their newborn daughter, Nargis. Born on 31 October 2011, Nargis was the symbolic seven billionth human presently on earth.

the birth rate is slightly more than one child per woman. Fourteen per cent of the Japanese population is younger than 15, and 21 per cent is older than 65. In contrast, 44 per cent of Nigeria's population is younger than 15, with only 3 per cent over 65. A typical Nigerian woman gives birth to six children over her lifetime.

In all these predictions, great uncertainties exist. Much depends on trends in infant mortality rates and attempts to curb the growth of AIDS in the developing world. In 2009, an estimated 2.6 million people were newly infected with HIV, and by the end of 2010, 32 million people were living with HIV, a 27 per cent increase since 1999 (UNAIDS, 2010). These numbers suggest that, from a global perspective, HIV/AIDS may not be having a significant impact on global populations, although it certainly will for some countries.

Population age structure is also important. If two countries have similar populations but differing age structures, that can have a dramatic impact on future population growth.

Environment in Focus

Box 1.2 Some Findings from the Millennium Ecosystem Assessment

- More land was converted to cropland in the 30 years after 1950 than in the 150 years between 1700 and 1850. Cultivated systems (areas where at least 30 per cent of the landscape is in croplands, shifting cultivation, confined livestock production, or freshwater aquaculture) now cover one-quarter of Earth's terrestrial surface.
- Approximately 20 per cent of the world's coral reefs were lost and an additional 20 per cent degraded in the last several decades of the twentieth century, and approximately 35 per cent of mangrove area was lost during this time.
- The amount of water impounded behind dams has quadrupled since 1960, and reservoirs hold three to six times as much water as natural rivers. Water withdrawals from rivers and lakes have doubled since 1960; most water use (70 per cent worldwide) is for agriculture.
- Since 1960, flows of reactive (biologically available) nitrogen in terrestrial ecosystems have doubled, and flows of phosphorus have tripled. More than half of all the synthetic nitrogen fertilizer (which was first manufactured in 1913) ever used on the planet has been used since 1985.
- Since 1750, the atmospheric concentration of carbon dioxide has increased by about 32 per cent (from about 280 to 376 ppm in 2003), primarily because of the combustion of fossil fuels and land-use changes. Approximately 60 per cent of that increase (60 ppm) has taken place since 1959.
- More than two-thirds of the area of two of the world's 14 major terrestrial biomes and more than half of the area of four other biomes had been converted by 1990, primarily to agriculture.
- Across a range of taxonomic groups, either the population size or range, or both, of the majority of species is currently declining.
- The distribution of species on Earth is becoming more homogeneous; in other words, the set of species in any one region of the world is becoming more similar to the set in other regions, primarily as a result of introductions of species, both intentionally and inadvertently in association with increased travel and shipping.
- The number of species on the planet is declining. Over the past few hundred years, humans have increased the species extinction rate by as much as 1,000 times over background rates typical over the planet's history. Some 10 to 30 per cent of mammal, bird, and amphibian species are threatened with extinction. Freshwater ecosystems tend to have the highest proportion of species threatened with extinction.
- Genetic diversity has declined globally, particularly among cultivated species.
- Most changes to ecosystems have been made to meet a dramatic growth in the demand for food, water, timber, fibre, and fuel. Between 1960 and 2000, the demand for ecosystem services grew significantly as world population doubled to 6 billion people and the global economy increased more than sixfold.
- Food production increased by roughly two and a half times, water use doubled, wood harvests for pulp and paper production tripled, installed hydroelectric capacity doubled, and timber production increased by more than half.

Source: Millennium Ecosystem Assessment (2005).

Environment in Focus

Box 1.3 Population and Exponential Growth

Population change is a result of the interaction between births and deaths. **The crude birth rate (CBR)** minus the **crude death rate (CDR)** will yield the **crude growth rate (CGR)**, all usually expressed as per thousand of the population per year. In this way, populations of different countries, regardless of their size, can be compared. The figures are known as 'crude' because they give no insights into factors such as age and sex ratios, figures that are very important for understanding future potential growth. If CBR and CDR are equal, a zero population growth will result if the effects of migration are excluded.

In 1798, a British clergyman, Thomas Malthus, pointed out that population growth was geometric or **exponential** in nature (i.e., 2, 4, 8, 16, 32, 64, and so on), whereas the growth in food supply was arithmetic (i.e., 1, 2, 3, 4, 5, and so on). This phenomenon would inevitably, said Malthus, lead to famine, disease, and war. Such a viewpoint was not popular in his day when population growth was considered very beneficial. For many years, the Malthusian view was ignored. The opening up of new lands for cultivation in North America and the southern hemisphere and later the development of Green Revolution techniques (Chapter 10) allowed food supplies to increase rapidly.

Increasing numbers of experts, watching the decline in food supplies per capita over the past few years (see Chapter 10) and the increase in population, particularly in less developed countries, now feel that the Malthusian spectre is quite real. Currently, 80 million people are being added every year to the population in less developed countries, compared to about 1.6 million in more developed countries. Figure 1.2 illustrates how global population has grown over the centuries and millennia.

On the other hand, some pundits, particularly economists, feel that more population simply furnishes more resources—human resources—upon which to build increases in wealth for the future. Indeed, there are concerns that some developed countries will start losing population in the future and that this will have a negative impact on their economies. For example, Japan is predicted to lose 20 per cent of its population by 2050, with declines also expected to take place in Germany, Russia, and Italy. In both the US and Canada, the trend is predicted to move in the opposite direction, largely as a result of immigration. By 2050, it is estimated that Canada's population will have increased by 14 per cent to 37 million through immigration, despite a fertility rate of 1.5 children per couple.

Political leaders in some of the less developed countries experiencing the most rapid population growth rates have argued that population growth per se is not a problem and that the main problem is overconsumption in the more developed countries. This distributive concern is echoed by women's groups—also wary of coercive birth control programs—who think that most progress can be achieved by improving the status of women. Women with more education have smaller, healthier families, and their children have a better chance of making it out of poverty. Yet two-thirds of the world's 876 million people who can neither read nor write are women, and a majority of the 115 million children not attending school are girls. Women who have the choice of delaying marriage and child-bearing past their teens also have fewer children than teen brides. Yet more than 100 million girls will be married before their eighteenth birthday during the next decade.

However, starting with the landmark 1994 International Conference on Population and Development (ICPD) at which 179 governments adopted a forward-looking, 20-year Program of Action, remarkable progress has been made on achieving consensus on approaches to population control. The ICPD Program of Action, sometimes referred to as the Cairo Consensus, recognized that reproductive health and rights, as well as women's empowerment and gender equality, are cornerstones of population and development programs. Furthermore, at the 2005 World Summit, the largest-ever gathering of world leaders reaffirmed the need to keep gender equality, HIV/AIDS, and reproductive health at the top of the development agenda.

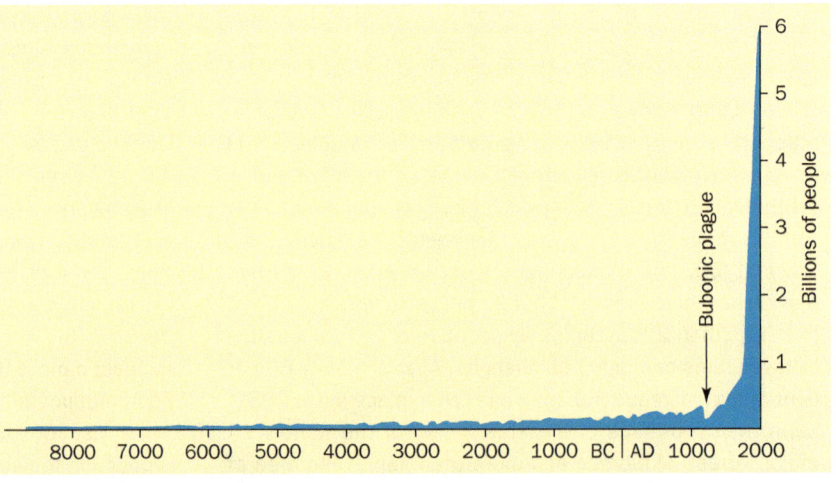

Figure 1.2 | The growth of human population over time.

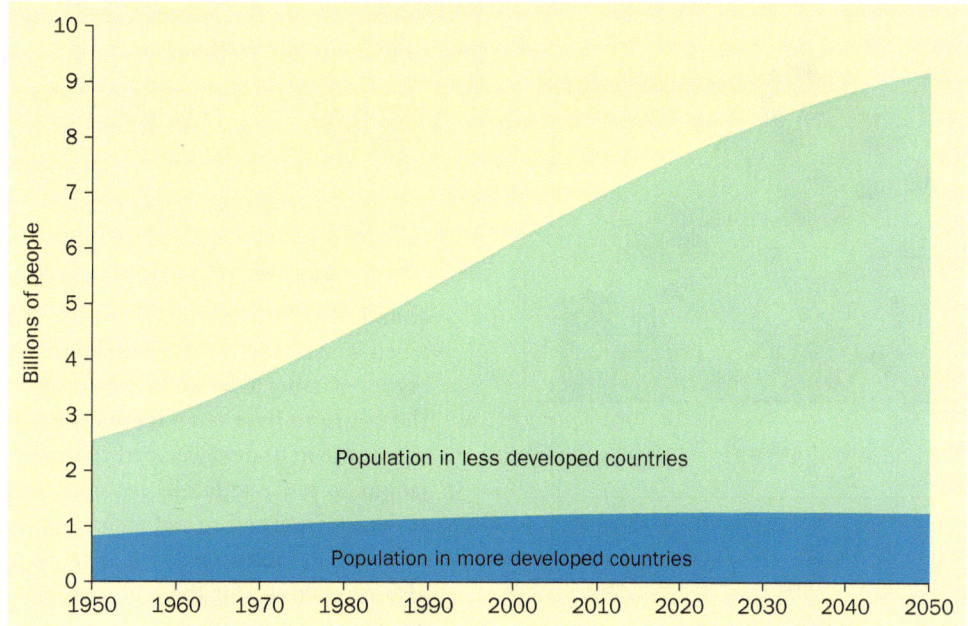

Figure 1.3 | World population growth and projections, 1950–2050. Source: *UN Population Division (2007b)*.

Population age structure is usually represented as a population pyramid (Figure 1.4), and the shape gives information about birth and death rates as well as life expectancy. It also reveals the number of dependants. Two groups are recognized: young dependants (aged below 15) and elderly dependants (those over 65). Dependants rely on the economically active for economic support. Many less developed countries have a high number of young dependants, while many developed countries have a growing number of elderly dependants.

A population with a high number of young dependants and a low life expectancy has a very triangular pyramid. A population with a falling birth rate and rising life expectancy is reflected in a population pyramid with fairly straight sides. Generally, the shape of a country's population pyramid changes from a triangular to a barrel-like shape with straighter edges as the country develops and passes through the demographic transition, discussed below. In fact, places experiencing an aging population and a very low birth rate may have a population structure that looks a little like an upside-down pyramid.

Human reproductive age is usually between 15 and 44 years of age. The larger the number of people in this age bracket, the greater the potential for population increase. **Total fertility rates** represent the average number of children each woman has over her lifetime. If the fertility rate is 2.0, then theoretically this will lead to stable populations as children replace their parents. If the rate is higher than 2.0, it will lead to population growth; if less, the population declines. However, because of infant mortality, particularly in less developed countries, the **replacement-level fertility** is calculated as higher than 2.0.

The largest generation in history—1.2 billion people—is now between the ages of 10 and 19, and the child-bearing decisions by this generation will be critical. If the world's women have on average just half a child more than predicted, then the population would swell to 15.8 billion by the end of the century. And although the average annual global population growth rate has fallen from more than 2 per cent, as it was from the 1950s to the 1990s, to less than 1.3 per cent today, that rate is being applied to a much larger and still increasing population. Furthermore, after early declines in fertility rates, the rates in many countries have now reached a plateau (Figure 1.5).

Women's group in Haryana
Empowering and educating women is one of the major steps towards lowering birth rates in many developing countries. Women's groups in some parts of India, such as here in Haryana, are having a major role in this transformation.

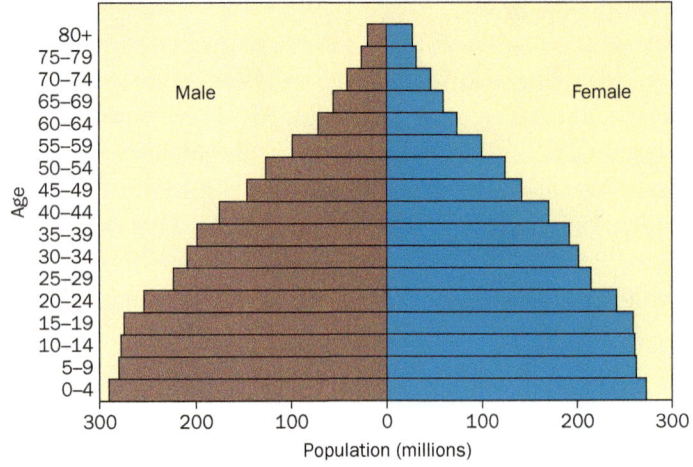

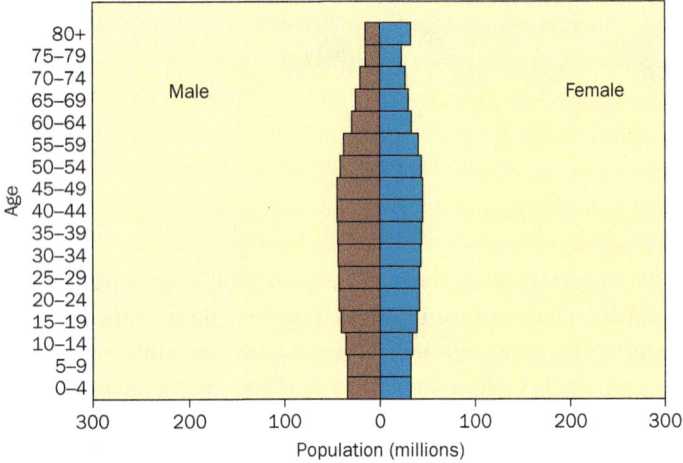

Figure 1.4 | Population pyramids for developing (top) and developed (bottom) countries. *Source: UN Population Division (2007b).*

Another important consideration is the speed of the **demographic transition** in each country. Demographers, those who study population structure and growth, have noted a relationship between economic growth and population that occurs in four main phases (Figure 1.6):

1. *High equilibrium.* Both death and birth rates are high, resulting in very little population growth. This situation usually occurs in pre-industrial societies.
2. *High expanding.* Advances in health care result in declining mortality rates but show no concomitant decrease in birth rates, leading to high population growth. This situation occurs in the early stages of industrialization when some benefits of technology and industrial society are starting to be felt but are insufficient to outweigh the desire to have large families. Large families are an advantage in underdeveloped countries, providing more labour to generate family income. Lacking the pension systems of more advanced societies, parents need to have someone to look after them as they age. Large families also compensate for the high rate of child mortality in pre-industrial societies.
3. *Low expanding.* Birth rates start to fall as the benefits of increased income begin to erode the advantages of having large families. In Western societies, where the cost of raising children is high, having large families is no longer an overall benefit.
4. *Low equilibrium.* Birth rates and death rates are in balance as a result of the decline in birth rates.

As with most simplified models, the model can be criticized because of its generality, largely based on European experience, and because it does not take a full range of cultural factors into account. Moreover, a fifth phase to the model is emerging in some nations, illustrated in Figure 1.6, as total populations fall.

Historically, the decline in death rates (the **epidemiological transition**) that occurred in most developed countries was relatively slow. Discoveries about the causes of disease and how they could be countered developed in conjunction with growing interest and investments in science. For example,

Children
What the future holds in terms of population growth depends on the reproductive decisions taken by today's children, such as these Maasai children in Loliondo, Tanzania, as they grow up.

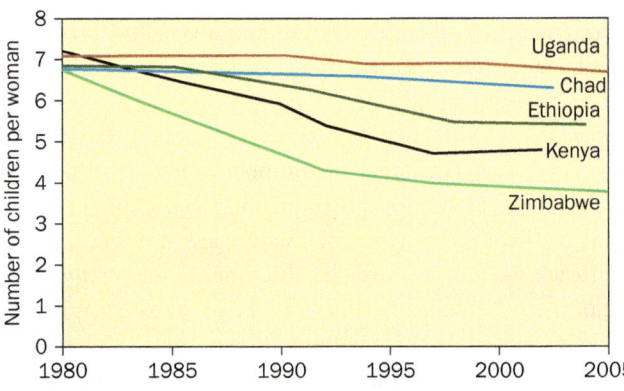

Figure 1.5 | Fertility trends in sub-Saharan Africa, 1980–2005. *Source: UN Population Division and national demographic and health surveys.*

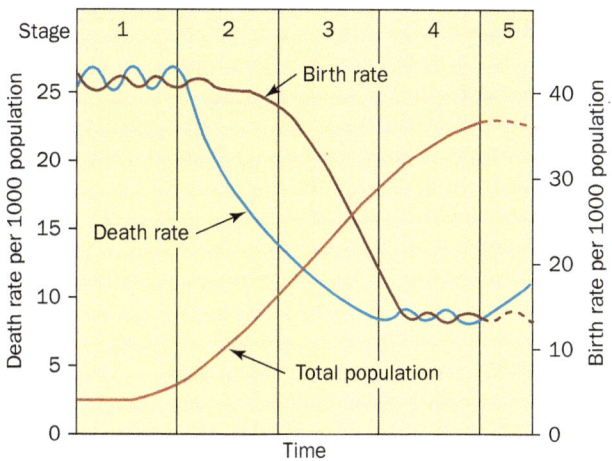

Figure 1.6 | The demographic transition.

Chapter 1 | Environment, Resources, and Society

is valued for both sexes, and there has been a latent demand for effective contraception and effective means to distribute contraceptive devices. These conditions often do not exist in many countries.

Recognition of the importance of the demographic transition in stabilizing population growth and of the role of economic development as a main driving force behind the transition was also a factor in the drive to industrialize the world promoted by many global organizations, such as the World Bank and the United Nations. Furthermore, a strong relationship was seen between some indicators of environmental degradation and economic growth. As economic growth increases, so does environmental degradation—until a threshold is reached. After that point, so it has been claimed, the wealth generated by increased industrial activity is sufficient to pay for environmental services (e.g., pollution control) that poorer countries cannot afford. Thus, the goal of development planning was to help countries reach that threshold so that they could enjoy the benefits of increased wealth while not succumbing to environmental degradation.

Unfortunately, the relationship among society, stages of development, and economic growth is more complex than these models allow. In particular, they ignore one of the most (and many scientists would argue the single most) important factor—the impact of consumption on the capability of planetary ecosystems to continue to provide life support services.

in the late 1800s, Louis Pasteur and others discovered the main infectious agents and the means by which they were transmitted. Vaccines were developed, and whole populations became immune to diseases such as typhoid and smallpox. By the 1930s, antibiotics such as penicillin were being developed. These medications led to cures for many other ailments. Sanitation improved, as did nutrition. However, these developments took time, and the decline in crude death rates was gradual in the developed world. In contrast, these innovations were made available in many less developed countries all at once, often leading to a precipitous decline in death rates without a corresponding decrease in birth rates.

Perspectives on the Environment

Life Expectancy Increases

The average life expectancy at birth in less developed countries rose from 41 years in 1950 to 66 years in 2007. The Middle East and North Africa region has experienced the largest increase in life expectancy since the late 1950s: from 43 years to 70 years. Since 1950, the greatest gains in life expectancy at birth occurred among women. In more developed countries, average life expectancy for women rose from 69 years in 1950 to 80 years in 2007, while the average for men rose from 64 years to 73 years.

—Population Reference Bureau (2007)

Some countries travel through this sequence more quickly than others, with rapid economic development followed by corresponding adjustments in birth rates. Thailand is a prime example, where a fertility rate of 6.4 in 1960 fell to 1.8 in 2009, with economists forecasting further reductions, leading to a future labour shortage. However, not all countries adjust as rapidly as Thailand. In Thailand, a Buddhist country, there are no religious obstacles to reducing family size, women play a major role in household decision-making, education

Consumption

The Earth's passengers do not all have the same impact on the life support system. Some passengers—those in first class—get special meals, three times a day, wine included; those in the economy section are lucky if they get one meal and must buy their own water, if it is even available. The richest 20 per cent of the world's population are responsible for more than 75 per cent of world consumption, while the poorest 20 per cent consume less than 2 per cent (World Bank, 2008). In terms of

Although much is being done throughout the world to curb population growth, as seen in this signboard in Vietnam encouraging couples to have only one child, consumption knows no bounds, and we are constantly exhorted to buy more.

metal use, for example, the 15 per cent of the global population living in the US, Canada, Japan, Australia, and Western Europe account for 61 per cent of aluminum use, 60 per cent of lead, 59 per cent of copper, and 49 per cent of steel. The average North American uses 22 kilograms of aluminum a year, the average African less than one kilogram. Rural populations are significantly poorer than urban populations, and one of the main drivers of future consumption will be the increased urbanization of global populations (Chapter 13).

Perspectives on the Environment
Urbanization

In 2008, the world reached an invisible but momentous milestone: for the first time in history, more than half its human population, 3.3 billion people, live in urban areas. By 2030, this number is expected to swell to almost 5 billion. Many of the new urbanites will be poor. Their future, the future of cities in developing countries, the future of humanity itself all depend very much on decisions made now.

—United Nations Population Fund (2007b)

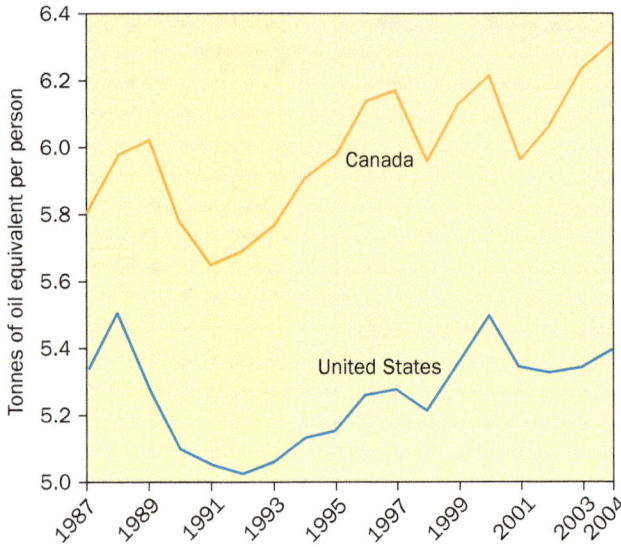

Figure 1.7 | Per capita energy consumption of Canada and the US, 1987–2004. Source: GEO data portal, compiled from IEA (2007) and UNPD (2007).

Energy consumption is also very unequally distributed, with the people in the wealthiest countries using 25 times more per capita than the world's poorest people. More than a third of the global population does not have access to electricity, but demands are growing. There are also large differences in energy consumption among developed countries, with the average Canadian and American consuming 2.4 times as much energy as the average person in Western Europe.

Canadians are among the top per capita consumers of energy in the world, with an even larger consumption rate than Americans (Figure 1.7). Each Canadian consumes as much energy as 60 Cambodians. Government policies encourage us to be wasteful by subsidizing energy production, and we as individuals normally do not resist. Energy is a good index of our planetary impact, reflecting our ability to process materials and disrupt the environment through pollution such as acid precipitation (Chapter 4) and the production of greenhouse gases (Chapter 7). However, as shown in Figure 1.8, there is no direct relationship between electricity consumption and human development. In other words, it is possible to have high standards of living without excessive energy consumption, as exemplified by many European countries. Canada has yet to make this transition.

Obviously, very different kinds of passengers share our planet, and the differences among them have grown rather than diminished as a result of increased wealth over the past 20 years. **Gross national product** (GNP) is an index used by economists to compare the market value of all goods and services produced for final consumption in an economy during one year. Over the past two decades, the planetary GNP has risen by $47 trillion, but only 15 per cent of this increase has trickled down to the 80 per cent of the passengers in the economy section of the spaceship. The rest has made the rich even richer.

Global poverty is a major challenge to planetary survival. Living in extreme poverty (less than $1 a day) means not being able to afford the most basic necessities. An estimated 8 million people a year die from absolute poverty. Moderate poverty, defined as earning about $1 to $2 a day, enables households to just barely meet their basic needs, but still forgo many of the things—education, health care—that many others take for granted. The smallest misfortune (e.g., a health issue, job loss) threatens survival.

However, progress is being made. Extreme poverty in developing countries fell from 46 per cent in 1990 to 27 per cent in 2005. Over the same period, however, the number of people in developing countries grew by 20 per cent to more than 5 billion, meaning 1.9 billion people live in extreme poverty. The economic downturn in 2010–11 is estimated to have put an additional 64 million people into extreme poverty. If developing countries can return to the previous rate of poverty eradication, it is estimated that less than 920 million people would be left in extreme poverty by 2015. These rates are led by a sustained period of economic growth in China and increasing growth in India. However, such success is not universal. In sub-Saharan Africa, the number of poor people has increased by a third.

Implications

The stresses on the planetary life support system are a consequence of overconsumption and the resulting pollution, as well as overpopulation and the resulting poverty. Together,

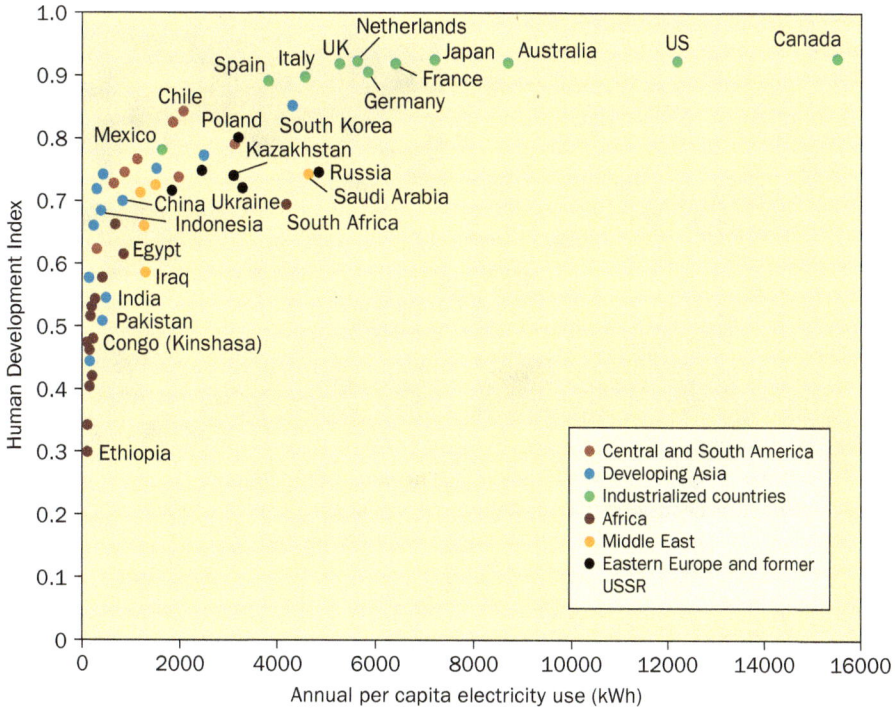

Figure 1.8 | The relationship between energy use and the Human Development Index. Source: UNDP (2005).

International Guest Statement
The Resource Dilemma: Environment, Livelihood, and Poverty Nexus in Developing Countries
Peter O. Adeniyi

Things have changed. The rains are not coming often and on time now, the land is no longer productive, nobody is available to take over our cocoa farms, our young men have all gone to the cities, and there is just nobody to help us. That's the story from a remote village in western Nigeria. You hear how in the past the village was green, with forests adorning the rolling hills surrounding the village. Then came the realization that the trees, some of which have been there over a hundred years, are articles of trade. One at a time the trees are ferried to sawmills (see photo, next page). No effort is made to replace them.

Indeed, things have changed. How could the poor villagers have realized that the rational economic behaviour of logging the trees could partly be responsible for the poor rain now being experienced in an environment where the local climate is controlled by meso-scale processes that depend on local forcing, especially vegetation? Nobody ever told them that. But would the villagers have kept the trees that provided 'just some additional income' for the trees' sake? This situation exemplifies the conflict among environment, development, and livelihood. Would or could the people survive at the expense of the environment, or should the environment be preserved from being exploited to meet human needs? This issue is one for which sustainable development does not provide much answer, but which the concept of sustainable livelihoods may resolve.

Oil exploitation started in Nigeria's Niger Delta during the late 1950s, and oil production presently accounts for over 90 per cent of export earnings and 80 per cent of government revenue. Yet, nowhere in the country is poverty as pervasive as in the Niger Delta. The creek dweller and wetland inhabitant who should have mattered the most in the profit-and-loss equation are excluded right from onset. Basic facilities are not available in rural villages, and a significant part of otherwise productive land and water is degraded, thereby compromising livelihoods. This situation is similar in another dimension to the Sydney Tar Ponds problem in that a science-based approach to management of resources was never fully implemented.

I think the most significant problem in most developing countries is not environment, but rather the poverty arising from poor leadership at all levels; poor institutions (where they exist); poor, selfish, and overly centralized legislative

Timber logs ferried to a sawmill in Ijebu-Igbo, Ogun State, southwest Nigeria.

The yam, the most popular staple root crop in Nigeria, may be under threat from climate change. Photo taken at Ilasa, Ekiti State, Nigeria.

mandates/administrative policies; and the arrogance of, and the disrespect for, indigenous knowledge by the elite. If sustainable livelihoods can be facilitated, then, maybe, we may have to worry less about the environment. But on the other hand, livelihoods for large populations are also tied to natural resource systems, creating a complex interconnectivity among people, livelihoods, environment, and poverty. Such interconnectivity also increases vulnerability to compounding and creeping disasters, including climate change effects such as drought, food insecurity, flooding, and land degradation.

In most developing countries, the vigorous young men and women are moving in large numbers to the cities, where conditions continue to deteriorate, in search of the perceived 'good life'. Government intervention in provision of basic amenities, including health and sanitation, in the face of poor planning cannot cope with the influx. Millions of school-age children are on the street, house yards and streets are treeless, and a 'green environment' is perceived as something from a past time and/or fit only today for the rich who have reached the 'good life'. Developmental efforts are hardly knowledge-based and science-driven, except where anchored by global organizations. The result is disorder in virtually every sector.

Whether the Sydney Tar Ponds, declining productivity in remote village of western Nigeria, or pollution and poverty in the Niger Delta, attitudes and approaches to environmental issues evolve and what may be viewed as appropriate behaviour at one time may be viewed as inappropriate later. The nexus among people, development, and environment is unassailable in all climates. A recurring theme is either institutional failure to act when necessary, which is common in developed countries, or the lack of institutional and legislative frameworks and unavailability of requisite data to address these issues in developing countries. Most developing countries do not even have accurate estimates of their resource stock. Because governance is weak and non-responsive, gains from resource exploitation are rarely translated effectively into physical and human development. This disconnect, pervading most resource-rich countries of Africa, seems to be the bane of our development.

Peter Olufemi Adeniyi, Ph.D., DES, is a Professor of Geography, University of Lagos, Nigeria, where he teaches and researches into the applications of Remote Sensing and Geographic Information Systems (GIS) for resource and environmental management, land-use planning, and change detection.

they create pressure on the **planetary carrying capacity** at all different scales. Although in the past many cultures violated the carrying capacities of their local environments with dire results, never before have we approached these limits at a global scale.

Clear evidence indicates that critical thresholds are being reached and surpassed. Rockström and his colleagues (2009) have looked at the scale of these changes in what has come to be known as the 'Anthropocene', the geological age when humanity drives most global biophysical processes. They propose that nine main planetary processes need to be taken into account (Figure 1.9). Change in these is often non-linear, and when certain thresholds are crossed there may be sudden and irreversible change with enormous consequences for the Earth as the home of humanity. Their work identifies thresholds for each of climate change; rate of biodiversity loss (terrestrial and marine); interference with the nitrogen and phosphorus cycles; stratospheric ozone depletion; ocean acidification; global freshwater use; change in land use; chemical pollution; and atmospheric aerosol loading.

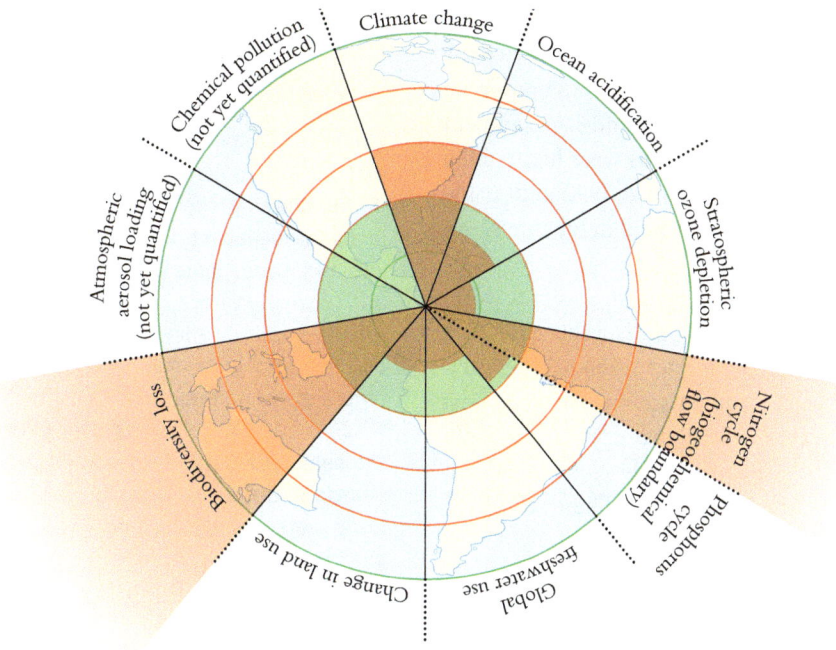

Figure 1.9 | Beyond the boundary. The inner green shading represents the proposed safe operating space for nine planetary systems. The red wedges represent an estimate of the current position for each variable. The boundaries in three systems (rate of biodiversity loss, climate change, and human interference with the nitrogen cycle) have already been exceeded. *Source: Rockström et al. (2009: 427).*

Three of these system processes have already exceeded the safe operating zones: rate of biodiversity loss (see Chapter 14), climate change (see Chapter 7), and interference with the nitrogen cycle (see Chapter 4), and the implications of these changes are discussed in more detail in the chapters indicated. What is interesting about this approach is the perspective of setting scientifically determined biophysical preconditions for human development and the need to stay within those boundaries. Violating these boundaries will result in a noted loss in the resilience of the Earth in its ability to produce the goods and services necessary to support humanity.

Jared Diamond, a geographer at UCLA, has written a fascinating book on why past societies collapsed and what we can learn from their experiences. Appropriately, the book is titled *Collapse* (2005), and we strongly recommend it. Diamond suggests that four main reasons explain why societies fail to make corrections to prevent societal collapse. They may not anticipate the problem; they may fail to appreciate the severity of the problem even though they are aware of it; they may appreciate the problem but neglect to address it; and they may perceive the problem as a serious threat, try to solve it, and fail. Diamond explores examples of all these situations. However, his main message is to alert us to how we can forestall such a collapse in modern society. At the heart of Diamond's analysis is the role of environmental degradation in causing societal collapse. One of his most sobering conclusions is that many of the societies that collapsed were very successful and collapse seemed impossible, yet it happened with frightening rapidity.

There have been many warnings about the impact of environmental degradation on society in the future, spanning

Families in Lesotho are still large, but the planetary impact of this entire family will be a fraction of that of one Canadian child.

back to Rachel Carson's *Silent Spring* (1962) of a half-century ago and including the famous *Limits to Growth* study of the early 1970s (Meadows et al., 1972) through to the report of the World Commission on Environment and Development in 1987 (WCED, 1987). Some of the trends were highlighted at the Earth Summit in Rio de Janeiro in 1992 and at the World Summit on Sustainable Development in Johannesburg 10 years later.

In 2000, at the United Nations Millennium Summit, world leaders agreed to a set of time-bound, measurable goals and targets for combatting poverty, hunger, disease, illiteracy, environmental degradation, and discrimination against women. Placed at the heart of the global agenda, the eight objectives, now called the Millennium Development Goals (MDGs), aim to improve human well-being. These goals are:

1. Eradicate extreme poverty and hunger
2. Achieve universal primary education
3. Promote greater gender equality and empower women
4. Reduce child mortality
5. Improve maternal health
6. Combat HIV/AIDS, malaria, and other diseases
7. Ensure environmental sustainability
8. Develop a global partnership for development

Under each of the MDGs, countries have agreed to targets to be achieved by 2015. Many of the regions facing the greatest challenges in achieving these targets coincide with regions facing the greatest problems of ecosystem degradation. At the 62nd General Assembly of the United Nations in 2007, the goals, targets, and indicators were revised. Table 1.1 lists the targets and indicators for goal 7, ensuring environmental sustainability. Progress is summarized in Chapter 15, while the use of indicators is discussed in greater detail later in the chapter.

The evidence for and causes of global climate change are discussed in more detail in Chapter 7, along with Canada's response. Canada has the second highest per capita emissions of greenhouse gases in the world. Canada agreed under the Kyoto Protocol to target a 6 per cent cut in emissions by 2012. Instead, emissions have increased by 29 per cent, putting the country 35 per cent above its Kyoto target. While greenhouse gas intensity has fallen, efficiency gains have been swamped by an increase in emissions from an expansion in oil and gas production. Net emissions associated with oil and gas exports have more than doubled since 1990. This compares with the UK's reduction of 23 per cent over the same time period and Germany's reduction of 22 per cent.

Scientists have suggested that an overall increase in temperatures in excess of 2°C will cause runaway environmental damage and that emissions need to be kept under 450 ppm Ce (carbon dioxide emission equivalents) to avoid this consequence. More recent modelling suggests that 2°C was

Perspectives on the Environment
The Imperative for Action on Climate Change

What we do today about climate change has consequences that will last a century or more. The part of that change that is due to greenhouse gas emissions is not reversible in the foreseeable future. The heat-trapping gases we send into the atmosphere in 2008 will stay there until 2108 and beyond. We are therefore making choices today that will affect our own lives but even more so the lives of our children and grandchildren. This makes climate change different and more difficult than other policy challenges.

. . . Even if we were living in a world where all people had the same standard of living and were impacted by climate change in the same way, we would still have to act. If the world were a single country, with its citizens all enjoying similar income levels and all exposed more or less to the same effects of climate change, the threat of global warming could still lead to substantial damage to human well-being and prosperity by the end of this century.

In reality, the world is a heterogeneous place: people have unequal incomes and wealth, and climate change will affect regions very differently. This is, for us, the most compelling reason to act rapidly. Climate change is already starting to affect some of the poorest and most vulnerable communities around the world. A worldwide average 3° centigrade increase (compared to pre-industrial temperatures) over the coming decades would result in a range of localized increases that could reach twice as high in some locations. The effect that increased droughts, extreme weather events, tropical storms, and sea level rises will have on large parts of Africa, on many small island states and coastal zones will be inflicted in our lifetimes. In terms of aggregate world GDP, these short-term effects may not be large. But for some of the world's poorest people, the consequences could be apocalyptic. In the long run, climate change is a massive threat to human development, and in some places it is already undermining the international community's efforts to reduce extreme poverty.

—UNDP (2007)

an optimistic threshold, with 1.5°C being more realistic. Nonetheless, to put emissions in perspective, of the overall planetary carrying capacity and using the suggested annual allowable emission ceiling of 14.5 Gt CO_2, if emissions were frozen at the current level of 29 Gt CO_2 to stay below the threshold, we would need two planets. However, emissions are not equally distributed. In Ethiopia, for example, the average per capita carbon footprint is 0.1 tonnes, compared to 20 tonnes in Canada. The per capita increase in emissions since 1990 for the United States (1.6 tonnes) is higher than the total per capita emissions for India in 2004 alone (1.2 tonnes). The overall increase in emissions from the United States exceeds sub-Saharan Africa's total emissions. If every person

Table 1.1 | Ensuring Environmental Sustainability: MDG Targets and Indicators

Targets	Indicators
Integrate the principles of sustainable development into country policies and programs and reverse the loss of environmental resources	• Proportion of land area covered by forest • Carbon dioxide emissions, total, per capita and per $1 GDP (PPP) • Consumption of ozone-depleting substances • Proportion of fish stocks within safe biological limits • Proportion of total water resources used
Reduce biodiversity loss, achieving, by 2010, a significant reduction in the rate of loss	• Proportion of terrestrial and marine areas protected • Proportion of species threatened with extinction
Halve, by 2015, the proportion of people without sustainable access to safe drinking water and basic sanitation	• Proportion of population using an improved drinking water source • Proportion of population using an improved sanitation facility
By 2020, to have achieved a significant improvement in the lives of at least 100 million slum dwellers	• Proportion of urban population living in slums

Source: UNs Millennium Development Goals.

living in the developing world had the same carbon footprint as the average for high-income countries, global CO_2 emissions would rise to 85 Gt CO_2, a level that would require six planets. With a global per capita footprint at Australian levels, we would need seven planets, rising to nine for a world with Canadian and US levels of per capita emissions. As a global community, we are running up a large and unsustainable carbon debt, but the bulk of that debt has been accumulated by the world's richest countries.

In terms of addressing the situation, a global perspective is obviously necessary, although responsibilities clearly differ. For example, a 4 per cent cut would be generated in global emissions if a 50 per cent cut were initiated in CO_2 emissions for South Asia and sub-Saharan Africa. A similar cut in high-income countries would reduce emissions by 20 per cent. Equity must also play a role. An average air-conditioning unit in Florida emits more CO_2 in a year than a person in Afghanistan or Cambodia is responsible for during his or her lifetime. A European dishwasher emits as much CO_2 in a year as the total carbon footprint of three Ethiopians. What kind of role do you think Canada and Canadians should assume to shoulder their share of global responsibility?

There are other important global perspectives on Canada that you should bear in mind. Our land is vast, about 13 million square kilometres, and our population small, about 34.2 million people (Figure 1.10). Population density is 0.04 people per hectare, compared to Bangladesh at eight people per hectare. Canada would have to have a population of more than 8 billion to equal this density. Immigration, rather than natural increase, is the most important factor in population growth in Canada. Migrants comprised 21 per cent of the total Canadian population by 2010. Are more people good for Canada when we consider that there is more to Canada than just the economy? Does Canada have a moral obligation to accept migrants from overcrowded countries elsewhere or from countries generating 'environmental refugees' as a result of environmental stress in their home countries? These are some of the important questions that policy-makers—and you—must consider.

In terms of numbers alone, Canada is not overpopulated compared to virtually any other country. Canada is the second largest country in the world in terms of area and includes 20 per cent of the world's wilderness, 24 per cent of its wetlands, 10 per cent of its forests, and 7 per cent of its renewable fresh water and has the longest coastline in the world. However, as discussed above, it is not simply numbers of people but rather the impact of those people that is critical. Canadians are among the world's top producers per capita

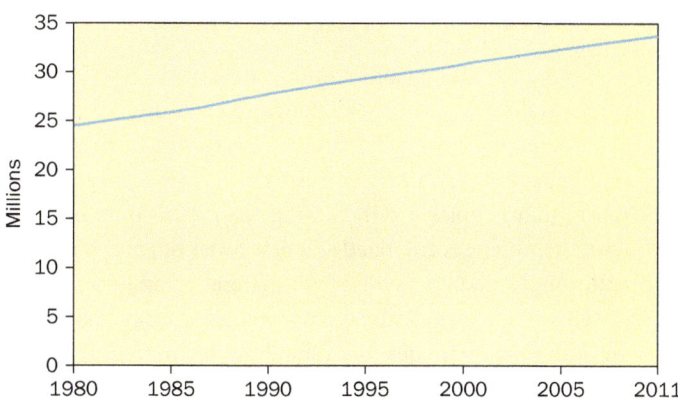

Figure 1.10 | Population growth in Canada, 1980–2011. Source: Statistics Canada, CANSIM database, Table 051-0001, at: cansim2.statcan.gc.ca.

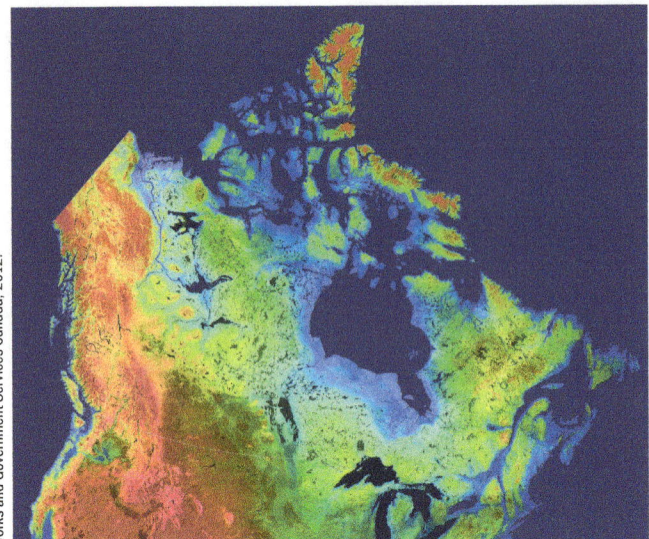

Satellite image of Canada (processed at the Canada Centre for Remote Sensing, Earth Sciences Sector, Natural Resources Canada).

of industrial and household garbage, hazardous wastes, and greenhouse gases (Box 1.4). Some point to the size of the country, the cold in winter, and the heat in summer as the reasons behind our remarkable energy consumption, but it is clear that Canadians can contribute substantially to reducing impacts on the planetary life support system. We offer suggestions on some of the ways that you can help contribute to the needed changes throughout the text, and we return to this theme in Chapter 15.

Changes in environmental directions for Canada require changes in policies and legislation and the strict implementation of those changes. It is therefore important for you to appreciate the jurisdictional arrangements for environmental management in Canada, the topic of the next section.

Jurisdictional Arrangements for Environmental Management in Canada

Under the Canadian Constitution, authority or responsibility for natural resources and the environment is divided between the federal and provincial governments, with territorial and municipal governments increasingly having a role. As well, Aboriginal peoples are increasing their role commensurate with their being recognized as a new order of government. In addition, Canada is involved in bilateral arrangements with the United States to address environmental problems such as air pollution and to deal with shared water bodies such as the Great Lakes and the Columbia River, as well as in multilateral arrangements with other nations or international organizations regarding resources such as fisheries, migratory birds and animals, and minerals on or under the ocean floor.

Federal, Provincial, and Municipal Roles

Canada is a federated state, with power and authority shared between federal and provincial governments and with municipal governments receiving their power and authority from provincial legislatures. Ownership and control of all Crown lands and natural resources not specifically in private ownership is given to the provinces, under section 92A of the Constitution Act, 1867, except for the Canadian North (north of 60 degrees latitude), where the federal government has proprietary rights to land and resources until the territories receive such power, and for resources found on or under seabeds off the coasts of Canada (some provinces, however, have challenged this right).

Legislative authority is mixed between the federal and provincial governments and often becomes a significant source of conflict. The federal government has jurisdiction over trade and commerce, giving it substantial authority over both interprovincial and export trading of resources (oil and natural gas, water). Alberta, Saskatchewan, and British Columbia, which have oil and natural gas, often object to the federal government becoming involved in setting prices and determining buyers, arguing that such matters are within provincial authority because of their responsibility for property and civil rights, as well as because these resources are under provincial control. The federal government has used its legislative authority for navigation and shipping and for fisheries to create water pollution regulations—even though water within provinces falls within the jurisdiction of the provinces. Thus, ambiguities and inconsistencies exist regarding jurisdiction over resources and the environment. One consequence is that it has been difficult to establish *national approaches* (combined federal and provincial) to deal with resource and environmental issues.

In the early to mid-1990s, many provincial governments began to download selected responsibilities, which they had traditionally held, to municipalities. The provinces argued that downloading was consistent with the principle of **subsidiarity**, which stipulates that decisions should be taken at the level closest to where consequences are most noticeable. While such an argument is rational, the primary motive for downloading often was the desire of provincial governments to shift the cost of many responsibilities to lower levels of government to reduce provincial debts and deficits. Whatever the motivation, the outcome has been that municipalities have become much more significant players in natural resource and environmental management, since in many instances provinces have withdrawn from related management activities.

Effective partnerships exist between provincial and municipal governments. Among the best and most enduring examples are the Ontario conservation authorities, watershed-based organizations established by statute in 1946 to manage many renewable resources within river basins. Individual

Environment in Focus
Box 1.4 Canada Facts

- We generate about 383 kilograms of solid waste per capita per year, ranking seventh in the world.
- We generate almost six tonnes of hazardous waste for each US$1 million of goods and services produced; Japan generates less than a quarter of a tonne.
- We have one of the highest per capita uses of water in the world, about 15 cubic metres per day, roughly three times that of Sweden and Japan.
- We are among the highest per capita energy consumers in the world.
- We use our cars nearly 10 per cent more than residents of other industrialized countries.
- We emit 2 per cent of the world's greenhouse gases with 0.5 per cent of the world's population and rank second in global production of greenhouse gases per capita.

---

authorities were established when two or more municipalities in a watershed petitioned the provincial government to establish one. When a majority of the municipalities in a watershed agreed that they would work collaboratively, the province established a conservation authority. Today, 36 authorities exist, primarily in the more settled parts of the province.

While the provincial government would not impose a conservation authority, it provided a strong incentive for local governments to form one by offering funds not available to municipalities on their own but available after a conservation authority was established. This cost-sharing arrangement was a powerful stimulus for municipalities to agree to establish authorities, and for many years the cost-sharing was 50/50 between the province and the municipalities. However, in the mid-1990s, the Ontario government significantly reduced its proportion of the funding to the conservation authorities as part of a drive to reduce government activities and costs. In Chapter 11, Barbara Veale's guest statement provides more insight about conservation authorities.

Another challenge in the reallocation of responsibility to municipal governments is the variable degree of competence to deal with resource and environmental matters. There is significant variation among municipalities and other local-level governments in technical expertise and required data, funding capability, leadership, community awareness and engagement, and ability to implement, monitor, and enforce solutions related to resource and environmental management. Often, the provincial governments have downloaded responsibility to local governments without having determined whether they had the necessary competence to take it on.

Monitoring Progress towards Sustainable Development

In 1997, the Office of the Auditor General began reporting on progress by 24 federal government departments and agencies regarding sustainable development. The Commissioner of the Environment and Sustainable Development observed in his 1997 report that Canadians expected governments to provide strong leadership and a clear vision and to lead by example through fostering a culture of environmental protection and sustainable development within federal organizations. The Commissioner also stated that few quick solutions were available for environmental problems and that progress would require persistence, patience, and focused effort. The challenge was characterized as a 'long journey' requiring systematic change if a real difference were to be realized for present and future generations.

Looking to the future, the Commissioner wrote that many environmental problems and sustainable development issues were 'difficult to manage' as well as 'scientifically complex' and involved long time frames. Furthermore, he noted, rarely do they fit tidily within one department's or government's mandate or jurisdiction. He highlighted three major aspects where improvement was necessary:

- Federal government agencies' performance often fell well short of stated objectives. As a result, an *implementation gap* existed, since policy direction too often was not translated into effective action. Issues surrounding implementation challenges are addressed in Chapter 6.
- Many pressing issues transcended departmental mandates and governmental jurisdiction. Consequently, *lack of co-ordination and integration* was frequent. The need was to manage 'horizontal issues', or those involving shared responsibility.
- There was often inadequate information about the benefits from environmental programs. Therefore, a strong need existed to resolve *inadequate performance review* processes so that both senior managers and parliamentarians could know what was being accomplished.

In his 2008 annual report, the Commissioner indicated that some progress was taking place in creating and implementing sustainable development strategies, but that in many more areas little or no progress had been made. He explained that where progress was not occurring satisfactorily, it was due to lack of commitment by senior managers and often insufficient funding. Examples of poor progress were found regarding protection of species at risk along with their habitat, control of aquatic invasive species, and restoration of heavily polluted areas in the Great Lakes. The Commissioner stated that a major concern was poor performance in the conduct of strategic environmental assessments at the time when policy and program proposals were prepared.

The Commissioner's fall report in 2009 focused on one aspect—ensuring high-quality information to design, implement, and monitor environmental management programs. In the words of the Commissioner, 'Informed decision-making is at the heart of sound policy-making. The environmental programs of the federal government need science-based environmental information that is timely, robust, and accessible in ways that both identify patterns of environmental degradation and help programs concentrate on the most urgent environmental problems.'

The Commissioner emphasized two major challenges: (1) individual environmental monitoring programs must accurately track environmental quality, and (2) the many environmental programs scattered among agencies 'can and should work in tandem to provide a composite or cumulative picture'. Having examined numerous individual programs, the Commissioner concluded that many do work as intended. Examples of effective monitoring programs include the Air Quality Health Index, which provides real-time air quality data; the Guidelines for Canadian Drinking Water Quality that track allowable levels of microbiological, chemical, and radiological contaminants in drinking water; and the national bio-monitoring program, which tests people for traces of chemicals.

In contrast, the Commissioner concluded that, 'Unfortunately, other systems are incomplete, out-of-date, or non-existent.' Examples offered were inability to monitor fish habitat protection, and the National Pollutant Release inventory, which tracks emissions of 347 different chemical and waste substances. With regard to the latter system, the Commissioner observed that 'Environment Canada does not have adequate systems to verify that all facilities required to report their emissions are doing so and that the information they report is accurate.' All of the Commissioner's reports can be found at: www.oag-bvg.ca/internet/English/oag-bvg_e_920.html.

Given the attention by the Commissioner to the need for timely, robust, and accessible environmental information, in the next section we examine practices for tracking environmental conditions, with particular attention to the role of indicators.

Measuring Progress

From the foregoing discussions, you will appreciate that the environmental situation is often so complex that many, including many decision-makers, give up on trying to make sense of it. However, if we do not understand the problem and whether it is getting better or worse, we cannot implement effective management strategies. One of the goals of science is to provide understanding of complex problems. And one way of doing this is through the use of indicators.

Ecological footprints enable us to make comparisons among various countries (Box 1.5). Some countries have ecological demands greatly in excess of their capabilities (Figure 1.11), and they import ecological capital from elsewhere to make up for this deficit. Although trade between nations is to be expected, this excess of **ecological footprint** over capacity allows some nations to live beyond their ecological means. Canada has one of the largest available

When you have your morning cup of coffee or tea with sugar, your ecological footprint is reaching out to the tropics where both drinks and sugar are grown. Tea plantation, Sri Lanka (top). Sugarcane fields, South Africa (bottom).

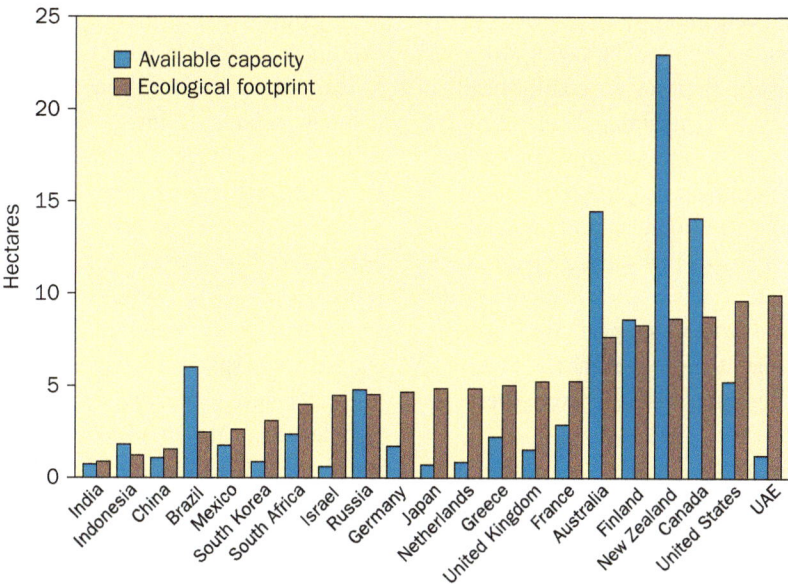

Figure 1.11 | Ecological footprint per person in selected nations, 1999. *Source: Worldwatch Institute (2004a).*

ecological capacities; we also have one of the largest ecological footprints per capita.

Indicators are not new. For many years, doctors have used body temperature, measured easily by thermometer, as one indicator of the health of the human body. Gross domestic product has been used as an indicator of economic performance, as has the Dow Jones industrial average. These indicators tell us something of the current state of a particular system,

Environment in Focus

Box 1.5 Your Ecological Footprint

Your 'ecological footprint' is the land and sea base required to provide your needs, including all energy and material requirements, and also to dispose of your wastes. In fact, most of a person's footprint comprises the space needed to absorb waste from energy consumption, especially carbon dioxide. Can nature provide enough of these services on an ongoing basis to meet the needs of an individual, community, or nation?

A 2004 study undertaken for the Federation of Canadian Municipalities (see www.fcm.ca/english/communications/eco.pdf) examined the ecological footprints of 20 municipalities—that is, the area of land and sea throughout the world required to produce the amount of food, energy, and other materials the citizens use. The Canadian average was 7.25 hectares per capita, third highest in world rankings. The lowest and highest footprints in Canada were both in Ontario. Sudbury was the lowest at 6.87 hectares, with York–Durham the highest at 10.33. One reason for Sudbury's success is the result of the efforts of the public works engineer in charge of heating and sewage. He has introduced many innovative programs, especially programs related to using local energy sources, such as the power of the wind, the sun, and the Earth's heat, that have reduced both the cost of heating and resource consumption. It is a striking example of how the initiative and energy of one person can make a significant improvement in planetary resource use.

York–Durham, on the other hand, uses 43 per cent more resources per capita than the average Canadian. This is mainly a result of the fuel consumption generated by the sprawling suburban developments near Toronto that are largely dependent on car transportation. Chapter 13 provides more detail about Canadian cities and their footprints.

On a global scale, only 1.6 hectares are available for each person, and this amount is shrinking every year, largely as a result of population growth. Yet the collective average global footprint is 2.2 hectares per person, with a North American average double that for Europeans and seven times greater than it is in Asia or Africa. Globally, humans are consuming the natural resources of 1.5 planets. To provide for everyone at Canadian standards would require three Earths, not just one. Estimates suggest that by 2030, less than 0.9 hectare will be available per person at a global scale.

but they do not help us to understand why the system is in that state. Over the past 20 years, there has been growing awareness of the need to develop indicators that would gauge the health of other aspects of societal well-being, including the environment. Indicators are often used to provide information on environmental problems that enables policy-makers to evaluate their seriousness, to support policy development and the setting of priorities by identifying key factors that cause pressure on the environment, to monitor the effects of policy responses, and to raise public awareness and generate support for government actions. Box 1.6 describes one framework that helps to develop causal linkages between indicators—the Drivers-Pressures-State-Impact-Response framework, or DPSIR.

One example of the DPSIR approach is the joint Environment Canada–US Environmental Protection Agency (Environment Canada, 2003b) series of indicators on the state of the Great Lakes, which is broken down into pressure, state, and response indicators. It is interesting to note that two of the richest nations in the world found that there was insufficient data on many of the 80 desired variables. On the basis of the 43 indicators used, the conclusion is that the overall trend is 'mixed'—i.e., some indicators show improving conditions, others deterioration.

Another, more recent example of the DPSIR framework, from the Commission on Environmental Cooperation (CEC), was established under the North American Free Trade Agreement to examine environmental challenges in North America. A recent report looks ahead to 2030 and tries to make assessments of how the drivers, pressures, states, and impacts will change (CEC, 2010). The results are not unexpected in that virtually all areas of the environment will come under increasing pressure. Three areas of prime concern are continued and accelerated warming, particularly in the Arctic, continued loss of terrestrial biodiversity, and persistence of elevated levels of ground-level ozone in urban areas. Again, though, particular attention is drawn to the lack of adequate information and understanding in many areas to be able to make assessments with confidence.

The federal government produces a suite of indicators, the Canadian Environmental Sustainability Indicators (www.ec.gc.ca/indicateurs-indicators/default.asp?lang=En&n=47F48106-1), to track progress towards meeting the goals and targets of the Federal Sustainable Development Strategy (FSDS). Information is provided in three main categories: air quality; freshwater quality and availability; and nature protection.

One issue difficult to resolve in reporting on environmental change is the degree of aggregation of information included in an indicator. An almost infinite amount of information could be collected on environmental systems (Figure 1.13). Much of this information might be useful for understanding the basic nature of the system, but it is not necessary for decision-making. Research scientists and line agencies may be involved in the routine collection of such data, and without such data, meaningful indicators cannot be constructed. At a higher level of sophistication, these raw data may form an *integrated database*, such as the integration of social and biophysical data as a basis for integrated watershed management planning. However, synthesis of these data into indicators is often most useful to decision-makers, and indicators themselves may show greater or lesser degrees of aggregation, especially in a spatial sense. The Canadian Environmental Sustainability Indicators are examples of the kind of indicators in which there is some spatial aggregation for the whole country.

Higher levels of thematic aggregation produce *indices*. Simple indices are composed mainly of similar indicators. The well-known Dow Jones industrial average, for example, combines changes in market processes for 30 blue-chip stocks listed on the New York Stock Exchange. The **Living Planet Index**, created by the World Wildlife Fund (WWF), is an example of a widely used index (www.wwf.panda.org/about_our_earth/all_publications/living_planet_report/living_planet_report_graphics/lpi_interactive/) that quantifies the overall state of planetary ecosystems. It tracks populations of 2,544 vertebrate species—fish, amphibians, reptiles, birds, mammals—from around the world. Separate indices are produced for terrestrial, marine, and freshwater species, and the three trends are averaged to create an aggregated index. Although vertebrates represent only a fraction of known species, it is assumed that trends in their populations are typical of biodiversity overall.

The index shows a 35 per cent reduction overall in the planet's ecological health since 1970. Figure 1.13 shows the temperate and tropical terrestrial Living Planet Indices for 1970 to 2007. The terrestrial index is a spatial aggregation of the temperate and tropical indices. The challenge with aggregation can be seen by comparing the two. The temperate index shows a slight improvement since 1970, which is one of the reasons why inhabitants of the temperate nations of the world have too optimistic a view of environmental conditions on the planet. On the other hand, the tropical index shows a clear drop, evident to anyone who has spent a lot of time in the tropics over the past 30 years. These differences are hidden when the two are aggregated.

Perspectives on the Environment
Humanity's Ecological Footprint

The Earth's regenerative capacity can no longer keep up with demand—people are turning resources into waste faster than nature can turn waste back into resources.

Humanity is no longer living off nature's interest but drawing down its capital. This growing pressure on ecosystems is causing habitat destruction or degradation and permanent loss of productivity, threatening both biodiversity and human well-being.

—World Wildlife Fund (2006: iv)

Environment in Focus

Box 1.6 The DPSIR Indicator Framework

The most widespread framework for classifying environmental indicators is the Drivers-Pressures-State-Impact-Response (DPSIR) framework developed by the Organization for Economic Co-operation and Development (OECD) and adopted by all European Union countries, the US, Canada, Australia, Japan, and many developing countries (e.g., Malaysia) and international organizations (such as the Commission for Sustainable Development of the UN, the United Nations Environment Programme, and the World Bank). The framework, as shown in the figure below, is popular because of its organization around key causal mechanisms of environmental problems.

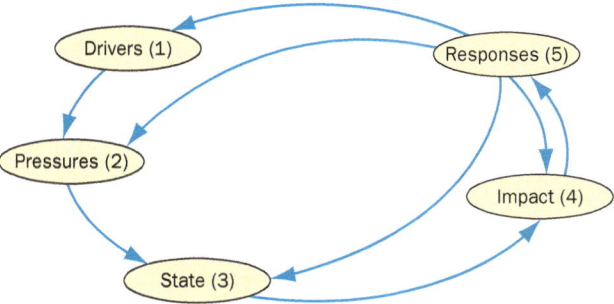

Figure 1.12 | The DPSIR Indicator's framework

Driving forces (1). Drivers are the underlying forces causing environmental change. They describe social, demographic, and economic developments in societies and corresponding changes in lifestyles, overall levels of consumption, and production patterns. Examples of drivers include population pressures and the demand for various consumer goods and services (e.g., cars, red meat, increasing travel).

Pressure indicators (2). These indicators are the pressures on the environment resulting from the drivers. Examples include emission of pollutants, use of resources, use of land for roads, water withdrawals, deforestation, and fisheries catch. Initial interest in these indicators focused further down the causal chain in the state indicators, described below. However, there is now widespread realization that state indicators are mere reflections of changes further up the chain, prompting much greater interest in drivers and pressures. One example of the link between drivers and pressures can be found in a driver such as the number of cars. Not only can we address the driver (by seeking to limit the number of cars through better public transportation, raising the price of gas, or imposing special taxes, licensing fees, or tolls on people who insist on driving their cars into downtown urban areas during the daytime), but we can also try to reduce the pressure by making vehicles more fuel-efficient and less polluting.

State indicators (3). State indicators describe the quantity and quality of physical phenomena (e.g., temperature), biological phenomena (e.g., fish stocks, extinctions), and chemical phenomena (e.g., CO_2 concentrations, phosphorus loading) and tell us the current state of a particular environmental system. They are often tracked over time to produce a trend.

Impact indicators (4). The changes in the state of the environment described by the state indicators result in societal impacts. For example, a rise in global temperatures (a state indicator) has an impact on crop productivity, fisheries value, water availability, flooding, and so on.

Response indicators (5). Response indicators measure the effectiveness of attempts to prevent, compensate for, ameliorate, or adapt to environmental changes and may be collective or individual efforts, both governmental and non-governmental. Responses may include regulatory action, environmental or research expenditures, public opinion and consumer preferences, changes in management strategies, and provision of environmental information. Examples include the number of cars with pollution control or houses with water-efficient utilities, the percentage of waste that communities and households recycle, use of public transport, and passage of legislation. Response indicators are critical to assessing the effectiveness of policy interventions but are often the most difficult to develop and interpret.

The WWF produces another global index, the ecological footprint, which shows the extent of human demand on global ecosystems. The index measures the area of biologically productive land and water needed to provide ecological resources and services—food, fibre, and timber, land on which to build, and land to absorb carbon dioxide (CO_2) released by burning fossil fuels. The amount of biologically productive area—cropland, pasture, forest, and fisheries—available to meet humanity's needs is the biocapacity. Since the late 1980s, the ecological footprint had exceeded the Earth's biocapacity. By 2007, the most recent year for which data are available, humanity's ecological footprint had grown to 50 per cent more than biocapacity.

The main components of the ecological footprint are shown in Figure 1.15, and again the aggregated index masks large global differences (Figure 1.16). When the ecological footprint is compared with biocapacity, the unsustainable direction becomes clear (Figure 1.18). The challenge is to

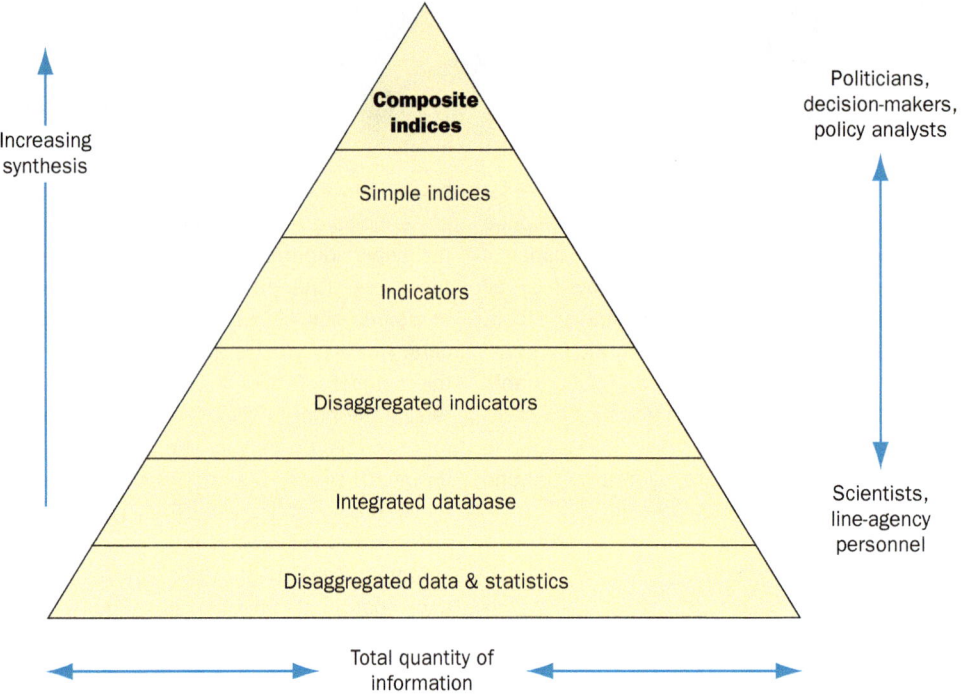

Figure 1.13 | Relationship among data, indicators, indices, and users.

devise effective strategies to reduce the footprint below biocapacity before the ecological debt becomes insurmountable. At the same time, it is important to take into account the very real needs reflected in the Millennium Development Goals discussed earlier. One way to summarize these is in the UN's Human Development Index, a comparative measure of life expectancy, literacy, education, and standards of living for countries worldwide. Developed countries have scores of 0.8 or more. Hence the goal is to provide for human development within the available global ecological footprint. As can be seen in Figure 1.17, this is a challenging task.

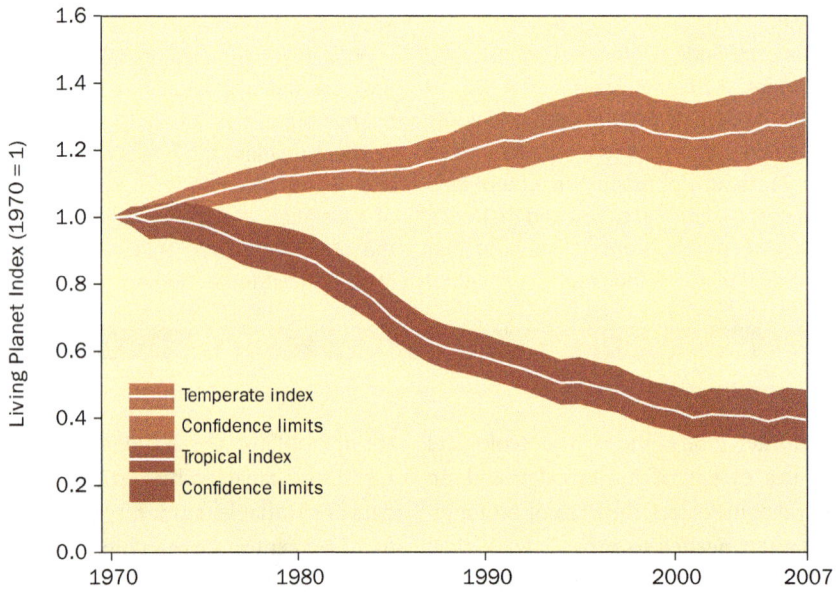

Figure 1.14 | The temperate LPI and the tropical LPI. The temperate index shows an increase of 29 per cent between 1970 and 2007. The tropical zone index shows a decline of more than 60 per cent over the same period (WWF/ZSL 2010). Source: WWF (2010: 9).

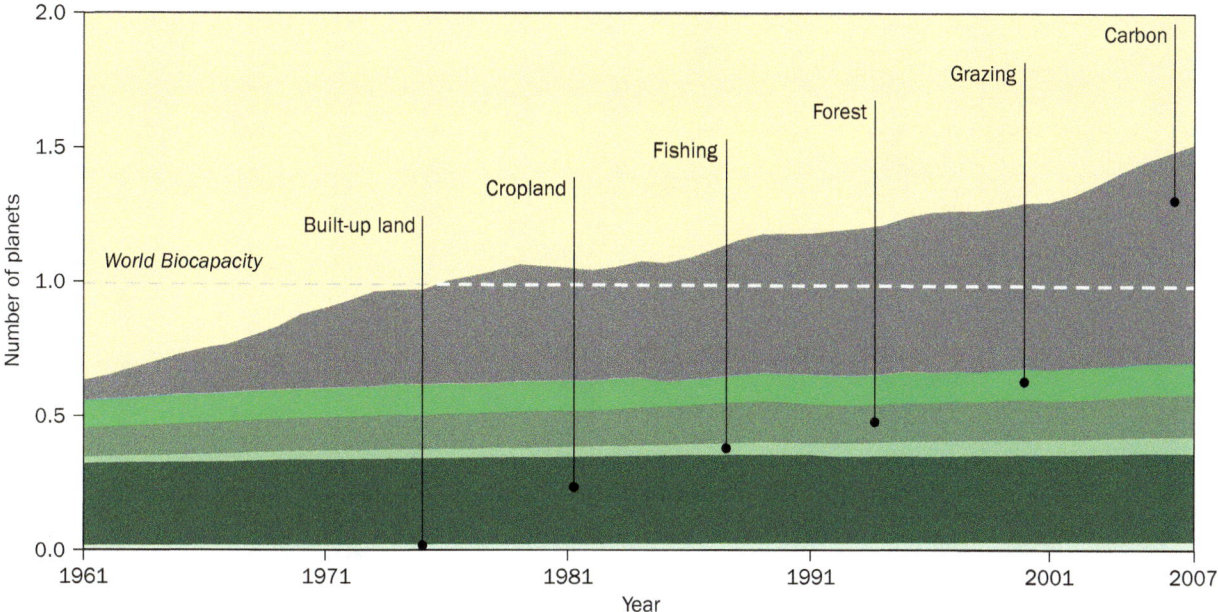

Figure 1.15 | Components of the ecological footprint, 1961–2007. One way of showing the ecological footprint is as the number of planets required to produce natural resources consumed in a single year. Total **biocapacity**, represented by the dashed line on the graph, is the equivalent to one planet Earth, although the biological productivity of the planet changes every year. In 2007, humanity was using the equivalent of one-and-a-half planets. Note that hydroelectric power is included in built-up land and fuelwood in the forest component. *Source: WWF (2010: 13), from Global Footprint Network.*

Composite indices, such as the ecological footprint, are often the most useful for decision-makers and represent the highest level of aggregation. They are few in number and incorporate many, often very different sub-variables. The Human Development Index, discussed above, created by the United Nations Development Programme (Figure 1.17), the Environmental Sustainability Index of the United Nations, and gross national product are other examples of aggregate indices.

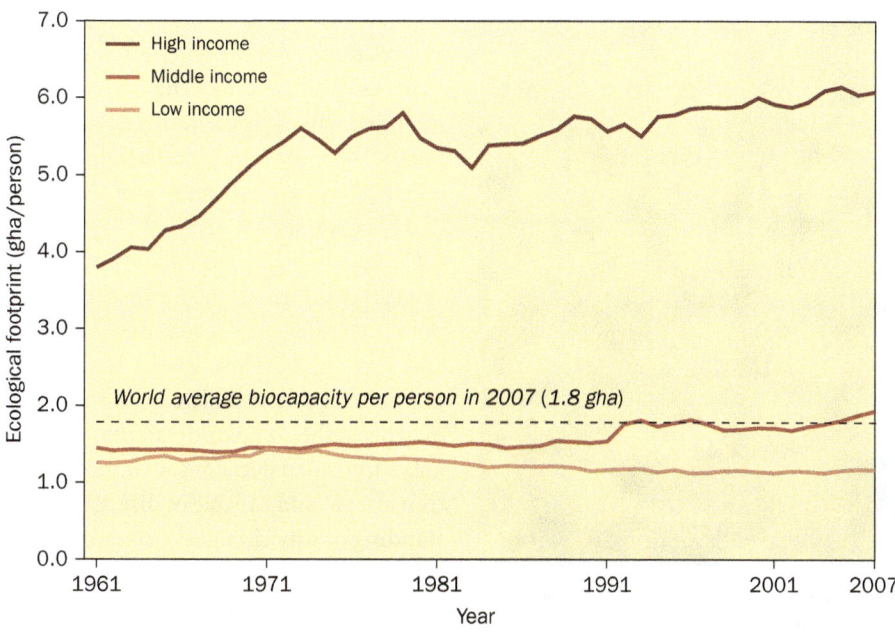

Figure 1.16 | Changes in ecological footprint per person in high-, middle-, and low-income countries, 1961–2007. The dashed line represents world average biocapacity in 2007. *Source: WWF (2010: 18), from Global Footprint Network.*

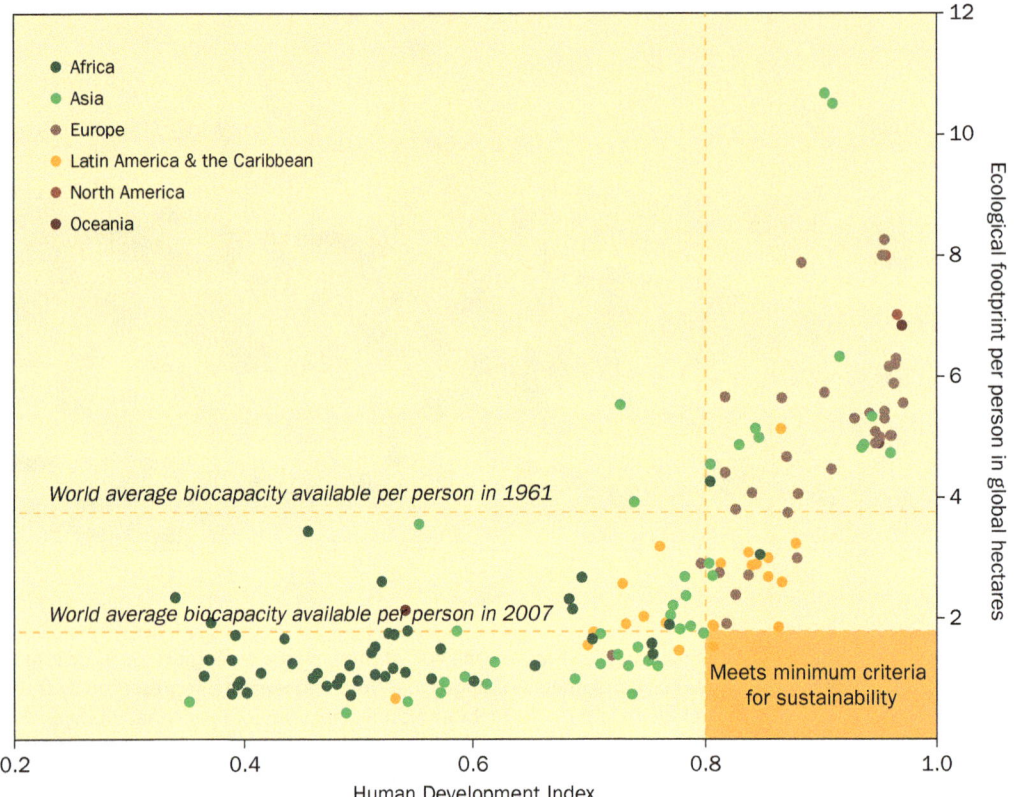

Figure 1.17 | Human Development Index correlated with the ecological footprint. *Source: WWF (2010: 21), from Global Footprint Network and UNDP.*

Even if we could determine the value of insect pollinators to Canada's agriculture adequately, money could never buy the services they provide.

In late 2011 the most recent Canadian Index of Wellbeing (CIW) was released, encompassing eight different categories of well-being: living standards, healthy populations, community vitality, democratic engagement, time use, leisure and culture, education, and environment. Based at the University of Waterloo, the CIW aims to produce a composite index with a single number that moves up or down like the TSX or Dow Jones Industrial, giving a quick snapshot of whether the overall quality of life of Canadians is getting better or worse.

Composite indices are highly attractive because they convey a lot of information and are useful for making macro-level policy decisions. However, these highly aggregated indicators also carry risks. They often tell us what is happening at the macro level but add little in terms of explaining why. They may mask the complex detail that decision-makers require to make informed decisions. Composite indices must be highly transparent and capable of disaggregation to facilitate understanding of why change is occurring.

Indicators provide some basis for assessing change and comparison among countries, but they raise questions about the role of science in environmental decision-making. This aspect is discussed in more detail in the next section.

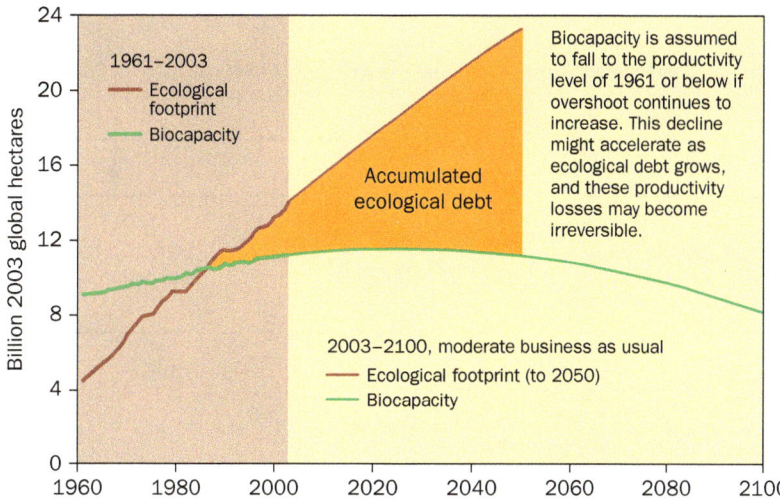

Figure 1.18 | Business-as-usual scenario and ecological debt. *Source:* WWF International, Living Planet Report 2006.

Implications

We are violating global thresholds related to the carrying capacity of the life support system of the planet. We have ceased to live off the interest and are often consuming the capital at such a rate that it threatens the future viability of the system. Many species reach such carrying capacity limits with their environment, overshoot them, and have their numbers drastically reduced by environmental factors, as discussed in Chapter 3. So far, we have been able to avoid this process because of human technological ability, which has increased carrying capacities. But can we continue to increase our numbers and our habits of consumption indefinitely? Or must even humans accept some limits to activities and numbers?

If the answer to the latter question is yes, then identification, in general terms, of the changes that need to be made is not that difficult. We need to balance birth and death rates, restore climatic stability, protect our atmosphere and waters from excessive pollution, curb deforestation and replant trees, protect the remaining natural habitats, and stabilize soils. The challenge, however, is charting a course to fulfill these objectives. Before its demise, the Soviet Union had possibly the most stringent and comprehensive environmental protection regulations in existence and yet still ended up as one of the most polluted environments on Earth. The regulations were simply not enforced. The secret is charting a course that not only addresses the goals mentioned above but is actually able to achieve these goals.

In this book, we aim to provide some background as to how this can be achieved, with particular reference to the Canadian situation. In this chapter, we started by discussing the characteristics of the 'environment' and 'resources'.

We also examined different approaches to understanding complex systems and considered issues related to the use of 'science' in decision- and policy-making. The case study of the Sydney Tar Ponds dramatically illustrates the complexity of environmental challenges, even at the local level. They are characterized by uncertainty, rapid change and conflict, and the need to appreciate both the scientific and technical aspects of a problem and the social dimensions. This book attempts to provide an introduction to both of these aspects. Part B outlines some of the main processes of the *ecosphere*, the basic functionings of the planetary life support system, and the ways in which we are disrupting them. Part C details some of the main planning and management approaches that have evolved within the Canadian context to address environmental challenges. Part D provides a thematic assessment of the challenges associated with particular activities such as fisheries, forestry, agriculture, wildlife use, water, energy production, and mineral extraction.

Figure 1.19 illustrates the relationship among these aspects. Natural systems form the basis of all human activity. A system is a recurring process of cause-and-effect pathways (Box 1.7). They range in scale from the giant atmospheric and oceanic circulation systems to the processes underway in a single living cell. The pollination of a flower by a bee, the melting of a glacier, and the biological fixation of nitrogen from the atmosphere are all parts of such systems. These systems are infinitely complex, and there is a great deal of uncertainty as to how they function. One of the goals of natural science is to try to understand this complexity. We do this by constructing simplified models of how we think they work (e.g., Figure 1.19, Box 1). The models presented in Chapters 2 and 4 on how energy flows through the biosphere and on the

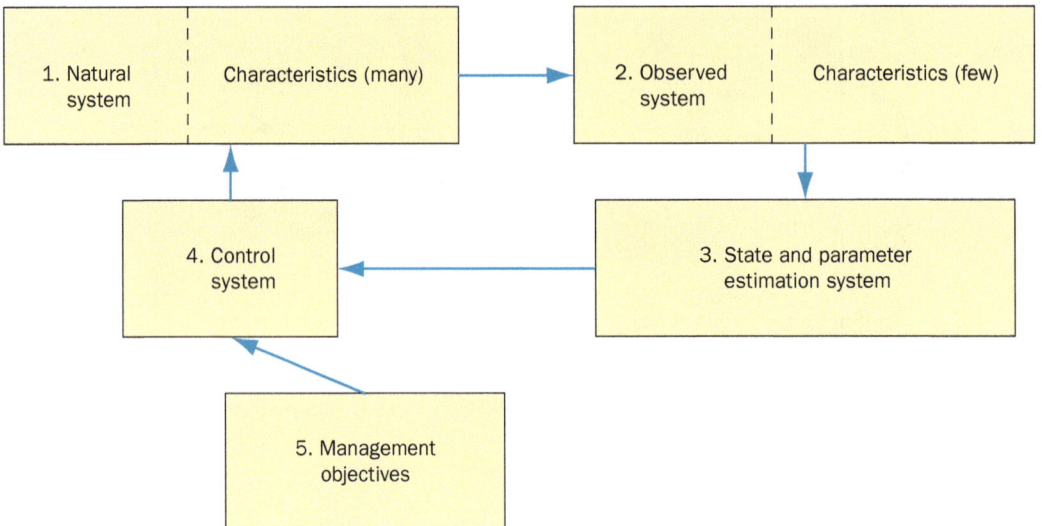

Figure 1.19 | Simplified model of interaction of biophysical and social systems in resource management.

nature of biogeochemical cycles are examples of these kinds of simplified representations of natural systems.

We do not know all the facts relating to these systems. We do not know all the components, let alone their functional relationships (Figure 1.19, Box 2). Many species, especially insects, still await discovery, even in well-explored temperate countries such as Canada. Of those we do know about, we have to select those we think are important and worth representing in our simplified models. Only recently, for example, have we become aware of the critical role played by various lichens in the circulation and retention of nitrogen in temperate rain forests (Chapter 4). Furthermore, not all characteristics are measurable, even if we are aware of their existence. Thus, our simplified models are fraught with uncertainty.

On the basis of these models, we estimate the status of a given system (Figure 1.19, Box 3). How many fish spawn in a certain river? What proportion of the landscape supports commercial tree growth? What soil characteristics are suitable to support a given crop? If we understand the current status of the system, we can also start to ask what the result will be if certain parameters are changed. What would happen to the system, for example, if we took a certain number of fish from the river before they spawned, if we removed tree growth from a portion of the landscape, or if we grew a given crop in the same soil for a particular time period? In other words, we try to assess the impact of various changes to the system. Formal processes of impact assessment have arisen in many jurisdictions. They assess not only the impact on natural systems but the impact on social systems as well, as described in Chapter 6.

On the basis of this understanding, we try to replace natural systems with control systems in which the main decision regulators are humans rather than nature (Figure 1.19, Box 4). Instead of natural forces determining the number of fish that reach the spawning grounds, or the age of trees before they are replaced by other trees, or what species will grow in a particular location, people make these decisions as we modify the environment to our own advantage. These

Environment in Focus

Box 1.7 Systems

You are in the educational system. You use the transportation system to go to your college or university, which is warmed by a heating system. What is a system?

Systems are composed of sets of things—e.g., educational institutions, buses, heating components—that are all related and linked together in some sort of functional way. Between these different components there is a flow of material, such as students, passengers, or heat, subject to some driving force—a thirst for knowledge, the need to get somewhere, or the need to get or stay warm. Systems are generalized ways of looking at these processes.

> ### Perspectives on the Environment
> #### When Science Meets Art
>
> The moral I labor toward is that a landscape as splendid as that of the Colorado Plateau can best be understood and given human significance by poets who have their feet planted in concrete—concrete data—and by scientists whose heads and hearts have not lost the capacity for wonder. Any good poet, in our age at least, must begin with the scientific view of the world; and any scientist worth listening to must be something of a poet, must possess the ability to communicate to the rest of us his sense of love and wonder at what his work discovers.
>
> —Abbey (1977: 87)

control systems are considered under topics such as forestry, water, energy, and agriculture in Part D of the book. And as control measures are introduced, we are increasingly appreciating that we may make natural and social systems less resilient, and therefore increase their vulnerability when significant or sudden changes occur.

Control systems are implemented on the basis of the social, economic, technological, and management constraints of a society (Figure 1.19, Box 5). These factors influence the demands for various outputs from the system and the speed of extraction. The environmental management strategies discussed in Part C of the book outline some main approaches to mediating between the social and economic demands of the society and the productive capacity of the system. As with the natural system, these strategies are characterized not only by complexity and uncertainty but also by *conflict* among different societal groups regarding the rate of outputs and distribution of benefits. Simply deciding which groups in society have a legitimate interest in a particular environmental issue is quite complex, as described in Chapter 5. Various dispute resolution mechanisms (Chapter 6) have emerged to address the conflicts arising from resource allocation decisions.

As we can already see, the challenges faced in environmental management are complex indeed, and consequently it is necessary to employ an integrated approach, such as the ecosystem approach described in Chapter 5, to understand this complexity. Both natural and social systems are fraught with uncertainty, making an adaptive approach (Chapter 6) a necessity, with strong adherence to the *precautionary principle*, as discussed in Chapter 5. Furthermore, our present predicament is largely the result of modification of natural systems before we had invested the time and effort—or had sufficient data or scientific expertise—to understand the consequences of our actions, especially related to the concept of resilience. The fisheries on both the Pacific and Atlantic coasts of Canada are in trouble (see Chapter 8) because our simplified system models were inadequate as a basis for decision-making. However, in some instances, even when the long-term implications of an activity on the future viability of a resource are understood, the activity continues for political and economic reasons. The overharvesting of timber (Chapter 9) is a good example.

Perhaps the most important message underlying the environmental challenges we face is the need for fundamental changes in the way we view our relationship with nature, as discussed earlier in this chapter regarding anthropocentric and biocentric/ecocentric perspectives, as well as later in Chapter 15. Changes in outlook and approach must take place at all levels, from international agencies such as the World Bank, through national and regional governments, to household and individual initiatives. Part of the goal of this book is to motivate the reader to become more involved in making these changes happen, both locally and globally.

Summary

1. The environment is the combination of the atmosphere, hydrosphere, cryosphere, lithosphere, and biosphere in which humans, other living species, and non-animate phenomena exist.

2. Some consider resources to be only those components of the environment of utility to humans. This view is considered to be anthropocentric; value is defined relative to human interests, wants, and needs. Another, contrasting perspective is that resources exist independently of human wants and needs. This view is ecocentric or biocentric; aspects of the environment are valued simply because they exist, and they have the right to exist.

3. Environmental and resource issues can be approached from disciplinary, multidisciplinary, cross-disciplinary, interdisciplinary, and transdisciplinary perspectives. Each provides a different basis or model for viewing the world. We should strive for interdisciplinary and transdisciplinary approaches in order to understand complex systems.

4. For a science-based approach to management of resources and the environment, we should (1) focus the science on key issues and communicate it in a policy-relevant form; (2) use scientific information to clarify issues, identify potential management options, and estimate consequences of decisions; (3) clearly and simply communicate key scientific findings to all participants; (4) evaluate whether or not the final decision is consistent with scientific information; and (5) be aware of the balance between scientists providing technical information and interpretation and being advocates for particular approaches or solutions.

5. The Sydney Tar Ponds case illustrates how attitudes and approaches towards environmental issues evolve and how what may be viewed as appropriate behaviour at one time can be viewed as inappropriate later.

6. There are at least three lessons from the Sydney Tar Ponds experience: (1) when basic science is not used from the outset to inform policy decisions related to environmental issues, the probability is high that effective solutions will not be found; (2) even when science is used, understanding can be incomplete, and decisions will have to be taken in the face of considerable uncertainty; and (3) when local stakeholders are not included in the process, challenges can be expected to proposed solutions. The third point reminds us that while science is important, it is often not sufficient by itself.

7. Having a vision of or sense of direction towards a desirable future condition is essential for planning and management. In that regard, sustainable development and sustainable livelihoods have been proposed as appropriate ideals to characterize what societies should aspire towards.

8. Resilience, or 'the ability of a system to absorb disturbance and still retain its basic function and structure', challenges some of the basic assumptions underlying sustainable development.

9. On the global scale, there is undeniable evidence of unprecedented environmental degradation as a result of human activities. Growing global population is a continuing challenge, as are the consumer demands of people in the wealthier countries.

10. An estimated 4.3 people are born every second around the world. By 2011, there were 7 billion people on Earth. The United Nations forecasts an increase to 9.3 billion people by 2050, representing more than 80 million additional people per year to feed.

11. Conditions continue to deteriorate in many poorer countries. The richest 20 per cent of the world's population is responsible for more than 75 per cent of world consumption, while the poorest 20 per cent consume less than 2 per cent.

12. The eight Millennium Development Goals adopted by the United Nations in 2000 aim to improve human well-being by reducing poverty, hunger, and child and maternal mortality; by ensuring education for all; by controlling and managing diseases; by tackling gender disparity; by ensuring environmental sustainability; and by pursuing global partnerships. Under each of the MDGs, countries have agreed to targets to be achieved by 2015.

13. In order to stay below the threshold level of what scientists predict would be runaway change and if emissions were frozen at the current level, we would need the absorptive capacity of two planets.

14. Canada is one of the most privileged countries, covering some 13 million square kilometres and with a population of some 34 million people. However, our environmental impacts are considerable. Our per capita consumption of water and energy is among the highest in the world. We also have some of the highest production per capita of waste products, including greenhouse gases.

15. Canada has the second highest per capita emissions of greenhouse gases in the world. Canada agreed under the Kyoto Protocol to target a 6 per cent cut in emissions. Instead, emissions have increased by 29 per cent, and the country is 35 per cent above its Kyoto target.

16. Responsibility for the environment and natural resources is divided between the federal and provincial governments, with the territories and municipalities taking on increasingly important roles. Aboriginal peoples also are much more involved. The shared responsibility often requires collaboration and partnerships, which can create tensions because of differing interests and perspectives.

17. Indicators are one way by which science can assist decision-makers to appreciate current trends. There are different kinds of indicators with different levels of complexity. Decision-makers often want few indicators that contain the most information to be able to understand the situation. Indicators that combine many different elements are called composite indices.

18. The Living Planet Index shows a 35 per cent reduction overall in the planet's ecological health since 1970. Ecological footprints show the extent of human demand on global ecosystems. These 'footprints' measure the area of biologically productive land and water needed to provide ecological resources and services—food, fibre, and timber, land on which to build, and land to absorb carbon dioxide (CO_2) released by burning fossil fuels.

19. A comparison of humanity's ecological footprint with biocapacity (the amount of biologically productive area—cropland, pasture, forest, and fisheries—available to meet humanity's needs) shows that since the late 1980s, the ecological footprint exceeds the Earth's biocapacity by more than 50 per cent.

20. The main challenge for humanity is to implement effective strategies to ensure that the ecological footprint falls below biocapacity as soon as possible.

Key Terms

anthropocentric view
biocapacity
crude birth rate (CBR)
crude death rate (CDR)
crude growth rate (CGR)
demographic transition
ecocentric (biocentric) values
ecological footprint
environment

epidemiological transition
estuary
exponential growth
global climate change
gross national product (GNP)
indicators
Living Planet Index
Millennium Ecosystem Assessment
planetary carrying capacity

population age structure
replacement-level fertility
resilience
resources
subsidiarity
sustainable development
sustainable livelihoods
total fertility rates

Questions for Review and Critical Thinking

1. What information is available in your municipality or province regarding environmental hazards? Could a Sydney Tar Ponds occur in your community?

2. Who should be responsible for dealing with environmental hazards resulting from earlier resource use and environmental standards no longer acceptable today?

3. If you were hired to provide recommendations related to the Sydney Tar Ponds, what information would you need to make a decision about the potential risk to health for people living in the community? How would you place a monetary value on any potential risk?

4. What is the distinction between sustainable development and sustainable livelihoods? How does the concept of resilience challenge the goals of sustainable development? Do you believe sustainable development or resilience makes a better basis for imagining a desirable future state?

5. Outline the main arguments for considering population growth a threat to global carrying capacity or as a building block for future economic growth.

6. What moral obligations, if any, do Canadians have to assist people in the developing world whose standards of living do not meet basic human needs?

7. Is population growth or environmental degradation the major problem in the less developed countries? Which is cause, and which is effect?

8. What are indicators used for? What are the main types of indicators? Are there any indicators used by your province or municipality that give insight into environmental changes?

9. What are some of the main initiatives of Canadian governments to address environmental problems?

10. What is a system? Outline the components of a system that you use on a regular basis.

11. What are the top three things you would do if you were the Prime Minister of Canada to make a contribution towards achieving global sustainability and resilience?

Related Websites

Canadian Environmental Sustainability Indicators
www.ec.gc.ca/indicateurs-indicators/default.asp?lang=En&n=47F48106-1

Canadian Index of Wellbeing
www.ciw.ca

Community-based indicators
www.sustainable.org

Fraser River Basin indicators
www.fraserbasin.bc.ca/programs/indicators.html

Living Planet Index
www.wwf.panda.org/about_our_earth/all_publications/living_planet_report/living_planet_report_graphics/lpi_interactive/

National Round Table on the Environment and the Economy (NRTEE)
www.nrtee-trnee.ca

Office of the Auditor General, Commissioner of the Environment and Sustainable Development, *Reports on the Environment and Sustainable Development*
www.oag-bvg.gc.ca

Resilience Alliance
www.resalliance.org

Selection of indicators for Great Lakes
www.binational.net

Sydney Tar Ponds Agency
www.tarpondscleanup.ca

United Nations Environment Programme
www.unep.org/geo/geo4/media/index.asp

United Nations Population Fund
www.unfpa.org

Worldwatch Institute
www.worldwatch.org

Further Readings

Note: This list comprises works relevant to the subject of the chapter but not cited in the text. All cited works are listed in the Bibliography at the end of the book.

Furimsky, E. 2002. 'Sydney Tar Ponds: Some problems in quantifying toxic waste', *Environmental Management* 30, 6: 872–9.

Guernsy, J.R., et al. 2000. 'Incidence of cancer in Sydney and Cape Breton County, Nova Scotia, 1979–1997', *Canadian Journal of Public Health* 91, 4: 285–92.

Joffres, M.R., et al. 2001. 'Environmental sensitivities: Prevalence of major symptoms in a referral center: The Nova Scotia Environmental Sensitivities Research Center Study', *Environmental Health Perspectives* 109, 2: 161–5.

O'Leary, J., and K. Covell. 2002. 'The tar pond kids: Toxic environments and adolescent well-being', *Canadian Journal of Behavioural Science* 34, 1: 34–43.

Worldwatch Institute. 2011. *Vital signs: The Trends That Are Shaping Our Future*. New York: W.W. Norton.

> The human brain now holds the way to our future. We have to recall the image of the planet from outer space: a single entity in which air, water, the continents are interconnected. That is our home.
>
> —David Suzuki

Part B
The Ecosphere

The first section of the book provided an overview of the global and Canadian situations regarding the relationship between humans and the Earth. As the Canadian communications theorist Marshall McLuhan noted, we are not passive bystanders in this interaction. We are members of the crew, helping direct what happens to our planet. As crew members, we need to have some idea of the workings of our 'Spaceship Earth', particularly of the nature of its life support system. Part B will provide this overview. Here we describe the natural systems through simplified models, following the conceptual framework outlined in Chapter 1 (Figure 1.12). This will provide the background you need to understand many of the environmental challenges facing society today.

Most of what follows is derived from just one form of environmental knowledge—natural science (see Box B.1). Natural science tries to find order in and to understand nature so that accurate predictions can be made regarding the outcome of given changes that may occur. An important underlying assumption is that patterns in nature can be discerned if we approach things in the right manner with the right tools. The *scientific method* lays the foundation for how scientists approach this task. An important assumption of the scientific method is that the same results will be obtained by different scientists if they repeat the experiment or observations in the same manner as the original observations. This is one reason why scientists must give detailed descriptions of their methodology and attempt to be as objective as possible. What use would a thermometer be, for example, if its readings of temperature varied according to who took the measurement?

Environment in Focus

Box B.1 Traditional Ecological Knowledge

Most of the concepts presented in this book are the result of the 'scientific approach' to understanding different phenomena. There are other approaches, however, and one gaining increasing attention is **traditional ecological knowledge** (TEK). The scientific community now understands that indigenous peoples often have detailed knowledge of their local environments, which is not surprising for peoples gaining their sustenance directly from that environment. Indigenous peoples tend to undertake the same kinds of tasks as Western scientists, such as classification and naming of different organisms and studies of population dynamics, geographical distributions, and optimal management strategies. Unlike Western science, however, this knowledge is rarely recorded in written form but is handed down orally from generation to generation.

Interest in TEK has been spurred by increasing industrial interest in northern regions and the potential impact of resource extraction on Native communities. Inevitably, this has given rise to discussions about which form of ecological knowledge, Western or traditional, is the 'best'. In reality, both have their advantages and disadvantages. Modern science is informed by developments around the world but is limited in its knowledge of changes over time in a particular place, an area where traditional knowledge is particularly rich. Scientists tend to concentrate on information that can be tested by replication and to ignore idiosyncratic and individual behaviour, which is given substantial weight by indigenous hunters.

A graphic example of these differences came to light in March 2008. Scientists from the Department of Fisheries and Oceans (DFO) claimed that the bowhead whale populations in the eastern Arctic were so low—about 5,000 animals—that the species was listed as threatened under the Species at Risk Act, discussed in Chapter 14. However, for many years the Inuit had claimed that there were far more animals and that the hunting quota of one whale every two years should be increased. As a result of new scientific information, this latter claim was found to be correct, and DFO scientists expanded their estimates to 14,400. That is nearly 300 per cent higher and roughly equal to the 11,000 whales thought to have frequented waters such as the Davis Strait and Lancaster Sound during the nineteenth century, when whales were vigorously hunted for their blubber as the main source of lamp oil. This example highlights that there is no 'best' form of knowledge and illustrates the potential for traditional knowledge and conventional science to complement one another. A similar dispute is now occurring over polar bear numbers, and it will be interesting to see how the issue is resolved over the next few years.

Management systems also differ between indigenous and scientific approaches. The traditional system is self-regulating, based on communal property arrangements. Conservation practices, such as rotation of hunting areas, were commonly practised. However, the system is not infallible, especially with the onslaught of outside influences and commercialization. There are examples around the world where indigenous peoples have hunted species to extirpation within their homelands. Similarly, the modern system of private property rights and state allocation of harvesting rights does not always work. This was recognized all too clearly with the complete collapse of the North Atlantic cod fishery: if scientists and policy-makers had given more credence to the local ecological knowledge of inshore fishers in Newfoundland outport communities, the destruction of the fishery might have been averted. Scientists and indigenous peoples are now realizing the benefits offered by the two systems of knowledge and management approaches and are trying to use both through co-management arrangements.

There is considerable dispute between scientists and indigenous peoples about the numbers of polar bears remaining.

Although the scientific method is designed to minimize the effects of scientists on the phenomenon they are studying, it is a myth to believe that science is 'objective'. Scientists work in a social environment and are greatly influenced by their peers and by society. The research topics selected and the way research questions are framed are key components of the scientific method and yet are heavily influenced by the individual scientist. Despite the fact that social biases and values influence science, it is important to maintain as value-free an approach as possible and ensure that biases are explicit and documented.

Scientists collect data or *facts* (observations widely accepted as truthful) about the environment and then try to make some order out of those facts. This order is called *theory*. The theory of natural selection, first outlined by Charles Darwin and Alfred Russel Wallace in 1858, is a good example. When there is universal acceptance of the theory, a *scientific law* may be established. A scientific law represents the most stringent form of understanding and lays down a universal truth that describes in all cases what happens in certain circumstances. In the next two chapters, you will be introduced to some of these laws. They are useful because they provide firm blocks upon which we can build our scientific understanding.

The scientific method links, with minimum error, the testing of *hypotheses* with the existing body of knowledge. It prescribes a series of steps that, over time, scientists have found are most likely to provide understanding about phenomena under study, irrespective of the observer undertaking the observations or experiment. The process starts with a question often derived either from observations of the environment or from existing literature. A hypothesis is generated to explain the phenomenon of interest, and experiments or observations are designed to disprove the hypothesis. If the hypothesis is not rejected, then additional experiments may be designed in further efforts to disprove it. If all efforts fail, the hypothesis becomes incorporated into theory and is accepted as a valid answer to the question. However, natural scientists do not have to follow one scientific method. Many important scientific insights have come about through quite irregular approaches. Nonetheless, it is wise to learn from previous experience and to know the kinds of procedures usually followed in any given area of inquiry.

That said, we should not feel that any of the hypotheses, theories, or even laws advanced by science are beyond question. The whole purpose of science is to ask questions, and science advances by continually changing and modifying previous knowledge. 'Be kind to scientists (and teachers!) but ruthless in your questions' is not a bad dictum. Debate and disagreement in science are normal. In the following chapters, for example, ideas on some topics are changing rapidly, and even some concepts (e.g., keystone species, climax vegetation) are being questioned by some scientists. This active debate is often misunderstood by those outside the academic community. It can also be used for political purposes to support inaction on measures that might be unpopular, such as limits on industrial emissions to reduce acidic precipitation. Debate and dispute, however, are signs of the vitality and strength of scientific thinking—and of the urgent need to get closer to truth—rather than the converse.

When we think of science, we often tend to think of it within the context of natural and physical sciences, and most of the next three chapters focus on this understanding of 'science'. However, social science is also critical to understanding and addressing environmental management. Just as natural scientists hope to provide greater understanding of the natural world, social scientists address the same need for social dimensions. How can we understand how individuals think, how to change their behaviour, how governments and the economy operate, and how legislation is formed and enforced? What can we learn from societies elsewhere and how they have interacted with their environmental surroundings in the past? These are the kinds of questions that we look to disciplines such as geography, anthropology, history, psychology, sociology, political science, and economics to address.

The methods of social science can also have much in common with the natural sciences, with hypothesis testing as a means of building theory and establishing laws that will help provide a generic understanding of seemingly disparate events. Yet the great variability in social systems, the difficulties with isolating and controlling many of the variables under study in a social context (as opposed to natural phenomena or laboratory work), and the challenges of repeatability (it is often impossible to duplicate the circumstances for repeated experiments) make adherence to strict scientific procedures impossible. These difficulties make it necessary for social scientists to apply a very broad range of approaches to gaining social understanding. Thus, although theory formation is a critical part of the social sciences, the formulation of universal laws is very rare.

Furthermore, although both natural and social scientists strive for *quantitative* (numerical) data because of the precision that numbers provide, *qualitative* (non-numerical) data are sometimes all that can be obtained or even all that can or should be sought, depending on the particular research question. A case in point: Over many years, federal fisheries scientists tended to dismiss the qualitative data from inshore Newfoundland fishers as 'anecdotal' and chose to rely largely on their own estimates of stock biomass, which were based on offshore data from the trawler fishery and from their own survey boats. As it turned out, the anecdotal observations and warnings based on generations of knowledge and experience were far more accurate than the quantitative data the scientists chose to believe. Both kinds of data are important. For example, if someone tells you that the weather was 'cool', that does not convey as much information as telling you that the temperature was −25°C. Data on the beauty of a certain landscape is less amenable to quantitative assessment, although some scientists have developed numerical ways of approaching such problems!

The three chapters in Part B provide a basic overview of the principal processes that maintain the planetary life support system. A simple model of the planet would look like the layers of an onion (Figure B.1). We are most concerned with the outer layer, the **ecosphere**, which consists of three main layers:

- The **lithosphere**, which is the outer layer of the Earth's mantle and the crust, and contains the rocks, minerals, and soils that provide the nutrients necessary for life.
- The **hydrosphere**, which contains all the water on Earth. Water, in a frozen state, is referred to as the **cryosphere**, a very important component for much of Canada.
- The **atmosphere**, which contains the gases surrounding the lithosphere and hydrosphere. It can be further divided into four main sub-layers. The innermost layer, or **troposphere**, contains 99 per cent of the water vapour and up to 90 per cent of the air and is responsible for our weather. Two gases, nitrogen (78 per cent) and oxygen (21 per cent), account for 99 per cent of the gaseous volume. This layer extends on average to about 17 kilometres before it gives way to the second layer, the **stratosphere**, wherein lies the main body of ozone that blocks out most of the ultraviolet radiation from the sun. At about 50 kilometres from the Earth's surface is the **mesosphere** and above that the **thermosphere**. As distance from Earth increases, the pressure and density of the atmosphere decreases as it melds into space.

These three layers combine to produce the conditions necessary for life in the ecosphere, which stretches from the depths of the ocean trenches up to the highest mountain peaks, a layer some 20 kilometres in width, no larger in scale than the peel of an apple, which contains some 30 million different kinds of organisms. The following chapters impart some idea of the main environmental processes of the ecosphere and describe how human activities interrupt these processes.

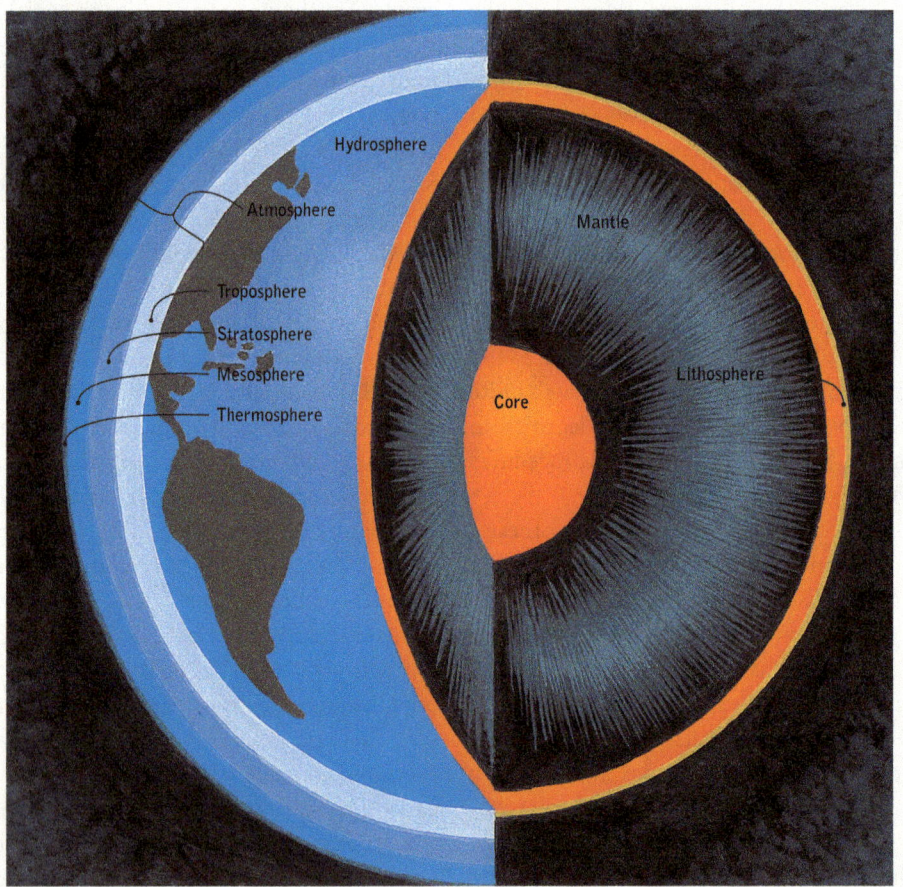

Figure B.1 | A simplified model of the Earth showing the ecosphere.

Key Terms

atmosphere
cryosphere
ecosphere
hydrosphere

lithosphere
mesosphere
stratosphere

thermosphere
traditional ecological knowledge (TEK)
troposphere

Chapter 2
Energy Flows and Ecosystems

Learning Objectives

- To know the nature of energy and the laws governing its transformation.
- To understand the way energy flows through the ecosphere and links ecosystem components.
- To outline the main influences on the structure and nature of ecosystems.
- To understand the nature and importance of biodiversity.
- To appreciate the Canadian context for biodiversity and approaches and challenges with biodiversity conservation in Canada.
- To understand the ecological implications associated with the loss of biodiversity.
- To learn the importance of reducing energy use in society.

Introduction

The annual arrival of the capelin to the beaches of Newfoundland to spawn on the first full moon of June had long been a bounty—not only for many animal species but also for the settlers who collected the fish for consumption and application to their gardens as fertilizer. The capelin is a small fish of the North Atlantic and an important food supply for many other species, including cod, salmon, halibut, mackerel, seals, various whale species, and many species of seabirds such as puffins and murres. In the late 1970s and early 1980s, the numbers of fish coming to spawn declined markedly.

Normally, 80 to 90 per cent of the Atlantic puffin's diet consists of capelin. In the early 1980s, scientists noticed capelin declined to 13 per cent of their diet, resulting in severe malnutrition of puffin chicks and subsequent declines in the population owing in part to starvation. Their numbers fell as a direct result of the removal of their food base, the capelin. The energy flow between the species had been interrupted by the opening of an offshore capelin fishery that had removed the capelin from the food chains that nourish many other marine species. The puffins were a noticeable victim of this appropriation, but other species feeding on the capelin suffered the same consequence. These species in turn would affect the abundance of other species at all levels in the food web, since the numbers of some species are controlled mainly by their predators. This example illustrates the importance of

The colourful bill is the most striking feature of the Atlantic puffin, which breeds among the rocks of sea islands.

Spawning capelin on a beach in Newfoundland.

understanding how energy links species and flows through ecosystems. Changing the energy available at one part of the food chain will have repercussions throughout the ecosystem.

Reading this book, taking notes in class, even snoozing at home all require energy. That energy comes ultimately from the **radiant energy** of the sun and is transformed into chemical energy in the form of food supplies before being converted to mechanical energy in the form of physical exertion and activity. In this chapter, you will gain an appreciation of energy in relation to such transformations, how energy flows through ecosystems, and the ecosystem consequences that result.

Energy

Energy is the capacity to do work and is measured in calories. A **calorie** is the amount of heat necessary to raise one gram or one millilitre of water one degree Celsius (°C), starting at 15 degrees. Energy comes in many forms: radiant energy (from the sun), chemical energy (stored in the chemical bonds of molecules), as well as heat, mechanical, and electrical energy. Energy differs from matter in that it has no mass and does not occupy space. It affects matter by making it *do* things—work. Energy derived from an object's motion and mass is known as **kinetic energy**, whereas **potential energy** is stored energy that is available for later use. The water stored behind a dam is potential energy that becomes kinetic energy as it pours over the dam. The gas in a car is potential energy before it is poured into the engine to create mechanical energy for propulsion.

Most of the energy available for use is termed **low-quality energy**, which is diffuse, dispersed at low temperatures, and difficult to gather. The total energy of all moving atoms is referred to as **heat**, whereas temperature is a measure at a particular time of the average speed of motion of the atoms or molecules in a substance. The oceans, for example, contain an enormous amount of heat, but it is very costly to harness this energy for use. They have high heat content but low temperature. On the other hand, **high-quality energy**, such as a hot fire or coal or gasoline, is easy to use, but the energy disperses quickly. Much of our economy and technology are now built around the transformation of low-quality energy into high-quality energy for human use. It is important that we match the quality of the energy supply to the task at hand. In other words, the aim is not to use high-quality energy for tasks that can be undertaken by low-quality supplies. Heating space, such as your house, for example, requires only low-temperature heat, yet many homes are heated through the conversion of high-quality energy sources that entail significant energy losses in generation, transport, and application. Nuclear energy, which involves high-quality heat at several thousand degrees converted to high-quality electricity transmitted to homes and used in resistance heating, is very inefficient. The most efficient way to provide space heating is to have super-insulated houses and passive solar heating.

All organisms, including plants, require energy for growth, tissue replacement, movement, and reproduction. To gain a comprehensive perspective on life, we must understand energy and how it is transformed and used. Box 2.1 provides an introductory definition of life.

Laws of Thermodynamics

Two laws of physics (or physical laws) describe the way in which trillions of energy transformations per second take place all over the globe. They are known as the *laws of thermodynamics*. The first one, the **law of conservation of energy**, tells us that energy can neither be created nor destroyed; it is merely changed from one form into another (nuclear is a form of potential energy—the energy is simply held in the nucleus of an atom). Organisms do not create energy; rather, they obtain it from the surrounding environment. When an organism dies, the energy of that organism is not 'lost'. It flows back into the environment and is transformed into different types of energy, the total sum of which adds up to the original amount. Similarly, we all know that cars obtain their energy from gasoline. As the fuel gauge goes from full to empty, this does not indicate that energy has been consumed; it has merely been transformed from chemical energy into other forms of energy, including the mechanical energy to move the car.

The second law of thermodynamics, the **law of entropy**, tells us that when energy is transformed from one form into another, there is always a decrease in the quality of usable energy. In any transformation, some energy is lost as lower-quality, dispersed energy that is dissipated into the surrounding environment, often as heat. The amount of energy lost varies depending on the nature of the transformation. In a coal-fired generating station, for example, 35 per cent of the coal's energy at most is converted into electricity. The rest is given off as waste heat to the environment.

In a car, only about 10 per cent of the chemical energy of the gasoline is actually converted into mechanical energy to turn the wheels. The remainder is dispersed into the environment. Put your hand onto the hood of a car that has just stopped running. The heat you feel is a result of this second law of energy or the *law of entropy*. **Entropy** is a measure of the disorder or randomness of a system. High-quality, useful energy has low entropy. As energy becomes dispersed through transformation, the entropy increases.

For organisms, the second law is particularly important because they must continuously expend energy to maintain themselves. Whenever energy is used, some of that energy is lost to the organism, creating a need for an ongoing supply that must exceed these losses if the organism is to survive. If losses exceed gains for an extended period of time, then the organism dies.

There are many other important ramifications of this law. It tells us, for example, that energy cannot be recycled. As it flows through systems, it is constantly degraded. We think of 'advanced' societies as being energy consumers. Large dams and nuclear power stations, for example, are visible signs of a modern economy. As we become more economically developed, we find new ways to transform energy. Cars, telephones, electric can openers, blenders, microwaves, hot tubs, computers, and iPods are all energy transformers. Yet as more energy is transformed, more is dispersed into the atmosphere because entropy increases. This dispersion can be likened to a bar of soap in a bowl of water. As the soap is used over time, it dissolves into the water, making it less and less useful. Similarly, as energy is used it gradually disperses into the atmosphere, becoming less useful.

Some of the principal transformations that have to take place to achieve a sustainable society are: to view high energy consumption as undesirable; to reduce energy waste; and to switch from the non-renewable sources of energy that now dominate (coal and oil particularly) to renewable sources, such as those discussed in Chapter 12. Until the Industrial Revolution, the speed of processing raw materials was limited by the energy available, supplied largely by human and animal labour combined with wood, wind, and water power.

For many people, such as villagers in India, biomass is the main form of energy. It can take many forms ranging from wood (a) through to dried buffalo feces (b) that are burned to cook food and, in some places, heat houses. These energy sources are ancient and depend on photosynthesis from the sun. Modern technology is now helping to capture the sun's energy in new and exciting ways, such as these solar cells in a remote village in western Thailand (c).

These sources were in turn limited by the input of solar energy over a relatively short time period. The use of coal, and later oil, to fuel steam engines removed these limitations and made accessible a vast storehouse of potential energy created by the sun over millions of years through the remains of compressed plants. Acid rain, climatic change, and many

Environment in Focus

Box 2.1 What Is Life?

We have said that energy is essential for all life, but what is life? Living organisms have a number of common characteristics, including:

- They use energy to maintain internal order.
- They increase in size and complexity over time.
- They can reproduce.
- They react to their environment.
- They regulate and maintain a constant internal environment.
- They fit the biotic and abiotic requirements of a specific habitat.

We think we have a fairly good idea of what constitutes life, but there is still a lot of debate as to how life developed on Earth. More than 80 years ago, two scientists proposed a theory, called the Big Bang Theory, explaining the origin of the universe as the result of a massive explosion that occurred some 15 billion years ago. The solar system came from the resulting matter. As the chunks of matter grew in size, they heated up. As the Earth cooled, warm seas formed, and precipitation occurred that helped to create a nutrient-rich environment. Over time, the constant bombardment of this nutrient-rich soup by high energy levels from the sun created chemical reactions producing simple organic compounds, such as amino acids. Scientists have managed to recreate several organic compounds necessary for life from inorganic molecules by bombarding them with energy.

Over billions of years, larger organic molecules came to be synthesized until the first living cells, probably bacteria, developed between 3.6 and 3.8 billion years ago. These cells passed through several stages over billions of years, with increasingly complex development. This activity took place in the ocean environment, protected from ultraviolet (UV) radiation. Between 2.3 and 2.5 billion years ago, a major change occurred when photosynthetic bacteria developed that emitted oxygen into the atmosphere as they manufactured carbohydrates from the carbon dioxide in the atmosphere. Over time, the oxygen reacted with the abundant and poisonous methane in the atmosphere, reducing levels of this gas and leading to the atmosphere we know today. Some oxygen was also converted to ozone in the lower stratosphere that protected evolving life from UV radiation and allowed the emergence of life from deeper to shallower waters and eventually onto land itself.

To simplify the vast array of life on Earth, biologists recognize five main kingdoms. The simplest kingdom (#1) consists of *monerans* that include bacteria and the photosynthetic blue-green bacteria and are single-celled micro-organisms. The genetic material of monerans is not contained within a nuclear envelope, and they are known as *prokaryotic* and were the first to evolve. The other four kingdoms (#2 to #5) all have nuclei and a high degree of internal structure and are known as *eukaryotic*. The *protists* (#2) comprise a large variety of unicellular and multi-cellular species such as algae, protozoans, slime moulds, and foraminifera. Kelp species, found in abundance around much of Canada's coastline, are multi-cellular algae from this kingdom. The kingdom consists of 14 phyla with more than 14,000 species described. *Fungi* have their own kingdom (#3), one that evolved relatively recently, some 400 million years ago. The kingdom includes both fungi, which are multi-cellular, and yeasts, which are unicellular. All fungi are heterotrophs and mostly digest dead organic matter or act as parasites. They are key components of the biogeochemical cycles described in Chapter 4. Many are asexual. Mushrooms are a good example.

The remaining two kingdoms (#4, #5) will be most familiar to you. The *plantae* (#4) are mostly photosynthetic, although there are some exceptions as described in Box 2.4. Unlike algae, plants are always multi-cellular. They dominate terrestrial ecosystems and contain two main groups, the *bryophtyes* and *vascular* plants. Bryophytes, such as mosses and liverworts, are restricted to moist environments because they lack a waxy cuticle to cover their foliage. They also lack vascular tissues. Vascular plants are very complex and have vascular tissues in their stems to convey water and nutrients. There are a further nine divisions within the vascular plants, including ferns, conifers, and flowering plants. It is estimated that there are more than 235,000 species of flowering plants. Other divisions, such as the ginkgo, have only one surviving species.

The final kingdom (#5) is the *animalia*, which are heterotrophic, multi-cellular organisms that have the ability to move. They ingest their food and digest it within their bodies. Most reproduce sexually. This is the largest kingdom, largely because of the vast numbers of insects. Insects are examples of the *invertebrates*, having no backbone. *Vertebrates*, on the other hand, have backbones and include amphibians, fish, birds, reptiles, and mammals. There are about 4,500 species of mammals. Mammals feed their young with milk, are homeotherms (i.e., they can regulate their body temperatures at a constant level), are hairy, and have a four-chambered heart. This book is mainly about the impacts of one mammal, humans, on the rest of life on this planet.

other environmental problems result directly from this transformation of the energy base of society from a renewable to a non-renewable one. In geological terms, we have released the energy input of millions of years in the blink of an eye—the past 250 years. Many current environmental problems are a result of this increase in entropy.

Energy Flows in Ecological Systems

Energy is the basis for all life. The source of virtually all this energy is the sun. More than 150 million kilometres away, the sun, a giant fireball of hydrogen and helium, constantly bombards the Earth with *radiant energy*. This energy, although it is only about 1/50 millionth of the sun's output, fuels our life support system, creates our climate, and powers the cycles of matter discussed in Chapter 4. About a third of the energy received is reflected by the atmosphere back into space (Figure 2.1). Of the remainder, about 42 per cent provides heat to the Earth's surface, 23 per cent causes evaporation of water, and less than 1 per cent forms the basis for our ecological systems. When we think of solar energy, it is important to remember not just the direct heat from the sun but also these indirect forms of energy created by heat input.

Figure 2.1 illustrates the law of conservation of energy. The total amount of energy received by the Earth is equal to the total amount lost. One of the changes caused by human activity is delaying the loss of heat to space by trapping it in the atmosphere through increased levels of heat-trapping gases such as carbon dioxide and methane. The mechanisms and implications of global climate change are discussed more fully in Chapter 7.

Producers and Consumers

The sun's energy is transformed into matter by plants through the process of **photosynthesis** (photo = light, synthesis = to put together). Through this process, plants combine carbon dioxide and water, using energy from the sun, into high-energy carbohydrates such as starches, cellulose, and sugars (Figure 2.2). Green pigments in the plants, called **chlorophylls**, absorb light energy from the sun. Photosynthesis also produces oxygen, some of which is used by plants in various metabolic processes. The rest goes into the atmosphere. Hundreds of millions of years of evolution have served to produce the oxygen in the atmosphere that we depend on for life.

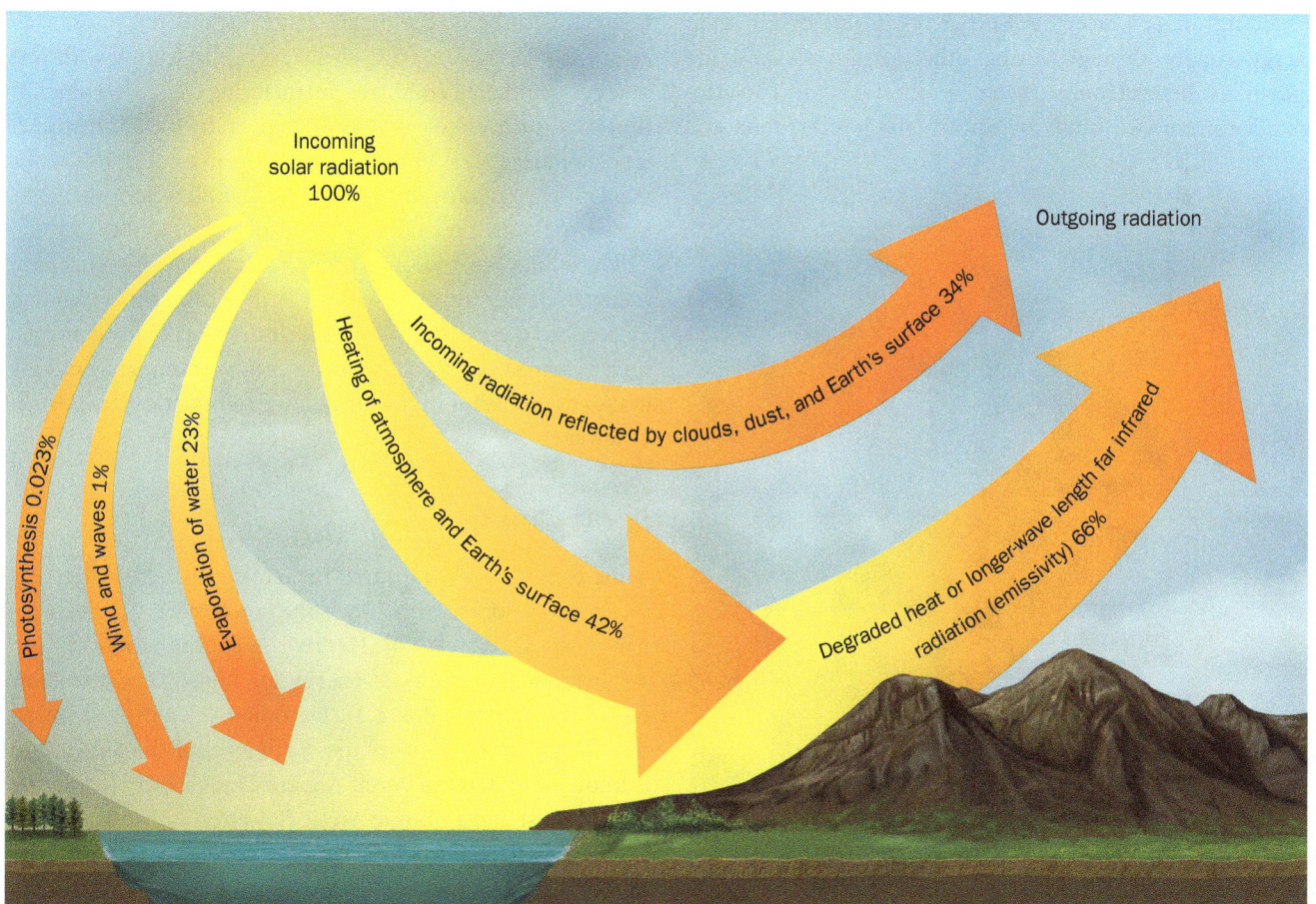

Figure 2.1 | The Earth's energy input and output, a good example of the first law of thermodynamics.

Plants that grow on the forest floor have differing strategies to obtain enough light to survive. Most, such as many ferns, can survive on relatively low light levels. Some, such as the devil's club shown here, grow very large leaves (over 40 cm wide for the devil's club) in order to expose as much photosynthetic surface as possible to the low light levels. The devil's club is a member of the ginseng family and well known among indigenous peoples in western North America for its medicinal properties.

Organisms with the ability to capture energy and manufacture matter are known as **autotrophs** (auto = self, trophos = feeding) or **producers**. All other organisms obtain their energy supply through eating other organisms and are known as **heterotrophs** (heter = different) or **consumers**. There are two kinds of autotrophs, **phototrophs** and **chemoautotrophs**. Phototrophs obtain their energy from light; chemoautotrophs gain their energy from chemicals available in the environment. Although most of us are aware of the critical role played by phototrophs (plants) in our life support system, the chemoautotrophs play an equally critical yet not so visible role (Box 2.2). Most of them are bacteria and play a fundamental role in the biogeochemical cycles, discussed in more detail in Chapter 4.

Phototrophs convert the light energy of the sun into chemical energy, using carbon dioxide and water to produce carbohydrates. The second law of thermodynamics instructs us that some energy will be lost in this transformation; indeed, the efficiency rate is only between 1 and 3 per cent. In other words, 97 to 99 per cent of the energy will be lost. Nonetheless, this conversion is sufficient to produce billions of tons of living matter, or **biomass**, throughout the globe.

Besides photosynthesis, *cellular respiration* is another essential energy pathway in organisms. In both plants and animals, this involves a kind of reversal of the photosynthesis process in which energy is released rather than captured. High-energy organic carbohydrates are broken down through a series of steps to release the stored chemical bond energy. In other words, the potential energy is now realized as kinetic energy in the way described above. This produces the inorganic molecules, carbon dioxide, water, heat (because of the law of entropy), and energy that can be used by the organism for various purposes, such as growth, feeding, seeking shelter, communicating with one another, producing seeds, and maintaining basic physiological functions such as constant body temperature and breathing. Since we are unable to obtain energy from photosynthesis or through chemotaxis, this is how we, and all other organisms unable to fix their own energy, get our energy supplies.

For cellular respiration to occur, most organisms must have access to oxygen or they will die. Such organisms are known as **aerobic** organisms. Some species, anaerobic organisms, such as some bacteria, can survive even without oxygen. This makes them useful in the breakdown of organic wastes, such as sewage.

Food Chains

Some of the energy captured by autotrophs is subsequently passed on to other organisms, the consumers, by means of a **food chain** (Figure 2.3). **Herbivores** eat the producers and are in turn the source of energy for higher-level consumers, or **carnivores** (Box 2.3). Decomposers will feed on all these organisms after they die. Each level of the food chain is known as a **trophic level**. A giant Douglas fir tree on the Pacific coast and a minute Arctic flower on Baffin Island are on the same trophic level—autotrophs. Herbivores, on the second level, range in size from elephants to locusts. The role in energy transformation, rather than the size of the organism, is the important factor in determining trophic level.

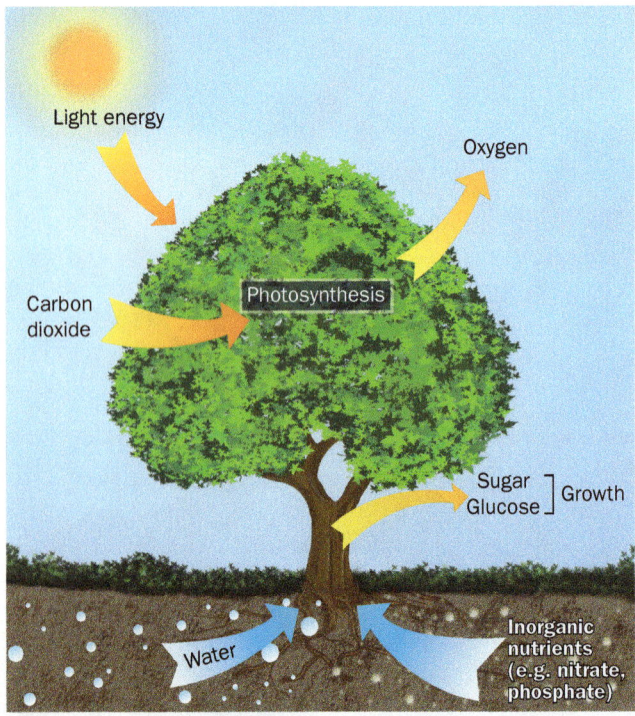

Figure 2.2 | The process of photosynthesis.

Animals as different as the grasshopper and the elephant are on the same trophic level.

Environment in Focus

Box 2.2 Deep-Sea Ecosystems

We think of the deep-sea floor as a biological desert. In the 1970s, however, scientists discovered that rich biological communities were supported at hydrothermal vents on the sea floor, mainly bacteria that derive their energy from sulphide emissions. Similar kinds of chemoautotrophic-based communities were discovered on whale skeletons found at depth, nourished by sulphides produced as the carcasses decay. Discoveries of fossils suggest that dead whales may have provided dispersal stepping stones for these communities for more than 30 million years. The question then becomes—what was the impact on these communities when whales were virtually eliminated from the oceans by whalers? Scientists do not yet have the answer to this question.

However, we do know that the biodiversity of the sea floor is much greater than imagined and may equal that of shallow-water tropical reefs. In addition to the vents, the sea floor contains many other habitats, including cold seeps, seamounts, submarine canyons, abyssal plains, oceanic trenches, and asphalt volcanoes, sure to contain a large number of **endemic species**, with total numbers perhaps as high as 10 million. Although seemingly far removed from human activities, deep seabeds are threatened by many activities, including pollution, shipping, military activities, and climate change. However, fishing is probably the greatest threat as new technologies penetrate the depths and new markets develop.

Deep-sea bottom trawling is a major concern and very damaging to seamounts and the coldwater corals they sustain. These habitats are home to several commercial bottom-dwelling fish species. Seamounts are also important spawning and feeding grounds for species such as marine mammals, sharks, and tuna, which makes them very attractive fishing grounds. Deep-sea fish are particularly vulnerable to large-scale fishing activities because of their long life cycles and slow sexual maturation. Lack of information on deep-sea environments and their species makes it difficult to establish whether sustainable fisheries can take place.

One way to protect such environments is by establishing marine protected areas (MPAs), as discussed in Chapters 8 and 14. Canada has designated Sable Gully, the largest underwater canyon in eastern Canada, as an MPA. The gully is located approximately 200 kilometres off the coast of Nova Scotia at the edge of the Scotian Shelf, where the sea floor suddenly drops by more than two kilometres. More than 70 kilometres long and 20 kilometres wide, this area is home to many interesting and unusual species. The gully is a productive ecosystem that supports a diversity of marine organisms. The world's deepest-diving whale, the bottlenose whale, is a 'vulnerable' species that lives in the gully year-round. Fin whales and northwest Atlantic blue whales, both also classified as 'vulnerable' by the Committee on the Status of Endangered Wildlife in Canada (Chapter 14), make use of the gully throughout the year. Deep-sea corals are a significant feature of the benthic fauna in the area, and nine species are confirmed to live in the gully. MPA regulations prohibit disturbing, damaging, destroying, or removing any living marine organism or habitat within the gully. The MPA contains three management zones, providing varying levels of protection based on conservation objectives and ecological sensitivities. The regulations also control human activities in areas around the gully that could cause harm within the MPA boundary.

Some organisms, such as humans, raccoons, sea anemones, and cockroaches, are **omnivores** and can obtain their energy from different trophic levels. When we eat vegetables, we are acting as **primary consumers**; when we eat beef, we are at the second trophic level, acting as **secondary consumers**; and when we eat fish that have derived their energy from eating smaller organisms, we may be **tertiary consumers** at the top of the food chain. The level at which food energy is obtained has some important implications, to be discussed later.

We tend to concentrate on these **grazing food chains**, but equally important are the decomposer food chains (Figure 2.4). Overall, some 80 per cent of the annually produced plant biomass cycles through the detritus chain rather than being consumed by herbivores. These chains are based on dead organic material or **detritus**, which is high in potential energy but difficult to digest for the consumer organisms described above. However, various species of micro-organisms, bacteria, and fungi are able to digest this material as their source of energy. Indeed, many large grazing animals such as cows and moose have such bacteria in their stomachs to help break down the cellulose in plant material. These decomposers (or saprotrophs) derive their energy from dead matter (sapro = putrid). They are joined by consumers such as earthworms and marsh crabs, known as detritivores, which may consume both plant and animal remains.

A **decomposer food chain** plays an integral role in breaking down plant and animal material into products such as carbon dioxide, water, and inorganic forms of phosphorus and nitrogen and other elements. For example, fungi that consume simple carbohydrates, such as glucose, first break down dead wood. Following this phase, other fungi, bacteria, and organisms such as termites break down the cellulose that is the main constituent of the wood. Were it not for these organisms, wood and other dead organisms would accumulate indefinitely on the forest floor.

Detritus plays a major role in ecosystem processes, as will be discussed in more detail in Chapter 4, as a source of nutrients and within and between ecosystem transfers of energy and matter. Overall plant biomass exceeds animal biomass by a factor of 10 and hence plant biomass has received a lot of attention in this regard. However, in certain ecosystems animal detritus is a main factor in nutrient supply and scientists are only recently beginning to understand the importance of dead salmon, for example, to the health of west coast rain forests in British Columbia (Hocking and Reynolds, 2011). When these nutrient flows are reduced, through large-scale fishing, for example, the repercussions are felt throughout the ecosystem as vegetative growth rates decline over the long term, as well as populations of species higher on the food chain, from insects through to bears.

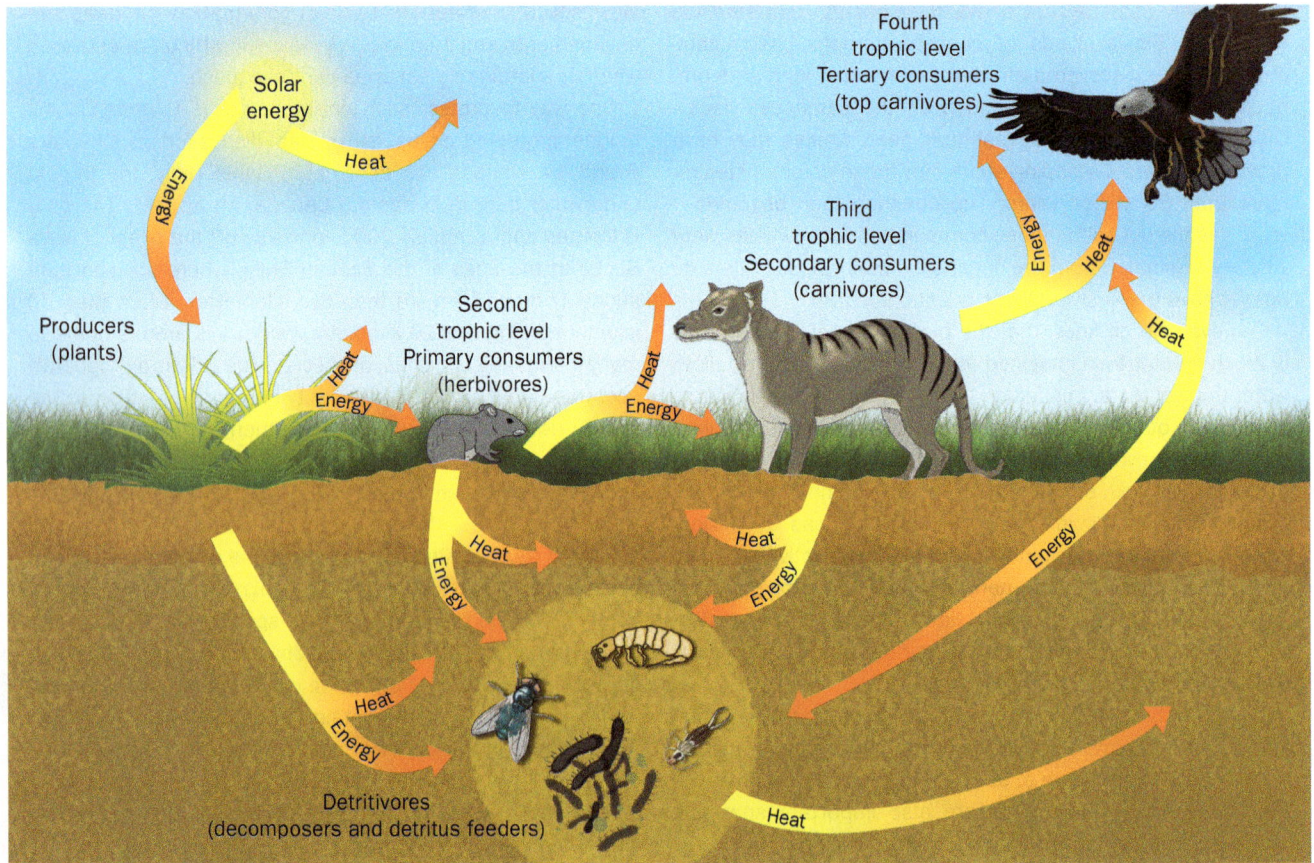

Figure 2.3 | A food chain.

Environment in Focus

Box 2.3 Carnivorous Plants

Not all plants are autotrophs. Carnivorous plants, such as the pitcher plant, the floral emblem of Newfoundland and Labrador, gain their energy from ingesting the bodies of insects that become trapped in their funnel-shaped leaves. The plant, which grows in boggy areas across Canada, has no photosynthetic surfaces. Instead, the leaves act as 'pitchers' to hold a soapy liquid from which a hapless insect cannot escape. The plant may be aided in the decomposition of dead insects by other insects that have developed immunities to the decomposing enzymes produced by the plant. The plant plays host to several insects that seem to thrive on the environment it provides. This is an example of **mutualism** in which both species benefit from a relationship.

The carnivorous pitcher plant, the provincial flower of Newfoundland and Labrador, grows in abundance in eastern Canada.

The relative importance of grazing and detrital food chains varies. The latter often dominate in forest ecosystems, where less than 10 per cent of the tree leaves may be eaten by herbivores. The remainder dies and becomes the basis for the detritus food chain. In the coastal forests of British Columbia, for example, there are some 140 different species of birds, mammals, and reptiles through which energy can flow. In contrast, more than 8,000 known, as well as many unknown, species are involved in breaking down the soil litter. The same is often true in freshwater aquatic systems, where there may be relatively little plant growth but abundant detritus from overhanging leaves and dead insects. However, the converse is true in marine ecosystems (see Box 2.4), where 90 per cent of the photosynthetic **phytoplankton** (phyto = plant,

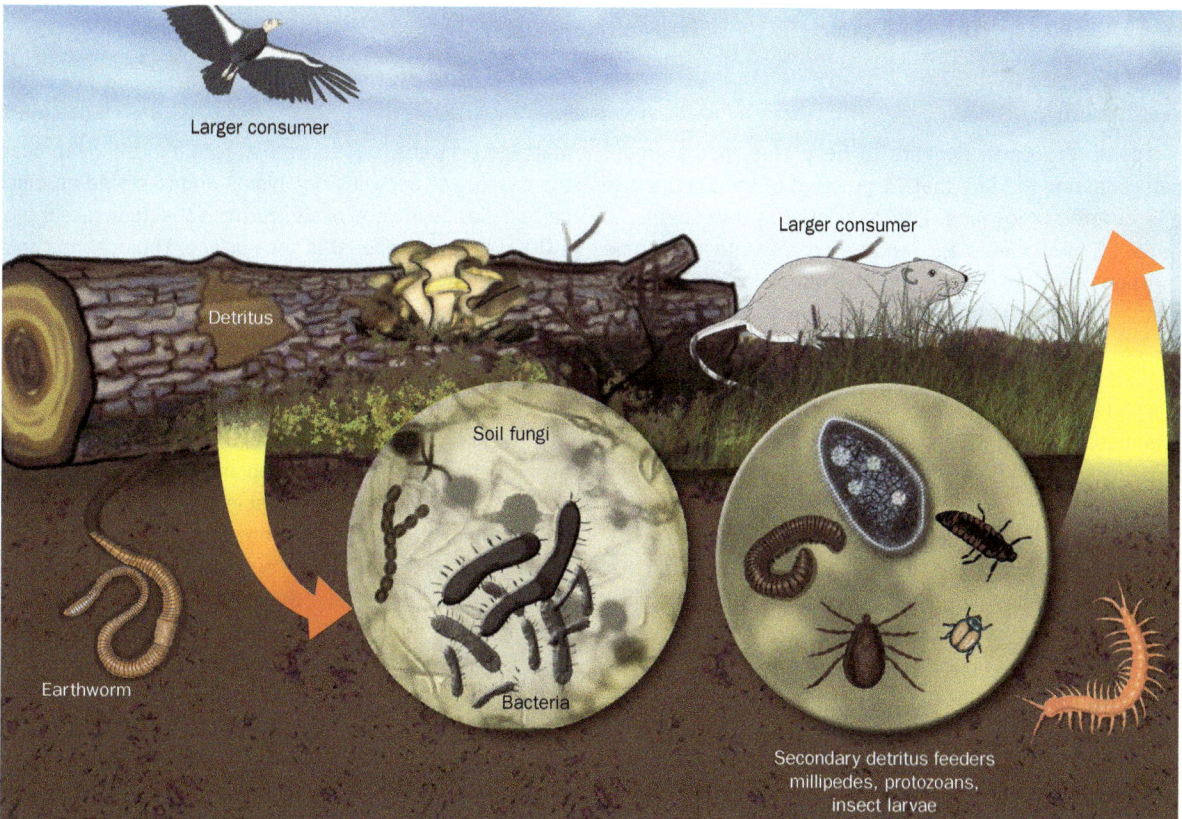

Figure 2.4 | Detritus-based food chain.

plankton = floating) may be grazed by the primary consumers, the **zooplankton**.

In general, ecological theory suggests that the more species in the ecosystem, the more alternative pathways are available for energy flow and the better able the ecosystem is to withstand stress and thereby be resilient. In the Arctic, for example, a simple food chain might be phytoplankton to zooplankton to cod to ringed seal to polar bear. All these species are heavily dependent on the species at the preceding trophic level. Were one of these species to be drastically reduced in number or made extinct, the chances of the role of that species being compensated for by other species is quite low, and the whole food chain might well collapse. This situation can be compared to that at the other extreme, such as a tropical forest, where there are many times more species and the chance of other species combining to fulfill the ecological role of a depleted one is much higher. This situation is sometimes referred to as **ecological redundancy** or **functional compensation** and it assumes that a given role in an ecosystem can be played by more than one species. A competing idea is that species can be likened to the rivets holding an airplane together. Just as the loss of species through extinction increasingly endangers an ecosystem—or the planet as a whole—so the likelihood of the plane in flight disintegrating increases as rivets are lost because no rivets can replace them.

In practice, however, many factors are involved, such as the relative degree of specialization of the various organisms. In some ecosystems there may be examples of functional compensation; others will tend more towards rivet-popping. In general, functional compensation will help build resilience, but care must be exercised before generalizing the theory to all ecological systems.

Rarely are food chains organized in the simple manner shown in Figure 2.3. Usually, there are many competing organisms and energy paths representing **food webs** rather than simple food chains (Figure 2.6).

The number of species increases from the poles to the tropics as conditions become more amenable for life (Figure 2.7). In the Arctic, for example, there are relatively few species and therefore relatively few alternative pathways for energy flow. If a prey species, such as the Arctic hare, decreases in number, then so will the organism dependent on it higher in the food chain, such as the lynx, because there are few other species upon which these organisms can feed. This gives rise to the familiar population cycles in the North as predator numbers closely reflect the availability of dominant prey species (Figure 2.8).

However, as discussed in Box 2.5, this relationship might not be quite so simple, and ecologists have long debated whether ecosystems are mostly controlled by predators

Environment in Focus

Box 2.4 Oceanic Ecosystems

From space, the Earth appears to be a blue, not a green, planet, reflecting the fact that 71 per cent of the Earth's surface is covered by oceans. Life originated in this blueness, perhaps 3.5 billion years ago, and only came onto land some 450 million years ago. Hence, much of our biological ancestry lies within these waters. Although we know about more different species on land than in the oceans, the number of phyla, distinguished by differences in fundamental body characteristics, is higher in the oceans. Of the 33 different animal phyla, for example, 15 exist exclusively in the ocean, and only one is exclusively land-based. We share the same phyla as the fishes, the chordata, characterized by a flexible spinal cord and complex nervous system.

Through their photosynthetic activity, the early bacteria that started in the oceans helped to create the conditions under which the rest of life evolved. Current photosynthetic activity is no less important to our survival.

Scientists estimate that the phytoplankton in the sunlit or **euphotic zone** of the oceans (10 metres to 200 metres in depth) produce between one-third and one-half of the global oxygen supply. In doing so, they also extract carbon dioxide from the atmosphere. Some 90 per cent of this is recycled through marine food webs, but some also falls into the deep ocean as the detritus of decaying organisms and is stored as dissolved carbon dioxide in deep ocean currents that may take more than 1,000 years to reappear at the surface. The oceans contain at least 50 times as much gas as the atmosphere and are playing a critical role in helping to delay the so-called greenhouse effect, discussed in more detail in Chapters 7 and 8.

These phytoplankton, so important to atmospheric regulation, are also the main autotrophic base for the marine food web. From tiny zooplankton through to the great whales, almost every marine animal has phytoplankton to thank for its existence. Phytoplankton flourish best in areas where ocean currents return nutrients from the deep ocean back to the euphotic zone. This occurs in shallow areas, such as the Grand Banks, where deep ocean currents meet the coast or where two deep ocean currents meet head on. Such areas are the most productive in what is generally an unproductive ocean, and they are the best sites for fisheries. Ninety per cent of the marine fish catch comes from these fertile near-shore waters. Unfortunately, these waters are also the sites of greatest pollution. The blueness of most of the rest of the ocean is a visible symbol of the low density of phytoplankton. That is why the sea is blue, not green.

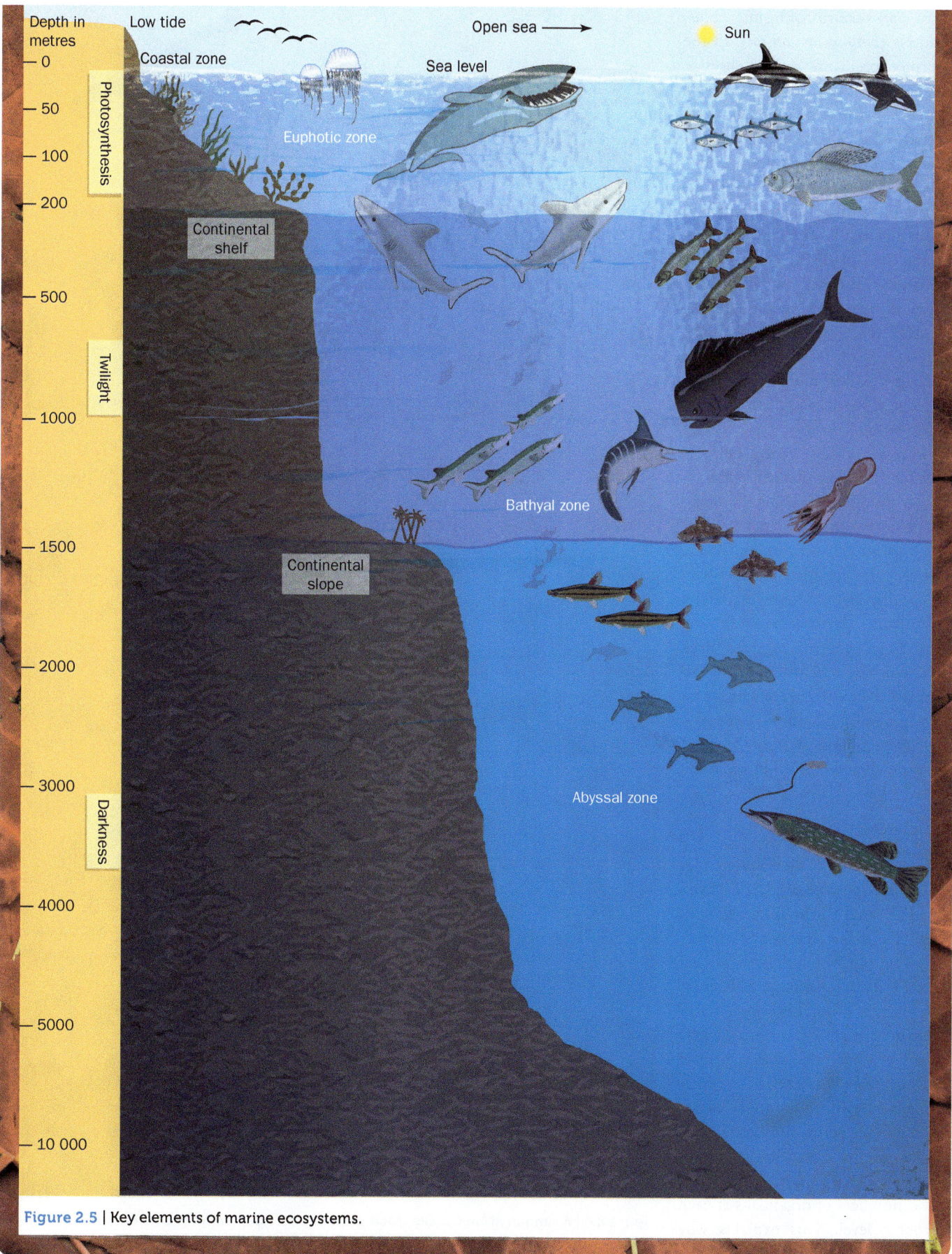

Figure 2.5 | Key elements of marine ecosystems.

(*top-down control*) or by prey populations (*bottom-up control*). In the former, predators restrict the size of the prey population. This seems to occur, for example, when wolves control deer, elk, or moose populations. Conversely, in some systems, the quality of available forage limits the number of herbivores, which in turn limits the number of predators. The predator numbers are essentially limited by the energy flow through the previous trophic levels. If herbivore populations fall as a result of disease or lack of forage, this will result in a drop in predator populations because of a lack of food. Ecosystems in which controls are dominantly bottom-up tend to have marked limits on plant productivity through abiotic factors, such as low nutrient supply, lack of water, and similar factors, or very close relationships between a specific plant, a herbivore, and a carnivore. Ecosystems reflecting top-down control typically lack these features. However, as with most ecological phenomena, these are general guidelines—most ecosystems contain elements of both top-down and bottom-up control.

Biotic Pyramids

The second law of thermodynamics describes how energy flows from trophic level to trophic level, with a loss of usable energy at each succeeding transformation. In natural food chains, the **energy efficiency**, or amount of total energy input of a system that is transformed into work or some other usable form of energy, may be as low as 1 per cent. In general, we expect about 90 per cent of the energy to be lost at each level (Figure 2.9). Similar losses may be experienced in biomass and numbers of organisms at each trophic level. This explains why there are fewer secondary than primary consumers and fewer tertiary

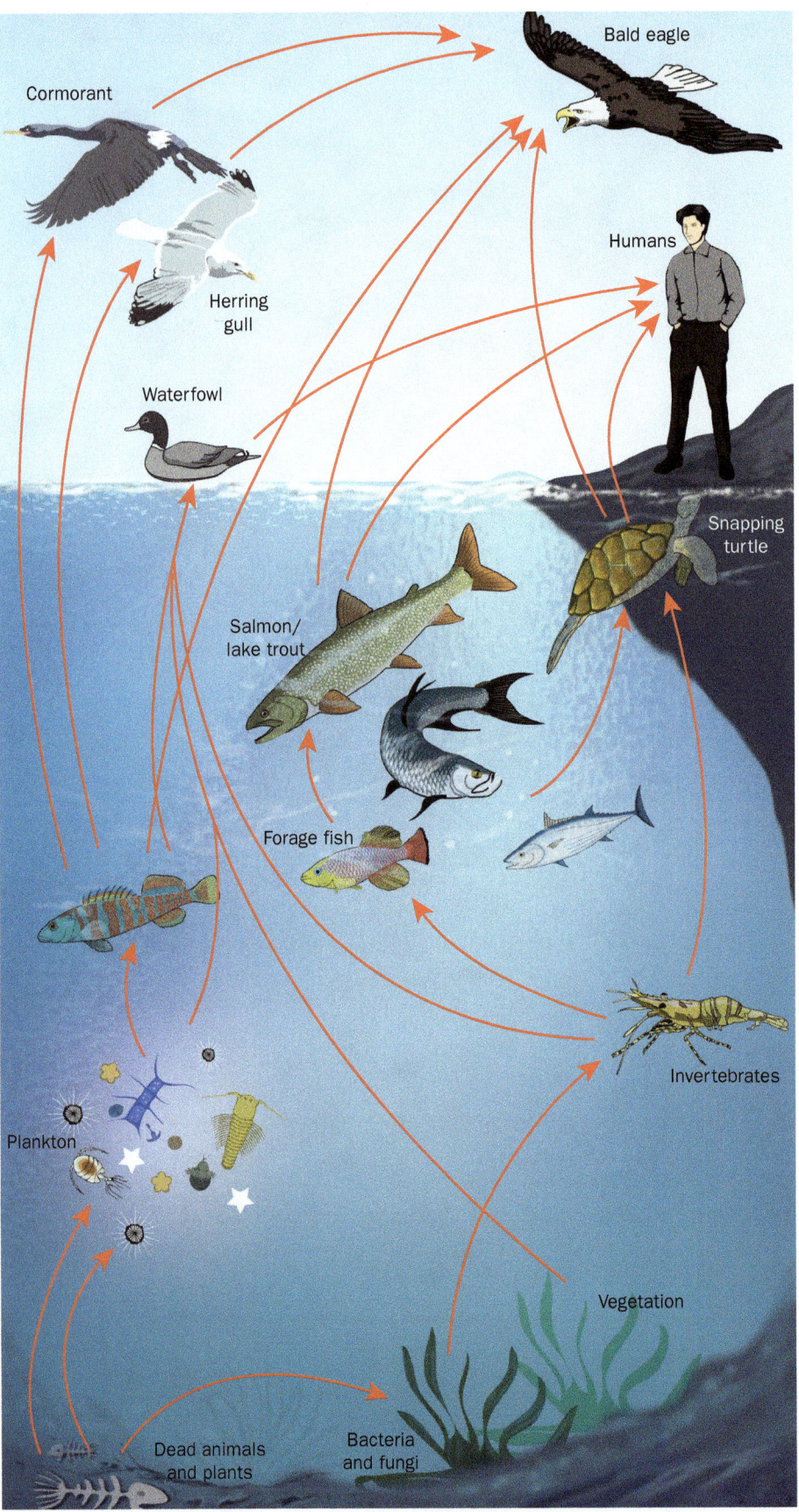

Figure 2.6 | A simplified Great Lakes food web. Source: Adapted from Environment Canada. 1991. *Toxic Chemicals in the Great Lakes and Associated Effects*. Toronto: Department of Fisheries and Oceans, Ottawa: Health and Welfare Canada.

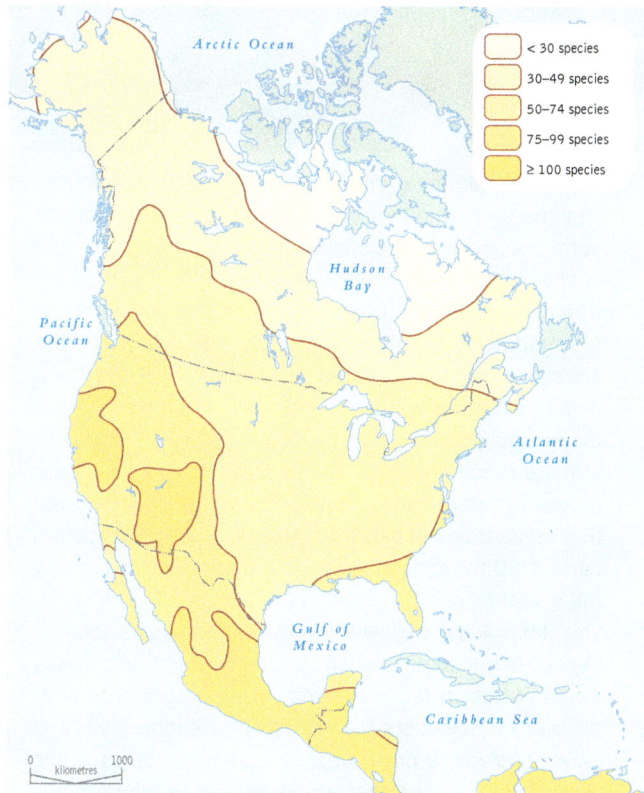

Figure 2.7 | The number of mammal species per latitude. *Source: After Simpson (1964).*

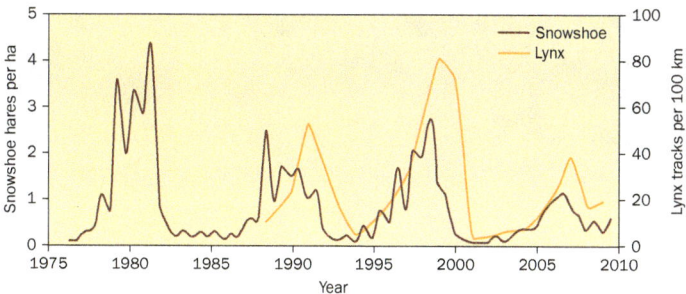

Figure 2.8 | Snowshoe hare and lynx cycles, boreal forest, Kluane, Yukon. *Source: Federal, Provincial, and Territorial Governments of Canada (2010: 101).*

than secondary. Carnivores must always have the lowest numbers in an ecosystem in order to be supported by the energy base below. The case of the Atlantic puffins, described in the introduction to this chapter, provides a good example. The biomass of carnivores (puffins) could no longer be supported by the energy from the preceding trophic level, the capelin.

Some ecosystems, however, may display an inverted **biomass pyramid**. In natural grasslands such as those found in southern Saskatchewan, the dominant species, such as grasshoppers, are small-bodied and do not have a large biomass. In contrast, many herbivores in this system, such as antelope and mule deer, are large-bodied and long-lived, with a large total biomass. The same situation exists in the oceans, discussed in more detail in Chapter 8. However, in both cases, the productivity of the plant base is much greater than that of the herbivores.

Several reasons exist for the low energy efficiencies of natural food chains. First, not all the biomass at each trophic level is converted into food for the next trophic level. Many organisms have developed characteristics to avoid getting eaten by something else. For example, many plant species have thorns or produce secondary chemicals to deter herbivores. Others have low nutritive levels. Generally, only between 10 and 20 per cent of the biomass of one trophic level is harvested by the next level. Furthermore, of that which is consumed, not all is digested. Humans, for example, are not well equipped to break down and consume the bones or fur of animals, nor are they equipped, compared to moose and other members of the deer family, to break down woody tissue. The proportion of ingested energy actually absorbed by an organism is the **assimilated food energy**. Finally, as cellular respiration occurs to liberate energy for the growth, maintenance, and reproduction of the organism, energy is further released as heat.

The longer the food chain, the more inefficient it is in terms of energy transformation, reflecting the second law of thermodynamics. An Arctic marine food chain that starts from the producers (phytoplankton) to primary consumers (zooplankton) that are subsequently grazed by the largest animals ever to exist on earth (whales) is very efficient because it is so short, with only three energy transformations in which energy is lost. Longer food chains involve a proportionately larger loss of energy because of the greater number of energy transformations. Entropy dictates that long food chains with five or six trophic levels, such as that supporting a killer whale, are very scarce.

The energy pyramid also has important implications for humans. For example, it takes between eight and 16 kilograms of grain to produce one kilogram of beef. This means that more land must be cultivated to provide people with a diet high in meat as opposed to a diet based on grains. Since humans are one of the species that can access food energy at several different trophic levels, in terms of energy efficiency it would be better to operate as low on the food chain as possible—that is, as primary consumers or vegetarians. This topic is discussed in more detail in Chapter 10.

Productivity

Productivity in ecosystems is measured by the rate at which energy is transformed into biomass, or living matter, and is usually expressed in terms of kilocalories per square metre per year. In terrestrial ecosystems, the large majority of production comes from vascular plants with much smaller amounts from algae, mosses, and liverworts. In the oceans,

Environment in Focus
Box 2.5 Arctic Population Fluctuations

Populations of many animals in the North show distinctive fluctuations in numbers over regular time periods. A three-to-four-year cycle is recognized for smaller animals, such as lemmings and meadow voles, and a longer one of nine to 10 years for larger animals, such as snowshoe hares and several species of grouse and ptarmigans. The predators of these herbivores show similar fluctuations in numbers as their energy source becomes critically depleted.

One of the themes that this book illustrates is the uncertainty surrounding many aspects of environmental management. In this case, there is a fair amount of certainty about the dates of fluctuations in these populations, but there is considerable uncertainty as to why these fluctuations occur. Four main ideas have been advanced:

1. The seasonality of the Arctic environment means that plants grow for only a few months every year, yet herbivores must eat throughout the year. Thus, when population levels rise, overexploitation of the food supply can occur rapidly and for a relatively long period, but the plants will take a long time to recover. The relationship between the herbivores and their food supply is the determining factor, and the predator numbers simply reflect those of the prey.
2. A second hypothesis builds on this idea but suggests that as populations of small animals such as lemmings increase, then more nutrients vital to plant productivity become tied up in this higher trophic level. This lack of nutrients causes reductions in plant productivity and quality, leading to starvation for the herbivores.
3. Other ideas postulate more of an interaction between the predators and prey. Keith et al. (1984) studied the ecology of snowshoe hares in northern Alberta to test a food supply–predation hypothesis. It suggested that food supply shortages halt populations of herbivores that are subsequently caught by predators until numbers fall enough to permit plant recovery. Keith et al. reported that malnutrition of hares was evident but that predators caused 80 to 90 per cent of deaths, thereby supporting the hypothesis.
4. The final idea suggests that food supplies play a negligible role in population cycles and that these cycles would not occur in the absence of predators (Trostel et al., 1987). Scientists have found that sudden drops in hare populations occur despite supplementary feeding programs, a finding that would appear to support this hypothesis (Smith et al., 1988).

In short, ecological systems are not easy to understand. In many cases, several factors may contribute to changes such as in population cycles.

---

most production comes from algae, although some vascular plants, such as sea grasses, have been found to have very high rates of production and sequestration of carbon. **Gross primary productivity** (GPP) is the overall rate of biomass production, but there is an energy cost to capturing this energy. This cost, *cellular respiration* (R), must be subtracted from the GPP to reveal the **net primary productivity** (NPP). This is the amount of energy available to heterotrophs.

All ecosystems are not equal in their ability to fix biomass. Light levels, nutrient availability, temperature, and moisture, among other factors, regulate the rates of photosynthesis. The most productive ecosystems per unit area are estuaries, swamps and marshes, and tropical rain forests (Figure 2.10). Recent data indicate that the temperate rain forests, such as those in the Pacific Maritime ecozone, are just as productive as the tropical forests. Other ecosystems are more limited because of deficiencies in one or more of the characteristics noted above. A desert, for example, lacks water, the Arctic lacks heat, and the ocean lacks nutrients.

Unfortunately, in Canada some of our most highly industrialized and polluted lands are adjacent to estuaries. There is not one sizable estuary on the east coast of Vancouver Island, for example, that is not used by the logging industry either as a mill site or for log storage. The estuary of the Fraser River has been extensively dyked, industrialized, and polluted. Paradoxically, relative few data exist on the effects of these intrusions on our most productive ecosystems. However, one estuary, the Musquash in the Bay of Fundy, should experience little further degradation. It has been declared a marine protected area under Canada's Oceans Act (Chapter 8), and one of the main characteristics noted for its designation was its high productivity. It is one of the last ecologically intact estuaries in a region that has seen extensive modification of its salt marshes.

Between 1985 and 2006, primary productivity increased markedly on over 22 per cent of Canada's vegetated surface and declined on only 1 per cent (Federal, Provincial, and Territorial Governments of Canada, 2010). The main growth areas are in the North where global climate change has been most strongly felt and where successional processes, as discussed in the next chapter, appear to be accelerating. Similar trends have been noted at the global level where increases in tropical productivity are being attributed to reduced cloud cover.

Humans take about 40 per cent of terrestrial NPP for their own use. The remainder supports all the other organisms on Earth, which in turn maintain the environmental conditions that keep us alive. The human population is projected to increase by about 40 per cent over the next 50 years. It is highly doubtful that the Earth's systems could withstand a

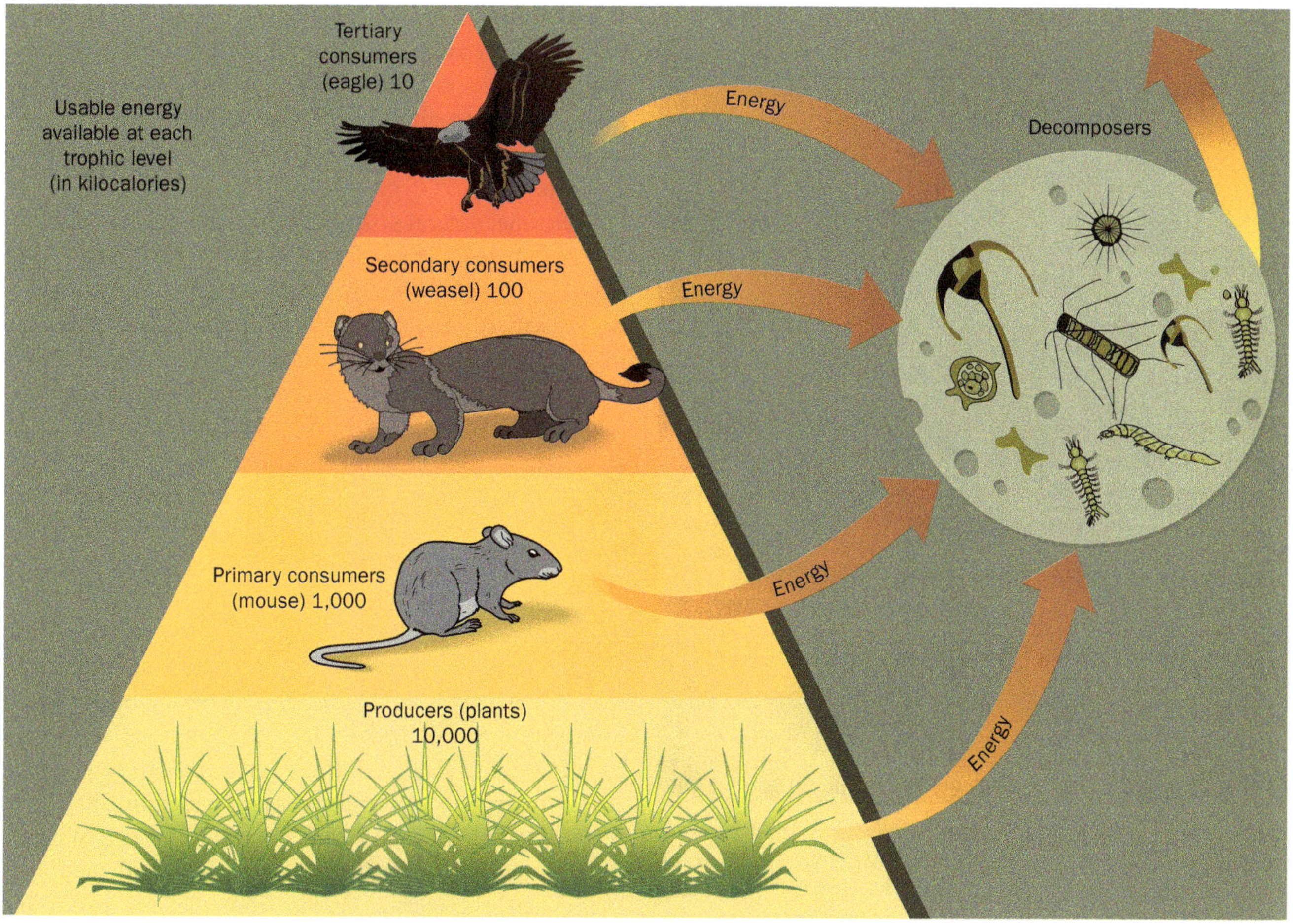

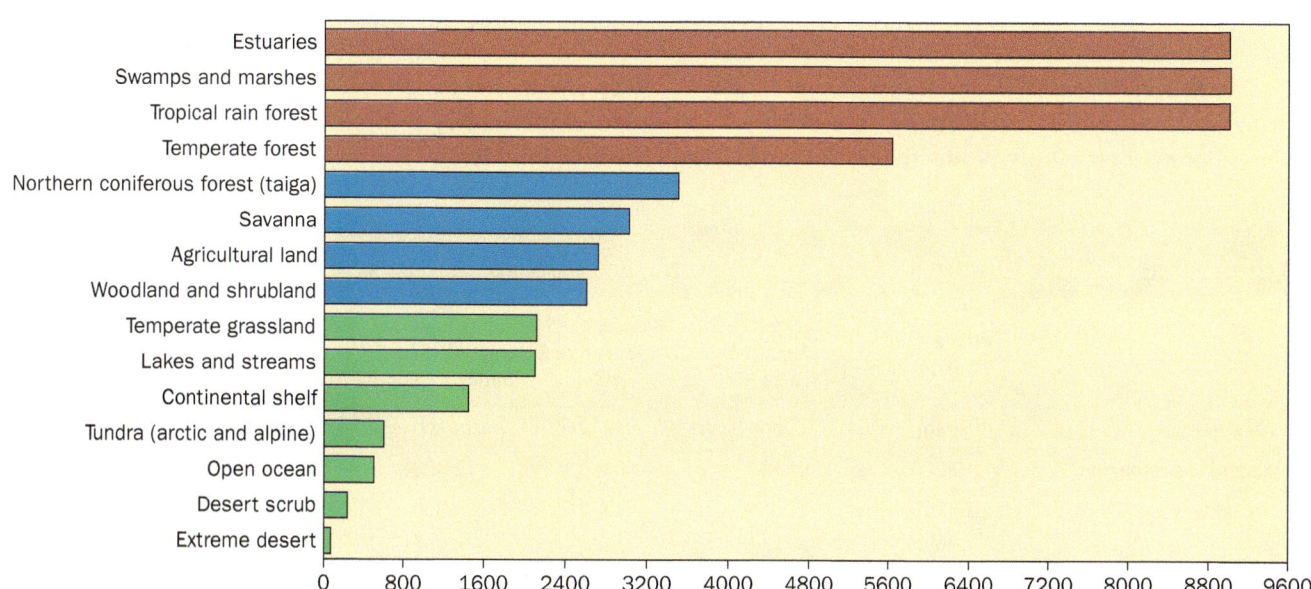

Figure 2.9 | Generalized pyramid of energy flow.

Figure 2.10 | Estimated annual average net productivity of producers per unit of area in principal types of life zones and ecosystems. Values are given in kilocalories of energy produced per square metre per year.

Estuaries are among the most productive ecosystems. Unfortunately, they are also very convenient sites for industrial activity, such as the log boom storage seen here in Campbell River, British Columbia, which inhibits productivity.

heterotrophic and autotrophic respiration. Measurements indicate that as communities mature, although GPP and NPP rise, an increasing proportion of the energy of the community is devoted to heterotrophic respiration (Table 2.1). In mature communities, the amount of respiration may be sufficient to account for all the energy being fixed by photosynthesis. There is thus no net gain, leading to the characterization of such communities by some foresters as 'decadent' because they are mainly interested in the productivity of the autotrophs.

Over time, natural systems mature towards maximization of NCP. On the other hand, humans are often concerned with maximizing NPP. This is an example of the 'decision regulators' discussed in Chapter 1. Natural forest system decision regulators may allow trees to achieve ages of several hundred to more than a thousand years before they die. The control system exerted by forest management determines that the life of the trees will be that which maximizes NPP, before considerable amounts of energy become devoted to heterotrophic respiration (Figure 2.11). The age of the trees in systems managed for forestry will hence be much younger than in natural systems.

Auxiliary energy flows allow some ecosystems and sites to be especially productive. For example, tidal energy in an estuary is a form of auxiliary energy flow that helps to bring in nutrients and dissipate wastes so that organisms do not have to expend energy on these tasks and can devote more energy to growth. Agriculture, as discussed in Chapter 10, relies extensively on the inputs of auxiliary energy in the form of pesticides, fertilizer, tractor fuel, and the like to supplement the natural energy from the sun to augment crop growth. In many cases, this subsidy, mostly derived from fossil fuels, exceeds the amount of energy input from the sun. Without this subsidy, productivity would be much reduced. There is a cost to the subsidy, however, in terms of high energy costs and the environmental externalities created as the subsidy disperses into the environment in the form of pollution.

concomitant increase in the amount of NPP being appropriated for human use—another illustration of the carrying capacity challenge we face.

In addition to primary productivity, measurements can be made of **net community productivity** (NCP), including

Table 2.1 | Production and Respiration as (kcal/m2/yr) in Growing and Climax Ecosystems

	Alfalfa field (US)	Young pine plantation (England)	Medium-aged oak-pine forest (NY)	Large flowering spring (Silver Springs, Florida)	Mature rain forest (Puerto Rico)	Coastal sound (Long Island, NY)
Gross primary production	24,400	12,200	11,500	20,800	45,000	5,700
Autotrophic respiration	9,200	4,700	6,400	12,000	32,000	3,200
Net primary production	15,200	7,500	5,000	8,800	2,500	2,500
Heterotrophic respiration	800	4,600	3,000	6,800	13,000	2,500
Net community production	14,400	2,900	2,000	2,000	little or none	little or none

Source: Adapted from: E.P. Odum. 1971. *Fundamentals of Ecology*. 3rd edn, p. 46. Toronto: Holt, Rinehart and Winston, CBS College Publishing. Copyright © 1971 by Saunders College Publishing.

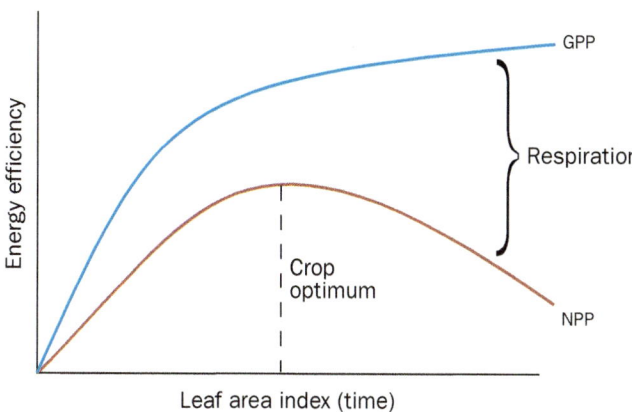

Figure 2.11 | Diagram to illustrate general relationship between productivity and time as a forest matures. Foresters might consider the optimal stage of the forest to be at maximum NPP, even though GPP continues to increase over time.

Ecosystem Structure

The energy flows described above are all part of the ecosphere. The ecosphere can be broken down in size to smaller units. At the smallest level is the individual **organism**. A group of individuals of the same species is a **population**. The populations in a particular environment are known as a **community**. The **ecosystem** is a collection of communities interacting with the physical environment. However, ecosystems represent a somewhat abstract conceptualization of the environment that can range greatly in scale. Because of the highly interactive nature of the relationship between organisms and their environment, it is often difficult to define precisely the boundary of an ecosystem. Furthermore, ecosystems are open systems in that they exchange material and organisms with other ecosystems. Ecosystems and communities thus provide useful abstractions for the study of the environment but should not be taken as precise categories that will be agreed upon by all scientists.

Similar ecosystems can be grouped together as ecozones, representing their dominant vegetation and animal communities. The main ecozones in Canada are shown Figure 2.12. In turn, these can be grouped into the largest classification of life forms, **biomes**, based upon dominant vegetation and adaptations of other organisms to that particular environment. Globally, six main biomes are recognized: marine, freshwater, forest, grassland, desert, and tundra. Canada, because of its size, has as many biomes as any country in the world. The main factors that control biome distribution are water availability and temperature. Figure 2.13 summarizes how these factors influence biomes on the global scale.

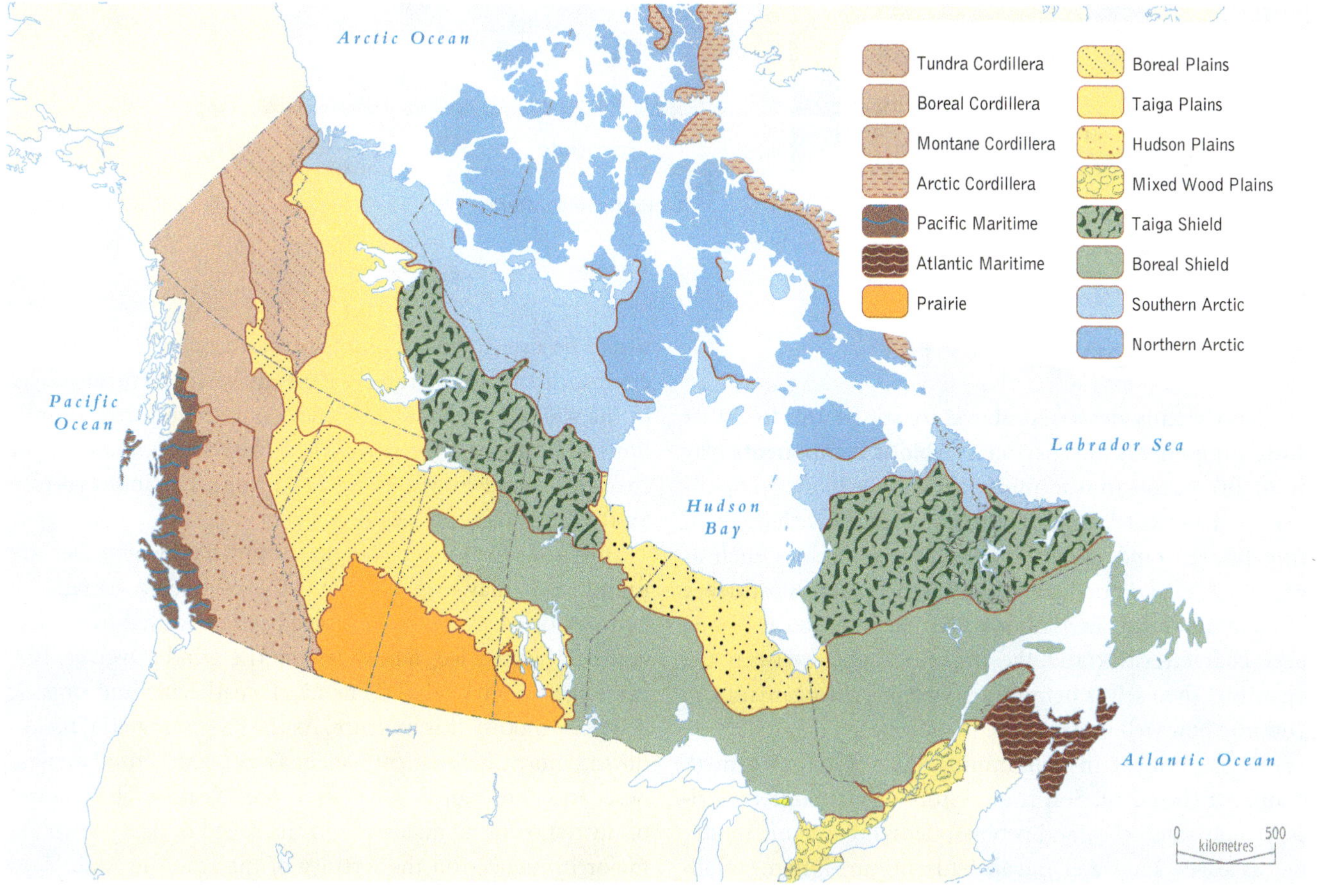

Figure 2.12 | Terrestrial ecozones of Canada. Source: *Terrestrial Ecozones of Canada,* http://sis.agr.gc.ca/cansis/nsdb/ecostrat/zones.gif, © Agriculture and Agri-Food Canada, 1995. Reproduced with the permission of the Minister of Public Works and Government Services, 2011.

Figure 2.13 | Influence of temperature and rainfall on biome.

Abiotic Components

The food chains described above constitute the living or **biotic components** of ecosystems. **Abiotic components** play an important role in determining how these biotic components are distributed. Important abiotic factors include light, temperature, wind, water, and soil characteristics such as pH, soil type, and nutrient status. All these factors influence different organisms in various ways. The interaction among these characteristics and the organisms and between the organisms themselves determines where each organism can grow and how well it may grow.

Soils are critical in determining the vegetation growth of an area (Box 2.6). Soil is a mixture of inorganic materials such as sand, clay, and pebbles, decaying organic matter such as leaves, and water and air. This mixture is home to billions of micro-organisms that are constantly modifying and developing the soil. In the absence of these organisms, earth would be a sterile rock pile rather than a rich life-supporting environment. Most of these organisms are in the surface layer of the soil, and one teaspoon may contain hundreds of millions of bacteria, algae, and fungi. In addition, many larger species—roundworms, mites, millipedes, and insects—play vital roles in this complex ecology.

Most soils form from the **parent material** where they are found. This may originate from the weathered remains of bedrock or where sediments have been deposited from elsewhere by water, ice, landslides, or the wind. Over centuries, ongoing physical and chemical weathering and organic activities modify this mixture. As the parent material breaks down, inorganic elements such as calcium, iron, manganese, and phosphorus (Chapter 4) are released. The amount of nutrients in the material and the speed of breakdown are major influences on the fertility of the resulting soil. Hard rocks, such as granite, break down slowly and yield few nutrients. Different soils will result, depending on the location.

These various processes result in different layers forming in the soil, called **soil horizons**. A view across these horizons is called a **soil profile**. Figure 2.15 shows a generalized profile. However, not all soils have all these different horizons.

Time is also a critical factor in soil development. Soils that have been exposed to millions of years of chemical and physical weathering, such as many tropical soils, have often lost their entire nutrient content. Conversely, where glaciation has scraped all the soil away, as in much of Canada as recently as 10,000 years ago, then the hard rocks, such as the granite of the Canadian Shield, have had little opportunity for weathering. They are also infertile. However, soils can be very fertile where ice sheets, as they retreated, deposited large quantities of clay rich in nutrients, as on the Prairies.

Environment in Focus

Box 2.6 Soils in Canada

Just as we can define ecozones, soil scientists can define soil zones that group together soils that are relatively similar in terms of their measurable characteristics. A glance at the soil map of Canada (below) will reveal a close resemblance to the ecozone map, since at this scale both tend to reflect the gross climatic and geological conditions of the region. The Canadian System of Soil Classification includes nine orders, the largest category of classification:

Brunisols cover 8.6 per cent and are brown soils found mainly under forests.
Chernozems cover 5.1 per cent, occur under grasslands, and are some of the most productive soils.
Cryosols are the dominant soils in Canada, covering some 40 per cent of the country's land mass, and are found in association with permafrost.
Gleysols only cover 1.3 per cent and are found in areas that are often waterlogged.
Luvisols cover 8.8 per cent and occur in a wide variety of wooded ecosystems. They have higher clay content than brunisols.
Organics cover 4.1 per cent of Canada, form in wetland ecosystems where decomposition rates are slow.
Podzols are found beneath heathlands and coniferous forests, are relatively nutrient poor, and cover 15.6 per cent of Canada.
Regosols cover less than 1 per cent of Canada and vary little from their parent material.
Solonets are saline soils, covering 0.7 per cent of Canada, and are found mostly in grassland ecosystems.

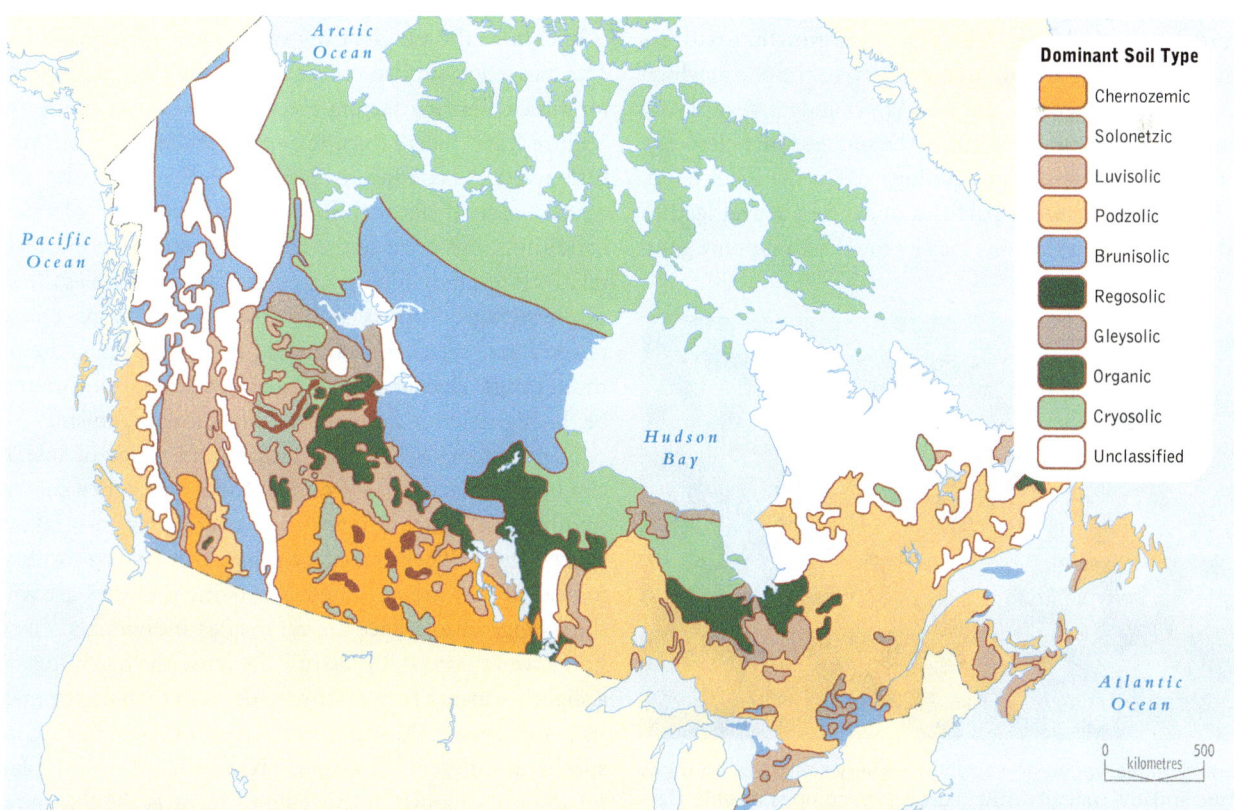

Figure 2.14 | Soil zones of Canada.

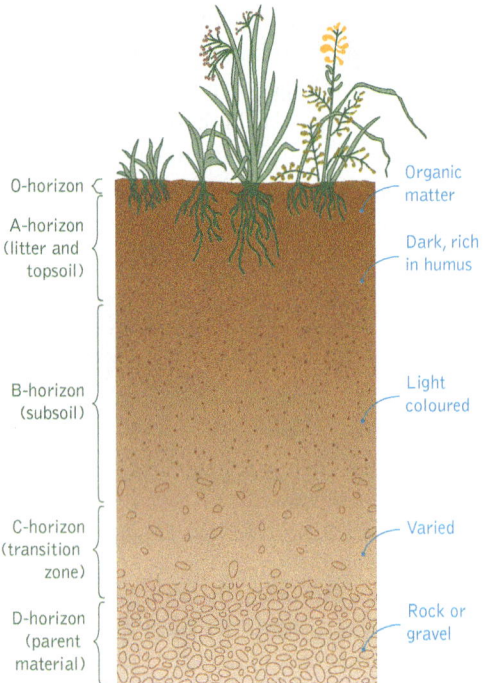

Figure 2.15 | Generalized soil profile.

Soils also differ in their texture, or sizes of different materials. Clay is the finest, followed by silt, sand, and then gravel, the coarsest. Soils that contain a mixture of all these with decomposed organic material, or **humus**, are called **loams** and often make the best soils for vegetation growth. Texture is a main determinant of **soil permeability**, or the rate at which water can move through the soil. Water moves very slowly through soils composed mainly of the smallest particles, clay, and the soil easily becomes waterlogged. On the other hand, the large spaces between particles of sand or gravel lead to rapid drainage, and the soils may be too dry to support good

In northern climates, vegetation structure is very simple, as both low temperature and low rainfall result in growth conditions in which few species can survive. In the short Arctic summer, however, areas of tundra are ablaze with brightly coloured flowers, such as these mountain avens and oxytropis on Victoria Island, Nunavut.

vegetation growth. Plants obtain their nutrient supply necessary for growth from ions dissolved in the soil water, and so permeability is critical.

Soil has many different chemical characteristics. One of the most important is the pH value (see Chapter 4), measuring the acidity/alkalinity of the soil, which helps to determine which minerals are available and in what form. Different plants have different mineral requirements. Farmers often try to change the acidity of their soils, for example, by adding lime if the soil is too acidic or sulphur if the soil is too alkaline.

Just as the laws of thermodynamics explain energy flows, some principles help us to understand how organisms react to different abiotic influences. The first of these is known as the **limiting factor**. This principle tells us that all factors necessary for growth must be available in certain minimum quantities if an organism is to survive. Thus, a surplus of water will not compensate for an absence of an essential nutrient or adequate warmth. In other words, a chain is only as strong as its weakest link. The weakest link is known as the **dominant limiting factor**. A major goal of agriculture is to remove the effect of the various limiting factors. Thus, auxiliary energy flows are employed to ensure that a crop has no competition from other plants (weeding), or that water supply is adequate (irrigation), or that the plant has optimal nutrient supply (fertilizer) (Chapter 10).

The corollary of the above is that all organisms have a range of conditions that they can tolerate and still survive. This is known as the **range of tolerance** for a particular species. This range is bounded on each side by a zone of intolerance for which limiting factors are too severe to permit growth (Figure 2.16). There may, for example, be too much or too little water. As conditions improve for the particular factor, certain individuals within the population can tolerate the conditions, but because the conditions still are not optimal, relatively few individuals can exist. This is known as the **zone of physiological stress**. Still further amelioration creates a range where conditions are ideal for that species, the **optimum range**. Here, in theory, barring other factors, there will be the highest population of the particular organism.

The concepts of limiting factors and range of tolerance can be illustrated by an example from Saskatchewan, where smallmouth bass are introduced every year into lakes for sport fishing. These hatchery-raised fish survive from year to year (if they are not caught), proving that they are within their range of tolerance for survival as individuals. They do not, however, breed successfully, because the hatchlings need a slightly warmer temperature to develop than do the mature fish to survive. Thus, different levels of tolerance exist for species at different life stages. The dominant limiting factor for the smallmouth in this environment is the low temperatures during hatching.

Water availability is often the critical factor that determines differences between communities. Where precipitation is in

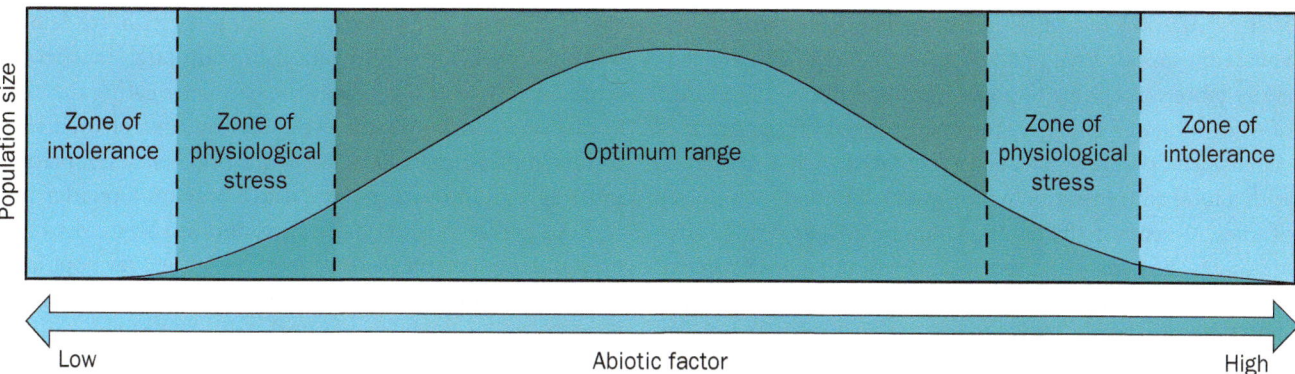

Figure 2.16 | Range of tolerance.

excess of about 1,000 millimetres per year, for example, trees will usually dominate the landscape if other factors are suitable. Below 750 millimetres, the range of tolerance for trees is exceeded, and grasses will dominate because they have a tolerance for water stress in the order of 100 millimetres per year. Below that level, even grasses run into their zone of intolerance, and cacti, sagebrush, and other drought-resistant species dominate.

We must keep in mind that organisms are not reacting to just one abiotic factor, such as water availability, but to all the factors necessary for growth. Sometimes the optimal range for one factor will not overlap with the optimal range for other factors, which would place the organism in the zone of physiological stress for that factor, and it would become the dominant limiting factor. Organisms may also be out-competed for a particular factor in their optimum range by another organism with a greater tolerance to that environmental factor and again be forced into a zone of physiological stress. In other words, the simple single-factor model represented in Figure 2.13 is more complicated because numerous abiotic and biotic influences must also be taken into account. The model does, however, provide a useful conceptual tool to help understand the spatial distribution of organisms.

Biotic Components

Other species also have an important role in influencing species distributions and abundance. Species interact in several ways, including competition for scarce environmental resources. Each species needs a specific combination of the physical, chemical, and biological conditions for its growth. This is known as the **niche** of that species. Where the species lives is known as the **habitat**.

The **competitive exclusion principle** tells us that no two species can occupy the same niche in the same area. Most species have a *fundamental niche*, representing the potential range of conditions that they can occupy, as well as a narrower *realized niche*, representing the range actually occupied. The physical conditions for growth exist throughout the fundamental niche, but the species may be out-competed in parts of this

The panda is a classic example of a specialist species.

Some species, such as the black bear, are very adaptable and have a relatively broad range of tolerance. On the Pacific coast, black bears are frequent scavengers of the intertidal zone, where many items are considered potential food. The year-round availability of an abundant and varied food source results in very large individuals.

area through the overlapping requirements of other species. **Specialist** species have relatively narrow niches and are generally more susceptible to population fluctuations as a result of environmental change. Many endangered species are specialists. The panda is a classic example of such a specialist, with a total concentration on one plant, bamboo, as a source of food. Whenever the bamboo supply falls, as it does after it flowers, this specialist species has few suitable alternative sources of food. Historically, when bamboo was abundant this did not particularly matter, because the pandas simply moved to a new area. However, as the animals have become increasingly restricted to smaller and more isolated reserves, it has become a major problem.

In Canada, specialist species include many of those discussed in later chapters, such as the burrowing owl and the whooping crane. **Generalist species**, on the other hand, like the black bear and coyote, may have a very broad niche, where few things organic are not considered a potential food item. Such generalist species have adapted most successfully to the new environments created by humans.

Competition

Intraspecific competition occurs among members of the same species, whereas **interspecific competition** occurs between different species. Both forms of competition are a result of demands for scarce resources. Intraspecific competition occurs particularly where individual species densities are very high. Interspecific competition occurs where species niches are similar. Competition may be reduced through **resource partitioning** in which the resources are used at different times or in different ways by species with an overlap of fundamental niches. Hawks and owls, for example, both hunt for similar types of prey but at different times, since owls are mainly nocturnal.

Intraspecific competition may lead to the domination of specific areas by certain individuals; the area is known as a **territory** and may be aggressively defended against intruders. Grizzly bears establish such territories, which may be as large as 1,000 square kilometres for dominant males, although the possibility of defending such a large territory from intruders at all times is remote. During the breeding season, male robins establish and defend nesting territories, the boundaries of which are advertised in song. This kind of behaviour aims to establish sufficient resources for breeding pairs to be successful. Ultimately, intraspecific competition contributes to regulation of population size in areas where favourable habitat is limited, since those individuals unable to defend territories are outcast to less favourable areas where their likelihood of success is limited.

Biotic Relationships

There are other kinds of relationships between species besides competition. In predation, for example, a **predator** species benefits at the expense of a **prey** species. The lynx eating the hare and the osprey eating the fish are familiar examples of this kind of relationship, although in a broader sense we should also consider the herbivore eating the plant. Predation is a major factor in population control and usually results in the immediate death of the prey species. A predator must be able to overwhelm and kill prey on a regular basis without getting hurt. Usually, predators are bigger than their prey and often target weaker members of the prey population to avoid getting injured. They may also hunt as a group to improve the likelihood of a kill and minimize the possibility of a debilitating injury.

One theory that addresses the relationship between the benefit of making a kill and feeding against the cost of the energy expended to make the kill is **optimal foraging theory**. The theory recognizes that there is a point of compensation between the benefit of obtaining the prey and the costs of doing so and that the predator's behaviour adjusts to optimize the benefits. It may be more worthwhile, for example, to hunt a smaller prey more often, even though it will result in less food intake, if the smaller prey can be dispatched with little fear of injury and eaten quickly so that another predator cannot steal it. Optimal foraging theory also suggests that as one type of prey becomes scarce, most predators will switch prey if they can. Several examples of this kind of behaviour are discussed within the marine context in Chapter 8.

Prey species have evolved many strategies to avoid being transferred along the food chain. Some plants develop physical defences such as thorns, while others may evolve chemical defences such as poisons to deter their predators. The chemicals manufactured by plants provide the raw material for many of our modern medicines, such as aspirin, which comes from willows. Animal species employ a wide variety of predator avoidance strategies ranging from camouflage, alarm calls, and grouping to flight (Box 2.7).

This rather stunned-looking red fox has just escaped from the clutches of a large female golden eagle that managed to lift it some three metres from the ground before one of the authors happened on the scene and unwittingly rescued one carnivore from becoming the prey of another one at the next trophic level.

Rubber is a natural product made by rubber tree to help deter herbivores. Depicted here is tree-tapping on Koh Lanta, Western Thailand.

These Maldivian anemone fish live in a mutualistic relationship with the anemone, which provides protection from potential predators with its stinging tentacles while benefiting from the food scraps brought by the fish.

A special kind of predator–prey relationship is **parasitism**, where the predator lives on or in its prey (or host). In this case, the predator is often smaller than the prey and gains its nourishment from the prey over a more extended time period that may lead to the eventual death of the host. This may cause the death of the parasite too, although some parasites, such as dog fleas and mosquitoes, can readily switch hosts. Tapeworms, ticks, lamprey, and mistletoe are all examples of parasites.

Not all relationships between species are necessarily detrimental to one of the species. **Mutualism** is the term used to describe situations in which the relationship benefits both species. These benefits may relate to enhanced food supplies, protection, or transport to other locations. The relationship between the nitrogen-fixing bacteria and their host plants, described in Chapter 4, is an example of such a relationship that results in enhanced nutrition for both species. Other examples include the relationship between flowering plants

Environment in Focus

Box 2.7 Canada's Olympic Champion: The Pronghorn Antelope

The pronghorn is the sole survivor of a family of North American antelopes, unrelated to the African antelopes, that at one time contained more than two dozen different species. Pronghorn are migratory, moving north in summer and south in winter, at which times they may gather together in large numbers. Their main predators (other than humans)—the plains wolf and grizzly—have become extirpated from their range, leaving the coyote and, farther south, the bobcat as predators. Unlike all other members of the bovid family (sheep, oxen, and goats), pronghorns shed their horns after the rut. Populations of pronghorn are now protected in Grasslands National Park in Saskatchewan.

Pronghorn are reckoned to be the fastest middle-distance runners on Earth, having been clocked at speeds of up to 98 kilometres per hour for more than 24 kilometres. They are superbly adapted for speed, with small stomachs and big lungs, long strides, small feet, flexible spines like the cheetah, and additional ligaments and joints, all designed to make this animal the fastest on the prairie. Speed is the logical defence mechanism for a vulnerable animal easily seen by predators on the open prairies. Confidence in their escape mechanism made them rather curious, and early settlers found them easy to attract and shoot. Initial herds estimated at between 20 and 40 million animals in North America were soon reduced to fewer than 30,000 animals as a result of hunting pressure and land-use change. There are now some half a million pronghorn remaining.

Guest Statement
Landscape Ecology
Chris Malcolm

The key to successful environmental resource management is an understanding of the ecological relationships between the myriad of environmental components that exist in space and time. These components are connected in a hierarchy from the micro-scale (energy and nutrient flows), through the meso-scale (organisms, populations, and communities), to the macro-scale (ecosystems and the biosphere). For example, in eastern North American deciduous forests the production of acorns by oak trees, termed masting, in one year can influence the prevalence of Lyme disease, spread by black-legged ticks, two to three years later. A good production of acorns attracts white-footed mice and white-tailed deer. The deer spend up to 40 per cent of their time in the forest in high mast years and only 5 per cent in low mast years, a spatial pattern of habitat selection based on differential acorn production on a temporal scale. The mice host a bacterium, *Borrelia burgdorferi*, which they spread to the ticks, which both increase in numbers as the mice population grows, and spread to the deer that spend more time in proximity to the mice. As the deer move about, in and out of the forest, they expand the range of the ticks. The ticks, carrying *Borrelia burgdorferi*, can cause Lyme disease if they come into contact with humans. Further complicating this relationship, the mice often eat the pupae of gypsy moths, which controls the moth population. If the moths outbreak, however, they feed on oak tree leaves, reducing acorn production and ultimately the potential for Lyme disease (Jones et al., 1998).

There are a number of approaches to understanding relationships across space and time at various scales. Various terms describe the meso-scale approach, including *landscape ecology*, *landscape connectivity*, *spatial ecology*, and *ecological integrity*. 'Landscape' is the common term at this scale, although it does not have a universal definition, which can make development of management policies with respect to landscape-scale phenomena confusing. However, those of us who study wildlife movements generally define 'landscape' as a heterogeneous area of land composed of a mosaic of habitat types. Landscapes come in the form of the patchwork of tundra, permafrost lakes, spruce krumholtz, bogs, and fens of the Subarctic, the glaciated network of kettle lakes, granite outcrops, and coniferous forests of the boreal Canadian Shield, or the transition of temperate rain forest to rocky tidal pools and kelp forests on British Columbia's west coast.

Landscape ecology is the science of studying and attempting to improve the relationships between spatial patterns and ecological processes on a multitude of spatial scales and organizational levels (Wu and Hobbs, 2007). **Landscape connectivity** is the degree to which the landscape facilitates or restricts movement between and among habitat patches (Taylor et al., 1993). Landscape connectivity can further be divided into **structural** and **functional** connectivity. Structural connectivity focuses solely on the physical relationships between habitat patches such as fragmentation, corridors, or distances between them, while functional connectivity includes the behavioural responses of organisms to structural connectivity. *Spatial ecology* has a decided geographical emphasis that examines how the spatial arrangements of organisms, populations, and landscapes influence ecological dynamics (Collinge, 2010). *Ecological integrity* describes a natural system in which the interconnected web of components and processes, from nutrient and energy flows to populations of species within complex communities, are intact and functioning (Wipond and Dearden, 1998). Ecological integrity tends to be an idealized concept, as gaining a holistic understanding of a landscape through one of the other approaches described above is extremely difficult!

As an ecological biogeographer, my students and I have been examining patterns of movement within fish populations at the landscape scale. We have been examining functional connectivity for northern pike, both in regard to natural and anthropogenic structure. In the southern portion of Riding Mountain National Park, Manitoba, Clear and South Lakes are separated by a narrow sand barrier bar several metres wide. Clear Lake is a large, deep, mesotrophic lake, while South Lake is small, shallow, and eutrophic. Northern pike spawning habitat in Clear Lake is rare and poor in quality, while South Lake provides prime spawning habitat. However, the pike can only enter South Lake if the spring melt breaks through the barrier bar, creating a temporary corridor. This does not occur every year. In the spring, prior to spawning, we placed VHF transmitters on 40 northern pike in Clear Lake and watched

Measuring pike from the Little Saskatchewan River.

as 39 of them entered South Lake when a corridor opened in the barrier bar. We placed micro-VHF transmitters in the oviduct of 19 of the 40 pike, which would be expelled with eggs. We relocated 15 of the micro-transmitters, all in South Lake. We were able to demonstrate that the northern pike population in Clear Lake is dependent on a natural, ephemeral connectivity to South Lake; one that requires conservation of the landscape in a manner to allow this process to continue.

As anthropogenic habitat fragmentation increases there is a pressing need to understand its impact on connectivity. The Little Saskatchewan River, in southwestern Manitoba, was divided into five disjunct stretches by a series of dams and weirs between 1820 and 1960. These barriers impeded upstream movement of fish. Between 1992 and 2004, fish ways were constructed around three of the dams. Again using VHF telemetry, we discovered that connectivity up and down the river is extremely important. Pike routinely climb fish ways and fall back down over dams, often more than once a year. They also show site fidelity outside of the spawning season, repeatedly returning to the same location within days or weeks, although it might require travelling back and forth around a dam. We even recorded two pike that swam 120 kilometres upstream! Landscape genetics, the study of how landscape features influence population genetics, has revealed to us that there is no significant genetic variability in northern pike within the river system. This connectivity within the Little Saskatchewan River would not have occurred during the period between dam and fish way construction.

Sometimes it isn't functional connectivity but the dynamic nature of limiting factors related to connectivity of habitat components that affects wildlife habitat selection. Back in Clear Lake, 30 metres deep, lives a small benthic fish, the slimy sculpin (*Cottus cognatus*). In early summer, like many other temperate lakes, Clear Lake stratifies into two thermal layers. The epilimnion, a warmer upper layer, remains connected to mixing processes at the surface, which help to maintain dissolved oxygen levels. The hypolimnion, a lower colder layer, does not mix with the epilimnion, becomes disconnected from the surface oxygen source, and dissolved oxygen levels decline over the summer. Dissolved oxygen is a limiting factor for fish presence, which in turn can be used to measure ecological health. In co-operation with Parks Canada, my students and I have discovered that during the summer dissolved oxygen levels in the hypolimnion of Clear Lake can decline to a level at which slimy sculpins must move out of their preferred habitat in the deepest water. Are these low levels of dissolved oxygen natural or anthropgenically enhanced? Have dissolved oxygen levels always lowered to levels that require slimy sculpins to move to areas of higher oxygen concentrations, or have humans contributed to lower levels by perhaps increasing eutrophic processes in Clear Lake?

One of the wonderful yet problematic aspects of ecological research at the landscape scale is the great expanse of unknown causal relationships. There's so much to learn, so many mysteries to solve! But at the same time, natural resource management is fraught with difficult decisions and controversy in the face of the unknown.

Chris Malcolm is an Associate Professor in Geography at Brandon University.

and their pollinators, which results in the transport of pollen to other plants, and the protection offered by ants to aphids in return for the food extracted from plants by the aphids. Interactions that appear to benefit only one partner but do not harm the other are examples of **commensalism**. The growth of **epiphytes**, plants that use others for support but not nourishment, is one example.

Keystone Species

Species with a strong influence on the entire community are known as **keystone species**. They are named after the final wedge-shaped stone laid in an arch. Without the keystone, all the other stones in the arch will collapse. In Canada, our national symbol, the beaver, is a good example of such a species (Box 2.8). Beavers can have a profound impact on their environments through the dams they build that raise and lower water levels. This, in turn, affects the limits of tolerance of other species in the community that may suddenly find themselves submerged under a beaver dam or facing lower water levels downstream. Different species will have different reactions to this change, depending on, for one thing, their range of tolerance relating to water. However, when a keystone species is removed, there is generally a cascading effect throughout the ecosystem as other species are affected. The same species may be a keystone in some communities and not in others, depending on the community composition in that particular locale.

It is especially significant when a keystone species is removed from an area, or **extirpated**, by human activity. Such changes may take some time before they become noticeably manifest. Changes to soil characteristics created by the extermination of major herbivores, such as bison from the prairie, may take centuries before they become noticeable and are generally not reversible. The same is true for the other large grazers, such as the great whales (discussed in Chapter 8), which have been decimated over the last couple of centuries.

Biodiversity

Over billions of years, interaction between the abiotic and biotic factors through the process of evolution, discussed in more detail in the next chapter, has produced many different life forms. **Biodiversity** is the sum of all these interactions, and high biodiversity is often taken as a surrogate for healthy ecosystems. The year 2010 was designated by the United Nations as the International Year of Biodiversity to celebrate life on Earth and recognize its value.

Biodiversity is usually recognized at three different levels:

1. **Genetic diversity** is the variability in genetic makeup among individuals of the same species and the ultimate source of biodiversity at all levels. In general, genetic diversity in a population increases the ability to avoid inbreeding and withstand stress.
2. Species are life forms that resemble one another and can interbreed successfully. **Species diversity** is the total number of species in an area and is also known as *species richness*.
3. **Ecosystem diversity** is the variety of ecosystems in an area. Some ecosystems are more vulnerable to human interference than others (Box 2.8). Estuaries and wetlands, for example, are highly productive but are often used for industrial and agricultural activities. As these ecosystems are replaced by human-controlled ecosystems, natural diversity at the landscape level is reduced.

Scientific knowledge of biodiversity is primitive (Box 2.9). There may be up to 100 million species, although most scientific estimates suggest between 5 million and 20 million, of which we have identified some 1.8 million (Figure 2.17). Some 56 per cent of these species are insects, 14 per cent are plants, and just 3 per cent are vertebrates such as mammals, birds, and fish. Even new mammals are still being discovered, such as the giant muntjac and saola discovered recently on the borders of Vietnam and Laos. However, most species awaiting discovery are probably invertebrates, bacteria, and fungi in the tropics. We also know relatively little about the ocean (Chapter 8). Only about 15 per cent of described species are from the oceans. Most biologists agree that there are fewer species to

Environment in Focus

Box 2.8 Carolinian Canada

Carolinian Canada is the wedge of land stretching from Toronto west to Windsor that contains 25 per cent of the country's human population. It also contains the highest number of tree species in the country as the mixing zone between the eastern deciduous forests to the south and mixed coniferous-deciduous forests to the north. Its location in the southernmost part of the country, with the mediating effects on climate of the southern Great Lakes, allows semi-tropical tree species such as the cucumber and sassafras to spread up into this land of ice and snow. After Vancouver and Victoria, Windsor ranks as the third warmest city in Canada, and it is the most humid city in the country. It is little wonder that in summer the humidity and southern vegetation can give the appearance of a much more southern location.

Besides the distinctive vegetation, the Carolinian zone also supports a noteworthy bird population. Point Pelee is one of the top birding spots in North America. Not only do migrating birds (exhausted from crossing Lake Erie on their migrations north in spring) rest here, but it is also part of the Carolinian forest and the nesting habitat for many species that are unusual for Canada, particularly warblers. Of the 360 bird species seen, about 90 stay to nest, and in spring there may be 25 to 30 different warblers spotted on a good day. In the fall, the birds are joined by thousands of monarch butterflies as they pause here before heading south on their 3,600-kilometre journey to the Gulf of Mexico for the winter.

Point Pelee is a national park. Most of the rest of the Carolinian forest is not so well protected and is heavily fragmented by agriculture and urban development. It is estimated that close to 40 per cent of Canada's rare, threatened, and endangered species are primarily Carolinian. The Carolinian Canada Program, started in 1984, has co-ordinated the efforts of government agencies and private landowners to try to protect the remaining forest. About half of the 38 targeted sites have some degree of protection, but biologists still worry that these fragments are too small and isolated to be capable of protecting this most diverse area of Canada.

A black-throated blue warbler male alights near the Lake Erie shoreline during the spring migration.

Environment in Focus
Box 2.9 Counting Critters

Since actually discovering and describing new species is a very slow process, scientists have devised several means to help estimate just how many species we might have on Earth.

Species-area curves are one of the most popular approaches. Generally, the larger the area, the larger the number of species found there, but the new species gradually level off as the same species are encountered repeatedly. This relationship has enabled scientists to construct species-area curves that show the number of species likely to be found in areas of different sizes and hence predict how many species might be found in larger unsampled areas.

Rain forest insect samples work from the idea that most of the species we have yet to describe are probably rain forest insects. Studies indicate that there are up to 1,200 beetle species in the canopy of a single tree in Peru. Since we know that 40 per cent of insects are beetles, this suggests that there may be more than 3,000 insect species in the canopy alone. About half this number will be found lower down and on the trunk, leading to estimates of 4,500 species on one tree. Given that there are more than 50,000 species of tropical tree, the numbers that may be found in total will be vast. Some researchers predict that there are more than 30 million insect species.

Ecological ratios can be used to predict the populations of little-known groups from their relationships with better-known groups. For example, the ratio of fungus to plant species in Europe is 6:1. If this holds worldwide, then there should be more than 600,000 species of fungus. At the moment, fewer than 70,000 have been described.

None of these approaches (and these are just three of the more popular ones) can deliver an absolute estimate of the numbers of species on the planet. Each one, however, is consistent in indicating that we have a long way to go before we can claim that we truly know the nature of life on this planet.

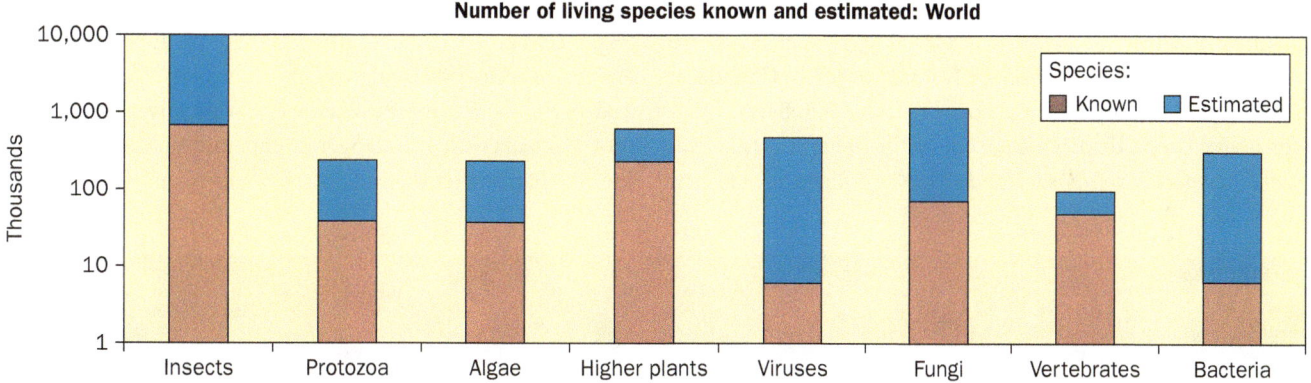

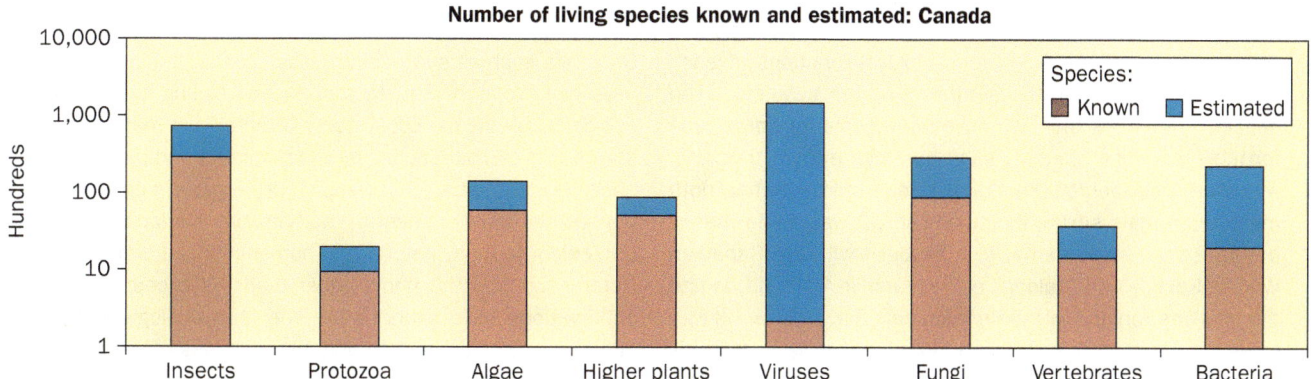

Figure 2.17 | Numbers of known and estimated living species in the world and in Canada. Source: B. Groombridge. 1992. *Global Biodiversity: Status of the World's Living Resources*, p. 17. London: Chapman and Hall. Reprinted with kind permission of Springer Science and Business Media B.V.

be found there than on land. On the other hand, there are 32 phyla in the oceans, compared with only 12 on land.

Species identification is only the first building block in biodiversity. We also need to understand the differences in genetic diversity within species and how species interact in ecosystems to really understand how the life support system of the planet works. However, even at the species level our knowledge is limited (Box 2.10), but we do know from many authoritative sources, such as the Millennium Ecosystem Assessment, that biodiversity is declining at unprecedented rates as a result of human pressures, and has been identified as the most stressed of all planetary systems by a team of international scientists (Rockstrom et al., 2009), as mentioned in Chapter 1. Extinction as an ecological process is considered in more detail in the next chapter and the main reasons behind these declines and possible solutions are considered in Chapter 14.

Biodiversity in Canada

Biodiversity is not evenly distributed around the world. Some biomes, mainly tropical forests, are extremely diverse (Box 2.11), but in temperate latitudes there is much less diversity. Overall, as discussed earlier, species numbers decline in a gradient from the tropics to the poles. Latin America, for example, is home to more than 85,000 plant species. North America has 17,000, of which only 4,000 occur in Canada. A similar gradient for birds is shown in Table 3.5. Several reasons have been advanced to account for the latitudinal gradient in species richness. Rohde (1992), following a review of the various reasons, concludes that many factors contribute at different scales but the primary cause appears to be the effect of solar radiation (i.e., temperature) that increases evolutionary speed at lower latitudes. For this reason most **biodiversity hotspots**, areas with high numbers of endemic species, are found mainly in tropical forest areas.

Estimates suggest that Canada has more than 140,000 different species, of which about half have been named. The taxonomic groups containing the most numbers of species are shown in Figure 2.18. Mosquin (1994) points out that although the groups represented in this graph are not as well known as other groups, such as birds and mammals, they undertake key functions in ecosystems, often functions that we are only just becoming aware of and that support the more familiar and larger organisms. Beneficial insects, for example, fertilize flowers and control pests; crustaceans provide food for fish; bacteria recycle nutrients; and bread, beer, and penicillin all come from fungi. The Canadian Endangered Species Conservation Council provides five-year assessments of the status of more than 7,000 species in Canada (Chapter 14).

Another important element of biodiversity is the concept of endemism. **Endemic species** are ones found nowhere else on Earth. In Canada, we have relatively few endemic species compared, for example, to southern Africa, where some 80 per cent of the plants are endemic, or southwest Australia, where 68 per cent are endemic. In Canada there are approximately 54 endemic species of vascular plants, mammals, freshwater fish, and molluscs. Examples include the Vancouver Island marmot (Canada's only endangered endemic mammal species), the Acadian whitefish, and 28 species of plants in the Yukon.

Environment in Focus

Box 2.10 Biological Uncertainty

Attention has been drawn to the high degree of biological uncertainty regarding the numbers of species in Canada and elsewhere, not to mention the ecological functions of each species. Few things in the natural world are absolute. Even gender differences blur. We have known this for some time for more primitive species, such as slugs and earthworms, which are hermaphrodites—that is, one individual has both male and female sexual functions. Still, it was somewhat of a surprise to Charles Francis, a biologist with the Canadian Wildlife Service undertaking field research in Malaysia, to find the first free-ranging wild male mammals that lactate. He was collecting Dayak fruit bats when he found that several males had breasts and milk. He speculates that this characteristic may have evolved among males that are monogamous. Another theory suggests that something in the bats' diet contains steroids that mimic female hormones.

There is also uncertainty regarding many of the basic ecological principles described in these last two chapters. Concepts such as succession, ecosystems, and communities have been found unsatisfactory for addressing many real-life ecological problems. Ecosystems often exhibit many different states that can be reached by various paths and that may produce no identifiable local climax. Furthermore, species distributions along environmental gradients may overlap broadly and be subject to many natural and human-induced disturbances. Heterogeneity is the norm rather than the exception. These observations have changed the way that ecologists look at ecological systems, recognizing the openness of systems, the importance of episodic events, and the numerous possibilities for intervention in ecological processes. Nonetheless, the classical concepts provide a useful background against which to organize this new, more flexible, 'non-equilibrium' approach.

Most of the world's species are insects, and many more await discovery.

Reasons for our low endemism include the recent glaciation over most of the country, which effectively wiped out localized species, and the wide-ranging nature of many of our existing species. In terms of protecting biodiversity, it is especially important that endemic species are given consideration.

Canada, as a requirement of the international **Convention on Biological Diversity**, produced the Canadian Biodiversity Strategy as a framework for action to conserve and sustainably use biodiversity, and the strategy was endorsed in 1996 by federal, provincial, and territorial ministers. Since that time, the Auditor General of Canada has undertaken three audits to assess progress, all of which pointed out significant weaknesses in the approach being taken (Auditor General of Canada, 2005b).

In response, in 2006 the federal, provincial, and territorial governments produced an outcomes-based framework that provides a more systematic approach towards national biodiversity priority identification, monitoring, and achievement. It seeks to produce a set of national outcomes relating to healthy and diverse ecosystems, viable populations of species, genetic resources, and adaptive potential and sustainable use of biological resources. The outcomes will be achieved by the framework based on an adaptive ecosystem-based approach (see Chapter 6) involving four steps, as shown in Figure 2.19.

As part of this process, in 2010 Canada produced the first assessment of biodiversity from an ecosystem perspective. Although there were some positive trends—for example, the amount of land in protected areas (see Chapter 14) has increased and populations of some marine mammals appear to be improving—the overall findings are not encouraging. In particular the report suggests action is urgently needed to address key findings:

> These findings include loss of old forests, changes in river flows at critical times of the year, loss of wildlife habitat in agricultural landscapes, declines in certain bird populations, increases in wildfire, and significant shifts in marine, freshwater, and terrestrial food webs. Some contaminants recently

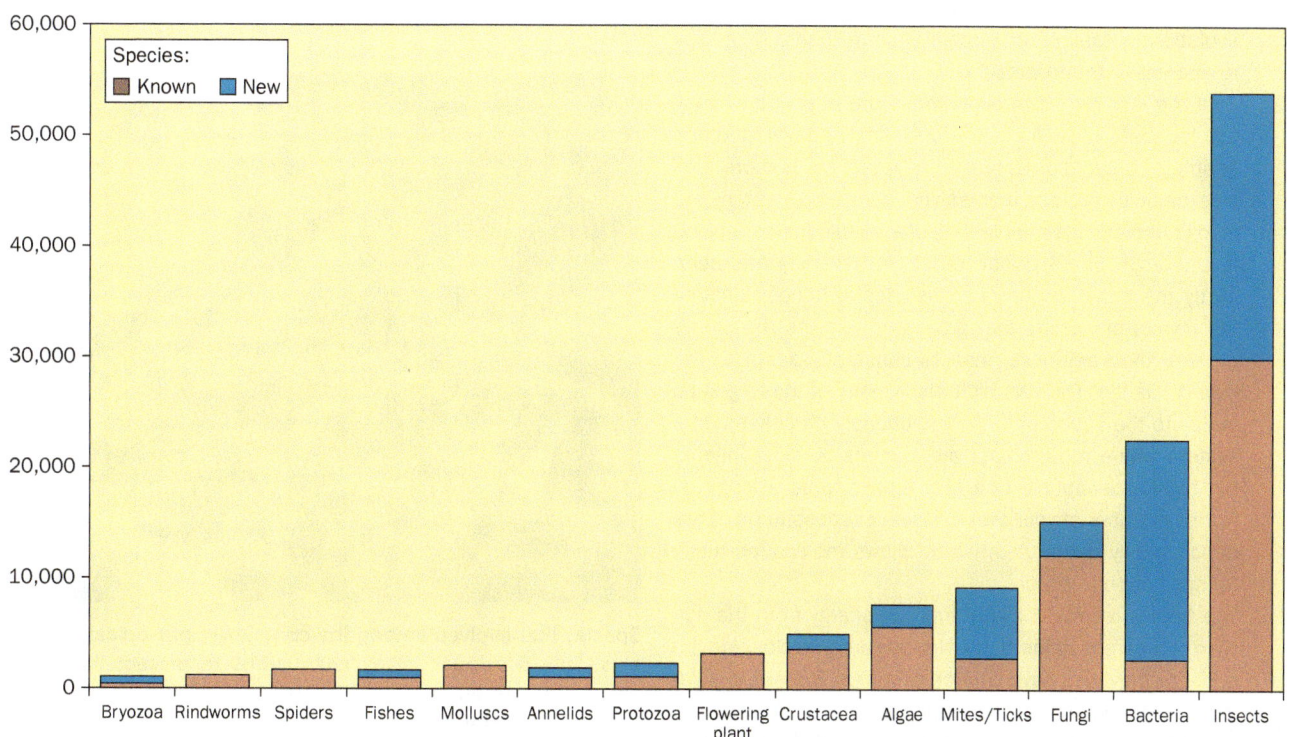

Figure 2.18 | Groups with the most species in Canada (excluding viruses). *Source: Mosquin et al. (1995: 58). Reprinted by permission of the Canadian Museum of Nature.*

Environment in Focus

Box 2.11 The Tropical Forests

Charles Darwin, who described the mechanisms of evolution in *On the Origin of Species* (1859), originated most of his ideas in the tropics. It was in the tropics—where life is speeded up through high energy inputs and abundant moisture, where adaptation is at its most complex and intricate, and where the struggle for survival is most dramatic—that evolution could most readily be appreciated.

The diversity of the tropical forests is astounding—estimates suggest that at least half of the world's species are within the 7 per cent of the globe's surface that the tropics cover. For example: in 100 square metres in Costa Rica, researchers found 233 tree species; one tree in Venezuela was home to at least 47 different species of orchids; there are 978 different species of beetles that live on sloths; and more than 1,750 different species of fish live in the Amazon basin. In general, the rain forests of South America are the richest in species, followed by Southeast Asia and then Africa. Several factors account for this abundance.

1. Tropical rain forests have been around for more than 200 million years, since the time of the dinosaurs and before the evolution of the flowering plants. It is thought that at that time there was just one gigantic landmass, before continental drift started to form the continents as we now know them. The vegetation of many areas was subsequently wiped out by succeeding glacial periods, which had minimal impact on the rain forests. Hence, evolutionary forces and speciation have had a long time to operate in the tropics.

2. Over the long period of evolution, there is a kind of positive feedback loop. As more species have developed and adapted, it has caused further adaptations as more species seek to protect themselves from being eaten and also to improve their harvesting of available food supplies. It is thought, in particular, that plant diversity has been partly the result of the need to adapt defences against the myriad of insects that graze on them. As the plants develop their defences, insects constantly adapt to the new challenge. The very high biodiversity of these groups is due to the speed of these evolutionary processes. In a system where most plants are immune to most insects but highly susceptible to a few, it pays to be a long way from a member of your own species. Successful trees are hence widely distributed, which allows more opportunity for speciation to occur.

3. The tropics receive a higher input of energy from the sun than other areas of the globe. Not only are they closer to the sun, but they also have little or no winter. The flux in solar input at the equator between the seasons is 13 per cent, but at latitude 50 degrees, the variation is 400 per cent.

4. Tropical rain forests receive a minimum of 2,000 millimetres of precipitation evenly distributed throughout the year. Moisture is therefore not a limiting factor, allowing for constant growth. There is a strong correlation between diversity and rainfall.

5. Tropical rain forests are the most diverse ecosystems that have evolved on Earth. They are also characterized by examples of coevolution and mutualism in which two species are absolutely co-dependent on one another. More than 900 species of wasp, for example, have evolved to pollinate the same number of fig trees. Each wasp has adapted to just one species of fig. Should anything destroy the food supply in such a finely tuned system, then the other species will also meet its demise.

While the evolutionary process has benefited from most of these characteristics, the soils have suffered. They have been exposed to weathering processes for a very long time, with no renewal and remixing from **glaciation**. The warm temperatures and abundant moisture are perfect for chemical weathering to great depths, and most tropical soils have long since had their nutrients washed out. A fundamental difference between tropical and temperate ecosystems is that in the tropics, unlike more temperate climes, most of the nutrients are stored in the biomass and not in the soils. When tropical vegetation is removed—by logging, for example—this removes most of the nutrients.

Species that evolved among the complexity of tropical forests have developed many adaptations to protect themselves. The camouflage of the leaf insect pictured here gives it some protection from predators.

detected in the environment are known to be increasing in wildlife. Plant communities and animal populations are responding to climate change. Temperature increases, shifting seasons, and changes in precipitation, ice cover, snowpack, and frozen ground are interacting to alter ecosystems, sometimes in unpredictable ways.

Some key findings identify ecosystems in which natural processes are compromised or increased stresses are reaching critical thresholds. Examples include: fish populations that have not recovered despite the removal of fishing pressure; declines in the area and condition of grasslands, where grassland bird populations are dropping sharply; and fragmented forests that place forest-dwelling caribou at risk. The dramatic loss of sea ice in the Arctic has many current ecosystem impacts and is expected to trigger declines in ice-associated species such as polar bears. Nutrient loading is on the rise in over 20 per cent of the water bodies sampled, including some of the Great Lakes where, 20 years ago, regulations successfully reduced nutrient inputs. This time, causes are more complex and solutions will likely be more difficult. Lakes affected by acid deposition have been slow to recover, even when acidifying air emissions have been reduced. Invasive non-native species have reached critical levels in the Great Lakes and elsewhere.

(Federal, Provincial, and Territorial Governments of Canada, 2010: 1)

The report notes that biodiversity and ecosystem monitoring is deficient in Canada, and that 'relevant ecosystem-level information is less available than decision-makers may realise' (ibid.). This lack of adequate biodiversity monitoring is also highlighted in an independent and comprehensive assessment of the state of biodiversity information in Canada (Hyde et al., 2010). Without an effective biodiversity information monitoring system Canada will be unable to respond to questions relating to ecosystem health, species at risk, invasive species, and changes in species distributions and abundance as they are affected by environmental changes such as climate change. Unfortunately, in Canada's official report to the UN Convention on Biodiversity, these significant deficiencies were not noted and instead the government emphasized that 'investment in building the biological data base has been significant across Canada' (Environment Canada, 2009c: 46). Obviously, much remains to be done.

Implications

The above discussion points to important implications for society and species distributions:

- All of the Earth's inhabitants are interlocked in environmental systems that depend on one another for survival. Perturbations in part of the system will have impacts on other parts of the system.
- The basic scientific laws that govern the transformation of matter and energy dictate that, sooner or later, society must transform itself from a throwaway society built on

The Vancouver Island marmot is Canada's only endangered endemic mammal species.

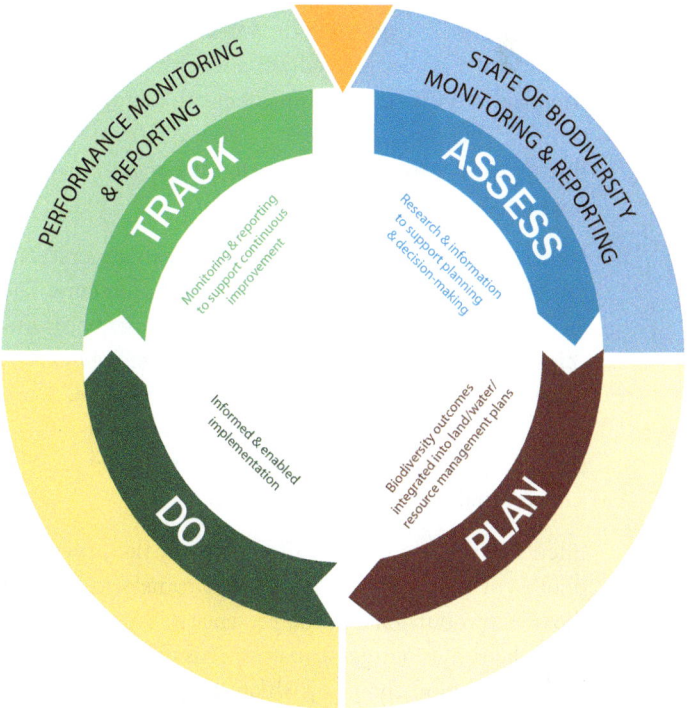

Figure 2.19 | Ecosystem approach and adaptive management used to achieve shared outcomes. *Source: Environment Canada (2009c: 43).*

processing ever-increasing matter and energy flows to one in which energy efficiencies are improved and matter flows are reduced.

- A species may have a wide range of tolerance to some factors but a very narrow range for others.
- Species with the largest ranges of tolerance for all factors tend to be the most widely distributed. Cockroaches and rats, for example, enjoy virtually global distribution.
- Many weed and pest species are successful because of their large range of tolerance. Eurasian water milfoil, a significant nuisance in many waterways in Canada, is an alien that can grow in conditions from Canada to Bangladesh.
- Response to growth factors is not independent. Grass, for example, is much more susceptible to drought when nitrogen intake is low.
- Tolerance for different factors may vary through the life cycle. Critical phases often occur when organisms are juveniles and during the time of reproduction.
- Some species can adapt to gradually changing conditions for some factors, up to a point. However, after this **threshold** of change is reached, the population will collapse.

The loss of biodiversity also has enormous implications, as discussed further in Chapter 14. Before the rise of biodiversity as a concept, the human-induced extinctions of species were looked on as tragic, isolated events. However, biodiversity has helped us reframe the problem and acknowledge the systematic nature of the process as well as the implications for ecological processes overall as genetic, species, and landscape impoverishment occurs at ever-increasing rates. Such is the concern over biodiversity loss that an international treaty, the Convention on Biological Diversity, is attempting to mobilize global responses to the problem. However, international progress, as well as that by the government of Canada on biodiversity protection, has been slow. We discuss this in more detail in Chapter 14.

From this discussion, it should be apparent that ecosystems are complicated. A complex set of interrelationships exists among organisms and between organisms and their environment. A change in part of this matrix will often result in corresponding changes throughout. Humans are now such a dominant influence on global environmental conditions at all scales that significant changes are underway as a result of human activities. There is considerable uncertainty as to how ecosystems and the entire life support system of this planet will react to these changes. Yet even under natural conditions, ecosystems are not static. The next chapter will focus on how ecosystems change over time.

Summary

1. Energy is the capacity to do work. Energy comes in many forms, including radiant energy (from the sun), chemical energy (stored in chemical bonds of molecules), and heat, mechanical, and electrical energy. Energy differs from matter in that it has no mass and does not occupy space.

2. Understanding energy flows is critical to an understanding of the ecosphere and environmental problems. The laws of thermodynamics explain how energy moves through systems. The first law states that energy can be neither created nor destroyed but merely changed from one form to another. The second law informs us that at each energy transformation, some energy is converted to a lower-quality, less useful form.

3. Energy is the basis for all life. Through the process of photosynthesis, certain organisms transform carbon dioxide and nutrients in the presence of radiant energy from the sun into organic matter. This matter forms the basis of the food chains by which energy is passed from trophic level to trophic level. At each transference, the second law of thermodynamics dictates that some energy is lost, typically as much as 90 per cent.

4. Productivity is a measure of the abilities of different communities to transform energy into biomass. The most productive communities are found in estuaries, wetlands, and rain forests.

5. The ecosphere is the thin, life-supporting layer of the Earth characterized by interactions between the biotic and abiotic components. It can be further subdivided into communities, ecosystems, and biomes.

6. The concepts of limiting factors and range of tolerance help us to understand the interaction between the biotic and abiotic components of the ecosphere.

7. Each species needs a specific combination of physical, chemical, and biological conditions for its growth. This is the niche of that species.

8. The principle of competitive exclusion tells us that no two species can occupy the same niche in the same area at the same time.

9. Species compete for scarce resources in any given habitat. However, there are many other forms of relationship between species, such as predation, parasitism, mutualism, and commensalism.

10. Species with a strong influence on the entire community are known as keystone species.

11. Biodiversity involves the variety of life at three different scales: genetic, species, and landscape. Estimates suggest that Canada has more than 140,000 species, of which about half have been named.

12. The progress of the government of Canada in implementing its biodiversity strategy has been very slow.

Key Terms

- abiotic components
- aerobic
- assimilated food energy
- autotrophs
- biodiversity
- biodiversity hotspots
- biomass
- biomass pyramid
- biomes
- biotic components
- calorie
- carnivores
- chemoautotrophs
- chlorophylls
- commensalism
- community
- competitive exclusion principle
- consumers
- Convention on Biological Diversity
- decomposer food chain
- detritus
- dominant limiting factor
- ecological redundancy
- ecosystem
- ecosystem diversity
- endemic species
- energy
- energy efficiency
- entropy
- epiphytes
- euphotic zone
- extirpated
- food chain
- food webs
- functional compensation
- generalist species
- genetic diversity
- glaciation
- grazing food chains
- gross primary productivity (GPP)
- habitat
- heat
- herbivores
- heterotrophs
- high-quality energy
- humus
- interspecific competition
- intraspecific competition
- keystone species
- kinetic energy
- law of conservation of energy
- law of entropy
- limiting factor
- loams
- low-quality energy
- mutualism
- net community productivity (NCP)
- net primary productivity (NPP)
- niche
- omnivores
- optimal foraging theory
- optimum range
- organism
- parasitism
- parent material
- photosynthesis
- phototrophs
- phytoplankton
- population
- potential energy
- predator
- prey
- primary consumers
- producers
- radiant energy
- range of tolerance
- resource partitioning
- secondary consumers
- soil horizons
- soil permeability
- soil profile
- specialist
- species-area curves
- species diversity
- territory
- tertiary consumers
- threshold
- trophic level
- zone of physiological stress
- zooplankton

Questions for Review and Critical Thinking

1. What are the main biotic and abiotic components of ecosystems?
2. How do the laws of thermodynamics apply to living organisms?
3. How do the laws of thermodynamics apply to environmental management?
4. What are chemoautotrophs, and what role do they play in ecosystem dynamics?
5. On what trophic level is a pitcher plant? Why? Are there plants on the same trophic level in your area? What are they, and where do they grow?
6. In what kinds of ecosystems do detritus food chains dominate?
7. What roles do phytoplankton play in maintaining ecospheric processes?
8. What are the management implications of recognizing concepts such as specialist, generalist, and keystone species? Can you think of any examples in your area?
9. What is optimal foraging theory?
10. What do you think the dominant limiting factors are for plant communities in your area?
11. Draw a cross-section across (E–W) and down (N–S) your province or territory, and show the main environmental gradients and the vegetational response.
12. What are some of the main transformations that have to take place in society to reflect the implications of the laws of thermodynamics and law of conservation of matter?
13. How does genetic diversity help to protect a species from extinction?
14. What is endemism, and why does Canada have relatively few endemic species?
15. What progress is Canada making on implementing its biodiversity strategy?

Related Websites

Biodivcanada.ca
www.biodivcanada.ca

Biodiversity in Canada
www.scib.gc.ca
canadianbiodiversity.mcgill.ca/english/species

Canadian Biodiversity Strategy
www.biodivcanada.ca/560ED58E-0A7A-43D8-8754-C7DD12761EFA/CBS_e.pdf; www.oag-bvg.gc.ca/internet/English/parl_cesd_200509_03_e_14950.html

Environment Canada, ecosystem information
ecoinfo.ec.gc.ca/index_e.cfm

Further Readings

Note: This list comprises works relevant to the subject of the chapter but not cited in the text. All cited works are listed in the Bibliography at the end of the book.

Hocking, M.D., and J.D. Reynolds. 2011. 'Impacts of salmon on riparian plant diversity', *Science* 331: 1609–12

Hodges, K.E., and A.R.E. Sinclair. 2003. 'Does predation risk cause snowshoe hares to modify their diets?', *Canadian Journal of Zoology* 81: 1973–85.

Krebs, C.J., et al. 2003. 'Terrestrial trophic dynamics in the Canadian Arctic', *Canadian Journal of Zoology* 81: 827–43.

Mills, E.L., et al. 2003. 'Lake Ontario: Food web dynamics in a changing ecosystem (1970–2000)', *Canadian Journal of Fisheries and Aquatic Sciences* 60: 471–90.

Predavec, M., C.J. Krebs, K. Dannell, and R.J. Hyndman. 2001. 'Cycles and synchrony in the collared lemming (*Dicrostonyx groenlandicus*) in Arctic North America', *Oecologia* 126: 216–24.

Chapter 3
Ecosystems are Dynamic

Learning Objectives

- To understand the nature of ecosystem change and its implications for society and environmental management.
- To understand the process of primary and secondary succession and the ways in which humans alter these processes.
- To appreciate the role of disturbance such as fires, insect infestations, and major storms as often being an integral and natural part of healthy ecosystem function.
- To explore the impact and management of invasive species.
- To recognize the main factors affecting species population growth.
- To appreciate the nature of evolution and extinction.
- To appreciate some of the implications of global climate change on species distributions and abundance.

Introduction

Communities and ecosystems change over time. The rate of change depends on factors driving change, the response of individual species, how species interact with one another, and how they respond collectively and individually to their abiotic environment from an ecosystem perspective. Part of the response of the species comprising these systems is related to their range of tolerance to such factors as the amount of light, nutrients, and soil type in the case of plants, discussed in Chapter 2. For animals, the response may be related to the type, distribution, and availability of food resources or potential for predation. Some changes are very rapid, such as a forest fire. Others, such as climate change under natural conditions, occur over long time periods and allow communities to adjust slowly to the new environment. Unfortunately, the speed of change now occurring as a result of greenhouse gas emissions is faster than any previously experienced, and many species will be unable to adapt at this speed. As vegetation communities change, so do the heterotrophic components dependent on plants for food. Similarly, if the components of the food web change, it may well cause a change in vegetation.

In this chapter, we examine aspects of change in ecosystems, starting with the process of ecological succession, and then discuss the concept of ecosystem function and its

Figure 3.1 | A general model of primary succession over time, from a bare rock surface to a forest community.

dynamic characteristics. Next we examine the role of a species population growth and how and why it varies. Last, we look at the role of longer-term change in the processes of evolution and extinction on biodiversity and ecosystem function. The impact of humans on these processes is often to alter their natural function relative to a time when humans were far less populous on Earth.

Ecological Succession

Ecological succession is a relatively slow process. It involves the gradual replacement of one assemblage of species by another as environmental conditions change over time. Some of these changes are created by the species themselves and others occur more indirectly. We divide succession into two basic types, with some additional variants. **Primary succession** is the colonization of a previously unvegetated surface, such as when a glacier retreats or a landslide removes all traces of the vegetation of the previous ecosystem (Figure 3.1). Little or no soil exists, and the first species to occupy the area, known as *primary colonizers*, must be able to withstand high variability in temperatures and water availability and highly limited nutrients. Few species can tolerate such conditions.

Only 10,000 years ago, most of Canada was covered in a thick layer of ice. The Kaskawulsh Glacier in Kluane National Park, Yukon, is a remnant of this time.

Lichens are typically the first colonizers because they can establish on bare rock surfaces that are virtually devoid of nutrients and have the ability to hold water (Box 3.1). Over time, lichens, in combination with other physical and chemical processes, break down rocks. Their biomass traps water and nutrients. Over centuries, their accumulating biomass and alteration of the environment make it possible for other species to colonize; mosses most often follow. Mosses grow faster than lichens, resulting in yet greater accumulation of biomass and the beginnings of soil. The lichens are eventually out-competed by the faster-growing mosses.

The next stage in successional advance is typically invasion by herbaceous plants such as grasses and species that we often think of as 'weeds'. Most of these species are annuals or biannuals. Such species are able to colonize a wide range of habitats and have reproductive strategies to disperse widely. Dandelions and fireweed are good examples. While some plant species physically disperse into the habitat patch, others are already present in the form of seeds lying dormant in the soil, sometimes for decades! These seeds germinate when environmental conditions, such as the availability of light, become favourable for growth. The seeds that lie 'in wait' are said be part of the soil **seed bank**.

Over time, these early herbaceous species create an environment conducive for the next successional stage to establish, which includes hardy shrubs and light-tolerant trees that in turn further ameliorate conditions until shade-tolerant tree species become established. Examples of light-tolerant ('sun-loving') trees are birch, oak, and trembling aspen. Species that can establish in the shade include western hemlock and western red cedar, which we typically find in old-growth forests. In areas where precipitation and temperature are

Environment in Focus

Box 3.1 Lichens

Some environments have such challenging growing conditions that virtually nothing can survive. However, lichens are one of the few types of organism that can be found in such places. Lichens are partnerships, part of an evolutionary mutualistic relationship between fungi and photosynthetic algae such that each benefits from the presence of the other (Box 2.1). The fungi are able to cling to rocks or trees with their filaments and to retain water. In turn, the algae produce food for both groups of species through photosynthesis. This may include the fixing of nitrogen from the atmosphere by cyanobacteria, as discussed in the next chapter. This combination is able to survive intense cold and drought and has been evolving for more than a billion years, making lichens one of the most primitive of living organisms. Individual lichens may be more than 4,000 years old. Over centuries, sufficient growth of lichens may occur so that the thinnest of soils is produced, allowing other species able to tolerate harsh conditions to colonize. Lichens are therefore very important primary colonizers.

More than 18,000 species of lichens have been described throughout the world, and undoubtedly many others have yet to be discovered. They have different life forms—dust, crust, scale, leaf, club, shrub, and hair. The best known in Canada include the leafy variety found growing on trees, an encrusting variety that grows on rocks and sometimes trees, and the so-called (and misnamed) reindeer mosses found throughout northern Canada.

Lichens are not only important agents of succession, but some species rely on them for food. A couple of species are found not in early successional environments but in old-growth forests, and these lichens constitute the main winter food supply for BC mountain caribou, a species precariously on the brink of extinction.

Besides providing an essential food supply for caribou, lichens have also been used by humans as flour (when dried and ground up) and as a dye for wool and other fabrics. They are now finding other uses as well. Lichens, because of their adaptive ability to absorb mineral requirements directly from the air, are very efficient accumulators of pollution. Unlike many other plant species, they concentrate pollutants to exceed their own tolerance levels and hence are excellent indicator species for air pollution, since lichens will be absent from heavily polluted areas. Wong and Brodo (1992) documented that of 465 species of lichen in Ontario, 52 are believed to be regionally extinct owing to pollution.

Orange lichens surround a pool in pre-Cambrian rock on Georgian Bay, south of Philip Edward Island, Ontario.

Fireweed, seen here growing along the shores of Bow Lake below Crowfoot Mountain in Banff National Park, Alberta, is a common herb in early successional sites throughout Canada.

The term 'treeline' is used to describe areas where vegetation communities dominated by trees give way to those dominated by other types of vegetation, such as herbs and grasses. Rarely, however, is there a sharp line; rather, there is usually an ecotone, where patches of both tree- and grass-dominated communities exist together.

adequate, trees typically dominate the final stage of this successional process, with fewer species in the understory. Each stage along the way is known as a **seral** stage.

In the first half of the past century, it was believed that vegetation would ultimately reach a well-defined, stable stage known as the **climax community** and that this final successional stage was in equilibrium with the environment. However, equilibrium conditions are rare and **disturbances** (such as fires, insect infestations, flooding, ice storms) are so common that most ecological systems never reach a dynamically stable climax stage even in large landscapes such as Banff National Park. More formally, disturbance can be defined as a relatively discrete event in time and space (such as a flood) that alters the structure and function of populations, communities, and ecosystems. This change occurs because the amount, kind, and distribution of resources is altered or the physical environment (including substrate availability) is significantly changed. Many agents of disturbance are a natural and integral part of the healthy functioning of ecosystems. This is contrary to our intuitive sense that phenomena like fires, flooding, and windstorms are harmful to ecosystem health.

An example of a major disturbance currently unfolding relates to the mountain pine beetle invasion affecting more than 16 million hectares of forest in British Columbia as well as Alberta, and this invasion is now poised to move further east across Canada. The beetles have killed and are killing millions of trees and thus are preventing these forests from achieving or maintaining a state of climax. As such, they are 'setting back the successional clock' to an environment represented by early successional states. The pattern of recovery following this disturbance will depend on the features of the species themselves, the nature of interactions among species, and many unpredictable factors. Thus, ecosystems and landscapes are dynamic, interacting in complex ways, often unpredictable, over large space–time scales.

Succession is not an inevitable linear progression. It is a guideline to help understand the changes that may take place in ecological communities. In some instances—in recently glaciated terrain, for example—very hardy species of trees, such as willows and alders, may become established in favoured sites with little previous colonization having occurred. **Cyclic succession** may also occur where a community progresses through several seral stages but is then returned to earlier stages by natural phenomena such as fire (Box 3.2) or intense insect attack. The different seral stages are not discrete but may blend from one into another. These blending zones tend to be the areas with the highest species diversity, since they contain species from more than one community. They are known as **ecotones** and occur as relatively richer zones between communities.

Environment in Focus

Box 3.2 Fire, Management, and Ecosystem Change

In many areas, fire is a natural occurrence that has a profound impact on plant and animal communities. In some communities, it may be the dominant agent of disturbance, and if suppressed by human interference, those communities may change significantly in species composition. Fire has been used as a tool by humans since earliest times to manipulate ecosystems to produce desired effects, such as removing forests to facilitate agriculture, burning grasslands to generate new grass growth, and herding animals so that they can be more readily hunted. Fire is used in forest management: hazard reduction for silviculture, insect and disease control, wildlife habitat enhancement, and range burning.

Fire has several important ecological and social implications:

- It favours the growth of certain species over others. Some species are fire-resistant (such as the Douglas fir), while the heat from fire may aid in the germination of other species. For example, lodgepole pine seeds can only be released from their cones when sufficiently high temperatures melt the resin that once held the cone tightly shut. The phenomenon is termed **serotiny**. Fire may result in the death of other species.
- At moderate levels of intensity and frequency, it tends to increase the diversity of species in a community. Fire releases nutrients from the biomass into the soil and atmosphere; some may be lost from the site, while the remainder help to stimulate growth of some species—for example, the pine seedlings mentioned above.
- It stimulates the growth of various grasses and herbs that provide fodder for herbivores, which may in turn increase carnivore populations.
- Soil temperatures are increased not only during the fire but also afterwards—the site has a lower **albedo** and is more open to the sun. This also influences chemical and biological properties of the soil, stimulating microbial activities and enhancing decomposition.
- Fires that are highly intense or very frequent may cause sufficient nutrient impoverishment of a site to preclude further growth of trees, and the vegetation may become dominated by grasses and low shrubs. Many of the heathlands of Northern Europe were created in this manner, and clear-cutting and fire in nutrient-poor black spruce forests in Canada can have the same effect.

Fire is also a highly emotive topic for management. Early concepts of forestry and conservation encouraged policies of total fire suppression, with little attention given to the role of fire in various ecosystems. One can see evidence of this in the message of Smoky the Bear: forest fires are 'bad'. This mindset led to unanticipated changes in some ecosystems. For example, in the absence of fires as a result of fire suppression, lodgepole pine cannot establish (as explained above) and thus establish what otherwise would be a forest dominated by lodgepole pine. Instead, these ecosystems with an altered fire regime may be dominated by species such as trembling aspen. Further, fire suppression results in the accumulation of organic debris such as dead trees. If and when fire occurs, this debris will help fuel a fire such that the fire will jump from the forest floor to the canopy. Often, these fires are so intense and spread over such a large area that they cannot

A 2009 road closure through the Saskatchewan Valley in Alberta's Banff National Park allows for a forest fire controlled burn.

be controlled. Managers of protected areas such as parks now realize that if fire is a natural part of an ecosystem, fire suppression policies are altering the ecosystem in unnatural ways. This has led to **prescribed burning** programs in many parks, such as Banff National Park (Chapter 14). A decision on whether fire should be suppressed or not should reflect knowledge of an ecosystem's natural fire regime. The regime includes factors such as the frequency, intensity, and spatial magnitude of this agent of disturbance. Such knowledge enables managers to mimic the regime, thus maintaining the natural state of the ecosystem. Some fires may be ecologically appropriate. Others may result from human carelessness or lack of ecological understanding. Furthermore, we cannot ignore the potentially destructive effects of fires on human livelihoods.

It is widely believed that global warming (see Chapter 7) will result in more frequent and intense fires. The burning of millions of tons of carbon that is biologically fixed in the biomass of the trees releases carbon dioxide, further exacerbating the buildup of greenhouse gases in the atmosphere. This is an example of a **positive feedback loop**. The hotter it gets, the drier it gets, the more fires we have, the more carbon dioxide is released, and the warmer it gets. Scientists predict that temperature rise associated with global warming will be in the order of 4–6°C within 40 years for the boreal forest biome. They also predict lower rainfalls. This will lead to greater drying of the land surface and, again, increased frequency and area of fire. Overall, forest ecosystems will show a high degree of disturbance not typical of that which they experience as part of their natural fire regime. Species dependent on old-growth ecosystems, such as woodland caribou, will be put under increasing pressure. Caribou are highly dependent on the forest for lichens, which form the major part of their winter diet and only grow in forests more than 150 years old.

Sand dune succession is another common form of primary succession in which the primary colonizers are not lichens but grasses that have the ability to withstand not only the high variability in temperature and water but also the constantly shifting sand. The grasses help to stabilize the sand until mat-forming shrubs invade. Later, conditions may become suitable for hardy trees such as pines that may in turn be replaced by other tree species such as oaks.

Climax is a relative rather than an absolute stage. Communities do not change up to the climax and then cease to change. However, the nature of the species assemblage is more constant over time once a **mature community** is established. Even in mature communities, future changes in pathogens, predation, and climate will generate ongoing changes.

The climax vegetation for most areas is strongly influenced by the prevailing climate and is therefore known as a **climatic**

Guest Statement
How Will Forests Respond to Rising Atmospheric Carbon Dioxide?
Ze'ev Gedalof and Aaron Berg

As humans we tend to think about global environmental change in terms of temperature and precipitation, or the frequency of hurricanes, or the persistence of drought. For plants, though, global environmental change includes the very composition of the atmosphere. Changing levels of ground level ozone [O_3], carbon dioxide [CO_2], and reactive nitrogen have the potential to affect all aspects of plant growth—from growth rates, to distributions, to reproductive success. Given the huge number of variables involved and the uncertainty regarding future greenhouse gas emissions and climate projections, the task of predicting these effects is extremely challenging. By necessity, most scientists focus on only one or two variables at a time and study short lived organisms growing in controlled environments. Understanding the effects of increasing CO_2 is especially important, as it is the most rapidly accumulating greenhouse gas and is involved directly in photosynthesis. Specifically, increasing CO_2 should increase the growth rates of trees due to two possibly complementary processes: First, direct CO_2 fertilization may occur because higher partial pressure of CO_2 increases the rate of CO_2 reactions with rubisco (a plant enzyme) during photosynthesis, thus inhibiting photorespiration. Second, increasing water use efficiency may occur due to reduced stomatal conductance, leading to greater drought tolerance.

Scientists have developed many tools for studying the effects of elevated CO_2 on plant growth. While much has been learned from these studies, the inferences that can be made about forests are limited. For example, while closed growth chambers allow for a high degree of control over environmental conditions, they can only be used to study small plants and seedlings, and there are typically few interspecific interactions, damaging agents, or climatic variations included in experiments. Open-top chambers allow for more natural conditions to be simulated, but are similarly restricted to studying small organisms. More recently, the development of the Free Air CO_2 Enrichment (FACE) sites has allowed large, natural ecosystems to be studied by providing a slow continuous supply of CO_2 from the upwind side of the site. The extreme high costs associated with FACE technology has meant that most of the 35 studies undertaken to date have focused on agriculturally important species, and only three have studied unmanaged forests. Furthermore, the FACE studies have been short in duration and, like virtually all CO_2 enrichment studies, have applied an abrupt change in CO_2 levels, rather than the gradual increase that has occurred over the past 150 years (Klironomos et al., 2005). Because many tree growth processes occur over a period of years to decades—including foliage retention in evergreen species, root versus shoot growth, reproduction cycles, and carbohydrate storage—trees could respond differently to abrupt increases in CO_2 than to gradual increases.

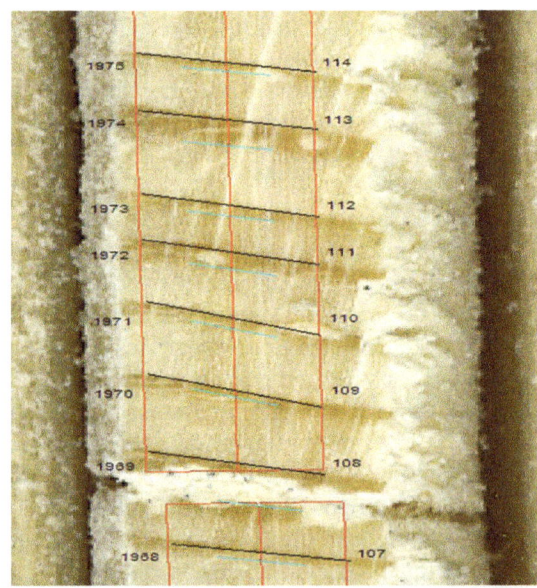

Figure 3.2 | A typical core sample from a conifer species. The black lines show annual ring boundaries, and the light blue lines show subseasonal anatomical differences. There is a small crack in the core in the middle of the 1969 growth ring. *Courtesy Ze'ev Gedalof and Aaron Berg.*

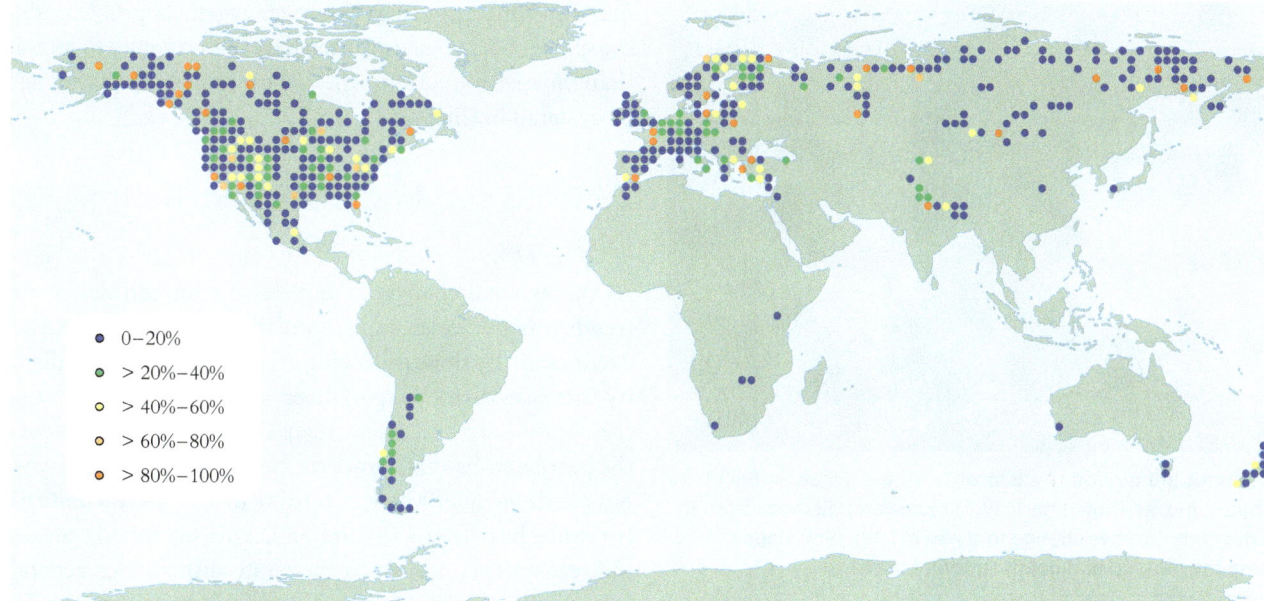

Figure 3.3 | The global distribution of the tree-ring data sites, and the proportion of sites (rounded to the nearest degree of latitude by longitude) showing unexplained increases in growth over the twentieth century. *Source: Adapted from Gedalof and Berg (2010).*

An alternative to these experimental approaches is to use natural history methods to examine how trees have responded to the observed increases in atmospheric CO_2 over the past century (Gedalof and Berg, 2010). In most temperate regions of the globe, trees produce annual growth rings that can be used to reconstruct the history of growth rates over the lifetime of the tree (Figure 3.2). While many factors contribute to the growth rates of trees, including site productivity, tree age, climatic variability, disturbance, and competition, most of these factors can be modelled mathematically, or averaged out by using many samples and many sites.

Using this approach, we asked the question: Is there an increasing trend in the growth of trees over the past century that can't be explained by these other competing explanations? To answer this question we used the International Tree Ring Data Bank (www.ncdc.noaa.gov/paleo/treering.html), a publicly accessible archive containing data on the annual growth rates of tens of thousands of trees worldwide. Using statistical models we removed the variability in growth that could be explained by factors unrelated to CO_2. While we can't control all causes of variability in growth rates, the large sample size we used suggests that these other effects should average out. While our analysis lacks the precision of the three natural FACE forest experiments, the fact that we analyzed over 2,300 sites allows even a small signal to emerge from the noise of the data. What we found is both surprising and interesting: approximately 20 per cent of trees worldwide show an unexplained increasing trend in growth (Figure 3.3)—about four times what one would expect by chance. There is no obviously discernable spatial pattern to the sites where growth is increasing, and no species is more likely than any other to show increasing growth rates. What this implies is that while CO_2 fertilization is clearly a locally important phenomenon, based on the CO_2 increases observed over the past century it is not universal.

This finding is important because it shows that forests cannot be relied on to accelerate their growth in response to rising atmospheric CO_2 and thereby slow down the rate of atmospheric accumulation. Second, those trees that are able to take advantage of rising CO_2 will have a competitive advantage over those that cannot—suggesting that future competitive interactions may be surprising. Finally, and most importantly, there is still a lot to learn about how rising atmospheric CO_2 will affect forests and forested ecosystems. It is an exciting time to work in the field of forests and global change.

Ze'ev Gedalof is an Associate Professor in the Department of Geography at the University of Guelph, and is director of the Climate & Ecosystem Dynamics Research (CEDaR) laboratory. **Aaron Berg** is an Associate Professor in the Department of Geography at the University of Guelph, and is a co-director of the University of Guelph Centre for Hydrogeomatics.

Sand dunes are a good place to observe the successional changes over time, as shown here. With increasing distance from the sea, the communities change to those in later seral stages representing the buildup and colonization of the sand.

climax. In some areas, other factors such as soil conditions may be more important than climate in determining community composition and structure. These are known as **edaphic climaxes** (Box 3.3).

In addition to the primary succession described above, successional processes also occur on previously vegetated surfaces such as abandoned fields or avalanche tracks or following a fire, where soil is already present. This process is known as **secondary succession**. The earlier soil-forming stages of primary succession are not repeated, so the process is much shorter, with the dispersal characteristics of invading species being a main factor in community composition. Annual weeds again dominate the community until perennial weeds, such as goldenrod, start to become established. Where conditions are suitable, the community will eventually be invaded by shrub and ultimately tree species. A major challenge for agriculture and forest managers is to prevent this natural recolonization taking place by species that may not yield the required products. As a result, chemical herbicides, as discussed in greater detail in Chapters 9 and 10, are often used to arrest secondary succession.

Similar kinds of processes also occur in aquatic environments. Here, the natural aging is called eutrophication (eu = well, trophos = feeding) as nutrient supplies increase over time with inflow and the growth and decay of communities. The process can be relatively rapid in shallow lakes, because the nutrients (one of the auxiliary energy flows discussed in Chapter 2) promote increased plant growth that leads to more biomass and nutrient accumulation. The lake becomes shallower over time, with less surface area of water, and the aquatic communities may eventually be out-competed by marsh and ultimately terrestrial plants. This process is another example of a positive feedback loop (the shallower the lake gets, the stronger the forces become to make it shallower), discussed in more detail in the next section. Eutrophication may also constitute a significant management problem, since the species being replaced often have higher values to humans than the species replacing them. This problem is discussed in more detail in Chapter 4.

Indicators of Immature and Mature Ecosystems

As successional changes take place in communities, several trends emerge. For example, annual net primary productivity declines as the slower-growing species establish, and diversity increases as more specialized species come to dominate the community and more finely subdivide the resources of the particular habitat. However, the increase in diversity will not continue indefinitely, according to the **intermediate disturbance hypothesis** (Figure 3.5). This hypothesis suggests that ecosystems subject to moderate disturbance generally maintain high levels of diversity compared to ecosystems that experience low levels of disturbance or those that have high levels of disturbance. Under low levels, competitive exclusion by the dominant species reduces diversity. With high disturbance, only those species tolerant of the stress can persist. Disturbance occurs at different scales, from small scale such as that associated with a gap created in a forest when a tree falls over from death or windthrow to large scale associated with widespread fire.

Certain differences between mature and immature systems are generic (Table 3.1). In general, mature ecosystems tend to have a high level of community organization among a large number of larger plants and have a well-developed trophic structure. Decomposers dominate most food chains, with a high efficiency of nutrient cycling and energy use. Net productivity is low. Immature ecosystems tend to have the opposite of these characteristics.

Effects of Human Activities

Humans influence ecological succession. Many activities are directed towards keeping certain communities in early seral stages. In other words, humans seek to maintain the characteristics of the immature ecosystems shown in Table 3.1 as opposed to those of the mature ecosystems that would result if natural processes were allowed to proceed. Agriculture, for example, usually involves large inputs of auxiliary energy flows to ensure that succession does not take place as weeds try to colonize the same areas being used to grow crops. The same can be said for commercial forestry. Maintaining ecosystems in early successional stages has several implications:

- The productivity of early successional phases is often higher than later phases.
- Nutrient cycling, discussed in more detail in the next chapter, is often more rapid in early stages. Trees, for example,

not only hold nutrients in their mass for a longer time than herbaceous plants, but they also maintain relatively low temperatures in soils. High temperatures result in more rapid breakdown of organic material and release of nutrients to the environment. Water uptake and storage by plants is also much reduced. Consequently, disturbance may result in a significant loss of nutrient capital from a site through losses in soil water to streams.

Environment in Focus

Box 3.3 Edaphic Climax: Table Mountain, Newfoundland

The west coast of Newfoundland (like most of the rest of the island) is dominated by the boreal forest (Chapter 9). In Gros Morne National Park (Figure 3.4), however, and at other locations on the west coast, this greenery (white spruce, paper birch, balsam fir) is punctuated by practically treeless orange-coloured outcrops that bear little if any similarity to the surrounding vegetation. These outcrops result from the distinctive chemical composition of the bedrock, known as 'serpentine'. Along with three other serpentine outcrops in western Newfoundland, the Table Mountain massif in Gros Morne was formed on the floor of the Atlantic Ocean millions of years ago and rafted up to its present position through the process of continental drift.

Serpentine is characterized by high levels of nickel, chromium, and magnesium and low levels of calcium. Most species of the surrounding forests cannot tolerate these conditions; if they grow at all, they are stunted. Instead, the serpentine is host to relict communities of tough arctic-alpine species that have survived since the retreat of the glaciers and have not been displaced through the process of succession like the arctic-alpines on the surrounding bedrock. These serpentine communities are edaphically driven, where the underlying geology is more important in determining plant cover than climate.

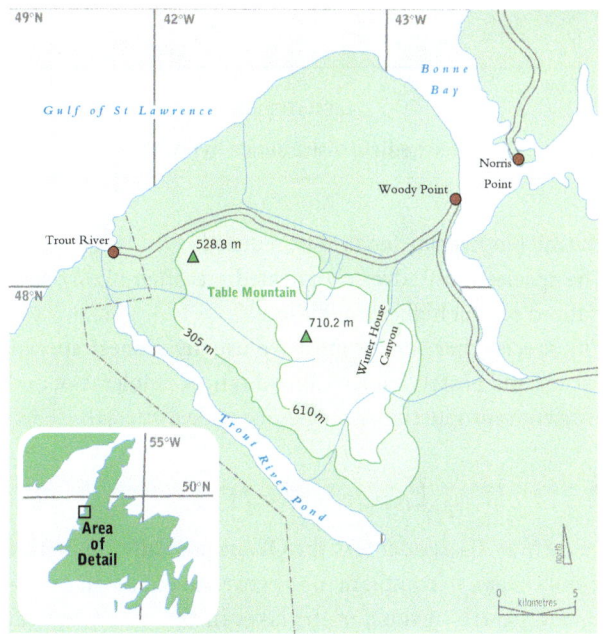

Figure 3.4 | Location of Table Mountain.

The difference between the dominant vegetation of the edaphic climax of outcrops in Newfoundland's Gros Morne National Park and the surrounding boreal forest can be clearly seen along the geological boundary.

The inhospitable soil chemistry has allowed rare species, such as this *Lychnis alpina*, to continue to grow in the area as relicts from the ice age.

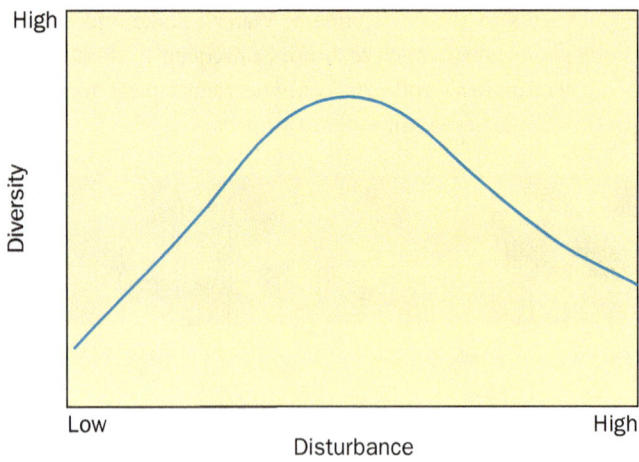

Figure 3.5 | The intermediate disturbance hypothesis.

- Overall biodiversity tends to be reduced.
- The species most adversely affected are often highly specialized ones at higher trophic levels.
- The species that benefit most are usually pioneer species (weeds and pests) that have broad ranges of tolerance and efficient reproductive strategies for wide dispersal.

Changing Ecosystems

In the early 1970s, residents of the Okanagan Valley in British Columbia began to complain to the government about excessive weed growth in some of the lakes in the valley. Several popular beaches were becoming unusable because of the weeds, and the extent of the invasion was growing rapidly. This was of considerable concern to the residents, not only because of the impact on their recreational activities but also because of the impact on the economy of this tourist area for which water-based recreation was the main attraction.

The culprit was Eurasian water milfoil, which arrived in the area in the 1970s and over the next couple of decades would spread not only to all the lakes in the Okanagan but also to many other lakes in southern BC and other provinces. The government spent significant amounts of money trying to control the spread of the species but to no avail. Originating in Eurasia, the milfoil had reached the eastern shores of this continent probably a century ago and since that time had spread across the continent, replacing native aquatic plants in many water bodies.

This ecological event, the spread of a Eurasian plant into North America, also illustrates the dynamic relationship among the biophysical, socio-economic, and management systems that is the main focus of this book. In BC, for example, the dependence of local economies such as that of the Okanagan Valley on water-based tourism triggered a strong response to milfoil that involved the use of the chemical 2,4-D. This created considerable conflict between different stakeholders regarding the relative impact of the plant versus that of the control mechanism. Critics claimed that management had failed to consider the broader perspectives that would have been included had they adopted an ecosystem-based approach to the problem and that management failed to adapt to the changing parameters of the situation. Chapter 6 discusses various approaches to these kinds of resource management issues in greater detail.

Situations such as this are common. We tend to think of ecosystems as having relatively constant characteristics, of being in a balance in which internal processes adjust for changes in external conditions. It is not a static state but one of **dynamic equilibrium.** James Lovelock (1988) postulated the **Gaia hypothesis,** which claims that the ecosphere itself is a self-regulating homeostatic system in which the biotic and abiotic components interact to produce a balanced, constant state (Box 3.4). This is an example of a highly integrated system in which there is a strong interaction between the different parts of the system. Other systems may not be so highly dependent on one another. Cells in a colony of single-celled organisms, for example, may be removed and have little effect on the remainder because of the low integration of the system.

Not all ecosystems are equal in their ability to withstand perturbations. **Inertia** is the ability of an ecosystem

Table 3.1 | Characteristics of Immature and Mature Ecosystems

Characteristic	Immature Ecosystem	Mature Ecosystem
Food chains	Linear, predominantly grazer	Web-like, predominantly detritus
Net productivity	High	Low
Species diversity	Low	High
Niche specialization	Broad	Narrow
Nutrient cycles	Open	Closed
Nutrient conservation	Poor	Good
Stability	Low	Higher

Source: Modified from Odum (1969). Copyright © 1969 by the American Association for the Advancement of Science.

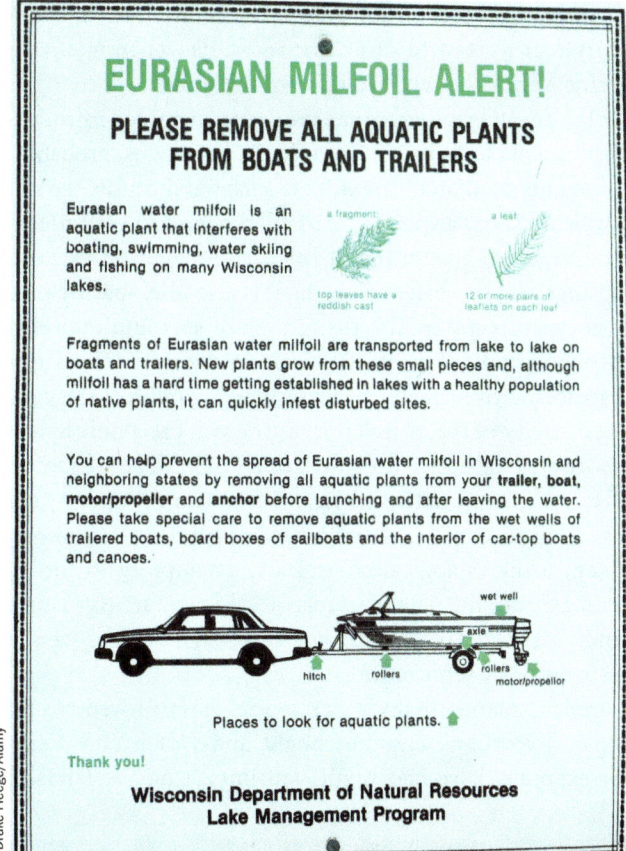

Signs similar to this one, warning of the spread of Eurasian water milfoil, were placed at boat-loading ramps throughout BC but did little to stem the colonization. In the tourist economy of the Okanagan Valley, where resorts rely on water-based activities to attract clientele, there was considerable conflict among different stakeholders regarding the most appropriate means of controlling the spread of milfoil.

to withstand change, whereas **resilience** refers to the ability to recover to the original state following disturbance. Ecosystems can have low inertia and high resilience or any combination thereof. In terms of human usage, it is best to work with systems that have both high inertia and high resilience. This means that they are relatively difficult to disturb and even when disturbed will recover quickly. Such systems are relatively stable. The best growth sites for forestry—alluvial sites in nutrient-rich areas at low elevations—would fit into this category. Many tropical and Arctic sites would fit into the opposite combination in which sites are readily disturbed and recover only very slowly, if at all.

Ecosystems are constantly subject to change, and equilibrium exists only in a dynamic form. In some cases, as with the milfoil described above, this is obviously true. The milfoil invasion involved the replacement of various native aquatic species with a mono-specific stand of the alien species. Similar effects are common with other non-native invaders, such as Scottish broom, purple loosestrife (Box 3.5), sea lamprey, and zebra mussels.

Invasive Alien Species

Organisms found in an area outside their normal range, such as Eurasian water milfoil and purple loosestrife, are considered **alien species**. The UN Convention on Biological Diversity defines 'alien species' as a species introduced outside its normal past or present habitat. Many species transported to a new environment do not survive. However, many others multiply rapidly, out-compete native species, change native habitats, and become **invasive** alien species. Characteristics that make a species more likely to be a successful invasive alien include:

Environment in Focus

Box 3.4 The Gaia Hypothesis

As humans became capable of probing deeper and deeper into space and thus able to view Earth from this unique perspective, it became increasingly clear that our planet was significantly different from all the millions of others. It seems a happy coincidence that of the vast range of temperatures that could be experienced, those on Earth are just right for life, between the freezing and boiling points for water, even though the energy output of the sun has increased by more than 30 per cent during the past 3.6 billion years. The gaseous composition of the atmosphere is also just what we need to breathe.

One hypothesis, the Gaia hypothesis, named for the ancient Greek goddess of the Earth by the originator of the idea, James Lovelock, postulates that the Earth acts as one giant self-regulating super-organism to help maintain these conditions necessary for life. Organisms act like cells in a body, with integrated functions to promote the health of that body, or in this case, optimum conditions for life. Active, automatic feedback processes among the atmosphere, lithosphere, hydrosphere, and ecosphere maintain this equilibrium.

There is no doubt that since the beginning of life on Earth, organisms have adapted to existing conditions and have also modified these conditions in ways that are beneficial to life. However, few scientists believe that Earth is a super-organism with all living things interacting to maintain an equilibrial environment. The historical record shows that organisms do not regulate natural cycles. Furthermore, it is wishful thinking to assume that Earth will adapt and compensate for all of the changes now being initiated by human activity in ways beneficial to humans.

being a fast-growing generalist with an ability to alter growth form to suit different conditions, being a fast reproducer able to reproduce both sexually and asexually with a good dispersal mechanism, and being associated with humans.

Invasive species are second only to habitat destruction as a leading cause of biodiversity loss. Globally, invasive non-native species are responsible for almost 40 per cent of all animal species extinctions for which the cause is known. On islands, they are often the main cause of extinctions, since there is little opportunity for the indigenous species to escape. Twenty-two per cent of species listed as endangered in Canada are in such a perilous state because of the effects of invasive species in their respective habitats (Venter et al., 2007).

In Canada, some 12 per cent of the 11,950 species assessed in *Wild Species 2010: The General Status of Species in Canada* are not native, and their numbers are increasing (Figure 3.6). Several of the invasive alien species on the World Conservation Union's list of 100 of the worst are established in Canada: Dutch elm disease, purple loosestrife, leafy spurge, Japanese knotweed, green crabs, spiny water fleas, gypsy moths, carp, rainbow trout, starlings, domestic (feral) cats, and rats (www.issg.org/database/species/search.asp?st=100ss). More than 500 species of alien plants in Canada have developed into agricultural weeds. They cost farmers millions of dollars every year to control, costs that we all pay when purchasing grains, vegetables, and fruit grown in Canada.

One example is the various species of knapweed introduced into Canada and the US from the Balkan states, probably in shipments of alfalfa. The diffuse knapweed causes the most problems; it has a wide range of tolerance and a very effective seed dispersal system that it has used to colonize vast areas of rangeland in western Canada. It is also **allelopathic**—that is, it can directly inhibit the growth of surrounding species through production of chemicals in the soil. The species displaces native species and considerably reduces the carrying capacity of the rangelands. Cattle will eat it only as a last resort, and the nutritive content is less than 10 per cent of that of the displaced native species. Initial control efforts relied on chemical sprays. A more integrated approach is now being taken, using biological control and attempting to limit its spread through stricter controls on vehicular access to rangelands, one of the main means of seed distribution as seeds get distributed by vehicle tires.

Besides plants, many other species have proved troublesome. Two fungi, chestnut blight and Dutch elm disease, for example, have had significant impact on the landscape

Environment in Focus

Box 3.5 Purple Loosestrife: Alien Invader

Purple loosestrife was inadvertently introduced to North America from Europe more than a century ago. Ocean-going ships typically carry 'ballast water'—that is, water to balance their cargo load in heavy seas—and it is taken on in the originating port. When the ship reaches calm water near its destination, this ballast and everything in it, including biological organisms, is discharged. An aggressive invader of aquatic systems, the purple loosestrife has spread through thousands of hectares of wetlands in Quebec and Ontario. In Manitoba, it is thought to be the most serious noxious weed. It is estimated that 190,000 hectares of wetland habitat in North America is invaded by purple loosestrife each year. After its woody root systems have become established, native plants and the animals that depend on them for food are forced out.

At the University of Guelph, experiments with the *Galerucella pusilla* beetle have showed promising results in controlling this invader plant. The beetle has a voracious appetite for purple loosestrife. They eat the metre-high plant at such a rate that the plant's capacity to produce seed (about 2.5 million per plant per year) is reduced by 99 per cent. Thus, use of the beetles to control purple loosestrife is promising, since previous control efforts that relied on physical removal, burning, mowing, and spraying produced negligible results. However, the beetles do forage on native plant species. Manitoba has initiated a biological control program using the highly host-specific weevil *Nanophyes marmorates* that is showing promising signs for controlling loosestrife.

Seen here growing along the banks of a stream, purple loosestrife grows aggressively in aquatic systems and has been a problematic invader of native species habitats in Ontario, Quebec, and Manitoba.

Sources: www.purpleloosestrife.org; www.ducks.ca/purple.

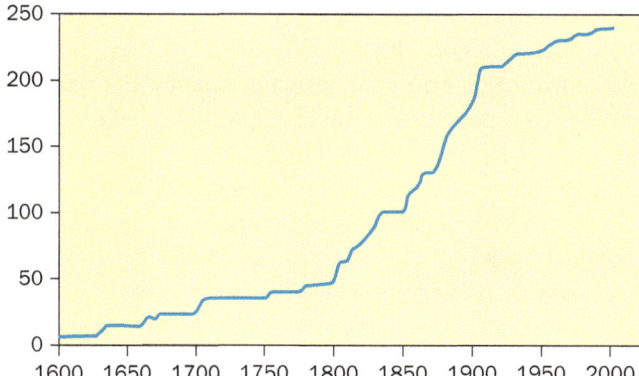

Figure 3.6 | Invasive non-native plants in Canada. Source: Federal, Provincial, and Territorial Governments of Canada (2010: 54).

of central and eastern Canada. Both attack native trees that at one time were conspicuous parts of the deciduous forests. The American chestnut was attacked by an Asian pathogenic fungus that was introduced on stocks of Japanese chestnuts during the past century and the elm by a European fungus transmitted between trees by beetles. More than 600,000 elm trees were killed in Quebec alone, and 80 per cent of Toronto's elms died within one year in the 1970s.

In Winnipeg, the boulevard elm trees are worth $307 million and contribute an estimated $160 million in property value. Annual budgets for control in Winnipeg have been as high as $2.25 million, with more than $30 million in control and research costs. Researchers at the universities of Manitoba and Toronto are using a natural toxin, *mansonone*, that occurs in some elms to breed seedlings more resistant to Dutch elm disease.

Another fungus, the white pine blister rust, illustrates the complexity of the impact of invasive species. The fungus, originating in Eurasia, attacks five-needled pines and causes extensive mortality. Whitebark pine is a key component of the subalpine ecosystems of the Canadian Rockies. It has a mutualistic relationship (Chapter 2) with Clark's nutcracker, a crow-like bird that caches the seeds for forage during the winter. Unlike those of many pines, the whitebark cones are not opened by fire but by animal activity. The seeds cannot be carried by wind and rely on the nutcracker for dispersal. The bird caches the seeds in forest openings for easy retrieval, creating perfect conditions for germination of the seed. However, the birds, while remarkable in their ability to remember hundreds of cache sites, invariably 'forget' some. These seeds may then germinate, resulting in the establishment of seedlings. Beyond the mutualistic relationship between these species in that both benefit, it is important to note that the seeds of the pine are too heavy to disperse very far, which means that the nutcracker is a keystone species, as discussed in Chapter 2. When keystone species are lost in an ecosystem, that system is subject to significant change. Stuart-Smith et al. (2002) measured mortality rates of the pine in excess of 20 per cent in some areas of the national parks as a result of fungus attack. There is concern that if mortality rates increase, it will lead to population declines of the Clark's nutcracker.

Often, invasive species have been deliberately introduced by humans for one reason or another and can have much the same impact as species introduced accidentally. One example is the introduction of Sitka black-tailed deer into Haida Gwaii as a food source for local people in the late nineteenth century. In the absence of predators such as wolves and cougars, the deer populations and distribution expanded rapidly. However, because of the nature of the archipelago, some islands were colonized early, others later, and others not at all. This created ideal conditions for scientists to study the impact of the deer over different time periods. Stockton et al. (2005) found that vegetation cover exceeded 80 per cent in the lower vegetation layers on islands without deer. This contrasted with 10 per cent for islands that had supported deer for longer than 50 years. Overall plant species richness was similar, but at the plot level it was reduced by 20 to 50 per cent on islands that had deer for more than 50 years. In general, these results show the potential of seemingly innocuous species to greatly simplify ecosystems.

Many of the most serious invasions occur in aquatic ecosystems. The Great Lakes, for example, are home to over 185 non-native reproducing species (Figure 3.7). One example is the spread of the zebra mussel. The mussel, named for its striped shell, joins a long line of aliens in the Great Lakes, including the sea lamprey, alewife, and rainbow smelt. The mussel, a native of the Black and Caspian seas, was introduced from the ballast of freighters in the mid-1980s. It was first found in 1988 in a sample of aquatic worms collected from the bottom of Lake St Clair at Windsor–Detroit, which connects lakes Erie and Huron. Evidence from Europe indicated that the species was an aggressive colonizer, able to displace most native species. In a short period of time, it displaced 13 species from Lake St Clair and caused the near-extinction of 10 species in western Lake Erie.

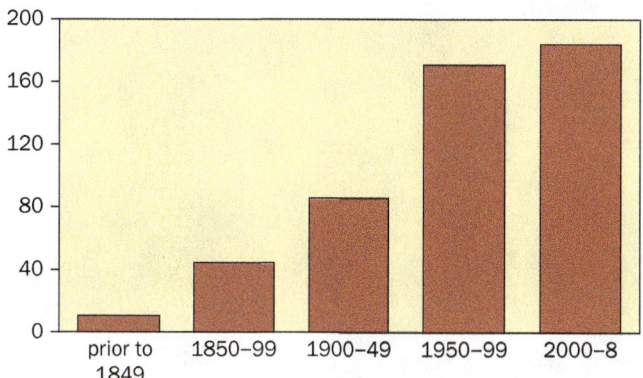

Figure 3.7 | Trends in non-native species in the Great Lakes. Source: Federal, Provincial, and Territorial Governments of Canada (2010: 52).

The mussels start their three- to six-year life as free-swimming larvae before attaching to a hard surface, usually in the top three to four metres of the water, although they can live as deep as 30 metres. By the end of 1988, the mussels had colonized half of Lake St Clair and two-thirds of Lake Erie at densities as high as 30,000 per square metre. Densities as high as 600,000 per square metre have been recorded as testament to the mussel's ability to grow on itself.

The mussel has now spread through the Great Lakes, where it appears capable of colonizing any hard surface. It has encrusted water intakes and discharges, severely reducing their efficiency and necessitating significant expense to remove them. Water intakes may be reduced by as much as 50 per cent. Many different approaches are being undertaken to screen out the mussels, but they appear to be able to pass through most physical barriers. At the moment, chlorination is the most common measure, but this raises problems related to the potential for formation of toxic organochlorines. Ontario Power Generation has spent more than $20 million on installing and maintaining chlorine applicators at its Great Lakes facilities and inland facilities and another $13 million on research to reduce chlorine use. Estimates of the damage to all Great Lakes utilities range from $200 to $500 million per year. The mussels also colonize spawning sites for other fish, with as yet undetermined impact on their populations or the $4.5 billion fishing and tourism industry in the region.

Impacts on the population levels of other species are likely to come about more indirectly through effects on food chains. The mussels are filter-feeders that remove phytoplankton from the water, thereby affecting all the species higher in the food chain, such as walleye, bass, trout, and perch. In the Great Lakes there was a marked reduction, for example, in the body size of whitefish following the colonization by the mussels. The linking factor seems to be the collapse of the deep-water amphipod *Dipoeria*, a major food source for whitefish. In some European locations, invasion by the mussels has led to clearer water as a result of the removal of phytoplankton. These changes may benefit some species, even fish species. Bottom-feeders, such as carp and whitefish, and invertebrates, such as crayfish, may benefit as more nutrients are returned to the lake bottoms, either in the form of mussels themselves or mussel feces.

However, the mussels do not remove all species of phytoplankton equally. This is creating problems with blooms of blue-green algae, such as the toxic *Microcystis aeruginosa*, that are not ingested by the mussels. Some scientists believe that the algae may be primarily responsible for Lake Erie's 500 to 1,000 km^2 dead zone, which had mostly been attributed to chemical pollutants.

It remains to be seen whether species higher in the food chain, such as scaup and other waterfowl, can help to control the spread of the mussel. Already, numbers of some of these species, which stop over to feed during their migration, appear to have risen considerably. Realistically, it appears that the ducks may have some impact, as they have in Europe, but that the infestation will be too large and the number of ducks too small for the problem to be controlled in this manner. Furthermore, once a species becomes established, it is difficult to prevent further spread. Despite major efforts in the US to stop the spread of zebra mussels, they were found for the first time in early 2008 in Lake Mead in the desert near Las Vegas, about 2,000 kilometres from the Great Lakes.

In 2005 another deadly invader, perhaps the worst to date, suddenly appeared: viral haemorrhagic septicaemia, or VHS, dubbed the Ebola virus for fish. Great Lakes fish have little immunity to it, and this has led to massive die-offs as they become infected. The virus is one of the world's most dreaded fish diseases, normally found only in salt water, and one of the first foreign pathogenic microbes to become established in the Great Lakes. The virus has been identified in 19 species, and in the St Lawrence River hundreds of thousands of round gobies have succumbed to the disease. Gizzard shad die-offs from VHS in Lake Ontario west of Rochester and in Dunkirk Harbor on Lake Erie also have been reported.

Zebra mussel infestations like this one have been clogging water intakes in the Great Lakes since the invasive species was first discovered in Lake St Clair in 1988.

Another threat—the Asian bighead carp—has appeared. The carp has a voracious appetite, eating up to 20 per cent its body weight in plankton every day and reaching almost a metre in length. The carp escaped from fish farms in the southern US in the 1990s and invaded the Illinois and Mississippi River systems. Only a canal in Chicago that connects to Lake Michigan, protected by an electric fence, prevents the fish from entering the Great Lakes. Biologists believe it is only a matter of time before the carp enter the lakes, which would lead to the demise of the entire fishery. One possible means of entrance has been through the live fish trade. Carp are brought live from fish farms in the US to Asian markets and restaurants in Toronto, and the water subsequently is discarded into the drainage system, along with any fingerlings. Since 2005, the importation of live carp into Ontario has been illegal but, despite the threat of large fines, some entrepreneurs continue to take that risk.

Many invasive aquatic species, including the zebra mussel, arrive in their new habitat courtesy of ocean freighters, which take on water for ballast in one part of the world and release it in another. More than 3,000 species are being transported around the world every day through this process. Given the magnitude of these introductions, it is inevitable that some of these species will not only find a tolerable home in their new location but also explode into great numbers.

In 2004, a new international convention to prevent the potentially devastating effects of the spread of harmful aquatic organisms carried by ballast water was adopted by the International Maritime Organization, the United Nations agency responsible for the safety and security of shipping and the prevention of marine pollution from ships. The convention requires all new ships to implement a Ballast Water and Sediments Management Plan. They will have to carry a Ballast Water Record Book and will be required to carry out ballast water management procedures to a given standard. Existing ships are required to do the same but after a phase-in period.

Canada has legislation and programs that ostensibly deal with the problem of invasive species, especially those that may damage agricultural and forest crops or pose a danger to human health. Under the terms of the United Nations **Convention on Biological Diversity**, discussed in Chapter 14, Canada is also committed to containing invasives that threaten biodiversity. However, audits undertaken by the Auditor General (2002, 2008) discerned little progress in addressing the problem (Box 3.6).

In 2006, the previously voluntary ballast measures became mandatory under the Ballast Water Control and Management Regulations. All ships arriving from beyond the exclusive economic zone (EEZ) and entering waters under Canadian jurisdiction must undertake one of the following: exchange their ballast water, treat their ballast water, discharge their ballast water to a reception facility, or retain their ballast water on board (Figure 3.8). In 2007, the government announced $4.5 million in funding to enforce the regulations for the ensuing five-year period. The funding was used to increase the number of marine inspectors enforcing ballast water regulations, to support the development of technologies to better address ballast water issues, and to equip marine inspectors with the proper tools to enforce the ballast water regulations. Increased aerial surveillance to combat marine pollution has also been undertaken. For example, during fiscal year 2006–7, Transport Canada conducted 1,649 hours of dedicated pollution patrol. This resulted in the detection of 98 marine pollution incidents from the 10,063 vessels that were flown over.

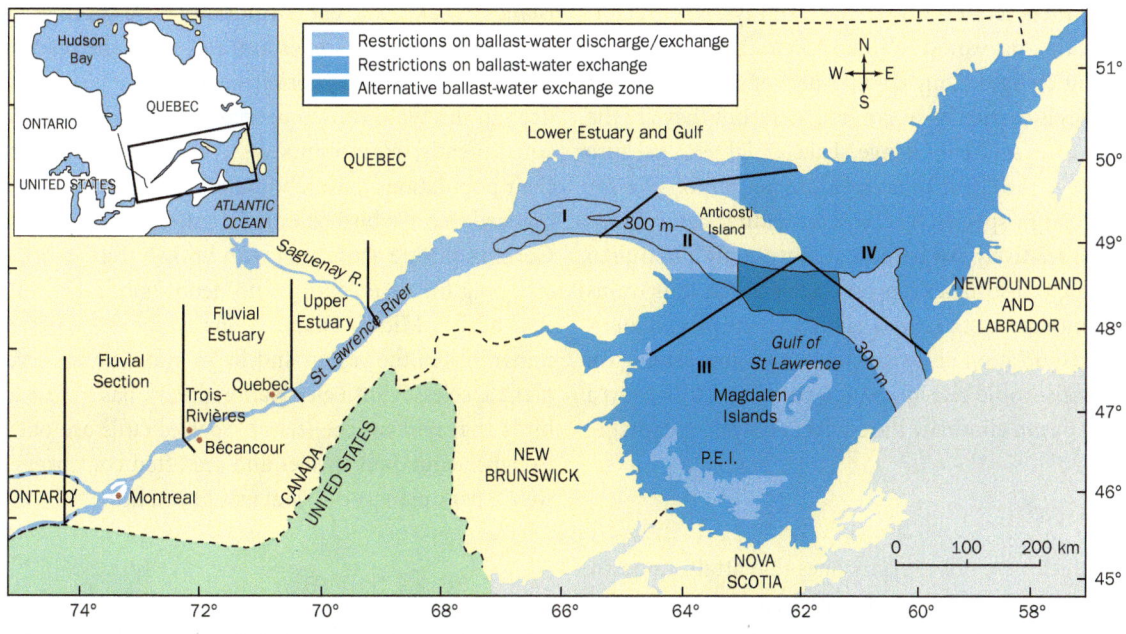

Figure 3.8 | Areas for ballast water control on the east coast.

Environment in Focus

Box 3.6 Summary of Findings of the Commissioner of the Environment and Sustainable Development on Control of Aquatic Invasive Species

- Fisheries and Oceans Canada has made unsatisfactory progress in implementing our 2002 recommendations. While the Department has identified some aquatic invasive species that pose major risks to Canada and put in place biological risk assessment guidelines, the rate at which new alien species are becoming established is exceeding the rate at which the Department is assessing risks. The Department has failed to assess economic and social risks, and priorities and objectives for prevention, control, or eradication of risks posed by aquatic invasive species have not been set.
- In addition, Fisheries and Oceans Canada does not have plans or mechanisms in place for early detection of, or rapid response to, aquatic invasive species and is therefore unprepared to prevent, control, or eradicate potential new aquatic invasive species. It has not monitored or reported how effective its efforts have been at preventing, controlling, or eradicating the aquatic invasive species it has identified.
- Transport Canada has made satisfactory progress in implementing one of our 2002 recommendations. Mandatory regulations for the control and management of ballast water came into force through the *Canada Shipping Act* in 2006. These regulations constitute a major step forward in addressing the issue of aquatic invasive species. Ships entering Canadian waters are now required by law to take steps to manage ballast water to reduce the likelihood of introducing aquatic invasive species.
- Transport Canada has made unsatisfactory progress on our 2002 recommendation related to monitoring and reporting on compliance. While Transport Canada has developed the tools it needs and is gathering information on compliance with its regulations for ships entering the Great Lakes, gaps remain in the Department's compliance monitoring and reporting at the national level.
- We found that risks posed by aquatic invasive species have not been adequately assessed or effectively managed. The federal government is not yet in a position to prevent, control, or eradicate invasive species that pose the greatest threat to Canada's aquatic ecosystems and economy. Much remains to be done to meet commitments made in the federal government's 1995 Canadian Biodiversity Strategy.

Source: Auditor General of Canada (2008: ch. 6, pp. 2–3).

Although regulations have slowly been tightened, the days of allowing ocean freighters into the Great Lakes might be numbered. If the approximately 500 ocean ships that now enter the Great Lakes each year were allowed no closer than the Port of Montreal and their cargo transferred to railways, lake freighters, or barges, the cost would be about $55 million (US) a year in extra freight charges. Only about 7 per cent of tonnage on the Great Lakes is carried by ocean freighters, but they are the main source of ecosystem change. Lake freighters, because they do not leave the Great Lakes area, are not responsible for bringing in foreign species. The cost of ending ocean shipping would be relatively low because it moves only a marginal amount of freight. Estimates suggest that as few as four trains could accommodate all the cargo carried by ocean freighters, while the transportation costs for companies now relying on the ocean vessels would rise by about 6 per cent. This seems a small price to pay to eliminate the current ecosystem damage.

Hyperabundance

Introduced species are not the only ones that attain undesirable numbers in some ecosystems. Native species may do the same. This often occurs where natural habitats have been disturbed and species, particularly predatory species, have been removed. Prey species previously controlled by natural factors may become hyper-abundant, becoming pest species and presenting considerable management challenges.

One example is the double-crested cormorant that nests on islands in Lake Erie. The cormorants are large, migratory water birds that nest in colonies and return to the place they were hatched to breed. Cormorants experienced a rapid population drop in the 1960s caused primarily by pesticides such as DDT; consequently, cormorants were targeted for protection and their populations have rebounded. Today, ecologists have recognized that the bird colonies are threatening rare vegetation. Cormorants are associated with broken tree branches, foliage stripping for nests, and guano deposits that threaten vegetation health. Middle Island is one of the few forested islands remaining in the region and in an effort to preserve the rare plant species, Point Pelee National Park has begun to cull the birds that nest on the island. Species culls are one response to hyper-abundant species and are often controversial. If you were a park manager—what would you do?

Species Removal

Just as the introduction of species to new habitats can disturb ecosystem function, so can the removal of species from food webs. The reduction of some species, the so-called keystone species discussed in the previous chapter, may be particularly disruptive. One well-known example relates to the extirpation of the sea otter from the Pacific coast.

When James Cook anchored at Nootka Sound on the west coast of Vancouver Island in 1778, he reported that the fur of the sea otter 'is softer and finer than that of any others we know of; and, therefore, the discovery of this part of the continent of North America, where so valuable an article of commerce may be met with, cannot be a matter of indifference.' Indeed, it was not. The British, seeking trading goods to barter with the Chinese in exchange for tea, discovered that sea otter pelts were in great demand in China and thus made every effort to ensure that the west coast became British (rather than Spanish or Russian!) Columbia.

The sea otter is a large sea-going weasel of the outer coasts, flourishing in giant kelp beds. They lack a protective layer of blubber but have a very fine fur that traps air and insulates them from the cold Pacific waters. They also need a lot of food (up to nine kilograms per day) to fuel the fast metabolism that counteracts energy loss to the environment. Favourite prey are sea urchins, crabs, shellfish, and slow-moving fish.

The otters were easy to catch, and Russian, American, and Spanish hunters, aided by local Native populations, finished off what the British had begun. Within 40 years, populations were reduced from more than half a million to 1,000 to 2,000. On the coast of British Columbia, it is likely that they were completely extirpated. However, relict populations remained around Monterey in northern California, and in the Aleutian Islands. Individuals from this latter population have now been reintroduced to the coast of British Columbia, where expanding populations thrive again.

Scientists discovered the otters' key role in maintaining ecosystems after studying two groups of islands off Alaska. They noticed that although the two groups were very similar in terms of location and physical conditions, one group had much more life—bald eagles, seals, kelp beds, and otters—than the other. Otters play a critical role in controlling sea urchin populations (Estes et al., 1989). Sea urchins are voracious eaters of kelp (large, brown seaweed), which may be the world's fastest-growing plant, with increments of up to 60 centimetres per day. Given the support of the ocean, kelp does not need to invest much energy in heavy support structures, leaving more energy for growth (another example of the auxiliary energy flow discussed in Chapter 2).

Kelp plays a major role in coastal communities. It provides food and habitat for many other species. Diatoms, algae, and microbes grow on the fronds of the kelp, along with colonies of filter-feeding bryozoans and hydroids. Predators abound. Fish come to feed off the colonists or to seek protection from open-water predators such as seals, sea lions, and killer whales. When overgrazed by sea urchins, this productive habitat disappears. The urchins eat through the holdfasts that anchor the kelp to the ocean floor, and the kelp is soon washed away into the open ocean or onto land. As the kelp disappears, so do the species dependent on it. On the two islands in Alaska, one island had managed to escape the fur rampage that eliminated otters elsewhere, and this one displayed the rich coastal community that should extend all along the outer coast of the North Pacific. Otter populations, through their control of the urchin population, are therefore critical to maintaining the productivity of the entire community, right up to bald eagle populations. The fact that the fashion tastes of Chinese mandarins 200 years ago, met by traders from the other side of the world who wanted to enjoy afternoon tea, is still reflected in bald eagle populations 7,000 kilometres away on the BC coast indicates the complex interactions between biophysical and human systems.

Feedback

Feedback is an important aspect of maintaining stability in ecosystems whereby information is fed back into a system as a result of change. Feedback initiates responses that may exacerbate (positive feedback) or moderate (**negative feedback**) the change. There is, for example, considerable debate regarding the role of feedback loops in global climate change, as discussed in Chapter 7. One positive feedback loop that may have a strong influence in Canada is the effect of increased temperatures in the North. It would increase the area of snow-free land in summer and is known as **polar amplification**. Snow has a high albedo; in other words, it reflects rather than absorbs much of the incoming radiation. As temperatures rise, the area covered in snow will be replaced by areas free of snow, uncovering rocks and vegetation with lower albedo values. This will cause more heat to be absorbed, which in turn will contribute to global warming. A similar situation with regard to forest fires was noted in Box 3.2.

Sea otters.

On the other hand, negative feedback loops may also be in operation and serve to counteract such positive feedback loops. One of them has to do with the possible role of phytoplankton in global warming. Phytoplankton produce a gas called dimethyl sulphide. When seawater interacts with the gas, sulphur particles formed in the atmosphere serve as condensation nuclei for cloud droplets. As the planet heats up, the productivity of the phytoplankton should increase, leading to an increase in the amount of gas and cloud droplets produced. This will have the effect of increasing cloud cover and reflecting away solar radiation, which could lead to cooling of the Earth. However, scientists feel that this cooling will be offset by the overall impact of global warming, as discussed in Chapter 7.

Almost all the examples in this chapter can be used to illustrate some aspect of feedback mechanisms. The allelopathic quality of the diffuse knapweed, for example, shows a positive feedback loop that promotes the spread of the species. The more the species spreads, the more conditions are created into which only it can spread. The sea otters produce a negative feedback loop on the sea urchin–kelp relationship. If the urchins become too numerous and overgraze the kelp beds, increases in otter populations will help to reverse this imbalance. When this negative feedback loop was removed from the system, there was nothing to maintain the dynamic balance of the system.

Similar examples of feedback loops occur at all scales, even down to the regulation of temperatures in individual organisms. Sometimes these feedback messages can be rapid, as in the case of organism thermoregulation. In other cases, there can be considerable delay between the stimulus for change and the resulting feedback response. Unfortunately, as the example of the positive feedback loop and snowmelt described above indicates, sometimes the delay between the stimulus and the response may be so long that we are not conscious of it. By the time we are aware, it may be too late to try to moderate the stimulus, and a powerful positive feedback loop may already, albeit slowly, have been set in motion. This is one reason why many scientists support immediate actions to reduce emission of greenhouse gases (see Chapter 7), even though we do not yet have a clear understanding of all the relationships involved.

We are also becoming more aware of the chaotic nature of many systems in which a slight perturbation becomes greatly enhanced by positive feedback. The so-called **butterfly effect**, for example, traces how the turbulence of a butterfly flapping its wings in South America might, through cascading effects on airflows, influence the weather in North America (see www.imho.com/grae/chaos/chaos.html). Further research has revealed the existence of similar phenomena in many different systems in which very small changes can have a great influence on outcomes. Chaos theory tries to discern pattern and regularity in such systems and allow for greater predictability.

Synergism

Synergism is another important characteristic that may influence change in ecosystems. A synergistic relationship occurs when the effect of two or more separate entities together is greater than the sum of the individual entities. One example is the problem of acid deposition, discussed in Chapter 4. The effects of acid deposition are often exacerbated by the presence of other pollutants, such as ground-level ozone. Individually, both these forms of pollution may cause a certain amount of damage to an ecosystem. In combination, however, their effects are magnified.

Ecological Restoration

Many ecosystem changes result from human activities, and these changes may have a negative impact on ecosystem components and function. However, this does not mean that severely damaged ecosystems have to stay that way, and restoration ecology has developed as a field of study and practice to help repair environmental damage. This book contains many examples, ranging from remediation of the Sydney Tar Ponds discussed in Chapter 1 and reclamation of the areas around Sudbury rendered treeless by acid rain (Chapters 4 and 13) to efforts to reintroduce endangered species into national parks (Chapter 14). The goals of these efforts vary enormously—from merely stabilizing an area with a self-maintaining cover of vegetation, such as the reclamation of many industrial sites, to efforts to restore areas to their pre-disturbance condition. One of the common difficulties in the latter situation is ascertaining the nature of the original ecosystem. Restoration ecologists are now concentrating more on trying to restore natural processes in an area rather than reintroducing components. However, **ecological restoration** is very challenging and costly, and there is widespread agreement that it is better to avoid degrading ecosystems in the first place rather than trying to restore them afterwards.

Population Growth

The number of individuals in a species is known as the **population**. When calculated on the basis of a certain area, such as the number of sea otters per hectare, it becomes **population density**. The number of organisms in a population is important, because low numbers will make a species more vulnerable to extinction. Changes in population characteristics are known as population dynamics.

Populations change as a result of the balance among the factors promoting population growth and those promoting reduction. The most common response is through adjustments in the birth and/or death rates to the factors shown in Figure 3.9, although emigration and immigration can be important factors in some species. Population change is calculated by the formula $I = (b - d) N$, where I is the rate of change in the number of individuals in the population, b is the average birth rate, d is the average death rate, and N is the number of individuals in the population at the present time. As long as births are more numerous than deaths, then a population will increase exponentially over time (Figure 3.10) until the environmental resistance of the factors shown in Figure 3.9 begins to have an inhibiting effect that will serve to flatten out the curve.

The **carrying capacity** of an environment is the number of individuals of a species that can be sustained in an area indefinitely, given a constancy of resource supply and demand. Most species will grow rapidly in numbers up to this point and then fluctuate around the carrying capacity in a dynamic equilibrium (Figure 3.11). The carrying capacity is not one fixed figure, however, but will vary along with changes in the other abiotic and biotic parts of the ecosystem. In the puffin–capelin example in Chapter 2, the carrying capacity of the North Atlantic waters to support the puffin population was severely reduced as a result of a reduction in their food supply caused by competition from another organism, humans. Management inputs, such as provision of supplementary feeding or other habitat requirements, are often used to change the capacity

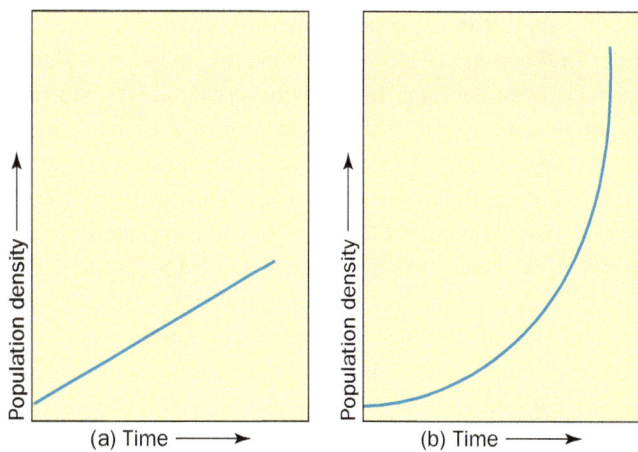

Figure 3.10 | Geometric (a) and arithmetic (b) growth patterns.

of an area to meet human demands. The concept of carrying capacity is now often applied to park management, with park managers establishing the number of visitors a park can sustain without either suffering ecological damage (known as *ecological carrying capacity*) or seeming too crowded (known as *social carrying capacity*).

Organisms that demonstrate the kind of S-shaped growth curve of Figure 3.11 are *density-dependent*, and as the population density increases, the rate of growth decreases. In other words, the larger the population, the lower the growth rate. This view is in accord with the equilibrium view of ecosystems discussed earlier but populations can still crash as a result of the dynamic nature of carrying capacities as discussed above. Some organisms, however, are *density-independent*, and the population operates with a positive feedback loop—the more

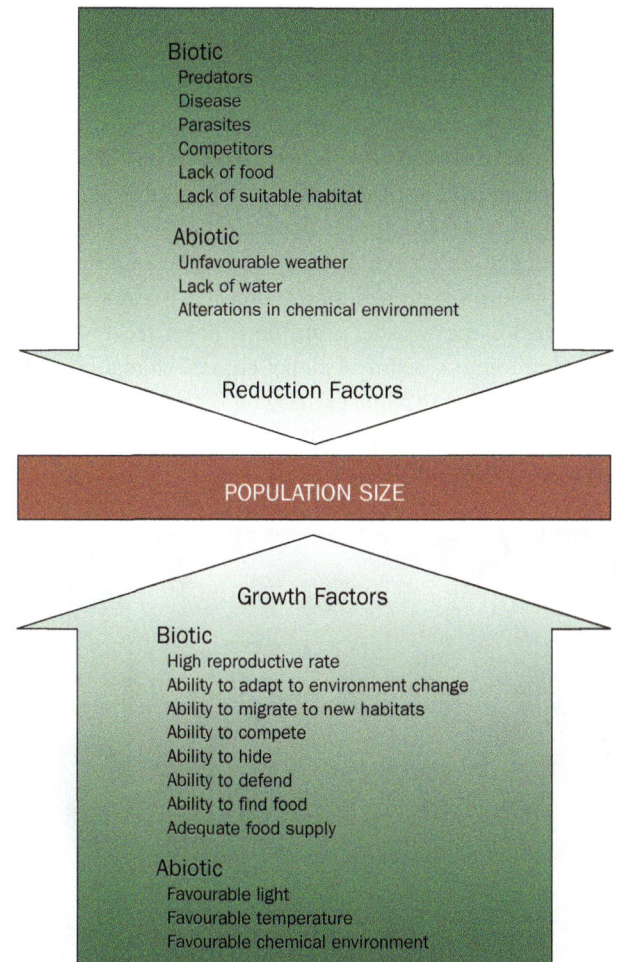

Figure 3.9 | Factors affecting population growth.

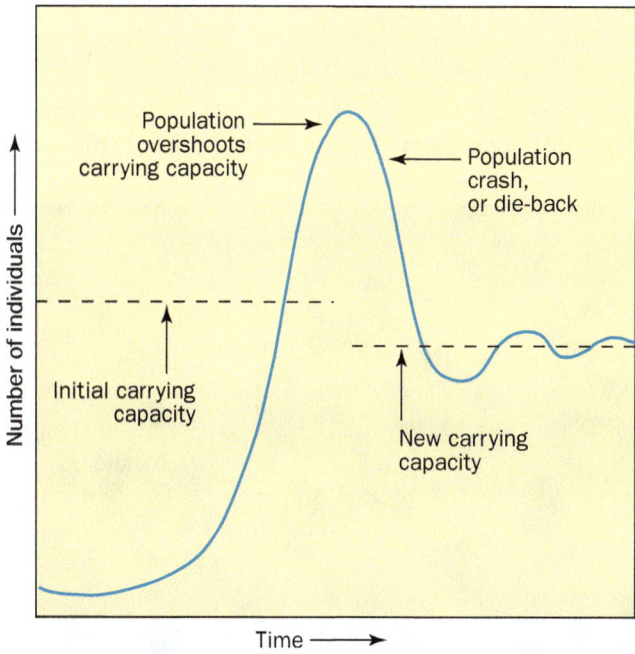

Figure 3.11 | Carrying capacity and population growth rates.

individuals in the population, the more that are born, and the population grows at an increasing rate to demonstrate a J-shaped curve. At some point, this population meets environmental resistance, causing the population to crash back to, or below, the carrying capacity. The algae blooms on ponds in the late spring or early summer are a result of this kind of growth. In reaction to the increased nutrient availability after winter, spectacular growth can occur until this food supply is exhausted and the population crashes. The spectacular and economically damaging increase in the numbers of the mountain pine beetle in BC is another example. Ecologists now accept that given the absence of cold winters because of climate change, the population will only reach its limit when it has exhausted its food supply (Chapter 9).

In some locations in Europe, this is what has happened with the zebra mussels discussed earlier. Mussel populations soared to a peak and stayed there for a few years before exhausting the food supply and crashing to between 10 and 40 per cent of the original numbers. However, in other locations, such as Sweden, the expected crash has yet to occur. Given the enormous food supplies of the Great Lakes and the low numbers of predators such as waterfowl, it may be a very long time before any natural population crash happens there.

The capacity of species to increase in number is known as their **biotic potential**, the maximum rate at which a species may increase if there is no environmental resistance. Different species, however, have different reproductive strategies. Some species, such as zebra mussels, are known as **r-strategists**, which produce large numbers of young early in life and over a short time period but invest little parental energy in their upbringing. Most of their energy is spent on reproduction, and they have few resources left to devote to maintaining a longer lifespan. Such species are usually small and short-lived and can respond to favourable conditions through rapid reproduction. They are opportunists, and their reproductive strategy is essentially based on quantity. Such species tend to dominate the early seral stages of the successional process.

K-strategists, on the other hand, produce few offspring but devote considerable effort to ensuring that these offspring reach maturity. Their strategy is based on quality. Individuals live longer and are usually larger. Populations of K-strategists often reach the carrying capacity of an environment and are relatively stable compared to r-strategists, which may experience large variations in population size. Table 3.2 summarizes the characteristics of these different strategists.

Examples of r-strategists include insects, rodents, algae, annual plants, and fish. A mature female codfish, for example, may produce more than 9 million eggs in one season. However, fewer than 5 per cent of these offspring may mature and last the first year. Most K-strategists are larger organisms, such as the larger mammals (including humans). Their lower biotic potential and ability to disperse often means that they are more restricted to the later seral stages of succession. Many endangered species (see Chapter 14) are K-strategists. The great whales (Chapter 8), with perhaps only one offspring every three years, are a good example. When the conditions to which they have become accustomed, and under which they evolved their reproductive strategy, change dramatically, such as with the introduction of new predators (humans), they have little capacity to respond in terms of increasing their reproductive rate. Muskox are another example (Box 3.7).

In addition to the factors outlined above, chance also plays an important role in determining population size. Severe winters, disease outbreaks, fires, droughts, and similar factors often have a major impact on populations. Peary caribou, for example, exist north of the 74th parallel by digging under the snow to feed on vegetation. In 1974–5, heavy snows and freezing rains led to high mortality as the herd starved to death, unable to reach their food source. Unfortunately, these are the very same weather conditions predicted to become more common as a result of global climate change. In 1993, there were more than 3,000 caribou on Bathurst Island in the High Arctic. By 1997, as a result of repeated bad winters, the number was down to 75.

Jellyfish and the Pacific white-sided dolphin are good examples of marine r and K species respectively.

Table 3.2 | Characteristics of K-Strategists and r-Strategists

K-Strategists	r-Strategists
Late reproductive age	Early reproductive age
Few, larger young	Many, small young
More care of young	Little care of young
Slower development	Rapid development
Later reproductive age	Early reproductive age
Greater competitive ability	Limited competitive ability
Longer life	Short life
Larger adults	Small adults
Live in generally stable environments	Live in variable or unpredictable environments
Emphasis on efficiency	Emphasis on productivity
Stable populations usually close to carrying capacity	Large population fluctuations usually far below carrying capacity

Environment in Focus

Box 3.7 Muskox

With its great shaggy head and flowing mane facing into an Arctic storm, the prehistoric-looking muskox seems to symbolize the determination of life to survive in even the most inhospitable environments. Muskox are superbly adapted to the rigours of the North. They have a long, coarse outer coat covering a fine underlayer of soft wool; all of their extremities (e.g., tail and ears) are covered by this coat. Their keen sense of smell enables them to find food in the winter darkness, they have low metabolic rates, and they minimize energy loss by remaining relatively inactive.

Muskox are not widely distributed throughout the Arctic—they concentrate in areas where vegetative conditions are best, where they can graze and browse on many Arctic herbs, grasses, and shrubs. In these areas, they stay in bands of females with a single bull that takes the leadership role in finding routes and repelling predators, mainly Arctic wolves. It is against such predators that they evolved their well-known defensive strategy of forming a circle around the calves with their massively horned heads pointing outward. Although effective against wolves, this strategy proved much less effective against humans with guns, who could stand at a distance and shoot this large target at will. More than 16,000 hides were shipped out of Canada between 1864 and 1916. In addition, Arctic explorers used them as a main source of meat. Consequently, with their low birth rate (one calf every second year) and a few hard winters, numbers fell drastically. They are now protected but may be hunted by licence and by indigenous hunters. About 8,000 muskox are left in the Canadian Arctic, ranging from the Arctic islands south to the Thelon game sanctuary and the coast of Hudson Bay.

Muskox.

Evolution, Speciation, and Extinction

When Charles Darwin first published his *On the Origin of Species* in 1859, he started a thought revolution that has seeped into virtually every realm of human ideology. His ideas challenged the static beliefs of many fields with the concept of evolution and the mechanisms by which change can occur. He postulated that over the long term, populations adapt to changing conditions through **evolution**, a change in the genetic makeup of the population with time. This can be achieved through mutations passed on to subsequent generations, eventually creating a new species. Within any population, some variation in the genetic composition also may predispose a certain segment of the population to adapt to certain conditions. If change occurs to favour those conditions, then the success of the part of the population genetically better adapted to the new conditions will be enhanced. In this way, over time, **natural selection** can lead to changes in the characteristics of a population.

Phyletic evolution is the process in which a population has undergone so much change that it is no longer able to interbreed with the original population and a new species is formed. This is the process of **speciation**. It can happen as a result of geographical isolation, when a single population becomes fragmented by a geographical feature, such as a mountain range or a water body, and the populations evolve separately from one another. If conditions differ in the respective environments of the different breeding groups, then natural selection will favour those individuals best suited to those conditions. Over time, they become so different that they can no longer interbreed. Another example occurs when part of a population adapts better to a new food source. It is better for these individuals to mate with similar individuals to enhance the ability to exploit that food source, and over time this process might create sufficient differences that they become a different species. Sometimes the effects of these influences may combine.

The evolution of the polar bear from the grizzly bear is one example. It is thought that the polar bear evolved as a separate species from the grizzly bear some 10,000 years ago. Bears with characteristics that helped them hunt seals on ice flows, such as lighter-coloured fur and greater strength, would be relatively more successful in the Far North rather than in the rest of the range, where a brown pelt and greater mobility are advantages. In this way, a single bear species became two bear species through adaptation to different environments and the process of natural selection. This process of local adaptation and speciation is known as *adaptive radiation*. However, instances of interbreeding between the two species have recently come to light.

Genetic diversity helps to protect species from extinction. The resilience of a species depends partly on the magnitude of the environmental change, how rapidly it takes place, and the capacity of the gene pool of the species to respond to these changes. In general, the broader the gene pool, the greater the capacity to adapt to change.

Changes in the abiotic environment are not alone in promoting evolutionary change. Species may also change through **co-evolution**, whereby changes in one species cause changes in another. Each species may become an evolutionary force affecting the other. A typical case is a prey species evolving to be more effective in avoiding a predator. In turn, the predator may evolve more efficient hunting techniques to detect the prey. Many such relationships have evolved in the tropical forests, especially between specific plants and animals, because of the long period of evolutionary change that has taken place in such environments. Canada has many examples as well, particularly relating to pollination in which various insects, birds, and bats have evolved to pollinate flowering plants and, in turn, a great diversity of plant shapes, sizes, and colours have developed as a direct response to the activities of the pollinators.

The processes of evolution described above have always been thought to take place very slowly. However, biologists have identified processes occurring much more quickly as a result of human activities in a process known as **contemporary evolution** (Stockwell et al., 2003). One of the main pressures for contemporary evolution is human harvesting of prey populations. In established fisheries, once fish enter targeted age classes, predation by humans occurs at rates two to three times higher than that of natural predators, often exceeding 50 per cent (Stokes and Law, 2000). Fisheries have hence selected for the survival of certain fish, usually small fish. For example, the average weight of groundfish on the Scotian shelf declined by 66 per cent between 1970 and 1995 (Leggett and Frank, 2008). These smaller fish produce fewer and less viable eggs, leading scientists to formulate the Big Old Fat Fecund Female Fish (BOFFFF) hypothesis. These old fish are irreplaceable in that they not only contain more eggs but also proportionately more viable eggs. As BOFFFFs are harvested, fish populations have increasingly greater difficulty in achieving replacement population and a negative feedback loop sets in.

Our predation of terrestrial vertebrates, especially ungulates, is also well developed. Fa et al. (2002) estimated that human predators kill more than 5 million tonnes of wild mammals in neotropical and Afrotropical forests per year. Gauthier et al. (2001) reported that half of the adult mortality of North American snow geese results from hunting, and Festa-Bianchet (2003) asserted that hunting accounts for most of the adult mortality of many of Europe and North America's large herbivores. Human predators induce

Many tropical orchids are products of coevolution. The flower has evolved to imitate the female wasp of the species that pollinates the flower. The male is deceived into thinking that it is a female, flies into the flower and in so doing picks up pollen that is subsequently taken to the next imitator, and pollination occurs.

micro-evolutionary change because we are replacing natural predators as dominant agents of selection.

Traditional harvesting strategies have often concentrated on taking the oldest and largest members of a population. For example, we shoot the largest ram and catch the largest fish, letting the others go so that they can grow larger and be harvested later. However, research is indicating that all individuals do not have the same capacity to become that large and that by eliminating the largest, we are systematically selecting for smaller individuals in the future. In general, individuals' size and growth rates decline, while reproductive investment increases and individuals become reproductively mature at smaller sizes and earlier ages. This is happening with many different species as average sizes continue to decline. Jachmann et al. (1995) reported an increase in tusklessness among African elephants. Tuskless males increased in the population from approximately 1 per cent in the early 1970s to about 10 per cent in 1993 and tusklessness among females rose from 10.5 per cent in 1969 to roughly 38 per cent in 1989 following intense poaching that targeted individuals with ivory-bearing tusks.

Extinction is the opposite of evolution and represents the elimination of a species that can no longer survive under new conditions. The fossil record suggests that perhaps close to 99 per cent of the species that have lived on Earth are extinct. The fact that we may still have up to 50 million species, more than have ever existed before, indicates that speciation has exceeded the extinction level. However, speciation takes time. Even with r-strategists, it may take hundreds and thousands of years; with K-strategists, it may take tens of thousands of years. Evidence suggests that in recent times, human activities have strongly tipped the scale in favour of extinction over speciation (Box 3.8), as discussed in Chapter 14. Table 3.3 gives some examples of species that at one time existed in Canada but are now extinct.

Extinction, like speciation, is not a smooth, constant process but one punctuated by relatively sudden and catastrophic changes. It appears that multi-cellular life, for example, has experienced five major and many minor mass extinctions. Scientists think that the age of the dinosaurs, a remarkably successful dynasty that relegated mammals to minor ecological roles for more than 140 million years, was brought to an end 65 million years ago by the impact of a large extra-terrestrial object. And then the mammals took over. Perhaps the dinosaurs, through the processes of evolution and speciation, managed to out-compete the mammals for a long period and, were it not for the chance impact of the asteroid, might still be the dominant animal life. However, this chance occurrence not only led to the demise of the cold-blooded dinosaurs but also favoured the survival of the rodent-like mammals with their smaller body size, less specialization, and greater numbers. Small body size was likely a sign of the mammals' inability to challenge the dinosaurs during the normal evolutionary process; however, small body size became a positive feature favouring survival under the new conditions.

Table 3.3 | Some Canadian Vertebrate Species Now Extinct

Species	Distribution	Last Recorded	Probable Causes
Great auk (*Alca impennis*)	Canada, Iceland, UK, Greenland, Russia	1844	Hunting
Labrador duck (*Camptorhynchus labradorius*)	Canada, US	1878	Hunting, habitat alteration
Passenger pigeon (*Ectopistes migratorius*)	Canada, US	1914	Hunting, habitat alteration
Deepwater cisco (*Coregonus johannae*)	Canada, US	1955	Commercial fishing, introduced predators
Longjaw cisco (*Coregonus alpenae*)	Canada, US (Great Lakes)	1978	Commercial fishing, introduced predators

There are other examples of the non-random impact of mass extinction on life. The features that make some species successful during ordinary times may be completely unrelated to the new conditions, making life's pathway somewhat chaotic and unpredictable rather than the smooth path that evolutionary theory might suggest.

Finally, as emphasized in Box 3.9, we should not think that evolution is fundamentally a story that demonstrates the

Environment in Focus

Box 3.8 Humans and Extinction

Extinction is a natural process. Scientists estimate the average rate of species extinction by examining the fossil record, which suggests that extinctions among mammals occur at the rate of about one every 400 years and among birds one every 200 years. Current extinction rates are difficult to estimate, because we do not have a full inventory of species and so we do not know what we are losing. Based on current rates of habitat destruction for tropical forests, estimates range as high as 100,000 extinct species per year. Many of these extinct species are likely to be undescribed arthropods, since they comprise the majority of species in tropical forests. The most recent and sophisticated assessments, based on detailed historical assessments rather than on models, suggest that extinction rates across all species groups range from 1 to 5 per cent becoming extinct since 1800 (Hambler et al., 2011). Some groups show rates that are double or triple this range, such as amphibians. If one group had to be selected as a proxy for overall species loss, it would be birds. Habitats that are home to most extinct species include wetlands, grasslands, and 'ancient' forests.

More than 18,000 species are listed as threatened on the Red List of the World Conservation Union (IUCN). Fewer than 10 per cent of the world's species have undergone status assessments, yet more than 30 per cent of amphibians, 23 per cent of mammals, and 12 per cent of birds are threatened, according to the IUCN. Species at risk of extinction are discussed in more detail in Chapter 14.

Humans have had a major impact on biodiversity for quite some time. Paul Martin (1967) was one of the first to suggest that humans may have been a major factor in causing the extirpation of several species of large mammals from North America at the end of the last ice age, some 10,000 years ago. During that time, at least 27 genera comprising 56 species of large mammals, two genera and 21 species of smaller mammals, and several large bird species became extinct. The extinctions included 10 species of horse, four species of camel, two species of bison, a native cow, four elephant species, the sabre-toothed tiger, and the American lion. Although this period was also a time of global climatic change, no such extirpations were associated with the same period in Eurasia. The period also saw a substantial in-migration of humans from the Asian continent, who began to prey on animals unfamiliar with and therefore not adapted to human hunting. This hunting, combined with the environmental stresses experienced through habitat alteration and repercussions through the food chain, was sufficient to extirpate the species.

Charles Kay (1994) has conducted research into the subsequent impact of Native Americans on ungulate populations before the onset of European influences. He concludes that even then, humans were the main limiting factor on ungulate populations in the intermountain West and that elk in particular were overexploited. The people had no effective conservation strategies and hunted to maximize their individual needs, irrespective of environmental impacts. Thus, the image of North America as a vast wilderness unaffected by human activities before the coming of the Europeans appears to be a myth. Even that mightiest symbol of the wild, the grizzly bear, was apparently under pressure from Aboriginal hunters in Alaska (Birkedal, 1993).

Other reviews agree that human predation was the central cause among a suite of forces responsible for Pleistocene extinctions (Barnosky et al., 2004; Brook and Bowman, 2004), and it is obvious that humans are super-predators who excel over all other species.

It seems difficult to believe that what we now consider primitive weapons, such as this spear and other hunting tools of the Orang Asli peoples of Malaysia, may have enabled humans to hunt many other species to extinction.

benefits of complexity over simplicity. Humans—the most complex organisms—are not the pinnacle of life's achievement or the most successful. For that accolade, we have to travel to the other end of the complexity spectrum. From the beginning of the fossil record until present times, bacteria have demonstrated the most stable presence. According to Stephen Jay Gould (1994: 87):

> Bacteria represent the great success story of life's pathway. They occupy a wider domain of environments and span a greater range of bio-chemistries than any other group. They are adaptable, indestructible, and astoundingly diverse. We cannot even imagine how anthropogenic intervention might threaten their extinction, although we worry about our impact on nearly every other form of life.

The number of *Escherichia coli* cells in the gut of each human being exceeds the number of humans that has ever lived on this planet.

Impacts of Global Change

Global climate change will have profound impacts on the numbers and distributions of species in the world. Overall, climate is the main determinant of the patterns of life. As temperatures and precipitation change (discussed in more detail in Chapter 7), the changes will have a profound effect on these patterns. In general, there will be a poleward shift of life zones. The east-to-west orientation of the main terrestrial ecozones of Canada, shown in Figure 2.12, is likely to be replaced by a predominantly north–south pattern. Prairie ecozones will grow and forested ecozones will contract as

Environment in Focus
Box 3.9 The Burgess Shales

Burgess Shales fossils in the foreground at the Burgess Shales World Heritage Site in Yoho National Park, British Columbia.

Stephen Jay Gould, the Harvard paleontologist, called the Burgess Shales in Yoho National Park in British Columbia the single most important scientific site in the world. The reason for this superlative is the extensive bed of fossils high on the flanks of Mount Wapta. They are fossils from the Cambrian era, some 530 million years ago, when there was a great flourishing of diverse life forms. The special feature of the site is that the fossils from this era are preserved in great detail, even down to the soft body parts, such as stomach contents.

The story revealed is one of great diversity at a time when all but one phylum of animal life made a first appearance in the geological record. The site also contains many body patterns for which there are no current counterparts. Thus, it seems as though life could be characterized as three billion years of unicellularity, followed by this enormously diverse Cambrian flowering in a brief five-million-year period and a further 500 million years of variations on the basic anatomical patterns set in the Cambrian period. Why, or how, this flowering took place is uncertain. It would seem to require a combination of explanations. First, there was literally an open field available for colonization—an environment ripe to support life but with little life in it. Therefore, species did not have to be particularly good competitors to survive. Virtually anything could survive. Since this time, even after mass extinctions, sufficient species have remained to make it pretty tough competition for any newcomers. Second, it seems as though the early multicellular animals must have maintained flexibility for genetic change and adaptability that declined as greater specialization arose and organisms concentrated on refining the successful designs that had already evolved. Furthermore, we have little idea why most of these early experiments in life died out and yet others remained. There seem to be no common traits shared by the survivors to indicate that they were the victors of Darwinian strife. Perhaps just the lucky ones survived.

Gould, in his fascinating book *Wonderful Life*, suggests that this challenges our established view of evolution as an inevitable progression over time from the primitive and few to the sophisticated and many. It also radically challenges our view of ourselves as being the logical end point of evolutionary change, the rightful inheritors of the Earth. In Gould's words: 'If humanity arose just yesterday as a small twig on one branch of a flourishing tree [of evolution], then life may not, in any genuine sense, exist for us or because of us. Perhaps we are only an afterthought, a kind of cosmic accident, just one bauble on the Christmas tree of evolution' (Gould, 1989: 44). In other words, we should be humble!

Archeologists work at a dig site of a fossilized dinosaur at Alberta's Dinosaur Provincial Park. More species of extinct dinosaurs have been found and identified at this World Heritage Site than anywhere else in the world.

precipitation levels fall. Species dependent on grasslands will increase their range; those dependent on forests will contract.

Already, there are many documented examples of species range changes in response to climate change (Parmesan, 2006). For example, egg-laying, flowering, and spawning are occurring earlier for many species, in some cases disrupting delicate cycles that ensure that insects and other food are available for young animals. Tree swallows across North America advanced egg-laying by as much as nine days from 1959 to 1991. Unfortunately, the hatchlings are now emerging before major insect hatches, and as a result populations of tree swallows are declining because of chick starvation. These types of changes are expected to continue and it is predicted that the drought in the Prairie Potholes region (southeastern Alberta and northeastern Montana to southern Manitoba and western Minnesota) will lead to significant reductions in the populations of 14 species of migratory waterfowl; 30 to 50 per cent fewer prairie ponds will hold water in spring by 2060, with an associated 40 to 50 per cent decline in the numbers of ducks settling to breed in the area.

Other changes are also taking place. Spring migration is occurring earlier and fall migration later for many species. For example, 25 migratory bird species are arriving in Manitoba earlier than they did some 60 years ago; only two are arriving later. However, as usual, there are complications. Short-distance migrants are migrating earlier because they can read the local cues that conditions are right for them in their destination. However, migrating birds in Costa Rica have few cues to tell them that conditions in Canada are right for their return. Consequently, they arrive at the normal time, only to find that the food supply has already waned (Wiley-Blackwell, 2008).

International Guest Statement
Life at the Crossroads: How Climate Change Threatens the Existence of the Maasai
Philip Osano

From 2008 to 2009, a devastating drought hit areas inhabited by Kenya's Maasai pastoralist community, destroying three-quarters of their cattle and two-thirds of their small stock. The drought was the worst the Maasai had experienced in decades, despite the fact that it followed in the wake of recurrent droughts brought about by climate change.

The Maasai live in Kenya's arid and semi-arid land areas (ASALs) in the southern part of Kenya, which are characterized by low and erratic rainfall with high evaporation rates and limited soil moisture—conditions that render the drylands fragile and unsuitable for rain-fed agriculture. In this tenuous ecosystem, pastoralism is considered the most suitable form of land use and livestock forms the principal source of livelihood for the Maasai.

The ASALs constitute 80 per cent of Kenya and are home to over 10 million people. Compared to the regions of Kenya with high potential for development, the ASALs have the lowest development indicators. Populations in the ASALs are characterized as being in abject poverty, where 60 per cent of inhabitants live on less than a dollar a day. In addition, environmental degradation, insecurity due to cattle rustling, climatic shocks, and diseases compound poverty rates in these areas. However, the ASALs support an estimated 60 per cent of Kenya's livestock. Nearly half of the total livestock herds in Kenya belong to the Maasai, and an estimated 75 per cent of the total household income in Maasai land is generated from livestock.

The Maasai have a reputation of rigid adherence to their traditional means of living. But in recent years, myriad changes in land tenure, land-use intensification, sedentarization, institutional changes, and climate change have forced the Maasai to abandon their old ways. Although the Maasai live in primarily rural areas and largely depend on livestock, the majority of their households are increasingly adopting agro-pastoralism, or shifting from natural resource-based livelihoods to non-farm activities, which often involve relocation to other regions.

Two major factors are fuelling the transformation of the pastoral livelihoods of the Maasai: fragmentation of once contiguously intact grasslands that reduces the scale of the pastoral landscape; and climate change, which is increasing the variability and frequency of rainfall perturbations in drylands. Climate change has been particularly devastating to the Maasai because of the negative effects manifested in recurrent drought, leading to increased food insecurity, starvation, and poverty.

Climate change is expected to occur at a faster rate, culminating in new weather patterns that are likely to result in increased suffering among the Maasai. It is anticipated they will be affected in several ways: climate change variability will induce droughts that will disrupt the traditional seasonal migration of herders, livestock, and wildlife to critical water and nutrient resource points. The disruption of livestock and wildlife migration patterns will constrain the space for coexistence between humans and wildlife, as competition for scarce resources increases. As this competition intensifies, the potential for violent conflicts between herders and farmers also will increase. Higher occurrence of droughts and floods also will limit capacities to diversify into crop farming because of increased risk of crop failure, and floods will limit abilities to relocate to other areas.

Similarly, negative factors induced by climate change, such as drought, erratic rainfall, and lack of access to watering points will result in poor nutrition for livestock because the quality and quantity of grass the animals feed on will be compromised. As well, conditions induced by climate change will worsen the severity and distribution of livestock diseases and pests.

Since 2003, the government has renewed its focus on improving livelihoods in the pastoral areas by taking note of the added dimension of climate change and its negative effects on poverty eradication. Key to this is the recognition of the existence of fundamental resources that could be used to develop the arid lands. Within Maasai territory, national wildlife reserves such as the world famous Maasai Mara and Amboseli in southern Kenya attract thousands of tourists each year, bringing in significant revenues. Tourist dollars could potentially be used in projects that improve living conditions for the Maasai.

The government has initiated short-term and long-term measures to combat climate change and mitigate its effects on pastoral communities. Short-term initiatives include emergency measures to tackle drought through the provision of food relief, water supplies, and emergency livestock off-take. The long-term, multi-year initiatives involve policy development and the creation of institutions for implementation. Some of the more innovative schemes include the following:

- Payments for ecosystem services (PES) provide incentives to pastoralists in exchange for administering their land for some ecological service that promotes environmental conservation. Pastoralists who participate in a PES scheme voluntarily agree not to cultivate, fence, or subdivide their land in return for a fee paid directly to them. In addition, the pastoralists also agree to keep the land open for livestock and wildlife grazing.

Climate change in southern Kenya is creating drought, erratic rainfall, and restricted access to watering points, and places pressures on the land and livestock that are critical for the traditional pastoralist ways of the Maasai people.

- A conservancy grass bank is a section of natural grassland that pastoral communities agree to reserve and not graze their animals on for the benefit of wildlife, and partly as a safety net during periods of drought. The pastoral communities apportion communal and private lands into respective zones for wildlife tourism, livestock herding, and community grass banks, which act as buffer zones during drought periods.

Most pastoral communities do not have access to insurance services despite the vulnerability posed by climate change to their livestock. The Index-Based Livestock Insurance (IBLI) was instituted in partnership with private industry and covers periodic drought that dries up the natural rangeland vegetation, leading to livestock mortality. Insurance payouts are made to herders who have bought annual insurance contracts.

There is no doubt that the dynamic ecosystems of the Maasai will be changing as a result of climate change, and now is the time to start implementing strategies that will lead to resilient livelihoods for the people in the future.

Source: First published in 2011 by the Africa Initiative (AI) and The Centre for International Governance Innovation (CIGI) on the Africa Portal at www.africaportal.org. Reproduced here with permission of the author, AI and CIGI.

Philip Osano is a Ph.D. candidate in the Department of Geography at McGill University, a Graduate Fellow at International Livestock Research Institute (ILRI), and a Research Associate with the African Technology Policy Studies Network (ATPS), in Nairobi, Kenya. He is investigating the impact of payments for ecosystem services (PES) on poverty among pastoral communities in East Africa's drylands. A longer version of this article was originally published in the Africa portal (www.africaportal.org).

Extinctions will take place because some species are incapable of adjusting at such rapid rates. Already, the US has listed the polar bear as threatened under the Endangered Species Act as a result of rapid melting of the sea ice that the bears depend on, and Canadian scientists are reporting similar findings (Stirling et al., 2008; and see Chapter 7). There are some interesting human dimensions to this situation. One implication is that the US has blocked the import of polar bear trophies into the US. This was followed by a similar ban by the European Union in 2008, effectively removing the incentive for US- and EU-based hunters. In 2010, a US proposal to place an international ban on polar bear trade through the Convention on International Trade of Endangered Species (CITES) was defeated. Canada opposed the proposal due to the potential impacts on Aboriginal economies. Many Inuvialuit people in western Arctic coastal communities depend on the hunt for their livelihood. Foreign hunters pay from $25,000 to $40,000 for the chance to shoot a polar bear, and the Northwest Territories estimates the total loss to the territorial economy would be around $3 million annually.

There are also some interesting political dimensions to the endangered species listing. The listing was long delayed by the US government and came about as a result of a lawsuit by an environmental group. During the delay, several additional exploratory licences for oil and gas drilling in Alaska were issued by the government. Because of the strict provisions of the Endangered Species Act, it is unlikely that these licences would have been issued after the listing. In announcing the listing, the government bemoaned the 'inflexible nature' of the Act that forced it to determine the listings on purely ecological grounds, irrespective of economic considerations (*Globe and Mail*, 15 May 2008, A6). On the other hand, environmentalists consider this 'inflexibility' one of the strongest aspects of the Act. There are no such inflexibilities in the Canadian Act. In Chapter 14, you will see that this feature of the US Act is in direct contrast to our own Species at Risk Act under which decisions on listing endangered species are made by politicians, not biologists, and can take into account economic and political considerations. Do you think this is a good or a bad approach?

Scientists predict that 9 to 52 per cent of all terrestrial species (up to 1 million plants and animals) will be on an irreversible path to extinction by 2050. These figures appear to be supported by past temperature changes. Mayhew et al. (2008) analyzed the fossil record for the past 520 million years against estimates of low-latitude sea surface temperature for the same period. They found that global biodiversity (the richness of families and genera) is related to temperature and has been relatively low during warm 'greenhouse' phases, while during the same phases the extinction and origination rates of taxonomic lineages have been relatively high. Calculations suggest that the world's biodiversity hot spots, the main engines of evolution, are particularly vulnerable to extinctions from climate change because of the high level of change predicted in these biomes and the low levels of out-migration (Malcolm, et al. 2006).

Climate change will also influence the functioning of ecosystems, the characteristic ways in which energy and chemicals flow through the plants, herbivores, carnivores, and soil organisms that comprise the living components of ecosystems, as described earlier in this chapter. Productivity will change; in some places it will increase, in others decline (see International Guest Statement). Food webs will be disrupted as predators and prey react differently to the changing conditions, as illustrated above with the example of the tree swallow. These changes will affect marine ecosystems as much as, if not more than, terrestrial ecosystems. Migratory whales, for example, will face shrinking crucial Antarctic foraging zones, which will contain less food and will be farther away. Levels of global warming predicted over the next 40 years will lead to winter sea-ice coverage of the southern ocean declining by up to 30 per cent in some key areas. Migratory whales may need to travel 200 to 500 kilometres farther south to find the 'frontal' zones that are their crucial foraging areas. The affected migratory whale species will include the blue whale, the Earth's largest living creature, and the humpback whale, only now coming back from the brink of extinction after populations were decimated by commercial whaling, mainly during the first half of the twentieth century.

Both species build up the reserves that sustain them throughout the year in the frontal zones, which host large populations of their primary food source, krill. Shrinking ice-covered areas affect krill production in two ways: sea ice is a refuge for krill larvae in winter and an area of intense algal blooms on which the krill feed in summer. Krill is so fundamental to the southern ocean ecosystem that the impact will not be confined to whales but will also affect seals, seabirds, and penguins as well as fisheries productivity. 'Frontal zones' are areas where water masses of different temperatures meet. They are associated with upwelling of nutrients supporting large plankton populations on which species such as Antarctic krill feed. As frontal zones move southward, they also move closer together, reducing the overall area of foraging habitat available. Since the krill depends on sea ice, less sea ice is also expected to reduce the abundance of food for whales in the feeding areas (Tynan and Russell, 2008).

Implications

This chapter emphasizes that ecosystems are dynamic entities that change over time. Without such change, we would not have evolved, and the dinosaurs would not have become extinct. The main implication is that we should accept and try to understand the nature of these changes and be able to distinguish between those essentially the result of natural processes and those that are the result of human activities.

Environment in Focus
Box 3.10 What You Can Do

It may seem that, with all the complexities of ecosystem change, an individual can do little to influence the situation. However, this is far from the case, since many changes are brought about by individuals and the sum total of their actions adds up to the tremendous changes described in this chapter. Specific actions that you can take include:

- Minimize your contribution to global warming. The many ways of doing this are outlined in Chapter 7.
- Avoid speeding up eutrophication processes by polluting waterways, using excessive fertilizer on your garden, and using phosphate-based detergents and other nutrient additives.
- Patronize organic farmers who do not use chemical additives.
- Do not introduce new species into the environment either deliberately by releasing them or inadvertently by, for example, transporting their seeds or fragments.
- Join a campaign to help eradicate an alien invasive species.
- Start your own campaign to eradicate an alien invasive species.
- Get to know the local flora and fauna in your area so that you can recognize alien invasive species.
- All change is not bad. Join a group that is trying to change an ecosystem for the better, sometimes called ecological restoration.

We cannot impose static management regimes on dynamic ecosystem processes without causing ecological disruption. A visible reminder of this was the fire-suppression policy characteristic of many national park services, which often ignored the natural role of fire in these ecosystems. When fires did start in such ecosystems, the buildup of fuel was often so great as to cause a major and very damaging fire, as happened in Yellowstone National Park in the US in 1988. Most park services have abandoned such practices for a more dynamic approach that tries to mimic the role of natural fires through prescribed burning programs.

Unfortunately, the temporal and spatial scales of ecosystem change are often so great that they are very difficult to observe in the human lifespan. Scientists are only now beginning to unravel the mysteries of some of these dynamic interactions between the different components of the ecosphere. There are complicated feedback loops and synergistic relations. In some cases, positive feedback loops are strengthened and accelerate undesirable changes that underlie some of the most serious environmental challenges facing humanity, such as global warming. Global climate change will place considerable stress on many species in terms of their limits of tolerance. This will lead to changes in range and abundance, and some species will become extinct. When faced with such dynamic ecosystem changes, we must use equally dynamic thinking to confront the challenges of the future.

Summary

1. Ecosystems change over time. The speed of change varies from very slow, over evolutionary time scales, to rapid, caused by events such as landslides and volcanic eruptions.

2. Ecological succession occurs as a slow adaptive process involving the gradual replacement of one assemblage of species by another as conditions change over time. Primary succession occurs on surfaces not previously vegetated, such as surfaces exposed by glacial retreat; secondary succession occurs on previously vegetated surfaces, such as abandoned fields. Fire is an important element in ecosystem change. Some ecosystems, such as much of the boreal forest, have evolved in conjunction with periodic fires. Fire suppression in such ecosystems can be detrimental to these natural processes.

3. Ecosystems tend towards a state of dynamic equilibrium in which the internal processes of an ecosystem adjust for changes in external conditions. Not all ecosystems are equal in their ability to withstand perturbations. Inertia is the ability of an ecosystem to withstand change; resilience is the ability to recover to the original state following disturbance. Both contribute to the stability of the system.

4. Important causes of ecosystem change include the introduction of alien species and the removal of native keystone species.

5. Feedback mechanisms exist in ecosystems that may either exacerbate (positive feedback loops) or mitigate (negative feedback loops) change.

6. Population change occurs as a result of the balance between factors promoting growth (e.g., increase in birth rates or reduction in death rates) and those promoting reduction (e.g., declines in birth or survival rates or increase in death rates).

7. Different species have different reproductive strategies. K-strategists produce few offspring but devote considerable effort to ensuring that these offspring reach maturity. In comparison, r-strategists produce large numbers of young starting early in life and over a short time period and devote little or no energy to parental care.

8. Populations adapt to changing conditions over the long term through evolution. Evolution results in the formation of new species as a result of divergent natural selection responding to environmental change. This is speciation. Extinction results in the elimination of species that can no longer survive under new conditions.

9. Although evolution can take thousands of years, scientists now detect evolutionary changes on the scale of tens of years as a result of humans acting as predators on a massive scale.

10. Global climate change will have a significant impact on the distribution and abundance of species. Some will flourish. Others will decline. Some will become extinct.

Key Terms

albedo	ecological succession	population density
alien species	ecotones	positive feedback loop
allelopathic	edaphic climaxes	prescribed burning
biotic potential	evolution	primary succession
butterfly effect	extinction	resilience
carrying capacity	Gaia hypothesis	r-strategists
climatic climax	inertia	secondary succession
climax community	intermediate disturbance hypothesis	seed bank
co-evolution	invasive	seral
contemporary evolution	K-strategists	serotiny
Convention on Biological Diversity	mature community	speciation
cyclic succession	natural selection	species
disturbances	negative feedback	synergism
dynamic equilibrium	polar amplification	
ecological restoration	population	

Questions for Review and Critical Thinking

1. What are different kinds of succession? Can you identify different seral stages in your area?

2. What is an edaphic climax? Can you find some local examples and identify the dominant limiting factor?

3. How does the concept of succession relate to environmental management?

4. How important was fire in the development of vegetation patterns in your region? Is there a fire management plan in your region? If so, what are its management goals?

5. Identify the main non-native plant and animal species in your region. What effect are they having on the local ecosystems? What are the implications for management?

6. Can you think of any other examples of negative and positive feedback loops in the ecosphere besides those mentioned in the text?

7. Are K-strategists or r-strategists most vulnerable to environmental change?

8. What is co-evolution? Can you think of any examples of co-evolution among species in Canada?

9. What place in Canada has been called the most important scientific site in the world, and why?

10. What are the implications of global climate change on species distributions and abundance?

Related Websites

Hinterland Who's Who, Invasive Alien Species in Canada
www.hww.ca/en/issues-and-topics/invasive-alien-species-in.html

Invasive species
www.invadingspecies.com; www.oag-bvg.gc.ca/internet/English/parl_cesd_200803_06_e_30132.html; www.imo.org/ourwork/environment/ballastwatermanagement/Pages/Default.aspx; globallast.imo.org; www.cwf-fcf.org/en/conservation/issues/issues-of-concern/invasive-species.html; www.invasivespeciesinfo.gov/international/canada.shtml

Invasive Species in Canada
www.invasivespecies.gc.ca

IUCN Red List of Threatened Species
www.redlist.org

Further Readings

Note: This list comprises works relevant to the subject of the chapter but not cited in the text. All cited works are listed in the Bibliography at the end of the book.

Balmford, A., R.E. Green, and M. Jenkins. 2003. 'Measuring the changing state of nature', *Trends in Ecology and Evolution* 18, 7: 326–30.

Butchart, S.H.M., et al. 2010. 'Global biodiversity: indicators of recent declines', Science 328: 1164–8.

Dearden, P. 1979. 'Some factors influencing the composition and location of plant communities on a serpentine bedrock in western Newfoundland', *Journal of Biogeography* 6: 93–104.

Environment Canada and US Environmental Protection Agency. 2009. *State of the Great Lakes 2009*. Ottawa and Washington: Governments of Canada and the United States of America.

Gould, S.J. 2002. *I Have Landed: The End of the Beginning in Natural History*. New York: Harmony Books.

Grigorovich, I.A., et al. 2003. 'Ballast-mediated animal introductions in the Laurentian Great Lakes: Retrospective and prospective analyses', *Canadian Journal of Fisheries and Aquatic Sciences* 60, 6: 740–56.

McNeely, J.A. 2001. *The Great Reshuffling: Human Dimensions of Invasive Alien Species*. Gland, Switzerland: World Conservation Union.

Martin, T.G., P. Arcese, and N. Scheerder. 2011 'Browsing down our natural heritage: Deer impacts on vegetation structure and songbird populations across an island archipelago', *Biological Conservation* 144: 459–69.

Noonburg, E.G., B.J. Shuter, and P.A. Abrams. 2003. 'Indirect effects of zebra mussels (*Dreissena polymorpha*) on the planktonic food web', *Canadian Journal of Fisheries and Aquatic Sciences* 60, 11: 1353–68.

Perrings, C. 2002. 'Biological invasions in aquatic systems: The economic problem', *Bulletin of Marine Science* 70, 2: 541–52.

Quammen, D. 1988. *The Flight of the Iguana: A Sidelong View of Science and Nature*. New York: Touchstone Books.

Chapter 4
Ecosystems and Matter Cycling

Learning Objectives

- To understand the nature of matter.
- To be able to describe why human intervention in biogeochemical cycles is a fundamental factor behind many environmental issues.
- To learn the main components and pathways of the phosphorus, nitrogen, sulphur, and carbon cycles.
- To be able to identify the main components of the hydrological cycle and the nature of human intervention in the cycle.
- To understand the causes, effects, and management approaches to eutrophication and acid deposition.

Introduction

The collapse in the Atlantic puffin population, described in Chapter 2, was a result of human interference with energy flow through the ecosystem. There are implications, however, for other aspects of ecosystem functioning. Puffins and most other seabirds play an important role in recycling nutrients, particularly phosphorus, from marine to terrestrial ecosystems. If these systems are disturbed, then the efficiency of the recycling mechanisms can be greatly reduced. Since the phosphorus cycle has very limited recycling capabilities from aquatic to terrestrial systems, the impact of interfering with it in this way could be substantial. This chapter explains how matter, such as phosphorus, cycles in the ecosphere and some of the implications of disturbing these cycles. The most critical environmental challenges facing the Earth, such as global warming, acid deposition, and the spread of dead zones in the ocean, result from cycle disturbance. Consequently, it is critical that you understand the nature of biogeochemical cycles if you are to fully appreciate the nature of these problems and their potential solutions.

This chapter is divided into three sections. First, it describes four biogeochemical cycles. Second, it outlines and explains the hydrological cycle. Finally, it examines the environmental consequences of human actions on these cycles and highlights a few of the important ways you can be a part of the effort to mitigate these changes.

Matter

Everything is either matter or energy. However, in contrast to the supply of energy, which is virtually infinite, the supply of matter on Earth is limited to that which we now have. **Matter**, unlike energy, has mass and takes up space. Matter is what things are made of and is composed of the 92 natural and 17 synthesized chemical elements such as carbon, oxygen, hydrogen, and calcium. *Atoms* are the smallest particles that still exhibit the characteristics of the element. Subatomic particles include *protons*, *neutrons*, and *electrons* that have different electrical charges. At a larger scale, the same kinds of atoms can join together to form molecules. When two different atoms come together, they are known as a **compound**. Water (H_2O), for example, is a compound made up of two hydrogen atoms (H) and one oxygen atom (O). Four major kinds of organic compounds—carbohydrates, fats, proteins, and nucleic acids—make up living organisms.

Matter also exists in three different states (solid, liquid, and gas) and can be transformed from one to another by changes in heat and/or pressure. At the existing temperatures at the surface of the Earth, we have only one representative of the liquid state of matter, water. We can also readily see water in its other two states as ice (solid) or vapour (clouds).

Just as the laws of thermodynamics explain energy flow, the **law of conservation of matter** helps us to understand how matter is transformed. This law tells us that matter can neither be created nor destroyed but merely transformed from one form into another. Thus, matter cannot be consumed so that it no longer exists; it will always exist but in a changed form. When we throw something away, it is still with us, on this planet, as matter somewhere (see International Guest Statement). There *is* no 'away'. All pollution stems from this law. The huge super-stacks on large smelters such as at Inco in Sudbury do not dispose of waste (see Chapter 12); they just disperse those wastes over a much larger area. The matter dispersed is the same and ultimately falls as acid deposition somewhere else. The same is true for all the wastes that we wash down our sinks. They do not disappear but collect in larger water bodies and create pollution problems.

According to the law of matter, emissions from stacks such as these do not simply disappear but end up somewhere else, often with undesirable consequences, such as acid deposition or global warming.

Water is the only substance that occurs in all three phases of matter at the ambient temperatures of the Earth's surface.

Biogeochemical Cycles

For millions of years, matter has been moving among different components of the ecosphere. These cycles are as essential to life as the energy flow described in Chapter 2. About 30 of the naturally occurring elements are a necessary part of living things. These are known as **nutrients** and may be further classified into **macronutrients**, which are needed in relatively large amounts by all organisms, and **micronutrients**, required in lesser amounts by most species (Table 4.1). About 97 per cent of organic mass is composed of six nutrients: carbon, oxygen, hydrogen, nitrogen, phosphorus, and sulphur. These nutrients are cycled continuously among different components of the ecosphere in characteristic paths known as **biogeochemical cycles.**

Perspectives on the Environment

We know from studies of chemistry that our bodies are reorganized star-dust, recycled again and again, so that, truly, our bones are of corals made.

—Rowe (1993)

International Guest Statement
Action-Oriented Research on Community Recycling in São Paulo, Brazil
Jutta Gutberlet

With more than half of the world's population already living in urban spaces, increased generation of solid waste is a serious global concern. Appropriate solid waste management is an effective measure for building more sustainable communities and for increasing urban resiliency. Improper forms of disposing solid waste cause harm to the environment and to the climate. Waste can also be a resource. By not recovering the resources embedded in waste the prevailing unsustainable exploitation of natural resources may continue. My research concerns are closely related to building more sustainable communities. I focus particularly on solid waste recovery and recycling as livelihood options to achieving this goal. Since 2005 the Participatory Sustainable Waste Management (PSWM) project—hosted by the Community-Based Research Laboratory at the University of Victoria—has helped recycling co-operatives in the metropolitan region of São Paulo, Brazil, to become more resilient, through empowering their participants, increasing their income, supporting inclusive public policies, and building environmental awareness.

Participatory sustainable waste management means 'solid waste recovery, reuse and recycling practices with organized and empowered recycling co-ops supported with public policies, embedded in solidarity economy and targeting social equity and environmental sustainability' (Gutberlet, 2009: 171). This approach facilitates the cyclical use of resources and spares virgin materials from being extracted. This form of resource recovery happens worldwide, particularly in the global South, where it generates work and employment among the most vulnerable populations. Improving the working conditions of the recyclers and expanding the resource recovery activity ultimately translate into stronger local economies and overall reduced social vulnerability, thereby leading to more sustainable communities.

In Latin America, Asia, and Africa, in particular, an extensive informal sector is involved in collecting and separating recyclable materials from the waste stream, providing an insight into their resourcefulness (Moreno-Sánchez and Maldonado, 2006). In Brazil approximately 800,000 to 1 million people engage in this activity, yet most of them remain extremely poor and marginalized.

The research that my graduate students and I undertake has revealed the diverse ways in which these informal collectors are redefining waste as a resource (Gutberlet, 2008; Gutberlet and Baeder, 2008; Tremblay et al., 2010). This research points to positive change through community-led recycling as a poverty reduction strategy that also improves environmental health.

We have recently investigated the potential for integrating organic waste management with urban agriculture, thereby further reducing the volume of waste going to landfills and returning valuable nutrients to depleted soils (Yates and Gutberlet, 2011). The majority of the organized recyclers in Brazil, and particularly their leaders, are women. The activity provides them with opportunities for income generation and allows for their capacity-building and collective engagement. Co-operative recycling thus makes an important contribution to the human development of those who are impoverished and socially excluded from society. The researchers that are part of the PSWM project engage in community outreach activities, help organize workshops or seminars, produce video documentaries, and conduct participatory and action-oriented research interventions in Brazil and in Canada. These activities have helped to increase the awareness among governments and communities about waste co-management issues and the need to decrease waste generation. Overall, the PSWM project illustrates the value of at-source separation in reducing waste. Lessons have been learned about how to inform policy-makers on sustainable and socially responsible waste management, lessons that need to be disseminated to further influence legislation and policy. The use of video in our research has played an important role to empower the participants and to inform the public and the local politicians about the work of the recyclers and their livelihood concerns (Tremblay et al., 2010).

A major threat to community recycling is the 'waste to energy' scheme, where solid waste is incinerated to extract energy (Gutberlet, 2010). Not only does this form of waste management cause environmental hazards, but it also dismisses the fact that resource recovery and recycling are more

Recyclers from Pacto Ambiental association in Diadema, Brazil.

socially and environmentally friendly, generate employment, and contribute to resource conservation. My current research aims to help build the arguments for more sustainable resources use and recovery. The findings will contribute to the design of solid waste policies that propel community building with zero waste, inclusive resource recovery, and more sustainable lifestyles.

Jutta Gutberlet is an Associate Professor of Geography at the University of Victoria. She directs the Community-Based Research Laboratory and undertakes research primarily in South America on community-based waste treatment.

Figure 4.1 shows a generalized model of such a cycle. Like all the subsequent diagrams of cycles in this chapter, it exemplifies the types of simplifying models that scientists construct to try to represent the vast complexity of Earth processes, as described in Chapter 1 and illustrated in Fig 1.19. Nutrients can be stored in the different compartments shown in Figure 4.1 for varying amounts of time. In general, there is a large, relatively slow-moving abiotic pool that may be in the atmosphere or lithosphere and is chemically unusable by the biotic part of the ecosystem or is physically remote. There is a more rapidly interacting exchange pool between the biotic and abiotic components. Nutrients move at various speeds from the biotic to the abiotic pools. For example, very rapid exchange takes place through respiration as carbon and oxygen move rapidly between the biotic and atmospheric components. The elements that now make up your body have undergone millions of years of recycling through these various compartments. You are a product of recycling!

Ecosystems also vary substantially in terms of the speed of cycling and the relative proportion of nutrients in each compartment. Some systems have nutrient-poor soils, for example, and have developed different mechanisms to store nutrients in other compartments. Tropical forest ecosystems are classic examples. Most of the nutrients are stored in the biomass as opposed to the soil system (Table 4.2). When leaves fall to the ground, they are rapidly mined for nutrients by plant roots before those nutrients have a chance to be leached out of the system. In contrast, many temperate

Table 4.1 | Relative Amounts of Chemical Elements That Make Up Living Things

Major Macronutrients (>1% dry organic weight)		Relatively Minor Macronutrients (0.2–1% dry organic weight)		Micronutrients (<0.2% dry organic weight)	
Name of Element	Symbol	Name of Element	Symbol	Name of Element	Symbol
Carbon	C	Calcium	Ca	Aluminum	Al
Hydrogen	H	Chlorine	Cl	Boron	B
Nitrogen	N	Copper	Cu	Bromine	Br
Oxygen	O	Iron	Fe	Chromium	Cr
Phosphorus	P	Magnesium	Mg	Cobalt	Co
		Potassium	K	Fluorine	F
		Sodium	Na	Gallium	Ga
		Sulphur	S	Iodine	I
				Manganese	Mn
				Molybdenum	Mo
				Selenium	Se
				Silicon	Si
				Strontium	Sr
				Tin	Sn
				Titanium	Ti
				Vanadium	V
				Zinc	Zn

Source: Kupchella and Hyland (1989).

Slash-and-burn agriculture is a common way to transfer nutrients from the biomass to the soil to increase agricultural productivity. It is a common agricultural practice in the tropics, where most of the nutrients are in the biomass and not in the soil. This photograph shows such fields cut in the forest in the Cardamom Mountains of Cambodia. The soils rapidly lose fertility owing to the burning of the biomass, and they are then abandoned for secondary succession to occur.

forests have soils of high fertility. Removal of the nutrients in the biomass, through logging for example, does not remove as high a proportion of the site nutrient capital as removal in tropical ecosystems. This is discussed in more detail in Chapter 9.

Speed of cycling may also change within a cycle, depending on the nutrient of concern and the time of year. For the carbon cycle, for example, there is greater uptake of CO_2 in spring and summer as deciduous trees grow leaves. In fall, there is a correspondingly greater release as the leaves fall off and decompose (see Box 4.1). On average, a carbon dioxide molecule stays in the atmospheric component of the cycle from five to seven years. This is known as the *residence time*. It takes, on average, 300 years for a carbon molecule to pass through the lithosphere, cryosphere, atmosphere, hydrosphere, and biotic components of the carbon cycle. By way of contrast, it may take a water molecule two million years to make a complete cycle. The speed of cycling is influenced by such factors as the chemical reactivity of the substance. Carbon, for example, participates in many chemical reactions. It also occurs as a gas. In general, a gaseous phase allows for a speeding up of a cycle, because gas molecules move more quickly than molecules in the other states of matter.

Cycles can be classified according to the main source of their matter. **Gaseous cycles**, as the name would suggest, have most of their matter in the atmosphere. The nitrogen cycle is a good example. **Sedimentary cycles**, such as the phosphorus and sulphur cycles, hold most of their matter in the lithosphere. In general, elements in sedimentary cycles tend to cycle more slowly than those in gaseous cycles, and the elements may be locked into geological formations for millions of years.

Under natural conditions, recycling rates between components achieve a balance over time in which inputs and outputs are equal. Human activity serves to change the speed of transference between the different components of the cycles. Many of our pollution problems result from a human-induced buildup in one or more components of a cycle that cannot be effectively dissipated by natural processes. It's similar to when you consume more alcohol than your body can effectively process and you wake up with a headache.

In addition to the biogeochemical cycles, some attention will be given in this chapter to the hydrological cycle. This cycle is critical to all other cycles, since water plays a major role in the mobilization and transportation of materials. The energy for this, as with all other aspects of the cycles, ultimately comes from the sun. Photosynthesis powers the biotic aspects of the cycles, and atmospheric circulation, fuelled by the sun's energy, controls the water power that is so important for weathering and erosion processes.

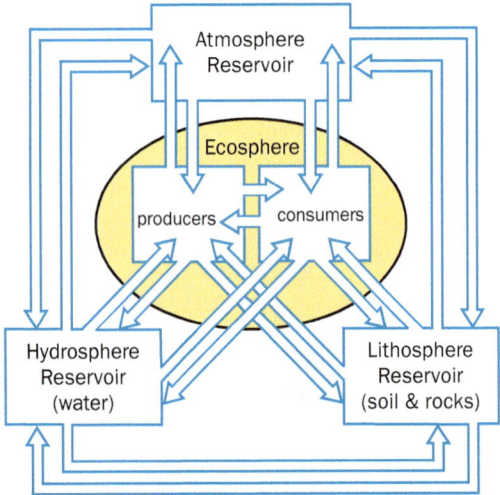

Figure 4.1 | Each nutrient is stored and released by components of the Earth's systems. Different nutrients follow slightly different paths through the systems and are stored and released at different rates.

Table 4.2 | Approximate Distributions of Carbon and Nitrogen in Temperate and Tropical Rain Forests

	Tropical Rain Forest	Temperate Rain Forest
Carbon in vegetation	75%	50%
Carbon in litter and soil	25%	50%
Nitrogen in biomass	50%	6%
Nitrogen in biomass above ground	44%	3%

Sedimentary Cycles

Sedimentary cycles mobilize materials from the lithosphere to the hydrosphere and back to the lithosphere. Some, such as sulphur, may involve a gaseous phase, while others, such as phosphorus, do not. These cycles rely essentially on geological uplift over long periods to complete the cycle. Human actions interfere with the speed at which many of these cycles occur and can result in serious environmental problems. The lack of a gaseous phase in the phosphorus cycle means that the cycle is missing one potential route for more rapid recycling, which can lead to problems when mobilization rates are increased through human activity. Phosphorus and sulphur will be discussed here, but other elements, such as calcium, magnesium, and potassium, follow similar pathways.

Phosphorus (P)

Phosphorus, a macronutrient incorporated into many organic molecules, is essential for metabolic energy use. It is relatively rare on the Earth's surface in relation to biological demand, so it is essential that phosphorus cycles efficiently between components. Many organisms have devised means of storing this element preferentially in their tissues, and phosphorus moves very readily within plants from older tissues to more active growth sites. Deciduous trees may recirculate up to 30 per cent of their phosphorus back to their more permanent components before the leaves fall in an effort to preserve the nutrient.

Under natural circumstances, phosphorus is a prime example of a nutrient held in a tight circulation pattern between the biotic and abiotic components. Replenishment rates through weathering and soil availability are limited;

Environment in Focus

Box 4.1 The Decomposers

In Chapter 2, attention was drawn to the importance of decomposer organisms and detritus food chains. These are the main means by which nutrients in the biotic component of the ecosphere are returned to the abiotic so that plants can once again use them. Photosynthesis has been described as the process of making a complicated product out of simple components; decomposition is the reverse process of making simple components out of that complicated product.

Decomposer organisms such as fungi may attack leaves while still on the plant; the fungi release products such as sugars, which are then washed to the ground by rainfall. Once leaves fall to the ground, they are broken down progressively by various groups of organisms. Larger organisms such as earthworms, slugs, snails, beetles, ants, and termites help to break up the leaf material initially. Many gardeners are fully aware of the ability of slugs, for example, to devour green leaves in great quantities.

Fungi and heterotrophic bacteria further break down the organic matter, releasing more resistant carbohydrates, followed by cellulose and lignin. The humus—the organic layer in the soil—is composed mainly of products that can resist rapid breakdown. A chemical process, oxidation, is mainly responsible for the decay of this material.

As everyone who has witnessed leaf decay in autumn knows, the process can occur quite rapidly. The speed varies depending on the environment. Warm environments tend to promote more rapid microbial activity. Leaf decay in the tropics takes place in a matter of weeks. In the boreal forest, however, where conditions are cold and the leaves, such as spruce and pine needles, quite resistant, **recycling** of the nutrients held in the leaves may take decades. Overall, the average recycling time for organic material in the wet tropics is five months; in the boreal forest, it is 350 years. The high amounts of lignin found in leafs of needle-leafed trees help to protect the trees against freezing conditions but offer little food value to decomposers, so decay is slow. In comparison, deciduous trees, such as maple, have a high reward for decomposers: high nitrogen levels and little protective lignin, so they decay very quickly. However, researchers are now discovering that the slow breakdown of organic matter in the boreal forest also plays a major role in global carbon storage. As the forests are removed through logging and other activities, not only does the supply of leaves disappear, but the built-up carbon is released either through burning associated with these activities or the higher ground temperatures resulting from the reduced amount of shade.

Slugs play an important role in breaking down vegetable matter. In the wet west coast forests the biomass of slugs is greater than that of any other animal in the ecosystem.

thus, the amount retained by the biomass is quite critical. The residence time of phosphorus in terrestrial systems can be up to 100 years before it is leached into the hydrosphere. Phosphorus is often the dominant limiting factor (Chapter 2) in freshwater aquatic systems and for plant growth in terrestrial soils. Agricultural productivity relies heavily on augmenting this supply (auxiliary energy flow) through fertilizer application. Gruber and Galloway (2008) suggest that both the nitrogen and carbon cycles in the ocean are ultimately controlled by phosphorus and, since these cycles are key to global warming response, the phosphorus cycle will be a main determinant of global futures.

Box 4.2 outlines the impact humans have on the phosphorus cycle.

The availability of phosphorus in the soil is influenced by soil acidity. Acidity is measured on the ph scale, which is discussed in more detail later in this chapter. Below pH 5.5, for example, phosphorus reacts with aluminum and iron to form insoluble compounds. Above pH 7, the same thing happens in combination with calcium. Obviously, things that change soil pH, such as acid precipitation, can have a critical effect on phosphorus availability. This is an example of the kind of synergistic reaction discussed in Chapter 3 in which the combination of either high or low pH values with low phosphorus availability, as a result of chemical reactions, can have a stronger effect than the sum of the two individually.

Rocks in the Earth's crust are the main reservoir of phosphorus (Figure 4.2). Geological uplift and subsequent weathering (Box 4.3) make phosphorus available in the soil, where it is taken up by plant roots. Phosphate ions are the main source of phosphorus for plants and are released from slowly dissolving minerals such as iron, calcium, and magnesium phosphates. Many higher plants have a mutualistic relationship with soil fungi, or mycorrhizae, which helps them gain improved access to phosphorus in the soil. Once incorporated

The Head-Smashed-In UNESCO World Heritage Site in southern Alberta is rich in phosphorous as Native peoples used the 11-metre-high cliff to kill stampeding bison by driving them off the edge. The decomposed bones left large deposits of this macronutrient behind.

into plant material, the phosphorus may be passed on to organisms at higher trophic levels.

Animal wastes are a significant source of phosphorus and return to the soil. All organisms eventually die, and the organic material is broken down by the decomposer food chains. This may take some time, since a considerable amount of the phosphorus is within animal bones. In the past, farmers have used concentrated sources of animal bones, such as the bison jumps used by indigenous peoples on the prairies, as a source of phosphate fertilizer.

Following breakdown in the soil, the phosphorus is then either taken up again by plants or removed by water transport. Bacteria mineralize the returned organic phosphorus into inorganic forms so that plants can take it up once more. Most of the water transport occurs in particulate form by

Environment in Focus

Box 4.2 Human Impacts on the Phosphorus Cycle

Humans intervene in the phosphorus cycle in several ways that serve to accelerate the mobilization rate:

- mining of phosphate-rich rocks for fertilizer and detergent production, creating excessive runoff into aquatic environments;
- biomass removal, leading to accelerated erosion of sediment and solutes into streams;
- concentration of large numbers of organisms such as humans, cattle, and pigs, creating heavy burdens of phosphate-rich waste materials;
- removal of phosphorus from oceanic ecosystems through fishing, with the phosphorus returned to fresh water again and ultimately the marine system through the dissolution of wastes.

The major implication of all these interventions is for excessive phosphorus accumulation in freshwater systems, resulting in eutrophication. Human activity is now estimated to account for about two-thirds of the phosphorus reaching the oceans. The environmental impact of this nutrient enrichment will be discussed in more detail later.

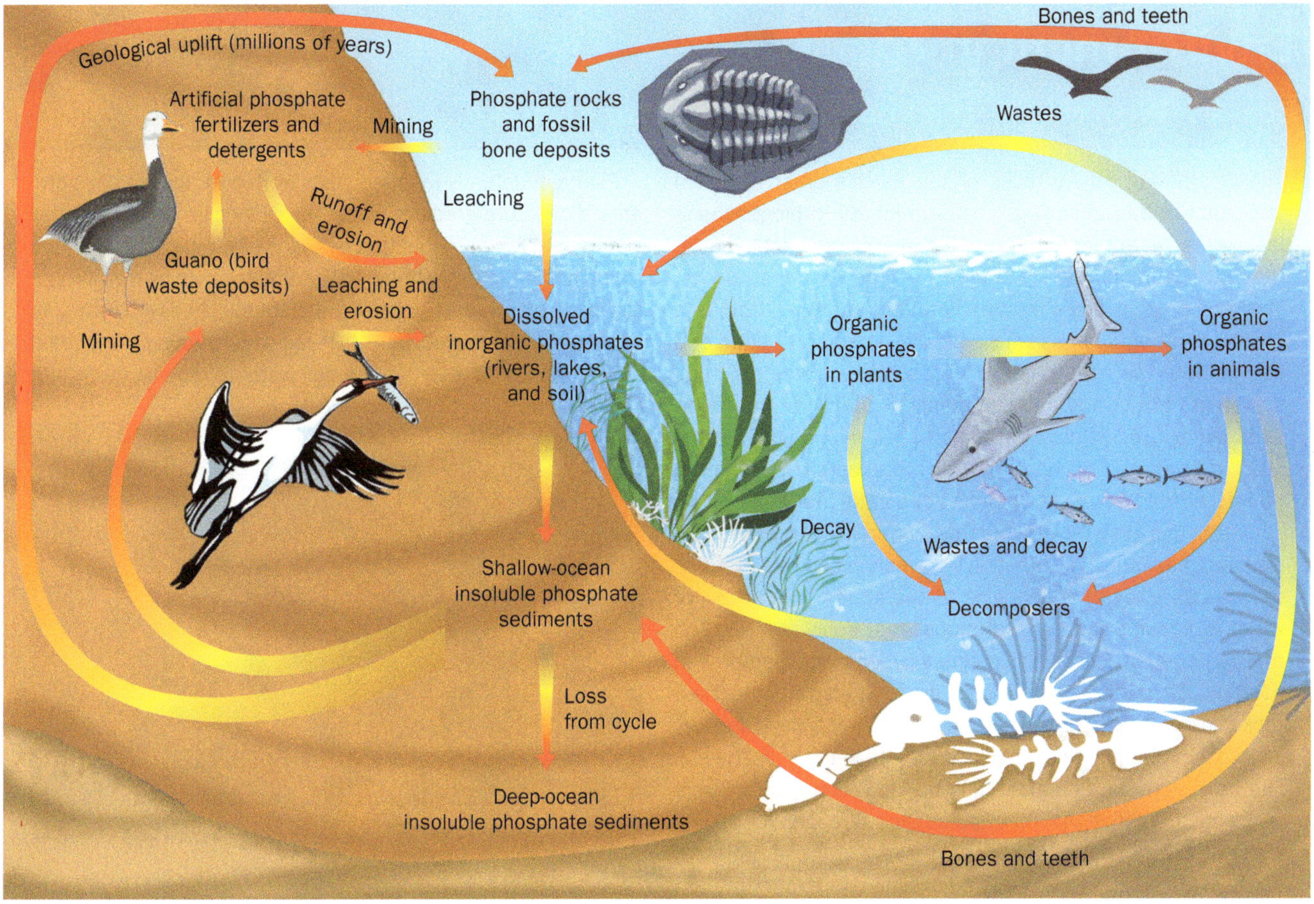

Figure 4.2 | The phosphorus cycle.

streams, which is one reason to be concerned about excessive sedimentation occurring through land-use activities such as logging and agriculture, as described in Chapters 9 and 10.

Stream transport ultimately ends up in the ocean. Estuaries have such high productivities, as discussed in Chapter 2, in part because of this nutrient input from upstream. The circulation patterns within estuaries tend to trap nutrients, but some phosphorus finds its way into the shallow ocean areas of the coastal zone. It may be fixed in biomass by phytoplankton or other aquatic plants in the euphotic (eu = well, photos = light) zone and once again incorporated into the food chain. The coastal zones, with this plentiful supply of nutrients and photosynthetic energy from the sun, cover less than 10 per cent of the ocean's surface but account for more than 90 per cent of all ocean species.

Beyond the coastal zone and the continental shelves, water depth increases into the open ocean. Phosphorus and other nutrients that have not been incorporated into food chains, plus elements from the death of oceanic organisms, filter through to the bathyal and ultimately the abyssal zones (see Figure 2.5). Here, uptake by organisms is extremely limited, and the nutrients must either be moved back to the euphotic zone by upwelling currents or wait to be geologically uplifted over millions of years to move into another component of the cycle. Where such upwellings occur, such as off the west coasts of Africa and South America, plentiful fisheries are found because of the combination of high nutrient and energy levels. Some fish species, such as salmon, are **anadromous**, spending part of their lives in salt water and part in fresh water, where they die after spawning. When they die, the nutrients they have collected during their ocean phase are returned to the freshwater system, resulting in a significant input of nutrients, including phosphates (Chapter 8).

The nutrients that have sustained this salmon are now being recycled.

Environment in Focus

Box 4.3 Weathering, the Rock Cycle, and Plant Uptake

The weathering of the rocks of the Earth's crust plays an important role in supplying long-term inputs to biogeochemical cycles. Weathering is part of the **rock cycle** whereby rocks that have been uplifted are eroded into different constituents. The rock cycle involves the transformation of rocks from one type to another, such as when volcanic rocks are eroded and washed into the ocean. Over millions of years, the resulting sediments are turned into sedimentary rocks. In turn, these sedimentary rocks may be compressed within the Earth's crust and altered by heat and pressure before once more being uplifted through the process of continental drift.

Weathering involves numerous different processes. In Canada, mechanical weathering involves the physical breakup of rocks as a result of changing temperatures. The action of water is important. Chemical processes, such as hydration and carbonation, further the process by removing elements in solution. Secondary clay minerals are produced from primary rock minerals by hydrolysis and oxidation. These clays are very important in terms of holding the nutrients in the soil. The soil can be thought of as a giant filter bed in which each particle is chemically active. As water percolates through, containing many different nutrients in solution, some of these nutrients are held by the clays and become available for plant uptake.

Plants constantly lose moisture from their leaves. This creates a moisture gradient within the plant that serves to draw water up to replace what has been lost. Water moves from the roots, and more nutrient-laden water is taken in by the roots. It is the job of the roots to keep the plant supplied with water. As nutrients are removed from the soil water around the plants, new nutrients move within the soil water to replace them.

These sedimentary rocks have been compressed and folded as part of the rock cycle.

Two other recycling mechanisms also occur. One is the biotic one described earlier as marine birds, such as puffins, cormorants, and other fish-eating birds, return phosphorus to land in the form of their droppings, representing the phosphorus that has concentrated through the marine food chain. This phosphorus, known as **guano**, constitutes the largest source of phosphorus for human use and is heavily mined for fertilizer production. A small amount of phosphorus is also returned to land through the atmosphere as sea spray.

Sulphur (S)

Like phosphorus, sulphur is a sedimentary cycle, but it differs from phosphorus in two important ways. First, it has an atmospheric component and therefore better recycling potential. Sulphur is not often a limiting factor for growth in aquatic or terrestrial ecosystems. Second, like most of the other cycles but unlike phosphorus, it has strong dependencies on microbial activity. Sulphur is a necessary component for all life and a building component of proteins.

Sulphur is not available in the lithosphere and must be transformed into sulphates to be absorbed by plants. Bacteria are critical here, changing sulphur into various forms in the soil (Figure 4.3). The exact form depends on factors such as the presence (**aerobic**) or absence (**anaerobic**) of oxygen, which is usually a reflection of the relationship of the particular site of transformation to the water table and the presence of other elements such as iron. From these microbial transformations by chemoautotrophs (discussed in Chapter 2), gases such as hydrogen sulphide (H_2S) may be released directly into the atmosphere (giving the familiar 'rotten egg' smell we associate with marshlands), or sulphate salts (SO_4^{2-}) are produced. Through their roots, plants can then absorb the sulphates, sulphur enters the food chain, and the same processes occur as in the biotic components of the other cycles.

The complexity of these cycles is illustrated further by some of the interactions that occur between cycles. For example, the phosphorus cycle benefits when iron sulphides are formed in sediments and phosphorus is converted from insoluble to soluble forms, where it becomes available for uptake. There are also important interactions with global climate change. For example, rising temperatures and reduced rainfall in many areas of the Canadian North are causing more frequent drying out of extensive peat beds. When these peat beds are re-wet they emit 3-4 times as much SO_2 as do continually wet peat beds and hence add to the acid deposition described later in this chapter.

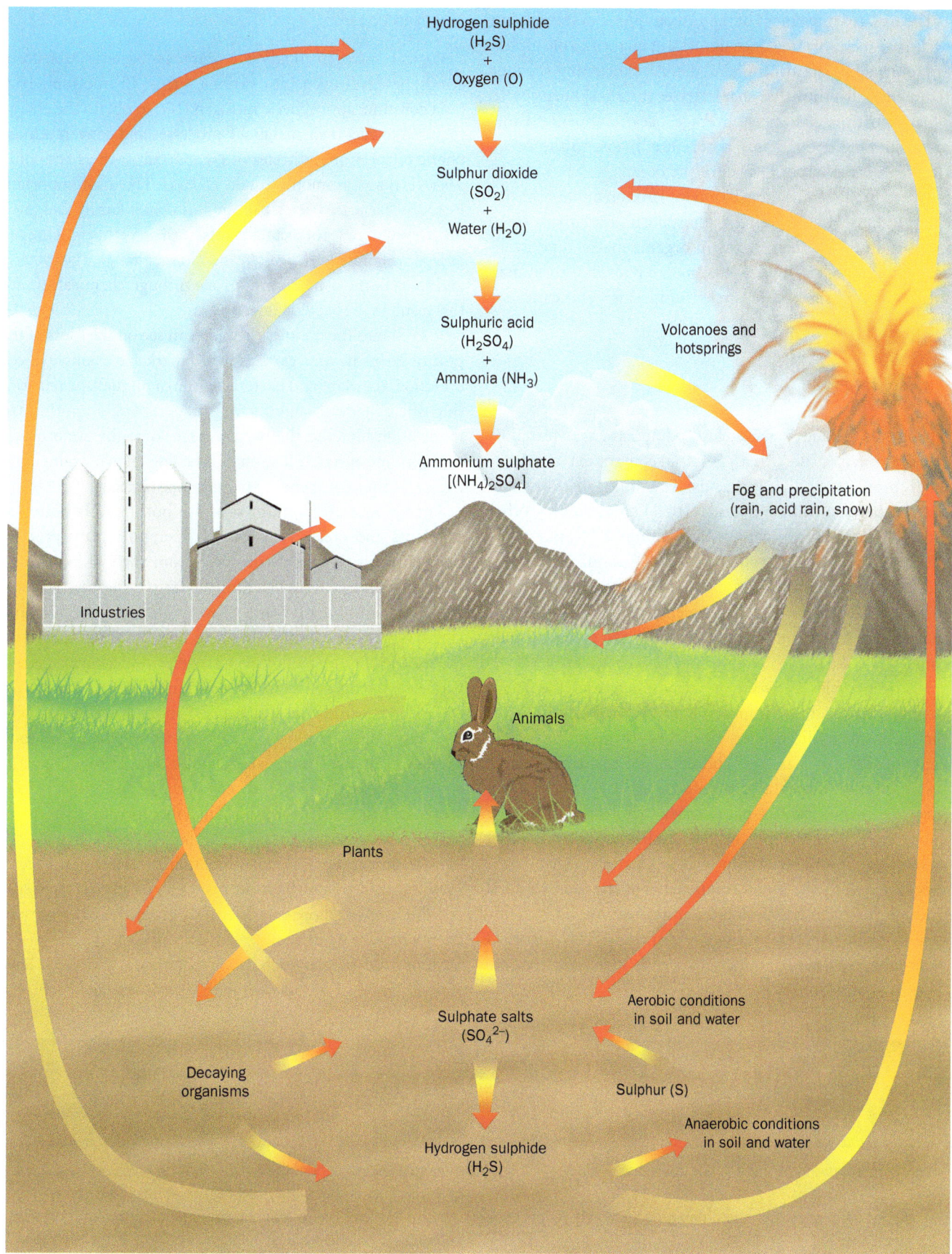

Figure 4.3 | The sulphur cycle.

The upward movement of the gaseous phase of the sulphur cycle is also important, since significant quantities of sulphur are returned to the atmosphere, thereby shortening the long sediment uplift time that characterizes the phosphorus cycle. This is fortunate, because average ocean residence times are quite long and sulphur is continually lost to the ocean floor. From the upper reaches of the oceans, sulphur can be returned to the atmosphere by phytoplankton or photochemical reactions. However, unlike phosphorus, a relatively small proportion of sulphur is fixed in organic matter, and availability is not usually a problem. As with phosphorus, human intervention in the sulphur cycle (Box 4.4) is significant.

Gaseous Cycles

Nitrogen (N)

Nitrogen is a colourless, tasteless, odourless gas required by all organisms for life. It is an essential component of chlorophyll, proteins, and amino acids. The atmosphere is more than 78 per cent nitrogen gas (N_2) and also contains other forms of gaseous nitrogen such as ammonia (NH_3), nitrogen dioxide (NO_2), nitrous oxide (N_2O), and nitric oxide (NO). Excess quantities of these other forms are involved in many of our most challenging environmental problems, such as acid deposition, ozone depletion, and global climate change (Box 4.5).

Nitrogen cycles between the atmosphere and the lithosphere with the most important interactions occurring at the atmosphere–lithosphere interface through biological activity. Nitrogen can also collect in the hydrosphere and result in environmental problems such as eutrophication. Most organisms cannot gain access to nitrogen from the atmosphere. The nitrogen is instead obtained from the soil as nitrates. The main way in which the atmospheric reservoir is linked to the biotic components of the food chain is through **nitrogen fixation** and **denitrification**, both mediated through microbial activity (Figure 4.4). The historical record shows a close coupling between the speed of these processes and atmospheric CO_2 levels.

Nitrogen Fixation

Biological nitrogen fixation occurs as bacteria transform atmospheric nitrogen into various forms. Chemotropic bacteria consume atmospheric nitrogen (N_2) to obtain the energy required to fuel their metabolic processes and convert atmospheric nitrogen into nitrates or compounds such as ammonia gas (NH_3) and ammonium salts (NH_{4+}). The most important nitrogen fixers are bacteria of the *Rhizobium* family that grow on the root nodules of certain plants, such as members of the pea or legume family (e.g., peas, beans, clover, alfalfa). The bacteria and roots of the plant communicate through chemical stimuli that result in the bacteria infecting root cells. Once infected, the cells swell into the nodules that you can see on the roots of the peas or beans in your garden. In a remarkable example of co-evolution, the plant and bacteria exist in a mutualistic relationship where the plant supplies the products of photosynthesis to the relationship, and the bacteria transform the atmospheric nitrogen into nitrates. It is one of the few known examples of two organisms co-operating to make one molecule.

Nitrates and ammonium salts are both readily absorbed by plants and create rich soils that support plant production. Nitrogen is quickly depleted from the soil and, along with phosphorus, is often a limiting factor in terrestrial soils, which explains why farmers grow crops such as alfalfa and clover as part of a crop rotation to help build up nitrates in the soil. About one-half of the nitrogen circulating in agricultural ecosystems comes from this source. The increasing cost of fertilizer worldwide has focused more attention on biological nitrogen fixation as a part of meeting the global food challenges of the future (see Chapter 10). Through genetic engineering, for example, it may be possible to inject other crops, such as cereals, with similar symbiotic unions between plants and nitrogen-fixing bacteria. However, it may not be that simple. For example, research indicates that species involved in nitrogen fixation may also be particularly susceptible to phosphorus deficiencies, given their high P and energy requirements.

Some wild species such as alder, lupines, and vetch have similar bacteria associated with them and hence play a

---

Environment in Focus
Box 4.4 Human Impacts on the Sulphur Cycle

Humans intervene in the sulphur cycle mainly through:

- the burning of sulphur-containing coal, largely to produce electricity;
- the smelting of metal ores that contain sulphates.

Almost 99 per cent of the sulphur dioxide and about one-third of the sulphur compounds reaching the atmosphere come from these activities. These sulphur compounds react with oxygen and water vapour to produce sulphuric acid (H_2SO_4), a main component of acid deposition, as discussed later in this chapter.

valuable role when they act as primary colonizers in the successional process or when they help to recolonize sites that have been logged (Chapter 9) or otherwise disturbed. These relationships are mutualistic in that both organisms gain. The plant receives enhanced nutrient supply, and the organisms find a home in which the plant supplies them with various sugars.

Other bacteria and algae that fix nitrogen are not attached to specific plants. These free-living nitrogen fixing microorganisms are particularly important in the Arctic and within the ocean. These free-floating relationships are not as efficient at fixing nitrogen from the atmosphere as vegetative relationships. Estimates suggest that in terrestrial ecosystems, about twice the amount of nitrogen is fixed by mutualistic relationships as by these free-floating relationships.

Nitrogen is also fixed through atmospheric fixation that occurs largely during thunderstorms. Lightning causes extremely high temperatures that unite oxygen and nitrogen to form nitric acid (HNO_3), which is subsequently carried to earth as precipitation and converted into nitrates (NO_3^-). These nitrates can then be taken up by plant roots. Estimates on the importance of atmospheric fixation vary, but 10 per cent of total fixation would be a maximum figure, and most estimates place it at about only 5 per cent.

Mineralization, or Why Compost Matters

Although nitrogen-fixing bacteria are an important source of nitrates within soil, most physical nitrogen (e.g., nitrates and ammonium salts) comes from the breakdown of existing biomass by decomposer food chains. In fact, nitrogen is tightly circulated in most ecosystems between the dead and living biomass.

Once fixed in the soil, nitrogen is incorporated into plant matter and then moved through the food chain. **Mineralization** is the process by which decomposing biomass (i.e., dead plants) is converted back to ammonia (NH_3) and ammonium salts (NH_{4+}) by bacterial action and returned to the soil. This process highlights the importance of compost in agricultural production and is explored more fully in Chapter 10. Primarily, mineralization does not produce nitrates directly, but rather indirectly through another process known as nitrification.

Nitrification and Denitrification

Chemotrophic bacteria, such as *Nitrosomonas* and *Nitrobacter*, convert ammonia and ammonium salts into nitrites and then into nitrates. Other bacteria—anaerobic bacteria—reverse the nitrogren-fixing process and convert nitrates into nitrogen gas, returning it to the atmosphere (Figure 4.4). Denitrification occurs in anaerobic conditions, especially where there are large amounts of nitrates available, such as on flooded agricultural fields.

Nitrates are highly soluble in water, and if not held tightly they may be lost to the ecosystem by surface runoff and become a major contributor to the problem of eutrophication, as discussed in more detail later in the chapter. Ammonia is also susceptible to loss by soil erosion, since it tends to adhere to soil particles. Like phosphorus, nitrogen is often a limiting factor for growth. When excessive concentrations occur in water, it is a major contributor to the process of eutrophication. Unlike phosphorus, however, nitrogen is not immobilized in deep-ocean sediments but has an effective feedback mechanism to the atmosphere from the ocean through microbial denitrification.

Environment in Focus

Box 4.5 Human Impacts on the Nitrogen Cycle

Humans disrupt the nitrogen cycle in many ways:

- Chemical fixation to supply nitrates and ammonia as fertilizer. The amount fixed is currently estimated to be about equal to that produced by natural processes. However, the Millennium Ecosystem Assessment (2003) predicts large increases in the future (Figure 4.5), and Gruber and Galloway (2008) predict that humans will double the turnover rates of the nitrogen cycle of the entire Earth. The main impacts are through runoff of excess fertilizer (contributing to eutrophication) and denitrification (contributing to climatic change). Agricultural fertilizers are implicated in both. Eutrophication will be discussed in more detail in the next section. Denitrification transforms nitrogen fertilizers into nitrous oxide, a greenhouse gas (Chapter 7), and is also involved in the catalytic destruction of the ozone layer. Health concerns also are related to excessive nitrate levels from fertilizers running into water supplies (Box 4.8).
- Removal of nitrate and ammonium ions from agricultural soils through the harvesting of nitrogen-rich crops.
- High-temperature combustion, which produces nitric oxides (NO) that combine with oxygen to produce nitrogen dioxide (NO_2), which reacts with water vapour to form nitric acid (HNO_3), a main component of acid deposition.

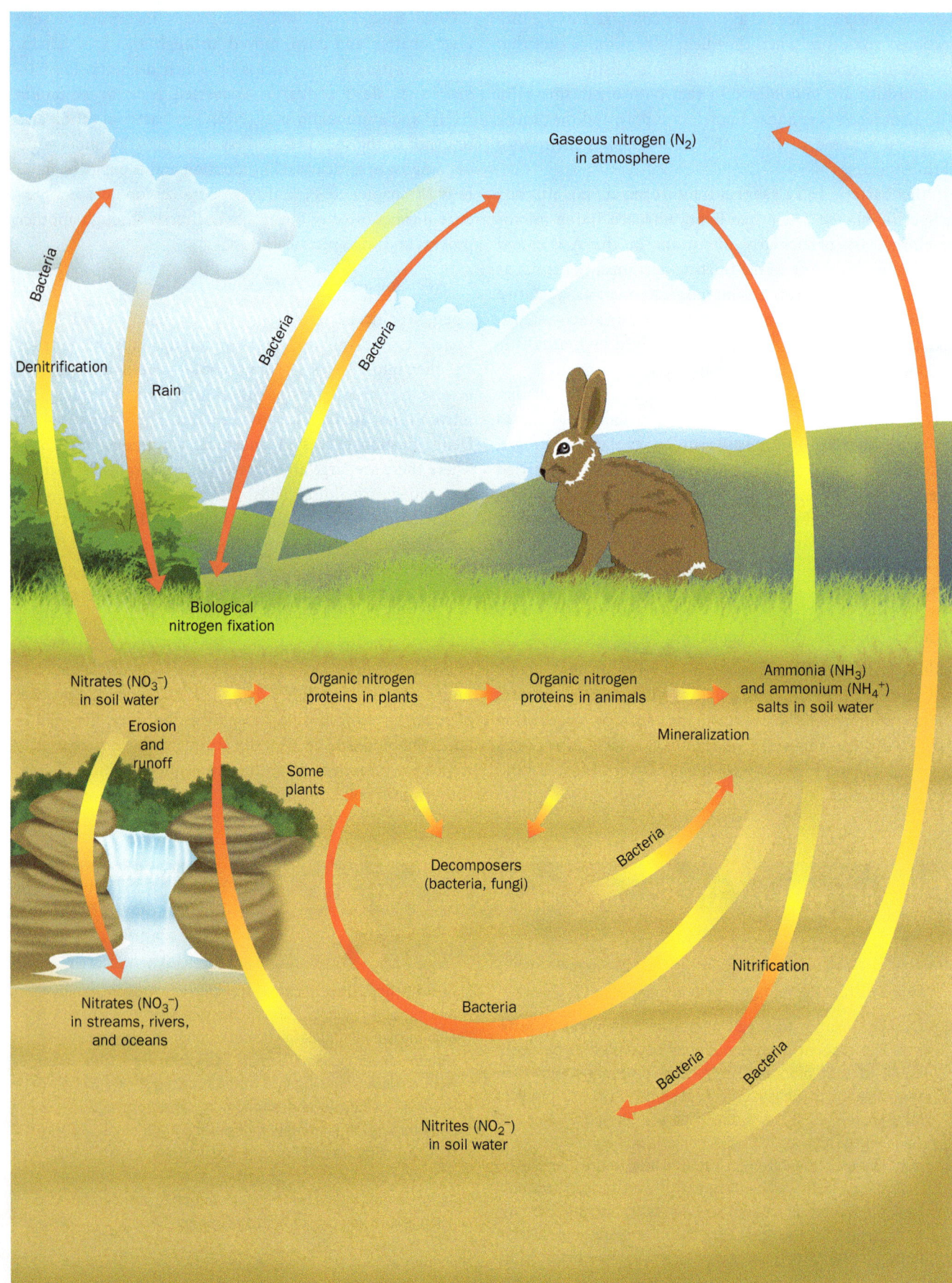

Figure 4.4 | The nitrogen cycle.

Chapter 4 | Ecosystems and Matter Cycling

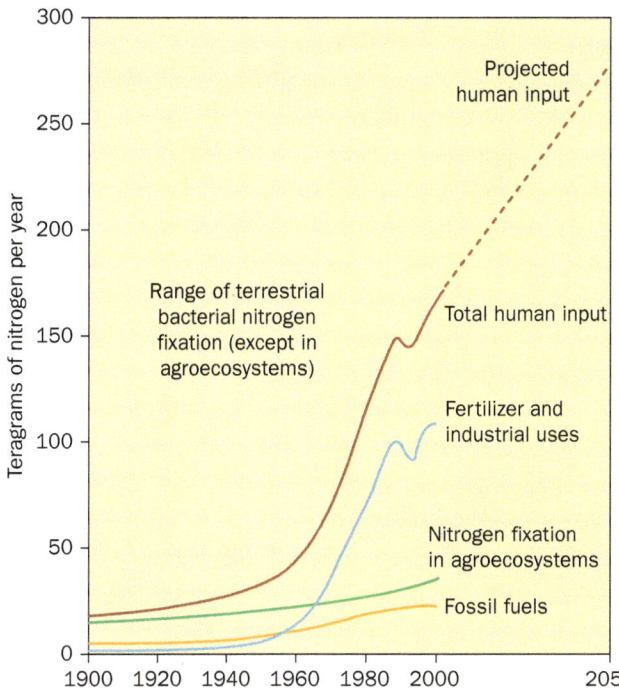

Figure 4.5 | Global trends in the creation of reactive nitrogen on Earth by human activities, with projections to 2050. Source: Millennium Ecosystem Assessment (2005).

Scientists are trying to understand the relationship between the nitrogen cycle and the major elements of global climate change, such as the carbon cycle. They know for sure, based on historical records, that the cycles are closely linked but are still unsure as to the direction and magnitude of changes that might be expected in the future and whether the nitrogen cycle will form a positive or negative feedback loop with rising atmospheric carbon levels (Figure 4.6). On the one hand, ocean acidification resulting from the ocean's taking up anthropogenic CO_2 might lead to an increase in the C/N uptake ratio of marine phytoplankton and enhanced nitrogen fixation. If this happens, the marine biosphere would act as a negative feedback for climate change, since the resulting enhanced fixation of carbon would draw additional carbon from the atmosphere, thus reducing the accumulation of anthropogenic CO_2 in the atmosphere.

On the other hand, current climate–carbon cycle models used for making projections of Earth's climate do not consider nitrogen limitation of the terrestrial biosphere but generally assume a strong CO_2 fertilization effect. In other words, the additional CO_2 available will act to stimulate further organic growth. However, the latter would require the availability of large quantities of nitrates, since nitrates are a necessary ingredient for all life, as discussed earlier. If they are not available, then nitrogen will become a major limiting factor (Chapter 3). Thus, nitrogen limitation will significantly affect the ability of the terrestrial biosphere to act as a CO_2 sink in the future.

Carbon (C)

Although carbon dioxide gas (CO_2) constitutes only 0.03 per cent of the atmosphere, it is the main reservoir for the carbon that is the building block for all necessary fats, proteins, and carbohydrates that constitute life. Plants take up carbon dioxide directly from the atmosphere through the process of photosynthesis and at the same time emit oxygen. The carbon becomes incorporated into the biomass and is passed along the food chain. Residence times can vary greatly, but older forests constitute a significant repository for carbon for centuries. Respiration by organisms transforms some of this carbon back into carbon dioxide (Figure 4.7), and the cellular respiration of decomposers helps to return the carbon from dead organisms into the atmosphere. Most of this is in the form of CO_2 but also methane (CH_4) in anaerobic conditions. Thus, the cycling of carbon and the flow of energy through food chains are intimately related.

Besides this relatively rapid exchange, some carbon can also be stored in the lithosphere for extended periods of time as organisms become buried before they decompose. This is particularly true under relatively inefficient anaerobic decay conditions such as in peat bogs. Through geological time, millions of years of photosynthetic energy have been transformed into fossil fuels by this process as a result of heat and compression. The highly productive forests and marine environments of the distant past have become the coal, oil, and natural gas

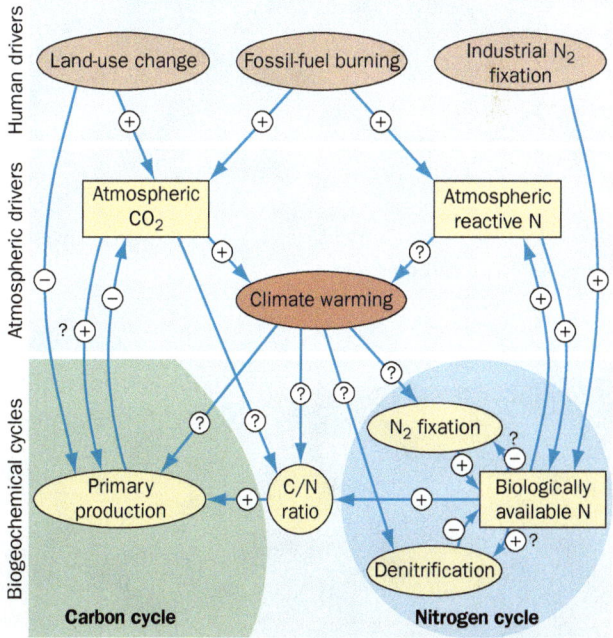

Figure 4.6 | Nitrogen–carbon–climate interactions. Source: Reprinted by permission from Macmillan Publishers, Ltd: N. Gruber and J.N. Galloway. 2008. 'An Earth-system perspective of the global nitrogen cycle.' Source: Gruber and Galloway (2008). Reprinted by permission from Macmillan Publishers Ltd. © 2008 Nature Publishing Group.

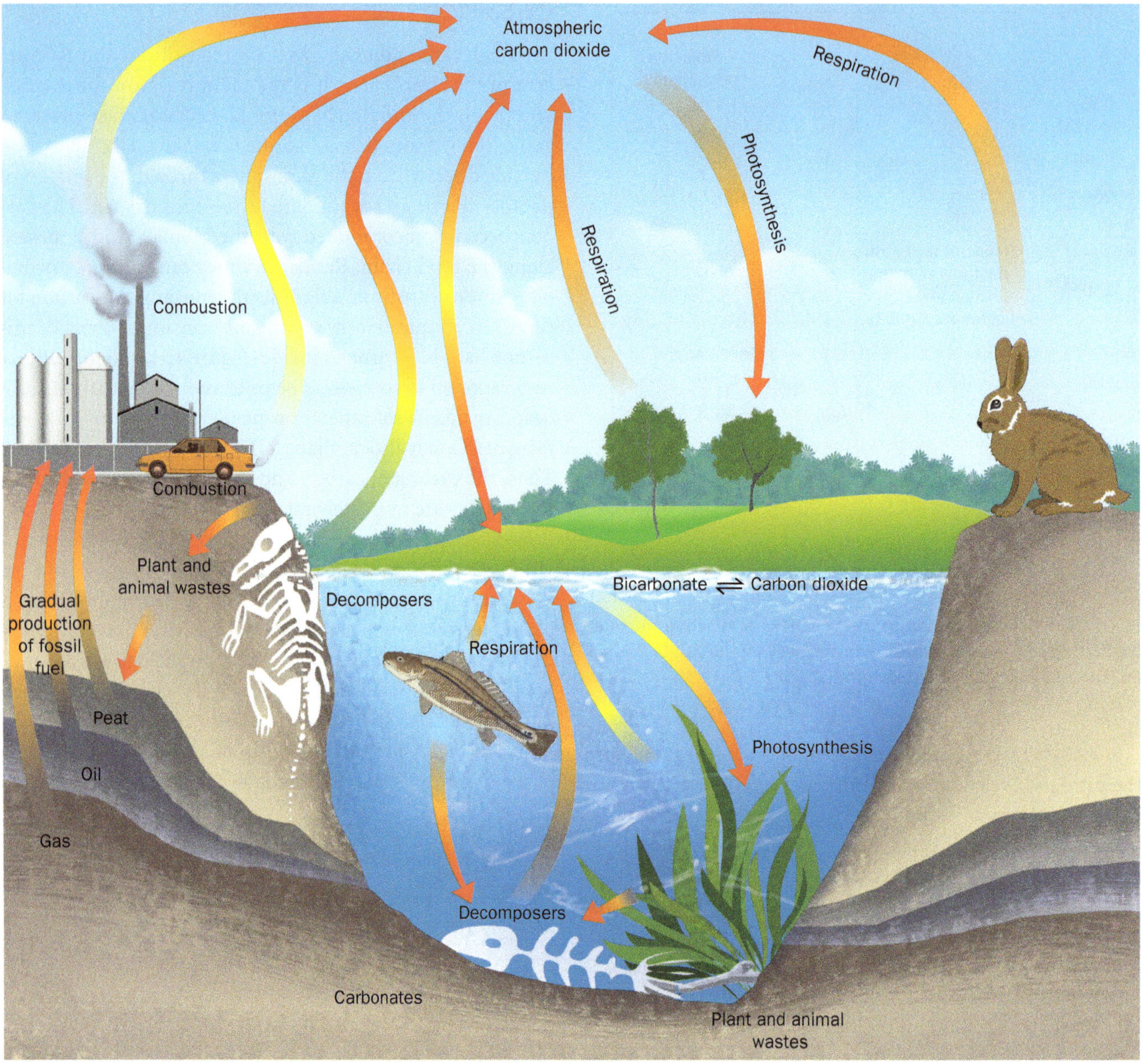

Figure 4.7 | The carbon cycle.

Large amounts of carbon are stored in the lithosphere, such as in these coal-beds, the product of millions of years of photosynthetic activity.

Coral reefs, such as this reef in the Andaman Sea at Koh Surin, Thailand, store large amounts of carbon from the remains of thousands of years of coral growth.

Environment in Focus
Box 4.6 Human Impacts on the Carbon Cycle

As human populations have increased, two major changes to the carbon cycle have occurred:

- Natural vegetation, usually dominated by tree growth, has been replaced by land uses, such as urban and agricultural systems, that have reduced capacity to uptake and store carbon.
- For the past 200 years or so, human activity, particularly industrial activity, has mobilized large amounts of fossil fuels from the lithospheric component of the cycle to the atmospheric component. Estimates suggest that this mobilization represents the release of one million years of photosynthetic activity every year. The concentration of carbon dioxide in Earth's atmosphere now exceeds 390 parts per million (ppm), which is more than 90 ppm above the maximum values of the past 740,000 years, if not 20 million years. Furthermore, increasing CO_2 has driven an increase in the global oceans' average temperature by 0.74°C and sea level by 17 centimetres.

that fuel the world's economy today (see Box 4.6). Scientists predict that there will be a positive feedback loop between increased atmospheric CO_2 and the terrestrial and marine elements of the cycle that will serve to further increase atmospheric CO_2 (Heimann and Reichstein, 2008).

Some of the carbon dioxide is dissolved into the shallower ocean before re-entering the atmosphere. Residence time is in the order of six years in these shallower waters but much longer (up to 350 years) when mixed with deeper waters. These residence times are now of considerable scientific interest because of the rising levels of carbon dioxide in the atmosphere and the potential for the oceans to absorb these increases (see Chapter 8). Recent predictions show that increased carbon dioxide within the atmosphere will also have a positive feedback loop with the carbon concentrations in the ocean and that the oceans' storage capacity for carbon may be decreasing.

Large amounts of carbon are stored for much longer periods in the ocean. When marine organisms die, their shells of calcium carbonate ($CACO_3$) become cemented together to form rocks such as limestone. Over millions of years, the limestone may be uplifted to become land and then is slowly weathered to release the carbon back into the carbon cycle.

The Hydrological Cycle

Water, like the nutrients discussed above, is necessary for all life. You are 70 per cent water. Although other planets such as Venus and Mars have water, only on Earth does it occur as a liquid. It also occurs in a fixed supply that cycles between various reservoirs driven by energy from the sun. By far the largest reservoir is the ocean, containing more than 97 per cent of the water on Earth. Most of the rest is tied up in the polar ice caps, with only a small amount readily available as the fresh water that sustains terrestrial life (Table 4.3). Water travels ceaselessly between these various reservoirs through the main processes of evaporation and precipitation known as the **hydrological cycle** (Figure 4.8).

Average residence times in the reservoirs vary greatly (Table 4.3). In the deep ocean, it may take 37,000 years before water is recycled through evaporation into the atmosphere, whereas once in the atmosphere, average residence time is in the order of nine to 12 days. These figures have special relevance with regard to the effects of pollution. Although many major rivers have suffered from critical pollution incidents, the flushing action of rivers, combined with the short residence time of the water, means that a relatively rapid recovery is often possible. This is not the case, however, with groundwater pollution, especially deep groundwater pollution.

The hydrological cycle involves the transport of water from the oceans to the atmosphere, through terrestrial and subterranean systems, and back to the ocean, all fuelled by energy from the sun. Eighty-six per cent of the water in the atmosphere is evaporated directly from the ocean surface. The remainder comes from evaporation from smaller water bodies, from the leaves of plants (**transpiration**), or from

Table 4.3 | Global Water Storage

Reservoir	Average Renewal Rate	Per Cent of Global Total
World oceans	3,100 years	97.2
Ice sheets and glaciers	16,000 years	2.15
Groundwater	300–4,600 years	0.62
Lakes (freshwater)	10–100 years	0.009
Inland seas, saline lakes	10–100 years	0.008
Soil moisture	280 days	0.005
Atmosphere	9–12 days	0.001
Rivers and streams	12–20 days	0.0001

Environment in Focus
Box 4.7 Some Important Properties of Water

Water has several properties that make it unique:

- Water is a molecule (H_2O) that can exist in a liquid, gaseous, or solid state.
- These molecules have a strong mutual attraction, promoting high surface tension and high capacity to adhere to other surfaces; these properties allow water to move upward through plants.
- Water has a high heat capacity, meaning that it can store a great deal of heat without an equivalent rise in temperature; this is the reason why the oceans have such a moderating influence on climate.
- It takes a lot of heat to change water from liquid to gaseous form; this is why evaporation results in a cooling effect.
- Few solids do not undergo some dissolution in water; this allows water to carry dissolved nutrients to plants, but it also means that water is easily polluted.
- Unlike other substances, when water passes from a liquid to a solid state, it becomes less rather than more dense; this is why ice floats on top of water and permits aquatic life to exist in cold climates.

Most of these properties spring from the fact that although the water molecule is electrically neutral, the charges are distributed in a bipolar manner. In other words, a positive charge is at one end of the molecule, and a negative charge is at the other. This means that water molecules have a strong attraction for each other and also explains why water is such a good solvent, since the charges increase the chemical reactivity of other substances.

the soil and plants (**evapotranspiration**). As it evaporates, water leaves behind accumulated impurities. The most common dissolved substance in the ocean is sodium chloride, or table salt, which also contains many other elements in trace amounts. Evaporation acts as a giant purification plant until further pollutants are encountered in the atmosphere.

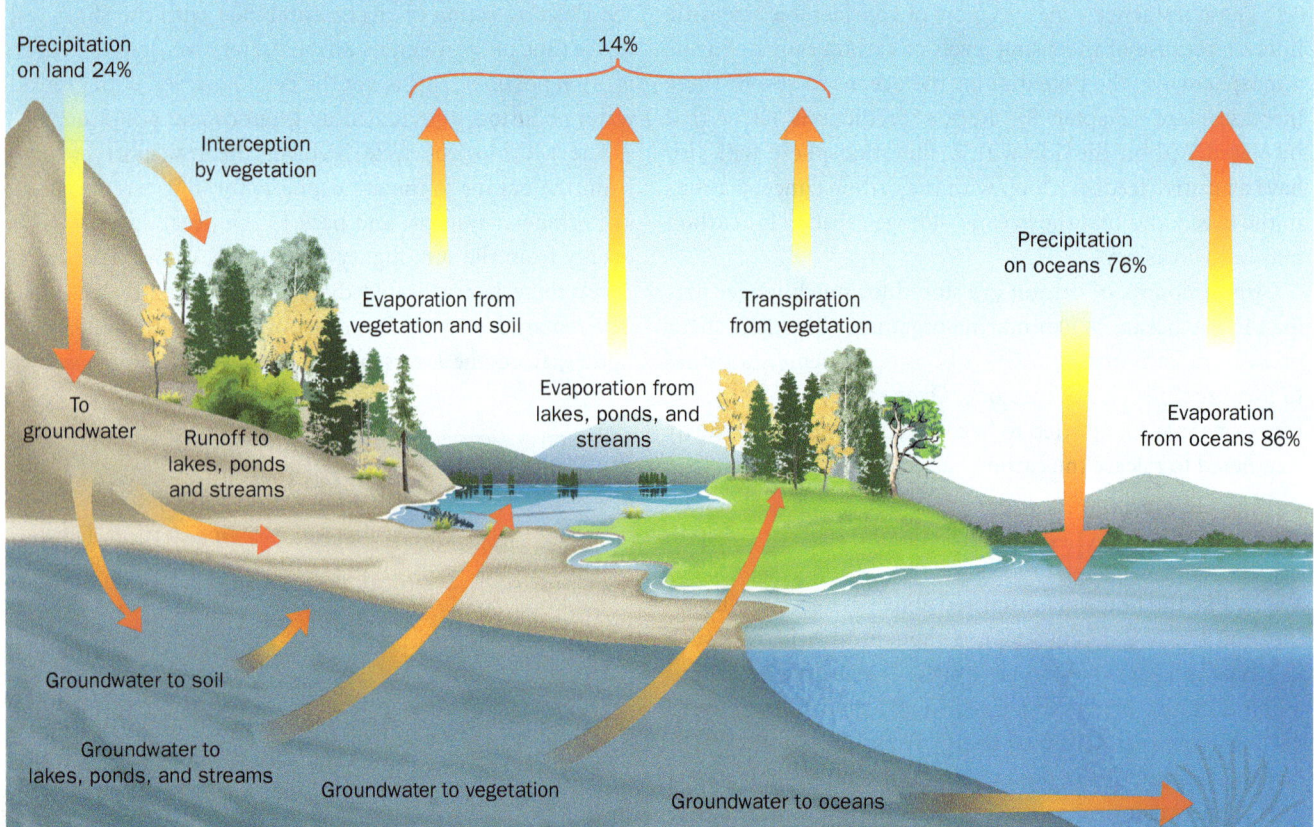

Figure 4.8 | The hydrological cycle. Water moves through the hydrological cycle as a liquid, as a vapour, and as snow.

Once in the atmosphere, the water vapour cools, condenses around tiny particles called **condensation nuclei**, forms clouds, and is precipitated to the earth as rain, snow, or hail. The warmer the air is, the more water it can hold. Moisture content can be expressed in terms of **relative humidity**, the amount of moisture held compared to how much could be held if fully saturated at a particular temperature. At a relative humidity of 100 per cent, the air is saturated, and cloud,

Environment in Focus

Box 4.8 Precipitation

Precipitation occurs in several forms: rain, snow, hail, dew, fog, and rime ice (frost). It occurs when the accumulated particles of condensed water or ice in clouds become large enough that they overcome the uplifting air currents and fall to earth as a result of gravity. Some of this precipitation may never reach the ground. Lower air layers may be warmer and drier, and re-evaporation may occur as the precipitates pass through these layers—an excellent example of the speed of some of these mini-cycles that occur as parts of the larger Earth cycles.

Distribution of precipitation is one of the main factors influencing the nature and location of global biomes and the ecozones of Canada. In Canada, precipitation varies from almost nothing in the Arctic to more than 3,000 millimetres annually on the west coast (Figure 4.10).

Differences in precipitation occur for various reasons. At the global scale, heating of equatorial regions causes air to rise. As it rises, it cools and condenses, clouds form, and precipitation occurs. As a result, equatorial regions tend to have consistent high rainfall. Where this air falls as it cools, over subequatorial regions, it tends to be dry, such as in the Sahara Desert.

In Canada, much of the precipitation comes from low-pressure systems, large cells of rising air that form along the boundary between warm and cold air masses. The main factor influencing relative precipitation levels in Canada is moisture-laden winds crossing the ocean and being forced to rise as a result of mountain barriers. The most extreme example occurs with westerlies coming across the Pacific and meeting the western Cordillera, thus creating the highest precipitation levels in the country. As the air warms up in its descent from the mountains, it can hold more moisture, and precipitation levels fall considerably to produce a **rainshadow effect** that accounts for the small amounts of precipitation across the prairies, with as little annual precipitation as 300 to 400 millimetres. Precipitation levels rise again in central Canada because of disturbances bringing moisture from the Gulf of Mexico and Atlantic Ocean. In southern Ontario, precipitation increases to 800 millimetres and more than 1,000 in the lee of the Great Lakes, a major source of moisture for downwind localities. In Atlantic Canada, exposure to maritime influences increases once more, with up to 1,500 millimetres of precipitation falling annually on the south coast of Newfoundland.

The Arctic is very dry because the prevailing winds from the north are very cold and therefore have little capacity for carrying moisture, they pass over terrain that has relatively few sources of water evaporation (sources often remain frozen for a proportion of the year), and there is an absence of low-pressure systems in winter. Topography is also an important determinant of whether the precipitation falls as snow or rain.

More localized precipitation may be produced by convection as warmer and lighter air rises and cooler, heavier air sinks. This mechanism is important for localized storms in summer when the air is heated by the warm ground and may send columns of moist, warm air to great elevations, resulting in thunderstorm activity.

Desert-like conditions occur in some areas of Canada, largely due to rainshadow effects. Although we associate the Fraser River with the high rainfalls of the west coast, the interior of British Columbia, like the Middle Fraser Canyon depicted here, is quite dry.

fog, and mist form. Clouds are moved around by winds and continue to grow until precipitation (Box 4.8) occurs and the water is returned to the earth.

About 76 per cent of precipitation falls into the ocean. The remainder joins the terrestrial part of the cycle in ice caps, lakes, rivers, groundwater, and transport between these compartments. Gravity moves water down through the soil until it reaches the water table, where all the spaces between the soil particles are full of water. This is the **groundwater** (Box 4.9). Lakes, streams, and other evidence of surface water occur where the land surface is below the water table. Surface water is a major factor in sculpting the shape of the surface of the Earth. At greater depths, the groundwater may penetrate to occupy various geological formations, known as aquifers.

As mentioned in Box 4.9, water is unique in that at the ambient temperatures and pressure of the Earth's surface, it is the only substance that exists in all three phases of matter (solid, liquid, vapour). Water is stored in all three forms within the hydrological cycle and moves among these forms by the processes shown in Figure 4.11. **Sublimation** is the process for direct transfer between the solid and vapour phases of matter, regardless of direction. This explains why on bright sunny winter days when the air is dry, snowbanks may lose size without any visible melting. About 75 per cent of the world's fresh water is stored in the solid phase, and it may stay in this phase for a long time. Measurements in the Antarctic, for example, indicate that some of the ice is more than 100,000 years old.

Although over the short term there are relatively constant amounts of water in the different storage compartments, over the long term these amounts can change markedly. For

Relative humidity is high most of the year in Atlantic Canada and produces some beautiful atmospheric effects.

example, large amounts of water are evaporated from the oceans and precipitated on land as snow during glacial periods. Over time, the snow accumulates and builds ice fields that may be more than a kilometre thick. This effectively removes water from the oceanic component, causing the sea level to fall. During glacial times, then, the area of land will increase relative to the ocean, and the area of Earth surface covered by ice may increase by up to 300 per cent.

In a more contemporary example, pumping rates of groundwater have more than doubled from 1960 to 2000, largely due to agricultural demands. Much of this water is evaporated and then precipitated and eventually reaches the ocean. Researchers have calculated that 25 per cent of the current rise in sea level is a result of this reallocation of water from the ground into the ocean (Wada et al., 2011). This illustrates well how human activities are making planetary cycles acyclic through the rapid mobilization of matter from one storage compartment to another.

In Canada, the solid phase of water is particularly important. Canada may have up to one-third of the world's fresh water but most of it is held in a solid state. Compared to the surface supply in Canada, the country's 100,000 glaciers contain more than 1.5 times the volume of fresh water. Furthermore, more than 95 per cent of the country is snow-covered for part of the winter. Spring melt is hence a critical part of the hydrological cycle in Canada as water moves from the solid to liquid phase. This creates a runoff regime for many Canadian rivers, characterized by low late-winter flows and a high spring melt flow that slowly diminishes over the summer into the winter lows as water becomes stored in the solid phase once more. This marked seasonality is one of the reasons why Canada has developed considerable expertise in the construction of water storage facilities. At

Water in the solid phase of ice may be in the Lowell glaciers of the St Elias range in western Yukon for thousands of years before melting and flowing to the ocean.

Environment in Focus
Box 4.9 Groundwater

Groundwater is found within spaces between soil and rock particles and in crevices and cracks in the rocks below the surface of the Earth. Above the water table is the unsaturated zone where the spaces contain both water and air. In this zone, water is called soil moisture. Groundwater moves the same way as surface waters, downhill, but rarely as quickly and not at all through impermeable materials such as clay. Permeable materials allow the passage of water, usually through cracks and spaces between particles. An **aquifer** is a formation of permeable rocks or loose materials that contains usable sources of groundwater. They vary greatly in size and composition. Porous media aquifers consist of materials such as sand and gravel through which the water moves through the spaces between particles. Fractured aquifers occur where the water moves through joints and cracks in solid rock. If an aquifer lies between layers of impermeable material, it is called a confined aquifer, which may be punctured by an artesian well, releasing the pressurized water to the surface. If the pressure is sufficient to bring water to the surface, the well is known as a flowing artesian well.

Areas where water enters aquifers are known as recharge areas; discharge areas are where the water once more appears above ground. These discharge areas can contribute significantly to surface water flow, especially in periods of low precipitation. Groundwater, of course, is a very significant part of the Canadian water supply. Dependencies range from 100 per cent in Prince Edward Island to 17 per cent in Quebec, with Ontario having the largest total consumption.

One example of the interaction between groundwater and biogeochemical cycle disruption is contamination of the groundwater of the Abbotsford Aquifer in the Lower Mainland of BC. Agriculture is the main land use. Changes in agricultural products over the past 25 years have led to high concentrations of nitrogen. These changes include a shift in animal production from locally fed dairy/beef cattle to poultry production based on outside feed. Crops have changed from grass and hay to raspberry production, which requires less nitrogen. Poultry manure is used to fertilize the raspberries. Nitrogen use is less than one-half of that produced. The remainder infiltrates to the groundwater. Of 2,297 groundwater samples taken over the years, 71 per cent have exceeded the Guideline for Canadian Drinking Water Quality of 10 milligrams per litre (mg/L), with some as high as 91.9 mg/L. This is cause for concern. These nitrates may be transformed to nitrites in the digestive systems of babies, which in turn may lead to an oxygen deficiency in the blood, known as methemoglobinemia, or blue-baby syndrome. High nitrate levels have also been linked to cancer. More information on this can be found at Environment Canada's 'ecoinfo' website, listed at the end of this chapter.

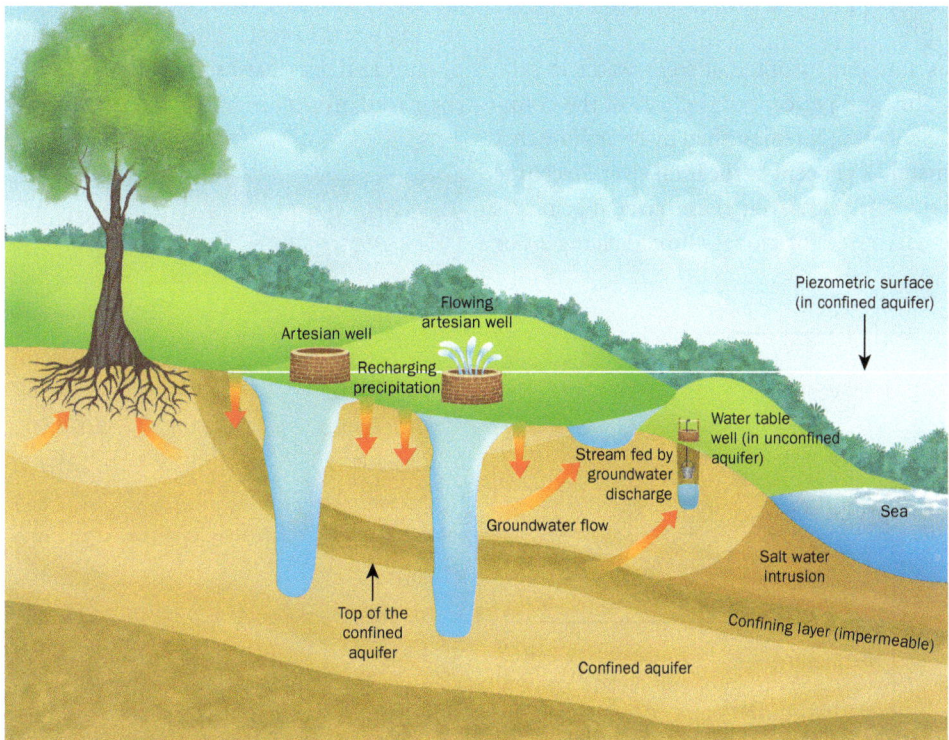

Figure 4.9 | Groundwater flow.

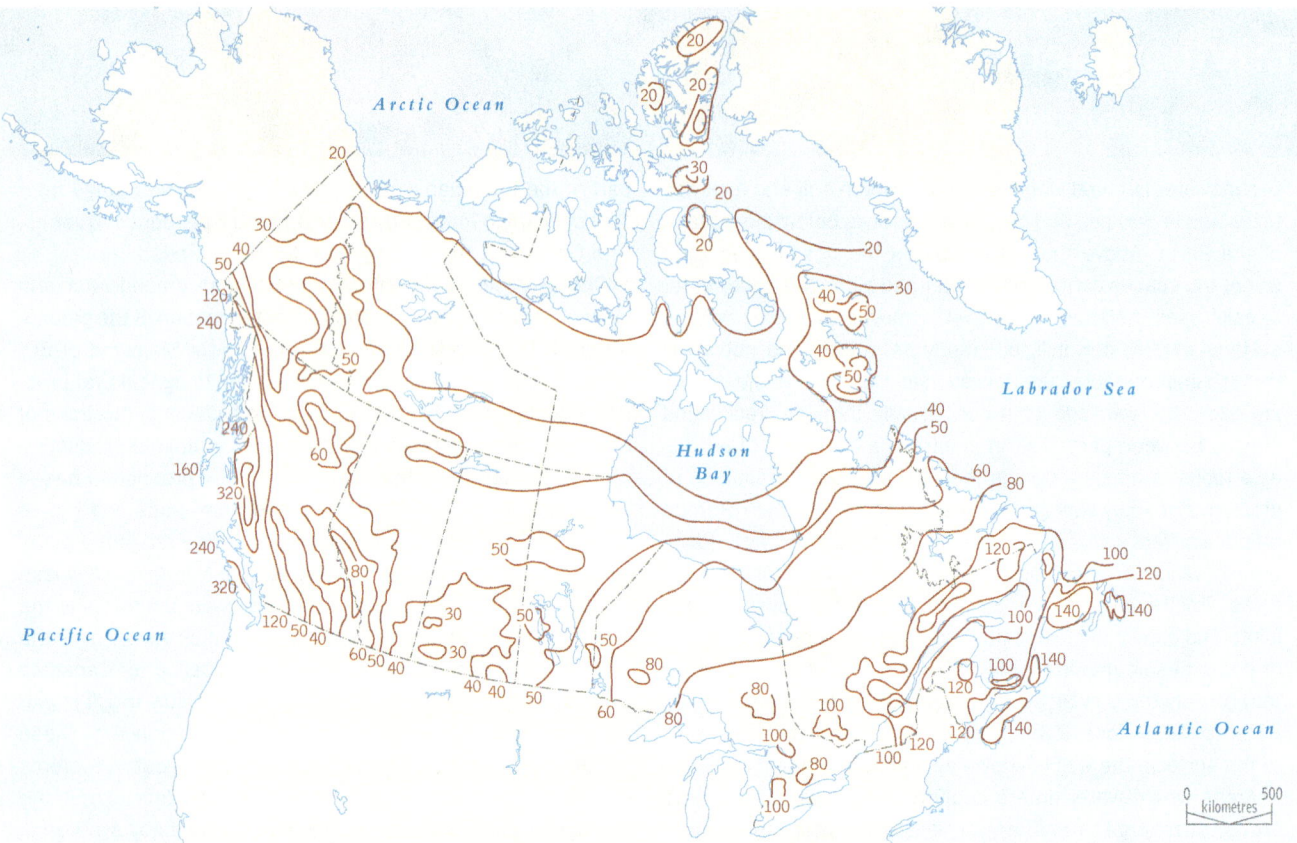

Figure 4.10 | **Average annual rain and snow for Canada (cm).** Source: Phillips (1990: 210).

the same time, human impacts on the hydrological cycle are significant (Box 4.10).

Canada also has abundant storage of fresh water in lake systems, covering almost 8 per cent of the area of the country. These lakes are constantly replenished by river flow that contains approximately 7 per cent of the total river discharge in the world (Table 4.4). However, these river discharges are also changing as a result of global climate change. For example, Environment Canada reports that of its monitoring sites, lowest annual flow increased significantly at 51 sites and decreased significantly at 27 sites. The sites with increased minimums occurred in northwest Canada and the

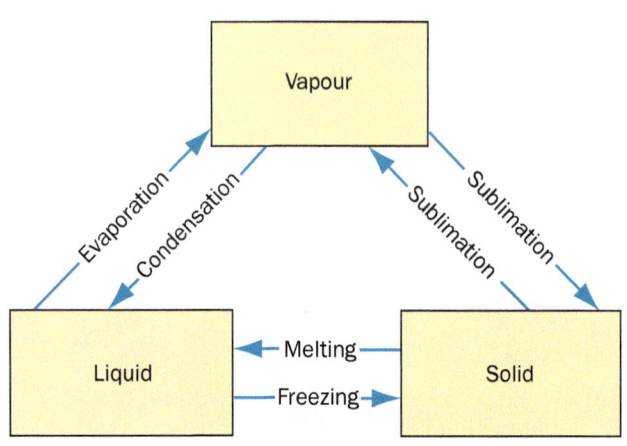

Figure 4.11 | **Changes of phase in the hydrological cycle.**

Table 4.4 | Mean Annual Stream Discharge to the Oceans for Selected Canadian Rivers

River	Watershed Area (km^2)	Discharge ($m^3 s^{-1}$)
Saguenay	90,100	1,820
St Lawrence	1,026,000	9,860
Churchill	281,300	1,200
Nelson	722,600	2,370
Albany	133,900	1,400
Koksoak	133,400	2,550
Yukon (at Alaska border)	297,300	2,320
Fraser	219,600	3,540
Columbia (at Washington border)	154,600	2,800
Mackenzie	984,195	10,800

Source: Briggs et al. (1993: 206).

Environment in Focus

Box 4.10 Human Impacts on the Hydrological Cycle

Human activities have also affected the hydrological cycle. Changes include:

- the storage and redistribution of runoff to augment water supplies for domestic, agricultural, and industrial uses;
- the building of storage structures to control floods;
- the drainage of wetlands;
- the pumping of groundwater;
- cloud seeding;
- land-use changes such as deforestation, urbanization, and agriculture that affect runoff and evapotranspiration patterns;
- climatic change caused by interference with biogeochemical cycles.

Arctic, whereas decreased flow sites were mainly in eastern and Atlantic Canada and across southern Canada from BC to the prairies. Aquatic biodiversity is affected by these changes in relation to spawning times and other aspects of the life cycle.

About 75 per cent of the rivers and 60 per cent of the discharge in Canada drains north to the Arctic Ocean (Figure 4.12), whereas 90 per cent of the Canadian population lives within 300 kilometres of the US border, creating the potential for water deficits in this water-rich country (see Chapter 11).

As water demands grow, we have become increasingly dependent on groundwater sources. Once they become polluted with agricultural biocides or industrial wastes, however, they may be unsuitable for human use for centuries. The importance of this is underscored by the fact that Canada is estimated to have 37 times the amount of water in underground sources as in surface sources and one-quarter of the Canadian population relies on groundwater for domestic use, while some communities, such as Fredericton, are almost totally dependent on groundwater.

Biogeochemical Cycles and Human Activity

Despite the apparent sophistication of human society, the humble fact remains that society could not exist without biogeochemical cycles and those unpretentious bacteria that make them work. Yet all the cycles are susceptible to perturbations by human activity. Such is the scale of human actions that the major transfers taking place between some of the reservoirs in the cycles are human-induced. Some of the most notable and difficult environmental challenges now faced by society spring from these transfers. The purpose of this section is to discuss two of them in more detail: eutrophication and acid deposition. A third example, global climatic change—largely resulting from disruptions in the carbon and nitrogen cycles—is of such significance that Chapter 7 is devoted to it.

Eutrophication

Eutrophication is a natural process of nutrient enrichment of water bodies that leads to greater productivity. In an assessment of safe limits for stressing global systems, an international team of researchers (Rockström et al., 2009) found that of the nine main systems, three were already stressed above these limits. The nitrogen cycle is one of these three and the phosphorus cycle is also rapidly approaching its planetary boundaries, although some scientists have pointed out that phosphorus is already well over its limit in freshwater ecosystems. Furthermore, a recent assessment of water-based pollutants in North America found that of the 256 pollutants released to water, just two—nitrate compounds and ammonia—comprised 90 per cent of the total discharge (Commission for Environmental Cooperation, 2011). Given this volume, it is little surprise that eutrophication of global water bodies is becoming an increasing challenge.

Phosphorus and nitrogen are often the two main limiting factors for plant growth in aquatic ecosystems. Systems with relatively low nutrient levels, **oligotrophic** ecosystems, have quite different characteristics from those with high nutrient levels (**eutrophic**), as summarized in Table 4.5. **Mesotrophic** bodies have characteristics in between these two extremes. Natural terrestrial ecosystems are relatively efficient in terms of holding nutrient capital. The progression from an oligotrophic to eutrophic condition, through the process of succession discussed in Chapter 3, may take place over thousands of years. This rate is influenced by the geological makeup of the catchment area and the depth of the receiving waters. Catchments with fertile soils will progress more quickly than those with soils lacking in nutrients. Depth is important, because shallower lakes tend to recycle nutrients more efficiently.

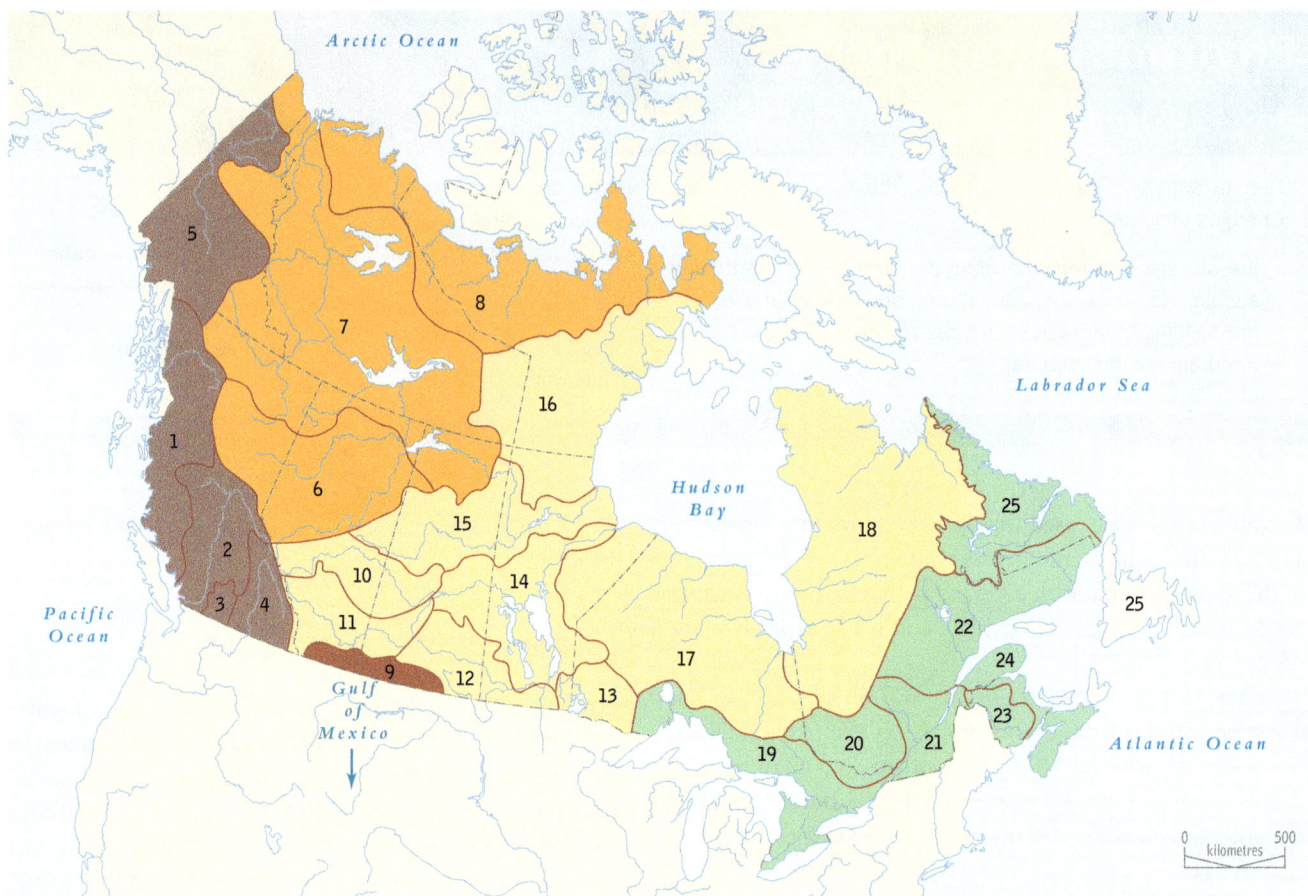

Ocean Basin Region		River Basin Region	Area in 000s km²
Pacific	1	Pacific Coastal	352
	2	Fraser-Lower Mainland	234
	3	Okanagan-Similkameen[a]	14
	4	Columbia[a]	90
	5	Yukon	328
Arctic	6	Peace-Athabasca	487
	7	Lower Mackenzie	1300
	8	Arctic Coast–Islands	2025
Gulf of Mexico	9	Missouri[a]	26
Hudson Bay	10	North Saskatchewan	146
	11	South Saskatchewan[a]	170
	12	Assiniboine-Red[a]	190
	13	Winnipeg[a]	107
	14	Lower Saskatchewan–Nelson	363
	15	Churchill	298
	16	Keewatin	689
	17	Northern Ontario	694
	18	Northern Quebec	950
Atlantic	19	Great Lakes[a]	319
	20	Ottawa	146
	21	St Lawrence[a]	116
	22	North Shore–Gaspé	403
	23	St John–St Croix[a]	37
	24	Maritime Coastal	114
	25	Newfoundland-Labrador	376
CANADA			9974

Figure 4.12 | Drainage regions of Canada. [a]Canadian portion only; area on US side of international basin regions excluded from total. Source: Environment Canada (1985: 35). Reproduced with the permission of the Minister of Public Works and Government Services Canada, 2011.

Table 4.5 | Characteristics of Oligotrophic and Eutrophic Water Bodies

Characteristic	Oligotrophic	Eutrophic
Nutrient cycling	low	high
Productivity (total biomass)	low	high
Species diversity	high*	low
Relative numbers of 'undesirable' species	low	high
Water quality	high	low

*Lakes that are extremely non-productive (e.g., high mountain lakes) will have low species diversity.

What Causes Eutrophication?

Cultural eutrophication (eutrophication caused by human activity) may speed up the natural eutrophication process by several decades, mainly through the addition of phosphates and nitrates to the water body. As lakes become shallower as a result of this input, nutrients are used more efficiently, productivity increases, and eutrophication progresses. This is a classic example of a positive feedback loop, with change in the system promoting even more change in the same direction. Additional phosphates and nitrates come from many different sources (Table 4.6), and in accordance with the law of conservation of matter discussed earlier, they do not simply disappear but accumulate in aquatic ecosystems. In total, between 8.5 and 9.5 tonnes of phosphorus finds its way into the ocean from the 20 million tonnes mined each year. This is approximately eight times the natural amount.

What Are the Effects?

This enrichment promotes increased growth of aquatic plants, particularly favouring the growth of floating phytoplankton over **benthic** plants rooted in the substrate. As the benthic plants become out-competed for light by the phytoplankton, they produce less oxygen at depth. Oxygen is critical for the maintenance of more diverse, oxygen-demanding fish species such as trout and other members of the salmonid family, which also start to decline in number. The oxygen produced by photosynthesis by the phytoplankton tends to stay in the shallower water, escaping back to the atmosphere rather than replenishing supplies at greater depths.

Oxygen depletion is further exacerbated by the decay of the large mass of phytoplankton produced. Dead matter filters to the bottom of the lake where it is consumed by oxygen-demanding decomposers. Once broken down, nutrients may be returned to the surface through convection currents and provide more food for more phytoplankton and algae. Blue-green algae replace green algae in eutrophic lakes, which further exacerbates the problem, since most blue-green algae are not consumed by the next trophic level, the zooplankton.

These effects of oxygen depletion in a water body also occur whenever excess organic matter is added. Under natural conditions, water is able to absorb and break down small amounts of organic matter, with the amount depending on the size and flow and the lower temperature of the receiving water body. The greater the size and flow are and the lower the temperature, the greater the ability to absorb organic materials and retain oxygen levels.

When organic wastes are added to a body of water, the oxygen levels fall as the number of bacteria rises to help break down the waste. This is known as the **oxygen sag curve** (Figure 4.13) and is measured by the **biological oxygen demand** (BOD), the amount of dissolved oxygen needed by aerobic decomposers to break down the organic material in a given volume of water at a certain temperature over a given period. At the discharge source, the oxygen sag curve starts to fall, and there is a corresponding rise in the BOD. As distance from the input source increases and the bacteria digest the wastes, then the oxygen content returns to normal, and the BOD falls.

Major sources of nitrates and phosphates, such as runoff from feedlots and sewage discharge, also contain large amounts of oxygen-demanding wastes. Heat is another source of oxygen stress. The overall result is a progression to a less useful and less healthy water body. The composition of the fish species changes to those less dependent on high oxygen levels, species that are generally less desirable for human purposes. Populations of waterfowl may fall as aquatic plants die off. The water becomes infested with algae, aquatic weeds, and phytoplankton, making swimming and boating unpleasant and giving off unpleasant odours. Water treatment for domestic or industrial purposes becomes more expensive.

What Can We Do about It?

The main way to control eutrophication is to limit the input of nutrients into the water body (Table 4.6). Domestic and animal wastes must be treated to remove phosphates. Advanced

Table 4.6 | Main Nutrient Sources Contributing to Cultural Eutrophication

Runoff from	fertilizers (N and P)
	feedlots (N and P)
	land-use change, such as cultivation, construction, mining natural sources
Discharge of	detergents (P)
	untreated sewage (N and P)
	primary and secondary treated sewage (N and P)
Dissolved nitrogen oxides	(from internal combustion engines)

treatment can remove up to 90 per cent of these wastes. More difficult problems occur with diffuse, **non-point sources**, such as runoff from urban areas and agricultural land. Such flows really have to be controlled at the source, since they enter the water body, by definition, in so many different locations. In the past, measures to control water pollution have been largely directed towards **point sources** of pollution, or single discharge points, such as effluent discharges from sewage plants or industrial processes. By and large, because of the high visibility of such sources, they are easy to identify and monitor, and pollution from such sources has fallen as a result. Increasing attention is now being directed towards the non-point sources (see Chapter 11).

Water quality is one of the three environmental sustainability indicators still measured by Environment Canada. Phosphorus levels in southern Canada exceeded limits set under the water quality guidelines for aquatic life more than half the time at 127 of 344 monitoring sites (Environment Canada, 2007b).

Eutrophication used to be considered a problem of smaller water bodies, but now entire areas of the world's largest water bodies, the oceans, are becoming so eutrophic they are being

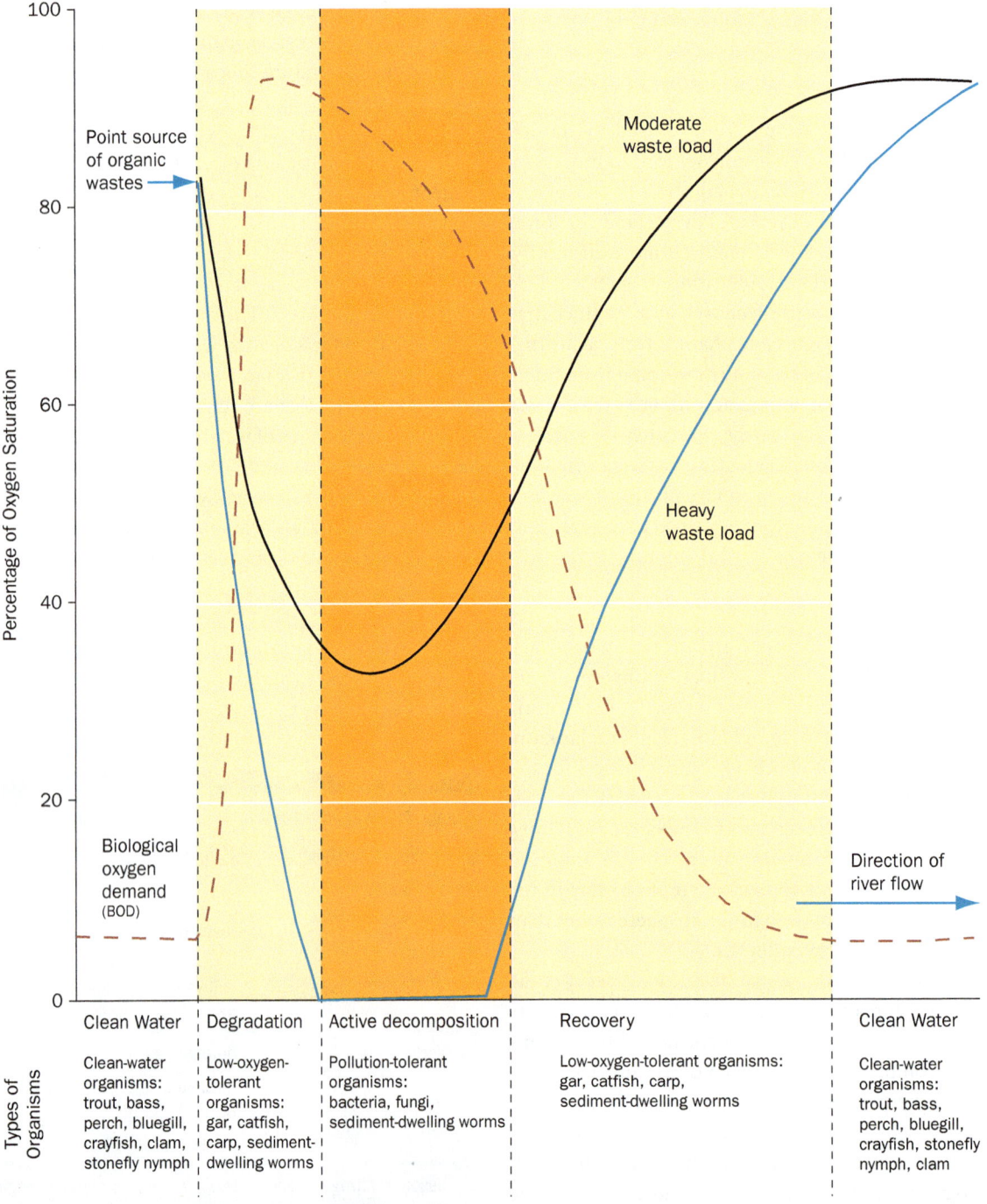

Figure 4.13 | Oxygen sag curve and biological oxygen demand (BOD).

described as 'dead zones'. More than 200 such zones have been recognized, some as large as 70,000 square kilometres in size. One of the largest and best-known areas occurs in the Gulf of Mexico, which receives all the excess fertilizers brought down from the Mississippi watershed. However, similar areas are now found in the waters off every continent, and researchers expect that their number and size will increase as global climatic change generates greater rainfall and greater runoff in many areas. Fearing that nutrient input control will not be totally successful, researchers in Sweden, where the Baltic Sea is heavily eutrophic, are now experimenting with pumping oxygen to the ocean floor using a large wind-driven pump. New ecosystems are created in the presence of oxygen that enable nature to deal with eutrophication through binding the excess phosphorus into ocean sediments and biomass.

It is important to continue to experiment with technological answers to environmental challenges, such as the one described above. However, experience suggests that it is unlikely that they will ever provide complete answers to modern-day environmental challenges and strong focus must be placed on addressing the root causes of the problems. Eutrophication was thought to be largely addressed through pollution control some 25 years ago. Now it is apparent that the challenge is still there, but at a larger scale. The next section explores one of Canada's most notable eutrophication challenges, Lake Erie, and points to some of the ways to address the problem.

Lake Erie: An Example of Eutrophication Control

In all parts of the country, there are many examples of eutrophic lakes. Lake Erie is a particularly well-known case because of its size and importance. Lake Erie is the second smallest and also the shallowest of the Great Lakes, which contain almost 20 per cent of the world's fresh water. Erie has experienced considerable changes in fish species composition since the early explorers described a highly diverse community. Gone are the lake sturgeon, cisco, blue pike, and lake whitefish as the human population in the basin increased and water quality declined. Some 11 million people live in the Erie drainage basin, and 39 per cent of the Canadian and 44 per cent of the American shore is taken up by urban development. There is also considerable industrial use around the shore and intensive agricultural use throughout the basin.

In the past, up to 90 per cent of the bottom layer of the central zone of the lake became oxygen-deficient in the summer. Huge algae mats more than 20 metres in length and a metre deep became common. Beaches were closed. The natural eutrophication that might have taken thousands of years was superseded by cultural eutrophication within the space of 50 years.

In 1972, Canada and the US signed the Great Lakes Water Quality Agreement to try to come to grips with this problem. The signing of the 1985 Great Lakes Charter, in which

Animal feedlots are a major source of nutrients, such as phosphates and nitrates, which speed up eutrophication.

the two countries agreed to take a co-operative and ecosystem-based approach to the lakes, further strengthened international efforts. Since the 1970s, phosphorus controls, implemented under the Canada Water Act, have led to significant reductions in the phosphorus concentration of the water (Figure 4.14). Phosphate-based detergents were banned and municipal waste treatment plants upgraded.

These measures have led to improvement of water quality, but significant problems still remain. The controls are largely on point-source pollution, discharges that have a readily identifiable source, such as waste treatment plants and industrial complexes. However, much of the remaining nutrient load comes from non-point sources, such as runoff from agricultural fields, lawn fertilizer, and construction sites that are much more difficult to regulate. Figure 4.14 shows that phosphorus levels are once more rising in the lake but remain below the target. The levels of nitrite and nitrate concentrations have also been rising, leading to increased eutrophication as a result of other nutrients.

Important synergistic effects between increasing phosphorus loads and invasive species can exacerbate the problem. For example, zebra mussels (Chapter 3) selectively feed on edible algae, giving opportunity for expanded blooms of the toxic blue-green algae *Microcystis*. The latter can be toxic to animals, including humans, and also is not a preferred food for zooplankton and therefore affect the base of the food chain, particularly for fish larvae (Vanderploeg et al., 2009).

Acid Deposition

In 1966, fisheries researcher Harold Harvey was puzzled to find that the 4,000 pink salmon he had introduced to Lumsden Lake in the La Cloche Mountains southwest of Sudbury, Ontario, the previous year had all disappeared. Their passage upstream and downstream of the lake had been blocked. To unravel the mystery, he began to take more

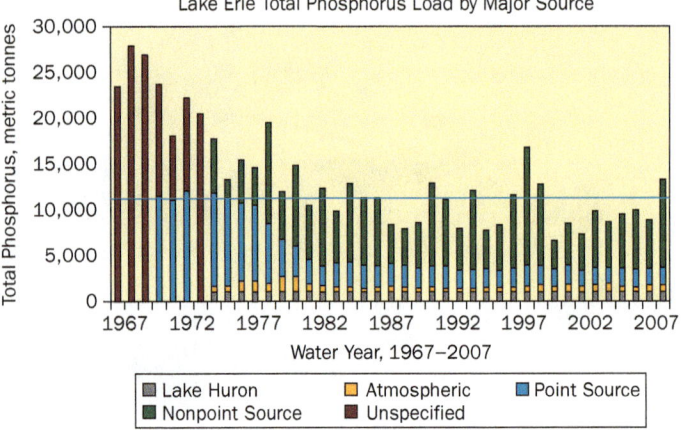

Figure 4.14 | Lake Erie total phosphorus load by major source, 1967–2007. Source: Ohio Environmental Council, Lake Erie Phosphorous Task Force: www.slideshare.net/OhioEnviroCouncil/ohio-lake-erie-phosphorus-taskforceresultslamp. Used with permission from the Ohio Environmental Protection Agency, Division of Surface Water.

measurements of the lake and look into its past history. The results were startling—not only had the salmon disappeared but many other species of fish indigenous to the lake had gone missing as well (Table 4.7). The reason soon became apparent. Between 1961 and 1971, Lumsden Lake had experienced a hundredfold increase in the acidity of its waters, as had many other lakes in the same region (Table 4.8). The changes had shifted the lakes outside the limits of tolerance of the species, as discussed in Chapter 2. The indigenous fish species, and many of the species upon which they depended for food, simply could not tolerate the new conditions and perished. They were victims of the effects of **acid deposition**.

What Is Acid Deposition?

Acids are chemicals that release hydrogen ions (H+) when dissolved in water, whereas a base is a chemical that releases hydroxyl ions (OH−). When in contact, acids and bases neutralize each other as they come together to form water (H_2O). Acidity is a measure of the concentration of hydrogen ions in a solution and is measured on a scale, known as the pH scale, that goes from 0 to 14 (Figure 4.15). The midpoint of the scale, pH 7, represents a neutral balance between the presence of acidic hydrogen ions and basic hydroxyl ions. The pH scale is logarithmic. A decrease in value from pH 6 to pH 5 means that the solution has become 10 times more acidic. If the number drops to pH 4 from pH 6, then the solution is 100 times more acidic.

Precipitation, either as snow or rain, tends to be slightly acidic, even without human interference, because of the chemical reaction as carbon dioxide in the atmosphere combines with water to form carbonic acid. Generally, a pH value of 5.6 is given to 'clean' rain. Acid rain is defined as deposition that is more acidic than this, and in Canada rainfall has been recorded with pH levels much lower. Acidic deposition is a more generic term that includes not only rainfall but also snow, fog, and dry deposition from dust.

What Causes Acid Deposition?

The increases in acidity reflected in the pH levels of the lakes in Table 4.8 are due to human interference in the sulphur and nitrogen cycles. The largest sources are the smelting of sulphur-rich metal ores and the burning of fossil fuels for energy. These processes change the distribution of the

Table 4.7 | Disappearance of Fish from Lumsden Lake

1950s	Eight species present
1960	Last report of yellow perch
1960	Last report of burbot
1960–5	Sport fishery fails
1967	Last capture of lake trout
1967	Last capture of slimy sculpin
1968	White sucker suddenly rare
1969	Last capture of trout and perch
1969	Last capture of lake herring
1969	Last capture of white sucker
1970	One fish species present
1971	Lake chub very rare

Source: H. Harvey, unpublished speech, based on Beamish and Harvey (1972). Reprinted with the author's permission.

Table 4.8 | Lake Acidification in the La Cloche Mountains, 1961–71

Lake	pH 1961	pH 1971
Broker	6.8	4.7
David	5.2	4.3
George	6.5	4.7
Johnnie	6.8	4.8
Lumsden	6.8	4.4
Mahzenazing	6.8	5.3
O.A.S.	5.5	4.3
Spoon	6.8	5.6
Sunfish	6.8	5.6
Grey (1959)	5.6	4.1
Tyson (1955)	7.4	4.9

Source: H. Harvey, unpublished speech, based on Beamish and Harvey (1972). Reprinted with the author's permission.

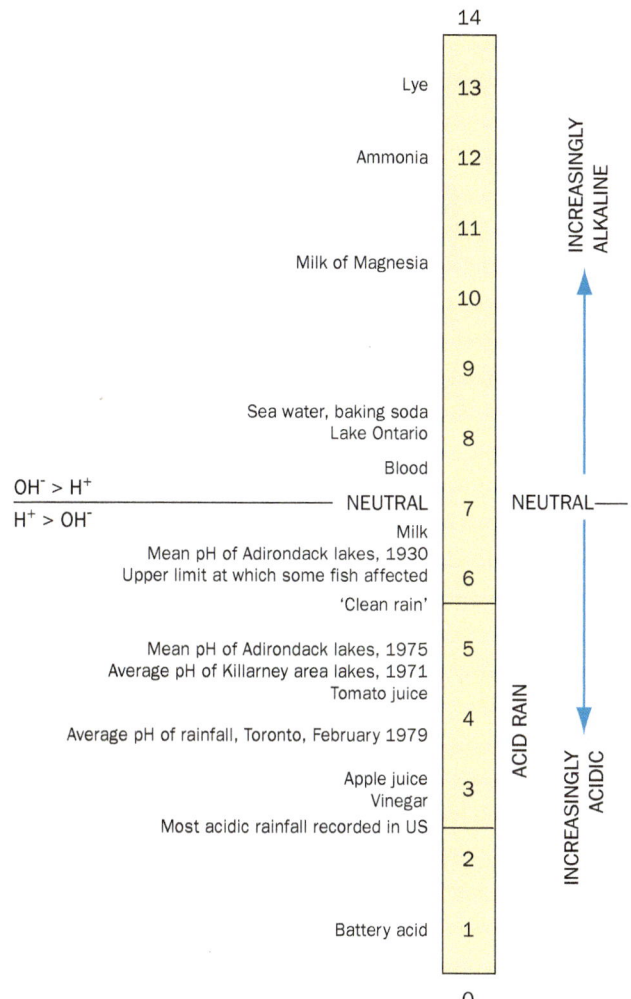

Figure 4.15 | The acid (pH) scale.

of these emissions became obvious fairly early on around smelting plants such as those at Inco in Sudbury and in Trail, BC. Trees were destroyed over large areas. Now there are very encouraging signs of rehabilitation, as discussed in Chapter 13.

These obvious signs of ecological damage were ignored for many years. However, as the ecological implications became better known, industry began to build higher stacks to eject the waste further into the atmosphere. Inco, for example, built a 381-metre 'Superstack' in 1972 (see also Chapter 13). At the local scale this improved matters somewhat but merely served to create problems elsewhere, especially in Quebec, as entire air masses became acidified and dropped their acid burdens over a larger area. Weather patterns are not random, and so acidified air masses tend to travel in the same kinds of patterns. In central and eastern Canada, as air masses travel from southwest to northeast, they bring a heavy pollution burden from the heavily industrialized Ohio Valley in the US, which falls mostly in Canada. It is estimated that approximately half of the sulphate falling in Canada originates in the US.

These point sources of pollution are, however, easier to monitor and control than the other main source of acids—nitrogen emissions—of which some 22 per cent comes from various means of transport as a result of high-temperature combustion (Figure 4.16). The remainder is split between emissions from thermoelectric generating stations and other industrial, commercial, and residential combustion processes.

What Are the Effects of Acid Deposition?

Aquatic Effects

The effects of acid deposition on the fish of Lumsden Lake and other aquatic ecosystems are one visible sign of some of the impacts of acid deposition. Other species are also affected as the pH of the water body declines. Indeed, as can be seen in Figure 4.17, fish are often not the most sensitive species and are really more of an indicator of the damage that has already occurred. As insects such as mayflies are eliminated, species higher in the food chain that feed on them become affected through food depletion. The same is true of fish-eating birds such as loons, whose young have been shown to have a lower chance of survival on acidified lakes because of starvation.

Unfortunately, some of these impacts are permanent. Examination of historical angling records in Nova Scotia, for example, indicates that of 60 main salmon rivers, 13 runs of salmon are extinct and a further 18 are virtually extinct. It is estimated that more than half of the total salmon stock has been lost as a result of declining pH levels. Atlantic salmon populations in rivers of the Southern Upland region of Nova Scotia continue to be severely negatively affected and

elements between the various sources shown in Figures 4.3 and 4.4, and consequently, natural processes are inadequate to deal with the buildup of matter. Increased amounts of sulphur and various forms of nitrogen are ejected into the atmosphere, where they may travel thousands of kilometres before being returned to the lithosphere as a result of depositional processes. Human activities account for more than 90 per cent of the sulphur dioxide and nitrogen oxide emissions in North America.

Excessive sulphur is produced when ore bodies, such as copper and nickel, are roasted (smelted) at high temperatures to release the metal. Unfortunately, such ores often contain more sulphur than metal, and the sulphur is released into the atmosphere as a waste product of the process. A similar effect is created when sulphur-containing coal is burned as the energy source in power plants. In Canada, the smelting of metal ores accounts for 69 per cent of SO_2 emissions, with power generation and other sources accounting for 30 per cent (Figure 4.16). In the US, electrical utilities are the largest source, accounting for 70 per cent of emissions. The effects

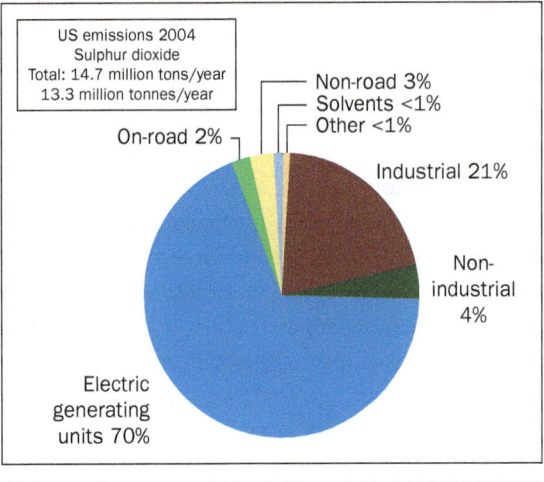
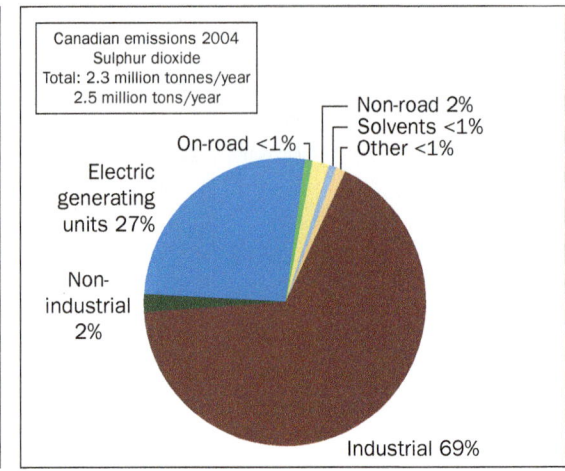
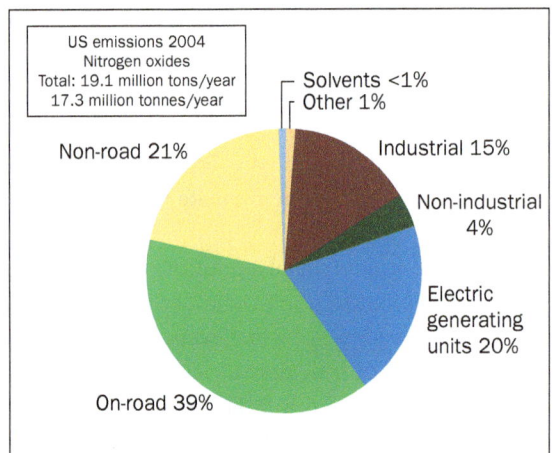
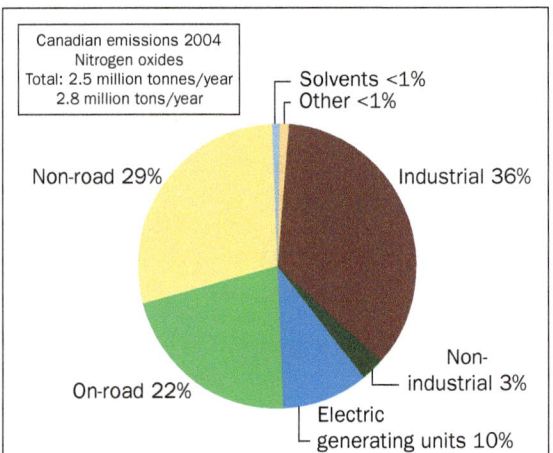

Figure 4.16 | Main sources of sulphur dioxide and nitrogen oxide emissions in Canada and US in 2004. *Source: Environment Canada (2006a).*

will likely become extinct if adult survival rates remain at current low levels and pH recovery continues to be delayed (Environment Canada, 2004). It has also been suggested that the endangered Acadian whitefish is threatened by declining pH levels.

Even where fish manage to survive, they may be grossly disfigured, with twisted backbones and flattened heads, because their bones have been deprived of the necessary nutrients for strength. Reproductive capacities may be sufficiently impaired that it leads to eventual population declines. Generally, the time of reproduction is the most sensitive part of the life cycle. The critical factor is often the lower pH level of the water as the snow melts. At this time, the buildup of acids over the winter can result in even higher acidity than experienced through the rest of the year. This pulse of acidity is called **acid shock**, and it may also be one of the causes of stress on amphibious creatures, such as frogs, that often use small temporary pools of water for breeding in the spring following runoff. At Lumsden Lake, for example, spring runoff produced a pH as low as 3.3, more than 100 times more acidic than the 1961 levels.

More acidic water and the subsequent food chain effects are not the only concern. Other chemical changes also occur. The increased acidity, for example, releases large amounts of aluminum from the terrestrial ecosystem into the rivers and lakes. Here it forms a toxic scum, lethal to many forms of aquatic life.

Terrestrial Effects

Terrestrial effects of acid deposition first became visible around emission sources, such as at Trail, BC, and Sudbury, Ontario, as trees began to die. Before joining the soil, the acids eat away at the sensitive photosynthetic surfaces of the leaves. Broad-leaved trees, such as sugar maples, the source of Quebec's $40 million annual maple syrup industry, are thought to be particularly susceptible because of the large surface area of their leaves.

Once in the soil, the acids leach away the nutrients required for plant growth, leading to nutrient deficiencies. The high levels of aluminum released by the acids also help to inhibit the uptake of nutrients. The bacteria so critical to the workings of many biogeochemical cycles are also adversely affected and cause changes in natural soil processes. Decomposition and humus formation are retarded. Soil contains many organisms involved in the critical ecological functions of biogeochemical cycling and energy flow. There may be more than 100 million bacteria and several kilometres of fungal hyphae in a single gram of healthy soil. Mycorrhizal fungi are very important for the growth of many plants because they help to transport nutrients from the soil water into the roots. Research on Jack pine, undertaken to understand the potential impact of increased acidity on these mutualistic relationships, found that changes in calcium-to-aluminum ratios caused by increased acidity influenced the succession of mycorrhizal fungi on tree root systems. The actual physical contact of the acids with the plant roots can also inhibit growth and lower resistance to disease. The long-term effects of these kinds of changes on tree growth, the ecological health of the community, and forest yields are still highly uncertain, but they are not going to be beneficial.

Eastern Canadian watersheds now exhibit releases of sulphur from soils in excess of deposition. Two internal catchment sources, sulphate desorption and release via decomposition of organic matter, are the likely causes of the budget imbalance. The release of this extra sulphur acts as an additional acid load for soils and downstream waters and may be partly retarding the recovery of surface waters in eastern Canadian forested watersheds (Environment Canada, 2004).

However, increased nitrogen deposits are not a problem, since most areas are nitrogen-deficient and the increased inputs act as fertilizer.

Damage from acid deposition is now visible over much wider areas than those surrounding sources of high emissions. Extensive areas of damage have been recorded in Europe and in the eastern United States. At these larger scales, it is often difficult, however, to single out one cause, such as acid deposition. It is likely that other factors—climate change and other pollutants, such as high levels of ozone brought about by excessive nitrogen oxides—are also important in placing stress on these communities in synergistic reactions.

In central and eastern Canada, 89 per cent of the high-capability forest land receives in excess of 20 kilograms per hectare of acid deposition annually, the figure originally defined by the Canadian government as being an 'acceptable' level of deposition. Severe impacts have been noticed. A survey of white birch around the Bay of Fundy, for example, found 10 per cent of the trees already dead and virtually all others showing signs of damage.

The impact on plant life is not restricted to natural ecosystems. Significant changes can also occur on agricultural lands as direct damage to crops or more indirect changes through changes to soil chemistry. The growth of crops such as beets, radishes, tomatoes, beans, and lettuce is inhibited, and biological nitrogen fixation is diminished at a pH of 4. In central and eastern Canada, 84 per cent of the most productive agricultural lands receive more than the 20 kilograms per hectare. In some areas, the application of lime to the soil has become a routine agricultural procedure in an attempt to neutralize the acids by adding more basic ions.

Heterotrophs can also be affected. As the forest cover diminishes, so does the habitat for many species. Toxic metals such as cadmium, zinc, and mercury, released from the

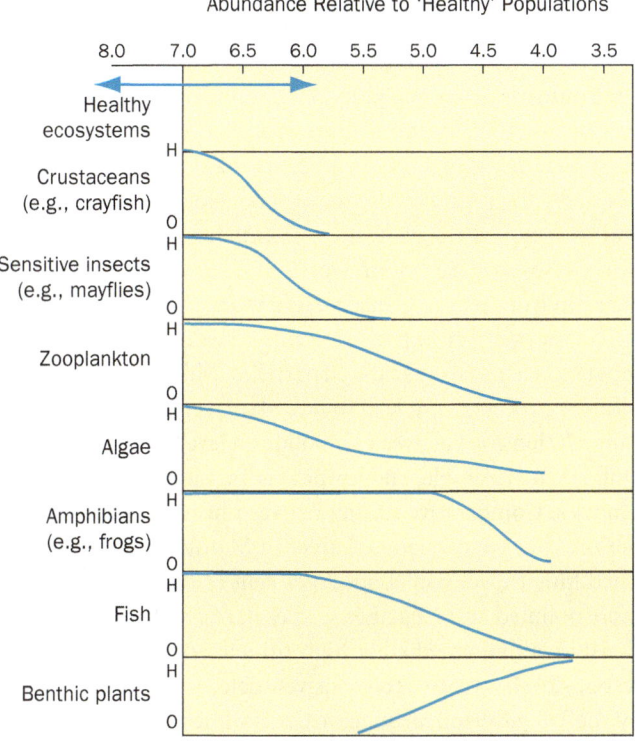

Figure 4.17 | Sensitivity of various aquatic organisms to pH level.
Source: Environment Canada (1991).

The smelter at Trail, BC, caused extensive damage to the surrounding vegetation.

Loons, with their beautiful coloration and haunting cries, are a wilderness symbol for many Canadians, but their breeding success has been reduced by the effects of acid rain.

soil by the acids, may be concentrated by certain species of plants and lichens and accumulated in the livers of species eating them, such as moose and caribou.

Ecosystem Sensitivity

Not all ecosystems are equally sensitive to the effects of acid deposition. The **critical load** is the maximum level of acid deposition that can be sustained in an area without compromising ecological integrity. Some areas have a high capacity to neutralize the excess acids because of the high base capacity of the bedrock and soils. For example, the prairies are not as sensitive to the effects of acid deposition, because underlying carbonate-rich rocks, deep soils, and other factors combine to provide high **buffering capacity**. On the other hand, areas with difficult-to-weather rocks with low nutrient content (e.g., granite) and with thin soils following glaciation often have very low buffering capacities. Much of central and eastern Canada and coastal British Columbia fit within this category (Figure 4.18). The provinces most vulnerable to acid deposition in terms of amount of area ranked as highly sensitive are Quebec (82 per cent), Newfoundland (56 per cent), and Nova Scotia (45 per cent). The spatial coincidence of low buffering capacities and high deposition rates explains why most attention in Canada has centred on the central and eastern parts of the country.

Socio-economic Effects

The environmental effects described above have socio-economic implications, and there are also direct effects of acid deposition on human health. In aquatic ecosystems, for example, declines in fish populations have implications for those involved in fishing, whether commercially or for sport. The impacts of acid deposition on tree growth and the forest industry are also of considerable concern. Some studies indicate that tree growth reductions of up to 20 per cent might be experienced. Preliminary estimates of the market value of lost wood production resulting from acid rain are in the hundreds of millions of dollars in Nova Scotia and New Brunswick alone (Canadian Council of Ministers of the Environment, 2006). The full effects, however, are likely to be much greater than this. Research indicates that a time lag of some 20 to 30 years is likely before the effects of acid rain are reflected in reduced tree growth.

Many values, however, are difficult to express in monetary terms. Thousands of Canadians, for example, maintain lakeside cottages. It is challenging, and some would say impossible, to ascribe a value to the changes that might occur as the lakes become devoid of life. Common loons, for example, are for many the quintessential symbol of the Canadian wilderness, yet populations have been found to be quite sensitive to lake acidification. Below pH levels of 4.5, loons do not seem able to find enough food to feed their young. However, as a result of emissions regulations in Canada and the US, overall improvements in the capacity of many lakes to support aquatic biota are being observed. For instance, a general increase in the number of breeding fish-eating water birds (such as loons) has been observed in lakes in Ontario, Quebec, and Newfoundland, particularly those in close proximity to reduced emission sources. At the same time, algae, invertebrates, and water-bird food chains in many lakes in this region continue to show acidification impacts (i.e., direct effects of acidification, metal toxicity, loss of prey species, and reduced nutritional value of remaining prey), particularly in lakes and rivers where fish communities have been affected.

The human-built environment, as well as the natural environment, is damaged by acid deposition. The acids eat away at certain building materials. The effects have been most damaging in Europe, with its many old monuments, but such effects can also be seen on the Houses of Parliament and nearby statues in Ottawa. Estimates suggest that acid rain causes over $3 billion worth of damage to the human-built environment in Canada every year.

The most direct impact on human health may arise from inhalation of airborne acidified particles, which can impair respiratory processes and lead to lung damage. There are significant relationships between air pollution levels in southwestern Ontario, for example, and hospital admissions for respiratory illnesses. Comparative studies between heavily polluted areas in Ontario and less polluted areas in Manitoba found diminished lung capacity in about 2 per cent of the children in the more polluted areas. Further studies have confirmed this relationship, although whether high sulphate or ozone levels are responsible has yet to be conclusively determined. Calculations by the US government suggest that southern Ontario is now saving more than $1 billion per year in health costs since 2010 as a result of efforts to reduce emissions in the US.

Humans can also be affected by ingesting some of the products of acid rain. Regarding drinking water, for example,

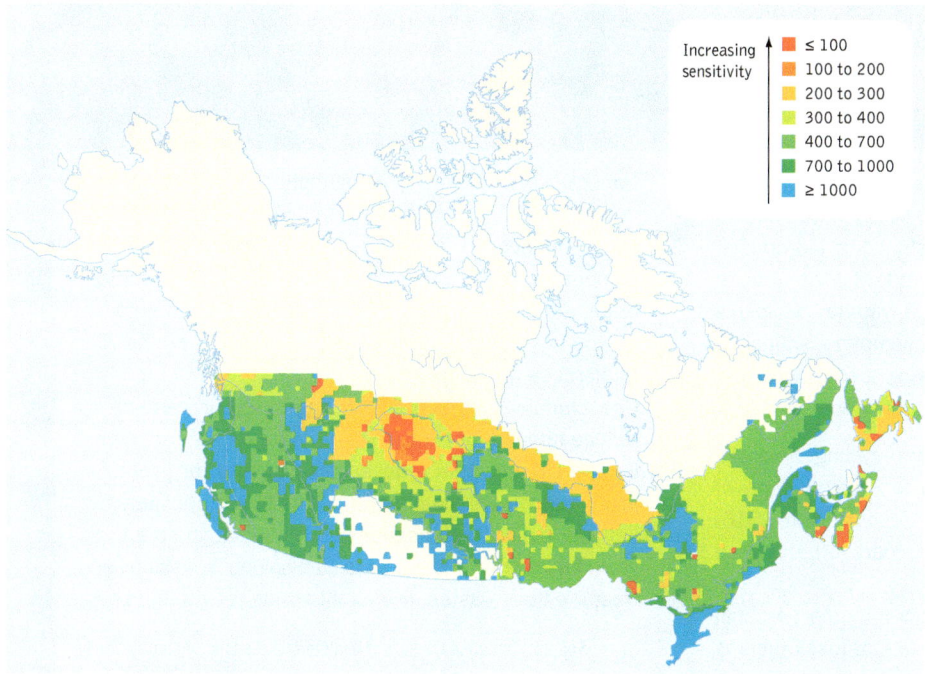

Figure 4.18 | Sensitivity of terrain to acidity. Critical load index, 2008; yellow through red categories are considered acid sensitive terrain. *Source: Federal, Provincial, and Territorial Governments of Canada (2010: 68).*

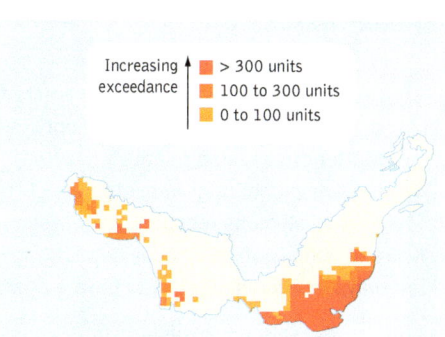

Figure 4.19 | Areas where the critical load has been exceeded in the Boreal Shield. Number of units above critical load, 2009. *Source: Federal, Provincial, and Territorial Governments of Canada (2010: 68).*

in some areas where older delivery systems are still in place, the increased acidity of water can corrode pipes and fittings and result in elevated levels of lead in the water supply. This is one reason why in certain areas, such as Victoria, it is recommended that schools flush their water fountains early in the morning to spill the water that has been held overnight and that might show elevated lead levels.

Acidified water may also hold other substances in solution that are deleterious to human health. Excess levels of aluminum have already been noted. When metals such as mercury, chromium, and nickel are leached from the substrate into water, they may be taken up and concentrated along the food chain and eventually cause a human health problem.

What Can We Do about It?

Perspectives on the Environment

Thresholds, Excesses, and Recovery: Key Finding 13

Thresholds related to ecological impact of acid deposition, including acid rain, are exceeded in some areas, acidifying emissions are increasing in some areas, and biological recovery has not kept pace with emission reductions in other areas.

—Federal, Provincial, and Territorial Governments of Canada (2010: 63).

One of the main challenges associated with acid deposition is that it is not limited to the areas generating the emissions and causing the problem. In Canada, more than half of the acid deposition originates in the US. Furthermore, the acidic fog that forms in the Canadian Arctic in spring is largely a result of emissions from Europe and Russia. To address such problems, international efforts are required, as is a greater concern on the part of individuals (see Box 4.11).

Canada has addressed the problem through national, bilateral, and multilateral efforts. In 1983, the Canadian Council of Resource and Environment Ministers agreed on an annual target deposition or critical load of 20 kilograms per hectare as an acceptable goal, taking political and economic costs into account. This is an important qualification. The **critical load** represents a **policy target value** (PTV) set by politicians. At the time scientists warned that a further 75 per cent reduction in SO_2 emissions would be needed to address the situation adequately. This is a **scientific target value** (STV), and subsequent experience demonstrates that the scientists were correct in their assessment. Many of the environmental problems you will read about in this text represent instances in which PTVs were established that conflicted with STVs. *Total allowable catch* in fisheries (Chapter 8) and *annual allowable cut* in forestry (Chapter 9), as well as designation of endangered species (Chapter 14), are good examples of this conflict.

In 1985, an agreement was reached among the provinces east of Saskatchewan (the area defined as 'eastern' Canada within the context of acid rain) that set specific emission reductions to reach this target. Progress was rapid. The area of eastern Canada receiving 20 kilograms per hectare or more

Environment in Focus
Box 4.11 What You Can Do

Many of the challenges and problems discussed in this book are international in scope and require the co-ordinated efforts of different levels of government, industry, and individuals. Matter cycles are a relatively easy way for individuals to reduce their environmental impacts through the day-to-day decisions that we all make regarding food, water consumption, shopping, and a host of other activities. The following are some of the ways individuals can try to have a positive influence.

1. Recycle your wastes. In BC, just by recycling beverage containers, consumers contributed to the reduction of 135,000 tonnes of carbon dioxide equivalent in 2010. This is equivalent to taking 39,000 cars off the road for a year.
2. Acid deposition is profoundly influenced by the personal decisions we all make regarding our use of fossil fuels in transport and electricity consumption. Think about your decisions and how you can minimize use.
3. Many consumer items, such as TVs, cell phones, computers, and other electronic products, contain materials such as lead and nickel that contain sulphur. As you buy and dispose of these items, you are helping to increase acid deposition.
4. Use chemical fertilizers sparingly on your gardens to reduce impacts of excessive nutrients on water bodies.
5. Eat less meat. This reduces the demand for livestock, and livestock are major contributors to eutrophication.
6. Let your political representatives know that you are in favour of mandatory measures to curb sulphur emissions and treat livestock wastes, even if this costs you more money.

Source: Canadian Council of Ministers of the Environment (2008).

---

of wet sulphate per year declined by nearly 59 per cent from 0.71 million square kilometres in 1980 to about 0.29 million in 1993.

In 1998, the provinces, territories, and federal government signed the Canada-Wide Acid Rain Strategy for Post-2000, committing them to further actions to deal with acid rain. Under this scenario, and assuming a further 50 per cent reduction in US emissions over and above those already committed to, the area receiving more than the critical load would be reduced to 247,000 square kilometres. By 2004, SO_2 emissions were down to 2.3 million tonnes, a 50 per cent reduction from the 1980 level of 4.6 million tonnes (Environment Canada, 2006a). Ontario, Quebec, New Brunswick, and Nova Scotia have announced a further halving of their provincial sulphur dioxide targets by 2010 to 2015, and Nova Scotia has a cumulative reduction target of 75 per cent by 2020 (Environment Canada, 2010).

Advances in science have allowed more comprehensive assessments of critical loads to be defined that combine sulphates and nitrogen oxides (NO_X) for both aquatic and terrestrial ecosystems. Sulphur and nitrogen have different atomic weights, so the combined critical load cannot be expressed as mass units (e.g., kilograms per hectare). They are now expressed as ionic charge balance in terms of equivalent/hectare/year, and the old load of 20 kilograms/hectare/year is represented by 416 equivalent/hectares/year. Current critical loads for the Boreal Shield are shown in Figure 4.19. Calculations show that between 21 and 75 per cent of eastern Canada is still receiving deposits in excess of the critical load. The very wide range represents the uncertainty regarding long-term effects of nitrogen and the degree of future absorption in soils (Environment Canada, 2006a).

Bilaterally, early efforts to convince decision-makers in the US to control emissions met with little success. However, in 1990, intense lobbying efforts came to fruition. The Canada–US Air Quality Agreement was signed to create a flexible and dynamic mechanism for bilateral environmental co-operation in reducing trans-boundary air pollution. The initial focus of the agreement was on reducing emissions of sulphur dioxide and nitrogen oxides, the major contributors to acid rain. Both Canada and the US have surpassed the emission reduction requirements in the agreement.

The US Clean Air Act was revised to cut by half the 1980 sulphur emission levels by the turn of the century, and significant progress has been made (Figure 4.20). Total emissions of SO_2 declined by approximately 40 per cent between 1980 and 2000 and are predicted to decline by approximately 38 per cent from 2000 levels by 2020. Overall, the amount of acid rain falling on eastern Canada originating in both countries is down an estimated 33 per cent since controls were introduced. More than half the eastern Canadian emission reductions have occurred at the smelters at Sudbury, Ontario, and Rouyn-Noranda, Quebec. However, since that time, reductions in Canada have been minimal, with western Canada showing increasing emissions, largely due to emissions from the Alberta oil industry.

Efforts to control the emissions of nitrogen oxides have not been as successful. Between 1980 and 1990, both Canadian and US emissions remained fairly constant, and Canadian emissions show an increase to 2000 (Figure 4.21). Mobile

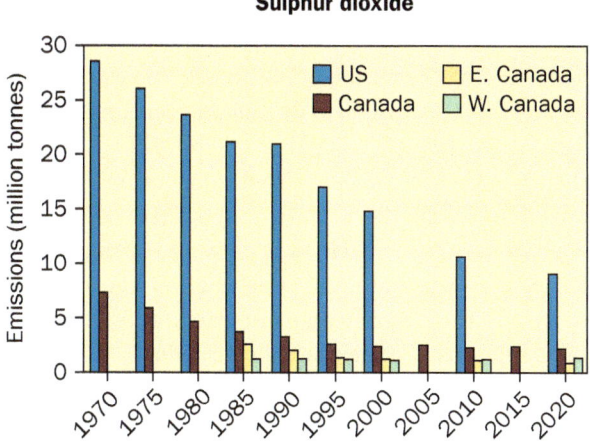
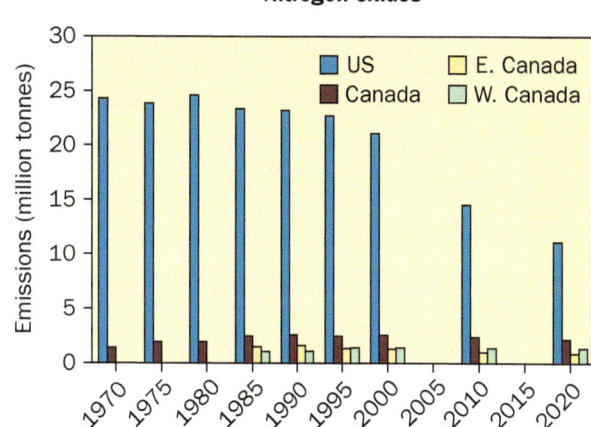

Figure 4.20 and 4.21 | Estimated emissions of sulphur dioxide in the US and Canada and estimated emissions of nitrogen oxides in the US and Canada. *Source: Environment Canada (2006a).*

emission reductions are being sought by introducing more stringent performance standards on exhaust emissions from new vehicles and are predicted to decline by approximately 17 per cent between 2000 and 2020. In eastern Canada, emissions of NO_X are predicted to decrease by approximately 39 per cent between 2000 and 2020. These declines are largely the result of power plant emissions reductions. Emissions from power plants in the Ontario portion of the Pollutant Emission Management Area were approximately 78 kilotonnes in 1990 and had decreased by almost half by 2004. In 2005, the Lakeview Generating Station closed, eliminating annual emissions of approximately 4,000 tonnes of NO_X and 15,000 tonnes of SO_2 upwind of the Greater Toronto Area. Ontario achieved a 12 per cent reduction in electricity use by the government in 2007. In 2010, Ontario's Long-Term Energy Plan was published, setting reduction targets of 6,300 MW by 2025. This plan also aims to double the supply of renewable energy by 2025. Further action in the province to achieve reductions includes agreements to purchase power from 19 new renewable energy projects, including three water power projects, three landfill gas and biogas projects, and 13 wind farms.

However, in the West, NO_X emissions are predicted to increase by approximately 5 per cent between 2000 and 2020. Emissions of NO_X in western Canada surpassed emissions in eastern Canada by the year 2000 and will continue to increase, driven by the oil sands developments in northern Alberta (see also Chapter 12). Greater reductions are predicted for the US, where NO_X emissions are predicted to decline by approximately 47 per cent from 2000 levels by 2020 (Figure 4.20).

An analysis of the US Acid Rain Program estimated annual benefits of the program to both Canada and the United States at $122 billion in 2010 and costs for that year at $3 billion (in 2000 dollars)—a 40 to 1 benefit/cost ratio. These benefits flow from such factors as improved air quality prolonging lives, reducing heart attacks and other cardiovascular and respiratory problems, and improving visibility.

Canada has met all its goals and commitments, some with considerable time to spare, in terms of acid deposition reductions. There has been a decreasing trend in lake sulphate levels in southeastern Canada in response to reductions in SO_2 emissions; however, many of these lakes are still acidified, and many do not meet a pH condition of 6, a key threshold for the sustenance of fish and other aquatic biota. Acid deposition remains a significant problem in Canada and will remain so for years to come. To some extent, the issue has been overshadowed by public and political interest in global climate change, but many causes for concern remain. Emission controls have focused largely on point-source control of sulphate emissions. It is more difficult to address the more diffuse nitrogen derivatives coming mainly from the transportation sector. The pattern of wet nitrate deposition has changed little over the last decade.

Automobile emissions reduce the air quality in many urban areas.

Concern has focused on eastern Canada, but pockets of acidity exist all across the country. The presence of acid-sensitive geology and increasing emissions of SO_2 and NO_X suggest that monitoring should expand into the western provinces to ensure that acid deposition does not damage ecosystems in that region. Large increases in emissions of SO_2 and NO_X from oil sands operations in northern Alberta are raising concerns that acid deposition could negatively impact the West. Between 2000 and 2020, emissions of SO_2 are predicted to decline by 21 per cent in eastern Canada and increase by 15 per cent in western Canada.

Furthermore, although some lake pH levels are showing signs of recovery, it is still too early to assess the biotic implications in terms of community health. Emission targets have been consistently set below levels required to ensure biological recovery. The pH levels are still low enough in many areas to have damaging effects on sensitive species. Similarly, low calcium concentrations in many southeastern Boreal Shield lakes could adversely affect important keystone species such as calcium-rich zooplankton (Jeziorski et al., 2008).

Implications

Understanding the nature of matter and the way that nutrients cycle in the ecosphere is fundamental to appreciating many of the more challenging environmental issues that society currently faces. Acid deposition, eutrophication, and global climate change all have their roots in disruption of biogeochemical cycles. Science is only just starting to unravel some of the secrets of these cycles, but we know enough to understand their significance. Research also clearly shows their complexity and their interconnectedness. We add nitrogen to soils to try to boost productivity and produce more food. However, this also results in eutrophication, depletion of the ozone layer (Chapter 7), and other problems. Clearly, we have to understand the basic science of this interconnectivity. However, this understanding has to be linked to our ability to manage the situation. The chapters in Part C discuss some of the main approaches in environmental planning and management.

Summary

1. Matter has mass and takes up space. It is composed of 92 natural and 17 synthesized chemical elements. The law of conservation of matter states that matter can neither be created nor destroyed but merely transformed from one form into another. Matter cannot be consumed.

2. Elements necessary for life are known as nutrients. They cycle between the different components of the ecosphere in characteristic paths known as biogeochemical cycles.

3. Humans disturb these cycles through various activities, resulting in environmental problems such as acid rain and global warming.

4. Cycles can be classified into gaseous or sedimentary, depending on the location of their major reserves.

5. Phosphorus is an example of a sedimentary cycle. The main reservoir of phosphorus is the Earth's crust. Phosphates are made available in the soil water through erosional processes and are taken up by plant roots and passed along the food chain. There is no atmospheric component to the cycle, making it especially vulnerable to disruption. The main human use for phosphorus is as fertilizer. It is a main cause of eutrophication.

6. Sulphur is also a sedimentary cycle but differs from phosphorus in that it has an atmospheric component. Like phosphorus, it is an essential component for all life. Bacteria enable plants to gain access to elemental sulphur by transforming it to sulphates in the soil. Sulphur is a main component of acid deposition.

7. Nitrogen is a gaseous cycle. Almost 80 per cent of the atmosphere is composed of nitrogen gas, yet most organisms cannot use it as a source of nitrates. Instead, various bacteria help to transform nitrogen into a form that can be used by plants. As with the other cycles, these nitrates are then passed along the food chain. Nitrates are used as fertilizers and contribute to eutrophication. Various nitrous oxides also contribute to acid deposition and the catalytic destruction of ozone.

8. Carbon dioxide constitutes only 0.03 per cent of the atmosphere, but it is the main source of carbon—the basis for life—through the process of photosynthesis. Carbon becomes incorporated into the biomass and is passed along the food chain. Respiration by organisms transforms some of this carbon back into carbon dioxide, and the cellular respiration of decomposers helps to return the carbon from dead organisms into the atmosphere. Carbon dioxide emissions from burning fossil fuels are a main contributor to global climatic change.

9. Water travels between the different components of the ecosphere by means of the hydrological cycle fuelled by energy from the sun. Ninety-seven per cent of water is in

the oceans. Less than 1 per cent is readily available for human use.

10. Canada has up to one-third of the world's fresh water. Most of this is held in the solid phase as ice. Canada also has high storage capacities for liquid water, with lakes covering an estimated 8 per cent of the country. These lakes are replenished by a river flow containing approximately 9 per cent of the total river discharge in the world.

11. Major pollution problems, such as eutrophication, acid deposition, and global warming, are the result of human disruption of biogeochemical cycles.

12. Eutrophication is the nutrient enrichment of water bodies over time. Although it is a natural process, human disruption of the phosphorus and nitrogen cycles has caused a marked acceleration in the rate of eutrophication. This promotes excessive plant growth that leads to oxygen depletion when the plants die and start to decay. Over time, this leads to changes in the composition of fish species and makes water treatment more expensive. The rate of eutrophication can be slowed down by limiting the inputs of nutrients into the water body.

13. Lake Erie is a classic example of eutrophication. Although phosphorus loadings have been reduced considerably, they are still above mandated guidelines and appear to be increasing.

14. Eutrophication is becoming a significant problem in oceanic ecosystems and some 200 'dead zones' are largely associated with terrestrial inputs of phosphates and nitrates.

15. Precipitation tends to be naturally acidic. However, as a result of disturbances in the sulphur and nitrogen cycles, the acidity has increased dramatically over much of Canada during the past few decades. The largest impacts are caused by the burning of sulphur-rich fossil fuels and the smelting of sulphur-rich metal ores. The resulting sulphur dioxide mixes with water in the atmosphere to produce sulphuric acid. Emissions of various nitrogen oxides as by-products of high-temperature combustion account for most of the remainder. The increase in acidity has a damaging effect on aquatic and terrestrial ecosystems as well as on human health.

16. Emission controls have been agreed upon to try to limit these impacts, and targets have been exceeded. However, concomitant improvements in reducing acid rain and the recovery of aquatic systems have not been seen. Many scientists believe that more stringent measures are called for.

Key Terms

acid deposition
acid shock
aerobic
anadromous
anaerobic
aquifer
benthic
biogeochemical cycles
biological oxygen demand (BOD)
buffering capacity
compound
condensation nuclei
critical load
denitrification
eutrophic
eutrophication
evapotranspiration
gaseous cycles
groundwater
guano
hydrological cycle
law of conservation of matter
macronutrients
matter
mesotrophic
micronutrients
mineralization
nitrogen fixation
non-point sources
nutrients
oligotrophic
oxygen sag curve
point sources
policy target value
rainshadow effect
recycling
relative humidity
rock cycle
scientific target value
sedimentary cycles
sublimation
transpiration

Questions for Review and Critical Thinking

1. Summarize some of the key differences and similarities between energy and matter.
2. Why is life dependent on biogeochemical cycles?
3. Explain why decomposer organisms are important in biogeochemical cycling.
4. What are some of the important implications of biogeochemical cycling for forestry and agricultural activities?
5. Outline the main characteristics of the hydrological cycle in Canada.
6. Which biogeochemical cycles are most responsible for eutrophication, and in what ways are they changed by human activities?
7. Are there any eutrophic lakes in your region? If so, what are the main inputs causing eutrophication, and where do they come from?
8. How can eutrophication be controlled? Discuss one example.
9. What is the pH scale, and what is it used for?
10. Which biogeochemical cycle is most responsible for acid deposition, and how do disruptions occur as a result of human activities?
11. What are the effects of eutrophication and acid deposition on aquatic ecosystems? What do you think might be the impacts of both of them together?
12. Are all areas equally sensitive to the impact of acid deposition? If not, what influences the relative vulnerability of different areas?
13. What are the main socio-economic impacts of acid deposition likely to be?

Related Websites

Canadian Council of Ministers of Environment
www.ccme.ca

Environment Canada, acid rain
www.ec.gc.ca/air/default.asp?lang=En&n=AA1521C2-1

Environment Canada, air
www.ec.gc.ca/Air/default.asp?lang=En&n=14F71451-1

Environment Canada, criteria air contaminants and related pollutants
www.ec.gc.ca/Air/default.asp?lang=En&n=7C43740B-1

Environment Canada, water
www.ec.gc.ca/eau-water/default.asp?lang=En&n=65EAA3F5-1

Nitrate levels
www.ecoinfo.ec.gc.ca/env_ind/region/nitrate/nitrate_e.cfm

Take Action for the Environment
www.ec.gc.ca/education/Default.asp?lang=En&n=E413CCE7-1

Further Readings

Note: This list comprises works relevant to the subject of the chapter but not cited in the text. All cited works are listed in the Bibliography at the end of the book.

Commission for Environmental Cooperation. 2011 *Taking Stock*. Montreal: Commission for Environmental Cooperation.

Dalton, H. 2003. 'Nitrogen: The essential public enemy', *Journal of Applied Ecology* 40, 5: 771–81.

Galloway, J.N., and E.B. Cowling. 2002. 'Reactive nitrogen and the world: 200 years of change', *Ambio* 31, 2: 64–71.

Janzen, H.H., et al. 2003. 'The fate of nitrogen in agroecosystems: An illustration using Canadian estimates', *Nutrient Cycling in Agroecosystems* 67, 1: 85–102.

Jeffries, D., et al. 2003. 'Monitoring the results of Canada/USA acid rain control programs: Some lake responses', *Environmental Monitoring and Assessment* 88, 1–3: 3–19.

Keller, W., J.H. Heneberry, and S.S. Dixit. 2003. 'Decreased acid deposit and the chemical recovery of Killarney, Ontario, lakes', *Ambio* 32: 183–7.

Magee, J.A., et al. 2003. 'Effects of episodic acidification on Atlantic salmon (Salmo salar) smolts', *Canadian Journal of Fisheries and Aquatic Sciences* 60, 2: 214–21.

Nasr, M., M. Castonguay, J. Ogilvie, B.A. Raymond, and P.A. Arp. 2010. 'Modelling and mapping critical loads and exceedances for the Georgia Basin, British Columbia, using a zero basecation depletion criterion', *Journal of Limnology* 69 (Suppl. 1): 181–92.

Scott, K.A., B.J. Wissel, J.J. Gibson, and S.J. Birks. 2010. 'Chemical characteristics and acid sensitivity of boreal headwater lakes in northwest Saskatchewan', *Journal of Limnology* 69: 33–44.

Vitousek, P.M., et al. 2003. 'Nitrogen and nature', *Ambio* 31, 2: 97–101.

Winter, J.G., et al. 2002. 'Total phosphorous budgets and nitrogen loads: Lake Simcoe, Ontario (1990 to 1998)', *Journal of Great Lakes* 28, 3: 301–14.

> Plans are nothing; planning is everything.
> —Dwight D. Eisenhower

Part C
Planning and Management: Philosophy, Process, and Product

The chapters in Part B focused mainly on the natural science pertinent to resources and the environment in Canada. In the following two chapters, attention is given to planning and management related to philosophies, processes, and products that can be applied in resource and environmental management. A key point needs emphasis. Often, it is inappropriate to think that humans 'manage' the environment or natural resources. Instead, we usually attempt to manage the *interaction* between humans and the environment. This is why resource and environmental management involves more than application of 'science' or 'technical expertise'. It also requires sensitivity to various—and often different—values, interests, needs, and wants.

Chapter 5 focuses on the 'philosophy' of planning and management relating to resources and environments, while Chapter 6 considers 'processes and products'. Together, they provide an overview of concepts, approaches, and methods to inform the subsequent discussions in Part D on management of various resources.

Chapter 5 begins by considering the concept of *best practice*. It then examines the importance of *context* for a problem-solving situation and the need to be able to design solutions to fit specific situations. The unique characteristics of a given place and time suggest that it is best to develop approaches and solutions specific to a situation. However, when this is done, a potential problem is that some people may perceive other people or regions receiving preferential treatment, since different arrangements are being applied to them. As a result, there often is pressure to use the same or a similar approach in all places, regardless of differences among them. Thus, debate arises over the merits of using a standard approach versus developing different approaches, depending on the circumstances in a particular place.

Best practice also stipulates that we need to understand the distinction among most probable, desirable, and feasible futures and why a *vision* or *direction* needs to be established to help in making choices. Often, those concerned with environmental problems give most of their attention to what is the most likely future and how to deal with it. Understanding what is likely to occur in the future is very important, but it is also important to recognize that the most probable future is not necessarily the most desirable future. Thus, we should have a clear sense of what kind of desirable future we aspire to so that we then can judge, knowing what is likely to occur, whether by intervening it is possible to move closer towards what is desired. A fundamental challenge, of course, is that societies are not homogeneous, and therefore at any given time there may be competing views about what kind of future is desirable. Thus, a major task is to identify, and achieve, a shared vision.

Various possibilities for what could constitute a desired future have been identified, and we focused in Chapter 1 on three of them: *sustainable development*, *sustainable livelihoods*, and *resilience*. In determining what might be a desirable future, it is important to appreciate that basic values shape perspectives. In that spirit, Chapter 5 examines the difference between *biocentric* and *anthropocentric* views. As part of this discussion, Dan Shrubsole draws attention to different interpretations about what constitutes a 'resource' and how that can influence thinking about what are appropriate decisions to take.

A basic concept associated with best practice is a *systems perspective* and how that can be applied as an ecosystem approach. In Chapter 5, discussion focuses on how the ecosystem approach can be translated from concept to practice, not always easy, especially since most often administrative or political boundaries do not reflect or respect ecosystem boundaries. Another best practice element is the need for thinking simultaneously in the short, medium, and long terms. Too often, our society expects instant results or gratification, which places emphasis on the short term. In addition, elections at municipal, provincial, and federal levels usually occur within five years and sometimes as frequently as every two or three years, which further drives attention towards the immediate and short term. The point is not that we should think only in the long term but that we should be thinking simultaneously at several time scales and also have the patience and understanding to recognize that results may take some time to emerge.

Planning and management continue to be the focus in Chapter 6, but the emphasis shifts to processes and products.

For example, if a systems perspective, noted in Chapter 5, is to be used, it requires considerable *collaboration* and *co-ordination* among governments and Aboriginal groups, public agencies, the private sector, and non-governmental organizations. Both are needed to overcome what is often termed the *silo effect*, or the propensity of organizations or individuals to focus only on their own interests and responsibilities and not to consider those of others. One means of achieving collaboration and co-ordination is through *stakeholder* and *participatory approaches* and by continuously seeking to enhance *communication* among participants. Chapter 6 examines alternative approaches for facilitating participation and communication. The challenge is to determine which mix of approaches to use so that the strengths of one can offset the limitations of another.

Best practice increasingly includes *adaptive management*, *impact and risk assessment*, and *dispute resolution*. These processes explicitly recognize that there are high levels of change, complexity, and uncertainty in environmental management, and that conflicts often occur. Therefore, the challenge is not to eliminate or avoid change, complexity, uncertainty, and conflict but to manage within the reality of their presence. *Adaptive management* involves monitoring experience in order to make systematic adjustments as a result of that experience. *Impact and risk assessment* encourages us to be proactive—that is, to look ahead and anticipate positive and negative results from management actions to enhance the positive and mitigate the negative. *Dispute resolution* methods and processes have been created to deal with the presence of different values, needs, interests, and behaviour, which need to be reconciled or at least managed.

A challenge in resource and environmental management is that strategies, programs, and plans have to gain credibility or legitimacy. This is normally realized by having a statutory

Programs such as this tour along Toronto's waterfront, near the Humber Bridge, are made possible with the help of public consultations during the planning process.

or legislative base, political commitment, or administrative endorsement. The more of these factors in place, the more visibility and authority there will be for strategies, programs, or plans. One way to enhance their credibility is to link them explicitly to other management tools that have a statutory base, such as *regional and land-use plans*. When connections are made with such tools or methods, the probability of initiatives being sustained and implemented usually goes up markedly.

Finally, strategies and plans need to be implemented, and there is recognition of factors or variables creating an 'implementation gap'. We need to understand the nature of this 'gap' in order to anticipate and be proactive to minimize the effect of factors that might hinder moving from plans to action.

The discussion of philosophies, processes, and products in Part C provides an overview of elements of best practice relating to resource and environmental management. The purpose of presenting these elements is threefold. First, being aware of them will enable you to examine ongoing initiatives to determine whether they reflect best practice. If they do not, you should consider what would have to be changed for best practice to be achieved. Second, understanding the elements of best practice will allow you to incorporate them into solutions you develop relative to emerging issues. Third, your awareness will enable you to consider how science can be used to provide a more solid foundation for best practices in planning and management.

Chapter 5
Planning and Management Philosophy

Learning Objectives

- To appreciate the significance of different planning and management approaches.
- To understand the importance of context for a problem-solving situation and the need to be able to design solutions to fit particular contexts.
- To appreciate the distinctions among 'government', 'governance', and 'management'.
- To distinguish among most probable, desirable, and feasible futures and understand why it is important to identify a vision or direction to help in making choices about what is the right thing to do.
- To know the difference between 'ecocentric' and 'technocentric' perspectives.
- To understand the significance of a systems perspective and how that can be applied as an ecosystem approach.
- To realize the importance of thinking simultaneously in the short, medium, and long terms.
- To understand the significance of 'social learning' as a foundation for resource and environmental management.
- To appreciate the implications of 'environmental justice' when making resource and environmental decisions

Introduction

Improved resource and environmental management is likely to be achieved if two principles are taken into consideration. First, as discussed in Chapter 1, using science to inform decision-making is desirable. Second, management and decisions should reflect the best planning and management approaches in terms of concepts, processes, and methods. In this chapter, we focus on *philosophy* and consider basic concepts widely accepted for addressing resource and environmental problems. In Chapter 6, attention will turn to *processes, methods,* and *products*, or those processes (such as public participation and community-based approaches), methods (impact and risk assessment), and outcomes (effective, efficient, equitable, implementable) recognized as leading to more effective environmental management.

Awareness of different approaches enables us to assess what is being done in any given situation and to judge whether the approach is likely to be appropriate. The ideas reviewed in Chapters 5 and 6 serve as a lens through which you can see whether the way a problem is being addressed reflects what many professionals around the world consider the most suitable approaches.

Planning and Management Components

Context

The **context**—i.e., the specific characteristics of a time and place—needs to be systematically considered when developing a strategy, plan, or approach for a resource or environmental management problem. Biophysical, economic, social, legal, and political conditions differ from place to place and from time to time, demonstrating that it is usually inappropriate to proceed as though one model or approach were sufficient for every situation. Instead, it is often necessary to custom-design solutions to conditions. In other words, one size does not fit all situations (Box 5.1). Considering context is especially important in a large country like Canada with a wide variety of ecosystems and a rich tapestry of differing cultural perspectives.

> ### Perspectives on the Environment
> #### Management Challenges
>
> Resource management is at a crossroads. Problems are complex, values are in dispute, facts are uncertain, and predictions are possible only in a limited sense. The scientific system that underlies resource management is facing a crisis of confidence in legitimacy and power. Top-down resource management does not work for a multitude of reasons, and the era of expert-knows-best decision-making is all but over.
> Some of the new directions that have been proposed include adopting learning-based approaches in place of set management prescriptions . . . , using a broader range of knowledge . . . , dealing with resilience and complexity . . . , and sharing management power and responsibility. . . .
>
> —Berkes et al. (2007: 308)

The importance of context reaffirms that it is advisable to include local people when developing a strategy or implementing initiatives, since they often have special insight into the conditions of a region or place (discussed further in Chapter 6). This is especially critical in situations involving Aboriginal populations. Understanding the significance of context also means understanding that context can change as conditions evolve, requiring the capacity and willingness to modify strategies and initiatives to ensure they remain relevant. This point is discussed further in Chapter 6 with regard to adaptive management.

While context is important, it is not unusual for public agencies to prefer a standardized approach to problem-solving. The rationale is that a standardized approach is easiest to defend or justify. If every area or region is treated the same way, no region can be perceived as receiving special or preferential treatment. But if a specific approach is designed for one region or place, people in other regions might charge that the region was favoured.

We suggest that despite possible criticism about **custom-designed solutions**, managers should recognize the specific conditions of a place and time and design accordingly. It is better to develop an approach specially designed for the needs of a place and accept criticism about perceived favouritism than to use a standardized approach that forestalls criticism but does not really suit the specific conditions or needs.

Context in the Big Picture

Management of natural resources and the environment involves many organizations and jurisdictions that frequently have overlapping and/or conflicting legal mandates and responsibilities, numerous and often conflicting interests regarding access and rights to environmental systems, and growing skepticism about the formal mechanisms of **government**—local, state, national, regional, international—to deliver services effectively, efficiently, and equitably.

Behind such challenges is the reality that **governance** of resources and the environment takes place in situations defined by high levels of *complexity* and *uncertainty*. Furthermore, managers often deal with rapid change and at the same time may need to become agents of positive change. Finally, managers increasingly deal with *conflict* because of the many different interests related to resources and the environment. As Homer-Dixon (2000: 1) has observed, 'the complexity, unpredictability, and pace of events in our world, and the severity of global environmental stress, are soaring.' In his view, such complexity, unpredictability, and rapid change create an **ingenuity gap**, or growing disparities between those who adapt well and those who do not. Homer-Dixon defines ingenuity as 'ideas for better institutions and social arrangements, like efficient governments and competent governments' (ibid., 3).

> ### Perspectives on the Environment
> #### Governance, Management, and Monitoring
>
> . . . governance is the process of resolving trade-offs and providing a vision and direction for sustainability, management is the realization of this vision, and monitoring provides feedback and synthesizes the observations to a narrative of how the situation has emerged and might unfold in the future.
>
> —Olsson (2007: 269)

Environment in Focus

Box 5.1 Custom-Designing to Deal with Agricultural Diffuse Pollution

Stonehouse (1994) reviewed Canadian experience regarding adoption and use of soil conservation practices, with implications for non-point-source pollution control. He concluded that both cultural and biophysical factors influence conservation decisions by farmers and that each set of factors operates at an individual farm or land-unit level and at a broader regional or national scale. As a result, he recommended that a targeted approach is the best way to apply public policies to help solve degradation problems, in contrast to the dominant universal application approach normally used in Canada. The latter approach assumes that farmers and landowners are homogeneous and that they contribute to degradation in more or less the same way. In contrast, he argued that farmers are heterogeneous regarding their personal characteristics, economic circumstances, and managerial abilities. Furthermore, farms vary in size, natural resource characteristics, and enterprise specializations. Such differences result in wide variations in conservation effort, leading to the need for differential treatment of farmers regarding remedial or preventive public policy measures.

Stonehouse suggested that farmers can be identified as innovators, early adopters, mainstream adopters, laggards, or apathetic types, in descending order of conservation effort. In a targeted approach, the type of farmer is matched with different policy instruments (ranging from voluntary to enforced compliance). For example, for innovators (those most likely to adopt new ideas), some form of public recognition of their stewardship and public leadership would be appropriate, along with, where necessary, information and extension assistance for newly emerging conservation practices. Early adopters, in contrast, would most likely be better assisted by information and extension assistance. Mainstream adopters would need some positive financial incentives, particularly for practices for which profitability was unclear or risk-loading was significant. For laggards, the people with a much lower inclination to adopt conservation measures, much stronger incentives would be required because of their lack of problem awareness, low management skills, economic problems, or other reasons. Thus, the most effective policy approach would be positive financial incentives and necessary information and extension support. Members of the final group, the apathetic, are often aware of degradation problems but are unwilling to take remedial action. For them, strong measures must be used, including regulation and control of farming activities associated with resource depletion, supported where needed by litigation, taxes, and penalties.

Key messages from Stonehouse's evaluation are the need to custom-design solutions, to apply a mix of policy instruments, and to use a targeted rather than a universal approach.

Experience during the past two decades suggests that four other contextual aspects are also important for understanding progress related to managing resources and the environment wisely. One is the preoccupation of many governments with debt and deficit reduction. Since the Rio de Janeiro Earth Summit in 1992, many governments have significantly reduced their allocation of funds to environmental infrastructure and services, which usually has had a serious negative impact on agencies responsible for resources and the environment.

In September 2007, for example, the CBC reported that budget problems at Environment Canada were threatening wildlife programs and services, freezing allotted money for some and reducing funds to zero for others. The Canadian Wildlife Service had its service budget frozen for the rest of that fiscal year, halting all of its scientific field and survey work. The budget for national wildlife areas, which protect bird and other animal habitats, was cut to zero. The Environmental Monitoring and Assessment Network, which tracks changes in ecosystems, lost 80 per cent of its budget. The Migratory Bird Program, which monitors the health of bird populations, had its budget cut by 50 per cent. Although the Environment Minister, John Baird, was not available for comment, his department reported that it was prioritizing spending for climate change initiatives (Chapter 7). This illustrates the way that governments sometimes announce new programs to address environmental challenges but are less forthcoming in detailing what programs might have been cut in order to support the new programs. In many cases, no new money is being allocated to the environment; it is merely shifted from other programs.

Second, and emerging from the concern about debt and deficit reduction, many national and state governments have been (1) downloading responsibilities for environmental services to lower levels of government, which usually do not have the human or financial resources to maintain levels of service; (2) commercializing such services; and/or (3) privatizing these services. Such decisions frequently are taken without rigorous analysis of the capacity of municipal governments or the private sector to take on such responsibilities.

Such initiatives are usually justified by referring to the principle of *subsidiarity* (allocating responsibilities to levels of government closest to where the services are used or

International Guest Statement
Water Governance for the Twenty-First Century
Ali Memon

Water governance is a major 'wicked problem' challenge of the twenty-first century. In order to address deep-seated water conflicts, globalization pressures, and climate change, it is imperative that water allocation and management policies are informed by scientific knowledge and analysis.

Traditionally, scientists have undertaken this function as independent experts in regulatory, and in some instances, 'top-down' or centralized government agencies, in the role of 'speaking truth to power'. However, one of the biggest disappointments with such water governance institutions is the extent to which water priorities are predominantly shaped by vested self-interests. A strongly regulatory planning approach also leads conflicting parties into mutual distrust and strategic posturing, as evident in adversarial science practices exemplified by the processes of New Zealand's Resource Management Act 1991.

A related concern from indigenous perspectives is that modernist water governance institutions in Western societies have been constrained by Eurocentric world views and practices. This arguably restricts the ability of indigenous peoples to manage their lands and resources on the basis of their own endogenous forms of knowledge and understandings. Thus, for example, in regard to water during the pre-colonial period in New Zealand, under *tikanga Māori* (Maori customary practices), the resource was allocated and used in specific ways that recognized its life-giving spiritual force and importance in sustaining aquatic life. These customary practices have been subsequently marginalized in the allocation and management of water resources for such activities as dairy farming.

In response to the above limitations experienced by regulatory water institutions, and reflecting wider societal shifts from *government* to *governance*, various scholars and practitioners have been exploring the potential for devolved, collaborative institutional arrangements for water management, with a reframed role for scientists. Civic science—also known as participatory, citizen, stakeholder, or democratic science—represents a collective ambition for increased public participation and representation in science policy issues. This model anticipates that successful decision and engagement processes should recognize the changed political, social, and scientific realities of contemporary policy-making and legitimates the science for non-science stakeholders (including indigenous peoples), who are also accountable to the wider society, by giving stakeholders responsibility for final policy decisions. The civic science model links scientists in a triangular interaction with both policy-makers and citizens in an attempt to restore public trust in science, to understand more fully the complexity of 'wicked' environmental problems, and to democratize science.

Despite the emphasis in the recent literature on the merits of practising civic science, arguably there is a research gap in our understandings of how well scientists are able to exercise a collaborative role in water governance in diverse scalar settings (local, regional, national), especially in the context of multicultural societies such as Canada and New Zealand. What are the key barriers multiple stakeholders encounter, and how may they be overcome in the global, national, and regional geopolitical settings of socially diverse Western societies? What can be done to assure a greater chance of science impact given current societal conditions and mistrust? Does it require broader adoption of collaborative civic science models as opposed to more traditional science and policy models? Or does the effectiveness of science impact vary less according to the particular 'model' being used and more to certain conditions in place in a particular water governance area?

To address such questions, my current research is focused on studying collaborative regional water governance institutions in New Zealand. We anticipate obtaining innovative insights into the dynamics of effective collaborative water governance.

Ali Memon received his Ph.D. from the University of Western Ontario and was appointed to a Personal Chair in Planning and Environmental Management at Lincoln University in New Zealand in 1999. Prior to that, he was the Director of the Postgraduate Planning Program at the University of Otago, also in New Zealand. His research interests focus on institutional arrangements for urban planning and natural resource management, and he has a distinguished research record.

received) or *efficiency* (providing services at least cost). However, these decisions are often ideological, reflecting a belief in market-based economies as the most effective way to allocate scarce societal resources and that less government involvement is desirable. Thus, while the rationale for the subsidiarity principle is sensible because it allows people closest to the outcome to participate directly in decisions that affect them, it also requires willingness to maintain, create, or enhance human and financial capacity at local levels to deal with resource and environmental management issues.

Guest Statement
Visioning: Wanted from Individuals and Beyond
Dan Shrubsole

If you don't know where you are going, any direction will take you there.

One important aspect for achieving sustainability is establishing a vision statement. Vision statements are desirable at the individual level as well as the government, organizational, and community levels (including the global community) because having an effective personal vision statement is a sign of leadership. If governments, organizations, and communities are to develop and implement a vision, it falls upon people, especially their leaders, to promote and convince others of the urgency and importance of fulfilling this vision. In *Interpersonal Skills for Leadership* (2005), Susan Fritz and her co-authors maintain that all leaders must have many qualities, including a clear and personal vision, because it defines what you really want out of life, gives your life direction and purpose, and allows you to experience life to its fullest potential. In the context of sustainability, if someone's personal vision is convincing, other people may be inspired to change their values and behaviours to become more sustainable. Thus, achieving sustainable development requires leaders who have compelling personal visions.

Developing personal vision statements is something many of us avoid or do not make the time to do. Part of the difficulty lies in not knowing how to develop a personal vision. Fritz and her co-authors suggest the following steps in creating a personal vision statement:

- List five things you are happy about in your life.
- List five things you are committed to in your life.
- List five things you are doing right now that require you to apply your talents.
- Consider your five most important roles in life right now (e.g., son/daughter, brother/sister, student, roommate, employee) and identify five adjectives that describe your behaviour in each one of these roles.
- Identify five roles you may have in the future (e.g., partner, parent, professional, activist, civic leader, volunteer) and identify five adjectives that describe your behaviour in each of these roles.
- List your five priority values or principles in life.
- List your 10 greatest strengths.
- List five things you would like to do to make a difference in the world.
- List five things you really enjoy doing.
- Write a paragraph that you would like someone to read at your funeral.
- Review all your responses above and summarize your 10 life-guiding principles (values) or core beliefs (e.g., fairness, hard work, respect).

Your personal vision statement can be about a page long and will address two key issues: (1) what you want out of life, and (2) what type of person you want to be.

In the context of sustainability, an effective leader, who has a compelling vision statement, who embraces some of the central values of sustainable development (e.g., reducing injustice and achieving equity; meeting basic human needs; a balance among social, environmental, and economic considerations; and increasing self-determination) and who also has the other personal attributes of a leader (e.g., excellent communicator; high level of integrity) is better equipped to engage in the demanding, participatory, and community-based process of establishing vision statements for government agencies, other organizations, or communities. In common with the personal vision statement, these will describe the desired future to be achieved and identify the underlying values to guide the group. In contrast, organizational and community vision statements are often but not always very brief, perhaps one to six sentences.

One of the most powerful personal vision statements was provided by Martin Luther King Jr on 28 August 1963, when he made his 'I Have a Dream' speech in Washington, DC. It can be seen at: www.dailymotion.com/video/x833ml_martin-luther-king-i-have-a-dream-s_news. The level of inspiration and commitment in that vision is similar to what may be required to make meaningful progress in achieving sustainable development. In the absence of this type of strong vision for sustainable development, any decisions we make might not get us there.

Dan Shrubsole is Professor and Chair, Department of Geography, University of Western Ontario. His research focuses on sustainable water management, flood hazards, and environmental planning.

When senior levels of government download their management responsibilities to local levels without regard for the capacity to handle those responsibilities or without consideration as to how the necessary capacity can be created, they may cite subsidiarity as the basis for decisions that in fact may have been driven more by financial cost-cutting considerations. Various ideological perspectives present quite different views of the environment and resources, and these differing views can affect planning and management dramatically.

Third, while many governments favour less government intervention, more reliance on the private sector and market forces to deliver products and services efficiently, acceptance of the value of globalization, and a 'business model' that emphasizes efficiency, results-based management, and tangible products, they demonstrate less interest in using systematic and thorough consultation processes regarding development and implementation of policy. Indeed, in some jurisdictions, transparent, accessible consultation has almost disappeared. In such cases, while the senior levels of government embrace subsidiarity as an appropriate principle to guide decisions, in practice their actions do not reflect adequate consideration of all the preconditions required for subsidiarity to function effectively.

Fourth, since the late 1990s, many governments have been steadily backing away from concern for or commitment to environmental issues and instead have been emphasizing strategies for economic growth. An often-cited example is the decision by President George W. Bush to have the United States withdraw from the Kyoto Protocol for reducing global warming (see Chapter 7) and to refocus energy policies in the US on supply solutions with much less attention to demand management strategies. Ironically, many polls indicate that the general public has become increasingly concerned about environmental issues over the past five years.

As Mascarenhas (2007) has explained, the above characteristics are usually associated with **neo-liberalism**, a political theory based on the belief that humans' well-being is best achieved by encouraging and facilitating individual freedom and minimizing the role of government. Defining features of neo-liberalism are strong private property rights, free markets, and free trade.

Winfield and Jenish (1998) reviewed the implications for resource and environmental management of the neo-liberalism that dominated in Ontario during the mid- and late 1990s. They concluded that neo-liberalism prompted changes that limited or constrained nearly all provincial legislation and regulations related to protection of the environment or management of natural resources, led to major reductions in the key agencies responsible for the environment and natural resources, resulted in a scaling back of opportunities for public participation in decision-making, and brought about a rearrangement of responsibilities for the environment among provincial and municipal governments and the private sector. Much more scope was allowed for voluntary compliance and self-monitoring within the private sector.

Thus, much of what is (or is not) happening in resource and environmental management around the world may be attributed to the shifting influence of differing ideologies. To be informed citizens, planners, or managers, individuals need to understand the basic values and assumptions of various ideologies so they can determine whether the arguments by governments realistically present the rationale for policies and actions. In Box 5.2, the environmental commissioner of Ontario, Gord Miller, identifies some basic challenges and poses fundamental questions that deserve our attention. While he refers to Ontario, his comments can be applied generally across the entire country.

Vision

Ends need to be distinguished from means. In other words, before deciding how to deal with resource and environmental management problems or opportunities, managers should determine what ends or desirable future conditions are sought (Box 5.2). This consideration is often referred to as a need to have a clear vision or sense of direction.

Perspectives on the Environment

You see things and you say 'Why?' But I dream of things that never were; and I say 'Why not?'

—George Bernard Shaw, Irish playwright

According to Nanus (1992: 8–17), a **vision** represents a realistic, credible, and attractive future for a region, community, or group. It is ideal to have a shared vision, one to which many people are committed. Achieving a shared vision is challenging, however, since many interests exist in a society, some of which are mutually exclusive. If a shared vision about a desirable future is to be achieved, it is important to involve stakeholders in the management process, a matter discussed further in Chapter 6.

In Chapter 1, sustainable development, sustainable livelihoods and resilience were examined as possible visions for the future. The examples below illustrate what different organizations or groups have developed in terms of vision.

- *Bay of Fundy Ecosystem Partnership*

    Promoting the ecological integrity, vitality, biodiversity, and productivity of the Bay of Fundy ecosystem, in support of the social well-being and economic sustainability of its coastal communities.

Environment in Focus
Box 5.2 Thinking beyond the Near and Now

When did we become so focused on the present moment and our immediate situation? It seems to me there was a time not so long ago when we were focused as a society on building a better future for our children and our grandchildren. We had a broader sense of the connections between the landscape, the communities of Ontario, and the economy. . . . And we valued that. We worked for a better tomorrow, but at the same time we didn't forget about the past, about those who had made sacrifices and worked hard to create the opportunities we enjoyed.

Somehow the awareness of the past and concern about what the future might become seem to be missing from current public discourse and decision-making. . . .

It's rather like speeding down a dark northern Ontario highway on a moonless June night with only your low beams on. You have that confident, comfortable feeling because there is no one on the road and you're making good time. But if you just click on your high beams, you'll see the moose standing there only a few seconds in front of you. . . .

The analogy of a speeding car works quite well in a discussion about the 'near and now', because as most drivers know, we have a tendency to focus on objects and surroundings just in front of the car. It takes training and discipline to bring your eyes up to the distant horizon where you become aware of events and objects far ahead—in both time and space. But only then can you acquire the capacity to anticipate, plan, and react to future hazards. And so it is with public decision-making that affects the environment. We all must keep our eyes on the horizon.

Source: Miller (2003: 4).

Facilitating and enhancing communication and co-operation among all citizens interested in understanding, sustainably using, and conserving the resource, habitats, and ecological resources of the Bay of Fundy.

—www.bofep.org/vision.htm

- *Fraser River Basin Charter*

The Fraser Basin is a place where social well-being is supported by a vibrant economy and a healthy environment.

—Fraser Basin Council (1997)

- *Toronto and Region Conservation Authority*

Our vision is for a new kind of community. The Living City, where human settlement can flourish forever as part of nature's beauty and diversity.

—Toronto and Region Conservation Authority (2006)

Perspectives on the Environment
A Vision for Your Area or Community

What would be the key features of a vision for resource and environmental management for your college or university, or community, region, or province? Does one exist now?

Visions can be developed in various ways. One is to ask three questions: (1) What is likely to happen? (2) What ought to happen? (3) What can happen? In much of resource and environmental management, the focus is on the first question. Answering it helps to establish the likelihood of some future state, assuming continuation of current conditions or estimating changes in them. However, this question does not to help determine whether the most probable or likely future is also the most desirable. To deal with the issue of desirability, we also need to ask what ought to happen and therefore consider what would be desirable future conditions for a society or place. These two questions (what is most likely, what is most desirable) reflect the difference between forecasting and backcasting, as explained by Tinker in Box 5.3. Finally, the third question imposes discipline by considering what is feasible or practical. We need to address all three of these questions instead of relying only on the first, which is the common practice.

Perspectives on the Environment
Forecasting versus Backcasting

Forecasting takes the trends of yesterday and today and projects mechanistically forward as if humankind were not an intelligent species with the capacity for individual and societal choice. Backcasting sets itself against such predestination and insists on free will, dreaming what tomorrow might be and determining how to get there from today. Forecasting is driving down the freeway and, from one's speed and direction, working out where one will be by nightfall. Backcasting is deciding first where one wants to sleep that night and then planning a day's drive that will get one there.

—Tinker (1996: xi)

Ethics and Values

To ensure that a shared vision is endorsed by a group or society, it must be consistent with and reflect basic ethics and values. Alternatively, a vision may outline a desirable future significantly different from the present situation and, if it is to be achieved, there will have to be a shift in fundamental values. The vision can help to clarify what different values must prevail for the desirable future to be realized.

As Matthews, Gibson, and Mitchell (2007: 337) have observed, an ethic is 'a set or system of moral principles or values that guides the actions or decisions of an individual or group'. Such principles or values help us to determine right from wrong and how to behave appropriately. At the same time, no set of ethics gives every needed direction, and sometimes ethical principles conflict. Furthermore, we normally do not understand or cannot predict all the consequences of our decisions or actions, and even when we can, many anticipated consequences contain 'shades of grey', making it difficult to know how much of a good or bad feature will be present.

It is desirable to have a clearly articulated foundation, based on ethical principles, from which we can make decisions. Ethical principles direct us regarding appropriate behaviour (e.g., always treating others as you like to be treated yourself) and also regarding process issues (e.g., striving for accountability, transparency, and equity). At a minimum, articulating such principles should help us in deciding among different options, assuming that we intend to act in accordance with them. In Chapter 11, we outline current thinking regarding a set of principles to guide decision-making for water.

We also need to be able to appreciate that the values in different societies usually reflect a mix of explicit principles as well as implicit and unstated principles, with the latter generally being understood as common knowledge among all members of the society. The presence of both explicit and implicit values and principles, of course, often raises challenges in cross-cultural situations, when someone from outside a society may not recognize or understand the significance of implicit values unless they have been exposed to that society for a significant length of time.

Two sets of values significant for resource and environmental management can be identified. At one end of a continuum are **ecocentric values**, which include a belief that a harmonious and balanced natural order governs relationships between living things, which humans tend to disrupt through ignorance and presumption. Other key values include reverence for, humility and responsibility towards, and stewardship of non-human as well as human nature. Those with ecocentric values are not against technology per se but favour application of low-impact technology, oppose bigness and impersonality in all forms, and advocate behaviour consistent with ecological principles of diversity and change. The ecocentric viewpoint is comparable to the **biocentric perspective** identified in Chapter 1.

In contrast, a **technocentric perspective** is based on the assumption that humankind is able to understand, control, and manipulate nature to suit its purposes and that nature and other living and non-living things exist to meet human needs and wants. While ecocentrics are concerned about choosing appropriate ends and using consistent means, technocentrics focus more on means because of their confidence in human ingenuity and rights, which makes them less concerned about the moral aspects of activities or consequences. Technocentrics admire the capacity and power of technology and believe that technology and human inventiveness will overcome possible resource shortages as well as remediate or rehabilitate environmental degradation. Technocentrics are similar to anthropocentrics, introduced in Chapter 1; both believe the environment and resources exist primarily to provide direct value for humans.

While recognizing that a spectrum of values is important if we are to understand why ideas are supported or opposed, we should not assume that the values held by a group or society can be neatly allocated into these two categories. Boundaries often are blurred and indistinct, and many people will support certain aspects of values in both categories, depending on the conditions. O'Riordan (1976) shares a story about a man who asked a socially conscious friend what he would do if he had two houses. The friend replied that he would keep one and give one to the state. The man then asked what he would do if he had two cows. Again, the friend answered that he would keep one and give one to the state. The man next asked what he would do if he had two chickens. The friend responded that he would keep them both. When asked why, the friend stated, 'Because I have two chickens.' This story highlights that circumstances (referred to earlier as 'context') can be very important in shaping outlooks about what is needed, important, and desirable. It is important for resource and environmental managers to be aware of dominant and secondary values in a society so they can determine how these values can be used or may have to be altered if a vision is to be defined and then achieved. In his guest statement, Dan Shrubsole provides further insight into the shades of meaning associated with perspectives reflecting different basic values.

Systems and Ecosystem Perspective

The Saskatchewan Round Table on Environment and Economy, in its *Conservation Strategy* published back in 1992, stated that 'ecosystems consist of communities of plants, animals, and micro-organisms, interacting with each other and the non-living elements of their environment.' In the same year, the final report of the Royal Commission on the

Future of the Toronto Waterfront argued that an ecosystem approach emphasizes that human activities are interrelated and that decisions made in one area affect all others. The Royal Commission concluded that 'dealing effectively with . . . environmental problems . . . requires a holistic or ecosystem approach to managing human activities.' These two statements remind us that people have been aware for some time of the value of an ecosystem approach for planning and management. Indeed, most researchers agree that the concept of ecosystem was formulated in 1935 by Arthur Tansley, who argued that organisms were affected by more than other organisms and that other key factors included soils, water, weather, and climate.

Such interpretations of ecosystems and an ecosystem approach reflect the ideas presented in Chapters 2, 3, and 4, which emphasized systems, interrelationships or linkages, energy flows, and ongoing change. In this section, we turn our attention to the characteristics, opportunities, and challenges presented by the systems or ecosystem approach for resource and environmental management.

Characteristics of the Ecosystem Approach

Slocombe (2010: 410) suggests that the ecosystem approach has a set of core characteristics, including systems concepts and analysis, ethical perspectives, stakeholder and public participation, a bioregional place-based focus, efforts to identify and develop common goals, and gaining a systematic understanding of the ecosystem of interest. Slocombe (ibid., 409–10) also argues that the ecosystem approach was developed to address a mix of problems frequently encountered in resource and environmental management. In his view, these problems include:

- people and their activities viewed as separate from nature;
- fragmentation of knowledge or disciplines, as well as of ecosystems, jurisdictions, and management responsibilities;
- single resource uses or economic sectors being emphasized, and conflicts over possible alternative uses being ignored;
- the many ways in which ecological and socio-economic systems are interconnected not being recognized;
- the propensity of biophysical and socio-economic systems to change, sometimes rapidly and unexpectedly, being ignored; and,
- rather than anticipating change and problems, being reactive and attempting to eliminate uncertainty by controlling complex, dynamic systems instead of adapting to them.

Viewed in the context of the characteristics and problems identified by Slocombe, it becomes apparent why the Royal Commission on the Future of the Toronto Waterfront stated that an ecosystem approach:

- includes the whole system, not just parts of it;
- focuses on the interrelationships among the elements;
- recognizes the dynamic nature of the ecosystem;
- incorporates the concepts of carrying capacity, resilience, and sustainability;
- uses a broad definition of environments—natural, physical, economic, social, and cultural;
- encompasses both urban and rural activities;
- is based on natural geographic units such as watersheds rather than on political boundaries;
- embraces all levels of activity—local, regional, national, and international;
- understands that humans are part of nature, not separate from it;
- emphasizes the importance of species other than humans and of generations other than the present;

Aerial view of Grasslands National Park, Saskatchewan.

Aerial view of the Toronto waterfront.

- is based on an ethic in which progress is measured by the quality, well-being, integrity, and dignity it accords natural, social, and economic systems.

Opportunities through the Ecosystem Approach

The above attributes present some challenges to contemporary environmental and resource management. First, in a Western, industrialized society such as Canada, many people believe they have a dominant role relative to nature and that the environment and natural resources exist to satisfy human needs and wants. This is an anthropocentric or technocentric perspective, in contrast to the ecocentric or biocentric world view described earlier. By emphasizing that humans are part of nature rather than separate from it and by recognizing the inherent value of non-human species and things, the ecosystem approach questions such a belief.

Second, by taking a holistic perspective focusing on interrelationships, the ecosystem approach reminds us of the need to consider management problems and solutions in the context of linked 'systems'. It forces us to appreciate that decisions made about one system, such as land, can have consequences for other systems, such as water or wildlife, and vice versa. In contrast, the conventional approach to environmental or resource management has often focused on systems in isolation from one another, as reflected by having one government agency responsible for forestry, another for wildlife, another for water, another for agriculture, and another for urban development.

Third, the ecosystem approach demands that the links between natural and economic or social systems be considered. This focus is also one of the basic thrusts behind sustainable development and resilience. When such linkages are recognized, it becomes apparent that certain thresholds normally exist in natural systems and that exceeding these thresholds leads to deterioration and degradation. For example, agricultural production can be increased by adding chemical inputs (fertilizers, pesticides, herbicides). However, the cumulative effects of agrochemicals may eventually make the product grown (e.g., fruit or vegetables) unsafe for human consumption. Other concerns revolve around introducing chemicals into adjacent environments, resulting in eutrophication and pollution, as discussed in Chapters 4 and 10. This reminds us that while sustainable development accepts the need for development to meet basic human needs, some kinds of growth are not sustainable because they lead to degradation of natural, economic, or social systems. And, they sometimes can make systems more vulnerable to sudden changes or 'flips' when critical thresholds are crossed, making systems much less able to withstand shocks.

Fourth, the holistic perspective reminds us that decisions made at one place or scale can have implications for other places or scales. If a community deposits untreated sewage into an adjacent river, people and communities downstream will bear costs related to that action. Or, if a community, province/state, or nation is unwilling to impose emission reductions on factories, some of the costs will be borne by people, provinces/states, or nations downwind, since air pollutants usually are carried well beyond the borders of the location in which the pollutants are generated, as is the case with acid deposition, discussed in Chapter 4.

Fifth, given the impact of decisions in one place for people and activities in other places, the ecosystem approach raises questions regarding the most appropriate spatial unit for planning and management. The conventional management unit usually has been based on political or administrative boundaries (e.g., municipal, regional, provincial, or national). In contrast, the ecosystem approach suggests that areas identified on the basis of other units, such as watersheds or airsheds, have more functional value. For example, in managing for migratory birds that travel from the Gulf of Mexico to the Canadian Arctic, national boundaries have little relevance. The management area in this situation comprises at least three nations (Canada, the United States, and Mexico).

Sixth, an ecosystem approach highlights that systems are dynamic or continuously changing. An ecosystem, whether a local wetland, prairie grassland, boreal forest, or urbanizing area, is not static. In addition to daily, seasonal, and annual variations, ongoing longer-term changes occur, as illustrated by the transition of natural grasslands to cultivated cropland or of farmland to urban land use. Climate change, discussed in Chapter 7, exemplifies how ecosystems may change and 'migrate' as result of different patterns of temperature, precipitation, and other climate variables. That is why managers and management strategies must be capable of adapting or adjusting to evolving situations. This imperative has led to an interest in *adaptive management*, discussed further in Chapter 6.

Perspectives on the Environment
Spatial Units for Ecosystem-Based Management

At the most basic level, problems derive from the definition of the units for which planning and management are undertaken. Management units often bear no relationship to the realities of ecological problems (even the home range of the species for which protection is sought), their connections to economic and social processes, or local people's cultural and political identity. Instead, they are arbitrary units defined by lines drawn on the map—lines often drawn by someone who has never been to the region and who, for example, decides a river would make a good boundary.

—Slocombe (1993: 616)

In summary, the ecosystem approach insists that humans are part of nature rather than separate from it, that interrelationships must be emphasized, and that critical thresholds exist. When these aspects are combined, it can be appreciated why two decades ago, the Royal Commission on the Future of the Toronto Waterfront (1992: 31–2) concluded that:

> the ecosystem approach is both a way of doing things and a way of thinking, a renewal of values and philosophy. It is not really a new concept: since time immemorial, aboriginal peoples around the world have understood their connectedness to the rest of the ecosystem—to the land, water, air, and other life forms. But, under many influences, and over many centuries, our society has lost its awareness of our place in ecosystems and, with it, our understanding of how they function.

The need and rationale for adopting an ecosystem approach were also spelled out some time ago by the Conservation Authorities of Ontario (1993: 2):

> The fundamental problem that exists in resource management . . . is not financial constraint. It is that the current body of legislation, agency structures, and mandates do not recognize the concept of ecosystem-based management. The overlapping of mandates between the Ministries of Natural Resources, Environment and Energy, Agriculture and Food, and Municipal Affairs, and Conservation Authorities and Municipalities is evident to everyone.

The situation has evolved as public agencies react to specific problems with specific solutions. This issue-by-issue approach results in a situation that, when viewed from an ecosystem perspective, borders on the ludicrous. Thus, many reasons can be identified as to why an ecosystem approach should be used more frequently in Canada for environmental management. However, as we have already pointed out in earlier chapters, implementing an ecosystem approach requires adjustments to arrangements for governance and management systems, aspects considered in more detail in later chapters.

Long-Term View

In resource and environmental management, it is important to have a long-term view (more than 15 years) while also being able to identify actions to be taken in the short term (less than five years) and middle term (five to 15 years). The rationale is that systems often change slowly and that a significant period of time may be required to shift values, attitudes, and behaviour. At the same time, some system changes can occur quickly and with little or no warning, so we need to be flexible and adaptable in our capacity to recognize problems and develop solutions.

Furthermore, many of our environmental problems have emerged after many decades, or even centuries, so it is unrealistic to assume that they can be reversed or 'fixed' in a few years. For example, changes in the aquatic ecology of the Great Lakes (see Chapters 3, 4, and 11) reflect decades of different kinds of land use (agriculture, industry, settlement) on lands adjacent to the lakes. Unfortunately, our society usually wants 'instant results' and does not often show the patience required to deal with problems that have been created over a long period.

There are many reasons for the short-term perspective. One is the relatively short time between elections. Politicians normally want to produce tangible results so they can show what has been accomplished during their term of office and why they should be re-elected. Conversely, they are reluctant to undertake projects that will require the commitment of public funds beyond their term of control. In contrast, those running for office for the first time emphasize how little has been accomplished by those previously elected and why they should be given an opportunity to demonstrate what they could do instead. This mindset drives decisions focusing on short-term, tangible results and usually results in low priority for long-term strategies involving intangible outcomes.

In many instances, changes are required in basic values, attitudes, and behaviour, such as Canadians modifying their basic patterns of activity so that the greenhouse gas emission targets can be achieved (see Chapter 7). There was much outcry from 2002 to 2011 against the targets the federal government committed to reach under the Kyoto agreement, based on fear that they would cause economic hardship or disadvantage. The former Premier of Alberta, Ralph Klein, was one of the most aggressive in criticizing the targets, since they could eventually lead to less reliance on a 'petro-economy', the backbone of the Alberta economy. Klein's criticism reflected the perspective that when changes occur, some gain and some lose. Those who might lose are understandably not usually enthusiastic about the changes.

The focus on tangible results also explains why one of the first casualties of budget cutbacks in the public education system usually is outdoor education and field trips, which are viewed as a luxury. And yet, if young students are to appreciate the role of natural systems and change their attitude towards them, what better way to accomplish that than by exposing the students directly to these systems through such programs? Unfortunately, the 'product' or result of these programs—changed values and attitudes—is difficult to document, and it often takes many years before behaviours are sufficiently influenced and changed that they have significant positive consequences for the environment.

Thus, while a long-term perspective accompanied by patience, perseverance, determination, and commitment is necessary, most people in our society are more preoccupied with shorter-term and visible results. This creates many difficulties for those who believe that solutions require a commitment of funds and human resources over an extended and sustained period of time.

Social Learning

As Diduck (2010: 500) observes, 'Social learning is an emerging framework in resource and environmental management. At its heart is the suggestion that learning is an idea that applies not only to individuals but also to social collectives, such as organizations and communities.' He and others argue that it is not enough to create opportunities for stakeholders to become engaged in resource and environmental management processes. Such processes should be designed so that individuals and organizations learn from their experience and thereby become more knowledgeable and effective in the future.

Based on the 'theory of action', those interested in the concept of **social learning** differentiate between 'single-loop' and 'double-loop' learning. The emphasis in **single-loop learning** is to ensure a match between intent and outcome. A metaphor often used is the thermostat. Reflecting single-loop learning, a thermostat is designed to monitor the temperature, and when the temperature becomes either too cold or too hot, it signals to turn the heat on or off. Thus, the thermostat receives information (about temperature in a room), takes corrective action (turns heat on or off), and in that way ensures an outcome consistent with what is intended or desired.

Double-loop learning, in contrast, addresses a different condition—when there is a mismatch between intention and outcome. Double-loop learning happens by challenging underlying values and behaviour rather than assuming that the prevailing values and behaviour are appropriate. Returning to the thermostat metaphor, if double-loop learning were to occur, the thermostat would question the basic underlying value (why is the desired temperature set for 22°C? why not for 20°C and adapt to a cooler temperature?) or question the prescribed behaviour (rather than requiring more heat, put on a sweater and/or close an open window).

In resource and environmental management, we should be drawing on both single- and double-loop learning—but especially the latter, because it encourages thinking 'outside the box'. In other words, single-loop learning emphasizes the right way to get something done, whereas double-loop learning focuses on the right thing to do.

In the next chapter, we will examine specific approaches to facilitating public participation and, when considering choices in that regard, you should recall the distinction between single-loop and double-loop learning and why we should strive to ensure that conditions are created that encourage both.

Environmental Justice

Environmental justice has been defined by the US Environmental Protection Agency (EPA) as 'the fair treatment and meaningful involvement of all people regardless of race, colour, national origin, or income with respect to the development, implementation, and enforcement of environmental laws, regulations and policies' (www.epa.gov/compliance.basics/ej.html). 'Fair treatment' means that no group of people should bear a disproportionate share of negative environmental consequences from industrial, commercial, or municipal operations, or from the implementation of federal, state, local or tribal policies or programs (EPA, 1997).

The concept of environmental justice was triggered by a protest related to a hazardous waste landfill site in Warren County, North Carolina. The protestors opposed a decision to establish a landfill site for PCB-contaminated soil to be removed from 14 different places in the state and transported for disposal near a small, low-income community whose residents were predominantly African-American. A follow-up study by the US General Accounting Office examined eight southern states to determine if an association existed between the location of locally unwanted land uses (LULUs)

Guiyu, China, has become that country's biggest e-waste destination. Such sites put health at risk through exposure to toxic pollutants as residents attempt to recover valuable metals like gold and copper from discarded electronic devices.

Environment in Focus
Box 5.3 What You Can Do

1. Ask elected officials and designated experts to explain how solutions to problems have been designed with regard to the particular conditions in the area of concern.
2. Identify a desirable future consistent with ideas presented in this chapter and book, and then ensure that your behaviour is consistent with this vision.
3. Urge your college or university to include ideas and practices consistent with ideas in this chapter in its curriculum and practices.
4. Strive to incorporate more ecocentric values into your personal outlook on living.
5. Expect public agencies to use a systems approach when dealing with resource and environmental problems, and challenge them when that does not seem to be the case.
6. Advocate both short-term and long-term thinking and action related to environmental problems.
7. Actively support an NGO that reflects your own values.
8. Seek to identify double-loop rather than single-loop solutions to environmental problems.
9. Advocate for environmental justice in your own community, province, and country.

and the racial and economic status of nearby communities. The report concluded that three of every four such landfills in the US were sited in or close to minority communities. Subsequent studies confirmed this pattern.

Environmental justice is not confined to 'local matters'. Due to increased restrictions on disposal of toxic wastes in developed countries, combined with increased concern about health problems associated with toxic waste sites, alternative disposal sites are often sought in other countries, often in less developed nations. The attraction for the latter is substantial financial compensation for becoming a destination for toxic wastes and opportunities to create employment in building and operating the waste sites.

A similar issue emerges related to climate change. Most developed countries grew their economies using fossil fuels, whose emissions are contributing directly to global warming. Leaders of developing countries reasonably question why they should be restricted in use of fossil fuels, as they now strive to achieve higher standards of living for their citizens. In brief, where is the social and environmental justice in such restrictions?

The attention to environmental justice reminds us that aspirations for sustainable development always need to incorporate social considerations. In addition, the emergence of environmental justice highlights that resource and environmental policy and management decisions often have public health implications, which can and do lead to reduced resilience of local communities.

Implications

The concepts and 'philosophy' that reflect best practice for management of natural resources and the environment are becoming more clearly identified. Important concepts include (1) recognizing the contextual factors that characterize a problem situation and being willing to design solutions that address specific attributes of the problem; (2) establishing a vision that identifies a desirable future so that appropriate means can be identified to achieve the desired end; (3) appreciating the strengths and limitations of potential desirable futures; (4) clarifying underlying values that influence attitudes and behaviour and developing ethical principles or guidelines consistent with the desired future; (5) adapting a systems perspective to ensure that the interactions of various environmental and human subsystems are considered; (6) looking beyond the present and immediate future to consider the longer term; (7) appreciating the significance of social learning; (8) addressing issues related to environmental justice; and (9) recognizing the importance of governance issues. In the spirit of transforming these ideals into action, Box 5.3 outlines initiatives that you can undertake as an individual or as part of a group.

Summary

1. Context refers to the specific conditions related to a time and place. Since context can vary significantly by place and time, it is desirable to design solutions to fit specific environmental problems. A challenge in doing so is that some people may conclude that a region or group is receiving special consideration.

2. Attention to context reinforces the importance of incorporating the experiential knowledge of local people with scientific knowledge when seeking to understand problems and develop solutions. Because they have lived in an area for many years, local people often have insight and understanding that scientists do not.

3. Important aspects of context include rapid change, high complexity and uncertainty, and significant conflict. We should not be surprised to encounter such aspects when dealing with environmental issues.

4. Neo-liberalism has become an increasingly important ideology shaping environmental problem-solving. Characteristics include reducing the scope of government, relying on market mechanisms to allocate scarce resources, and accepting globalization. This ideology has driven many governments to a preoccupation with deficit and debt reduction and to commercialize or privatize management functions previously accepted by governments.

5. The subsidiarity principle, which stipulates that people closest to problems should participate directly in decisions affecting them, encourages involving local people in decisions. It also is used to justify decisions to download or privatize responsibilities.

6. A vision represents a realistic, credible, and attractive future for a region, community, or group. Achieving a shared vision is challenging because many interests—some of which are mutually exclusive—compete in a society. Without a vision or a well-defined end point, managers, their political masters, and society itself have difficulty in determining the most appropriate means for achieving the desired ends.

7. Explicit ethical principles should be identified in order to provide guidance when choosing among options.

8. Ecocentric and technocentric world views reflect different basic values and interests. Their existence is one explanation for the conflicts that arise when decisions must be made regarding the environment or resources.

9. Ecosystems consist of communities of biotic and abiotic elements interacting with each other. Their management requires a systems or holistic perspective.

10. A long-term view (15 years or more) needs to be maintained, but at the same time, there must be a capability to deal with immediate or short-term issues. Too often, only a short-term perspective is taken.

11. Social learning emphasizes the importance of appreciating that both individuals and organizations can and should learn from engagement in resource and environmental management decision-making.

12. Single- and double-loop learning models remind us of the difference between 'doing the thing right' and 'doing the right thing right'.

13. Environmental justice reminds us that some communities or areas are bearing a disproportionate cost related to locally unwanted land uses and that attention must be given to matters of equity as well as of efficiency.

Key Terms

biocentric perspective
context
custom-designed solutions
double-loop learning
ecocentric values
environmental justice
ingenuity gap
LULUs
neo-liberalism
single-loop learning
social learning
technocentric perspective
vision

Questions for Review and Critical Thinking

1. Can science be objective and provide unbiased input to inform management decisions related to natural resources and the environment? Before responding, consider what is meant by the word 'objective'. What are the reasons for your position on objectivity and unbiased evidence?

2. Why do many believe that it is desirable to custom-design solutions to the specific conditions of a problem rather than to design approaches applied uniformly throughout a region?

3. What is the 'ingenuity gap'?

4. Explain why the principle of subsidiarity generates debate.

5. What are the implications of neo-liberalism for resource and environmental management?

6. Why do many consider that establishing a vision is important for resource and environmental management?

7. What might be a vision for resource and environmental management for your community?

8. Do you believe that you are primarily ecocentric or technocentric in your values and behaviour? What are the implications of this view for the way you interact with the natural environment?

9. What are the major obstacles to effective implementation of an ecosystem-based approach? What things must first be changed if implementation is to improve?

10. If there were a model of an ideal ecosystem approach, what characteristics would it have?

11. Why is there a predisposition in governments to favour short-term rather than long-term strategies regarding resources and the environment? What is the best example of a long-term strategy in your community or province?

12. What should be done to facilitate social learning by individuals and organizations?

13. What is the distinction between 'single-loop' and 'double-loop' learning, and what is its significance for resource and environmental management?

14. What are the interconnections between LULUs and environmental justice? What criteria should be used to assess resource and environmental decisions to determine if environmental justice has been achieved?

Related Websites

Environmental Commissioner of Ontario
www.eco.on.ca

Environmental justice
www.epa.gov/compliance/environmentaljustice

Envirolink: The Online Environmental Community
www.envirolink.org

International Union for Conservation of Nature
cms.iucn.org/index.cfm

National Round Table on Environment and Economy
www.nrtee-trnee.ca

Sustainable Business
www.sustainablebusiness.com; www.sustainable.org.nz/cms

Sustainable Livelihoods
www.livelihoods.org/index.html; www.undp.org

United Nations Conference on Sustainable Development 2012, Rio+20
www.uncsd2012.org

Further Readings

Note: This list comprises works relevant to the subject of the chapter but not cited in the text. All cited works are listed in the Bibliography at the end of the book.

Bernstein, S., and B. Cashore. 2000. 'Globalization, fourth paths of internationalization and domestic policy change: The case of ecoforestry in British Columbia, Canada', *Canadian Journal of Political Science* 33: 67–99.

Brown, L.M. 2000. 'Scientific uncertainty and learning in European Union environmental policy-making', *Policy Studies Journal* 28: 576–97.

Bryant, R. 2007. *Environmental Advocacy: Working for Economic and Environmental Justice*. Ann Arbor, Mich.: Bunyan Bryant.

Bullard, R.D. 2005. *The Quest for Environmental Justice: Human Rights and the Politics of Pollution*. Berkeley: University of California Press.

Coward, H., R. Ommer, and T. Pitcher. 2000. *Just Fish: Ethics and Canadian Marine Fisheries*. Social and Economic Papers no. 23. St John's: ISER Books.

Diduck, A., J. Moyer, and E. Briscoe. 2005. 'A social learning analysis of recent flood management initiatives in the Red River basin, Canada', in D. Shrubsole and N. Watson, eds, *Sustaining Our Futures: Reflections on Environment, Economy and Society*. University of Waterloo Department of Geography Publication Series no. 60. Waterloo, Ont.: University of Waterloo, 126–64.

Draper, D., and B. Mitchell. 2001. 'Environmental justice considerations in Canada', *Canadian Geographer* 45: 93–8.

Dwivedi, O.P., et al., eds. 2001. *Sustainable Development and Canada: National and International Perspectives*. Peterborough, Ont.: Broadview Press.

Keen, M., V.A. Brown, and R. Dyball, eds. 2005. *Social Learning in Environmental Management*. London: Earthscan.

Krajnc, A. 2000. 'Wither Ontario's environment? Neo-conservatism and the decline of the Environment Ministry', *Canadian Public Policy* 26: 111–27.

Leeuwis, C., and R. Pyburn. 2002. 'Social learning in rural resource management', in C. Leeuwis and R. Pyburn, eds, *Wheelbarrows Full of Frogs*. Assen, Netherlands: Koninklijke Van Gorcum, 85–105.

Pahl-Wostl, C. 2002. 'Towards sustainability in the water sector: The importance of human actors and the processes of social learning', *Aquatic Sciences* 64: 394–411.

Peterson, B. 2007. 'Using scenario planning to enable an adaptive co-management process in the Northern Highlands Lake District of Wisconsin', in D. Armitage, F. Berkes, and N. Doubleday, eds, *Adaptive Co-management: Collaboration, Learning, and Multi-level Governance*. Vancouver: University of British Columbia Press, 286–307.

Rechtschaffen, C., and E. Gauna. 2002. *Environmental Justice: Law, Policy and Regulations*. Durham, NC: Carolina Academic Press.

Schusler, T.M., D.J. Decker, and M.J. Pfeffer. 2003. 'Social learning for collaborative natural resource management', *Society and Natural Resources* 15: 309–26.

Stewart, J.M. 1993. 'Future state visioning—A powerful leadership process', *Long Range Planning* 26: 89–98.

Zimmermann, E. 1964 [1933]. *Introduction to World Resources*. New York: Harper and Row.

Chapter 6
Planning and Management: Process, Method, and Product

Learning Objectives

- To understand the strengths and limitations of key methods and processes related to resource and environmental management.
- To know the distinction between collaboration and co-ordination.
- To appreciate the benefits and limitations of participatory approaches.
- To understand the obstacles inhibiting communication of scientific results and conclusions to policy-makers and the public.
- To examine why adaptive management has been designed to expect uncertainty and surprise.
- To appreciate the concept of adaptive co-management.
- To understand the connection between risk and impact assessment.
- To understand the role of strategic environmental assessment and sustainability assessment.
- To identify alternative approaches to resolving conflicts.
- To realize the relationship between regional and land-use planning and resource and environmental management.
- To recognize implementation barriers that must be overcome.

Introduction

In Chapter 5, we argued that planning and management can reflect basic philosophies. In this chapter, attention turns to various processes, methods, and products with regard to resource and environmental management. This chapter provides a lot of information. To make it less abstract, we suggest that after reading each section you pause and consider the implications of what you have just read for the Sydney Tar Ponds case study in Chapter 1, as well as what this reading might mean relative to an issue or problem in your own community.

Collaboration and Co-ordination

In Chapter 5, we noted that the systems or ecosystem approach emphasizes that different components of resource and environmental systems are interconnected and, therefore, that a holistic perspective is required. However, for practical reasons, public agencies often focus on a subset of resources or the environment. Hence, we have departments of agriculture, environment, forestry, water, and so on. If a holistic approach is to be taken, there must be **collaboration** among the various agencies and with other stakeholders.

Perspectives on the Environment

Collaboration and Co-ordination

Collaboration: working together
Co-ordination: harmonious adjustment

For Himmelman (1996: 29), collaboration involves exchanging information, modifying activities in light of others' needs, sharing resources, and enhancing the capacity of others to achieve mutual benefit and to realize common goals or purposes. Selin and Chavez (1995) argue that collaboration also involves joint decision-making to resolve problems, with power being shared and stakeholders accepting collective responsibility for the outcomes.

Collaboration is needed within, between, and among organizations. Once collaboration is agreed to, then **co-ordination**—the effective or harmonious working together of different departments, groups, and individuals—can be sought. For example, interdepartmental committees or task forces often provide a way to co-ordinate the activities of different agencies. Such mechanisms provide the means through which effective collaboration can be achieved. Different public participation processes, addressed in the next section, also can be used to facilitate collaboration.

Collaboration increasingly is accepted as desirable, because the complexity and uncertainty associated with resource and environmental issues create a challenge for any individual or organization in terms of having sufficient knowledge or authority to deal with them. Indeed, this is why multi-, inter-, and transdisciplinary research teams are often used, as explained in Chapter 1. Furthermore, differing values and interests contribute to conflict, another reason for various stakeholders to come together to determine how they can meet their respective needs. Done well, collaboration involves sharing information and insight to achieve multiple goals. When accomplished, collaboration can contribute to more open, participatory processes and to solutions to which different stakeholders feel committed. Thus, there is greater likelihood that acceptable solutions leading to effective implementation will be found.

However, collaboration is not always accepted by everyone. Some may decide not to reach out to other groups because they are determined to satisfy their own interests, regardless of what others want. And when such stakeholders are powerful, they may single-mindedly pursue their own interests. Another potential disadvantage of collaboration is that all parties may compromise principles or interests to reach a 'common denominator' that may not always represent a good, long-term decision. Therefore, we should remember that collaboration and co-ordination are means to an end, not an end in themselves. Box 6.1 outlines a notable success in using collaboration to develop a shared vision.

Stakeholders and Participatory Approaches

The Manitoba Round Table on Environment and Economy (1992) argued some time ago that one of 10 basic principles for sustainable development should be *shared responsibility* and that one of six guidelines to achieve sustainable development should be *public participation*. The Manitoba Round Table concluded that it was essential to 'encourage and provide opportunity for consultation and meaningful participation in decision-making processes by all Manitobans' as well as to ensure 'due process, prior notification, and appropriate and timely redress for those affected by policies, programs, decisions, and developments'.

More recently, Manitoba Water Stewardship (2006) declared that *public participation* was one of 13 principles guiding its programs and activities. Specifically, it explained that 'Watershed management and planning will provide forums that will encourage and provide opportunity for consultation and meaningful participation in decision-making processes by Manitobans. Transparent processes and opportunity for information sharing and open communication will be provided.'

As another example, the Nova Scotia Department of Environment and Labour (2002: 7) identified the following principle related to the provincial drinking water strategy: 'the province will foster **partnerships** among individuals, organizations, governments, communities, and businesses for the purpose of developing water management programs. Partnerships enhance communication and support and will make the strategy more effective and more affordable.' And regarding its *Water for Life* strategy, the government of Alberta stated that one of three core areas would be partnerships and that 'citizens and stakeholders will have opportunities to actively participate in watershed management on a provincial, regional, and community basis' (Alberta Environment, 2005).

Here we focus on public participation as a means of reallocating *power* among participants and on alternative ways of facilitating *empowerment* of people relative to the environmental management process.

Degrees of Sharing in Decision-Making

In Chapter 1, the basic characteristics of sustainable development, sustainable livelihoods, and resilience were outlined.

Environment in Focus

Box 6.1 Boundary Waters Treaty, 1909

Canada and the United States share many lakes, rivers, aquifers, and wetlands along their 8,840-kilometre border. Indeed, more than half of the border passes through water bodies. Examples include the Saint John and Saint Croix rivers in New Brunswick, the Great Lakes–St Lawrence River system in Ontario and Quebec, the Souris River in Saskatchewan and Manitoba, the Red River in Manitoba, the St Mary and Milk rivers in Alberta, and the Columbia River in BC.

The two countries signed the Boundary Waters Treaty in 1909 to provide a framework for a collaborative approach and resolution of disputes over water resources shared by the two countries.

The International Joint Commission, or IJC, has six commissioners (three Canadian; three American). The IJC has power to adjudicate regarding issues associated with water development proposals in one country that could affect water levels or flows in the other country. The two countries signed the Boundary Waters Treaty in 1909 to provide a framework and to serve as a commission of inquiry when matters are referred to it by the two federal governments.

The six commissioners are expected to consider what would be best for both countries rather than being advocates for their own country. The record indicates that they have effectively done exactly that.

Several characteristics—achieving inter- and intragenerational equity, increasing self-determination—provide the rationale for sharing power in environmental management. However, redistribution of power for environmental management from government agencies and private firms to members of the public can challenge vested interests and/or undermine regulatory authority. As a result, such redistribution is not always easily or readily accepted. Furthermore, power-sharing raises challenging questions about accountability and responsibility for decisions.

Perspectives on the Environment

Tensions Exist from Use of Participatory Approaches

At a conceptual level, the domains of science and of democratic politics have different goals, standards of merit, norms of participation, and procedures for resolving differences. At a practical level, desired knowledge is often unavailable, and available relevant knowledge is often not adequately used. Knowledge is often inadequate to give high confidence in the consequences of decisions, and decisions sometimes cannot be delayed until high confidence is obtained. Uncertainty is thus unavoidable and pervasive . . . .

—Parson (2000: S128)

Arnstein (1969) provided a perspective still relevant regarding the issue of power redistribution by identifying 'rungs' on the ladder of citizen participation (Table 6.1). Even though the ladder has been modified in many subsequent writings, the essential steps remain.

As long ago as the late 1960s and early 1970s in Canada, public involvement programs had moved up to Arnstein's rungs of informing, consultation, and placation. However, because of a belief that public agencies were accountable for resource allocation decisions and expenditure of public funds, the position usually taken by public agencies was that information and advice received through public participation was only one of several sources to be considered and that the public agency would retain decision-making authority.

Resource and environmental managers believed that they had the legal mandate, responsibility, and power to decide which trade-offs best reflected societal needs and interests and to make final decisions. This viewpoint was usually reinforced because not one public interest but many different interests existed, and frequent conflicts among them were common. Giving responsibility for decision-making to citizens frequently was viewed by public agencies as dangerous, since it could too easily evolve into a form of anarchy in which there would be an absence of government and laws and no one would be responsible or accountable for decisions or behaviour.

During the 1980s, dissatisfaction with many resource and environmental management decisions arose. Growing numbers of Canadians rejected the idea that 'technically correct' answers could always be found. Instead, a prevailing view emerged that such decisions ultimately depended on weighing conflicting goals, aspirations, and values. In these situations, technical or scientific expertise was seen as a legitimate input, but only one of several.

From these considerations arose the idea that 'stakeholders' had a right to participate in decisions. **Stakeholders** are those who should be included because of their direct interest. These stakeholders include: (1) any public agency with prescribed management responsibilities; (2) all interests significantly affected by a decision; and (3) all parties who might intervene in the decision-making process to facilitate, block, or delay it. Because more and more decisions of an

Table 6.1 | Rungs on the Ladder of Citizen Participation

Rungs	Nature of Involvement	Degrees of Power
8. Citizen control		
7. Delegation	Citizens are given management responsibility for all or parts of programs.	Degrees of citizen power
6. Partnership	Trade-offs are negotiated.	
5. Placation	Advice is received from citizens but not acted on.	Degrees of tokenism
4. Consultation	Citizens are heard but not necessarily heeded.	
3. Information	Citizens' rights and options are identified.	
2. Therapy	Power-holders educate or 'cure' citizens.	Non-participation
1. Manipulation	Committees rubber-stamp political decisions.	

Source: Adapted from Arnstein (1969).

increasingly complex nature had to be made, the traditional forum for public participation—the political process with elected representatives reflecting constituents' views—no longer seemed adequate. As a result, various individuals and non-governmental organizations (NGOs) began pressing for public involvement to move higher up on Arnstein's ladder. While few politicians and public servants believed that total 'citizen control' was feasible or desirable, expectations were that 'partnerships' and 'delegated power' that gave effective power to the public were desirable and achievable.

Partnerships among governments, private companies, and the general public have become increasingly popular. The partnership concept has been implemented through **co-management** initiatives and other approaches that reflect a genuine reallocation of power to citizens and away from elected officials or technical experts. Co-management arrangements have been developed particularly with Aboriginal peoples regarding management of forests, fish, and wildlife. In these situations, power is allocated to Aboriginals or other local people. Some of the challenges associated with power-sharing with Aboriginals are highlighted in the context of marine resources in Chapter 8.

Perspectives on the Environment
Co-management in Northern Canada

With diverse institutional structures, integration of traditional knowledge, partnering with local indigenous groups, and successful implementation, northern Canada co-management institutions are building and strengthening cross-scale interaction.

—Trainor et al. (2007: 638)

Communication

At an international conference focused on 'climate change communication', the organizers observed that communication has three main purposes: (1) raise awareness; (2) confer understanding; and (3) motivate action (Scott et al., 2000: iii). Furthermore, at the same meeting, Andrey and Mortsch (2000: WP1) argued that 'communication is thought to be effective only when these changes in awareness and understanding result in attitudinal adjustments and/or improve the basis upon which decisions are made.'

Carpenter (1995) identified important aspects to consider regarding how communication of scientific understanding should be conducted. They continue to be relevant, and include the following:

- Much of the public does not understand science or how scientific research is conducted.
- Other than for weather and gambling, much of the public does not understand 'probability', and the idea of risk as part of life is rejected by many persons. Association and causation are often confused or assumed to be the same.

Stakeholders from various governmental organizations (federal, territorial, and First Nations/Aboriginal) involved in Nunavut take part in a Natural Resources Canada workshop on geomatics products intended to support sustainable development in Canada's North.

- The media do not deal well with the 'ebb and flow' of scientific research, which progresses by new research disproving or challenging existing understanding. This creates confusion and doubt about the authority or credibility of scientists when different views exist.
- In the courts, expert witnesses appear for different sides in a case and present conflicting testimony. Diverging opinions should be understood as a characteristic of scientific 'findings'.

To overcome communication challenges, we must recognize the range of target audiences, such as other scientists, planners and managers, elected decision-makers, and the general public. Consequently, we should be sure that messages are crafted with regard to who the recipients will be and what their level of understanding can reasonably be assumed to be, with as full as possible an appreciation for their mistaken assumptions. To guide the preparation of messages, Carpenter recommended attention to four complementary questions, regardless of audience:

1. What do we know, with what accuracy, and how confident are we about our data?
2. What don't we know, and why are we uncertain?
3. What could we know, with more time, money, and talent?
4. What should we know in order to act in the face of uncertainty?

Perspectives on the Environment
Communicating Uncertainty: Catch-22

If environmental professionals are candid about uncertainties, the client/recipient will likely be disappointed and perhaps berate the professionals for not being helpful and just simply telling them what is going to happen and what to do. If the environmental professionals ignore or obscure the uncertainties and give unambiguous predictions and advice, events may very well show them to be substantially wrong. Then their credibility will be gone, and they will not be consulted in the future.

—Carpenter (1995: 129)

The important message in this section is that while it is important to achieve understanding of natural and human systems and their interactions, it also is important to determine how this knowledge and insight can be shared with others who may not have the same scientific background but are key stakeholders in terms of taking, facilitating, or thwarting action. To test yourself, consider the four questions above in relation to the Sydney Tar Ponds case study in Chapter 1. How would you answer each question and how would you craft your answer as the target audience changes from scientists, to elected decision-makers, to the general public?

Adaptive Management
Surprise, Turbulence, and Change

Trist (1980) observed some 30 years ago that there is no such thing as *the future* but instead there are *alternative possible futures*. Which future actually emerges depends very much on choices made and on actions taken to implement those choices. He argued that 'the paradox is that under conditions of uncertainty one has to make choices and then endeavour actively to make these choices happen rather than leave things alone in the hope that they will arrange themselves for the best.' His conclusion continues to be relevant today.

A challenge in making choices and taking initiatives is what Trist referred to as the *turbulent conditions* that have become increasingly prevalent. For example, energy plans in Canada became obsolete in the early 1970s when OPEC countries rapidly quadrupled oil prices. Such an increase in prices had not been included in the forecasts and assumptions on which energy plans had been based. It is interesting to reflect on whether we have learned much from that experience, considering the rapidly increasing but fluctuating oil prices between 2007 and 2011 (see Chapter 12). Would anyone have been taken seriously in 2007 if they had suggested that crude oil would be selling for an all-time high of $145 a barrel in June 2008, would fall to $30 a barrel by December of the same year, and then would reach $103 in the last week of February 2011? The spike in early 2011 was triggered by unrest in Libya as citizens began the eventual overthrow of Muammar Gaddafi. During the 1980s when decisions were being made about the east coast fishery, there was little expectation that in the early 1990s the cod fishery would be effectively closed and thousands of people in Atlantic Canada would lose their jobs (see Chapter 8). Few anticipated the terrorist attack on the World Trade Center in New York City in mid-September 2001. Today, to what extent is the possibility of terrorist attacks on basic infrastructure such as oil pipelines or water supply reservoirs incorporated into strategies and plans (see Chapter 11 for more detailed discussion)? These and other events, such as the economic 'meltdown' in the autumn of 2008 and the increasing melting of sea ice in the Arctic and Antarctic over the past decade, came as surprises to many people. Such events create bewilderment and anxiety, and raise doubts about the capability of science, planning, planners, managers, and policy-makers, because decision-makers apparently could not anticipate or adapt to rapid change.

Adaptive Environmental Management

Awareness of the need for **adaptive environmental management** was popularized by Holling (1978, 1986) and his colleagues. They concluded that policies and approaches should be able to cope with the uncertain, the unexpected, and the unknown. Holling and his co-workers observed that the customary way of handling the unknown is through trial and error. Errors or mistakes provide new information so that subsequent activity can be modified. Thus, 'failures' generate new information and insight, which in turn lead to new knowledge. However, effective trial-and-error management has preconditions. The experiment should not destroy the experimenter. Or, at a minimum, someone must remain to learn and benefit from the experiment. The experiment also should not cause irreversible, negative changes in the environment. Furthermore, the experimenter should have the will and ability to learn and to begin again.

In Holling's view, a major challenge for the use of adaptive environmental management is that it can be difficult to satisfy the preconditions. For example, the concern about climate change discussed in Chapter 7 reflects the fear that we may not be able to reverse such change before serious problems have occurred. Moreover, even when errors are not irreversible, the magnitude of the original capital investment, as well as personal and political investment in a particular decision or course of action, often makes reversing the process unlikely. Many people simply do not like to admit to or to pay for mistakes.

Adaptive Co-management

Earlier in this chapter, we introduced the concept of co-management, which we now link to adaptive management. Armitage, Berkes, and Doubleday (2007: 1) differentiate between co-management and adaptive management in the following way:

- *Co-management* is primarily concerned with user participation in decision-making and with linking communities and government managers.
- *Adaptive management* is primarily concerned with learning-by-doing in a scientific way to deal with uncertainty.

In their view, when the two concepts are connected, the outcome is a combination of the iterative *learning dimension*

Perspectives on the Environment
Commentary: The Nature of Adaptive Management

Adaptive management is an approach to natural resource policy that embodies a simple imperative: policies are experiments; learn from them. In order to live we use the resources of the world, but we do not understand nature well enough to know how to live harmoniously within environmental limits. Adaptive management takes that uncertainty seriously, treating human interventions in natural systems as experimental probes. Its practitioners take special care with information. First, they are explicit about what they expect so that they can design methods and apparatus to make measurements. Second, they collect and analyze information so that expectations can be compared with actuality. Finally, they transform comparison into learning—they correct errors, improve their imperfect understanding, and change action and plans. Linking science and human purpose, adaptive management serves as a compass for us to use in searching for a sustainable future.

—Lee (1993: 9)

from adaptive management with the *linkage dimension* of collaborative management through which rights and responsibilities are shared jointly. Consequently, several characteristics become prominent, including: (1) learning-by-doing; (2) integration of different knowledge systems; (3) collaboration and power-sharing among community, regional, and national levels; and (4) management flexibility. Plummer and FitzGibbon (2007) have also argued that **adaptive co-management** and social learning (discussed in Chapter 5) complement each other and can be used to mutual benefit. Box 6.2 outlines key features of adaptive co-management.

The concept of visions and visioning was introduced in Chapter 5, and the first point in Box 6.2 refers to the role of a shared vision in adaptive co-management. Olsson (2007: 269–70) argues that visioning processes should promote two

Environment in Focus
Box 6.2 Features of Adaptive Co-management

1. Shared vision, goal, and/or problem definition to provide a common focus among stakeholders and interests.
2. A high degree of dialogue, interaction, and collaboration among participants.
3. Distributed or joint control across multiple levels, with shared responsibility for action and decision-making.
4. A degree of autonomy for different stakeholders at multiple levels.
5. Commitment to the pluralistic generation and sharing of knowledge.
6. A flexible and negotiated learning orientation with explicit recognition of uncertainty.

Source: Armitage et al. (2007: 6).

important qualities: *adaptability*, or the capacity to sustain a system on a desired trajectory or trajectories in the context of changing conditions and disturbances, and *transformability*, or the capacity to create a fundamentally new system when evolving ecological, economic, social, and political conditions make an existing system untenable or unsustainable. The feature of transformability helps achieve resilience, as explained in Chapter 1.

Olsson explains how these two basic ideas (adaptability; transformability) underlay the vision for a wetland area in Sweden: 'to preserve and develop the natural and cultural values of the Kristianstads Vattenrike wetland area, while at the same time making careful and sustainable use of these values, and thus set a good example that can help promote the region' (ibid., 272). To achieve this vision, attention had to be given to building a sense of place and trust among stakeholders, identifying common interests, facilitating learning, encouraging both horizontal and vertical collaboration, and managing conflict.

Among various lessons learned from Kristianstads Vattenrike, Olsson (ibid., 280) highlighted that creating a vision does not ensure successful adaptive co-management. There is always a risk that creating a vision may stimulate other stakeholders to articulate a different vision. Those involved in the Swedish wetland understood this challenge. They dealt with it by initially inviting a small number of people to become engaged and then gradually expanding the number as others became aware of the positive aspects of the vision.

Another lesson learned relates to the power of 'transformational leadership'. The Kristianstads Vattenrike initiative moved forward effectively because one creative and committed individual helped to define the desired direction, aligned people to the vision, and motivated and inspired them to become committed and engaged. All of these accomplishments were important for the ultimate goal of generating movement or change in perceptions, attitudes, and values related to the wetland area. Much of the accomplishment was due to the remarkable leader's ability to recognize windows of opportunity as well as to identify and modify constraints or limitations (such as conflicts of interest, values, and opinions). Of course, the downside of reliance on a transformational leader is that the initiative becomes vulnerable if that individual retires or leaves the region.

In summarizing, Olsson provides valuable insights regarding the role of a vision in achieving adaptive co-management. He observed that it:

> helps develop values and builds motivation for ecosystem management among actors by envisioning the future together and developing, communicating, and building support for this vision. Key factors for the success of this process are dialogue, trust-building, and sense-making. The vision defines an arena for collaboration, frames the adaptive co-management processes, and fosters the development of social networks and interactions among actors, including those dealing with conflict resolution. Visioning processes can therefore be as important as implementation of rules and regulations for framing and directing adaptive co-management. (Ibid., 281–2)

Impact and Risk Assessment

Environmental impacts, intended and unintended, positive and negative, are common to all development initiatives. The use of environmental impact assessment (EIA), starting in the early 1970s, has been a response to the increasing size and complexity of projects, greater uncertainty in predicting impacts, and growing demands by the general public and special interest groups to become more involved in planning and decision-making processes (Lawrence, 2003).

Various definitions of **environmental impact assessment** have been put forward. For example, some time ago, Beanlands and Duinker (1983: 18) defined it as 'a process or set of activities designed to contribute pertinent environmental information to project or programme decision-making'. Twenty years later, Lawrence (2003: 7) provided a more in-depth interpretation when stating that EIA is a process for:

> determining and managing (identifying, describing, measuring, predicting, interpreting, integrating, communicating, involving, and controlling) the potential (and real) impacts (direct and indirect, individual and cumulative, likelihood of occurrence) of proposed (or existing) human actions (projects, plans, legislation, activities) and their alternatives on the environment (physical, chemical, biological, ecological, human health, cultural, social, economic, built and their interrelations).

Initially, EIA emphasized the physical and biological resources that might be affected by a project, with attention to reducing negative consequences. Partly in response to public pressure, the focus gradually broadened to incorporate social concerns, leading to the concept of social impact assessment (SIA). A major contribution regarding inclusion of social considerations into impact assessments came from Canada in the mid-1970s, when Thomas Berger completed his landmark report focused on the proposed Mackenzie Valley pipeline (Berger, 1977). In addition, more attention has been given to basic policy questions, such as establishing the appropriateness of the objectives a project is designed to meet, considering alternative projects that also could meet the same objectives, and examining how compensation could be provided for impacts or losses that cannot be mitigated. Elements of best practice for EIA are highlighted in Table 6.2.

Table 6.2 | Best Practices for Environmental Impact Assessment

1. A strong legal foundation establishes EIA as a mandatory and enforceable process, and one that provides clarity, certainty, fairness, and consistency.
2. A broad definition of the environment and related processes stipulates requirements to ensure EIA is applied to all environmentally significant undertakings.
3. The EIA process identifies the best options rather than merely acceptable proposals, and requires critical examination of purposes and comparative evaluation of alternatives to the initiative as well as of alternative means to undertake the proposal.
4. The EIA process limits ministerial discretion.
5. The EIA process is open and fair, provides a significant role for the public, and contains provisions for public notice, comment, access to information, and participant funding.
6. The EIA process has enforceable terms and conditions for approval of an initiative.
7. The EIA process explicitly addresses monitoring and other post-approval follow-up to ensure terms and conditions are met.
8. The EIA process ensures assessment work is connected to a larger context, including establishment of overall biophysical and socio-economic impacts.

Source: Sinclair and Doelle (2010: 464).

Risk assessment underlies impact assessment, since it focuses on determining the probability or likelihood of an environmentally or socially negative event of some specified magnitude, such as an oil spill, and the costs of dealing with the consequences. Of course, since risks have to be estimated, our calculations may be incorrect. Thus, at the Earth Summit of 1992, the **precautionary principle** was endorsed. This principle states that in order to protect the environment when there are risks of serious or irreversible damage, lack of full scientific certainty regarding the extent or possibility of risk should not be used as an excuse for postponing cost-effective measures to prevent environmental degradation. In other words, decision-makers and managers should estimate risks on the side of caution. The Canadian Environmental Protection Act of 1999 commits the federal government to follow the precautionary principle, as do other statutes, such as the Canada Marine Conservation Area Act of 2002.

Challenges in Impact Assessment

People conducting impact assessment have to balance technical matters and value judgements. Often, there are no right or wrong answers but rather different answers, depending on the starting point for the assessment and the assumptions made.

The Types of Initiatives to Be Assessed

In Canada and most other countries, impact assessments have primarily been conducted for development and waste management projects, especially major capital projects such as dams and reservoirs, nuclear or other types of power plants, oil or natural gas drilling or pipelines, waste disposal facilities, major highways, and runway expansions at major airports. The rationale has been that such development usually has the potential for significant environmental and social impact and that readily identifiable stakeholders—proponents, people, communities—would be affected.

It has been argued for some time, however, that impact assessments could and should be completed for policies and programs. The argument in favour of this approach, called **strategic environmental assessment** (SEA), is that projects often are simply the means of implementing policies and programs. Waiting to conduct an impact assessment until a policy or program evolves into a project means that the assessment may come too late. At the policy or program level, decisions may already have been taken to preclude or eliminate possible alternatives, and the project may become virtually irreversible. Partidario and Clark (2000: 4) define strategic environmental assessment as follows:

> SEA is a systematic, ongoing process for evaluating, at the earliest appropriate stage of publicly

Bruce Nuclear Power Plant, Ontario.

accountable decision-making, the environmental quality, and consequences of alternative visions and development intentions incorporated in policy, planning, program initiatives, ensuring full integration of relevant biophysical, economic, social, and political considerations.

An example of the need to consider policy issues before assessing a project occurred when BC Hydro proposed a pipeline across the Strait of Georgia to provide natural gas to Vancouver Island for generating power. The EIA was to be conducted on the specific impact of the pipeline on the environment. However, environmental groups argued for a much broader EIA that would examine the need for the power and the alternatives available rather than assessing mitigation strategies for one option (i.e., a pipeline).

While it is easy to accept the logic of having impact assessments completed for policies and programs as well as for projects, such assessments can be difficult. Policies can be general or specific, stated or implicit, incremental or radical, independent or linked to other policies. In that regard, strategic environmental assessment is most applicable in three types of situations: (1) sectoral policies, plans, and programs (e.g., mineral extraction, energy generation, waste management); (2) regional or area-based policies, plans, and programs (e.g., land-use plans, development plans); and (3) indirect policies, plans, and programs (e.g., fiscal, trade, or science and technology policy) (Thérival, 1993).

Other challenges arise regarding which activities should be subject to impact assessments. One issue is particularly difficult. On the one hand, society expects government regulations to be reasonable and efficient. In that regard, it is normal to have some lower limit or threshold below which assessments are not required. For example, it is unlikely one homeowner building an outdoor barbecue in her backyard would adversely affect air quality in the community. However, if every homeowner built an outdoor barbecue, air quality could be affected. The issue, then, is how to balance reasonableness and efficiency against the dilemma that many small developments *in aggregate* might have significant implications. This problem has been described as one of *cumulative effects*.

When Impact Assessments Should Be Done

The final report by the federal Environmental Assessment Panel (1991: 2) that reviewed the Rafferty–Alameda Dam and reservoir projects in southeastern Saskatchewan stated that 'environmental impact assessment should be applied early in project planning. That is the intent of both provincial and federal processes.' The panel concluded that the Rafferty–Alameda projects were 'well advanced, however, when both the first and this panel became involved. This put some limits on the usefulness of the review.' Unfortunately, too often in Canada, developments are well advanced before environmental impact assessments are conducted.

EIA should be used jointly with other analyses to determine the appropriateness of development proposals and to help in the design of mitigating measures. However, as with the Rafferty–Alameda Dam, environmental assessments have often been conducted after the basic decision has already been taken to allow the project to proceed. As a result, impact assessment rarely has been used to determine whether a project should be approved; rather, it has become a tool for establishing which mitigation measures could be used to reduce or soften negative impacts.

Determining the Significance of Impacts and Effects

A difficult challenge is to determine the significance or implications of impacts. This issue is not solely scientific or technical. For example, scientific data may be able to determine whether water is too polluted to support fish, but not whether the absence of fish is humanly significant because that is influenced by the prevailing values in a society at a specific place and time. Furthermore, people from different cultural backgrounds or of different value systems and ideologies in the same place and at the same time might have quite

Highway 407 construction, Rouge Valley, Ontario.

Guest Statement
Cumulative Environmental Effects: Thinking beyond the Project in Environmental Assessment
Bram Noble

Each disturbance on a landscape can represent a high marginal cost to the environment—this is the basic concept behind **cumulative environmental effects**, which are the combined effects of an action with the effects of other past actions and that have implications for the present and future. Not only are cumulative effects *additive* because individual actions contribute to incremental levels of change, they are *synergistic* when the total effect becomes greater than the sum of the individual actions. For example, in a typical Canadian watershed a variety of sources of stress affect river system health, including industrial effluent, seepage from municipal sewers, nutrient enrichment due to agricultural runoff, sedimentation caused by forest clearing, and changes to water chemistry due to reservoir creation for hydroelectric generation. When a new mining project is proposed, the effects of that project on the health of the river system act in combination with the effects of all other activities in the watershed.

Environmental assessment is the primary tool in Canada for assessing and managing the effects of development actions. When a project is proposed and the need for an environmental assessment determined, the potential impacts of that project are assessed. A decision is then made as to whether the project might cause significant adverse effects, and whether such effects can reasonably be mitigated. Although many jurisdictions have some requirement to assess the cumulative effects of project proposals, project-based environmental assessment is ill-equipped to deal with cumulative change.

First, many sources of cumulative effects are not subject to environmental assessment. Large-scale and long-term disturbances, such as those associated with agriculture or urban systems, typically do not require environmental assessments. Many small projects, such as natural gas wells and their access roads, often are deemed 'insignificant' and exempt from environmental assessment. Collectively, the effects of many small disturbances can be significant. The Great Sand Hills in Saskatchewan, for example, a 1,940 km^2 area of native grassland, including many rare species, is home to 1,500 natural gas wells and 3,175 km of access roads! None of these developments was deemed individually significant and only five environmental assessments were completed. An independent assessment in 2007 identified considerable cumulative change in the Great Sand Hills, including adverse effects to range health and to biodiversity due to habitat fragmentation.

Second, when project proponents are required to assess the cumulative effects of their projects, experience indicates that they have done a poor job. This is not necessarily the fault of the project proponent. Cumulative effects are complex and often beyond the control of a project proponent.

Consider the above example of a proposed mining operation in an already stressed watershed. Can we truly understand the cumulative effects of mining by applying environmental assessment at the project level? Can the mining proponent reasonably be expected to assess the cumulative effects of all activities in the watershed—past, present, and future—and propose ways to manage them?

The traditional approach to environmental assessment has been to focus on individual projects, mitigating impacts until they are deemed acceptable, rather than also to grapple with cumulative change. However, after decades of 'not seeing the forest for the trees', progress is occurring. In 2009, the Canadian Council of Ministers of the Environment released 'Regional strategic environmental assessment in Canada: principles and guidance', establishing a framework for cumulative effects assessment and to provide direction for regional planning and development decision-making beyond that possible at the project scale. Regional strategic environmental assessment is designed to evaluate the sources of cumulative effects associated with multi-sector land uses and disturbances under different future scenarios, providing a context for project development, reducing uncertainty, and capturing the total effects of actions that individually or collectively may not be subject to environmental assessment. Efforts to advance regional and strategic forms of cumulative effects assessment are ongoing, including initiatives in Canada's western watersheds and in the Beaufort Sea for offshore energy development.

Extensive work is needed to advance the practice of cumulative effects assessment and management. Cumulative effects assessment requires collaboration among scientists, planners, regulators, and development proponents to understand cumulative environmental change, identify disturbances associated with development actions, establish thresholds to guide development, and ensure support for ongoing monitoring and evaluation. Managing cumulative effects requires a focus on ecologically significant boundaries, such as watersheds or eco-regions, and not political boundaries or those of individual projects. We have all the tools needed to make cumulative effects assessment viable. The real challenge lies in changing institutional mindsets and developing the capacity to act.

Bram Noble, Ph.D., is a Professor in the Department of Geography and Planning and the School of Environment and Sustainability at the University of Saskatchewan. His research focuses on environmental assessment, resource planning, and development.

different perspectives regarding what is important to protect or preserve. Accepting that judgements about significance are not strictly technical or scientific matters strengthens the rationale for extending partnerships in environmental management and for ensuring that key stakeholders participate in planning and assessments.

One of the major challenges in determining 'significance' is that the issues in a dispute often do not lend themselves to a monetary valuation; rather, they are characterized by intangible features. What is the value of a wetland that is to be drained to allow the building of a subdivision or to make it possible for a farmer to increase food production? What is the cost to wildlife of disturbance to their habitat resulting from a mine or a pipeline? What is the ecological value of a stand of old white pine or of Douglas fir? Questions such as these pose major challenges for scientists in designing appropriate metrics, which is why they are raised in impact assessments. Usually, no single answer suffices to resolve impact dilemmas; people of differing views and vested interests may come to quite different conclusions about their significance and what action is appropriate.

Inadequate Understanding of Ecosystems

The scientific understanding of ecosystems may be incomplete, as discussed in Chapter 2. Furthermore, information about a specific ecosystem may not be adequate to permit estimates about what might be the effects of human intervention.

Even the most basic ecological concepts are not without problems. For example, some time ago, McIntosh (1980) commented that ideas such as community, stability, succession, and climax were causing (and still do cause) major disagreements among ecologists. In some instances, scientists

On 20 April 2010, an explosion on the *Deepwater Horizon*, a marine oil-drilling platform being operated by British Petroleum (BP), killed 11 workers, injured 17 others, and triggered an oil gusher on the ocean floor that released approximately 4.9 million barrels of crude oil over three months before it was contained. The event, which took place in the Gulf of Mexico near the Mississippi River Delta, is to date the largest oil spill in the petroleum industry's history. The accident had a negative impact not only on wildlife and their habitats, but also on the area's fishing and tourism industries as 6,500 to 180,000 km^2 of the Gulf were affected at various points during the oil's dispersal. Although originally slated for a full recovery by 2012, a spokesperson for the American National Oceanic and Atmospheric Agency (NOAA) has indicated that the area may feel the spill's effects beyond that date. It is also expected that residual oil will remain until it is deteriorated by natural processes such as weathering, evaporation, and microbial activity (www.guardian.co.uk/environment/2011/apr/20/deepwater-horizon-key-questions-answered). The above photos show an aerial view of the Gulf spill (left) and an employee of Plaquemines Parish, Louisiana, using a portable vacuum to collect oil from the water's surface (right).

disagree over terminology and definitions, even after more than three-quarters of a century of research. Some of these difficulties were considered in Chapters 2, 3, and 4. With uncertainty and disagreement over basic ecological concepts, it is understandable why predictions are difficult to make with confidence or why the same data result in different scientific interpretations, especially when such data are incomplete. This situation partly explains why it is not unusual for proponents and opponents at environmental hearings to have their own scientific experts who have reached opposite conclusions about impacts or risks, as noted earlier in the discussion on communicating scientific understanding.

To address in part the problem of incomplete information and understanding, Nakashima (1990) argued that more use should be made of **indigenous knowledge** or traditional ecological knowledge (TEK). He illustrated his argument by discussing the ecological understanding of seabirds by the Inuit in the coastal communities of Inukjuak and Kuujjuaraapik, Quebec, and Sanikiluaq, Nunavut, along southeastern Hudson Bay. He noted that Inuit hunters are positioned to make a contribution to the protection of the environment, since they have accumulated excellent information and insight about the range and behaviour of seabirds as a direct outcome of their hunting lifestyle. Such knowledge can be invaluable in estimating and assessing the potential impact of an oil spill on wildlife.

Incomplete understanding and inadequate data will remain a challenge for those involved in impact assessments in Canada. For this reason, the adaptive environmental management approach is attractive, given its emphasis on learning by trial and error and its acceptance of uncertainty. Such problems also highlight the valuable contribution that indigenous knowledge can make to environmental management.

The Nature of Public Involvement

Public involvement has at least three functions in impact assessment:

1. Including the public helps to make the assessment process fair. Furthermore, decision-making that is accessible to the public enhances the credibility of the process. However, there is always a danger that the public or individuals will be 'co-opted' through such processes.
2. Public involvement helps to broaden the range of issues and potential resolutions, and allows the public to share in devising mitigation measures. In this manner, citizens share in establishing the conditions under which a proposal will be approved.
3. Public participation can contribute to social change. That is, by participating in the decision-making process, the public in a particular place becomes more aware of conditions in their own environment. Such enhanced awareness can lead to new initiatives within the community to identify and to begin to address other problems. This point is often referred to as 'social learning', a concept discussed in Chapter 5.

The Development of Monitoring

Chapters 2, 3, and 4 highlight that many ecological processes unfold over *decades* and that, therefore, the effects of some ecosystem changes may emerge only slowly. Because of incomplete ecological science, it is often difficult to predict which changes may occur in ecosystem structures or processes as a result of development. If we are to improve our knowledge of the resiliency and recuperative powers of ecosystems, monitoring is essential. Monitoring can help to ensure that recommended mitigating measures have actually been implemented. It can track public concerns or fears regarding a project and thereby help to ensure that they are recognized and addressed.

Too often in environmental and resource management in Canada and in other countries, such monitoring is not conducted. It is usually time-consuming and expensive, and

Students participating in the western pond turtle monitoring project, Oregon. It has been estimated that western pond turtle populations have declined by 80 per cent. The intent is to gather information on the quality and quantity of the remaining turtles as a first step to restoration and population recovery.

the results may not become useful until after many years of monitoring. When financial and human resources are scarce, managers may be tempted to reduce or eliminate monitoring and redirect resources to new development activity, as happened with the Experimental Lakes Area in northwestern Ontario (Box 6.3). However, by following that route, we reduce opportunity to learn from past experience and thus make it difficult to apply the adaptive management approach.

Sustainability Assessment

Earlier, we noted growing support for strategic environmental assessment. A more recent innovation is a concept known as sustainability assessment, an extension of sustainable development reviewed in Chapter 1.

Gibson (2007: 73–4, 80–7) reminds us that environmental assessment emerged to change decision-making by ensuring that environmental values are considered along with economic and technological matters, and that sustainability assessment is a more recent innovation. In his words, its core focus is on 'efforts to apply some form of sustainability analysis, appraisal, or assessment or otherwise to adopt sustainability objectives as core guides for evaluations and decisions' (ibid., 81). He further notes that the proliferation of sustainability initiatives can be surveyed readily through any reputable Internet search engine, and we encourage you to do that and determine whether any such activities are taking place in your province, region, or community.

To stimulate your thinking about sustainability assessment, Table 6.3 presents what Gibson refers to as the basic sustainability requirements in the form of criteria for decision-making that can guide sustainability assessments.

If you were to apply the sustainability requirements in Table 6.3 to present or future development proposals in your area, how well would they rate against these criteria, and what changes would you suggest for them? If the same requirements were to be applied in the Sydney Tar Ponds experience, what conclusions would you draw, and what recommendations would you offer? What is your view about the Fish Lake decision, outlined in Box 6.4?

Dispute Resolution

Conflicts and disputes occur for many reasons. They may emerge because of clashing or incompatible values, interests, needs, or actions. In an environmental context, conflicts may arise as a result of either substantive or procedural issues, or both. At a substantive level, disputes may arise about the effects of resource use or project development, about multiple uses of resources and areas, about policies, legislation, and regulations, or over jurisdiction and ownership of resources. At a procedural level, conflicts may occur regarding who should be involved, at what times, and in what ways.

Conflict is not necessarily undesirable. Conflict can help to highlight aspects of a process or system that hinder effective performance. It also can lead to clarification of differences stemming from poor information or misunderstandings. Approached in a constructive manner, conflict can result in creative and practical solutions to problems. On the other hand, conflict can be negative if it breeds mistrust or misunderstanding or reinforces biases. It can be negative if it is ignored or set aside, leading to the escalation of a problem or to the creation of stronger obstacles that must be overcome later.

Environment in Focus

Box 6.3 The Experimental Lakes Research Area

Generating enough electricity to fuel society's demands while minimizing environmental impacts is one of the main challenges we face today. Hydroelectric power has generally been regarded as preferable to coal- and oil-generating stations, which are linked to acid deposition, and to nuclear power plants with their associated difficulties in waste disposal. So it was somewhat of a surprise when researchers at the Experimental Lakes Research Area (ELA) in northwestern Ontario discovered that reservoirs created for hydroelectric generation were responsible for releasing large amounts of carbon dioxide and, which is more troubling, of methane into the atmosphere. The emissions occurred as a result of bacterial decomposition of flooded peat and forest biomass.

This was not the ELA's first finding of global significance. Since its founding in 1968, ELA researchers have made significant contributions to the understanding of eutrophication and acid deposition. Fifty-eight small lakes and their watersheds are used as experimental sites to track the impact of various environmental perturbations. Research on eutrophication at the ELA was instrumental in developing the phosphorus control strategies in the Great Lakes Water Quality Agreement. In 1987, research on acidification was initiated and contributed to new estimates regarding damage to aquatic ecosystems. Again, the results were used as the basis for international accords to limit emissions. These lakes have provided further valuable information as changes in environmental conditions, such as global warming, are tracked over time. In view of the national and global importance of the research at the ELA, it is unfortunate that government funding has been sharply cut back.

Table 6.3 | Sustainability Requirements as Criteria for Sustainability Assessment

1. Socio-ecological system integrity	Build human–ecological relations to establish and maintain the long-term integrity of socio-biophysical systems and protect life-support functions on which human as well as ecological well-being depends.
2. Sufficiency and opportunity	Ensure that everyone and every community has enough resources for a decent life and that everyone has opportunities to seek improvement in ways that do not compromise future generations' possibilities for sufficiency and opportunity.
3. Intragenerational equity	Ensure that sufficiency and effective choices for all are pursued in ways that reduce dangerous gaps in sufficiency and opportunity (and health, security, social recognition, political influence, etc.) between the rich and the poor.
4. Intergenerational equity	Favour present options and actions most likely to preserve or enhance the opportunities and capabilities of future generations to live sustainably.
5. Efficiency	Provide a larger base for ensuring sustainable livelihoods for all while reducing threats to the long-term integrity of socio-ecological systems by reducing extractive damage, avoiding waste, and reducing overall material and energy use per unit of benefit.
6. Democracy and civility	Build the capacity, motivation, and inclination of individuals, communities, and other collective decision-making bodies to apply sustainability requirements through more open and better-informed deliberations, greater attention to fostering reciprocal awareness and collective responsibility, and more integrated use of administrative, market, customary, and personal decision-making practices.
7. Precaution and adaptation	Respect uncertainty, avoid even poorly understood risks of serious or irreversible damage to the foundation for sustainability, plan to learn, design for surprise, and manage for adaptation.
8. Immediate and long-term integration	Attempt to meet all requirements of sustainability together as a set of interdependent parts, seeking mutually supportive benefits.

Source: Gibson (2007: 84).

Environment in Focus

Box 6.4 Fish Lake, BC: Proposed Mine near Williams Lake, British Columbia

In early November 2010, the federal Minister of the Environment announced that the proposed $800 million Prosperity copper–gold mine about 125 km southwest of Williams Lake in the Chilcotin Region of British Columbia was not approved 'as proposed', although the BC provincial government previously had approved it. The minister stated that significant adverse environmental impacts were not acceptable, and mitigation measures could not overcome the anticipated 'severe damage'.

The proposed mine had been viewed by some in BC, and especially in Williams Lake, as a 'lifeline' for the local economy, which depends on the forestry resource and has been struggling due to slumping markets for lumber and the negative effects from the extensive pine beetle infestation (see Chapter 9). Several sawmills in Williams Lake had been closed, causing a significant loss of jobs.

Fish Lake is recognized for its 85,000 rainbow trout. Many other species, including moose, mule deer, grizzly bear, and long-billed curlew, depend on the lake. The Tsilhqot'in Aboriginal people have fished, hunted, and trapped in the area for centuries, and for them the lake is a sacred place.

Taseko Mines proposed to drain Fish Lake. Subsequently, up to 700 million tons of tailings and other waste material containing arsenic, mercury, lead, and cadmium would be placed in a new tailings pond, created by flooding an adjacent smaller lake and associated creek.

The review by the provincial government, completed in January 2010, concluded negative environmental impacts would be 'offset' by significant numbers of new jobs, tax revenue to local and provincial governments totalling millions of dollars, and spinoff benefits.

In contrast, the federal environmental assessment, completed in July 2010, concluded the mine would cause significant damage to fish and fish habitat in Fish Lake. In addition, concern was expressed about negative consequences for 'potential or established Aboriginal rights or title'.

The decision by the federal government to stop development of the copper–gold mine is noteworthy. The Prosperity mine proposal was only the third project, in almost 20 years of environmental assessments under the Canadian Environmental Assessment Act, to be rejected by the federal government. The other two were the Kemess Mine in northern BC and the Whites Point Quarry in Nova Scotia.

Conflicts over resource management can become high-profile news issues in Canada. Some of these conflicts are highlighted in subsequent chapters, such as the conflicts over the seal hunt, Native whaling, and the lobster fishery in New Brunswick, discussed in Chapter 8. One of the largest conflicts over resource management focused on the temperate rain forests of Clayoquot Sound on the west coast of Vancouver Island (Box 6.5). A more localized but still significant conflict at the national level is the dispute between Six Nations people and developers in Caledonia, a community adjacent to the Grand River in southern Ontario. In 1784, the 'Haldimand Tract' was granted to the Six Nations by the British government in appreciation for their support during the US War of Independence and as compensation for having lost their territory south of the Great Lakes to the United States. The original Haldimand Tract, just under 20 kilometres (12 miles) wide along the Grand River from the mouth at Lake Erie for a considerable distance upriver, covered a total of 384,450 hectares. It included the present-day cities of Kitchener, Waterloo, Cambridge, and Brantford.

Over time, much of the land in the tract was sold. For example, in 1798, the tract lands in what is now the Regional Municipality of Waterloo were sold by Joseph Brant, the Mohawk chief. However, with regard to situations such as this, Six Nations traditional chiefs have since argued that Brant did not have the right to sell the land and/or that some of the land was purchased from Native leaders who did not appreciate what the sale involved. As a result, the Six Nations traditional government has registered a land claim for the tract lands and maintains that it has jurisdiction over them. The Six Nations traditional chiefs have not claimed ownership over all the land but insist on being consulted before any new developments occur.

The conflict erupted on 28 February 2006 when a group of Six Nations protestors occupied the Douglas Creek Estates subdivision, a 40-hectare housing development in Caledonia, a small community southwest of Hamilton. The Native protestors moved onto the property and insisted that they would not leave until the land claim was resolved. In 2007, the provincial government bought the property from the developer and awarded $1 million as compensation for businesses in the Caledonia area that had been adversely affected by the protest.

At the one-year anniversary of the occupation in February 2007, Janie Jamieson, a spokesperson for the Six Nations, stated that the occupation represented too much for them to back down. She was quoted as saying that the protestors had

Environment in Focus

Box 6.5 Clayoquot Sound, Vancouver Island

On 13 January 1993, a full-page advertisement appeared in the New York Times with the question: 'Will Canada do nothing to save Clayoquot Sound, one of the last great temperate rainforests in the world?' It was paid for by eight major international conservation groups. Six months later, Greenpeace International in London produced a 17-page colour booklet, *British Columbia's Catalogue of Shame*, outlining the background to the Clayoquot decision and demanding 'an end to the all clear-cut logging in Clayoquot Sound, full inventory of all plants and animals to be carried out, and outstanding Native land claim issues to be settled'. Robert Kennedy Jr flew in as a lawyer representing the Natural Resources Defense Council of the US and promised the support of his organization. A resource and environmental management issue that had made nightly headline news in Canada also resulted in the arrest of more than 800 protestors and captured worldwide attention.

At stake was the future of one of the greatest remaining temperate rain forests of the world and the largest remaining tract of old-growth forest on Vancouver Island, which was slated to undergo forest harvesting. The area was also under Aboriginal land claim negotiations, coveted by the mining industry, an arena for conflict between the rapidly expanding aquaculture industry and fishers and recreationists, and included part of a national park.

The attention drawn to Clayoquot Sound as a result of the protests resulted in several innovations. The area planned for complete protection was expanded. Part of the Sound is now included in an international biosphere reserve recognized by UNESCO. The reserve helps to plan a more sustainable future. Some logging operations are now joint operations between logging companies and Aboriginal groups and are addressing the chronic poverty in some Aboriginal communities. Tourism has continued to increase in the area, again helping to combat poverty and attracting more support for further preservation.

However, conflict is looming once again. Despite booming tourism in the area and the rapid growth of Tofino, few of these economic benefits have been felt by Aboriginal communities. Iisaak Forest Resources, an Aboriginal company, has applied for permits to build roads and for 'helicopter log-drop zones' to prepare for logging in different intact watersheds on Flores Island. Environmental conservation organizations such as the Sierra Club are galvanizing support to oppose the application. There are also differing opinions within Aboriginal communities about the logging, but one factor is common to all sides: the need to do a much better job of distributing tourism benefits if these conflicts are to be resolved long-term.

made it through a cold winter in 2006 and would remain on the land for several more winters if they had to.

In terms of jurisdiction, the Ontario government's position has been that the federal government, through Aboriginal Affairs and Northern Development Canada, has the responsibility and authority to deal with land claims. The federal minister for Aboriginal affairs has maintained that the Haldimand Tract land claim was one of the oldest in Canada, involved many challenging issues, and thus required time and patience to resolve. Negotiations have been underway between the Six Nations and the federal government for many years.

The provincial government and police made a conscious decision by not forcibly removing the Six Nations occupiers from the Caledonia site for fear such action would precipitate

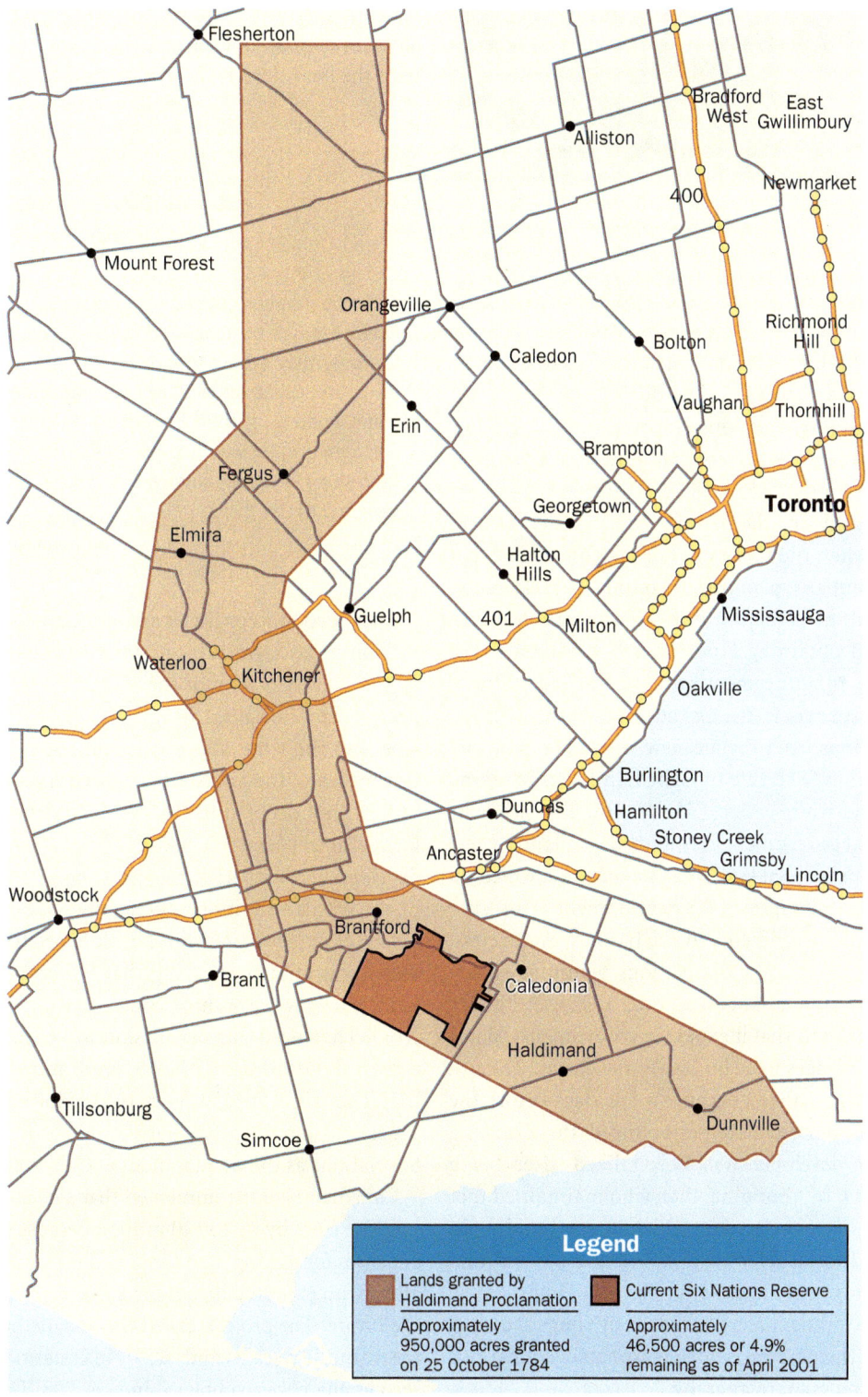

Figure 6.1 | Haldimand Tract.

Caledonia protesters at blockade.

violence. Violence at Oka, Quebec, in 1990 left a police officer dead during a conflict between the Mohawks and the Quebec provincial police (during which the Canadian army was called in), and violence erupted during the Ipperwash Provincial Park dispute when one Native protestor, Dudley George, was shot and killed by an Ontario Provincial Police officer.

Tensions escalated in the Grand River watershed in September 2007 when the traditional chiefs announced that any person or company planning to build on Haldimand Tract land would need approval from the Six Nations. Approval required obtaining a permit and paying a fee of up to $7,000. The announcement prompted a lead editorial in one of the newspapers in the Grand River basin with the headline 'Six Nations can't dictate new laws'. The editorial began by stating that 'The rule of Canadian law is breaking down in the Grand River valley' (*The Record*, 2007). For its part, the provincial government maintained that landowners were governed by provincial land title system rules, which do not require Aboriginal approval for development. This view was shared by local governments in the basin. Nevertheless, some developers had paid the Six Nations planning agency fees ranging from $3,000 to $7,000 for each approval. The Six Nations also announced that it expected governments to pay fees for three new bridges to be built over the Grand River.

In part, concerns escalated because a few days before the Six Nations of the Grand River announced the planning permit system, a developer was hospitalized after being assaulted by a group of Aboriginal men who maintained that they had authority over the land on which he was building in Caledonia. The editorial referred to above commented that 'This kind of outrage should have no place in Ontario. The demand for a development fee, backed up by thugs accountable to no one but themselves, is not a reasonable request; it is blackmail and a precursor to anarchy.'

In November 2009, a couple whose property was adjacent to the Douglas Creek Estates was in court, seeking damages through a civil lawsuit against the Ontario Provincial Police for not having protected their family and property during the protest. An out-of-court settlement was made (Blatchford, 2010). At the end of February 2011, on the fifth anniversary of the occupation, three protestors continued to occupy the land, and scuffles broke out when residents of Caledonia organized a rally to mark the anniversary. The provincial government had not taken a final decision about what to do with the land (ibid.).

Perspectives on the Environment
$20 Million Settlement in Caledonia

In early July 2011, the Government of Ontario paid $20 million to about 440 homeowners and 200 businesspeople as compensation related to the occupation of the Douglas Creek Estates development in Caledonia. Attorney General Chris Bentley stated that while the payment was intended to compensate individuals for losses, the basic land claim underlying the occupation had not been resolved. Furthermore, he was quoted as stating that 'Let's not pretend there are easy answers and solutions to challenging claims such as this.'

—Edwards and Talaga (2011)

The Caledonia conflict is not unique. In August 2007, a $275 million wind farm project started in 2006 near Shelburne in the northern part of the Grand River basin was stalled because an Aboriginal group claimed that it owned both the land and the wind above the land (Burt, 2007a, 2007b). In this instance, the Six Nations served notice of its land claim rather than occupying the site. The developer of the wind farm, Calgary-based Canadian Hydro, planned a two-phased development, with 45 turbines in the first phase. The second phase, with an additional 88 turbines, was challenged.

One of the spokespersons for the Six Nations, Kahentinetha Horn, was quoted as saying that 'We've been robbed of our land and now we want it back. It is our land and it is our wind. They need our permission to use it. We are very concerned about the use of the air, land, and water' (Burt, 2007a: B1). The wind farm project is also located on Haldimand Tract land and hence was subject to the same challenge by the Six Nations as the development at Caledonia.

Canadian Hydro commented that a delay of one year would increase costs by up to $10 million because of the incremental expenses for storing, handling, and transporting the turbines, along with expected higher prices for materials and services in the future. The project, called the Melancthon wind farm and the first utility-scale wind facility in Ontario, continued and the second phase was completed in late 2008. The project contains

133 wind turbines that generate 200 megawatts of power, sufficient to meet the energy needs of about 70,000 average households. In Chapter 12, claims and counterclaims related to health problems resulting from wind farms are presented.

Whether conflict is positive or negative, it is often present because people see things differently, want different things, have different beliefs, and live their lives in different ways. Such differences can be exacerbated by other factors. These factors include lack of understanding of other people or groups; using different kinds and sources of information; differences in culture, experience, or education; and differing values, traditions, principles, assumptions, experiences, perceptions, and biases. In fact, conflicts are a normal part of life, and we need to devise ways to deal with them.

Approaches to Handling Disputes

Disputes usually centre on three main issues: *rights*, *interests*, and *power*. The traditional means of dealing with societal disputes are political, administrative, and judicial. The latter is the most familiar and involves court action. In litigation, the main issues of concern are *fact*, *precedent*, and *procedure*. Attention focuses on establishing a winner or on punishing an offender.

The judicial or litigation approach uses a process that has evolved over centuries. Standards for procedure and evidence are well established. Accountability is ensured through appeal mechanisms and the professional certification of lawyers. However, the judicial process is often viewed as unduly adversarial, time-consuming, and expensive. An adversarial and adjudicative process may also encourage participants to exaggerate their private interests, conceal their 'bottom lines', withhold information, and try to discredit their opponents. In addition, the courts do not always provide a level playing field between, for example, a small group of private citizens or a First Nation opposing a resource project and a large multinational corporation or government agency as the proponent. Because of its greater financial resources, the latter can afford expensive legal expertise and invest in gaining access to and influence within the corridors of political power.

One alternative to the judicial approach is alternative dispute or conflict resolution. **Alternative dispute resolution** (ADR) emphasizes the interests and needs of the parties involved. And while the judicial approach ends by declaring a winner and loser or identifying a party to be punished, the focus of ADR is on reparation for harm done and on improving future conduct. Another key distinction between the two is that the judicial approach emphasizes *argument*, while the ADR approach stresses *persuasion*.

Attributes of Alternative Dispute Resolution

At least six strengths or advantages of ADR can be identified (Shaftoe, 1993):

1. emphasis on issues and interests rather than on the procedures;
2. an outcome normally resulting in a greater commitment to the agreement;
3. attainment of a long-lasting settlement;
4. constructive communication and improved understanding;
5. effective use of information and experts;
6. increased flexibility.

These strengths highlight some limitations of the judicial or court-based approach. However, we should not conclude that legally based approaches are never appropriate for dealing with environmental issues. No approach is perfect, and ADR is no exception. Thus, while in some circumstances ADR

Turbine construction.

The Melancthon wind facility in Shelburne, Ontario.

may be more effective than litigation, the key is to recognize the strengths and weaknesses of each approach and to determine which one would be most effective.

Types of Alternative Dispute Resolution

The features of specific types of ADR approaches include public consultation, negotiation, mediation, and arbitration. Various aspects of *public consultation* were reviewed earlier in the context of the ideas of partnerships and stakeholders. Public participation has been used explicitly since the late 1960s for resource and environmental management in Canada. Initially, such consultation focused on having the public help in identifying key issues and in reviewing possible solutions. This public input was one of many inputs considered by the managers who ultimately determined which trade-offs were appropriate and then made the final decisions. Members of the public had no real power or authority in the management process.

By the mid-1980s, this approach had been modified as public participation moved towards the concepts of partnership and delegated power. Some co-management initiatives illustrated the shift towards giving real power to the public. However, public consultation is not normally considered one of the emerging types of alternative dispute resolution, because all decisions related to the dispute are the exclusive domain of stakeholders in any case.

Negotiation is one of the two main types of ADR. Parties involved in a dispute come together in a voluntary, joint exploration of issues with the goal of reaching a mutually acceptable agreement, and participants can withdraw at any time. Through joint exploration, the parties strive to identify and define issues of mutual concern and to develop mutually acceptable solutions. The normal procedure is to reach an agreement by consensus.

Mediation is the second main type of ADR. Its distinguishing feature is a neutral third party (called a mediator) who helps the disputants overcome their differences and reach a settlement. The third party has no power to impose any outcome. The responsibility to accept or reject any solution remains with the stakeholders in the dispute. In addition, the third party has to be acceptable to all parties in the conflict.

Mediators play various roles. They assist the parties to come together and in this role act as facilitators. They can also help the parties with fact-finding. The mediators do not necessarily have the expertise to provide the needed information, but they can help to identify essential information and then assist in finding it. Mediators help each stakeholder understand the interests and objectives of the other stakeholders, find points in common, and settle differences through negotiation and compromise. Key roles for the mediator include maintaining momentum in the negotiations, keeping the parties communicating with each other, and ensuring that proposals are realistic.

Arbitration differs from negotiation and mediation because it normally involves stakeholders accepting a third party with the responsibility to make a decision on the issue(s) in conflict. In mediation, the third party has no power to impose a settlement. In arbitration, the arbitrator's decision is usually binding on the parties. However, in instances known as 'non-binding arbitration', the arbitrator makes a decision on the conflict, but the stakeholders may accept or reject it. In some ways, the judicial or court-based approach is similar to binding arbitration, since a judge reaches a decision that is imposed on those involved in the dispute. The main difference is that in arbitration, the stakeholders usually have a voice in selecting who the arbitrator will be. In the judicial procedure, the participants play no role in determining which judge will hear their case.

As already noted, public consultation or public participation has long been used for environmental management in Canada, so there has been considerable experience with it. Judicial or court-based approaches have been used ever since the country was settled by people from Europe. Negotiation and mediation are newly emerging approaches, although aspects of negotiation have been used as part of public consultation and judicial approaches. Some of the case studies in Part D provide further details as to how negotiation and mediation are used to address conflicts over the environment. At this point, you might consider how ADR methods could be used to deal with the different interests represented in the Sydney Tar Ponds situation or in the dispute involving the Six Nations, developers, the community of Caledonia, the Ontario Provincial Police, and the provincial and federal governments.

Perspectives on the Environment
Consensus

Traditional [indigenous] society was based upon the principle of consensus for government. . . . Consensus must be the most perfect form of democracy known because it means that there is no imposition of the rule of the majority. Everyone has input and no one is excluded.

—Ovide Mercredi, in Mercredi and Turpel (1993: 115)

Regional and Land-Use Planning

Regional and land-use planning represents a process and a product, and ideally the product (or plan) reflects a vision of desirable development in a region. Resource and environmental managers should be aware of arrangements for regional and land-use planning for several reasons. First, work undertaken to create or update a plan is often directly relevant to resource and environmental management because it can provide valuable information and insight. Second, regional and land-use plans frequently have a statutory or legislative basis and thus govern activity in an area. In contrast, many resource and environmental management plans, such as a watershed plan, usually do not have a legal basis, and this often creates difficulty for effective implementation. Once a resource management plan has been created, it is common for various agencies to have responsibility for implementing specific recommendations. However, such agencies usually have other responsibilities as well, and they must determine what priority recommendations from resource management plans should have. As a result, if resource managers can link their plans to official regional land-use plans, there is greater likelihood that their recommendations will be acted upon.

The same argument can be made with regard to environmental impact assessment. As with regional and land-use plans, environmental impact assessment has a legal basis at both the federal and provincial levels. Thus, the likelihood of action being taken on recommendations in a resource management plan should be higher if they are related to associated impact assessment statements and to regional and land-use plans. Put another way, if such links are not created, it becomes too easy for decision-makers to overlook or ignore resource management plans because they usually do not have any legal underpinning.

In the section on system and ecosystem perspectives in Chapter 5, we noted that ecosystems are dynamic and therefore continuously changing. One implication is that resource management plans need to be updated to ensure their relevance. The notion of monitoring and modifying plans in light of changing circumstances, new knowledge, and lessons learned is also embodied in adaptive environmental management, as already discussed. In most jurisdictions, provision is made for reviewing, updating, and modifying regional and land-use plans. This is another reason why it is important to link resource management plans to them, since both sets of plans can be reviewed and updated at the same time.

Implementation Barriers

Many challenges accompany moving from plans to implementation. While attention has been allocated to determining how to overcome what is referred to as the **'implementation gap'** (Joseph et al., 2006), simple solutions rarely exist. In this section, we identify key matters requiring attention to achieve effective implementation (Mitchell, 2009). Specifically, attention needs to be given to:

1. Recognizing the *context*, and developing a custom-designed solution.
2. Maintaining a *long-term perspective*, usually including initiatives phased over time.
3. Identifying a *vision* regarding the desired future condition to be achieved.
4. Establishing *legitimacy* and *credibility* for the proposed direction, and means to achieve it. This is normally achieved through a mix of legislation, policy, administrative arrangements, and sufficient funding.
5. Ensuring one or more *leaders* or *champions*, especially to keep moving forward when obstacles, setbacks, and disappointments emerge.
6. Facilitating willingness to *share* or *redistribute* power.
7. Establishing a *multi-stakeholder* approach to obtain commitment and buy-in by diverse stakeholders.
8. Building capacity for *adaptability* and *flexibility*.
9. *Monitoring* and *assessing* outputs and outcomes so adjustments can be made.
10. Emphasizing effective *communication* to stakeholders regarding ends, means, and achievements.
11. Using *demonstration projects* to highlight tangible evidence of progress.
12. Profiling and celebrating *accomplishments*, and openly acknowledging those who facilitated the accomplishments.

Addressing the above 12 points does not guarantee effective implementation. Other points could be added and you are encouraged to consider such possibilities. However, experience indicates that attention to these points often does lead to improved implementation.

Implications

The approaches discussed in this chapter represent what many would view as ideals for resource and environmental management regarding processes, methods, and products. Some pose fundamental challenges to us as individuals and society. For example, basic values in Western industrialized societies (and many others) include self-interest and competition. It is often assumed or believed that scarce resources—natural and/or human—are allocated most efficiently and equitably through competition. It is thought that as each agent—individual citizens, interest groups, corporations, resource managers—pursues his or her or their own best interests, social balance will be ensured as governments mediate the process in the greater public interest. However, the concepts of *collaboration*, *co-ordination*, *partnerships*, and *stakeholders* suggest a different paradigm, one based on a willingness to recognize the legitimacy of many interests and needs and to try to accommodate them. In a similar manner, *alternative dispute resolution* is predicated on the idea of groups working together for mutual gain. In a different way, *adaptive management* rejects a belief that humans can

International Guest Statement
Spatial Planning and Disaster Risk Reduction in Indonesia: Mediating Long-Term and Short-Term Interests
Bakti Setiawan

As a professor in a graduate program of urban and regional planning, I frequently ask whether planning, particularly spatial planning, contributes significantly to solve major problems faced by Indonesia.

Planning has been practised in Indonesia since the 1960s. As a result, one would expect that it has contributed to improving the lives of the more than 240 million people in this archipelago country. The reality, however, tells a different story. Many cities and regions in Indonesia have critical environmental problems; living conditions of millions of urban residents are poor; and, communities seem to be becoming more vulnerable to natural disasters.

Mount Merapi (a) and the destruction left in the wake of its November 2010 eruption (b)–(e).

Questions about the role of planning are becoming more relevant in the context of natural disasters that have brought unprecedented losses and damages. For example, on the morning of 26 December 2004 in Aceh province, a massive earthquake (9.0 Richter) hit the region and was followed by a massive tsunami that devastated the human population living on the coastline of Aceh province and parts of North Sumatera province. The impact was huge: more than 200,000 people died and more than 700,000 were displaced.

In May 2006, a big earthquake (6.3 Richter) rocked Yogyakarta, taking the lives of 6,716 people and destroying more than 200,000 houses in the region. Another earthquake occurred in West Sumatera in October 2009 (7.6 Richter), causing more than 1,000 deaths and destroying at least 135,000 houses and public buildings. A final example was the disaster that hit Yogyakarta in November 2010, resulting from the eruption of nearby Mount Merapi, its worst eruption since 1870. Continuing eruptions from the volcano destroyed six villages and took the lives of at least 277 people and created 71,579 refugees. How to explain such intensive and massive natural disasters? And what should be the contribution from urban and regional planning to mitigate such destruction? And particularly for planners, to what extent do they consider such disaster potential when developing spatial arrangement of land uses for cities and regions?

Spatial planning is concerned with the efficient use of spaces. It deals with the physical structures and forms of the city and regions, particularly arrangements for land use and zoning. Planners believe that a 'blueprint' for our future, including our cities and regions, can be formulated and realized. Such a view, while not totally wrong, could be misleading as it could result in a Utopian approach to planning. That the environment we are living in now is complex, full of uncertainties, not always free from hazards, and cannot be precisely predicted suggests we need to adopt a flexible and adaptive planning approach—one that may give us more room to manoeuvre and survive in changing and turbulent conditions.

In regard to natural disaster or hazard issues, a more flexible and adaptive planning approach may be integrated with the concept of *disaster risk reduction* (DRR). The *Hyogo Framework for Actions 2005–2015: Building Resilience of Nations and Communities to Disaster*, provides a bold and clear agreement among international communities to substantially reduce disaster losses, in lives, and in the social, economic, and environmental assets of communities and countries.

What is the potential connection between spatial planning and the DRR ideas? As proposed by Fleischhauer (2008), four possible roles can exist when linking spatial planning with DRR: (1) making decisions about land use or zoning plans; (2) keeping vulnerable areas free from development; (3) following recommendations for legally binding land use or zoning plans; and (4) undertaking hazard modification. Such options are, of course, ideal, and are not always easy to implement. Several conditions may hinder achieving integration between DRR and spatial planning. First, data and information on hazard-prone areas are not always available. Second, for generations, communities may have already occupied the hazard-prone areas, and therefore it may not be realistic to resettle them in other, safer areas. Third, safer areas for resettling millions of people already occupying hazard-prone areas are not always available.

In this context, planners frequently face difficulties in arranging land use of cities and regions, or in integrating them with DRR concepts. Planners are not always able to identify and allocate sites or areas most suitable for settlement. Nor do they always have enough information on whether such areas are vulnerable from a certain natural hazard. In such situations, planning cannot readily produce a long-term, static 'blueprint' for a certain city or region. Instead, planning should be seen more as a dynamic and flexible process to help achieve sustainable outcomes.

The tsunami in Aceh, the earthquake in Yogyakarta and West Sumatera, and the recent Mount Merapi eruption near Yogyakarta have provided lessons for planners. The planners' tasks are not only to produce long-term, static spatial plans (in the planning studios). Equally crucial is to determine how to integrate capacity-building of the community into the planning process itself. By involving communities in the planning process, planners and communities may learn together how to systematically build resilience relative to hazards.

Dr Bakti Setiawan is Professor and Head of the School of Architecture and Planning at Gadjah Mada University in Yogyakarta, Indonesia. He completed his Ph.D. at the University of British Columbia. His research focuses on the urban environment, housing, and community development.

completely understand and control natural systems. Instead, this approach accepts that our understanding will always be incomplete and limited, resulting in ongoing surprises that will require us to adapt and modify our policies and practices. *Impact* and *risk assessments* further reflect acceptance of uncertainty with regard to resource and environmental systems and encourage us to strive to anticipate and monitor outcomes—intended and unintended, desirable and undesirable—so that we can determine where and when adjustments are needed.

In Chapter 5, we stressed the importance of a systems approach, and the discussion in this chapter about *regional and land-use planning* highlights that resource and environmental management does not occur in a vacuum. Other management processes occur in parallel, which means that it is important to watch for how connections can be made with these processes, particularly when they have a legal basis. Furthermore, if we want to facilitate positive change it is necessary to pay attention to factors that contribute to *implementation gaps* or *failure*.

Finally, *communication* reminds us that it is critically important that stakeholders share information, insights, needs, and priorities with each other in problem-solving situations. Indeed, inadequate communication often is the variable that undermines otherwise well-conceived management initiatives.

What can you do as an individual? First, by being aware of the desired characteristics of processes, methods, and products, you can critically examine any initiative related to resource or environmental problems and determine whether these approaches and their inherent values have been included. If any elements are absent, you can draw attention to the need to incorporate them. Second, you can reflect on whether the values you have been acculturated to believe—such as self-interest and competition—are appropriate.

Alternative approaches exist, and we should always approach problems questioning why and how we tend to deal with them in particular ways. Third, you can pay attention to related approaches, such as regional and land-use planning and risk and impact assessment, to see how they can support or advance resource and environmental management. Fourth, you can do your best to share the information and understanding you have with others not only to improve our understanding of structures and processes related to natural and human systems but also to enhance appreciation of the almost inevitable range of underlying values, assumptions, and attitudes that will shape behaviour. Awareness of all these processes, methods, and products, as well as sensitivity to them, will help to generate diverse ways to define, frame, and solve problems.

Summary

1. If a systems approach is to be used, collaboration and co-ordination are required.

2. Collaboration involves exchanging information, modifying activities in light of others' needs, sharing resources, enhancing the capacity of others in order to achieve mutual benefit and to realize common goals or purposes, and joint decision-making to resolve problems during which power is shared and stakeholders accept collective responsibility for their actions and the outcomes.

3. Participatory approaches aim to incorporate insight from individuals and groups affected by decisions or with responsibility for issues, and also aim for a sharing power and authority. Not all groups welcome power-sharing, since relinquishment of power may undermine their role in environmental management.

4. Stakeholders should be involved in decision-making because of their direct interest, and should include (1) any public agency with prescribed management responsibilities; (2) all interests significantly affected by a decision; and (3) all parties who might intervene in the decision-making process or who might block or delay the process.

5. Co-management involves sharing power through giving responsibility and authority to local people for certain aspects of resource and environmental management.

6. Effective communication of science and local knowledge is essential, and communications should be prepared for the target audience and its level of understanding.

7. Effective communication should address four complementary questions: (1) What do we know, with what accuracy, and how confident are we about our data? (2) What do we not know, and why are we uncertain? (3) What could we know, with more time, money, and talent? (4) What should we know in order to act in the face of uncertainty?

8. The concept of adaptive environmental management accepts that (1) surprise, uncertainty, and the unexpected are normal; (2) it is not possible to eliminate them through management initiatives; and (3) management should allow for them. Thus, management is viewed as an experiment, requiring systematic monitoring so we can learn from experience.

9. Adaptive co-management includes (1) learning by doing; (2) integration of different knowledge systems; (3) collaboration and power-sharing among community, regional, and national levels; and (4) management flexibility.

10. Environmental impact assessment identifies and predicts the impact of legislative proposals, policies, programs, projects, and operational procedures on the biophysical environment and on human health and well-being. It also investigates and proposes means for their management.

11. Risk assessment focuses on determining the probability or likelihood of an event of some specified magnitude, as well as the likelihood of the associated consequences. Since risks have to be estimated, our calculations may be incorrect. For this reason, the precautionary principle is used, which states that in order to protect the environment when there are risks of serious or irreversible damage, lack of full scientific certainty should not be used as an excuse for postponing cost-effective measures to prevent environmental degradation.

12. Sustainability assessment focuses on applying some form of sustainability analysis, appraisal, or assessment or otherwise adopting sustainability objectives as core guides for evaluations and decisions.

13. Disputes usually centre on three main issues: rights, interests, and power. The traditional means of dealing with societal disputes are political, administrative, and judicial. Increasingly, attention is being drawn to alternative dispute resolution in which information and understanding are shared and efforts are made to find solutions that address the needs of all stakeholders.

14. Regional and land-use plans often have a statutory or legislative basis and govern activity in an area. In contrast, many resource and environmental management plans do not have a legal basis, and this often creates difficulty for effective implementation. By connecting resource and environmental management plans to regional and land-use plans, statutory authority can be obtained.

15. Implementation failure is a non-trivial issue when moving from a plan to action, and must receive ongoing attention.

Key Terms

adaptive co-management
adaptive environmental management
alternative dispute resolution
arbitration
collaboration
co-management
co-ordination
environmental impact assessment
impact assessment
'implementation gap'
indigenous knowledge
mediation
monitoring
negotiation
partnerships
precautionary principle
risk assessment
stakeholders
strategic environmental assessment
sustainability assessment

Questions for Review and Critical Thinking

1. Explain the difference between collaboration and co-ordination. Why are both needed in resource and environmental management?

2. What does the word 'stakeholder' imply for resource and environmental management? How would you go about identifying stakeholders in a specific problem-solving situation?

3. What are the underlying principles of co-management?

4. Why is it often difficult to communicate results from scientific research to the public? What can be done to improve such communication?

5. Do you believe that scientists are objective when they conduct research?

6. What motivated people to develop the concept of adaptive management? What are its strengths and weaknesses? How might it be applied to an environmental problem in your community or province?

7. What are the main characteristics of adaptive co-management?

8. Why is the precautionary principle considered to be important? Has it been effective in practical terms?

9. What is the difference between strategic environmental assessment and sustainability assessment?

10. What are the distinctive features of alternative dispute resolution? In what kinds of situations might it be a better way to deal with conflicts than the judicial approach?

11. What is the benefit of connecting or relating resource and environmental management to regional and land-use plans and to environmental impact assessments? Is that being done in your community or province?

12. To make the transition from a plan to action, which variables do you believe are most important to consider to ensure effective implementation?

13. To what extent did collaboration and co-ordination occur during the process used to deal with the Sydney Tar Ponds problem discussed in Chapter 1? What initiatives should be taken to achieve a more collaborative and co-ordinated approach in the future?

14. Remediation of the Sydney Tar Ponds experienced various false starts. In your view, what were the main reasons causing difficulty in implementing a remediation plan?

Related Websites

Adaptive co-management, resilience alliance
resalliance.org/2448.php

Clayoquot Sound
forests.org; www.focs.ca; www.clayoquotbiosphere.org

Co-management
www.pcffa.org; www.iasc-commons.org/about

Conflict management:
　The Conflict Resolution Information Source
　www.crinfo.org

　Public dispute resolution
　www.sog.unc.edu/programs/dispute

Environmental Impact Assessment:
　Canadian Environmental Assessment Agency
　www.ceaa.gc.ca

　Environment Canada, Environmental Assessment Program
　www.ec.gc.ca/ese-ees/default.asp?lang=En&n=BA0E21A9-1

　Experimental Lakes Research Area
　www.dfo-mpo.gc.ca/regions/central/science/environmental-science-environnement/ela-rle-eng.htm; www.umanitoba.ca/institutes/fisheries

　Parks Canada, EIA Overview
　www.pc.gc.ca/nature/eie-eia/index_e.asp

Fraser River Estuary Management Program
www.bieapfremp.org/main_fremp.html

International Joint Commission
www.ijc.org

Kristianstads Vattenrike Biosphere Reserve
www.vattenriket.kristianstad.se/eng/biosphere.php

Manitoba Round Table for Sustainable Development
www.gov.mb.ca/conservation/susresmb/mrtsd/index.html

Monitoring:
　Community Based Environmental Monitoring Network, Atlantic Canada
　www.envnetwork.smu.ca/

　Ecological Monitoring and Assessment Network
　www.ec.gc.ca/faunescience-wildlifescience/default.asp?lang=En&n=B0D89DF1-1

　Monitoring and Reporting on Ontario's Forests
　www.mnr.gov.on.ca/en/Business/Forests/2ColumnSubPage/STEL02_166359.html

NatureWatch
www.naturewatch.ca/english

Stewardship Canada
www.stewardshipcanada.ca

Further Readings

Note: This list comprises works relevant to the subject of the chapter but not cited in the text. All cited works are listed in the Bibliography at the end of the book.

Armitage, D. 2005. 'Adaptive capacity and community-based natural resource management', *Environmental Management* 35: 703–15.

———. 2005. 'Collaborative environmental assessment in the Northwest Territories, Canada', *Environmental Impact Assessment Review* 25: 239–58.

——— and R. Plummer. 2010. *Adaptive Capacity: Building Environmental Governance in an Age of Uncertainty*. Heidelberg: Springer.

Dalal-Clayton, B., and B. Sadler. 2011. *Sustainability Appraisal: A Sourcebook and Reference Guide to International Experience*. London: Earthscan.

Gibson, R.B., et al. 2005. *Sustainability Assessment: Criteria and Processes*. London: Earthscan.

Hahn, T., P. Olsson, C. Folke, and K. Johansson. 2006. 'Trust-building, knowledge generation and organizational innovations: The role of a bridging organization for adaptive co-management of a wetland landscape around Kristianstad, Sweden', *Human Ecology* 34: 573–92.

Hanna, K.S. 2005. 'Planning for sustainability: Experiences in two contrasting communities', *Journal of the American Planning Association* 71: 27–40.

———., ed. 2009. *Environmental Impact Assessment: Practice and Participation*, 2nd edn. Toronto: Oxford University Press.

Nasen, L., B.F. Noble, and J. Johnstone. 2011. 'Environmental effects assessment of oil and gas lease sites in a grassland ecosystem', *Journal of Environmental Management* 92: 195–204.

Noble, B.F. 2004. 'Integrating strategic environmental assessment with industry planning: A case study of the Pasquai-Porcupine Forest Management Plan, Saskatchewan, Canada', *Environmental Management* 33: 401–11.

———. 2005. 'Integrating human health into environmental impact assessment: Case studies of Canada's northern mining resources sector', *Arctic* 58: 395–405.

———. 2010. *Introduction to Environmental Impact Assessment: A Guide to Principles and Practice*, 2nd edn. Toronto: Oxford University Press.

———. 2010. 'Applying adaptive environmental management', in B. Mitchell, ed., *Resource and Environmental Management in Canada: Addressing Conflict and Uncertainty*, 4th edn. Toronto: Oxford University Press, 434–61.

——— and J. Birk. 2010. 'Comfort monitoring? Environmental assessment follow-up under community–industry negotiated environmental agreements', *Environmental Impact Assessment Review*. At: doi:10.1016/j.eiar.2010.05.002.

Olsson, P., C. Folke, and F. Berkes. 2004. 'Adaptive co-management for building resilience in social-ecological systems', *Environmental Management* 34: 75–90.

Plummer, R., and J. FitzGibbon. 2004. 'Some observations on the terminology in co-operative environmental management', *Journal of Environmental Management* 70: 63–72.

Sadler, B., R. Aschemann, J. Dusik, T.B. Fischer, M. Partidario, and R. Verheem. 2011. *Handbook of Strategic Environmental Assessment*. London: Earthscan.

Schultz, L., C. Folke, and P. Olsson. 2007. 'Enhancing ecosystem management through social-ecological inventories: Lessons from Kristianstads Vattenrike, Sweden', *Environmental Conservation* 34: 140–52.

Seitz, N., C. Westbrook, and B.F. Noble. 2010. 'Bringing science into river systems cumulative effects assessment practice', *Environmental Impact Assessment Review*. At: doi:10.1016/j.eiar.2010.08.001.

Storey, K., and P. Jones. 2003. 'Social impact assessment, impact management and follow-up: A case study of the construction of the Hibernia offshore platform', *Impact Assessment and Project Appraisal* 21, 2: 99–107.

Therivel, R. 2011. *Strategic Environmental Assessment in Action*, 2nd edn. London: Earthscan.

Trist, E. 1980. 'The environment and system-response capability', *Futures* 12, 2: 113–27.

Warner, J., Ed. 2007. *Multi-stakeholder Platforms for Integrated Water Management*. Aldershot, UK: Ashgate.

Wollenberg, E., D. Edmunds, and L. Buck. 2000. 'Using scenarios to make decisions about the future: Anticipatory learning for the adaptive co-management of community forests', *Landscape and Urban Planning* 47: 65–77.

Our environmental performance has been mixed, with improvement in some areas and stagnation or deterioration in others. Canada's overall 'C' grade, however, reveals that Canada is not taking the necessary steps toward environmental sustainability.
—Conference Board of Canada, 2009–10 Report Card on the Environmenti

Part D
Resource and Environmental Management in Canada

Parts B and C focused on science and management related to resources and the environment. In Part D, we discover how ideas and methods from science and management can be applied in practical problem-solving situations.

Several comments about the structure of this section should be helpful. It may appear contradictory to argue for a systems or ecosystem approach in Parts B and C and then organize Part D on the basis of specific 'resources' or attributes of the environment, such as agriculture, forestry, or wildlife. Indeed, it would be inappropriate to isolate components of resource or environmental systems and examine them on a sector-by-sector basis, as the chapter headings in Part D might suggest. However, it is appropriate to use one resource or environmental aspect as the starting point for a discussion, as long as we give attention to other elements of the ecosystem and their linkages. This is the approach used in this section.

Four types of resources or environments are considered in the following eight chapters. The first, addressed in Chapter 7, is the atmospheric system, with emphasis on climate change. The second, the focus of Chapters 8 to 11, comprises various renewable resources—oceans, forestry, agriculture, and water systems. The third type, considered in Chapter 12, deals with non-renewable resources—minerals and fossil fuels that are created (or renewed) over the course of a geological time span rather than human lifespans. The fourth type, place-based, focuses on the environmental opportunities and challenges in urban areas (Chapter 13), in which four of five Canadians now live, and on endangered species and protected areas (Chapter 14).

Interest centres on examining how science can be used to inform analysis and management and how elements of best practice in management have been or could be applied. For example, after examining the nature of weather and climate, Chapter 7 turns to assessing the scientific evidence related to climate change and to the role of climate models in aiding our understanding of changes. While solid scientific understanding of environmental systems and how they may be changing is of critical importance, it is equally important to communicate that understanding to non-technical specialists. Consequently, this chapter addresses the challenges encountered by researchers when they seek to communicate scientific insight about climate change to policy-makers and the general public. These challenges too often are not systematically addressed by scientists, who are frequently more interested in the 'purity' of their scientific work and in communicating their findings to peers. As in other chapters, we attempt to relate scientific understanding to 'real-world' situations.

Case studies give you opportunities to learn about how science and management can be connected. For example, Chapter 8 on oceans looks at the challenges involved in the depletion of the Pacific salmon fishery, the debate over the seal hunt off the east coast, and the closure of the groundfish fishery in Atlantic Canada. In later chapters, case studies cover the environmental impact of the James Bay hydroelectric project in northern Quebec, the development of diamond mines in the Northwest Territories and Nunavut, the rapid growth of the Alberta oil sands, the implications of the Walkerton, Ontario, experience for water security, the lessons learned about managing natural hazards from flooding of the Red River in Manitoba, and the opportunities presented by and issues to be overcome in the development of wind and other renewable sources of energy production.

The issues and examples considered here raise fundamental questions about humans' relationships with the environment and resources. We see a range of attitudes towards the environment and other living things, covering the spectrum from humans dominating nature to humans striving to live in harmony with nature. The extinction of species, such as those reported in Chapter 14, reminds us that we have been (and can still be) incredibly arrogant in believing that it is acceptable for us to eliminate some species forever. In contrast, as also shown in Chapter 14, conscious decisions are being taken to protect valued areas, in some instances because we believe it is important to protect examples of different biomes. Notwithstanding significant and positive accomplishments, current lifestyles in Canada continue to contribute to global climate change in a major way, as discussed in Chapters 7 and 15. The changes associated with climatic warming may confound attempts to identify and protect examples of biospheres if these biospheres change in the future because of different climatic conditions.

Finally, the case studies and examples illustrate how pervasive conflict can be in resource and environmental management. This reinforces our belief that resource and environmental management must recognize, identify, and incorporate different values and interests. For this reason, managers often spend significant time trying to resolve conflicts. Given this reality, scientists must also develop a greater appreciation that their work will usually be used in situations in which values and emotions can be as important as, or of greater importance than, theories, models, and quantitative evidence.

The mix of case studies and examples in Part D and throughout this book will help you to appreciate that change, complexity, uncertainty, and conflict are integral parts of resource and environmental management.

Chapter 7
Climate Change

Learning Objectives

- To understand the difference between weather and climate.
- To know the difference between climate change and global warming.
- To appreciate why the science of climate change is characterized by complexity and uncertainty.
- To understand the nature of scientific evidence regarding climate change.
- To understand the scientific explanation for climate change.
- To realize the implications of climate change for natural and human systems.
- To appreciate the challenges of sharing information and insight related to climate change.
- To comprehend the Kyoto Protocol, as well as subsequent international summits seeking to renew or replace it.
- To understand the strategies and tactics of 'climate change deniers'.
- To appreciate Canada's role in the global context as a contributor to both climate change challenges and solutions.
- To appreciate the importance of including both mitigation and adaptation in a strategy for reducing vulnerability to climate change.
- To understand the implications of 'geo-engineering' initiatives related to climate change.
- To discover what you can do as an individual to minimize the impact of climate change.

Introduction

Climate is naturally variable. It is never exactly the same from one period to another. Sometimes it can shift dramatically within a few hundred or thousand years, as it does when ice ages begin and end. Usually it varies within much narrower limits. For most of the past 1,000 years, for example, the world's average temperature has remained within about half a degree of 14°C.

Over the past 100 years or so, however, the world's climate has changed noticeably. The world's average temperature was approximately 0.6°C warmer at the end of the twentieth century than it was at the beginning, and the 1990s was the hottest decade in 140 years of global climate records. In 2010, the globalized average land temperature was 0.96°C above the average for the twentieth century, making it tied in second place with 2005 as the second warmest year—2007 was the warmest, at 0.99°C above the twentieth-century average. For Canada, in 2010 the national average temperature was 3°C above the 1961–90 average, making it the warmest year on record since 1948. The same year also had the driest winter since 1948 (NOAA, 2010). Such changes may seem trifling, but the difference between global temperatures now and at the peak of the last ice age is a mere 5°C. Evidence of earlier climates suggests that global temperatures warmed more during the twentieth century than in any other century during the past 1,000 years (Canadian Council of Ministers of the Environment, 2003: 5). At the same time, much uncertainty and complexity are encountered in seeking to understand the significance of such changes, requiring knowledge about both science and societies.

The 2009 United Nations Climate Change Conference, also known as the Copenhagen Summit, failed to produce any legally binding agreement between participating countries on actions to combat climate change.

Perspectives on the Environment
Uncertainty Related to Climate Change

Climate change is now a scientifically established fact. The exact impact of greenhouse gas emissions is not easy to forecast, and there is a lot of uncertainty in the science when it comes to predictive capability. But we now know enough to recognize that there are large risks, potentially catastrophic ones. . . .

Prudence and care about the future of our children and their children requires that we act now. This is a form of insurance against possibly very large losses. The fact that we do not know the probability of such losses or their likely exact timing is not an argument for not taking insurance. We know the danger exists. We know that the damage caused by greenhouse gas emissions is irreversible for a long time. We know it is growing with each day of inaction.

—Human Development Resource Office (2007: v)

Nature of Climate Change

The condition of the **atmosphere** at any time or place—that is, the **weather**—is expressed by a combination of several elements, primarily (a) *temperature* and (b) *precipitation* and *humidity*, and, to a lesser degree, (c) *winds* and (d) *air pressure*. These are called the elements of weather and climate because they are the ingredients out of which various weather and climatic types are compounded. The weather of any place is the sum total of its atmospheric conditions (temperature, pressure, winds, moisture, and precipitation) for a *short* period of time. It is the momentary state of the atmosphere. Thus, we speak of the weather, not the climate, for today or of last week.

Climate, on the other hand, is a composite or generalization of the variety of day-to-day weather conditions. It is not just 'average weather', because the variations from the mean, or average, are as important as the mean itself.

Perspectives on the Environment
Weather and Climate

Canada is perceived to be a cold, northern country, and the average annual temperature of −3.7°C compared to the global average of 15°C supports this view. However, as usual, averages mask a lot of variation, especially in a country as big as Canada. Vancouver, for example, has an average temperature of 10°C, Toronto 8°C, and Halifax 6°C, compared with Alert in the Arctic at −18°C. Precipitation also varies widely, with more than 3,200 millimetres along parts of the west coast and less than 200 millimetres in the Arctic. This variation, spanning more than 40 degrees of latitude between the northern and southern extremities, is reflected in the ecozones described in more detail in Chapter 3.

—After Trewartha (1954: 4)

A distinction should also be made between climate change and global warming. Climate represents average day-to-day weather conditions as well as seasonal variations for a particular place or region. In that context, **climate change** is defined as 'a long-term shift or alteration in the climate of a specific location, a region, or the entire planet' (Hengeveld et al., 2002: 1). A shift is measured for variables associated with average weather conditions, such as temperature, precipitation, and wind patterns (velocity, direction). A change in variability of climate is also included as climate change.

In contrast, **global warming**, often mentioned by the media, addresses changes only in average surface *temperatures*. It does not address whether conditions are becoming wetter or drier, for example. A frequent misunderstanding is that global warming means uniform warming throughout the world. An increase in average global temperatures drives alterations in atmospheric circulation patterns, which can contribute to some areas warming at higher rates, others at lower rates, and others even to become cooler.

What are the 'causes' of climate? The Earth's surface and atmosphere are heated differentially by short-wave radiation from the sun. The differences in heat and pressure between the poles and the tropics fuel the global circulation system as heat and moisture are redistributed around the world. The temperature balance of the Earth is maintained through the return of the continually absorbed solar radiation back to space as infrared radiation, consistent with the first law of thermodynamics (Chapter 2). Long-term temperature changes result from shifts in the amount of energy received or absorbed. These shifts may be caused over long cycles (100,000 years) by factors such as the shape of the Earth's orbit around the sun, wobbles of the Earth's axis, and the angle of tilt. Such a 100,000-year cycle of glaciation can be traced over 600,000 years, using evidence from sources such as glacier ice and the chemical characteristics of marine sediments. It is more difficult, however, to explain some of the shorter-term fluctuations.

Perspectives on the Environment
Greenhouse Effect

The **greenhouse effect** describes the role of the atmosphere in insulating the planet from heat loss, much like a blanket on our bed insulates our bodies from heat loss. The small concentrations of greenhouse gases within the atmosphere that cause this effect allow most of the sunlight to pass through the atmosphere to heat the planet. However, these gases absorb much of the outgoing heat energy radiated by the Earth itself and return much of this energy back towards the surface. This keeps the surface much warmer than if they were absent. This process is referred to as the 'greenhouse effect' because, in some respects, it resembles the role of glass in a greenhouse.

—Hengeveld et al. (2002: 2)

Natural events, such as the eruption of large volcanoes, and changes in ocean currents, such as **El Niño**, have an influence on climate. Volcanoes, when erupting, eject large quantities of dust and sulphur particles into the atmosphere, which reduce the amount of solar radiation reaching the surface of the Earth. Changes in ocean currents can also be influential (Chapter 8). El Niño represents a marked warming of the waters in the eastern and central portions of the tropical Pacific as westerly winds weaken or stop blowing, usually

These coral fossils, from the Canning River in the Arctic National Wildlife Refuge, Alaska, illustrate how climates have changed in the past.

two to three times every decade. In normal years, the trade winds amass warm water in the western Pacific. As the winds slacken, this water spreads back eastward and towards the pole into the rest of the Pacific. This triggers weather changes in at least two-thirds of the globe, causing droughts and extreme rainfall in areas along the Pacific and Indian oceans, including Africa, eastern Asia, and North America.

However, it is increasingly apparent that climatic change may occur more rapidly than ever before as a result of human activities. We examine these factors below.

Scientific Evidence Related to Climate Change

In the context of the distinction among weather, climate, global warming, and climate change, the following statements are supported by solid scientific evidence:

1. The world has been warming, with the average global temperature at the Earth's surface having increased by

about 0.6°C, with an error range of plus or minus 0.2°C, since the late nineteenth century (Figure 7.1). This increase is a global average, and in some areas, especially over continents, warming has been several times greater than the global average (IPCC, 2007c).

2. The increase in the average temperature of the northern hemisphere during the twentieth century was the largest of any century in the past 1,000 years (Figure 7.2).

3. **Greenhouse gas** concentrations, especially those of carbon dioxide, methane, nitrous oxide, and tropospheric **ozone**, have been rising for several decades (Figure 7.3). Carbon dioxide and methane are at higher concentrations now than at any time over the past 420,000 years (Figure 7.4).

4. In most parts of the world since 1980, glaciers have lost more mass than they have gained. Furthermore, most mountain glaciers have been retreating during the past 100 years. Exceptions are found in Norway and New Zealand, where some glaciers have been advancing. The explanation for these exceptions is that they are associated with remarkable increases in precipitation rather than with decreases in temperatures.

5. In many areas of the world, reduced snow cover has been documented, as well as earlier spring melting of ice on rivers and lakes. For example, snow cover in the northern hemisphere has decreased by approximately 10 per cent since 1996. The cold temperatures and large amounts of snow in the winters of 2004 and 2008 across most of Canada were an exception to this general pattern. Regarding Arctic sea ice cover, two scientific organizations, one in Germany and one in the US, reported

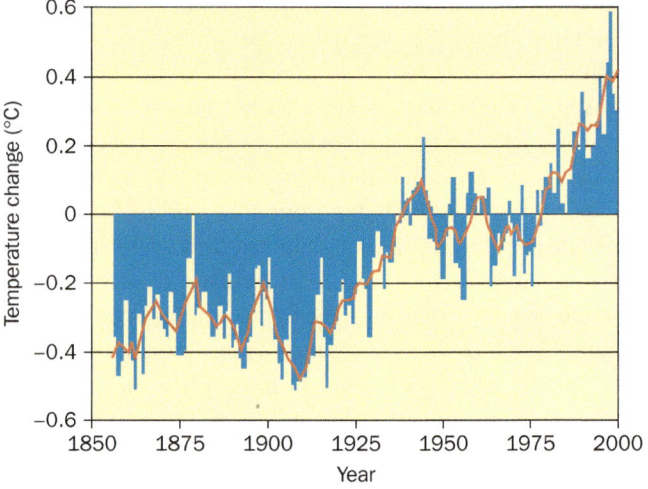

Figure 7.1 | Variation in global average surface temperature between 1856 and 2000. Source: Miller (2002: A-2), prepared by Professor Danny Harvey, Department of Geography, University of Toronto, using data in electronic form available from the UK Meteorological Office website www.meo.gov.uk.

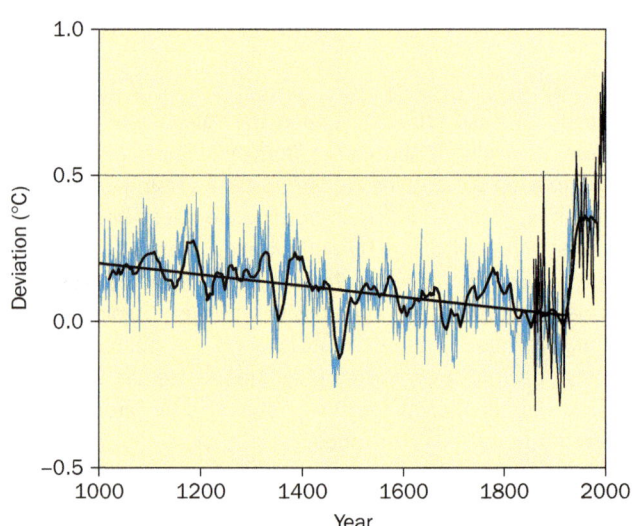

Figure 7.2 | Variation in northern hemisphere average surface temperature. Based largely on ice core, tree ring, and coral reef data (blue line). Also shown is the 20-year running mean of the annual paleoclimatic data and the 1000–1900 trend line (thick black lines) and the directly observed temperature variation of Figure 7.1 (thin black line). Source: Miller (2002: A-2), prepared by Professor Danny Harvey, Department of Geography, University of Toronto, using paleoclimatic and historical data from UK website (see Figure 7.1) and data from the US National Oceanographic and Atmospheric Administration paleoclimatology website: www.ngdc.noaa.gov/paleo.

Following a volcanic eruption, large amounts of sulphur dioxide (SO_2), hydrochloric acid (HCl), and ash spew into the Earth's stratosphere. In most cases, HCl condenses with water vapour and is rained out of the volcanic cloud. SO_2 from the cloud is transformed into sulphuric acid. The sulphuric acid quickly condenses, producing aerosol particles, which linger in the atmosphere for long periods of time.

in the autumn of 2011 that coverage was about as low as it had been since satellite data started to be collected in 1978. Specifically, the National Snow and Ice Data Center in Colorado reported that the ice coverage in the Arctic had almost reached the all-time low measured previously in 2007. By 2011 there was approximately two-thirds ice coverage relative to the average for the period 1979 to 2000.

6. Measurements show that permafrost is warming in many regions.
7. The Intergovernmental Panel on Climate Change (IPCC, 2001c) reported that the average rate of sea-level rise increased from 0.1 to 0.2 millimetres per year during the past 3,000 years to one to two millimetres per year in the twentieth century, a tenfold increase. Since 1961, global average sea levels have risen at an average rate of 1.8 (1.3 to 2.3) millimetres annually and since 1993 at 3.1 (2.4 to 3.8) millimetres annually (IPCC, 2007c, 1). Caution is needed in interpreting such findings, however, since it is recognized that in some areas, land is still rebounding from the weight of the last glaciation. Thus, data from tidal gauges must be interpreted in light of the combination of sea-level rise and land rebound, which could mask the increase in sea-level rise.

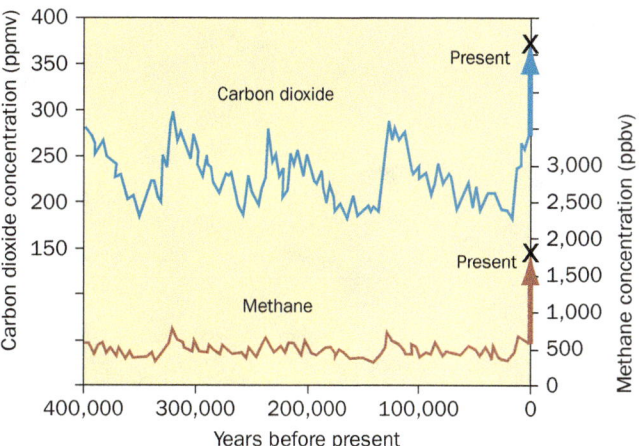

Figure 7.4 | Variation in atmospheric concentrations of carbon dioxide and methane to 400,000 years before present. Measured in the Vostok ice core in Antarctica (thin lines) and during the past 200 years (heavy line). *Source: Miller (2002: A-1), prepared by Professor Danny Harvey, Department of Geography, University of Toronto, using data in electronic form obtained from the US National Oceanographic and Atmospheric Administration paleoclimatology website http://www.ngdc.noaa.gov/paleo.*

Hengeveld (2006: 28) summarized the above findings:

> The debate about climate change within the science community is gradually shifting away from whether it is happening and how serious it will be to how to deal with it. The general consensus is that global warming is already occurring, that fossil fuels combustion is the primary cause, and that the negative impacts of weather and climate events will get much worse.

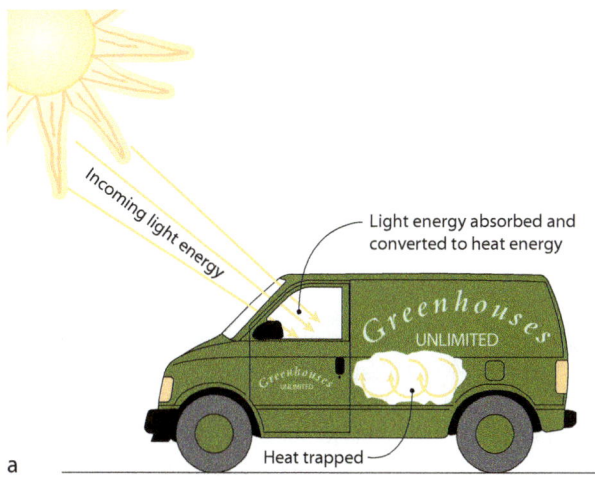

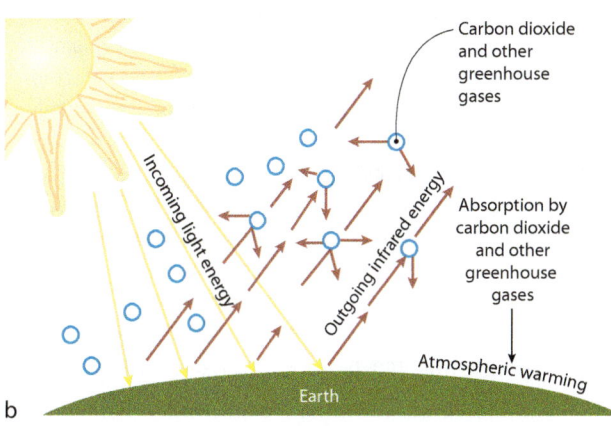

Figure 7.3 | The greenhouse effect.

The forest industry is a major contributor to the rising levels of carbon dioxide in the atmosphere, not only through deforestation, but also through emissions from processing plants.

Some countries, such as the Maldives in the Indian Ocean, are so low-lying that they could be mostly flooded as early as 2050 if global sea levels rise as predicted. Indeed, in late 2008 the new president of the Maldives announced an emergency plan whereby revenue from tourism is to be set aside in a special fund for the purchase of land in Sri Lanka, India, or Australia so the nation of 300,000+ people, in a worst-case scenario, can continue to exist in another location.

It is not just the scientific community that has reached this consensus. The world's largest oil company, Exxon, which up to 10 years ago was questioning the veracity of global climate change, in its 2011 annual, *Outlook for Energy*, concludes that the scientific community is being unrealistically optimistic in its projections. Exxon expects carbon emissions to rise a further 25 per cent over the next 20 years as demands for power will rise by 40 per cent over the next 40 years, leading to annual increases of 0.9 per cent emissions per year. These figures far surpass any predictions from the scientific models, discussed in more detail in the next section. Unfortunately, so also do the empirical data now being collected on climate change, suggesting that the Exxon may well be right.

One attribute of 'good science' is the use of cross-checking data sources to ensure that findings are not unduly influenced by measurement error or limitations of any single data source. The findings above related to temperature, greenhouse gas concentrations, glaciers, snow cover, river and lake ice breakup, permafrost, and sea-level rise all indicate that climate change is occurring. In the next section, we turn our attention to the reasons for this change. Once the pattern and causes of climate change are understood, we have a foundation for considering the implications for designing possible policies and actions.

Perspectives on the Environment
Implications of Sea-Level Rise

Rising sea levels threaten familiar shoreline environments. Coastal wetlands, which are important ecosystems and barriers against shoreline erosion, gradually disappear. Bluffs and beaches are more exposed to erosion by waves, groundwater is more likely to become contaminated by salt water, and low-lying coastal areas may be permanently lost. In addition, wharves, buildings, roads, and other valuable seaside property face a greater risk of damage as a result of flooding from storms.

—Canadian Council of Ministers of the Environment (2003: 13)

Perspectives on the Environment
Glaciers and Climate Change in Canada

Since 1950, the greatest warming in Canada has occurred in the West and Northwest. Most glaciers in these regions are also shrinking rapidly. The 1,300 or so glaciers on the eastern slopes of the Rockies, for example, are now about 25 per cent to 75 per cent smaller than they were in 1850. The area of warming also covers many of the High Arctic islands in Nunavut, where glaciers such as the Melville Island South Ice Cap have been shrinking gradually since at least the late 1950s. In eastern Nunavut, however, the situation is more complex: some glaciers are shrinking, while others are growing.

The melting of glaciers is a concern for Alberta, Saskatchewan, and Manitoba. Farmers depend on glacier-fed rivers like the Saskatchewan and the Bow for irrigation water, and cities like Edmonton, Calgary, and Saskatoon rely on them for municipal water supplies and recreation. At The Pas in Manitoba, reduced flows on the Saskatchewan could interfere with the native fishery and hydroelectric power generation.

—Canadian Council of Ministers of theEnvironment (2003: 20)

A view of the Moose River from Moose Factory Island. Traditional environmental knowledge is vital to understanding the complex changes that occur in regions such as Moose Factory and Moosonee. Climate change has not only affected average temperatures but also animal and bird migration, weather patterns, and the freeze/thaw cycles of the Moose River (image from the Cree Village Ecolodge website).

Modelling Climate Change

The uncertainty associated with global climate change has led scientists to explore different ways of assessing past and future climates (Box 7.1). One approach is **climate modelling**. While concern about climate change due to greenhouse gas (GHG) emissions is relatively recent, climate modelling is not. The earliest global climate models date back to the 1950s, far ahead of when scientists became concerned about carbon dioxide emissions and their effect on the atmosphere. However, more recent concerns about global warming have propelled the science of climate modelling to the forefront.

Climate Models

All climate models consider some or all of five components in order to predict future climates:

- radiation—both incoming (solar) and outgoing (absorbed, reflected);
- dynamics—the horizontal and vertical movement of energy around the globe;
- surface processes—the effects of the Earth's surface (snow cover, vegetation) on climate (i.e., albedo, emissivity);
- chemistry—the chemical composition of the atmosphere and its interactions with other Earth processes (i.e., carbon cycling);
- time step and resolution—the time step of the model (minutes or decades) and the resolution, or spatial scale of the model (your backyard or the entire globe).

The nature of the Earth's climate and its complexity make comprehensive climate modelling difficult. The many components, interactions, and feedback loops in the global climate cannot be entirely represented by any mathematical model, and therefore all models simplify certain aspects of climate. There are four main types of climate models, each increasing in complexity. They are outlined in Box 7.2.

Unlike energy balance models (EBMs), **general circulation models** (GCMS) attempt to examine all of the climatic elements and processes, making these models very complex. GCMs model the Earth's atmosphere and oceans under certain climate change scenarios, the most popular being $2 \times CO_2$. In this situation, the Earth's climate is modelled to indicate the changes that would occur if atmospheric concentrations of carbon dioxide were doubled from pre–Industrial Revolution levels, which many scientists believe will occur by 2050.

In a GCM, the Earth's surface is divided into a grid, in which case a larger grid results in a simpler model and a smaller grid requires more calculations. For each grid, a series of equations are solved at the surface of the grid (sea level) and for several layers of the atmosphere and subsurface layers (the vertical dimension). The equations deal with:

- conservation of momentum;
- conservation of mass;
- conservation of energy;
- ideal gas law.

Beginning with present-day or known values, the solutions for these equations are solved and repeated at each time

Environment in Focus

Box 7.1 Measuring Climate Change

An essential step in attempting to assess climatic change is to see how present variations in climate compare with those of the past. Current data are largely instrument-based weather observations—i.e., the instrumental record. Even here there are difficulties. More modern and accurate data—for example, data on the upper atmosphere gathered by satellite—are available only for the past three and a half decades or so. This period also coincided with the greatest human impact on climate and does not provide any type of control for climate change in the absence of industrialization.

Former climates are reconstructed by scientists using proxy information from many different sources. For example, examination of historical records of climate-influenced factors such as the price of wheat in Europe over the past 800 years, the blooming dates of cherry trees in Kyoto, Japan, since AD 812, the height of the Nile River at Cairo since AD 622, the number of severe winters in China since the sixth century, examination of sailors' and explorers' logs, and other such sources all contribute to a picture of past climates.

Scientists also use climate-sensitive natural indicators such as tree rings and glacial ice. Cores obtained from ice in Greenland and Antarctica that go back tens of thousands of years have been analyzed using the ratio of two oxygen isotopes that indicate the air temperature when the original snow accumulated on the glacier surface. Tree rings are also very useful. Outside the tropics where there are noted differences in seasons, the width and density of tree rings reflect growth conditions, including climate. Some species—red cedar in coastal BC, for example—may live for well over 1,000 years and can provide valuable indicators as to past climates. The same kinds of rings also characterize the growth of many long-living corals in the tropics, which may be more than 800 years old, and provide valuable evidence about previous El Niño events.

step of the simulation, and then the results are interpolated between the grid points to cover the Earth's entire surface. Most models operate at spatial resolutions of a few degrees latitude and longitude and at time steps of less than one hour. The vertical dimension is often divided into 10 layers, with two subsurface layers. Because of these simplifications, GCMs are best used for global or overall climate modelling, not for regional representations of climate change.

Other aspects of the global climate are simplified as well, thus limiting the predictive capabilities of many GCMs. For example, known or present-day values are required to run many models, but in some areas of the world, these values are unavailable or scarce for some variables (temperature, sea-ice cover, cloud cover). Therefore, assumptions are made to fill in the missing values, which may not be accurate. In addition, many complicated feedbacks cannot all be accounted for in GCMs, partly because of their complexity and uncertainty about how they react under given circumstances. While the relationship between greenhouse gas emissions and temperature is a relatively straightforward positive feedback loop (where a positive change in one variable results in a positive change in the other), the relationship between increased temperature and cloud cover relies on many other variables. Finally, many of the climatic interactions at the Earth's surface are difficult to model and are under-represented in many GCMs. For example, ocean layers and interactions are difficult to model, but their effect on regional and global climate can be significant.

In summary, while GCMs are becoming increasingly sophisticated, many complex aspects and interactions of the global climate need to be understood more fully, and computational facilities need to be better developed.

Perspectives on the Environment
Intergovernmental Panel on Climate Change

The Intergovernmental Panel on Climate Change (IPCC) was established in 1988 by the World Meteorological Organization and the United Nations Environment Programme.

The IPCC was created to assess scientific, technical, and socio-economic information related to understanding the risks from human-induced change to climate. The IPCC does not conduct original research, nor does it monitor climate data. Its assessments are based on peer-reviewed and published scientific literature.

The IPCC has three working groups and a task force. Its *First Assessment Report*, published in 1991, had an important role related to the UN Framework Convention on Climate Change, adopted at the Earth Summit at Rio de Janeiro in 1992.

Its *Second Assessment Report*, published in 1995, became a significant input into negotiations that resulted in the Kyoto Protocol in 1997. The Kyoto Protocol is discussed later in the chapter.

The *Third Assessment Report*, produced in 2001, was the product of the work of more than 2,000 scientists from many disciplines from all around the world.

A *Fourth Assessment Report* was published in 2007 and is discussed later in this chapter.

Environment in Focus
Box 7.2 Four Types of Climate Models

1. **Energy Balance Models (EBMs)**
   EBMs can be either non- or one-dimensional. In the first case, the Earth (or any point on the Earth) is treated as a single entity, and only incoming and outgoing radiation are modelled. In one-dimensional EBMs, temperature is modelled as a function of latitude and radiation balance.

2. **One-Dimensional Radiative-Convective (RC) Climate Models**
   In this model, the one dimension is altitude. One-dimensional RC models take into account incoming and outgoing solar radiation, as well as convective processes that affect the vertical distribution of temperature. These models are useful for examining the vertical distribution of solar radiation and cloud cover and are very useful for examining the effects of volcanic emissions on temperature.

3. **Two-Dimensional Statistical-Dynamic (SD) Climate Models**
   This model takes into account either the two horizontal dimensions or one horizontal dimension and the vertical dimension. The latter are most frequently modelled, thus combining the latitudinal EBMs with the vertical RC models. These models can examine wind speed, direction, and other horizontal energy transfers.

4. **General Circulation Models (GCMs)**
   While the first three types of climate models are still used for various purposes in climate research, since the 1980s, general circulation models (GCMs) have largely taken over the field of climate modelling, and most model development is devoted to them. It is by far the most complex type of model, since the GCM takes into account the three-dimensional nature of the Earth's atmosphere, oceans, or both.

Limitations of GCMs

While GCMs provide overall indications of future climates, their limitations for policy and planning need to be appreciated. Scientists recognize that the coarse spatial resolution, poor predictive capacity for precipitation, relatively weak simulation of oceans, lack of baseline data, and many other limitations cause GCM outputs to be highly variable. Some researchers have become discouraged by the dominance of GCMs in climate change sciences and by disregard of other climate models (i.e., EBM, RC, SD, or combinations). Others caution against the misinterpretation that GCMs are accurate and realistic models of global climate and stress that much more improvement is needed so that GCMs may best represent the complex nature of the Earth's climate.

Scientific Explanations

In its 2007 report, the Intergovernmental Panel on Climate Change concluded that, as noted earlier in this chapter, worldwide trends in the twentieth century consistently and strongly reveal an increase in global surface temperature. There is now virtually no scientific debate about this finding. Furthermore, as also noted above, greenhouse gas concentrations in the atmosphere have increased significantly. There is strong scientific consensus that the increase in greenhouse gases has been caused by human activities. The basis for this conclusion comes from data showing that (1) the rate of increase of greenhouse gases in the atmosphere over the past century closely matches the rate of human-driven emissions into the atmosphere; (2) atmospheric oxygen has been falling at the same rate that fossil-fuel emissions of carbon dioxide have been increasing; and (3) significantly, the change in proportions of carbon isotopes in the atmosphere is evidence that the atmosphere is being enriched with carbon from fossil-fuel sources rather than from natural sources. Regarding the third point, the IPCC (2007b: 2) concluded that 'The global increases in carbon dioxide concentrations are due primarily to fossil fuel use and land-use change, while those of methane and nitrous oxide are primarily due to agriculture' (see Chapter 4).

Natural and human variables both contribute to climate change. It is difficult to determine their relative contribution, since they usually operate at the same time. Such variables alter the balance of incoming and outgoing energy in the Earth's atmospheric system and are usually referred to as 'radiative forcings'. Positive forcings produce warming; negative forcings lead to cooling. Greenhouse gases, aerosols, variations in solar output, and volcanic eruptions are viewed to be the most influential radiative forcings.

Figure 7.5 shows the relative contribution of different radiative forcings. Greenhouse gases, in particular carbon dioxide, have been the main forcing factor. In comparison, the effect of variation in solar output has been minor. Working in the opposite direction, sulphate aerosols, emissions from volcanic eruptions, biomass aerosols, and depletion of the stratospheric **ozone layer** have contributed to cooling at the Earth's surface. Figure 7.5 is derived from a high level of scientific understanding about most greenhouse gases, based on excellent data regarding their concentrations in the atmosphere and good insight about their radiative forcing characteristics.

What about the role of humans and their activities regarding climate change? The reports from the IPCC (2007a, 2007b, 2007c) are explicit that most warming since the mid-twentieth century is associated with human activities. This view is based on climate modelling showing that when only the influences of variation in solar output and emissions from volcanic eruptions are included, simulated temperature changes do not match the observed changes very well. In contrast, when the models include the forcing factors related to human activities (greenhouse gases, stratospheric ozone depletion, sulphate aerosols), better matches are found between the observed and simulated patterns of temperature change. Not surprisingly, the best match occurs when both natural and human forcing factors are combined.

Implications of Climate Change

The implications of climate change are potentially widespread and significant. In this section, we briefly outline some of the impacts on different resource-based systems. A general overview of the impacts (Figure 7.6) was developed by the National Round Table on the Environment and the Economy.

Perspectives on the Environment
Global Reach of Climate Change

In Canada and across the globe, we are already seeing the effects of warming temperatures and changing climate conditions. As climate change persists, we can expect, for example, further melting of glaciers and sea ice, rising sea levels, earlier springs, shifts in the distribution of animals and plants, and increasingly volatile weather. No region and no aspect of our geography will be immune; but impacts will vary in time and intensity.

—NRTEE (2010: 013)

Terrestrial Systems

It is conceivable that within your lifetime, many terrestrial systems, along with the associated fauna and flora, will change dramatically. For example, on the Canadian Prairies, boreal

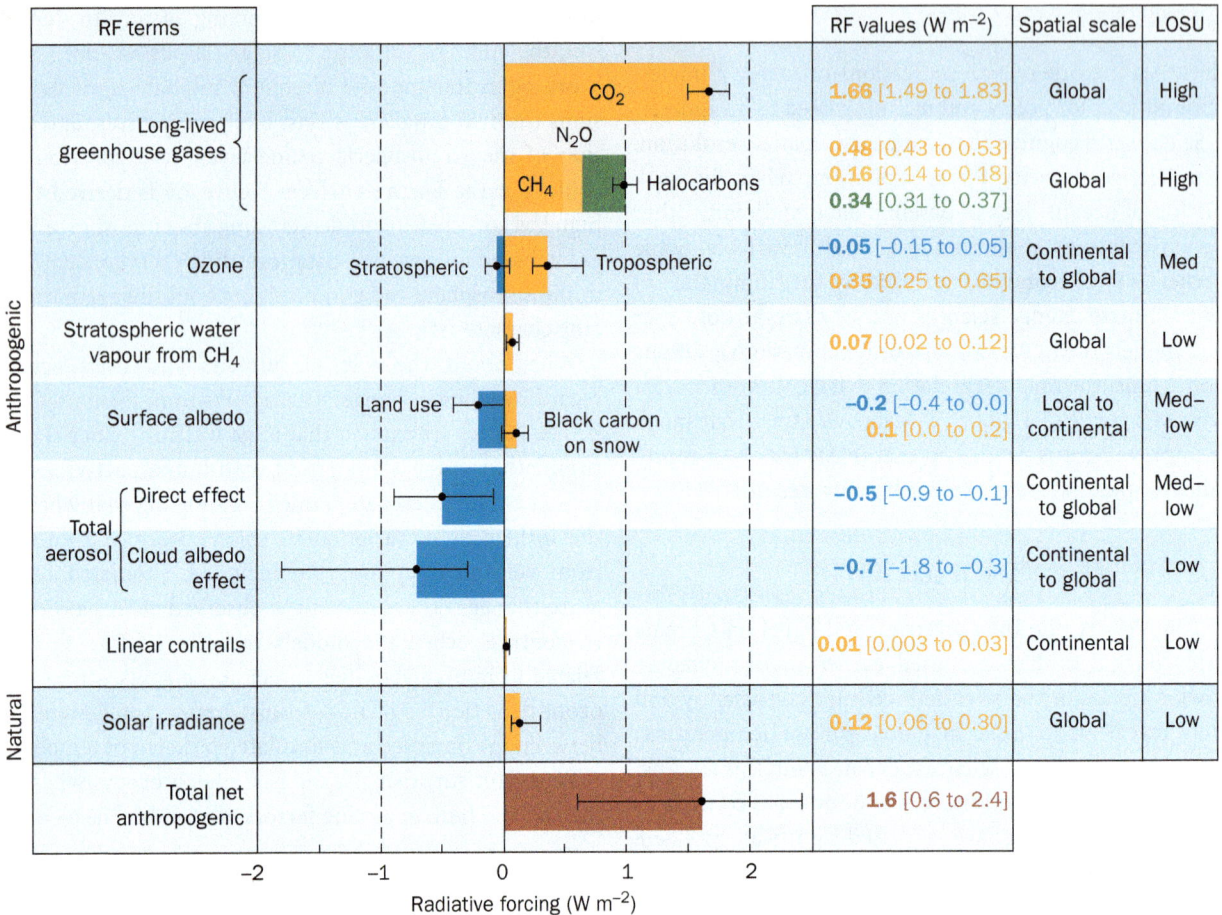

Figure 7.5 | Global average radiative forcing (RF) estimates and ranges in 2005 for anthropogenic carbon dioxide (CO_2), methane (CH_4), nitrous oxide (N_2O), and other important agents and mechanisms, together with the typical geographical extent (spatial scale) of the forcing and the assessed level of scientific understanding (LOSU). The net anthropogenic radiative forcing and its range are also shown. They require summing asymmetric uncertainty estimates from the component terms and cannot be obtained by simple addition. Additional forcing factors not included here are considered to have a very low LOSU. Volcanic aerosols contribute an additional natural forcing but are not included in this figure because of their episodic nature. The range for linear contrails does not include other possible effects of aviation on cloudiness. Source: IPCC (2007b: 16).

forests may shift anywhere from 100 to 700 kilometres to the north, to be replaced by grasslands and more southern forest species. In the Arctic, the southern permafrost border could move 500 kilometres northward, and the treeline could move from 200 to 300 kilometres to the north. These shifts are illustrated in Figure 7.7. Boreal forests in particular would also be affected by increases in insect infestation, disease, and fires.

The consequences of change to terrestrial systems could be dramatic. For example, polar bears may no longer remain to breed in Wapusk National Park in Manitoba, yet the park was created in 1996 to protect the habitat of polar bears. At the other extreme, the hoary marmot in the Rockies and other areas in the Western Cordillera is likely to thrive, since changed climate leads to more avalanches, which will expand its preferred habitat of open meadow. These examples indicate that the rationale for national and provincial parks, created to protect representative ecosystems (Chapter 14), may dramatically change as the distinctive ecosystems currently protected by such parks evolve into something totally different.

Agriculture

One of the major limitations on agricultural activity in most areas of Canada is our cold climate. In southern Canada, the frost-free growing season is about 200 days, and in the Far North the growing season is normally just a few weeks. Furthermore, early frosts or severe winters can damage even dormant vegetation.

Given the above conditions, the scientific consensus is that, on balance, Canada would be one of the countries to benefit from global warming, since it would extend the growing season and reduce the damage from severe cold or frosts. For example, some scenarios indicate that by 2050, growing

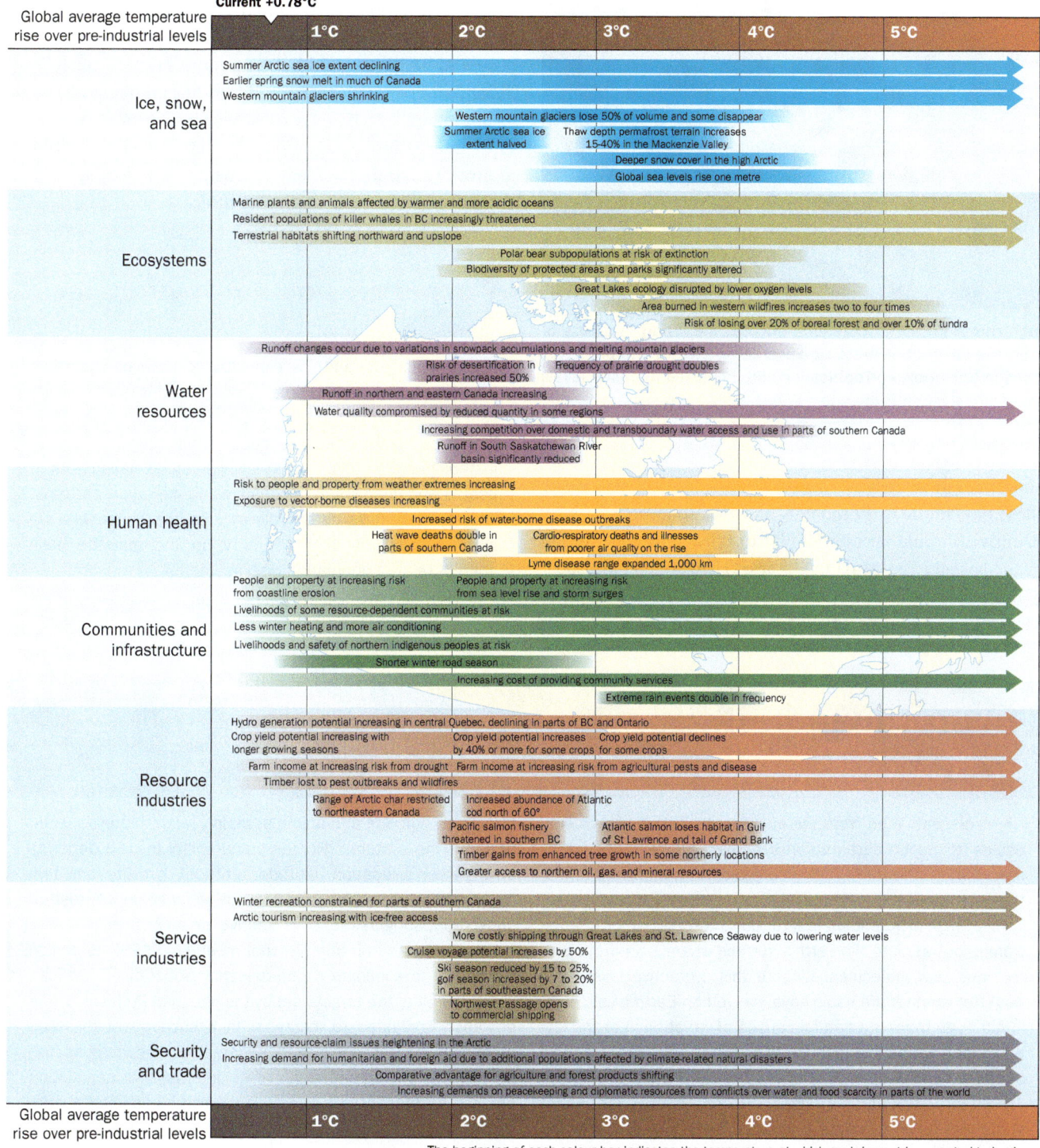

Figure 7.6 | Summary of the impacts of climate change expected in Canada over the twenty-first century. Source: NRTEE (2010: O15).

Although this rural scene may look bucolic, the rising numbers of cattle on the Earth contribute significantly to two of the problems discussed in this book: eutrophication and increased methane levels leading to global climatic change.

temperatures increased appreciably or water became limited, crops could be adversely affected. In the next subsection, we look at the effect of climate change on aquatic systems.

Moreover, climate change may have significant effects on food production in other regions of the world, and the first people to be hurt would be the poorest farmers. Such conditions and consequences could lead to significant increases in migration, which could cause regional instability, as well as in international migration. The latter would contribute to growing numbers of 'environmental' refugees for whom Canada could be a destination of choice.

Perspectives on the Environment

Climate Change and Environmental Refugees

Future climate change is expected to have considerable impacts on natural resource systems, and it is well established that changes in the natural environment can affect human sustenance and livelihoods. This in turn can lead to instability and conflict, often followed by displacements of people and changes in occupancy and migration patterns. Therefore, as hazards and disruptions associated with climate change grow in this century, so, too, may the likelihood of related population displacements.

—McLeman and Smit (2003: 6)

conditions in Whitehorse and Yellowknife would approximate those now found in Edmonton, and that conditions in New Brunswick would become similar to those now experienced in the Niagara Peninsula in Ontario (Hengeveld et al., 2005: 37). However, as in most situations, challenges also may arise. Many plants are vulnerable to heat stress and drought, and if

Environment in Focus

Box 7.3 Ozone Depletion

Ultraviolet radiation from the sun causes some oxygen molecules to split apart into free oxygen atoms. These may recombine with other oxygen molecules to form ozone (O_3) in the outer layer of the atmosphere, known as the stratosphere. This layer of ozone helps to filter out ultraviolet (UV) radiation from penetrating to the Earth's surface where it destroys protein and DNA molecules. Without this protective layer, it is doubtful whether life could have evolved on Earth at all.

Although there are natural causes of variation in ozone levels, observations suggest that this layer is being broken down by the emission of various chemicals from the Earth. Since 1979, the amount of stratospheric ozone over the entire globe has fallen by about 4 to 6 per cent per decade in the mid-latitudes and by 10 to 12 per cent in higher latitudes. These decreases have led to average increases in exposure to ultraviolet-b (UV-b) of 6.8 per cent per decade at 55°N and 9.9 per cent in the same latitude in the southern hemisphere. In general, penetration of UV-b radiation increases by 2 per cent for every 1 per cent decrease in the ozone layer. UV-b radiation is responsible for various potentially negative health effects on humans and animals, mainly related to eyes, skin, and immune systems. Human vulnerability to UV-b depends on a person's location (latitude, altitude), duration and timing of outdoor activities, and precautionary behaviour (use of sunscreen, sunglasses, or protective clothing).

Canada hosted an international meeting in 1987 to design a program to eliminate ozone-depleting substances. Despite achievement of the targets set by the **Montreal Protocol**, ozone depletion continued to increase. In a subsequent follow-up meeting in 2007, hailed by a UN undersecretary-general as 'the most important breakthrough in an international negotiation process for at least five or six years', also hosted by Canada, more ambitious targets were set, more in accord with the scientific target values (STVs). The new agreement saw developed countries capping production of ozone-depleting hydrochlorofluorocarbons (HCFCs) at 2009–10 levels by 2013, replacing the earlier date of 2016. Agreement was also reached to reduce production by 75 per cent by 2010 and by 90 per cent by 2015. Significantly, developing countries, including India and China, agreed to end production by 2020 rather than by 2030.

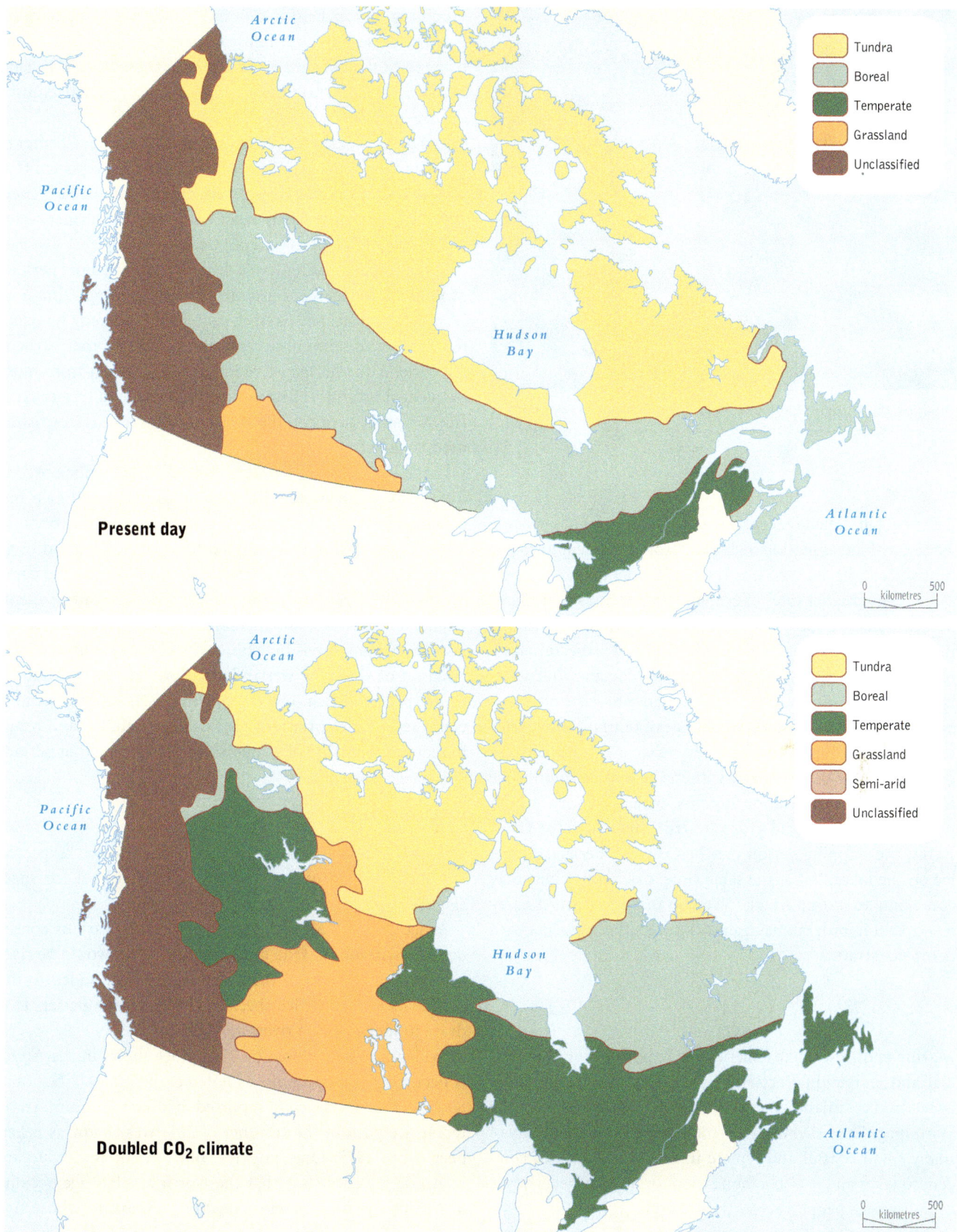

Figure 7.7 | Changes in forest and grassland boundaries resulting from a typical doubled CO_2 climate. The montane region of western Canada is 'unclassified' because altitude changes result in various terrestrial systems. *Source: Adapted from Hengeveld 1991, 44, and Thomas Curran, 'Forests and Global Warming' (Ottawa: Government of Canada/Science and Technology Division, 1991), found at http://publications.gc.ca/Collection-R/LoPBdP/BP/bp254-e.htm.*

Polar bear on the Hudson Bay coast near Wapusk National Park.

Freshwater Systems

Every part of Canada except the southern Prairies has become wetter, with precipitation increasing from 5 per cent to 35 per cent since 1950. At the same time, generally higher temperatures cause higher rates of evapotranspiration. What might be the outcomes? On the west coast of British Columbia, increased cloud cover and more rain can be expected. As a result, water supplies should be relatively secure, but tourism may be adversely affected if potential tourists avoid an area already known for its abundance of rainfall, sometimes referred to as 'liquid sunshine'. In other areas, agriculture operations may become more vulnerable, leading to pressure for expansion of irrigation systems, which may place high pressure on surface and groundwater systems. On the Great Lakes, the shipping season may be extended because of less ice on the lakes, but at the same time, drier conditions may contribute to a drop in lake levels so that lakers must carry less freight in order to navigate locks and other passages with depth constraints.

Fisheries

Marine and freshwater fisheries are important for commercial and recreational activity on the west, east, and Arctic coasts and on inland lakes such as the Great Lakes and Lake Winnipeg. The fishery is a key component of the economy of many small coastal and remote interior communities adjacent to lakes and is also a foundation for many aspects of life of indigenous peoples across the country.

Fish are vulnerable to changes in temperature, precipitation, wind patterns, and chemical conditions in or related to aquatic systems. Hengeveld et al. (2005: 39) have noted that climate change may be an important cause of the decline in salmon stocks in British Columbia and that water temperature increases in the Atlantic Ocean have most likely contributed to the decrease in flounder in those waters. In fact, changing oceanic conditions as a result of global climate change may well underlie the failure of the Atlantic cod stocks to recover from overfishing, as discussed in Chapter 8. On the other hand, in Arctic waters it has been reported that pink and sockeye salmon are being found well beyond their normal range, and this is most likely due to warmer water conditions in northern waters.

In the future, if water levels drop or there are more periods of lower water levels, the mortality of spawning salmon in BC rivers and streams is likely to increase, thereby reducing the number of salmon successfully completing their spawning cycle. A further negative impact could result from more frequent short, intense rainstorms, which could trigger flash floods that in turn could damage the gravel beds in streams used by salmon as spawning beds.

For freshwater systems, warmer water would enhance conditions for warm-water fish such as sturgeon and bass but create additional stress for cold-water fish such as trout and lake salmon. Fish better adapted to warm-water conditions could migrate into waters that have become warmer and compete with and perhaps even prey on species already present. Higher temperatures and increased evapotranspiration could also lower water levels in lakes, and one result would be degradation of shoreline wetlands that provide key habitat for some species of fish. On the positive side, warmer winters could reduce ice cover during the winter months and result in reduced mortality of some fish species during that period.

Cryosphere

Warmer temperatures in higher latitudes are expected to cause melting of ice, such as the Greenland ice sheet (Figure 7.8).

As ice cover in Arctic latitudes is reduced, various consequences will follow. One of the most obvious would be rises in sea levels because of the incremental water added to the oceans from melting ice sheets, as well as from glaciers that drain to the ocean via rivers and lakes.

Evidence indicates that melting of ice sheets in the Arctic is occurring. For example, as noted earlier, the US National Snow and Ice Data Center reported that the minimum Arctic Ocean ice pack in the summer of 2011 was as low as it had been since 1979. One 'positive' outcome of the reduction in Arctic ice cover was that the Northwest Passage became nearly ice-free and therefore readily navigable for about five weeks in the late summer. For some Canadians, however, easier passage raises concerns about Canadian sovereignty in this area, since non-Canadian ships without special ice-breaking capacity could navigate the passage.

Great Lakes freighter leaving Duluth harbour, Minnesota.

The record in the Arctic is paralleled in the Antarctic, where since 1950 the northwestern peninsula of Antarctica has warmed more than nearly anywhere else on the Earth. Winter temperatures have increased by about 6°C, and annual temperatures are up by 2.8°C. Ninety per cent of Antarctic glaciers are in retreat. Sea ice adjacent to the western Antarctic peninsula today remains for almost three months fewer annually compared to the extent in 1979.

Perspectives on the Environment

Glaciers Melting at Increasing Rate

In March 2008, the United Nations Environment Programme reported that data from about 30 glaciers in nine mountain ranges around the world indicated that these glaciers were thinning and melting at a rate that had more than doubled between 2004–5 and 2005–6. This finding reflected the pattern of accelerated ice loss from glaciers over the previous two and a half decades.

Achim Steiner, UN undersecretary-general and UNEP executive director, commented that 'There are many canaries emerging in the climate change coal mine. The glaciers are perhaps among those making the most noise, and it is absolutely essential that everyone sits up and takes notice.'

—UNEP (2008)

Melting of ice has not been confined to the polar regions. In the Rockies, glaciers less than 100 metres thick could disappear by 2030. Such glaciers are relied upon to provide water to the rivers that drain from the eastern slope of the Rockies across the Prairie provinces, causing concern about water security in the mid and long term.

Another consequence will be degradation of permafrost in both alpine and high-latitude regions. One result will be changes in the hydrology of aquatic systems in northern regions as well as stability of land. As Hengeveld et al. (2005: 41) have observed, 'decaying permafrost would also destabilize coastal land areas, increasing the risks of landslides and compounding the coastal erosion caused by rising sea levels and reduced ice cover.'

Another effect of reduced ice cover and degradation of permafrost would be a reduction of the expansive and stable ice cover that northerners count on for winter travelling as well as for hunting and other traditional activities. There also would be a reduced season and lower load capacity for 'winter roads' constructed over snow and ice and relied on for the delivery of goods to and from remote communities in Canada's Arctic and Subarctic. Winter roads also are important for transporting equipment, fuel, and other goods to the various mines across northern Canada and in helping to move the resources from these projects to southern markets.

The shrinkage of sea ice contributes to a decline in polar bear populations (Regehr et al., 2007). Furthermore, some communities, such as Arviat in Nunavut, have developed measures to protect the community from polar bears coming into the community while they wait for the delayed freeze-up of sea ice. The Nunavut government and the World Wildlife Fund have provided two large metal shipping containers in which Arviat residents, on western Hudson Bay, can store seal and caribou meat, while smaller metal bins are used to store food for dog teams. And dog teams are encircled by electric fencing. The intent is to proactively minimize the attractiveness of such a community to polar bears looking for food, especially since these predators are known to stalk and kill humans.

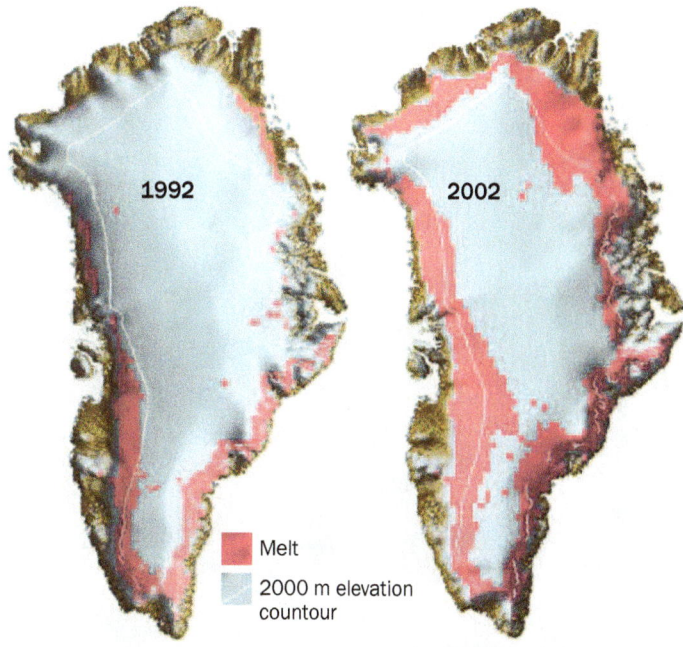

Figure 7.8 | Greenland ice sheet melt, 1992 and 2002. *Source: Walsh et al. (2004: 205).*

Environment in Focus

Box 7.4 Sea Ice in the Arctic and Implications for Cruise Tourism

Records show that surface air temperatures are increasing in the Canadian Arctic. The extent of northern hemisphere sea ice has also been decreasing since 1979. One conclusion is that these two patterns should lead to an increase in cruises in Arctic waters, which currently extend from about the end of July until mid-September.

Stewart et al. (2007: 377) observed, however, that the record of sea ice provides 'little to no evidence' in support of claims that climate change has affected sea-ice conditions in the Canadian Arctic to the extent that it would allow easier movement of ships through the Northwest Passage. In their words:

> While some increases in open water have been recognized, the navigable areas through the Northwest Passage actually have exhibited increases in hazardous ice conditions; navigation choke points remain and are due primarily to the influx of multi-year ice into the channels of the Northwest Passage. Thus, cruise operators working in the Northwest Passage face considerable uncertainty. In the future, rather than widespread accessibility, as some have claimed, there is likely to be much more variability of ice conditions across this region.

Whether such analysis is optimistic or pessimistic depends on one's point of view, but the fact is that cruise lines have dramatically increased their Arctic business over the past decade, including transiting the Northwest Passage—not always successfully. In 2010, one ship ran aground and had to be towed by a Canadian icebreaker to Kugluktuk, Nunavut, from where the adventure tourists were flown south.

Ocean and Coastal Systems

Both sea temperatures and sea levels will increase. The IPCC (2007b: 4) concluded that since 1961, the average global water temperature has increased to depths of at least 300 metres and that the ocean has been absorbing more than 80 per cent of the additional heat to the climate system. The IPCC also noted that this warming causes sea water to expand, which contributes to sea-level rise. The latter result will affect coastal communities, the severity depending on the nature of the coastline and the amount of increase. However, it is also clear that wave action may change, becoming more severe in areas previously covered by ice for part or most of the year. Wave action on shorelines will contribute to erosion and to changes in wetland complexes, both enhancing and damaging them.

Infectious Diseases

Given the predictions about climate change in North America—warmer temperatures, more rainfall—Health Canada (2007) has indicated that Canadians can expect to experience a greater incidence of disease.

Greer et al. (2008: 716) have noted that most people recognize the close association among climate, environment, and infectious disease in developing countries, but this association is not so obvious in nations like Canada because people normally enjoy clean drinking water, lower exposure to insect vectors, and higher-quality housing. All of these advantages reduce the vulnerability of residents. Nevertheless, Greer et al. conclude that predicted climate changes can be expected to increase both the incidence and burden of infectious diseases. Examples include:

1. Lyme disease (a tick-borne borreliosis), currently uncommon in Canada, is found mainly in southern Ontario and British Columbia. Temperature is the key factor limiting the northern extent of Lyme disease, and anticipated warmer temperatures may lead to expansion of the range for ticks 200 kilometres north by 2020. If this were to occur, Lyme disease could appear in Alberta and Saskatchewan.
2. Other diseases, such as dengue fever, could spread northward as the habitat for mosquitoes expands.

Glacial retreat of the Athabasca Glacier, 1992 to 2005.

3. Earlier springs will also facilitate expansion of West Nile virus, which first appeared in Canada in 2002.
4. More diseases may be imported as a result of Canadians travelling abroad and being exposed to disease vectors in subtropical and tropical countries. Dengue fever and malaria are expected to increase in Latin America, the Caribbean, Asia, and Africa, so it is highly probable that a greater number of returning travellers will carry these diseases into Canada.

Opportunities

While the above discussion identifies many challenges and potential problems, climate change may also create opportunities, as highlighted in Box 7.5. On balance, what conclusions do you draw about opportunities and challenges created by climate change?

A pair of trucks drive the Ice Road in the Northwest Territories. Truckers drive the frozen ice highway on the Mackenzie River in the few short months it is safe for use.

Environment in Focus

Box 7.5 Global Warming's Silver Lining? Northern Countries Will Thrive and Grow, Researcher Predicts

Move over, Sunbelt. The New North is coming through, a UCLA geographer predicts. As worldwide population increases by 40 per cent over the next 40 years, sparsely populated Canada, Scandinavia, Russia, and the northern United States will become formidable economic powers and migration magnets, according to Laurence C. Smith in *The World in 2050* (2010). . . .

While wreaking havoc on the environment, global warming will liberate a treasure trove of oil, gas, water, and other natural resources previously locked in the frozen North, enriching residents and attracting newcomers, according to Smith. And these resources will pour from northern rim countries—or NORCs, as Smith calls them—precisely at a time when natural resources elsewhere are becoming critically depleted, making them all the more valuable.

'In many ways, the New North is well positioned for the coming century even as its unique ecosystem is threatened by the linked forces of hydrocarbon development and amplified climate change', writes Smith, a UCLA professor of geography and of earth and space sciences.

Other tantalizing predictions:
- New shipping lanes will open during the summer in the Arctic, allowing Europe to realize its 500-year-old dream of a direct trade route to the Far East, and resulting in new access to and economic development in the North.
- Oil resources in Canada will be second only to those in Saudi Arabia, and the country's population will swell by more than 30 per cent, a growth rate rivalling India's and six times faster than China's.
- NORCs will be among the few places on Earth where crop production will likely increase due to climate change.
- NORCs collectively will constitute the fourth largest economy in the world, behind the BRIC countries (Brazil, Russia, India, and China), the European Union, and the United States.
- NORCs will become the envy of the world for their reserves of fresh water, which may be sold and transported to other regions.

An Arctic scientist who has consistently sounded alarms about the approach of global warming, Smith is best known for determining the role of climate change in the disappearance of more than a thousand Arctic lakes over the last quarter of the twentieth century. . . .

According to Smith, in the best-case scenario, climate change can be expected to raise temperatures an average of 3°C by century's end, a change greater than the difference between a record cold and a record hot year in New York. At worst, temperatures will rise twice as much. And don't expect relief from wind, solar, and hydrogen technologies, he warns. By 2050, they still won't satisfy global energy needs. In fact, growing water shortages may force societies to choose dirtier power sources, such as coal, over cleaner, water-intensive sources like hydroelectric power.

Smith paints a picture of wet regions of the globe getting wetter, parched regions becoming drier, and increasingly erratic and dangerous weather events.

As a result of these and other threats, wildlife will suffer the greatest rate of extinction since the disappearance of the dinosaurs 65 million years ago, he writes. Climate change will push wildlife that manage to survive northward and into higher elevations, with increasing hybridization between northern and southern species.

Hardest hit by the changes, Smith predicts, will be megacities in the developing world, which are already struggling to accommodate an influx of rural migrants, 'despite being hell on earth'. But not so in the New North. New prosperity awaits communities that lie north of the 45th parallel as global warming diminishes winter's severity and the world's energy appetite increasingly turns to natural gas and unconventional oil, he writes.

'In many ways, the stresses that will be very apparent in other parts of the world by 2050—like coastal inundation, water scarcity, heat waves and violent cities—will be easing or unapparent in northern places', Smith said. 'The cities that are rising in these NORC countries are amazingly globalized, livable and peaceful.'

Cities expected to increase in size and prominence over the next 40 years include Toronto, Montreal, Vancouver, Seattle, Calgary, Edmonton, Minneapolis–St Paul, Ottawa, Reykjavik, Copenhagen, Oslo, Stockholm, Helsinki, St Petersburg, and Moscow, he writes.

Of the 10 'ports of the future' cited by Smith, only three—Alaska's Prudhoe Bay, Canada's Churchill, and Iceland's Reykjavik—will sound familiar. Future beneficiaries of increased Arctic traffic will also include Nuuk in Greenland; Hammerfest, Kirkenes, and Tromsø in Norway; and Archangelsk, Dudinka, and Murmansk in Russia.

Although they will be facing severe threats to their traditional culture, northern indigenous communities can be expected to share in the wealth, Smith predicts. In the northern US, Canada, and Greenland, these societies are expected to trade harpoons for briefcases, as increasingly common self-determination agreements allow them to exploit natural resources just as climate change is making them more accessible.

'Northern aboriginal people don't like being portrayed as hapless victims of climate change', Smith said. 'They want the power and resource revenues to save themselves, and at least in North America, it looks like they'll have it.'

Research for the book in no way abated Smith's concern about the prospects of climate change, but it did leave him optimistic in a lot of ways. . . . 'There's a new part of the world that's emerging, with vast continents and a harsh geographical gradient but also resource and immigration bonanzas. Humanity will increasingly look north in response to the four global pressures of rising population, resource demand, globalization and climate change.'

Source: Adapted from *Science Daily*, 5 Sept, 2010, at: www.sciencedaily.com/releases/2010/09/100904164915.htm.

Adaptation to Climate Change

Five types of **adaptation** are usually recognized: (1) prevent the loss by adopting protective measures that reduce vulnerability; (2) tolerate the loss by doing nothing and absorbing the cost of losses when they happen; (3) spread or share the loss by distributing the costs over a larger population, such as through insurance; (4) change the affected activity by ceasing to do certain things or by shifting to other activities; and (5) change the location of the activity by moving to a less vulnerable location. These adaptations, except for the option of doing nothing, are discussed in three groups below: protective measures, accommodation, and retreat.

Protection usually involves structural measures to protect property, buildings, and infrastructure. These measures can involve individual initiatives or major public works projects, such as sea walls, revetments, and groynes designed to trap sediment and to protect coastal areas.

Accommodation usually involves a mix of approaches—redesign of structures to reduce their vulnerability, zoning to guide appropriate land use involving low capital investment in vulnerable areas, and other measures such as rehabilitating coastal dune systems, renewing wetlands, nourishing beaches, and replacing causeways with bridges. However, 'stabilizing' natural systems can undermine the natural functioning of ecosystems and can be counterproductive.

Retreat, the third general approach, seeks to avoid vulnerability. It usually involves recognizing the high risk or vulnerability of a place and taking a conscious decision to relocate buildings, other capital works, or infrastructure away from hazardous places. The initial cost of relocating is normally very high, but in the long term the costs are usually much lower than what would have been spent on rebuilding or repairing properties after each damaging natural event.

Communicating Global Change

In Chapter 6, we identified communication as one of the chief attributes of best practice related to resource and environmental management. Andrey and Mortsch (2000) have highlighted several challenges for communicating information or understanding about global change.

1. *Global change is a complex issue.* The global climate system is enormously complex, mainly because of the many linkages and feedback mechanisms in the atmospheric system. Furthermore, the associated socio-economic system is complex and continuously changing.

    A related complication is that while many people have heard about 'global warming' or 'global change', the level of in-depth understanding is usually poor. Polls consistently show that many Canadians have a poor understanding of the meaning, causes, or effects of global change. In addition, few see the connections between energy use and deforestation and climate change. Another complication is that the media often provide misleading or incorrect information. In that regard, Hengeveld (2006: 29) identified a particular problem for scientists related to media reporting:

> Scientists need to challenge journalists on their tendency to seek 'balanced reporting' by presenting opposing views of a topic with equal weight, without considering or reporting the credibility or marginality of these views. Such reporting can create a significant bias in communication—a bias that some argue is particularly apparent in high-profile North American media.

A further complication is that scientists and the general public often do not speak the same language when they talk of global change. As Chalecki (2000: A2, 15) observed, 'Scientists often examine small pieces of larger environmental problems in great detail within the limits of their discipline, while most non-scientists have a somewhat fuzzy understanding of the larger issues, often fed by outdated knowledge and half-formed opinions.'

2. *Uncertainties exist regarding almost every aspect of the global change issue, and these uncertainties increase when moving from natural to human systems.* There are four main sources of **uncertainty**: (1) statistical randomness, or the variability in nature; (2) lack of scientific understanding of the processes involved; (3) lack of or inadequate data; and (4) imprecision in risk assessment methods because of varying protocols for conducting research. All of these are relevant in global change research. They collectively contribute to uncertainty, which encourages a 'wait-and-see' attitude on the part of some policy-makers because they are skeptical about the information and understanding provided by scientists (Fraser Institute, 1999). Ehrlich and Ehrlich (1996) refer to such a view as 'brownlash' because the intent is to 'minimize the seriousness of environmental problems' and 'help to fuel a backlash against "green" policies'.

Erosion caused by high surf destroyed this home.

3. *The impacts of global change will be disproportionately heavier on people in less developed countries and on future generations.* Human-induced climate change impacts will fall mainly on future generations. Furthermore, areas at greatest risk are those with limited fresh water, prone to drought, along coasts, and generally in less developed nations. One consequence is that many people in developed nations, confronted by the various issues and problems in their lives, will give less attention to global change challenges. Most give priority to issues with some immediacy or urgency, and global change does not fall into that category. As Andrey and Mortsch (2000) observe, the consequences of climate change are diffuse rather than concentrated, indirect rather than direct, unintended rather than intended, and affect statistical or anonymous people rather than identifiable individuals.

4. *The basic causes of global change are embedded in current values and lifestyles.* In the developed world, including Canada, relatively high standards of living and materialistic lifestyles depend on extensive use of energy based on **fossil fuels**. Much of this use is devoted to residential heating and cooling and personal transportation by car. It is the cumulative effect of billions of people going about their normal lives that contributes to global warming. Thus, it is easy for any one individual to conclude that a change in his or her lifestyle will make virtually no difference, and the 'tragedy of the commons' is played out at a global scale. This creates a dilemma for any individual, city, province, or country, because the scale of the challenge requires unprecedented collaboration (see Chapter 6). In short, individuals believe that they are helpless to make a difference on their own, while for many people, more immediate issues compete for attention and resources.

Climate Change Deniers

Communication about climate change has become increasingly important, given that Hoggan (2009) has shown that **climate change deniers** have skilfully used communication tactics to question the science underlying climate change. As Anderegg (2010: 655) observes, 'Hoggan provides a sobering perspective of how public relations strategies can be used to propagate uncertainty and politically motivated messages into public understanding of climate change, even as the science itself accumulates evidence and certainty.'

Hoggan identifies various approaches or tactics used by climate deniers to highlight serious uncertainty associated with the findings and conclusions of climate change science. First, he draws parallels to tactics used by the tobacco industry when it argued that there was no direct link between cancer and smoking. Various variables other than anthropocentric actions are identified as causes. The purpose is to suggest that, because cause-and-effect relationships are difficult to establish, until clarification is achieved it is inadvisable to develop regulations. The real intent is to create doubt in the minds of citizens and policy-makers, with the result that introduction of regulations will be postponed.

A second tactic is based on 'Astroturf campaigns', in which well-funded organizations create what appear to be grass-roots criticism of climate change science. The reality is that the campaigns are facilitated by organizations with a vested interest, such as the fossil-fuel industry. The intent is to raise uncertainty about whether climate change is occurring. A key element in this tactic is use of 'experts' who state that they disagree with or challenge the science supporting climate change. The intent is to create doubt in the minds of the public. The media are often complicit in disseminating such views, given their commitment to 'balanced' reporting. As a result, the media usually provide comments from supportive and dissenting scientists, but rarely provide independent assessment about the proportion of researchers supporting or opposing the science underlying climate change. Nor do they often assess the qualifications of the spokespeople.

Third, a close cousin of the Astroturf campaigns is the tactic of assembling results of surveys of researchers critical of climate change science. Again, the dissenting commentators often are not scientists actively involved in climate change research, and the target audience of the survey results is not other scientists. Instead, the target is the general public and policy-makers, and the latter are often not scientists. As Hoggan observed, the ultimate goal is to convince both the general public and policy-makers that because there is much uncertainty and confusion associated with climate change science, the best course would be to wait. Thus, the climate 'deniers' become climate 'delayers'.

Perspectives on the Environment
Implications of Climate Denier Arguments

. . . the continuing popular belief in the existence of a scientific debate can only stand as a tribute to the success of the disinformation effort.

The upshot of that success is twofold. First and most obviously, it has prevented, in North America at least, the implementation of any serious policy initiatives to mitigate climate change. People who believe scientists may still be arguing over the details are not going to demand urgent action from their government, and—quite clearly—governments that are in thrall of (or in debt to) private interests that are profiting from inaction are not about to impose policies that will discomfit their voters and offend their financial backers.

—Hoggan (2009: 219)

Given the arguments presented above, what should you do when you hear someone arguing that the science underlying climate change does not provide a solid basis to develop actions? We suggest you ask yourself the following questions posed by Hoggan (2009: 231):

- Does the 'expert' have relevant credentials? That is, does he or she have a credible academic background, such as degrees in physics, chemistry, etc.
- Is the 'expert' actively engaged in scientific research related to climate change? And, if so, is he or she publishing regularly in peer-reviewed journals? Or, are their written statements found only in 'op-ed' commentaries or letters to the editor in newspapers or magazines?
- Is the 'expert' being paid by groups with a direct interest in climate change research and policies, such as petroleum companies or associations, or is the person affiliated with a think-tank with an ideology opposing government being involved in solutions to challenges created by climate change?

Notable Climate Denier Events

Two examples highlight how deniers or skeptics create doubt and uncertainty about the science associated with climate change.

One example relates to the reports from the IPCC in 2007, already referenced in this chapter. The thousands of pages in the multi-volume report made it likely, notwithstanding a peer-review process, that some errors would be included. And, indeed, that happened. Particular attention was drawn to a statement that glaciers in the Himalayas were likely to disappear by 2035. It was acknowledged later by those responsible for the report that the statement about the Himalayan glaciers was an error. However, that acknowledgement did not stop citation of this error time and time again across the Internet and in many other media, with no comparable acknowledgement that in nearly all parts of the world there were significant reductions in the extent and magnitude of glaciers. The main message from the deniers was simple: given such an error, why should other conclusions in the IPCC reports be trusted? On the other hand, other statements in the IPCC reports, which could be claimed to be underestimates or understatements of the rate of change of climatic conditions, never received commentary.

Another example was the leaking of e-mail messages and other documents from scientists at the University of East Anglia Climatic Research Unit during November 2009. Comments in those messages that appeared to indicate the scientists were actively trying to suppress evidence contradicting results from their research were emphasized. The release of this information, a few weeks before the United Nations-sponsored climate change summit in Copenhagen during December 2009 (discussed later in this chapter), was viewed by some as a deliberate act to undermine the Copenhagen meetings. Indeed, some commentators remarked that the release of the e-mails from the research institute helped to energize the climate change skeptics and also contributed to destabilizing the summit.

The controversy created became known as '**Climategate**'. It led to the head of the research institute being placed on temporary leave and four reviews being conducted. The final review report, published in July 2010 and authored by Sir Muir Russell, a former civil servant, concluded there was no evidence of dishonesty or corruption by the scientists. However, he also commented that the scientists in the institute should have done a better job in sharing their data with critics of their work.

Perspectives on the Environment

Climate Change Deniers and Resistance to Change

In the long run, the climate-deniers will be a footnote to history. But by delaying action, they will have helped prevent us from taking the steps we need to take while there's still time.

If we're going to make real change while it matters, it's important to remember that their skepticism isn't the real root of the problem. It simply plays on our deep-seated resistance to change.

That inertia is what gives the climate cynics ground to operate. That's what we need to overcome, and at bottom that's a battle about data, but also about courage and hope.

—McKibben (2010: A9)

What is your view about the position and tactics of the climate deniers and skeptics? What are your conclusions about how communication can and should be used to help people understand that uncertainty is a core element of science? How can we move forward, when we will rarely have complete understanding of the natural and human environments, and the way in which they interact?

Kyoto Protocol

During December 1997, representatives from more than 160 countries met in Kyoto, Japan. The outcome was an agreement, referred to as the **Kyoto Protocol**, with targets for 38 developed nations as well as the European Community to ensure that 'their aggregate anthropocentric carbon dioxide equivalent emissions of the greenhouse gases [e.g., carbon dioxide (CO_2), methane (CH_4), nitrous oxide (N_2O), hydrofluorocarbons (HFCs), perfluorocarbons (PFCs), sulphur hexafluoride (SF_6)] . . . do not exceed their assigned amounts . . . with a view to reducing their overall emissions of such gases by at least 5 per cent below 1990 levels in the commitment period 2008 to 2012' (Kyoto Protocol, 1997: Article 3).

Delegates from more than 160 countries at the conference in Kyoto, Japan, that led to the Kyoto Protocol, December 1997.

The Protocol would become legally binding when ratified by at least 55 countries accounting for at least 55 per cent of the developed world's 1990 emissions of carbon dioxide. Table 7.1 shows the targets for selected countries. Developing countries, including China and India, were not included in the targets because their per capita emissions were much lower than those of developed countries. Another reason was that their economies were judged to be much less able to absorb the costs of changing to cleaner fuels, since the main source of greenhouse gas emissions is carbon dioxide from use of fossil fuels.

Table 7.1 | Greenhouse Gas Emission Reduction Targets under the Kyoto Protocol for Selected Countries

Country	Reduction commitment as percentage of base year (1990)
Australia	108
Canada	94
European Community	92
France	92
Germany	92
Iceland	110
Japan	94
Netherlands	92
New Zealand	100
Norway	101
Russian Federation	100
Sweden	92
Ukraine	100
United Kingdom	92
United States	93

Source: Kyoto Protocol (1997: Annex B).

Canada was to reduce greenhouse emissions to 6 per cent below 1990 levels by between 2008 and 2012. Canada ratified the Protocol in December 2002. However, at the end of March 2001, shortly after taking office, US President George W. Bush stated that he opposed the Kyoto agreement, the US would not agree to it, and the US would develop its own approach. He argued that it was inappropriate for China and India, countries with the largest populations, not to be included in the Kyoto targets. Bush was quite correct that these two countries have the largest populations, but he ignored the fact that their per capita emissions of greenhouse gases are much lower than those of the United States, which has the worst record. With only 4 to 5 per cent of the world's population, the US accounts for about 25 per cent of the global emissions of greenhouse gases.

Bush explained in February 2002 that the US would use a 'voluntary approach' related to greenhouse gas emissions, with the purpose of reducing 'greenhouse gas intensity' by 18 per cent over 10 years, a general approach later endorsed by Canadian Prime Minister Stephen Harper. Greenhouse gas intensity is the ratio of greenhouse gas emissions to economic output. Unlike the Kyoto Protocol, which requires an absolute reduction in greenhouse gas emissions, the American approach would result in emissions continuing to increase as its economy grows but at a slower rate than without this arrangement.

The Bush approach was based on determination to protect the US economy in the short term, ensure that jobs were not lost because of the costs associated with reducing emissions, and maintain its international economic competitiveness. Because the US is such a dominant player in the global economy, its position has been cited by people in other countries, including Prime Minister Harper in Canada, who believe it would be economically foolish to accept the Kyoto Protocol targets when the nation with the largest economy has decided not to do so.

Against this background, we can now turn to some of the specific aspects of the Protocol and then to Canada's approach.

Specific Features of the Kyoto Protocol

Legal Basis

Unlike the Framework Convention on Climate Change signed at the Earth Summit in Rio de Janeiro in 1992, which committed countries only to 'aim' to stabilize emissions at 1990 levels by 2000, the Kyoto Protocol commitments are legally binding on nations under international law.

Assigned Amounts

For the period 2008–12, the Protocol states that overall average emissions are to be 94.8 per cent relative to 1990 levels. 'Assigned amounts' are identified for each developed nation. While the targets were set for allowed emissions with reference to population, gross national product, and carbon intensity of economies, the final targets were determined politically. Canada was to reduce its emissions by 6 per cent. Australia was allowed to increase emissions by 8 per cent, and Iceland could increase by 10 per cent.

Greenhouse Gases

The Protocol identified six greenhouse gases. Three are viewed as the main greenhouse gases produced by human activity: carbon dioxide, nitrous oxide, and methane. The other three—hydrofluorocarbons, perfluorocarbons, and sulphur hexafluoride—are released in small quantities but are long-lasting and significant contributors to climatic change.

Exclusion of Most Forest and Soil Sinks

The assigned emission amounts for most nations are a percentage of gross emissions in 1990. Gross emissions are the anthropocentric (human-caused) greenhouse gas emissions from energy, industrial processes, agriculture, and waste. However, they do not include carbon fluxes from forests, soil, and other carbon reservoirs.

When a nation calculates whether it is in compliance with its target emissions, it must count emissions and carbon flux changes due to afforestation, reforestation, and deforestation since 1990. In Canada, the view of the federal government after signing the Protocol was that it could interpret the Protocol to include loss of carbon from agricultural soil in calculating the balance between emissions and carbon flux removal. Indeed, the target for Australia of 108 per cent of 1990 emissions was partly based on arguments that it had positive net emissions related to land-use change and forestry in 1990.

Because of methodological challenges in measuring emissions from land-use change and forestry, some observers were concerned that countries would use forest and soil sinks to claim credits that are difficult to verify. There are also problems in reaching agreement about the meaning of key terms such as reforestation, afforestation, and deforestation. In Canada, reforestation is normally interpreted to mean replanting and natural regeneration after logging, and afforestation to mean planting trees in areas traditionally forested. In contrast, the Intergovernmental Panel on Climate Change defines reforestation as planting of forests on lands historically forested but later converted to another use and afforestation as planting of new forests on lands historically not forested. These different interpretations are important. If reforestation is interpreted to include planting trees after harvesting, as Canada does, then emissions from harvesting are not counted. As a result, only the credit from planting would be used in calculating the carbon reservoir credit.

At a conference in Milan during December 2003, the signatories to the Kyoto Protocol reached agreement on how industrialized countries can earn credit towards their emission targets by preserving or establishing forests.

Clean Development Mechanism

Emission reduction commitments can be fulfilled through a clean development mechanism, allowing **emission credits** in countries not given targets through the Protocol to be used by countries included in the Protocol targets. Initiatives are certified as satisfying the clean development mechanism when they involve voluntary participation by each party; real, measurable, and long-term benefits for **mitigation** of climate change; and emission reductions in addition to those that would have occurred without the initiative. Such initiatives are also intended to help the host nation. The Protocol allows countries with targets to meet their commitments for the 2008–12 period by claiming certified clean development emission reductions achieved between 2000 and 2007.

The major concern about the clean development mechanism is that emission credit may be given for projects that would have occurred without such a mechanism in place. The Protocol only requires reductions in emissions that would be in addition to what would have occurred without an initiative.

Emissions Trading

Under the Protocol, a country can meet its emission commitments by acquiring from other countries 'emission reduction units' related to initiatives that cause a reduction in emissions or enhancement of sinks incremental to what would otherwise have occurred without the initiative. When a nation buys some emission reduction units, they are added to its allowable emissions and subtracted from the allowable emissions of the selling country.

However, because developing countries are not given emission targets under the Protocol in order to allow them to develop their economies, they cannot sell emission credits to developed countries, even when such sales could benefit them economically. Developing countries also cannot agree to voluntary emission targets, which could benefit some if they could introduce low-cost emission reductions and then sell emission credits.

The theory of **emissions trading** is based on the belief that it is more efficient for one country to purchase emission credits from another country that can generate credits in a less costly manner. In such a way, overall compliance goals can be reached at a reduced overall cost.

Canada's Initial Approach to Implementing the Kyoto Protocol

On 10 December 2002, the House of Commons voted 195 to 77 to ratify the Kyoto Protocol. The Liberal majority led those supporting the motion, while the Canadian Alliance and Progressive Conservative parties opposed it. Through this vote, Canada committed itself to cut average greenhouse gas emission levels to 6 per cent below 1990 levels by 2008–12.

Prime Minister Jean Chrétien signed Canada onto the Kyoto Protocol on 16 December 2002.

Canada ratified the Kyoto Protocol without a clear plan on how it would be implemented. The Prime Minister argued that details would be worked out. This position was supported by the Minister of the Environment, David Anderson, who compared the challenge of global warming to World War II. He was often quoted as saying that 'If Winston Churchill had said in 1939 that we are not going to challenge the Nazis until we know exactly how much it will cost and how long the war will last, then we would have never won the war.' In contrast, Stephen Harper, leader of the then Canadian Alliance, argued that it was inappropriate to ratify the Protocol without providing a clear plan as to how it would be achieved. His view was that 'I don't think we have any idea what the government is going to do' and that the implementation would turn into a great disaster for Canada and Canadians.

The federal government allocated $2 billion in its 2003 budget for Kyoto Protocol initiatives over a five-year period. One specific initiative, the creation of Sustainable Development Technology Canada, was identified. This arm's-length foundation was to receive $250 million to support technologies so new that they were not yet commercially viable. The remaining $1.75 billion was to be allocated to other initiatives once the details were worked out, with the budget indicating only that the money would be 'to support climate change science, environmental technology, and cost-effective climate change measures and partnerships in areas such as renewable energy, energy efficiency, sustainable transportation, and new alternative fuels'.

The components of a plan for implementing the Kyoto Protocol included requiring major industrial emitters to reduce their greenhouse gas emissions, taxes on private vehicles such as sports utility vehicles, minimum requirements for fuel alcohol, subsidies to install energy-efficient windows, and an emissions trading framework.

The federal government compiled a list of large industrial emitters of greenhouse gases, such as oil and gas, mining, and pulp and paper companies, each of which generates large amounts of carbon dioxide per unit of product. Each industry was to be required to reduce emissions extensively.

Accomplishments in Reducing GHG Emissions in Canada

In the early fall of 2004, the federal deputy minister of natural resources, speaking at a conference in Australia, stated for the first time that Canada would not meet its commitments under Kyoto and would be unlikely to realize even two-thirds of the reductions of GHGs that Canada had committed to achieve under Kyoto. Indeed, by December 2004, data indicated that emissions of GHGs in Canada had increased by 20 per cent relative to the base year of 1990.

A Change of Federal Government and a Change of Course

In January 2006, a federal election led to a minority Conservative government. As leader of the opposition, Harper, the new Prime Minister, had consistently opposed the Kyoto Protocol, arguing that it would hurt economic growth in Canada. In addition, Harper had frequently expressed doubt about the credibility of the scientific conclusions that climate change was a significant issue.

Once in office, Harper indicated that Canada's Kyoto commitment was unrealistic and unachievable. He also noted that if Canada were to remain economically competitive with its largest trading partner, the United States, it would be unwise to aggressively seek to reduce GHG emissions unless and until the US had also accepted a binding target to reduce GHGs. Furthermore, he argued that it was not reasonable that countries such as China and India were not included in the agreement with binding targets.

In 2007, Prime Minister Harper attended an APEC (Asia Pacific Economic Co-operation) summit of 21 Pacific Rim countries in Australia. The main outcome was a joint statement endorsing a long-term but unspecific target to cut GHG emissions by all participants, including the US and China. This unspecified non-binding target was characterized as an **aspirational approach**. This was the first time that the United States and China had agreed to sign a climate change agreement. In addition, other developing countries had agreed to participate, something they had not been required to do under the Kyoto Agreement. Harper was reported as having said at the summit, 'Kyoto divided the world into two groups, those that would have no targets and those that would reach no targets. The reality is that the world is now making efforts toward a new protocol post-2012.' The Prime

The role of the oceans in helping to mitigate the impacts of global warming through absorption of carbon dioxide is still uncertain, as is the oceanic response to warmer temperatures. Scientists are already detecting larger wave swells in many parts of the world that may be linked to these changes.

Minister acknowledged that much more work needed to be done but that having China and the United States 'on board' was a noteworthy accomplishment.

UN-Sponsored Climate Change Conference, Bali, December 2007

In December 2007, a major UN climate change conference took place in Bali, Indonesia, with 192 countries attending over a two-week period. The purpose of the **Bali Conference** was to start a process for creating a new framework to replace the Kyoto Protocol when it expires in 2012. At the conference, UN Secretary-General Ban Ki-moon remarked that 'The situation is so desperately serious that any delay could push us past the tipping point, beyond which the ecological, financial, and human costs would increase dramatically.' Global attention to climate change had increased after the 2007 Nobel Peace Prize was awarded in mid-October jointly to former US Vice-President Al Gore for his climate change activism, including the film *An Inconvenient Truth*, and to the UN Intergovernmental Panel on Climate Change in recognition of the significance of their work and the seriousness of climate change for worldwide stability and peace.

Throughout the Bali Conference, federal government representatives from Canada and the United States argued that GHG reduction commitments should be required for all countries but that specific numerical targets should not be identified. At the last moment during negotiations, the US government shifted its position regarding explicit targets after a delegate from Papua New Guinea made the following remark directed at the US representatives: 'We seek your leadership. But if for some reason you are not willing to lead, leave it to the rest of us. Please get out of the way.'

Canada's position at the conference was labelled as that of a 'climate hypocrite' by Yvo de Boer, the head of the UN climate change agency, because Canada was calling for binding targets on developing countries but refused to accept them for itself under Kyoto. Rajendra Pachauri, the head of the IPCC, suggested that Canada's position made it clear that the Canadian government was not prepared to take significant action on climate change. During the conference, Canada received numerous 'Fossil of the Day' awards created by non-government organizations.

Canada also lobbied to change the base year of 1990 established for the Kyoto Protocol. The base year of 1990 has been the accepted international standard since the Kyoto Protocol was created in 1997, and any decision to move that date forward would reward nations that had been slow in curbing GHG emissions. In that regard, the Conservative government's approach to the baseline year has been to refer not to 1990 but to 2006, the year that the Conservatives were elected to power.

The outcome at Bali was mixed. The participating nations agreed to continue meeting to determine what to do about climate change and in particular how to reduce GHG emissions.

Environmentalists dressed as polar bears demonstrate in front of the conference centre where the negotiations for a post-Kyoto agreement took place during the UN Climate Conference in Nusa Dua, Bali, Indonesia, in December 2007.

The negotiations were to be completed to allow time for a ratified agreement by individual nations before the Kyoto Protocol ended in 2012.

Perspectives on the Environment

China and Climate Change

Although China is often criticized for emphasizing economic development without regard for accompanying pollution, Gorrie (2008) has argued that China is moving aggressively to cut all types of GHG emissions and is going well beyond what Canada and the US have done in terms of regulations and targets for car emissions, renewable fuels, carbon storage, forest renewal, energy efficiency, and industrial pollution.

In terms of fuel efficiency standards for automobiles, Gorrie notes the following: China: 36 miles per gallon (current), 43 by 2009; Canada: 27 miles per gallon (current average, no standard), an unspecified standard, starting in 2011; US: 25 miles per gallon (current average, with 35 proposed by 2030); California: 25 miles per gallon (current), 36 proposed by 2016; Europe: 40 miles per gallon (current), 48.9 proposed by 2012; Japan: 40 miles per gallon (current), 48.9 proposed by 2015.

Post-Bali: Copenhagen (2009) and Cancún (2010)

Copenhagen

The two-week **Copenhagen Summit** has been viewed as either a total failure or a modest success. Those viewing it as a failure believed giant strides forward were essential and noted that (1) the 192 countries represented at the conference were unable to develop a new legally binding agreement to replace the Kyoto Protocol; (2) no firm national targets for GHG emissions were included in the declaration (the Copenhagen Accord) and instead nations agreed to set their own emission reduction targets; (3) no deadlines were specified for future action; and (4) there was a deep divide between developed and developing countries, with developed nations not prepared to take actions to reduce GHGs unless developing countries did the same, and developing countries unwilling to curtail GHG emissions if that would dampen economic growth needed to overcome huge poverty problems.

Perspectives on the Environment

An Asymmetrical Climate Deal

The reason no deal was possible is that public opinion in the developed countries is still in denial about the fact the final climate deal must be asymmetrical. Until the general public grasps that, especially in the United States, there will be no real progress.

. . . The developed countries must cut their emissions deeply and fast, and give the developing countries enough money to cover the extra cost of growing their economies with the clear sources of energy that they must use instead of fossil fuels. That's the deal, but most voters in the United States don't understand it yet.

—Dyer (2009: A11)

Those who concluded the conference was a success stressed that expectations for a legally binding treaty were unrealistic, and that 'baby steps' forward represented a positive outcome. Specific achievements viewed as positive included (1) an agreement between both developed and developing nations that GHGs need to be limited and that global temperatures need to be stopped from rising by more than 2°C; (2) the first time major developing countries such as China, India, and Brazil agreed their GHGs needed to be curtailed; and (3) establishment of a Copenhagen Green Climate Fund, with the intent to have primarily developed nations contribute $100 billion (US) by 2020 to help developing nations cut carbon emissions, and developing countries agreeing to make their records open for international scrutiny regarding emissions plans.

At the conclusion of the Copenhagen Summit, Prime Minister Harper stated that the agreement reached was 'comprehensive and realistic' and thus was supported by Canada. In contrast, media commentators suggested that the Copenhagen Summit had been a 'public relations nightmare' for Canada. At the very popular and public Fossil of the Day award ceremonies each day at the Summit, Canada was invited to accept the award, on its own or as part of a group of countries, 10 times, more than any other nation. In conferring the award to Canada, it was stated that the target by Canada for reducing GHG emissions was among the worst in the industrialized world, and its plan was so weak that its modest targets were unlikely to be achieved. Canada also received criticism from developing nations and environmentalists for refusing to make concessions to help reach an agreement, and thereby lost its traditional role as a progressive player on the global stage.

The level of commitment, or not, to the Summit by the Canadian government was suggested by some to have been demonstrated when the Prime Minister attended a dinner hosted by the Queen of Denmark on the evening when Canada made its official presentation. The Minister of Environment, Jim Prentice, argued in a three-and-a-half minute speech that the Kyoto Protocol needed to be replaced, that all major emitters had to be signatories to any agreement, and that Canada's climate change strategy reflected the reality that Canada had a large and diverse land mass and its energy sector was important for meeting global demand. Prentice also noted that Canada's approach reflected its strong economic ties with the United States, meaning a need to align Canada's approach to that of the US, and that Canada would contribute to the proposed climate-aid fund but he would not speculate on what amount Canada would provide. (US Secretary of State Hillary Clinton earlier at the Summit had committed the US to provide $100 million over a decade to help the most vulnerable nations cope with climate change).

Prentice's remarks, and Harper's choice to attend a royal dinner rather than his minister's Summit presentation, led many to conclude that the Canadian government viewed climate change only as a cost, not an opportunity. As one

A journalist reads a draft of the Copenhagen Accord at the December 2009 climate conference in Copenhagen, Denmark. Some feel the conference was a total failure while others argue that the talks inched participating countries closer to action.

columnist wrote, 'Canada has had a rough ride at the Copenhagen climate change summit. It is unloved, even despised, by the scientists, the environmental groups and most developing countries. It will leave the summit with a sullied image—the arrogant rich country that is part of the problem, not part of the solution' (Reguly, 2009: A16).

Cancún, December 2010

The **Cancún Summit** followed one year after the Copenhagen Summit, with 193 nations represented. As at Copenhagen, some modest gains were achieved, but major issues were left unresolved. In terms of positive outcomes, the Cancún agreements included: (1) a general framework to assist developing nations in reducing their carbon output and dealing with negative effects of climate change; (2) commitment by developed countries to provide $30 billion (US) to support climate action in the developing world up to 2012, with the intent to raise $100 billion by 2020; (3) technology transfer from developed countries to developing countries to enhance adaptation and mitigation; and (4) endorsement of a system to compensate developing nations for not cutting down trees in rain forests. Furthermore, for those who support **geo-engineering** initiatives, there appeared to be willingness to examine them as one option to complement other mitigation strategies.

However, various major issues were not resolved. Specifically, delegates at the conference postponed determining how developed and developing nations will work collaboratively to make significant reductions in GHG emissions over the next 10 to 15 years. Thus, no progress was made at either Copenhagen or Cancún on this issue. In addition, the specific contributions developed nations would make to the green climate fund were never clarified, nor was it determined whether the Kyoto Protocol would be extended after 2012 if a new agreement is not reached. These limitations from Cancún were captured nicely by one commentator's words that 'the planet is being scorched by climate change, and the best the international community can do is shoot back with a squirt gun' (*The Record*, 2010).

At Cancún, the Canadian government maintained its positions from Copenhagen. And, as at Copenhagen, Canada was not well regarded for its stance or performance. To illustrate, the NGO Germanwatch conducted a survey of 190 climate experts regarding efforts by nations to address climate change. The survey considered actual measurements of each country's GHG emissions, whether they had been increasing or decreasing, and their national policies. Of the 57 countries identified in the survey, Canada was ranked fifty-fourth, ahead of only Australia, Kazakhstan, and Saudi Arabia. Canada placed last among the top 10 carbon dioxide emitters and second to last among developed nations. The United States was ranked fifty-first, while the countries ranked best overall were Brazil, Sweden, and Norway, in that order.

The next climate change conference sponsored through the United Nations Framework Convention on Climate Change was scheduled for South Africa from 28 November to 9 December 2011. The ultimate objective was to 'stabilize greenhouse gas concentrations in the atmosphere at a level that will prevent dangerous human interference with the climate system'. There is some distance to travel on the journey to this destination, with the probability high that there will be many twists and turns on the road.

Domestic Approach to Climate Change

As shown from the discussions at the international conferences in Bali, Copenhagen, and Cancún, the federal Conservative government's position has been consistent: (1) the Canadian approach to climate change has to be harmonized with the United States, given the importance of trade with the US; (2) Canada will not unilaterally take a leadership role at a global scale because that would hurt Canada's economic competitiveness and thereby negatively affect the standard of living for Canadians; (3) all countries, especially China and India, must be part of any international agreement, targets, and timelines; (4) the Kyoto targets are unrealistic, and 'intensity-based' emission regulations should be used instead; (5) 2006 rather than 1990 should be used as a baseline against which to measure GHG emissions; and (6) a sector-by-sector approach will be used to recognize the needs of different industrial groups.

Perspectives on the Environment
The Canadian Government's Lack of a Climate Change Strategy

About 20 years ago, the federal government acknowledged that the impacts of climate change would pose significant, long-term challenges throughout Canada from more frequent and severe storms in Atlantic Canada to changes in the amount of rain available to farmers. And, today, the federal government still lacks an overarching federal strategy that identifies clear, concrete actions supported by coordination among federal departments.

—Commissioner of the Environment and Sustainable Development (2010: 2)

Reflecting the above views, the Conservative majority in the Senate defeated the Climate Change Accountability Act by a vote of 43–42 in mid-November 2010. Marjorie LeBreton, the leader for the government in the Senate, stated that Conservative-appointed senators voted against the bill because it would have hurt Canada's economy, the view of the federal government when the bill was passed by the House of Commons in May 2010. The proposed legislation called for cuts to GHGs of 25 per cent below 1990 levels by 2020. In contrast, the Conservative government argued that a cut of 17 per cent relative to emission levels in 2005 to be achieved by 2020 was appropriate, and was the same target identified by the Obama administration for the United States. By recalibrating its GHG reduction target, Canada became the only country to reduce its aspiration level after the Copenhagen conference.

Opposition parties have argued that, by directly connecting the Canadian approach to that in the US, the federal government is forfeiting Canadian sovereignty. The opposition also has noted that with Republicans in control of the US Senate starting in January 2011, little progress could be expected regarding climate change in the United States for at least two years. Indeed, President Obama remarked in early 2011 that he did not expect any major climate change legislation to pass in the following two years. In response, the new federal Minister of Environment, Peter Kent, was quoted in late January 2011 as saying that 'We've made a decision to proceed on the basis of continental realities, with regulation. We know that will work.' This position was reiterated by the Conservative government in the campaign leading to the federal election in May 2011, in which the Conservatives achieved a majority.

Kent's comments were made when he rejected a proposal from the National Round Table on the Environment and the Economy (2011a) to create a 'made-in-Canada' strategy based on a cap-and-trade market to encourage GHG cutbacks by emitters. In contrast to Kent's view, in its conclusions the National Round Table (ibid., 129) remarked that:

> The federal government has indicated that Canadian climate policy will be harmonized with US climate policy as much as possible, given the integrated nature of our two economies. Overall, this is a sensible and realistic approach.
>
> But how we pursue that goal matters just as much. As the US struggles internally with its domestic climate policy, Canada must protect its own interests, both environmentally and economically.... Canada faces some economic competitiveness risks in moving too far ahead of the US, but also faces both environmental and economic risks by simply waiting. Delay leads to rising carbon emissions each year, and a higher financial and economic cost in ultimately acting to meet our stated GHG emission targets for 2020 or beyond. Neither outcome is desirable or inevitable.

Provincial governments do not necessarily share the view of the federal government. In June 2008, the premiers of Quebec and Ontario, Jean Charest and Dalton McGuinty, announced that their two governments had agreed to develop and implement a 'cap-and-trade' plan to reduce carbon emissions because they did not believe that the federal Conservative government's strategy to develop 'intensity-based' emissions targets would be effective.

A joint cap-and-trade strategy means that companies producing fewer GHG emissions than permitted under their caps can sell their unused quota on an open market as credits to other companies exceeding their caps. In the view of Charest and McGuinty, a cap-and-trade system would lead to an absolute reduction in GHG emissions from central Canada relative to 1990 as a base year, the year identified in the Kyoto Protocol. Their view was that the federal intensity-based emissions, in contrast, would lead to a steady increase in GHG emissions and that the Conservative plan to use 2006 as its base year results in a much higher level of baseline emissions.

By the fall of 2010, the provincial governments of Ontario, Quebec, Manitoba, and British Columbia, along with seven

Perspectives on the Environment

A Made-in-Canada Climate Plan?

And the real challenge for Canada's prime minister is whether he is able to create a made-in-Canada climate plan—or leave it to American lawmakers to decide our climate and, therefore, economic policy.

The rationale [in waiting for the US to create an American climate change policy] is that Canada needs to harmonize its policy with the US because of our strong economic ties—yet [Environment Minister] Prentice denied that Canada was abandoning its sovereignty by noting Canada is developing its own climate policy. If there is a made-in-Canada approach, can we please see it?

—Marlo Reynolds, Pembina Institute (2009)

Perspectives on the Environment

The Commissioner of the Environment and Sustainable Development on Government Climate Change Initiatives

The Government of Canada has committed to addressing climate change by reducing its national greenhouse gas (GHG) emissions in various plans and agreements since 1992. Since this time, however, national GHG emissions have risen and were 24 per cent higher in 2008 than in 1990 and 31 per cent higher than Canada's Kyoto target.

In 2010, the federal government committed to a new GHG emission reduction under the Copenhagen Accord.

... However, the 2010 plan does not contain measures with GHG emission reductions sufficient to achieve the level required to meet the obligations of the Kyoto Protocol or the Kyoto Protocol Implementation Agreement. Furthermore, expected emission reductions reported in the plans have been revised downward by more than 90 per cent between 2007 and 2010.

—Commissioner of the Environment and Sustainable Development (2011: 47–8)

US states, agreed to collaborate in developing a regional cap-and-trade system by 2012 because they believed there would be no significant climate change initiatives from either Ottawa or Washington. In addition, Saskatchewan, Nova Scotia, New Brunswick, and Yukon would be involved as observers. Notably, the two major oil-producing provinces, Alberta and Newfoundland and Labrador, are neither participants nor observers. The Ontario Minister of Environment, John Wilkinson, commented in November 2010 that 'It is important for us to signal to the federal government that provinces representing the vast majority of the economy of Canada are prepared to enter into a cap-and-trade system' (Babbage, 2010).

The provincial initiative has precedents in the United States. While the US federal government has not been a leader, some states have taken leadership roles to move away from reliance on fossil fuels. For example, in January 2010 California approved an implementation plan for Low Carbon Fuel Standards, which impose a stringent new pollution standard on imported fuel used for transportation. Effective in 2011, the standard sets a threshold of 96.88 grams of carbon dioxide equivalent per megajoule of fuel, a standard challenging for even corn-derived ethanol to meet. One implication is that this standard could result in California not accepting petroleum from Alberta's oil sands. Eleven other states in December 2009 agreed to work towards a regional fuel standard, prompted by the California initiative.

Another option is a **carbon tax**. The purpose of such a tax, which was introduced by the BC government in 2008, is to modify human behaviour in favour of activities that generate lower GHG emissions. When the then federal Liberal leader, Stéphane Dion, proposed a carbon tax in 2008, Prime Minister Harper stated that it was 'crazy economics' and 'crazy environmental policy' and that it would 'shaft' all Canadians. In the federal election held in October 2008, the Conservatives were returned as a minority government, and many commentators suggested that one of the problems for the Liberals during the campaign was a difficulty in explaining the carbon tax in easily understood terms.

The differing positions taken by the federal government, the opposition parties, and the provincial governments highlight that policy-makers have a range of choices, each of which has strengths and weaknesses. What are your views on the relative merits of 'intensity-based', carbon tax, and 'cap-and-trade' approaches?

Policy and Action Options

Regarding climate change, strong agreement exists on several matters related to policy and action. (1) International collaborative action will be required, since climate change is a shared problem. No one country can take unilateral action to resolve it. The challenge, of course, is that each national government may be reluctant to take the 'first step' for fear that it might become less economically competitive with other nations. (2) A mix of strategies will be required, including both mitigation and adaptation.

In the discussion of the Kyoto Protocol and the process to create a successor framework, we noted the importance of nations working collaboratively to reduce GHG emissions. We also saw that some of the most developed countries, including the United States and Canada, have argued that they will not take a leadership role because of a concern about hindering economic growth and impairing short-term livelihoods. One consequence is that it has been very difficult to develop any meaningful international strategy to address the fundamental causes of climate change. Work will have to continue on creating an international approach. Central elements to any approach will be mitigation and adaptation.

Mitigation

Mitigation involves reducing emissions of GHGs, which in turn will limit future temperature changes. Many scientists agree that a critical threshold would be an upward change of 2°C in temperature, although more recent research suggests that 1.5°C is a more accurate figure. Ensuring that global warming does not pass 2°C will require emissions of GHGs not exceeding two times current CO_2 levels, or about 560 parts per million. Attaining such an emissions target will not be a perfect solution, but will allow time for other adaptations to be introduced. To achieve this goal, emissions would have to be reduced significantly below current levels, and about 75 per cent of energy production would have had to be based on 'carbon-free' sources. To reach 75 per cent carbon-free energy production will require development and implementation of new technologies.

In contrast, if CO_2 emissions continue at present levels, then incremental warming of between 2°C and 6°C is likely over the next four centuries. If production of all GHGs were stopped immediately, it is most likely that because of past emissions, the Earth would still experience a warming of 1°C.

In terms of mitigation strategies, a mix of options exists (after Hengeveld, 2006: 30–2). They include a carbon tax, new technologies, carbon sequestration, and geo-engineering.

Carbon tax. A tax would be levied on countries based on their generation of GHGs. The measure could also include creating a market for GHG emissions, with some countries not able to meet their targets purchasing emission capacity from other countries well below their limits (see the accompanying box on BC's carbon tax).

Rivers and Sawyer (2008) have outlined the rationale for a carbon tax. They argue that a primary cause of global warming is that producers or users of goods do not have to pay any charge for the associated carbon emissions. As a result, all Canadians 'pay the price' through a degraded environment.

Perspectives on the Environment

Carbon Tax in British Columbia

As of July 1st, 2008, the provincial government in British Columbia introduced a carbon tax on fuels—gasoline, diesel, coal, propane, natural gas, and home heating fuel. The rationale was a conviction that consumers respond to price signals, and would reduce their use of fuels and thereby overall emissions.

The carbon tax started at a relatively low rate to allow time to adjust. The initial rate was $10 per tonne of associated carbon, or carbon-equivalent, emissions, or about 2.4 cents a litre at the gas pump and 2.8 cents a litre for diesel and home heating oil. The rate per tonne increased by $5 a year for each of four years to reach $30 a tonne (or 7.2 cents per litre of gasoline) by 2012. The tax is levied on a per volume basis, and therefore is not related to the actual selling price of the fuel.

The tax is intended to be revenue neutral, with the collected revenue returned to residents of the province through lower personal and business income taxes. A feature was a Low Income Climate Action tax credit paid quarterly, included so the tax would not unfairly affect low-income British Columbians.

—Adapted from British Columbia Ministry of Finance, (2011a and b)

In contrast, they note that most residents and businesses in cities accept a charge to send waste to a local landfill site. In their words, the bottom line is that Canadians need to stop using the atmosphere as a free dumping ground.

Rivers and Sawyer (ibid., 6) also argue that the necessary shift to clean, renewable energy is hindered by unfair competition from the fossil-fuel sector because it does not have to include the cost of GHG emissions in its pricing. They further state that 'a carbon price can help fix the imbalance between clean energy and carbon-intensive fossil fuels in two ways. First, putting a price on carbon will introduce the true cost of carbon emissions into the equation, and second, the substantial revenue generated by a carbon price can be used to pay for the massive increase in home efficiency retrofits and the deployment of green energy technology.'

Technologies. One key technology involves finding alternatives to fossil-fuel combustion for heating buildings, running manufacturing and industrial equipment, and powering vehicles, aircraft, and ships. A specific example is fuel switching through which fossil-based fuels are replaced by other fuels, ranging from hydro-generated electricity to ethanol fuels. Wind-based energy is another alternative, discussed in more detail in Chapter 12. Other options include renewable biomass and developing a 'hydrogen economy'.

Carbon sequestration. Carbon can be sequestered in biological sinks. Land-use practices that encourage agricultural

Global Policy Challenges
Barry Smit

Climate change is ubiquitous, connecting with human societies and economies globally, through our influence on climate via greenhouse gas emissions and through the effects of climate change on our livelihoods and lives. Managing climate change is particularly challenging, in part due to the wide range of stakeholders and interests, but especially because of the spatial scales (from local to global) and the time frames (from hours to centuries) over which climate changes and their effects are experienced.

Climate change is a global process, as emissions of greenhouse gases in one place affect not only that place but also all other places. Climate change is a classic case of the 'tragedy of the commons' in that the atmosphere is a 'common resource' open to all. There is no immediate disincentive for a country (or an industry) to emit gases, and if they were to cut back, others, it is assumed, will continue to pollute the atmosphere to their short-term economic advantage. Yet all people in all countries experience the impacts of climate change, and many of the most vulnerable in developing countries have contributed little to global emissions.

The 'global commons' feature of climate change has prompted attempts to reach an international accord to reduce greenhouse gas emissions (known as 'mitigation'). Some progress has been made on principles (e.g., the UN Framework Convention on Climate Change, Kyoto), but the narrow, short-term self-interests of countries (and their dominant economic sectors) have stymied any effective global agreement.

Policy on climate change is also constrained by its interconnectedness with most aspects of our lives and economies. The Montreal Protocol on ozone depleting substances was largely successful because it dealt with only CFCs and the relatively small industrial sector employing them. However, greenhouse gases are numerous and are emitted from almost every facet of human living. Policy initiatives to reduce greenhouse gas emissions are necessarily undertaken through measures relating to energy, transportation, land use, forestry, urban development, agriculture, etc. In all these areas, numerous policy drivers exist. Climate change is just one, and usually not the most pressing to those directly involved.

Communities reliant on irrigation for crop production in northern Chile are having to adapt to reduced water supplies as the alpine glaciers shrink.

Families in coastal Nigeria are having to adapt to loss of food sources and livelihoods as fish stocks have moved with changing ocean temperatures.

Market measures are increasingly promoted to reduce emissions. Carbon taxes are employed in many jurisdictions and several carbon markets exist, with values in the billions of dollars, within which emissions (typically over a 'cap') have a monetary cost, and reductions or credits have a price and can be traded. These approaches accommodate the interconnectedness of climate with economic sectors and human behaviour, but they still require defined targets or caps, ideally with some global consistency. To date they focus on carbon, omitting many other important greenhouse gases.

A key issue in the debates over climate change policy relates to the costs of taking action versus the costs of not acting. Of course, 'costs' can be assessed in terms of physical environments, species, ecosystems, resources, livelihoods, societies, cultures, and lives. From a solely economic perspective, eminent economist Sir Nicolas Stern concluded that the cost of 'mitigating' climate change would be in the order of 2 per cent global GDP, a very large sum, yet the cost of not mitigating was between 5 and 20 per cent global GDP.

The climate change issue is highly politicized, reflected in the campaign to discredit climate change science and hence prevent or limit public policies that might constrain interests in the fossil-fuel sector. This public relations campaign employs the same strategies (and some of the same people) as the tobacco companies did in their challenge of the science linking smoking to lung cancer.

Notwithstanding mitigation efforts, emissions will continue to rise for decades, and even if emissions slow, lags in the atmosphere would ensure that changes in climate would continue. Hence, there is a need to adjust to the effects of a changing climate (known as 'adaptation'). Internationally, the main issue is funding of adaptation in the less developed and vulnerable countries—who pays? how much? who receives? for what activities? who administers? Adaptation is already being addressed in many countries, rich and poor. In Canada, for example, BC and Ontario have climate change adaptation programs.

Vulnerability research has provided insights into the practicalities of adaptation at the community level in a wide range of countries. Community livelihoods are particularly vulnerable to climate-related extremes, and there is limited awareness that these may shift with climate change. Initiatives to adapt to climate are invariably undertaken as part of ongoing risk and resource management strategies, rarely as stand-alone strategies to mitigate climate change.

Dr Barry Smit is Professor of Geography and Canada Research Chair in Global Environmental Change at the University of Guelph. He has been active in the science and policy of climate change in Canada and internationally for 30 years, and was a co-recipient of the 2007 Nobel Peace Prize as a member of the IPCC.

crops and forest systems with the capacity to sequester carbon have been endorsed as a legitimate way for nations to achieve GHG targets under the Kyoto Protocol. A challenge with this option is that biological sinks are not permanent, and hence attention must be given to ways of preventing the sequestered carbon from entering the atmosphere.

Geo-engineering. Geo-engineering is sometimes considered to be a third tool, in addition to mitigation and adaptation, but here we include it under 'mitigation' since the intent is to reduce GHG emissions and thereby reduce global warming through systematic large-scale manipulation of the Earth's climate. Examples of geo-engineering can involve **carbon sequestration** by *direct* (capture of carbon dioxide air) or *indirect* (iron fertilization of oceans) approaches. Another approach is management of solar radiation, such as by producing stratospheric sulphur aerosols, or using space mirrors and enhancement of cloud reflectivity. So far, no large-scale geo-engineering approaches have been initiated outside of laboratory experiments. Smaller-scale geo-engineering approaches include 'cool roof' projects and tree planting.

Advocates of geo-engineering approaches argue that climate change has already passed key 'tipping points', and as a result future reductions in GHGs will not be sufficient to reverse climate change. They claim that geo-engineering could reverse, even if temporarily, some negative aspects of climate change and thereby provide more time and opportunity to reduce GHG emissions through various mitigative measures. Critics express concern about the safety and appropriateness of geo-engineering methods, and worry about unintended side effects at a global scale.

Adaptation

As noted above, even if all GHG emissions were stopped tomorrow, there are enough GHGs already in the atmosphere to generate significant climate change. The change will have impacts, some positive and some negative. Given that there will be negative impacts, we need to develop adaptation strategies so that adversely affected activities and regions can create capacity for resilience.

In developing countries in particular, the tendency has been to prepare strategies that emphasize recovery following a disaster, such as preventing starvation as a result of a loss of crops because of flooding, hail, drought, disease, or pests. Today, social scientists argue that more attention should be directed towards creating capacity to deal with disasters.

For example, climate change is likely to have an impact on agricultural production. And as Hengeveld (2006: 32) observed, some choices—such as increasing production through greater crop specialization, more water-intensive crops, more tile drainage and water competition, and less grazing—will make agriculture more vulnerable to climate extremes. In the resource-based recreational sector, warmer winters are likely to adversely affect downhill skiing operations, so an investment in snow-making equipment now would be prudent as an alternative to waiting for government support after a series of lower-than-average snowfall years.

Choices are always available, and both policy-makers and individuals in a resource-based sector can opt for choices that allow greater scope for adaptation in the future when climate conditions are quite likely to be different from what they are today. However, to promote more adaptive choices, governments will have to review current systems of incentives, such as crop insurance and agriculture support programs. Any proposed changes would probably meet vocal resistance, since some people profit from such programs and can be expected to oppose any changes to them. Currently, more attention is being given to adaptation. Walker and Sydneysmith (2008), for example, have laid out some of the adaptive mechanisms that might be appropriate for British Columbia.

It is unlikely, however, that either mitigation or adaptation strategies on their own will suffice, as highlighted by Barry Smit in his guest statement. Both are needed, and each one creates many opportunities for those able and willing to look beyond 'business as usual' practices.

What Else?

Clearly, considerable uncertainty remains about the precise effects of anticipated climate change. Furthermore, concerted and co-ordinated initiatives by provinces, states, and nations will be necessary to reduce the projected negative impacts. Some of the multilateral initiatives underway have been identified in this chapter, but it is obvious that some nations, especially the United States and Canada, have been unwilling to participate in co-ordinated global activities if they could create economic disadvantage for them in the short term. But what can individuals do?

Hengeveld et al. (2005: 51) provide thoughtful guidance:

> How can the individual Canadian citizen influence the outcome of a global environmental issue that is already challenging the wisdom and resources of the world's governments and international agencies? The answer, simply put, is that it is the individual citizens who must create the environment of opinion which will encourage governments to act.
>
> And it is the individual citizens who can take actions themselves to reduce their personal emissions and who can support the policies that an effective response to the risks of climate change will demand.

They go on to observe that it is not the governments of countries that release the estimated 23 billion tonnes of CO_2 annually into the atmosphere through use of fossil fuels, but rather the actions of the almost 7 billion people on the Earth. In their view:

> Each time we turn on a hot water tap, or open a refrigerator door, we add to the problem. By changing our attitudes and lifestyles, by becoming more knowledgeable about the issue and rethinking our attitude to the environment, we can make a difference.

Box 7.6 presents some options. A change in one person's behaviour will not have a significant impact. However, cumulatively, many individual actions can be significant, as

Environment in Focus
Box 7.6 What You Can Do

1. Much of global climate change is influenced by our society's love affair with the automobile. Find something else to love, such as a bicycle or a bus. Simply aim to reduce your consumption of fossil fuels.
2. Use energy more efficiently in your home and in all other aspects of life. Make sure your appliances, lighting, and heating are energy-efficient.
3. Encourage power utilities to invest more effort in renewable energy resources, as discussed in Chapter 12.
4. Consider getting by without air conditioning in your car and at home, or if/when you use it, do so sparingly. Plant shade trees to reduce the summer heat levels in your house.
5. Consider purchasing carbon offsets related to transportation or energy use (Box 7.7).
6. Plant trees to help absorb atmospheric carbon dioxide.
7. Let your political representatives know that you are in favour of mandatory measures to curb emissions, even if it costs more money in the short term.

Environment in Focus
Box 7.7 Carbon Offsets

Carbon offsets involve a financial transaction to compensate for the generation of greenhouse gas emissions through our activities, such as travel or use of electricity. Offsets are measured in metric tons of CO_2 equivalent, with one carbon offset equalling one ton of CO_2 or equivalent greenhouse gas.

There are two categories of carbon offsets. One involves governments and companies purchasing carbon offsets in order to comply with caps placed on their GHG emissions. The Kyoto Protocol approved offsets as one option for governments and companies to meet their emission targets. In that context, companies providing energy from wind, solar, small hydro, geothermal, and biomass all create offsets by substituting or displacing fossil fuels and their associated GHGs.

The other category is voluntary and involves individuals and companies who purchase carbon offsets to mitigate the GHG emissions generated by activities such as air travel. The next time you purchase an airline ticket, check to see whether you can purchase carbon offsets. The funds generated from such purchases are often used to plant trees that absorb GHGs.

For further information, see the David Suzuki Foundation (2007).

---

emphasized in Chapter 15. The challenge will be to decide whether we are prepared to make such modifications to our behaviour, since in many instances we will not be the direct beneficiaries—more likely, people one or two generations in the future, such as your grandchildren, will reap the benefits. In thinking about why we should change our behaviour, you should consider the following statement from the National Round Table on the Environment and the Economy (2011b: 15, 119):

Climate change costs for Canada could escalate from roughly $5 billion per year in 2020—less than 10 years away—to between $21 billion and $43 billion per year by the 2050s. The magnitude of costs depends on a combination of two factors: global emissions growth and Canadian economic growth and population growth. The highest costs result from a refusal to acknowledge these costs and adjust through adaptation.

Summary

1. The *weather* of any place is the sum total of its atmospheric conditions (temperature, pressure, winds, moisture, and precipitation) for a *short* period of time. It is the momentary state of the atmosphere. *Climate* is a composite or generalization of the variety of day-to-day weather conditions. It is not just 'average weather', since the variations from the mean, or average, are as important as the mean itself.

2. Scientific evidence confirms that the world has been warming, with the average global temperature at the Earth's surface having increased by about 0.6°C, with an error range of plus or minus 0.2°C, since the late nineteenth century.

3. The increase in the average temperature for the northern hemisphere during the twentieth century was the largest of any century in the past 1,000 years.

4. Evidence showing increases in greenhouse gases in the atmosphere, loss of mass in glaciers, reduction in permafrost and snow cover, and rises in sea level are all consistent with global temperature increases.

5. There are four basic climate models, the most commonly used being the general circulation model. GCMs are best used for global or overall climate modelling rather than for regional representations of climate change.

6. Coarse spatial resolution, poor predictive capacity for precipitation, relatively weak simulation of oceans, lack of baseline data, and many other limitations make GCM outputs variable.

7. The Intergovernmental Panel on Climate Change is explicit that most warming since the mid-twentieth century is associated with human activities.

8. At a global scale, records document that mean sea level has been rising at a rate of 0.1 to 0.2 metres per century during the past 100 to 200 years.

9. There are five ways to adapt to the hazards caused by climate change: prevent the loss, accept the loss, spread the loss, change behaviour, and change the location for activity.

10. Climate change has implications for terrestrial systems, agriculture, freshwater systems, fisheries, the cryosphere, ocean and coastal systems, and infectious diseases.

11. Climate change negatively affects the poorest people the most and can trigger out-migration, which can lead to 'environmental refugees' for whom Canada will be one destination.

12. The challenges of communicating information or understanding about global change include: (1) global change is a complex issue; (2) uncertainties exist regarding almost every aspect of the global change issue, and they increase when moving from natural to human systems; (3) the impact of global change will be disproportionately heavier on people in less developed countries and on future generations; (4) the basic causes of global change are embedded in current values and lifestyles.

13. Climate deniers aim to raise doubt in the minds of the public and decision-makers about the science of climate change, with the intent to delay decisions about changes needed to reduce greenhouse gas emissions.

14. During December 1997, representatives met in Kyoto, Japan, and reached an agreement, popularly referred to as the Kyoto Protocol, with targets for 38 developed nations as well as the European Community to reduce their overall emissions of greenhouse gases by at least 5 per cent below 1990 levels by the period 2008–12.

15. Canada, which ratified the Protocol in December 2002, committed to reduce greenhouse emissions to 6 per cent below 1990 levels by between 2008 and 2012. In contrast, US President George W. Bush stated that the US would not agree to the Protocol and would develop its own approach.

16. Despite the commitment to the Kyoto Protocol, Canada has been unsuccessful in reducing GHG emissions.

17. Canada has received the 'Fossil of the Day' award from NGOs at various international climate change conferences because of its poor performance in reducing GHG emissions.

18. At the Bali conference in December 2007, an effort was made to develop a new climate change agreement to replace the Kyoto Protocol. At that event, Canada argued against specific targets and wanted the timeline extended from 2020 to 2050.

19. Canada was labelled a 'climate hypocrite' at the Bali conference.

20. Climate summits at Copenhagen (2009) and Cancún (2010) made only incremental progress to develop an international accord to replace the Kyoto Protocol.

21. Both mitigation and adaptation must be used in policies and strategies to deal with climate change.

22. Geo-engineering involves deliberate and systematic large-scale manipulation of the global climate system to reduce the impact of GHG emissions on temperature increases.

Key Terms

- adaptation
- aspirational approach
- atmosphere
- Bali Conference
- Cancún Summit
- carbon offsets
- carbon sequestration
- carbon tax
- climate
- climate change
- climate change deniers
- 'Climategate'
- climate modelling
- Copenhagen Summit
- El Niño
- emission credits
- emissions trading
- fossil fuels
- general circulation models (GCMs)
- geo-engineering
- global warming
- greenhouse effect
- greenhouse gas
- Kyoto Protocol
- mitigation
- Montreal Protocol
- ozone
- ozone layer
- uncertainty
- weather

Questions for Review and Critical Thinking

1. Explain the difference between weather and climate.

2. What are some of the key natural and human causes of climate change?

3. How credible is the evidence for climate change or global warming? For which aspects is there the most uncertainty?

4. Have you noticed any indication of climate change in the area in which you live? If so, what are they, and how confident are you that they are valid and reliable indicators of climate change?

5. What would be the best sources of traditional ecological knowledge about climate change in your area?

6. Why have general circulation models (GCMs) become dominant in the modelling research focused on climate change? What are their limitations?

7. What is viewed as the main cause of increased carbon dioxide emissions to the atmosphere in the twentieth and twenty-first centuries?

8. Which of the negative impacts of climate change should be of greatest concern to Canadians? Will the priority shift depending on which Canadian region is considered?

9. What role should Canadians have with regard to the challenge of environmental refugees?

10. What are the main strategies normally used to adapt to natural hazards caused by climate change?

11. What are the basic communication challenges related to climatic change, and what, in your view, should the first steps to overcome them be?

12. Do you agree or disagree with the views of climate change deniers?

13. What was the significance of 'Climategate'?

14. What are the strengths and limitations of the Kyoto Protocol? If it did not exist and you were to start with a 'blank sheet', what type of protocol would you propose?

15. Why has the Canadian Conservative government opposed the Kyoto Protocol? What has it proposed as an alternative? What are the positions of provincial and territorial governments?

16. Why do people argue that climate change will only be addressed successfully if there is international collaborative effort? In that context, do you agree or disagree that developing nations such as China and India have to be signatories to any international agreement?

17. What has been the significance of the Bali (2007), Copenhagen (2009), Cancún (2010), and South African (2011) climate change conferences?

18. How should mitigation measures—carbon tax, technology, carbon sequestration, geo-engineering—be used together to reduce GHG emissions?

19. There have been different perspectives in Canada regarding the best way to reduce greenhouse gases. What do you think is the best strategy to resolve differences about climate change among federal, territorial, and provincial governments and among the provincial governments as a group?

Related Websites

Centre for International Climate and Environmental Research, Oslo
www.cicero.uio.no/home/index_e.aspx

Climate Action Network Canada
www.climateactionnetwork.ca

The Climate Group
www.theclimategroup.org

David Suzuki Foundation, Climate Change: Impacts and Solutions
www.davidsuzuki.org/issues/climate-change

Environment Canada, climate change site
ec.gc.ca/default.asp?lang=En&n=2967C31D-1

Environment Canada, 'Climate Trends and Variations Bulletin'
www.ec.gc.ca/adsc-cmda/default.asp?lang=En&n=4A21B114-1

Environment Canada, Environmental Indicators
www.ec.gc.ca/indicateurs-indicators

European Union
ec.europa.eu/environment/climat/home_en.htm

Federation of Canadian Municipalities, Partners for Climate Protection
www.sustainablecommunities.fcm.ca/Partners-for-Climate-Protection

Government of Canada, ecoACTION
www.ecoaction.gc.ca

Intergovernmental Panel on Climate Change
www.ipcc.ch

Meteorological Service of Canada
www.ec.gc.ca/meteo-weather/default.asp?lang=En&n=FDF98F96-1

Natural Resources Canada
www.nrcan.gc.ca/earth-sciences/climate-change/11610

Union of Concerned Scientists, Citizens and Scientists for Environmental Solutions
www.ucsusa.org/global_warming

United Nations Framework Convention on Climate Change
unfccc.int/2860.php

United States Environmental Protection Agency
www.epa.gov/climatechange

World Meteorological Organization
www.wmo.ch/pages/index_en.html

Yukon College, Northern Climate Exchange
www.taiga.net/nce/index.html

Further Readings

Note: This list comprises works relevant to the subject of the chapter but not cited in the text. All cited works are listed in the Bibliography at the end of the book.

Boucher, O., J.A. Lowe, and C.D. Jones. 2009. 'Implications of delayed actions in addressing carbon dioxide emission reductions in the context of geo-engineering', *Climatic Change* 92, 3 and 4: 261–73.

Gearheard, S., M. Pecernich, R. Stecoart, J. Sangnya, and H.P. Huntington. 2010. 'Linking Inuit knowledge and meteorological station observations to understand changing wind patterns at Clyde River, Nunavut', *Climatic Change* 100, 2: 267–94.

Gleick, P., et.al. 2010. 'Climate change and the integrity of science', *Science* (7 May): 689–90.

Gore, A. 2006. *An Inconvenient Truth: The Planetary Emergency and What We Can Do about It*. Emmaus, Penn.: Rodale Books.

Gould, J., and J. Church. 2010. *The Climate: Science and Impacts*. London: Earthscan.

Grunster, S. 2010. 'Self-interest, sacrifice and climate change: (Re)-framing the British Columbia carbon tax', in M. Maniates and J.M. Metyer, eds, *The Environmental Politics of Sacrifice*. Cambridge, Mass.: MIT Press, 187–215.

Hansen, L.J., and J.R. Hoffmann. 2010. *Climate Savvy: Adapting Conservation and Resource Management to a Changing World*. Washington: Island Press.

Hay, L.E., and G.J. McCabe. 2010. 'Hydrologic effects of climate change in the Yukon River basin', *Climatic Change* 100, 3 and 4: 509–23.

Hulme, M. 2009. *Why We Disagree about Climate Change: Understanding Controversy, Inaction and Opportunity*: Cambridge: Cambridge University Press.

Hunt, A., and P. Watkiss. 2011. 'Climate change impacts and adaptation in cities: A review of the literature', *Climatic Change* 101, 1: 13–49.

Jacques, P., R. Dunlap, and M. Freeman. 2008. 'The organization of denial: Conservative think tanks and environmental skepticism', *Environmental Politics* 17: 349–85.

Laidler, G.J., J.D. Ford, W.A. Gough, T. Ikummaq, A.S. Gagnon, et al. 2009. 'Travelling and hunting in a changing Arctic: Assessing Inuit vulnerability to sea ice change in Igloolik, Nunavik', *Climatic Change* 94, 3 and 4: 363–97.

McBean, G. 2010. 'Climate change, adaptation, and mitigation', in B. Mitchell, ed., *Resource and Environmental Management in Canada: Addressing Conflict and Uncertainty*, 4th edn. Toronto: Oxford University Press, 122–53.

Millerd, F. 2011. 'The potential impact of climate change on Great Lakes international shipping', *Climatic Change* 104, 3 and 4: 629–52.

National Round Table on the Environment and the Economy. 2009. *True North: Adapting Infrastructure to Climate Change in Northern Canada*. Ottawa: NRTEE.

Oreskes, N., and E.M. Conway. 2010. *Merchants of Doubt: How a Handful of Scientists Obscured the Truth on Issues from Tobacco Smoke to Global Warming*. London: Bloomsbury.

Parker, P., and T. Gliedt. 2010. 'Integrated energy resource management', in B. Mitchell, ed., *Resource and Environmental Management in Canada: Addressing Conflict and Uncertainty*, 4th edn. Toronto: Oxford University Press, 154–85.

Parkinson, C.L. 2010. *Coming Climate Crisis? Consider the Past, Beware the Big Fix*. Lanham, Md: Rowman & Littlefield.

Richter, B. 2010. *Beyond Smoke and Mirrors: Climate Change and Energy in the 21st Century*. Cambridge: Cambridge University Press.

Scott, D.N., et al. 2000. *Climate Change Communication: Proceedings of an International Conference*. Waterloo, Ont.: University of Waterloo; Hull, Que.: Environment Canada, Adaptation and Impacts Research Group.

Simpson, J., M. Jaccard, and N. Rivers. 2007. *Hot Air: Meeting Canada's Climate Change Challenge*. Toronto: McClelland & Stewart.

Stern, N.H. 2007. *The Stern Review of the Economics of Climate Change*. Cambridge: Cambridge University Press.

Stewart, E.J., and D. Draper. 2006. 'Sustainable cruise tourism in Arctic Canada: An integrated coastal management approach', *Tourism in Marine Environments* 3, 2: 77–88.

Chapter 8
Ocean and Fisheries

Learning Objectives

- To understand the nature of oceanic ecosystems and their similarities to and differences from terrestrial ecosystems.
- To know the main challenges facing the oceans.
- To learn about some of the global management responses to these challenges.
- To understand the reasons behind the collapse of Canada's east coast groundfish fishery.
- To appreciate the background to Aboriginal use of marine resources and to examine some current conflicts.
- To gain an understanding of Canada's main strategies for ocean management.
- To be aware of some of the challenges regarding aquaculture.

Introduction

The major challenges faced by society today are the global and transnational. They are very difficult to solve not just because of their scale—the changing composition of the atmosphere or pollution of the world's oceans—but because they need nations to act in ways that, over the short term, may yield little direct advantage to them. Can governments act for the long-term good of the majority when they may incur the wrath of voters at home?

Canada does not have a good record in this regard, especially recently and related to ocean resources. Canada would not join the worldwide moratorium on bottom trawling suggested by then US President George Bush, is a noted laggard in its international commitments to establish marine protected areas where fishing is not allowed, and refused to support an international treaty to protect the bluefin tuna.

The Atlantic bluefin tuna is the most valuable fish in the sea, and these fish regularly fetch over $100,000 each at market in Tokyo, with the record close to five times that amount. Obviously, the fish is highly sought after, and that demand has resulted in an 80 per cent reduction in stocks over the last 100 years. As a result, in 2010 Monaco proposed a temporary fishing ban, supported by the US and many other governments signatory to the UN Convention on International Trade In Endangered Species (CITES). The ban was opposed by Japan and Canada, among others.

The management body for the tuna, the International Commission for the Conservation of Atlantic Tunas (ICCAT), has consistently set quotas well above the catches suggested by their own scientists. In 2008, for example, the suggested quota was 8,500 to 15,000 tonnes, but members agreed on a

Perspectives on the Environment
Atlantic Bluefin Tuna at Risk

As far as science is concerned, it is not a controversial issue. The species is in great danger, it is not rebuilding. Every year, one is more worried that it will lead to a total collapse. The quota should be zero. . . . The species is going to be lost in the next decade simply because of stupidity.

—Daniel Pauly, quoted in Leeder (2010)

A vast quantity of tuna awaiting inspection and auction at Tokyo's Tsukiji fish market, the world's largest daily fish market.

22,000-tonne limit. There is also virtually no recourse for catching over the limit. In 2007 France was allocated 5,500 tonnes but reported catching double that amount. Unfortunately, given the value of the fish, these legal quotas are far exceeded by the illegal and unreported catch. Obviously, the measures put in place for management are not working.

Canada only has a small allocation of the fishery, the rod–and-reel fishery based in PEI, and this fishery is well managed. However, in voting to keep this fishery going, Canada is contributing to the overall demise of the species. If all nations continue to act solely in their own best interests in the short term, as Canada has done in this case, then the very severe global challenges faced by the oceans will never be solved.

Oceanic Ecosystems

Perspectives on the Environment
High-Intensity Stressors and Extinction

Not only are we already experiencing severe declines in many species to the point of commercial extinction in some cases, and an unparalleled rate of regional extinctions of habitat types (e.g., mangroves and seagrass meadows), but we now face losing marine species and entire marine ecosystems, such as coral reefs, within a single generation. Unless action is taken now, the consequences of our activities are at a high risk of causing, through the combined effects of climate change, overexploitation, pollution and habitat loss, the next globally significant extinction event in the ocean.

It is notable that the occurrence of multiple high-intensity stressors has been a prerequisite for all the five global extinction events of the past 600 million years

—Barnosky et al. (2009)

The oceans and their well-being are integral to sustaining life on this planet. They are key components in global cycles and energy flows (Chapter 2). Marine ecosystems are home to a vast array of organisms displaying greater diversity of taxonomic groups than their terrestrial counterparts. Marine organisms help feed us, and they are also the source of many valuable medicinal products. The value of marine ecological goods and services is estimated at $21 trillion annually, 70 per cent greater than that of terrestrial systems (Costanza et al., 1997). We use the seas to dump our waste products and to transport most of our goods around the world. The oceans enrich our cultures, and nations draw strength and inspiration from their links to the vital life-giving nature and awesome power of the seas.

One of the major difficulties working against sustainable human use of the oceans is our lack of understanding of oceanic ecosystems. In 2003, a $1 billion, 10-year expedition announced that it had described 150 new species of fish and another 1,700 plants and animals in just the first three years of its travels and anticipated that as many as 5,000 species of fish were waiting to be discovered. By 2010, the expedition had described more than 250,000 marine species with approximately 750,000 remaining to be described (see Census of Marine Life: www.coml.org/coml.htm). Not all new discoveries are small and in distant locales. For example, one discovery was a new species of squid more than nine metres in length in the Gulf of Mexico; another was a giant jellyfish in the heavily studied waters off Monterey, California. If we know so little about oceanic ecosystem components, it is even more difficult to understand the functional relationships among them.

Yet the general principles that govern life and energy flows and matter cycles, discussed in Chapters 2 and 4, also hold true for oceanic ecosystems (see Box 2.4). But there are important differences. On land, water is the most common limiting factor for life. In the oceans, this is obviously not the case. Here it is nutrients. While the oceans cover more than 70 per cent of the Earth's surface, they account for only 50 per cent of global primary productivity. Much of the ocean's surface is the marine equivalent to a desert, with productivity limited to the areas where nutrients are abundant. Generally, nutrient concentrations increase with depth because of the decomposition of organisms falling from the surface layer.

The most productive areas are coastal zones and in areas of the ocean where upwellings from the deep ocean return

The oceans sustain an amazing variety of life. Here are two fish with very different life strategies. The whale shark, the world's largest fish, cruises the world's oceans feeding on plankton. The stone fish rarely moves and sits on coral reefs, disguised as a stone, until it spots prey to ambush. It is highly poisonous to humans.

nutrients to the surface layers and where photosynthetic activity occurs. In terrestrial ecosystems, productivity generally increases from the poles to the tropics. In the ocean, this is not true. Although there are productive areas in the tropics (Box 8.1), some of the most highly productive marine areas in the world are situated off the coast of Canada, such as the Grand Banks and in the Arctic Ocean, where nutrient upwellings occur. These upwellings promote large phytoplankton populations, especially in the Arctic where there is virtually unlimited light in the summer. Many species of whales and birds migrate to these waters to take advantage of this abundance. The largest whale, the blue whale, and the bird with the largest wing span in the world, the albatross, are two good examples.

Besides nutrient availability, the other major ecological influences on marine life are temperature and light. Both temperature and light decrease with depth. This means that surface waters in the euphotic zone (Box 2.4) are warmer and have higher light levels, resulting in higher productivity. There is usually a sharp transition in temperature between the warmer surface waters and the cooler waters underneath. This is known as the *thermocline* and generally occurs at a depth of 120 to 240 metres, depending on latitude and ocean currents.

The deepest part of the ocean is more than 9,000 metres deep, but more than three-quarters is between 4,000 and 6,000 metres in depth. Most productivity is on the continental shelves at a depth of less than 200 metres, and especially within the top 100 metres. Most fisheries are concentrated in these areas. In the 1970s, however, scientists discovered that rich biological communities, mainly made up of bacteria that derive their energy from sulphide emissions, were centred around hydrothermal vents on the sea floor. Rich communities of tube worms, clams, and mussels have been documented at depths exceeding 2,000 metres at more than 100 sites worldwide, including off the west coast of Canada. Scientists speculate that life on Earth originated in such hydrothermal vent systems, based on chemosynthesis. Similar kinds of chemoautotrophic-based communities are found on whale skeletons at depth, nourished by sulphides produced as the carcasses decay, as discussed in Box 2.2.

Another difference between terrestrial and marine ecosystems is in the shape of the biomass pyramids discussed in Chapter 2. In terrestrial ecosystems, you will recall, the pyramids stand upright. There is a broad base of primary producers and a reduced biomass at each subsequent trophic level. This arrangement is reversed in oceanic pyramids because the biomass of the primary producers, the phytoplankton, alive at any one time is quite small compared to their predators. The food chains still depend on a broad energy base, but the turnover of biomass at the first trophic level is rapid. This continuously replenished, short-lived base supports long-living predators, such as the blue whale, that store energy over a much longer period.

The **carbon balance** in oceanic ecosystems is the subject of a lot of scientific research because of the crucial importance of the oceans in absorbing carbon dioxide, a greenhouse gas, and the role it may play in mitigating the impact of global climate change (Chapter 7). The ocean surface takes up about 2 billion metric tonnes of carbon per year by gas exchange, equivalent to one-third of annual anthropocentric emissions. This uptake is driven by wind exchange and the imbalances between the amount of carbon in the atmosphere and the oceans. Marine primary producers get their carbon from the dissolved CO_2 in the water as bicarbonate. There is a balance between the amount of CO_2 in the atmosphere and bicarbonate in the water. If there is too much in either compartment, a gradient is created, and the carbon migrates along the gradient between the ocean's surface and the atmosphere. However, carbon is also continuously being moved out of

Environment in Focus

Box 8.1 Coral Reefs—The Rain Forests of the Sea

Coral reefs are in many ways oceanic analogues of the tropical rain forests. Found throughout tropical and subtropical seas, they are among the most diverse and productive ecosystems on Earth. Like the rain forests, they have an ancient evolutionary history, having first appeared more than 225 million years ago, with some living reefs perhaps as old as 2.5 million years. With solar radiation the primary source of energy, these habitats are found predominantly within 30° north and south of the equator at depths of less than 50 to 70 metres. Coral reefs are categorized into three main types: (1) fringing reefs that are found along the shorelines of continents and offshore islands; (2) barrier reefs that have been separated from land by lagoons; and (3) atolls that appear as their own coral islands, having developed around (now submerged) lagoons.

Coral reefs are made up of countless numbers of individual **coral polyps** and the calcium carbonate skeletons deposited by prior generations of corals and other reef-associated organisms (e.g., coralline red algae and molluscs). These limestone secretions serve as a substrate for live coral polyps to grow and flourish.

Many coral species are involved in symbiotic relationships with unicellular algae, or **zooxanthellae**, that live inside the coral's protective skeleton. These photosynthetic algae produce carbohydrates that serve as the primary food source for the corals in which they live. When water temperatures get too warm, the zooxanthellae are often expelled, leading to the eventual death of the corals. This is called **coral bleaching** and has been recorded over large areas of reef throughout the world over the past decade. In tropical shallow waters, a temperature increase of only 3°C by 2100 may result in annual or biannual bleaching events of coral reefs from 2030 to 2050. Even the most optimistic scenarios project annual bleaching in 80 to 100 per cent of the world's coral reefs by 2080. This is likely to result in severe damage and widespread death of corals around the world.

Acting as sources of food and refuge to an incredible diversity of sea life, coral reefs are vitally important habitats. These complex ecosystems provide a number of critical ecosystem services, including the regulation of environmental disturbances, the treatment of organic wastes, the production of food, and the creation of recreational opportunities—for example, for scuba diving. In monetary terms, coral reef habitats have been estimated to generate approximately US$375 billion annually (Costanza et al., 1997).

Despite their value, coral reefs are now highly threatened worldwide, having been plagued by the effects of destructive fishing practices, coastal erosion, marine pollution, and irresponsible tourism activities. Southeast Asia is the world's epicentre for coral reef diversity, and more than 80 per cent of its reefs are considered threatened, with 50 per cent in the high-risk category. Over the past two decades, coral in the Indian and Pacific oceans, home to 75 per cent of the world's reefs, has disappeared at five times the rate of Earth's rain forests. A recent survey spanning four decades and 2,600 reefs found that more than 3,000 square kilometres of living coral reef is lost each year. Conservationists had previously believed that accelerated declines began in the 1990s, but the researchers found reports of widespread loss dating back to the 1960s, when pollution, deforestation, and overfishing trends began. Reefs vanished at an annual rate of 1 per cent during the 1980s, with declines climbing through the 1990s to the current rate of 2 per cent—nearly five times the pace of rain forest elimination (Bruno and Selig, 2007).

Globally, less than 2 per cent of coral reefs are protected from extraction, poaching, and other major threats. Local, national, and international initiatives have been launched to counter the effects, but consumers can help, too. Coral has

Healthy coral reefs are very complex structures that take centuries to develop and may last for thousands of years.

A coral reef that has been dynamited for aquarium fish in the Sulu Sea, Philippines.

become a fashionable accessory; avoid purchasing jewellery with coral. You can also help by not purchasing coral as a souvenir or for your aquarium. If you keep tropical fish, ensure that they have been bred in captivity and in a sustainable way. As fish species are removed from the reefs, the whole community becomes less stable and more likely to collapse.

The Banggai cardinal fish, a reef species, is an example of what can happen to fish populations if they are overexploited for the aquarium trade. The fish is found only in the Banggai Archipelago near Sulawesi in Indonesia, and its iridescent beauty and rarity has made it a favourite of aquariumists. The fish has experienced an 89 per cent drop in population since the aquarium trade began in 1995, and in 2007 it was declared an endangered species by the World Conservation Union. The fish has no dispersal capability, low fecundity, long gestation periods, high early mortality, and a very restricted habitat. These are conditions that make some species more vulnerable to extinction than others, as discussed in Chapter 14. Breeding programs have been undertaken, but unless unsustainable exploitation can be controlled, ultimately they will not work.

An even bigger challenge is now looming for coral reefs. Excess carbon dioxide in the atmosphere, a main source of global warming (Chapter 7), is also predicted to lead to the extinction of coral reefs. The CO_2 is absorbed by the oceans and produces carbonic acid, making the oceans more acidic. The hydrogen ions released from this acid lower the pH (like acid rain; see Chapter 4). The oceans have absorbed nearly half of the fossil-fuel CO_2 emitted into the atmosphere since pre-industrial times, causing a measurable reduction in seawater pH and carbonate saturation. If CO_2 emissions continue to rise at current rates, upper-ocean pH will decrease to levels lower than have been experienced for tens of millions of years and, critically, at a rate of change 100 times greater than at any time over this period. Scientists predict that by the end of the century, nowhere on Earth will water chemistry still permit the growth of coral reefs. Their calcium skeleton will simply be melted away by the increased acidity, and the concentration of available carbonate ions will be too low for marine calcifiers, such as coral reefs, molluscs, crustaceans, and some algae, to build their shells and skeletons. Ocean **acidification** will have a very visible impact on coral reefs, but its impact will be felt not only throughout the oceans but throughout planetary life support systems. It is one of the greatest challenges we now face.

Cold-water corals are expected to be particularly vulnerable to acidification due to their very slow growth. Such corals were only recently discovered off Canada's coasts. Scientists on a research ship in 2003 discovered a reef between Cape Breton Island and Newfoundland. Unfortunately, the reef, made up mainly of **Lophella pertusa**, had already been badly damaged by fishing activities, i.e., bottom trawling. Cold-water corals are found in waters ranging from 3 to 14°C and have now been recorded in 41 countries (Freiwald et al., 2004). They occur where cold, clear, nutrient-rich waters are present. Living mostly in perpetual darkness, cold-water corals do not possess symbiotic, single-celled algae like warm-water corals and rely solely on zooplankton and detritus for sustenance. Some species, such as *Lophelia*, can form large, complex, three-dimensional reef structures several metres high. Other soft corals living in colder waters, such as *Gorgonia*, do not form reefs but large 'gardens' covering vast areas—for example, around the Aleutian island chain in the North Pacific. The ecological functions of such reefs and gardens in the deeper waters are similar to those of tropical reefs: they are biodiversity hotspots and home, feeding, and nursery grounds for a vast number of other organisms, including commercial fish and shellfish species.

Coral reefs are home to some of the most complex and colourful organisms on Earth. Here are the Banggai cardinal fish from Indonesia, an endemic nudibranch from the Andaman Sea, and a clam and coral from the Maldives.

the surface layers of the ocean into deeper water, where it is stored in dead organisms, ocean sediments, and coral reefs (Figure 8.1). The more that is stored at depth, the greater will be the gradient pulling carbon out of the atmosphere and into the surface waters to compensate for these losses. Scientists estimate that the oceans have absorbed at least half of the anthropocentric CO_2 emissions occurring since 1750. However, the lag time in these movements is considerable and too slow for us to hope that the process can compensate for all the extra CO_2 released into the atmosphere by human activities. The Intergovernmental Panel on Climate Change (IPCC, 2007c) predicts that ocean acidity will increase by 150 per cent by 2100 (see Box 8.1).

The carbon-saturated water does not stay where it is but moves around the globe, mainly as a result of differing water densities. This is known as **thermohaline circulation** and involves warm surface water that is cooled at high latitudes sinking into deeper basins with water close to 0°C. The key to this circulation is the high salt content of the sea water that allows the water density to increase before it freezes and sinks at certain sites. The main sites for conversion are in the North Atlantic, the Arctic Ocean, and the Weddell Sea in the Antarctic. When the water sinks, it carries with it large quantities of carbon. This sinking is the main mechanism for the removal of atmospheric carbon by the oceans. The cold water is then carried along the ocean floor until it mixes with surface waters and is transported back by wind-driven currents to the conversion areas (Figure 8.2)

The system is in fact a series of interlinked and variable currents that serve to mediate the Earth's climate through the transport of heat and water around the globe (Figure 8.2). Its scale is vast, with the flow estimated at more than 100 Amazon Rivers and the heat delivered to the North Atlantic being about one-quarter of that received directly from the sun. One of the main concerns with global warming (Chapter 7) is the impact it may have on thermohaline circulation. For example, with increased temperatures, there will be increased freshwater melting from the polar ice caps. Fresh water is less dense and will not sink to the same depth as super-cooled salty water, and therefore less carbon will be sequestered for shorter periods of time.

This is a classic example of a positive feedback loop (Chapter 3): the warmer the atmosphere gets as a result of increased CO_2 emissions, the more ice will melt, resulting in less CO_2 being absorbed by the ocean, which will result in increased atmospheric warming . . . and more melting. The cycle will perpetuate itself unless equally strong negative feedback loops come into play. One example of a potential negative feedback loop in this context is the plankton that produce dimethyl sulphide gas. Given enough nutrient

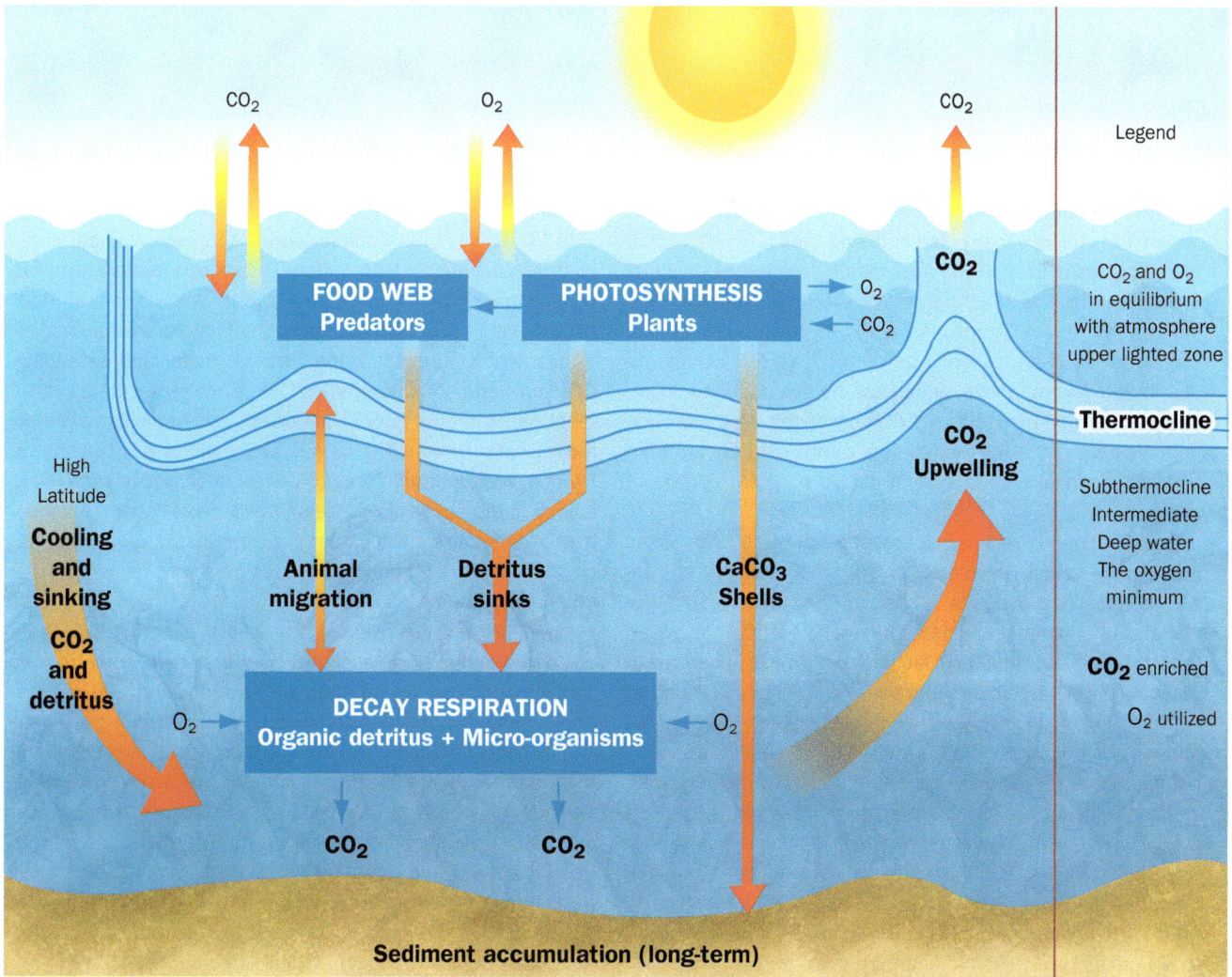

Figure 8.1 | The ocean–atmosphere carbon cycle. Source: Field et al. (2002: 13). Copyright © 2002 Island Press. Reproduced by permission.

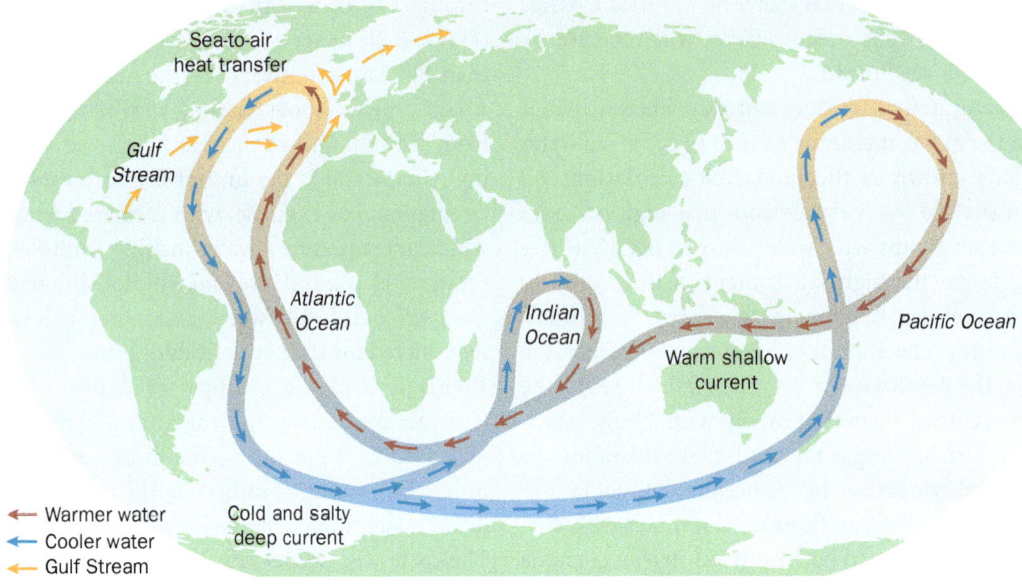

Figure 8.2 | The global ocean conveyor.

Environment in Focus

Box 8.2 Regional Scale Nodes and the Canadian Neptune Project

The Regional Scale Nodes (RSN) program is a component of the US National Science Foundation's Ocean Observatories Initiative. Formerly known as NEPTUNE, RSN is a cabled ocean observatory being constructed in the northeast Pacific Ocean off the coasts of Oregon and Washington. Led by the University of Washington, the RSN program will extend continuous high-bandwidth (tens of Gigabits/second) and power (tens of kilowatts) to a network of many different types of instruments distributed across, above, and below the sea floor and ocean associated with the southern portion of the Juan de Fuca tectonic plate. NEPTUNE Canada is a complementary program building a similar cabled system on the northern portion of the plate.

Such Internet-linked, sensor-robotic networks offer novel approaches to human interaction with many remote or dangerous oceanic processes intrinsic to the habitability of our planet. Continuous, real-time information flow from the ocean via electro-optical cables will launch rapid growth in our understanding of (1) the habitats and behaviour of known and novel life forms; (2) many climate-changing processes; (3) erupting underwater volcanoes; (4) the migration of charismatic marine mammals; (5) the assessment and management of living and non-living marine resources; (6) the timing and intensity of major under-sea earthquakes; (7) the mitigation of natural disasters; and (8) a host of new discoveries continuously unfolding within the 'Inner Space' of the global ocean. Researchers, educators, and members of the public have open web access to all imagery and information in their own laboratories, classrooms, and living rooms as the technologies and visualization software required to operate these in situ, submarine sensing networks become more sophisticated.

In 2009, NEPTUNE Canada (www.neptunecanada.ca) completed installation of its regional-scale observatory with an 800-kilometre cable loop from the shore station at Port Alberni, Vancouver Island, connecting five observatory nodes in coastal, continental slope, abyssal plain, and spreading ridge environments. Ironically, in 2011 one of the main instrument nodes on the floor of the ocean was badly damaged by a deepwater trawler. The area is supposed to be off limits to trawling and illustrates the difficulties in enforcing such regulations in the oceans. This illegal trawler was caught because it was detected by the cameras of the instrument platform. However, in the vast majority of such cases, no cameras exist and illegal trawling continues.

VENUS (www.venus.uvic.ca) is a coastal observatory in waters near Victoria and Vancouver. The first four-kilometre line was installed in Saanich Inlet, with the node at 100 metres depth near the oxic/anoxic transition zone within the fjord. The second 40-kilometre line with two nodes extends from the Fraser River Delta across much of the Strait of Georgia. The observatories investigate ocean and biological processes and delta dynamics in waters of up to 300 metres in depth. Real-time data and imagery are relayed from Saanich Inlet through the VENUS website. VENUS is used in an interactive mode for researchers to trigger experiments remotely and for educators to involve students in on-line studies.

The University of Victoria owns and operates the observatories and has established Ocean Networks Canada as a not-for-profit agency to manage NEPTUNE Canada and VENUS as national facilities.

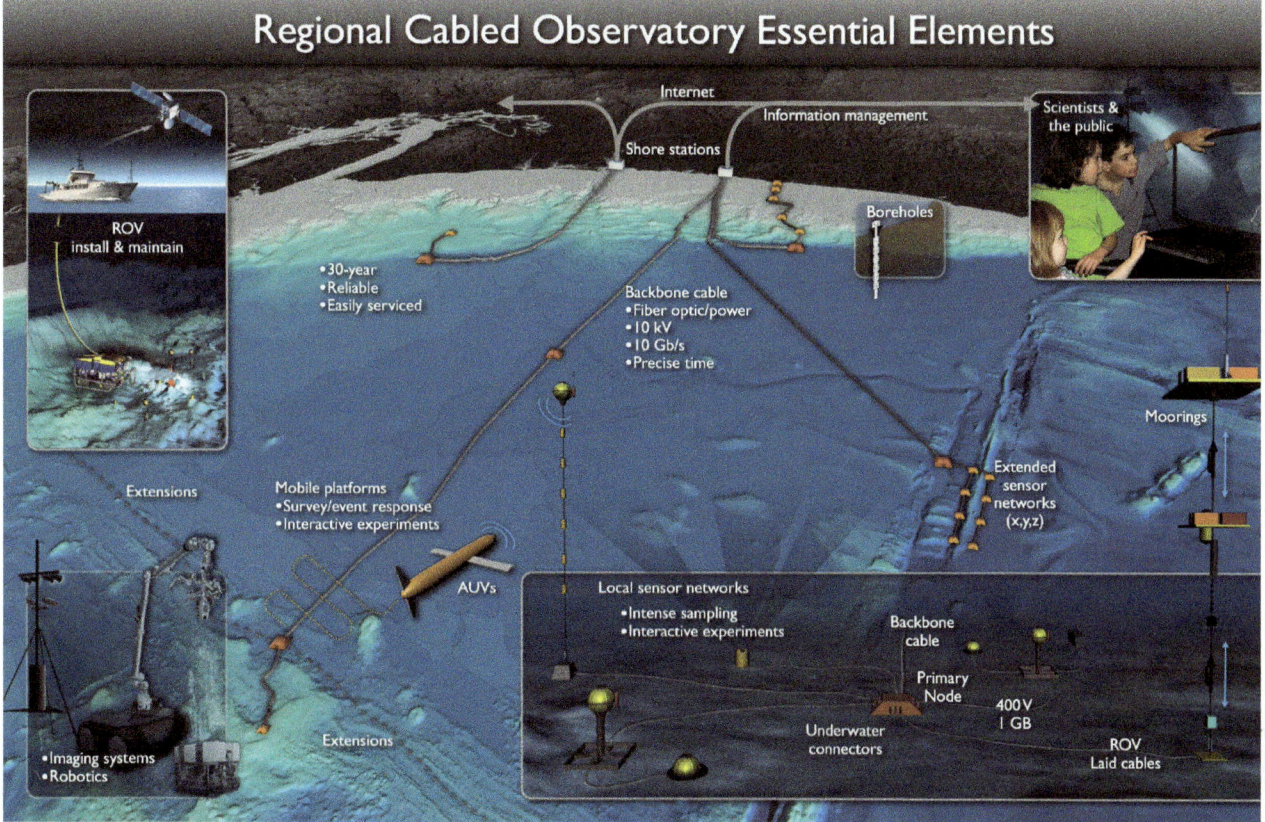

Figure 8.3 | Regional cabled observatory essential elements. *Source:* OOI Scale Nodes Program and the Center for Environmental Visualization, University of Washington.

supply, a warmer climate should produce a greater abundance of plankton. The gas is important in enhancing the formation of atmospheric sulphate aerosol particles and cloud condensation nuclei that would tend to screen out sunlight and hence produce a negative feedback loop for the effects of global warming. Unfortunately, scientists do not feel that the effects of this negative feedback loop will be enough to counteract the effects of the numerous positive feedback loops likely being triggered by global warming.

There is much scientific uncertainty about these global systems. Scientists know they exist and that they are of crucial importance in determining global climatic conditions. They also know that these are dynamic systems that change over time. However, because of a lack of good baseline data, it is often difficult to assess the dimensions of natural change and the underlying mechanisms. Large spatial changes happen only rarely and slowly, requiring the collection of very long data sets of frequent measurements in order to understand them. Ocean scientists are now trying to establish these kinds of monitoring systems so that we can understand observed changes and reduce uncertainty (Box 8.2). Meanwhile, the physical evidence to support global climatic change continues to grow. In 2007, late-summer Arctic sea ice cover was at the lowest extent ever recorded, in 2010 the third lowest extent was recorded, and 2011 saw the second lowest coverage since satellite imagery began tracking the sea ice cover in the late 1970s.

Ocean Management Challenges

Oceanic ecosystems have been providing humans with sustenance since time immemorial. Coastal zones, including the continental shelf, occupy about 18 per cent of the Earth's surface, supply about 90 per cent of the global fish catch, and account for roughly 25 per cent of global primary productivity. Approximately 50 per cent of the world's population now lives in the coastal zone, i.e., within 100 kilometres of a coast. By 2100, nearly 75 per cent of the world's population will live within the coastal zone, mostly clustered into 'mega-cities'.

The oceans are crucial to the way in which planetary ecosystems work and to the functioning of human society. Recent global research indicates that virtually no region of the ocean is untouched by humanity and more than 40 per cent is heavily affected. In 1883, Thomas Huxley, the great nineteenth-century biologist, voiced the common opinion 'that probably all the great sea fisheries are inexhaustible.' This has proven to be far from the truth. This section reviews some of the main management challenges facing ocean ecosystems as a result of fisheries and other human activities (Figure 8.4).

Fisheries

The most important fishing grounds in the world are found on and along continental shelves within fewer than 200 nautical miles (370 kilometres) of the shore. The distribution of these fishing grounds is patchy and localized. More than half of the marine landings are caught within 100 kilometres of

Small-scale fishers on the east coast of Sri Lanka sort the morning's catch.

the coast in depths generally less than 200 metres covering an area of less than 7.5 per cent of the world's oceans, and 92 per cent of marine landings are caught in less than half of the

International Guest Statement
The Rise and Fall of Industrial Fisheries
Daniel Pauly

Industrial fishing started in 1880, when the first steam trawlers were deployed along the English coast. This form of fishing spread rapidly in the North Atlantic, as it did in the North Pacific when Japan developed similar fishing methods. However, only after World War II did industrial fisheries begin their conquest of the world ocean. The growth of these fisheries was particularly rapid in the 1950s and 1960s, in terms of both input into the fisheries (invested capital, vessel tonnage, etc.) and output (tonnage or ex-vessel values of the landings). This period was also a time when fisheries appeared to behave like any other sector of the economy, with increased inputs leading to increasing outputs.

The 1950s and 1960s also saw the first massive fisheries collapses, as stocks that sustained entire fishing fleets, processing plants, and thousands of workers and their families, such as the California sardine fishery, disappeared practically overnight. Other fisheries were rebuilt after a few years; such as the Peruvian anchoveta fishery, which first experienced a massive collapse in 1972. The Peruvian example illustrates an approach already prevalent in the heyday of the California sardine fishery: blame the environment. Thus, in Peru it was El Niño that did it, never mind the fact that the catch in the year prior to the collapse was about 16 million tonnes, rather than the 12 million tonnes that were officially reported, which

itself hugely exceeded what the best experts of the time had recommended as sustainable.

Various concepts have been deployed to apprehend these events. One of these is Garrett Hardin's 'Tragedy of the Commons', which can be used to explain why the pathologies mentioned above were likely to occur in the largely unregulated fisheries then prevalent. These pathologies—which are still with us—have other aspects, notably: (1) not monitoring the fisheries, which results, among other things, in catches that are generally under-reported; (2) ignoring scientific advice aimed at restricting the catch and buildup of fishing effort; and (3) blaming 'the environment' for the fisheries collapse that inevitably follows. These and related pathologies existed long before the overfishing became widespread. However, when generalized overfishing became undeniable, a battery of new terms had to be coined to deal, at least conceptually, with the new developments. Hence the words 'bycatch' (fish caught but not targeted by fishing operations) and 'discards' (bycatch that is thrown overboard, mostly dead), and the emergence of concept of 'IUU' (illegal, unregulated, and unreported) fisheries.

In 1975, catches peaked in the North Atlantic, before going into a slow decline continuing to the present. This was accentuated when the giant stock of northern cod off Newfoundland and Labrador collapsed, plummeting thousands of families and an entire Canadian province into dislocation and economic hardship, and setting off a frantic search for something to blame (hungry seals, cold water, etc.) other than the out-of-control fishing industry.

The relatively well-documented freshwater and coastal fisheries of ancient times had the capacity to induce severe decline in and even extirpation of vulnerable species of marine mammals, fish, and invertebrates, as documented by a variety of sources. However, only since the onset of industrial fishing has the successive depletion of inshore stocks, followed by that of more offshore stocks, become routine. Thus, in the North Sea, it took only a few years for the accumulated coastal stocks of flatfish and other groups to be depleted and for the newly deployed English steam trawlers to be forced to move on to the central North Sea, then further, all the way to Iceland. A southward expansion soon followed, towards the tropics and through the development of industrial fishing in the nascent Third World, often through joint ventures with European (e.g., Spanish) or Japanese firms. Obviously, this expansion created new resource access conflicts and/or intensified earlier ones, and hence the protracted 'cod war' between Iceland and Britain, or the brief 'turbot war' of March 1995 between Canada and Spain. At the close of the twentieth century, the bottom fish resources of all large shelves of the world, all the way south to Patagonia and Antarctica, had been depleted, mainly by trawling, along with those of seamounts and oceanic plateaus. Indeed, from 1950 to 1980, industrial fisheries expanded their reach by about one million km^2 per year; the increase was to 3–4 million km^2 per year in the 1980s, then declined. By 2000, the geographic expansion was essentially over, and the emphasis turned to two other forms of expansion.

The second dimension of the expansion of fishing was bathymetric (i.e., offshore), which affected both the open (pelagic) waters and the sea floor. In the pelagic realm, the exploitation of tuna, billfishes, and sharks by longlines and similar gear has strongly modified oceanic ecosystems, which now have much reduced biomass of large predators. This is intensified by the use of fish aggregating devices (FADs), which, starting around the Philippines, have spread throughout the inter-tropical belt and have made accessible to fisheries small tuna and other fish that could not be captured before, thus representing an additional expansion of sorts. In the demersal or sea-floor realm, trawlers were deployed that can reach down to depths of several kilometres. They yield a catch increasingly dominated by slow-growing, deep-water species with low productivity, which cannot be exploited sustainably. Therefore, given that the High Seas (the waters outside 200-mile exclusive economic zones) are legally unprotected against such depredations, their oceanic plateaus and seamounts are subjected to intense localized fishing pressure, with subsequent collapse of the resources; the same is then repeated on the adjacent plateau or seamount. This fishing mode is no more sustainable than tropical deforestation.

Finally, the two expansions described above, which deplete traditional species, also cause previously spurned fish species to be caught and processed, thus generating a 'taxonomic' expansion. This is the reason why North American and European fish markets increasingly display unfamiliar seafood.

Overall, these three expansions—geographic, bathymetric, and demersal—and the massive import of seafood products by the global North from the global South are the reasons why global fisheries appear sustainable. They are also the reason why they are not.

Daniel Pauly is a French citizen who received his doctorate from the University of Kiel in Germany and for many years taught and did research at the International Center for Living Aquatic Resources Management in the Philippines. He is a Professor of Fisheries at the University of British Columbia and is the Principal Investigator of the Sea Around Us project, which aims at documenting and communicating global fisheries impacts on marine ecosystems (see www.seaaroundus.org).

total ocean area. These areas of the ocean play a major role in sustaining global populations.

Global annual per capita fish consumption has risen from 9.9 kg in the 1960s to over 17 kg by 2009, supplying 3 billion people with over 15 per cent of their protein requirements. Fisheries supply some 20 per cent of the world's annual animal protein supply, but in some areas of Asia and Oceania, fish provide virtually all the protein supply. More than one billion people rely on fish as their primary source of protein. This supply is tapped by fishers ranging from villagers using homemade canoes trying to feed their families through to multi-million dollar offshore factory ships owned by multinational corporations. Given this great scale and variation, it is only recently that fisheries scientists have begun to understand more about what is happening in global fisheries

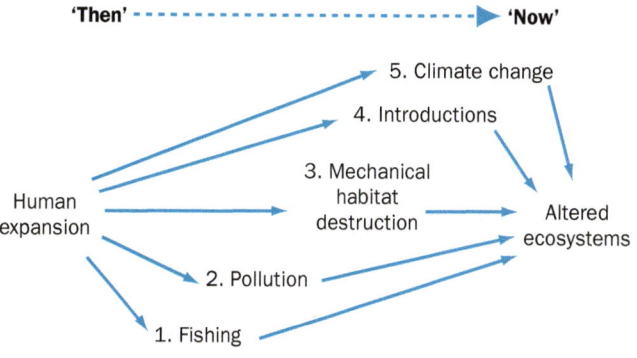

Figure 8.4 | Historical sequence of human impacts on marine ecosystems. Source: Roberts (2003: 170), from Jackson et al. (2001). Reprinted with permission from AAAS: www.sciencemag.org/cgi/content/abstract/293/5530/629.

as discussed by Daniel Pauly in the International Guest Statement on page 247.

Clear evidence, obtained from bottom sediment cores going back hundreds of years, shows that fish populations fluctuated widely as a result of changing environmental conditions, even before the advent of modern fishing. However, the scale and speed of current changes are unprecedented. As stated earlier, more than 80 per cent of global fisheries are now fully utilized or overexploited (Figure 8.5). Because of advances in technology and subsidies, fishing capacity is now estimated at as much as 2.5 times what is needed to harvest a sustainable yield from the world's fisheries. Figure 8.6 shows a gradual levelling off in more recent years, but many scientists argue that this masks a major unprecedented collapse of ocean fisheries. Scientists at the Fisheries Research Centre at the University of British Columbia, for example, revealed that the catch data shown in the graph may be up to 25 per cent above actual catch data. This may be the result of the routine reporting of inflated catches by China (Watson and Pauly, 2001). The problem is that permitted catch levels are based largely on historical catch information. If the latter is inflated, it leads to unsustainable catch levels in the future. Some fisheries scientists have predicted the global collapse of all taxa currently fished by 2048, if current levels of exploitation continue (Worm et al., 2006), although others (e.g., Branch, 2008) dispute this finding.

The fish being caught are also substantially smaller than the ones caught in the past (Figure 8.7). Research suggests that for top predators, current sizes are one-fifth to one-half what they used to be (Myers and Worm, 2003). Modern fisheries management often encourages fishers to select the large individuals of targeted stocks, either by using size-selective gear or by releasing small individuals back to the water. The reasoning has been that this allows smaller, younger individuals to grow up to reproductive age, thereby sustaining the stock. Recent research, however, shows that removing the larger, older individuals of a population may actually undermine stock replenishment. Some researchers have proposed that maintaining old-growth age structure can be important for replenishing fished populations. It is termed the Big Old Fat Fecund Female Fish (BOFFFF) hypothesis. Research shows that removing BOFFFF and other large adults can result in evolutionary changes in populations (Longhurst, 2002).

Changes in fish populations are not immediately obvious, since scientists tend to look only at the most recent data rather than comparing them with historical catches. This problem is known as a **shifting baseline** in which scientists have no other option than to take the current degraded state as the baseline rather than the historical ecological abundance. Removal of virtually all the large predatory fish from oceanic ecosystems will have significant implications for the structure and functioning of these ecosystems.

Some authors (e.g., Jørgensen et al., 2007) now suggest that the impact of commercial fishing is so great that 'evolutionary impact assessments' should be undertaken, since fishing causes changes to occur in decades rather than millennia, as would happen under normal conditions. The scientists note that fisheries-induced mortality is now reckoned to exceed natural mortality by 400 per cent and that increased mortality generally means that fish will reach sexual maturity at a younger age and increase levels of reproduction, but these younger fish produce smaller fish, as explained in the BOFFFF hypothesis. These trends are further exacerbated when fisheries selectively target larger or more mature individuals.

Many fish species show changes in their size and age at maturation, reduced annual growth, and loss of genetic diversity. Such fisheries-induced evolution may be slow to reverse, or even could be irreversible, and will affect many other species through trophic and ecological interactions. Managing fisheries from an evolutionary perspective will almost certainly require a variety of strategies. For some fisheries, we may need no-take zones such as **marine protected areas** (MPAs), as discussed later in the chapter, where the full range of sizes and ages of a given species can thrive. For those that target

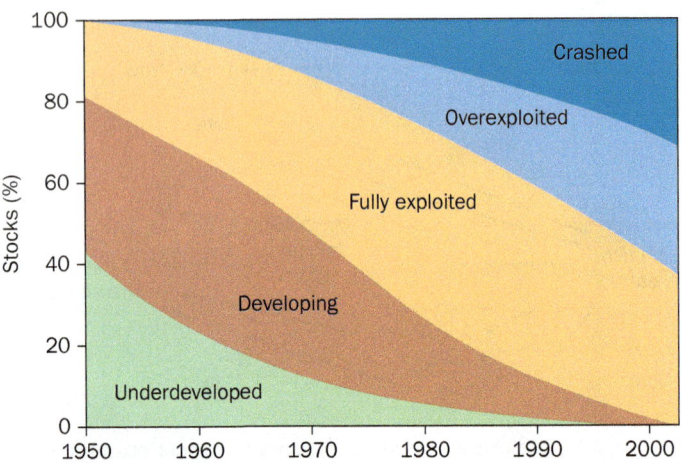

Figure 8.5 | State of the world's fishery stocks. *Source: Nellemann et al. (2008: 17).*

Perspectives on the Environment

Fish Stocks in Decline

. . . persistent changes in size-at-age may be a harbinger of declining condition, reproductive capacity, or the ability to withstand additional anthropogenic or environmental stressors. Declines in size-at-age have often been precursor indicators of impending population collapse. In several Pacific and Atlantic biogeographic units, this review noted long-term declines in size-at-age that may be indicators of significant changes in population productivity and resilience which should be a cause for concern and investigation.

—Fisheries and Oceans Canada (2011)

migratory species, setting both maximum and minimum size restrictions might work better.

The total catch is made up of many different species, which creates several problems. In theory, fishing should be self-regulating. As the catch of the target species declines, this should result in a reduced fishing effort as it becomes unprofitable. Unfortunately, this is not what happens. Instead, fishing fleets increase their efforts towards the target species and then turn to the next most profitable species until that, too, is depleted. Then they pursue the next most profitable species and so on. This is a familiar foraging behaviour in ecology, known as **prey switching**. In the fishery, it leads to **serial depletion** in which one stock after another becomes progressively depleted even if the total catch remains the same. This phenomenon has been well documented in whales (Box 8.3). Unfortunately, after switching the target species, many fisheries take some of the depleted species as bycatch, making it even more difficult for the stocks to recover.

The shift in target species is not the only change in the world's fishing activities. We are progressively exploiting lower and lower trophic levels to derive our catch. Fisheries in many areas have now focused on invertebrates, like the Atlantic crab. Large-sized fish no longer exist. This is known as **fishing down the food chain**. The gains in fish catch shown in Figure 8.6 were mainly the result of reaching further down the food chain to previously underexploited trophic levels, as well as accessing fish from greater and greater depths (Figure 8.8).

Eating lower on the food chain is not a bad idea. Few of us habitually eat wolves, tigers, and other top terrestrial predators, but in the oceans we target the top predators. In terms of protein production, the seas might not be as overexploited as we think. For example, it takes close to 60 million metric tons of potentially edible fish per year to feed the three million metric tons of the three major tropical tuna species we harvest annually because of the second law of thermodynamics, discussed in Chapter 2. If we could replace some of our tuna sandwiches with the anchovies, sardines, squid, and other species the tuna eat, we would open up a substantial supply of protein that could feed millions more.

Besides the impact on target species, fishing activities have many other ecological repercussions. Of particular concern is the impact on non-target organisms, or bycatch. Estimates suggest that 25 per cent of the world's catch is dumped because it is not the right species or size. Virtually

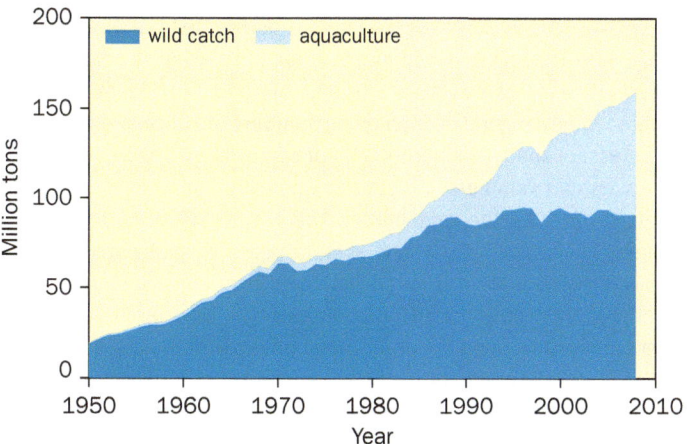

Figure 8.6 | World seafood production, 1950–2008. Source: Theobold (2009).

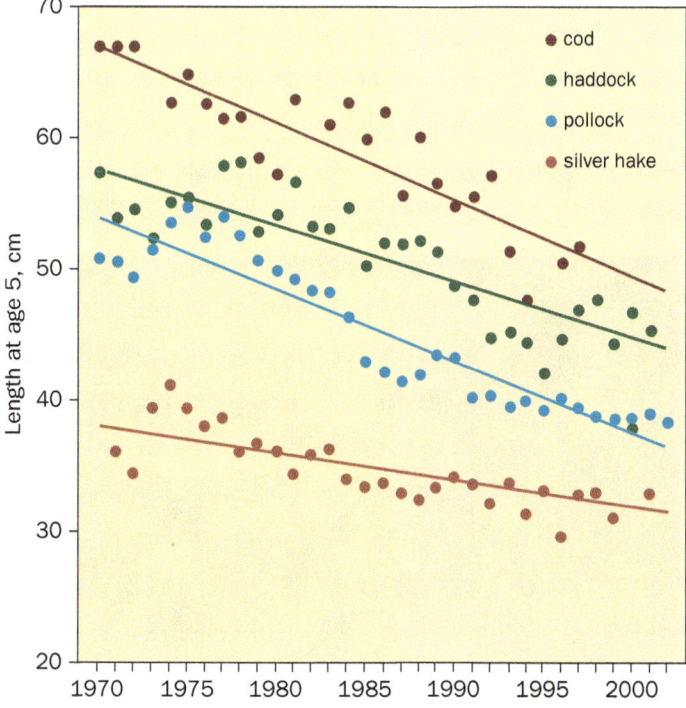

Figure 8.7 | Fish size decline, 1970–2000. Source: Fisheries and Oceans Canada 2003. Source: DFO (2010: 25). Reproduced with the permission of the Minister of Public Works and Government Services Canada, 2011.

This shark was caught by a 'ghost net' snagged on the bottom but still doing the deadly function it was designed for.

Environment in Focus

Box 8.3 Leviathan

> The mammoth bones of the California Gray lie bleaching on the shores of those silvery waters, and are scattered along the broken coasts, from Siberia to the Gulf of California; and ere long it may be questioned whether this mammal will not be numbered among the extinct species of the Pacific.
>
> —Captain Charles Scammon, 1874

Canada has the longest coastline in the world, as well as some of the richest waters. Cold oceanic waters from the north and deep upwelling currents mix with warmer waters from the tropics to create an abundance of life. Rich nutrient supplies, accompanied by shallow seas and long daylight hours in the summer, have created some of the richest waters off the east coast. Plankton flourish and provide the base for a diverse food web that supports three main groups of marine mammals—the Odontoceti or toothed whales, the Mysticeti or baleen whales, and the Pinnipedia or seals and walruses. The baleen whales all feed on plankton, small fish, and marine algae by means of plates of baleen in their mouths that serve to filter these organisms from the water. From these small food items, the baleen whales have evolved as the largest creatures on this planet. The blue whale is the largest of the baleen whales; the largest recorded was a female more than 30 metres in length and weighing 140 tonnes. A calf is typically seven metres long at birth and may weigh almost three tonnes. Blue whales, as well as other baleen whales such as the fin and sei whales, were once found in abundance off the east coast, as was the largest toothed whale, the sperm whale. These whale populations were all decimated by hunting and are only now starting to recover.

In the late fifteenth century, Europe found itself increasingly short of the oil needed to light its lanterns. Marine mammals, with a thick layer of blubber, were the solution. Exposed to high heat, the blubber can be rendered down to oil. The Basques—a great seafaring people from northern Spain—discovered the rich whaling grounds off the east coast of Canada, and the slaughter began. The abundance of whales in these waters at that time is difficult to imagine. Whales could be harpooned from shore. Early mariners complained of whales as a navigational hazard because they were so numerous; one missionary in the Gulf of St Lawrence reported that the whales were so numerous and loud that they kept him awake all night! The limiting factor was not the number of whales but the ability to process them, and as increasing numbers of shore stations were established, another toehold of colonization began.

As human numbers increased, whale numbers declined. First to go was the one hunted the most, the black right whale. The black right whale was targeted because it was slow, it floated when it was killed, and one whale could be rendered into more than 16,000 litres of oil. Besides the oil, the baleen was used for other indispensable purposes such as clothing supports, brush bristles, sieves, and plumes for military helmets. A Basque shipowner could pay off his ship and all his expenses and still make a good profit in one year from such whales. Although the whales gained some respite when England destroyed the Spanish Armada in 1588, other nations finished off the job.

Other whales—sperms, humpbacks, blues, fins, seis, and minkes—soon joined the right whale as commercially, if not biologically, extinct. This is a classic example of serial depletion. By the early 1970s, there were no commercially viable populations of large whales remaining. A global moratorium on whaling was announced in 1987 and still stands today, despite pressure applied by Japan and Norway, both of which continue to kill whales for 'research purposes'. The appeal of 'scientific

Humpback whales were once common off both the Pacific and Atlantic coasts of Canada.

Whale harpoon in the Fisheries Museum of the Atlantic, Lunenburg, Nova Scotia.

whaling' is not only that it bypasses any internationally agreed catch limits but that it also circumvents all other rules regarding protected species, closed areas, killing of juveniles, killing methods, and so on.

Iceland became the third country left whaling when it resumed the practice in 2006 and by 2011 had killed 280 endangered fin whales and 186 minke whales. In late 2007, Japan's whaling fleet left port for the South Pacific with orders to kill up to 50 humpback whales—the first known large-scale hunt for the whales since a 1963 moratorium put them under international protection. International protests and growing awareness have led to decreased demand for whales globally. Following an unsuccessful hunt in 2010, in February 2011 Japan suspended its annual whale hunt in Antarctica due to harassment by anti-whaling vessels. Iceland has also felt the lack of demand and in 2011 postponed the start of its fin-whaling season, although there remains a self-allocated quota of 154 whales for the year. The International Whaling Commission estimates the global population of fin whales to be between 1,400 and 7,200 animals and the World Conservation Union lists the species as endangered. Norway continues whaling despite shrinking markets, and set itself a quota of 1,286 minke whales for 2011.

The removal of such a large biomass from the top of the food chain obviously has ecological repercussions for other organisms. One possible implication is an increase in other krill eaters that would benefit from removal of these large and efficient competitors. For example, it has been suggested that increases in the populations of krill-eating seals (e.g., fur and crab-eating seals) in the Antarctic came about as a result of whaling. More breeding sites are required to support the higher populations of seals, so seals colonize areas previously used for nesting by birds such as the albatross. Did declining whale numbers also lead to a decline in the albatross? We cannot say for sure, but the example does illustrate the complexities of changes in food webs.

Populations for most species have been slow to recover because of the slow reproductive rates of these K-strategists (Chapter 3). Whaling has now been controlled throughout most of the world, and on all of Canada's coasts a resurgence is occurring, albeit very slowly for some species such as the right whales. The largest whale, the blue, is also in trouble. Although numbers still exist, finding a mate is difficult for this wide-ranging species. Scientists have now found evidence that the blues are interbreeding with the more common fin whales, the second largest whale in the world, and fear that this hybridization will result in a loss of genetic identity for the blue, which could then disappear as a species. Unfortunately, the rapacious killing by our ancestors denied all future generations the spectacle of our seas full of these mighty creatures. Are our actions today denying future generations similar opportunities?

all organisms dumped overboard die. The world's largest turtle, the leatherback turtle, is rated as critically endangered on the IUCN (World Conservation Union) Red List (see Chapter 14). In 1980, some 91,000 leatherbacks were left in the Pacific; there are now fewer than 5,000. The Atlantic population is stronger and the turtles spend a significant amount of time foraging off the Scotian Shelf. They are mainly the victims of the **longline** and gill-net fisheries. Longline fishing in all the world's deep oceans kills some 40,000 sea turtles each year, along with 300,000 seabirds and millions of sharks. Bycatch of albatross, petrels, and shearwaters in longline fisheries is one of the greatest threats to these seabirds. All 21 of the world's species of albatross are now considered at risk of extinction, along with 57 species of sharks and rays.

Some innovative ways to address bycatch are starting to be developed, in some cases helped along by cash prizes offered by NGOs. The World Wildlife Fund, for example, launched the International SmartGear Competition in 2004

As large, long-lived fish high on the food chain are depleted, we seek smaller species. *Reprinted by permission of Adrian Raeside.*

Figure 8.8 | Fishing down the food chain.

to encourage innovative, practical, and cost-effective gear designs that safeguard marine life while enabling fishers to better target their intended catch. One winner, knowing that most turtles are hooked at shallow depths, proposed a system for setting baited hooks deeper than 100 metres, thereby minimizing encounters with sea turtles while maintaining the catch of target fish. Turtle mortality can be further reduced by changing the types of hooks used and using fish rather than squid as bait. Another winner proposed placing magnets on fishing lines to scare away sharks, which are particularly sensitive to magnetic fields.

One of the most destructive means of fishing is **bottom trawling**, in which heavy nets are dragged along the sea floor, scooping up everything in their path. This is a common method for catching shrimp, and the ratio of shrimp to other organisms caught is generally around 10 per cent (i.e., one shrimp caught for every 10 organisms caught unintentionally). Trawling has been estimated to be as damaging to the seabed as all other fishing gear combined. The damage extends to more than half of the seabed area of many fishing grounds and is worse in the inner and middle parts of the continental shelves, with particular damage to small-scale coastal fishing communities.

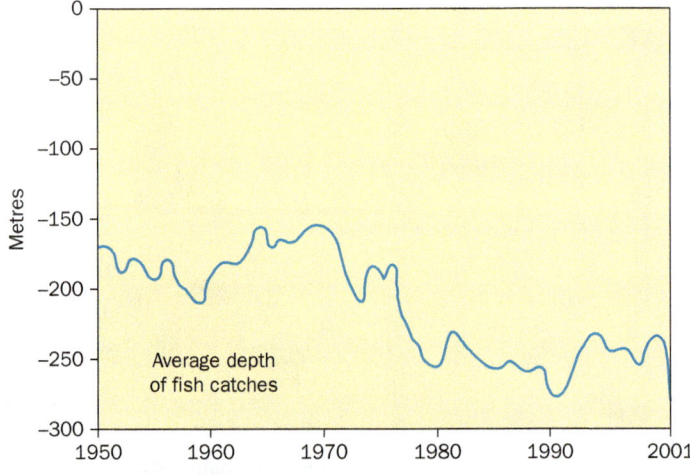

Figure 8.9 | Trend in mean depth of catch since 1950. *Source: Millennium Ecosystem Assessment (2005).*

Rich ocean waters attract many pelagic (open ocean) bird species, such as this adult waved albatross.

One noted victim of bottom trawling was the ancient sponge reefs found off the coast of BC. The sponge reefs, at depths between 165 and 240 metres, are more than 9,000 years old and may reach as high as a six-storey building. Previously, they were known only from fossils in Europe dating from 146 to 245 million years ago, and they were thought to be extinct before their discovery in 1988. Sponge reefs are widely believed to be the first multi-cellular animals on earth, appearing almost 570 million years ago. The reefs thrive in silica-rich water and are the only siliceous sponge reefs on the planet. Some live in shallower water but are limited to individuals rather than reefs. The federal government called for *voluntary* trawl restrictions, but in 2002, after documentation of extensive damage to the most pristine reefs by trawling, mandatory closures for groundfish trawling were finally implemented to protect the northern reefs. In 2006 the original boundaries were expanded and closures included shrimp trawling. Smaller reefs in the Strait of Georgia remain unprotected, although a 2010 Federal Strategy provides hope for their protection and outlines preservation plans for sponge reefs in BC by 2015—if they still exist by then!

In addition to the direct destructive effects of fishing on non-target organisms, it has indirect effects through food chain relationships. Steller's sea lion, for example, was once abundant in the North Pacific, with more than 300,000 animals recorded in 1960. By 1990, this number had fallen to 66,000, and the US declared the sea lion endangered. The main reason for the decline is thought to be the decline in pollock, their chief food source. As the harvests of cod diminished elsewhere, the demand for pollock increased, leading to unprecedented catches and a subsequent decline in sea lion numbers.

Scientists have recently suggested that another factor may have contributed to this decline. Earlier industrial whaling activities removed the bulk of the killer whales' main prey. In response, the killer whales started 'fishing down the food chain', eating sea lions in much greater numbers. Reaching even further down the food chain, killer whales are now implicated in the decline of sea otters in the North Pacific (Chapter 3). Other scientists (e.g., Trites et al., 2007) have questioned this relationship. But sea lions and sea otters are not the only marine mammals to experience declining numbers; two-thirds of all marine mammals are now classified as endangered on the IUCN Red List (Chapter 14).

Another challenge related to ecosystem dynamics is the rapid growth in jellyfish populations in many parts of the world. Over 2,000 species of jellyfish have been found to be increasing in number and also appearing earlier every year as a result of warming ocean temperatures. Warmer ocean temperatures favour flagellate-dominated food chains (zooplankton) preferred by jellyfish rather than diatom-based food chains (phytoplankton) favoured by fish. In effect, evolution is running backwards to the Precambrian period

Model of a box trawl in the Fisheries Museum of the Atlantic, Lunenburg, Nova Scotia. The large weights are dragged along the sea floor, destroying everything in their path.

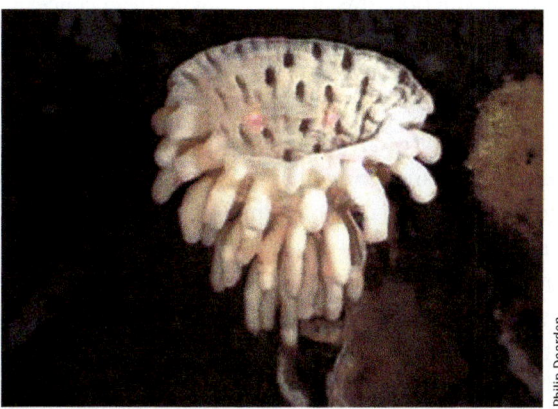

One of the deep sponges found off the BC coast.

Steller's sea lions, the world's largest sea lions, are now highly endangered in the North Pacific. These are at Cape St James, off the southern tip of the Haida Gwaii (formerly the Queen Charlotte Islands), also the windiest spot in Canada.

when jellyfish were more dominant than fish in the world's oceans. The current growth in jellyfish populations is related to several factors, including increased nutrient supplies from agricultural runoff and warmer waters, but a main factor is thought to be the reduction in predation on jellyfish as fisheries have depleted the larger fish and turtles that prey on jellyfish.

This increase in jellyfish has many implications. Jellyfish, for example, eat vast quantities of plankton, which thus reduces the base of the food chain for many fish species. Ocean bacteria are critical in absorbing and recycling nutrients such as phosphorus, carbon, and nitrogen. Unlike the dead bodies of fish, bacteria cannot effectively recycle the higher carbon levels found in jellyfish biomass; instead, the jellyfish exhale increasing amounts of carbon dioxide, which in turn adds to global warming. Scientists have detected a 40 per cent decline in global phytoplankton levels since 1950. Since phytoplankton are responsible for half of the annual oxygen production, this decline is obviously of concern.

Pollution

As the recipient of all the polluted water that flows off the land as well as airborne contaminants, the world ocean is the ultimate sink for many pollutants. The scale of global pollution is now astounding—even the oceans are being rendered eutrophic (Chapter 4), in some cases from farming practices thousands of kilometres away. For example, scientists suspect that iron particles picked up into the atmosphere in Africa and blown across the Atlantic are a major factor in increasing the growth of smothering algae on the coral reefs of the Caribbean.

Marine pollutants take various forms, originate from many different sources, and have a wide range of effects. About 80 per cent of ocean pollution comes from activities on land. The remaining 20 per cent comes from activities at sea, such as waste disposal, oil spills, vessel traffic, oil and gas exploration, and mining. Although we are all familiar with major oil spills, such as that of the *Deepwater Horizon* explosion and wellhead blowout in the Gulf of Mexico, we think little of the many diffuse sources of oil, such as leaking car engines, that eventually seep into the oceans. In fact, the total amount from non-point sources of pollution is considerably greater than from point-source pollution. During the past 20 years, many governments have made considerable progress in monitoring and regulating point-source pollution, such as effluent discharge from factories. However, addressing non-point-source pollution is a much more challenging task, since it requires a change in behaviour by billions of people.

Chemical pollutants take two main forms: toxic materials and nutrients. We live in a chemical society, with more than 100,000 chemicals used in manufacturing and released into the environment every year. Many of these chemicals are harmful to life; when they end up in the ocean, they may cause instant death if released in sufficient quantity, or they may have sub-lethal effects such as inhibition of reproduction. The chemicals are also subject to *bioconcentration*, discussed in more detail in Chapter 10. Synthetic organic chemicals and toxic metals tend to concentrate along two main interfaces in the ocean: the boundary between the seabed and water and the boundary between the water and the atmosphere.

One rapidly emerging impact relating to pollution is **endocrine disruption**. The endocrine system consists of glands and hormones that control many bodily processes such as sex, metabolism, and growth. Many chemicals in everyday use have been found to mimic these processes and may

This dovekie was killed by an oil spill. Oil penetrates through the feathers to the layer of down beneath and decreases the effectiveness of the feathers' insulating properties.

stimulate, replace, or repress the natural processes. More than 50 such endocrine disrupters have been positively identified, and scientists suspect that there are hundreds more. Many of them are in commonly used products such as soaps and detergents. The major effect of these chemicals on marine life detected so far is the feminization of various aquatic species. Hermaphroditic fish are appearing all over Europe and also in the Great Lakes. All the fish in the River Aire in England, for example, show signs of feminization. Increasing numbers of reports from the Arctic are reporting hermaphroditic polar bears. Many endocrine disrupters, such as the pesticide DDT and other organochlorines, are vulnerable to both the *grasshopper effect* and *biomagnification* (as described in Chapter 10) and would logically find their way to the top Arctic predator. More research is now underway on this issue, since it is feared that these so-called 'genderbenders' may be at least partly behind the documented fall in human male sperm counts in industrialized societies over the past 50 years.

Oxygen depletion occurs as a result of nutrient enrichment, as described in Chapter 4, and leads to large dead areas within the oceans. These areas (**hypoxic** or oxygen-deficient areas) increased from 149 in 2003 to more than 400 in 2011. Their growth could also promote the development of far more male fish than female, threatening some species with extinction. Shang, Yu, and Wu (2006) found that low levels of dissolved oxygen decreased the activity of certain genes that control the production of sex hormones and sexual differentiation in embryonic zebra fish. As a result, 75 per cent of the fish developed male characteristics versus 61 per cent of those raised under normal oxygen conditions. This gender shift decreases the likelihood that they will be able to reproduce in sufficient numbers to maintain sustainable populations. The study raises new concerns about vast areas of the world's oceans, known as 'dead zones', that lack sufficient oxygen to sustain most sea life. Fish and other creatures trapped in these zones often die. Those that escape may be more vulnerable to predators and other stresses. This study suggests that dead zones potentially pose a third threat—the inability of offspring to find mates and reproduce.

Global warming will further promote the growth of dead zones. As water warms it is capable of holding less oxygen. Furthermore, the ocean receives its oxygen from the atmosphere and from photosynthesizing algae floating at the top. The oxygen is distributed to the deeper ocean as the water sinks. However, as noted earlier, global warming will lead to lighter surface water since it will contain more freshwater from melting ice. It will also become lighter as thermal expansion occurs. The net effect will be to reduce mixing of surface waters and further reductions in oxygen content in the ocean, especially in deeper waters. The drop in oxygen levels will further promote the growth of jellyfish, which can store oxygen in their jelly, over oxygen-demanding fish.

Danish researchers have calculated that, even if we reduce carbon emissions to zero by 2100, the effects of global warming will cause oxygen levels to fall by 30 per cent over the next thousand years. It seems that a low oxygen-level ocean is going to be the reality of the future.

Plastics are also an increasing source of concern, even in supposedly more isolated areas such as the Canadian Arctic. Researchers have found that 84 per cent of fulmars have plastic in their stomachs. Fulmars are particularly vulnerable since they skim the top of the water searching for food, but other species are also starting to show significant accumulations: 11 per cent of thick-billed murres from five Arctic colonies, for example, were found to contain plastics in their stomachs. Unlike fulmars, murres dive deep to find food. Ingestion can lead not only directly to death but also to more subtle effects such as loss of appetite, stunted growth, and exposure to pollutants that can leach out of the plastics.

In the North Sea, seven European countries have jointly implemented an aggressive program to reduce the amounts of plastic in the oceans and, consequently, in bird populations. Canada as yet has no such program, even though the levels evident now in the Arctic are in excess of those set as acceptable by the European partners. Already the problem is seen as worsening. UNEP estimates that the average North American uses around 100 kg of plastic a year and predicts that this figure will rise to 140 kg by 2015. Furthermore, with an increasingly ice-free Arctic Ocean there will be greater boat traffic and increasing potential for more plastic disposal unless measures are put in place.

Energy

World demand for energy, particularly oil and gas, continues to rise (Chapter 12). Every time you jump into your car or go on an airplane, you are sending a financial message to industry and the government that you support further development of fossil-fuel sources. All of the world's cheap and accessible terrestrial sources have now been tapped. Many of the world's main oilfields, such as the North Sea in Europe and Hibernia off the coast of Newfoundland, are situated in sedimentary basins under the oceans. More than 60 per cent of current global production comes from these sources. Offshore oil rigs are a source of chronic, low-level pollution caused by the disposal of drilling mud and drill cuttings, which smother the local environment and are often contaminated with oil or chemicals. Oil is pumped to the surface and loaded on tankers or piped ashore, where it is often stored close to the ocean before being refined and further distributed. There is potential for spillage at every stage. Many seabirds and marine organisms are highly vulnerable to oil pollution. Following the wreck of the *Exxon Valdez* off Alaska, for example, more than 750 sea otters were killed. Peterson et al. (2003) argue, however, that the long-term effects of the spill, including mortalities of species such as sea otters, were much greater because of ongoing impacts such as contaminated food chains. Oil is lethal to marine life by virtue of its physical effects and its chemical composition. Cleanup attempts following oil spills can also damage many species.

Major gas and oil discoveries have been made in Atlantic Canada. The Hibernia field on the eastern edge of the Grand Banks, some 300 kilometres east–southeast of St John's, was the first field put into production. In 2002, it was joined by the Terra Nova field, with a third, White Rose, coming into production in 2005. Chapter 12 provides more detail about these oil and gas operations. However, most accessible ocean oil basins have already been developed, which means that exploration and development are pushing into increasingly challenging and fragile environments, such as the Arctic Ocean and North Pacific, and are seeking oil at ever-greater depths, as was the case with the *Deepwater Horizon* disaster. The most effective way to reduce the effects of oil exploration and development in these regions is by reducing demand for oil products.

Recent research has highlighted the potential for seabed-based methane hydrates to meet some energy demands. Methane hydrates are ice-like deposits found in the top few hundred metres of sediment in certain deep ocean areas of the continental margins. The methane gas is actually trapped in ice cages and can be easily extracted from it, but removing the hydrates from the seabed has proved problematic. Russian energy experts have tried using antifreeze to remove the methane from hydrates, and research has focused on trying to pipe warm surface water down to melt the hydrates and then piping the gas to the surface using a parallel set of pipes. However, melting the hydrates to release methane may cause the seafloor to become unstable and could have untold ecological impacts. In addition, if methane is lost to the atmosphere during the process, it could add to the global warming phenomenon, since methane is a potent greenhouse gas.

Coastal Development

Twenty-one of the world's 33 mega-cities are coastal. With half of the world's population living in coastal regions, and that proportion expected to grow significantly during the twenty-first century, these regions also contain the highest concentrations of supporting infrastructure, industrial plants, energy use, and food production in the world, and they are the focal points of global tourism. Half of the world's coastal wetlands have been filled in to support these developments. Meanwhile, as in Canada, environmental decision-making and management within the coastal zone are often highly fragmented among many different agencies. These extreme pressures on fragile ecosystems with ineffective management control have often led to highly degraded coastal environments in many countries. Half of the countries in the world do not have any coastal legislation in place to address the situation.

Climate Change

Global average air temperatures are expected to warm by anywhere from 1.4 to 5.8°C this century. In 2007 the IPCC estimated that this would raise the global sea level by 18 to 59 centimetres. This figure was revised to over a metre by the time of the Copenhagen talks in 2009, and in 2010 scientists from the Arctic Monitoring and Assessment Program raised the estimates to between 0.9 and 1.6m. The main causes of sea-level rise include thermal expansion of sea water and the melting of ice in land-based glaciers. Over-pumping of groundwater supplies that eventually run off into the ocean is another cause. The rate of sea-level rise has increased significantly over the past few years. Since 1961, global average sea levels have risen at an average rate of 1.8 (1.3 to 2.3) millimetres annually, and since 1993 this rise has been 3.1 (2.4 to 3.8) millimetres annually (IPCC, 2007c). Sea-level rise is not, however, uniform. For example, the Bering Sea between Alaska and Russia rose 0.5mm per year since 1993 whereas the Indonesian Seas rose 6.4mm. Higher sea levels increase the impact of storm surges, accelerate habitat degradation, alter tidal ranges, exacerbate flooding, and change sediment and nutrient circulation patterns. Rising levels will displace approximately 1 billion people, many of them among the poorest in the world.

Estimates suggest that an increase in mean sea-surface temperature of only 1°C could cause the global destruction of coral reef ecosystems. The waters off British Columbia have already experienced a rise in sea-surface temperature of between 0.3 and 0.9°C over the past 50 years. Increases in temperature may slow or shut down the thermohaline circulation

Coastal mega-cities, like Mumbai, India, place marine ecosystems under increasing pressures, resulting in issues such as coastal erosion, intrusion of sea water into freshwater supplies, loss of habitat for birds, fish, and other marine wildlife, depletion of fishery resources, and marine pollution.

(Figure 8.2), causing widespread climatic change, changes in the geographic distributions of fisheries, and increased risk of hypoxia in the deep ocean. Scientists have already detected changes in the distribution patterns of many marine species as a result of changes in ocean temperatures attributed to global warming. Early breakup of sea ice is thought to be one of the main contributing factors in the increased stress experienced by polar bear populations (Rode et al., 2010; Durner et al., 2011).

The Arctic Ocean is one of the most sensitive indicators of global climate change. Significant changes are taking place rapidly. In addition to the well-documented loss of perennial ice cover as a whole, the amount of oldest and thickest ice within the remaining multi-year icepack has declined significantly. The oldest ice types have essentially disappeared, and 58 per cent of the multi-year ice now consists of relatively young two- and three-year-old ice, compared to 35 per cent in the mid-1980s. Scientists now are predicting an ice-free Arctic in another 40 years.

Global Responses

International Agreements

Strong international action can help to address these problems, and there are many examples of international treaties concerning the oceans. The overall international legal framework is provided by the United Nations Convention on the Law of the Sea (UNCLOS), which entered into force in 1994. Activities under its jurisdiction include protection and preservation, navigation, pollution, access to marine resources, exploitation of marine resources, conservation, monitoring, and research. One of the most important provisions was the agreement for coastal nations to establish **exclusive economic zones** (EEZs). The Convention also established 45 per cent of the seabed as common property. Canada ratified the Convention in November 2003.

More specific agreements include the 1972 London Dumping Convention, which has led to a gradual decline in the amount of sewage sludge and industrial waste dumped into the oceans. In terms of fisheries, the UN Moratorium on High Seas Driftnets was quite successful. The same kind of approach now needs to be applied to reduce the impact of other highly damaging types of fishing gear, especially longlines and industrial trawlers with high rates of bycatch. The United States has already taken some steps to protect embattled marine species by closing the west coast to longlining and by prohibiting the Hawaii longlining fleet from fishing for swordfish. There are also temporal and spatial closures for gill-netting and moves to ban bottom trawling, which may be the most destructive fishing method of all.

Some strong ecologically-based marine agreements have emerged over the past decade, including the UN Convention on Straddling and Highly Migratory Fish Stocks and the Global Program of Action for the Protection of the Marine Environment from Land-Based Activities. However, some of these new agreements have not yet been fully implemented. For example, 10 of the world's top 15 fishing nations have not yet ratified the Straddling Stocks agreement, although Canada has done so.

In addition to these international agreements, international programs are aimed at creating a more comprehensive, global approach to oceans management. At the 2002 World Summit on Sustainable Development (WSSD) in Johannesburg, governments committed to restore most of the major global fisheries to commercial viability by 2015.

However, Daniel Pauly, of the Fisheries Centre at the University of British Columbia, suggests that to achieve real sustainability, we need to rethink how fishing is undertaken. Policy-makers have concentrated on promoting industrial fisheries under the mistaken belief that they catch the vast majority of fish. Pauly (2006) argues, however, that not only are small-scale inshore fisheries (artisanal fisheries) as productive as industrial fisheries but they are much more

Table 8.1 | Comparative Benefits of Large- and Small-Scale Fisheries

Benefits	Large-Scale Fishery	Small-Scale Fishery
Number of fishers employed	About ½ million	More than 12 million
Annual catch of marine fish for human consumption	About 29 million tonnes	About 24 million tonnes
Capital cost of each job on fishing vessels	US$30,000–$300,000	US$250–$2,500
Annual catch of marine fish for industrial reduction to meal, oil, etc.	About 22 million tonnes	Almost none
Annual fuel oil consumption	14–19 million tonnes	1–3 million tonnes
Fish caught per tonne of fuel consumed	2–5 tonnes	10–20 tonnes
Fishers employed for each US$1 million invested in fishing vessels	5–30	500–4,000
Fish and invertebrates discarded at sea	10–20 million tonnes	Little

Source: Pauly (2006: 17).

efficient and less damaging, and provide more support for local communities (Table 8.1). A wealth of knowledge has built up around the world that reflects the intimate relationship that artisanal fishers have with the resource and the more geographically-based management schemes that have evolved to reflect this relationship.

Pauly and others say that the quickest way to effect a change in fishing methods is to end the subsidies that support commercial fisheries, estimated at over $15 billion per annum. Subsidies make up about 25 per cent of the total landed values from the catch of these fleets, whereas the reported profit per landed value is no more than 10 per cent. Were the subsidies removed, the fisheries would die as uneconomical (Sumaila and Pauly, 2007). Since more than half the subsidies are for fuel, eliminating fuel subsidies would be simple and effective, while contributing to a reduction in fossil-fuel consumption.

Although not an agreement, another international movement that may show some promise for improving fishery management is through certification, just as forestry certification programs have begun to influence that industry (Chapter 9). The Marine Stewardship Council, based in London, is an independent global assessment body that sets standards and co-ordinates efforts in this area. In a recent ruling the Council certified BC's spiny dogfish fishery as the world's first sustainable shark fishery, which means it had to meet requirements for healthy fish stocks, minimal ecosystem impacts, and an effective management system. The certification lasts for five years and there is an annual audit. The dogfish takes 35 years to become sexually mature and has a two-year pregnancy period, meaning that extreme care must be taken to avoid any overfishing, since recovery times would be very long. However, if monitored and implemented rigorously and with public awareness about the importance of eating only certified products, then certification schemes may help improve fisheries management.

Marine Protected Areas

Compared to the terrestrial environment, where ecological communities are often associated with areas defined by geographical features (such as mountains and rivers), precise boundaries of distinct communities or processes in oceans are rare. Geographic scales are large, and biological processes are not self-contained within a given area. The water overlying the seabed is a mobile third dimension that provides nourishment for much of ocean life. These characteristics of the ocean environment mean that surveys take time and are costly, and because of the high variability in sizes of marine populations, survey results are typically uncertain. Until recently, these difficulties prevented scientists from being able to assess the effectiveness of marine protected areas (MPAs). However, in 2001, following two and a half years of rigorous scientific review, 161 of the world's top marine scientists signed a consensus statement (www.nceas.ucsb.edu/Consensus) that marine reserves:

- conserve both fisheries and biodiversity;
- are the best way to protect resident species;
- provide a critical benchmark for the evaluation of threats to ocean communities;
- are required in networks for long-term fishery and conservation benefits;
- are a central management tool supported by existing scientific information.

The establishment of MPAs has lagged substantially behind their terrestrial counterparts—currently just over 1 per cent of the ocean has been designated as protected. However, with increasing political awareness of the importance of the oceans and their highly degraded state, MPAs are starting to receive some attention. The 2002 World Summit on Sustainable Development set 2012 as the target date for completion of an effectively managed, ecologically representative network of marine and coastal protected areas within and beyond areas of national jurisdiction. The latter occur in the 64 per cent of the oceans beyond the 200-nautical-mile (370 km) limit of the EEZs of coastal states. This area is known as the High Seas, and because of a lack of an established legal framework for their allocation and management other than UNCLOS, it will be necessary to take concerted international action to achieve their protection. Researchers have calculated the impact on fisheries of establishing such areas outside of territorially defined waters and conclude that closing 20 per cent of waters on the High Seas to fisheries might lead to a loss of only 1.8 per cent of the current global reported marine fisheries catch (Sumaila et al., 2007).

Canada's Oceans

Canada has the longest coastline of any country and the second largest continental shelf, equal to 30 per cent of Canada's land mass (Figure 8.10). There are some 1,200 species of fish and many globally important populations of marine mammals. Unfortunately, several of these species are also on Canada's list of species at risk (see Chapter 14), including the beluga, bowhead, northern right, and Georgia Strait killer whales. More than 7 million Canadians live in coastal communities.

The federal government largely holds jurisdiction for the marine environment below the high-water mark. Twenty-seven different federal agencies and departments are involved. The lead agency is the Department of Fisheries and Oceans (DFO; since 2008, formally titled Fisheries and

Figure 8.10 | Canada's coastline and continental shelf.

Oceans Canada), which has traditionally focused its efforts on commercial fishery management but is also responsible for all marine species (except seabirds, which come under Environment Canada). DFO is also the lead agency for Canada's Oceans Strategy, discussed later. Provincial governments have responsibility for shorelines, some areas of seabed, and some specific activities, such as **aquaculture**. However, in BC the responsibility for aquaculture has recently been transferred to the federal government. Municipal governments also influence the coastal zone, since they have responsibility for many land-based activities affecting the oceans. There are quite a few areas of overlap between these different levels of jurisdiction, and, inevitably, conflicts have arisen. As a result, coastal and marine resource management is typically fragmented and often ineffective. In addition, the very size of Canada's coastal zone, ranging from the Mediterranean climate of BC's Gulf Islands to the High Arctic, means that a wide range of conditions and activities exists. The surface area of Canada's ocean estate is 7.1 million km^2.

Fisheries

Canada has been blessed with some of the most productive marine environments in the world, and these environments were critical in sustaining populations of Aboriginal peoples and attracting European attention on both coasts. Unfortunately, the squandering of this rich biological heritage stands as one of the sharpest reminders of our inability to manage ourselves in a way that sustains resources over a long period of time. Since 1992—the year of the cod fishery closure in Atlantic Canada—commercial landings have declined by 19 per cent overall, including cod by 162 per cent, turbot by 59 per cent, redfish by 122 per cent, and hake by 65 per cent. Bear in mind, too, that these percentages reflect a shifting baseline, as discussed earlier. The number of vessels in Canada's commercial marine fishing fleets has decreased by 37 per cent, but the total income of the industry has increased. By 2009, sea fishery landings accounted for $1.6 billion. The average income earned within fleet sectors has doubled. In Atlantic Canada, for instance, the average vessel earned $18,400 in 2002, more than double the average income of $8,700 in 1992. Overall, the catch has doubled in value over the past decade. This is mainly due to the growth in shellfish catches, up by 77 per cent, spurred by the closure of the Atlantic cod fishery. Similarly, in freshwater fisheries, although catch levels continue to decline, the value of the catch has increased because of higher prices.

The freshwater fishery accounts for about 3 per cent of the total value of commercial fishing in Canada, with 88 per cent of this coming from Ontario and Manitoba. Overall, the landed volume of freshwater fish increased by 1 per cent between 2008 and 2009, the last year for which data are available (Fisheries and Oceans Canada, 2010).

The Fisheries Act is federal legislation dating back to Confederation and was established to manage and protect Canada's fisheries resources. It applies to all fishing zones, territorial seas, and inland waters of Canada and is binding on federal, provincial, and territorial governments. As federal legislation, the Fisheries Act overrides provincial legislation when the two conflict. The Act, most recently revised and passed in 1985, is badly outdated, and although a new Act was formulated, it was not approved by Parliament. An important change proposed in the new Act would have reduced the discretion of the minister in several important matters that led to many of the problems described below in the collapse of the east coast fishery.

Case Study: East Coast Fisheries

The marine fishery has been an essential component of the economy and culture of Atlantic Canada for centuries. After 1977, when Canada declared a 200-nautical-mile fishing limit off its coasts, cod were the mainstay for more than 50,000 fishers and 60,000 fish-plant workers in Atlantic Canada. In Newfoundland and Labrador alone, about 700 communities depended entirely on the cod fishery, which had a 1991 value to fishers of more than $226 million.

However, between July 1992 and December 1993, decisions to cut back on the rate of harvesting resulted in employment losses for 40,000 to 50,000 people in Newfoundland, the Maritime provinces, and Quebec. How could this dramatic collapse of a renewable resource occur in such a relatively short period of time? Why did fisheries scientists fail to anticipate the collapse? Why was action not taken earlier to avoid degradation of the fishery and economic disruption? Is it possible for the fishery to rebound and become a mainstay of the regional economy once again?

The Nature of the Collapse

There are four main areas for this fishery in Atlantic Canada: the Scotian Shelf, the Gulf of St Lawrence, the Grand Banks, and the Labrador coast (Figure 8.11). The two areas most affected by the harvesting cutbacks have been the Scotian Shelf, which extends from the mouth of the Bay of Fundy to the northern tip of Cape Breton Island, and the Labrador coast.

These areas supported two different kinds of fishery. The fishery on the Scotian Shelf is readily accessible to the inshore fishers along the coast of Nova Scotia and New Brunswick and included a wide mix of species, including cod, haddock, flounder, pollock, hake, herring, redfish, crab, scallop, and lobster. In contrast, the Labrador coast fishery was dominated by the northern cod stock, extending east of the Labrador coast and north and east of Newfoundland. The northern cod traditionally yielded about half of Atlantic Canada's cod catch and one-quarter of all groundfish landings in the region. The northern cod has formed the backbone of the Atlantic fishery. This explains why stock depletion has been such a blow to regional economies, where fishing has provided a significant percentage of the jobs—and almost all jobs in small outport communities. These communities were further affected by the international attention and subsequent harvesting cuts to the seal fishery (Box 8.4)

The northern cod were caught by larger inshore vessels and especially by offshore draggers or factory trawlers, multi-million dollar boats that drag huge nets across the bottom of the ocean and stay on the fishing grounds for extended periods. However, the northern cod migrate to the shores of Newfoundland in the summer and thus also supported an inshore fishery that relied on much smaller boats using traps, hooks, and nets. The inshore fishery has been an important one. Until the late 1950s, the inshore catch was typically higher than 150,000 tonnes. By 1974, as a result of overfishing, the inshore catch had fallen to 35,000 tonnes. After Canada declared its exclusive fishing zone in 1977 and banned fishing by foreign draggers in that area, the inshore catch increased. It peaked at 115,000 tonnes in 1982, but by 1986 the catch had fallen to 68,000 tonnes, and the fish caught were very small. Local fishers had identified the first signs of serious problems.

Unfortunately, the models used by the fishery scientists indicated that stocks were still abundant, so these early warnings were not heeded. Total catches of northern cod increased, reaching 252,000 tonnes in 1986, almost twice what they were in 1978. However, by 1989, on the basis of new scientific advice, the minister reduced the total allowable catch (TAC) for northern cod to 235,000 tonnes. By 1991, it was clear that the stock was in trouble and the TAC was reduced to 120,000 tonnes, and in July 1992 a moratorium on northern cod until May 1994 was announced. Ottawa agreed to provide $500 million (later rising to $912 million) to compensate the 20,000 fishers and plant workers expected to lose their jobs (Figure 8.12).

In 1993, the government banned cod fishing in five more areas and sharply reduced quotas for other valuable species. The result was a total loss of 35,000–40,000 fisheries jobs in Atlantic Canada since the closures began in 1992. In Atlantic Canada, there were 17,200 groundfish licence holders in 1992

Environment in Focus

Box 8.4 The Seal Hunt

Since the mid-eighteenth century, the harp seal had been the target of hunting, mainly for pelts. Between 1820 and 1860, for example, about half a million harp seals were killed every year. In 1831, more than 300 ships and 10,000 sealers pursued the hunt; 687,000 pelts were taken. However, publicity over the hunt in the early 1980s led to bans on the importation of sealskins into Europe. Celebrities such as French actress Brigitte Bardot appeared on television across the world as they tried to protect helpless white-coated seal pups from being clubbed to death. Eventually, following bans on the import of sealskins by the US and the European Union, the Canadian government banned the hunt in 1987, and the number of seals is estimated to have tripled. Now, however, a new hunt has begun, ostensibly to help in the recovery of the endangered cod stocks.

Today most scientists agree that seals do not substantially alter the cod recovery and politicians are now stating that the seal hunt must continue to provide economic revenue for Inuit communities. International environmental groups have suggested that the real motive for reinstituting the seal hunt is not related to seal predation on cod but rather to the need for a political scapegoat in economically depressed areas and the demand for seal penises on the Asian market.

By 2009 the European Union had implemented a complete ban on the imports of seal pelts, dramatically reducing Canada's market. In response, Canada has challenged the ban and continued to develop relationships with Asian markets. In 2011 the Canadian government developed an arrangement with China to open up the seal market. China is the world's leading consumer of seafood and offers an alternative to the now non-existent European market.

In 2006, DFO introduced a five-year management plan (2006–10). The **total allowable catch (TAC)** of seals was set on an annual basis to allow for adjustments to changing environmental conditions and changes in harvest levels in Arctic Canada and Greenland. A one-year TAC of 330,000 harp seals was set for 2010, up from the 280,000 allowed in 2009. The harvest of seals did not meet the TAC for several years. In 2010, 67,327 harp seals were taken, down from the 74,581 taken in 2009 and a significant drop from the harvest in 2008, which exceeded 210,000 animals. Markets for seal pelts are subject to significant variation from one year to the next. For example, the landed value of the harp seal hunt in 2006 was $33 million. The average price per pelt received by sealers was $97, an increase of 77 per cent over the 2005 average value of $55. In 2009, the price was down to $15 per pelt, partly due to the EU ban. By 2010, the value had begun to recover with an average pelt price of $20–$25. Income from sealing may account for 35 per cent of total annual income for some coastal families, and many of their communities have unemployment rates more than 30 per cent higher than the national average.

In 2011 a new seal hunt was proposed, this time in the southern Gulf of St Lawrence, but this time for a different quarry, the grey seal. Seal numbers have been increasing in the area and catches of the southern cod falling so the Fisheries Resource Conservation Council, made up of scientists and fishing industry representatives appointed by the minister, have proposed killing 140,000 seals in the area as part of an experiment to see what happens. Independent scientists have been highly critical of the proposal, both as an experimental design and because of the overall impacts on the ecosystem.

A beater, like this harp seal pup, is one that has moulted its white fur.

and 10,783 in 2000, representing a decline of roughly 35 per cent. In total, the government spent $3.9 billion for income support, industry adjustment measures, and economic development assistance programs for the Atlantic fishing industry between 1992 and 2001.

In 2003, almost 11 years after the moratorium on cod fishing, Fisheries Minister Robert Thibault announced the closure of what remained of the cod fishery in Newfoundland, the Maritime provinces, and Quebec. The Atlantic cod was officially listed as endangered as the Committee on the Status of Endangered Wildlife in Canada (COSEWIC; see Chapter 14) estimated a 97 per cent decline in cod off the northeast coast of Newfoundland and Labrador over the previous 30 years.

Some Reasons for the Collapse

At the time of the collapse, a single cause for the depletion of the groundfish stocks seemed unlikely. It was suggested that changing environmental conditions, creating colder, less hospitable water temperatures for a period during the 1990s, had driven the cod away, while growing seal populations had devoured entire stocks of both cod and capelin, the favourite food source for cod. However, research shows that environmental factors played only minor roles in the disappearance of the fish. With study after study, it has become increasingly likely that 'the northern cod stock had been overfished to commercial extinction' (Steele et al., 1992: 37). The politicians and bureaucrats running Canada's Atlantic fisheries created opportunities for overfishing through the provision of inappropriate incentives for processing plants and lucrative subsidies (unemployment insurance) to all fishers and plant workers involved in the fishery. Similar perverse subsidies are still a significant contributor to overfishing around the world.

Foreign Overfishing

Once Canada established the 200-nautical-mile fishing limit in 1977, foreign fleets were required to fish outside that boundary or to fish inside the boundary for that portion of the domestic quota not taken by Canadian vessels. Foreign fishing fleets were monitored by the Northwest Atlantic Fisheries Organization (NAFO), and during the 1970s and early 1980s foreign fleets for the most part followed the rules for harvesting. However, in 1986, Spain and Portugal entered the European Community (EC), and that year the EC unilaterally established quotas considerably higher than those set by NAFO. Furthermore, the EC boats harvested fish well beyond the EC limits. The EC then raised the quota the following year, and again the NAFO quota was exceeded by the actual catch. In 1988, just half of the EC target was achieved, even though it was 4.5 times higher than the target recommended by NAFO. The EC, now known as the European Union (EU), later rejected a NAFO northern cod moratorium. In 1993, however, the EU finally accepted all NAFO quotas,

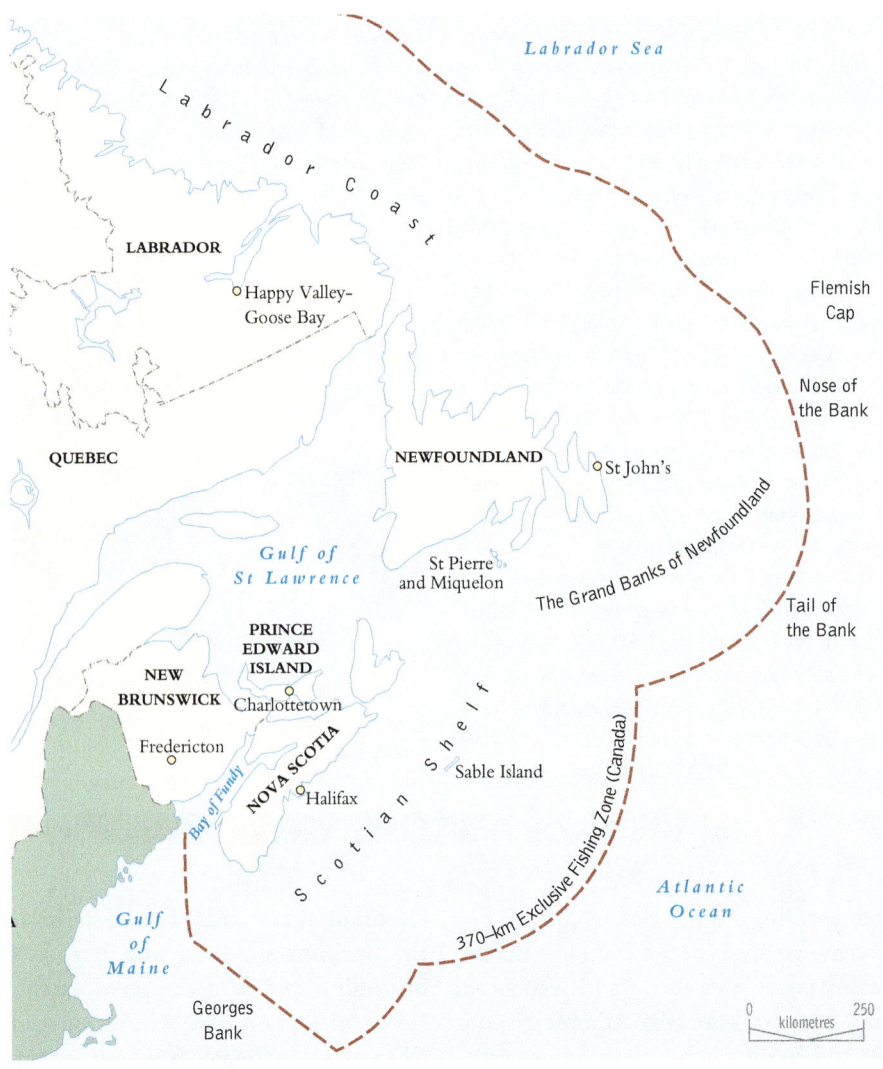

Figure 8.11 | Major fishing areas in Atlantic Canada. *Source: Adapted from Cameron (1990: 30).*

Chapter 8 | Ocean and Fisheries   263

Small outport communities in Newfoundland have always relied heavily on harvesting marine products, from seals to fish.

after having set its own quotas at a much higher level since the mid-1980s.

Thus, strong evidence exists that foreign vessels, especially those from Spain and Portugal, were overfishing at least during the mid- and late 1980s. Since cod migrate towards the coast in summer and then move offshore in winter to spawn in deeper waters, the fish are vulnerable to foreign fishing.

Domestic Overfishing

Despite the pressure placed on the stocks by foreign fishing vessels, most of the principal fishing grounds have been under Canadian control since the 200-mile limit was set. Two fisheries—both inshore and offshore—must be managed, which has been and continues to be a challenge.

For hundreds of years, the inshore fishery consisted of many fishers (particularly from Newfoundland) relying on small wooden boats, lines, traps, and nets to catch cod during the spring and summer months when the cod move close to shore. Until the mid-1950s, the inshore fishery, combined with limited offshore fishing by Canadian boats, resulted in annual landings of 200,000 tonnes or more. Foreign fishers were harvesting another 30,000–50,000 tonnes each year. Such harvesting did not appear to adversely affect the then estimated breeding stock of 1.6 million tonnes in the North Atlantic.

In the mid-1950s, the introduction of large offshore trawlers that operated year-round in the North Atlantic was a significant change to this pattern. Initially, catches were very high, but the spawning stocks were placed under great pressure. In the 1970s, yields reached a high of 800,000 tonnes per year before they started to drop. Until 1977, foreign trawlers did most of the offshore fishing. Following the establishment of the 200-mile limit, the Canadian offshore fleet expanded, and Canadian-based offshore trawlers became the main harvesters of northern cod. By the time the moratorium was placed on the fishery in the summer of 1992, Newfoundland was the base for some 55 large and 30 medium-sized offshore trawlers. Thus, Canadian offshore draggers, operating on a year-round basis, placed considerable pressure on groundfish stocks.

Critical in this regard are the ecology and behaviour of the northern cod. Initially, the harvesters caught a mix of ages

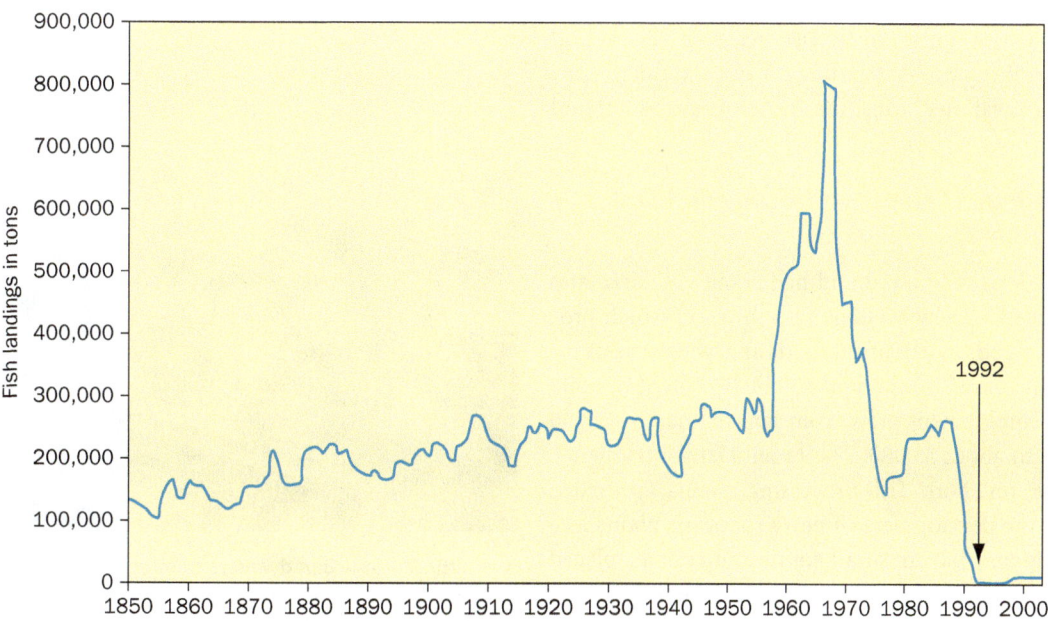

Figure 8.12 | Collapse of Atlantic cod stocks off the east coast of Newfoundland, 1992. Source: Millennium Ecosystem Assessment (2005).

and sizes of fish. However, market demand and net mesh sizes led to a focus on larger, older fish. Cod swim in groups or schools of similar ages primarily because the larger cod will eat the smaller and younger cod. The emphasis on larger fish had two consequences. First, the northern cod normally do not reach maturity and begin to spawn until seven years of age. Second, older fish produce more eggs. As the larger fish became scarce, the fishery then concentrated on fish in the five-to-seven-year age range. The result was that by the early 1990s, most of the older fish had been overharvested and attention had shifted to pre-adolescent fish, which were being caught before they were old enough to spawn. The consequence was a dramatic decline in the fish stock.

A further complication was that many domestic fishers overharvested. Estimates suggest that up to 50 per cent more fish were being landed than were being reported.

Imperfect Science and Management

Fishery scientists did not anticipate the collapse of the Atlantic fishery, especially the northern cod stocks. One reason was that sampling procedures do not provide sufficient ecological information about fish stocks. Until 1991, the total allowable catch was based on the assumption of a biomass of 1.1 million tonnes of cod. However, in 1991, the sampling from DFO research ships indicated only 600,000 tonnes. Sampling in distant areas did not reveal that the cod had migrated to other areas. Significant numbers of diseased or dead fish had not been found. The scientists simply did not know what had happened. They had been receiving warnings for a number of years from the inshore fishers that the fish being caught were fewer in number, smaller in size, and lower in weight. However, the scientists, who put much greater credence in the quantitative data gathered by DFO research vessels and from what were likely under-reported offshore landings, considered the observations of inshore fishers as anecdotal and less credible.

Inappropriate Incentives for Processing Plants and Fish Workers

By the early 1990s, Newfoundland had about 100 large and small fish-processing plants, some two-thirds of which processed northern cod. In Atlantic Canada as a whole, the number of plants increased from about 500 in 1977 to nearly 900 in 1988, and employment grew from approximately 25,000 full-time jobs to about 33,000. The provincial governments, which license on-shore fish-processing plants, provided incentives for the development of new processing plants as a way of creating new jobs in small communities. This placed political pressure on DFO to keep increasing the total allowable catch.

Another incentive for people to enter or stay in the fishing industry was the federal unemployment insurance program. After working for 10 weeks, fish-plant workers were permitted to collect unemployment insurance for the other 42 weeks of the year. This arrangement resulted in several individuals in a community sharing one job but all qualifying for separate benefits. For fishers, the unemployment benefits were based on the sale value of fish caught during the May–November season (26 weeks), which created the potential for receiving unemployment benefits for the other 26 weeks of the year.

Perspectives on the Environment
Destruction of Northern Cod Stocks

The state of our ignorance is appalling. We know almost nothing of value with respect to behaviour of fish. We don't even know if there's one northern cod stock, or many, or how they might be distinguished. We don't know anything about migration patterns or their causes, or feeding habits, or relationships in the food chain. I could go on listing what we don't know. . . . Our technology has outstripped our science. We have underestimated our own capacity to find, to pursue, and to kill.

—Leslie Harris, chairman of the Northern Cod Review Panel, 1990, quoted in Cameron (1990: 35, 29)

This program, which was intended to provide a social safety net, encouraged more people to become involved in the Atlantic fishery than could be justified economically. There was little incentive to consider other types of work, and the program also helped to reinforce an outlook in which little value was placed on education. By the time the fishery was closed, 50 per cent of Newfoundland's 19-year-olds were

It is difficult for urban dwellers to imagine the close relationship that built up over the centuries between the people in the outports of Newfoundland and the sea. Virtually every family would be involved in some way with fishing. When the fish were exposed to such fishing pressure that they could no longer be caught in any numbers, it was not just the economy that suffered but a whole way of life.

already on unemployment insurance, and 80 per cent of the fishery workers did not have a high school diploma. Thus, the fishery involved more people than realistically could be supported over the long term, yet unemployment insurance programs provided little incentive for individuals to consider alternatives. This made the trauma of the 1992 moratorium even greater than it might otherwise have been.

Changing Environmental Conditions

One theory for the depletion of groundfish in the North Atlantic fishery is based on the idea of environmental change. Records show that in 1991 the ocean temperatures off Newfoundland were the coldest ever measured. The water warmed slightly in 1992, then cooled again in 1993. It is possible that colder waters, combined with overfishing in the previous two decades, prevented or inhibited the shrunken stocks from regenerating. However, since relatively little is known about the migratory patterns of the northern cod, it is difficult to determine what the specific implications of changing water temperatures might have been. Furthermore, water temperatures began to heat up and stabilize around 1998, restoring a theoretically favourable habitat for groundfish. Despite this warming trend, no significant recovery has been observed within the affected groundfish stocks.

Predators

At the time of the fisheries collapse, it was also popular to blame predation. Seals in particular were identified because of their 'voracious appetites' and their growing numbers because of the closure of the seal hunt in the early 1980s (Box 8.4). There is no scientific evidence to support this view.

However, there is some evidence to suggest that seal predation may be a factor in the slow recovery of the east coast groundfish stocks. The seal population has more than doubled in the past three decades, and although cod represents only a small percentage of seals' diets, they are consuming more northern cod than fishers are catching. In fact, the total allowable catch for northern cod in 1999 was only 9,000 tonnes, or less than 20 per cent of corresponding predation by harp seals in that year (Doubleday, 2000).

The science of seals and cod is inconclusive and will remain inconclusive for the foreseeable future. Seals are a significant source of mortality for northern cod in the Northwest Atlantic and the Gulf of St Lawrence. Nevertheless, reducing the abundance of harp seals may or may not lead to recovery of depleted cod stocks. Even if a reduced seal population resulted in an increased number of fish in the ocean, there are other predators in marine ecosystems. Any increase in the size of a commercially important fish stock could well be eaten by these other predators before being caught by fishers. Furthermore, seals eat predators of commercially important fish, and so fewer seals could actually mean fewer fish for fishers.

Perspectives on the Environment
Climate Change versus Overexploitation

We reject hypotheses that attribute the collapse of the northern cod to environmental change. . . . We conclude that the collapse of the northern cod can be attributed solely to overexploitation [by humans]. . . .

—Hutchings and Myers (1994)

Fishing down the food chain has been discussed earlier and is exemplified in Atlantic Canada with the replacement of cod fisheries by shrimp and crab fisheries. Recent observations of the poor physiological condition of many predator fish in the area also suggest an overall lack of prey, forcing predator species such as the Atlantic cod to the same alternative as human fishers: 'fishing down the web'. After the cod were fished down, fishing pressure increased on shrimp. 'Cod feed on shrimp. If you remove the shrimp, how will the cod ever recover?' asks Daniel Pauly.

Lessons

The collapse of the Atlantic groundfish fishery highlights how some contemporary resource management practices may encourage resource liquidation. In theory, fisheries managers around the world and in Canada seek to manage the fishery according to three priority areas: ecological sustainability, economic goals, and social outcomes. In reality, these goals are often in conflict, at least in the short term, and the second goal often becomes paramount because of political interference. The case study illustrates that fisheries management requires scientific understanding of the biophysical resource system, a greater appreciation of traditional or local ecological knowledge, and parallel understandings of the history, culture, economy, and politics of the region, as well as federal and provincial fisheries and regional development policies. The Atlantic fishery also provides an excellent example of how inexact science often is and the extent to which complexity and uncertainty are paramount. It demonstrates conflict among different values and interests (Chapter 5) and the manner in which conditions can change dramatically over a relatively short time period.

Perspectives on the Environment
Fishery Collapse: A Bureaucratic Tragedy

The disaster in the cod fishery is now worse than anyone expected. . . . It may be a generation before we see a recovery of the cod. That a five-hundred-year-old industry could be destroyed in fifteen years by a bureaucracy is a tragedy of epic proportions.

—Ransom Myers, quoted in Harris (1998: 332–3)

The situation is readily comparable with the framework introduced in Chapter 1. Not only were our simplified models of the complex biophysical system inadequate for a proper understanding of the east coast fisheries, but our resulting attempts to assess the status of the system were inadequate and there was a lack of clarity regarding societal expectations and management directions. Davis and Wagner (2006) have provided a useful analysis of some of the social dimensions of fisheries regulation. Some of the approaches suggested in Chapter 6, relating to identification of stakeholders, resolving conflict, and taking more ecosystem-based and adaptive approaches, could usefully be applied to the resolution of these problems.

However, one of the chief proponents of adaptive management, Dr Carl Walters of Simon Fraser University, warns in a 2007 paper that this approach (Chapter 6), which was designed to address some of these problems, has not been applied successfully in the vast majority of cases. He traces the failures to three institutional problems: first, a lack of management resources for the expanded monitoring needed to carry out large-scale experiments; second, the unwillingness of decision-makers to admit and embrace uncertainty in making policy choices; and third, lack of leadership in the form of individuals willing to do all the hard work needed to plan and implement new and complex management programs.

Aboriginal Use of Marine Resources

One of the most challenging aspects of fisheries management in Canada is allocation of catch, especially when catches are declining. This is particularly difficult when allocation involves Aboriginal communities. There is a patchwork of treaties with Aboriginal peoples in different regions of Canada, and the rights to sustenance from fishing were often written into these treaties. However, it has never been clear to which regulations Aboriginals should be subject and how broad a range of activities the concept of sustenance might cover. Over the past decade, many important court cases have helped to clarify some of these issues. Nonetheless, high-profile conflicts still occur on both coasts, as detailed in the examples below.

In the fall of 1999, a violent and complex dispute erupted between Native and non-Native fishers in Miramichi Bay in northeastern New Brunswick. Lobster traps were cut and damaged, threats were exchanged, boats were rammed, and multiple shots were fired.

The crux of the dispute lay in the Supreme Court's 1999 decision in the case of a Nova Scotia Mi'kmaq, Donald Marshall Jr, who had caught and sold eels out of season and claimed protection under a 1760 treaty. The Court's ruling upheld the 1760 treaty, which effectively gave Mi'kmaq, Maliseet, and Passamaquoddy bands the right to earn a 'moderate livelihood' from year-round fishing, hunting, and gathering. As news of this decision spread, it spurred an immediate reaction among the Natives of the Burnt Church band of Miramichi Bay to resume catching and selling species like lobster, even though the season was formally over. Within days, the local Native bands had more than 4,000 traps in the water. This infuriated the non-Native fishers, who were confined to the short summer season and who believed that this continued harvest would lead to the destruction of the lobster fishery.

Despite the rising tension and conflict between the two groups, the Department of Fisheries and Oceans was hesitant to intervene, stating that Natives now had a right to fish that had been denied for more than two centuries. However, DFO did promise a peaceful solution to resolve the ambiguity of the *Marshall* decision. The *Marshall* case specifically indicated that the federal government still retained the right to regulate the fishery but was equally firm that Ottawa's authority to regulate treaty rights was limited to those actions that could be 'justified'. This led to an examination of the term 'justification' and, more specifically, of whether or not the DFO's limit on the number of lobster traps and length of fishing season was reasonable and 'justifiable' according to the rights of the Mi'kmaq in their 1760 treaty.

After several years of research into the affected lobster stocks, commissioned in order to 'justify' the decision scientifically, the Minister of Fisheries and Oceans announced an agreement with the Burnt Church band. The $20 million agreement included enhanced commercial fishery access for Native fishers, including additional lobster licences and extra boats and gear. However, a quota was set for the Native fall fishery of 25,000 pounds of lobster for food and 5,000 pounds of lobster for ceremonial use. Furthermore, the fishery would be limited to six weeks or until the quota was filled, and the sale of lobster would be strictly prohibited at all times.

Since the federal decision, there has been a significant decrease in violent conflicts on the water, and an improved relationship has emerged among all parties. The non-Native lobster fishers and many fisheries scientists still maintain that any kind of fall fishery needs to end because, they believe, it exploits the lobster at their most vulnerable stage. Yet they recognize that while the negotiated agreement does not represent a definitive solution to the conflict, it is a vital step towards the government providing a more appropriate degree of oversight and accountability in the management of the lobster fishery.

The Atlantic Integrated Commercial Fisheries Initiative (AICFI) was announced in 2007. This program aims to ensure that Mi'kmaq and Maliseet First Nations (MMFNs) in the Maritimes and Gaspé region of Quebec have the capability to manage and maximize access to the integrated commercial fishery provided through the Marshall Response Initiative. Individuals who want to participate in the program must undergo a rigorous application process to gain access to some of the $2 million made available over three years through AICFI. An evaluation conducted in 2010 showed that 71 per cent of MMFN representatives were either 'satisfied' or 'very satisfied' with services provided by AICFI. Service providers reported similar levels of success, indicating that 73 per cent of services were meeting specified objectives.

On the west coast, Native communities have unemployment rates of between 65 and 85 per cent, and the fishery plays a critical role. There is an Aboriginal right, established by the courts, to fish 24 hours a day, seven days a week, wherever Aboriginal people wish to, for food and ceremonial purposes. In the past, this right was sometimes abused, with fish being caught for commercial use. This prompted DFO to establish a special Natives-only commercial salmon fishery for some areas. In 2003, the BC Appeal Court, following complaints by non-Native fishers, struck down the Natives-only commercial salmon fishery. The judge said that it amounted to 'legislated racial discrimination' and was against the Charter of Rights. DFO then cancelled the program, but Native fishers vowed to continue catching and selling the salmon as they always have. This decision was the second court setback for the Natives-only fishery. A few months earlier, another judge had given absolute discharges to 40 non-Native fishers who had cast their nets during a Natives-only fishery on another part of Vancouver Island. However, many critics point out that Natives are being allocated fishing rights not because of their ethnicity per se but because their fisheries were wrongfully appropriated in the first place. In July 2008, a Supreme Court decision ruled that such redistributive justice was necessary. Others have argued that non-Native fishers should not bear the costs for wrongs perpetrated in the past by society as a whole. What do you think?

In many ways, the story of the BC coastal Aboriginal cultures is a story of the sea in general and of salmon in particular (Box 8.5). The bounty of the sea allowed these peoples to establish a more sedentary lifestyle than that of other Aboriginal peoples in North America. Consequently, nowhere else did hunter-gatherer societies develop such complex social structures, rigid hierarchies, and dense populations in permanent winter villages. From these villages, the people developed complex and effective hunting practices for whales, sea lions, seals, sharks, tuna, wolf eels, sole, oolichan, greenlings, herring, halibut, crabs, clams, mussels, skate, sturgeon, and, above all, salmon. The salmon fishery was managed effectively; no stocks crashed. And the salmon was venerated through myth and legend among the coastal peoples.

Conflict and sometimes tenuous resolution will continue to arise as Canada strives to achieve equitable solutions to fish resource allocation problems involving Aboriginal peoples. The clock cannot be rolled back to pre-treaty times, yet there must be some recognition of the centrality that fish and fishing have played in the societies of many Aboriginal peoples in Canada and of their intimate knowledge of coastal ecosystems (Box 8.6). Roughly three out of four British Columbians told a federal government-commissioned pollster that they oppose special commercial fishing rights for Aboriginal Canadians, and Prime Minister Stephen Harper stunned Aboriginal groups, federal bureaucrats, and the BC government by declaring that his government is opposed to 'racially divided' fisheries. Yet the federal government has since gone ahead with west coast treaties that include exclusive rights to the commercial fishery and has developed a BC First Nations Fisheries Action Plan that will allow greater involvement of First Nations in commercial fisheries.

A First Nations fisher uses a dip net to intercept a chinook salmon at Moricetown Falls, Bulkley Valley, British Columbia.

Environment in Focus

Box 8.5 Salmon: The Stories They Tell

From the shores of Japan to almost 2,500 kilometres up the Yukon River, a tangible thread exists—the Pacific salmon. Every year, millions of salmon make their way back from the other side of the Pacific Ocean to the streams of their birth. The five species of Pacific salmon—chum, coho, chinook, pink, and sockeye—are anadromous—that is, they spend part of their lives in fresh water and part in salt water. They depend on a wide range of conditions that link the mountains to the seas: the amount of snowpack to feed the streams, the lack of floods to wash away spawning gravel, unpolluted rivers and estuaries, the right temperature for entry into the marine environment, avoidance of predators, and avoidance of fishing nets. If any of these myriad factors go awry, then higher mortality rates can drastically reduce the numbers of fish returning to spawn in subsequent years. These factors are the links in a chain reflecting the limiting factor discussed in Chapter 2. It also means that salmon are good indicators of the overall health of our environment and our resource management practices.

What have these indicator species been telling us? The story is not a good one. Salmon in their millions sustained populations of coastal Aboriginal peoples. Early descriptions of the Fraser River by explorers talk about a river that could be crossed on the backs of the salmon. But early logging and mining practices, along with wasteful fishing practices, soon made a considerable dent in these numbers by negatively affecting spawning grounds. Habitat destruction and overfishing led to the virtual closing of the fishery in many areas by the 1990s. Scientists estimate that the salmon biomass has been diminished by half from pre-commercial fishing levels. Some stocks have been declared extinct, while others are now on the official endangered species list in both Canada and the US. The weight of salmon landings fell by more than 50 per cent between 1997 and 2006. In 2008, the Pacific Salmon Commission proposed cutting the chinook harvest by a third until 2018 in an effort to rebuild stocks.

The salmon have also been trying to tell us something that scientists are only now starting to realize. Salmon spend anywhere from two to seven years in the ocean environment before returning to spawn and die. When they die, the nutrients they have collected over this sojourn do not disappear (law of conservation of matter) but are released into the surrounding environment. The salmon provide food and nourishment not only for the plankton and insects that feed the next generation of fish and propel their journey to the sea but also for the terrestrial riverine environment. When the fish die, they provide a feast for many other species, including eagles, raccoons, and bears. As the fish are digested, their nutrients are distributed throughout the forest as feces, providing the rich fertilizer on which some of the tallest trees in the world depend. Tom Reimchen of the University of Victoria estimates that BC's bears could be transferring 60 million kilograms of salmon tissue into coastal forests each year. When we take away the fish, we take away this fertilizer, and forest growth suffers. Scientists have found that up to 40 per cent of the nitrates in the old-growth forests in coastal BC originate from marine environments. Rivers with barriers to salmon, such as waterfalls, have noticeably poorer forest growth.

Scientists have now discovered another aspect of this linkage. Not only do salmon collect nutrients from the ocean, they also collect pollutants and concentrate them within their

Grizzly bear tracks alongside a coastal river in the Great Bear Rainforest, British Columbia. Research now shows that the nutrients bears carry back from the oceans and rivers are central to promoting rich forest growth along many coastal streams.

bodies (Chapter 10). Scientists in Alaska have found that when the fish die in their millions after spawning, there is a sevenfold increase in the concentration of PCBs in remote, pristine, freshwater lakes. Lakes with the highest numbers of spawning salmon also have the highest concentrations of PCBs.

The biggest challenge, however, may be global warming and the changes this will bring to all aspects of the salmon habitat. From the amount of snowpack in the mountains that controls the water in the rivers, through river temperatures, to changes in oceanic currents and predator–prey relationships, the salmon will be very vulnerable. Already, some species are at the edge of their temperature tolerance range (Chapter 2) for fresh water, and the most productive salmon river in the world, the Fraser, may soon be too warm: the water temperature of the Fraser has risen over 2°C during the past 50 years. Over the same time period sea-surface temperatures have risen between 0.3°C and 0.9°C on the BC coast.

In response, the annual cycles of phytoplankton bloom that nourish the entire food chain are occurring earlier, raising the possibility of a growing mismatch between food supplies and emergence of salmon smolts into the ocean.

The salmon have an eloquent and tragic story to tell about how we are treating their environment, and they will be one of the best indicators of the impact of global change on coastal and marine environments. The story is not an easy one to unfold. In 2010, some 31 million sockeye returned to the Fraser, a run not equalled in size in the last century. Theories abound as to the cause, ranging from the fertilization of the northern ocean by an Alaskan volcano that produced increased nutrients and an exceptional plankton bloom when the plankton were feeding, through to enhanced survivorship of smolts because of favourable ocean conditions when the salmon first entered salt water. This is a mystery story that will continue to puzzle scientists for decades to come.

Environment in Focus

Box 8.6 So How Many Whales Was That? Traditional Ecological Knowledge (TEK) and Science in Canada's Arctic

DFO scientists believed that bowhead whales numbered only in the hundreds and were divided into two separate populations. Since 1996, their figure of 345 bowhead whales was used to determine an Inuit bowhead whale quota in Nunavut of about one every two years. But the scientists' new, much higher bowhead whale estimate—showing a population that could run as high as 43,105—supports an annual hunt of between 18 and 90.

Inuit have said for years that the eastern Arctic's stock of bowhead whales are part of one large and healthy population. The first sign that bowhead whales were more numerous than scientists first thought came from a study of Inuit bowhead knowledge completed in 2000. That study was based on interviews with 252 Inuit hunters and elders in 18 communities. In those interviews, most Inuit informants said they see far more bowhead whales now than in the 1950s. It took more than seven years for the DFO's science to catch up. DFO estimates of the bowhead population jumped from 345 in 2000 to about 3,000 in 2003, then to 7,309 in 2007, and to 14,400 in 2008. The DFO's most recent stock assessment based on surveys conducted between 2002 and 2004, says this latest number is only a 'partial estimate' and that there are between 4,800 and 43,105 bowhead whales in the eastern Arctic.

The DFO also concedes that rather than two populations of bowhead whales in the eastern Arctic, there is one. Based on tagging and genetic studies, scientists now say that bowhead whales off Canada and west Greenland share the same summering grounds along the east coast of Baffin Island and in the Canadian High Arctic and the same wintering grounds in Hudson Strait. The DFO stock assessment says the bowhead whale population, downlisted in 2005 from 'endangered' to 'threatened' under the federal Species at Risk Act, may have completely regained its health. This assessment suggests that, based on the most recent numbers, an annual hunt of 18 bowhead whales is realistic and 'conservative'.

However, the message to take from the reassessment is not that one side was right and the other wrong. It is that population estimates in remote locations are very difficult to make and all sources of information should be considered fairly. It is also appropriate to take a precautionary approach, especially when endangered species are being considered.

Pollution

The main sources of marine toxic pollution in Canada originate with the deposition of airborne pollutants from fossil-fuel combustion, agricultural runoff, inadequately treated sewage, and by-products or waste materials from refining processes (e.g., effluent from pulp and paper mills). Some chemicals, known as POPs (persistent organic pollutants), including PCBs (polychlorinated biphenyls) and DDE (the breakdown product of the now-banned pesticide DDT, discussed in Chapter 10), can take decades or even centuries to degrade and tend to bioaccumulate in the fatty tissues of organisms over time. The concentrated contaminants are then passed along through the food chain (biomagnification—see Chapter 10) and can reach very high concentrations in the tissues of animals in the top trophic levels (such as polar bears, whales, and humans).

A study by the Department of Fisheries and Oceans concluded that the killer whales of the Strait of Georgia are among the most contaminated mammals on the planet (Ross et al., 2000). Although PCBs have been the main concern, levels of a toxic flame retardant (PBDE) are growing so quickly that they are expected to surpass PCBs as the leading contaminant in endangered southern resident killer whales. At the current rate of increase, PBDE levels are predicted to exceed those of PCBs in killer whales by 2020. Unlike PCBs, largely used as coolants in industrial transformers before being banned 30 years ago, PBDEs are widely used as flame retardants in polymer resins and plastics and found in consumer products such as furniture, TVs, stereos, computers, carpets, and curtains. PBDEs find their way into the marine environment through the air or through runoff and effluent and pose a risk to the endocrine system, reproductive health, and the immune system. PBDE levels in harbour seals in Puget Sound in the state of Washington have increased steadily: 14 parts per billion in 1984, 281 in 1990, 328 in 1993, 644 in 1996, and 1,057 in 2003. Killer whales carry 10 times the contaminants of harbour seals, which means an increase in PBDEs in seals is an immediate cause for concern about the whales.

The dangers of pollutants are further exacerbated by the long-range polar transport of toxins in the atmosphere, and many Aboriginal people in Canada's North have bioaccumulated extraordinarily high levels of toxins in their bodies because of their dietary reliance on marine mammals. Some of these persistent toxins are also endocrine disrupters, which have been linked to severe growth, development, and reproductive problems in wildlife populations, as discussed earlier in this chapter. However, even toxic substances that are not persistent or bioaccumulative (such as benzene) can have significant harmful effects on the health of the marine environment.

In recent years, Canada has made progress in reducing emissions from a number of marine toxic pollution sources. Agricultural industries have been forced to develop and use more environment-friendly pesticides and fertilizers and to increase conservation tillage to reduce runoff pollution. There has also been a significant decrease in the amount of toxic pollutants coming from other industries such as pulp and paper, petroleum refining, and aluminum. For example, discharges of dioxins and furans from Canada's forest products industry have decreased by 99 per cent since 1988 (Environment Canada, 2003d) because of new regulations under federal and provincial legislation.

Already, some of these reductions are apparent in ocean life. Temporal trends of persistent, bioaccumulative, and toxic (PBT) chemicals were examined in beluga whales from the St Lawrence estuary. Blubber samples of 86 stranded adult belugas were collected between 1987 and 2002 and analyzed for several regulated PBTs, including PCBs, DDT and its metabolites, chlordane (CHL) and related compounds, HCH, HCB, and Mirex. Concentrations of most of the PBTs examined had exponentially decreased by at least a factor of two in belugas between 1987 and 2002, while no increasing trends were observed for any of the PBTs measured (Lebeuf et al., 2007).

Although the concentration of toxins in the Canadian environment has declined, they have not disappeared. Existing toxic residues will be recycled and dispersed throughout ecosystems for some time. In addition, toxic substances from sources outside Canada continue to enter our ecosystems through oceanic and atmospheric transport. One study of contaminant concentrations in the eggs of double-crested cormorants shows a significant decrease in the past 30 years (Canadian Wildlife Service, 2003). However, the lack of further declines, despite the banning of these chemicals in Canada, leads scientists to believe that it may be the result of long-range transport of POPs used outside of Canada, as well as the slow release of contaminant residues from bottom sediments and dump facilities.

The population of the southern pods of killer whales has fallen 20 per cent over the last 10 years and is now considered endangered.

In Canada, federal law and policy have declared the management and reduction of toxic substances in the environment 'a matter of national priority'. Under the Canadian Environmental Protection Act, the Minister of Environment is mandated to virtually eliminate the production of POPs and manage the discharge of other pollutants and wastes into the environment. Canada has also been active on an international scale and was the first nation to ratify an international treaty, known as the Stockholm Convention on Persistent Organic Pollutants, which aims to identify problematic substances for which comprehensive global action is required.

Organic pollution is also of concern in some areas (the overall process of organic decay was outlined in Chapter 4). Given the immense volume of water in the ocean, it might be thought that an infinite adsorption capacity exists for receiving and breaking down organic wastes. However, where there are dense populations and waste is deposited in a site with low adsorptive capacity, even the ocean can become polluted. On the east coast of Canada, for example, 52 per cent of all towns and cities lack any sewage treatment. The problem became quite obvious in Halifax, where sewage has been deposited directly into the harbour since 1749 and toilet paper, tampon applicators, and condoms have become a familiar sight. In 2003, Halifax made a historic move after 30 years of delay, announcing that it would install a network of treatment plants for the 181 million litres of raw sewage pumped out every day.

At the other end of the country, Victoria has taken a different approach. The city pumps out 100 million litres of raw sewage into Juan de Fuca Strait every day through two deep-sea pipes that extend more than a kilometre offshore and are more than 60 metres deep. Victoria's situation is rather different from that of Halifax in that the large volume of fast-moving water in the Strait breaks down and disperses the sewage very quickly. The plume from the discharge never reaches the surface in summer and only 1 per cent of the time in winter. Marine biologists have monitored the situation for years and have not been able to detect virtually any negative impacts. Most biologists are satisfied that the dilution of organic waste is acceptable. However, there is much greater concern over the non-organic wastes that are disposed of illegally through the sewage system. Over a two-year period, estimates suggest that these wastes include 2,920 kilograms of oil and grease, 17,400 kg of zinc, 9,000 kg of copper, 2,560 kg of cyanide, and 1,360 kg of lead. The regional government has introduced educational programs on waste disposal, since the most effective way to deal with these substances is to halt their entry into the sewage system rather than trying to treat them once they are there (see the CRD ad, below).

In 2005, Victoria commissioned an independent study to review liquid waste practices, which suggested that an increasing population would soon require increased treatment. A plan was devised and is now being implemented at a cost of more than $782 million. Many scientists and health professionals question whether this is the wisest investment for that amount of money. There is a strong consensus among health professionals that current disposal practices pose no medical risk, and marine biologists have failed to find any significant biological changes over time. One of their main points of contention is that no study has been undertaken on the impact of the land-based treatment and disposal that is now being planned. Despite strong opposing positions, the sewage treatment plant is under development. Ongoing research is informing finalized plans for land acquisition, program development, and facility planning. The treatment plant is expected to be completed and operational by 2016.

Ad from CRD (Capital Regional District, BC) campaign to control source pollution.

Some Canadian Responses

Canada's Oceans Strategy

The Oceans Act was passed in 1998 to provide a comprehensive and co-ordinated approach to marine resource management in Canada. One of the main requirements of the Act was for the Minister of Fisheries and Oceans to develop a national

Oceans Strategy. This strategy established three principles to guide *all* ocean management decision-making:

1. *Sustainable development* 'recognizes the need for integration of social, economic, and environmental aspects of decision-making and that any current and future ocean resource development must be carefully undertaken without compromising the ability of future generations of Canadians to meet their needs.'
2. *Integrated management* 'is a commitment to planning and managing human activities in a comprehensive manner while considering all factors necessary for the conservation and sustainable use of marine resources and shared use of ocean spaces.'
3. The *precautionary approach* is defined in the Oceans Act as 'erring on the side of caution'.

As underlying principles, they provide the essential litmus test against which all ocean management decisions should be judged and to which the federal government is accountable. Unfortunately, this has not proved to be the case. Globally important glass-sponge reefs off the coast of BC have been heavily damaged as a result of a lack of protection from fishing—a failure to apply the precautionary principle and poor consideration of values other than economic ones.

The Oceans Strategy also identified three desired policy objectives or outcomes:

- understanding and protecting the marine environment;
- supporting sustainable economic opportunities;
- providing international leadership.

These are laudable objectives but meaningless unless progress can be shown and resources are devoted to them. Over a decade after the passage of the Oceans Act, the Oceans Directorate of DFO has virtually no resources, and very little progress has been made on realizing any of the lofty objectives of the Act or the subsequent strategy. Canadians think that progress is being made because of periodic policy announcements, but in fact it is largely business as usual. A public poll undertaken in Atlantic Canada and New England on behalf of a consortium of conservation organizations found that most respondents think that between 20 and 23 per cent of their offshore waters are protected (Seaweb, 2003). In reality, the figure is less than 1 per cent.

Marine Protected Areas in Canada

Canada, like the rest of the world, has paid little attention to protecting the marine environment through marine protected areas (MPAs), especially when compared to the attention and protection given the terrestrial environment (see Chapter 14). Depending on the definition applied, estimates suggest that Canada has more than 790 protected areas, with a marine component covering approximately 4.6 million hectares, or 0.66 per cent of Canada's ocean area (Environment Canada, 2011). However, almost all of this area is in terrestrial protected areas that happen to be coastal. This figure compares with more than 9.8 per cent of the terrestrial environment under protection.

In response to this situation, three programs have been created at the federal level to establish MPAs (Table 8.2). The first is an MPA program established under the Oceans Act within

Table 8.2 | Federal Statutory Powers for Protecting Marine Areas

Agency	Legislative Tools	Designations	Mandate
Fisheries and Oceans Canada	Oceans Act	Marine protected areas (MPAs)	To protect and conserve: • fisheries resources, including marine mammals and their habitats; • endangered or threatened species and their habitats; • unique habitats; • areas of high biodiversity or biological productivity; • areas for scientific and research purposes.
	Fisheries Act	Fisheries closures	Conservation mandate to manage and regulate fisheries, conserve and protect fish, protect fish habitat, and prevent pollution of waters frequented by fish
Environment Canada	Canada Wildlife Act	National wildlife areas	To protect and conserve marine areas that are nationally or internationally significant for all wildlife but focusing on migratory birds
		Marine wildlife areas	To protect coastal and marine habitats that are heavily used by birds for breeding, feeding, migration, and overwintering
Parks Canada	National Parks Act; National Marine Conservation Areas Act	National parks; national marine conservation areas (NMCAS)	To protect and conserve for all time marine conservation areas of Canadian significance that are representative of the 29 natural marine regions identified on Canada's coasts, and to encourage public understanding, appreciation, and enjoyment

the Department of Fisheries and Oceans. The purpose of these MPAs is to conserve commercial and non-commercial fisheries, protect species at risk, and conserve unique habitats—i.e., areas of high biodiversity or biological productivity. Five Atlantic MPAs have been designated: The Gully near Sable Island; Basin Head in the Gulf of St Lawrence; Eastport in Bonavista Bay; Gilbert Bay in the Labrador Sea; and Musquash Estuary in the Bay of Fundy. In addition, the Endeavour Hydrothermal Vents and Bowie Seamount MPAs have been created in the Pacific as well as Tarium Niryutait in the Arctic. DFO committed to establishing nine MPAs between 2005 and 2010. Besides the five Atlantic MPAs designated within this period, another eight areas of interest are under consideration: Manicouagan (Quebec), St Lawrence Estuary (Quebec), Race Rocks (BC), Hecate Strait/Queen Charlotte Sound (BC), Laurentian Channel (Quebec–Atlantic Canada), St Anns Bank (Cape Breton, NS), Shediac Valley (NB), and American Bank (Quebec). A new designation, called 'protected marine areas', has also been introduced. At present, the Scott Islands at the north end of Vancouver Island is the only site being considered for protection under this Environment Canada designation.

Second, through the Canadian Wildlife Service (CWS), Environment Canada has several programs that may include designation of marine sanctuaries such as national wildlife areas and migratory bird sanctuaries. Although some of the CWS sanctuaries are large, especially in the Arctic, they are designed primarily to protect specific species (particularly seabirds) rather than ecosystems, and they have no minimum standards to control extractive activities.

The third program involves national marine conservation areas (NMCAs), developed by Parks Canada. These areas differ from terrestrial national parks in that they are managed

Environment in Focus
Box 8.7 The Endeavour Hydrothermal Vents Marine Protected Area

The Endeavour Hydrothermal Vents Marine Protected Area lies in water 2,250 metres deep, 250 kilometres southwest of Vancouver Island. As part of the Juan de Fuca Ridge system, the Endeavour segment is an active sea floor–spreading zone where tectonic plates diverge and new oceanic crust is extruded onto the sea floor. In this zone, cold sea water percolates downward through the crust, where it is heated by the underlying molten lava, eventually emerging through the sea floor as buoyant plumes of particle-rich, superheated fluid. The five known vent fields on the Endeavour Segment are separated from one another by about two kilometres along the ridge. Their associated plumes rise rapidly about 300 metres into the overlying water column.

Hydrothermal vents in the Endeavour area consist of large, hot, black smokers (chimney-like structures) and surrounding lower-temperature sites. The fields span a wide range of hydrothermal venting conditions characterized by different water temperatures and salt content, sulphide structure morphologies, and animal abundance. Temperatures associated with black smokers are typically in excess of 300°C. Formation of the large polymetallic sulphide chimneys takes place when dissolved minerals and metallic ions carried upward by the smokers precipitate on contact with the cold sea water. Cooler waters below 115°C on the sea floor and along the flanks of the chimneys support an abundance of flora and fauna. This rich ecosystem is supported by microbes whose life processes are fuelled by the chemical energy from the emerging fluids in the hydrothermal vents in the process of chemosynthesis (Chapter 2).

Hydrothermal venting systems host one of the highest levels of microbial diversity and animal abundance on Earth. The deep ocean near the Endeavour area normally only supports sparse animal abundance of about 20 worms and brittlestars per square metre. In the diffuse vent flows around the sulphide structures, these abundances can range up to half a million animals per square metre. An amazing abundance of life is in concentrated areas around such vents, surrounded by a veritable desert in the deep oceans.

Globally, hydrothermal venting systems foster numerous unique species of animals. Some 60 distinct species are native to the Juan de Fuca Ridge. Many of these species were first identified in this area. Hydrothermal vents at Endeavour are home to 12 species that are not known to exist anywhere else in the world.

The Endeavour Hydrothermal Vents Marine Protected Area was designated to ensure the protection of the vents and the unique ecosystems associated with them. Damage to or removal, disturbance, or destruction of the venting structures or the marine organisms associated with them is prohibited. This is important because hydrothermal vents are a major focus for future mineral mining. High temperatures cause the sea water to react with the rocks and minerals are formed, such as copper, gold, and silver. One Canadian company, Nautilus Minerals, is at the forefront of deep-sea ocean mining and projects that several billion tons of these minerals are produced every year. The UN has created the International Seabed Authority to regulate these activities. However, conservation is not currently one of its concerns. Scientists are now working to have conservation included from the onset of exploration activities to help protect at least some of the hydrothermal vents from mining damage.

for sustainable use. NMCAs are larger than the MPAs established by DFO, selected to represent Canada's ocean heritage, and contain an explicit mandate for recreation and education. NMCAs contain zones with special protection measures, such as no fishing. The location and size of these zones is decided through consultation among fishers, scientists, conservationists, government agencies, and other stakeholders. These decisions are crucial. Setting aside small fragments in unproductive areas will produce few benefits.

Apart from no-take zones, commercial and recreational fishing will continue in NMCAs, although additional conservation measures may be stipulated. Some activities, such as exploration or exploitation of hydrocarbons, minerals, aggregates, or any other inorganic material, are prohibited. Dumping is not allowed. Conservation interests sought to have bottom trawling, dragging, and fin-fish aquaculture prohibited as well because of their destructive impact on ocean ecosystems, but such prohibitions were not included in the National Marine Conservation Areas Act.

The goals of NMCAs are to conserve areas representative of the ocean environment and the Great Lakes and to foster public awareness, appreciation, and understanding of our marine heritage. The interpretation aspect of these areas may be their greatest contribution and a unique Parks Canada mission. Canadians are poorly informed about the marine environment. Creating ocean literacy among Canadians, who must

Guest Statement
Public and Political Will Needed to Protect Our Oceans
Sabine Jessen

In 2003, scientists at Dalhousie University reported that only 10 per cent of the largest fish still remain in the world's oceans. In their 10-year study of global fisheries they found that 90 per cent of tuna, sharks, swordfish, and cod have been fished out. Overfishing, removal of top predators, and fishing down marine food webs are having huge impacts on ocean ecosystems. Scientists point out that there are a variety of solutions to reducing fish mortality and ensuring the health of ocean ecosystems: reducing quotas, reducing overall effort, cutting subsidies, reducing bycatch, and creating networks of marine reserves where no fishing is allowed.

With all the scientific evidence available to show the decline in ecosystem health and the various tools to address this problem, why has it been so difficult for most countries to make these needed changes?

Part of the answer lies in the serious disconnect between what scientists are telling us and what the public believes is happening in the oceans. When the public is asked to name the greatest threat to ocean ecosystems, their answer is usually pollution. But scientific studies clearly show that the greatest threat is overfishing—the most serious problem is in what we are taking out of the oceans, not what we are putting in. Couple this with the results of another poll of Americans, asking them what the most important decision was that they make every day. You might be shocked to hear that deciding what to wear was considered the day's most difficult decision for one in every 10 people.

Why have I highlighted these points in talking about ocean management? Simply, without public understanding of the issue, and public demand for change, it is difficult to persuade politicians to make the difficult decisions that will lead to the fundamental changes required to better manage our oceans. Political support determines the priorities and the resources allocated to address these issues. Ocean management programs and marine protected areas are receiving very few resources in Canada, and the government has just announced plans to cut these resources.

While Canada has made many commitments to better oceans management, including marine protected areas, we still have reached only about 1 per cent protection of Canadian waters. And the MPAs that have been established rarely exclude fishing. While California has almost completed a network of fully protected marine reserves and Australia is moving ahead with a national network of marine protected areas, Canada lags behind these efforts.

A group of 14 scientists in Canada recently developed guidelines for MPAs and MPA networks in Canada to help the government and the public do it right. MPA networks in Canada must include areas that are fully protected from human uses, especially fishing, if ocean ecosystems are to recover and be more resilient in the face of climate change.

Until concrete steps are taken, we will continue to witness the ongoing destruction of the blue frontier. And by the time the public really understands and demands change, it could be too late.

Sabine Jessen is the National Manager for Oceans and Great Freshwater Lakes at the Canadian Parks and Wilderness Society, and one of the foremost activists pushing for increased conservation of Canada's ocean environment.

support public policies for the sustainable use and protection of Canada's marine environment, will pose unique challenges to interpreters.

One of the main challenges with all these programs is to actually designate areas. It is essential that local communities support these conservation measures, and gaining support can be time-consuming. Prime Minister Jean Chrétien made a commitment at the World Summit in Johannesburg in 2002 to create five NMCAs by 2007. By 2011 only one site, Haida Gwaii in BC, had been protected under the Act. Some progress has been made at sites in Lake Superior, the southern Strait of Georgia in BC, Lancaster Sound in the Arctic, and the Magdalen Islands in the Gulf of St Lawrence. Parks Canada has a marine system plan analogous to its terrestrial system plan (Chapter 14), with 29 marine regions. The goal is to have representation within each of these regions. Canada has international treaty commitments under the Convention on Biodiversity to complete the system by 2012. Clearly this commitment will not be met.

Aquaculture

One response to the declining catch in wild fisheries is to produce more seafood through farming or aquaculture. Aquaculture is the fastest-growing food production sector in the world and accounts for nearly half of the fish produced worldwide. It is expected that by 2030 it will be the dominant source of fish and seafood. Aquaculture could play a critical role in reducing world hunger. However, the global market is dominated by the production of salmon and shrimp, energy-intensive species for expensive markets rather than species designed to feed the poor.

Canada has been part of this growth and currently ranks twenty-sixth in the world in aquaculture production, although DFO predicts that Canada has the potential to be among the top three. In 2008, wild harvest and aquaculture production totalled more than 159 million tonnes. Farmed fish and seafood production was valued at $800 million in 2009, comprising approximately 25 per cent of all Canadian fisheries and aquaculture. Salmon is the predominant farmed species in Canada, producing 100,220 tonnes in 2009 and generating more than $598 million. Ninety-seven per cent of Canada's farmed fish and seafood is exported to the United States. Some 14,000 people are employed in aquaculture, and it is predicted that employment levels will quadruple over the next 15 years.

BC has Canada's largest output, worth $413 million in 2009 (Figure 8.13), followed by New Brunswick, PEI, and Newfoundland. Most of BC's production comes from salmon. Salmon production contributes 88 per cent of all cultured seafood in BC. The 2009 cultured salmon harvest of 68,000 tonnes was down 14 per cent from 2006. The increasing importance of cultured salmon to BC fisheries production can be seen in Figures 8.13 and 8.14. British Columbia is the fourth largest producer of farmed salmon in the world after Norway, Chile, and the United Kingdom.

A typical salmon farm consists of 10 to 30 cages, each 12 or 15 metres square, and contains on average 20,000 fish. The cages are made of open nets that allow water to flow through and antibiotics, uneaten food, feces, and chemicals used to prevent excessive marine growth on the cages to flow out.

There are more than 140 fish farms on the BC coast, mainly concentrated in three small areas. In addition, many applications await approval to expand operations to other areas along the coast. BC's salmon farming industry employs approximately 3,000 people in full-time, year-round jobs either directly (on farms) or indirectly (in processing). More than 92 per cent of the direct jobs are located in coastal communities outside of Greater Victoria and Vancouver. These economic opportunities can be lifesavers for some remote communities, especially Aboriginal communities. Nonetheless, there are several concerns about aquaculture.

Escapement. Salmon farms in BC mainly raise Atlantic salmon, primarily because the Norwegian-dominated industry had more experience with and well-developed markets for Atlantic salmon. The Atlantic salmon are also more efficient in converting feed into flesh, are less aggressive, and tolerate crowded conditions. The farming of Atlantic salmon is an environmental concern because escapement from farm fish cages is a regular occurrence and escapes often occur in high numbers. It has been estimated that up to now, more than 1 million Atlantic salmon have escaped. In July 2008, more than 30,000 ready-for-harvest Atlantics escaped after a net anchor was apparently displaced by high tides. There is irrefutable proof that Atlantic salmon are now spawning wild in Pacific rivers. DFO maintained for many years that this was impossible, until scientists proved otherwise (Volpe et al.,

The federal and Ontario governments have agreed to create an NMCA on the north shore of Lake Superior. It would be the largest freshwater protected area in the world.

2000). There are concerns that these escapers, an invasive species (Chapter 3), may displace the native salmon. Atlantic salmon have been found in more than 80 rivers on the BC coast, and spawning has occurred.

If salmon farming were confined to native Pacific species alone, it might not help. At the moment, there is no evidence that Atlantic and Pacific salmon interbreed. However, if Pacific salmon were to be found both wild and in farms, there would undoubtedly be interbreeding. This genetic introgression could have a devastating effect on wild stocks, as determined by McGinnity et al. (2003) on the east coast. There is considerable scientific uncertainty entailed in all these issues. And although DFO's stated policy is to 'err on the side of caution', the department has failed to do so with respect to the dangers associated with escapement.

Disease. The high stocking levels of fish in netted areas promote rapid spread of infectious diseases and parasites. Since fish farms are along migration routes for wild salmon, diseases and/or parasites can be passed along easily, with a detrimental impact on wild populations. In Clayoquot Sound on the west coast of Vancouver Island, a viral disease, infectious hematopoietic necrosis (IHN), swept through fish farms in the fall of 2002. As a result, the main operator in the region—Pacific National Aquaculture—lost some US$5.7 million in 2002 and closed down its processing plant in Tofino.

Lice. Strong scientific evidence indicates that pink salmon smolts in certain areas of the coast are being weakened by excessive sea lice coming from farms near their migration routes. One study found 90 per cent mortality among populations of juvenile pink salmon and predicted that stocks in some rivers will be extinct within 10 years if no mitigating actions are taken (Krkošek et al., 2007).

Pollution. To combat the diseases mentioned above, farmed salmon are treated with antibiotics. More antibiotic per weight of livestock is used by salmon aquaculture than by any other form of farming. Antibiotics can harm other marine organisms, since they are released directly into the ocean. As well, excess food and feces create a large amount of organic

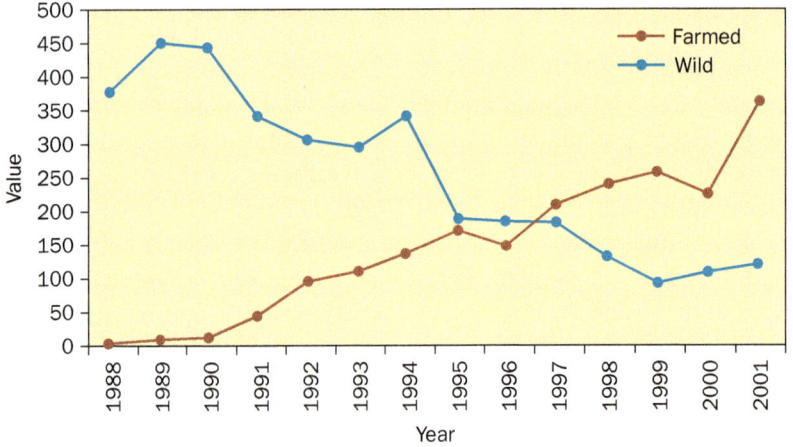

Figure 8.13 | Value of BC salmon exports in millions of dollars. *Source:* Adapted from BC Salmon Market Database, at: www.bcsalmon.ca/database/export/summary/sumvlpd.htm.

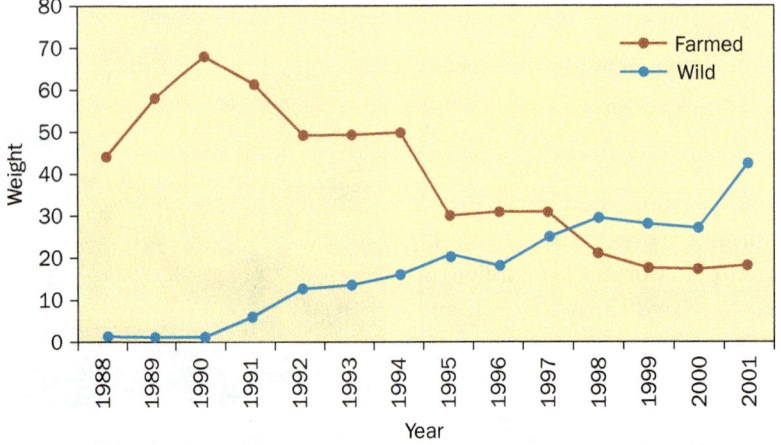

Figure 8.14 | Weight of BC salmon exports in millions of kilograms. *Source:* Adapted from BC Salmon Market Database, at: www.bcsalmon.ca/database/export/summary/sumvlpd.htm.

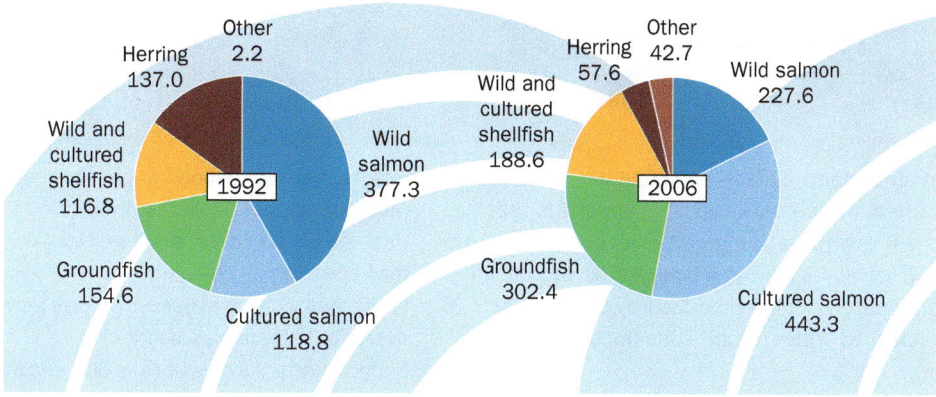

Figure 8.15 | BC seafood wholesale value species shares, 1992 and 2006 ($ millions). *Source: Copyright © Province of British Columbia. All rights reserved. Reprinted with permission of the Province of British Columbia. www.ipp.gov.bc.ca*

pollution. The substances build up on the ocean floor, depleting oxygen levels, releasing noxious gases (as a by-product of decomposition), and smothering benthic organisms. On a daily basis, the aquaculture industry in BC dumps the same amount of sewage in the ocean as a city of half a million people. Because this occurs in relatively protected coves and inlets, the waste is not readily dispersed as is, for example, that of the city of Victoria.

Predator control. Predators such as seals and sea lions are one of the main problems for the farmers, since they literally eat profits. Farmers are permitted to shoot animals that rip nets open. In 2001, farmers reported killing more than 400 seals and sea lions. Many observers think that these killings are grossly under-reported.

Energetics. Unlike the herbivorous fish produced in the vast majority of fish farms around the world, salmon are carnivorous. As a result, farmed salmon are mainly fed other fish, in pellet form, such as anchovies and mackerel that are caught as far away as South America. As dictated by the second law of thermodynamics (Chapter 2), only one kilogram of farmed salmon is produced for every three to four kilograms of feed fish. This is a poor use of fish protein and leads to the reduction of fish stocks elsewhere.

Social dimensions. Most profits from production go to five multinational companies that control 80 per cent of the industry in BC. As a result, a high percentage of the economic benefits attached to salmon aquaculture are exported out of the province. Increased mechanization is leading to lower employment figures, further limiting the economic benefits accruing to local communities. It is also feared that further growth in the industry will be detrimental to the wild fishery and reduce the health of communities dependent on wild fish. Increased supply of farmed salmon may continue to depress the price of BC's wild salmon.

Human health. To turn the white flesh pink, farmed salmon are fed artificial colouring. The most commonly used colourants are synthetic astaxanthin and canthaxanthin. In 2003, the European Union reduced the amount of canthaxanthin that can be fed to salmon by two-thirds because of concerns over retinal damage caused by ingesting too much of the chemical. One study found that farmed salmon contained 11 times the amount of toxic contaminants found in wild salmon (Hites et al., 2004).

Most of these problems are not insurmountable. Salmon can be produced in closed, land-based systems that all but eliminate some of the problems. One large grocery chain on Vancouver Island, Thriftys, now buys salmon from closed-pen systems for approximately 20 to 30 per cent more than for salmon from net-pen fish but sells them at roughly the same price in order to promote more sustainable practices. The main factor in why more environmentally and socially sound farming techniques are not being more widely adopted is the consumer. If people were willing to pay more for salmon produced using techniques that avoided the problems outlined above, then producers would not be so resistant to adopting these more sustainable systems (Box 8.8). At the moment, however, the environment pays those extra costs.

Small fish farm on the Broughton archipelago, Vancouver Island, British Columbia.

Environment in Focus
Box 8.8 What You Can Do

1. Fish are an important dietary component for many people and a healthy one. However, it is important that the fish you eat are not endangered or caught with a method that involves killing other species as bycatch. Use Canada's Seafood Guide produced by Sustainable Seafood Canada to inform your consumption (www.seachoice.org).
2. Buy only certified brands where they are available, such as dolphin-free tuna.
3. If you buy farmed seafood, consider paying a little more for products that have been produced using low-impact methods. For example, Thrifty Foods in BC sells farmed salmon produced using land-based, closed-system methods.
4. Ensure that you dispose of any toxic materials in the correct manner, not down the drain.
5. Use natural cleaners, such as vinegar and water, rather than commercial cleaners.
6. Using less water for your own needs leaves more water in rivers for fish such as salmon.
7. Support NGOs, such as Oceans Blue and the Canadian Parks and Wilderness Society, which are working for ocean conservation and the development of marine protected areas.
8. Whenever you do something that involves carbon emissions, whether travelling or buying a product, your actions are leading to ocean acidification. The car you drive and the coral reef in the South Pacific are intimately connected.

Perspectives on the Environment
'Extinction Vortex'

The combined impact of hybridization and competition means that when a large number of farm salmon spawn in a river, the number of adult salmon returning to the river and the potential offspring production in the next generation are reduced. . . . As repeated escapes are now a common occurrence in some areas, a cumulative effect is produced generation upon generation, which could lead to extinction of endangered wild populations as a result of this 'extinction vortex'.

—McGinnity et al. (2003)

In response to these concerns and vocal opposition from many communities, the BC government announced in 2008 that salmon aquaculture would not be allowed to expand to the north coast of the province. The moratorium remains in place today, although production has increased and the industry is pressuring governments to allow for major expansion. The main challenge to the aquaculture community is how to resolve the problems raised above so that salmon aquaculture can play an important role in meeting the food needs of the future.

Implications

Ocean health is now a major concern. Scientific efforts have intensified, and understanding has increased, yet much remains to be done. The very visible collapse of fishing stocks around the world and on the east and west coasts of Canada has helped to direct a little more political attention to oceans in general and fisheries in particular. Commitments have been made at both international and national levels to adopt more sustainable ocean practices, including encouraging and enabling sustainable fisheries, limiting pollution, and establishing systems of marine protected areas. At the moment, most of these measures are in the embryonic stage. Some plans, such as Canada's Oceans Strategy, have shown little progress. Only time will tell whether international and national commitments will be successful in turning around the trends described in this chapter.

Summary

1. Throughout history, the resources of the oceans have been thought of as vast and undiminished. The past decade has furnished conclusive proof that this view is far from correct. More than 70 per cent of global fisheries are now at or over their maximum exploitation levels.

2. Oceanic ecosystems are controlled by the same general principles that influence terrestrial ecosystems, but their manifestations may be different. There may be up to 5,000 species of fish still awaiting discovery.

3. The carbon balance of the oceans is of great interest because of its relationship with global climatic change.

4. Ocean fisheries supply about 20 per cent of the world's annual animal protein. Catch statistics showed very large increased catches over the past 50 years, but they have now levelled off considerably.

5. The oceans are the ultimate sink for many pollutants, and about 80 per cent of ocean pollution comes from activities on land.

6. More than 60 per cent of global oil production originates under the oceans. Exploration, drilling, transporting, and processing this oil is a major source of contamination.

7. Half of the world's population lives within 100 kilometres of the coast, a proportion that is expected to increase to 75 per cent by 2100.

8. Global climatic change will lead to increases in sea level of between 0.9 and 1.6 metres during this century. This will create severe challenges for many coastal communities.

9. There are many international agreements and programs on ocean management. Most have yet to fulfill their potential in improving oceanic conditions.

10. Marine protected areas have been endorsed by the scientific community as necessary to improving ocean conservation, but establishment at both international and national levels is a long way behind targets.

11. Canada has the longest coastline of any country and the second largest continental shelf, equal to 30 per cent of Canada's land mass.

12. Since 1992, commercial fishery landings in Canada have declined by 19 per cent and the number of vessels by 31 per cent. However, the value of the catch has doubled.

13. The east coast fisheries have experienced profound changes over the past couple of decades with the total collapse of the northern cod stocks.

14. Management of Aboriginal use of marine resources is an important concern on all coasts.

15. Exploitation of offshore hydrocarbons in Canada has taken place over the past two decades, mainly off the east coast. Increased attention is now being given to the Pacific and Arctic coasts.

16. Pollution levels of many substances have declined over recent years. However, a recent study concluded that the killer whales of Georgia Strait in BC are among the most polluted animals on the planet.

17. Canada passed a comprehensive Oceans Act in 1998, but it has been ineffectual because of a lack of political support and funding.

18. Three federal programs establish marine protected areas (MPAs) in Canada, yet less than 1 per cent of the area of Canada's marine environment is currently protected.

19. Aquaculture accounts for almost 30 per cent of the volume and 39 per cent of the value of global fish landings. Aquaculture is the fastest-growing food production sector in the world.

20. BC has Canada's largest share of the total value of aquaculture production, focused mainly on salmon. Although economically important to some communities, salmon farming also raises concerns over escapement, disease, proliferation of sea lice, killing of predators, energetics, and pollution.

Key Terms

- acidification
- aquaculture
- bottom trawling
- carbon balance
- coral bleaching
- coral polyps
- endocrine disruption
- exclusive economic zones (EEZs)
- fishing down the food chain
- longline
- marine protected areas (MPAs)
- prey switching
- serial depletion
- shifting baseline
- thermocline
- thermohaline circulation
- total allowable catch (TAC)
- zooxanthellae

Questions for Review and Critical Thinking

1. In what ways are oceanic and terrestrial ecosystems the same, and in what ways do they differ?
2. What are the most biologically productive areas of the ocean?
3. What is thermohaline circulation, and why is it important?
4. Give an example of a positive feedback loop related to global climate change and the oceans. Are there any negative feedback loops?
5. Explain the concepts of shifting baselines, serial depletion, and fishing down the food chain.
6. Give an example of the destructive effects of bottom trawling.
7. What are the two main forms of chemical pollutants in the oceans, and what are their main effects?
8. What are some of the main international conventions concerning ocean management?
9. What are the jurisdictional arrangements for ocean management in Canada?
10. Discuss the principal reasons behind the collapse of the Atlantic groundfish stocks and some of the lessons to be learned from this experience.
11. Outline some of the challenges involving the Aboriginal use of marine resources in Canada.
12. Discuss the differing approaches of Halifax and Victoria to ocean pollution resulting from sewage.
13. What are the main principles underlying Canada's Oceans Strategy?
14. Outline the three federal programs for creating marine protected areas in Canada and their similarities and differences.
15. Discuss the positive and negative aspects of aquaculture production.
16. What are the main interactions between global climate change and the oceans?

Related Websites

Canadian Parks and Wilderness Society (CPAWS)
www.cpaws.org

David Suzuki Foundation, oceans, fishing, aquaculture, MPAs
www.davidsuzuki.org/issues/oceans

Endeavour Hot Vents Marine Protected Area
www.dfo-mpo.gc.ca/oceans/marineareas-zonesmarines/mpa-zpm/pacific-pacifique/endeavour-eng.htm

Fisheries and Oceans Canada, Aquaculture
www.dfo-mpo.gc.ca/aquaculture/aquaculture-eng.htm

Fisheries crisis
www.fisherycrisis.com

Marine protected areas, Canada
www.dfo-mpo.gc.ca/oceans/marineareas-zonesmarines/mpa-zpm/index-eng.htm; www.pc.gc.ca/eng/progs/amnc-nmca/index.aspx

Marine Protected Areas Research Group (UVic)
mparg.wordpress.com/

Notes from Sea Level
www.jonbowermaster.com/

Save Our Seas
www.saveourseas.org

Seafood consumption choices
www.seafoodchoices.com; www.oceantrust.org; www.seachoice.org/page/resources; www.legalseafoods.com; www.montereybayaquarium.org/cr/seafoodwatch.aspx

United Nations Food and Agriculture Organization, Fisheries
www.fao.org/fi/default_all.asp

University of British Columbia Fisheries Centre
www.fisheries.ubc.ca

Value of marine protected areas
www.nceas.ucsb.edu/ecology/marine

Watershed Watch Salmon Society
www.watershed-watch.org

World Wildlife Fund Canada, Oceans
wwf.ca/conservation/oceans

Further Readings

Note: This list comprises works relevant to the subject of the chapter but not cited in the text. All cited works are listed in the References at the end of the book.

Baum, J.K., and B. Worm. 2009. 'Cascading top-down effects of changing oceanic predator abundances', *Journal of Animal Ecology*. doi: 10.1111/j.13652656.2009.01531.x.

Beaugrand, G., M. Edwards, and L. Legendre. 2010. 'Marine biodiversity, ecosystem functioning, and carbon cycles', *Proceedings of the National Academy of Science* 107: 10120–4.

Cosandey-Godin, A.C., and B. Worm. 2010. 'Keeping the lead: How to strengthen shark conservation and management policies in Canada', *Marine Policy* 34, 5: 995–1001.

Dearden, P., and R. Canessa. 2008. 'Marine parks', in P. Dearden and R. Rollins, eds, *Parks and Protected Areas in Canada*, 3rd edn. Toronto: Oxford University Press, 403–31.

Forbes, D.L., ed. 2011. *State of the Arctic Coast 2010—Scientific Review and Outlook*. International Arctic Science Committee, Land-Ocean Interactions in the Coastal Zone, Arctic Monitoring and Assessment Programme, International Permafrost Association. Geesthacht, Germany: Helmholtz-Zentrum Geesthacht Centre for Materials and Coastal Research.

Hocking, M.D., and J.D. Reynolds. 2011. 'Impacts of salmon on riparian plant diversity', *Science* 331: 1609–12.

Hutchings, J.A., C. Minto, D. Ricard, J.K. Baum, and O.P. Jensen. 2010. 'Trends in the abundance of marine fishes', *Canadian Journal of Fisheries and Aquatic Sciences* 67: 1205–10.

Pauly, D., et al. 1998. 'Fishing down marine food webs', *Science* 279: 860–3.

———. 2001. 'Fishing down Canadian aquatic food webs', *Canadian Journal of Fisheries and Aquatic Sciences* 58: 51–62.

———. 2003. 'The future for fisheries', *Science* 302: 1359–61.

Pinskya, M.L., O.P. Jensen, D. Ricard, and S.R. Palumbi. 2011. 'Unexpected patterns of fisheries collapse in the world's oceans', *Proceedings of the National Academy of Science*. doi: 10.1073/pnas.1015313108/

Ricketts, P.J., and L. Hildebrand. 2011. 'Coastal and ocean management in Canada: Progress or paralysis?', *Coastal Management* 39: 4–19.

Rogers, A.D., and D.d'A. Laffoley. 2011. *International Earth system expert workshop on ocean stresses and impacts*. Summary report. Oxford: IPSO.

Rose, G.A., and R.L. O'Driscoll. 2002. 'Capelin are good for cod: Can the northern stock rebuild without them?', *ICES Journal of Marine Science* 59: 1018–26.

Safina, C. 1997. *Song for the Blue Ocean*. New York: Henry Holt.

———. 2002. *Eye of the Albatross: Visions of Hope and Survival*. New York: Henry Holt.

Smedbol, R.K., and J.S. Wroblewski. 2002. 'Metapopulation theory and northern cod population structure: Interdependency of subpopulations in recovery of a groundfish population', *Fisheries Research* 55: 161–74.

Vaquer-Sunyer, R., and C.M. Duarte. 2010. 'Thresholds of hypoxia for marine biodiversity', *Proceedings of the National Academy of Science* 105: 15452–7.

Chapter 9
Forests

Learning Objectives

- To understand what the boreal forest is, its significance to Canada, and the main threats it is facing.
- To describe the eight main forest ecozones of Canada.
- To discuss the economic and non-economic values of Canada's forests.
- To appreciate the management arrangements and different approaches for harvesting Canada's forests.
- To understand some of the environmental and social aspects of forest management practices.
- To discuss the theory and practice of 'new forestry'.
- To describe current directions for forest use in Canada.

Canada's Boreal Forest

When the combined effects of climate warming, acid deposition, stratospheric ozone depletion, and other human activities are considered, the boreal landscape may be one of the global ecoregions that changes the most in the next few decades. Certainly, our descendants will know a much different boreal landscape than we have today.

—David W. Schindler (1998)

The Pew Environment Group, an American non-profit conservation think-tank, recently released a report (Wells et al., 2010) suggesting that Canada's **boreal forest** produces over $700 billion of ecosystem services to the world every year and calling for enhanced protection of this globally important landscape. Stretching 3,800 kilometres from the eastern tip of Newfoundland to western BC, the **Boreal Shield**, which is coterminous with the geological formation known as the Canadian or Precambrian Shield, houses roughly one-quarter of the world's remaining original forests. The Boreal is Canada's largest ecozone, covering almost 58 per cent of the country's land mass and stretching through all provinces except PEI, Nova Scotia, and New Brunswick. The forests are home to a wide diversity of terrestrial and aquatic wildlife, and 30 per cent of North America's bird population relies on the Boreal for breeding. Many Aboriginal people depend on the resources of the forests for subsistence, and more than 600 Aboriginal communities retain their roots in the forest.

'Borealis', a term that literally means 'of the North', comes originally from the Greek god of the north wind, Boreas. The term is now applied to many northern phenomena, perhaps the most famous being the aurora borealis, or northern lights. Many animal and plant species that live in the North have 'borealis' as part of their Latin name, such as the delicate twinflower, *Linnaea borealis*, which is found all across the country. It is also the name used to characterize the great northern forests that stretch not only across Canada but all across the northern hemisphere.

The boreal forests also support commercial activities such as logging, wood fibre and sawlog production, pulp and paper mills, and fibreboard production. Its wealth of minerals supports prospecting, mining, and smelting activities. There are large-scale hydroelectric developments, and the abundant fish and wildlife resources support subsistence, sport, and commercial harvesting activities, as well as a growing tourism industry. Recreation-related activities, such as canoeing, hiking, and birding in the Boreal, contribute more than $4 billion to the economy every year. In addition, more than 186 billion tonnes of carbon are stored in the Boreal's trees, soils, water, and peat—equivalent to 913 years' worth of greenhouse gas emissions in Canada. The global Boreal is the largest terrestrial carbon 'bank account' on the planet with values at least double those of tropical forests (Schindler and Lee, 2010). In fact, Anielski and Wilson (2009) show that the total market value of boreal resource extraction is only about 7 per cent of the value of the ecosystem services.

Almost 50 per cent of the boreal forest is currently allocated to industry and is open to harvesting. Although the Boreal Shield is the largest of Canada's 15 terrestrial ecozones, it has one of the lowest proportions of land (6 per cent) dedicated to protected areas in which all forms of industrial activity are prohibited. Approximately 4 per cent of the ecozone includes additional protected areas where forestry, mining, and other activities may be permitted.

The Canadian Boreal Forest Agreement was signed in 2010 by 21 of Canada's largest forestry companies and nine national environmental organizations to ensure a more protected and sustainable boreal forest and a stronger, more competitive forest industry. Over 72 million hectares of land between Yukon and Newfoundland and Labrador have been included in the agreement. Forestry companies have committed to practise sustainable harvesting that will preserve large tracts of old-growth forest, a habitat necessary for the woodland caribou, an endangered species that has symbolized this agreement. In return, environmental organizations will end their campaigns against Canadian forest products.

Virginia Falls, Nahanni River, NWT.

Environment in Focus

Box 9.1 Canadian Boreal Forest Agreement

The Canadian Boreal Forest Agreement has six strategic goals that encompass the needs for both conservation and economic development:

- Complete a representative network of protected areas that can serve as ecological benchmarks of the boreal forest.
- Protect species at risk.
- Reduce greenhouse gas emissions through the entire life-cycle of a forest product.
- Develop sustainable forest-management based on ecosystem principles (see Chapter 5), active management (see Chapter 14), and third-party verification.
- Improve the prosperity of the forest sector and associated communities.
- Recognize the importance of consumers and investors.

Source: The Canadian Boreal Forest Agreement, at: canadianborealforestagreement.com.

The six-goal agreement is shaping the forest industry to cater to environmentally conscious buyers and is the first of its kind in the world (Box 9.1).

The Canadian Boreal Forest Conservation Framework is another initiative supported by more than 1,500 scientists and thousands of others to gain permanent protection of at least 50 per cent of the Canadian Boreal and application of strict protective management standards in any other areas where development will occur. The framework has received significant support, including from more than 25 Aboriginal groups and some major forest companies, such as Domtar and Tembec. It is hoped that it will lead to some mitigation of the vast range of challenges now facing the Boreal, and some progress is already being made (Box 9.2).

Large areas of the Boreal are now experiencing a number of serious environmental stresses. In many ways, these stresses are no different from those experienced elsewhere in Canada, and they epitomize the challenges of developing strategies for the management of Canada's sustainable forest ecosystems. This chapter outlines the main challenges and some of the strategies developed to address them.

Environment in Focus

Box 9.2 Boreal Forest Conservation Framework

The Boreal Forest Conservation Framework is a shared vision to sustain the ecological and cultural integrity of the Canadian boreal forest in perpetuity.

The Boreal Forest Conservation Framework calls for conservation of at least 50 per cent of Canada's boreal forest in a network of interconnected, protected areas, and application of state-of-the art ecosystem-based resource management practices across the remaining landscape. It was developed by the Boreal Leadership Council (BLC), an unusual partnership of leading conservation organizations, resource companies, and Aboriginal groups, who joined together to promote the conservation and sustainable use of Canada's boreal forest region. Members of the BLC, convened by the Canadian Boreal Initiative, recognize that all who depend on the forest must come together to plan for its ecological, cultural, and economic future. The Framework is based on the best available principles of conservation biology and land-use planning, and has been endorsed by 1,500 international scientists, 25 Canadian Aboriginal communities, international conservation groups, and major businesses with annual sales totalling over $30 billion. Significant progress has been made towards the goals set out in the Framework. Recent key land protection actions include the following:

- In September 2010, the Ontario legislature passed a bill protecting 44,515,420.646 ha of pristine boreal forest and wetlands in the northern half of the province. The Far North Act is one of the largest wilderness protection efforts in the history of the province. It mandates that the entire area undergo conservation planning, and puts a minimum of 22 million ha permanently off limits to development.
- In Quebec, Premier Jean Charest pledged in March 2009 to protect at least 50 per cent of the area covered by the Plan Nord; this commitment totals more than 645,000 km^2.
- In the Northwest Territories, over 120,000 km^2 has been slated for protection since 2007; in April 2010, 33,000 km^2 was set aside for creation of a new national park around the East Arm of Great Slave Lake, the tenth largest lake in the world.

These and other land protection actions represent a significant commitment to the future of Canada's boreal forest, although much remains to be done to ensure equal treatment of conservation, sustainable development, and Aboriginal rights across the region.

Source: Wells et al. (2010).

An Overview of Canada's Forests

Canada is a forest nation. The symbol on our national flag is a maple leaf. Along with our northern latitude, the forests have provided the historical context for our national identity. Canada has one-tenth of the world's forests, and these forests cover almost half of the nation's land area (Figure 9.1) and a much higher proportion of southern Canada, where most Canadians live. Canada has one-quarter of the world's temperate rain forests and more than one-third of the world's boreal forests. Furthermore, estimates suggest that more than half of Canada's forest area consists of as yet undisturbed tracts of more than 50,000 hectares in size. More than one-third of Canada, however, is naturally treeless, and most of it occurs in the North. Together, Quebec, the NWT, Ontario, and British Columbia account for almost two-thirds of the country's boreal forest. Canada clearly has a major international role to play in forest conservation and management.

Deforestation is the permanent conversion of forests to other land uses. In Canada the main process is conversion to agricultural land, with conversion to oil and gas use being the next main factor and the most rapidly growing. Forest degradation by unsustainable harvesting practices is another issue. Overall, the annual rate of deforestation is falling, dropping from 64,000 hectares in 1990 to around 45,000 ha by 2009 (Natural Resources Canada, 2011).

The use of wood products is an integral part of the livelihood of many Canadians, as illustrated by this local boat-building in Newfoundland.

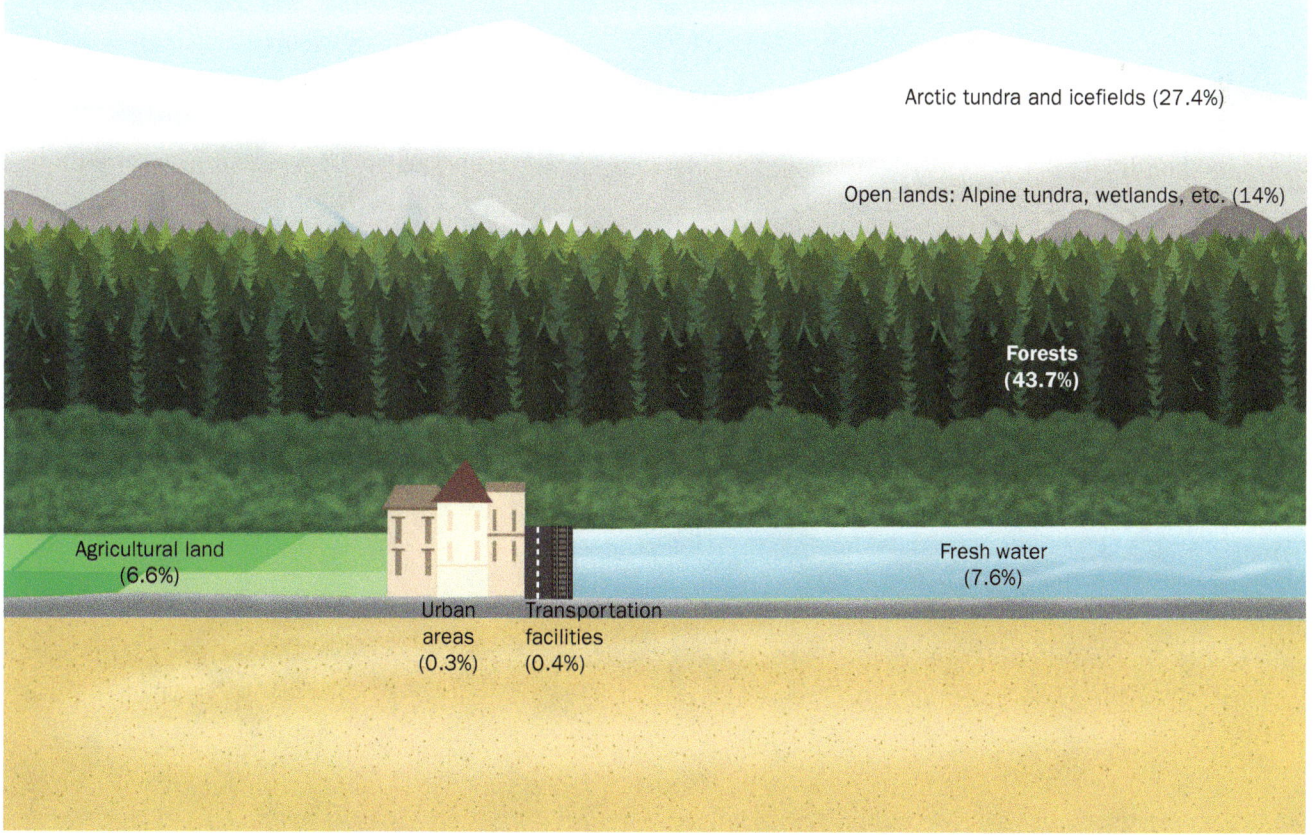

Figure 9.1 | Canada's physical makeup.

Canada's Ecozones

Although all the provinces are dominated by forest land, forest types differ significantly. Canadian forest ecosystems are a mixture of forest, woodlands, wetlands, lakes, glaciers, and rock, providing habitat for a great variety of plants, animals, insects, fungi, and other organisms. Of the estimated 140,000 species in Canada, approximately 66 per cent occur in forests, (Chapter 14). There are 15 terrestrial ecozones in Canada (Figure 9.2), although the majority of Canada's forests lie within the eight discussed below.

The **Boreal Cordillera** ecozone, to the south and west of the Tundra Cordillera, is found in northern BC and southern Yukon. This zone is made up of mountains in the west and east, separated by intermontane plains. A wet climate gives rise to more tree growth, hence the boreal designation. Much of the terrain along the western side of the ecozone is covered by permanent ice and snow. In parts of the ecozone located in BC, south-facing slopes are covered with meadows, while north-facing slopes feature boreal forest. Vegetative cover ranges from dense to sparse on a large part of the plateaus and valleys. Tree species include trembling aspen, balsam poplar, white birch, black and white spruce, alpine fir, and lodgepole pine. Higher elevations have large areas of rolling alpine tundra, with sedge meadows and stone fields colonized by lichens. Mammals characteristic of the Boreal Cordillera ecozone include woodland caribou, moose, Dall's sheep, mountain goats, black and grizzly bears, martens, lynx, hoary marmots, and Arctic ground squirrels. Representative bird species include willow, rock, and white-tailed ptarmigan, spruce grouse, and a range of migratory songbirds and waterfowl. The rich resources contained within this ecozone have fostered mining, forestry, and tourism industries, in addition to hydroelectric development and localized agriculture.

Stretching along the entire BC coast, the **Pacific Maritime** ecozone is influenced by the Pacific Ocean. Westerly winds sweeping across the ocean pick up considerable amounts of water. Most is precipitated on the windward side of the coastal mountains, producing the highest annual rainfall figures in Canada (up to 3,000 millimetres). The maritime influence also results in the warmest average temperatures in the country, with mild winters and relatively cool summers. These climatic characteristics provide ideal growth conditions and at lower elevations give rise to the most productive forests in the country. Some of the tree species are long-lived and grow to impressive heights. For instance yellow cedars can live more than 1,000 years, Douglas firs 750 years, and Sitka spruces 500 years. Douglas fir and Sitka spruce can grow to more than

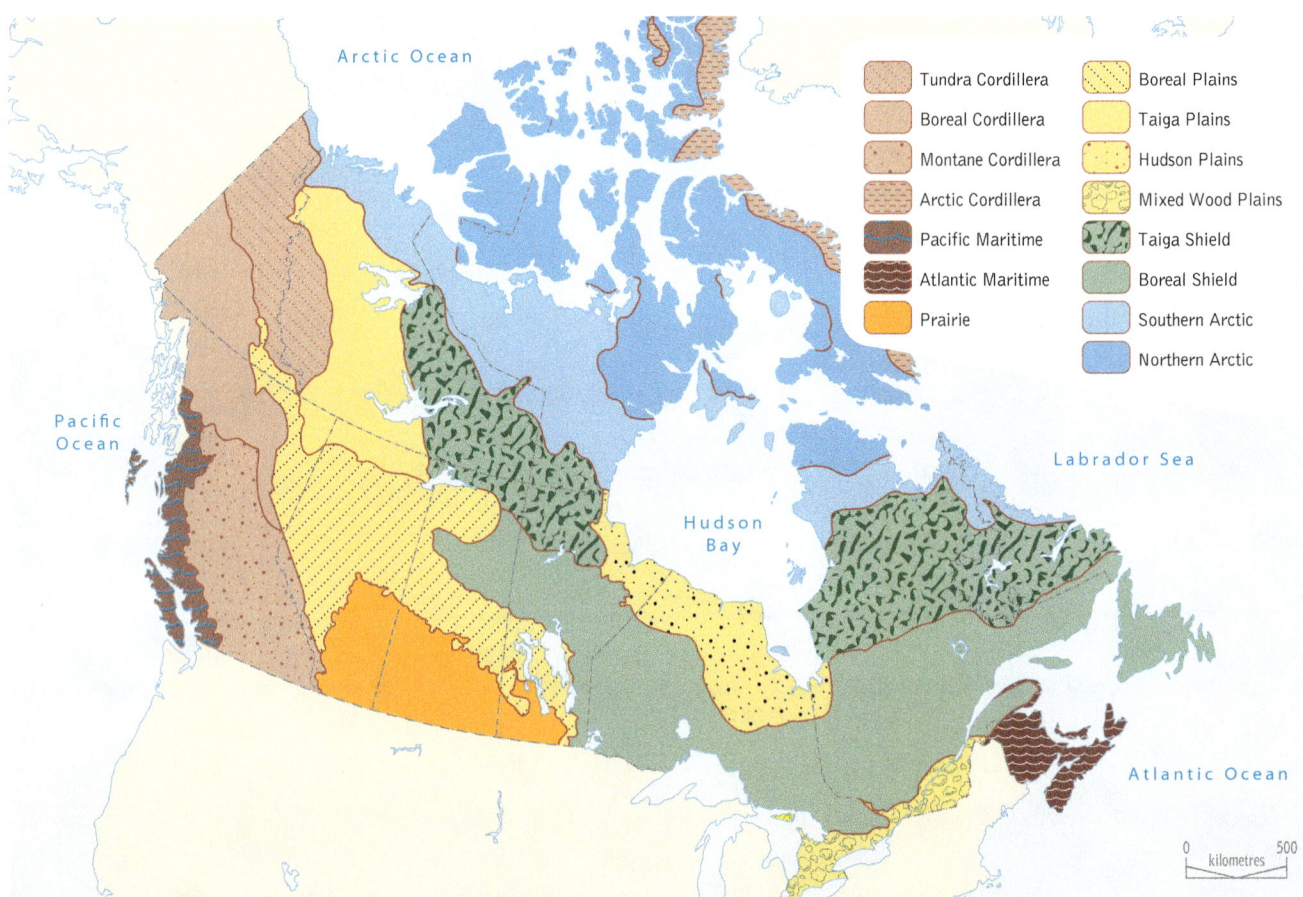

Figure 9.2 | Terrestrial ecozones of Canada. Source: Wiken (1986).

70 metres in height. The most common tree species in this ecozone are western red cedar, western hemlock, amabilis fir, Sitka spruce, and yellow cedar. The Pacific Maritime ecozone is also home to a variety of mammals and birds, including black-tailed deer, elk, otters, black and grizzly bears, cougars, fishers, wolves, American black oystercatchers, mountain and California quails, blue grouse, tufted puffins, pygmy owls, Steller's jays, and northwestern crows.

Forestry is the dominant industry of the ecozone. Douglas fir and western red cedar form the backbone of the BC lumber industry, but are no longer of infinite supply as they were once assumed to be. Approximately 25 per cent of the coastal temperate rain forest within BC has been logged and reforested and an additional 3 per cent of the total area has been logged and urbanized. Approximately 50,000 hectares of coastal temperate rain forest are logged annually.

The great altitude variation in the **Montane Cordillera** ecozone leads to considerable contrast among the summits of the snow-bound peaks, the high montane valleys, rolling plateaus, and deeply entrenched desert-like conditions in the BC Interior. The climate is similarly varied but is generally characterized by long, cold winters and short, warm summers. Annual precipitation ranges from highs of more than 1,200 millimetres along the mountain summits to as little as 205 millimetres in valleys in the rain shadow. Vegetation varies according to these conditions and can be thought of as a series of vertical zones, changing with altitude. At the summits, the vegetation is alpine, characterized by lichens, herbs, and small shrubs. In the lower subalpine environment, trees such as alpine fir, Englemann spruce, and lodgepole pine become more common. Below this zone, there is considerable variation, depending on local conditions. Ponderosa and lodgepole pines, Douglas fir, and trembling aspen are found towards the north, and in moister conditions to the southeast, western hemlock, red cedar, and Douglas fir are found. Characteristic mammals include black and grizzly bears, woodland caribou, mule deer, moose, mountain goats, California bighorn sheep, wolverines, and fishers. Common bird species include the pileated woodpecker, Clark's nutcracker, and red crossbill.

British Columbia has some of the largest trees in the world. This is Cathedral Grove, a stand of large Douglas fir and western red cedar on Highway 4 en route to Pacific Rim National Park Reserve.

Many Canadians are familiar with the spectacular mountain landscapes of the Montane Cordillera ecozone, which includes Banff, Canada's first and most-visited national park.

Trembling aspen is characteristic of the Boreal Plains.

Commercial forest operations have been established in many parts of the Montane Cordillera ecozone, particularly in the northern interior. Other significant industries include mining, oil and gas production, tourism, and agriculture. The southern valleys are well-known for their orchards and vineyards.

The **Boreal Plains** ecozone extends from the southern part of the Yukon in a wide, sweeping band down into southeastern Manitoba. The underlying glacial moraine and lacustrine deposits give a generally flat to undulating surface similar to the Prairie zone to the south. Climatic differences, making it wetter and cooler than the Prairie zone, produce vegetation dominated by trees rather than grasses. Coniferous trees include tamarack, Jack pine, and black and white spruce, although deciduous trees such as trembling aspen, white birch, and balsam poplar are common, especially at the transition zone into the true Prairie. Mammals found in the Boreal Plains include woodland caribou, white-tailed deer, bison, wolves, black bears, mule deer, and elk. Characteristic bird species include boreal and great horned owls, blue jays, various warblers, grouse, red-tailed hawks, cormorants, gulls, herons, and terns. One of Canada's most famous endangered species, the whooping crane, nests in the wetlands of Wood Buffalo National Park at the northernmost part of the ecozone. A number of species considered at risk of extinction can be found in this ecozone, including woodland caribou, wolverines, grizzly bears, and wood bison. The southern and northwestern areas of the ecozone have been transformed by agricultural development and timber, mining, and oil production.

Taiga, a Russian word, is used to describe coniferous forests in that country. In North America, it describes that portion of the boreal forest lying between the southern boundary of the tundra and the closed-crown coniferous forest to the south. Topography in the **Taiga Plains** is gently rolling, with a high proportion of surface water storage, wetlands, and organic soils. The climate is cold and relatively dry, with as little as 200 millimetres of precipitation annually in the northern sections. Canada's largest watercourse, the Mackenzie River and its tributaries, is located in the Taiga Plains. Winters are long and cold, with mean daily January temperatures ranging from −22.5 to −30°C. These conditions, plus the topography, give rise to large areas of wetlands dominated by species such as Labrador tea, willow, dwarf birch, mosses, and sedges. On better-drained localities and uplands are found mixed coniferous–deciduous forests containing white birch, trembling aspen, balsam poplar, lodgepole pine, tamarack, and black and white spruce. Large mammals are found in this ecozone, including moose, caribou, black bear, wolf, and wood buffalo. Common bird species include the bald eagle, the peregrine falcon, and the osprey. The Mackenzie Valley forms one of North America's most travelled migratory corridors for waterfowl breeding along the Arctic coast. Economic activity in the area is based on subsistence hunting, trapping, and fishing, in addition to a few industrial activities, such as mining and oil extraction.

The Boreal Shield is the largest ecozone in Canada, stretching along the Canadian Shield from Saskatchewan to Newfoundland. It is also part of one of the largest forest belts in the world, the boreal forest, extending all across North America and Eurasia, encompassing roughly a third of the Earth's forest land and 14 per cent of the world forest biomass. It is the belt that generally separates the treeless tundra regions to the north from the temperate deciduous forests or grasslands to the south. Winters are cold and summers warm to hot, with moderate precipitation. In Canada, the zone is influenced by cold Hudson Bay air masses that yield relatively high precipitation, ranging from 400 millimetres annually in the east to more than 1,000 millimetres in the west. Average January temperatures range from −10 to −20°C, with a July range of 15 to 18°C. The terrain is characteristically rolling, with bedrock outcrops, glacial moraine, and many lakes dominated by coniferous forests of black and white spruce,

The boreal forest in Newfoundland has bedrock outcrops, lakes, and muskeg.

The Mixed Wood Plains ecozone has the highest tree diversity in Canada.

balsam fir, tamarack, and Jack pine but with significant cover by aspens and white birch. This forest cover is interrupted by large areas of wetlands, particularly moss-dominated bogs. Overall, the boreal forest is characterized by its lack of diversity of tree species; large areas are covered in just one or two species, particularly the spruces. Balsam fir becomes more dominant in areas of heavier precipitation towards the east. Further south, hardwoods such as white birch, yellow birch, and trembling aspen are more common, along with softwoods such as red pine, Jack pine, and eastern white pine.

Characteristic mammals include the woodland caribou, moose, white-tailed deer, black bear, wolf, marten, snowshoe hare, striped skunk, Canadian lynx, and bobcat. Characteristic bird species include the common loon, boreal owl, great horned owl, and evening grosbeak. The rich natural resource base of the Boreal Shield ecozone supports various industries, including mining, forestry, energy, and tourism, in addition to commercial and subsistence hunting, trapping, and fishing.

The **Mixed Wood Plains** ecozone is the most urbanized and densely populated in Canada, spreading from the lower Great Lakes north through the St Lawrence Valley. The topography is gentle, mainly resulting from the lacustrine, marine, and morainic deposits. The climate is continental, with warm, humid summers and cool winters. Precipitation ranges from 720 to 1,000 millimetres per year. These conditions have produced the most diverse tree coverage in Canada, with more than 64 species. However, few intact areas of natural vegetation remain. In the northern part of the ecozone, the mixed coniferous–deciduous forest is dominated by red and white pine, oak, maple, birch, and eastern hemlock. Further south, the warmer zones contain deciduous species such as sugar maple, beech, basswood, and red and white oak. The white elm is also native to this zone. Unfortunately, many of them have been devastated by an imported fungus that causes Dutch elm disease. First noticed in 1944, the fungus is spread from tree to tree by beetles. It is a good example of the problems created by alien organisms, discussed in more detail in Chapter 3.

White-tailed deer and black bears were once common in this ecozone, but small mammals such as the raccoon, striped skunk, black squirrel, groundhog, and eastern cottontail rabbit now dominate. Common bird species include the whippoorwill, blue jay, red-headed woodpecker, great blue heron, cardinal, and northern oriole. Most of the deciduous forest has been cleared for agriculture and urban development, with small patches of forest cover found scattered throughout the zone. Two tree species, eastern white pine and eastern hemlock, were overharvested and are now under-represented. Service industries and the manufacturing sector are the largest employment sectors.

The **Atlantic Maritime** ecozone, stretching from the mouth of the St Lawrence River across New Brunswick, Nova Scotia, and Prince Edward Island, is heavily influenced by the Atlantic Ocean, which creates a cool, moist maritime climate. However, conditions vary considerably between the upland masses of hard crystalline rocks, such as the Cape Breton and New Brunswick highlands, through the coastal lowlands that support most of the population. Mean annual precipitation is as high as 1,425 millimetres on the coast but falls to fewer than 1,000 millimetres further inland. Temperatures are also moderated by the ocean, with mean daily January temperatures of −2.5 to −10°C and a mean July daily temperature of 18°C. Forests are generally mixed stands of deciduous and coniferous species, such as balsam fir, red spruce, yellow birch, sugar maple, eastern hemlock, and red and

Philip Dearden

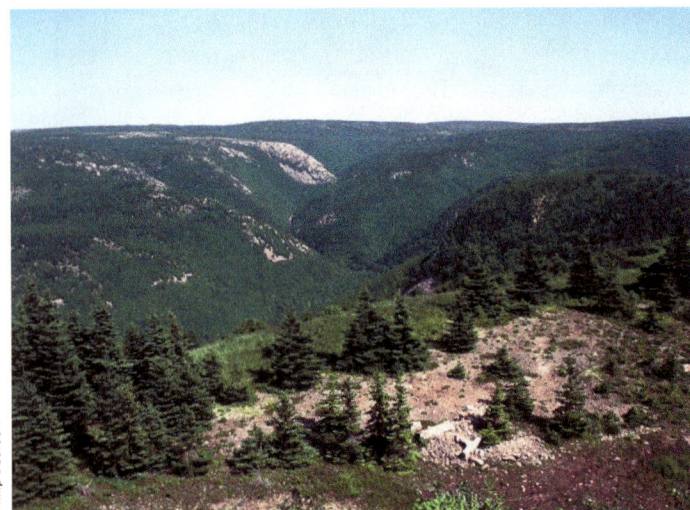

Cape Breton Highlands.

white pine, mixed in with boreal species such as black and white spruce, white birch, Jack pine, and balsam poplar. Characteristic animal species include the white-tailed deer, moose, black bear, bobcat, snowshoe hare, wolf, eastern chipmunk, mink, whippoorwill, blue jay, eastern bluebird, and rose-breasted grosbeak.

Major land-oriented activities include forestry, agriculture, and mining. Forest landscapes in the region are undergoing significant change because of overharvesting on private woodlots and spruce budworm outbreaks, which have increased dramatically in frequency, extent, and severity.

Forest Ecosystem Services and Products

Canada's forest ecosystems provide an array of beneficial services arising from ecological functions such as nutrient and water cycling, carbon sequestration, and waste decomposition. For example, plant communities are important in moderating local, regional, and national climate conditions. Biological communities are also of vital importance in protecting watersheds, buffering ecosystems against extremes of flood and drought, and maintaining water quality. The contributions of forest lands to the maintenance of ecological processes (Chapters 2 and 4) within Canada are substantial. However, the sheer scale of Canada's forests means that they are significant contributors on a global scale. It is estimated that 20 per cent of the world's water originates in Canada's forests. The forests are also major carbon sinks, with an estimated 50,000 million tonnes stored and a yearly accumulation of some 72 million tonnes.

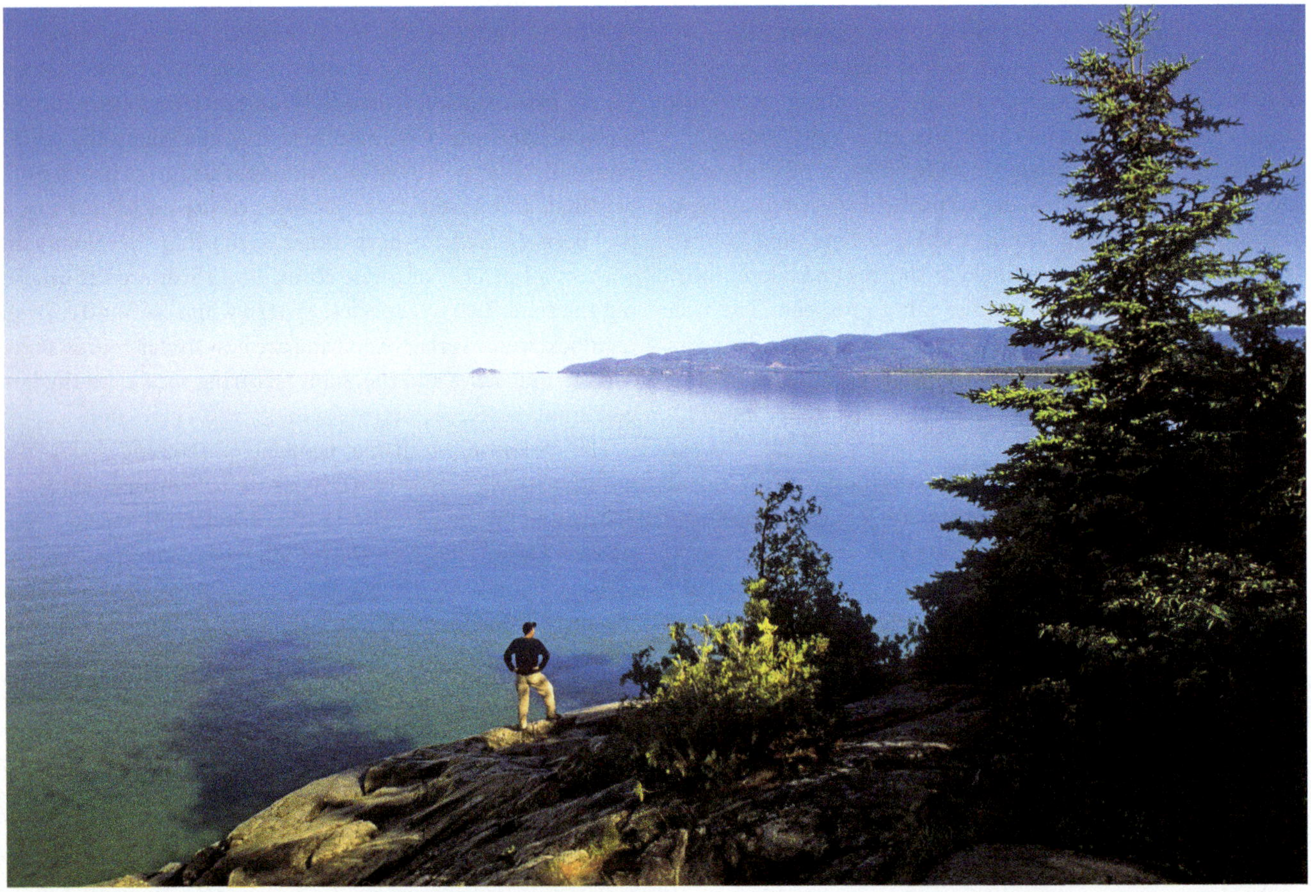
© All Canada Photos/Ethan Meleg

Many people find forest lands a source of recreational and spiritual fulfillment.

Forests are also places of exceptional scenic beauty, and millions of Canadians travel each year to participate in nature-related recreational activities such as wildlife viewing in parks and protected areas, nature walks, and bird watching. The monetary value of these activities can be significant. For example, more than 20 million Canadians participate in nature-related recreational activities, creating approximately 245,000 jobs and contributing more than $12 billion to Canada's GDP (NRTEE, 2003b).

Non-Timber Forest Products

In addition to the important 'services' that forest ecosystems in Canada provide, forests are also a valuable source of commodities. Wild rice, mushrooms and berries, maple syrup, edible nuts, furs and hides, medicines, ornamental cuttings, and seeds—collectively known as **non-timber forest products** (NTFPs)—are typical examples. Their total value is unknown, but these products have the potential to generate $1 billion per year for the Canadian economy (Natural Resources Canada, 2010a). In 2009 Canada produced upward of 41 million litres of maple products valued at $354 million. Similarly, 1.8 million Christmas trees generated $39 million while blueberries contributed $127 million in sales.

Some NTFPs are harvested commercially and are allocated by licence, while others are freely available and contribute significantly to recreational values, including tourism. These commodities are also important in sustaining First Nations communities. With careful management, NTFPs are renewable. Some inspection agencies have expressed concern about a lack of regulation over the harvesting, safety, and economic contribution of these products. In response, the Quebec NTFP association recently initiated a training program for harvesters to provide instruction on product identification, ethical harvesting, bush safety, and food storage and safety. A useful reference website on NTFPs is maintained by the Centre for Non-Timber Resources at Royal Roads University in Victoria, BC (cntr.royalroads.ca)

Historically, non-timber products and services of Canada's forests have received little attention. However, as timber harvest levels have increased and the public has become more aware of and vocal about declines in these other forest values, forestry companies are being required to take these values into account in their cutting plans. In other words, they are having to take a more ecosystem-based approach, as described in Chapter 5. NTFPs are also seen as bringing diversification to rural economies and can yield valuable economic returns. Some First Nation bands, for example, have succeeded in harvesting and marketing such forest products as mushrooms and wild rice, and the maple syrup industry in Quebec, which produces more than 90 per cent of the Canadian total, has become a significant business. In 2009, a record year, Quebec operators produced 8.3 million gallons of maple syrup at a value of nearly $305 million (www.honeycouncil.ca/documents/Honey%20and%20maple%20production%202009.pdf). However, Canada has a long way to go to catch up to most other countries in introducing this kind of product diversification to the forest land base.

NTFPs may be wild or managed and may come from both natural and managed forests. It is important to understand these differences if NTFPs are to play a fuller role in forest

Environment in Focus

Box 9.3 Canada's 'Button' Mushroom

Button, or matsutake, mushrooms.

Pine mushrooms (also know as button mushrooms, matsutakes, or *Tricholomamagnivelare*) are found in the Pacific Northwest of North America, some northern parts of Europe, and select regions in northern Asia. Button mushrooms are mycorrhizal fungi that have a symbiotic relationship with nearby trees, making them difficult to produce outside of the forest. They are a highly priced delicacy in parts of Asia and northern Europe and symbolize fertility and happiness. Their intense flavour and good omen mean these mushrooms have been known to generate upwards of $400 a kilogram in Japan. The industry has been estimated at $49 million annually but the development of a substantial black market can skew approximations. Limited wild production and a high market value can drive tension between harvesters: some who enjoy mushroom picking and others who are driven by the high value of this small forest product.

valuations and decision-making. For example, the harvesting of some wild stocks, such as mushrooms, from managed forests may conflict with timber production activities. However, harvesting some managed NTFPs can be encouraged alongside timber production and raise the overall level of return from the land. Such an approach has been adopted for blueberries in some areas of Quebec. It is an example of symbiotic use between the different resources in which both kinds of resource use can benefit. We often see this approach used in agro-forestry ecosystems in the tropics but rarely in Canada.

Other kinds of relationships include complementary, competitive, and independent resource use. *Complementary* relationships occur when NTFPs and timber are extracted from the same land base in non-conflicting ways. Craftspeople, for example, may get improved access to their raw materials (e.g., tree bark, boughs) because of the development of logging roads. In contrast, *competitive* relationships often involve mutually exclusive uses. Logging **old-growth forests** in western Canada, for example, would devastate the lucrative pine mushroom industry. Finally, *independent* systems develop when the two uses operate on different units of land—for example, in commercial and non-commercial forests.

Besides tangible non-timber forest products, forests fulfill a host of less tangible values related to cultural and spiritual fulfillment and knowledge and understanding. Such values are difficult to assess, let alone manage. The aesthetic value of forests in particular has become prominent over the past decade. Our forested areas include some of our most scenic landscapes. Not just an attraction for tourists, they also offer recreational and spiritual satisfaction. Most provinces have introduced procedures for including assessments of aesthetic quality into harvesting plans. Unfortunately, many of these procedures still leave it up to the people in charge of timber extraction to decide what interests should be considered in assessing scenic value and what harvesting regime might follow. As a result, modifications to cutting plans to take aesthetics into consideration often tend to be minimal.

Timber Forest Products

Despite recognition from federal and provincial governments that Canadian forests provide a broad range of values (wilderness, recreation, wildlife habitat, etc.), forest management paradigms over the past century have focused on the management of Canadian forests to supply wood. The economic benefits arising from timber products are substantial. For some 200 communities, the forest sector makes up at least 50 per cent of the economic base. Direct timber industry employment in 2010 totalled 222,500. Employment is concentrated in Quebec (77,900), British Columbia (54,400), and Ontario (46,700). The total number of people directly employed by the industry has been declining for over a decade and dropped 20 per cent between 2008 and 2010 (cfs.nrcan.gc.ca/pubwarehouse/pdfs/32683.pdf), in part a reflection of the global economic downturn that began in 2008 and a weak American housing market. Between 2003 and 2009, Canada's forest sector lost upwards of 130,000 jobs and experienced 455 mill closures (cfs.nrcan.gc.ca/pages/182). Despite this decrease, Canada remains the world's leading forest product exporter. In 2010, Canada's forests contributed a net $22.5 billion to national GDP (Table 9.1). The forestry industry is the largest single contributor to Canada's balance of trade, with exports totalling over $25 billion in 2010. British Columbia accounted for $9.0 billion (34.7 per cent) of this total, Quebec $7.5 billion (28.9 per cent), and Ontario $4.1 billion (15.8 per cent). The United States is by far the largest buyer of Canadian forest products, purchasing just under 71 per cent of all exports in 2009.

As the Canadian dollar strengthened against the American dollar in 2007 and 2008 and with severe pressure on US housing markets, this reliance on the US market led to a collapse in the forest industry in much of Canada. By early 2011 the industry was seeing clear signs of improvement. Some mills reopened, more jobs were created than lost, and new markets were developed, particularly in China where increases in the value of timber exports of 9 per cent per year until the year 2015 have been predicted. By the year 2015, China is predicted to have an annual timber deficit of some 182 million cubic metres. This amount can be compared with BC's 2010 harvest of 27 million cubic metres. Predictions suggest that China could supplant the US as the major importer of BC lumber by 2013.

The volume of wood produced per unit area differs across the country, rising to highs in excess of 800 cubic metres per hectare on the most productive sites in coastal British Columbia, where mild temperatures, deep soils, and abundant rainfall create some of the most productive growing sites in the world. Volumes harvested also vary by province. For example, in 2010, British Columbia harvested over double the volume of any other province, while less area was

Table 9.1 | Canada's Forests

Total land	882.1 million ha
Total forest	347.7 million ha
Commercial forest	294.8 million ha
Managed forest	1.4 million ha
Harvested forest (2009)	0.6 million ha
Value of exports	$26.0 billion
Contribution to the balance of trade	$16.6 billion
Contribution to the GDP	$22.5 billion
Direct employment	222,500
Annual allowable cut (2009)	207 million m^3
Harvest (2009)	118 million m^3

Source: Natural Resources Canada (2011).

Environment in Focus

Box 9.4 Canada's Unique Forest Industry

Canada is the world's largest exporter of forest products. In the international marketplace, Canada has a number of assets, including:

- Of Canada's forest land, 93 per cent is publicly owned.
- The federal government is responsible for trade, the national economy, and federal lands.
- Eighty per cent of Aboriginal peoples live in forested areas.
- The federal government has constitutional, treaty, political, and legal responsibilities for Aboriginal peoples.
- Over 8 per cent of the total forest area is protected by legislation.
- Less than 1 per cent of forests is harvested every year.
- By law, all harvested forests must be successfully regenerated.
- Over half of the energy used by the forest energy is bioenergy.
- Canada leads the world in exports of softwood lumber, newsprint, and wood pulp.
- By 2011, 150 million hectares of forest were certified as being sustainably managed.

Source: Natural Resources Canada (2011).

---

involved in BC than in either Quebec or Ontario (Natural Resources Canada, 2011).

The Canadian forestry industry is also a frequent flashpoint of conflict. Names such as Carmanah, Temagami, and Clayoquot became well known across the country in the 1990s as they appeared in newspaper headlines and on national news broadcasts. All these conflicts revolved around questions of whether particular areas should be logged or preserved. These conflicts reflected the increasing appreciation of the many values provided to society by forests besides the usual economic benefits. Few of these non-economic values are easy to calculate in monetary terms and compare against the financial returns of the forest industry. However, these values are gaining increasing recognition on the part of the public and decision-makers as the process of converting old-growth forests across Canada into managed forests continues.

The sheer scale of the industry has also attracted international attention. Concern over the destruction of the tropical forests and resulting impacts on the biosphere and on biodiversity has spread from countries such as Indonesia and Brazil to temperate forest nations, Canada in particular, which has been dubbed, rightly or wrongly, 'the Brazil of the North'. Governments in Canada have launched large public relations missions around the world to combat this perception. They have also started to look more closely at the basis for these accusations, and substantial changes are underway in many aspects of the forest industry in Canada.

When considering the allocation of any resource to different uses, it only makes good sense to evaluate the relative values that will result from the various allocation decisions. However, resources are commonly allocated in society with little appreciation of their true value. Anielski and Wilson

'Brazil of the North' sign visible as two protestors block the path of logging trucks during the 1993 protests (left) in the region of Clayoquot Sound (right).

(2009) undertook a study to assess the total economic values of Canada's boreal forest and arrived at an annual figure of over $700 billion. They found that the value of the non-timber forest products outweighed that of the timber products by a ratio of 2.5 to 1, yet these values are rarely if ever taken into account in decisions about how to allocate the boreal forests among competing uses.

Forest Management Practices

The provincial governments are responsible for 77 per cent of the nation's forests, with the federal and territorial governments responsible for 16 per cent. The remaining 7 per cent of forests are managed by 450,000 private landowners. These figures are likely to change over the next decade as an increased number of land claims by Aboriginal peoples are settled and more land comes under their control. Areas under current land claims account for about one-quarter of large, intact forest landscapes in Canada. Currently, forest management is primarily a provincial responsibility, with governments managing forest resources on behalf of the public through agreements with private logging companies. Different forms of tenure exist, but generally all involve the submission of plans by the logging company that outline where it intends to cut, the details of the harvesting process (including the location of roads), and **reclamation** plans. The governments provide regulations and guidelines for these practices and have the authority to ensure that they are followed.

Of the total of 347.7 million hectares of forest land in Canada, some 119 million hectares are currently managed for timber production. On these lands, forest ecosystems are being transformed from relatively natural systems to controlled systems, as described in Chapter 1, in which humans, not nature, influence the species that will grow there and the age that they will grow to. Over the past decade, increasing awareness of and concern about the environmental impact of forest harvesting has prompted questions about the environmental sustainability of forestry and about the different kinds of management approaches that might lead to sustainability. Key questions relate to the amount of forest protected from logging, the amount of fibre harvested over a specific period, the way by which it is logged, and what happens to the land after harvesting.

Guest Statement

Forest Ownership, Forest Stewardship, Community Sustainability

Kevin Hanna

Forest tenure refers to the conditions that govern forest ownership and use. Tenure is an important and fundamental element in determining forest policy and management. As this book points out, significant challenges face Canadian forest management and how we approach ownership, and must play a role in realizing a more innovative and flexible forest industry and a greater sense of stewardship. But because tenure is a difficult and thorny subject, governments, industry, labour, and environmental groups have tended to ignore it. I would like to say this is beginning to change, but it's not.

Right now about 95 per cent of Canada's forest land is owned by governments, largely the provinces. About 80 per cent of Canada's private forest land is located east of Manitoba, most of it in the Maritimes. British Columbia has the highest level of provincial forest ownership, at 96 per cent. Compared to other major forest nations, Canada's tenure profile is quite unique. Our biggest competitors in the wood product export market are Finland and Sweden. Finland's forests are mainly privately owned. Individuals hold about 62 per cent of forest land; timber companies about 6 per cent, and the national government 31 per cent (most of which is in the far north). There are about 280,000 private forest holdings with an average size of 37 hectares. These smallholdings are very productive and intensively managed. They supply the majority of Finland's domestic production and about 80 per cent of the stumpage income and 80 per cent of annual growth and cut.

In Sweden, small-scale landowners have about 50 per cent of the forests, while the state and forest companies each have about 25 per cent. There are about 240,000 private forests in Sweden; and about 30 per cent of these are less than 50 hectares. They also provide a major part of Sweden's timber needs. In Canada, while production on private lands has grown, the great majority of timber still comes from provincial forests, and it always will.

Swedish and Finnish forestry investment levels (regeneration, tending and harvesting techniques, and worker training) are relatively high, much higher than in Canada. The woods also hold a special place in the psyche of each nation. Public access, what the Swedes call *allemansrätten*, is an old concept simply meaning the right of access to the land. Except for an area nearest to a landholder's house and cultivated areas, anyone is free to traverse and enjoy the land of another, even to camp overnight. Intrinsic to this tradition is an appreciation and expectation of mutual respect. Stewardship is not only a policy concept, it is a cultural one. This is not to say that Scandinavian forestry is without its problems and controversies, but a strong culture of stewardship has endured and their forest sectors have created firms that are larger than their Canadian equivalents.

Many who work in Canada's forest industry would say that a culture of stewardship also exists here. Workers care about sustaining the resource on which their livelihoods depend. They might also suggest that it's really the short-term vision of companies and governments that limit the potential for such a culture to really flourish. But industry counters that tenures are unstable, the times too short, or conditions too uncertain to see a significant increase in investment levels, to develop more non-timber forest products, or to follow forestry practices that clearly acknowledge the services that forests provide beyond timber. And despite a tenure model that has supported large companies, Canada has not been able to create the large, globally dominant firms that the Scandinavians have.

A solution offered at various times in Canada has been to emulate the Scandinavian model. Large companies might be more willing to invest in innovative forestry practices if they owned the land. But there would not be much public support for selling provincial forests, certainly not to large companies, nor would many large firms necessarily want to buy forests or have the financial resources to do so. Another option is to increase the role of communities through community-held tenures. Yet another possibility would be to offer better tenure opportunities for small firms and individuals, perhaps creating many small private forests even if only leased from the province. But would smallholdings, community forestry, or corporate ownership result in better forestry practices?

Experience from the US and Canada shows that some companies with private forests are not always good forest managers. Some have logged their lands quickly for short-term profit and made few investments in regeneration. But others treat their forests as the foundation of their long-term survival. Community tenures may also result in a more stable, long-term vision of forest management. Alternatively, some communities will support short-term timber production as the way to realize immediate employment and prosperity. Some individuals may also log their lands quickly for the sake of short-term profit. Small private holdings would require capacity-building and stable investment sources, but they offer the greatest promise and may be the best hope for reforming the industry and enhancing community sustainability. The corporate model has not worked. In Scandinavia, governments act to blunt some negative tendencies by regulating private forestry, requiring forest plans, setting cutting rates, and financing forestry renewal and management, all while supporting a strong private forest context.

In Canada tenure reform will require a careful consideration of the lessons learned from other places. While we can look to other jurisdictions for information and experience, Canada's forests, geography, history, and culture are distinct. As part of addressing Canada's forest management problems, tenure reform will require innovative approaches and leadership. The answers ultimately lie in creating a context that provides new opportunities for individuals, small firms, and stronger and more effective community-based tenures—all conditions that can encourage stable long-term business investments and support a more complex vision of what forests provide. Tenure reform must be part of realizing more sustainable forests and sustaining and growing our forest communities.

Kevin Hanna teaches in the Geography and Environmental Studies Department at Wilfrid Laurier University in Waterloo, Ontario. His main research interest is community-based resource management.

Rate of Conversion

The rate of conversion of natural to managed forests is one of the most controversial issues in Canadian forestry. Each provincial government establishes an **annual allowable cut** (AAC), which is based on the theoretical annual increment of merchantable timber, after taking into account factors such as quantity and quality of species, accessibility and growth rates, and amounts of land protected from harvesting because of other use values, such as parks and wildlife habitat (Table 9.2). The AAC should reflect the **long-run sustained yield** of a given unit of land, or what that land should yield in perpetuity. This target is ultimately limited by the growth conditions, the biological potential of the site, and how that potential can be augmented by silvicultural practices. It is not sustainable to have an AAC that consistently exceeds this biological potential. Economists, however, often argue for the need to maximize the monetary return of the first cut in order to invest in other wealth-producing programs and to provide social services. The dominance of this line of thought has led to rates of conversion significantly higher than can be supported biologically.

To calculate AACs, it is also necessary to know the rotation period for each forest type. This is the age of economic maturity of the tree crop and varies widely but usually falls within the 60- to 120-year range in Canada. Foresters call this the **culmination age**.

The AAC will also vary substantially depending on the proportion of old-growth to second-growth timber included in

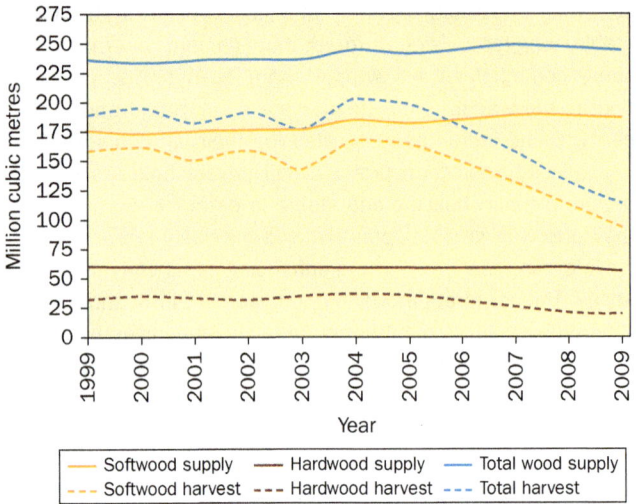

Figure 9.3 | Annual harvest versus supply deemed sustainable for harvest. Source: *Natural Resources Canada, (2011: 33)*.

the proposed cutting unit. Old-growth forests have very high timber volumes—for example, up to 800 cubic metres per hectare in BC's coastal forests. However, at the culmination age for **second growth** on these sites, volumes will be much lower, in the region of 500 cubic metres. This is known as the **falldown effect** and results in AACs up to 30 per cent lower as old-growth forests are eliminated.

For Canada, the total AAC is calculated by adding together all the provincial and territorial AACs where these figures are available. In 2009, the AAC stood at 207 million cubic metres, yet only 98 million cubic metres of timber were harvested. The long-term harvesting trends can be seen in Figure 9.3 and show the overall reductions in harvesting since 2004 as a result of the slowdown in the US housing market. Canada's wood supply is estimated by combining provincial AACs with the wood supply estimated for private, federal, and territorial lands. Harvest levels on private, federal, and territorial lands are not regulated by legislation, although the managers of these lands often set harvest targets. This means that wood supply from these lands can only be estimated based on the sum of these targets and, for lands where targets have not been set, the average of past harvest levels.

Silvicultural Systems

Silviculture is the practice of directing the establishment, composition, growth, and quality of forest stands through a variety of activities, including harvesting, reforestation, and site preparation.

Harvesting Methods

Perhaps no aspect of resource or environmental management has created as much conflict in Canada as the dominant forest harvesting practice of **clear-cutting**, and this method is consistently used on much of the forest lands harvested in Canada (Figure 9.4). The size of the clear-cuts varies widely, from approximately 15 hectares to more than 250 hectares, and in some cases clear-cuts extend for many thousands of hectares. Although openings of larger size have few ecological advantages, it should not be assumed that more but smaller clear-cuts are necessarily superior to fewer, larger clear-cuts. More clear-cuts create more fragmentation and less undisturbed forest area, to the detriment of 'interior' forest species.

Not only are clear-cuts aesthetically unappealing to many Canadians, but their environmental impact, especially cumulatively as they spread across the landscape, can be substantial. For example, Nova Scotia has implemented guidelines for the establishment of wildlife corridors to reconnect habitat. If clear-cuts are in excess of 50 hectares,

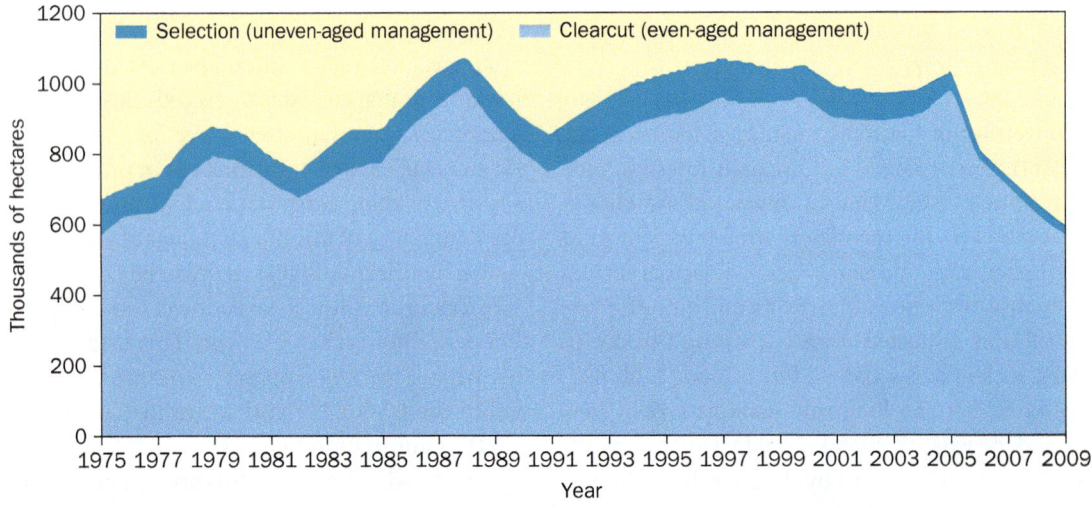

Figure 9.4 | Silviculture: Area harvested, 1975–2009. Source: *National Forestry Database, Silviculture—National Tables, at: nfdp.ccfm.org/silviculture/national_e.php.*

The rate of conversion from natural to managed forests has alarmed many environmentalists, who claim that forestry companies have been allowed to extract too much wood too quickly. This will result not only in environmental problems but also in a lack of adequate fibre for industrial use in the future.

the guidelines recommend that at least one corridor be created, with irregular borders and a minimum width of 50 metres. Furthermore, in 2010, Nova Scotia announced that whereas clear-cutting currently accounted for 95 per cent of the harvested area, this would be reduced to 50 per cent over the next five years. These changes are a step in the right direction, but it is the implementation of such guidelines, not just their specification, that is important. British Columbia tried to take a tougher approach by legislating forest practices in detail through the Forest Practices Act. Yet current government policy has seen considerable weakening in the implementation of the legislation as economic conditions have become less favourable to the forest industry.

Clear-cutting is the preferred means of harvest on vast areas. It is the most economical way for the fibre to be extracted and also allows for easier replanting and tending of the regenerating forest. In certain types of forests, it may mimic natural processes more closely than selective or partial cutting systems. This is especially true where natural fires have created even-aged stands of species such as lodgepole and Jack pine, black spruce, aspens, and poplars. Researchers working in the eastern boreal forest, however, have been questioning this assumption (Bergeron et al., 2001; Lesieur et al., 2002). They have found a dramatic decrease in fire frequency in this area over the past 150 years as a result of climatic change, and they suggest that this trend will continue into the future, leading to a higher proportion of old-growth forest in the landscape. In turn, this will lead to natural species replacement in these forests, with deciduous and mixed stands replaced by balsam fir and Jack pine by black spruce. Clear-cutting would counteract these changes rather than mimic natural processes and would lead to a dramatic decrease in stand diversity at the landscape level. These authors suggest that, in the future, forest management will have to employ mixed harvesting systems (Table 9.2) if forest management is to more closely emulate natural systems.

Reforestation

Until 1985, Canada's forests were considered to be so extensive that little effort was given to reforestation. Sites, once logged, might be burned to facilitate rapid nutrient return to the soil but then were abandoned in the hope that they would be satisfactorily recolonized by seeds from the surrounding area. Sometimes this was successful, but often it was not. Thus, with increased harvesting levels, the amount of land that no longer supported trees that could be harvested in the future gradually grew. In 2009, 0.6 million hectares of land were harvested, 15.2 million hectares were defoliated by insects, 3.2 million hectares were burned by fire, and yet less than 450,000 hectares were planted and reseeded (Figure 9.5), and the success of these plantings cannot be guaranteed. Given this annual deficit between what is cut and what is replanted, many conservationists have difficulty understanding further allocation of old-growth timber to the forest industry.

Site Preparation—Biocide Use

Biocides are used on forest lands in Canada to reduce competition for seedlings on replanted sites and to protect seedlings from insect damage. Sites regenerating from forest harvesting return to an earlier successional phase (Chapter 3), and under natural conditions a vigorous and diverse secondary succession takes place. However, for many years the community will not be dominated by the commercial species desired by foresters. Chemicals are used to suppress early successional species, to compress the successional time span, and to maximize the growth potential of the more commercially desirable species, usually conifers. Chemical use is generally quicker, easier, and more effective than using mechanical

A clear-cut area on Vancouver Island.

Table 9.2 | Main Characteristics of Common Silvicultural Systems Practised in Canada

Clear-cutting

The most commonly applied silvicultural system in Canada, clear-cutting involves the removal of all trees in a cutblock, in one operation, regardless of species and size. Some trees are left along riparian zones to protect streams. The objective is to create a new, even-aged stand, which will be regenerated naturally or through replanting.

Advantages	Disadvantages
• Cost-effective. Clear-cut areas are easily accessed for site preparation and tree planting. • Stands of even-aged trees are created, producing wood products with more uniform qualities. • Newly planted seedlings quickly take root and grow in the sunlight reaching the ground. This can benefit certain animal species. • In some respects, clear-cutting simulates natural disturbances such as wildfire and insect disease outbreaks. • Safest harvesting method with least risk of worker injury.	• Nutrients stored in the bodies of trees are removed from the ecosystem. • Loss of habitat for some wildlife species. Loss of biodiversity. New vegetation does not maintain the complexity and stability of mature forests. • Clear-cutting in sensitive ecosystems can cause soil erosion, landslides, and silting, which can damage watersheds, lead to flooding, and inhibit successful fish reproduction. • Large gaps are opened up, fragmenting the forest and exposing more area to the edge effect. • Aesthetically unattractive. Clear-cutting can conflict with other forest values. • No timber products for a long period of time (e.g., 50 to 70 years).

Seed tree

Method of clear-cutting in which all trees are removed from an area in a single cut, except for a small number of seed-bearing trees, which are intended to be the main source of seed for natural regeneration after harvest. Tree species that have been managed under this system in Canada include western larch, Jack pine, eastern white pine, and yellow birch.

Advantages	Disadvantages
In addition to the above: • Next to clear-cutting, this system is the least expensive to implement. • The system can result in improved distribution of seedlings and a more desirable species mix, since the seed source for natural regeneration is not limited to adjacent stands.	In addition to the above: • Regeneration can be delayed if seed production and/or distribution are inadequate.

Shelterwood

Mature trees are removed in a series of two or more partial cuts. Residual trees are left to supply seed for natural regeneration and to supply shelter for the establishment of new or advanced regeneration. The remaining mature cover is removed once the desired regeneration has been established. Thirty to 50 per cent canopy removal on the first cut is common. In Canada, this system has mainly been applied to conifers (e.g., red spruce in the East, white pine in Ontario, interior Douglas fir in the West).

Advantages	Disadvantages
• Trees left after the first cut grow faster and increase in value. • For eastern white pine, this system can be an effective management tool against white pine weevils, which are more attracted to pine shoots under full light exposure than under shade. • Visually more appealing than clear-cutting.	• Complex and costly to plan and implement. • Young trees can be damaged during removal of mature trees. • Windthrow is a serious concern. Uprooted stems can displace significant amounts of soil.

Selection

Involves the periodic harvest of selected trees of various ages in a stand. Trees are harvested singly or in groups as they reach maturity. Valuable, mature trees, along with poorly shaped, unhealthy, crooked, and leaning trees and broken or damaged trees, are selected for removal. The objective of this method is to create and maintain an uneven-aged stand. Small gaps created by harvesting leave room for natural seeding.

Advantages	Disadvantages
• This method is often favoured in areas where recreation or scenic values are important, since the harvested area is less visually offensive. • The method results in a continuous, regular supply of mature trees over time. Overall stand quality should improve after each harvest cycle. • Biodiversity loss is minimized.	• This system can only be successfully applied to stands containing shade-tolerant tree species (e.g., sugar maple, western red cedar, red spruce, balsam fir, eastern/western hemlock). • Requires skilled workers to implement successfully. • Can require the maintenance of more roads and skid trails per unit area. • Complex and costly system to plan and implement. • In some instances, landowners take the best trees in a forest, leaving poorly shaped and unhealthy trees to provide seed for the next generation, resulting in forest degradation over the long-term.

Source: National Forestry Database (2005); Alberta Food and Rural Development (2001); Canadian Institute of Forestry (2003).

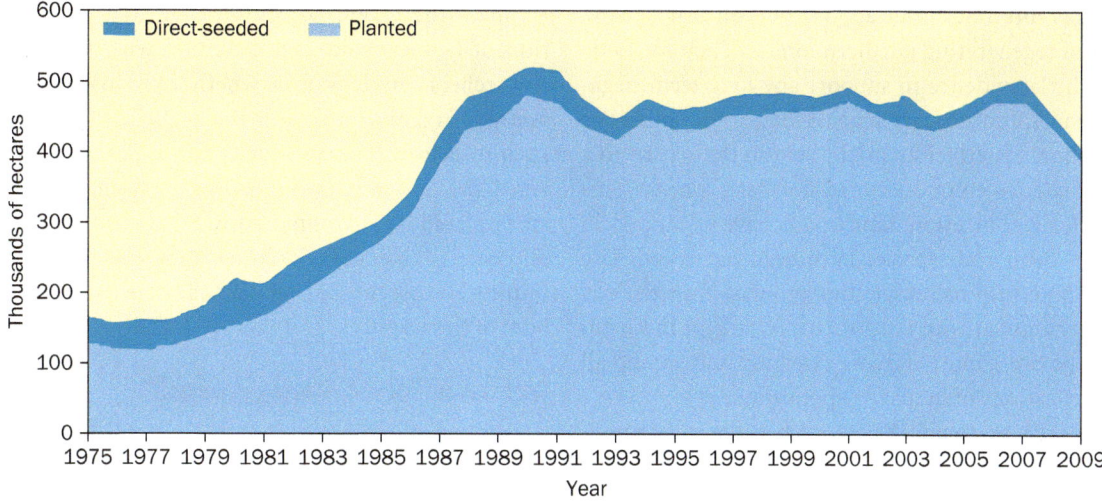

Figure 9.5 | Area planted or direct-seeded, 1975–2009. Source: National Forestry Database, Silviculture—National Tables, at: nfdp.ccfm.org/silviculture/national_e.php.

alternatives for weed suppression. Three herbicides (2,4-D, glyphosate, hexazinone) are registered for forest management in Canada. Glyphosate (or 'Roundup'), the most widely used, affects a broad spectrum of plants but degrades quickly and is relatively non-toxic to terrestrial animals.

Early colonizers often compete more effectively for soil nitrogen than conifers. Where nitrogen is the most limiting factor, as it is in the boreal forest, this can inhibit conifer growth over the short term. However, the law of conservation of matter (Chapter 4) tells us that these nutrients have not disappeared—they are simply being held by different species. As these species die and decay, the nutrients will be returned to the soil and become available for uptake. Furthermore, by holding nutrients in this way, early colonizers often slow down the loss of nutrients from the site that might otherwise occur from leaching. They act as a biological sponge over the short term. Herbicide application may also eliminate species that are ecologically advantageous, such as nitrogen fixers like the red alder, exacerbating nutrient loss from logged sites.

The balance between these effects needs to be evaluated over long periods and probably differs from site to site. Lautenschlager and Sullivan (2002) reviewed the literature

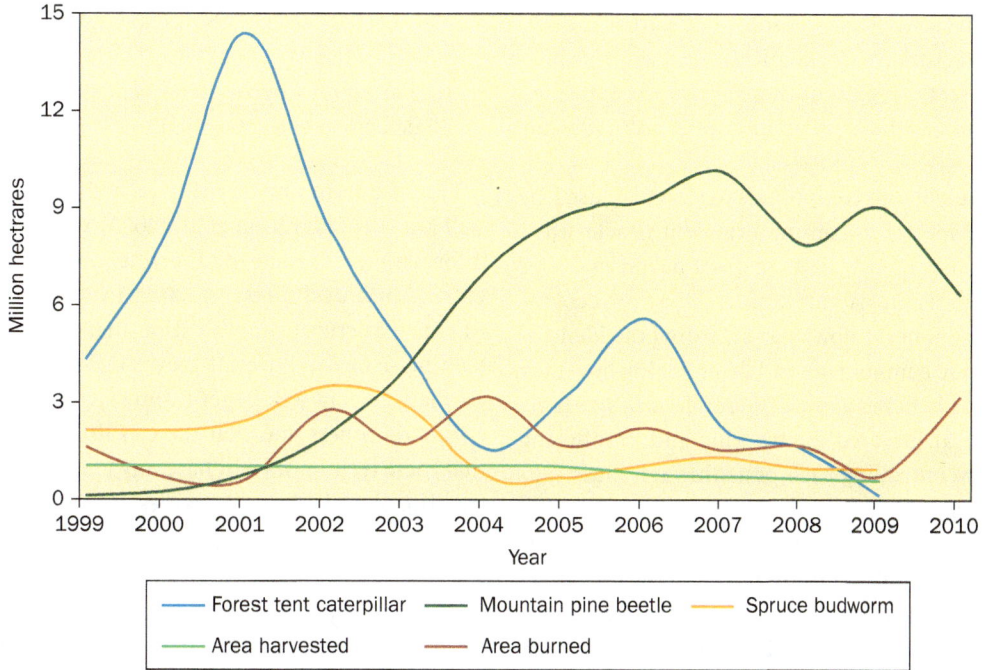

Figure 9.6 | Area of forest disturbed by fire, insects, and harvesting. Source: Natural Resources Canada (2011: 26).

on the effects of forest herbicide applications on major biotic components of regenerating northern forests. They conclude that there is little evidence to support any long-term negative consequences. Indeed, they suggest that, at the landscape scale, application of herbicides might help in the return to a more natural species composition in northern forests. This is because since colonization, hardwoods have expanded in these areas as coniferous species have been harvested. Use of herbicides may help to reverse this process. Nonetheless, Lautenschlager and Sullivan caution that large gaps in knowledge need to be overcome before we can truly understand all the biotic implications of herbicide application.

Insecticides are used to attack pests such as the spruce budworm, Jack pine budworm, hemlock looper, mountain pine beetle (see Box 9.7), gypsy moth, and forest tent caterpillar. The amount sprayed varies, depending on the population dynamics of these insects, which changes in response to environmental factors (Figure 9.6). Most spraying has occurred in eastern Canada; the spruce budworm spraying program in the Maritime provinces is the best-known incidence of such spraying and has caused considerable controversy (Box 9.5). However, the total area treated with forest chemicals in Canada has steadily declined since 1990 (Figures 9.7 and 9.8).

The health and ecological concerns arising from widespread application of chemicals (discussed further in Chapter 10) have led to several high-profile confrontations. Attention is being increasingly directed towards the replacement of synthetic insecticides with biological control agents such as *Bacillus thuringiensis* (Bt). Bt now accounts for almost all the insecticide used in Canadian forests and is non-toxic to humans and most wildlife, although it does affect moth and butterfly larvae of some non-target species. Plants also manufacture many chemicals themselves as protection against insects. Several of these chemicals appear to be good prospects for the development of insecticides for forestry use.

Pheromones are volatile compounds, or 'scents', used by insects of a given species to communicate with each other. Various types of pheromones serve different purposes—e.g., alarm, aggregation, territorial marking, tracking or recognition, and sex pheromones. Knowledge regarding the identification of pheromone components and determining how they are used by insects is being applied to developing synthetic pheromones that can be used for tracking and monitoring insect pests or for controlling such pests through mass captures or mating disruption.

Installing a few pheromone traps as opposed to systematically sampling trees substantially reduces the time spent on tracking and monitoring. This technique also makes it possible to increase the area covered at a cost lower than or equivalent to that of traditional sampling. Tracking is used to determine whether an insect population has exceeded a critical threshold, whereas monitoring is designed to estimate the extent of the damage that the next generation of larvae could cause. Armed with this information, researchers can determine whether phytosanitary measures are required and can select the best time for implementing such measures. The most common types of traps are sticky traps, the capacity of which is limited to the size of the sticky surface, and large-volume traps. Sticky traps are inexpensive and easy to use. Large-volume traps have a receptacle containing an insecticide that kills the insects that gather at the base of the trap.

Intensive Forest Management

After a new crop of trees has been established and reaches a free-growing condition, future timber resource values can be further enhanced by intensive silvicultural practices. Activities undertaken by foresters to improve stand growth include:

- pre-commercial thinning (homogenizes the stand, increases mean tree size, and lowers the age at which the stand can be harvested);
- commercial thinning (attempts to recover lost volume production in a stand as a result of competition-induced mortality);
- scarification (physical disturbance of the forest floor to create improved seedbeds for natural regeneration);
- prescribed burning (removes slash and woody debris, sets back competing vegetation, provides ash as fertilizer, and increases nutrient mobilization and availability through increased soil temperatures);
- pruning and shearing (increases the value of individual trees by prematurely removing the lower branches so that clear wood, free of knots, is laid down around an unpruned knotty core);
- timber stand improvement (cutting down or poisoning all deformed and unwanted trees within older stands).

The long-term impacts of these activities are generally not well known. Thompson et al. (2003) reviewed the published information on some of the likely impacts on vertebrate wildlife in boreal forests in Ontario. One important impact was the structural simplification that occurs with increased tending. However, overall they conclude that intensive forest management will benefit some species and have negative effects on others and that there is little evidence to suggest that these techniques will have any greater impact than more traditional silvicultural activities.

Fire Suppression

In certain areas, fire is a frequent occurrence and necessary to the reproduction of forest tree species. Fire is part of the long-term dynamics of these ecosystems. Fire initiates secondary successions, renewing vegetation through regeneration involving a complete change of species, regeneration of

Environment in Focus
Box 9.5 The Spruce Budworm Controversy

Since 1952, aircraft have been dousing the forests of eastern Canada, particularly New Brunswick and Nova Scotia, with an array of chemicals in the competition to see who would harvest the area's lumber—humans or the eastern spruce budworm. A native in Canada, the budworm feeds primarily on balsam fir but will also eat white spruce, red spruce, and, to a lesser extent, black spruce. Its range extends wherever balsam fir and spruce are found, from Atlantic Canada to the Yukon. Damage varies considerably from one stand to another. Mortality rates are related to stand composition and age, as well as site quality (soil, water, climate). Ordinarily, the influence of the insect goes unnoticed in a forest, but its impact occasionally reaches epidemic levels, with massive damage to the host trees. Outbreaks last between six and 10 years and have been documented for the past two centuries as the product of a long-term budworm–fir ecological cyclic succession. However, spruce budworm epidemics are occurring with increasing frequency as a result of human intervention related to forest harvesting. Human intervention has reduced the natural diversity of the forest through removal of preferred species such as white pine, creating a less diverse forest composed of large areas of mature balsam fir, the budworm's preferred food. Extensive mortality occurs in stands that have suffered defoliation for several years. In 1975, at the height of infestation, 54 million hectares of forest were defoliated, resulting in serious economic losses for the forestry industry.

Population levels of the spruce budworm have declined, following a downward trend that started in 2003 and a precipitous drop in 2004. The area affected also declined, from 755,325 hectares in 2004 to 696,483 in 2005. For the past several years outbreaks have been relatively small and localized. Manitoba had approximately 100,000 hectares infected from 1997 until a sharp drop in 2011. The spruce budworm is still present in central Canada, and in 2009 resulted in the defoliation of about 1 million ha (Figure 9.6). A 2006 outbreak in Quebec had expanded to more than 600,000 hectares by 2010 and was expected to grow. This trend of increasing infestations is expected to continue over the next five to 10 years as habitat and environmental conditions become favourable.

The long-term ecological stability of the budworm–fir system does not match the shorter-term demands of the economic system dependent on the forests for products. As a result, the forests have been extensively sprayed to limit defoliation and mortality. When spray programs started in 1952, DDT was the chemical of choice, and some 5.75 million kilograms were sprayed in New Brunswick alone between 1952 and 1968, when use was suspended. Other chemicals replaced DDT, such as phosphamidon, aminocarb, and fenitrothion, until questions were raised about their ecological and health impacts. Phosphamidon, for example, is very toxic to birds. Up to 1985, 118.5 million hectares (mostly in New Brunswick) were sprayed, some areas on an annual basis. Continual spraying appears necessary once natural controls are disrupted.

The organophosphate fenitrothion then became the most popular chemical. It also became the source of heated controversy regarding its health and ecological impacts. In particular, there was concern over the link between the chemical and Reye's syndrome, a rare and fatal children's disease. As a result, and because of unfavourable reviews of the ecological impact of the chemical, particularly on songbirds, the use of fenitrothion in aerial applications was cancelled by 1998. The biological control *Bacillus thuringiensis* (Bt) is now being used more extensively. In some areas, it is supplemented by an insecticide called Mimic that kills the budworm through interruption of the moulting process, starving the larvae to death. Although Mimic is not a broad-spectrum biocide, it has the same lethal effect on the larvae of butterflies and moths. Nonetheless, it has been used in some areas, such as Manitoba, since 1998.

Attention is also being devoted to other less toxic approaches, such as species and landscape diversification, so that large stands of mature balsam fir do not dominate the landscape. Biological control is also being investigated with a parasitic wasp that attacks the larvae of the budworm.

Western provinces also use chemicals to try to limit damage from the spruce budworm. Manitoba, for example, uses dimethoate and malathion, and Bt is commonly used in Saskatchewan.

Management of spruce budworm outbreaks provides a graphic example of the challenges presented by the conflict between longer-term ecological cycles and shorter-term economic dependencies.

Spruce budworm.

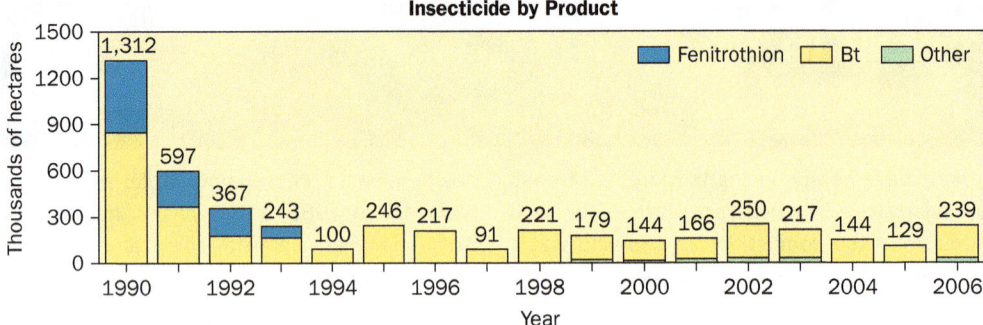

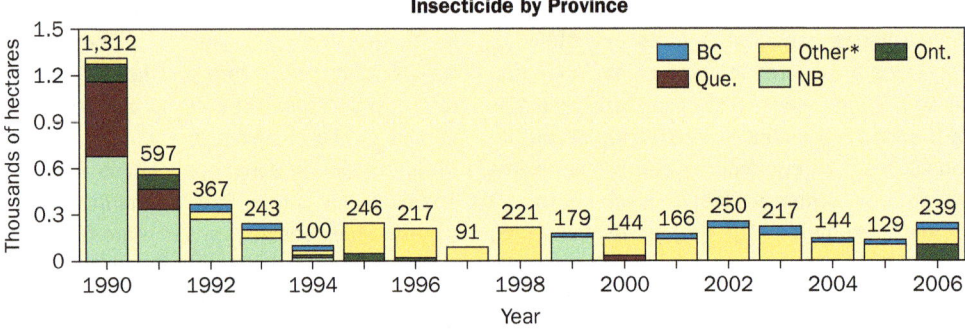

Figure 9.7 | Forest insecticide use by product and province, 1990–2006. Source: National Forestry Database.

the same species, or diversification of the species. Jack pine, birch, and trembling aspen are common in areas where fires have recently occurred. Fire suppression, viewed as essential to protect lives, property, and commercially valuable timber, has resulted in ecological changes not characteristic of fire-dominated ecosystems. For example, it has contributed to the very dense regeneration of almost pure Douglas fir in old-growth ponderosa pine stands in the interior of BC. Old-growth ponderosa pine stands are maintained by low-intensity, naturally occurring surface fires that burn brush and prevent maturation of the more shade-tolerant Douglas fir.

In the absence of recurring fires, ground fuel may accumulate, increasing the risk of a major wildfire event. The area affected by wildfires in Canada each year is immense: during the 1990s, an average of 8,248 fires burned 3.2 million hectares annually, including more than 700,000 hectares of commercial forest land, or 74 per cent of the annual area harvested (Neave et al., 2002). In 2009 there were 7,167 fires, burning 0.8 million hectares, the fifth lowest year for area burned since 1970. The area burned was not evenly distributed, with BC and Yukon accounting for 60 per cent (Figure 9.6). However, in 2010 nearly 3 million ha burned, almost double the 10-year average (Natural Resources Canada, 2011). Over half of this total was in Saskatchewan, with 90 per cent north of Churchill River.

Environmental and Social Impacts of Forest Management Practices

Change in species and age distributions arising from forest management practices has a major impact on ecological processes such as energy flows, biogeochemical cycles and the hydrological cycle, and the habitat for other species. We have only a rudimentary knowledge of how forest ecosystems function. It is therefore difficult to be precise about the possible impacts of wholesale conversion from complex natural to more simple human-controlled systems. In addition, important differences occur among forest ecosystems. Some, such as the boreal forest, have naturally evolved with periodic disturbances such as fire or insect attack that stimulate forest renewal. Others, such as the rain forests of the west coast, have little history of disturbance. The difference between disturbance through forestry and natural processes (Box 9.6) must be considered against this background. One essential difference is that natural disturbances such as fire or insect attack do not result in the physical removal of the biomass from the site; it is merely converted from one form to another at that site, consistent with the law of conservation of matter (Chapter 2). In contrast, logging results in the physical translocation of nutrients from the site. The closer that forest

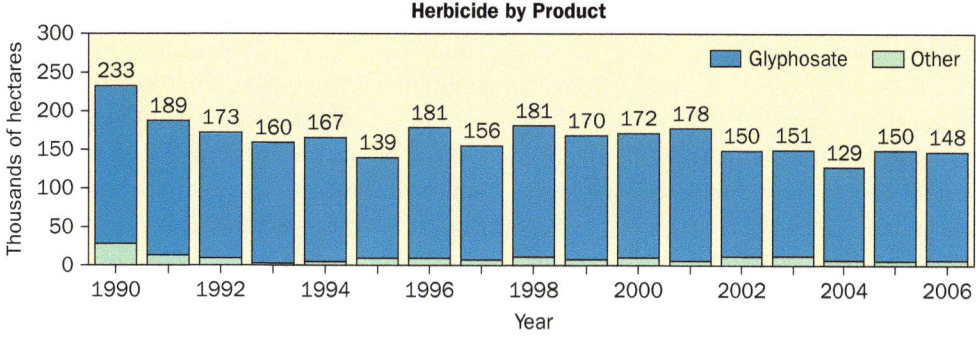

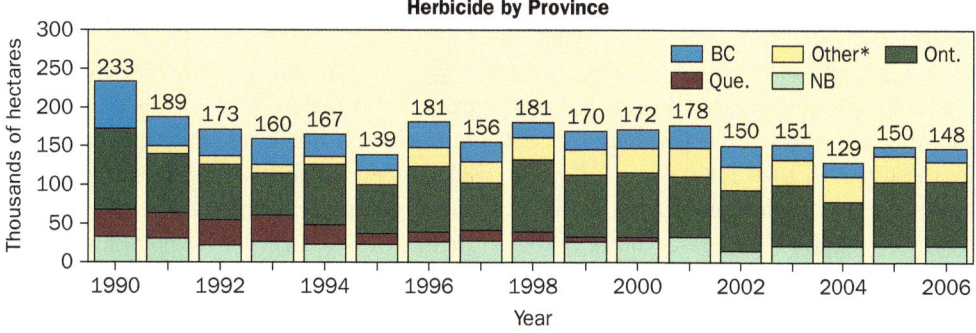

Figure 9.8 | Forest herbicide use by product and province, 1990–2006. *Source: National Forestry Database.*

harvesting approximates the conditions of natural perturbations, the less disturbing it will be to ecosystem processes.

Chapter 2 described how energy flows through ecosystems and is stored in various compartments. Trees, for example, represent energy stored in the autotrophic component of the ecosystem; deer represent storage in the herbivorous component. Thus, a forester may wish to maximize the energy storage in trees and minimize energy losses to herbivores, whereas a wildlife or recreation manager may wish to move the energy storage further up the food chain to support more wildlife. Conflicts therefore arise about where energy should be stored in ecosystems to optimize societal values. This section describes some of the environmental and social implications of forestry activities.

Forestry and Biodiversity

Most natural forest land is dominated by forests with old-growth characteristics, although the age of the trees and degree of structural and compositional attributes described for old growth vary greatly across Canada. Old-growth forests have ecological attributes that tend to be absent from forests that have been harvested.

Various definitions of old-growth forests have emerged. This one was suggested by the Forest Land Use Liaison Committee of British Columbia: 'Old-growth forests include climax forests but do not exclude sub-climax or even mid-seral forests.' The age structure of old growth varies significantly by forest type and from one biogeoclimatic zone to another. The age at which old-growth forests develop their characteristic structural attributes varies according to forest type, climate, site characteristics, and disturbance regime.

New forestry would devote much more attention to stand tending on commercial sites, such as this birch stand, where limbing and thinning enhances growth.

However, old growth is typically distinguished from younger stands by several of the following attributes:

- large trees for species and site;
- wide variation in tree sizes and spacing;
- accumulations of large dead, fallen, and standing trees;
- multiple canopy layers;
- canopy gaps and understorey patchiness;
- decadence in the form of broken tops or boles and root decay.

Old-growth forests typically contain trees that span several centuries. Based on age, approximately 18 per cent of Canada's forests can be classified as old-growth. In Canada, the longest-lived tree species are yellow cedar on the west coast and eastern white cedar in central Canada. Trees of both these species may live more than 1,000 years.

Old-growth forests supply high-value timber, contain large amounts of carbon, contain a large reservoir of genetic diversity, provide habitat for many species, regulate hydrologic regimes, protect soils and conserve nutrients, and have substantial recreational and aesthetic values.

Logged forests undergo a number of changes, including modifications of the physical structure of the ecosystem and changes in biomass, plant species mixtures, and productivity. These changes affect biodiversity directly and indirectly by changing the nature of the habitat. Since more than 90,000 species depend on forest habitats in Canada, this is obviously of concern. About 65 per cent of species considered at risk of endangerment in Canada are forest-related species (Natural Resources Canada, 2005b). The largest concentrations of these species are in the Carolinian forests of southern Ontario and the coastal forests of BC.

Direct changes arising from forestry practices include the effects on the genetic and species richness of a biotic community. Most species contain a wide range of genetic variability that helps them adapt to changes in the environment. As natural forests are replaced by plantations, this natural variability is reduced, since most plantation-grown trees are specially selected from the same genetic base to have desirable characteristics. This makes them more susceptible to pest infestations and disease and less able to adapt to future environmental changes (Box 9.7). For example, it has been suggested that the emergence of the spruce forest moth in central and eastern Canada since 1980 is at least partly due to the establishment of white spruce plantations. More than 85 per cent of the forest harvesting in Canada is done by clear-cutting. This does not destroy the ecosystem per se in that an ecosystem still exists on that unit of land, but it does dramatically alter the attributes of that ecosystem. For vegetation, changes include removal of the previously dominant trees and their ecological influence, followed by the vigorous growth of other assemblages of plants (where herbicides have not been applied) as the successional process starts again.

Early successional species dominate in the immediate post-harvesting phase but are much reduced over time as the canopy closes. Tree species such as alder, birch, cherry, Jack pine, poplar, and aspen often fit into this category, along with semi-woody shrubs such as elderberry and blackberry, annual and short-lived perennial herb species such as members of

Environment in Focus

Box 9.6 Forest Disturbance: Natural versus Clear-Cut

Some forests are more susceptible to disturbance than others, and different disturbances have differing impacts. There are similarities between some disturbances and clear-cutting, but important differences also exist. Some of the differences between the effects of fire and clear-cutting are:

- Openings created by fire are generally irregular in shape, with high perimeter-to-edge ratios that facilitate natural reseeding. Boundaries tend to be gradual rather than the abrupt edge of a clear-cut.
- Fires leave standing vegetation in wet areas that continues to provide habitat for wildlife and acts as a natural seed source. Clear-cuts remove all trees.
- Fires tend to kill pathogens; clear-cutting allows many pathogens to survive.
- Fire releases nutrients into the soil; clear-cutting removes nutrients in the bodies of the trees.
- Fire helps to break up rock that aids in soil formation; clear-cutting tends to physically disturb the site, leading to compaction and erosion.
- Fire stimulates growth of nitrogen-fixing plants that help to maintain soil fertility; clear-cutting does not.
- Fire encourages the continued growth of coniferous species in many areas through stimulating cone opening; clear-cutting often leads to dominance by shade-intolerant hardwoods. Ultimately, this changes species composition, as found in studies in Ontario where, in 1,000 sampled boreal clear-cut sites, regenerating poplar and birch had increased by 216 per cent and spruce had fallen by 77 per cent (Hearnden et al., 1992).

the aster family, and various grasses and sedges. Other species that existed in low abundance in the original forest may survive the harvest and, freed from the competitive suppression of the harvested trees, may dominate the community for some time. Trees such as red maple, yellow birch, and white pine often fall into this category in central and eastern Canada. As the canopy closes, most of these species will eventually be out-competed. Some species that survive the harvest or invade from surrounding areas may be found through all stages of succession. They have low light compensation thresholds, allowing survival under heavy shading. Balsam fir, hemlock, sugar maple, and beech fall into this category.

Once clear-cuts have had the opportunity to start regenerating, species diversity increases rapidly and usually results in a plant community that is more species-rich and diverse than the harvested community. The exception is where replanting takes place, which involves few species and where steps are taken to reduce competition for plantation trees. Herbicides such as Roundup are commonly applied to achieve this result, and may be applied consistently until the planted trees become established. Under these circumstances, an artificial lack of diversity is created, just as a farmer creates a similar system to maximize the amount of energy stored in the particular component that he or she wishes to harvest (Chapter 10).

Environment in Focus

Box 9.7 The War against the Mountain Pine Beetle—Who's Winning?

The outbreak of mountain pine beetles in the interior forests of BC is one of the most dramatic changes ever to happen to BC's forest landscape. The mountain pine beetle epidemic has spread throughout BC's range of lodgepole pine forests as a result of a combination of natural beetle population cycles, continuous mild winters, and an abundance of uniformly mature pine stands. Since 1997, mountain pine beetles have infested more than 9.2 million hectares of pine forests in BC, totalling some 726 million cubic metres of wood by 2011. Defying predictions, the beetles have spread as far east as Great Slave Lake in Alberta, and by 2007 an estimated minimum of 3 million trees in Alberta had been affected, up from 20,000 in 2005. The extent in BC is more than double that predicted by the models in 2001, and there is no sign of a slowing down of the invasion. By 2013, more than 80 per cent of the mature pine will have been consumed, and it is likely that the epidemic will only be over once it has infested most of the mature pine in BC. There is no doubt who is winning the war in BC. By 2010 more than 700 million cubic metres had been killed, and expectations are that this will reach the 1 billion mark before the infestation subsides (Figure 9.6). This amounts to three-quarters of BC's pine trees and as much as $30 billion lost in forest products. However, an even greater cost may be the impacts on carbon release and the contribution to global warming. In the period 2000–2020, it is estimated that the affected forests will release 270 Mt carbon per year to the atmosphere, equivalent to 75 per cent of the average annual forest fire emissions from all of Canada in 1959–1999 (Kurz et al., 2008).

There is now concern that the beetles may be able to spread all across the boreal forest. As the beetle spread eastward from central BC, it adapted to Jack pine from its main host, lodgepole pine. Jack pine is a main component of the boreal forest, creating the potential for the problem to become nationwide.

Bark beetles are small, cylindrical insects that attack and kill mature trees by boring through the bark and mining the phloem (the layer between the bark and wood of a tree). The mountain pine beetle and other bark beetles are native species and natural and important agents of renewal and succession in the forest of the BC Interior. However, when the beetles reach epidemic levels, natural predators like woodpeckers cannot reproduce quickly enough to maintain the insect population at manageable levels.

The dead and dying trees, visible from the air as huge swaths of red and grey forest throughout central BC, are the source of a hotbed of controversy as government agencies try to combat the rate of infestation through increases in the annual allowable cuts for some areas, reductions in environmental regulations and planning requirements on treatment units, and reduced stumpage (fees paid for logging on public lands). Beetle-infested trees retain economic value and can be salvage-harvested, but since wood quality and value decline with time since attack, there is significant pressure to cut beetle-infested timber promptly to maximize the economic value of salvaged trees. The recommended size of clear-cut areas is set at 60 hectares in the BC Interior, but exemptions for salvage loggers have resulted in cutblocks as large as 1,300 hectares, more than 20 times the recommended maximum. The salvage clear-cuts can be placed adjacent to previously logged areas that have not yet regenerated, creating potentially very large contiguous openings. Furthermore, little attention has been devoted to the hydrological implications of suddenly deforesting large areas in the Interior, and now greater flood damage is predicted as a result.

Accelerated cutting of large attacked areas can prevent the decline of timber value in the short term, but a number of social and environmental costs are associated with this approach. Increases in AACs in response to insect infestations can disrupt forest plans, oversupply markets, cause

Mountain pine beetle larvae and adult in pine tree bark, Smithers, British Columbia.

Forest in central British Columbia infested by mountain pine beetles.

short-term decreases in timber value, and affect employment levels when workers are no longer needed to support the temporary increase in harvest levels.

Research in forest ecology has increasingly recognized the essential role that natural disturbances play in shaping forest structure and maintaining ecosystem processes. Small infested areas create patchiness at the stand level, increasing local diversity, and generate habitat and food for a large number of species. In addition, outbreaks accelerate forest succession (Chapter 3) to other tree species in some areas, as well as creating the conditions for stand-replacing fires that maintain the dominance of lodgepole pine.

The approach of the BC Ministry of Forests has received considerable attention from environmental groups, which view the response of government agencies as drastic, particularly in light of a government announcement that BC parks and protected areas would not be exempt from the war against pine beetles. Environmentalists generally advocate natural control methods. Cold winters or fires will kill mountain beetle larvae, reducing the size of beetle populations. However, winters are now warmer than ever before, and many scientists point to the beetle outbreak as an indicator of the kinds of disturbance that will happen more commonly in the future (Woods et al., 2010).

In a further implication, in 2011 the US accused Canada of violating the 2006 softwood timber agreement by allowing lumber companies to sell off vast amounts of timber at cut-rate prices. Lumber that prior to the infestation would be sold to mills for as much as $18 per cubic metre was sold for as little as 25 cents. This resulted in falling lumber prices all across North America. Although Canada claims the sales were only of damaged timber, the US claims that many good logs were included in the sell-off and their mills could no longer compete. Despite nearly a century of actively managing the mountain pine beetle, efforts to suppress the outbreaks across BC and other parts of North America have been largely unsuccessful. Why do you think this is? What approach would you recommend, given the value of lodgepole timber stands? Do you think provincial parks should be open to clear-cutting as a response to the beetle infestation?

Naturally regenerating clear-cuts may also produce a higher biomass of herbivorous species such as white-tailed and mule deer that require brushy habitats for at least part of the year. Both the quantity and quality of browse is greater in regenerating clear-cuts up to the time that the canopy starts to close and is usually optimal during the first eight to 13 years. In many parts of their range, white-tailed deer are more abundant than they were before European colonization when the landscape was covered mostly in mature forest. However, even these species benefit most from a pattern of small clear-cuts, since they prefer edge habitats where the protective cover of the forest is not too far distant. Optimal clear-cut size for deer in southern Ontario, for example, is around two hectares.

Unharvested forests in most areas across Canada comprise a patchwork of forest stands of different ages and different diversities regenerating from the effects of various natural disturbances such as fire and insect attack. Clear-cutting at a certain rate and scale may not be inappropriate in some of these ecosystems. However, some forests, such as the coastal forests of BC and the mixed deciduous forests of southern Ontario and Quebec, are heavily influenced by the pattern of death of individual canopy trees. For example, more than half of the coastal rain forest is more than 250 years in age, and much smaller interventions are required to mimic natural processes.

Some animal and bird species require forests with old-growth characteristics—for example, ample lichen growth for woodland caribou or the presence of dead trees to provide nesting cavities for birds such as woodpeckers (Table 9.3). If forest harvesting takes place at a rate and scale that eliminates stands with these characteristics, these species will decline in numbers and may become extirpated. The case study presented in Box 9.8, the spotted owl, is a good example of the difficulties associated with maintaining populations of species dependent on old-growth habitat. Other species that have suffered in this regard include the woodland caribou, American pine marten, and marbled murrelet. There are 348 forest-associated species currently on the at risk list in Canada (see Chapter 14).

The pressure on high-profile species such as the grizzly bear (Box 9.9) is not the only concern in considering the destruction of old-growth forests; less well-known species, even unknown ones, and the kinds of ecological functions undertaken by these species are also at risk as the old-growth forests are cut away. Some 85 per cent of the species in some forests are arthropods such as insects and spiders. Only recently is the richness of this fauna being realized. There may be more than 1,000 species of invertebrates within a single forest stand. Many of these species are new to science, and we have little idea of their ecological role in maintaining healthy ecosystems. Specific spider species are associated with different stages of the successional process, and at least 30 years is required for the spider community to recover from clear-cutting. Studies of the upper branches of coastal forests in BC are revealing very complex

Table 9.3 | Examples of Forest-Dwelling Species at Risk

Mammals	Birds	Plants	Reptiles
Endangered			
American marten	Acadian flycatcher	American ginseng	Blue racer (snake)
Woodland caribou	Kirkland's warbler	Bashful bulrush	Night snake
Wolverine (eastern population)	Northern spotted owl	Blunt-lobed woodsia	Rocky Mountain tailed frog
Vancouver Island marmot	Prothonotary warbler	Cucumber tree	
	Western yellow-breasted chat	Deltoid balsamroot	
	White-headed woodpecker	Drooping trillium	
		Heart-leaved plantain	
		Large whorled pogonia	
		Nodding pogonia	
		Prairie lupine	
		Purple twayblade	
		Red mulberry	
		Seaside centipede (lichen)	
		Small whorled pogonia	
		Spotted wintergreen	
		Tall bugbane	
		Wood-poppy	
Threatened			
Ermine haidarum subspecies	Hooded warbler	American chestnut	Black rat snake
Pallid bat	Marbled murrelet	Deerberry	Blanding's turtle
Wood bison	Queen Charlotte goshawk	Goldenseal	Eastern Massasauga rattlesnake
		Kentucky coffee tree	Jefferson salamander
		Lyall's mariposa lily	Pacific giant salamander
		Phantom orchid	
		Purple sanicle	
		Round-leaved greenbrier	
		Scouler's corydalis	
		White wood aster	
		White-top aster	
		Yellow montane violet	

Sources: Natural Resources Canada (1995); Alberta Food and Rural Development (2001); Canadian Institute of Forestry (2003).

predator–prey relationships among many different species. These relationships are not replicated in managed forests, suggesting that continued elimination of old-growth habitat will lead to species extinctions, a decrease in genetic diversity in these communities, and removal of natural controls on forest pests.

One kind of impact on biodiversity that is often overlooked results from human intrusion to undertake silvicultural and other activities. For example, after looking at the impact of forestry on grizzly bears in the Selkirk Mountains of BC, Wielgus and Vernier (2003) reported that the main impact seemed to be disturbance from roads rather than clear-cuts or young forests.

Environment in Focus

Box 9.8 Case Study: The Northern Spotted Owl

In the latter part of the 1980s, the northern spotted owl in the western United States became the focal point of high-profile conflicts between conservationists and logging interests. The owl was accorded threatened status throughout its entire range in the US under the US Endangered Species Act. A 'threatened status' designation means that the owl is likely to become an endangered species within the foreseeable future throughout all or a significant part of its range. This designation requires that critical habitat be identified and a recovery plan implemented. Because the owl depends on old-growth forests, this decision led to severe conflicts with forest harvesting activities. In 1994, the Clinton administration established reserves on more than 4 million hectares where harvesting would be severely restricted.

The spotted owl also is found in the old-growth forests of southwestern British Columbia. This northern extension of the range is important, since individuals at the extremes of a species' range are often the most important to protect because they may have the genetic diversity best suited for future adaptability. Furthermore, just as in the US, the old-growth forest on which the owl depends was allocated for harvest. The spotted owl therefore provides a good case study of the impact of forest harvesting on biodiversity.

Biology and Range
The northern spotted owl is found from northern California to southwestern BC. Scientists estimate that, historically, 500 pairs of spotted owls once inhabited the old-growth forests of southwestern BC. Between 1985 and 1993, some 39 active spotted owl sites were recorded, totalling a minimum of 71 adult owls. By 2003, research reported only 25 breeding pairs left in the province; by 2011, this number was down to six. The historic range of the species in the province is probably not that much different from the current range, although the distribution of the owl within that range has changed significantly, primarily because of destruction of its prime habitat, old-growth forests.

Superior habitat for the owls has old-growth characteristics, such as 'an uneven-aged, multi-layered multi-species canopy with numerous large trees with broken tops, deformed limbs, and large cavities; numerous large snags, large accumulations of logs, and downed woody debris; and canopies that are open enough to allow owls to fly within and beneath them' (Dunbar and Blackburn, 1994: 19). Main prey for northern spotted owls are small mammals such as the northern flying squirrel and dusky-footed and bushy-tailed wood rats. Both prey species are abundant in old-growth forests.

Threats
The greatest threat to the spotted owl is the logging of old-growth forests, leading to loss of suitable habitat. Estimates suggest that probably less than 50 per cent of the old-growth habitat that once covered the Lower Mainland of BC is suitable habitat, and much of it is highly fragmented. The provincial government has approved logging in at least three and as many as six of the 10 areas in which the owl was detected in 2003. Meanwhile, even British Columbia's largest logging companies have agreed to a voluntary halt to logging in spotted owl habitat.

Provincial and regional parks provide some protection in two main blocks, covering some 110,000 hectares in total. Unfortunately, these blocks are about 85 kilometres apart, with little interconnecting habitat. It is extremely unlikely that the two populations will be able to interbreed. Provincial forests managed for timber production contain much larger amounts of suitable habitat, but much of this land is scheduled to be harvested over the next 100 years. The northern spotted owl is listed as endangered by the Committee on the Status of Endangered Wildlife in Canada (COSEWIC), but nothing in BC law 'precludes approval of a forest development plan if there is any element of risk to a forest resource, even where that forest resource is an endangered species' (*Vancouver Sun*, 2003).

Management Options
Concern over the future status of the spotted owl resulted in the formation of a Spotted Owl Recovery Team in 1995 to develop a recovery plan for the species. The primary goal was to outline a course of action for stabilizing the current population, which in turn could lead to an improvement of the status of the species so that it could be removed from the endangered species category. Sixteen management

options, corresponding with the different categories of abundance status outlined by COSEWIC, are discussed in detail in Chapter 14. The options range from banning all timber harvesting that could degrade suitable owl habitat within the entire range of the owl to no owl management beyond existing parks and protected areas. Several attempts have been made to assist the recovery of the species, including various captive breeding projects. Although two were successfully born in the United States in 2008, Canada's breeding facilities have failed to have any fertilized eggs hatch. This lack of success raises a gloomy picture for the future of the species. Unfortunately, the spotted owl is not alone in its predicament. Of the forest-related species reassessed by COSEWIC between 1999 and 2007, 17 per cent were placed in a higher category of endangerment, and only 4 per cent were assigned to a lower category or removed from the list. For instance, a re-evaluation of 32 species in 2011 found only four to be less at risk than during their initial assessments 10 years earlier. Although these reassessments may in part result from new information, the results are not what is desired.

The northern spotted owl.

From this brief review of the impact of forest harvesting activities on biodiversity, we can conclude that substantial changes can occur. These changes are beneficial to some species and detrimental to others. It is critical, therefore, to understand the complex ecological relationships involving forests if these changes are to be fully evaluated and taken into account in designing harvesting activities. And it is essential to leave some areas in their original state to maintain landscape biodiversity. Currently, about 8 per cent of Canada's forest area is protected by legislation.

Cameron (2006) has reviewed some of the interactions between protected areas and the so-called 'working forest' in Canada and suggests:

- retaining or restoring natural climax forest species composition;
- reducing edge contrast between working forest and protected areas;
- maximizing protection of watercourses draining into protected areas;
- planning road networks to minimize undesirable effects on nearby protected areas.

Many conservationists do not like using the term 'working forest' to differentiate between logged and unlogged forests. They maintain that the unlogged forests are in fact working the hardest for society through means such as water storage and carbon sequestration, and to imply that they are not working does not reflect modern scientific understanding of the ecosystem value of unlogged forests.

When enlightened forestry companies realize how whole forests work, they come up with innovative solutions. For example, in 2007 the Canadian Parks and Wilderness Society (CPAWS) and the forestry corporation Tembec negotiated a minimum 50-year halt on logging in an area used extensively by woodland caribou on the east side of Lake Winnipeg. The area is further protected under the Canadian Boreal Forest Agreement to ensure adequate conservation measures can be taken to protect the woodland caribou. Habitat protection is key to maintaining populations of this threatened species, since they are extremely sensitive to human developments. Estimates suggest that the Manitoba woodland caribou population has decreased by 50 per cent since 1950, and in 2006 the species was listed as threatened under the Manitoba Endangered Species Act. The 26,000-hectare area where harvesting is deferred is the 'winter core zone' of the Owl Lake woodland caribou herd—in other words, the lands the herd uses most during Manitoba's cold months, the most critical time of year.

Forestry and Site Fertility

Forest harvesting removes nutrients from the harvested site (Figure 9.9). The amount of nutrients removed depends

Environment in Focus

Box 9.9 Forestry and Grizzlies

The grizzly bear once extended across most of North America. Unlike its smaller cousin, the black bear, however, it is not able to tolerate human disturbance, and as human populations grew, the grizzly populations shrank. Some of the densest remaining populations are in the lowlands of the coastal valleys in BC, where the bears are attracted not only to the annual salmon runs but also to the abundant berry-producing shrubs and other nutritious vegetation found on the flood plains. These nutrient-rich sites also constitute some of the most productive forestry sites.

Studies over the past 10 years have documented declines in grizzly populations following logging activities. One of the main problems was the changes in vegetation as a result of logging. The favourite forage foods of the grizzly are those that compete with the re-establishment of trees as secondary succession takes place (Chapter 3), and because the re-establishment of trees has meant profit for logging companies and governments down the road, the mixed vegetation has often been suppressed by the use of chemicals. In turn, this helps to create dense stands of conifers, which have the long-term effect of shading out preferred forage for the grizzly. The result is a critical shortfall in grizzly food over an extended period.

Concern about these changes has led to new approaches. For example, lower replanting densities for conifers and more open space between tree clusters are being implemented in an effort to mimic the natural environment. The use of chemicals is restricted to the tree stands. A new adaptive management approach has been formulated for silviculturists dealing with grizzly bear habitats. It involves not only the measures outlined above for new logging activities but also revisiting past sites to undertake remedial action. The approach not only is adaptive but also exemplifies an ecosystem approach, as discussed in Chapter 5, which takes a more holistic view of resource management by accounting for the limits of tolerance (Chapter 4) of one of the most spectacular animal species in Canada. Although it is too early to assess the success of these efforts, in combination with the absolute protection of some of these sites, such as a grizzly reserve established in the Khutzeymateen Valley, they should help to ensure that grizzly bear populations survive in the coastal valleys of BC.

The grizzly bear is not very tolerant of human disturbance and has suffered large declines in number and distribution since European colonization.

on the kind and extent of harvesting. Selective tree-length harvesting removes relatively few nutrients compared to large clear-cuts of complete-tree (above- and below-ground biomass) harvesting. The latter maximizes the short-term yield of biomass from the forest but may compromise the potential of that site over the long term to produce further harvests. **Complete-tree harvesting** is rare in Canada; however, **full-tree harvesting**, where trees are felled and transported to roadside with branches and top intact, is the most common system in use. Alternatively, **tree-length harvesting** involves felling, delimbing and topping the trees in the cut-over area.

To judge the potential effects of forest harvesting on site fertility, it is necessary to consider the size of the soil nutrient pool, the amount of nutrients being removed, the net accretions and depletions of nutrients in the forests, and the

(a) Unmanaged Forest Ecosystem

Precipitation →
Dust, gases →
Rock weathering →
→ Sediment losses
→ Nutrient leaching
→ Gaseous losses (CO_2, NH_3, etc.)

(b) Managed Forest Ecosystem

Fertilizers →
Other inputs (precipitation, dust, gases, rock weathering) →
→ Harvest removal
→ Other outputs (leaching, sediments, denitrification, etc.)

Figure 9.9 | Nutrient inputs and outputs from managed and unmanaged forest ecosytems.

---

ways in which these variables interact. The process of site impoverishment over time as a result of harvesting is shown in Figure 9.10. On some sites, the proportion of nutrient capital removed in the biomass will be relatively minor, while on others it may be substantial. This depends to a large degree on the existing nutrient capital of the site. Areas with high soil fertility will be less affected. Some sites will recover quickly from harvesting and can sustain relatively short rotations. Other sites will not recover adequately between rotations, and site nutrient capital will fall, making tree regeneration difficult and in some cases impossible. Thus, nutrient-deficient sites should have long rotations with just stem harvesting in order to maintain productivity.

The amount of nutrients removed by harvesting is influenced by tree species, age, harvesting method, season of harvesting, and other factors. Older trees contain larger amounts of nutrients—such as nitrogen, phosphorus, potassium, calcium, and manganese—than younger trees. There is also considerable variation among species in the amount of nutrients organically bound and the nutrients preferentially held by different species. The differences between whole-tree and stem-only harvests can also be significant. In a study of black spruce in Nova Scotia, Freedman (1981) found an almost 35 per cent increase in biomass take for whole-tree harvesting. The loss amounted to 99 per cent for nitrogen, 93 per cent for phosphorus, 74 per cent for potassium, 54 per cent for calcium, and 81 per cent for magnesium, derived mainly from the nutrient-rich foliage and small branches. When deciduous trees are harvested, the loss of nutrients can be reduced significantly by cutting during the dormant period when leaves are not present.

Nova Scotia has announced that it will ban whole-tree harvesting, except for Christmas trees, as concerns intensify about the long-term impacts of intense biomass removal on site nutrient capital and the long-term productivity of the site. These concerns are not related solely to soils, as a recent review (Berch et al., 2011) details the potential impacts on biodiversity. As demands for biofuels increase in future (Pare et al., 2011), the links among soil nutrient capital, productivity, and biodiversity will need much greater understanding.

Forest harvesting can also lead to dramatically increased rates of nutrient loss through **leaching** (the downward movement of dissolved nutrients) to the hydrological system. The amount of loss varies according to the intensity and scale of the harvest and the particular ecosystem. Loss of nitrate is of most concern, since it is not only often a dominant limiting factor (see Chapter 2) but also the nutrient lost most often in large quantities. One reason for this is disturbance in the nitrogen cycle (Chapter 4) by logging, which results in an increase in the bacterial process of nitrification turning ammonium to nitrate. Nitrate is highly soluble, resulting in significant losses of nitrogen site capital in some ecosystems, particularly if the soil is not too acidic. Other factors, such as warmer soil temperatures, decreased uptake by vegetation, and abundant decaying organic matter on the forest floor, also contribute to increased losses of soluble nutrients following logging. In addition, younger stands (less than 145 years old) support a smaller biomass of lichens with nitrogen-fixing abilities compared to old-growth trees on the west coast of BC.

It is also important to consider nutrient inputs. Precipitation adds substantial amounts of nutrients over time. For a maple–birch stand in Nova Scotia, for example, Freedman et al. (1986) calculated that it would take 96 years of precipitation to replace the nitrogen lost through whole-tree removal, 83 years for potassium, 166 years for calcium, and 41 years for magnesium. Other nutrient inputs occur through dry deposition of gases and particulate matter, the weathering of minerals, and the fixation of atmospheric dinitrogen. Soil mycorrhizae are particularly important for the fixation of atmospheric dinitrogen (Chapter 4), but land treatments after clear-cutting, such as slash burning and pesticide use, may adversely affect fixation rates.

Forestry and Soil Erosion

Besides its direct influence on nutrients, forest harvesting can have a substantial impact on soil through erosion, especially on steep slopes in areas of heavy precipitation. Such losses also contribute to loss of site fertility, remove substrate for further regrowth, and contribute to flooding and the destruction of fish habitat. Poor road design and maintenance are often key factors behind accelerated soil erosion. Cutting roads across steep terrain exposes large banks of unprotected topsoil. Compacted road surfaces encourage overland flow with high erosive power. Many jurisdictions are implementing much stricter regulations on road construction. Other regulations to minimize soil erosion losses

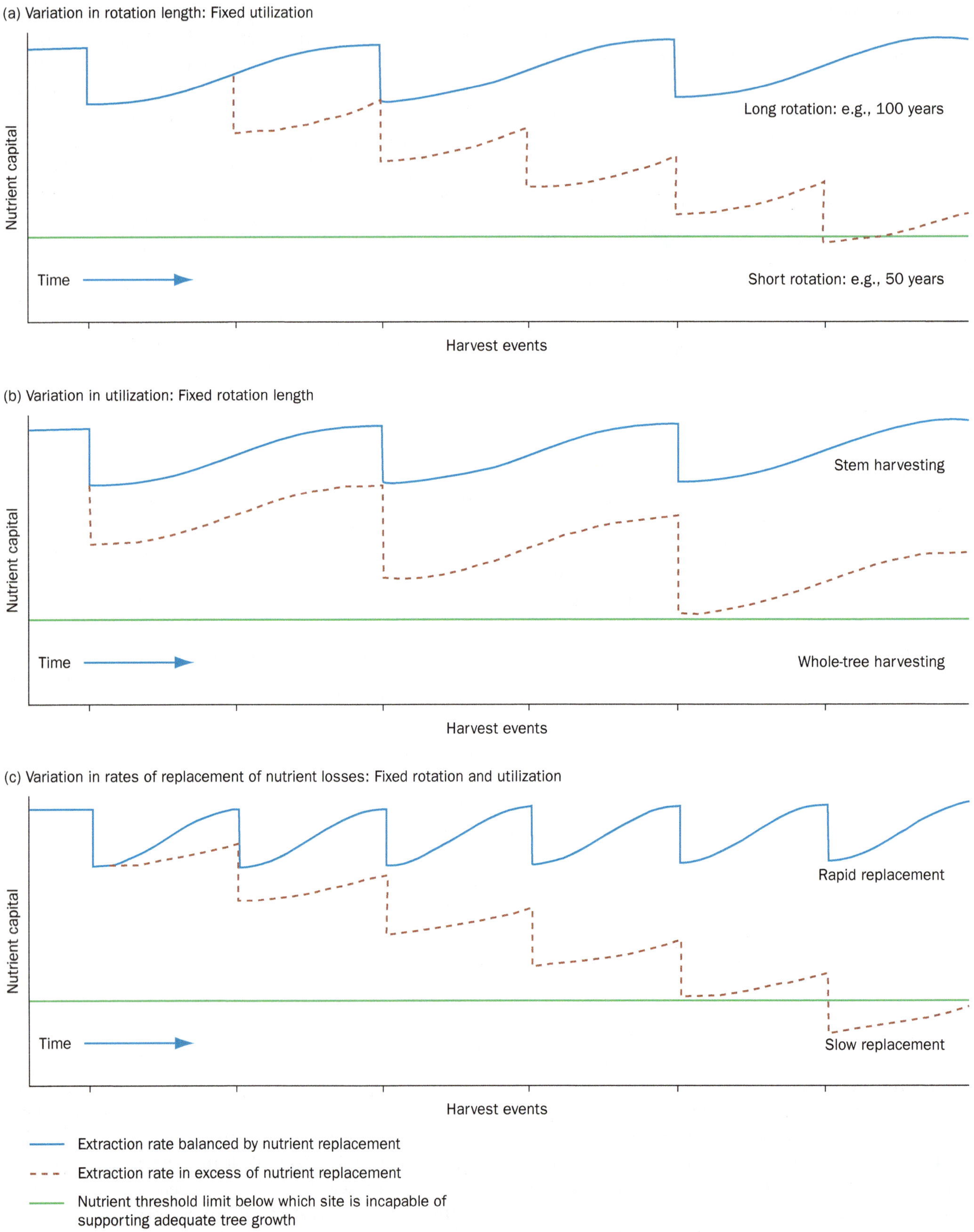

Figure 9.10 | Site impoverishment as a result of forest harvesting. Source: Adapted from Kimmins (1977).

When an old-growth tree such as this giant Sitka spruce on the west coast of Vancouver Island dies, the nutrients it contains recycle to fuel new tree growth. When logging removes the whole tree, these nutrients are lost to the ecosystem.

Logging has resulted in severe changes in the morphology of many coastal streams in BC, with large amounts of logging debris accumulating following flooding.

include leaving buffer strips along watercourses, using minimal impact logging techniques, and pursuing selective harvesting rather than clear-cutting.

Forestry and Hydrological Change

Large-scale forest harvesting can have a significant impact on hydrology. Under natural conditions, large amounts of water are returned into the atmosphere by the trees through the process of transpiration. Evapotranspiration includes transpiration plus evaporation from non-living surfaces. These processes are normally greatest during the summer. Removal of the trees significantly reduces this mechanism and other storage capacities, releasing large amounts of water into stream flow and often resulting in flooding during high-discharge periods. Furthermore, without the delaying mechanism of the trees, and with compacted soils from harvesting, the speed of flow is often increased, again raising the potential for flooding as well as increased erosion. In turn, sediment from erosion can damage fish spawning beds.

Forestry and Climate Change

Perspectives on the Environment

Canada's Forest and Peatland Ecosystems

As stewards of one of the largest biological carbon stores on the planet, Canadians have an opportunity to contribute to climate change mitigation at an international scale through improved land management. The development of sound policies for maintaining the role of Canada's forest and peatland ecosystems in climate regulation has been elusive, however, in part due to the rapidly evolving, and at times complex, science that is involved.

—Carlson et al. (2010: 439)

Forests are a carbon sink. They take in carbon dioxide and convert it to wood, leaves, and roots. They are also a carbon source. They release stored carbon into the atmosphere when they decompose or burn. Because of this ability to both absorb and release huge amounts of carbon dioxide (a major greenhouse gas), forests play a major role in the global carbon cycle (Chapter 4), the exchange of carbon between the atmosphere and the biosphere. More carbon is stored in forest biomass (trees and other living plants), dead organic matter, and soil than is contained in the atmosphere. That is why forests are a key part of the global carbon cycle.

Large changes in forest carbon sinks and sources, whether due to human or natural causes, can affect the climate by altering the amount of carbon dioxide in the atmosphere. As the climate changes, forest carbon storage will be affected. A warmer climate can speed up vegetation growth, which means more carbon storage. However, it can also accelerate decomposition, resulting in more carbon emissions, and boost the risk of drought, pest outbreaks, and fire, all of which can significantly reduce carbon storage. The extent of these effects is also influenced by the amount and/or timing of precipitation changes.

A rapidly changing climate has important implications for the forest sector and the communities whose livelihood is closely associated with forests. One example is the effects on timber supply. Growth and yield databases used in timber supply forecasting will need to be re-evaluated because of changing tree growth and productivity. Long-term timber supply planning may also need to take into account changes in species composition over time. More frequent large-scale disturbances will cause timber supply fluctuations and result in more salvage-harvesting of trees killed by disturbances, which affect fibre quality.

Because wood continues to store carbon even after it is made into products (such as lumber and paper), only a fraction of the carbon removed from the forest is actually emitted into the atmosphere. As well, some of the wood-waste from product manufacture is burned to produce energy, offsetting fossil-fuel use. After harvest, 40 to 60 per cent of the carbon remains in the forest in the roots, branches, and soil and decomposes slowly, providing nutrients for the newly regenerating forest. Natural disturbances such as forest fires and insect infestations release large amounts of carbon dioxide into the atmosphere, although the areas affected and emission levels can vary considerably from year to year. The area of forest burned each year is on average 2.5 times the area harvested and is projected to increase under warmer, drier climate conditions. A significant difference between fires and harvesting is that, with harvesting, much of the carbon ends up being stored in long-lived products, but with fires the carbon goes into the atmosphere.

Analysis using the Canadian Forest Service's Carbon Budget Model indicates that Canada's managed forests were a net carbon sink in most years between 1990 and 2009. But in some years they were a source, mostly because of wildfires (Figure 9.11). In 2009 Canada's forests sequestered 130 million tonnes of CO_2. Annual amounts ranged from a large *sink* of 174 million tonnes of carbon dioxide equivalents (CO_2e) in 1992 to a large *source* of 171 million tonnes of CO_2e in 1995, mostly because of wildfires. There has also been a slight increase in carbon losses resulting from forest harvesting over the years. International accounting rules dictate that

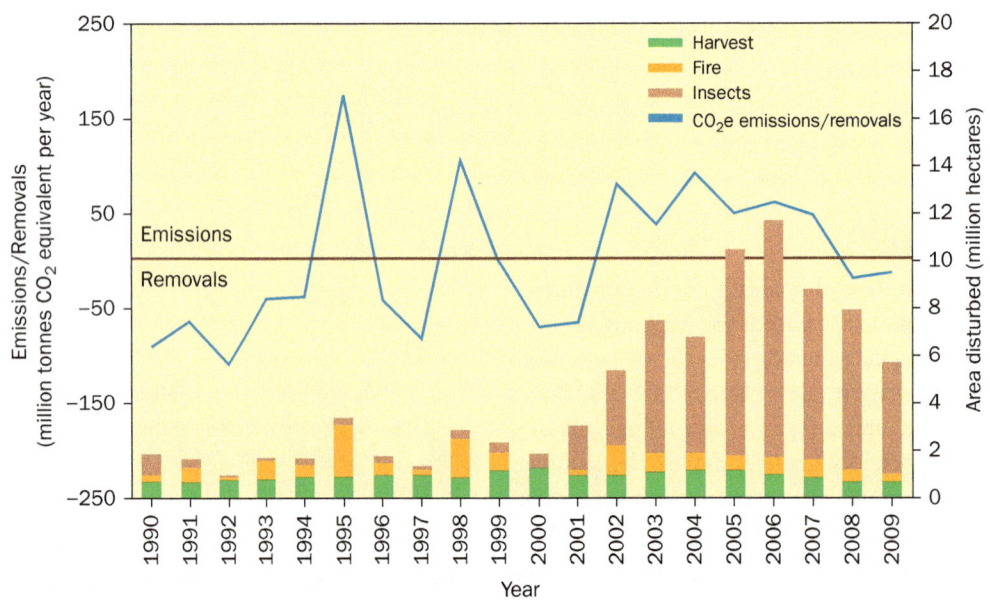

Figure 9.11 | Carbon emissions/removals in Canada's managed forests. *Source: Natural Resources Canada (2011: 29).*

they be included with other carbon losses, although in fact the carbon will remain stored in timber products for many years. The impact of the mountain pine beetle infestation in western Canada has continued to increase. In 2009 BC Ministry of Forests and Range surveys detected 16.3 million hectares under attack by the mountain pine beetle. This infestation will affect the sink/source balance and trends for decades to come.

Deforestation, the permanent conversion of land from forest use to other uses such as agriculture and urban and industrial use, has slowed over the years although it remains a subject of major international concern and negotiations (Chapter 7). Worldwide, deforestation creates about 20 per cent of human-generated greenhouse gas emissions, more than is produced by the transportation sector. In Canada, deforestation accounts for less than 3 per cent of national emissions, a figure that is declining. In 2009, an estimated 44,800 hectares of forest were converted to other land uses, equivalent to 18 million tonnes of CO_2e, down from 70,000 hectares (29 million tonnes) in 1990. However, this is still considerably in excess of the sink amount of roughly 1 million tonnes of CO_2 per year created by afforestation (planting forests on land previously used for other purposes, usually agriculture).

Considering carbon sequestration in timber management is different from any concept previously applied, but it might become a major factor in how we manage our forests. Neilson et al. (2007), for example, point out that hardwood stands in New Brunswick typically contain 10 to 20 per cent more carbon per hectare than a similar volume of softwoods of similar age. Forest managers might well change their management practices as global climate change becomes more severe and, thus, shift to growing multi-aged hardwood stands that promote carbon sequestration instead of the single-aged softwoods that now dominate their thinking.

New Forestry

Concerns over the impacts and sustainability of forest practices have given rise to calls for what has been termed **new forestry**, which involves new ways of looking at the management of forest ecosystems. Current approaches usually emphasize economic maximization over the short term through intensive forest management subsidized by auxiliary energy flows, leading to a simplification of forest biology. This entails genetic simplification through the exclusion of non-commercial species from regrowth areas and genetic manipulation to homogenize the species grown. Intensive forestry emphasizes production of a young, closed-canopy, single-species forest, usually the least diverse of all successional stages. Moreover, the strength and reliability of the wood produced from such forests have been questioned, and these plantations are susceptible to **windthrow**, insect infestations, and gradual nutrient depletion. Structural simplification also takes place as the range of tree sizes and growth forms is reduced, snags and fallen trees are removed, and trees are regularly spaced to optimize growth. At the landscape scale, simplification occurs as old growth is removed and the irregularity of wind- and fire-created openings is replaced by the regularity of planned clear-cuts. Successional simplification also takes place, since intensive management aims at eliminating early and late successional stages from the landscape.

New forestry embraces an approach that mimics natural processes more closely, emphasizing long-term site productivity by maintaining ecological diversity. This includes rotation periods sometimes longer than the minimum economic periods, reinvesting organic matter and nutrients in the site through snag retention and stem-only harvesting, minimizing chemical inputs, and diversifying the range of tree species and other forest products. Growth of traditionally non-commercial species such as alder and other early successional species, particularly nitrogen fixers, is permitted, and all stages of the successional process are accommodated. Old-growth big-leaf maple, for example, provides excellent growth sites for many epiphytes (plants that use other plants for physical support but not nourishment), which supply valuable nutrient accumulation and water retention. Riparian or riverbank habitats receive special attention; litter from streamside vegetation provides the primary energy base for the aquatic community, and management of coarse woody debris is particularly important for the structure of smaller streams. Needless to say, large woody debris cannot be produced by a forest that no longer contains large trees.

New forestry also emphasizes the maintenance of non-timber parts of the forest community. Special attention is given to the impact of the size, shape, and location of forest patches on wildlife and how these patches can be connected to sustain populations. The ecological complexities of forests are only just starting to be revealed. Recent research, for example, suggests that the younger the forest, the less conifer seed production. Species that rely on these seeds, such as crossbills, also experience a decline. In western Canada, five species of crossbill have evolved, each specializing on a different species of conifer and even different varieties of the same species. Protection of this diversity of crossbills will require protection of old-growth stands and an increase in rotation ages throughout the range of each conifer. Similar consideration must be given to the entire range of forest biodiversity if it is to be maintained into the future.

The kinds of changes suggested by new forestry make it unlikely that the dominant practices of today, such as extensive clear-cuts, can continue, and indicate that other harvesting systems (Table 9.2) will play a larger role as an ecosystem-based perspective becomes more widespread. New forestry may not suit the need for short-term economic return on the part of the large corporations now dominating the industry, and it is likely that more, smaller, and community-based companies will emerge (Box 9.10). Monetary returns over the short term will probably fall as the amount of wood fibre extracted from the forest declines. Proponents of new forestry argue, however, that these changes will have to occur anyway. Continuing the old approaches will simply lead to an abrupt decline in the amount of timber available and consequently will diminish future prospects. This way of thinking is gaining wider acceptance. The development of Canada's National Forest Strategies, described in the next section, is one result.

Environment in Focus

Box 9.10 New Forestry in Action

The ideas of new forestry must be put in motion if change is to occur. Several examples of alternatives to the dominant way of managing our forests are already in operation, both regarding individual woodlots and management of more extensive areas by communities.

At the individual scale, one well-known example is Merv Wilkinson and his 55-hectare woodlot, Wildwood Forest, on southern Vancouver Island. From 1936 until his death at the age of 97 in the late summer of 2011, Wilkinson practised **sustained yield** forestry, and despite the removal of more than 4,000 cubic metres of timber, his woodlot still contains as much wood as it did when it was first assessed in 1945. His practice involved removal of forest products by cutting in five-year rotations. The straightest, most vigorous trees with good foliage and abundant cone production are left as seed trees, including some estimated to be as old as 1,800 years. There is no clear-cutting, slash burning, or use of chemicals. The canopy is left intact to shield seedlings but thinned a little to promote good growth. Sheep are used for brush control. Wilkinson's model may not apply everywhere, but it worked for him and has offered a good example of how a forest can be maintained while still retaining its essential ecological characteristics.

At a regional scale, attention has focused on 'community forests'. Decisions on forest use are often made in boardrooms at the dictates of international capital. Such decisions may not be to the benefit of the local communities dependent on the forests for their livelihoods. Concern over this situation has prompted interest in how to manage forests to maximize the benefits to local communities.

Many types of community forests exist in Canada, with different forms of land tenure and administrative arrangements. However, they are all aimed at achieving benefits for the community and encouraging local involvement in decision-making. In BC, for example, the provincial government amended the Forest Act to create Community Forest Agreements, a new form of tenure designed to enable more communities and First Nations to participate directly in the management of local forests. In 1998, the Ministry of Forests launched a pilot project involving a special form of the tenure, called a community forest pilot agreement (CFPA). To date, 12 agreements have been issued to a range of communities and Aboriginal groups. The pilot agreements are located throughout the province and range in size from about 400 hectares to more than 60,000 hectares. The BC government has a number of objectives for the program:

- provide long-term opportunities for achieving a range of community objectives, including employment, forest-related education and skills training, and other social, environmental, and economic benefits;
- balance the use of forest resources;
- meet the objectives of government with respect to environmental stewardship, including the management of timber, water, fisheries, wildlife, and cultural heritage resources;
- enhance the use of and benefits derived from the community forest agreement area;
- encourage co-operation among stakeholders;
- provide social and economic benefits to British Columbia.

Community forests are typically managed to reflect community goals, including:

- supporting the local economy by hiring and buying supplies locally and selling timber to local timber-processing facilities;
- diversifying the local economy by making small volumes of wood fibre available to new and existing niche markets such as small-scale, value-added manufacturers and artisans;
- maintaining and enhancing local recreational opportunities;
- protecting drinking water, viewscapes, wildlife, and other environmental attributes;
- providing a source of income to support local community initiatives.

Under the Forestry Revitalization Plan announced in 2003, the BC government committed to significantly increasing the volume of timber allocated to community-based forest tenures. New community forest pilot agreements are no longer issued. Instead, 'probationary' agreements, carrying a five-year initial term, allow both communities and the ministry an assessment period. If successful, the agreement-holder is offered a long-term community forest agreement, which has a 25- to 99-year term and is replaceable every 10 years. Existing pilot agreements, if successful, can also be rolled over into long-term agreements. Burns Lake Community Forest was offered the first long-term (25-year) licence in 2004. As of June 2010, approximately 1.17 million hectares in BC were being managed as community forests.

More partnerships are now being made with Aboriginal communities in BC. The government currently has harvesting agreements with 172 Aboriginal groups, covering 55 million cubic metres and producing more than $243 million in revenue annually.

Alternative approaches to conventional forestry do exist. It is essential, however, to specify the goals of forestry activities before the most appropriate approach can be chosen, as emphasized in the framework in Chapter 1. Current models have evolved to maximize economic returns over the short term; the alternatives described above have different goals, more consistent with the demands of today. However, as Bullock and Hanna (2008) point out, community forestry is not a panacea for resolving conflict in forest management.

Perspectives on the Environment

Community Forests

The belief that community forests will enable local control, sustainable community development, and multiple-use management plans that reflect diverse values and interests has heightened expectations from community forests. As Duinker et al. (1991) warn, we cannot assume that all communities are suited to or have the capacity for this approach to forest tenure and management. The community forest is not a panacea for all the difficulties associated with industrial forestry or the failures of existing management systems, but it does hold promise.

—Bullock and Hanna (2008: 84)

Canada's National Forest Strategies

During the 1980s, it became increasingly clear that forestry in Canada could not continue as it had in the past; new ways had to be found to develop more sustainable management practices. This realization resulted in the formation of a National Forest Strategy (NFS): Sustainable Forests: A Canadian Commitment (1992–7). Canada was the first country in the world to develop such a national strategy. It was revised and extended in a second strategy covering 1998 to 2003. The National Forest Strategy Coalition (NFSC), composed of 52 governmental and non-governmental agencies, was formed to oversee implementation of the strategy. The strategy made 121 commitments under nine strategic directions to move Canada along the road to a more sustainable use of forest ecosystems.

Such an approach is only useful, however, if the commitments are meaningful and progress is made to achieve them. An independent evaluation concluded that there had been substantial progress on 37 commitments, some progress on 76 commitments, little progress on six commitments, and no progress on two commitments. The evaluation commended Canada for showing international leadership but suggested a simplification of any future strategy, as well as the inclusion of clear targets and timetables.

In 2003, a further five-year strategy was adopted, along with a commitment on the part of the NFSC partners to work towards its completion. However, Alberta, Quebec, and the Forest Products Association of Canada did not sign the strategy. Eight themes were outlined, each of which specifies an objective and action items to be undertaken. Two noteworthy themes were the urban forest, which emphasized the need to engage more of Canadian society in forest questions, and recognition of the importance of private woodlots to sustainability. The latter is particularly important in the Maritime provinces, where more than half of the forests are in private hands and not subject to provincial forestry regulations. Overcutting and neglect of these lands have been problems in the past, and the action items in the strategy included

Perspectives on the Environment

Different Views of the National Forest Strategy

The current NFS (2003–8) fulfills an important role of informing interested and concerned Canadians and the international community about the state of Canada's forests and forest practices.

—NFSC (2005)

After three years of working to implement the National Forest Strategy from within the Nation Forest Strategy Coalition, the Sierra Club of Canada has decided to disengage.

—Sierra Club of Canada (2006)

What cannot be determined in any traceable way is the process of taking the civil society inputs and translating them into final texts in the NFS. Thus, the processes of generating NFSs have not been entirely transparent, and in the case of the current NFS the outcomes were not entirely satisfactory.

—KBM Forestry Consultants (2007)

When we look at the international milestones in sustainable forest management, Canada's leadership stands out.

—Natural Resources Canada (2007b)

Table 9.4 | A Vision for Canada's Forests: 2008 and Beyond

Published by the Canadian Council of Forest Ministers, *A Vision for Canada's Forests: 2008 and Beyond* takes a broader and more aspirational approach than previous national forest strategies.

Overall Vision Statement

Be the best in the world in sustainable forest management and a global leader in forest sector innovation.

- Take a non-prescriptive approach and increase awareness of forest issues at home and abroad.
- Encourage domestic and international engagement; promote partnerships among traditional and non-traditional forest interests.
- Reflect the collective ambitions of Canadians for their forests and communities; open opportunities for all to draw on each other's strengths.
- Mobilize the talents of all Canadians in the search for creative solutions.

Commitment to Sustainable Forest Management

Demonstrate commitment to the sustainable management of forests by continuing to implement the principles of sustainability (including ecosystem-based and integrated landscape management) and embrace the principles of stewardship, innovation, partnership, transparency, and accessibility.

- Recognize the evolving nature of sustainability; for example, give increasing consideration to the effects of climate change, the provision of environmental good and services, and the importance of urban forestry to inhabitants of towns and cities.
- Strive to improve the forestry sector's environmental record while considering the social implications of sustainable forest management practices and expanding the contribution of forest resources to wealth and prosperity.
- Maintain the variety, quality, and extent of forest types; conserve biological diversity and soil and water resources; enhance the resilience of forests by managing carbon balances and adopting innovative forest-protection strategies.
- Innovate to facilitate the establishment of partnerships among the various forest sector members to maintain and expand the contribution of renewable forest resources, bioeconomy, and new market mechanisms for environmental good and services to Canada's economy.
- Include communities who depend on forest resources for their well-being and livelihood in helping decide how forest resources are managed.
- Diversify Aboriginals' role in development and implementation of sustainable forest management to include development of non-timber forest products and tourism initiatives.
- Consider urban populations' influence on the management of Canada's forests, including the need for urban forests that reduce the negative effects of airpollution, conserve energy, reduce soil erosion, provide wildlife habitat, and offer a place for recreation and spiritual renewal.

Priorities of National Importance

1. *Forest Sector Transformation: Ensure a prosperous and sustainable future for Canada's entire forest sector.*
   - Develop new ideas, technologies, processes, and markets through a systematic engagement of science and technology organizations in collaborative research and public–private partnerships.
   - Maximize economic value from forest resources through the diversification of uses.
   - Make products from Canada's forests recognizable as an environmentally and socially responsible choice for consumers around the world.
   - Take the provision of environmental goods and services into account in the sustainable management of Canada's forest resources, including through the development of new markets.
   - Ensure meaningful Aboriginal participation in an innovative forest sector, including use of their insights and expertise.
   - Have highly skilled workers contribute to the expansion of knowledge-based forest industries through education and training.
   - Put into place creative public policies that facilitate forest sector transformation.

2. *Climate Change: Become a world leader in innovative policies and actions to mitigate and adapt to the effects of climate change on our forests and forest communities.*
   - Include climate-change considerations in all aspects of the sustainable management of Canada's forests.
   - Recognize, harness, and manage the economic value of carbon in trees, forests, and wood products.
   - Identify and address the knowledge gap in the impacts of climate change on forests, industries, and communities.
   - Ensure policies and institutions provide means for forests, industries, rural and urban communities, and private woodlot owners to adapt to changing conditions and mitigate the effects of climate change.
   - Develop, share, and implement innovative adaptation and mitigation practices, including those that integrate Aboriginal knowledge.
   - Involve many forest-resilient communities, including Aboriginal ones, in the development, sharing, and implementation of forest mitigation and adaptation strategies.
   - Enable innovators and entrepreneurs, through institutions and creative policies, to take advantage of transformative and sustainable bioenergy opportunities that contribute to broad climate-change objectives.

Realizing the Vision through Partnerships

To be effective, the Vision needs to be embraced by Canada's entire forest sector, including future players.

1. *Canadian Council of Forest Ministers (CCFM)*
   - Champion the Vision to generate greater public awareness of forest issues and communicate the Vision's goals and outcomes.
   - Convey progress on the Vision, including advancement of sustainable forest management in Canada and progress on priority issues.
   - Facilitate the consolidation of progress reports from forest sector partners and ensure coordination with national initiatives and activities of other ministerial councils, and with its own initiatives and strategies.
   - Support selected events, including conferences and workshops that advance the Vision.
   - Host a national workshop every three years to showcase the initiatives that advance the Vision and to help inform the review process.

2. *Federal, Provincial, and Territorial Governments*
   - Act individually to promote sustainable forest management and bring focus to priority issues.
   - Use a variety of arrangements to consult and involve forest sector members and the general public.

3. *Private Owners and Forest Companies*
   - Continue to manage owned forests through stewardship, working within a framework of norms, regulations, and incentives defined by governments.
   - Continue to manage public lands with oversight from public agencies and innovate within new approaches that give regional authorities or communities more responsibilities for managing public forest.

4. *Aboriginals*
   - Continue to maintain a relationship with the forests that represent a source of food, medicine, and economic activity as well as provide opportunities for recreation, social interaction, and spiritual growth.
   - Play a role in forest economy, including involvement in the development of sustainable forest management practices, notably through the application of their knowledge and practices.

5. *Other Forest Sector Members*
   - Have consultants, land-use planning experts, environmental and conservation groups, communities, academics, professional associations, and other organizations play a role in helping to develop and deliver effective sustainable forest management policies and practices.

6. *International Partners*
   - Collaborate with those at home and abroad on science and technology issues and share expertise and resources to advance global forest development.
   - Remain engaged in multilateral and bilateral international forestry initiatives and conventions.
   - Remain engaged in international dialogues dedicated to international cooperation to further the sustainable management and development of forest resources.

Assessing and Communicating Progress

Communicating progress on the Vision requires coordinating reporting activities and sharing information.

- Use CCFM's Criteria and Indicators Framework as an evaluation tool to gauge improvements in the sustainable management of Canada's forests.
- Use federal and provincial State of the Forests reports in addition to collaborative tools available on the Internet.

Source: CCFM (2008).

---

providing more incentives and support for landowners to manage their woodlots on a sustainable basis.

The objectives of the 2003–8 NFS were highly laudable and seemed to herald a new and progressive approach to forest management in Canada. A self-evaluation in 2005 concluded that 'National Forest Strategy Coalition members and other partners across Canada's forest community are leading change, as highlighted in their two-year accomplishments report.' As Jean Cinq-Mars, the NFSC chair, further noted: 'Collectively, the achievements show a strong commitment to work together to face the challenges along the path towards a sustainable forest, nationally and globally' (NFSC, 2005: iii). However, others were less convinced. The Sierra Club withdrew in 2006, concerned about the lack of progress, lack of transparency, and efforts to select inappropriate indicators for assessing progress. The Sierra Club published its own set of indicators and an evaluation report (Sierra Club of Canada, 2006). The Ontario Association of Anglers and Hunters also withdrew, feeling that inadequate attention was being devoted to non-timber values in the forests and that the situation was really 'business as usual'.

An evaluation on the ecosystem-based forest management aspects of the NFS was undertaken in 2007 by an independent team of evaluators, KBM Forest Consultants, who found many trends to be stable. The evaluating team took this outcome to be a positive finding in that things were no longer deteriorating. However, they were generally critical of the NFS. They found an overall lack of engagement and financial support. For example, only 15 out of 66 coalition members attended the 2007 annual meeting. They were also highly critical of the evaluation system designed by the NFS:

- 'The action items were formulated either to represent activities already underway in Canada's forest sector...or written so vaguely and ambiguously that it is virtually impossible to determine what the action item actually means.'
- 'Many of the indicators seem (a) disconnected from the action items, and (b) difficult to measure with confidence.'
- 'No one has set any desirable levels or targets for the indicators. Without targets, how can one determine adequacy of performance?' (KBM Forestry Consultants, 2007: 48)

The report identifies some positive trends in Canadian forestry, but 'these trends would no doubt continue with or without an NFS' (ibid., 50). The main conclusion is that 'the NFS is successful at communicating a consistent pattern of behaviour of Canada's forest jurisdictions for the benefit of its citizens and its trading partners around the world' (ibid.). Unfortunately, what this statement seems to indicate is that rather than the NFS creating real progress in forest management in Canada, it is mainly a communication vehicle. Understandably, NGOs interested in generating real change may feel a little co-opted by such a process. The Sierra Club (2006: 7) advised other NGOs: 'Our experience has shown us that, overall, it's not worth our while to work within the NFS. While the NFS umbrella itself is good, efforts spent working outside of the NFS proved more fruitful than the inside approach.'

Both the Sierra Club and the independent evaluation indicate that the future trend of the government seems to be to limit inclusivity and return more to a top-down process of input to forest land decision-making. The 2008–18 strategy confirms this approach, with the Canadian Council of Forest Ministers reassuring Canadians that they will lead the process 'on behalf of all Canadians' (CCFM, 2008: ii) and therefore will not need the participatory approach adopted previously. The new strategy has two themes: transforming the forest sector and mitigating and adapting to climate change. There seems to be little appetite for re-engaging in the more detailed aspects of forest management change that emanated from the previous strategy.

The Model Forest Program

One commitment from Canada's 1992–7 National Forest Strategy that has borne fruit is to develop a system of model forests in the major forest regions. The objectives of the program are:

- to increase the development and adoption of sustainable forest management systems and tools within and beyond model forest boundaries;
- to disseminate the results of and knowledge gained through Canada's Model Forest Program at local, regional, and national levels;
- to strengthen model forest network activities in support of Canada's sustainable forest management priorities;
- to increase opportunities for local-level participation in sustainable forest management.

Proposals were solicited for areas between 100,000 and 250,000 hectares where partners would develop a management structure to facilitate co-operation and include a vision and objectives to balance a variety of values, as well as actions to demonstrate sustainable forest management. Key attributes of model forests include:

- a partnership that includes principal land-users and other stakeholders from the area;
- a commitment to sustainable forest management, based on an ecosystem-based approach;
- operations at the landscape or watershed level;
- activities that reflect stakeholder needs and values;
- a transparent and accountable governance structure;
- commitment to networking and capacity-building.

Eleven model forest agreements in six major forest regions across the country were initiated, each with a unique management structure designed to address the particular situation. Each model forest is an independent, not-for-profit organization, and in 2006 the Canadian Model Forest Network became an independent, not-for-profit organization, although Natural Resources Canada remains a key partner and supporter. Core issues relate to **ecosystem-based management**, Aboriginal participation, public participation, science and innovation, and the integration of non-market values into decision-making.

Model forests have a deliberate strategy of intra- and inter-site demonstration and networking. This strategy has expanded to the international context, with sites in Mexico, Chile, Argentina, China, the US, Japan, Indonesia, Thailand, Myanmar, the Philippines, and Russia as part of an International Model Forest Program. These initiatives have been supported by Canadian aid programs totalling more than $11 million, with additional support from other donors of more than $7.5 million. This support reflects recognition that the programs constitute a tangible demonstration of the value of co-operatively working together towards sustainability. They illustrate many of the approaches outlined earlier in Chapters 5 and 6.

The Model Forest Program terminated in 2007, but the Canadian government announced the Forest Communities

Program, patterned on the Model Forest Program. The new $25 million, five-year program is intended to help urban, rural, and Aboriginal communities make the most of new resource-based economic opportunities. As a part of the Canadian government's economic action plan, three $1 billion programs were implemented to aid in the economic diversification and development. The Community Development Trust was established in 2008, followed by a Community Adjustment Fund in 2009. A separate Pulp and Paper Green Transformation fund provides incentive for companies to improve the environmental standards of their operations.

Global Forest Strategies

Most of this chapter has concentrated on the Canadian situation, but as outlined in Box 9.11 challenges also exist at the global level. Between 2000 and 2010, the area covered by forests in the world shrank by 1.3 per cent, or 520,000 square kms, an area roughly the size of France (FAO, 2011). Forests now occupy some 31 per cent of the Earth's land surface. There is some good news in that the rate of deforestation has declined from 0.20 per cent a year in the 1990s to 0.13 per cent in the first decade of the century. The challenge is to reverse that figure so that, overall, forests are increasing and not declining. However, it should be noted that these figures are based on self-reported statistics and that they also include all lands growing trees, such as oil palm plantations, that might add to forest cover, but do little for biodiversity.

Significantly, some of the most influential and far-reaching social movements in the world have arisen out of local concerns regarding unsustainable forestry practices by major logging interests. The Chipko Movement—also known as the 'Hug the Trees' Movement; *chipko*, in Hindi, means 'embrace' or 'cling'—began in the Himalayan forest region of northern India in the 1970s when local people sought to protect local forests and in protest encircled trees to stop the logging that was destroying local ecosystems and ways of life, as well as causing erosion and flooding. Similarly, the Green Belt Movement in Kenya has fought against multinational interests by involving a half-million schoolchildren, thousands of farmers, and thousands of local women in tree planting and the creation of hundreds of tree nurseries for reforestation in the effort to halt desertification and provide for the livelihood needs of entire communities. The leader of this movement, Wangari Maathai, received the Nobel Peace Prize in 2004 (see, e.g., Knight and Keating, 2010: 257–60).

One of the major disappointments of the United Nations Conference on Environment and Development (UNCED) in 1992 was the failure to establish an international convention on forests. However, a set of non-binding forest principles was adopted, and these principles informed the development of Canada's national forest strategies. Some legally binding outcomes related to forestry did emerge from UNCED, in particular the Convention on Biological Diversity, which committed signatories to prepare and adhere to a national biodiversity strategy, including the designation of representative samples of their forest lands as protected areas and ensuring that forest management does not impair biodiversity. Chapter 14 delves into these commitments.

Forests were also included in Agenda 21, the non-binding principles that emerged from UNCED and set an agenda for development in the twenty-first century. In 2000, the United Nations Forum on Forests (UNFF) was created, with the objective of promoting the forest principles focusing on sustainable management contained in Agenda 21. In 2006, at its sixth session, the UNFF finally agreed on four shared Global Objectives on Forests:

- reverse the loss of forest cover worldwide through sustainable forest management (SFM), including protection, restoration, afforestation, and reforestation, and increase efforts to prevent forest degradation;
- enhance forest-based economic, social, and environmental benefits, including by improving the livelihoods of forest-dependent people;
- increase significantly the area of sustainably managed forests, including protected forests, and increase the proportion of forest products derived from sustainably managed forests;
- reverse the decline in official development assistance for sustainable forest management and mobilize significantly increased new and additional financial resources from all sources for the implementation of SFM.

In 2007, another breakthrough was achieved with agreement on the Non-Legally Binding Instrument on All Types of Forests. This was the first time that member states had agreed to an international instrument for sustainable forest management, and it is expected to have a major impact on international co-operation and national action to reduce deforestation, prevent forest degradation, promote sustainable livelihoods, and reduce poverty for all forest-dependent people. A stronger initiative has emerged in the EU where a legally binding agreement is under development, setting objectives for sustainable forest management.

Other global initiatives are expanding. Approximately 8 per cent of the world's forest area is certified (see Box 9.12 for the implications of certification), a fivefold increase since 2000. The two largest **certification** organizations worldwide, with 28 and 26 per cent of global certified forest area respectively, are the Forest Stewardship Council (FSC), a membership organization dedicated to sustainable development principles, and the Pan-European Forest Certification Framework (PEFC), a voluntary initiative led by the forest industry to promote an internationally credible certification framework. Both organizations develop principles and criteria for SFM using stakeholder participation and accredit third-party auditors to verify compliance through annual audits. Certifiers may issue a Forest Management Certificate for forest stewards or a Chain-of-Custody Certificate for forest product manufacturers and distributors. Consumers can then identify certified wood products through a certification logo.

Environment in Focus

Box 9.11 Forests: A Global Perspective

- Some 40 per cent of the land surface of the Earth supports trees or shrubs while 30 per cent is fully forested.
- Five countries—Canada, Russia, the US, Brazil and China—contain more than 50 per cent of the world's forests while 10 countries have no forest at all.
- Each year, about 13 million hectares of the world's forests are lost to deforestation, but the rate of net forest loss is slowing down, thanks to new planting and natural expansion of existing forests.
- From 2000 to 2010, the net forest loss was 5.2 million hectares per year—an area the size of Costa Rica.
- Deforestation accounts for up to 20 per cent of the global greenhouse gas emissions that contribute to global warming.
- Primary forests comprise over a third of all forests but lost over 40 million hectares since 2000 through deforestation or selective logging.
- Plantation forests are established at a rate of 5 million hectares per year.
- Plantations cover more than 264 million hectares (7 per cent of total forested area).
- The 10 countries with the largest net forest loss per year between 1990 and 2010 (Brazil, Indonesia, Sudan, Myanmar, Nigeria, United Republic of Tanzania, Mexico, Zimbabwe, Democratic Republic of the Congo, Argentina) had a combined net forest loss of 7.9 million hectares per year.
- The 10 countries with the largest net forest gain per year between 1990 and 2010 (China, Spain, Vietnam, United States, Italy, Chile, India, France, Finland, Philippines) had a combined net forest gain of 3.4 million hectares per year as a result of afforestation and natural expansion of forests.
- Estimates suggest that more than 80 per cent of the world's terrestrial species are found in forests. The tropical forests are our richest terrestrial biome. Tropical forests occupy only 7 per cent of the world's land area, but they contain more than half of the world's species.
- Developing countries consume more than 80 per cent of their wood as fuel; in developed countries, only 16 per cent goes to fuel, with the rest being processed as wood products. Approximately 1.5 billion tonnes of wood is harvested for fuel annually worldwide.
- Most wood products (85 per cent) are used domestically.
- Wood products are valuable, worth over US$100 billion annually between 2003 and 2007.
- Global forests provide wage employment and subsistence equivalent to 60 million work-years annually worldwide, 80 per cent of which is in developing countries. More than 1.6 billion people depend to varying degrees on forests for their livelihoods (e.g., fuelwood, medicinal plants, and forest foods).
- People in developing countries consume much less wood products (30 cubic metres per 1,000 people) and paper (12 tonnes per 1,000 people) than people in developed countries (300 cubic metres of wood products per 1,000 people and 150 tonnes of paper per 1,000 people).
- Thirty per cent of the world's forests are designated for production, with just 8 per cent to protection and 12 per cent to conservation.
- Worldwide, an estimated 460 million hectares of forested land are designated for the protection of biological diversity. However, of 200 areas of high biological diversity, 65 per cent are threatened by illegal logging. Illegal logging is estimated to cost governments approximately $15 billion annually.

Can tropical forests, such as these in Sri Lanka, sustain the needs of both the people and animals that depend on them?

Source: FAO (2010).

Although the early proponents of forest certification hoped to target tropical deforestation, the temperate and boreal forests of industrialized countries account for 87 per cent of all certifications. Some argue that this is because certification is not conducive to forest management schemes involving communities or small enterprises, which are typical in developing countries. However, tropical forest certifications in the developing world are continuing to grow.

Perhaps the biggest factor that might affect the way we look at global forests is their role in mitigating the impact of global climate change. In the accompanying International Guest Statement, someone actively involved in conservation through carbon trading discusses the issue.

Implications

Forestry is 'at a watershed' in Canada in terms of how forests, their values, and their management are viewed. The next decade will be crucial in determining whether Canadians will still consider themselves a forest nation in another 50 years. Although society in general and government and industry in particular have a much greater appreciation of the changes

Environment in Focus
Box 9.12 The Seal of Approval

Increasingly, customers for wood products around the globe are asking for guarantees that the products they buy come from forests managed and logged according to ecologically responsible standards. The trend towards responsible consumerism is supported by certification and labelling, a process in which an independent audit of a forestry company is conducted to assess whether it meets internationally and/or nationally recognized guidelines for responsible forest management. Certification is designed to enable consumers and participants to measure forest management practices against approved standards and also provides forest owners with an incentive to maintain and improve forest management practices.

In Canada, there are three main certification systems:

- Forest Stewardship Council (FSC);
- Canadian Standards Association (CSA);
- Sustainable Forestry Initiative (SFI).

Important differences exist among these certification systems related to standards, policies, procedures, and on-the-ground results. The FSC is an international system with strict standards; the others are largely generated and controlled by the forest industry. Canada boasts that it has the world's largest area of certified forest (Natural Resources Canada, 2007b), but only a small proportion of this certified forest meets international standards specified by the FSC. One independent report (ÉEM Inc., 2007) found that the FSC had the only system prohibiting the use of genetically modified trees, preventing the conversion of natural forests to plantations, and requiring a precautionary approach to the management of areas with high conservation value. Although it found that certification systems often mentioned similar requirements, the study raised concerns that, under some systems, it was left to the individual forest manager to decide what to do on the ground. The study identified the FSC as generally more rigorous in its performance requirements, and this conclusion has been subsequently supported by a further audit (Masters et al., 2010).

Forest certification systems are designed to link environmentally and socially conscious consumers with like-minded producers, retailers, and distributors and typically involve:

- independent third-party auditing;
- **chain-of-custody** procedures (verification of compliance from the forest through to the final product);
- on-the-ground inspections of forested areas to determine whether they are managed according to established sets of environmental and social standards;
- certified product labelling;
- multi-stakeholder involvement.

The range of issues considered in defining responsible forest management include wildlife habitat protection, endangered forests identification and maintenance, riparian and water utility protection, indigenous peoples' rights, and the equitable sharing of benefits with forestry-dependent workers and communities. For example, the FSC advocates that all functions of a forest ecosystem remain intact after an area is logged. This requires that a mix of tree species of different ages still remains standing after the forest is logged and that the functions of trees and other plant species also remain intact.

Forest products given the seal of approval should give consumers confidence that the products they purchase are derived from responsibly managed forests. As of 2009, 142.8 million hectares in Canada had been certified under one or more of the main certification systems listed above (Figure 9.12).

More and more, Canadian companies are trying to meet international standards. One company operating mainly in the boreal forest of northeastern Ontario has earned the FSC logo for voluntarily meeting its high standards for forest management. Clear-cutting once dominated the 2-million-hectare forest managed by Tembec Inc., but today considerable patches of trees are left standing, large tracts of old-growth forest are being protected, and selections of all forest types are being set aside to serve as wildlife habitat. Tembec is a leader in environmental forestry and now has almost 10 million hectares of FSC-certified forest in Canada.

Tembec has demonstrated that it is possible to dramatically improve forest management practices to attain the coveted FSC logo. In a climate in which consumers are more concerned about environmental and social issues surrounding primary resource extraction, securing a seal of approval from a highly rated certification program will become increasingly important.

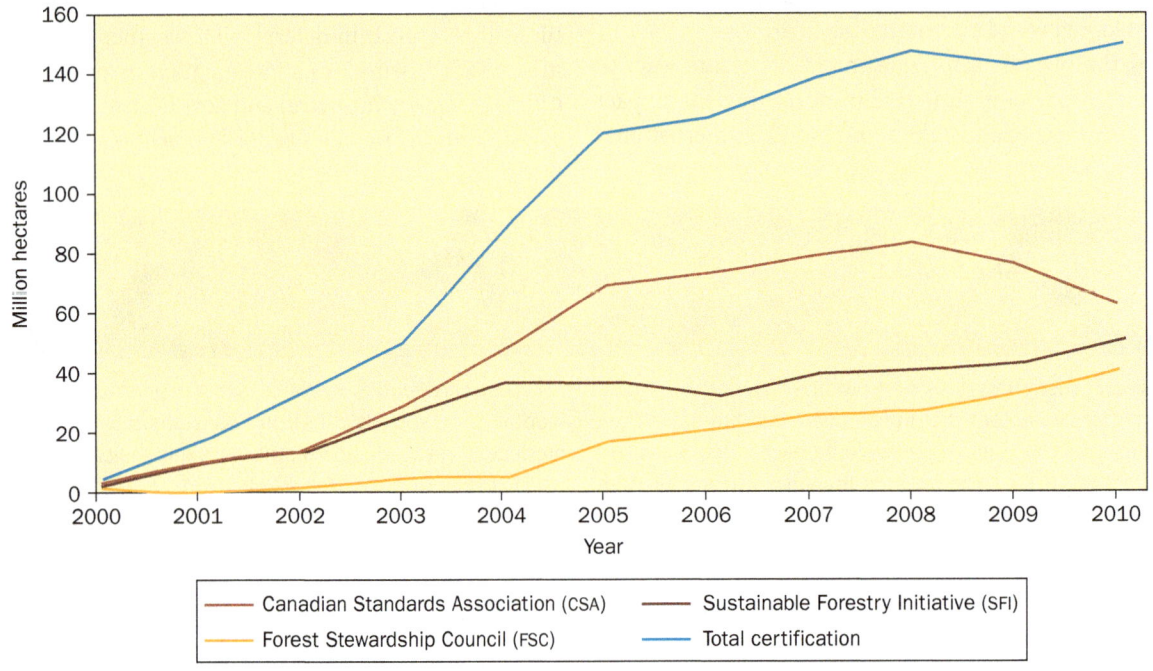

Figure 9.12 | Forestry certifications in Canada, 2000–2010. *Source:* Natural Resources Canada (2011: 34).

that need to be made in the industry to move towards more sustainable practices, actually making these changes will take some time. However, the world is watching, and the power of consumers to effect change is well in evidence. Box 9.13 offers some suggestions on things you can do in this regard. Large companies such as IKEA and Home Depot have reacted to consumer pressure by agreeing not to sell products from forestry operations not certified as sustainable. One can only hope that similar changes in other areas of forest product use will occur.

Forestry will no longer be the main economic backbone for many communities, as it was in the past, because mills continue to close. Some places will be hit harder than others. In BC, which accounts for over 50 per cent of Canada's lumber production, the annual timber harvest is expected to drop by one-third over the next decade. This decline is partly due to the mountain pine beetle, but much is also due to the failure of forest management to heed warnings about overcutting during previous decades. It is to be hoped that changes will arise as a result of this situation and help to generate a more sustainable industry in the future.

Some changes are already evident. The United Steelworkers Union, representing thousands of forestry workers, is now joining with environmental groups such as the Western Canadian Wilderness Committee—new allies that were bitter foes 10 years ago—in demonstrations to draw attention to the situation. Communities like Ucluelet, Port Alberni, and Port Hardy on Vancouver Island, which in the past strongly resisted attempts to preserve forest lands for other values such as recreation, are now looking at forests with a much greater appreciation of the multiple values they contain. Similar changes are occurring across the country. Forestry will continue to play a role in many economies, but the days of the one-industry town are gone.

International Guest Statement
Can the Private Sector Help Slow Deforestation?
Gijsbert Nollen

The private sector is usually looked at as being a main cause of deforestation, and in the past it has been. The price mechanism dictates that if a forest is worth more money cut down than standing, then the chances of it being cut down are very high. However, how forest value is calculated may now be changing so that higher values are achieved through leaving the trees standing. Of particular importance is the attention now being given to carbon sequestration to help mitigate the effects of global warming, as discussed in Chapter 7.

More than 20 per cent of total annual carbon emissions are emitted by land-use change, most of which is tropical deforestation, and approximately 14 million hectares of forest are lost annually (Kanninen et al., 2007). Inclusion of Reduced Emissions from Deforestation and Forest Degradation (**REDD**) in any future global climate change agreements is a must for mitigating the impacts of global climate change. Although for policy-makers next year may seem to be in the near future, for our forests every minute counts. By the time you have read this statement, approximately 250 ha of forest have been lost. Currently, proper platforms are lacking for payments for ecosystem services, such as carbon and water, and there is a global failure to account for all costs—specifically environmental costs—in supply chain and production processes. At the same time, policy-makers are unwilling (or unable) to address these two issues, leaving the protection of ecosystems, including forests, at a huge disadvantage in marketplace economics and making their liquidation more attractive.

Ideally, an ecosystem approach would develop a framework to address the deforestation drivers, integrating ecosystem services into a profitable structure with a conservation goal, and going hand-in-hand with sustainable development. This approach incorporates various methods, including ecosystem service dependency and impact assessment, valuations of ecosystem services, land-use scenarios, development requirements, and policies to support the transaction. This approach could also act as a catalyst to change contemporary economic land use and revenue generation from resource extraction to sustainable land use and sustainable development and provision of ecosystem services (based on resource conservation).

Therefore, if we want to address carbon emissions stemming from land-use change effectively and curb unsustainable land-use change (such as deforestation), returns from REDD should be more attractive than the alternative. Inclusion of payments for other ecosystem services, such as water, siltation, and biodiversity, could add to the viability of such a project. However, given the complexities of structuring these transactions, a deal should be viable and attractive on carbon trade alone.

Estimates of the potential global value of REDD payments vary, depending on the underlying assumptions. Assuming a conservative carbon value of $10 per ton of CO_2e, estimates include a net present value of $150 billion (Chomitz et al., 2007) and annual revenue of $2.3 billion to $12 billion (Ebeling, 2006; El Lakany et al., 2007). These large values make an interesting market for financial institutions, emerging forested countries, investors, and transaction developers. As such, the private sector has already moved into the market, and deals have been struck in Indonesia (Aceh by Merrill Lynch), Guyana (Iwokrama by Canopy Capital), Brazil (Amazonas by Marriott Hotel Chains), and Bolivia (Noel Kempff, traded on the Chicago Climate Exchange).

Much research is still required to develop rigorous scientific assessments and to trade forest carbon effectively. However, time is not on our side; the drivers for deforestation are winning, and conservation is losing. If deforestation continues at current rates, approximately another 52 million hectares will be lost over the period 2009–12. Therefore, it is imperative that we do not wait until the policy-makers finally draw up a framework. Under the current regulated carbon forestry market within the Kyoto Protocol, the Clean Development Mechanism (CDM), and Joint Implementation (JI), REDD is not an approved methodology, and only one reforestation/afforestation project was registered under the CDM by the end of 2007. Even though REDD is already being traded on the voluntary markets (approximately 5 per cent of total voluntary market value), the future of this market depends on what happens post-2012. Therefore, if REDD were to be traded on a lead platform through a post-Kyoto United Nations Framework Convention on Climate Change, the regulator has to be capable of dealing with these transactions effectively and efficiently. So far, the CDM has been too expensive and bureaucratic to be a trading platform addressing land use, land-use change, and forestry project proposals effectively and leading to successful carbon return-based land management transactions.

Policy-makers should lower the transaction costs, simplify the verification and monitoring of the transactions, increase benefits to developing nations trading carbon return-based land management transactions, and address the demand for unsustainable forest products and biofuels at home. An effective framework should be simple for developing countries willing to look at carbon trading from land management transactions, as well as for developers of the transactions, investors, communities in the transaction area, and purchasers of the carbon.

The private sector is taking a lead in this, and several projects are already being developed and traded on a voluntary basis. It will take visionary entrepreneurs, nations, and policy-makers to curb current deforestation rates and combat climate change in the land-use change arena. Let us hope that policy-makers follow the trend by drawing up an effective framework and trading platform, allowing REDD transactions to enter the mainstream of economic transactions in the future. The market has been a key force in creating global deforestation; it can also play a key role in preventing future deforestation.

Courtesy Gijsbert Nollen

Gijsbert Nollen has been based in Asia for over fourteen years, where he has provided project management services to develop, finance, manage, and implement sustainable development projects. In recent years, he has focused on developing projects that effectively address the poverty-environmental nexus. He has twelve years of experience in advising clients from both the private and public sectors on sustainable development projects. He specializes in comprehensive problem solving, designing and setting up management schemes, analyzing and developing institutional frameworks, project management, and advising on transactions. Gijsbert is inspired by a wide range of varying interests and opinions that drive his passion to advance sustainable development causes. He is now based in Bangkok and can be reached at gijsbert@nollengroup.com.

Environment in Focus

Box 9.13 What You Can Do

A number of environmental stresses threaten the health of Canada's forests, but individuals can help.

1. Reduce your consumption of paper products. For example, use all available space on your paper (write in the margins and on both sides); carry fabric with you and use it instead of paper tissues, towels, and napkins; resist the temptation to print materials from the Internet—read on-line instead; insist that no unsolicited flyers be delivered to your home.
2. Recycle paper products, and purchase recycled paper products.
3. Choose unbleached paper products whenever possible. Unbleached paper is less harmful to the environment, since it does not require the toxic chemicals used to whiten paper.
4. Purchase certified wood products. Make an effort to purchase wood that has been certified by at least one certification agency. This will help to ensure that the wood you purchase comes from a logging company that has introduced measures to promote long-term ecological, social, and economic sustainability.
5. Reduce the risk of human-caused fires. Many forest fires are started by human carelessness. Obey fire restrictions when visiting parks and protected areas, and do not discard fire accelerants (e.g., cigarettes) along highways.
6. Join one of the NGOS that support sustainable use of forest resources.
7. Write or phone your MP or provincial or territorial representative, and/or e-mail a letter to the editor of your local newspaper. Canadians own Canadian forests. If you are unhappy about how forests in your province or territory are being managed, voice or otherwise publicize your discontent.

Summary

1. Some 40 per cent of the land surface of the Earth supports trees or shrubs. Four countries—Canada, Russia, the US, and Brazil—contain more than 50 per cent of the world's forests. Each year, about 13 million hectares of the world's forests are lost to deforestation, but the rate of net forest loss is slowing down, thanks to new planting and natural expansion of existing forests. From 2000 to 2005, the net forest loss was 5.2 million hectares per year—an area the size of Costa Rica.

2. Canada is a forest nation. Not only do we have 10 per cent of the world's forests, but we are also the largest exporter of forest products in the world. In 2010, Canada's forests contributed $22.5 billion to national GDP. The forestry industry is the largest single contributor to Canada's balance of trade. The forests, along with the North, are dominant elements in the history and culture of the nation.

3. There are 15 terrestrial ecozones in Canada. Most of Canada's forests are within eight of these ecozones: Boreal Cordillera, Pacific Maritime, Montane Cordillera, Boreal Plains, Taiga Plains, Boreal Shield, Mixed Wood Plains, and Atlantic Maritime. They differ with respect to climate, geology, genetic diversity, and level of human development. The Boreal Shield, Canada's largest ecozone, is threatened by a number of environmental stresses, including logging, stand replacement, mining, hydroelectric development, acid precipitation, and climate change.

4. Ecosystems provide an array of beneficial services arising from ecological functions such as nutrient and water cycling, carbon sequestration, and waste decomposition. Forests are also places of exceptional scenic beauty, and millions of Canadians participate in nature-related recreational activities each year.

5. Forests are also a valuable source of commodities. Non-timber forest products contribute millions of dollars to the Canadian economy and are also an important aspect of Aboriginal peoples' subsistence economies. Despite government recognition that Canadian forests provide a broad range of values, forest management paradigms have traditionally focused on the management of forests to supply wood.

6. The provincial and territorial governments are responsible for 77 per cent of the nation's forests and the federal government for 16 per cent on behalf of the owners, the people of Canada. The remaining 7 per cent is owned privately. Governments enter into contract arrangements with private companies in which they can specify the forest management practices to be followed.

7. Approximately 143 million hectares are currently managed for timber production. On these lands, forest ecosystems are being transformed from relatively natural systems to control systems in which humans, not nature, influence the species that will grow there and the age to which they will grow.

8. The rate of conversion from natural to managed forest is controlled by provincially established annual allowable cuts (AACs). In theory, the AAC should approximate what the land should yield in perpetuity. It is not sustainable to have an AAC that consistently exceeds this biological potential. In most provinces, the AAC is greater than the harvest, but in some regions, the AAC is approaching or exceeds the harvest for softwoods.

9. In 2009 Canada harvested 118 million cubic metres of timber, which was 89 million cubic metres, or 40 per cent, less than the allowable cut for 2009.

10. Silviculture is the practice of directing the establishment, composition, growth, and quality of forest stands through harvesting, reforestation, and site preparation.

11. Clear-cutting is the dominant harvesting system in Canada. It is the most economical way for extracting fibre for short-term profit and also allows for easier replanting and tending of the regenerating forest. In certain types of forests, it may mimic natural processes more closely than selective or partial cutting systems. However, clear-cutting may not be the most appropriate way to harvest timber in some areas. Clear-cuts are aesthetically unappealing to many Canadians, and their environmental impact can be substantial.

12. Biocides are used in forestry to control populations of vegetation and insect species that compete with or eat commercial species. Several high-profile conflicts have arisen over application of chemicals. Concern over spraying to control the spruce budworm in the Maritime provinces is one of the most significant. The biological control agent *Bacillus thuringiensis* is increasingly being used against insect attacks in Canada.

13. Intensive forest management techniques are used to further enhance future timber resource values. Intensive silvicultural practices include pre-commercial thinning, commercial thinning, scarification, prescribed burning, pruning and shearing, and timber stand improvement. The long-term impacts of these activities are not well understood.

14. Fire suppression has resulted in ecological changes not characteristic of fire-dominated ecosystems, and has led to a gradual increase in the area burned over the past 30 years.

15. Various environmental impacts are associated with current forestry management systems, including changes to ecosystem, species, and genetic diversity; changes to biogeochemical and hydrological cycles; and soil erosion.

16. Timber harvesting can significantly alter species composition and abundance as the proportion of forest with old-growth characteristics is reduced. Species such as the woodland caribou and marten that depend on old-growth characteristics decline in abundance. Other species, such as deer, may increase as regenerating cut areas produce more forage for them.

17. The spotted owl is perhaps the best-known example of the impact of logging on biodiversity. The spotted owl requires old-growth forests to maintain populations, but logging in BC's old-growth forests continues to threaten this endangered species. Management options to maintain populations are currently being considered.

18. Forest harvesting removes nutrients from the site. The significance of this for future growth varies, depending on the nutrient capital of the site and type of harvesting system used. Sites with abundant capital and/or selective harvesting systems that leave branches behind will suffer less growth impairment of future generations than nutrient-poor sites or sites that are clear-cut with complete-tree removal.

19. Forest harvesting may also contribute to increased soil erosion and water flows.

20. Forests are a carbon sink. They take in carbon dioxide and convert it to wood, leaves, and roots. They are also a carbon source. They release stored carbon into the atmosphere when they decompose or burn. Because of this ability to both absorb and release huge amounts of carbon dioxide (a major greenhouse gas), forests play a major role in the global carbon cycle.

21. Canada's managed forests were a net carbon sink in most years between 1990 and 2010 But in some years, they were a source, mostly because of wildfires. In 2009, Canadian forests sequestered 130 million tonnes of CO_2.

22. Carbon sequestration was not considered in timber management in the past, but it might become a major influence on how we manage our forests. Hardwood stands in New Brunswick, for example, typically contain 10 to 20 per cent more carbon per hectare than a similar volume of softwoods of similar age. Forest managers might well change management practices in the future as global climate change becomes more severe and, therefore, grow multi-aged hardwood stands that promote carbon sequestration instead of the single-aged softwoods that now dominate thinking.

23. Forests produce many values for Canadians. In the past, attention focused almost exclusively on the monetary returns from forest harvesting. However, as the amount of forest brought under management has increased and as the public becomes increasingly aware of the changes occurring in Canadian forests, more attention is being devoted to the assessment and management of other values besides timber production. An ecosystem perspective is being adopted.

24. Concern over the impact and sustainability of forest practices has given rise to calls for what has been termed 'new forestry'. Such an approach embraces an ecosystem and adaptive management perspective that seeks to mimic natural processes more closely and give greater attention to the full range of values from the forests.

25. Management of Canada's forests is directed by a National Forest Strategy, developed by provincial and territorial forest ministers and the Canadian Minister of Natural Resources. The aim is to develop and implement more sustainable management practices. Independent reviews of the NFS have found many weaknesses. Despite this, Canada continues to boast of its international leadership in sustainable forest management.

26. Canada's Model Forest Program is one commitment arising from Canada's 1992–7 National Forest Strategy. Eleven model forest agreements were developed in six forest regions across the country, but the program was terminated in 2007. Core issues relate to ecosystem management, Aboriginal participation, public participation, science and innovation, and the integration of non-market values into decision-making.

27. Canada's Boreal Forest Agreement committed environmental groups and forestry companies to innovative approaches for managing 72 million hectares of Canadian boreal forest. The agreement aims to develop a strong, sustainable forest industry while better protecting ecosystems.

28. Internationally, the Convention on Biological Diversity commits signatories to prepare and adhere to a national biodiversity strategy, including the designation of representative samples of their forest lands as protected areas and ensuring that forest management does not impair biodiversity.

29. The United Nations Forum on Forests has agreed on four shared Global Objectives on Forests relating to reversing the loss of forest cover worldwide through sustainable forest management, enhancing forest-based economic, social, and environmental benefits, increasing the area of sustainably managed forests, and reversing the decline in official development assistance for sustainable forest management.

30. Approximately 8 per cent of the world's forest area is certified—a nearly fivefold increase since 2000. Certification is designed to enable consumers and participants to measure forest management practices against approved standards and also provides forest owners with an incentive to maintain and improve forest management practices.

Key Terms

annual allowable cut (AAC)
Atlantic Maritime
biocides
Boreal Cordillera
boreal forest
Boreal Plains
Boreal Shield
certification
chain-of-custody
clear-cutting
culmination age
DDT (dichlorodiphenyltrichloroethane)
ecosystem-based management
falldown effect
forest tenure
long-run sustained yield
Mixed Wood Plains
Montane Cordillera
new forestry
non-timber forest products (NTFPs)
old-growth forests
Pacific Maritime
pheromones
reclamation
REDD
second growth
silviculture
sustained yield
taiga
Taiga Plains
windthrow

Questions for Review and Critical Thinking

1. Outline some of the ways in which forests are important to Canada.

2. Compare and contrast the physical and biological attributes of the Boreal Shield and Pacific Maritime ecozones. What major economic activities affect the health of these ecozones?

3. How is forestry an ecological process?

4. What is an AAC?

5. Outline some of the advantages and disadvantages of clear-cutting.

6. Is Canada reforesting all lands that are harvested? What are some of the issues associated with current replanting schemes?

7. Outline some of the pros and cons of using chemical sprays to control insect infestations in Canada's forests.

8. List all the different values that society realizes from forests. What do you think the priorities should be among these different and sometimes conflicting uses?

9. Name some species that might increase in abundance as a result of forest harvesting and others that might decline. What are the characteristics of these species that would encourage this response?

10. What are the impacts of forest harvesting on site fertility, and how do they differ between sites?

11. What attributes of old-growth forests appear to explain their use by spotted owls?

12. What are the implications of global climate change for forest management?

13. How is forest management administered in Canada? What are the main strengths and weaknesses of this approach? What alternatives might you suggest? Do examples of such alternatives exist in your region?

14. What is 'new forestry'?

15. What tools are used to evaluate the sustainability of Canadian forests? Are these tools adequate to the task?

16. Natural Resources Canada (2007b) states that 'When we look at the international milestones in sustainable forest management, Canada's leadership stands out.' Do you agree?

Related Websites

A Vision for Canada's Forests, 2008 and Beyond
www.ccfm.org/english/coreproducts-nextnscf.asp

BC Ministry of Forests
www.for.gov.bc.ca/hfp/sof

Canadian Boreal Forest Agreement
canadianborealforestagreement.com

Canadian Boreal Initiative
www.borealcanada.ca; www.borealbirds.org

Canadian Forest Service
cfs.nrcan.gc.ca/centres/read/pfc

Canadian Forest Service, the State of Canada's Forests
cfs.nrcan.gc.ca/series/read/90

Canadian Institute of Forestry
www.cif-ifc.org

Canadian Model Forest Network
www.modelforest.net

Canadian Parks and Wilderness Society (CPAWS)
www.cpaws.org

Certification Canada
www.certificationcanada.org

David Suzuki Foundation, forests
www.davidsuzuki.org/cgi-bin/mt1/mt-search.cgi?IncludeBlogs=1&tag=forests&limit=10

Environment Canada What you can do
www.ec.gc.ca/eco/main_e.htm

Food and Agriculture Organization of the United Nations
www.fao.org/forestry/fra/fra2010/en/

Foothills Research Institute
foothillsresearchinstitute.ca/pages/home

Forest Stewardship Council
www.fsccanada.org

Global Forest Watch
www.globalforestwatch.ca

Government of Manitoba
www.gov.mb.ca/conservation/forestry/health/index.html

Government of Saskatchewan
www.gov.sk.ca/news?newsId=f31337b1-85f1-4100-b2ad-fc1b07990e9e

Industry Canada
www.ic.gc.ca/eic/site/fi-if.nsf/eng/h_fb01340.html

Model Forest Program
www.modelforest.net/index.php?option=com_k2&view=item&layout=item&id=27&Itemid=28&lang=en

National Forest Information System
www.nfis.org/

National Forestry Database Program
nfdp.ccfm.org

Natural Resources Canada, forests in Canada
cfs.nrcan.gc.ca/pages/64

Natural Resources Canada, silvicultural terms in Canada
nfdp.ccfm.org/silviculture/quick_facts_e.php

Parks Canada, Species at Risk
www.pc.gc.ca/eng/nature/eep-sar/index.aspx

Sierra Club
www.sierraclub.ca/national/programs/biodiversity/forests/index.shtml

United Nations Forum on Forests
www.un.org/esa/forests/index.html

Wilderness Committee
www.wildernesscommittee.org

Further Readings

Note: This list comprises works relevant to the subject of the chapter but not cited in the text. All cited works are listed in the Bibliography at the end of the book.

Benkman, C.W. 1993. 'Logging, conifers, and the conservation of crossbills', *Conservation Biology* 7: 473–9.

Bradshaw, C.J.A., I.G. Warkentin, and N.S. Sodhi. 2009. 'Urgent preservation of boreal carbon stocks and biodiversity', *Trends in Ecology and Evolution* 24: 541–8.

Buchert, G.P., et al. 1997. 'Effects of harvesting on genetic diversity in old-growth eastern white pine in Ontario, Canada', *Conservation Biology* 11, 3: 747–58.

Canadian Boreal Initiative. 2005. *The Boreal in the Balance: Securing the Future of Canada's Boreal Region*. Toronto: Canadian Boreal Initiative.

Cyr, D., et al. 2009. 'Forest management is driving the eastern North American boreal forest outside its natural range of variability', *Frontiers in Ecology and the Environment* 7: 519–24.

Elgie, S., G.R. Mccarney, and W.L. Adamowicz. 2011. 'Assessing the implications of a carbon market for boreal forest management', *Forestry Chronicle* 87, 3: 367–81.

Hanna, K. 2010. 'Transition and the need for innovation in Canada's forest sector', in B. Mitchell, ed., *Resource and Environmental Management n Canada: Addressing Conflict and Uncertainty*, 4th edn. Toronto: Oxford University Press, 269–300.

McCarty, J. 2005. 'Neoliberalism and the politics of alternatives: Community forestry in British Columbia and the United States', *Annals, Association of American Geographers* 96: 84–104.

McLachlan, S.M., and D.R. Bazley. 2003. 'Outcomes of long-term deciduous forest restoration in southwestern Ontario, Canada', *Biological Conservation* 113: 159–69.

McRae, D.J., et al. 2001. 'Comparisons between wildfire and forest harvesting and their implications in forest management', *Environmental Reviews* 9: 223–60.

National Forest Strategy Coalition (NFSC). 1992. *Sustainable Forests: A Canadian Commitment*. Ottawa: Canadian Council of Forest Ministers.

———. 2003. *National Forest Strategy (2003–2008), A Sustainable Forest: A Canadian Commitment*. Ottawa: Canadian Council of Forest Ministers.

———. 2005. *Boreal Futures: Governance, Conservation and Development in Canada's Boreal*. Ottawa: NRTEE.

Pimm, S.L., N. Roulet, and A. Weaver. 2009. 'Boreal forests' carbon stores need better management', *Nature* 462: 276.

Rayner, J., and M. Howlett. 2007. 'The National Forest Strategy in comparative perspective', *Forestry Chronicle* 83: 651–7.

Tarnocai, C., J.G. Canadell, E.A.G. Schuur, P. Kuhry, G. Mazhitova, and S. Zimov. 2009. 'Soil organic carbon pools in the northern circumpolar permafrost region', *Global Biogeochemical Cycles* 23, GB2023. doi:10.1029/2008GB003327.

Timoney, K.P. 2003. 'The changing disturbance regime of the boreal forest of the Canadian Prairie provinces', *Forestry Chronicle* 79: 502–616.

Wulder, M.A., J.C. White, and N.C. Coops. 2011. 'Fragmentation regimes of Canada's forests', *Canadian Geographer* 55: 288–300.

Chapter 10
Agriculture

Learning Objectives

- To understand the environmental and social impacts associated with the growth of agriculture.
- To appreciate the global food situation and some of the factors that influence it.
- To understand the nature and importance of biofuels and some of their advantages and disadvantages.
- To gain an understanding of the role of energy inputs in agriculture and the Green Revolution.
- To realize the main trends in Canadian agriculture and Canada's contribution to the global food supply.
- To know some of the main environmental implications of agriculture in Canada.
- To appreciate the contributions of agriculture to global climate change.
- To understand some of the main problems arising from the use of agricultural chemicals.
- To analyze the implications of a diet with a high level of meat consumption.
- To discover some of the changes that have to be made to move towards more sustainable modes of agricultural production.

Introduction

The food we grow and eat represents the most intimate interaction between humans and the natural world. For those of us lucky enough to have sufficient food, we will eat three or more times a day: consuming food and, along with it, all of the energy, chemicals, and organisms that have gone into producing it. Like much of the world, our agricultural system has shifted dramatically over the past centuries, altering the way we produce, consume, and think about our food. Unfortunately, many people think about food all the time, because they are not getting any. As this chapter is being written over 10 million people are starving, with 3.5 million people in danger of starving to death in Somalia as a result of drought. Canada has committed $50 million in food aid to the situation, the World Bank $500 million, and the UN is calling for an extra $1.6 billion; however, the rebel forces that control most of the country will not allow food aid to be distributed. The famine is the worst the region has experienced in 60 years, but global climate change patterns and political instability mean that this situation is going to be a reoccurring challenge. Clearly, questions of food security are a matter of urgency.

The origins of agriculture date back 9,000–11,000 years to a few regions where societies domesticated both plant and animal species. Through domestication, such desired traits as increased seed/fruit concentration and fleshiness, reduced or increased seed size, controlled seed dispersal, and improved taste could be achieved. Various agricultural practices—seedbeds, improved animal nutrition, and water management—also were devised. In turn, the increased availability of food, feed, and fibre provided the impetus for societies to prosper and support a larger non-farming population. Societies around the globe flourished by improving their capacity to expand agricultural production (Box 10.1).

The domestication of plants and animals continues today but under a much different set of social, economic, and environmental conditions than existed even a century ago. Agriculture is a dominant influence on the global landscape outside the major urban centres, if not the dominant influence.

Historically, agricultural output has been increased by bringing more land into production. For example, the global extent of cropland increased from around 265 million

Children play in the United Nations High Commissioner for Refugees Ifo extension camp in eastern Kenya. In summer 2011, the camp registered more than 1,000 newcomers a day, refugees displaced by the drought in the Horn of Africa, which triggered the worst famine the region had seen in a generation.

Environment in Focus

Box 10.1 Social Implications of the Development of Agriculture

Agriculture has had a profound influence on society, which in turn has further implications for ecosystems.

- More reliable food supplies permitted growth in populations.
- A sedentary life became more possible as a result of these food supplies and the ability to store food; this allowed the establishment of larger, permanent settlements.
- Permanent settlements allowed greater accumulation of material goods than was possible under a nomadic lifestyle.
- Agriculture allowed food surpluses to be generated so that not all individuals or families had to be involved in the food-generating process and specialization of tasks became more clearly defined. One end result is that only some 4 per cent of Canada's population is directly involved in food production today, permitting the rest of the population to direct their energies to other tasks, historically, the processing of raw materials into manufactured goods, thereby increasing the speed of flow-through of matter and energy in society. As indicated in Chapter 4, this high rate of throughput is at the core of many current environmental problems.
- The creation of food surpluses and more material goods promoted increasing trade between the now sedentary settlements. This led to the development of road and later rail connections to facilitate the rapid transport of materials, involving the consumption of large amounts of energy.
- Land and water resources became more important, leading to increased conflict between societies regarding control over agricultural lands.
- Aggregation of large numbers of people together in sedentary settlements also served to concentrate waste products in quantities over and above those that could be readily assimilated by the natural environment. Today we call this pollution.

Huge monuments, such as Angkor Wat in Cambodia, could not have been constructed, nor the large cities that surrounded them, without the development of agriculture.

hectares in 1700 to around 1.2 billion in 1950, predominantly at the expense of forest habitats, and now stands at more than 1.5 billion hectares. However, the opportunity for further geographic expansion of cropland is small because of the comparatively limited amount of land well-suited for crop production, the increasingly concentrated patterns of human settlement, and growing competition from other land uses (Box 10.2). Global climate change will increase the area suitable for agricultural production in some countries, especially in northern latitudes, but also lead to declines elsewhere, particularly in the tropics where future food demand will be the highest. South America may lose anywhere from 1 per cent to 21 per cent of its arable land area, Africa 1–18 per cent, Europe 1–17 per cent, and India 2–4 per cent (Zhang and Cai, 2011).

The intensification of production—obtaining more output from a given area of agricultural land—is a key development strategy. World grain production has almost tripled since 1961 (Figure 10.1), mainly because farmers are harvesting more grain from each hectare. In fact, only slightly more land today is planted in grain: 671 million hectares in 2002 compared with 648 million in 1967 (Worldwatch Institute, 2003b), with little change to 2009 (FAO, 2009). Worldwide, intensification of production, dubbed the 'Green Revolution', has doubled the average harvest of grain from a given hectare from 1.24 tonnes in 1961 to 2.83 tonnes in 2009 (FAO, 2009). The amount of grain produced per hectare grew from 285 kilograms in 1961 to a peak of 376 kilograms in 1986, falling to 330 kg/ha by 2011 (FAO, 2011a). On average, humans get about 48 per cent of their calories from grains, a share that has declined just slightly, from 50 per cent, over the past four decades. Grains, particularly corn, in conjunction with soybeans, also form the primary feedstock for industrial livestock production.

World grain production in 2011 is estimated at 2.313 billion tonnes, although with growing demand, the global cereal stocks remain low (Figure 10.2). However, future capacity to deliver agricultural outputs also depends on the continuing ecological viability of agro-ecosystems, and the Green Revolution carries with it a number of significant long-term negative environmental impacts that will ultimately lead to long-term productivity losses. Despite past successes, affordably feeding the current world population—and the more than 70 million people per annum by which that population will continue to grow over the next 20 years—remains a formidable challenge (Box 10.3). The International Food Policy Research Institute (IFPRI) suggests that between 1995 and 2020, global demand for cereals will increase by 40 per cent, while meat demand is projected to grow by 63 per cent and the demand for roots and tubers by 40 per cent. Between 2000 and 2006, world demand for cereals rose by 8 per cent; over the same period the price increased by 50 per cent, and prices more than doubled by early 2008 compared to 2000. Slow-growing supply, low stocks, and supply shocks at a time of surging demand for feed, food, and fuel have led to drastic price increases, and these high prices will not fall soon, according to the IFPRI (von Braun, 2007).

Grain prices reached a record high in 2008 and although still high, experienced significant declines in 2009. The global trade declined by 7 per cent in 2009 and continued to fall into early 2010 due to high stocks and good crop prospects. Drought and export closures in Russia pushed grain prices up again by mid-2010. The global food market expected stable prices and sufficient supplies into 2011, but worrisome outlooks drove prices to their highest in decades. Maize production dropped to a critical low in the US. As the primary producer and exporter of maize, the US crops largely influence global price fluctuations. Despite much turbulence in other crops, rice has remained stable in both production and price. As crops improved, Russia lifted its export ban in July 2011 and prices are expected to stabilize (Worldwatch Institute, 2011).

International food aid agencies also purchase their supplies on the world market and have been forced to scale back. In response to a 35 per cent increase in the cost of agricultural commodities, as well as the rising costs of fuel for shipping, the volume of aid provided through the largest assistance program in the United States, Food for Peace, has dropped by nearly half since 2005, to 2.4 million tonnes (USAID, 2011). The combination of rising food costs and declining aid can be fatal for the millions of people worldwide who experience hunger on a regular basis.

All regions will face difficulties in meeting the growing demand for agricultural products while also preserving the productive capacity of their agro-ecosystems. Global climate change will be an additional stress. In more than 40 developing countries, mainly in sub-Saharan Africa, cereal

Intensification has resulted in a doubling of grain production over the last 40 years. One of the main goals has been to increase the amount of grain produced per plant, as with these short-statured rice plants in the middle Himalayan region.

Environment in Focus
Box 10.2 Urbanization of Agricultural Land

Despite the fact that Canada is the second largest country in the world and one of the biggest exporters of foodstuffs worldwide, only 676,000 square kilometres, or about 7 per cent of Canada's overall land mass, is used for agricultural production. The amount of arable land free from severe constraints on crop production is even smaller, totalling less than 5 per cent of the land base. Limitations such as climate and soil quality reduce the amount of land that can be relied on for agricultural use; consequently, about 40 per cent of agricultural activities occur on marginal or poorer-quality land, which may not be dependable for long-term agricultural activity (Statistics Canada, 2001).

Most of the prime agricultural land is in southern Canada, where 90 per cent of Canadians live. In Ontario, for example, more than 18 per cent of Class 1 farmland is now being used for urban purposes. This juxtaposition of prime agricultural land and the main urban centres has meant that suburban expansion invariably leads to losses in agricultural land. By 2001, half of Canada's urbanized land was located on dependable agricultural land, and 7.5 per cent was on our very best agricultural land (Hofmann et al., 2005).

Between 1966 and 1986, 301,440 hectares of rural land were converted to urban use. Furthermore, some 58 per cent of the converted land was of prime agricultural capability. To replace the productivity of this land would require bringing twice as much land under cultivation on the agricultural margins. One study found that urbanization had consumed 15,200 km^2 of surrounding lands between 1971 and 2001, an increase of 96 per cent in the total amount of urban land over the period. In Ontario alone, urbanization increased by 4,300 km^2, a growth of almost 80 per cent (ibid.).

Urbanization of agricultural land affects specialty crops that have a limited ability to flourish in Canada. These crops often represent an important resource to local economies (e.g., the fruit belts in the Niagara and Okanagan regions). Cities also affect the use of surrounding lands in indirect ways—golf courses, gravel pits, and recreational areas are often located on agricultural land in areas adjacent to urban areas, and as a result, the effects of urban areas extend beyond their physical boundaries.

In an effort to slow the rates of conversion, several provinces have enacted legislation regarding the protection of agricultural lands. In 1972, for example, British Columbia enacted the Agricultural Land Reserve (ALR). At the time of its inception, some 6,000 hectares of prime agricultural land were being lost to urbanization each year. The annual loss has now fallen to less than one-seventh of that amount. The ALR initially covered 4.7 million hectares (5 per cent of BC); despite boundary changes over the decades, its current area remains approximately the same. A similar program exists in Quebec. In 1978, the Quebec government introduced the Agricultural Land Preservation Act, which now protects more than 63,000 square kilometres of prime agricultural land.

However, these programs do have their problems. Often, farmers whose lands fall within the program and cannot be sold for non-agricultural uses cannot compete with cheap agricultural imports from elsewhere. Hence, they cannot make a decent living in agriculture and yet cannot sell their lands for non-agricultural purposes. And as a nation, should we be concerned about becoming increasingly dependent on other countries for our food? What do you think?

Most Canadian cities of any size are surrounded by good agricultural lands. As the cities increase in size, they invariably encroach on the surrounding lands, as seen here in Quebec.

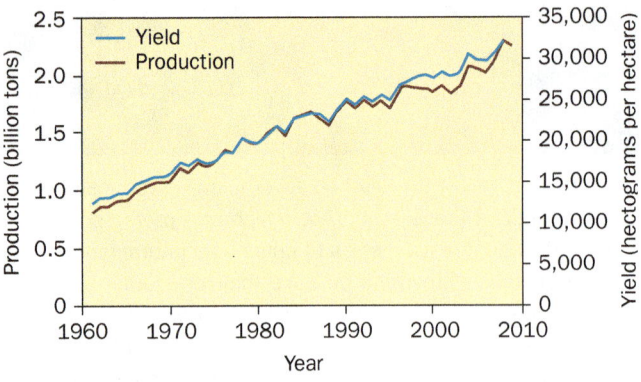

Figure 10.1 | World grain production and yield, 1961–2009. Source: Worldwatch Institute (2011: 56).

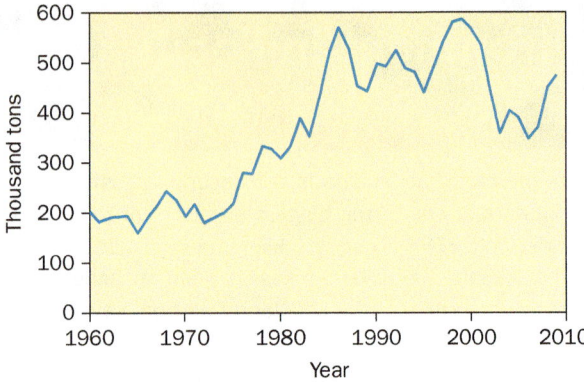

Figure 10.2 | World grain stocks, 1960–2009. Source: Worldwatch Institute (2011: 56).

yields are expected to decline, with mean losses of 15 per cent by 2080 (Fischer et al., 2005). Grain yields are likely to decline by 10 per cent for every 1°C increase over 30 years. Increased carbon dioxide and higher temperatures may promote lush growth, but they are deadly at the pollination stage, reducing some yields by 30 per cent. The International Rice Research Institute has found that the fertility of rice flowers falls from 100 per cent at 34°C to virtually zero at 40°C (Sanchez, 2001). In addition, warm, wet weather promotes diseases like blight, and pests survive warmer winters, meaning that longer growing seasons lead to an increased incidence of pest attacks.

World agricultural GDP is expected to fall by an average of 16 per cent as a result of climate change. However, these impacts are not evenly distributed, with a 20 per cent decline in less developed countries compared to a 6 per cent decline in industrial countries (Cline, 2007). Agricultural prices also will be affected, with price increases of up to 40 per cent expected if temperature increases more than 3°C (Easterling et al., 2007). Distribution will remain a major problem. For example, output per person for grain varies dramatically by region, standing at roughly 1,230 kilograms per year in the United States, most of which is fed to

Environment in Focus

Box 10.3 Hunger

Agriculture provides approximately 94 per cent of the protein and 99 per cent of the calories consumed by humans. On average, people need about 2,500 calories per day, although this amount varies from person to person depending on weight, age, level of activity, and other factors. North Americans and Europeans on average consume 3,400 calories per day. Conversely, over 1 billion people consume under 1,800 calories, and are classified as chronically hungry. Calories come mainly from carbohydrates, such as potatoes and rice, but to remain healthy, humans also need protein, vitamins, fatty acids, and minerals. Fatty acids are essential for hormonal control, and minerals are necessary for bone formation and growth.

People are considered undernourished if their caloric intake is less than 90 per cent of the recommended level for their size and level of activity. When people meet only 80 per cent of their recommended caloric intake requirement, they are considered severely undernourished. Malnourished people have adequate caloric intake but are deficient in other requirements. Protein deficiencies in children are obvious in their pot-bellied, skinny-limbed, and undersized appearance. TV news coverage of famines, mainly in Africa, have made such images very familiar to us. Vitamin deficiencies lead to many diseases, depending on the vitamin that is lacking. The deficiency can be fatal, such as the vitamin B deficiency that causes beriberi disease. Other effects include blindness, loss of hair and teeth, and bow-leggedness. Malnourishment also increases susceptibility to other diseases. Vitamin A deficiency, for example, impairs the immune system and is a significant contributor to the 1.5 million deaths from diarrhea each year. Over 3 billion people, half of the world's population, suffer some from some form of micro-nutrient deficiency.

The effects of malnutrition cross generations as well. Infants of malnourished, underweight women are likely to be small at birth and more susceptible to disease and death. Overall, 60 per cent of women of childbearing age in South Asia, where half of all children are underweight, are themselves underweight. In Southeast Asia, the proportion of underweight women is 45 per cent, in sub-Saharan Africa 20 per cent. Globally, the proportion of underweight children under five declined by one-fifth over the period 1990–2005. However, if current trends continue, the Millennium Development Goals target of halving the proportion of underweight children will be missed by 30 million children, largely because of slow progress in South Asia and sub-Saharan Africa.

Although global food supply could provide adequate nutrition for the entire global population, more than one billion people are currently malnourished. More than half of the undernourished people (61 per cent) live in Asia, while sub-Saharan Africa accounts for almost a quarter (24 per cent). Hunger is most pronounced where there are natural disasters and where warfare is ongoing. Every year, more than 3 million children die as a result of starvation. In stark contrast, the World Health Organization (WHO) reports that more than 300 million people are clinically classified as obese, and about half a million people die from obesity-related diseases every year. Obesity rates have tripled since 1980. The gross inequity between starvation and obesity rates suggests that hunger is a problem of distribution. Regardless of increasing global production levels, rising food prices and heavy agriculture subsidies result in food remaining inaccessible to many of the world's poorest people.

The 1974 World Food Summit promised to eradicate hunger within the next decade. The 1996 summit was a little more realistic and promised to cut the number of hungry people in half by 2015, and this was the target adopted in the United Nations Millennium Development Goals. Meeting the goal would require a reduction in the number of hungry people by 22 million each year; the actual number so far has been less than 6 million. The time frame to meet the goal has now been extended to 2060.

World population is expected to increase by 2.3 billion over the next 40 years, from 7.0 billion in 2011 to 9.3 billion in 2050. Almost all growth (97 per cent) is projected to occur in the less developed regions, where a large majority of the world's people live today. This will pose significant challenges

Hunger is a daily reality for many of the world's children. This food program in Malawi, Africa, provides phala (a maize porridge) for hungry children in the village of Buli.

for global food production systems. In many countries, more than half of family income goes to food, and with a doubling of food prices, people will have few options.

No single solution can address the challenge of increasing the quantity and quality of affordable foods. The solution will involve a variety of approaches, including: exploration of new marine and terrestrial food sources; continued research to increase yields of existing crops; improvements in the efficiency of natural resource use; family planning programs aimed at reducing population growth rates; elimination of global agricultural tariffs; more efficient food distribution systems to address chronic hunger; and a moderation in demand on the part of the already overfed countries.

Sources: FAO (2007b); von Braun (2007); Pappas (2011); *Science Daily* (2011).

---

livestock, compared to 325 kilograms in China and just 90 kilograms in Zimbabwe.

The very large subsidies involved in global agriculture are a major confounding factor in examining food production capabilities. Industrial nations collectively pay their farmers more than $300 billion each year in subsidies, and this amount goes to the largest farmers in the richest countries, promotes chemical dependencies, inhibits change, and discriminates against producers in less developed countries. However, in 2008, World Trade Organization members agreed to significant commitments. One obliges the European Union to eliminate agricultural subsidies, but a time frame was not set, and France's agriculture minister estimated that European export subsidies might not be finally eliminated until 2015 or 2017.

As the world becomes more crowded and as pressures on biological systems and global biogeochemical cycles mount (Chapter 3), it is no longer sufficient to ask whether we can feed the planet. We need to ask ourselves more difficult questions. Can the world's agro-ecosystems feed today's planet and remain sufficiently resilient to feed tomorrow's hungrier planet? Will intensive production systems cause some agro-ecosystems to break down irreversibly? Presuming we can maintain current food production, are we paying too high a price in terms of the broader environmental effects of agriculture?

Agriculture is fundamentally an ecological process as solar radiation is converted through one or more transformations into human food supplies. Rapid growth in human population has entailed increasing disruption of natural systems in order to feed burgeoning populations and, particularly in developed countries, growing appetites. This chapter provides some context for this ecological process before considering some of the environmental challenges facing Canadian agriculture.

Agriculture as an Ecological Process

Agriculture is a food chain, with humans as the ultimate consumers. Energy flows through this food chain in a manner similar to the way it does in natural food chains. Thus, the second law of thermodynamics is also important to agricultural food chains—the longer the food chain, the greater the energy loss (Chapter 2). This fact is one of the arguments for a vegetarian diet. By eating at the lowest level on the food chain as herbivores, humans will maximize the amount of usable energy in the food system. There are, however, other aspects of food production at higher levels of the food chain that should also be considered.

Food meets more than energy requirements alone. Important protein and mineral demands must also be satisfied. Animal products, by and large, are the main suppliers of these proteins and minerals, including such elements as calcium and phosphorus. Areas currently used as rangeland to support animal production often cannot be used as cropland. They may be too dry or otherwise ecologically marginal, and severe problems, such as excessive soil erosion, may have arisen in the past when humans tried to bring such lands under cultivation. Thus, it is not always correct to presume that rangelands can produce more food under tillage. Furthermore, grazing animals in many parts of the world provide not only food supplies but also other products and services, such as their energy as draft animals and their hides and other animal parts for clothing.

However, there are many valid arguments from an ecological, health, moral, and spiritual point of view for reducing meat consumption in favour of a diet with a higher vegetable content, especially in industrialized nations. For instance, while the Canadian Food Guide recommends that the average adult consume 54 kilograms of protein per year (including meat, fish, nuts and seeds, as well as the protein acquired from plant products), Canadians consume 90 kilograms of meat per capita each year (Statistics Canada, 2010). Meat consumption still varies widely by region and socio-economic status. For instance, in the developing world, people eat about 32 kilograms of meat per year, compared to an average of 85 kilograms per person in the industrial world.

The domestication of plants and animals thousands of years ago led to profound changes to the global land base. Complex natural systems that once dominated the landscape have been replaced by relatively simple control systems in which humans are in command of the species and numbers that exist in a given area. However, unlike the process that took place during the Industrial Revolution, these changes occurred over an extended time period, allowing greater potential for adaptation. Furthermore, until the past 150 years or so, energy inputs were largely limited to photosynthetic energy from the sun, the energy of domesticated draft animals such as oxen, camels, and horses, and human energy input. It was not until the Industrial Revolution unlocked past deposits of photosynthetic energy in the form of coal and later oil that industrial agriculture began and energy inputs and environmental impacts increased dramatically. Indeed, only during the latter half of the twentieth century did some of the most damaging impacts of agriculture come into play.

One of the most significant impacts is in the concentration of greenhouse gases (GHGs) in the atmosphere. Research on atmospheric trace gas concentrations has revealed that even early agriculture had a significant impact (Ruddiman, 2003). An increase in atmospheric carbon dioxide (CO_2) concentration of 20 to 25 parts per million between 8,000 and 2,000 years ago appears to be related to increased forest clearing for agriculture. About 5,000 years ago, there was an increase of 250 parts per billion in atmospheric methane (CH_4) concentration, coinciding with the adoption of 'wet rice' farming in Asia. These changes demonstrate that even with low populations, small changes over long periods can be as important as large changes caused by high populations over a short time.

Since cropland cultivation began in Canada, an estimated 1 billion tonnes of soil organic carbon has been lost. Early cultivation would have removed most of it, and further losses continued as a result of intensive tillage, biomass burning, and removal of residues. Canada also has extensive grasslands, much of which is used for seasonal grazing. If pastures are overgrazed, they store less carbon. Paradoxically, grazing stimulates growth up to a point; grazing that does not reach that threshold can shorten vegetative growth, producing the same effect as overgrazing. Since cultivation began, up to 30 per cent of the carbon originally present in the surface soil layer has been lost. However, the rate of loss appears to be

Horse power is one form of auxiliary energy used in agricultural production, but the environmental impacts of this system of ploughing with horses in British Columbia are much less than those resulting from the fossil-fuelled, mechanical processes favoured by most Canadian farmers.

falling because of smaller amounts of land being converted into cropland, decreases in summer fallow, increases in no-till farming, and increased fertilizer use in the Prairie provinces (Hengeveld et al., 2008).

Of course, CO_2 is not the only GHG affected by land-use change (Chapter 7). Fertilizers have a significant impact, as discussed in Chapter 4, as do the decomposition of crop residues and soil organic matter and the transport of nitrogen off-farm through leaching, runoff, and evaporation. Livestock is also a significant source, particularly of CH_4 and nitrous oxide (N_2O). Up to 10 per cent of feed energy is lost in ruminant animals through belching and other gaseous contributions, and additional amounts come from manure. The agricultural contribution of these gases has been increasing, and by 2005 the sector accounted for 66 per cent of Canada's N_2O emissions and 25 per cent of CH_4 emissions (Desjardins et al., 2008). These changes are driven mainly by increases in livestock numbers, discussed later in the chapter.

Opportunities to reduce the contributions of agriculture to global warming are many, but an obvious one is to reduce the numbers of livestock by eating less meat. Carbon sequestration in soils can be improved through such measures as no-tillage cultivation and reducing summer fallow, as discussed later in the chapter. Planting perennial crops such as trees also helps. On grasslands, grazing intensity can be reduced and productivity increased by adding nutrients and water. However, fertilizer has to be applied more precisely than in the past to ensure that the amounts added are taken up by the crops and will not denitrify into N_2O.

The complexity of finding the optimum solution for reducing GHG emissions from agriculture is illustrated by the fact that where soils are fine and it is relatively humid, such as in eastern Canada, no-tillage agriculture may result in increased N_2O emissions as a result of higher surface temperatures (Rochette et al., 2008). However, reducing no-tillage agriculture conflicts with the desire to maintain soil carbon levels through no-tillage systems. On the prairies, because of the drier environment and heavier soils, no-tillage is a very appropriate way to reduce overall GHG emissions from agriculture. Solutions obviously have to take such geographical differences into account.

Modern Farming Systems in the Industrialized World

The Green Revolution

Dramatic changes in food production systems have occurred through a variety of technological advances that were in turn influenced by changes in demographics (e.g., increases in population densities), social structure (e.g., urbanization, social stratification), and economic conditions (e.g., global trade). Early food production systems were small in scale, and were largely dictated by fixed environmental conditions—the quantity and quality of food produced depended heavily on existing local climatic conditions (e.g., annual rainfall), native vegetation (i.e., plants indigenous to the region), and availability of human and/or animal labour. Soil fertility was maintained or enhanced by using locally available, natural elements such as manure, bones, and ashes.

Today, local conditions are manipulated to improve both the quantity and quality of outputs. The amount of food produced per unit of land has increased dramatically as a result of a variety of technological advances, including genetic engineering, greater mechanization (e.g., tractors for ploughing and seed sowing, mechanized food processing), and the creation of auxiliary energy flows, as discussed in Chapter 2. The package of inputs or agricultural techniques, which together are referred to as the **Green Revolution**, includes the introduction of higher-yield seeds (e.g., shorter maturation, drought resistance) and a reliance on auxiliary energy flows.

The development and commercialization of higher-yielding seeds through **hybridization** led to significant gains in grain yields throughout the world. In the 1940s, scientists developed a 'miracle wheat seed' that matured faster, producing wheat that was shorter and stiffer than traditional breeds and less sensitive to variation of daylight. India more than doubled its wheat production in five years with the new technology, and other Asian and Latin American countries recorded similar productivity increases. Miracle rice seeds and high-yield maize (corn) were developed by scientists soon afterwards, and their use has diffused rapidly around the world.

Were it not for these developments, there would no doubt be many more people in the world suffering chronic food shortages. However, they also pose challenges. Most of the new hybrid seeds grew better than their native counterparts only if fertilizers and biocides were applied with sufficient frequency and in sufficient quantity. As these chemicals became more expensive, many poorer farmers were unable to take advantage of the high-yield seeds. The Green Revolution also encouraged a narrowing of the genetic base of the crop, and with each farmer growing exactly the same strain, any disease or pest that managed to adapt to the strain had an almost unlimited food supply.

Today, critics of the Green Revolution point to stagnation in scientific progress—the gains that can be made through technology have already been made, and the damage done to agro-ecosystems by modern production systems will limit the ability of farmers to increase yields significantly. For example, macronutrients in the soil, such as nitrogen, phosphorus, and potassium, can be replenished by fertilizers, but many of the micronutrients required in trace amounts cannot be replaced. As we take out more crops from the soil, these nutrients may become exhausted, leading to greatly diminished returns in the future. Signs of such declines are already appearing in some areas. The increasing cost of fossil fuels also leads to increasing fertilizer costs.

However, some advances are still being made. In 2001, for example, a new rice strain, called NERICA, was produced from hybridization of African and Asian rice varieties. According to the United Nations Development Programme (UNDP), the project's main supporter, NERICA produces 50 per cent more yield, uses less fertilizer, is richer in protein, and is more resistant to drought, disease, and pests. It was also developed in full consultation with farmers and consumers and is now being successfully grown in a wide range of African countries. In Guinea, thanks to the success of NERICA varieties, farmers are now able to gross US$65 per hectare with minimal inputs and $145 per hectare with a moderate level of inputs. The country saved more than US$13 million on rice imports in 2003. NERICA is now being used as a 'success story' for the Millennium Development Goals (www.mdgmonitor.org/goal1.cfm#).

The development of these 'miracle seeds' relied only on genetic combinations found in nature. The development of **genetically modified organisms** (GMOs), on the other hand, involves combining genes from different and often totally unrelated species. Such genetic manipulation is now a multi-billion dollar industry, and Canada is one of the world's leading participants (Box 10.4).

The reliance on large auxiliary energy flows in modern industrialized agricultural systems is one of the main differences between natural and agro-ecosystems. Auxiliary energy flows include natural and chemical fertilizers, biocides (insecticides, herbicides, fungicides), fossil fuels, and irrigation systems. These energy subsidies have significantly increased crop yields over the past century, particularly over the past three decades. **Subsistence farming**, in which the production of food is intended to satisfy the needs of the farm household, relies on natural energy supplies and may produce 10 food units for every unit of energy invested. However, in the most energy-intensive food systems, such as those in Canada and the US, 10 times as much energy on average needs to be invested through auxiliary energy flows for every unit of food produced.

Improved water management, a key component in Green Revolution technologies, helped to boost worldwide productivity or output of 'crops per drop' by an estimated 100 per cent since 1960 (FAO, 2003b). Agriculture accounts for 70 per cent of fresh water withdrawn from natural sources for human use and as much as 90 per cent in many developing countries. The water needs of humans and animals are relatively small—the average human drinks about four litres a day. But producing the same person's daily food can take up to 5,000 litres of water. For example, it takes 1,000 tonnes of water to produce one tonne of grain.

Although agriculture is not Canada's largest user of water in terms of withdrawals, it is its largest consumer. Withdrawal is the amount of water removed from a source for a particular use. Consumption is the difference between the withdrawal and the amount returned to the source. Agriculture removes significant quantities of water from the landscape, tying it up in agricultural products or evaporating it back into the air rather than returning it directly to streams or groundwater; therefore, consumption is high. Agriculture relies on a reliable supply of good-quality water for growing crops, raising livestock, and cleaning farm buildings. Approximately 75 per cent of all agricultural water withdrawals in Canada occur in the semi-arid prairie region. More than 780,000 hectares of cropland in Canada are under irrigation, and the amount is increasing. Alberta accounts for 63 per cent of the national

Rice fields on the east coast of Sri Lanka. Rice yields, where sufficient water is available, have tripled in some locations.

The agricultural system of these hill tribe people in northern Thailand, based on dry rice cultivation with livestock such as pigs and chickens, may seem primitive but yields a much higher return than modern farming systems in terms of the energy budget.

Environment in Focus

Box 10.4 Genetically Modified Organisms

Yields of many crops in the developed world will not increase significantly with conventional agricultural techniques, even those related to the Green Revolution. Instead, farmers are turning to biotechnology, or the genetic modification of crops, to increase production. The technology already dominates the production of a few crops in several countries, yet other countries ban its use. Some nations require labelling of foods that have been genetically modified, while others, such as Canada, have rejected this openness. Why are there such differences in approach towards this new 'genetic revolution'? Before we answer that question, it is necessary to explain what GMOs entail.

Advances in understanding of DNA have shown that the basic building blocks of life are all very similar. Traits controlled by single genes can be transplanted from one species to another. Transgenic crops are produced when a single species contains pieces of DNA from at least one other species. One of the most common examples is the implantation of genes containing the toxin from *Bacillus thuringiensis* (Bt). For years, farmers have sprayed this naturally occurring toxin on their fields as an insecticide. Geneticists were able to isolate the toxic gene in the bacteria and are now inserting it into crops such as corn and soybeans. The crops then produce their own toxin. The other common genetic modification is to produce crops with genes that make them resistant to a particular herbicide. When a farmer sprays the crop, all competing plants are killed.

The growth in use of such transgenic crops has been very rapid. At the time of the Earth Summit in Rio de Janeiro in 1992, production of transgenic crops had not yet begun, but it now stands at almost 150 million hectares. In 2010, the US, followed by Brazil, Argentina, India, and Canada, continued to be the principal adopters of biotech crops globally, with 66.8 million hectares planted in the US (Figure 10.3). Although in 2006 the majority of the area devoted to biotech was in industrial nations, by 2010, 90 per cent of all biotech farmers were in developing countries.

Biotech soybeans remained the principal biotech crop in 2010, occupying 73.3 million hectares (50 per cent of global biotech area), followed by maize (28.8 million hectares at 19 per cent), cotton (16.1 million hectares at 11 per cent), and canola (7.0 million hectares at 5 per cent). About 47 per cent of the total area planted with these crops is now devoted to genetically modified varieties, and two traits—insect resistance and herbicide tolerance—dominate. Throughout the world, several thousand GMO field tests have been conducted or are underway, and many more crop–trait combinations are being investigated, with greater focus on virus resistance, quality, and in some cases tolerance to abiotic stresses.

In Canada, scientists at the University of Victoria have developed a potato that resists bacteria and fungi, allowing it to be stored for 10 times longer than 'normal' potatoes. In Guelph, Ontario, scientists have engineered a pig that produces manure 20 to 50 times lower in phosphorus content than that of 'regular' pigs. Since phosphorus is a main cause of eutrophication (Chapter 4), the number of pigs that can be kept in a given area is often limited. The new pig can be stocked in higher densities.

The global value of the biotech crop market was projected at more than $11.2 billion for 2010. There is no doubt that GMOs hold great potential for the future. However, development has

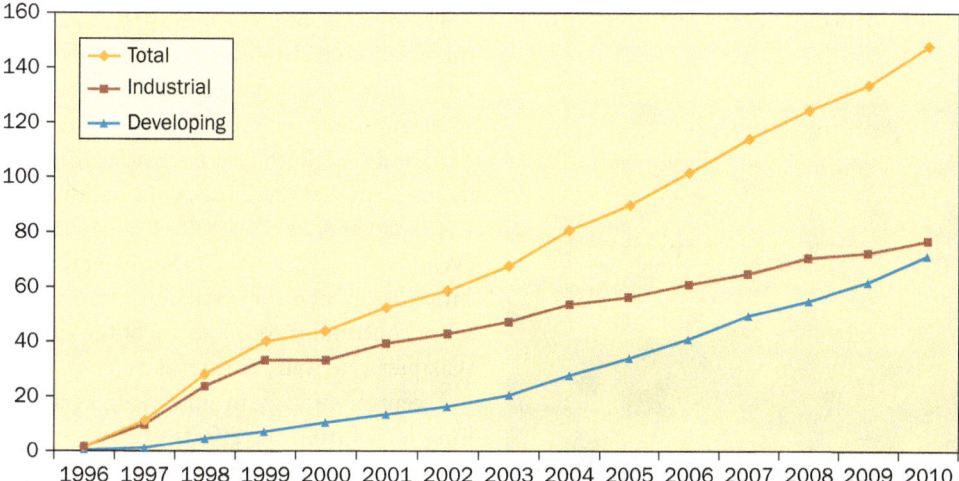

Figure 10.3 | Global area of biotech crops, industrial and developing countries (million hectares).
Source: James (2010).

been very rapid, and several areas of uncertainty regarding their effects remain:

- *Pleiotropic effects*. These are unexpected side effects that might be suffered by the target organism as a result of incorporation of the new gene. For example, there may be a change in the toxins produced or the nutrient content.
- *Environmental effects*. There will be impacts on natural processes, such as pollination and biogeochemical cycles, as a result of creating new crops. It is also feared that there may be unanticipated gene flow to other organisms and perhaps interbreeding with wild relatives. This has the potential to produce 'superweeds', as has happened in Canada with canola, where the biocide-resistant superweeds are now growing in wheat fields. The only way to get rid of them is to resort to broad-spectrum herbicides, those that will kill everything.
- *Unintentional spread*. Pollen and seeds from transgenic crops may spread onto lands where they are not intended to grow. This is already a problem for organic canola and honey producers on the prairies, who can no longer guarantee that their products are transgenic-free.

There is little consensus among governments, consumers, farmers, and scientists concerning the benefits and risks associated with biotechnology or genetic engineering. A lack of perceived benefits for consumers and uncertainty about their safety have limited their adoption in some countries, while others have developed legislation requiring mandatory labelling on genetically modified foods. Canada has chosen not to do so. The government voted down a bill that would have required mandatory labelling for all foods containing more than 1 per cent genetically engineered ingredients, even though polls suggested that a large majority of Canadians would prefer to have a choice. What is your view about whether GMO food products should be explicitly labelled?

total. Alberta has the largest irrigated field crop and irrigated hay and pasture areas. Most of the vegetable irrigation is in Ontario, with 45 per cent of the total irrigated vegetable area. British Columbia applies most of the irrigation to keep fruit areas, such as the orchards in the Okanagan Valley, thriving. As a whole, the province accounts for 52.8 per cent of total irrigated fruit area in the country.

Agricultural practices also have a negative impact on water quality. Box 10.5 details some of agriculture's impact on water resources and aquatic ecosystems.

Fertilizers are also important inputs to modern farming systems. New strains of crops, such as wheat, will only produce superior yields if fertilized adequately. Despite predictions that between 2008 and 2011 fertilizer production would come close to demand, the economic downturn of 2008 saw a rapid fall in demand. Several fertilizer factories were temporarily closed to cope with this downturn but have now reopened as the global economy recovers. Regardless, the International Fertilizer Association expects a significant surplus in fertilizers by 2013.

Large increases in fertilizer inputs, particularly nitrogen, have occurred over the past few decades in Canada (Figure 10.4). Western Canada, for example, saw a fivefold increase in fertilizer applications between 1970 and 2000, and the three Prairie provinces now account for 69 per cent of Canada's total fertilizer sales, up from 29 per cent in 1970. By 2006, fertilizer was applied to over 25 million hectares of Canadian farmland. In recent years rising prices have caused a steep reduction in fertilizer consumed worldwide, although fertilizer application remains widespread (Worldwatch Institute, 2011).

The *amount* of fertilizers applied *per hectare* (kg/ha) has increased considerably since the early 1970s, although the rate of increase has slowed since the mid-1980s. In 1970, 1980, 1990, and 2000, fertilizers were applied at a rate of 18.4, 42.4, 45.1, and 54.2 kg/ha, respectively. These rates are not high by international standards, with the United States, Australia, and Japan applying 103.4, 151.7, and 301.0 kg/ha, respectively (Worldwatch Institute, 2003a). Nonetheless, fertilizer application is of environmental concern because fertilizers are a main contributor to the speed of the eutrophication process (Chapter 4) as well as to groundwater pollution in some areas.

A central strategy in improving agricultural output is to limit losses from the effects of pests and diseases and from weed competition. Since the mid-1990s, the approach to crop protection has relied increasingly on the use of biocides (insecticides, nematocides, fungicides, and herbicides). Pesticides should be referred to as biocides, since their application often affects more than just the target species. Biocides

Irrigation water is critical for producing crops on much of Canada's landscape. This irrigation system is located in Saskatchewan.

are applied to boost yields. Yields are improved by reducing the amount of energy flowing to the next trophic layer of the food chain through the respiration of heterotrophs (often insects) and by eliminating non-food plants that compete with the crop plants for available growth resources.

Biocide use continues to increase dramatically, indicating that farmers find biocides cost-effective from a production perspective, particularly where alternative forms of crop protection are labour-intensive and labour costs are high, as they are in Canada. Canada and the United States lead the world in the consumption and use of biocides, accounting for 36 per cent of world biocide use (UNEP, 2002). The biocide market has been growing at approximately 6 per cent per year since 1990 (ibid.), with significant increases in Canada (Figure 10.4). These large increases in the application of agricultural biocides over the past 20 years have profound environmental implications, as will be discussed in more detail later.

The Biofuel Revolution

The rising cost of crude oil and the availability of improved processing technologies have made biofuel production more

Environment in Focus

Box 10.5 Agricultural Impacts on Canada's Water Resources and Aquatic Ecosystems

Canadian farms have employed modern technology to increase production, yet some of these practices have contributed to environmental degradation, including the decline of water quality. The main pollutants of water coming from farmland are nutrients, pesticides, sediment, and bacteria. Agriculture has also changed the physical presence of water across the Canadian landscape through the construction of dams and reservoirs, distribution of irrigation water, drainage of wetlands, and sedimentation of streams and lakes.

Nutrient losses. In certain parts of Canada, the use of additional nutrients in the form of mineral fertilizer, manure, compost, and sewer sludge to increase crop productivity has led to a nutrient surplus in the soils, with the potential loss of nutrients to surface and groundwater. Nutrient addition to aquatic ecosystems promotes eutrophication (Chapter 4). Other indirect consequences include changes in the abundance and diversity of higher trophic levels (e.g., benthic invertebrates and fish), increased abundance of toxic algae, and fish kills caused by loss of oxygen from the water. High nitrate concentrations contribute to the decline in population numbers for 17 of Canada's 45 frog, toad, and salamander species.

Pesticide losses. The use of pesticides for disease, weed, and insect control has resulted in pesticide losses to the atmosphere and subsequent deposition from the atmosphere to surface water and non-agricultural lands, as well as runoff and leaching to surface and groundwater. Depending on the compound and the concentrations involved, pesticides introduced into surface waters can kill fish and other aquatic organisms; cause sub-lethal effects on reproduction, respiration, growth, and development; cause cancer, mutations, and fetal deformities in aquatic organisms; inhibit photosynthesis of aquatic plants; and bioaccumulate in an organism's tissues and be biomagnified through the food chain. Pesticide concentrations in surface waters sometimes exceed Canadian water quality guidelines for irrigation water or protection of aquatic life.

Sedimentation. Agricultural practices such as tillage and allowing livestock access to streams increase erosion and the movement of soil from farmland to adjacent waters. Alterations to soil conditions caused by tillage and cropping patterns may cause soil degradation, which can lead to less infiltration of water into soil and increased runoff and movement of nutrients and pesticides to surface waters. The introduction of soil into aquatic ecosystems can increase turbidity and thereby reduce plant photosynthesis, interfere with animal behaviours dependent on sight, impede respiration (by gill abrasion) and feeding, degrade spawning habitat, and suffocate eggs.

Pathogens. Pathogens from farm livestock can migrate into groundwater or be carried in runoff into surface waters. Pathogen contamination of irrigation waters, drinking waters, or shellfish areas can create significant risks to human food supplies and can also pose a threat to aquatic ecosystems and biodiversity. Regions of the country with high livestock densities are reporting fecal coliform counts that exceed Canadian water quality guidelines for both drinking and irrigation water. The interaction between agriculture and water quality came to the attention of Canadians in 2000 through the tragic deaths at Walkerton, Ontario, discussed in Chapter 11.

Wetland drainage. Drainage of wetlands has increased the area of agriculturally productive soils, but it has also modified local ecosystems and changed the pattern of water partitioning between evaporation, stream flow, and infiltration. Wetland drainage destroys the habitat of a number of species of arthropods, reptiles, amphibians, birds, mammals, and fish. More than 85 per cent of the decline in Canada's original wetland area has been attributed to drainage for agriculture.

Source: Chambers et al. (2001).

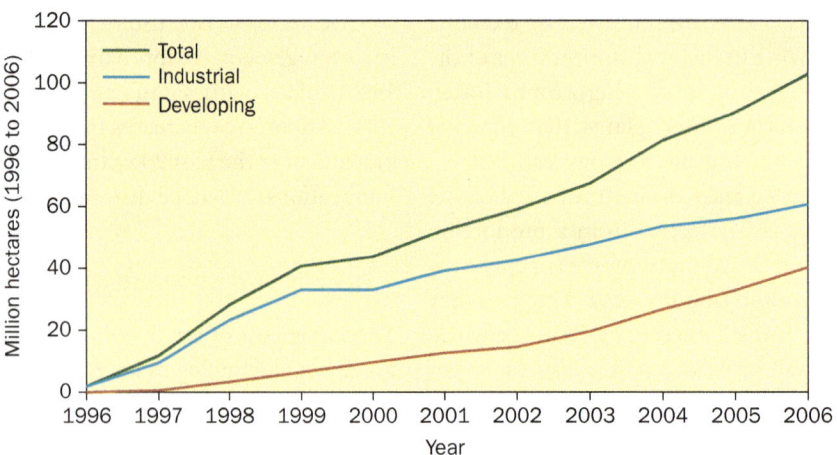

Increase of 13%, 12 million hectares (30 million acres) between 2005 and 2006.

Figure 10.4 | Chemical input usage in farming in Canada, 1971–2005. Source: Statistics Canada and AAFC calculators.

economically feasible. Some people characterize biofuels as an environmental saviour; others see them as one of the main challenges to environmental sustainability. What are biofuels, and why do views on them diverge so radically?

Biofuels are solids, liquids, or gases that have been derived from recently dead biological material and are processed into an oil that acts as a petroleum replacement. As they are derived from plants and other organic material, they seem to have great potential to help curb global greenhouse gas emissions. About one-quarter of GHG emissions are produced by the transport sector, and biofuels, including ethanol and biodiesel, are the only existing renewable fuels compatible with the current transportation infrastructure. Biofuels generate a fraction of the pollutants of traditional petroleum-based fuels, and the plants that produce them remove carbon, a climate-altering GHG, from the atmosphere. Biofuels also have the potential to reduce foreign oil dependency, lower fuel prices, increase income for farmers, and provide a host of new jobs. Nations that develop domestic biofuel industries will be able to purchase fuel from their own farmers rather than spending scarce foreign exchange on imported oil.

These advantages led experts at an African Union conference to conclude that 'Promotion of biofuels industry in developing countries has the capacity to propel such countries to achieve the MDGs [Millennium Development Goals] through poverty reduction (especially job creation and economic enhancement), health impact, and climate change.'

In response, biofuel production is growing at a rate of roughly 15 per cent per year, more than 10 times that of oil, although oil still accounts for 95 per cent of the global transportation fuel market. World ethanol production reached 88 billion litres in 2011, a 3 per cent increase over production in 2010. The United States produces almost 90 per cent of global ethanol, followed by Brazil, Europe, China, and Thailand. Biodiesel production, which started from a smaller base, has experienced a growth rate of over 20 per cent since 2005 and is predicted to reach 6 billion barrels by 2015. Europe is currently the leading producer of biodiesel, which is processed from vegetable oils derived from soybeans, oil palm, and canola, among other crops.

In 2005, the US pledged to nearly double ethanol production by 2012, and the European union announced that biofuels will meet 10 per cent of its transportation fuel needs by 2020. By 2009, the global production of ethanol was up over 452 million barrels per year, while biodiesel was produced at a rate of 308,000 barrels per day. Some experts feel that over the next 15 to 20 years, biofuels may supply up to 25 per cent of the world's energy needs.

Canada is not a large producer of biofuels in the global context, and the federal government introduced the ecoAGRICULTURE Biofuels Capital Initiative (ecoABC), a $200 million

Not all crops require the same level of chemical inputs. Tomatoes are particularly demanding compared with, for example, cereal crops.

four-year program to stimulate production that ended in 2012. The program provided repayable funding for the construction or expansion of transportation biofuel production facilities. Funding was conditional on investment in the biofuel projects by agricultural producers and the use of agricultural feedstock to produce the biofuel.

In view of these advantages, why are there so many critics of biofuels? One question relates to the inputs used to grow many of the crops used for biofuels. Fossil fuels are invested in fertilizers, pesticides, machinery, and processing, and the second law of thermodynamics (Chapter 2) warns us that energy is lost in each of these energy transformations, which contributes to global warming. As explained in Chapter 2, this energy is largely given off as heat, which contributes to global warming. Furthermore, nitrogen, a key fertilizer applied to biofuel crops, turns into nitrous oxide (N_2O), a greenhouse gas that is over 296 times more powerful than CO_2. Some scientists report that to produce ethanol from maize will cause 0.9 to 1.5 times as much GHG emissions as an equivalent proportion of fossil fuels (Crutzen et al., 2007).

Concerns around the effects of biofuels on global warming are compounded by the land conversion associated with expanding biofuel crops. Biofuel crops displace existing agriculture, which in turn requires the conversion of natural ecosystems to maintain food production levels. Scientists have noted that when rain forests, peatlands, savannahs, or other biologically diverse habitats are converted for biofuel production, the cost of this land conversion will result in a total contribution of CO_2 that is between 17 and 420 times higher than fossil fuels (Fargione et al., 2008). Indonesia and Malaysia account for 86 per cent of the world's palm oil and accelerating demand is prompting the conversion of tropical rain forests, contributing to GHG emissions and threatening biodiversity.

However, a comparative study (Farrell et al., 2006) suggests that, overall, using corn ethanol instead of gasoline could reduce GHG emissions by some 13 per cent. For sugarcane, the reduction could be even greater, as high as 87 to 96 per cent, according to some sources (www.unica.com.br/i_pages/files/pdf-ingles.pdf) because the crop can be grown at higher yields with fewer fossil-fuel inputs. Furthermore, ethanol derived from cellulosic material, including crop residues and woody plants, offers the potential of reducing GHG emissions by more than 100 per cent, particularly if bio-energy is also used in producing the fuel.

Another problem is that with the increased profit to be made from land used for growing biofuel inputs, land is being taken away from food production. The cars of the wealthy may end up consuming the food of the poor. The International Food Policy Research Institute projects that a drastic increase in biofuel production in sub-Saharan Africa will result in an 8 per cent decline in calorie availability (von Braun, 2007). In 2005, it took 13 per cent of the US corn harvest to displace less than 3 per cent of fuel needs (www.ethanolrfa.org/resource/facts/agriculture). Figures like that raise concern around the global capacity to produce both biofuels and food sufficient to meet growing demand. In 2007, interest in biofuels was high and corn prices doubled, causing social unrest in Mexico, where corn tortillas are a dietary staple. Unrest continued in many countries as food prices remained high into 2011. Overall, deeper analysis is needed to understand the global and local impact of expanded biofuel demand on food prices.

One possible answer to both the challenges of energy inputs into biofuel production and competition with food crops is to produce ethanol from non-food sources. Cellulosic ethanol is made from a wide variety of plant materials, including wood wastes, crop residues, and grasses, some of which can be grown on marginal lands not suitable for food production. The process for converting these materials to fuel is often more efficient, because plant material rather than fossil fuels can be used to provide heat and power. As a result, cellulosic ethanol has an energy yield at least four to six times the energy expended during production and can reduce greenhouse gas emissions by 65 to 110 per cent relative to gasoline (Worldwatch Institute, 2007b).

With the continued improvement of cellulosic technology, the ethanol feedstock supply is expected to shift from edible crops such as corn and canola oil to fast-growing grasses such as switchgrass and miscanthus, agriculture and forestry residue, and even the organic portion of municipal solid waste. The cellulosic conversion process is still relatively expensive, but the costs of cellulosic biofuel production are declining, and one Canadian company, Iogen Corporation (www.iogen.ca), is already selling cellulosic ethanol commercially.

A further challenge for biofuels is that large-scale biofuel production can threaten biodiversity, as seen recently with oil palm plantations in Indonesia and sugarcane plantations in Cambodia that are encroaching on forests. In Brazil, the Cerrado, a vast landscape of biologically rich forests, brush, and pasture just south of the Amazon, is coming under pressure as sugarcane cultivation expands. One response to this threat has been to convince decision-makers that areas rich in biodiversity may yield greater long-term returns by being preserved for ecotourism rather than bulldozed for sugarcane. For example, the government of Uganda had been supporting a proposal for a 7,000-hectare sugarcane concession in the supposedly protected Mabira Forest Reserve near Lake Victoria. However, a study by the conservation group NatureUganda showed that the financial benefits of protecting the forest and encouraging ecotourism vastly outstripped the potential of biofuel crops. The commercial value of tourism and carbon capture in Mabira was estimated at more than $316 million a year, whereas sugarcane production would be worth less than $20 million. A similar rationale has led the Cambodian government to restrict

Aerial view of sugarcane plantations abutting rain forest near Ribeirao Preto, Sao Paulo State, Brazil.

the number of concessions for sugarcane production in the fabled Cardamom Mountains in the southwest of the country to see whether ecotourism can produce sufficient yields to address poverty in the area. However, it should be remembered that tourism can also generate additional stresses on the environment, such as the GHG emissions generated by flying long distances.

Despite associated challenges, it seems that biofuels will be part of a portfolio of options to deal with global warming that also includes dramatic improvements in vehicle fuel economy, investment in public transportation, better urban planning, and many other aspects discussed in Chapter 13. The long-term potential of biofuels is probably in the use of non-food feedstock. Following the model of Brazil's sugarcane-based biofuels industry, cellulosic ethanol could dramatically reduce the carbon dioxide and nitrogen pollution that results from today's biofuel crops.

Modern industrial cropping systems that rely on auxiliary energy flows and other inputs are responsible for the production of a large percentage of the cereals, pulses, oil crops, roots and tubers, fruits, and sugar crops produced and consumed worldwide. Together, these crops represent only 60 per cent of the total value of output from the world's agro-ecosystems. Livestock production is responsible for the remaining 40 per cent.

The Livestock Revolution

Hundreds of years ago, livestock (e.g., cattle, sheep, goats) raised for local consumption were permitted to graze on surrounding natural vegetation. Stocking densities were dictated by surrounding environmental conditions—i.e., the availability of water and food supplies on a given unit of land and the ability of the local environment to assimilate animal wastes. As a result, farms were typically small, with fewer than 100 animals. However, as the industrialized world's appetite for meat continues to grow exponentially, traditional livestock production systems are being replaced with industrial technologies and intensification. Worldwide, meat consumption has doubled since 1977, and over the past half-century it has increased fivefold. Production of beef, poultry, pork, and other meats has risen to nearly 40 kilograms per person per year, more than twice as much as in 1950.

The **Livestock Revolution** has led to a number of changes with respect to how animals are brought to market. In industrialized countries such as Canada, the desire to supply the phenomenal growth in demand for protein has led to a reliance on industrial feedlots, which now produce more than half of the world's pork and poultry and 43 per cent of the world's beef (Worldwatch Institute, 2003b). Industrial systems of livestock production depend on outside supplies of feed, energy, and other inputs. Thirty per cent of the Earth's ice-free land surface is used for livestock production, in sharp contrast with the 8 per cent used for crops grown to feed humans (Worldwatch Institute, 2011). Livestock production requires high water inputs and accounts for over 30 per cent of the water used by the agricultural sector globally. Technology, capital, and infrastructure requirements are based on large economies of scale, and production efficiency is high in terms of output per unit of feed. As the world's main providers of eggs, poultry, beef, and pork at competitive prices, intensive farm operations (or 'factory farms') meet most of the escalating demands for low-cost animal products.

Agricultural production of livestock has grown across Canada, while the number of farmers has declined and the size of the average farm has increased. New farms are often capital-intensive operations with very large numbers of livestock. Farms with 3,000 or more pigs or 1,200 cattle are increasingly common, and some farms in Ontario and Quebec house in excess of 10,000 animals. For pigs, herd size grew by 150 per cent between 1996 and 2006, and average herd size is now 1,308 pigs per farm. Over the same time period, dairy cattle herd size increased by 44 per cent and beef by 35 per cent (Agriculture and Agri-Food Canada, 2008). As the livestock industry expands and becomes more intensive, health and environmental concerns over livestock manure are growing, particularly when livestock are produced in large numbers under confined conditions such as beef feedlots and intensive hog and poultry barns. In feedlots, animals are typically fed grains, which have undergone extensive energy and chemical inputs. Furthermore, at such high densities, animals easily acquire and transmit diseases and require extensive antibiotic treatments. Again, the second law of thermodynamics suggests that this may not be an efficient system. The social and environmental impacts associated with intensive farming operations will be discussed in greater detail later.

Agriculture's Impact on the Global Landscape

Various impacts are associated with the development of agriculture and more specifically with the development and spread of modern farming systems:

- Humans, rather than natural selection, have become the primary influence on the number and distributions of species. In modern agriculture, the dominant mechanism for production is through **monoculture cropping**, where crops are often made up of a single species with each individual having exactly the same genetic code. New species have been created for the purpose of maximizing the output of food for humans, while native species have been displaced. Domesticated plants and animals greatly increase in number and range, but wild species are drastically reduced.
- Energy flows are increasingly directed into agricultural as opposed to natural systems. It is estimated that humans now appropriate some 40 per cent of the net primary productivity of the planet.
- Biogeochemical cycles are interrupted as natural vegetation is replaced by domesticates that are harvested on a regular basis. Auxiliary energy flows in the form of fertilizer are used in an attempt to replenish some of the nutrients extracted through harvesting.
- Auxiliary energy flows used in modern agricultural systems to supplement the natural energy flow from the sun are often in excess of those derived from natural sources.
- In many areas, agriculture involves supplementing rainfall with irrigation to provide adequate water supplies. This has led to large-scale water diversions and to changes in groundwater, soil characteristics, precipitation patterns, and water quality.
- Soils are altered not only chemically through fertilizer and biocide inputs but also physically through ploughing. There is no natural process that mimics the disturbance created by ploughing.
- Natural food chains are truncated as humans destroy and replace natural consumers and predators at higher trophic levels.
- Natural successional processes are altered to keep agricultural systems in an early seral stage; auxiliary energy flows in the form of herbicides and mechanical weeding are often used to accomplish this.
- The stocking densities of domesticated herbivores are often much higher than that of natural herbivores, leading to a reduction in standing biomass and changes in the structure and composition of the primary production system.
- The industrial system of livestock production acts directly on land, water, air, and biodiversity through the emission of animal waste, use of fossil fuels, and substitution of animal genetic resources. It also affects the global land base

Agriculture has had a profound impact on the distribution of species. Large areas of the agricultural landscape are dominated by monocultures—in this case sunflowers—in which each plant has the same genetic makeup.

indirectly through its effect on the arable land needed to satisfy its feed concentrate requirements. The industrial system requires the use of uniform animals of similar genetic composition, contributing to within-breed erosion of domestic animal diversity, as discussed in Box 10.6.

Trends in Canadian Agriculture

Approximately 7 per cent of Canada's total land area (68 million hectares) is agricultural land, of which 46 million hectares are cropland, pasture, or summer fallow. The total area of farmland in Canada has remained relatively constant over the past 50 years. The Prairie provinces contain 81 per cent of the agricultural land base, while Quebec, Ontario, and BC account for 17 per cent, with the remaining 2 per cent in Atlantic Canada.

The agriculture and agri-food sector is a $130 billion industry, exporting more than $28 billion in products annually (Statistics Canada, 2007a; Agriculture and Agri-Food Canada, 2010a). Total government support to the agriculture sector rose to $10.1 billion in 2009 (UN Statistics Division, 2009). The number of farms has declined since the early 1970s, with almost 230,000 farms in Canada in 2006, a decline of 7 per cent since 2001. At the same time, average farm size has been increasing (Figure 10.5). The largest farms are in Saskatchewan (586 hectares), an increase of 13 per cent since 2001, and the smallest in Newfoundland (65 hectares).

Wheat is still the dominant crop in Canada, although the area dedicated to wheat production is declining (Figure 10.6). Saskatchewan grows 51 per cent of Canada's wheat. Other traditional grains, such as barley and oats, are also declining. The production of pulses (e.g., dry field peas, lentils, field beans, soybeans) has increased since the late 1970s and

Environment in Focus

Box 10.6 Conserving Canada's Breeds at Risk

Modern livestock production systems rely on a few specialized breeds that produce high yields. Consequently, many less productive but genetically valuable traditional breeds are threatened or have disappeared. According to the Canadian Farm Animal Genetic Resources Foundation:

- The genetic base for future breeding of dairy cattle is threatened by reliance on a single breed. Within the Holstein breed, 80 per cent of the cows are bred to 20 sires or their sons.
- The poultry breeding industry is limited to a few large breeding organizations, all of which appear to have similar genetic material. About a dozen such breeding companies provide stock for the world's poultry industry. The farm birds of the 1950s are now seen only as 'fancy' flocks.
- The Canadian swine industry is based predominantly on crossbreds of four breeds.

Loss of farm animal diversity is not just a Canadian problem. The United Nations Food and Agriculture Organization (FAO) estimates that of the 6,500 breeds of domesticated mammals and birds worldwide, 1,350 are at risk of extinction, 119 are officially confirmed as extinct, and 620 are reported to be extinct. The number of mammalian breeds at risk of extinction has risen from 23 per cent to 35 per cent since 1995, while the total percentage of avian breeds at risk of being lost increased from 51 per cent in 1995 to 63 per cent in 1999 (FAO, 2000). A recent global assessment suggested that one livestock species per month had become extinct over the previous seven years (FAO, 2007a). Since the mid-twentieth century, a few high-performance breeds, usually of European descent—including Holstein-Friesian (by far the most widespread breed, reported in at least 128 countries and in all regions of the world) and Jersey cattle; Large White, Duroc, and Landrace pigs; Saanen goats; and Rhode Island Red and Leghorn chickens—have spread throughout the world, often crowding out traditional breeds.

This progressive narrowing of genetic diversity is largely complete in Europe and North America and is now occurring in many developing countries, which have so far retained a large number of their indigenous breeds. The loss of genetic diversity in farm animals may represent a serious food security risk. Modern production systems are very efficient, but they are not well prepared for unexpected challenges, such as new diseases, feed crop failures, or sustained changes in consumer demand (e.g., the desire for low-fat foods and issues of animal welfare and environmental sustainability).

---

now accounts for almost 8 per cent of national crop area. Soybeans, the second largest oilseed crop grown in Canada after canola, are a major field crop in eastern Canada, along with grain corn. The area under soybean production, more than 1.2 million hectares, has grown eightfold since 1976, mostly as a result of strong demand for the crop and of breeding that has produced more cold-tolerant varieties with short growing seasons. The high-protein, high-oil beans are used as food for human consumption, animal rations, and edible oils as well as in many industrial products. They are also nitrogen fixers and help to replenish the soil (Chapter 4).

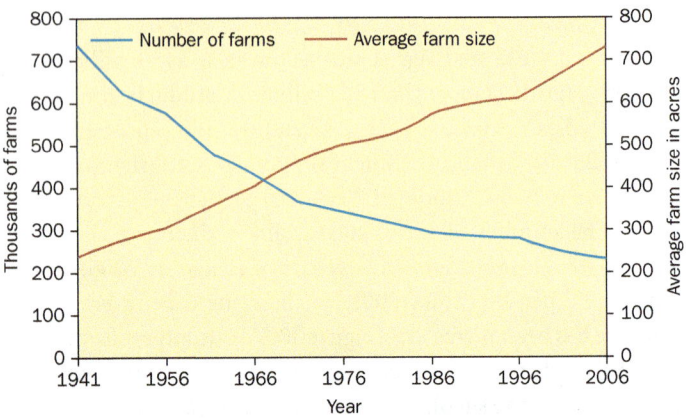

Figure 10.5 | Number and size of farms in Canada, 1941–2006. Source: Statistics Canada. Census of Agriculture, various years.
Note: It is interesting that although Canada officially uses metric units, Agriculture and Agri-Food Canada continues to use imperial units (acres) even in recent publications such as this graph. In a sense, this can be seen as symbolic of the inertia in Canada's agricultural sector.

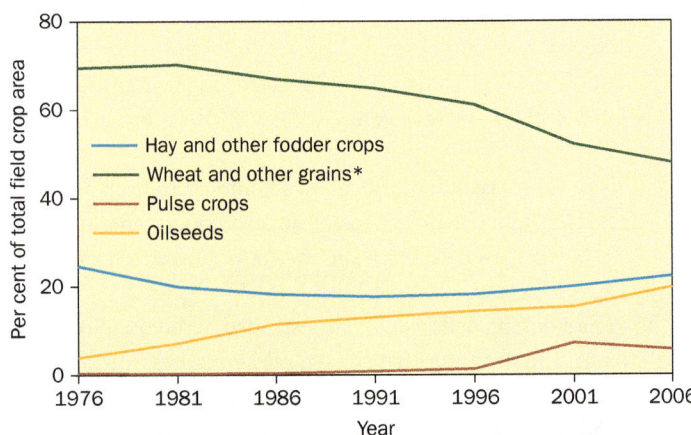

* Other grains include oats, barley, mixed grains, corn for grain, rye, and buckwheat.

Figure 10.6 | Crop area allocation, 1976–2006. Source: Statistics Canada. Census of Agriculture, various years.

Big Muddy Valley, about 200 km south of Regina, Saskatchewan. The Prairie provinces contain 83 per cent of the agricultural land base of the country.

Western Canada produces most of Canada's second largest crop, hay, which is required to keep up with the increasing numbers of livestock to be fed. The livestock industry is an important part of Canadian agriculture, accounting for $17.9 billion in farm cash receipts in 2009 (Statistics Canada, 2010). The steady growth in numbers of cattle and hogs on Canadian farms met a sudden downturn in 2009 due to a strong Canadian dollar, concerns about livestock diseases, and a global reduction in demand. The number of hogs dropped by 2.8 per cent between 2008 and 2009, to 11.8 million animals. Similarly, the number of cattle decreased by 1.3 per cent to a total of 13 million by the start of 2010 (ibid.). In the poultry sector most of the 126 million hens are produced in Quebec and Ontario.

Many Canadians have access to a ready supply of fresh produce year-round. Innovations include a thriving greenhouse sub-sector, as well as some of the most advanced storage technologies in the world. Other technologies have increased the availability of fresh food through production techniques that improve yields. Yield-enhancing technologies include mechanization, fertilizers and biocides, and genetic research. Canada is a main producer of food for the global market (Box 10.7). Unfortunately, many of the innovations and technologies currently employed by Canadian farmers have negative implications for ecosystem health. These implications will be discussed in more detail in the next section.

Environmental Challenges for Canadian Agriculture

Land Degradation

Land degradation reduces the capability of agricultural lands to produce food. As agricultural activities have intensified with increased cultivation and addition of agricultural chemicals to produce better yields, so has pressure on the soil resource.

Soil Erosion

Soil erosion is a natural process whereby soil is removed from its place of formation by gravitational, water, and wind processes. In Canada soil erosion is a serious land degradation problem and is estimated to cause up to $707 million in damage per year in terms of reduced yields and higher costs. In some parts of southwestern Ontario, erosion has caused a loss in corn yields of 30 to 40 per cent. Further costs are incurred off the farm when sedimentation blocks waterways, impairs fish habitat, lowers water quality, increases the costs of water treatment, and contributes to flooding.

Under natural conditions in most ecozones, soil erosion is minimal, since the natural vegetation tends to bind the soil together and keep it in place. Agricultural activities may totally remove this natural vegetation and replace it with intermittent crop plantings, thereby exposing the bare soil to erosive processes, or keep the land under full vegetation for grazing purposes. The latter approach provides much better protection for the soil but may still result in erosion, particularly under conditions of high livestock density.

The rate of soil formation varies as a function of different environmental factors. Because of Canada's latitude, soil formation is slow, with an average annual rate of 0.5 to 1.0 tonne per hectare. Any soil erosion above this amount will result in some loss of productive capacity. Losses in excess of 5 to 10 tonnes per hectare per year may lead to serious long-term problems. These figures have often been exceeded, and 30 tonnes per hectare has been recorded in BC's Fraser River Valley under row crops, while 20 tonnes per hectare is not uncommon in Prince Edward Island. Wind erosion is more difficult to measure but is a significant problem in the Prairie provinces, where high wind speeds, dry soils, and cropping practices often leave the soil unprotected. One study in Saskatchewan detected a net output of soil of 1.5 tonnes per hectare on a near-level field as a result of wind erosion, whereas a field with a greater incline (three degrees) was found to lose 6.6 tonnes per hectare, with water and wind erosion combined.

Soil Compaction

Soil compaction occurs from frequent use of heavy machinery on wet soils or from overstocking with cattle. Compaction serves to break down the soil structure and inhibit the throughflow of water. Crop yields can be reduced by up to 60 per cent in such conditions. Soil compaction is a problem mainly in the lower Fraser River Valley in BC and in parts of central and eastern Canada. One estimate puts the annual cost of compaction in Canada at between $68 million and $200 million.

Environment in Focus

Box 10.7 Food Production and Consumption in Canada

Food Production

- In Canada's early years, agriculture employed more than 80 per cent of the population. Today, only 3 per cent of Canadians are directly occupied in farming.
- Canada exports a wide range of products to more than 200 trading partners around the world. Value-added and processed goods, together with prime-quality meats, live animals, bulk grains, oilseeds, and vegetables are Canada's top agricultural exports. Other important export foods include milk products, fish and seafood, maple syrup and honey, organic, natural, and health foods, and confectionaries and beverages.
- Between 1990 and 2000, canola production in Canada more than doubled and increased again by one-third between 2000 and 2006. These increases are driven by consumers' growing awareness of canola's health benefits as well as demand for biodiesel. Saskatchewan and Alberta are the largest producers.
- There are about 16,500 vegetable growers in Canada, producing close to 7 million tonnes of vegetables worth approximately $2 billion annually. The area under glass used to produce vegetables (10.6 million square metres) exceeded that used to produce flowers (9.3 million square metres) for the first time in 2003. This increase in greenhouse area for vegetables offsets the reduction in area for growing the same vegetables, mostly tomatoes and cucumbers, outdoors.
- Potatoes, along with sweet corn and green peas, are the most extensively grown vegetables in Canada. Approximately 4.5 million tonnes of potatoes (valued at more than $700 million) are grown each year. More than half of them are processed, mostly into french fries.
- The total area in fruit climbed by 5.3 per cent between 2001 and 2006, to 110,069 hectares across the country. Fruit production is dominated by blueberries, which accounted for 46.6 per cent of the total acreage in 2006, an increase of 16.7 per cent since 2001. Quebec's area under blueberries increased by 24.5 per cent between 2001 and 2006, and the province is now Canada's leading producer, followed by Nova Scotia and New Brunswick.
- There are more than 12,000 maple syrup producers in Canada. Canada accounts for more than 85 per cent of world production.
- Canada's red meat and meat products industry includes beef, pork, lamb, venison, and bison. From 1990 to 2003, red meat and live animal exports increased in value from $1.9 billion to $5.3 billion.
- Canada's commercial chicken and turkey meat production totals 1.18 billion kilograms, an increase of about 100 million since 2001, and the average Canadian eats almost double the amount of chicken than in 1976.
- Poultry production and processing are among the most highly mechanized sectors in agriculture. One person can operate a unit of 50,000 broiler chickens. Poultry processing plants in Canada are effectively mechanized, which allows for the slaughter and preparation of 25,000 broiler chickens for market per hour.

Food Consumption

- In Canada, food and non-alcoholic beverages cost about 10 per cent of the average person's disposable income, making our food among the least expensive in the world.
- The level of food energy consumed per Canadian, which remained relatively stable from the mid-1970s to the early 1990s, increased by 17 per cent between 1991 and 2001.
- Red meat consumption totalled 27.1 kilograms per person in 2002.
- Consumption of poultry surpassed 13 kilograms per person in 2002, up 23 per cent over the past decade. In 2006, egg consumption amounted to more than 13 dozen eggs per person.

Sources: Agriculture and Agri-Food Canada (2003, 2008); Statistics Canada, Agricultural Division (2003); Statistics Canada (2003a, 2007a).

Soil Acidification and Salinization

Acidity in soils can occur naturally but can also be augmented by fallout from acid precipitation (see Chapter 4) and the use of fertilizers. Nitrogen fertilizers undergo chemical changes in the soil that result in production of H+ ions, causing greater acidity. In the Maritime provinces, where significant declines in soil pH have been measured, it is estimated that 60 per cent of the change can be attributed to fertilizer use and 40 per cent to acid precipitation. In the prairies, concern over acidity is relatively recent because the substrate is generally alkaline. However, increased use of fertilizer has now led to acidification in some areas. Excess acidity reduces crop yields and leads to nutrient deficiencies and the export of soluble elements such as iron and aluminum into waterways. The yields of crops such as barley and alfalfa fall sharply at soil pH of less than 6. Liming is a common agricultural practice to combat the effects of acidity.

Salinization is the deposition of salts in irrigated soils. Soil salinization is a major problem in many areas of the world where irrigation is common as it leaves soil unfit for growing most crops. As water evaporates, it leaves behind dissolved salts. Over time, these salts can accumulate in sufficient quantities to render the land unusable. Ancient civilizations that

Large areas of tilled soil are particularly vulnerable to erosion.

designed complex irrigation systems were unable to counter the effects of salinization, contributing to their eventual decline. Estimates suggest that 50 to 65 per cent of irrigated croplands worldwide are now less productive due to salinization.

Alkaline soils occur naturally in areas of western Canada that have high sodium content and shallow water tables. Salinization can also be exacerbated through cropping practices that remove natural vegetation and increase the rate of surface evaporation, leading to greater salt concentration at the surface. Summer fallow has this effect. **Summer fallow** is a practice common on the prairies in which land is kept bare to minimize moisture losses through evapotranspiration. On the prairies, crop yields have been reduced by 10 to 75 per cent as a result of salinization. Despite the increased use of fertilizers, it is estimated that in some regions, salinization is increasing in area by 10 per cent every year. However, summer fallow decreased by 25 per cent between 2001 and 2006. The economic need to keep arable land productive, along with diversified and extended crop rotations, improved seeding and tilling methods, and proper use of herbicides, have all contributed to the reduction in summer fallow.

Organic Matter and Nutrient Losses

Cultivation involves a continuous process of removing plant matter from a field. In so doing, both the organic and nutrient content of the soil are reduced. Organic matter is critical for maintaining the structure of the soil, influencing water filtration, facilitating aeration, and providing the capacity to support machinery. It also helps to maintain water and nutrient levels.

On the prairies, current organic matter levels are estimated to be 50 to 60 per cent of original levels, representing a probable annual loss of about 112,000 tonnes of nitrogen (Figure 10.7). This nitrogen is replaced by the addition of

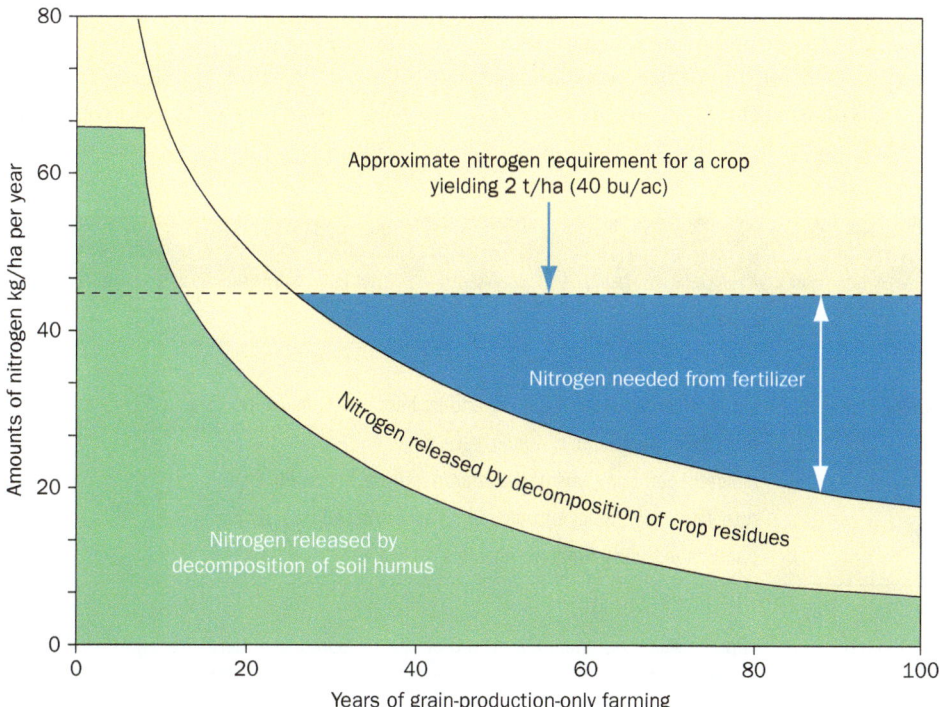

Figure 10.7 | Diagrammatic illustration of approximate sources of nitrogen needed to maintain grain yields of about two tonnes per hectare (40 bushels per acre) of barley under a system of continuous grain production in the prairie region.
Note that this diagram illustrates plant requirements, not supply—i.e., the amount of fertilizer nitrogen applied would normally be greater than the plant requirements because of losses due to denitrification and/or leaching. *Source: Bentley and Leskiw (1985).*

synthetic fertilizers, which in turn contribute to the problem of acidification. An alternative way to replace the nitrogen is the growth of leguminous crops to enhance biological nitrogen fixation (BNF), discussed in Chapter 4. It is estimated that before the increased use of fertilizers began in the 1960s, nitrogen exports from prairie grain exceeded fertilizer applications by more than tenfold and phosphate removals exceeded inputs by threefold. Current estimates still show depletion of the soil but with nitrogen now reduced to double the exports over inputs and phosphorus inputs to about 50 to 60 per cent of the export.

Soil degradation, including erosion and nutrient depletion, is undermining the long-term capacity of many agricultural systems worldwide. According to scientists at the International Food Policy Research Institute, nearly 40 per cent of the world's agricultural land is seriously degraded. This degradation will likely have serious implications for future generations, since the production of food in sufficient quantity and quality requires a healthy natural resource base.

Biocides

Biocide use affects almost all Canadians—biocides are used to produce and preserve the food we eat, and homeowners use them to control weeds in their lawns, insects in their gardens and homes, and parasites on their pets. Since the publication of Rachel Carson's classic book *Silent Spring* in 1962, which outlined some of the environmental problems associated with the use of biocides, there has been considerable controversy over the risks of using chemicals to control pests. Much of the controversy stems from the fact that biocides are designed to be toxic and are deliberately released into the environment.

Biocides have helped to boost yields throughout the world to meet the food demands of rising populations (many more people would be starving without their use), and biocides have also saved countless lives throughout the world by assisting in the control of various diseases through attacking vectors, such as malaria-carrying mosquitoes. But scientific evidence indicates that many chemicals have profound negative impacts on ecosystems. The possible environmental and health impacts may be delayed, in some cases for decades. Despite various risks such as pest resistance, non-selectivity, chemical persistence, biocide mobility, biomagnification, and bioaccumulation, biocide use continues to grow on a global scale.

Resistance

Part of the scientific debate on crop protection relates to the ability of pests, weeds, and viruses to develop resistance to biocides. When a population of insects, for example, is sprayed with a chemical, individuals within the population will react in various ways. If the biocide is effective, most of the population will be killed, but it is likely that a small number of individuals will have a higher natural resistance and survive the chemical onslaught. This remnant resistant population may then grow rapidly in numbers, a result of the lack of competition from all the dead insects. Seeing a resurgence of the pest insect, the farmer sprays again and is again successful in killing a proportion of the population but not as high a proportion as before, since the natural resistance has been passed on to a larger proportion of the population. As this process repeats itself, the use of the chemical creates

Table 10.1 | Major Types of Pesticides

Type	Examples
Insecticides	
Chlorinated hydrocarbons	aldrin, chlordane, DDT, dieldrin, endrin, heptachlor, mirex, toxaphene, kepane, methoxychlor
Organophosphates	malathion, parathion, dizainon, TEEP, DDVP
Carbamates	aldicarb, carbaryl (Sevin), carbofuran, propoxur, maneb, zineb
Botanicals	rotenone, nicotine, pyrethrum, camphor extracted from plants
Microbotanicals	bacteria (e.g., Bt), fungi, protozoans
Fungicides	
Various chemicals	captan, pentachorphenol, methyl bromide, carbon bisulphide
Fumigants	
Various chemicals	carbon tetrachloride, ethylene dibromide (EDB), methyl bromide (MIC)
Herbicides	
Contact chemicals	atrazine, paraquat, simazine
Systemic chemicals	2,4-D, 2,4,5-T, daminozide (Alar), alachlor (Lasso), glyphosate (Roundup)

a population that will ultimately be quite resistant to it. This results in a constant need to develop new biocide products (or pest-resistant plant varieties) to keep one step ahead of biological adaptation.

Perspectives on the Environment
Silent Spring

It was a silent spring without voices. On the mornings that had once throbbed with the dawn chorus of robins, catbirds, doves, jays, wrens, and scores of other bird voices, there was now no sound; only silence lay over the fields and woods and marsh.

—Rachel Carson, *Silent Spring* (1962)

The 'biocide treadmill' has led to biological adaptations resistant to most commercially available biocides. One estimate suggests that in Canada and the United States, more than 900 major agricultural pests are now immune to biocides, including some 500 insects and mites, 270 weed species, and 150 plant diseases (UNEP, 2002). As a result, more frequent applications are needed today to accomplish the same level of control as in the early 1970s. Across Canada, farmers have experienced increasing difficulty in controlling pests through spraying. In New Brunswick, Colorado potato beetles that used to be killed with one spraying a season now must be sprayed five or six times, and they still cause substantial damage. New approaches to controlling beetle damage have involved hybridization with a naturally resistant wild variety, illustrating the value of maintaining as wide a spectrum of wild species as possible. In British Columbia, the pear psylla, an aphid-like insect that feeds on pears, has become resistant to the five main synthetic pyrethroids registered for use against it, and another one is being developed. One of the main beneficial impacts that biocides have had is in controlling disease-bearing organisms such as mosquitoes. Some of the world's deadliest diseases, including malaria, are spread through mosquito bites. More than 1 million people are infected every year, and one person dies from malaria every 30 seconds. The numbers of infections and deaths are rising. Fifty years ago, medical experts predicted that malaria would be eliminated because of the control of mosquitoes with DDT. Unfortunately, the mosquitoes soon developed resistance to DDT, and since that time have become resistant to virtually all control mechanisms. Given global warming trends, it is possible that tropical diseases such as malaria could invade Canada. The advent of West Nile virus may well be a forerunner of what is to come in the future.

Non-Selective

Many biocides are popular because they are broad-spectrum poisons. In other words, there is no need to identify the specific pest, because a broad-spectrum poison will kill most insects. Unfortunately, they tend to eliminate not only the pest species but also other, valuable species, including some that may act to control the population of the pest. This may result in a population explosion of the resistant members of the pest population after spraying due to the reduced abundance of their predators. Similarly, the lack of predators may allow new pest problems to develop that were previously kept in check by natural predators.

Many biocides are also extremely toxic to species other than those directly targeted, such as soil micro-organisms, insects, plants, mammals, birds, and fish (Box 10.8). For example, in PEI more than 20 instances of fish kills since 1994 have been attributed to pesticides, with up to 35,000 dead fish collected in each incident (Commissioner of the Environment and Sustainable Development, 2003). The government instituted new buffer zone regulations as a result of excessive kills, but two further large kills occurred in 2007. Non-target organisms poisoned by biocide use may be beneficial to agriculture or other human economic activities, and as part of biodiversity, they are valued by society for recreational, cultural, ethical, or other reasons.

Mobility

The purpose of biocides is to reduce the impact of a particular pest species (or several species) on a particular crop in a particular area. However, the effects of the chemical application are often felt over a much wider area, sometimes spanning thousands of square kilometres, because of the mobility of the chemicals in the Earth's natural cycles, particularly the hydrological cycle, and the manner in which chemicals are applied (Figure 10.9). The US Department of Agriculture estimates that aerial spraying of insecticide results in less than 2 per cent reaching the target and for herbicide applications,

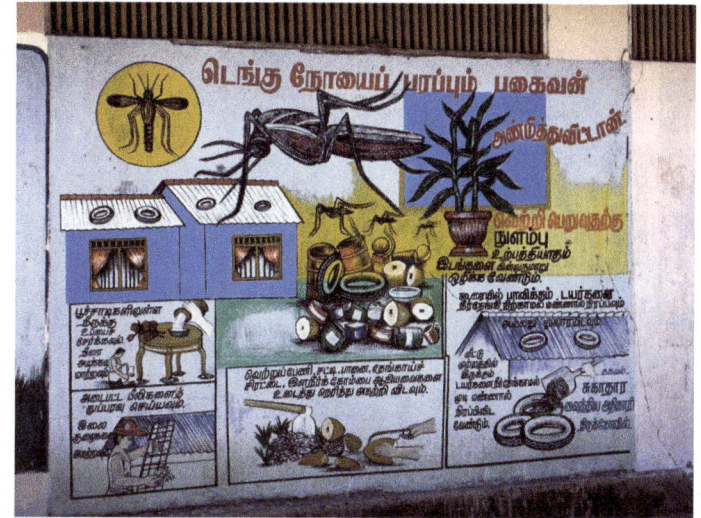

Malaria kills many people a year but receives little funding because it has little impact on the richer countries of the North. This is an educational sign in Sri Lanka showing villagers how to minimize the risks.

less than 5 per cent. The remainder finds its way into the ecosystem where it may contaminate local water supplies or be transported by atmospheric processes to more distant sites (Box 10.9).

Places with well-developed agricultural sectors and frequent use of chemicals might find that the entire environment is becoming contaminated. This seems to be the case in PEI, where more than 8 kilograms of biocides per person

Environment in Focus
Box 10.8 Carbofuran and Birds

Carbofuran is a carbamate (Table 10.1) that was registered for agricultural use in Canada. It was available either as a liquid or in granular form on particles of grit and was often applied in the latter form during seeding to protect recently germinated seedlings from insects. Considerable amounts of the chemical (up to 30 per cent of that applied) are often exposed on the soil surface following application. These granules are highly attractive to many birds that ingest grit in their gizzards to grind seeds. The chemical is highly toxic to birds, and the consumption of just one grain can be fatal to small seedeaters. Carbofuran also contaminates invertebrates such as earthworms, which in turn will poison organisms higher on the food chain. Flooded fields are also highly dangerous, since the chemical goes into solution. This is particularly serious in acidic fields, where the breakdown of carbofuran is very slow. There have been many documented bird kills as a result of the use of carbofuran, including 1,000 green-winged teal in flooded turnip fields in BC in 1975 and more than 2,000 Lapland longspurs in a canola field in Saskatchewan in 1984.

Of particular concern was the impact of carbofuran on one of Canada's designated threatened species, the burrowing owl. The owl nests on the prairies in abandoned mammal burrows, feeding on small mammals and insects. Two-thirds of the Canadian breeding population nests in Saskatchewan, where studies suggest that breeding numbers declined by 50 per cent in the south-central region between 1976 and 1987. Carbofuran spraying for grasshopper infestations is suspected to be the main cause. There were significant declines in nesting success and brood size with increasing proximity of carbofuran spraying to the nests. A high percentage of adult owls also disappeared after spraying. There were particularly severe infestations of grasshoppers in the early 1980s, and in Saskatchewan alone more than 3 million hectares was sprayed in 1985, 40 per cent of the area with carbofuran.

The registration of carbofuran was cancelled in 1996. The burrowing owl remains on Canada's endangered species list. One evaluation suggested that the voluntary habitat stewardship program (see Chapter 14), Operation Burrowing Owl, is having some success in protecting the remaining habitat (Warnock and Skeel, 2004), and 52 pairs were reported in the 2010 annual Saskatchewan census. However, this is significantly down from the 79 pairs reported in 2009.

The burrowing owl is an endangered species that has been particularly threatened by the use of agricultural biocides.

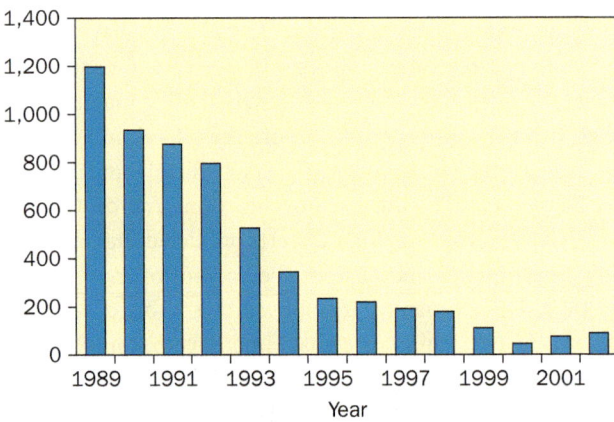

Figure 10.8 | Estimated pairs of burrowing owls reported by Alberta and Saskatchewan landowners. *Source: Nature Saskatchewan. Bird Species at Risk, 2010 Report. Habitat Stewardship for Bird Species at Risk in Saskatchewan,* www.naturesask.ca/docs/assets2/BSAR_Executive_Summary_11Apr11_1_.pdf

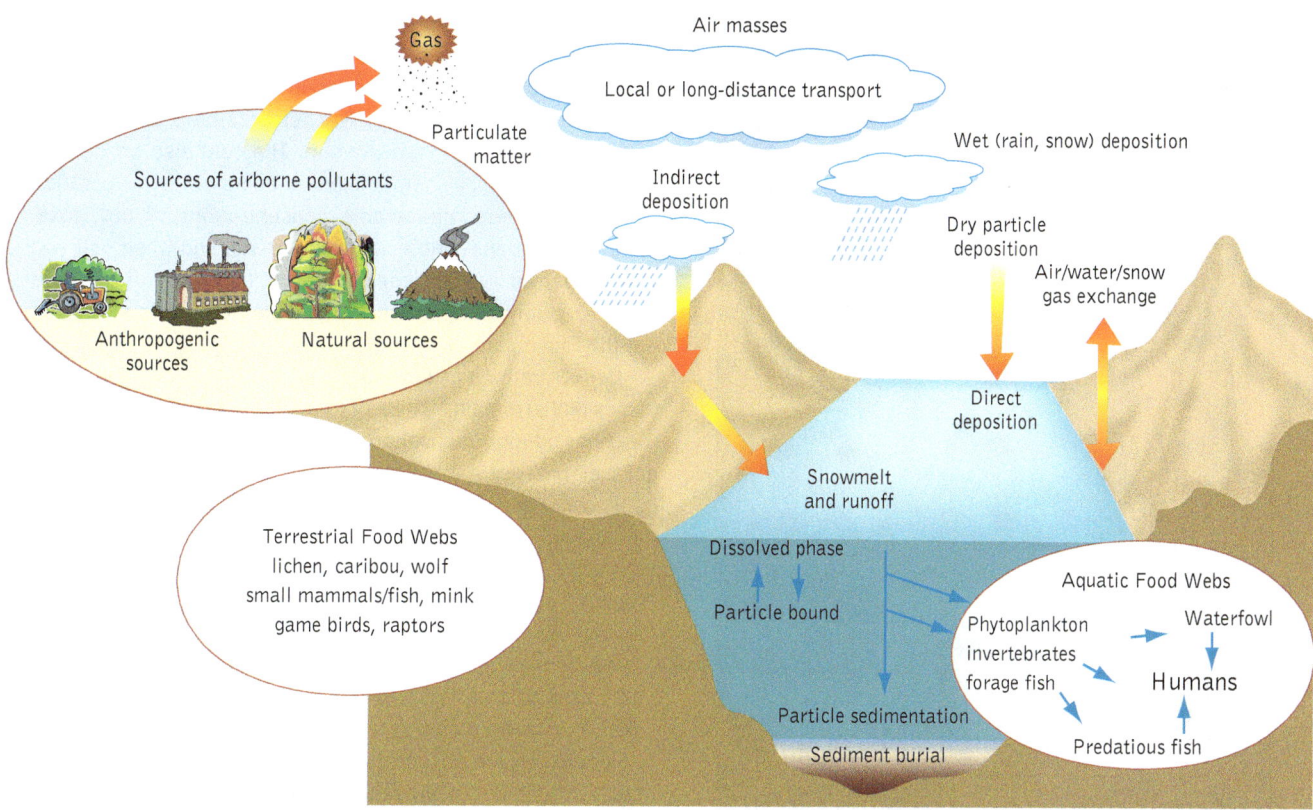

Figure 10.9 | Pesticide transportation in the environment. *Source: Adapted from Indian and Northern Affairs Canada (1997b). Reproduced with the permission of the Minister of Public Works and Government Services, 2004.*

Environment in Focus

Box 10.9 The Grasshopper Effect

Imagine for a moment, if you will, the emotions we now feel: shock, panic, grief—as we discover that the food which for generations nourished us and keeps us whole physically and spiritually is now poisoning us. You go to the supermarket for food. We go out on the land to hunt, fish, trap, and gather. The environment is our supermarket. . . . As we put our babies to our breasts, we feed them a noxious chemical cocktail that foreshadows neurological disorders, cancers, kidney failure, reproductive dysfunction. That Inuit mothers—far from areas where POPs are manufactured and used—have to think twice before breast-feeding their infants is surely a wake-up call to the world.

—Sheila Watt-Cloutier, President, Inuit Circumpolar Conference (Canada), 2000

We tend to think of the Arctic as 'pristine wilderness'. However, research has indicated that this is far from the truth. At Ice Island, for example, a floating ice-research station 1,900 kilometres above the Arctic Circle, concentrations of a family of pesticides called hexachlorocyclohexanes (HCHs) have been measured that are twice as high as those in agricultural southern Ontario. Yet there is not a single pesticide-dependent product grown in the North! How did the chemicals get there?

The so-called **grasshopper effect** is one reason behind Arctic pollution. Indeed, atmospheric transport and deposition is a major pathway of contamination. After chemicals are introduced into an environment, they are absorbed into the soils and/or plant tissues or deposited into rivers, lakes, and wetlands. Persistent and volatile pollutants evaporate into the air in warmer climates and travel in the atmosphere towards cooler areas, condensing out again when the temperature drops. The cycle then repeats itself in a series of 'hops' until the pollutants reach climates where they can no longer evaporate. Chemicals released in southern Canada, for example, may go through the grasshopper cycle several times and take 10 years to reach the Arctic. Extremely volatile chemicals will travel farther and recondense in greater concentrations. For example, biocides such as lindane and HCH

People have the perception of the Arctic and Rockies as pristine wilderness areas. Marketers use this image to sell products such as water. In reality, these cold environments can have high toxic burdens due to the grasshopper effect.

are quite volatile compared to DDT and therefore usually reach the North in greater quantities.

However, if only air transport were responsible, then the pollutants should have a fairly even distribution across the Arctic. Researchers found this not to be the case, with marked concentrations in some areas, particularly near bird colonies. These birds, such as fulmars, are at the top of the food chain and serve to concentrate the chemicals before excreting them on the local landscape, where concentrations can be 60 per cent higher than in the surrounding landscape.

The implications of this long-distance transport are serious. Arctic ecosystems are more vulnerable to toxic chemicals because they last longer in the North. Degradation processes are inhibited by low temperatures and reduced ultraviolet radiation from the sun. The cold also condenses the toxins, keeping them locked up and slowing evaporation rates. There are measurable concentrations of DDT, toxaphene, chlordane, and PCBs in the Arctic, and when fish and other species ingest these chemicals, they travel up the food chain, accumulating in the fatty tissue of animals at the top of the food chain.

Concentrations of several persistent organochlorine pesticides (POPs) remain high in many aquatic food webs in Canada. This has serious implications for the Inuit in particular because of their high consumption of wildlife. More than 80 per cent of Inuit consume caribou, almost 60 per cent consume fish, and almost 40 per cent consume marine mammals. Because of bioconcentration, the consumption of traditional foods places the Inuit at greater risk for developing several ailments related to toxic chemical exposure, including endocrine disruption, reproductive impairment, and cancer. These concerns are not restricted to the Arctic, since large concentrations of toxic chemicals have also been found in the mountains of British Columbia. Fish in alpine lakes have chemical levels that make them toxic to eat in large quantities.

Since POPs travel great distances, a global approach to tackle the issue is required. Various initiatives have been undertaken to control or eliminate POPs, including the Global POPs Protocol, signed in Johannesburg in December 2000. Signatories, including Canada, agreed to a global ban of 12 chemicals, including the pesticides aldrin/dieldrin, endrin, DDT/DDE, HCH/lindane, chlordane, heptachlor, chlordecone, mirex, and toxaphene. Canada is also a signatory to the United Nations Economic Commission for Europe (UNECE) POPs Protocol, which lists 16 chemicals for phase-out. But while governments around the globe have a significant role to play in reducing Arctic pollution, consumers also share the responsibility. Chemicals used in consumers' everyday environment can end up polluting some of the most 'pristine' environments on Earth. Consumers must make an effort to phase out domestic use of toxic chemicals.

are used every year, compared to the average of 1.3 kilograms used in the US. Tests on airborne pollution found every sample to be contaminated. Even tests taken at the end of a wharf, far from any farms, showed chemicals. One of the most heavily used chemicals on the island, chlorothalonil, has been identified by the US government as a potential carcinogen, and its effects can be detected in air samples two hours after spraying has taken place.

Persistence

Not only do biocides spread over vast areas, they also continue to contaminate through time, as many are very persistent.

DDT is one of the best-known insecticides. First synthesized more than 100 years ago, it was not until the 1940s that DDT became widely used, first in health programs in World War II to control disease vectors and later as an agricultural chemical. Production peaked by 1970 when 175 million kilograms was manufactured. By that time, the environmental effects of DDT were becoming better understood, and its use, but not manufacture, was banned in the US in 1972. It was not until 1985 that registration of all DDT products was discontinued in Canada. More than 7 million kilograms of DDT were sprayed on forests in New Brunswick and Quebec between the early 1950s and the late 1960s. DDT is extremely persistent. Even now, there are still considerable residues of DDT and its main breakdown product, DDE, in the environment. Because it is soluble in fat, DDT may also gradually accumulate over time in the tissues of organisms. This is known as bioaccumulation (Figure 10.10).

Organisms with long lifespans are particularly susceptible to bioaccumulation. In British Columbia, for example, an insecticide banned in the late 1970s called hexachlorocyclohexane (HCH), once used as a timber preservative and agricultural spray, is still being found in geoducks, a type of clam, off the west coast of Vancouver Island and in Puget Sound. These large clams are filter-feeders that may live as long as 140 years. They are therefore very susceptible to bioaccumulation, and concentrations have been sufficient to have the clams refused by processing plants. In another example of the persistence of biocides in aquatic environments, researchers have detected high levels of toxaphene in trout in Bow Lake in Banff National Park, Alberta. The toxaphene had been applied in 1959 to rid the lake of what were then seen as undesirable fish species. DDT and other chemicals have been detected in trout in many of the lakes in Waterton, Banff, Jasper, and Yoho national parks.

Biomagnification

The Arctic has generally been viewed as one of the few unpolluted regions of the world. But a closer look reveals a different story. High concentrations of persistent organic pollutants (POPs) such as organochlorine biocides have been detected in top predators of the Arctic food chain, including indigenous peoples. The ability of biocides and other toxic chemicals to traverse long distances, combined with the ability of biocides to concentrate and accumulate in the lipids of **pelagic** marine organisms (e.g., Atlantic cod, whales, seals), is to blame for the introduction of contaminants into the Arctic food web (Box 10.9). Concentrations of POPs multiply five- to tenfold with every step in the food chain. This process is illustrated in Figure 10.11, showing the concentration of DDT and its derivatives along a food chain in the North Pacific Ocean. The relatively low concentrations at the lower end of the food chain are magnified many times by the time they reach top fish-eating predators. Most of the visible effects of POPs on animals are related to the ability to conceive and raise young. Malformations in reproductive organs, fewer young, and even complete failure to reproduce are some of the detrimental

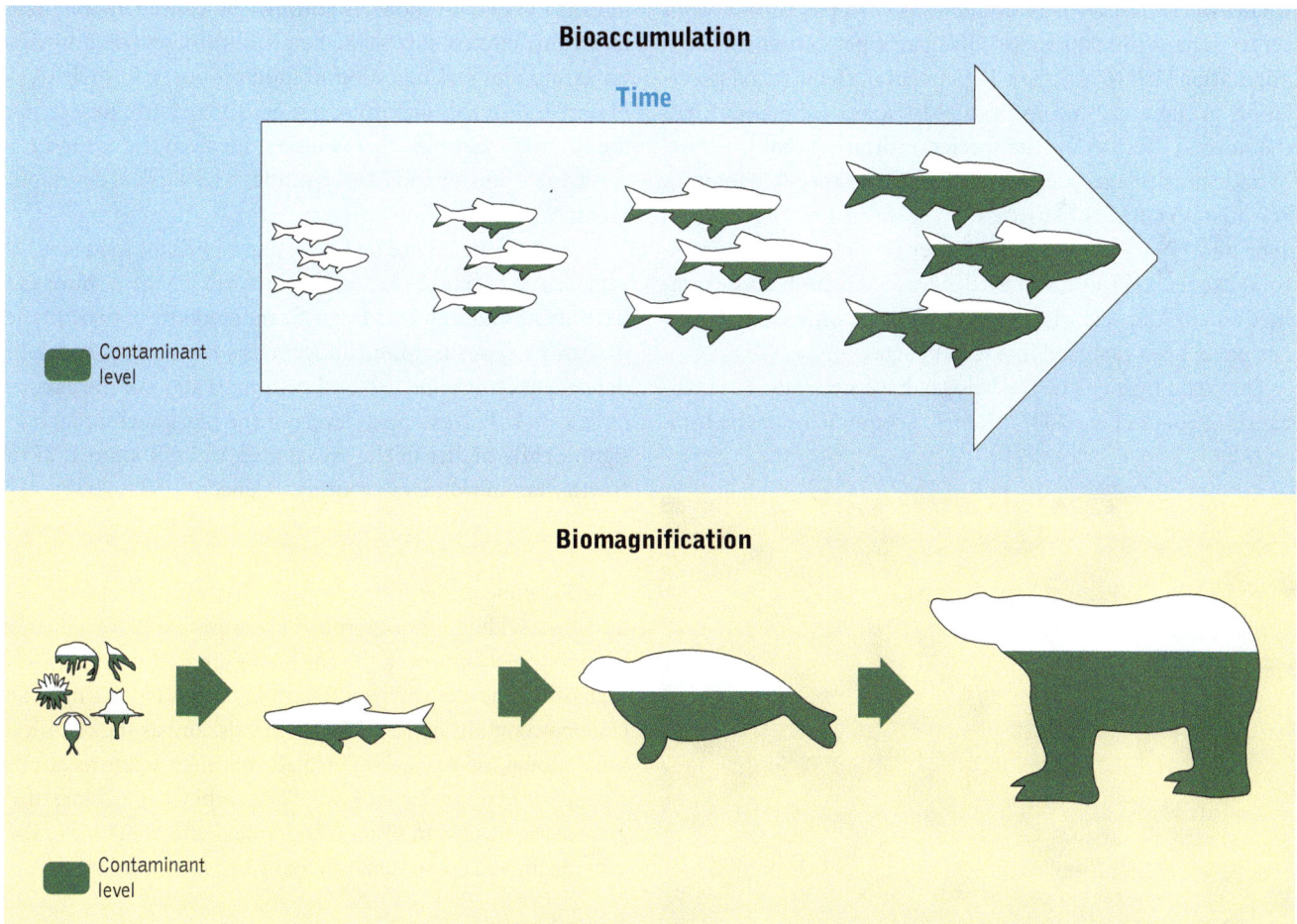

Figure 10.10 | Bioaccumulation and biomagnification. *Source: Adapted from Indian and Northern Affairs Canada (1997a). Reproduced with the permission of the Minister of Public Works and Government Services, 2004.*

Although Prince Edward Island is known for its beautiful rural landscape, the intensity of agricultural production has left large biocide residues, even in ocean sediments.

signs of high contaminant levels. Animals with a high load of organic contaminants are also more susceptible to infections, and POPs are suspected of being responsible for increased rates of malignant tumours in wildlife.

Some species build up very large concentrations. The beluga whales of the St Lawrence estuary, for example, showed concentrations of 70,000 to 100,000 parts per billion of DDT. Populations fell to less than 10 per cent of the original population in the area, and individual lifespans were about half the normal lifespan for the species, indicating that this level of toxic burden exceeded their level of tolerance (Chapter 2). Scientists examined 73 carcasses that washed ashore between 1983 and 1994 and found that 20 per cent of them had intestinal cancer. Of the 1,800 whales washed ashore and examined in the US, scientists found cancer in only one. Cancer has never been reported in Arctic belugas.

The toxic burdens of the belugas have now fallen significantly (Lebeuf et al., 2007). A major component of the toxic burden for the belugas was mirex and its by-products. Mirex is a biocide, now banned, which was never produced along the St Lawrence. Biologists think that the source was Lake Ontario, where American eels accumulate the chemical. During their downstream migration, these eels constituted a significant part of the whales' food supply.

Biomagnification was largely responsible for the drastic population reductions of many birds during the 1960s and 1970s. Birds of prey such as ospreys, peregrine falcons, and bald and golden eagles and fish eaters such as double-crested cormorants, gannets, and grebes were particularly affected by the widespread use of insecticides. Some birds were killed directly through bioaccumulation and biomagnification, while many others were unsuccessful in breeding. DDT affects the calcium metabolism of these species, resulting in thinner eggshells and leading to breakage and chick mortality. The banning of DDT and similar chemicals has led to a recovery of many of these species in temperate countries (see Chapter 14). However, the continued use of the chemicals in some tropical countries still affects populations in these areas as well as the populations of migratory species such as peregrines.

Together, biomagnification and bioaccumulation are often known as **bioconcentration**. Humans are exposed to the harmful effects of biocides through bioconcentration. They may come into contact with chemicals through contaminated water supplies and ingestion of food products, in their workplaces, and/or through domestic use (Box 10.10). Researchers found, for example, that women involved in farming in Ontario's Windsor and Essex counties have a risk of developing breast cancer nine times greater than non-farm women.

A greater dependence on fish or game birds for food may elevate the risk of biocide exposure, since fish and game birds may have already concentrated significant amounts of toxic matter in their fat deposits. Many indigenous communities are highly dependent on marine fish and mammals and may be particularly at risk. For example, levels of the biocide chlordane are significantly higher in the breast milk of Inuit women in the North than in women in southern Canada (UNEP, 2002).

Synergism

When chemicals are tested for their harmful effects, they are tested individually in controlled situations. When applied on farmers' fields, however, the chemicals are free to interact with each other and the environment in myriad ways. A single biocide may contain up to 2,000 chemicals, and as the chemicals break down, new ones are created that may again react with each other in unpredicted ways. The combined effects are often greater than the sum of their individual effects. This is called synergism, and it can result in many unanticipated effects.

Biocide Regulation

The concerns related to the use of biocides—resistance, non-selection, mobility, persistence, bioaccumulation,

The beluga whale population has declined rapidly in the St Lawrence, and the toxic burden from biocides appears to be one cause.

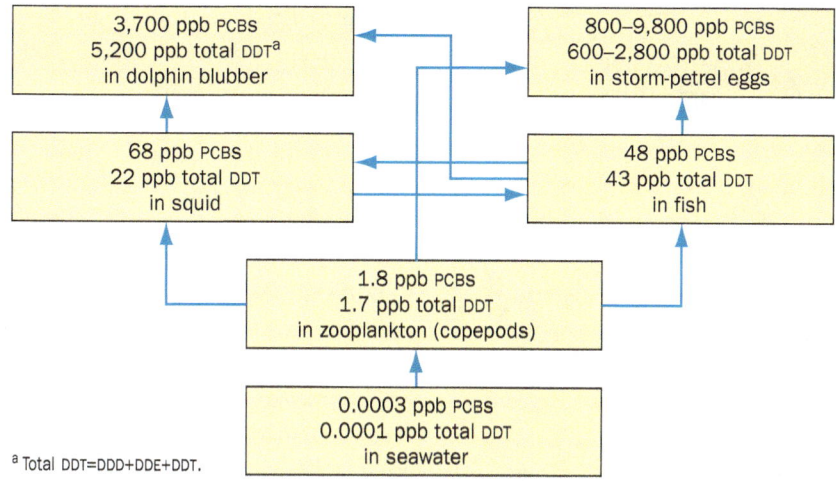

Figure 10.11 | Organochlorines in a North Pacific food chain. *Source: Nobel (1990).*

biomagnification, and synergism—are significant. However, the economic value tied to the use of biocides is also very significant. Because the use of biocides will likely continue to grow in the foreseeable future, there is a pressing need to strengthen the regulatory and enforcement mechanisms governing biocide import, production, and use.

In Canada, an estimated 1,000 new chemicals are introduced annually, adding to the more than 20,000 already in industrial, agricultural, and commercial use. New chemicals cannot be introduced onto the market without undergoing scientific tests regarding their capacity to cause cancer, birth defects, and mutations. The Pest Management Regulatory

Environment in Focus

Box 10.10 Biocides and You

Although most biocides in Canada are used by commercial producers, large amounts are also used domestically in homes and on gardens and lawns to control unwanted organisms. How people use, store, and dispose of these chemicals is very important in terms of minimizing environmental damage. Here are a few tips.

- Use chemicals only as a last resort. Ask yourself why you need to kill the organism. If it is just for aesthetic reasons, such as dandelions on your lawn, then maybe you need to change your perceptions rather than automatically reaching for a chemical solution. For each pest or weed, there are usually several other approaches you can take as part of your own integrated pest management strategy. You can find out about more specific strategies from government ministries such as the ministry of agriculture in your area.
- Use the safest chemicals available in the minimum quantities. Many plant nurseries now sell products that are less toxic than traditional biocides. Often, they need more skill in application, but they are less environmentally damaging than regular chemicals. Examples include the 'Safer Soap' line of products.
- Apply all chemicals in strict accordance with the manufacturer's instructions.
- Store unused chemicals so that they do not leak and cannot be accidentally upset.
- Dispose of chemicals and containers in a safe manner. Contact your local ministry of environment to see what programs are in place in your province for safe disposal. Some provinces, for example, have specific sites where biocides and other toxic chemicals can be disposed. If your province does not have such a program or acceptable alternative, start lobbying for one!

You should also protect yourself against the risk of ingesting chemicals that have been applied to food.

- Grow your own food; don't use chemicals.
- Buy organically grown produce whenever possible.
- Fruit and vegetables that look perfect often do so because they have had heavier applications of fertilizers and pesticides. Choose products that show more natural blemishes; this is a sign that chemical use has not been as high.
- Carefully wash all produce in soapy water.
- Remove the outer leaves of vegetables such as cabbage and lettuce, and peel all fruit.

Biocides and human health.

Agency (PMRA), a branch of Health Canada created in 1995, has the primary responsibility for regulating biocides. Other Health Canada branches and other federal departments and agencies that play important roles in biocide management include Agriculture and Agri-Food Canada, the Canadian Food Inspection Agency, Environment Canada, Fisheries and Oceans Canada, and Natural Resources Canada. The federal government shares the responsibility for managing biocides with provincial, territorial, and in some cases municipal governments (Box 10.11). About 5,000 biocides are currently registered for use in Canada.

To predict the effectiveness of biocides and their risks to human health and the environment, the PMRA relies on the expertise and judgement of the agency's scientists and managers. Evaluation of a new biocide ends with the approval of the biocide label, which describes the biocide's hazards and its proper use. This process, however, has not been foolproof.

The House of Commons Standing Committee on Environment and Sustainable Development reviewed biocide management in Canada and published a report, *Pesticides: Making the Right Choice*, in 2000. The report expressed strong concern over practices and urged a much stronger regulatory approach based on the precautionary principle. This was followed in 2003 by a report titled *Managing the Safety and Accessibility of Pesticides*, released by the Commissioner of the Environment and Sustainable Development (Box 10.12). This report also documented the adequacy (or lack thereof) of regulatory practices in Canada to approve biocide use, as well as the health and environmental standards relating to compliance, the government's commitment to research, and monitoring. The report identified several instances of poor overall compliance. To predict occupational exposure to pesticides and pesticide residues on food, PMRA evaluators

Environment in Focus

Box 10.11 Cosmetic Use of Pesticides

The proportion of households using pesticides on their lawns and gardens dropped slightly from 31 per cent in 1994 to 29 per cent in 2005. Municipalities also use substantial amounts of pesticides. Figures are difficult to obtain, but in Ontario in 1993, the amount was estimated at 1.3 million kilograms, about one-quarter of the amount used in agriculture in that province.

Because of potential negative impacts on human health, the Standing Committee on Environment and Sustainable Development (2000) recommended a ban on the use of biocides for cosmetic purposes—i.e., lawn care—but the federal government refused to endorse the recommendation (See cartoon, above). The government preferred to take a voluntary, educational approach to reducing the cosmetic use of pesticides and launched a 'Healthy Lawns Strategy' to address the issue.

However, the government's educational approach did not satisfy all Canadians, particularly residents of a small community in Hudson, Quebec. Residents were worried about the health consequences of lawn and park applications of herbicides and insecticides, particularly on children. The community wanted the municipal government to enact a bylaw that would ban the cosmetic use of pesticides, but chemical companies won the first battle, arguing that municipal governments lacked the authority to introduce such bylaws. The community took the matter to court, and in June 2001, Canada's Supreme Court unanimously ruled that towns and cities have the right to enact bylaws banning the purely cosmetic use of pesticides. The Supreme Court ruling grants municipalities across the country the right to impose similar pesticide restrictions. In 2002, Halifax became the first large city to place a ban against pesticide use. Today, more than 130 communities in Canada have bylaws banning the use of chemicals for cosmetic lawn purposes, including the entire province of Quebec. Toronto's Board of Health has endorsed a similar bylaw, initiating steps to phase out the use of pesticides on lawns for cosmetic purposes. In 2009 the province of Ontario followed suit and instigated a strict ban against cosmetic pesticide use, banning over 250 products for sale and 95 for use on lawns, gardens, patios, driveways, cemeteries, parks, and schoolyards.

Environment in Focus

Box 10.12 A Report Card on Canada's Use of Pesticides

In 2003, the Commissioner of the Environment and Sustainable Development released a report card on Canada's use of pesticides. The document, titled *Managing the Safety and Accessibility of Pesticides* (Commissioner of the Environment and Sustainable Development, 2003: ch. 1), detailed the shortcomings of the Pest Management Regulatory Agency (PMRA), a branch of Health Canada responsible for the evaluation of new pesticides, monitoring and compliance of existing pesticides, and the re-evaluation of old pesticides to ensure compliance with modern standards. Overall, the Commissioner concluded that 'the federal government is not managing pesticides effectively.' A few of the main findings in the report are highlighted here.

Some pesticides are approved based on inadequate information. As a result, many pesticides are used before they have been evaluated fully against current health and environmental standards. (See sections 1.36–1.39.)

Key assumptions are not tested, and some are not valid. Agency evaluators must make a series of assumptions to link the laboratory studies they receive to the possible impacts of the pesticide's use. Such assumptions include how large a crop area will be treated, how much treated food Canadians will eat, and how the pesticide will be applied. Despite the uncertainties in all of the different assumptions evaluators make, they have not determined how reliable their predictions of the risks are. In addition, agency staff have unrealistic assumptions about user behaviour. They assume that pesticide users will follow label instructions, even though the agency's own compliance reports show that they may not. (See sections 1.40–1.45.)

The agency manages a legacy of older pesticides. Many pesticides have been registered in Canada based on evaluations that did not apply the more stringent methods and standards used today. The agency has implemented re-evaluation programs. However, many re-evaluations do not consider new information about the pesticide's effectiveness resulting from new research, and as a consequence, opportunities may be missed to reduce the rate of application. (See sections 1.51–1.53.)

The agency does not know to what extent users are complying with pesticide labels. Few inspections are undertaken and inspections do not determine systematically whether the label requirements are being met. Problems with unclear labels also make it difficult to ensure compliance. Ambiguous labels mean that enforcement action cannot be taken for some possible violations of the Act. (See sections 1.79–1.82.)

Methods for measuring pesticide residues on food are not up-to-date. The Canadian Food Inspection Agency (CFIA) conducts an extensive chemical sampling program that includes testing each year for pesticide residues on food. However, methods for measuring pesticide residues are not inclusive or thorough. (See sections 1.90–1.91.)

Critical information on pesticide use and exposure is still missing. In 1994 and again in 2000, the federal government committed to developing a program of mandatory reporting on adverse effects of pesticides, but it has yet to do so. The lack of reliable information on pesticide use, exposure, and impacts is a major hurdle that continues to interfere with the agency's ability to regulate pesticides. (See sections 1.94–1.100.)

Federal research on the health impacts of pesticides has not been a priority. Health Canada has very limited dedicated funding for research on human exposure to pesticides or the resulting health effects. Three researchers are working on current pesticides, and they rely primarily on outside funding. (See sections 1.101–1.102.) (Commissioner of the Environment and Sustainable Development, 2003: ch. 1)

The Commissioner published a follow-up audit in 2008, concluding that the PMRA procedures had improved, as had procedures for re-evaluating chemicals, although there was still a considerable backlog and no action plan to deal with it. The Canadian Food Inspection Agency had also broadened the range of residue tests on fresh fruits and vegetables (Commissioner of the Environment and Sustainable Development, 2008: ch. 2).

---

assume that agricultural users will follow good practices—i.e., users will follow label instructions. But evidence suggests otherwise. In 2001, evaluators collected soil samples from 20 onion growers in Ontario. Of those, 14 (or 70 per cent) had violated the Act by using pesticides not registered in Canada. Even though this situation and that described in Box 10.12 have since improved, it is still worth noting, since Canadians place trust in their government to safeguard their health and that of the environment.

Lack of compliance is partly due to problems with pesticide labels. Some agricultural pesticides may have 30 or more pages of directions in fine print, while other label instructions are difficult to follow. For example, labels are often ambiguous, and application therefore depends on the applicator's interpretation. Ambiguous, vague terms used on pesticide labels include:

- *Appropriate* buffer zones should be established between: treatment areas and aquatic systems, and treatment areas and *significant* habitat.
- Do not apply in areas where soils are *highly* permeable and groundwater is *near* the surface.

- Do not apply *near* buildings inhabited by humans or livestock.
- Do not apply where fish and crustaceans are *important* resources.

Failure to follow label instructions could increase the risks to consumers and the environment, but it is difficult to apply pesticides appropriately when the directions are unclear. Many poisonings have been attributed to inappropriate application, and this is one reason why farmers in Ontario have welcomed mandatory biocide safety courses dealing with the use, mixing, handling, and transportation of biocides as well as laws governing their use. The certificate from these courses must be renewed every five years, and it must be presented in order to buy agricultural chemicals.

The federal Pest Control Products Act came into force in 2006. Its objective is to better protect Canadians and the environment from the risks of pesticide use. The Act requires that all pesticides be re-evaluated every 15 years against the most current health and environmental standards.

Intensive Livestock Operations

Livestock farming can have significant and far-reaching environmental implications. The production of livestock manure has both environmental benefits and drawbacks. Although manure is a valuable fertilizer for crop production, it can also become a source of pollution if not managed properly.

Manure consists of a variety of substances, including nitrogen, phosphorus, potassium, calcium, sodium, sulphur, lead, chloride, and carbon. Manure also contains countless micro-organisms, including bacteria, viruses, and parasites. Some of these micro-organisms are pathogenic, and therefore direct consumption or recreational use of water containing these organisms can lead to a variety of illnesses and even

Guest Statement
Canada Feeding the World?
Peter Schroeder

The myth woven into agricultural policies and subliminally embedded in all agricultural thinking throughout the developed countries is that agricultural output must continually increase, at all costs, or humanity will face the consequences of global starvation. This myth formed the bases for the theology driving the Green Revolution and is the primary reason stated for the global rush to accept and adopt genetically modified organisms in all aspects of modern commercial farming systems. The 'need to feed the world' myth would not be a bad thing if in some form modern agriculture's hyper-productivity actually helped to reduce global hunger, but there is little, if any, evidence of that. There is ample evidence that maximizing production from every hectare has caused more problems than it has solved by ignoring some of history's most important lessons.

Since its inception, agricultural production has been based on the 'natural fertility' of the land on which it is practised. As farmers became more adept at farming, they began to understand what depletes as well as what enriches their soil's inherent fertility. By paying close attention to these factors, they began to manipulate their environment to increase yields and harvest reliability. In a relatively short time, they had a thorough understanding of the direct linkage between sustainable yield and sustainable fertility. Left to themselves, farmers are loath to destabilize this critical balance because they have nothing but the land to sustain them.

The great civilizations throughout antiquity rose and fell according to how well their farming and herding people were able to balance the land's sustainable fertility with the harvest output. The fundamentals of that technology and science have not changed because, now as then, agriculture is all about repackaging solar energy into calories that can be digested by humans. This includes the use of all food animals that convert plant material not digestible by humans. Since so much of this planet is more pasture land than 'farm' land, food animals form an important link in the solar energy reprocessing system that is agriculture.

Modern agriculture, which began in the 1950s and 1960s, is based on cheap petroleum energy and a surplus of nitrogen production left over from World War II explosives manufacturing. These industrial surpluses were all redirected at agricultural production and marketing. Without fossil fuel in all its forms to power modern agricultural equipment and provide the fertilizer as well as transportation, the contemporary commercial agricultural models could not exist.

It was at this point that the Canadian government turned prairie agriculture into a full-blown exporting component of the Canadian economy. Grain, specifically wheat, was produced beyond any possible local demand and shipped around the world. Everybody involved in growing grain on the prairies had a fuzzy warm feeling; they were working hard, making a decent living, and feeding the world. What could be better?

The long-term truth is that the wheat was sold to the highest bidder, often disrupting local markets and bankrupting farmers as far away as India and other developing countries. The cheap Canadian wheat was often used to lower local grain prices, forcing massive disruptions in the local agricultural economies. Canadian wheat was cheap in part because Canadians had learned to use modern farming systems but also because it was subsidized by cheap fuel and transportation.

Intensive livestock production can only happen when the supply of grain in the world is too large for the human population to consume. In the overstimulated production environment as in North American and Western Europe, this artificially created surplus of grain is used as livestock feed for everything from poultry to pigs, cattle, and sheep. In addition to producing the cheapest food the world has ever seen, this farming model brings with it environmental and biological challenges.

For almost 100 years, government subsidies have directed agricultural production in western Canada. The best example is how the Canadian taxpayer paid the bulk of the freight costs to get western wheat and barley from the Canadian prairies to deepwater ports on Canada's west coast or the Great Lakes. The same story is now being repeated with canola, minus the transportation subsidy. Today's incentives come in the form of tax breaks and risk management programs. A similar scenario is also being played out in Manitoba with its massive export-dependent hog industry and across the prairies where the beef industry expanded to supply a growing demand in the US.

Today, Canadian and American beef and hog producers are suffering economically as US energy policy diverts corn from feed grain to ethanol production. Because of the feed grain shift to ethanol, Manitoba hog producers are killing newborn piglets and brood sows. Diverting even a small part of the feed grain stocks has also caused food riots in the poorest countries, because it takes only a very small supply shortfall to cause dramatic price increases.

From a farmer's perspective, modern agriculture has nothing to do with feeding the world's hungry and everything to do with not going broke. The world's hungry cannot afford to pay the freight for the food it would take to solve their problem, never mind cover the production costs. At the same time that the world's starving poor are rioting, people in developed nations are fighting an epidemic of obesity.

The Green Revolution has been underway for more than 40 years, and one has to wonder how it has changed the world for the better. There are fewer farmers on the land in Canada's prairies every year. Those who are left earn less per acre and animal unit than their parents did before the Green Revolution started. In their struggle, they have abandoned many of the fundamentals that sustained the generations on the land before them. Many of today's farming practices now contribute to ecological damage, with chemical and nutrient runoff contributing to contaminating waterways and lakes instead of keeping the land productive and healthy. Likewise, the Green Revolution has not been kind to farmers. The net beneficiaries of the Green Revolution are the people with shares in the farm service and food processing sectors.

Some farmers in North America have opted to step off the industrial farming treadmill. They distinguish themselves as grass farmers, holistic ranchers, natural and organic farmers. The common link among all of them is that they are striving to achieve a balance between the 'natural' fertility, sustainable yield, and their own economic needs without resorting to the farming stimulants: chemical fertilizers, herbicides, pesticides, and genetically modified seeds. These people are a minority, but their numbers are growing.

In my own farming experience covering 25 years of sheep ranching, we had our best economic return after we adopted the grass farming principles and applied them. Simply put, grass farming is all about using livestock to repackage the solar energy stored in plants for human consumption. Grass farming requires the rancher to put the needs of pasture ahead of those of livestock. It also requires the rancher to forget about maximizing production in favour of sustainable yields. The credo for many grass farmers is the 80:20 rule: 'Take the easy 80 per cent nature is offering for free, and forget about the last 20 per cent.' In my experience, the net return to the enterprise increased substantially, and there was a marked improvement in lifestyle with less labour and lower stress levels.

The biggest threat facing food production today is that by well-intentioned regulatory means, the traditional subsistence and small-scale commercial farms could be eliminated in the drive to accommodate intensive agri-business operations as they continue to perpetuate the myth that they are feeding the world's hungry.

Peter Schroeder farmed the Canadian prairies for many years and now shares his experience as a columnist with the *Winnipeg Free Press*.

death. The contamination of drinking water with E. coli that killed seven residents of Walkerton, Ontario, in May 2000 was related to livestock manure. Pathogens from manure that have reached watercourses also have the potential to spread disease to livestock. The spread of bovine spongiform encephalopathy (BSE), commonly referred to as 'mad cow disease', is an example of an inter-species disease transmission. BSE is thought to cause Creutzfeldt-Jakob disease among humans.

Other risks to human and ecosystem health arise from air pollution. Odour and air pollution are identified as serious environmental and human health concerns related to **intensive livestock operations** (ILOs). High concentrations of

Aerial view of mixed breeds of cattle in pens at a large, modern beef feedlot with a 12,500-head capacity in Alberta.

noxious gases such as methane, hydrogen sulphide, carbon dioxide, and ammonia are often found in manure pits and confinement barns. Pigs and poultry, for example, excrete some 65 and 70 per cent of their nitrogen and phosphorus intake, respectively. Nitrogen, under aerobic conditions, can evaporate in the form of ammonia (Chapter 4). Ammonium nitrate and ammonium sulphate emitted to the air from animal housing can be harmful to human and animal health. Foul odours emitted by ILOs are a significant problem for neighbours, and studies have shown an increase in chronic respiratory diseases reported by people who live in close proximity to a large animal farm. Ammonia can also have toxic and acidifying effects on ecosystems. Ammonia in high concentrations in the air can have a direct effect on plant growth by damaging leaf absorption capacities, but its indirect effect on soil chemistry is even more important—ammonia acidifies the soil, interfering with the absorption of other essential plant elements.

Issues also arise from the storage and use of livestock manure. There are two basic manure storage types in operation in Canada: liquid manure storage (concrete enclosures; steel tanks, either open or covered; earthen basins; lined/unlined lagoons) and solid manure storage (e.g., manure stored indoors with bedding or as a pack in the barn; manure stored outside as a pile on the ground). Problems arise when storage systems are inadequately built or when they are sited too close (less than 30 metres) to water supplies. Liquid manure stored in lagoons, for example, may overflow during periods of heavy rainfall, or the lagoons can fail to prevent the leaching of organic and inorganic materials into the surrounding environment. The health and environmental impacts associated with manure spills are similar to those arising from the improper use of manure as fertilizer.

Raw, well-rotted manure is often spread onto farm fields as fertilizer, a reasonable environmental practice as long as farmers have sufficient cropland to absorb the manure of their livestock. However, new large-scale farms produce vast quantities of manure and often do not have correspondingly large areas of farmland. In 2001, Canadian livestock produced an estimated 177.5 million tonnes of manure (Hofmann and Beaulieu, 2006). The hogs in Ontario currently produce as much raw sewage as the province's people.

If manure and commercial fertilizers are misused, spilled, or applied in excessive quantities, the result is contamination of soil and water by nitrogen, phosphorus, and bacteria. Although crops take up the bulk of added nutrients, a portion—the nutrient surplus—remains in the field. For all agricultural land in Canada, annual inputs of nitrogen and phosphorus from commercial fertilizers and livestock manure exceed annual outputs. There is a national surplus of approximately 0.3 million tonnes of nitrogen and 56,000 tonnes of phosphorus, or 8.4 kilograms per hectare of nitrogen and 1.6 kilograms per hectare of phosphorus. Fertilizers, whether organic or synthetic, are viewed as essential to maintain crop yield and soil health, but their application in excess of what crops can utilize can have significant implications for human and environmental health.

Nutrient loading can result in runoff to streams, rivers, lakes, and wetlands, spurring additional growth of algae and other aquatic plants. Accelerated eutrophication results in loss of habitat and changes in biodiversity (Chapter 4). For example, long-term exposure to elevated nitrate concentrations has contributed to the recent decline in frog and salamander populations in Canada. Concentrations of nitrate greater than 60 milligrams per litre in water kill the larvae of many amphibians. From 1988 to 1998, a total of 274 manure spills were reported in Ontario, of which 53 resulted in fish kills, primarily because of the ammonia in liquid manure. Reporting on fish kills from accidental spills/discharges of nutrient-related compounds is currently voluntary, so they are believed to be widely under-reported.

Factors influencing the effect of manure and commercial fertilizers on the environment include soil type, climate, precipitation, topography, and the quantities of manure produced. Most of these factors are beyond the farmer's control. However, manure management practices also influence the magnitude and extent of ecological impacts. Unfortunately, the environmental laws governing manure management were created when small operations were the norm, and therefore they fail to address the environmental and health risks that come with more intensive livestock operations. For example, in Ontario there are no legally binding standards for constructing manure storage facilities or for the application of manure. Nor are there any monitoring mechanisms to ensure that farmers use best practices for managing manure.

Sustainable Food Production Systems

While Canadians continue to enjoy a wide variety of high-quality foodstuffs year-round, concern over the vulnerability of the productive capacity of our agro-ecosystems mounts. The future capacity to deliver agricultural outputs depends on the continuing ecological viability of agro-ecosystems, yet significant stresses are imposed on them by intensification. The challenge is to foster agro-ecosystem management practices that will meet growing food, feed, and fibre needs while providing more environmental protection.

Improving agro-ecosystem management so that all levels of agricultural production can be associated with better environmental performance requires new knowledge and better skills, which can be achieved by improvements in technology, natural resource management systems, and landscape planning, as well as by policies and institutional arrangements that help to integrate environmental values into agricultural investment and management decisions. Examples include integrated pest management, integrated plant nutrient systems, no-till and conservation agriculture, and permaculture. These approaches seek to meet the dual goals of increased productivity and reduced environmental impact.

Integrated Pest Management

Integrated pest management (IPM) seeks to avoid or reduce yield losses caused by diseases, weeds, insects, mites, nematodes, and other pests while minimizing the negative impacts of pest control (resistance, non-selection, mobility, persistence, bioconcentration, etc.). Originally used to reduce excessive use of pesticides while achieving zero pest incidence, the concept has broadened over time. The presence and density of pests and their predators and the degree of pest damage are monitored, and no action is taken as long as the level of pest population is expected to remain within specified limits.

IPM considers the crop and pest as part of a wider agro-ecosystem, promoting biological, cultural, and physical pest management techniques over chemical solutions to pest control. Combinations of approaches are used, including:

- bacteria, viruses, and fungi (pathogens);
- insects such as predators and parasites (biological management);
- disease and insect-resistant plant varieties;
- synthetic hormones that inhibit the normal growth process;
- behaviour-modifying chemicals and chemical ecology products (such as pheromones, kairomones, and allomones).

If pesticide use is deemed essential to pest control, only pesticides with the lowest toxicity to humans and non-target organisms are applied.

The adoption of IPM practices has economic and other benefits for farmers, but it requires more expertise than simply applying chemicals. For this reason, Ontario has established a formal system of IPM for 21 of the province's 95 agricultural commodities. Producers can obtain expert and current advice on these products, their pests, and optimal courses of action by phone. Ontario has also introduced a formal accreditation program for companies and facilities wishing to be IPM-certified. Ontario has cut the use of pesticides by 52 per cent since the program's introduction in 1987. Overall pesticide use on fruit and vegetable crops has decreased by 20 per cent over the past five years alone. Specifically, insecticide use in fruit-growing has declined by 57 per cent, while fungicide use for both fruit and vegetables has been reduced by 54 per cent since 1998. Increased adoption of IPM and alternative pest control strategies, such as border sprays for migratory pests, mating disruption, alternate row spraying, and pest monitoring, are major reasons for these large declines. Internationally, some countries have developed aggressive IPM programs; Indonesia, for example, has managed to cut pesticide use by as much as 90 per cent, while Sweden has adopted a similar aggressive approach, reducing pesticide use by 50 per cent.

Integrated Plant Nutrient Systems

Agricultural production removes plant nutrients from the soil, reducing its organic and nutrient content. Imbalances in nutrient availability can lead to excessive depletion of nutrients that are in short supply, with corresponding reductions in crop yield. The goal of **integrated plant nutrient systems** (IPNSs) is to maximize nutrient use efficiency by recycling all plant nutrient sources within the farm and by using nitrogen fixation by legumes (Chapter 4) to the extent possible. Soil productivity is enhanced through a balanced use of local and external nutrient sources, including manufactured fertilizers. Fertilizers supplied in excess can pollute soils and waters, as discussed in Chapter 4, so IPNSs also seek to minimize the loss of nutrients through the judicious use of external fertilizers. IPNSs aim to optimize the productivity of the flows of nutrients passing through the farming system during a **crop rotation** (Scialabba, 2003). The quantities of nutrients applied are based on estimates of crop nutrient requirements—i.e., knowledge of the quantities of nutrients removed by crops at the desired yield level.

No-Till/Conservation Agriculture

To destroy weeds and loosen topsoil to facilitate water infiltration and crop establishment, agricultural land is ploughed, harrowed, or hoed before every planting. Topsoil disturbance of this magnitude and frequency destabilizes the soil structure, leading to soil erosion and soil compaction, negatively affecting productivity and sustainability. The economic and ecological costs associated with conventional tillage systems are becoming more apparent, leading farmers to search for alternative land preparation techniques, such as **no-till/conservation agriculture** (NT/CA). NT/CA (or zero, minimum, or low tillage) protects and stimulates the biological functioning of the soil while maintaining and improving crop yields.

Essential features of NT/CA include: minimal soil disturbance restricted to planting and drilling (farmers use special equipment to drill seeds directly into the soil instead of ploughing); direct sowing; maintenance of a permanent cover of live or dead plant material on the soil surface; and crop rotation, combining different plant families (e.g., cereals and legumes). Crops are seeded or planted through soil cover with special equipment or in narrow cleared strips. Soil cover inhibits the germination of many weed seeds, minimizing weed competition and reducing reliance on herbicides. Soil cover also reduces soil mineralization, erosion, and water loss, builds up organic matter, and protects soil micro-organisms. Crop sequences are planned over several seasons to minimize the buildup of pests or diseases and to optimize plant nutrient use by synergy among different crop types, and these sequences involve alternating shallow-rooting crops with deep-rooting ones to utilize nutrients throughout various layers of the soil. Other advantages associated with NT/CA include: increased yields in the order of 20 to 50 per cent higher than with conventional tillage practices; reduction in the variability of yields from season to season; significant reductions in labour costs; and lower input costs, particularly for machinery (e.g., smaller tractors can be used, reducing fuel costs).

Despite these advantages, conventional tillage-based agricultural systems continue to dominate worldwide. There are several reasons for this—a reluctance to change (conventional approaches have been working for decades), a lack of knowledge of the damage to soil systems associated with plough-based agriculture, and the complex management skills required for a successful transition. The short-term economic cost associated with the transition from conventional tillage to NT/CA is also a deterrent. During the transition years, there are extra costs for tools and equipment, higher weed incidence may increase herbicide costs initially, and yields will improve only gradually.

In Canada, farmers appear to recognize the advantages of NT/CA, with more than 63 per cent employing some or all of the elements of NT/CA. No-till increased to 46.4 per cent of the area tilled in 2006 from 29.7 per cent in 2001. The area worked with conventional tillage, which had historically been the most popular tillage method, dropped to 28 per cent in 2006 from 40.5 per cent in 2001. Conservation tillage—the midpoint between conventional and no-till, dropped from 29.8 per cent of tilled area to 25.6 per cent over the same five-year period.

Figure 10.12 shows the percentage of farms in Canada using various soil conservation methods. Crop rotation is the most common and is an important means of recharging soil nitrogen through use of legumes such as alfalfa and clover. Grassed waterways are used to control overland flow of runoff, thereby controlling the formation of gullies on exposed soil surfaces. **Contour cultivation** involves cultivating the soil parallel to the contour of the slope, which serves to reduce the speed of runoff by catching soil particles in the plough furrows. More than 16 per cent of PEI and Saskatchewan cropland is protected in this way. **Strip cropping** is a similar technique in which different crops may be planted in strips parallel to the slope. While one crop may be harvested, leaving bare soil, the other crop serves to provide some protection. This technique is commonly used against wind erosion and is most prevalent in western Canada. The soil surface can also be protected in winter through growth of a winter cover crop. This is effective not only for wind erosion in winter but also to protect the soil from intense rainfall in the spring.

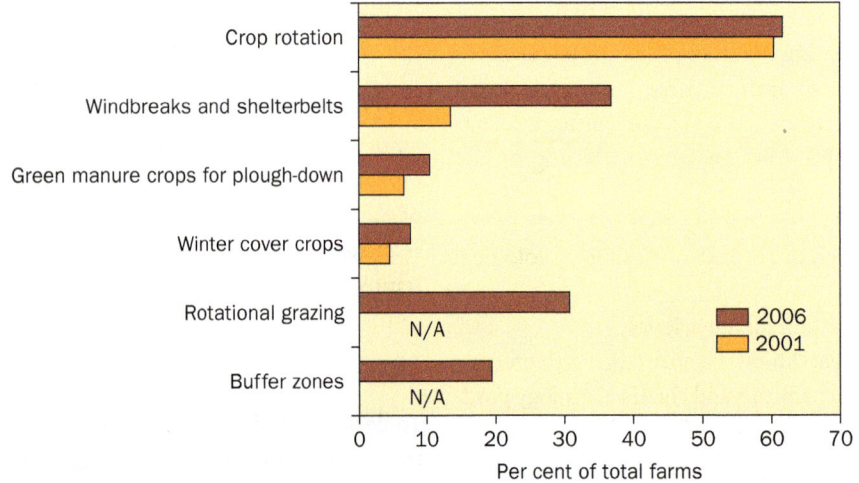

Figure 10.12 | Soil conservation practices, 2001 and 2006. Source: Statistics Canada (2001, 2007a).

Strip farming in Saskatchewan.

Organic Farming

Approaches to sustainable agriculture such as IPM, IPNS, and conservation tillage consider only one aspect of the farming system components—pest ecology, plant ecology, and soil ecology, respectively. Organic agriculture, however, combines these and other management strategies into a single approach, focusing on food web relations and element cycling to maximize agro-ecosystem stability. Organic agriculture is a production management system that aims to promote and enhance ecosystem health. It is based on minimizing the use of external inputs and represents a deliberate attempt to make the best use of local natural resources while minimizing air, soil, and water pollution. Synthetic pesticides, mineral fertilizers, synthetic preservatives, pharmaceuticals, genetically modified organisms, sewage sludge, and irradiation are prohibited in all organic standards (FAO, 2001; Canadian General Standards Board, 2003).

Organic agriculture encompasses a range of land, crop, and animal management procedures designed to:

- enhance biological diversity within the whole system;
- increase soil biological activity;
- maintain long-term soil fertility;
- recycle wastes of plant and animal origin in order to return nutrients to the land, thus minimizing the use of non-renewable resources;
- rely on renewable resources in locally organized agricultural systems;
- promote healthy use of soil, water, and air as well as minimize all forms of pollution that may result from agricultural practices;
- handle agricultural products with emphasis on careful processing methods in order to maintain the organic integrity and vital qualities of the product at all stages;
- become established on any existing farm through a period of conversion, the appropriate length of which is determined by site-specific factors such as the history of the land and the type of crops and livestock to be produced.

For example, organic practices that encourage soil biological activity and nutrient cycling include manipulation of crop rotations and strip cropping, the use of **green manure** and organic fertilizer (animal manure, compost, crop residues), minimum tillage or zero tillage, and avoidance of pesticide and herbicide use. Organic agriculture significantly increases the density of beneficial invertebrates, earthworms, root symbionts, and other micro-organisms essential to maintaining soil health. For example, the biomass of earthworms in organic systems is 30 to 40 per cent higher than in conventional systems.

Enhanced soil fertility is the cornerstone of organic agriculture, but there are other benefits. For example, organic fields in excess of 15 hectares contain a greater diversity and abundance of flora and fauna than conventional fields, including endangered varieties (e.g., organic grasslands contain 25 per cent more herb species than conventional grasslands). Diverse crops have been shown to be less vulnerable to diseases and droughts, increasing crop yields and decreasing the amount of biocides necessary to meet production targets.

Organic farming systems are also more energy-efficient per unit crop than conventional farming techniques, since organic systems resemble closed or semi-closed nutrient cycles. Organic land management permits the development of a rich weed flora, and a versatile flora attracts more kinds of beneficial insects. Organic farming systems are also better at controlling erosion, since organic soil management techniques improve soil structure. Organically grown foods also benefit human health, since they contain fewer pesticide residues than foodstuffs grown under intensive farming methods.

Like most industrial countries, Canada has national organic standards, regulations, and inspection and certification systems that govern the production and sale of foods labelled as 'organic'. In 2007, federal regulations were introduced for organic certification, and were revised in 2009 to ease the way for organic trade of produce between provinces and primarily, with other countries. The global organic products industry was worth over US$50 billion in 2008 and included over 35 million hectares of certified organic land, with Australia, Argentina, Italy, Canada, and the United States having the largest areas of organic farmland (Agriculture Agri-Food Canada, 2009). In 2000, agricultural land under certified organic management averaged 0.25 per cent of total agricultural land in Canada, compared to 2.4 per cent of total agricultural land in Western Europe, 1.7 per cent in Australia, and 0.22 per cent in the United States (Scialabba, 2003). In most developing countries, agricultural land reported under certified organic production is less than 0.5 per cent, although the extent of non-market, non-certified organic agriculture may be considerable. For example, an estimated one-third of West African agricultural produce is produced organically (ibid.).

More than 15,000 farm operations in Canada (6.8 per cent) reported at least one type of organic product in 2006. Non-certified organic production is most common, with the most likely application being animal products. Field crops such as buckwheat, rye, and caraway dominate certified Canadian organic production. Canada is among the five top world producers of organic grains and oilseeds. By 2010, organic food sales contributed $2 billion a year to the Canadian economy. The number of certified producers increased by 5.4 per cent between 2008 and 2009; however, the certified growers in Alberta increased by 23 per cent during the same time period. The domestic organic market is strongest in British Columbia, Alberta, Quebec, and Ontario. Nevertheless, Saskatchewan has the highest proportion of certified organic producers, with 38 per cent of the national total.

Consumer health and food quality concerns (e.g., concerns about growth-stimulating substances, genetically modified [GM] food, dioxin-contaminated food, and livestock epidemics such as bovine spongiform encephalopathy and foot-and-mouth disease) continue to drive demand for organic products in Canada and around the globe. Organic food production systems are considerably more respectful of the environment than chemically intensive farming practices, but chemical-intensive farming systems still dominate.

In the absence of governmental support for the expansion of organic production, farmers may be reluctant to convert to organic farming for several reasons. Conversion from conventional, intensive systems to organic production causes a loss in yields, the extent of which varies depending on the biological attributes of the farm, farmer expertise, the extent to which synthetic inputs were used under previous management, and the state of natural resources. Yields can be 10 to 30 per cent lower in organic systems, and it may take several years (e.g., three to five) to restore the ecosystem to the point where organic production becomes economically viable. In addition, production costs per unit of production (e.g., labour, certification and inspection fees) and marketing expenses can be higher with organic produce, but once produce qualifies as *certified* organic, some costs can be offset by price premiums. In developed countries, retail organic products can command 10 to 50 per cent more than conventional prices for the same commodity. In many places, rising fuel costs and depleting supplies have increased the costs of agricultural inputs, making organic agriculture a more economically viable alternative. In Canada, demand for most organic products continues to exceed supply despite the higher prices charged for certified organic produce. Consumers in industrialized countries are willing to pay a premium for organic food because they perceive environmental, health, or other benefits.

In Canada and elsewhere, cases are emerging in which organic production is constrained or no longer feasible because of the advent of GM crops. Organic farmers in Canada can no longer grow organic canola (i.e., oilseed rape) because of GM canola contamination in Saskatchewan.

While organic agriculture has become prominent, governmental support does not reflect the growing demand. In 2010, the Canadian government invested $170,000 in the organic sector, which pales in comparison to the $10.1 billion provided in 2009 to non-organic farmers in agricultural subsidies (UN Statistics Division, 2011).

In many places, demand for organic foods cannot be met with the current supply. As a result, supermarkets such as Wal-Mart are beginning to offer organic produce, increasing the availability of these foods for the general population. In fact by 2010, 54 per cent of organic food products were sold by mass-market retailers. Many mass-market retailers rely on large quantities of food products from a high production source—increasing the incentive for intensified organic agriculture. With higher premiums for organics, the only way to maintain low prices is through subsidies. By 2009 the US subsidized the agricultural sector with more than $124.5 billion (ibid.). Most small farmers cannot compete with the impossibly low prices offered by intensified organic agriculture, and are threatened by these developments. If you walk through the organic section in a supermarket, how many products are available from your local area? From Canada? Should we have more local products available?

Local Agriculture

Canada currently imports over 70 per cent of its food. The majority of domestic products that feature a 'Made in Canada' label contain imported ingredients (CFIA, 2010). The global food system allows us to eat bananas while there is snow on the ground, but this also dramatically increases our food miles. Food miles measure the distance your food must travel from where it was produced to reach your plate. As food miles increase, greenhouse gas emissions, rural unemployment, and local food insecurity grow correspondingly. For instance, the average food item in Toronto travels close to 4,500 kilometres before it is consumed (Toronto Public Health, 2007). The development of local food systems can contribute to reducing the impact of agriculture on the environment and local agricultural economies.

Local agriculture often consists primarily of small farms, which are characterized by diverse crops, low capital, and, consequently, low energy inputs and high levels of human labour. This is in stark contrast with the high energy inputs common to producing a single crop in most industrial agricultural practices. Many small farms have a greater yield per hectare than large farms, in part due to the human labour involved (Cornia, 1985). The amount of food produced and consumed in a region can influence the food security of the area, which is highly variable both spatially and temporally (Morrison et al., 2011).

Various countries collaborated at the World Food Summit in 1996 and stated that 'Food security exists when all people, at all times, have physical and economic access to sufficient, safe, and nutritious food to meet their dietary needs and food preferences for an active and healthy life' (www.who.int/trade/glossary/story028/en/). As can be imagined, various constraints limit the ability of Canadian communities to be food secure, including a lack of arable land, water shortages, unsuitable climate conditions, and land-use change. Toronto is surrounded by arable land, yet in 2005, Ontario imported $4 billion more in food than it exported, resulting in less than three days of fresh food in the city at any one time. And, a study conducted in 2011 found that currently in British Columbia, local food production is not sufficient to meet local nutritional needs (Morrison et al., 2011).

While there are challenges associated with local food consumption, various solutions have emerged as people aim to decrease their impact on the environment and increase their food security. For example, urban agriculture is being practised throughout Canada in the form of community gardens, rooftop gardens, or growing peri-urban crops. In Montreal, a group of eager gardeners has collaborated to offer city-wide tours of urban agriculture ventures featuring the unique practices employed by each. One practice used in many urban agriculture centres is **permaculture**. Permaculture agricultural designs are based on ecological relationships with the fundamental principle of minimizing wasted energy. In these systems, the wastes of one component become the inputs for another. Some rooftop permaculture catches rainwater to feed plants that are strategically positioned to ward off pests, capture nutrients, and provide shelter.

Implications

Agricultural modification is arguably the main impact that humans have had on natural ecosystems. It is, however, also one of the oldest and one that is basically a modification of ecological systems to benefit humans. Over centuries, natural and human-modified agricultural landscapes have existed and transformed from one state to the other with little lasting damage to planetary life-support systems. However, as additional auxiliary energy flows were applied to boost the productivity of agriculture, the differences between these two ecological systems became more distinct, and the impacts of agriculture on natural ecosystems increased. Agricultural production (certainly in Canada's commercial agricultural sector) is now more similar to industrial production than to the natural ecosystems from which agriculture was derived.

This industrialization has led to many environmental challenges for agriculture. Yields are declining in some areas as crops become less responsive to fertilizer input, biocides continue to eliminate many natural enemies of pests, and soils are eroded, salinized, and compacted. In response, researchers are suggesting that a fundamental restructuring is required in how agriculture is undertaken, with the emphasis changing from maximizing productivity to ensuring sustainability.

Achieving sustainability and resilience will require greater attention to the agro-ecosystem and particularly to the soil base that sustains agriculture. In addition, the socio-economic and regulatory dimensions will need to be integrated into systemic change. Organic farming will not be successful, however, unless customers are willing to pay for the produce, not only as a benefit to themselves but also to sustain and nourish healthier ecosystems overall. In addition to buying organic produce, we can do a number of things as individuals to ensure that we at least do not exacerbate the challenges now facing agricultural systems (Box 10.13).

Finally, it is important to remember that millions of people still experience hunger on a daily basis, as highlighted by the situation in Somalia during 2011. Programs designed to address these problems have always failed to meet their goals. Food prices soared rapidly in 2008, and in countries where more than half of family income (as opposed to less than 10 per cent in Canada) goes to feeding families, significant civil unrest took place. Hungry families are angry families,

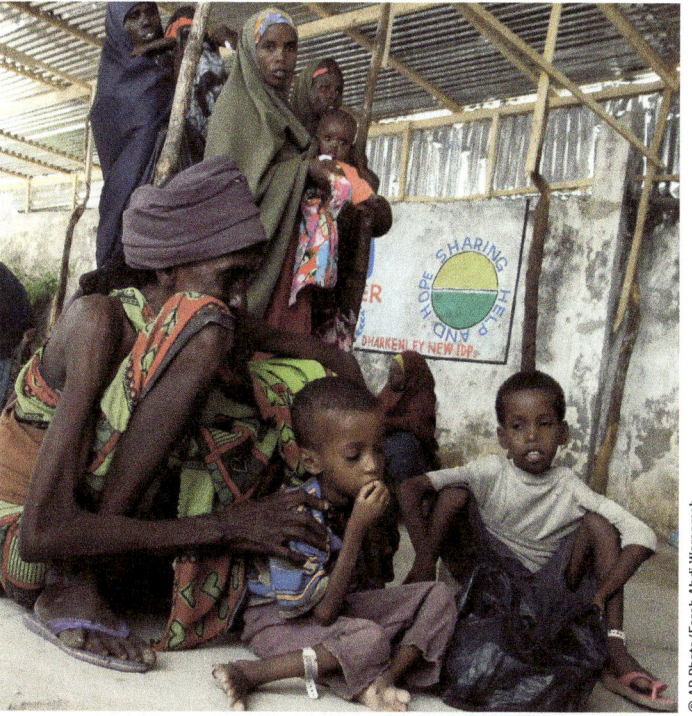

A Somali woman waits with her malnourished children at a medical centre in a refugee camp in Mogadishu, Somalia. The World Food Program estimated that only 20 per cent of the 2011 famine victims were receiving aid owing to the rebel groups controlling large portions of the country.

Environment in Focus
Box 10.13 What You Can Do

Although the challenges facing agriculture at the global and national levels are immense, there are still some ways in which individuals can help.

1. Eat less. This entails finding out about good nutritional habits so that we consume only the food that we really need.
2. Eat lower on the food chain. Most North Americans eat far too much meat. Eating more vegetables will benefit not only the global food situation but your own health.
3. Feed your pet lower on the food chain. Dogs and cats will also be healthier if fed on balanced grain pet foods rather than meat.
4. Waste less food. Studies indicate that as much as 25 per cent of food produced in North America is wasted.
5. Grow at least some of your own food. If Canadians were to devote a fraction of the time and resources on growing food that they spend on their lawns, it would allow more food for others elsewhere.
6. Support local food growers and food co-ops. This helps to protect agricultural land in Canada from being transformed to other uses.
7. Learn what foods are in season in your area, and try to build your diet around them.
8. Ask the manager or chef of your favourite restaurant how much of the food on the menu is locally grown, and then encourage him or her to source food locally. Urge that the share be increased. People can do the same at their local supermarket or school cafeteria.
9. Take a trip to a local farm to learn what it produces.
10. Host a harvest party at your home or in your community that features locally available and in-season foods.
11. Produce a local food directory that lists all the local food sources in your area, including farmers' markets, food co-ops, restaurants emphasizing seasonal cuisine and local produce, and farmers willing to sell directly to consumers year-round.
12. Buy extra quantities of your favourite fruit or vegetable when it is in season and experiment with drying, canning, jamming, or otherwise preserving it for a later date.
13. Join one of the NGO groups that specializes in rural development in less developed countries.

---

and ways must be found to ensure greater food security for the world's poor if this situation is not to deteriorate into widespread violence.

In 2011, food prices reached record highs, jumping 36 per cent between April 2010 and April 2011, with millions more people being pushed below the poverty line. In response, the World Bank called for a targeting of social assistance and famine relief at the poorest elements of society, removing grain export restrictions, and relaxing government requirements to achieve biofuel production mandates when certain threshold food prices are reached, as well as more restrictions on commodity speculators—speculation in commodity markets has increased the volatility of the food market. The interlocking of so many environmental, political, social, and financial factors, compounded by the increasing challenges presented by global climate change, suggest that the production and distribution of adequate food supplies is going to be a major issue in the years to come.

Summary

1. Agriculture originated at least 9,000 years ago, when societies domesticated both plant and animal species. Societies around the globe flourished by improving their capacity to expand agricultural production.
2. Over much of history, agricultural output has been increased by bringing more land into production. Today, intensification of production—obtaining more output from a given area of agricultural land—has become a key development strategy in most parts of the world to meet the increased demand for foodstuffs. The future capacity to deliver agricultural outputs depends on the continuing ecological viability of agro-ecosystems.
3. Agriculture is a food chain, with humans as the ultimate consumers. The second law of thermodynamics dictates that the shorter the food chain, the more efficient it will be.

4. The Green Revolution and the Livestock Revolution have led to profound changes to the global land base. Complex natural systems have been replaced by relatively simple control systems in which humans are in command of the species and numbers that exist in a given area.

5. The Green Revolution relies on auxiliary energy flows, such as fertilizers, biocides, fossil fuels, and irrigation systems, to increase yields. Yields are falling in many of the poorest parts of the world because of soil exhaustion and the rising price of fertilizers. Industrial systems of livestock production also depend on outside supplies of feed, energy, and other inputs to satisfy growing worldwide demand for meat.

6. Biofuels have great potential to be part of adaptation to reducing GHGs and creating higher values for agricultural products in many underdeveloped countries, but they also create challenges in terms of competition with food supply and encourage greater transformation of biodiversity-rich forests into agricultural land. Cellulosic production seems to have the greatest potential to address these problems in the future.

7. Agriculture is an important industry in Canada, accounting for 8.5 per cent of gross domestic production, 10 per cent of employment, and about 6 per cent of total merchandise export earnings. Many of the innovations and technologies employed by Canadian farmers to produce agricultural products have extraordinary implications for ecosystems.

8. Large increases in fertilizer inputs have occurred in Canada over the past three decades.

9. Between 1970 and 2000, the value of biocides applied to agricultural crops in Canada increased threefold. More than 30 million hectares were treated with at least one biocide in 2000, compared to fewer than 10 million hectares in 1970.

10. Land degradation includes a number of processes that reduce the capability of agricultural lands to produce food. One study suggests that such processes cost Canadian farmers more than $1 billion per year.

11. Soil erosion is estimated to cause up to $707 million in damage per year in terms of reduced yields and higher costs. Soil formation in Canada is slow. A rate of 0.5 to 1.0 tonnes per hectare (t/ha) per year may be considered average. Losses of 5 to 10 t/ha are common in Canada, and figures of 30 t/ha have been recorded in the Fraser River Valley.

12. Increasing acidity as a result of the application of nitrogen fertilizers and acid deposition is also a problem that reduces crop yields. Salinization occurs where there are high sodium levels in the soils and shallow water tables, such as in the Prairie provinces. One estimate suggests that salinization causes economic losses four to five times as great as losses due to erosion, acidification, and loss of nitrogen.

13. Cultivation involves a continual process of removing plant matter from a field. In the process, both the organic content and nutrient content of the soil are reduced. On the prairies, current organic matter levels are estimated to be 50 to 60 per cent of the original levels.

14. Biocides are applied to crops to kill unwanted plants and insects that may hinder the growth of the crop. They have boosted yields throughout the world and helped feed many hungry mouths. There is also clear scientific evidence that they have serious negative impacts on ecosystem health.

15. Biocides promote the development of resistance among target organisms. Over the past 40 years, more than 1,000 insects have developed such resistant populations. They are non-selective and tend to kill non-target as well as target organisms. They may also be highly mobile and move great distances from their place of application. In addition, they may persist for a long time in the environment and accumulate along food chains. Such biomagnification has resulted in drastic reductions in the populations of some species at higher trophic levels, such as ospreys and bald eagles.

16. Chemicals, and their constituents as they break down, may interact synergistically.

17. The government must register chemicals before they can be used. Registration involves the chemical company providing evidence that the chemical is not carcinogenic or oncogenic. Many chemicals have been registered but subsequently fail these tests.

18. Attention is now being devoted to sustainable food production systems that maintain or enhance environmental quality, generate adequate economic and social returns to all individuals/firms in the production system, and produce a sufficient and accessible food supply. Integrated pest management is now becoming more popular, and several provinces have such programs.

19. Organic farming is growing rapidly but is still a relatively small part of overall agricultural production.

Key Terms

- bioaccumulation
- bioconcentration
- biofuels
- biomagnification
- contour cultivation
- crop rotation
- genetically modified organisms (GMOs)
- grasshopper effect
- green manure
- Green Revolution
- hybridization
- integrated pest management (IPM)
- integrated plant nutrient systems (IPNSs)
- intensive livestock operations (ILOs)
- Livestock Revolution
- monoculture cropping
- no-till/conservation agriculture (NT/CA)
- organic farming
- pelagic
- permaculture
- salinization
- *Silent Spring*
- soil compaction
- soil erosion
- strip cropping
- subsistence farming
- summer fallow

Questions for Review and Critical Thinking

1. How do the laws of thermodynamics apply to agriculture?

2. Indicate how the concepts of 'limiting factors' and 'range of tolerance', discussed in Chapter 4, can be applied to agriculture.

3. Considerable interest is being directed towards ecosystem management. What is the relevance of this concept, if any, for agriculture?

4. Do you think that biofuels have an important role to play in agricultural systems in the future? How would you try to mitigate some of the challenges presented by biofuels?

5. What do you think might be the main challenges facing the global agricultural supply in five years?

6. In this chapter, mention was made of the resilience of farm systems. In Chapter 6, in relation to adaptive management, attention was also given to resilience. What value do the concepts of resilience and adaptive management have regarding agriculture?

7. How would you go about identifying and assessing the impacts (environmental and social) of agricultural policies and practices in Canada? What ideas from Chapter 6 might be helpful in this exercise?

8. In Chapter 6, attention focused on the ideas of partnerships and stakeholders. What relevance do such concepts have regarding agriculture if partnerships are to be formed among the federal and provincial governments, agribusiness, and family farms?

9. It has been suggested here that conflict or tension can exist between the needs of urban areas and adjacent rural areas. To what extent do you think alternative dispute resolution ideas might be applied to deal with such tensions?

10. If you were a commercial farmer in Canada, to what extent would it be important for you to consider the implications of climate change for your farming operations?

Related Websites

Agriculture and Agri-Food Canada
www.agr.gc.ca/site_e.phtml#env

Canadians against Pesticides
www.caps.20m.com

Consultative Group on International Agricultural Research (CGIAR)
www.cgiar.org

Environment Canada
www.ec.gc.ca

Food and Agriculture Organization of the United Nations (FAO)
www.fao.org

Government of Canada, agriculture and the environment
www.agr.gc.ca/policy/environment/pubs_sds_e.phtml

Pesticides and wild birds
www.hww.ca/en/issues-and-topics/pesticides-and-wild-birds.html

World Hunger: We Feed People
sustag.wri.org

World Resources Institute, Earth Trends
earthtrends.wri.org

Worldwide Fund for Nature, chemicals and wildlife
panda.org/downloads/toxics/causesforconcern.pdf

Further Readings

Note: This list comprises works relevant to the subject of the chapter but not cited in the text. All cited works are listed in the Bibliography at the end of the book.

Burke, M., D. Lobell, and L. Guarino. 2009. 'Shifts in African crop climates by 2050, and the implications for crop improvement and genetic resources conservation', *Global Environmental Change* 19, 3: 317–25.

Ericksen, P.J. 2008. 'What is the vulnerability of a food system to global environmental change?', *Ecology and Society* 13, 2: 14.

Food and Agriculture Organization (FAO). 2003. *World Agriculture: Towards 2015/2030. An FAO Perspective*. Rome: FAO.

Koning, N., and M.K. Van Ittersum. 2009. 'Will the world have enough to eat?', *Current Opinion in Environmental Sustainability* 1: 77–82.

Nass, L.L., P.A.A. Pereira, and D. Ellis. 2007. 'Biofuels in Brazil: An overview', *Crop Science* 47: 2228–37.

Chapter 11
Water

Learning Objectives

- To appreciate the need for credible science to inform decision-making about water.
- To recognize the water endowment in Canada.
- To understand the hydrological cycle.
- To know the environmental and social impacts associated with water diversions.
- To appreciate various perspectives related to water export from Canada.
- To understand the significance of point and non-point sources of pollution.
- To learn about the concept of 'water security'.
- To appreciate the growing importance of 'water terrorism'.
- To gain an understanding of the concept of a 'multi-barrier approach' to drinking water protection.
- To understand the challenges and opportunities regarding water security on Aboriginal reserves.
- To appreciate the distinctions among supply management, demand management, and the soft path approach.
- To understand the concepts of 'virtual water' and 'water footprint'.
- To realize that water is both resource and hazard.
- To appreciate the difference between structural and non-structural approaches to flood damage reduction.
- To understand the significance of droughts.
- To appreciate the significance of heritage related to protection of aquatic systems.
- To understand the importance of hydro-solidarity and integrated water resource management (IWRM).
- To appreciate the evolving ideas related to 'water ethics'.
- To understand perspectives related to the issue of 'water as a human right'.

Introduction

Statistics Canada (2003b) reports that in terms of water, although Canada has just 0.5 per cent of the world's population, Canadians have access to almost 20 per cent of the global stock of fresh water, and Canada has 7 per cent of the total flow of renewable water. This apparent natural bounty is often taken for granted, it seems, since Canadians are among the highest consumers of water in terms of per capita water use, second only to citizens of the United States. This high use was characterized some time ago by Foster and Sewell (1981: 7) as due to a 'myth of superabundance'.

With or without a myth of superabundance, Canada has a relatively and absolutely generous endowment of water. Indeed, as O'Neill (2004: xi) has observed, 'There can be no question that Canada's freshwater supply is an immensely valuable national resource. Estimates of water's measurable contribution to the Canadian economy range from $7.5 to $23 billion annually, values comparable to the gross figures for agricultural production and other major economic components.' In contrast, countries in the Middle East and Sahelian Africa usually experience significant water deficits that are a major impediment to overcoming poverty and facilitating development. Furthermore, at a forum at the Munk School of Global Affairs at the University of Toronto to mark World Water Day on 22 March 2011, experts cautioned that the likelihood is high for violent conflict between nations due to shortages of fresh water in the near future. Rapid population growth, climate change, and food shortages could combine to exacerbate tensions related to scarcity of fresh water.

The significance of water is recognized by many Canadians. The fourth annual Canadian Water Attitudes Study, commissioned by RBC and Unilever in 2011, revealed that the majority (55 per cent) of Canadians identified fresh water to be the most important natural resource for Canada (RBC, 2011). Despite such awareness, the 2011 survey also revealed that 61 per cent of households reported they had no idea how much their household pays for water use, but still concluded the unknown price is high enough to ensure water is treated as a valuable resource. In addition, 72 per cent indicated they flushed items down their toilets that could be disposed in other ways.

Hydrological Cycle

Water or aquatic resources are one component of a system that includes the atmosphere, cryosphere, biosphere, and terrestrial components. Evaporation from surface water (rivers, lakes, wetlands) and transpiration from plants release water vapour into the atmosphere that condenses and forms clouds while moving upward. The tiny droplets of water in clouds eventually fall to the Earth as rain, fog, hail, or snow. After reaching the surface, the water evaporates back into the atmosphere, moves into rivers, lakes, or oceans, or percolates into the soil to become groundwater. Chapter 4 provides a more detailed discussion of the hydrological cycle.

About 12 per cent of Canada (1.2 million km^2) is covered by lakes and rivers, with only 3 per cent of that area located in inhabited regions. There are more than 2 million lakes, with the largest being the Great Lakes shared between Canada and the United States. Other large lakes are Great Bear Lake and Great Slave Lake in the Northwest Territories and Lake Winnipeg in Manitoba. Lake water represents about 98 per cent of the surface water available for human use. Canada has

Perspectives on the Environment
An Abundance of Water?

The perception that Canada is blessed with an abundance of fresh water has led to misuse and abuse of the resource; from household toilets that use 18 litres per flush where 6 litres would do, to industrial plants—and some municipalities—that use water bodies as convenient sewers.

—Environment Canada (2009: 1)

Perspectives on the Environment
Future Conflict over Freshwater Shortages?

Water resources in themselves have rarely been the sole source of conflict or war. Unfortunately, our global water situation is changing rapidly and may soon no longer resemble anything that has existed on Earth before. The tensions and conflicts over water of the kind that have typically occurred in the past will soon represent only one of many emerging explosive hydro-climatic issues that are likely to bring sovereign nations into internal and external discord that could erupt in violence. Humanity's numbers appear to be the greatest threat to water security globally. We have created a hydro-climatic bomb, and that bomb has started to tick. But we don't know how big the bomb is or where and when it will go off.

—Bob Sandford, water policy expert (quoted in Perkel, 2011).

Perspectives on the Environment
Water Availability in 2025

By 2025, the greater part of the Earth's population will likely live under conditions of low and catastrophically low water supply. Approximately 30–35 per cent of the world population will have catastrophically low freshwater supply (less than 1,000 m^3 per year per capita). At the same time . . . high water availability can be found in northern Europe, Canada, and Alaska, almost all of South America, Central Africa, Siberia, the Far East, and Oceania.

—Shiklomanov (2000: 28)

Perspectives on the Environment
The Need to Manage Water in Canada

Water, by its very nature, presents managers with three issues that are difficult to resolve: competition between users of water resources; vertical coordination between the multiple scales at which water is used and managed; and mismatch between geopolitical and administrative boundaries. . . .

—Bakker (2009: 18)

Wetlands of the Hudson Bay Lowland—Mansel Island.

more than 8,500 named rivers, and the Mackenzie River, with an average surface flow of 8,968 m³/second, has the highest volume. There are more than 1,000 named *glaciers*, and they are an important source of fresh water for rivers and lakes.

Various types of **wetlands** exist, all being hybrid aquatic and terrestrial systems. They are a key habitat for waterfowl and also store and gradually release water, thus serving as an important 'sponge' to aid in reducing flooding. Wetlands, found in the greatest number and extent in the Prairie provinces and in northern Ontario, cover about 14 per cent of the land area in Canada. Canada has about 25 per cent of the wetlands in the world, the largest amount of any country.

Groundwater is a key source of water for rivers and lakes and is created by surface water passing into the ground and becoming contained in sand and gravel, as well as in pores and cracks in bedrock. During dry periods, many rivers receive much of their water via base flow from groundwater aquifers.

Sprague (2007: 23–5) has observed that the belief of many Canadians about an abundance of water most likely stems from the apparent large volume of fresh water contained in lakes across the country, totalling about 20 per cent of the water in all the lakes of the world. He emphasizes, however, that water contained in lakes is not the same as what is considered a 'renewable' supply. The latter is based on precipitation that falls, then runs off into rivers, often being held in lakes before draining to the ocean or moving downward into aquifers. The flows associated with precipitation or snowmelt should be identified as the renewable supply. As Sprague (ibid., 24) states: 'To use a financial analogy, the water sitting in lakes and aquifers is comparable to a capital resource of money that can be spent only once. The rivers running out of the lakes would represent interest and dividends that could be used every year for an indefinite time.'

More specifically, Sprague highlights that the apparent bountiful abundance of water in Canada is due to several characteristics of the landscape and the hydrological cycle, including: (1) a few very large lakes and many shallow, small lakes; (2) a cool climate; and (3) low evaporation of water. Calculations suggest that Canada ranks between third and sixth in terms of **renewable water supply**. The leaders are Brazil with more than 12 per cent of the global renewable supply, followed by Russia with 10 per cent. After that, Canada (6.5 per cent) is in a virtual tie with Indonesia (6.5 per cent), the United States (6.4 per cent), and China (6.4 per cent). The next five on the list, ranging from 5 to 2 per cent of the global renewable water supply, are Colombia, Peru, India, and the Democratic Republic of Congo.

Some 60 per cent of Canada's water flows northward to Arctic and Subarctic areas in which few people live. Such northward-flowing water is generally not available to southern Canada, which is where most people live and work. Consequently, when considering the renewable water supply available to southern Canada, the proportion drops from 6.5 to 2.6 per cent of the global supply. Sprague argues that it is the 2.6 per cent figure that citizens, managers, and political leaders should keep in mind, and remarks that 'notably, the figure [of 2.6 per cent] is tenfold lower than the frequently used and mythical 'one-quarter of the world supply.'

Perspectives on the Environment

Water Diversions in Canada

Interbasin diversion projects are found in almost all provinces, and the total flow of water diverted currently between drainage basins is enormous—approximately 4,500 cubic metres per second. No other country diverts nearly as much water or concentrates so much flow for a single function—hydroelectric power generation.

—Quinn et al. (2004: 3)

Human Interventions in the Hydrological Cycle: Water Diversions

Given that water is often not in the right place at the right time, humans modify aquatic systems to store, divert, or modify flows. There are more than 900 large dams in Canada and about 60 large inter-basin diversions. Quebec has 333 large dams, followed by Ontario with 149 and British Columbia with 131 (Environment Canada, 2010c). **Diversions** are completed for one or more of the following reasons:

- To increase water supplies for a community or in a region, as illustrated by the St Mary Irrigation District in Alberta. While diversions for irrigation are important in the southern Prairies, this type of diversion is not as typical of the Canadian experience as it is for countries such as India and the United States.

- To deflect watercourses away from or around areas to be protected, such as the Portage Diversion in Manitoba. Here, the purpose is not to move water to a place of need but to protect a community from flood damage. Other reasons are to drain land to allow agricultural production or to drain a mine site.
- To enhance the capacity of a river so that it can be used to support activities such as floating logs or to allow passage of ships, disposal of wastes, or sustaining of fish. For example, dams on the Ottawa River were designed partly to facilitate the moving of logs downriver to sawmills.
- To combine or consolidate water flows from several sources into one channel or route in order to facilitate hydroelectric generation, such as the James Bay Project in northern Quebec. Canada is a global leader in water diversions for hydroelectricity generation, and diversions for hydropower purposes dominate overwhelmingly in both number and scale of diversions in Canada (Day and Quinn, 1992: 10–11).

While diversions can create positive capacity, they also can cause negative environmental impacts and impose costs on people or regions not benefiting directly from them. The James Bay Cree and their homeland in northern Quebec represent a case in point.

The James Bay Hydroelectric Project

Governments and private corporations have pursued many **megaprojects** in Canada to meet energy demands, and virtually every region in the country has experienced such megaprojects (Figure 11.1). Perhaps the most massive, and one that has garnered a great deal of national and international attention, has been the **James Bay Project** in Quebec. Other huge hydroelectric developments include Churchill Falls in Labrador, the Nelson–Churchill in Manitoba, and the Columbia and Nechako rivers in British Columbia. Nuclear power plants in Ontario, the development of oilfields off the coast of Newfoundland, the Sable Island natural gas exploration off Nova Scotia, and the exploitation of the oil sands in northern Alberta are among other major Canadian energy projects and are discussed in detail in Chapter 12.

Background

In 1971, Quebec Premier Robert Bourassa proposed hydroelectric development using the rivers on the eastern side of James Bay. The purpose was to satisfy future electricity needs in Quebec. The cost was estimated at $2 billion. The decision was to develop La Grande River basin to double the flow in that river by diverting water from adjacent catchments (Figure 11.2). Other river systems north and south of La Grande were to be developed in later phases.

Environment in Focus

Box 11.1 Hydroelectricity and Politics on the Lower Churchill River

In November 2010, Premier Danny Williams of Newfoundland and Labrador announced a $6.2 billion project to develop hydroelectricity on the Lower Churchill River, in partnership with Nova Scotia, and subject to ratification by the Labrador Innu. The planned project will include $2.9 billion to build a power generating facility at Muskrat Falls with capacity to produce 824 megawatts of electricity, $2.1 billion for a subsea transmission link from Labrador to Newfoundland, and $1.2 billion for a 180-km subsea link from Newfoundland to Nova Scotia. The decision to plan for the costly underwater transmission route was driven by the ongoing conflict between Newfoundland and Labrador and the Quebec government over the transmission of power to southern markets from the Churchill Falls hydroelectric project (built 1967–71) via Hydro-Québec lines. The 1969 agreement favoured Quebec, and when Hydro-Québec balked at offering a significantly better deal for transmitting Muskrat Falls power, the Newfoundland and Labrador government opted to move in a different direction. Nova Scotia will receive 170 megawatts of electricity annually for 35 years, which is 8 to 10 per cent of its total power needs, with additional capacity transmitted to New England markets.

At the end of June 2011, the Innu people of Labrador ratified the New Dawn Agreement, which will provide them with benefits and compensation, both from the new project and from the Churchill Falls project of more than 40 years ago, and opens the way for the Muskrat Falls development. Political—and intergenerational—conflict also was reflected in the New Dawn ratification. Younger people, a generation removed from Innu traditional lands and lifestyle, voted overwhelmingly for the agreement, and for the money it would provide. Community elders, however, having seen what was lost as a result of the Churchill Falls project, did not want to see more of their traditional lands inundated and altered by the new project (CBC News, 2011). This was epitomized in the opposing views of federal Conservative cabinet minister Peter Penashue and his mother, Elizabeth Penashue, a respected elder, speaker, and widely known environmental activist.

378 Part D | Resource and Environmental Management in Canada

Figure 11.1 | Hydroelectric megaprojects in Canada. Source: Adapted from Day and Quinn (1992: 16).

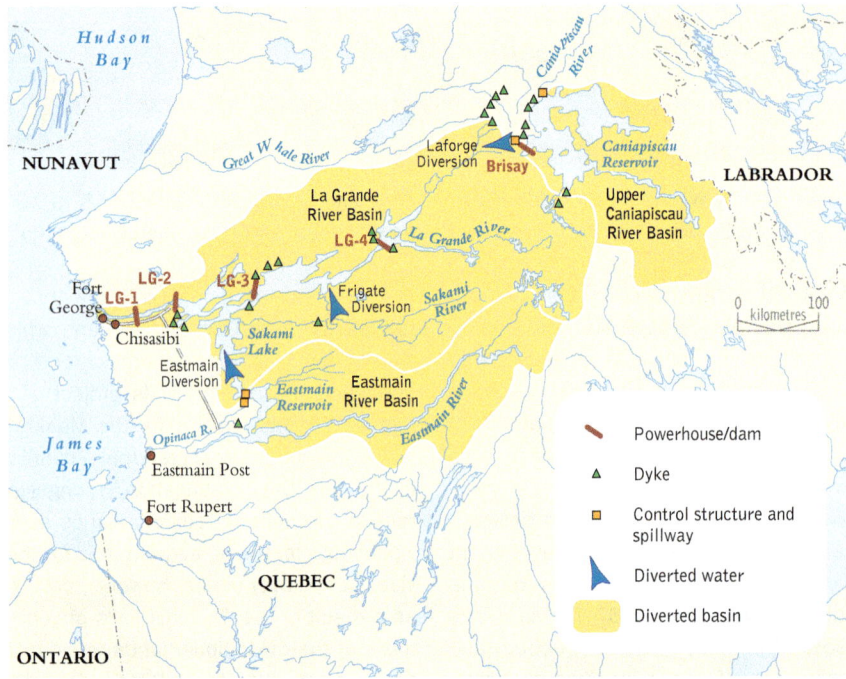

Figure 11.2 | La Grande River hydroelectric development project, Phase 1. Source: Day and Quinn (1992: 134).

Two major diversions channelled water into La Grande basin. These diversions together added an average of 1,635 m^3/sec to La Grande, almost doubling the natural flow in that river. Over a 15-year period, the cost increased to $14.6 billion, compared to the $2 billion estimate in 1971.

In Phase I of the development, three hydroelectric plants (LG2, 3, 4) with a combined 10,283 megawatt (MW) capacity were built. The first electricity was generated from LG2 in 1979, and LG4 was completed in 1986. LG1 and other dam construction were deferred to Phase II.

The scope and magnitude of the James Bay development has been described as 'breathtaking'. It produces electricity from rivers flowing in a 350,000-km^2 area of Quebec, more than one-fifth of the province or an area equivalent to France. The provincial government and Hydro-Québec justified the James Bay development on the grounds of the jobs to be created, the industrial growth to be attracted to the province, and the stability to be generated. However, in the enthusiasm over the perceived benefits from hydroelectricity, little regard was given to the fact that it was the homeland for about 10,000 Cree and Inuit whose people had lived and hunted in the region for centuries.

James Bay and Northern Quebec Agreement

The **James Bay and Northern Quebec Agreement** is the first 'modern' Aboriginal land claims agreement in Canada. However, when Premier Bourassa first announced the construction of the hydroelectric megaproject, no systematic environmental or social impact assessments had been completed. The Cree people in northern Quebec soon organized themselves to fight the project. The outcome was the James Bay and Northern Quebec Agreement, signed on 11 November 1975 and subsequently approved by the government of Canada and Quebec's National Assembly.

The agreement, although complex and often ambiguous, provided for land rights and guaranteed a process to deal with future hydroelectric developments. The agreement included provisions for environmental and social impact assessment for future developments, monetary compensation, economic and social development, and income security for Cree hunters and trappers.

James Bay II

When Premier Bourassa announced Phase II in 1985, he explained that the development would (1) generate revenue for Quebec through exports of electricity to the United States under long-term contracts and (2) attract energy-intensive industries (such as aluminum and magnesium smelters) as a result of competitively priced electricity. James Bay II involved completion of development in La Grande basin, particularly the building of LG1, as well as new hydroelectric development in the Great Whale and Nottaway–Broadback–Rupert river systems. During 1986, the Cree agreed to the completion of the development in La Grande basin but opposed the projects in the adjacent basins.

The projects in the Great Whale basin would provide just under 3,000 MW of new power by diverting several adjacent rivers. One outcome would be a reduction by 85 per cent in the flow of the Great Whale River at the community of Whapmagoostui (Figure 11.3). The Nottaway–Broadback–Rupert development would produce 8,000 MW of additional power and, as with development on La Grande, would involve inundation of land as a result of dam construction.

During the construction period from 1974 to 1984, various concerns had emerged. These concerns included the relocation of Fort George to a new site at Chisasibi, the quality of the drinking water in the new community, problems in maintaining traditional hunting activity in areas that had become accessible from the new roads built for the construction of dams, and, because of the altered patterns of ice breakup on the lower river and estuary as a result of the release of relatively warmer water from the reservoirs in winter and early spring, the difficulty hunters faced in travelling to the northern coastal area across the river from Chisasibi.

At the community level, other concerns emerged. For example, increased erosion along the banks of La Grande, the result of fluctuating water levels in the river caused by releases from the upstream reservoirs, threatened the site of the new community at one point. The newly built road exposed the community to other people and values, contributing to problems such as alcohol abuse for some individuals.

Following completion of the first three dams on La Grande, the major problem became the very high levels of mercury in fish caught in the reservoirs or connecting rivers. As a result, by the end of 1985, the Cree completely stopped fishing in the LG2 area. Another problem was that hunters from Chisasibi purchased vans to travel to distant inland hunting grounds. However, after construction of LG4 was completed, maintenance of the road network east of LG4 was stopped, and the vans could no longer be used.

Against this changing mix of issues and concerns, we look next at the challenges of estimating impacts regarding some specific issues. These changing issues and concerns reinforce the arguments for an adaptive management approach, discussed in Chapter 6, and highlight the presence of uncertainty, complexity, and change in terms of both science and management.

Mercury in reservoirs. No environmental impact assessment studies had predicted the appearance of mercury in reservoir fish. Evidence about elevated mercury levels in fish was available from earlier hydroelectric projects at the Smallwood Reservoir in Labrador and from Southern Indian Lake in Manitoba but such impacts apparently were dismissed as being of only short duration and not significant for La Grande.

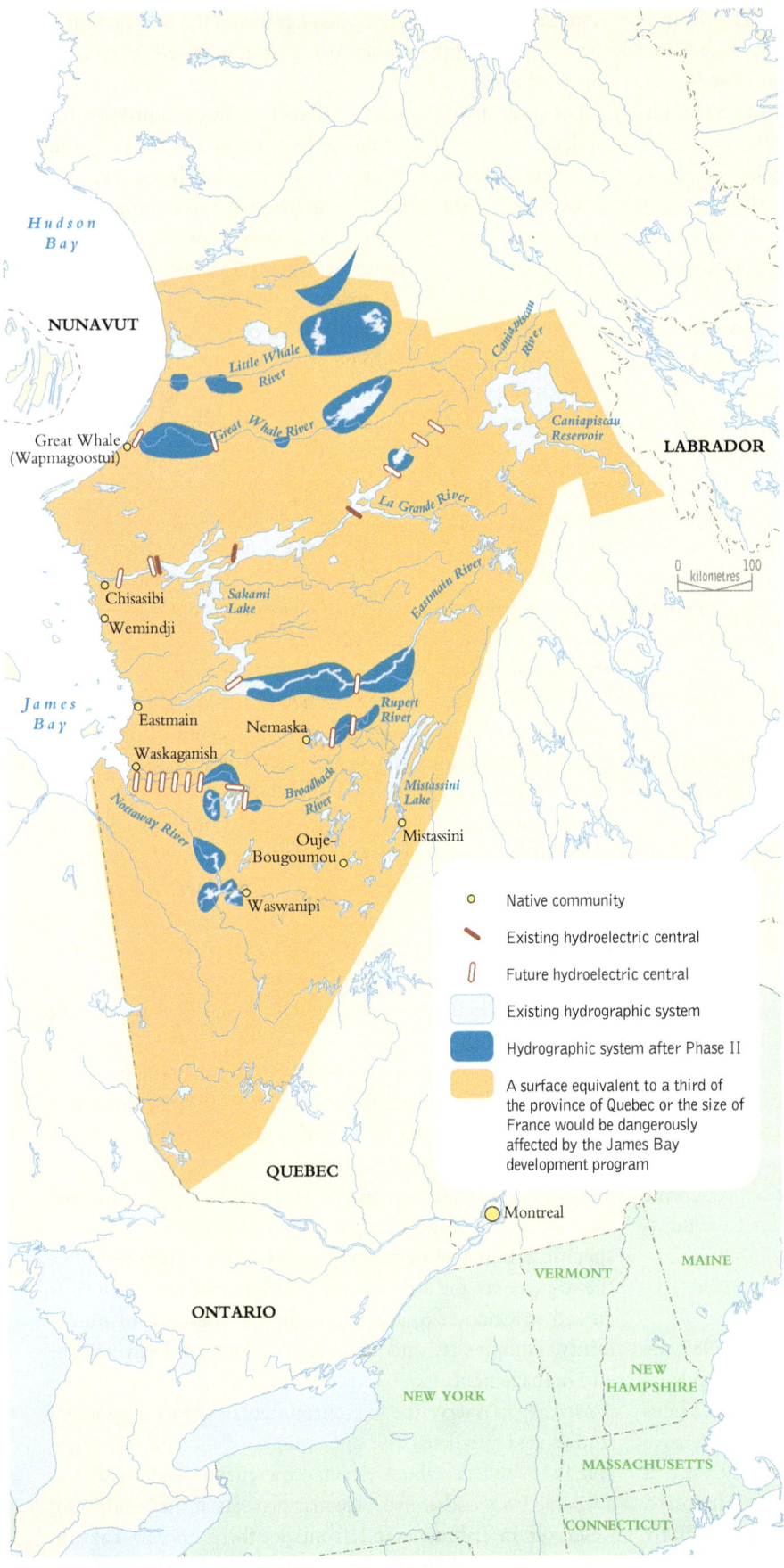

Figure 11.3 | The Great Whale project. *Source*: Diamond (1990: 32).

Mercury is common in rocks throughout the North (Gorrie, 1990: 27–8) in an insoluble form. However, when such rocks are inundated by a reservoir, bacteria associated with the decomposition of organic material in the reservoir water transform the insoluble mercury into methyl mercury that vaporizes, is released into the atmosphere, and returns to the water. Once in the water, the mercury enters the food chain and through biomagnification reaches the highest trophic levels in fish species. Such predator fish had been an important source of high-quality protein food for the local people. Berkes (1988) indicated that in most years, about one-quarter of the total community wild food harvest came from fishing, averaging about 60 kilograms per year for every man, woman, and child.

In new reservoirs, a burst of decomposition often accelerates the release of mercury. In the La Grande river system, initially few trees were removed prior to the flooding of the reservoir area, so there was a lot of organic matter to decompose. (In later stages of the massive project, tree removal was done—with much of the timber going to Cree-operated sawmills.) Downstream from the dams on La Grande, levels of mercury in fish climbed to six times their normal levels within months of completion of the dams. By the sixth year following the impoundment, concentrations of mercury were four to five times higher in all species sampled. A 1984 survey of the Cree at Chisasibi showed that 64 per cent of the villagers had unsafe levels of mercury in their bodies.

It was expected that as time passed and the drowned vegetation completely decomposed, the release of mercury would return to normal (and safe) levels. Monitoring, as reported by Chevalier et al. (1997) and Dumont et al. (1998), focused on the species most often consumed by the Cree people. The results indicated that 15 years after the impounding of the LG2 reservoir, the concentrations of methyl mercury were higher than in natural lakes but also that they were decreasing in both predatory and non-predatory species.

Monitoring revealed that mercury concentrations in the Cree people had decreased and were stabilized and that for most of the population, these concentrations 'do not present a health risk' (Chevalier et al., 1997: 79). Dumont et al. reached a more qualified conclusion. While mercury levels had dropped in all communities between 1988 and 1993–4, in their view, 'the significance of these mercury levels for the health of the Cree population remains unknown.' Their reasoning was that the decrease in mercury concentrations could be due to the Cree having changed the type of fish eaten from contaminated to less contaminated fish or from having decreased their total fish consumption, or both. As a result, Dumont et al. (1998: 1444) concluded that the 'present low mercury levels in the Cree communities may not be permanent and must not lull health authorities into believing that the mercury problem has been definitively resolved.'

La Grande estuary fish. The pre-construction impact assessments indicated that the estuarine fishery in La Grande was unlikely to survive the development of the dams and reservoirs. The impact study predicted that when the reservoirs began to fill behind the dams, with the resulting absence of ice cover to dampen the impact of ebb and flow of the tidal water from James Bay, salt water would move farther up into the river. The consequence would be the elimination of the freshwater overwintering fish habitat for species important for the local fishery. On the other hand, if the river water flow were reduced *after* the formation of ice cover, then saltwater intrusion would be impeded and a critically important pocket of fresh water could be maintained in the key habitat area.

Partly as a result of pressure exerted by local fishers, the river flow was not cut off until after an ice cover had formed on the river. Monitoring revealed that this action did result in the creation of the necessary freshwater pocket, which remained in place throughout the winter. The outcome was that the predicted fish kill did not occur, and subsequent fish populations were about the same as in the pre-construction period.

> ## Perspectives on the Environment
>
> ### Limiting Factor Principle
>
> In Chapter 2, we discussed the **limiting factor principle**, which indicates that all factors necessary for growth must be available in certain quantities if an organism is to survive. We also noted that the weakest link is known as the dominant limiting factor. Are these ideas helpful in understanding the impact of interrupting river flow, and changing patterns of ice cover, on the overwintering fish habitat in La Grande estuary?

Improving Impact Assessment

The James Bay experience reinforces the viewpoint that surprises should be expected in impact assessment and other aspects of environmental management, a point made in Chapters 1 and 6. Some time ago, Berkes (1988) concluded that little mystery surrounds some sources of uncertainty. Impacts do not occur simultaneously but over an extended period and are the result of decisions taken. Furthermore, impacts do not happen in isolation; in many instances, they combine to form *cumulative impacts*, which is particularly important in such projects as James Bay where development is contemplated over decades.

Water Exports, Diversions, and Other Options

Growth and development in the US Southwest has the potential to lead to water scarcity in that region. A 'supply management' solution would be to look to Canada as a source and to import water in bottles, by ship, or via pipeline. Canada exports oil and natural gas to the United States, so why not water?

As Day and Quinn (1992: 41–2) and Quinn (2007) have explained, those who support the export of water to the United States argue that water is just another resource with value and can be exchanged on the open market, Canada has more water than it needs to meet its foreseeable needs, substantial income could be earned from selling water to the US and elsewhere, jobs would be created through the necessary major construction projects, such as pipelines, and some of the diverted water could be sent to regions in Canada facing shortfalls. These advantages, identified nearly 25 years ago, continue to be promoted by those advocating export of water (Boyer, 2008; Katz, 2010).

Those opposing the export of water emphasize that the scale or magnitude of the proposed water diversion projects would be much larger than any previous project, creating significant risks and uncertainty. Once the taps are 'turned on', it would be virtually impossible to turn them off, since receiving areas would have become dependent on the diverted water. In addition, negative environmental and social impacts could be significant and irreversible (Nikiforuk, 2007; Lasserre, 2009).

During the 1960s, various southern state governments in the US considered the feasibility of large-scale water diversions through pipelines from one or both of the Columbia and Mississippi river systems. The northern states that would be the source of these diversions strongly opposed the proposals. One outcome was that some states and private companies began to consider whether water from Canadian river or lake systems could be diverted southward. Two examples illustrate the scope and magnitude of these proposals (Day and Quinn, 1992: 40).

North American Water and Power Alliance (NAWAPA)

The NAWAPA project was the most publicized major diversion project. Conceived by the Ralph M. Parsons Company in California, NAWAPA was proposed to store the headwaters of the Yukon, Skeena, Peace, Fraser, and Columbia rivers in the Rocky Mountain Trench in eastern British Columbia. The stored water would be diverted to both the Canadian Prairie provinces and western states by pipeline. The Canadian federal and provincial governments strongly opposed the possibility of major water diversions from Canada to the US, rejecting Parsons's basic assumption that water in Canada was a 'continental resource'.

Grand Recycling and Northern Development (GRAND) Canal

Thomas Kierans, a Canadian engineer, proposed a major diversion involving a dyke across James Bay, creating a reservoir in the bay, and then pumping water up 300 metres to move it into the Great Lakes basin from which it could be moved by pipeline to the southwestern states. The eight states (Minnesota, Wisconsin, Michigan, Illinois, Indiana, Ohio, Pennsylvania, NewYork) and two provinces (Ontario and Quebec) in the Great Lakes basin all opposed this proposal.

Other Options

Exports of water do not have to take place only via pipelines or major diversions. Some exports come about between communities adjacent to one another on either side of the Canada–US border. In such situations, water supply systems are often shared between communities, such as between St Stephen, NB, and Calais, Maine, or Coutts, Alberta, and Sweetgrass, Montana. In these arrangements, water often flows in both directions, the volumes are modest, and mutual accommodation is achieved.

Another type of diversion involves movement of water between national and boundary waters. For example, in northern Ontario, diversions from Ogoki Lake and Long Lac move water from Ontario rivers into Lake Superior in order to increase power-generating capacity. In the United States, a channel was constructed at Chicago to divert water from Lake Michigan to the Illinois–Mississippi river system to facilitate navigation southward from Chicago. While the amounts of water involved are modest in the Chicago Diversion, there is concern that it could be regarded as a precedent for larger-scale diversions out of Lake Michigan for other purposes. There is also always a risk that diversion channels between watersheds can lead to introduction of invasive species from one to the other.

Tanker shipment represents another option. It would involve containerized vessels taking fresh water from coastal rivers to destinations as close as California or as distant as Middle East nations. One arrangement would use ships bringing petroleum to Canada from the Middle East to transport fresh water on the return journey after suitable cleaning of holds. Another option would be to use tanker ships dedicated to carrying only water. A third option, still under development, would be floating bags or membranes towed behind a ship. The volumes of water would be relatively small, and the cost of collection and transportation would have to be competitive with alternatives available at the destinations, such as desalinization.

El Ayoubi and McNiven analyzed possible export of water by tanker from the Annapolis Valley in western Nova Scotia to Brownsville, Texas. They concluded that the project was unprofitable: 'Pricing policies and alternate technologies will probably mean that the utility of tanker-based projects will remain only a future possibility. The unprofitability of such projects is likely the reason why there have been so few examples of bulk water exports in the world and not because of the political opposition to them' (El Ayoubi and McNiven, 2006: 14).

Export of water in bottles or similar containers is a possibility. It does not raise concerns on the same scale because the quantities involved would not be significant and such trade could be stopped or modified with minimal consequences, unlike turning off the water passing through a major pipeline.

Reactions and Responses

The situation in the US is evolving. For example, Lasserre (2007: 152) observed that 'water demand has been stagnating in the United States for the past two decades.' Several reasons account for this pattern. First, although still relatively low, water prices in the US have been steadily rising. Second, water-short cities are relatively wealthy compared to the agricultural sector and have been able to drive reallocation of water from agriculture to urban areas. Third, agriculture in the US is facing growing competition from producers in Mexico and Asia, leading to reduced crop production. As a result, Americans appear more likely to look to other solutions than to Canada for bulk imports.

In Canada and in the Great Lakes states, however, concern has persisted. This concern was reflected during 1985 when the Ontario and Quebec premiers joined the governors of the eight Great Lakes states in signing the Great Lakes Charter. The Charter is a non-binding agreement, but those signing agreed to notify and consult each other regarding any possible diversion. In addition, they agreed that no jurisdiction would start a new or increase an existing diversion involving more than 5 million gallons per day without seeking the consent of all Great Lakes states or provinces that would be affected.

The Canadian government has taken various actions to curtail water export, drawing on its authority for international and interjurisdictional matters. The first major initiative was the federal water policy introduced in November 1987. In that policy, the government prohibited 'large-scale' export by inter-basin water diversions to the United States.

A different challenge emerged in the late 1980s when 'free trade' negotiations between Canada and the US were at their final stage. Concern was expressed that provisions for free

Perspectives on the Environment

Water Shortages in the Great Lakes Basin

The US Geological Survey has concluded that the Great Lakes basin could experience localized water shortages due to climate shifts and/or growing demand (Reeves, 2010). For example, its five-year study shows that groundwater levels have dropped approximately 300 metres over the past several hundred years in the Chicago–Milwaukee area due to pumping of water from aquifers to provide municipal water, and it estimates a further drop of 30 metres over the next three decades if withdrawal rates increase as expected.

---

trade would make it impossible for Canada to prohibit export of water to the United States. Reflecting this concern, in June 1988 the government of Ontario introduced a bill entitled the Water Transfer Control Act to prohibit all forms of water export from Ontario, except for bottled water, and it became law in 1989.

Then in 1998, the Ontario government of Premier Mike Harris awarded a licence to the Nova Group to export water from Ontario. The Nova Group proposed to ship water from Sault Ste Marie to Asia. There was an immediate uproar across Ontario and in the Great Lakes states, which led the Ontario government to reverse its decision. The Ontario government's original willingness to approve water exports raised doubts about the capacity of the Great Lakes Charter to prevent such decisions.

One outcome was that the federal government in 1999 announced a strategy to address water exports from Canada. There were three components to the strategy. First, the **International Joint Commission** was asked to provide a legal opinion regarding exports from the Great Lakes basin. Second, the provincial governments were urged to pass laws that would ban water exports. This was viewed as necessary, since provinces own the water within their boundaries. And third, a federal law was to be designed to restrict exports.

With the exception of New Brunswick, all provinces subsequently passed legislation to prevent bulk water diversions or exports outside their borders. The door was still left open for other types of exports, such as bottled water. The risk is that if one province allows what might be viewed as significant water exports, then under the North American Free Trade Agreement (NAFTA), all provinces might be compelled to treat water as a 'tradable good'.

As Lasserre (ibid.) explained, rather than passing a law, the federal government chose a different approach. The choice was based on a feature of the river basins associated with the international border. Nearly all the large basins are located either entirely in Canada or in the US, with only a few systems, such as the Great Lakes and the Columbia River, straddling the border. Given this physical reality, if provinces prohibited transferring water out of their jurisdiction, that would effectively stop water exports to the US. This arrangement could be reinforced by the federal government restricting inter-basin transfers because of negative environmental impacts. Thus, Canada did not need to have a direct policy or law prohibiting water exports to the United States, either of which might be challengeable under NAFTA.

The federal government also approved an amendment to the International Boundary Waters Treaty Act during December 2002. The amendment prohibits bulk water exports from river basins if they exceed 50 m^3/day. It was also intended to deter the diversion of water out of the Great Lakes basin, since the International Joint Commission's legal position in response to Canada's request for guidance was that diversions from boundary waters could occur only if they were authorized by the appropriate governments with jurisdiction for the water.

In the US, some initiatives have been undertaken to restrict large-volume diversions from the Great Lakes. In 2000, the US government stipulated that any diversion of water from the Great Lakes system by any state or federal agency or private organization for use outside of the Great Lakes basin is prohibited unless the governors of each of the eight Great Lakes states give approval. Further support for this policy came in December 2005 when the eight governors and two premiers, through the Council of Great Lakes Governors, agreed on a set of principles to review proposals for transfers of water from the Great Lakes. Any proposal would be authorized only if:

- no reasonable alternative to the proposed transfer exists;
- withdrawals are limited to reasonable volumes for specified uses;
- all withdrawn water, after an allowance for consumptive use, is returned to the Great Lakes basin from which it was removed;
- for 'major' proposals, an explicit conservation plan has been prepared.

The 2005 agreement by the Council of Great Lakes Governors effectively banned major water exports or diversions from the Great Lakes basin but left open the option for low-volume transfers that might be anticipated between adjacent communities on either side of the border.

Water Quality

Humans can adversely affect water quality in numerous ways, and water quality has been selected as one of the three main Environmental Sustainability Indicators by Environment Canada (Environment Canada et al., 2006). Water quality is assessed according to an index to measure the ability of surface waters to protect aquatic life at selected river and lake locations within Canada (Environment Canada, 2009b).

Environment Canada (2010b: 7) has reported three key insights related to its water quality indicator based on information gathered up to 2006:

- Fresh water at the 379 monitoring stations located in southern Canada was rated as 'good' or 'excellent' at 48 per cent of the sites, 'fair' at 30 per cent, and 'marginal' or 'poor' at 22 per cent.
- Freshwater quality at the 32 monitoring sites in northern Canada was rated as 'good' or 'excellent' at 66 per cent of the sites, 'fair' at 28 per cent, and 'marginal' at 6 per cent.
- The St Lawrence River basin, which includes the Great Lakes, had the highest percentage of sites with water quality rated as 'poor' or 'marginal' (28 per cent). Phosphorus was the biggest problem in this region.
- In contrast, the Maritime and Arctic drainage basins, each with only seven measuring stations, reported the highest percentages (71 per cent) of stations with 'good' or 'excellent' freshwater quality.

Notwithstanding the data provided above based on federal government monitoring programs, we need to be careful in using such information. As the Commissioner of the Environment and Sustainable Development (2010: 3) reported, 'Environment Canada is not adequately monitoring the quality and quantity of Canada's surface water resources. . . . The Department is not monitoring water quality on the majority of federal lands and does not know whether other federal departments are doing so.'

The most important issue is pollution from various sources but especially from industrial and other urban wastes and from agricultural runoff. The first two are easier to identify because they are usually associated with *point sources*, such as manufacturing plants or sewage treatment plants. Agricultural runoff is more challenging, since it is usually diffuse pollution from *non-point sources*, such as fertilizers, herbicides, and pesticides from farm fields. Some types of urban runoff, such as oil and salt from road surfaces, also are characterized as non-point, since they cannot be identified with specific places.

Perspectives on the Environment

Municipal Waste Water

Municipal waste water can result in increased nutrient levels, often leading to algal blooms; depleted dissolved oxygen, sometimes resulting in fish kills; destruction of aquatic habitats with sedimentation, debris, and increased water flow; and acute and chronic toxicity to aquatic life from chemical contaminants, as well as bioaccumulation and biomagnification of chemicals in the food chain.

—Environment Canada (2003b: 34)

Point Sources

Urban waste water can be treated up to three levels of treatment: (1) *primary*, which removes only insoluble material; (2) *secondary*, which removes bacterial impurities from water previously having received primary treatment; and (3) *tertiary*, which removes chemical and nutrient contaminants following secondary treatment.

In 1999, 78 per cent of Canadians living in municipalities serviced by sewers had secondary and/or tertiary treatment, an increase from the 56 per cent with such service in 1983. Nine per cent of all Canadians had their wastes go to waste stabilization ponds (also known as sewage lagoons), simple treatment systems that provide the equivalent of secondary treatment. In 1998, 3.46 million Canadians living in larger municipalities were not connected to wastewater systems and used either septic tanks or other arrangements to handle their wastes. Another 4.88 million people lived in smaller

Industrial point-source pollution on the Calumet River, Chicago.

Point-source, end-of-pipe type of water pollution, Great Lakes Basin.

municipalities, and almost 57 per cent of them had no wastewater treatment facilities. The percentages reported here for the late 1990s remain relevant at the start of the second decade of the twenty-first century.

A growing concern is that many wastewater treatment facilities are old and need expensive maintenance, upgrading, or replacement. The Canadian Water and Wastewater Association calculated that $5.4 billion in new investment would be required *each year* between 1997 and 2012 to modernize and upgrade all existing water and wastewater treatment facilities, as well as to provide such facilities to communities currently without them. Such investment did not occur.

Industry is an important source of wastes deposited into water bodies. Tables 11.1 and 11.2 identify the most prevalent chemicals and the water bodies receiving more than 500 tonnes of pollutants in 2001. These data are from the National Pollutant Release Inventory (NPRI), which in 2006 tracked the release of more than 300 substances into the environment.

Ninety per cent of the 9,000 facilities that reported chemical releases in 2006 were in Alberta (3,717) and Ontario (2,379) (www.ec.gc.ca/pdb/npri/npri_factsheet06_e.cfm).

Ammonia and nitrogen represented more than 94 per cent of the total releases to water. Other chemicals such as mercury are released in much smaller amounts but have serious negative impacts on human and aquatic system health. Mercury bioaccumulates and biomagnifies (see Chapter 10) in the liver, kidneys, and muscles of affected organisms, and chronic exposure can result in brain and kidney damage. Mercury levels in the Canadian environment continue to rise. The main sources are metal mining and smelting, waste incineration, and coal-fired power plants.

Runoff from urban areas either flows directly into water bodies from roads and other non-point sources or can be channelled by stormwater systems. Stormwater can contain various contaminants such as suspended solids, sediment, and grit; nutrients, including different forms of phosphorus and nitrogen; toxic metals, including copper, lead, and zinc; hydrocarbons, including oil, grease, and polycyclic aromatic hydrocarbons; trace organic contaminants, including pesticides, herbicides, and industrial chemicals; and fecal bacteria. As a result, stormwater should be treated in municipal wastewater plants. Unfortunately, such treatment does not always occur.

A third important source of wastes into water bodies is from agricultural activity, but that is more appropriately discussed under the category of non-point sources.

Non-Point Sources

As discussed in Chapter 10, crop and livestock production has increased significantly as a result of more effective farm machinery, new genetic crops, agrochemicals, and irrigation.

Table 11.1 | Top Releases of Chemicals to Water, 2001

Chemical	Releases (tonnes)
Ammonia (total)*	26,106
Nitrate ion in solution at pH equal to or greater than 6.0	22,450
Manganese (and its compounds)	1,157
Methanol	697
Zinc (and its compounds)	308

*Total includes both ammonia (NH_3) and ammonium ion (NH_4^+) in solution.
Source: Statistics Canada (2003b: 18).

Table 11.2 | Water Bodies Receiving More Than 500 Tonnes of Pollutants, 2001

Water Body	Total Release (tonnes)	Dominant Release	Share of Total Release (%)
Fraser River	9,168	Ammonia*	49.2
Lake Ontario	8,877	Ammonia*	41.6
Bow River	8,264	Nitrate ion	90.8
Ottawa River	3,066	Ammonia*	76.6
North Saskatchewan River	2,953	Nitrate ion	61.3
Red River	2,766	Ammonia*	72.7
Hamilton Harbour	1,516	Ammonia*	70.6
South Saskatchewan River	1,275	Nitrate ion	62.4
St Lawrence River	1,086	Nitrate ion	43.6

*Total includes both ammonia (NH_3) and ammonium ion (NH_4^+) in solution.
Source: Statistics Canada (2003b: 18).

The latter two also contribute to environmental impacts, especially through fertilizers, pesticides, and herbicides being carried in runoff from farm fields, which ends up in streams, rivers, and lakes. In this section, we examine the experience with diffuse pollution in the Great Lakes basin.

Diffuse pollution is a policy issue in the Great Lakes basin. Since the early 1960s, interest has evolved from concern about sedimentation from soil erosion and eutrophication from phosphorus and nitrate loading to persistent toxic chemicals. As the definition of the problem has evolved, so have ideas regarding appropriate responses.

What has been learned about the strategic implications of how the problem has been defined? A key lesson is that diffuse pollution represents a 'layered' problem—it is much like an onion. Too often, attention does not go beyond the first layer. At the first layer, concern about diffuse pollution focuses on *environmental degradation* and the *economic costs* imposed on downstream users. The motivation for defining the problem in this manner appears to be that people will see the connection between diffuse pollution and loss of economic production or increased costs for economic production. A second layer, receiving greater attention, is the link between diffuse pollution and negative impacts on *ecosystem health* or *integrity* and especially on *human health*. It is believed that making the connection to human health should create a powerful image in the minds of both policy-makers and residents regarding diffuse pollution. A third layer is the interpretation of diffuse pollution as a problem touching on human *values, beliefs, attitudes, and behaviour*. From this perspective, the fundamental problem is behaviour by individuals and groups, driven by inappropriate values, beliefs, and attitudes. If attention is focused on this third level, then the prescription to resolve diffuse pollution is certainly different from what it would be if attention were limited to the first level.

What mechanisms have been effective in achieving recognition of diffuse pollution as a policy issue? First, there is a need for *credible science* to document the nature of the problem.

Credible Science and Institutional Commitment

Appreciation of diffuse pollution as a policy issue has been encouraged in Canada and the Great Lakes basin by a combination of science, institutions, and individuals. Several initiatives by the International Joint Commission (IJC), a bilateral institution created in 1909 to manage interjurisdictional water issues between Canada and the United States, have been significant.

During the 1960s, the media declared that 'Lake Erie is dying', a reference to the highly eutrophic state of that lake (see Chapter 4). In 1972, the governments of Canada and the United States entered into an agreement to restore and enhance water quality in the Great Lakes. Initial attention focused on reducing phosphorus loading from municipal sewage treatment plants and other point sources. Initiatives in that regard were effective, but it was suspected that non-point sources might also be significant. However, data were not available to indicate the importance of these sources.

Under the 1972 agreement, the IJC was asked 'to conduct a study of pollution of the boundary waters of the Great Lakes System from agricultural, forestry, and other land use activities'. Subsequently, the International Reference Group on Great Lakes Pollution from Land Use Activities, otherwise known as PLUARG, examined two major pollution problems in the basin: eutrophication from elevated nutrient inputs, and increasing contamination by toxic substances. PLUARG studied the pollution potential from various land uses, including agriculture, urbanization, forestry, transportation, and waste disposal, as well as natural processes such as lakeshore and riverbank erosion.

PLUARG concluded that the 'combined land drainage and atmospheric [non-point] inputs to individual Great Lakes ranged from 32 per cent (Lake Ontario) to 90 per cent (Lake Superior) of the total phosphorus loads (excluding shoreline erosion). Phosphorus loads in 1976 exceeded the recommended target loads in all lakes' (International Reference Group on Great Lakes Pollution from Land Use Activities, 1978: 4–5). The PLUARG study findings, the first credible science to document the contribution of non-point sources to phosphorus loading, were difficult to ignore. The report also stated that toxic substances such as PCBs (polychlorinated biphenyls) were entering the Great Lakes system 'from diffuse sources, especially through atmospheric deposition. Through land drainage, residues of previously used organochlorine pesticides (e.g., DDT) are still entering the boundary waters in substantial quantities.' In terms of the sources, it was reported that 'intensive agricultural operations have been identified as the major diffuse source contributor of phosphorus.' In addition, 'Erosion from crop production on fine-textured soils and from urbanizing areas, where large-scale land developments have removed natural ground cover, were found to be the main sources of sediment. Urban runoff and atmospheric deposition were identified as the major contributors of toxic substances from non-point sources' (ibid., 6).

Regarding necessary actions, PLUARG concluded:

> The level of awareness about pollution from non-point sources among Great Lakes Basin residents is inadequate at present. Control of non-point sources will require all basin residents to become involved in reducing the generation of pollutants through conservation practices. Improved planning and technical assistance are prerequisites to long-term solutions of land drainage problems. (Ibid., 10–11)

The PLUARG report, and other analyses, led to renewal of the Great Lakes Water Quality Agreement in 1978 and to the signing in 1987 of a protocol amending the 1978 agreement. The 1987 amendments extended the scope of the agreement, but diffuse pollution was still recognized as a priority problem. Specifically, Annex 13 of the protocol focused exclusively on 'pollution from non-point sources' and identified 'programs and measures for abatement and reduction of non-point sources of pollution from land-use activities' in order 'to further reduce non-point source inputs of phosphorus, sediments, toxic substances, and microbiological contaminants contained in drainage from urban and rural land, including waste disposal sites, in the Great Lakes System' (Canada and Ontario, 1988: 55).

The PLUARG study commissioned by the International Joint Commission, along with the IJC's prestige and watchdog role, was significant in helping elected officials and the public to understand the severity of the diffuse pollution problem in the Great Lakes. Without such a credible voice to draw attention to the issue, it is unlikely that action would have been forthcoming.

Agricultural Non-point Source Pollution

After the PLUARG studies were completed, Canada and the United States agreed to deal with the issue of high phosphorus loadings from rural non-point sources. In Canada in 1987, the federal and Ontario governments created a cost-shared program—the Soil and Water Environmental Enhancement Program, or SWEEP. The purpose was to meet by 1990 the target reduction for Canada of 200 metric tons per year of phosphorus loading in Lake Erie from non-point sources.

SWEEP consisted of various programs. The first focused on technology evaluation and development. It was intended to stimulate adoption of soil management and cropping practices to improve water quality and to reduce soil erosion and degradation. A second thrust focused on pilot watershed programs, local demonstrations, and technical assistance at the farm level. The pilot watershed program tested the effectiveness of state-of-the-art conservation practices on working farms. The effects of using different practices on all farms in three experimental watersheds were compared with conventional practices in three control watersheds regarding water quality, hydrology, soil quality, crop production, and economics. A third component involved information services, with attention to informing the public about the nature and consequences of soil and water quality problems and about the SWEEP objectives. Cressman (1994: 421), whose consulting firm was actively involved in the pilot watershed studies, later remarked that 'It was during this time that interest in, and the practice of, conservation tillage grew significantly among Ontario farmers.'

In parallel with SWEEP, the Ontario Land Stewardship Program was introduced by Ontario. This $40 million program provided financial incentives for first-time adoption of conservation measures on farmland. Financial assistance was provided for practices to protect soil structure, build structures to ameliorate soil erosion, purchase conservation equipment, and obtain technical training. Funds also were dedicated to research projects related to stewardship practices. The SWEEP program overlapped with the National Soil Conservation Program, a $150 million shared-cost program. When SWEEP and the National Soil Conservation Program both terminated in the early 1990s, a newly elected federal government introduced another program—the Green Plan—which provided funds for soil conservation and diffuse pollution control. However, by that time attention had shifted from soil erosion, sedimentation, and eutrophication to toxic substances. One observer remarked to one of the authors that in his view, the three programs had been developed with little consultation among key agencies, resulting in duplication and overlap and a time frame not long enough to allow measurement of program impacts. Indeed, perceived results and the provision of support to farmers for production improvements took precedence over ameliorating environmental degradation problems.

Some positive results were achieved nevertheless. In the Lake Erie and Lake Ontario watersheds, the loads from phosphorus have been reduced significantly. The main initiatives for non-point sources helped farmers to modify how their land was cropped, especially by encouraging conservation tillage to reduce erosion and thereby reduce sediment and toxics placed into aquatic systems. The main actions regarding point sources involved upgrading municipal sewage treatment plants, and regulations were established to reduce phosphorus in laundry detergents.

In terms of a summary of the environmental status of the Great Lakes, the St Lawrence River, and the St Clair River–Lake St Clair–Detroit River ecosystem in 2006, Environment Canada and the US Environmental Protection Agency (2007: 3) concluded that 'the overall status of the Great Lakes ecosystem was assessed as mixed because some conditions or areas were *good* while others were *poor*. The trends of Great Lakes ecosystem conditions varied: some conditions were *improving* and some were *worsening*.'

Water Security: Protecting Quantity and Quality

A central concern in water management is to ensure a sufficient quantity of water of adequate quality for human use. By the end of the first decade of the twenty-first century, more than one in six people on the Earth did not have access to safe water supplies, and two out of five did not have access to adequate sanitation, notwithstanding substantive efforts during the United Nations International Drinking Water Supply and Sanitation Decade throughout the 1980s to improve conditions.

In terms of current per capita water use, the range extends from as little as 20 litres to more than 500 litres each day. Only 4 per cent of the world's population use water in the range of 300 to 400 litres per person per day, with people in the United States, Canada, and Switzerland being the highest per capita users. In contrast, about two-thirds of the global population use fewer than 50 litres for each person daily.

Most humans become thirsty after losing only 1 per cent of their bodily fluid and are in danger of death once the loss approaches 10 per cent. The minimum water requirement to replace loss of fluid for a normal healthy adult in an average temperate climate is about three litres each day. In tropical or subtropical conditions, the minimum amount becomes about five litres per person per day.

Most Canadians receive their drinking water from the 4,000 municipal water treatment plants across the country, but a significant number depend on private wells or other arrangements. About 9 million Canadians, most living in small towns or rural areas, draw on groundwater for their drinking water.

The relative abundance of water in Canada, the high levels of water use, and the myth of superabundance, all referred to at the beginning of this chapter, made most Canadians complacent about the adequacy and safety of their water supplies. For many, this all changed in mid-May 2000 when the town of Walkerton in southwestern Ontario, population about 5,000, experienced contamination of its water supply system by deadly bacteria, *Escherichia coli* O157:H7, or **E. coli**, and *Campylobacter jejuni*. Children under five years of age and the elderly are at greatest risk from these bacteria. Seven people died, and more than 2,300 became ill. It is anticipated that some individuals who became sick in Walkerton, especially children, may experience effects for the rest of their lives. For example, 10 years after the 'Walkerton event', a medical team at the University of Western Ontario reported that adults from Walkerton who developed acute gastroenteritis in May 2000 and had been monitored between March 2002 and August 2008 showed higher probabilities of developing hypertension, kidney problems, or cardiovascular disease compared to adults who had not become ill or who had become mildly ill (Clark et al., 2010).

Perspectives on the Environment
Groundwater Quality

Groundwater sources generally provide water that is safe to drink. This is especially true if the well field is protected from pollutants. Aquifers that are close to the surface are more prone to contamination by pollution, which partially explains the poor water quality that is characteristic of many shallow wells in Canada.

—Statistics Canada (2003b: 24)

The concern generated by the Walkerton experience was reinforced during March 2001 in North Battleford, Saskatchewan, a community of 14,000, where thousands of residents suffered from contamination of the municipal water system by the parasite *cryptosporidium*. The parasite got into the water supply system over three weeks following routine maintenance at the treatment plant. Residents were under a boil-water order for three months. An inquiry ordered by the provincial premier concluded that the Saskatchewan government had not been effective in safeguarding drinking water in the province.

The Walkerton Inquiry

A public inquiry by Justice Dennis O'Connor (2002a, 2002b) established that:

1. The E. coli, contained in manure spread on a farm near one well of the Walkerton water supply system, entered the system through that well.
2. The farmer who spread the manure followed proper practices and was not at fault.
3. The outbreak would not have occurred if the water had been treated. The water was not treated because the chlorination equipment was being repaired.
4. The provincial government's approvals and monitoring programs were inadequate.
5. In addition to lack of training, the operators of the well system had a history of improper operating practices.

Environment in Focus
Box 11.2 Bottled Water

In his book, *Bottled and Sold: The Story Behind Our Obsession with Bottled Water*, Peter Gleick (2010) remarks that:

- More than 1,000 bottles of water are sold every second of every day in the US, and at the same time another 4,000 bottles are sold in other countries around the world.
- In North America, quality standards for bottled water are often lower than for tap water.
- Marketing information about bottled water can be misleading. Examples: 'Everest' water is from Texas and 'Arctic Wolf Springs' is from Philadelphia.

Walkerton water memorial.

6. When people began to fall ill, the general manager of the water system withheld from the public health unit critical information about adverse water quality test results. This resulted in delay of a boil-water advisory.
7. Budget reductions by the provincial government had led to closure of government laboratory testing services for municipalities, and private laboratories were not required to submit adverse test results to the Ministry of the Environment or to the Medical Officer of Health.

Walkerton: Lessons and Recommendations

Justice O'Connor offered recommendations to ensure the safety of drinking water across the province. Many are relevant to all regions of Canada. Overall, he recommended a **multi-barrier approach** to drinking water safety. In his words, 'Putting in place a series of measures, each independently acting as a barrier to passing water-borne contaminants through the system to consumers, achieves a greater overall level of protection than does relying exclusively on a single barrier (e.g., treatment alone or source protection alone). A failure in any given barrier will not cause a failure of the entire system' (O'Connor, 2002b: 5). He argued that the first barrier involves selecting and protecting reliable, high-quality drinking water sources. He thus recommended 'a source protection system that includes a strong planning component on an ecologically meaningful scale—that is, at the watershed level' and said 'the Province [should] adopt a watershed-based planning process' (ibid., 6, 3). Within a watershed-based approach, he recommended:

1. A comprehensive approach for managing all aspects of watersheds was needed.
2. A framework for developing watershed-based source protection plans should be established.
3. To ensure that local considerations are fully taken into account and to create goodwill in and acceptance by local communities, source protection planning should be undertaken as much as possible at a local (watershed) level by those most affected (municipalities and other affected local groups).

The multi-barrier approach has subsequently been endorsed by other federal and provincial governments. At a national level, Environment Canada, Health Canada, the Canadian Council of Ministers of the Environment, and the Committee on Environmental and Occupational Health collaborated to prepare a multi-barrier approach. Details about this initiative can be found on the website of the Canadian Council of Ministers of the Environment (www.ccme.ca/sourcetotap/) under 'Source to tap—Protecting our water quality'.

Another positive outcome was that in 2005 the Walkerton Clean Water Centre was officially opened. It operates from a facility completed in 2010 at a cost of $8.3 million. By the tenth anniversary of the Walkerton water incident in 2010, more than 21,000 drinking water system owners and operators had attended mandatory and elective training courses at this centre. In addition, mobile units provide training to remote communities in the province. On-line courses also are provided.

In late May 2011, the Ontario provincial government announced it had paid out $72 million to victims of the tainted water. The Attorney General explained that more than 99 per cent of the over 10,000 claims for compensation had been resolved.

First Nations Water Security

The events at Walkerton and North Battleford drew attention to structural and human resource issues related to reliable and safe water supply systems in Canada, systems that most citizens had previously taken for granted. Events in October 2005 at the Cree First Nations community of Kashechewan close to the western shore of James Bay, however, highlighted the Third World conditions on many First Nation reserves across the country. This situation was verified by the Commissioner of the Environment and Sustainable Development, who observed that 'When it comes to the safety of drinking water, residents of First Nations communities do not benefit from a level of protection comparable to that of people who live off reserves' (Auditor General of Canada, 2005a: ch. 5, p. 1).

The Commissioner reported that the Department of Indian and Northern Affairs indicated that in 2004 about 460,000 First Nations people were living on some 600 reserves across the country. In 1995, data showed that 25 per cent of reserves had water systems that 'posed potential health and safety risks to the people they served'. In 2001, when Indian and Northern Affairs completed its next assessment, 75 per cent of the water systems on First Nations reserves posed 'significant risk to the quality of the safety of drinking water' (ibid., 11). More recently, Health Canada (2010) reported that as of October 2010, 116 First Nation communities out of a total of 615 across Canada were under a drinking water advisory. This situation reinforces a conclusion by Harden and Levalliant (2008: 7): 'The fact remains that unsatisfactory access to safe drinking water persists for many First Nations people despite numerous reports and policies.'

This serious situation regarding poor water quality has been attributed to inconsistent testing and inadequate training of equipment operators. Others, while agreeing with this conclusion, note that many non-reserve communities in rural Canada also have poor water supply services. Indeed, in December 2006, Health Canada reported to the Senate that 1,174 boil-water advisories were in place at that time, a remarkably high number for a nation that usually perceives itself to be 'developed'.

Perspectives on the Environment
Boil-Water Advisories in Canada

In 2008, the Canadian Medical Association reported that 1,760 boil-water advisories were in place across the country, in addition to those in place on 93 First Nations reserves. The number of advisories by province, in descending order, were: Ontario 679; British Columbia 530; Newfoundland and Labrador 228; Saskatchewan 126; Nova Scotia 67; Quebec 61; Manitoba 59; Alberta 13; New Brunswick 2; Northwest Territories 1; Prince Edward Island, Nunavut, and Yukon 0.

The advisories included those for communities, commercial facilities, and trailer parks. Some had been in place for at least five years.

The Federation of Canadian Municipalities estimates that about $31 billion is needed to upgrade water and waste water treatment facilities across the country.

—CBC and CTV News, 7 and 8 April 2008

Kashechewan, a community of 1,900 people located on the Albany River some 400 kilometres north of Timmins,

Established in 2004 as an agency of the Ontario government, the Walkerton Clean Water Centre provides education and training on drinking water systems and advises on research needed to maintain high-quality, safe drinking water.

Ontario, became a flashpoint for the poor water services in many remote and distant communities. Kashechewan was established by the federal government in 1958 on the flood plain of the Albany River near the shore of James Bay because supply barges were unable to travel through rapids on the Albany River to a site further upstream preferred by the Cree people.

The present water treatment plant was built in 1996, with the intake located downstream from large sewage lagoons that leach continuously into the creek containing the water supply intake. Furthermore, tidal action from James Bay pushes waste-laden water into the creek, past the water intake. A further challenge was that the plant was run by local operators who did not have sufficient background or expertise to recognize serious problems.

A contributing factor to the contamination of Kashechewan's drinking water in October 2005 was the failure of a chlorine pump. An emergency backup system should have taken over but did not because it had not been connected. In most modern plants, an emergency paging system alerts operators to a system malfunction, but such a system had not been installed. Thus, the sequence of events leading to contamination was similar to that at Walkerton, where chlorine equipment was not working and operators did not have the necessary training to realize the implications.

On 12 October 2005, Health Canada discovered unacceptable levels of E. coli in the treatment system but did not alert the band office for two days. Once a qualified contractor arrived at the treatment plant, it took about six hours to have the equipment operating properly, and safe potable water was being provided by 22 October. Nevertheless, nearly 1,000 community members experienced negative side effects. All 1,900 residents needed vaccinations for hepatitis A and B. Furthermore, many residents had scabies and impetigo because of the ongoing poor water quality, conditions exacerbated by high chlorine levels.

A further concern was that in 2003, the Ontario Clean Water Agency had alerted the government to problems with the water treatment system at Kashechewan. The next year, the provincial Minister of Health and the Minister of Community Safety visited the community, but no action was taken. The view of the provincial government was that because First Nation reserves are under the jurisdiction of the federal government, it was the responsibility of federal departments, specifically Health Canada and Indian and Northern Affairs.

The short-term solution was to evacuate residents needing treatment and care. The evacuation began on 26 October, and by 2 November 815 people had been removed to Ottawa (245), Sudbury (251), Cochrane (206), Timmins (50), Attawapiskat (43), and Moosonee (20). The cost of the evacuation was estimated at about $16 million. The provincial Minister of Natural Resources, David Ramsay, attributed the delay in evacuating people to 'a jurisdictional misunderstanding as to who should have the lead on this'. In addition to the evacuation, other interim steps were taken to ship bottled water to the community, send certified water treatment operators to assess and repair the treatment plant, assess the quantity and quality of the water in the river from which the treatment plant takes water, and assess the state of the sewage treatment lagoon.

Subsequently, the federal government examined various long-term solutions, including a $200 million package over five to seven years to reinforce a dyke to reduce flood damage vulnerability as well as to construct better drainage systems to protect low-lying areas; a $200 million project to relocate the community to Timmins; or a $500 million initiative to move the entire reserve to a new location on higher ground. At the end of July 2007, the federal Minister of Indian and Northern Affairs announced that his government would rebuild and redevelop the low-lying reserve on its present location, since relocation was too expensive.

Perspectives on the Environment

Flooding in Kashechewan in 2008

On 25 April 2008, Kashechewan was again evacuated, this time because of flooding. Hundreds of the most vulnerable residents were evacuated by plane and were housed in motels, hotels, or other types of accommodation in Cochrane, Greenstone, Kapuskasing, Hearst, Sault Ste Marie, and Thunder Bay. Subsequently, about 1,900 people were evacuated from Kashechewan and Fort Albany, with 1,000 sent to Stratford and nearby communities such as St Mary's, Mitchell, and Milverton. The plan was to accommodate them for up to three weeks, but many were able to return in less than a week.

This was the fourth evacuation caused by flooding for Kashechewan since 2004 and the first for Fort Albany.

It should be noted that as a result of the Walkerton experience in 2000, the federal government had initiated a First Nations Water Management Strategy in 2003, with a budget of $600 million spread over five years. The main purpose was to improve the quality and safety of drinking water on First Nation reserves through developing comprehensive policies, guidelines, and standards; educating on-reserve residents about drinking water issues; clarifying roles and responsibilities; building and upgrading water systems to meet standards; improving operation and maintenance; providing training to operators; and expanding water testing. The federal budget in 2008 included over $330 million for a renamed First Nations Water and Wastewater Management Action Plan for two years, and in the 2010 budget the plan was extended for two more years. Given what happened at Kashechewan and continuing conditions on many other First Nations reserves, it appears that much more needs to be done.

Perspectives on the Environment

Monitoring of Fresh Water on Reserves

. . . there are unacceptable gaps in the federal monitoring of fresh water—notably, that Environment Canada has water quality monitoring stations on only 12 of some 3,000 First Nations reserves.

—Commissioner of the Environment and Sustainable Development (2010: 2)

Supply Management, Demand Management, and Soft Path

Water security can be achieved through various approaches. The best known are supply management and demand management, but an emerging approach is called soft path. More details can be found in Gleick (2003), Brooks (2005), and Brandes and Brooks (2005).

Perspectives on the Environment

Uncertainty and Choices

Water problems (scarcity, flooding, pollution) have no single solution: options to deal with scarcity include supply augmentation through the mobilization of more resources through capital-intensive projects; efforts to conserve water; or redefining allocation to users. All these options have political and financial implications. They all come with risks, costs, and benefits, private or public, which strongly shape what solutions particular stakeholders are likely to push for.

—Molle (2007: 361)

Supply Management

Supply management is the traditional approach. When a water shortage is anticipated, the solution is to develop a new source of supply, normally accomplished through either augmenting an existing supply (e.g., raising the height of a dam in order to be able to impound more water) or developing a new supply (e.g., a new dam and reservoir, new wells, a pipeline to a new source [a lake or river], a desalinization plant). The rationale is that populations will grow and the economy will expand and each requires additional water supplies. Without incremental water, human well-being and development will be impeded.

Supply management has served societies well when it has resulted in an adequate supply of suitable-quality water available to meet demands. There are also some downsides. If people believe that additional supplies will always be found to meet demands, little incentive exists to avoid wasteful water use or to adopt water conservation measures. One consequence is that a society may invest more money than would

Tributary joining the Albany River at Kashechewan, Ontario.

Dyke and drain control between the Albany River and Kashechewan, Ontario.

New water treatment plant beside the dyke, Kashechewan, Ontario.

otherwise be required. Such additional costs may be significant, since the low-cost sources have usually already been developed. Another consequence is that the construction of new dams and reservoirs or pipelines may have significant environmental or social impacts at local and regional scales.

Given the above mix of benefits and limitations, it is worthwhile to consider other approaches that can be used in combination or in place of supply management.

Demand Management

The basic approach in supply management is to manipulate the natural system to create new sources of supply. The approach in **demand management** is to influence human behaviour so that less water is used.

Various methods can be used to influence human behaviour in water use. The most basic is pricing. It signals to users that water has a cost and that by using less water, people can save money. Volume-based pricing can be designed so that as consumers use more water, they pay an increasingly higher per unit charge. For this system to work, all use, whether in homes, offices, institutions, manufacturing plants, or farms, has to be metered. However, significantly higher prices for water could disadvantage the poorest members of a society, who may find it difficult to pay for the minimal amounts of water to meet basic needs. To avoid inequities, other social policies must be in place to ensure that vulnerable people are not in jeopardy. This is particularly important with regard to water, because there is no alternative to water for meeting basic needs.

Another incentive is to offer price rebates for the purchase and installation of water-saving devices such as low-flow showerheads or toilets. Low-flow toilets, for example, use on average about 75 per cent less water per flush than a 'regular' toilet. Since showers and baths account for 35 per cent and toilet flushing 30 per cent of water used in Canadian households, reducing volumes of water for them can lead to significant reductions in water used. Further savings can be realized if previously used water, or grey water, is used for waste disposal, rather than using drinking-quality water for toilet flushing.

A third incentive is to restrict outside water use (watering of lawns, flower beds) by regulation during the hottest months when water use peaks. If the high peaks of water use in summer can be reduced, there would be no need to invest in additional capacity only required during a few months each year.

A fourth measure in demand management is to inform and educate water users so that, over time, they reduce water use. For example, people can be educated to turn off the faucet when brushing their teeth except when rinsing the toothbrush or to turn off the water in a shower except when initially soaking or rinsing off soap. Other practices include running a dishwasher or washing machine only with a full load.

Perspectives on the Environment
How to Reduce Water Use in the Home

So, where do we start? The first step is to identify where we use water in the home. Then we need to decide on what to do to reduce the amount of water we use, either by eliminating wasteful practices and habits, or improving the efficiency of our water using fixtures and devices. Since we waste so much, this should be a relatively easy and painless process. The prime area to target is the bathroom, where nearly 65 per cent of all indoor water use occurs. . . . Based on the three rules of water conservation—reduce, repair, and retrofit—a typical household can reduce water consumption by 40 per cent or more, with no effect on lifestyle.

—Environment Canada (2009a: 10–11)

Demand management is not a new concept, but it has been introduced often as a second-level approach after decisions focused on supply management have been taken. An ideal approach would combine the two in a well-integrated system.

Soft Path

The **soft path** approach extends demand management. Soft path aims to improve water use efficiency by challenging basic patterns of consumption (Brandes and Brooks, 2005; Brooks and Holtz, 2009; Brooks et al., 2009). While demand management emphasizes the question of 'how', or how to do the same with less water, the soft path approach addresses the question of 'why', or why water is even used for a function.

Several examples illustrate the significance of asking 'why'. In the words of Brandes and Brookes (2005: 9):

- Why . . . do we use water to carry away our waste? Demand management would urge low-flow toilets, but waterless systems are available—perhaps not for homes (because of the need for regular maintenance) but certainly for larger buildings.
- Why do we use half the potable water piped to a house in the summer for watering lawns and gardens—and sidewalks? Demand management would urge more efficient sprinklers with automatic shut-offs, maybe even water restrictions. The soft path goes further: recycling water from bathtubs and washing machines or, better yet, drought-resistant greenery that requires little or no watering once it is established.

The soft path approach is based on four basic principles (ibid., 10–13):

1. *Water is treated as a service rather than as an end.* In the soft path, water is not viewed as the final product, other than for a few human uses (drinking, washing)

and for support of ecosystems. Instead, water is viewed as a means to accomplish specific functions, including sanitation, farm production, and yard maintenance. This principle moves managers to think about alternative ways that services traditionally supported by water might be achieved. For example, the end is not to flush toilets or irrigate crops but to dispose of wastes or to grow food.

2. *Ecological sustainability is fundamental.* Ecosystems are viewed as legitimate users of water and also as one foundation of economies. Consequently, ecosystem health and ecosystem resilience must be considered when calculating the cost–benefit ratio of solutions to meet water demand. One result is that environmental needs are identified from the outset, and the amount of water required to satisfy such needs is subtracted from what is available to meet human needs.

3. *Quality of delivered water is matched to an end-use requirement.* While high-quality water is necessary for human consumption, the quality may vary significantly for other uses. The soft path seeks to match water quality to what is needed to accommodate an end use. One implication is recognition of 'cascading water systems': waste water from one use becomes the supply for another use needing less stringent quality. Examples include using water from a washing machine on a garden or shower or bath water for toilet flushing.

4. *Determine the desired future condition, and plan back to the present.* Conventional planning for water takes the present as its starting point. Future needs are projected, and then decisions are taken to meet these needs. The soft path focuses not on the most probable future but instead on the most desirable future, normally characterized by sustainability characteristics or criteria. Once the attributes of the desirable future are defined, decisions are taken about the most appropriate means to meet desired ends—without assuming that the way water is used now will be the same in the anticipated desirable future. This approach is referred to as backcasting, discussed in Chapter 5.

Brooks and Holtz (2009: 164–6) examine how soft path analysis was applied in Nova Scotia's Annapolis Valley. They concluded that current water use practices mean that available surface water will be inadequate to meet annual demand at least once every 12 years, and nearly every second year during summer seasons. Also, groundwater will not meet annual demands in two years out of every five. Based on that assessment, it was proposed that a mix of demand management measures, such as high-efficiency technologies (low-flush toilets; crop, golf course, and lawn irrigation; industrial procedures), drip irrigation or high-efficiency sprinklers, repair of leaks in municipal water mains, and capping of artesian wells, along with soft path measures, such as waterless technologies or practices (toilets, cooling systems, industrial systems), rainwater/runoff storage, and water/wastewater recycling and reuse, be introduced. They also offer the following insight:

> The analysis underlying soft path planning does not generally yield a single, best path. Different policy and program combinations will lead us to the desired future. Soft path analysis can identify possible paths, describe the advantages and disadvantages (where quantifiable, the benefits and costs) and determine the likely social appeal. It is up to the society at large, with decision making that employs community consultation and participation, to choose the path most appropriate to its values. (Ibid., 163)

Water security, both quantity and quality, should be achieved through a mix of supply management, demand management, and soft path approaches. The goal should be to use more of the tools of demand management and soft path approaches as part of an integrated strategy that includes supply management. The key is to use creative and innovative ways of moving beyond reliance on supply management.

Virtual Water

Growing water scarcity, whether shortages of quantity or inadequate quality, may damage economic development, human livelihoods, and well-being. A relevant concept is **virtual water**, which has grown out of recognition of the importance of water for agriculture and food production. Aldaya et al. (2010: 942) have explained the virtual water concept in the following manner:

> The virtual water content of a product (a commodity, good or service) refers to the volume of water used in its production . . . virtual water 'trade' represents the amount of water embedded in traded products. A nation can preserve its domestic water resources by importing water intensive products instead of producing them domestically. . . . Thus, virtual water 'import' is increasingly perceived as an alternative source of water as well as an opportunity to preserve environmental flows in water-stressed nations, and is slowly changing the prevailing paradigms of water and food security.

By pursuing a virtual water strategy, national governments in water-stressed countries can plan to meet food security needs even if their nation has a limited water endowment. For example, the net virtual water import by Egypt, as a percentage of its own water resources, has been calculated to be about 23 per cent (El-Sadek, 2010: 2445). Given that a water shortage within that country has been a barrier to expanding cropland, Egypt benefits by importing crops that require significant amounts of water to grow. However, as El-Sadek comments, before adopting an explicit virtual water approach, 'Egypt needs to be assured that it can have fair and secure trade with water-abundant nations' (ibid.).

Perspectives on the Environment
Singapore and Virtual Water

Singapore has only 5 per cent of the water it needs. Yet there is no hint of water shortages, nor of the constrained economic development that many feel ought to be inevitable for a seriously water-short island economy. Ninety per cent of the total water need is brought in through trade in food commodities. The other non-native 5 per cent has until recently been imported across the straits from Malaysia. This dependence is now being reduced by investment in desalinization, an increasingly affordable technology.

—Allan (2011: 53)

Perspectives on the Environment
Limitations of the Virtual Water Concept

Estimates of 'virtual water flows' are helpful in generating public awareness regarding the volume of water required to support production and consumption activities. However, the true policy relevance is gained only by considering information regarding the scarcity of water . . . in a given region or country. . . . The policy relevance . . . will be greater where scarcity values (opportunity costs) are substantial.

—Wichelns (2010: 2204)

A virtual water strategy does have the potential to ameliorate water shortages at a national level. Nevertheless, as with all concepts or strategies, and as already indicated above, it has limitations. First, if a nation decided to cut back significantly on domestic production of crops demanding a large amount of water and import them from other countries, many local farmers would likely lose their livelihoods. One outcome could be migration of poor rural people into urban areas, where they would probably become part of a growing marginalized group. Second, a relatively poor nation would not likely have the foreign currency necessary to purchase food products requiring significant water inputs. Third, a nation might be reluctant to become dependent on other countries for food needs. Liu et al. (2007: 86) have commented that with regard to China: 'For food security, the government pays more attention to food self-sufficiency than to water use efficiency. Food self-sufficiency is overwhelmingly favoured by the Chinese government, which regards reliance on international food markets as a threat to domestic security.' And fourth, use of virtual water might mask or obscure the reality of in-country water shortages and lead to delays or inaction regarding policy changes providing environmental and social as well as economic benefits.

Nevertheless, 'virtual water' provides an option for consideration. For example, a virtual water assessment was conducted by Brown et al. in British Columbia to understand the virtual water requirements and contents for crops and livestock within watersheds of wet and dry regions to create a foundation for water conservation management strategies. The Okanagan Basin was chosen as a dry region and the Lower Fraser Valley as a wet region. The virtual water content in both areas related to fruits was higher compared to the average for Canada, and was 50 per cent higher compared to global averages. Regarding grain and field crops, 55 per cent were above Canadian averages and 68 per cent were above global averages. Brown et al. (2009: 2694) concluded that 'Some major decisions will need to be made on how to reduce water consumption in order to accommodate future anticipated growth. The data generated is a first step in providing science based information to assist decision makers in strategic choices of reallocation and conservation of water use.'

Perspectives on the Environment
Professor Tony Allan and Virtual Water

Professor John Anthony Allan of King's College, London and the School of Oriental and African Studies received the 2008 Stockholm Water Prize for his pioneering work in developing the concept of 'virtual water'. The Stockholm Water Prize, established in 1990, is awarded annually by the Stockholm Water Foundation. It is conferred for 'outstanding water-related activities'. The recipient receives US$150,000 and a crystal sculpture. Professor Allan received the prize from King Carl Gustaf XVI of Sweden.

What are your views on 'virtual water'? Does it offer promise to address water security challenges? Or does it have the potential to disadvantage people already vulnerable and benefit those who enjoy a relatively high standard of living?

Water Footprints

If 'virtual water' relates to the volume of water used to produce a commodity, good, or service, a **water footprint** serves as an indicator of water consumption by tracking both direct and indirect water use by a consumer or a product (Water Footprint Network, 2011). As Hockstra and Chapagain (2007: 36) observed, the water footprint is analogous to the ecological footprint, discussed in Chapter 1. More specifically, they commented that 'The water footprint of a nation is defined as the total volume of freshwater . . . used to produce the goods and services consumed by the people of the nation. Since not all goods consumed in one particular country are produced in that country, the water footprint consists of two parts: use of domestic water resources and use of water outside the borders of the country.'

Hockstra and Chapagain calculated that the global water footprint is 7,450 Gm^3/year (where G stands for Gigameters, or 10^9). In terms of direct factors determining a water footprint, they identify (1) volume of consumption (related to gross national income), (2) consumption patterns (such as high versus low meat consumption), (3) climate, and (4) agricultural practice (water use efficiency). The relative importance of the four factors varies from country to country. To illustrate, at the start of the twenty-first century, the footprint of the US was high (2,480 m^3/capita/year) due to high meat consumption and high consumption of industrial products. Iran also had a relatively high footprint (1,624 m^3/capita/year), mainly due to low crop production yields and high evapotranspiration rates.

In terms of water footprint measured by total water use, India (987 Gm^3/year, or 13 per cent), China (883 Gm^3/year, or 12 per cent), and the US (696 Gm^3/year, or 9 per cent) are the largest consumers of global water resources; Canada's total footprint is 62.80 Gm^3/year. However, if the measure is water use per capita, tracking both direct and indirect water use by a consumer or a product, the largest footprint is that of the US, as noted above, followed by Italy (2,332 m^3/capita/year) and Canada (2,049 m^3/capita/year). In contrast, India's per capita footprint is 980 m^3/capita/year and for China it is 702 m^3/capita/year, both relatively small footprints.

Various options exist to reduce water footprints. Improved technology that reduces the amount of water needed per unit of product is one. A second is to change to consumption patterns that require less water, such as through reduced consumption of meat. Third, behaviour can be altered through a mix of pricing, raising awareness, labelling of products, or other incentives to encourage behaviour that uses less water. And, fourth, production can be shifted away from low productivity per unit of water to high productivity, through altering trading patterns and thereby achieving global water use efficiency.

Water Terrorism

In the context of water security, Gleick (2006: 481) notes that contemporary society relies on complex and sophisticated infrastructure systems to provide safe and reliable supplies and to remove waste water, especially in large cities. Such infrastructure, while vital for human welfare and economic development, is 'vulnerable to intentional disruption from war, intrastate violence, and, of more recent concern, terrorism'. He further observed that water infrastructure systems are attractive terrorist targets because 'there is no substitute for water.' Communities can also suffer if a dam and reservoir are destroyed, leading to flooding immediately downstream.

Terrorists can disrupt water infrastructure or related services by directly attacking the infrastructure or by contaminating water. Damage can come in the form of serious health implications, unusable water, or the destruction of infrastructure facilities and equipment. As Gleick (ibid., 482) commented, 'the typical scenario for a terrorist attack on domestic water supplies involves putting a chemical or biological agent into local water supplies or using conventional explosives to damage basic infrastructure such as pipelines, dams, and treatment plants.'

While water infrastructure is vulnerable, the difficulties of disrupting it through **water terrorism** are considerable. For example, most biological pathogens do not easily survive in water, and most chemicals have to be present in very large volumes to contaminate the supply. Treatment systems normally deal with biological agents or chemicals. In addition, most water systems contain redundancy in order to deal with equipment failure or routine maintenance. Notwithstanding such redundancy, the bombing of a major pipeline supplying water to Baghdad in 2003 illustrates that serious disruption and hardship can be caused. Another possible threat is using remotely controlled computers to interrupt the functioning of valves, pumps, and chemical processing equipment regulated by computer-based control systems. Such an attack is referred to as cyber-terrorism. During April 2003, a man used a computer along with a radio transmitter to gain control of a district wastewater system in Queensland, Australia, and released sewage into parks and rivers and onto property (Gleick, 2006: 492).

Managers must consider two matters: probability and consequences of a terrorist attack. Both require examination to calculate degree of risk. While vulnerability can never be eliminated, mitigation measures can reduce the probability of an attack as well as the consequences.

To reduce the probability of an attack, various initiatives can be undertaken. The most fundamental is to reduce or remove the basic motivation. Such an option is normally well beyond the scope or authority of resource managers, but it should always be kept in mind. However, attention normally focuses on protection and detection initiatives.

Protection involves limiting or denying access to facilities or at least to key or vulnerable parts of the facilities. This often involves 'gates, lights, and guards'. However, such actions are not practical for pipelines or aqueducts that extend over many kilometres, and other options must be considered for them. Detection can involve surveillance cameras or motion detectors, often at significant expense. By themselves, such tools cannot stop intruders, but they act as a deterrent.

In addition to protection and detection, society can prepare by ensuring that various emergency measures are in place. These measures can include public advisories, temporary shutdown of systems, information about alternative services or supplies, treatment for contamination (e.g., boil-water advisories), health interventions, and emergency response teams. A central element is to have a plan in place for an emergency response, as well as training sessions for key stakeholders, including water utility staff, law enforcement officers, community leaders, and the media.

A six-step process has been proposed to enhance security related to water (Ping, 2010: 59):

1. *Vulnerability assessment.* Evaluate the security related to people, equipment, and facilities.
2. *Risk management.* Estimate risk based on the probability of an attack occurring and methods and tools that could be used to attack a water system.
3. *Strengthening existing infrastructure.* Add layers of protective rings for a specific facility: e.g., *inner ring*: access controls for entrances and exits; *middle ring*: perimeter security, including anti-climb fences, motion detection sensors, alarms; *outer ring*: active guard patrols; *outermost ring*: early-warning monitoring systems.
4. *Designing security into future infrastructure.* Incorporate security considerations when designing new facilities rather than adding them as an afterthought.
5. *Contingency and response plans.* After assessing vulnerability and identifying risk, prepare responses related to anticipated emergency situations.
6. *Exercises.* Arrange pre-planned events to test alertness and readiness of staff and equipment, including simulated attacks on a facility by a group playing the role of terrorists.

Water as Hazard

Flooding

Humans settle adjacent to rivers and lakes for many reasons. Proximity provides access to potable water, a place to dispose wastes, and sometimes, a source of power and a means of transportation. The relatively flat land beside many rivers and lakes facilitates construction of roads, homes, and places of business. The aesthetic quality—serenity, beauty, natural appeal—of a river or lake view also often means that waterfront lots command a premium price. However, the term 'flood plain' exists for a reason. From time to time, rivers and lakes extend beyond their normal limits to cover adjacent areas or flood plains. Flooding is a normal hydrological function. Indeed, many species of flora and fauna depend on flooding to survive and flourish. In addition, humans often benefit from flooding, as when flood plains are enriched by the deposit of silt that then supports agriculture.

When flooding occurs, the result is often only a minor inconvenience, and this was usually the case when settlements were relatively small. However, as population concentrations on flood plains increase, the potential of flood damage goes up. Examples of major floods and serious associated damages in Canada include the floods in the lower Fraser River Valley in BC (1948); Manitoba (1950, 1997 [an estimated $300 million in damage], 2009 [an estimated $40 million in damage], 2011 [an estimated minimum of $550 million in damage]); Toronto (1954 [Hurricane Hazel, causing more than 80 deaths and millions of dollars in damage]); Fredericton (1973); Cambridge, Ontario, Maniwaki, Quebec, and Montreal (1974); the Saguenay River Valley, Quebec (1996 [10 deaths and $800 million in damage]); and eastern Ontario and Quebec as well as the St John River in New Brunswick in the spring of 2008. The latter flood was of similar magnitude to the one in 1973 and threatened 1,300 homes.

During 2011, serious flooding was experienced across Canada. Manitoba had the worst flooding since 1976, and in mid-May the provincial government deliberately breached a dyke adjacent to the Assiniboine River. This action was taken to reduce flood damage for many properties by inundating the farmland of relatively few people. The Premier announced that the province would be spending $60 million in aid to homeowners for damage to property and lost wages, and $115 million to repair infrastructure such as roads and dykes across the province. In Quebec, the worst flooding in 150 years occurred on the Richelieu River, and it was estimated that more than 1,000 people evacuated their homes and the province would spend $40 million in support. Finally, in late June and early July, the Souris River caused major flood damage in southern Saskatchewan, with downstream damage also occurring in North Dakota and Manitoba.

Internationally, the worst flooding in decades in the state of Queensland in Australia, starting in late November 2010 and continuing into January 2011, highlighted the danger and damage from flooding, especially when sudden flash floods occur. Another example of the danger was in mid-January 2011 in Brazil, when washed-out mountainsides resulted in mud slides that caused over 500 deaths.

Humans have various ways of reducing flood damage potential. *Structural approaches* modify the behaviour of the natural system by delaying or redirecting flood waters. Common methods are upstream dams and storage reservoirs, protective dykes or levees, and deepening or straightening river channels to increase their capacity. All these measures provide protection. However, because they are designed and built with a standard in mind, such as the magnitude of flooding that may occur once in 100 years, a flood of greater magnitude will eventually occur (such as a flood that occurs once every 200 or 500 years). If people perceive the structural measures as 'protecting' the flood plain, resulting in more development on it, then when the inevitable flood event greater than the design capacity of the structural measures does happen, potential and actual flood damage will be greater.

A *non-structural approach* focuses on modifying the behaviour of people. Methods include land-use zoning to restrict or prohibit development in flood-prone areas, relocation of existing flood-prone structures, information and education programs to alert people to the hazard of occupying flood plains, and insurance programs to help people deal with the costs of flood damage. The best strategies use a mix of structural and non-structural approaches. In the following discussion of the 1997 flood of the Red River in Manitoba, you will see how both structural and non-structural approaches are used.

Red River Flood, 1997

In late April and early May 1997, the Red River 'experienced a catastrophic regional flood that far surpassed any previous flood in the historical record' (Todhunter, 2001: 1263). Areas in Minnesota, North Dakota, and Manitoba had what the media called 'the Flood of the Century'. Total damages in the US part of the basin were calculated at US$4 billion, nearly US$3.6 billion of which was within the Grand Forks (North Dakota) and East Grand Forks (Minnesota) metropolitan area. Per capita, this event was the most costly flood for a major metropolitan area in the United States. Downstream, southern Manitoba experienced $300 million in damage. The Winnipeg area just avoided a catastrophe, with only 54 homes flooded. While there were no deaths, 28,000 rural residents had to evacuate their homes (Figure 11.4).

The Red River originates in the United States, in southern North Dakota/Minnesota, and flows northward, draining into Lake Winnipeg. It is 880 kilometres long, and the watershed is 290,000 km^2, including the Assiniboine River basin (163,000 km^2), which joins the Red River at the 'Forks' in Winnipeg (Brooks and Nielson, 2000). In Manitoba, the modern Red River is on the bed of glacial Lake Agassiz in an eroded valley up to 15 metres deep and 2,500 metres wide.

The river has a long history of flooding. The largest recorded flood occurred in 1826. Records also document major floods in 1852 and 1861. Subsequently, significant floods happened in 1950, 1979, 1996, 1997, and 2009. Various factors combined to cause the major flood of 1997: (1) high precipitation in the autumn of 1996, saturating the soil by the time of the winter freeze-up; (2) near-record levels of precipitation during the winter; (3) a long and unusually cold winter, leading to a high water content of the snowpack into the spring; and (4) after spring melting had started, a major blizzard over much of the basin on 5 April, with snow accumulations of up to 50 centimetres. These factors combined to produce ideal conditions for flooding.

During the flood, an area of about 2,000 km^2 and up to 40 km wide was covered by water from the Canada–US border to the southern edge of Winnipeg (Figure 11.4). For most rivers, flood waters are confined to the river valley, even when they spread out onto the flood plain. However, extensive flooding occurs in the Red River Valley because of the low capacity of the river to accommodate extreme flows as a consequence of the low valley gradient and shallowness of the valley. This situation is exacerbated because areas adjacent to the river are broad and flat, allowing the flood waters to spread out over many kilometres on each side of the river. The flood peaked at Emerson, the most southern community along the river in Manitoba, on 27–8 April and then at the Floodway inlet in Winnipeg on 3–4 May. This pattern was typical for the Red River, where floods usually rise and fall slowly over a period of several weeks.

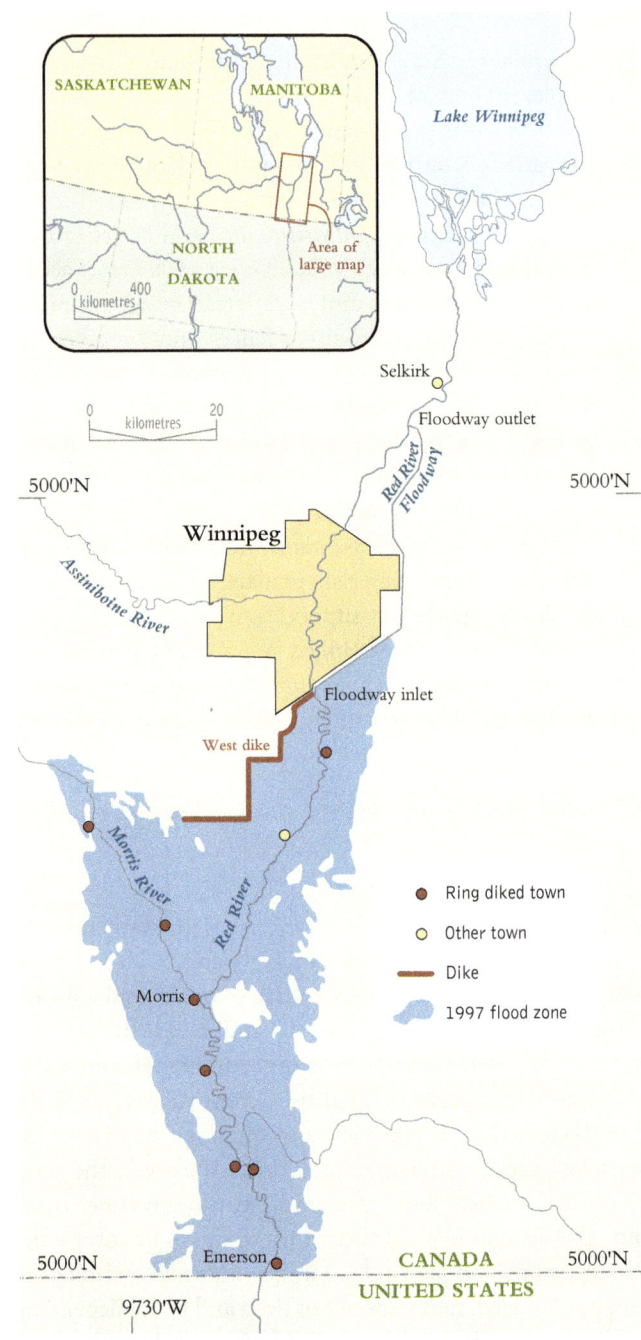

Figure 11.4 | Extent of Red River flooding, 1997. *Source: Brooks and Nielson (2000: 307).*

Flooding on the Red River in 1979.

However, the Floodway operated at maximum capacity (a once-in-160-years flood) during the 1997 flood, indicating that a flood of larger magnitude would have caused serious damage.

2. *Portage diversion.* This diversion is located at Portage La Prairie, 84 kilometres west of Winnipeg, and is an excavated channel, 29 kilometres long and 54 to 366 metres wide, that can divert up to 700 m^3/sec from the Assiniboine River to Lake Manitoba. Completed in 1970, the diversion reduces the flow of the Assiniboine River into Winnipeg during high water in the Red River.

3. *Shellmouth Dam and Reservoir.* Completed in 1972, this storage reservoir is able to hold water for later release at times of high flows.

4. *Earth dykes.* A system of dykes has been built along the Red, Assiniboine, and Seine rivers within Winnipeg. One hundred twenty kilometres of primary dykes have been built within the city, as well as the 32-km West Dyke that extends west and south from the Floodway.

5. *Ring dykes.* South of Winnipeg, large dykes encircle various small towns on the Red River flood plain. Smaller ring dykes surround individual buildings, such as homes and barns.

6. *Elevated roads and railway beds.* Roads and railway beds are raised one to two metres above the valley. This elevation protects transportation infrastructure from flooding, and the roads and rail lines also serve as dykes.

After the devastating flood in 1950, when a large part of Winnipeg was inundated, structural measures were undertaken to protect the city, smaller towns to the south, and transportation routes. These measures included:

1. *Red River Floodway.* Completed in 1968, this **floodway** is an excavated channel (48 kilometres long, 210 to 305 metres wide, and 9.1 metres average depth) that can divert up to 1,700 m^3/sec of the flow of the Red River around the eastern side of Winnipeg to rejoin the Red River downstream of the city. During the 1997 flood, the Floodway kept flooding in Winnipeg to a minor level.

There were both positive and negative outcomes from the 1997 Red River flood (Burn and Goel, 2001; Shrubsole, 2001; Todhunter, 2001). First, the Winnipeg Floodway prevented an estimated $6 to $10 million in damage in Winnipeg. Second, flood-fighting by local and provincial people resulted in

Flooding on the Red River in 1997.

Control gates for the Red River Floodway.

more than 8 million sandbags being filled in Winnipeg, more than 600,000 m³ of clay excavated to provide material for new temporary dykes, and 50 temporary dykes modified to become permanent structures. Third, in addition to filling sandbags, thousands of volunteers looked after children and provided food and other support to their neighbours. Fourth, non-government organizations provided critical assistance, such as $10,000 grants from the Red Cross to residents whose homes were damaged beyond repair.

Balancing these positive outcomes were some negative aspects. First, operation of the gates at the Winnipeg Floodway caused the upstream water level to rise, exacerbating flooding by 0.64 metres in some smaller upstream communities (south of Winnipeg). This led some rural upstream residents to conclude that their well-being had been sacrificed to protect the residents of Winnipeg. Second, the peak flow at the Floodway was underestimated by 1.5 to 1.7 metres. Furthermore, the flooding of the upstream community of Ste Agathe was a surprise, since the town had never been flooded before and thus a ring dyke had not been built. Eight other small communities on the flood plain but within ring dykes were not flooded. Third, some rural municipalities hesitated to spend their funds on fighting the flood until arrangements were clarified with the provincial government, since provincial law does not allow municipalities to have operating deficits. Fourth, some Aboriginal communities experienced problems because of confusion regarding agencies responsible for them.

The International Joint Commission (2000) conducted an inquiry and offered the following conclusions and recommendations: (1) although the 1997 flood was natural but rare, floods of the same or greater magnitude are possible; (2) both people and property in the Red River Valley will remain at risk to flooding until comprehensive, integrated, and binational solutions are developed; (3) a mix of structural and non-structural approaches is needed; (4) specific communities in both Canada and the US need to take flood damage initiatives; and (5) aspects of the ecosystem need more explicit attention, specifically, hazardous material needs to be more carefully controlled, and banned substances need to be removed from flood-prone locations.

Since the 1997 flood, the Canadian and Manitoba governments have spent more than $130 million on structural measures. Much of this amount ($110 million) has been for protective adjustments for 16 rural communities. In December 2003, the two governments announced an agreement to allocate $240 million for expansion of the Red River Floodway around Winnipeg. This expansion will allow an increase in the capacity of the existing 49-km channel to meet a flood event comparable to the one in 1826, calculated as a once-in-280-years occurrence. This will be realized by increasing the width of the diversion channel by about 110 metres and the depth by up to two metres. Upon completion

The inlet control structure during the spring of 2005.

The inlet control structure is situated on the Red River downstream from the Floodway channel entrance, June 2007.

An integral component of the Manitoba Floodway protection system, the outlet structure, located downstream of Lockport, Manitoba, was designed to dissipate energy from flood water as it returns to the Red River. (Photo from 2004.)

All three photos courtesy the Manitoba Floodway Authority.

of the structural work in 2009, protection was provided against a once-in-700-years flood. Work over several following years focused on upgrading bridges over the river and construction of the inlet control structure

The expansion was challenging, since it involved upgrading and improvements to 12 bridge crossings, the Floodway outlet, the West Dyke, utilities, and drainage services. There was potential to disrupt the groundwater table during the excavation, which moved more than 30 million m^3 of earth. An environmental impact assessment began in 2004 to determine adjustments to the design. This project, needing a decade to complete, was managed by the Manitoba Floodway Expansion Authority. Up-to-date details can be found at its website (www.floodwayauthority.mb.ca/floodfacts.html).

The spring 2009 flood was a reminder of the continuing natural hazard posed by the Red River. The 2009 event was the second-worst flood of the century in the southern portion of the Red River valley, while in the portion north of Winnipeg it was the worst flood of the century. The federal and provincial governments provided an estimated $40 million in aid, with the province spending additional money to buy out flood-prone properties.

As with the 1997 flood, conditions combined to cause the serious flooding: (1) a wet autumn in 2008 saturated the clay soil in the valley; (2) heavy snowfall in the upper (North Dakota) portion of the valley generated a lot of melt water flowing into Manitoba; and (3) a very cold spring in 2009 created severe ice jams north of Winnipeg, causing the river to back up and spill over its banks. Some homes were damaged by large chunks of ice driven into them.

Canadians will continue to live and work on flood plains for the reasons mentioned at the beginning of this section. In that context, the comments of Shrubsole (2001: 462) deserve consideration:

> The present practice of flood management in Canada is characterized by at least three realities. First, it is impossible to provide absolute protection to people and communities. Second, a mix of structural and non-structural adjustments that cover the entire range of protection, warning, response, and recovery is needed to effectively protect lives and property. Third, implementation of flood adjustments requires the effective participation of all levels of government and the public.

Droughts

If flooding represents situations with too much water, droughts represent the opposite problem—insufficient water. Flooding is immediate and apparent. The beginning or end of a **drought** is more difficult to determine, since droughts are a function of a lack of precipitation, temperature, evaporation, evapotranspiration, capacity of soil to retain moisture, and resilience of flora and fauna in dry conditions.

Consequently, as Gabriel and Kreutzwiser (1993) have noted, a significant challenge when seeking to identify 'drought-prone' areas is to define what is meant by a drought. As they noted, interpretations are based on causes and effects. Regarding those based on *causes*, a meteorological drought is due to a prolonged deficiency of precipitation, which reduces soil moisture. This type of drought can trigger a second type (hydrological drought, an effect that then becomes a cause), manifested by reduced stream flows and lowered **water table** and/or lake levels. In terms of *effects*, an agricultural drought results in reduced crop yields because of a lack of moisture. An urban drought happens when there is insufficient water, because of lower stream flows or water tables, to support all demands in the community.

Droughts, as Gabriel and Kreutzwiser (ibid., 119) explain, reduce the amount of water for use by depleting soil moisture and groundwater reserves as well as by lowering stream flows and lake levels. These reductions can start a 'depletion cycle': less than normal rainfall leading to low soil moisture, triggering demand for irrigation development, in turn depleting non-recharging surface and groundwater supplies. The high evapotranspiration rate associated with hot, dry periods also contributes to depleting soil moisture and surface water supplies, which are not restored to normal levels without unusually high rainfall.

What has been the experience with drought in Canada, where the Prairies, the interior of British Columbia, and southern Ontario are most vulnerable?

Droughts are most usually associated with the Prairie provinces, especially in that area of southern Alberta and Saskatchewan and extreme southwest Manitoba known as **Palliser's Triangle** (Figure 11.5). This area is named after Captain John Palliser, sent by the British government and the Royal Geographical Society to explore the territory between the Laurentian Shield and the Rocky Mountains between 1857 and 1860 and to determine the nature of the soil, its capacity for agriculture, the quantity of its timber, and the presence of coal or other minerals.

Palliser divided the area into two sections—a fertile belt and a semi-arid area. He considered the southern or semi-arid area unfit for settlement. Palliser commented that this area 'has even early in the season a dry patched look. . . . The grass is very short on these plains, and forms no turf, merely consisting of little wiry tufts. Much of the arid country is occupied by tracts of loose sand, which is constantly on the move before the prevailing winds' (Mackintosh, 1934: 11). Palliser concluded: 'There is no doubt that the prevalence of a hard clay soil derived from the cretaceous strata which bakes under the heat of the sun, has a great deal to do with the aridity of these plains, but it is primarily due more to want of moisture in the early spring' (ibid., 34). Thus, over

150 years ago, Palliser identified the drought-prone nature of the southern Prairies. His 'heads-up' was reinforced during both 2001 and 2002 when large areas of the Prairie provinces experienced drought conditions.

Figure 11.6 illustrates that 'in 2002, drought-stricken areas covered over three-quarters of the Prairies (including the northeastern part of British Columbia)' (Statistics Canada, 2003b: 13). While there were many impacts, the most pronounced was inadequate water to support agriculture. As the data in Table 11.3 show, yields of spring wheat, barley, and canola fell significantly during 2002 relative to the average yields between 1991 and 2000, which were non-drought years.

Livestock were also affected negatively. The greatest impact was in Alberta, where the inventory dropped by 605,000 cattle, a decrease of 10.4 per cent between January 2002 and January 2003. At the same time, declining supplies of cattle feed, a result of the drought conditions, pushed up feed prices and led many ranchers to reduce herds.

Another indicator of the drought conditions was the drying up of many dugouts (small human-made ponds), potholes (small natural ponds), and sloughs. By September 2002, 80 per cent of Prairie farms were in regions in which dugouts were half empty, and 20 per cent reported their dugouts were completely dry. The drying up of potholes and sloughs not only affected agriculture. They are also critically important habitat for migratory wildfowl, which were adversely affected (ibid., Table 10.4, regarding the effect of drought on wetlands).

While droughts represent an extreme condition related to water shortages, we also should be aware that the historical record indicates for some regions of Canada that water availability, even over many previous decades, may not be a good guide regarding what should be expected as 'normal' water availability in natural systems. For example, Wolfe et al. (2011) examined the record for more than 5,200 years at Lake Athabasca and came to some startling conclusions, as noted in the accompanying box.

Perspectives on the Environment

Historical Perspective on Natural Water Availability in the Athabasca River System

. . . a new 5,200-year record of Lake Athabasca water-level variations, which serves as a sensitive gauge of past changes in alpine-sourced river discharge, reveals that western Canadian society has developed during a rare period of unusually abundant water 'subsidized' by prior glacier expansion. As the 'alpine water tap' closes, much drier times are ahead. Future water availability is likely to become similar to the mid-Holocene when Lake Athabasca dropped 2–4 metres below the twentieth century mean. Regions dependent on high elevation runoff (i.e., western North America) must prepare to cope with impending water scarcity of a magnitude not yet experienced since European settlement.

—Wolfe et al. (2011: 1)

In Ontario, drought can occur in any season but is most likely in the summer when demand for water is usually the highest. Southwestern Ontario is most vulnerable, especially in the summer and early autumn. Extended dry periods for more than a month are unusual, but shorter droughts are not uncommon. For example, dry periods of at least seven consecutive days occur at least once a month during the agricultural growing season in southern Ontario, and short-term (10 to 20 days) dry spells occur every year. Longer droughts (more than four weeks) happen once in three years. One of the most severe and extensive droughts occurred in 1966 when during a 41-day period between mid-June and the end of July, most of southern Ontario received less than a quarter of normal rainfall. Another significant drought occurred in the summer of 1988 when southwestern Ontario received less than 40 per cent of the average precipitation from early May to mid-July.

Lake levels are affected by dry periods, and the Great Lakes illustrate this effect. The variation between minimum and maximum lake levels is 1.2 metres on Lake Superior, 1.8 metres on Lakes Huron and Erie, and two metres on Lake

Figure 11.5 | Palliser's Triangle. Source: Adapted from Bone (2005: 410).

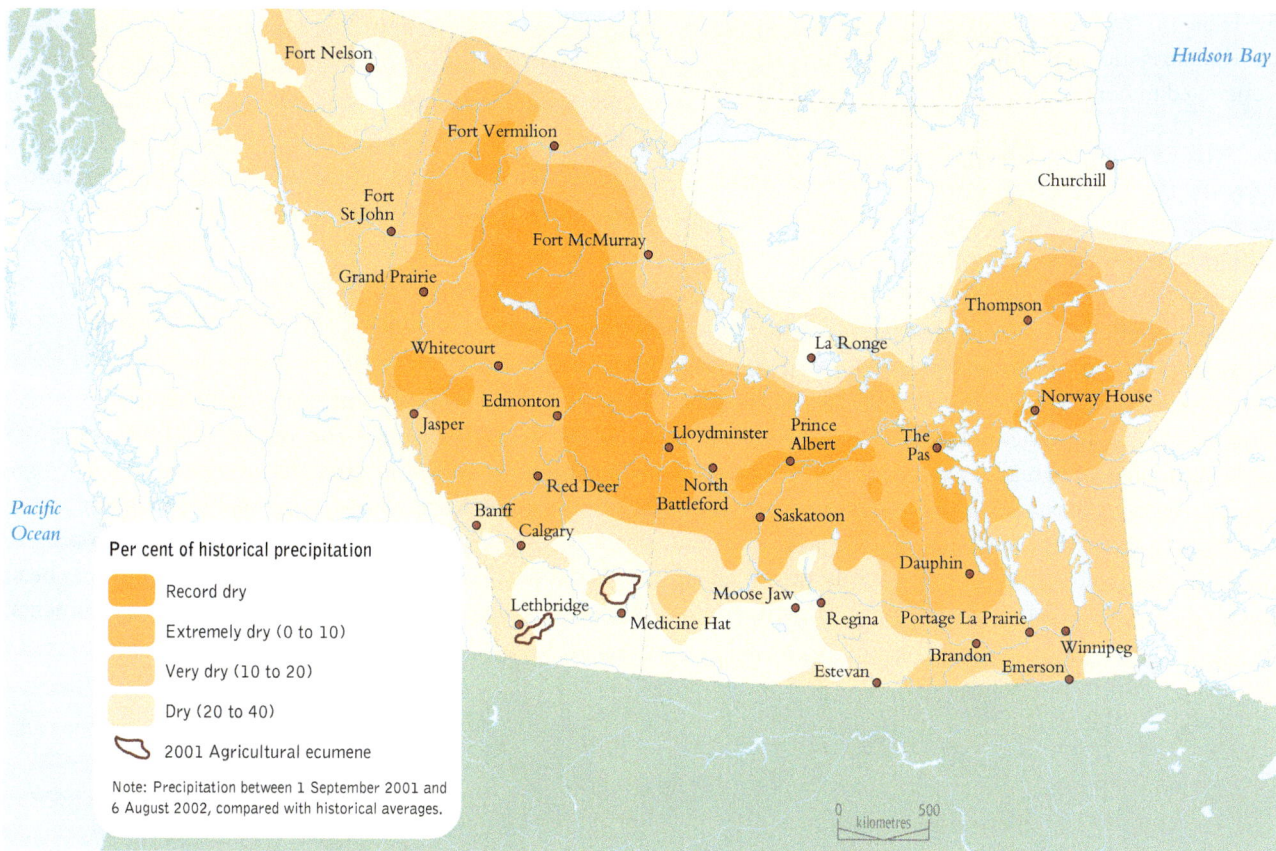

Figure 11.6 | Precipitation below historical averages, 2002. *Source*: Statistics Canada (2003b: 13).

Ontario. Low levels affect shipping, especially in terms of cargo tonnages, which have to be reduced so that ships do not run aground. To reduce the draft by only 2.5 centimetres requires a reduction of up to 90 tonnes on most ships and more than 180 tonnes on ships between 244 and 305 metres in length.

In Ontario, many streams are almost totally supplied by groundwater discharges during low rainfall periods. In average conditions, groundwater discharge provides 20 per cent of the water for streams and rivers in most of Ontario. For some rivers or streams, the contribution can be up to 60 per cent, extending up to 100 per cent in the summer months. Thus, depletion of groundwater reserves resulting from drought can have a serious impact on surface flows, especially for smaller streams. The lowering of water tables due to drought can also lead to the drying up of wells dependent on shallow aquifers.

Heritage Rivers

The mission of the Canadian Heritage Rivers Program (CHRP) is to 'develop a river conservation program that is nationally valued, internationally recognized, and reflects the significance of rivers in the identity, and history of Canada; and ensure that the natural, cultural, and recreational values for which rivers are designated are managed in a sustainable manner'. The CHRP is overseen by a board, with representatives from the federal, provincial, and territorial governments.

Table 11.3 | Crop Yields and Insurance Payments, 2002 Percentage Variation from 1991–2000 Average, Prairie Provinces

Province	Spring Wheat	Barley	Canola	Crop Insurance Payment
Alberta	−29.4	−26.8	−13.0	399.1
Saskatchewan	−32.0	−34.1	−21.4	224.1
Manitoba	2.6	7.6	3.9	69.9

Source: Adapted from Statistics Canada (2003b: 13).

The French (French River Provincial Park in Ontario) and Alsek (Kluane National Park in Yukon) were the first two rivers designated under this program in February 1986. By 2009, 41 rivers had been nominated, totalling nearly 11,000 kilometres, and 37 had been formally designated, totalling just over 9,000 kilometres (Figure 11.7).

For almost a decade after the beginning of the program, designated rivers were located in federal or provincial parks, the territories, or in areas within provinces with relatively few people. Designation thus involved rivers primarily on Crown land, avoiding the complication of having to deal with private landowners and municipalities, often suspicious of the Heritage Rivers Program, viewing it as possible intrusion into property or municipal rights.

However, in 1994, the entire Grand River basin, located in southern Ontario and with most of its land in private ownership, was designated, and that experience is explained by Barbara Veale in her guest statement below. Other rivers in highly settled areas have since been designated, such as the Humber River in Toronto and the Thames River and the Detroit River in southwestern Ontario.

In the Fraser River basin in British Columbia, a Charter for Sustainability, which outlines what is needed to achieve sustainability for the Fraser River and its watershed, has been signed by federal, provincial, and municipal governments,

Perspectives on the Environment

Rivers as Heritage

Rivers teach valuable lessons about renewal. It is said, and it is true, that you cannot enter a river at the same spot twice; because, of course, the river 'rolls along'. This very character of rivers generates health and well-being. It also connects one part of the waterway to another.

—Harry Collins, chairperson, Canadian Heritage Rivers Board, 2002

Figure 11.7 | Heritage Rivers in Canada, 2009. Source: Based on Canadian Heritage Rivers Board (2010: 7).

Grand River, Cambridge, Ontario. A former factory has become a park setting adjacent to the river, making the river accessible to the public while retaining a sense of heritage.

as well as by First Nations and other organizations. A Fraser Basin Council, a not-for-profit organization, was created and provides oversight for activities to achieve the future identified by the charter.

These examples illustrate the types of initiatives being taken as part of the management strategies for designated heritage rivers. All are oriented to protecting the integrity and health of the ecosystems, ranging from biophysical to cultural components. In September 2007, a 10-year Canadian Heritage Rivers System strategic plan was approved, and the plan will be in place until March 2018. Four priorities are in this plan: (1) build a comprehensive and representative system that recognizes Canada's river heritage; (2) conserve the natural, cultural, and recreational values and integrity of designated Canadian heritage rivers; (3) engage communities and partners to maximize the full range of benefits associated with the Canadian Heritage Rivers Program; and (4) foster excellence in river management.

Guest Statement

How Becoming a Heritage River Can Influence Water Management

Barbara Veale

The Grand River is located in the heart of southern Ontario. Its rich diversity illustrates key elements in the history and development of Canada, and many of the river-related heritage resources remain intact today. These resources include the historical buildings and settlements in the watershed, the land use associated with different ethnic groups, such as Mennonite farmers still using horse-drawn ploughs, and the Six Nations area near Brantford. In addition, the river and its tributaries provide a broad range of excellent recreational opportunities.

In 1987, the Grand River Conservation Authority spearheaded a participatory process to have the Grand River and its major tributaries declared a Canadian heritage river. This was achieved in 1994.

The designation of the Grand River as a Canadian heritage river marked the beginning of a second generation of **heritage rivers**. Prior to 1990, almost all nominated rivers either were within protected areas or were short sections of larger rivers. In contrast, the Grand River is located in one of the most densely populated and fastest-growing parts of Canada where almost all lands are privately owned and managed within a complex multi-agency, multi-jurisdictional setting. The management plan presented to the Canadian Heritage Rivers Board—The Grand Strategy—deviated from past management plans for heritage rivers in that it provided a framework for an ongoing, community-based watershed approach sustained by consensus, co-operation, and commitment. It was based on a common vision, beliefs, values, principles, and goals, and the designation of the Grand River set a precedent for the Canadian Heritage Rivers Board to accept other rivers in highly settled areas of Canada where river management is complex and shared among all levels of government, First Nations, and non-government entities.

Since 1994, The Grand Strategy has evolved into a shared management approach for integrated watershed management. Management partners, including federal and provincial governments, watershed municipalities, First Nations, non-government groups and organizations, and educational institutions, jointly identify critical resources issues, develop creative solutions, pool resources, implement actions, monitor results, and evaluate progress on an ongoing basis.

Under the umbrella of The Grand Strategy, the Grand River Conservation Authority works with its partners to address a wide array of existing and emerging resource issues and to identify priorities for action. The philosophy is that everyone who shares the resources of the Grand River watershed is encouraged to be a part of a collective effort to address key watershed issues:

- reducing non-point sources of pollution in rural areas;
- pursuing excellence in wastewater treatment;
- slowing increases in water use and advocating the wise use of water;
- protecting groundwater resources;
- developing long-term water quality and water budget/supply plans;

- maintaining the water control system;
- implementing a watershed-wide fisheries management plan;
- developing community-based plans that advance forest management, wildlife management, and natural heritage management;
- developing community riverfront plans;
- developing the watershed's potential for outdoor recreation, cultural education, and eco-tourism;
- building a sense of community around the river and celebrating successes.

In keeping with Canadian Heritage Rivers System requirements, the first Ten-Year Monitoring Report for the Grand River was completed in 2004. This assessment was undertaken within the participatory structure created by The Grand Strategy and provided participants with the opportunity to revisit and reaffirm the vision, values, principles, and goals of The Grand Strategy. It revealed that great strides had been made in increasing awareness of the river as a community asset, strengthening heritage conservation, and developing new outdoor recreational opportunities. The general level of concern about the river had increased, resulting in more stewardship activities taking place.

Despite many gains, it was noted that a continuing challenge was nurturing the process of collaborative research, planning, education, action, and monitoring, especially given significant staff turnover in partner agencies and organizations. Participants voiced concern about emerging water quality issues and degradation of ecosystem health.

Today, mounting stresses on water resources place even more emphasis on the need to grow partnerships and strengthen collaborative management efforts. According to the 2006 census, the Regional Municipality of Waterloo was the fourth fastest-growing urban area in the province, with a growth rate of nearly 9 per cent between the years 2001 and 2006. The Growth Plan for the Greater Golden Horseshoe, released by the government of Ontario in 2006, anticipates rapid population growth in the watershed's five cities (anywhere from 52 to 60 per cent between 2001 and 2031) and associated land intensification. The Greenbelt Act, passed in 2005, imposes rigid planning policies limiting growth in the Greater Toronto Area. The greenbelt area runs almost entirely along the eastern boundary of the Grand River watershed. There is concern that the greenbelt will cause 'leapfrogging' of development from the Toronto–Hamilton region into the Grand River watershed, creating even more demands on land and water resources.

The Clean Water Act of 2006 adds another layer of complexity to the governance of water in Ontario. The Clean Water Act creates source water protection authorities charged with developing plans for source protection areas, which are defined on a watershed basis. In most cases, source protection areas coincide with the boundaries of conservation authorities and are combined to create larger source protection regions. Within each region, one conservation authority is

Kayaking on the Eramosa River at Rockwood (left); historic bridge, Irvine Creek at Elora Gorge (above). An updated water management plan for improving river water quality in the Grand River watershed provides the dual benefits of supporting healthy aquatic ecosystems and ensuring that the river's key recreational and cultural resources are appreciated and enjoyed for years to come.

appointed as the source water protection authority. The Grand River Conservation Authority is the authority for the Lake Erie region. A Source Protection Committee consisting of members from watershed municipalities and various sectors, including agriculture, business, industry, environment, and health, has been selected to guide the development of the source water protection plans for each watershed within the Lake Erie Source Protection Region. Once completed, the plans will be implemented through municipal planning documents.

While the watershed has recovered from years of abuse and misuse through the efforts of the Grand River Conservation Authority and its partners, the river system is now showing signs of stress. Detrimental impacts on water quality, river flow, and water availability are visible. The continued good health of the watershed is at risk. The collective challenge is to find effective solutions and to stretch limited dollars in support of actions to maintain river health and resiliency. Finding suitable means for tackling emerging water issues requires renewed effort.

In late 2009, a multi-agency steering committee was formed to oversee the development of a plan to update the Grand River Basin Water Management Plan, 1982. This revised plan will give specific consideration to sustainable water supply, reduced flood damage potential, and improvements to water quality in order to maintain and improve the health of the Grand River and Lake Erie. Expected to be completed by 2013, the new plan will provide a road map for the future and align the efforts of all partners by:

- identifying the capacity of the Grand River to accommodate more wastewater and non-point pollution, while improving water quality;
- identifying priority areas for water quality improvements and recommending rehabilitation and restoration works;
- identifying priority areas for groundwater recharge protection;
- determining water sources and water needs by sector, including ecosystems (taking into account the impacts of climate change, growth, and cumulative impacts), recommending opportunities for reducing water demand, and identifying infrastructure requirements;
- identifying options to improve river hydrology and health;
- confirming operating procedures for reservoir operations to address flooding, water supply, and wastewater assimilation under changing climatic conditions and ecosystem needs;
- identifying options to reduce flood risk;
- understanding the socio-economic and environmental trade-offs associated with the recommended actions.

The Grand Strategy establishes a firm foundation for collaboration from which new partnerships and approaches can evolve within the Grand River watershed and illustrates that strong science and technical tools must be combined with an ongoing participatory approach that allows people to regard water management within the broadest societal context. In the face of unprecedented growth and increasing governance complexity, this integration is essential for motivating collective action in a timely fashion to resolve current and anticipated water issues. Focusing on measures that improve water quality and sustain water supply in addition to actions that conserve and interpret the watershed's natural and human heritage will ensure that the Grand River and its major tributaries will continue to deserve their national status as Canadian heritage rivers.

Barbara Veale is co-ordinator of policy planning and partnerships, Grand River Conservation Authority, Cambridge, Ontario.

Canada has not been alone in showing interest in heritage rivers. To illustrate experience in one other country, Shuheng Li provides a perspective from China related to the Grand Canal. The situation she outlines in China offers an interesting contrast to that of the Grand River in Ontario.

Hydrosolidarity

The term **hydrosolidarity** refers to an approach that recognizes interconnections among aquatic, terrestrial, and other resource systems, leading to management that is integrated, participative, collaborative, co-ordinated, and shared, whether at local, state, national, or international levels. The challenges of achieving hydrosolidarity increase when moving from local to international situations. Notwithstanding such difficulties, hydrosolidarity is most often evoked as necessary in international settings (International Water Resources Association, 2000). Hydrosolidarity stands in contrast to more traditional approaches to international rivers or lakes in which sovereign states claim control over the water within their boundaries without regard for the implications for countries sharing the resource, especially as downstream users.

Elements of best practice associated with hydrosolidarity reflect the reality of aquatic systems, including that: (1) water flows downhill, leading to differing upstream and downstream interests in a river basin or catchment; (2) interconnections exist between water and land systems, meaning that land-based activities can have significant implications for water quantity and quality; and (3) the multiple uses that water can serve range from drinking, to crop and industrial production, to supporting migratory bird habitats and recreation. As a result, best practice regarding hydrosolidarity includes management by (1) using river basins or watersheds as the spatial unit for planning and management; (2) ensuring attention to upstream–downstream issues; (3) recognizing the interrelationships among water, land, and other resource systems; (4) engaging a range of stakeholders in a collaborative and participatory manner; and (5) acknowledging the needs of biophysical systems as well as of humans.

Consistent with hydrosolidarity is the concept of **integrated water resource management** (IWRM), which the Global Water Partnership (2000: 22) defined as 'a process which promotes the co-ordinated development and management of water, land, and related resources in order to maximize the resultant economic and social welfare in an equitable manner without compromising the sustainability of vital ecosystems'.

The motivation for IWRM is to overcome the challenges arising from various groups having interest in and organizations being responsible for water and related resources, along with the reality that they each often focus only on their own interests and responsibilities. One result, for example, can be a ministry of agriculture providing support for farmers to drain wetlands in order to expand land in agricultural production, while a ministry of natural resources in the same jurisdiction provides support to farmers to expand wetlands

International Guest Statement
The Grand Canal of China
Shuheng Li

The Grand Canal of China (also known as the Jing-Hang Grand Canal) is the longest artificial waterway in the world. Its length is 1,794 kilometres, extending from Beijing in the north to Hangzhou, Zhejiang Province, in the south, via Tianjin, Hebei, Shandong, and Jiangsu provinces (Figure 11.8). It connects five large rivers: the Haihe, Yellow, Huaihe, Qiantang, and Yangtze. In a country dominated by west–east-flowing rivers, the Grand Canal provides a north–south connection among the river systems.

The canal we see today was built section by section in different areas and dynasties. The oldest part dates back to 486 BC. A canal, called Han Gou, was first cut near Yangzhou, Jiangsu Province, to guide the waters of the Yangtze River north to the Huaihe River by using existing waterways, lakes, and marshes. It was lengthened significantly during the Sui Dynasty. The shift of China's economic centre from the north to the south prompted Emperor Sui Yang, between AD 605 and 610, to extend the canal to link the productive southern region to the nation's capital, Luoyang, in the north. During the Yuan Dynasty of the thirteenth century, the capital was moved to Beijing. The canal was rebuilt and became what is now known as the Grand Canal.

With a history and culture dating back about 2,500 years, the Grand Canal has served as the major transportation artery between north and south China. In terms of improving communication between the north and south and promoting economic and cultural exchanges, the canal, like the Great Wall, was an astonishingly huge project in the history of Chinese civilization. Once the canal opened, many cities along the waterway developed quickly. Communities along the

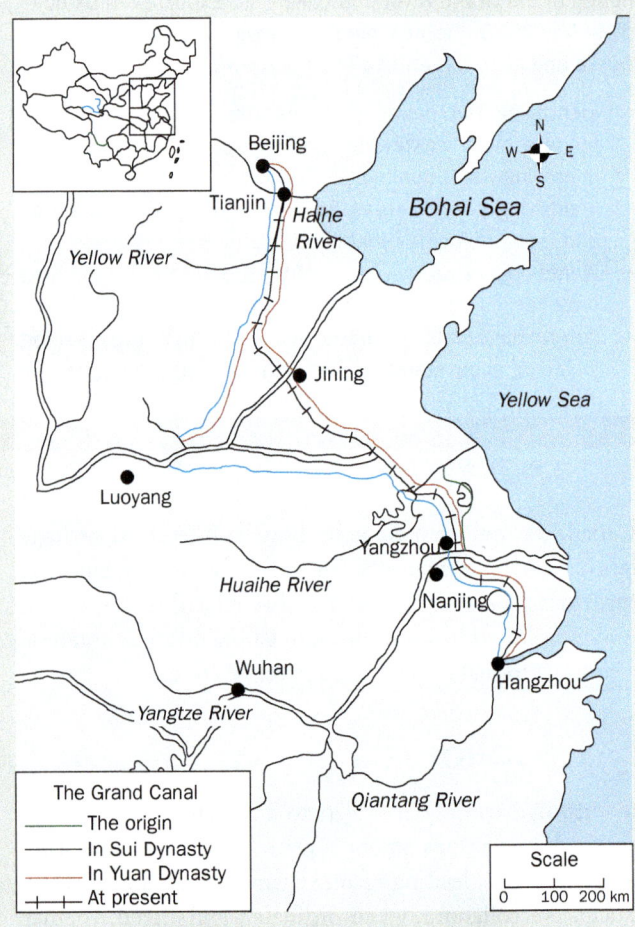

Figure 11.8 | Development of the Grand Canal.

Community of Xitang adjacent to the Grand Canal.

Barges on the North Jiangsu section of the Grand Canal.

canal grew into the richest agricultural areas in the country. In addition to its contribution to business and the economy, the canal also helped culture to flourish. It helped different ideas, beliefs, folk customs, and architecture styles to collide and blend. After more than 2,500 years, unique culture and folk customs associated with the canal formed, including countless magnificent cultural relics.

However, because of inadequate water quality and the narrow width, most of the canal's northern section (from Beijing to Jining) is inaccessible to boats. Currently, the navigable sector of the Grand Canal is 883 kilometres in length, stretching from Jining to the Sanbao Shiplock in Hangzhou. The accessible part still plays an important role in regional transportation, irrigation, and flood control, despite the decline during the Qing Dynasty and 'Republic of China'. The canal connects major coal deposits in the north and serves as a cheap and convenient channel to transport coal to the fast-developing Yangtze River Delta. More than 100,000 vessels use the canal and carry 260 million tonnes of goods every year, three times as much as the Beijing–Shanghai Railway. It also functions as part of a huge project for water diversion from the south to the north as the main channel of the Eastern Route of the South North Water Transfer Project. By using the canal, the project can save a great deal of investment, labour, and land. Moreover, the lakes and wetlands it connects also help to reduce flooding.

The historical, cultural, social, and economic significance of the canal has attracted great attention both in China and abroad. Protecting the canal is vitally important not only for maintaining its great historical and cultural values but also for promoting its significant economic and social values. Since 2005, the Chinese government has sought to have the Grand Canal included on the world heritage list. Led by vice-chairperson Chen Kuiyuan and organized by the History Study Committee, a group of experts toured the Grand Canal in May 2006 for a field study of the canal's preservation and utilization. A seminar was held in Hangzhou on preservation of the canal and application for world cultural heritage site recognition for the canal. At the seminar, a declaration was adopted to promote the preservation and application efforts. The State Administration for the Preservation of Cultural Relics has also placed it on a revised list of the country's candidates for world cultural heritage statuses. The canal was selected as the key cultural relics site under the state protection in May 2006. The Grand Jing-Hang Canal Museum, set up at a cost of nearly 100 million yuan (about $16 million, Canadian), was opened in Hangzhou in October 2007 with a floor space of 10,000 square metres.

However, the preservation and application task is challenging, and some thorny problems exist because of inadequate attention paid by the public to the importance of the task. Over the past 10 years, various preservation and maintenance projects have been initiated under different regional administrations, without integrated efforts and results. There also has been no unified plan for preservation or a mechanism for overall co-ordination and management. A result is that the canal as a whole has been seriously degrading. The section north of Jining has been dry since the lower part of the Yellow River changed its channel in 1855. Some sections are seriously polluted. Traffic congestion on the canal is frequent along various sections. Some sections of the ancient waterway have been vandalized repeatedly. Some ancient water control projects on the canal and important ancient structures related to the canal have been dismantled to make room for new urban projects. Intangible cultural heritage, along the canal, such as folk lifestyles and arts, has been deteriorating.

Thus, China needs to double its efforts to protect the Grand Canal and make all the appropriate preparations for successfully getting World Cultural Heritage status for the canal.

Shuheng Li received her Ph.D. from Nanjing University in 2008. She is a lecturer in the School of Urban and Environmental Science at Northwest University in Xi'an, China. Her research focuses on *environmental change* and water resource management.

in order to extend migratory bird habitat and enhance capacity to slow the release of flood waters. As Dale and Newman (2007: 59–60) have observed, an integrated approach seeks to overcome the silos, stovepipes, and solitudes that often characterize the approaches of governmental organizations responsible for water, land, and other related resources.

Ideally, IWRM drives managers to consider integration at several levels, including:

1. Integration of various dimensions of water, such as quantity and quality, surface and underground, and upstream and downstream. Use of river basins or watersheds for planning and management is recognized as the most appropriate spatial unit to achieve this.
2. Integration of water considerations with those for terrestrial and other related resources. This involves recognizing that many water problems, such as pollution and flooding, originate from or are exacerbated by land-based activities. Here again, the river basin or watershed is viewed as the most appropriate unit to facilitate attention to a range of interconnected resource systems.
3. Integration of water, as part of the environmental system, with aspects related to economic and social systems. Regional land-use planning, environmental impact assessment, and strategic sustainability assessment are used to connect environmental, economic, and social considerations.

Perspectives on the Environment

Great Lakes System

The Great Lakes—Superior, Michigan, Huron, Erie, and Ontario—constitute the largest system of fresh surface water on earth, containing about 20 per cent of the world's surface water. Less than one per cent of the system's water is renewed annually by precipitation. In addition to the natural beauty, ecological complexity, and significance of the lakes, the region is home to nearly 40 million Americans and Canadians, supports the culture and life ways of tribes and First Aboriginal Peoples, provides drinking water to millions of people, and is the foundation for billions of dollars in shipping, trade, agriculture, fishing, and recreation.

—Agreement Review Committee (2007: 5)

The rationale for IWRM is intuitive: water is but one subsystem, and therefore managers need to take an ecosystem approach to ensure linkages and connections are addressed. Such connections occur at scales ranging from a local sub-catchment to a major interjurisdictional river basin or catchment. However, the challenges to implementing IWRM can be formidable, because human behaviour is often competitive rather than co-operative, individuals focus on narrow interests without concern for the consequences for others today or in the future, and understanding complex natural and human systems can be very difficult (Biswas, 2004, 2008; Rahman and Varis, 2005; Lenton and Muller, 2009; Butterworth et al., 2010).

The following example illustrates the opportunities and challenges in implementing a hydrosolidarity approach.

Great Lakes Water Quality Agreement

In 1972, the governments of Canada and the United States signed the Great Lakes Water Quality Agreement (GLWQA). The agreement was amended in 1978, 1983, and 1987. It focuses on the Great Lakes ecosystem, defined to include the interacting components of air, land, water, and living organisms, including humans. The purpose is to restore and maintain the chemical, physical, and biological integrity of waters in the Great Lakes ecosystem.

The objectives of the 1972 agreement were to reduce nuisance conditions and discharge of substances toxic to humans, animals, and aquatic life, as well as to reduce phosphorous loadings in Lakes Erie and Ontario. During 1978, the agreement was amended to specify that an ecosystem approach would be used, with attention directed to both human health and environmental quality. Furthermore, amendments indicated that the intent was to virtually eliminate persistent toxic substances.

In 1983, further amendments stipulated increased effort to reduce phosphorous inputs into each lake. During 1987, a key change was to identify 'areas of concern', near-shore areas that had been significantly degraded and for which rehabilitation efforts would be pursued. In addition, new initiatives were to be undertaken to address non-point pollution sources and to create Lakewide Management Plans, as well as to deal with contaminated sediment, airborne toxic substances, and contaminated groundwater.

Despite impressive progress, the Agreement Review Committee (2007: 8) remarked that 'there are still serious threats to the physical, biological, and chemical integrity of the Ecosystem. Many scientists have voiced concern that the Great Lakes are exhibiting symptoms of stress from a variety of sources and impacts, including nutrient loadings, toxic contaminants, invasive species, and land use changes.' The drivers behind some of these challenges were viewed to include climate change, urbanization, long-range transport of toxics, and inadequate approaches to stop invasive species. Regarding the latter, the International Joint Commission (2004: 13) has stated that new aquatic alien species are introduced into the Great Lakes at a rate of one every eight months from ocean-going ships or from bait fish, aquarium fish, aquaculture, or connecting tributaries.

Perspectives on the Environment
Different Views on the Great Lakes Water Quality Agreement

Today, the Agreement remains one of the most farsighted international agreements and is a model of cooperative environmental research and ecosystem management.

—International Joint Commission (2004: 1)

Often cited as one of the most forward-thinking diplomatic achievements for the environment, the Agreement has served as a model for other international agreements to protect and restore environments elsewhere in the world. Its strengths include the establishment of common objectives and commitments for protecting and restoring the waters of the Great Lakes Ecosystem, the facilitation of information sharing, and cooperation on research and monitoring.

—Agreement Review Committee (2007: 6–7)

In some respects it could be argued that leadership in the Great Lakes governance regime forgot how to learn and adapt. . . . Matters of accountability, transparency, distributed governance, and shared decision-making are all absent from the current GLWQA, leading to the threat that if not addressed in a new agreement, an implementation deficit is nearly certain.

—Krantzberg and Manno (2010: 4274, 4275–6)

In terms of future needs, the Agreement Review Committee (2007) identified the following:

- clarification related to the ecological scope of the agreement to establish whether the focus is only on water quality or on a full ecosystem approach;
- attention to the geographical scope to clarify whether the focus is on open waters or on a combination of open waters, near-shore waters, inland areas, tributaries, and watersheds within the basin;
- clarification regarding whether groundwater resources are included as part of the Great Lakes ecosystem (the agreement considers groundwater only as a source of contamination, not as a component of the overall ecosystem);
- clarification regarding whether the entire St Lawrence River system should be included as part of the ecosystem (at the moment, the agreement includes only that part of the river down to the international boundary at Cornwall, Ontario);
- determination of how to apply management concepts such as watershed planning, adaptive management, pollution prevention, biodiversity initiatives, and airshed management;
- determination of how best to deal with chemicals, such as pharmaceuticals, flame retardants, and personal care products, not addressed in the agreement;

Perspectives on the Environment
Invasive Species

. . . the flow of new invasive species to the Great Lakes has not been stopped. In 2001, scientists estimated that 162 invasive species had entered the lakes from all pathways. Today, some scientists have raised that estimate to more than 170 non-indigenous fish, invertebrates, plants, algae, protozoa, and parasites. . . .

—International Joint Commission (2004: 15)

- in addition to the two national governments, clarification of the critical role and essential participation of other levels of government, such as states and provinces, local governments, and First Nations.

The above needs capture some of the challenges of achieving a hydrosolidarity approach, in spite of the high regard earned by the Great Lakes Water Quality Agreement through collaboration involving two national governments and numerous state/provincial and municipal governments.

Perspectives on the Environment
New Pathogens

Some experts believe that the massive and largely unregulated use of antibiotics in agriculture and aquaculture, coupled with the increasing number of antibiotic-resistant pathogens found in nature, may present the greatest risk to the aquatic environment and to public health. Antibiotic-resistant bacteria have been spread in the environment through the indiscriminant use of antibiotics in human and animal health.

—International Joint Commission (2004: 30)

The spirit of hydrosolidarity is that upstream and downstream jurisdictions work together collaboratively to ensure that initiatives in upstream parts of an aquatic system do not cause significant damage in downstream parts. The IJC was created for exactly that purpose and is viewed as a model of how transjurisdictional resource issues can be addressed. Certainly, when a nation agrees to participate in mechanisms and processes to deal with cross-border problems, it acknowledges that it is sacrificing some autonomy in order to reach decisions that benefit all nations.

Water Ethics

Water is a necessity for human life. There is no substitute. Furthermore, many needs and interests compete for their share of water. Given this situation, it is puzzling that more attention has not been given to development of **water ethics** to establish principles on which water management decisions could be based. UNESCO has published some relevant reports (Selbourne, 2000; Priscoli et al., 2004), but study focused on water ethics has been paltry.

Matthews et al. (2007) have addressed the matter of water ethics, and the following comments are based on their work. They remind us that an ethic normally is a statement of principles or values to identify appropriate behaviour by individuals or groups. At the same time, a set of ethics cannot provide all the answers needed or resolve all dilemmas. Frequently, uncertainty and complexity make outcomes difficult to predict, contributing to fuzzy understanding about what might happen. Also, different ethical principles, each desirable, can sometimes conflict, making it unclear as to which path to pursue.

Notwithstanding the above difficulties, Matthews et al. (ibid., 350–3) offer six 'imperatives' for a new water ethic. They are:

- Meet basic human needs to enhance equity today and for the future.
- Safeguard ecosystems by allocating sufficient water resources.
- Encourage efficiency and conservation of water resources.
- Establish open and participative decision-making processes.
- Respect system complexity and emphasize precaution.
- Seek multiple sustainability benefits from water-centred initiatives.

We encourage you to think about the six imperatives presented above, and to decide whether they provide a reasonable foundation for a 'water ethic'. If you conclude that they do not, then decide what should be changed or added. For help, you may wish to review the reports by Armstrong (2009), Graenfeldt (2010), and Sandford and Phare (2011).

Perspectives on the Environment

Water as a Human Right

. . . consideration should be given to the creation of a human right to water in Canada. It might come as a surprise to learn that no such right exists within Canada, or indeed within the main UN conventions on human rights (although other elements of international law do suggest such a right exists). The international campaign for a human right to water has gained considerable momentum over the past few years, but the Canadian federal government has been one of the most consistently outspoken opponents of this right in the international arena.

—Bakker (2009: 20)

Regarding the first imperative of the proposed water ethic (meet human needs to enhance equity today and for the future), the United Nations Human Rights Council has sought to have water and sanitation recognized as a basic human right and to establish an international monitoring organization to track actions of nations. The Council met for three weeks in March and April 2008 to address this and other matters. The occasion was the third time in six years that the United Nations had attempted to have human rights to water and sanitation recognized. Canada opposed the resolution for **water rights**, which had been proposed by Germany and Spain. Russia and the United Kingdom also did not support it. Nevertheless, it was generally agreed that Canada led the opposition.

Canada has been consistent in its opposition. During a Human Rights Council meeting in 2002, Canada was the only one of 53 nations voting against a motion to appoint a special rapporteur on water. And in October 2006, Canada voted against a resolution to have the Human Rights Council conduct a study on the right to water.

Canada's lack of support appears to be driven by a belief on the part of federal politicians that Canadian sovereignty over its own water is ambiguous under NAFTA. In that trade agreement, water could be viewed as a commodity or service comparable to any other. As a result, if Canada were to support the principle that water is a basic human right, some believe that it could become vulnerable to claims from the United States that since Canada has more water than it needs and places such as Atlanta and the US Southwest face water shortages, then Canada has a moral duty to share its water through bulk water transfers or other means. In contrast, others argue that the Human Rights Council resolution explicitly excluded transborder water issues and therefore Canadian sovereignty over its water would not be threatened.

Another reason undoubtedly has influenced the federal government. If it were to support the concept of a human right to water, liability issues could emerge related to the many small communities across Canada under boil-water advisories. If water were prescribed to be a basic human right, could the federal or provincial governments become liable to lawsuits when sufficient quantities of potable water are not provided to communities? The federal government would be especially liable regarding Aboriginal communities, since under the Canadian Constitution, the federal government is responsible for their reserves.

As you think about the matter of water as a human right, consider that Article 25.1 of the Universal Declaration of Human Rights, adopted by the United Nations General Assembly on 10 December 1948, states that 'Everyone has the right to a standard of living adequate for the health and well-being of himself and his family, including food, clothing, housing, and medical care and necessary social services. . . .'

If food, clothing, and housing are viewed as basic human rights, why could not 'water' be included in this statement?

In 2002 the United Nations Committee on Economic, Social and Culture Rights (CESCR) issued 'General Comment No. 15', or GCN15, in which the following words appear: 'The human right to water entitles everyone to sufficient, safe, acceptable, physically accessible and affordable water for personal and domestic uses. An adequate amount of safe water is necessary to prevent death from dehydration, reduce the risk of water-related disease and provide for consumption, cooking, personal and domestic hygienic requirements.' The CESCR is explicit that this statement is an 'interpretation' rather than a 'treaty', and as result is not legally binding on member states. However, it reflects a powerful moral position (Debreuil, 2006: 8).

On 28 July 2010, the UN General Assembly adopted a resolution recognizing access to clean water and sanitation as a human right. The vote was 122 nations for, none against, and 41 abstentions. Bolivia's representative, who introduced the resolution, argued that the resolution was needed because a human right to water was not fully recognized despite being referred to in various international instruments. Countries abstaining, which included Canada, the US, the UK, Australia, Denmark, Greece, Ireland, Israel, Japan, Kenya, Netherlands, New Zealand, Poland, and Sweden, took the position that this matter was being examined by the Geneva-based Human Rights Council, and, as a result, such a resolution was premature until that Council's work, which is intended to clarify the scope of such a right, is concluded.

What is your position related to water as a human right? As you consider this question, you may find the insights of Debreuil (2006), Jayyousi (2007) and Khadka (2010) helpful.

Implications

Canada has a relative abundance of high-quality water, even though some areas experience scarcity in terms of quantity and/or quality. As a society, however, we have placed stress on aquatic systems, sometimes degrading them significantly. As the discussion in this chapter has shown, considerable scientific understanding can be drawn upon to assist in the management of water systems. And there have also been some significant improvements, confirming that individuals, groups, communities, and societies can reverse degradation and deterioration.

Perspectives on the Environment
A Public Policy Matter

Ultimately, if we are to sustain Canada's water supplies, we all have to better understand how we use water as individuals, as communities and as consumers. Water management is not just a government problem or an industry problem. Rather it is one of the most important public policy issues facing the world today—with the key word in that phrase being 'public'.

—Taylor (2009: 47)

Each of us has an opportunity to modify our basic values and change behaviour to help protect our water resources and to maintain the integrity and health of aquatic ecosystems. Table 11.4 identifies action that individuals and/or governments can take to further contribute to improved water resources in Canada.

Table 11.4 | What You Can Do: Ten Water Conservation Initiatives

Focus	Problem	Solution	Challenges	Savings
1. Educate	Lack of understanding about need for and potential benefits from water conservation	Introduce outreach and education programs beyond information dissemination to change behaviour	Engaging community members in meaningful education that changes view of 'water abundance'	Many experts agree 50 to 80 litres of high-quality water per capita per day is needed for a good standard of living
2. Design communities for conservation	Municipal-level decisions too often having negative impacts on watersheds	Limit urban sprawl, reduce green lawn syndrome, promote 'green' infrastructure, expect land-use decisions to be judged for their impacts on watersheds	Belief that water-sensitive urban design is much more expensive than standard approaches	Conservation-oriented urban design can save 50 per cent of outdoor water use
3. Close urban water loop	All municipal water treated to drinking-water quality but more than two-thirds used for non-drinking functions	Use reclamation, reuse, and recycling to ensure better match of water quality to end uses	Possibility that lower water prices may make reuse and recycling less financially attractive; risks associated with reused and recycled water	Up to 50 per cent water savings can be achieved by reusing or recycling water for toilets and outdoor irrigation

Continued

4. Use rainwater	Rainwater not usually viewed as source of water for homes, businesses, etc.	Use decentralized infrastructure to harvest rainfall and create xeriscaped landscapes that rely on rainfall	Building and plumbing code restrictions; financial cost of rainwater harvesting infrastructure for homeowners and businesses	Rainwater harvesting and xeriscaping can lead to 50 per cent savings in outdoor water use and up to 40 per cent savings for indoor use for toilet flushing and clothes washing
5. Plan for sustainability	Too many conservation programs viewed as short-term solutions until next supply source can be developed	Plan with 10- to 50-year time horizon, involve all stakeholders, and place ecological health in central position	Looking beyond the three- or four-year electoral cycle and investing in programs that provide long-term returns; engaging the community	Effective water conservation plans can lead to water savings ranging from 20 to 50 per cent
6. Adopt appropriate pricing	Normal water pricing rates encourage wasteful use	Use 'full cost' pricing with volume-based pricing structures	Need to implement metering and gain political support; need to ensure full accessibility to meet basic needs for water	Effective pricing can lead to 20 per cent reduction in water use over the long term
7. Link conservation to development	Present arrangements for funding urban water infrastructure promote neither conservation nor innovation	Connect conservation to development by requiring water infrastructure funding and development permits to depend on use of demand management	Local resistance to conditional funding arrangements; capacity for enforcement and follow-up on conditions	Water savings from 20 to 30 per cent can be achieved by using 'off the shelf' technologies and modest water pricing reforms
8. Make managing demand part of regular business	Demand management approaches often not comprehensive or part of daily business in most communities	Implement permanent water conservation measures and hire full-time staff with appropriate skills	Reluctance of utilities to commit financial resources to hire demand management professionals and implement long-term demand management programs	Depending on how aggressive and creative demand management programs are, 'the sky is the limit' in terms of potential savings
9. Stop flushing the future	Inefficient fixtures and appliances common in most homes	Install efficient toilets, faucets, and showerheads as well as water-saving dishwashers and washing machines	Permissive building and plumbing codes and lack of incentives and resources to promote efficient technologies	Efficient fixtures and appliances can achieve 33 to 50 per cent indoor water savings with payback within two years in most instances
10. Fix leaks and reduce waste	Significant water loss from leaks, often due to old infrastructure	Detect and repair leaks by regular water audits and maintenance programs	Financial challenges on utilities of up-front costs for integrated metering, detection, maintenance, and monitoring programs	Fixing leaks can easily result in 5 to 10 per cent water savings, and up to 30 per cent savings are possible with older infrastructure

Source: Based on Brandes et al. (2006: 6–42); see also Casselman (2011).

Summary

1. While comprising only 0.5 per cent of the world's population, Canadians have access to almost 20 per cent of the global stock of fresh water and 7 per cent of the total flow of renewable water. About one-fifth of the Canadian population relies on groundwater for daily water needs.

2. Canada's per capita demands on water resources are the second highest in the world and have been calculated at about 329 litres per person per day at home.

3. Canada is a global leader in terms of water diversions for hydroelectric generation, and diversions for hydro power dominate overwhelmingly in both number and scale of diversions.

4. Canada has used many megaprojects to meet energy demands, and virtually every region in the country has energy megaprojects. One of the most significant is James Bay in Quebec, while others are Churchill Falls in Labrador, the Nelson–Churchill in Manitoba, and the Columbia and Nechako rivers in British Columbia.

5. Various major projects have been proposed, such as NAWAPA and the GRAND scheme, to transfer large volumes of water from Canada to the United States. They all have significant financial costs and environmental implications.

6. Free trade agreements, such as stipulations under NAFTA, trigger concern that water could become a tradable commodity.

7. Pollution involves point and non-point sources, with the latter being the most challenging to manage.

8. Progress regarding diffuse pollution has been associated with *credible science* to document the nature of the problem.

9. The governance of the Great Lakes is evolving, and key aspects include a shift from an exclusive command-and-control emphasis to voluntary measures and from top-down management to environmental partnerships.

10. At the start of the second decade of the twenty-first century, one in six people worldwide did not have access to a safe water supply and that two out of five people did not have access to adequate sanitation.

11. Many Canadians have been complacent about the adequacy and safety of their water supplies. For many, this changed in mid-May 2000 when Walkerton, Ontario, experienced contamination of its water supply system by deadly bacteria. Seven people died, and more than 2,300 became ill.

12. The judge who conducted a public inquiry into the Walkerton tragedy recommended a 'multi-barrier approach' to drinking water safety.

13. The water crisis at Kashechewan in 2005 highlighted the often unsatisfactory water supply infrastructure on First Nation reserves as well as in many other rural communities in remote regions. Water supply systems in such places are often more like those in developing countries.

14. Increasing attention is being given to determining the most appropriate mix of supply management, demand management, and soft path approaches to provide water to communities. The soft path approach shifts attention from questions about *how* to meet needs to questions about *why* needs are met by using water.

15. The concept of 'virtual' water encourages nations to determine whether it would be more sensible to import water-intensive crops than to grow them in their own countries. This sometimes leads to conflict between two desirable goals: food self-sufficiency and efficient water use. The concept of 'water footprint' complements the idea of virtual water, and helps to educate people and communities about the impact of their use of water.

16. Awareness is growing about the vulnerability of water infrastructure to terrorism. No infrastructure system can ever be made totally invulnerable, but the risk can be reduced.

17. Floods along rivers and lakes are normal occurrences. Humans adapt to flooding through some mix of structural and non-structural approaches.

18. Droughts are a function of precipitation (or lack thereof), temperature, evaporation, evapotranspiration, capacity of soil to retain moisture, and resilience of flora and fauna in dry conditions.

19. The most drought-prone areas in Canada are south-central British Columbia, the southern Prairie provinces, and southern Ontario.

20. The Canadian Heritage Rivers Program recognizes the significance of rivers in the identity and history of Canada and is intended to ensure that their natural, cultural, and recreational values are protected.

21. Hydrosolidarity and integrated water resource management promote an ecosystem approach to water management in which attention is given to connections among water, land, and other resources, relationships between upstream and downstream parts of a basin, and linkages between surface and groundwater.

22. The Great Lakes Water Quality Agreement contains many elements of best practice related to hydrosolidarity.

23. Attention is being drawn towards developing principles related to a 'water ethic' to guide decisions regarding water allocation, development, and use.

24. International dialogue continues related to the idea of 'water as a human right'. Canada has consistently opposed this proposition.

Key Terms

- demand management
- diversions
- drought
- E. coli
- floodway
- heritage rivers
- hydrosolidarity
- International Joint Commission
- integrated water resource management (IWRM)
- James Bay and Northern Quebec Agreement
- James Bay Project
- limiting factor principle
- megaprojects
- multi-barrier approach
- Palliser's Triangle
- renewable water supply
- soft path
- supply management
- virtual water
- water ethics
- water footprint
- water rights
- water table
- water terrorism
- wetlands

Questions for Review and Critical Thinking

1. Does the 'myth of superabundance' adequately account for the high per capita water use by Canadians?
2. Why do some believe that Canada has 20 per cent of the world's renewable water supplies while others suggest that this number is 2.6 per cent?
3. What are the main reasons for water diversions in Canada?
4. Why is the likelihood of pressure emerging in the United States to import bulk water supplies from Canada relatively low?
5. Explain the significance of the 'limiting factor principle'.
6. Explain the significance of bioaccumulation and biomagnification of chemicals in the food chain and what role water plays in it.
7. Why is diffuse pollution from non-point sources a challenge for managers?
8. What is meant by 'water security'?
9. What are the implications of the experience with contaminated drinking water in Walkerton, Ontario, and North Battleford, Saskatchewan?
10. What is the significance of the failure of the water supply system in Kashechewan?
11. Explain the distinctions among supply management, demand management, and soft path approaches.
12. What is the relationship between 'virtual water' and a 'water footprint'?
13. How serious is the threat of water terrorism, and what options are available to reduce the risk of terrorist actions?
14. What is the significance of structural and non-structural approaches for reducing flood damages?
15. What lessons were learned from major floods in Canada in the past five years?
16. How do we know when a drought begins and ends?
17. What are the most significant flood and drought-prone areas in Canada?
18. Which criteria and indicators should be used to identify the natural, cultural, and recreational value of rivers?
19. Are there river systems in your region that could be candidates for nomination as heritage rivers?
20. Explain the relationship between the concepts of 'hydrosolidarity' and 'integrated water resource management'.
21. What is the significance of the Great Lakes Water Quality Agreement in terms of best practice associated with hydrosolidarity?
22. What principles should underlie a 'water ethic' to guide decisions about allocation and use of water?
23. What are arguments for and against the proposition that water should be a basic human right? What is your view about this proposition?

Related Websites

Atlantic Coastal Action Program
www.acapsj.com/Home.html

Canada's Water Quality Guidelines
www.ec.gc.ca/eau-water/default.asp?lang=En&n=F77856A7-1

Canadian Council of Ministers of the Environment
www.ccme.ca

Canadian Global Change Program
www.globalcentres.org/cgcp

Canadian Pollution Prevention Information Clearinghouse
www.ec.gc.ca/cppic

Canadian Water and Wastewater Association
www.cwwa.ca

Canadian Water Resources Association
www.cwra.org

Clean Annapolis River Project
www.annapolisriver.ca

Environment Canada's Water Website
www.ec.gc.ca/eau-water

Experimental Lakes Area
www.umanitoba.ca/institutes/fisheries

Fisheries and Oceans Canada, Freshwater Institute
www.dfo-mpo.gc.ca/regions/central/pub/fresh-douces/01-eng.htm

Great Lakes Information Network
www.great-lakes.net/lakes

International Joint Commission
www.ijc.org

National Water Research Institute
www.ec.gc.ca/inre-nwri/Default.asp?lang=En&n=0E7169DE-1

Northern River Basins Study
www3.gov.ab.ca/env/water/nrbs/nrbs.html

Protect Your Watershed, Royal Canadian Geographical Society
www.canadiangeographic.ca/watersheds/map/index.aspx?path=english/

Statistics Canada, Human Activity and the Environment Annual Statistics
www.statcan.ca/bsolc/english/bsolc?catno=16-201-X&CHROPG=1

St. Lawrence Plan for Sustainable Development
www.planstlaurent.qc.ca/index_e.html

UNESCO water portal bimonthly newsletter
www.unesco.org/water/news/newsletter

WaterCan
www.watercan.com

Water Footprint Network
www.waterfootprint.org

Water Survey of Canada
www.ec.gc.ca/rhc-wsc/

Further Readings

Note: This list comprises works relevant to the subject of the chapter but not cited in the text. All cited works are listed in the Bibliography at the end of the book.

Armstrong, A. 2006. 'Ethical issues in water use and sustainability', *Area* 38, 1: 9–15.

Bakker, K., ed. 2007. *Eau Canada*. Vancouver: University of British Columbia Press.

——— and C. Cook. 2011. 'Water governance in Canada: Innovation and fragmentation', *International Journal of Water Resources Development* 27: 275–89.

Botts, L., and P.K. Muldoon. 2005. *Evolution of the Great Lakes Water Quality Agreement*. East Lansing: Michigan State University Press.

Bourassa, R. 1985. *Power from the North*. Scarborough, Ont.: Prentice-Hall Canada.

Brelet, C. 2004. *Some Examples of Best Ethical Practices in Water Use*. Paris: UNESCO.

British Columbia Ministry of Environment. 2008. *Living Water Smart*. Victoria: British Columbia Ministry of Environment.

Brooymans, H. 2001. *Water in Canada: A Resource in Crisis*. Edmonton: Lone Pine Publishing.

Canada. 2007. *Canadian Water Sustainability Index: Project Report*. Ottawa: Policy Research Initiative.

Chiefs of Ontario. 2008. *Water Declaration of the First Nations in Ontario*, Oct.

Conservation Ontario. 2010. *Navigating Ontario's Future: Integrated Watershed Management—Summary Report*. Newmarket, Ont.: Conservation Ontario.

Diamond, B. 1985. 'Aboriginal rights: The James Bay experience', in M. Boldt and J.A. Long, eds, *The Quest for Justice: Aboriginal People and Aboriginal Rights*. Toronto: University of Toronto Press, 265–85.

Environment Defense. 2007. *Up to the Gills: Pollution in the Great Lakes*. Toronto: Environment Defense.

Falkenmark, M., and C. Folke. 2002. 'The ethics of socio-ecohydrological catchment management: Towards hydrosolidarity', *Hydrology and Earth System Sciences* 6: 1–10.

Falkenmark, M., and J. Rockstrom. 2005. *Balancing Water for Humans and Nature: The New Approach to Ecohydrology*. London: Earthscan.

Glendon, R. 2010. *Unquenchable: America's Water Crisis and What To Do About It*. Washington: Island Press.

Griffiths, M., A. Taylor, and D. Woynillowicz. 2006. *Troubling Waters, Troubling Trends*. Calgary: Pembina Institute.

Hellegers, P., et al. 2008. 'Interactions between water, energy, food and environment: Evolving perspectives and policy issues', *Water Policy* 10 (supplement 1): 1–10.

Maas, T. 2001. 'Water footprints: Exposing the invisible corporate risk', *Water Canada* 10, 1: 12–16.

McCammon, A. 2011. 'A long and winding road: Integrated watershed management is a journey, not a destination', *Water Canada* 11, 1: 32–5.

McCutcheon, S. 1991. *Electric Rivers: The Story of the James Bay Project*. Montreal: Black Rose Books.

Mainville, R. 1992. 'The James Bay and Northern Quebec Agreement', in M. Ross and J.O. Saunders, eds, *Growing Demands on a Shrinking Heritage: Managing Resource Conflicts*. Calgary: Canadian Environmental Law Association, 176–86.

Mascarenhas, M. 2007. 'Where the waters divide: First Nations, tainted water and environmental justice in Canada', *Local Environment* 12: 565–77.

Mysiak, J., C. Pahl-Wostl, C. Sullivan, J. Bromley, and H.J. Henrikson, eds. 2009. *The Adaptive Water Resources Management Handbook*. London: Earthscan.

Nanos, N. 2009. 'Canadians overwhelmingly choose water as our most important natural resource', *Policy Options* (July–Aug.): 12–15.

Perkel, C.N. 2002. *Well of Lies: The Walkerton Water Tragedy*. Toronto: McClelland & Stewart.

Phare, M.-A. 2009. *Denying the Source: The Crisis of First Nations Water Rights*. Surrey, BC, and Caster, Wash.: Rocky Mountain Books.

Pollution Probe. 2008. *A New Approach to Water Management in Canada*. Toronto: Pollution Probe.

Schindler, D.W., and W.F. Donahue. 2006. 'An impending water crisis in Canada's western Prairie provinces', *Proceedings of the National Academy of Sciences of the United States of America* 103: 7210–16.

Swain, H., S. Louttit, and S. Hrudey. 2006. *Report of the Expert Panel on Safe Drinking Water for First Nations*, vol. 1. Ottawa: Department of Indian Affairs and Northern Development. At: www.eps-sdw/gc/ca/reprt/index-e.asp.

United Nations World Water Assessment Programme. 2006. *Water: A Shared Responsibility*. The United Nations World Water Development Report. Paris: UNESCO; New York: Berghahan Books.

Walkem, A. 2007. 'The land is dry: Indigenous peoples, water, and environmental justice', in K. Bakker, ed., *Eau Canada*. Vancouver: University of British Columbia Press, 303–19.

Chapter 12
Minerals and Energy

Learning Objectives

- To understand the characteristics of non-renewable resources relative to the renewable resources discussed in previous chapters.
- To appreciate the significance of minerals and energy for Canada.
- To understand the management issues associated with non-renewable resources in general and minerals and energy in particular.
- To identify the relative importance of different minerals for the Canadian mining industry, as well as Canada's importance in global mining trade.
- To discover how science is used in environmental assessments, illustrated by the Ekati Diamond Mine in the Canadian North.
- To appreciate that energy resources can be both renewable and non-renewable.
- To learn of the potential of alternative, renewable energy sources, particularly wind and solar power.
- To understand the significance of non-renewable energy resources, including offshore petroleum and natural gas, the Athabasca oil sands, and nuclear power.
- To know what you can do to have a lighter 'footprint' related to use of minerals and energy.

Introduction

Previous chapters focused on **renewable or flow resources**, those renewed naturally within a relatively short period of time, such as water, air, animals, and plants (Rees, 1985: 14). Other renewable resources are solar radiation, wind power, and tidal energy. Given this mix, a distinction is often made between renewable resources not dependent on human activity (e.g., solar radiation) and those that renew themselves as long as human use allows reproduction or regeneration (e.g., fish).

Figure 12.1 highlights that flow or renewable resources can exist in critical or non-critical zones. Those in the critical zone can be harvested or exploited to exhaustion. The most vulnerable depend on biological reproduction for renewal. Whether through overhunting, overfishing, polluting, or destroying habitats, humans can create conditions such that renewable resources cannot replace or replenish themselves. Indeed, the 'collapse' of the northern cod in the Northwest Atlantic (see Chapter 8) is a classic case of overharvesting leading to depletion.

In this chapter, emphasis is mostly on **non-renewable or stock resources**, which take millions of years to form. As a result, from a human viewpoint, such resources are for practical purposes fixed in supply and therefore not renewable. However, Figure 12.1 indicates that non-renewable or stock resources are not homogeneous. Some are consumed through use, whereas others can be recycled. Those consumed by use are best illustrated by fossil fuels (coal, oil, natural gas). Once used, they are effectively not available to humans, even though they do not really disappear but are changed into another form, often pollutants. In contrast, stock resources such as metals can be recycled many times, so the stock of resources in the ground is not the only source. However, recycling often requires significant amounts of energy, so the recycling of one type of stock resource (e.g., aluminum) may hasten the depletion of another (coal, oil, or natural gas) stock resource.

Our attention here focuses on both minerals and energy. And to emphasize that non-renewable resources are not homogeneous, the discussion of energy examines wind and solar power, usually viewed as types of renewable or flow resource. In addition, attention is given to non-renewable types of energy resources, such as offshore petroleum and natural gas, the Athabasca oil sands, and nuclear power.

Mineral and energy resources are important in and for Canada. Canada is the sixth largest user of primary energy (commercially traded fuels) in the world. Such a high level of use is attributed to Canada being a large country with long travel distances, as well as its cold climate, an energy-intensive industrial base, relatively low energy prices, and a high standard of living. Regarding minerals, Canada is one of the major global exporters. More details about energy and minerals will be provided later in the chapter, but the key message here is that both are important for regional and national economies in this country and that in their extraction and use, they can have significant environmental impacts.

STOCK			FLOW	
Consumed by Use	Theoretically Recoverable	Recyclable	Critical Zone	Non-critical Zone
OIL GAS COAL	ALL ELEMENTAL MINERALS	METALLIC MINERALS	FISH FORESTS ANIMALS SOIL WATER IN AQUIFERS	SOLAR ENERGY TIDES WIND WAVES WATER AIR

Flow resources used to extinction

Critical zone resources become stock once regenerative capacity is exceeded

Figure 12.1 | A classification of resource types. *Source:* Rees (1985: 13). Reproduced with permission of Taylor & Francis Group LLC.

Framing Issues and Questions

The challenge for renewable resources is to manage them so that they remain sustainable and resilient. For non-renewable resources, however, extraction usually results in absolute depletion in any time frame other than a geological one. Given the characteristics of non-renewable resources, the management issues are usually different from those associated with renewable resources. Specifically, concerns focus on:

1. how to use the proceeds from resource extraction to generate new wealth, to benefit generations today and in the future;
2. how to conserve mineral or fossil-fuel assets to extend the longevity of reserves and how to identify substitutes for use in the long run;
3. how to minimize negative environmental impacts at each stage in the life cycle of use: exploration, extraction, transformation, consumption, recycling, and final disposal;
4. how to create improved socio-economic relationships with stakeholders, especially the local communities located in the mining area, after the resource has been exhausted;
5. how to manage recyclable non-renewable resources—i.e., many metals and minerals—as a renewable or flow resource.

What would motivate mining and fossil-fuel firms to engage in environmental management, given that their priority is to maximize profits and remain competitive in international markets? As Hilson (2000: 203) noted, if done systematically and correctly, enhanced 'environmental management practices and extended social responsibility almost always generate some kind of economic return on investment for business, although usually over the long term. A documented reduction in effluent discharges, for example, leads to a reduction in costly government inspections and auditing practices.'

The main environmental issues for the mining and energy sectors include **acid mine drainage**, **sulphur dioxide emissions**, and **metal toxicity**.

1. *Acid mine drainage*. Most non-ferrous metals exist as sulphides and usually are accompanied by iron sulphides. When ore minerals are separated from minerals without economic value, significant quantities of waste rock and tailings are created, and they contain iron sulphides that can readily oxidize to become sulphuric acid. When exposed to precipitation (rain or snow), sulphuric acid can dissolve residual metals, leading to acidic drainage, which can continue for centuries. Liabilities in the Canadian mining industry related to acidic drainage are estimated to range between $2 billion and $5 billion.
2. *Sulphur dioxide emissions*. One outcome of smelting sulphide ores is the release of huge quantities of sulphur, mainly in the form of dioxides, into the atmosphere, thereby creating acid precipitation as discussed

in Chapter 4. The Canadian mining sector is the main contributor to sulphur dioxide emissions in Canada. The burning of fossil fuels is also a major source of atmospheric emissions of sulphur dioxides, creating pressure for alternative sources of energy.
3. *Metal toxicity*. The mining industry is being challenged about the toxic effects of metals on human and ecosystem health. For example, many uses of asbestos are now not acceptable because of connections established between it and cancer. Lead is also toxic. Emissions from smelting and steelmaking processes can also threaten health.

To these three issues can be added challenges related to energy:

- disruption of remote ecosystems due to exploration, test drilling, and operation of oilfields or gas wells, ranging from habitat degradation to alteration of nesting, denning, and migration patterns of birds and animals;
- disturbance to aquatic ecosystems from escape of waste heat produced from nuclear energy production;
- threat to human and ecosystem health from radioactive waste associated with nuclear energy production over a period of thousands of years;
- alteration to ecosystems from building hydroelectric dams and generating stations.

These issues, individually and collectively, provide a strong rationale for increased attention to environmental aspects and management in the mining and fossil-fuel sectors.

'Best practice' related to environmental management for mining and fossil-fuel firms in Canada should include a combination of basic scientific research to ensure understanding of natural and social systems that can be affected by operations and design of appropriate mitigation measures, environmental impact assessments and reporting, environmental audits, corporate policies that explicitly include environmental aspects, environmental management systems, and life-cycle assessments.

Non-Renewable Resources in Canada: Basic Information

According to Natural Resources Canada (2009a), Canada was severely affected by the global economic and financial crisis in 2008–9, with negative consequences for the mining and mineral processing industries. As the global economic downturn intensified, demand for many commodities declined, leading to lower prices, quantities, and value of production, reduced investment in exploration, and decreased workforces. To illustrate, in 2008 the gross domestic product for Canadian mining and mineral processing industries was almost $40 billion, or 3.2 per cent of Canada's total GDP. In 2009, the GDP for mining and mineral processing industries fell to just under $32 billion, or 2.7 per cent of the total Canadian GDP. Another indicator of the negative impact of the global recession is the 'annual rate of capacity utilization' in the mining industry, which fell from just under 78 per cent in 2008 to slightly over 55 per cent in 2009, the 'lowest annual utilization rate on record' (ibid., 6). A third indicator—the value of production of the mining industry (metallic minerals as well as non-metallic minerals and coal)—decreased from $47 billion in 2008, a record high, to just over $32 billion in 2009. This was the first annual drop since 2001. At the same time, however, production increased in some regions, notably British Columbia.

Notwithstanding the challenges associated with the 2008–9 financial crisis, Canada maintained its position as a global leader in exports of minerals and metals. Canada is ranked in the top five countries in the world as a producer of aluminum, diamonds, nickel, platinum, group metals, potash, uranium, and zinc. During 2009, the total trade (exports minus imports) of minerals and metals was over $121 billion, representing 17 per cent of total trade for Canada. Minerals and metals were exported to more than 200 countries, with the main destinations being the United States (55 per cent), the European Union (16 per cent), China (5 per cent), and Japan (4 per cent).

After significant growth between 2003 and 2007, the drop in value of metal produced in Canada reflected mostly a drop in commodity prices rather than decreased output. Exceptions to this pattern were gold and uranium. The increase in value of gold primarily reflected higher prices. The increase in

Perspectives on the Environment

Diversification in Export of Minerals and Metals

The most notable recent trend in the export of minerals and metals has been the diversification away from the United States. In 2000, the vast majority (78 per cent) of Canada's export of minerals and metals were destined for the United States. However, the dominance of the United States has declined in the past few years. In 2009, the United States accounted for 55 per cent of Canada's exports of minerals and metals.

Canada's exports of minerals and metals to other countries, particularly China, have grown rapidly over the past decade. Over the period 2000–2009, Canada's exports of minerals and metals to China grew at an average annual compound rate of an astonishing 17 per cent, reaching $3.9 billion in 2009. Minerals and metals accounted for 31 per cent of Canada's total exports to China—the world's fastest growing and second-largest economy—and are by far Canada's most valuable export to China.

—Natural Resources Canada (2009a: 8)

value of uranium reflected a significant increase in volume of production (150 per cent) along with higher prices. Despite a strong performance by the uranium sector, Canada dropped to second place behind Kazakhstan as the leading global producer of uranium in 2009. It is not yet known what impact, if any, the earthquake and subsequent tsunami in Japan in March 2011, and the associated damage to four nuclear power plants, especially the Fukushima Daichi nuclear complex in northeastern Japan, might have on global uranium markets, and therefore on demand for Canadian uranium.

Potash in Saskatchewan

The potash industry began in Saskatchewan during the early 1960s and expanded steadily during the 1970s and 1980s. The potash had been discovered in the 1940s during exploratory drilling for petroleum. Canada now is the largest producer and exporter of potash in the world. Furthermore, Saskatchewan's potash industry has been judged the most productive worldwide, accounting for about 25 to 30 per cent of world potash production (Saskatchewan, 2008), but there can be significant variability from year to year. For example, the value of potash production fell from $7.7 billion in 2008 to $3.4 billion in 2009 because of a steadily dropping price for potash. However, value of sales rebounded in 2010 and 2011.

Potash is a generic term, which covers different kinds of potassium salts. The most important type is potassium chloride. Potassium, a necessary ingredient for plant growth, has become one foundation of modern fertilizers. Some 95 per cent of potassium production is used in fertilizers. In Saskatchewan, potassium is located at depths of more than 1,000 metres beneath the surface in southern Saskatchewan. The estimated reserves are sufficient to meet global demand for several hundred years, assuming current levels of use. In addition to massive supply, the quality of the potash is very high. Situated in flat beds, the potash can be mined efficiently, and the mines in Saskatchewan are considered to be among the most efficient in the world.

The potash industry in Saskatchewan attracted significant attention during the second half of 2010 as a result of a takeover bid by the Anglo-Australian mining giant BHP Billiton to purchase Saskatchewan's Potash Corporation, the world's largest producer of potash, for a price of $38.6 billion. The takeover bid, considered a hostile one, would have been the biggest takeover in Canadian history. The provincial government, led by Premier Brad Wall, argued that jobs and revenues in Saskatchewan would be at risk if a foreign owner controlled Potash Corp.

In early November 2010, federal Industry Minister Tony Clement decided that the takeover bid 'did not likely present a net benefit to Canada' but did not elaborate. In mid-November, Billiton withdrew its takeover bid. The decision by Clement caused much speculation as to the 'real reasons' for rejection of the bid, given that the federal Conservative government had been explicit earlier in 2010 that a key principle for it was commitment to free trade and Canada being open to foreign investment. Observers commented that if potash was such a strategic natural resource, why had the federal government allowed the foreign takeover of other natural resource companies such as Alcan, Falconbridge, and Inco?

Some suggested that, with a federal election likely in 2011, with 13 of the 14 federal seats in Saskatchewan held by the Conservatives, and with fierce and emotional opposition from Saskatchewan to the takeover, the decision reflected basic partisan politics. The political concern for the Conservatives was undoubtedly heightened by support for Saskatchewan from the premiers of Alberta (Conservative), Manitoba (NDP), and Quebec (Liberal). This example highlights that many aspects often influence decisions related to resource and environmental management. However, the amount of the rejected bid—over $38 billion—is a clear indicator of the importance of fertilizer in the future as the nations of the world will struggle to feed their human populations with increased pressures on agricultural systems, as discussed in Chapter 10.

Canadian mineral production is not only exported but is also used within the country. For example, coal and uranium are the basis for one-third of electricity production. Alberta, which produces nearly half of the coal mined in Canada, depends on coal for almost 50 per cent of its electrical power (Alberta Energy, 2008).

In the following sections, we look at how science has been incorporated into initiatives to remediate landscape degradation associated with mineral extraction and how it has been used in understanding and mitigating environmental impacts when a new mining venture is being designed. Finally, we consider the role of science in exploring alternatives to fossil-fuel energy sources.

The exterior of the Potash Corp. in Rocanville, Saskatchewan. The company expected a record year for potash shipments in 2011, driven by international agricultural demand.

Perspectives on the Environment
Recycling Growing in Importance

Natural Resources Canada (2006a) has estimated that trade in recyclable ferrous and non-ferrous metal, including waste, scrap, ash, and residue, reached 6.7 metric tons, valued at $4.8 billion. Most of Canada's import and export of recycled metals is with the United States. In 2005, for example, Canada exported 87 per cent of its recycled metal to the US, while imports from the US represented nearly 99 per cent. More than 3,000 metal recycling companies operate in Canada, employing some 15,000 people.

Developing a Diamond Mine: Ekati, NWT

In 1991, following more than a decade of geological detective work, two Canadian geologists discovered minable diamonds underneath Lac de Gras in the Northwest Territories, leading to the opening in 1998 of the first diamond mine in Canada by BHP Diamonds Inc., after an investment of $700 million. Called Ekati, the mine is located 200 kilometres south of the Arctic Circle. A second diamond mine, Diavik, began producing in 2003 after a $1.3 billion investment by Rio Tinto/Aber Resources.

In 2003, the Nunavut Impact Review Board conditionally approved Canada's third diamond mine and the first in Nunavut. The mine was operated by Vancouver-based Tahera Corporation, and its Jericho mine, located 420 kilometres northeast of Yellowknife began commercial production in July 2006 (Tahera Diamond Corporation, 2008). However, while more than 786,000 carats of gem-quality diamonds were produced from the Jericho mine between 2006 and the winter of 2008, Tahera closed the mine in February 2008 and put the mine up for sale in early 2010. It cited the high operating costs, especially due to the high value of the Canadian dollar and increasing energy costs, which made the mine unprofitable. De Beers, a South African-based mining multinational, began producing diamonds from a fourth mine at Snap Lake, NWT, in January 2008, 220 km northeast of Yellowknife.

In addition to exploration and diamond production in the NWT and Nunavut, major exploration is underway in Alberta, Saskatchewan, Manitoba, Ontario, and Quebec, and has begun to yield results. The Victor diamond mine, another De Beers operation, is located about 500 km north of Timmins in the Hudson Bay Lowland of Ontario and began production in 2008. One of the largest bodies of diamond-bearing ore in the world has been found in the Fort á la Corne forest in central Saskatchewan. The executive vice-president of exploration of Shore Gold has indicated that diamonds could be mined from its holdings in Saskatchewan by 2016, with expectation that the mine would operate for 12 years. Some predictions indicate that, in 20 years, Canada could be the global leader in terms of the value of diamond production.

The story of Ekati is now part of Canadian mining lore. In 1980, exploration geologists Chuck Fipke and Stewart Blusson noticed alluvial traces of pyrope garnet, ilmenite, and chrome diopside—all indicator minerals associated with diamonds—while working near the border of Yukon and the Northwest Territories. These indicator minerals had been dispersed during the last ice age 10,000 years ago, and Fipke and Blusson realized that their presence did not mean diamonds were close by. Analysis of paleoglaciation and drainage patterns led them to believe that they would have originated in **kimberlite pipes** somewhere in a 65-million-square-kilometre area of tundra to the east.

They began an extensive program of exploration, collecting thousands of alluvial samples and examining each one for traces of indicator minerals. By 1983, they observed a promising pattern. As they moved east, the concentration of indicator minerals increased, and more significantly, the crystals had less alluvial wear. By 1989, the trail of indicator minerals had led them some 640 kilometres to the east, and other geologists also were searching. In that year, although no diamonds had been found, Fipke and Blusson staked claims on 1,800 km² of tundra. Later that year, travelling by helicopter, they noticed that Point Lake was circular and much deeper than other lakes in the area. They sampled along its shoreline and discovered a chrome diopside crystal with no alluvial wear, suggesting a kimberlite pipe was close by, perhaps even underneath the lake. Verification required expensive core drilling.

A partnership was arranged with a major Australian mining company, BHP World Minerals, and in 1991, core samples from Point Lake yielded the first diamonds found in Canada. The subsequent public announcement, required under Canadian

Perspectives on the Environment
Kimberlite

Diamonds are a crystalline type of carbon, stable at depths of 150 km or more beneath the Earth's surface. Kimberlite is a rare igneous rock found at the same or greater depth. Eruptions of kimberlite can transport diamonds to the surface of the Earth. Diamond content can be highly variable in a carrot-shaped kimberlite pipe, ranging from nothing to economic concentrations. Because of glaciation in Canada's North, the top part of the diamond-yielding kimberlite pipes often have been scoured out, leaving a circular depression. Glacial alluvium and water then fill the depression, creating a circular lake over the kimberlite pipe, as was the case at what has become the Ekati mine. There are 136 known kimberlite pipes in the NWT.

—Rylatt (1999: 39); Voynick (1999); Couch (2002: 267)

law, triggered a mineral rush, with 260 companies from eight nations staking claims totalling 194,000 km². Following an environmental impact assessment review spanning 1994 to 1996, Canada became a major producer of diamonds in October 1998 when the Ekati diamond mine started up. The Ekati mine is expected to produce a gross value of $9.5 billion over its projected life of 25 years. By 2009, production had grown to be about 4 per cent of the world's total diamond supply, based on more than 4.5 million carats of rough diamonds annually (Natural Resources Canada, 2009b: 1).

The Environmental Context

A claim was staked by BHP for an area of 3,400 km² situated some 300 kilometres northeast of Yellowknife (Figure 12.2). The mining activity is located mainly in the Koala River watershed, which drains into Lac de Gras and then northward into the Coppermine River and on to the Arctic Ocean. The mine is in the Low Arctic Ecoclimate region in which the average annual temperature is −11.8°C. The temperature range is large, with daily temperatures in summer reaching 25°C and winter temperatures often falling below −30°C. Precipitation is low, averaging only 300 millimetres, most as snow.

Perspectives on the Environment
Meaning of 'Ekati'

'Ekati' is a Dene word, meaning 'fat lake', but it has been said that it actually refers in this context to the white granite rock outcrops that resemble caribou fat.

—www.ccrs.nrcan.gc.ca/resource/tour/43/index_e.php

The BHP claim area is in the tundra region, 100 kilometres north of the treeline. About one-third of it is covered by some 8,000 lakes, and the landscape has continuous permafrost, with permanently frozen subsoil and rock up to

Figure 12.2 | Location of NWT Diamonds Project. *Source: Canadian Environmental Assessment Agency (1996: 6) Reproduced with the permission of the Minister of Public Works and Government Services, Canada, 2005.*

250 metres deep, and an overlayer of about one metre that thaws during summer. The main vegetation includes stunted shrubs and grass tussocks, with willows and scrub bush appearing in low areas. Wetlands include water sedges and sedge-willow communities.

The area supports the Bathurst caribou herd and grizzly bears. The caribou herd has approximately 350,000 animals and moves around a range of about 250,000 km². They spend the winter south of the treeline, then in the spring start their northward migration to calving grounds near Bathurst Inlet on the Arctic Ocean. The grizzly bear, because of low numbers, density, and reproduction rates, has been designated as vulnerable (see Chapter 14).

Economic and Social Context

The economic aspects of the Ekati diamond mine are significant. The total project capital cost is estimated to be $1.2 billion, the contribution to the Canadian gross national product will be $6.2 billion, and the direct, indirect, and induced benefits to the NWT will be $2.5 billion (60 per cent being wages and benefits). The mining company's policy has been to first hire NWT Aboriginal people, then non-Aboriginal NWT residents, and finally other Canadians. When Aboriginal people do not have the necessary skills, the company provides appropriate education and training. The company also committed to give preference to businesses owned by Aboriginal people for contracting, and to establish scholarship programs, on-the-job training programs for Aboriginal students, and cross-cultural training in the workplace

Assembling Data Related to Environmental Impacts

The mining company, BHP, collected baseline data during 1992, with systematic and intensive field sampling begun in 1993. The sampling program addressed biological, cultural, and socio-economic issues. Through the environmental

Panda Pit.

Barren-ground caribou.

assessment process, various federal government departments provided comments about the field sampling design. The most significant comment focused on the need for BHP to make additional effort to incorporate traditional knowledge (see the Introduction to Part B for a discussion on TK) into the collection of conventional scientific data.

In response, BHP observed that it faced serious challenges in its efforts to include traditional knowledge into its research program. First, the Treaty 8 and Treaty 11 Dene groups were in the midst of land claim negotiations and were reluctant to release traditional knowledge into the public domain because the knowledge was important for their negotiation strategy. Second, Aboriginal people expressed concern about traditional knowledge being used outside the context of the cultures and broader system of knowledge that gives it meaning. Third, there was not one set of traditional knowledge, since the Inuit, Métis, and Dene each have their own traditional knowledges, which do not always coincide. Fourth, traditional knowledge was viewed by Aboriginal people as their intellectual property, and therefore its use and management had to remain within their control. And fifth, there was no documented baseline of traditional knowledge, nor were there any generally accepted standards or methods to guide traditional knowledge research.

Mining Tailings

Management of mining tailings is necessary because of the potential impact on downstream water quality. During the mining operations, 35 to 40 million tonnes of waste rock are excavated each year. The ore is crushed, and diamonds are separated through physical means. The crushed rock or tailings are being placed in the Long Lake tailings impoundment basin for the first 20 years and then in one of the mined-out pits for the final five years of the project.

The capacity of Long Lake was increased by building three perimeter dams. Each dam has a central core of frozen soil saturated with ice and bonded to the natural permafrost. The core is surrounded with granular fill to ensure both stability and thermal protection. The use of a frozen core and permafrost foundation ensures that no water can escape through the dams as long as the soil remains saturated with ice

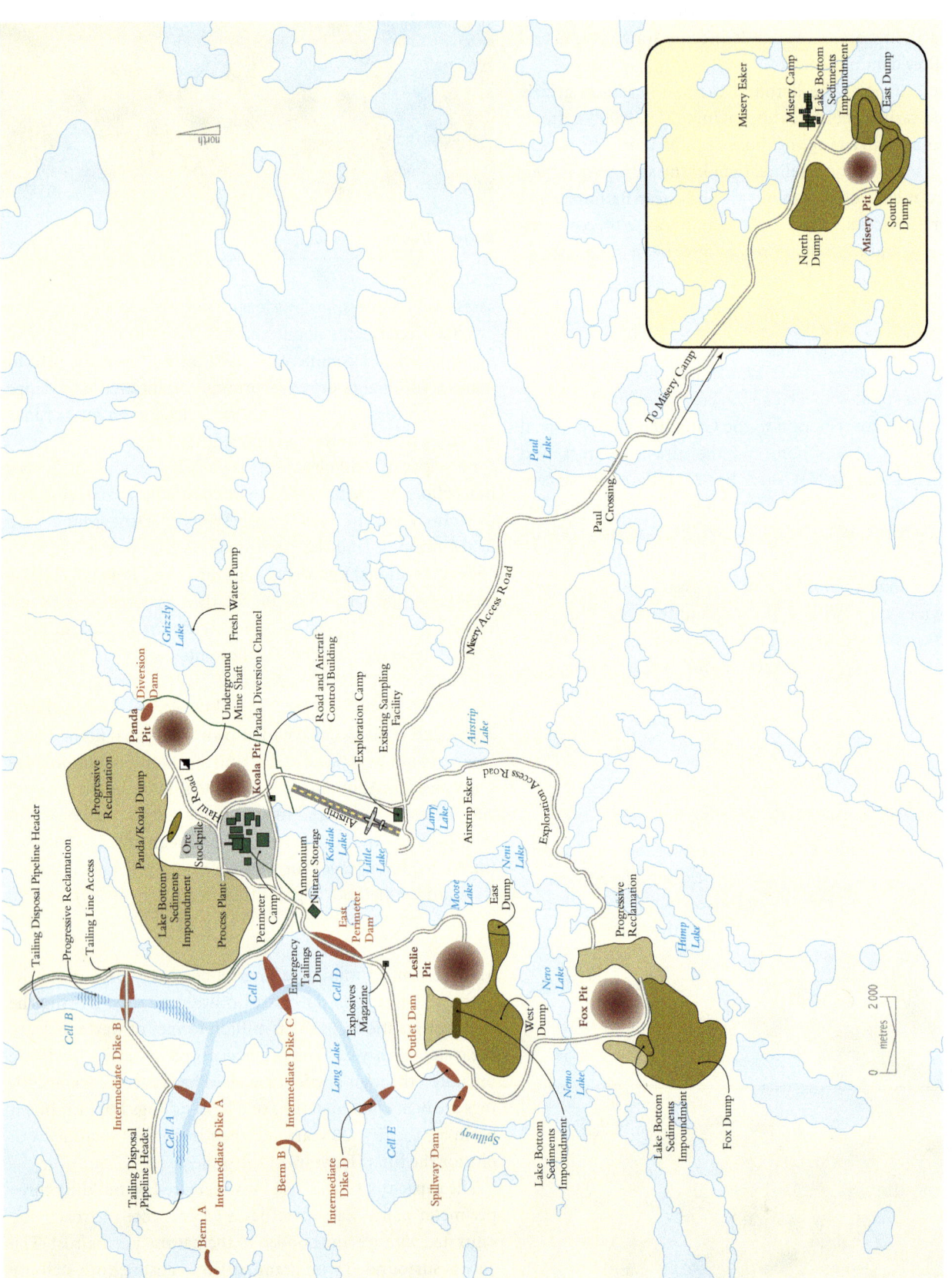

Figure 12.3 | Development plan area, NWT Diamonds Project. Source: Canadian Environmental Assessment Agency (1996: 7). Reproduced with the permission of the Minister of Public Works and Government Services, Canada, 2011.

(Figure 12.3). The frozen core dam design was chosen because of the lack of impervious fill needed for construction of a conventional dam, because the climate is conducive to a frozen core design, and because previous experience with frozen core designs in Canada and Russia could be drawn upon.

The design has the tailings gradually consolidating and becoming permafrost. Once a frozen crust has formed over a cell, it will be covered with waste rock and then topped with fine granular soil. Such a covering will be thick and moist enough to facilitate creation of a new active layer in the new permafrost system. Subsequently, the soil will be revegetated, with the ultimate purpose being to create a wetland.

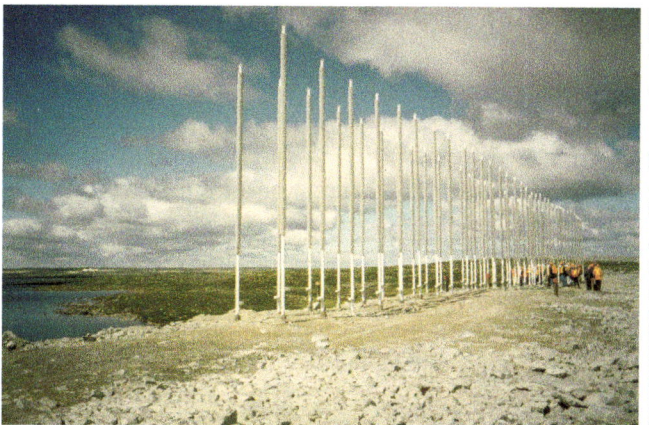

The Panda Pit frozen core dam.

Reproduced with the permission of Natural Resources Canada 2011, courtesy of the Geological Survey of Canada (Photo 2001-260 by Lynda Dredge).

Migratory Caribou

BHP conducted and supported research on the Bathurst caribou herd, the largest in the NWT (Figure 12.4). Baseline data were collected in 1994 and 1995 to determine the numbers of animals using the Lac de Gras area during migrations, the location of migration corridors, and the use of habitat. The migration patterns differed during the two baseline years, which was consistent with natural variability in caribou migration and use of habitat. The government of the Northwest Territories agreed that the ability to predict, on an

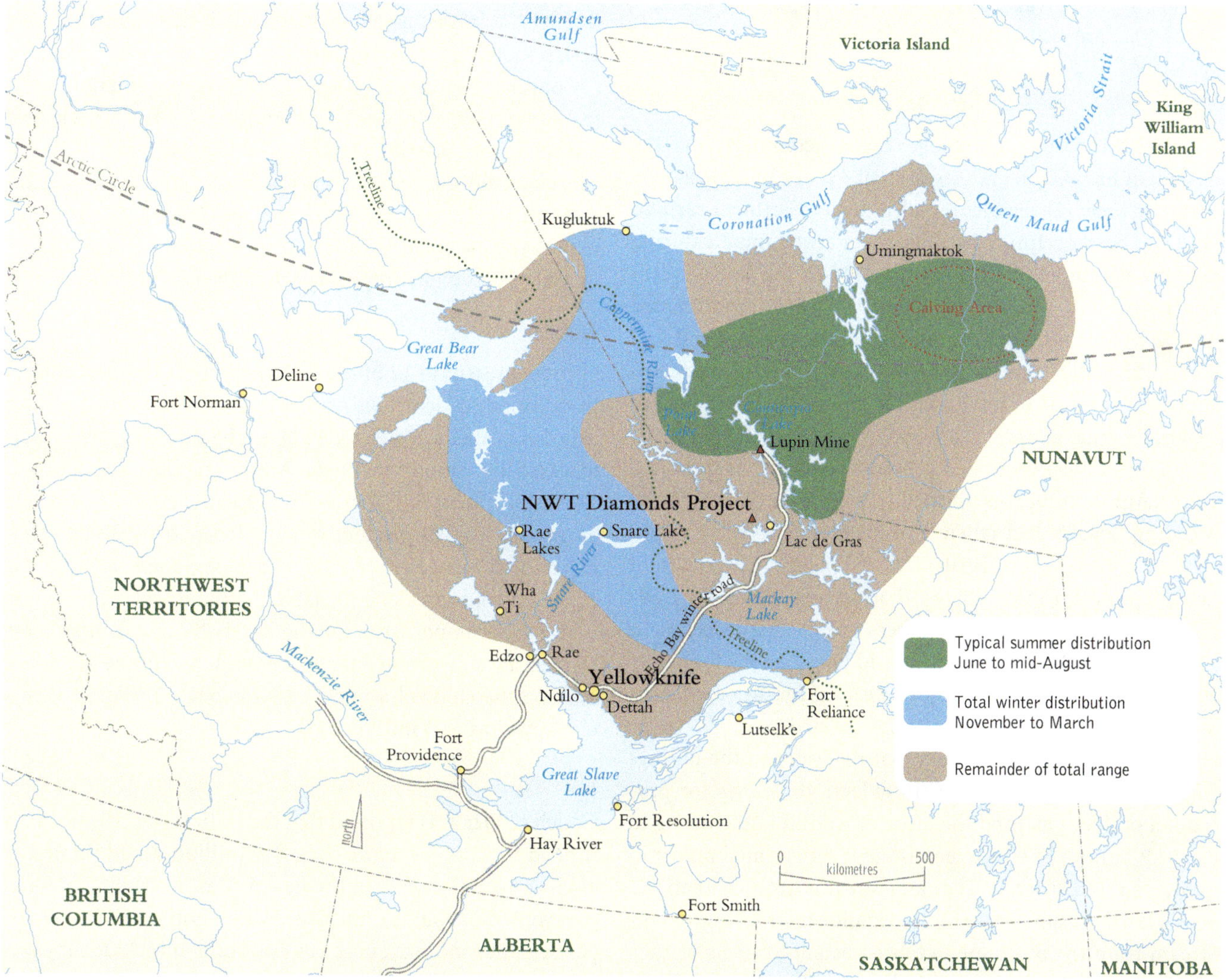

Figure 12.4 | Distribution of Bathurst caribou herds. *Source: Canadian Environmental Agency (1996: 40). Reproduced with permission of the Minister of Public Works and Government Services, Canada, 2011.*

annual basis, the timing and numbers of caribou in the vicinity of the proposed mine was low (Canadian Environmental Assessment Agency, 1996: 39).

Since the caribou herd does not follow the same migration route each year and the areas affected by mining activity represent less than 0.01 per cent of the range of the herd, it was believed that mining would have a very small impact. Attention was also given to the possible effects of roads and the new airport landing strip, either through collisions of caribou with vehicles or as a barrier to migration. It was concluded that these developments would not cause problems.

Perspectives on the Environment
Bathurst Caribou Herd

Among environmental issues, the potential effect of the Project on the health, numbers, and migratory patterns of the Bathurst caribou was the most important concern raised. . . . The GNWT told the Panel that the cultural value of the herd could not be estimated but that the dollar value of the harvest, based on meat replacement costs, was $11.2 million annually.

—Canadian Environmental Assessment Agency (1996: 39)

Notwithstanding the above boxed comment by the Canadian Environmental Assessment Agency in 1996, a 2006 survey of breeding females revealed that the Bathurst caribou herd had been declining by about 5 per cent annually during the past decade (NWT, 2006). In 2006–7, the Independent Environmental Monitoring Agency (IEMA) commented on the lack of progress on understanding the regional and cumulative effects on caribou and recommended increased action on the part of territorial governments. The results of a photographic survey conducted during June 2009 on the calving grounds by the NWT Department of Environment and Natural Resources showed the estimated number of breeding females had fallen from 55,600 in 2006 to 16,000 (plus or minus 4,500 animals). The Minister of Environment and Natural Resources stated that caribou herds do traditionally cycle, but the low numbers and dramatic decline in 2009 highlighted intervention would be necessary to ensure recovery by the herd. However, the reasons for the precipitous fall in the herd size could not be determined.

Finally, modelling indicated that the water in the tailings impoundment would be within federal guidelines for protection of livestock (and therefore of wildlife). According to the BHP Billiton Environmental Agreement annual report in 2001, 'the undisturbed lakes and streams around Ekati are very clean.' Results from water quality monitoring downstream from the mine site indicated balanced levels of zooplankton and phytoplankton, indicators of healthy lakes and streams. The slight fluctuations in pH and nitrate levels in the Koala River watershed are not viewed as a threat to fish.

Water Issues

Water flow changes would be caused by draining the lakes to facilitate open-pit mining as well as by diversion of flows around the pits and by the infilling of Long Lake with the tailings. In total, 15 lakes will be affected. Drainage of the lakes prior to open-pit mining will be managed so that flows will not be greater than 50 per cent of the mean annual flood levels in any downstream water system containing fish. As a result, the main consequence of draining the lakes will be to extend the peak spring flows for a longer period of time. Because the connecting channels between the lakes are both wide and braided, the effects of the extended period of higher flow were judged to be negligible.

The potential impact of mining operations on water quality attracted considerable attention. A primary concern was that contaminants from the mining operation could affect downstream consumers of fish and drinking water in the Coppermine River watershed. The main issue was whether the tailings impoundment in Long Lake would ultimately release water of acceptable quality. Analysis focused on three water quality variables (suspended solids, total nickel, and total aluminum). During the impact assessment process, it was agreed that the design should meet all regulatory standards for water quality.

Concern also arose about possible contamination from toxicity of kimberlites, acid generation from waste rock, and nitrogen from blasting. Analysis led to agreement that such contamination would be controlled satisfactorily. Effects on groundwater also were addressed. Baseline data were collected, and a long-term monitoring program was established so that effects on hydrogeology could be tracked.

Fish

Fish are present in 12 of the 15 lakes affected by the mining. In addition, 43 connecting streams, outflow streams, and inflow streams are affected. The main species is lake trout, followed by round whitefish, Arctic grayling, and burbot. The Department of Fisheries and Oceans (DFO) has a policy of 'no net loss' of productive capacity of fish habitat. That means that whenever fish habitat is degraded or lost, DFO expects a counterbalancing habitat replacement.

BHP has compensated for the lost fish habitat in streams by creating a diversion channel between two of the key lakes, making the channel a quality fish habitat. The cost of this initiative was $1.5 million.

Costs

Couch (2002: 274) reported that the initial scientific research funded by BHP cost more than $10 million. In addition, the environmental assessment review process cost the Canadian government about $1 million, with another $255,000 for participant funding. These amounts do not include the costs incurred by various federal departments, as well as by the government of the Northwest Territories. In Couch's view, 'in comparison with the Project's capital cost, the anticipated profits to BHP Diamonds Inc., and the tax revenue to governments, this outlay was very small' (ibid.).

Environmental Assessment Process

Beginning in 1992, BHP began research to understand the impact of the proposed mining activity and to develop mitigation measures. Company representatives visited all communities in the project area. BHP made public presentations, organized field trips, held community meetings and open houses, facilitated cultural exchanges and workshops, and sent a group of Aboriginal people to its mines in New Mexico, where 76 per cent of its employees were Native Americans.

In 1994, the Minister of Indian Affairs and Northern Development referred the mining project for an environmental assessment, and in 1995 BHP submitted its environmental impact statement. From late January to late February 1996, an Environmental Assessment Panel appointed by the Minister of Environment held public meetings, and its report was submitted to the federal government in June 1996. In 1997, the federal government gave its formal approval, and construction started. In January 1999, the first diamonds from Ekati were sold in Antwerp, Belgium.

Agreements and Arrangements

Emerging from the process outlined above were several agreements and arrangements.

Announcement from the Minister of Indian Affairs and Northern Development. In August 1996, the minister announced his acceptance of the Environmental Assessment Panel's report and gave participants 60 days to work out detailed agreements. Given that land claim negotiations were also underway, potential existed for much debate and disagreement. The tight time frame put pressure on all participants to work out details.

Environmental Agreement. The Environmental Agreement is legally binding and requires BHP to: (1) prepare a plan for environmental management during the construction and operation of the diamond mine; (2) submit annual reports related to the environmental management plan; (3) prepare an impact report every three years related to the impact of the project; (4) establish a monitoring program for air and water quality and for wildlife; (5) submit a reclamation plan for approval; (6) establish a security deposit ($11+ million) for potential land impacts and a guarantee of $20 million for potential water impacts; and (7) incorporate traditional ecological knowledge into all environmental plans and programs.

In addition, an Independent Environmental Monitoring Agency (IEMA) was established by the two governments and BHP as a public watchdog. The IEMA: (1) prepares annual reports on the project's environmental implications; (2) reviews impact reports; and (3) provides a public document repository at its Yellowknife office. This was an innovative feature, since it had not been recommended by the assessment panel.

Perspectives on the Environment
Independent Environmental Monitoring Agency

The Monitoring Agency was created to ensure that the Ekati Diamond mine meets the conditions set forth for its environmental approval . . . the Government of Canada, the government of the Northwest Territories, and BHPB Diamonds Inc. . . . wrote a legally binding Environmental Agreement. . . . The Environmental Agreement called for the establishment of the Independent Environmental Monitoring Agency, and provided its mandate. . . . The agency is a non-profit organization. . . .

—Independent Environmental Monitoring Agency (2010a)

In the IEMA report published in 2010, the chair of the agency stated that he 'was pleased to report that BHP Billiton (BHPB) has continued to do a reasonably good job of protecting the environment at EKATI Diamond Mine.' He also added, however, that the agency 'still has concerns that need to be addressed so that this good performance can last. These deal mostly with water quality downstream from the Long Lake Containment Facility . . . and with wildlife (especially caribou)' (IEMA, 2010b: 1). Various observations were made, including:

- The three diamond mines in the NWT should work together on wildlife monitoring to ensure cumulative effects are tracked.
- A long-term management plan is needed for the tailings operations in the Long Lake Containment Facility.
- How BHPB has incorporated traditional knowledge into its environmental programs at Ekati needs to be clarified. Regarding projects such as showing how to make traditional drums or providing donations for cultural gatherings, 'it is not clear how they improve environmental management' (ibid., 9).
- The Wildlife Effects Monitoring Program (WEMP), 'while covering a lot of information, does not really focus on what is most important. It still gives findings that BHPB admits are weak at best, often based on poor data and sample sizes. Parts of the report do not give a big-picture point of view' (ibid., 11)
- Regarding closure planning for the Ekati mine, and the Interim Closure and Reclamation Plan (ICRP), an important issue is 'BHPB's proposal not to restore the pit lakes for fish use or travel. . . . BHPB says it made an agreement with Fisheries and Oceans Canada (DFO) and paid money for the right to destroy fish habitat. So, BHPB says it does not have to create fish habitat. The company is alone in this view' (ibid., 20).
- The monitoring agency concludes that BHPB 'runs the Ekati Mine in an environmentally sound way' but 'there is, however, always room to improve.' In addition, it 'found the Environmental Impact Report 2009 to be not good enough' (ibid., 25).

Guest Statement
Accountability in Resource Management: Independent Oversight of Proponents and Government
Patricia Fitzpatrick

As environmental problems become increasingly characterized by uncertainty, complexity, and conflict, there is growing interest in how to manage resource-based activities to be ecologically sound, socially responsible, and fiscally prudent, as well as to ensure that the best interests of the public(s) are achieved. Both aspects traditionally fall under the purview of government. However, some government departments have conflicting mandates; for example, Natural Resources Canada is charged with enhancing the development of Canada's resources, but through its Major Projects Management Office it also facilitates the regulatory process designed to protect the environment. Budget cuts to the federal civil service also create concerns about the capacity of departments to complete commitments. Finally, catastrophic failures by government and industry to protect the environment, especially when resource development promises tax revenues and huge profits, result in public uneasiness surrounding government priorities. These issues, among others, have led to the creation of different models to ensure project-specific accountability.

One model with growing resonance over the last 20 years involves private contracts. These contracts or agreements, between the proponent and different parties (including some combination of government and non-governmental organizations), are designed to ensure a proponent fulfills its environmental, social, and economic commitments. Agreements address many issues, including the development of environmental management programs, reporting requirements, reclamation plans, security deposits, local training commitments, health and social services programs, local business development initiatives, and, as discussed by Michael Hitch (see page 432), social investment by corporations for the benefit of local indigenous groups and organizations. Independent oversight is one way to address accountability when private contracts are used.

Independent oversight involves the creation of an institution or board that is autonomous or semi-autonomous from government and/or the proponent. Oversight bodies may fill several potentially overlapping functions, including:

- providing surveillance of the environmental and social system affected by the development;
- offering verification of environmental impacts;
- facilitating a technical review of management systems used by the proponent and/or government involved in the development;
- creating a venue for public participation in management decisions;
- functioning as a vehicle for public communication about a specific development.

Independent oversight has been created or proposed for mineral, oil, and gas developments in various stages of production (proposed, operating, post-closure) around the world. Examples include the Independent Environmental Monitoring Agency for the Ekati diamond mine in the Northwest Territories; the Prince William Sound Citizens' Advisory Committee for the ongoing operation of the Alyeska Pipeline Transfer Station in Alaska, and the proposed oversight for the Giant Mine Remediation in the Northwest Territories.

The literature on independent oversight bodies is small, but growing. Studies largely focus on case studies involving an oversight body or a group of oversight bodies in a geographic area (e.g., diamond mining in the Northwest Territories). However, four aspects related to best practice emerge.

Giant Mine is an abandoned gold mine on the edge of Yellowknife, Northwest Territories. Intervenors in the environmental assessment of the proposed remediation project are studying the need for and potential models of independent oversight.

1. *Clear mandate.* Since independent oversight can serve various purposes, the agreement must specify the roles and responsibilities of the institution or board.
2. *Independence.* Although some oversight agencies include representatives from the federal and territorial governments and the proponent, the more successful ones comprise representatives from local non-governmental organizations and local and Aboriginal governments.
3. *Adequate, long-term funding.* Gaining adequate funding to fulfill tasks in the mandate is a struggle for most oversight agencies. Funding ranges from nothing up to $3.3 million annually, in the case of the Prince William Sound Citizens' Advisory Group. Beyond sufficient funding, best practice involves multi-year funding arrangements so that long-term programs can be established.
4. *Experience.* Organizational learning and experience are important. Assessments during the first decade of operation focus on aspects for improvement, while literature reviewing agencies in existence for 20 years or more (e.g., the Prince William Sound Citizens' Advisory Group) focus on organizational strengths or achievements. More research is needed. Specifically, to what degree do different contractual arrangements influence the long-term success of an oversight agency? Are positive, long-term reviews more favourable because, in fact, the structure and function of the agency are more effective and efficient than those of younger agencies? Or, alternatively, are experience and learning necessary aspects for success?

Independent oversight is an important tool in efforts to facilitate environmental sustainability in resource development. It is important to ensure that each oversight agency reflects the circumstances of the project for which it was created, and employs best practices, so resources are not wasted.

Patricia Fitzpatrick, Ph.D., is an Associate Professor in the Department of Geography at the University of Winnipeg. Her research is concerned with the changing nature of resource management. She is currently exploring the relationship between public and private regulation of environmental and social sustainability in the minerals sector.

Socio-economic Agreement (SEA). This agreement between BHP and the government of the Northwest Territories focuses on commitments beyond existing statutory requirements. The concern was economic benefits and social impacts related to all NWT residents, not just traditional users of the project area. The agreement covered matters such as preferential hiring of NWT residents (with a target of 62 per cent northerners and 31 per cent Aboriginal peoples), criteria to guide recruitment, employment targets, employment of local contractors, training programs, and employment support. Targets were specified for awarding contracts to and purchases from northern businesses, as well as for employment of Aboriginal and northern residents.

Although not part of the agreement, a noteworthy initiative has been the establishment of diamond-cutting and polishing businesses in Yellowknife. The traditional centres for polishing are Antwerp, Tel Aviv, New York City, and India. In 1999, a small Vancouver-based diamond-polishing company opened a facility in Yellowknife, recruiting a South African diamond cutter from Antwerp. Shortly afterwards, another company opened, with cutters recruited from Armenia. Other firms have opened facilities as well, and local people are learning the trade under the guidance of cutters from Europe, Israel, and Africa.

Impact and Benefit Agreements. In 1994, BHP began negotiations with the four Aboriginal groups. Each group was involved in land claim negotiations, and BHP did not want to get entangled in those processes. **Impact and benefit agreements** (IBAs) are one tool for addressing community and industry relations in mining or other extractive resource activities. They are voluntary agreements, beyond formal impact assessment requirements, and are intended to facilitate extraction of resources in a way that contributes to the economic and social well-being of local people and communities. IBAs create opportunities for communities to realize direct economic benefits from natural resource development projects as well as to participate in the management, monitoring, and mitigation of impacts. All of these matters were addressed in the IBA between BHP and the four Aboriginal groups. In the guest statement below, Michael Hitch provides further insight regarding IBAs in the context of the concept of corporate social responsibility.

Managing Change and Conflict

Mining activity often generates conflicts relative to other land uses. Two examples illustrate the challenges.

Ring of Fire

Some 500 km north of Thunder Bay a pristine wilderness area of just over 5,000 km² is the traditional homeland of the Marten Falls Aboriginal people. Recent geological exploration has revealed massive deposits of chromite, used in making stainless steel. The deposits are believed to be sufficient to maintain mining for 150–200 years. This area has been labelled the 'Ring of Fire', named by a mining company

Guest Statement

Corporate Social Investment

Michael Hitch

Corporate social responsibility is the new face of the sustainability discourse. Over the past decade, we have seen the enrichment of the definition of this amorphous concept called sustainability. Each of us has our own personal notion of what sustainability is or can be, and some may align themselves with the structure outlined in Figure 12.5.

Two critical elements of sustainability are the mutual understanding between developer and affected parties under the banner of social licence. Social licence is often achieved through negotiated impact and benefit agreements, as described above. Adjacent, and possibly part of the same process, is laying the groundwork and supporting local capacity for social and economic continuation beyond the life of the mining project. The impetus for both of these is usually some form of financial support from the proponent developer and administered as **corporate social investment** (CSI). CSI can be defined as contributions or actions by a company to assist the affected community in achieving its own development priorities in ways sustainable and supportive of both parties' objectives.

In some sustainability circles, the notion of corporate morality is becoming *de rigueur*. This seems to be especially so in mining. The idea of corporations having a sense of morality at first glance seems counterintuitive to the corporation's rationale for existing: to increase in value and hence to ensure that shareholders prosper. The argument can be made that the corporate entity itself cannot be moral but rather should operate to minimize negative biophysical and social consequences, at the same time employing tools that foster community development beyond the operating life of the mine. What is the difference between a new mining operation and a new auto assembly plant when it comes to this idea of corporate morality? Is the auto plant required to comply with the same demands as a mining operation in terms of community development or capacity development? Perhaps there is an enhanced sense of expectation or maternal role the mine will satisfy, and how the developer addresses it will make or break its acceptance in the community. Mining companies are sensitive to this and are normally eager for a strategic alignment with their overall business goals.

When social investment is aligned with corporate goals, a strong business case exists for its continuance. Essential to this is an overall acceptance and buy-in by management and a focus on an understanding and engaged workforce. For the most part, focused social investment programs are viewed as an element of the cost of doing business rather than as voluntary philanthropy.

The alignment of the mineral developer's and the affected community's goals and aspirations is an essential element in achieving the condition of 'shared value'. Such achievement consists of policies and practices that enhance a company's competitiveness while simultaneously advancing the economic and social conditions in the communities in which it operates.

The application of CSI also comes with some dangers. First is a tendency for the company's financial involvement to replace or become a substitute for programs under the purview of an existing government body (e.g., community health care, primary education, appropriate governance). Similarly, CSI may exacerbate existing tensions between groups within the community. I have seen cases where certain segments of a population are excluded from the benefits from impact and benefit agreements. This is particularly the case in Nunavut, where only signatories of the Nunavut Comprehensive Land Claims Agreement receive any CSI-oriented benefit from mineral development in the territory. Those not parties to the agreement experience shortages in housing and labour, limited business opportunities, overstressed social services, and the complicating effects of two governance structures in

Figure 12.5 | Core elements of sustainability.

Grandview Gold Ltd developing an inoculation program in northern Peru.

Grandview Gold Ltd introducing drought-resistant corn seed and organic fertilizers in northern Peru.

a hamlet of fewer than 2,000 people (e.g., Cambridge Bay/Iqaluktuuttiaq). Another negative aspect of CSI is that community dependency can develop, replacing one dependency for another.

The key to successful CSI programs is to sustain their positive impact. To achieve this, an exit strategy must be incorporated into its design. These programs cannot be open-ended: just as a mine has a finite life, so does a CSI program. CSI programs need to build on existing assets, capacities, and plans the community identifies with. There also needs to be a careful recognition of the community's responsibility for its own success and the importance of making choices and setting priorities.

CSI programs are as much about community empowerment as financial assistance and capacity-building. By sharing the wealth from a mine, the community is in a better position to assert itself, emerge from the bonds of dependency on any single industry, and soar on the wings of its own strengths.

Michael Hitch, Ph.D., a professional geologist and engineer, is an assistant professor in the Norman B. Keevil Institute of Mining Engineering, University of British Columbia. Beyond his 23 years of industry experience in 165 countries, his research interests include sustainable mining in Canada, small-scale mining and poverty alleviation in Peru, CO_2 sequestration in mine wastes in India, and energy-efficient bulk materials handling systems in Indonesia.

executive who also is a fan of Johnny Cash. If the deposits are developed, massive change will occur in the area, ranging from a new 350-km railway, a processing plant, jobs for Aboriginal people for several generations in an area with few employment opportunities, and significant tax revenue to the Ontario provincial government.

Complexity and conflict exist because environmental groups are concerned about negative impacts on a sensitive ecosystem, and Aboriginal leaders are determined that any development will create local jobs and contribute to the local economy. This determination was demonstrated during two months in the winter of 2010 when two Aboriginal communities blockaded their local airstrips to prevent companies from landing.

The world's supply of chromite primarily comes from South Africa and Zimbabwe, which have 70 per cent of the world's reserves. Its discovery in the Hudson Bay Lowland was accidental. Geologists were examining the area near Marten Falls and McFaulds Lake for diamonds when they discovered massive deposits of copper, nickel, and platinum. During the follow-up exploration, chromite was discovered in an area of wetlands and bush. Its value has been estimated to be $30 billion.

Falling prices for chromite, the Aboriginal blockage of airstrips, and concerns about opposition from environmental groups, such as the Canadian Parks and Wilderness Society, slowed exploration and development activity during 2010. On the other side, a provincial government with a large deficit and debt is vigorously examining ways to diversify the provincial economy, and a long-term mining operation in the North of the province has many attractions.

Horn Plateau

The Horn Plateau, in the Northwest Territories, is an extensive region of boreal forest, uplands and wetlands, and headwaters for three river systems. It provides habitat to woodland caribou and wolverines, both endangered species, as well as for migratory birds. The area also has spiritual and cultural significance for various Aboriginal peoples.

Beginning in 1992, discussions began to protect the area. In 2007 the federal government agreed that 25,000 km² would be off limits for staking mineral claims. Subsequently, that area was reduced to 14,000 km², after discussions with industry, Aboriginal peoples, and environmental stakeholders. However, in 2010, without consultation, the federal government opened the entire area for mineral claim staking. A legal challenge was subsequently submitted by the Deh Cho First Nation on the basis that the federal decision would jeopardize the possibility of the area becoming a national wildlife area within the NWT Protected Areas Strategy.

The Ring of Fire and Horn Plateau highlight the multiple dimensions often associated with a proposed mining operation. They emphasize that, in addition to addressing technical issues, decisions-makers normally have to resolve conflict and uncertainty for and among stakeholders. What would you recommend for these two situations, based on the experience of the Ekati diamond mine, as well as the planning concepts discussed in Chapters 5 and 6?

Perspectives on the Environment
Significance of Energy

... energy is society's critical master resource: when it's scarce and costly, everything we try to do, including growing our food, obtaining other resources like fresh water, transmitting and processing information, and defending ourselves, becomes far harder.

—Homer-Dixon (2007: 12)

Energy Resources

Energy resources are classified as renewable and non-renewable. Renewable resources are those that can be replenished in a relatively short time period. Figure 12.6 identifies three renewable energy sources, and one of these, gravity, is ongoing and widespread but remains as a potential unless associated with significant motion, such as tides or river flow. Geothermal heat also is persistent and widespread but at great depths below the surface. Manifestations of geothermal heat at the surface or shallow depths, as is the case, for example, in Iceland and in areas in New Zealand, are much more limited and are usually associated with the heat being carried by water or steam, so the renewability of geothermal heat depends on a reliable and ongoing supply of water. Solar supplies come from continuous emission of radiation from the sun, but it arrives discontinuously on the surface of the Earth because of diurnal and seasonal variation as well as cloud cover. As a result, renewable, solar-based energy supplies are intermittent and often cyclic, meaning that they usually must be supplemented by other sources.

Biomass energy sources are frequently used in rural areas in developing countries and can take the form of millions of people and their draft animals doing subsistence work. Metabolic energy (muscle power) is supplemented by heat created from burning firewood, from crop and animal wastes in basic biogas converters, and from direct sunlight used to dry and preserve agricultural or marine products (e.g., dried fish). Biomass energy is renewable as long as the rate of use and capacity to produce biomass are balanced.

Non-renewable sources cannot be replenished over a period of time short enough to support humans. These sources result from geological processes over millions of years, which lead to solid (coal) and liquid (oil) fuels, natural gas, and nuclear fuels. While they all share the characteristic of offering high energy content per unit of weight or volume, they also differ. Solid fuels are mined, which is labour-intensive and requires expensive infrastructure. For efficient transport, they must be carried in bulk or batch containers, such as rail cars or ships. When burned, solid fuels release gaseous and particulate matter in large quantities. In contrast, oil and natural gas can be produced with facilities requiring relatively little labour but capital-intensive refineries or processing plants. Once processed, the product can be transported continuously through pipelines or in batches (trains, ships, trucks). Nuclear fuels contain the highest content per unit of weight but require sophisticated facilities and highly skilled human resources. They are used only to generate electricity and demand careful handling in processing and waste

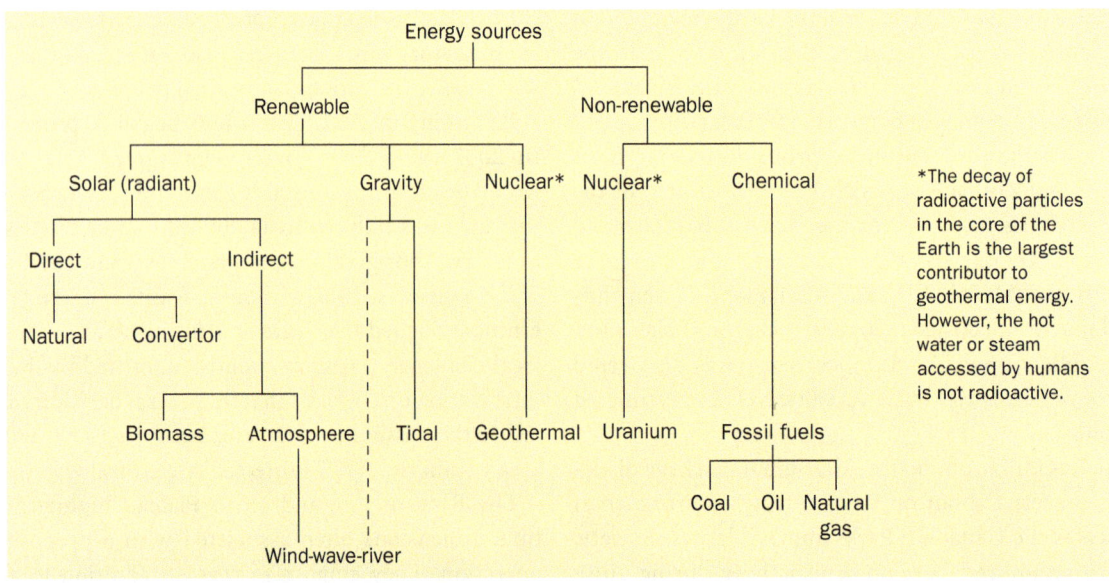

Figure 12.6 | Energy sources. Source: Chapman (1989: 4).

disposal. Given these different attributes, the most appropriate sources of energy will vary depending on circumstances. Box 12.1 highlights the different variables that need to be considered when making a choice.

Energy Use and Issues in Canada

According to Statistics Canada (Ménard, 2009: 1), Canada is 'a huge consumer of energy'. The main explanations are a growing population and economic growth, while other important factors are long, cold winters; large travel distances; and an economy reliant on high energy-consuming industries (mining, forestry, pulp and paper, petrochemicals, aluminum smelters, refining and steel manufacturing). The region with the greatest increase in energy use has been Alberta, attributable to high population growth and an economy based on energy-consuming industries. The major source of energy has been fossil fuels (refined petroleum products, natural gas, and coal). Fossil fuels also are the energy source that releases the greatest amount of greenhouse gases.

Perspectives on the Environment

Petajoule

The energy content of a 30-litre tank of gasoline is about one gigajoule. One million gigajoules is equal to one petajoule. On average, Canada consumes about one petajoule of energy every 50 minutes for all uses.

—Ménard (2009: 2)

Fossil fuels are the main type of energy consumed by Canadians (Figure 12.7), and one of the largest consumers of energy is the transportation sector, which depends overwhelmingly on fossil fuels. Between 1990 and 2007, energy use in the transportation sector increased by 38 per cent, from 1,878 petajoules (PJ) to 2,595 PJ (Natural Resources Canada, 2010c). In 2007, passenger transportation consumed 54 per cent of energy; freight, 42 per cent; and off-road vehicles, 4 per cent. Freight was the fastest-growing subsector related to transportation, accounting for 62 per cent of the increase for the entire sector between 1990 and 2007.

Some fuel efficiencies in automobiles occurred between the early 1970s and the early 1980s, but since then there have been no significant improvements. Instead, there has been an increased use of less fuel-efficient vehicles, such as light-duty trucks and sport utility vehicles. A clear message is that Canada is an intensive energy-using nation and that despite growing efficiencies, energy use is climbing. Given our dependency on fossil fuels, it is appropriate to examine alternative energy supplies. However, it is important to recognize that alternative sources, usually renewable (solar, geothermal, hydro, tides, wind), are not problem-free from an environmental perspective. For example, the impacts of the hydropower development at James Bay, examined in Chapter 11, included increased levels of mercury in aquatic systems, with consequences for both humans and other living species in the area, and changes in fish populations. Since hydro power was addressed in Chapter 11, it is not addressed in this chapter.

Environment in Focus

Box 12.1 Choosing among Energy Sources

1. *Occurrence*. Many energy sources are confined to specific environments and locations and are only available at other locations when transport systems exist. Even physically present sources may not actually be available because of technical, economic, or other constraints.
2. *Transferability*. The distance over which an energy source may be transported is a function of its physical form, energy content, and transport technology.
3. *Energy content*. This is the amount of usable energy by weight or volume of a given source. Low-energy-content sources are inadequate when demand is large and spatially concentrated.
4. *Reliability*. Uninterrupted availability gives one source an advantage over one that is intermittent.
5. *Storability*. To meet interruptions of supply or peaks of demand, a source that can be stored has an advantage over one that cannot.
6. *Flexibility*. The greater the variety of end uses to which a given source or form may be put, the more desirable it is.
7. *Safety and impact*. Sources that may be produced or used with low risk to human health and the environment will be preferred over less benign sources.
8. *Cleanliness and convenience*. The cleaner and more convenient source will be preferred over the dirty and the cumbersome.
9. *Price*. The less expensive source or form will be preferred over the more expensive.

Source: Chapman (1989: 5).

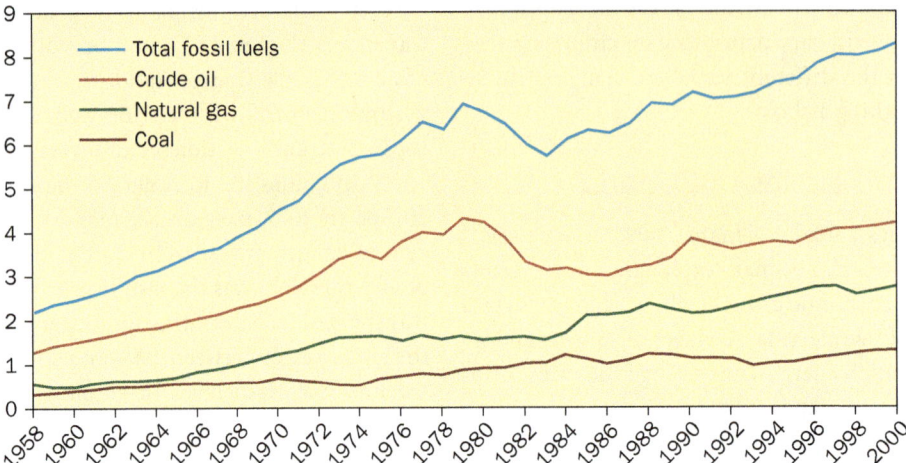

Figure 12.7 | Fossil fuel consumption in Canada (exajoules). *Source:* Environment Canada (2003a). Adapted in part from Statistics Canada Energy Division. Reproduced with the permission of the Minister of Public Works and Government Services, Canada, 2011.

Perspectives on the Environment

Energy and Climate Change

In June 2006, the [National] Round Table [on the Environment and the Economy] released its advice to the federal government on a long-term strategy on energy and climate change. It noted that significant greenhouse gas emission reductions could take place in Canada in mid-century only if energy is used more efficiently and if it is produced while emitting less carbon. It pointed to the need to increase energy efficiency, to perfect carbon capture and storage, and to transform energy generation to clean coal technology, co-generation, and renewable energy, particularly wind power.

—Auditor General of Canada (2006)

Wind Power

Natural Resources Canada (2006c) explains that wind energy converts kinetic energy available in wind to forms of energy more useful to humans, such as mechanical energy or electricity. Furthermore, wind energy is 'a pollution-free, sustainable form of energy. It doesn't use fuel; it doesn't produce greenhouse gases, and it doesn't produce toxic or radioactive waste.'

Humans have used wind energy for centuries, beginning with windmills to provide mechanical energy for pumping water and grinding grain. Sailing ships also depend on the power of the wind. Frequent contemporary uses of wind energy are electricity production and water pumping.

Capacity to generate power from wind depends on several variables, the most important being wind speed. Wind turbines are located in the windiest areas, and within such areas they usually are situated on high spots, since wind speed increases with elevation above the surface. Exceptions are 'wind tunnel areas' at lower elevations.

The expansion of **wind power** has been impressive. Kitasei (2011: 26) noted that between 2001 and 2009, the wind energy industry experienced an annual rate of growth of 31 per cent at a global scale. In 2009, more than 38 gigawatts (GW) of new wind energy installations were installed, bringing the world total to over 158 GW. The United States maintained its status as global leader with 40 per cent of installed capacity. China and Germany were ranked second, and Canada eleventh. Within Canada, Ontario was the leader, with 1,168 MW installed in 2009, followed by Quebec (659 MW), Alberta (590 MW), and New Brunswick (171 MW). In 2009, British Columbia completed its first wind power installation, resulting in every province having an operating wind farm. Wind energy accounted for 1.1 per cent of total energy production in Canada in 2009, a tenfold increase in capacity in six years.

A continuing issue is some public opposition to wind farms because of noise and aesthetics, and concern about impacts on health. As an example, in April 2010, some 300 protestors assembled in front of the provincial legislature in Toronto to support a resident whose home is near a wind farm between Kincardine and Port Elgin. She stated that she has had migraines, insomnia, high blood pressure, and stress ailments that she attributes to the presence of the wind turbines. Another concern was raised regarding social divisions within communities near wind farms because some landowners receive payments for having a turbine on their land, while nearby neighbours receive no financial benefit.

In January 2011, a resident of Prince Edward County in eastern Ontario took the provincial government to court, claiming that the 550-metre setback specified between wind turbines and homes was not based on medical evidence. In

Oppose Belwood Wind Farm Association (OBWF) is a community effort to prevent the installation of industrial wind turbine projects until all long-term effects on health of residents living near such installations have been studied and addressed.

early March, three judges in the Ontario divisional court dismissed his claim, but observed that the setback provision and related regulations under the government's Green Energy Act could be challenged before the Environmental Review Tribunal. In February 2011, the provincial government declared a moratorium on all offshore wind farms until their impact was better understood. A government spokesperson stated that 30 to 40 years of research existed for onshore wind turbines, but little research had been completed regarding offshore turbines.

In September 2011, a family in Thamesville in southwestern Ontario initiated a lawsuit related to the Kent Breeze wind farm, which they allege is causing pain and suffering, and loss of enjoyment of the normal use of their property. The family is suing for $1.5 million for damages, and asked the courts to close the wind farm.

Physiologically, the deleterious effect of wind turbines on human health is fairly simply explained, but at this point it is not yet known or fully understood by the medical profession, by industry, by policy-makers, or by those who have suffered symptoms. Briefly, low-frequency sound, especially of a continuous nature, saps energy and wellness, and can cause a wide range of symptoms—physical and psychological—that, at the lower range of disability, has been called chronic fatigue syndrome. So, living within hearing range of a wind farm may be harmful—possibly extremely so for some.

Thus, debate and conflict can be expected to continue. In sum, wind energy has advantages and disadvantages.

Advantages of Wind Power

- Wind power does not require fuel, create greenhouse gases (GHGs), or produce toxic or radioactive wastes.
- The production of wind energy is quiet and not a significant hazard to birds or other wildlife.
- When large wind farms are established, containing many wind turbines, they require 2 per cent of the land area, making the balance available for farming, livestock, and other uses.
- Payment is made to landowners, which provides another source of income.

Disadvantages of Wind Power

- Wind is not constant, meaning that there will be times when no power is generated.
- When wind turbines are built, conflict often arises because landowners view them as a negative feature on the landscape.
- With a large wind farm containing many wind turbines, noise from the turbines may be intrusive for nearby landowners.

These advantages and disadvantages can be considered with reference to the 'criteria' for sustainable energy options introduced at the beginning of the chapter.

Environmental Impacts of Wind Turbines

With regard to the significance of wind turbines in terms of birds, habitat, noise, safety, and aesthetics, Dillon Consulting Ltd (2000) and Kuvlesky et al. (2007) examined the possible impacts and reached the conclusions outlined below.

Wildlife

Kuvlesky et al. concluded that most research has focused on the impact of wind farm development on birds and bats, with emphasis on mortality due to collisions with turbines. A key aspect of their conclusions can be summarized with the

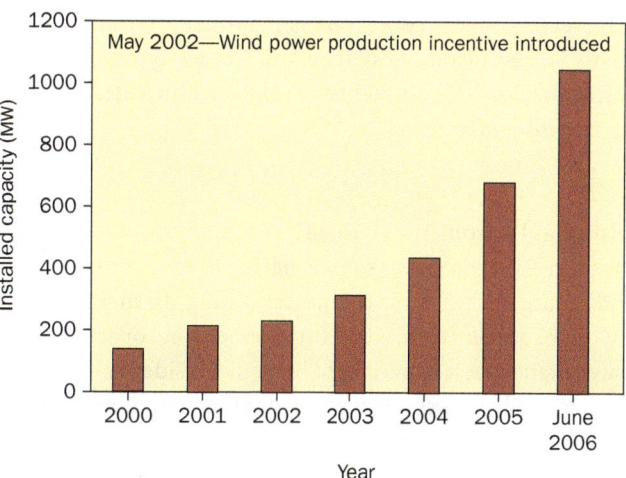

Figure 12.8 | Total installed wind power capacity in Canada. Total installed wind power capacity in Canada is growing; depicted is nationwide capacity as of November 2011. The installed capacity changes frequently as new wind farms are established. *Source:* Canadian Wind Energy Association (CanWEA).

words, 'it depends'. That is, layout of a wind farm, specific attributes of turbines, topography, weather conditions, and the specific types and numbers of birds and their behaviour all affect impact.

Most research has focused on passerines, especially nocturnal migrants, which suffer the highest mortality, regardless of the type of habitats on which wind farms are constructed. Nevertheless, they still concluded that, 'generally collision fatalities are not thought to be substantial enough to impact bird populations' because few birds collide with turbines' (Kuvlesky et al., 2007: 2488). In contrast, they note justifiable concern about impact on raptor populations because substantial raptor fatalities occur due to wind farms. Despite such concerns, their overall view was that raptor populations were not affected by collisions with wind turbines.

Another finding, with significance for waterfowl, is that their collision rates are higher within offshore wind farms than with those on terrestrial sites. In addition, offshore facilities have been shown to divert migration routes of sea ducks from traditional migration paths, but the consequences are not clear.

The above findings indicate that research results are mixed, but in general it appears as if wildlife populations are not significantly adversely affected by wind farms.

Habitat Loss and Change

Loss of habitat due to wind farms is a greater threat to bird, mammal, and herpetofauna populations. Wind farms often render habitats unsuitable for birds. The fragmentation of habitats due to wind farms or related infrastructure (roads, electric transmission lines) can create challenges for wildlife. Another negative impact associated with related road construction and maintenance is the introduction and range expansion of exotic species, as well as increased probability of mortality from collisions with vehicles.

Nevertheless, the conclusion is that disruption from wind farms is significantly less than from other types of energy extraction, such as oil and gas exploration or extraction, or surface mineral mining.

Noise

Disturbance from noise is influenced by many variables, including distance from source, nature of background noise, and nature of the source (frequency, time pattern, intensity). All noise levels from wind turbines during operation are lower than what is experienced in a quiet residential area and similar to what is experienced inside an average home. Dillon Consulting (2000) concluded that given the normal background noise in an average suburban residence, the noise from a wind turbine would be inaudible at a distance of 260 metres. However, it should be noted that Dillon's research focused primarily on volume (decibels). The frequency (hertz) of sound, or pitch, is another key variable, and can affect health and well-being.

Safety

The main safety concern is ice thrown from turbine blades or falling off the tower. Proactive steps can be taken to ensure public safety. First, setback criteria can be used to ensure that people are kept at a reasonable distance from a wind turbine tower and the rotating blades. Second, temperature sensors as well as sensors to monitor the balance of blades can provide early information about ice buildup. Once ice accumulation is detected, the wind turbine can be shut down and not restarted until operators determine that conditions are safe.

Aesthetics

Some people may view a wind turbine or wind turbine field as an unwelcome visual intrusion on the landscape, especially if they feel the turbines are not in keeping with an area's historical, cultural, or natural values. On the other hand, others may enjoy the look of a wind turbine, appreciating its modern, futuristic appearance as well as the symbolism and educational role of a visible environmentally benign technology. The challenge, as Dillon Consulting (2000: 44) observed, is that 'Given the conformity of view that windmills are a good thing but that they should be placed "somewhere else" and not "here", and the recognition that everyone's "somewhere else" is someone else's "here", a balanced answer is needed.'

Research regarding wind turbines or windmills in Europe and North America has indicated that prior to their construction, nearby residents usually have some concerns. However, after the wind turbines are operating, their views normally became either neutral or positive. If the homes receive electricity from the turbines, attitudes are likely to be more positive.

Summary

Evidence indicates that wind turbines have minimal adverse environmental effects. However, issues of health and well-being for those living in proximity remain to be resolved and, compared to conventional fossil-fuel energy sources, wind turbines are still relatively expensive, but that could change as the technology becomes less costly and/or fossil-fuel supplies become more expensive. Their increased use in the future will require governments to be proactive and create requirements or incentives for energy suppliers to include renewable sources in their mix of sources. Furthermore, in weighing the cost of alternative sources against that of conventional sources, the total costs of each source should be considered, including the costs entailed in emissions into the atmosphere. If such comprehensive costing were done, the gap between conventional sources and renewable energy sources would not be as large as it seems to be at the moment.

Solar Power

Solar power is another renewable energy option. Energy generated by the sun travels to the Earth as electromagnetic radiation. The solar energy available at any place on the Earth is a function of several variables, the most important being how high the sun is in the sky and cloud cover. There are three general categories of use for solar energy: heating/cooling, production of electricity, and chemical processes. The most widespread uses are for heating of space and water.

The end use for solar power varies from country to country. For example, in China, Taiwan, Japan, and Europe, the main use is for heating water and space. In contrast, the dominant use in the US and Canada is for heating swimming pools. Europe has the most diverse and sophisticated market for solar power, with end uses ranging from heating water, space heating for single- and multi-family houses and hotels, and large-scale plants for district-scale heating, as well as air conditioning of homes along with cooling and industrial uses.

Based on total installed solar power capacity, the leading countries are Germany, Italy, Japan, and the United States. Canada ranks twenty-ninth. Photovoltaic prices decreased on average by 4 per cent annually between 1995 and 2010. The main reasons have been a steady improvement in conversion efficiencies along with enhanced economies of scale in manufacturing processes. The steady drop in cost is important, as it has been suggested that for the two million people around the world without access to electricity, solar photovoltaics would be the most appropriate source if they could afford it—which at the moment they cannot.

Perspectives on the Environment

Solar Power Potential in Canada

The potential for solar energy varies across Canada. The potential is lower in coastal areas, due to increased cloud coverage, and is higher in the central regions. . . . In general, many Canadian cities have a solar potential that is comparable internationally with that of many major cities. For instance, about half of Canada's residential electricity requirements could be met by installing solar panels on the roofs of residential buildings.

—National Energy Board (2009a: 5)

In considering the prospects for solar energy, Natural Resources Canada (2005b) compiled a list of its advantages and disadvantages relative to conventional energy sources.

Advantages Relative to Conventional Energy Sources

- Energy from the sun is effectively free after the initial cost of infrastructure is recovered.
- The sun is a virtually unlimited source of solar energy.
- Depending on how the energy is used, the payback time can be very short.
- Solar energy systems can be 'stand alone', making them attractive for isolated areas or facilities, since they do not have to be part of a power or natural gas grid.
- Use of solar energy as an alternative to conventional energy sources normally leads to a proportional decrease in GHG emissions.

Disadvantages Relative to Conventional Energy Sources

- Cloudy conditions significantly reduce capacity to produce energy from solar systems.
- Large surface areas of land are required to install enough solar panels to produce significant quantities of energy, with potential negative disruption to farming as well as concern about aesthetic impact.

Offshore Petroleum

About three-quarters of the surface of the Earth are covered by oceans. As land-based reserves of petroleum and gas become depleted, exploration has moved to offshore locations. The outcome is that about three-fifths of production of petroleum around the world is from offshore facilities in waters adjacent to more than half of the world's coastal nations. The extraction of offshore petroleum is also increasingly occurring in highly challenging environments, including at greater depths (2,500 metres below the surface in the Gulf of Mexico off the coast of Louisiana) or in extreme climate conditions (storms in the North Sea; hurricanes in the Gulf of Mexico; high winds and waves, cold temperatures, ice and icebergs, and fog in the Northwest Atlantic east of Newfoundland).

With Pacific, Atlantic, and Arctic coastlines, Canada is an offshore producer. Offshore petroleum production started in 1992 southwest of Sable Island, off the coast of Nova Scotia. Production continued there until 1999, with a total output of 44.5 million barrels of light crude. The next major production started in mid-November 1997 at the Hibernia field on the Grand Banks, about 315 kilometres south–southeast of St John's in Newfoundland. The fixed production platform used at Hibernia is anchored on the seabed at a depth of 80 metres. Because of the prevalence of icebergs, the outer edge of the platform is serrated. A support vessel is always stationed near the production platform, and one task is to tow small and medium-sized icebergs away from the platform. Tankers take the petroleum from the production platform to an inshore storage terminal near an oil refinery at Come-by-Chance.

Located 350 kilometres east-southeast of St John's and discovered in 1984, the Terra Nova project is Canada's third field and began production in January 2002. The Terra Nova field, the second largest on the east coast after Hibernia, is estimated to hold 440 million barrels of recoverable petroleum. Producers use a floating facility with capacity for production, storage, and off-loading. The floating facility design was

A tug positions itself near the base of the Hibernia platform in Bull Arm, Trinity Bay, Newfoundland, on 22 May 1997 as the massive rig is prepared to be towed out to the Grand Banks.

chosen in light of the harsh environment, and it can be disconnected relatively quickly from its mooring system and moved off-location in case of an emergency. The hull of the facility was designed to withstand the force of an iceberg weighing up to 100,000 tonnes or sea ice covering up to 50 per cent of the ocean surface around the platform. Other protective measures include sub-sea wells within 'glory holes' (excavations on the seabed) to protect the wellheads from icebergs that scour the ocean bottom and flexible pipes to take oil from the wells so that oil can be flushed out of them and replaced by sea water if an approaching iceberg might damage the pipes.

Hibernia originally had an expected production life of 25 years and Terra Nova 15+ years. In 2006, the Canada–Newfoundland Offshore Petroleum Board revised its estimate of Hibernia's recoverable reserves at 1,244 billion barrels, an increase of 379 million barrels from the previous estimate. This upward revision means that Hibernia is expected to be in production until about 2030 rather than until the early 2020s.

In 2005, a fourth field containing both petroleum and gas, named White Rose, was brought into production. It is located about 50 kilometres from the Hibernia and Terra Nova fields, placing it on the northeastern part of the Grand Banks. The White Rose field extends over 40 square kilometres at a depth of 120 metres and is estimated to contain 250 million barrels of recoverable oil. A floating production and storage facility is used here, similar to the one at Terra Nova.

White Rose received regulatory approval in 2001 following an environmental impact assessment from which the federal Minister of Environment concluded significant negative environmental effects were unlikely as long as mitigation measures were used. The environmental assessment report, completed in 1997, concluded that extreme weather and ice regimes at the production site would be the most serious challenges (Canada–Newfoundland Offshore Petroleum Board, 1997), a concern demonstrated in late November 2011 when a Husky Energy supply ship, in rough seas, collided with a drilling platform at the White Rose site, tearing a five-metre hole in one of the platform's supporting legs. Although no oil spill or environmental damage was reported, and damages to both the rig and the supply vessel were reported to be above the water line, such an event shows the environmental and human risks involved in open-sea oil extraction. The Environmental Assessment Panel believed that a floating production system, by allowing avoidance strategies in the face of extreme conditions, reflected the *precautionary approach* that should underlie all aspects of the project. Notwithstanding its confidence in the capacity for avoidance of possible environmental dangers to the production platform, the panel recommended continuing effort to improve operational forecasting capacity regarding both weather and iceberg trajectories.

The panel also observed that the developers of White Rose could not be held responsible for the effects of subsequent development projects on the Grand Banks. The real possibility of future projects, however, made it clear to the panel that significant difficulties and uncertainties exist in terms of calculating *cumulative effects* from a number of offshore projects. As one step to respond to this dilemma, the panel recommended a systematic and peer-reviewed monitoring system as a foundation for *adaptive management* decisions in the future.

Perspectives on the Environment

Iceberg Hazards at White Rose Oil Field

In early April 2008, oil production from the White Rose field was shut down because of a threat from nearby icebergs.

Production was stopped because thick sheets of ice along with large icebergs were close to the offshore production facilities. Six vessels were used to tow icebergs away, in addition to water cannons to deflect the icebergs' trajectories.

The White Rose oil rig, along with its 109 crew members, was towed to an ice-free area some 40 kilometres southwest of the oil field.

Beyond the possibility of a major oil spill, the panel noted that discharges of oil-based drilling mud, various chemicals, and product water (water used in processes to extract and produce oil, as well as general cleaning) into the ocean were the project's biggest environmental hazards. If a major oil spill were to occur, the panel believed that mitigative measures were unlikely to be effective because of the fragile environment. Consequently, the panel argued that it was 'absolutely essential' that prevention be the top priority. In the view of the panel, the project operators had to 'adopt a zero-tolerance for oil spills of any kind'.

The possible impact of light oil on seabirds was recognized, as well as the feasibility of mitigative measures if there was a sound monitoring program. Another risk to seabirds would arise when oil is moved from the production site to the shore refinery. An oil spill close to the shoreline could threaten the large seabird colonies on the Avalon Peninsula. The panel recommended development of a systematic coastal zone management regime for the Avalon Peninsula shoreline.

Regarding natural gas, the coastal continental shelf adjacent to Nova Scotia contains significant gas fields. In the 1970s, recoverable reserves were discovered in various locations near Sable Island, some 100 kilometres from the Canadian mainland. During 1979, a drill rig successfully identified a commercial field. At that time, however, the cost of developing the field, combined with low natural gas prices, did not make extraction viable. By the mid-1990s, improvements in drilling technology and increased prices for natural gas made commercial extraction feasible. A consortium of oil and gas extraction companies began developing the gas fields in 1996, and production began from the Sable Island Project during 1999. This was the first offshore natural gas project in Canada.

The project has two components. The initial one focused on extracting gas from six fields near Sable Island and constructing a pipeline to take the product to a plant for further processing near Goldsboro, Nova Scotia. The second component involved building the Maritimes and Northeast Pipeline to move processed gas from the Goldsboro plant to a transfer point at the border between Canada and the United States.

Given the possible environmental impact of these projects, various federal and provincial departments collaborated on an environmental impact assessment process (Canadian Environmental Assessment Agency, 2003). A five-member assessment board was created in 1996, and its report was made public in 1997. Subsequently, all appropriate regulatory agencies gave approval, subject to adoption of the recommendations in the assessment report. It has been estimated that the Sable Island Project will produce for up to 25 years, with royalty payments to the province ranging between $1.6 and $2.3 billion.

Atlantic Canada has been the major source of offshore oil and gas production so far. And by 2005, Newfoundland was producing 12 per cent of Canada's crude oil and equivalent. However, significant discoveries of oil and gas have been made in the Beaufort Sea off the coast of the Northwest Territories. Production will likely begin there once a Mackenzie Valley pipeline has been built to move the oil and gas to southern markets. There also are estimates of significant oil and gas reserves off the coast of British Columbia, but a federal moratorium on exploratory drilling has been in place there since 1972.

Perspectives on the Environment

Estimated Oil and Gas Reserves in BC Offshore Waters

The Geological Survey of Canada has estimated that in the offshore sedimentary basins, there may be as much undiscovered hydrocarbon reserves as 9.8 billion barrels of oil (BBL) and 25.9 trillion cubic feet (TCF) of gas in the Queen Charlotte Basin, 9.4 TCF in the Tofino and Winona Basins, and 6.5 TCF in the Georgia Basin. Such estimates can only be confirmed by drilling, which must be preceded by and based on sophisticated scientific and technical surveys, which in turn will only be carried out if permitted by the governments of British Columbia and Canada.

—Scientific Review Panel (2002: 2–3)

The moratorium affecting the waters off the British Columbia coast reflects at least the following concerns: jurisdictional uncertainty regarding whether the federal or provincial government owns the seabed, Aboriginal land and related ocean claims, and environmental risks. The oil spill from the tanker *Exxon Valdez* in 1989 in Alaskan waters highlighted the vulnerability of BC coastal waters to environmental risk. Some 10.8 million US gallons of unrefined crude oil were released into Prince William Sound from the tanker,

A cormorant killed by an oil spill.

the largest oil spill to that time in North American waters. The oil eventually covered more than 1,900 kilometres of rocky shoreline and caused the death of tens of thousands of birds, a thousand sea otters, several hundred seals, and unknown numbers of fish and other sea life. Exxon had 10,000 workers on site in the summer of 1989 for the cleanup work, which ultimately cost US$2.2 billion. Additional costs were a US$1 billion fine payable to the US and Alaskan governments and several billion dollars for damage experienced by fishers, property owners, and others.

Environment in Focus

Box 12.2 The BP Oil Spill, Gulf of Mexico, April 2010

The risk of offshore oil extraction was highlighted on 20 April 2010 when an explosion occurred on an offshore drilling rig, the *Deepwater Horizon*, operated by British Petroleum (BP) in the Gulf of Mexico, killing 11 rig workers and resulting in an uncontrolled wellhead blowout and the worst offshore oil spill in the deep ocean in North American history. The flow of oil into the Gulf affected states from Florida to Texas. Major negative economic impacts were caused to fishing (especially shrimping) and tourism industries. It will take decades to understand the long-term effect on the Gulf ecosystem. Another consequence was a moratorium placed on deepwater offshore drilling in US waters.

After many unsuccessful tries, on 15 July a temporary cap was placed over the wellhead, nearly four kilometres beneath the surface of the Gulf of Mexico, and on 19 September 2010, five months after the blowout began, a permanent cap had been installed. It was estimated that the costs for cleanup, government fines, lawsuits, and damage claims will be well over $40 billion. Some experts suggested the final costs could be up to $200 billion.

A seven-member US presidential commission reported that the oil well blowout was caused by cost-cutting and time-saving business decisions by BP and its partners, Halliburton and Transocean (National Commission on the BP *Deepwater Horizon* Oil Spill and Offshore Drilling, 2011). In the view of the commission, without significant reform in business practices and government policy, more such spills are likely.

Some commentators also observed that we are all responsible for the accident, due to our collective excessive reliance on fossil fuels and reluctance to make significant behavioural changes regarding our use of energy.

The *Deepwater Horizon* oil rig burning after an explosion in the Gulf of Mexico, off the southeast tip of Louisiana on 20 April 2010.

Tourists look on as a worker cleans oil from the sand along a strip of beach in Gulf Shores, Alabama, two months after the initial *Deepwater Horizon* explosion.

The potential for offshore oil and gas (estimated to be up to 25 per cent of the globe's undiscovered oil and gas) and mineral deposits has led the 'Arctic Five' countries—Canada, Denmark, Norway, Russia, and the United States—to make political claims to seabed resources in the Arctic. For example, in August 2007, a remote-controlled Russian mini-submarine was used to plant a Russian flag on the seabed at the North Pole. This action was a symbolic gesture by Russia to claim rights to the Lomonosov Ridge, an underwater mountain range extending 1,995 kilometres, as part of Russia's continental shelf. One month later, Russia stated that initial analysis of samples collected by one of its scientific teams proved that the mountain range beneath the Arctic Ocean was an extension of its continental shelf. The consequence of Russia asserting this area as part of its economic zone would be to give Russia rights to the resources on the ocean floor in that area. Shortly after that flag-planting event, Danish scientists travelled to the Arctic to search for evidence supporting Denmark's claim that the Lomonosov Ridge is an extension of the continental shelf of Greenland, which would place it under the control of Denmark.

Canada's Prime Minister declared in 2007 that the Northwest Passage was 'Canadian internal waters' rather than an 'international waterway'. In international law, foreign ships in general can travel through an international waterway that is within a nation's territorial waters without permission, but foreign warships must obtain permission before sailing in such waters. The complication is that the Northwest Passage consists of a number of linked channels situated between Canadian-owned Arctic islands, and some countries do not recognize Canada's jurisdiction over these waters. For example, the United States has never accepted the Canadian claim that the Northwest Passage consists of internal or territorial waters, instead arguing that it is 'international waters', meaning US warships can use the Northwest Passage without seeking Canada's permission. As Dyer (2007: A9) commented, 'There is a scramble for the Arctic, but it is not military. It's about laying claim to potentially valuable resources.' This situation highlights the need to combine understanding of scientific, technical, management, and policy aspects in resource and environmental management. More recently, the Arctic Five, sometimes in co-operation, have been involved in scientific surveys of the Arctic Ocean seabed in preparation for presenting their cases for extended economic zones, beginning in 2013, to the United Nations Commission on the Limits of the Continental Shelf, part of the UN machinery related to the Law of the Sea Convention.

Athabasca Oil Sands

Background

Extensive and intensive development is occurring in northeastern Alberta, focused on the oil sands located in an area north of Fort McMurray along both sides of the Athabasca River. The Athabasca oil sands are one of three deposits, the others being in the Peace River and Cold Lake areas (Figure 12.9 and Table 12.1). Together, the three deposits extend under an area of 149,000 square kilometres, almost one-quarter the area of Alberta and larger than the state

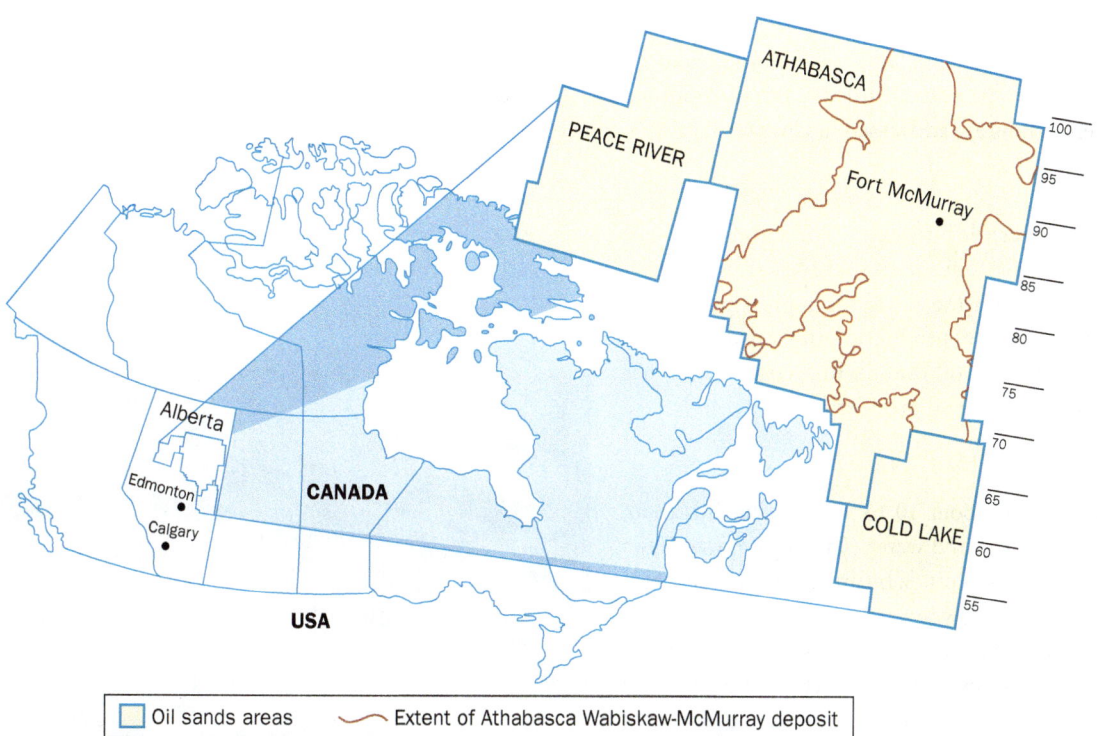

Figure 12.9 | Athabasca oil sands region. *Source: ERCB ST98-2011: Alberta's Energy Reserves and Supply/Demand Outlook*

Table 12.1 | Area and Bitumen Resources in the Oil Sand Deposits of Alberta

Deposit	Initial Volume of Crude Bitumen in Place (Barrels)	Land Area (km^2)
Athabasca (in situ + surface minable)	1.484 trillion	93,000
Cold Lake	182 billion	18,000
Peace River	136 billion	29,000
Total oil sands	**1.803 trillion**	140,000

Source: Woynillowicz et al. (2005: 2). Reprinted by permission of the publisher.

of Florida. The oil sands make Canada second only to Saudi Arabia in terms of global oil reserves. Given the relatively stable political situation in Canada, the oil sands are attracting billions of dollars in investment.

Development of the oil sands started during the mid-1960s, but it was not until the mid-1990s that it became financially viable as a result of improving technology, preferential financial arrangements (low provincial royalties and federal tax concessions), and strong demand as well as rising prices for oil. Between 1995 and 2008, production from the oil sands more than doubled, reaching an average daily output of 1.3 million barrels. With oil prices expected to remain high and demand to grow, it is anticipated that between now and 2030, investment in the oil sands will be $100 billion and daily output will reach 5 million barrels. This level of output is bringing major economic benefits to Alberta and to Canada. Nevertheless, the development also has implications for the integrity of the northern boreal forest as well as for air quality, water in the Athabasca River, and land-based resources. Each of these aspects is considered here, while the implications for greenhouse gas emissions are addressed in Chapter 7.

Context

In 2001, production of **crude bitumen** (a thick and heavy oil) surpassed production of conventional crude oil in Alberta. By 2005, production from the oil sands represented about 50 per cent of total crude oil production in Canada. And by 2015, it is expected that the oil sands will account for 75 per cent of total Canadian crude oil production.

Extracting Oil from Oil Sands

The oil sands consist of about 10 to 12 per cent **bitumen**, mixed with sand, silt, clay, and water. The oil removed from the oil sands is referred to as crude bitumen, and because it is thick and heavy, it cannot flow towards a well. As a result, two different methods are used to extract bitumen. If the bitumen is not more than 100 metres below the surface, it is removed through surface or strip mining from open pits. Subsequently, the mined oil sands are mixed with hot water, which washes the bitumen out of the sand. The other method, used at depths greater than 100 metres, is referred to as **in situ recovery**, and the specific technique is called **steam-assisted gravity drainage**, or SAGD. This method is used for more than 90 per cent of the oil sands. The usual approach is to inject high-pressure steam into the oil sands to separate the bitumen from the sand, silt, and clay. Once exposed to the steam, bitumen can flow to a well from which it can be pumped to the surface.

For surface or strip mining, Woynillowicz, Severson-Baker, and Raynalds (2005) have explained that:

- To produce one barrel, or 159 litres, of synthetic crude oil, about four tonnes of material (two tonnes of soil and rock above the deposit, then two tonnes of oil sands) must be removed. On average, every two days the mining operations move oil sands equivalent to what would fill Toronto's Rogers Centre.
- To extract one barrel of bitumen, two to five barrels of fresh water and 250 cubic metres of natural gas are needed. The amount of natural gas is equivalent to what would be needed to heat an average Canadian home for one and a half days.

Athabasca oil sands, north of Fort McMurray.

The surface mining method results in about 90 per cent of the bitumen being recovered. In contrast, to extract a barrel of bitumen, in situ oil sands extraction requires:

- 2.5 to 4 cubic metres of steam;
- 1,000 cubic metres of natural gas.

Through the SAGD method, 60 to 80 per cent of the bitumen in the oil sands is recovered.

Perspectives on the Environment
Evolution of Tar Sands Technology

Technology will evolve related to extraction of bitumen from tar sands. For example, E-T Energy is developing an 'electro thermal dynamic stripping process' that heats bitumen in the ground with electrodes. Once the bitumen has liquefied, it can be pumped out relatively inexpensively with relatively minor environmental impacts.

—Reichel (2011: A5)

After recovering the bitumen, subsequent processing stages involve producing the final synthetic crude oil from the bitumen and then transportation to final destinations in Canada or the US. The removal of the bitumen from the oil sands and the production of synthetic crude oil require significant energy inputs. The energy-intensive methods have led to consideration of nuclear power as a source of energy to produce the final product.

Pipelines move most of the oil to markets in Ontario and BC and in the state of Washington as well as in Rocky Mountain and Midwest states. A pipeline has also been proposed from Alberta to a new marine terminal in Kitimat, BC, to export the synthetic oil to China and other Asian markets as well as to California. For that to happen, however, the moratorium restricting oil tanker traffic along the BC coast would have to be lifted. The pipeline is also strongly opposed by 61 First Nations, including 50 over whose territories the pipeline would cross. The demand in the US for oil from the Alberta oil sands has led Calgary-based TransCanada Corp. to propose a pipeline, the $7 billion Keystone XL pipeline, from Alberta to the Gulf coast in Texas, where the raw bitumen would be upgraded and refined. Public protests occurred in Canada in the fall of 2011 related to the Keystone XL pipeline. In contrast, Prime Minister Harper and Natural Resources Minister Joe Oliver argued that such a pipeline would be beneficial for the Canadian economy through the profits earned from selling oil to the US as well the creation of construction jobs. They also argued that oil from Canada would be a secure and reliable source of energy for the US. In November 2011, after large-scale protests centred on the possible negative impacts of the pipeline route through Nebraska, where it would cross a large aquifer that provides water to several Great Plains states, the US State Department announced that TransCanada would be required to examine rerouting the pipeline, and indicated that the review of alternative routes was anticipated to be completed in the first quarter of 2013. President Obama supported the decision by the US State Department, and was quoted as saying the decision 'could affect the health and safety of the American people as well as the environment'.

Environmental Impacts

The oil sands operations have impacts on the boreal forest system and on water and water levels in the Athabasca River and Lake Athabasca, which feeds into the Mackenzie River system, as well as on air quality and wildlife. The consequences of the development are also significant in relation to *cumulative impact assessment*, discussed in Chapter 6.

Boreal Forest and Wetlands

A major impact of oil sands development is fragmentation of the boreal forests. Fragmentation is significant because boreal ecosystems, involving a mix of forest and wetlands, are the habitat for many species of wildlife and also support the highest diversity of breeding bird species in North America. Furthermore, the boreal forest system is valuable in the context of global change because it influences climate and is a source for storage of carbon. Fragmentation occurs from the removal of forests and wetlands through either surface or strip mining or in situ removal of bitumen, as well as the building of roads and above-ground pipelines. The outcome is the breaking up of continuous areas of extensive woodland and wetlands into smaller and separated patches. This means reduced habitat for wildlife, as well as constraints on movement of wildlife from patch to patch. Such changes have serious negative implications for birds that nest in interior forest systems and for larger carnivores that range over large territories.

Reclamation programs have been designed, but they have limitations. For example, the eventual reclaimed landscape, as proposed by the oil sands industry, will be different from the mix of forest and wetlands being altered by mining. It will consist mainly of dry, forested hills, more lake area arising from the end-pit lakes used in oil sands production, and absence of peatlands. The latter take thousands of years to develop and so cannot be replicated by reclamation. After reclamation, an estimated 10 per cent of the wetlands in the original boreal ecosystem in the region will be gone forever. The loss of wetlands will have several consequences because they:

- provide habitat for rare plants and wildlife;
- regulate surface and groundwater flow through retaining snowmelt and summer storm flows;
- recharge aquifers;
- serve as natural filters, removing contaminants from waters that flow through them.

In addition, the river needs sufficient water to support various fish species. The natural flow in the river fluctuates seasonally, with lows occurring in the winter and highs in the spring. Fish, such as northern pike, walleye, and burbot, that occupy the river during the winter are vulnerable if flows drop below their minimum needs, and the natural low flow in winter combined with extraction of water for oil production is a threat to their well-being. Given the low temperatures associated with a boreal ecosystem, the low water temperatures result in low reproduction rates and many species not beginning to reproduce until they are six to 10 years old. The challenge, then, is to determine what is termed 'instream flow needs', or the minimum threshold for water flow needed to sustain the health of an aquatic ecosystem.

Perspectives on the Environment

Abstraction of Water from the Athabasca River

The ecological integrity of the river is threatened during the winter months in years when low precipitation rates in the Athabasca River basin lead to low flow conditions. Industrial water withdrawals must be limited during these brief periods to protect the health of the river. These withdrawal limits must be based on the precautionary application of current scientific knowledge and must acknowledge the impacts to the Athabasca River's resilience arising from active oil sands mining operations within its basin.

—Woynillowicz and Severson-Baker (2006: 2)

Athabasca River north of Fort McMurray, with the town of Fort McMurray in the background. The road leads to the oil sands operations.

Overall, 'it is likely that the reclaimed landscape will lack the biodiversity of its pre-disturbance state, and it is acknowledged that it will be a major challenge to re-establish self-sustaining ecosystems' (ibid., 38).

Athabasca River

Both surface mining and in situ extraction of oil sands have significant implications for aquatic systems. Impacts can occur from draining or removing wetlands, dewatering aquifers, withdrawal of water from the Athabasca River, and storing tailings.

The Athabasca River is about 1,540 kilometres in length, starting in Jasper National Park and emptying into Lake Athabasca through the Peace–Athabasca Delta in Wood Buffalo National Park, the largest boreal delta on the planet and a major nesting and staging area for migratory birds. The national park has been designated as a World Heritage Site (discussed in more detail in Chapter 14). Given the huge demand for water for oil sands production, large quantities of water are being removed from the Athabasca River. Such withdrawals pose a potential threat to the Peace–Athabasca Delta, which requires minimum flows from the river as well as natural fluctuations.

Woynillowicz and Severson-Baker (2006: 4) note that withdrawal of water from the Athabasca River to support production from the oil sands is further challenging because such operations 'return very little water to the Athabasca River'. They also highlight that oil sands operations are by far the largest withdrawers of water from the river, representing some 65 per cent of withdrawals in 2006 (about two times the amount needed to meet the annual water demand of Calgary with its population of more than 1 million). Future anticipated oil sands production will increase the withdrawals to about one and a half times the amount now being removed.

Water quality issues also have been identified. David Schindler, an ecologist at the University of Alberta, and a group of colleagues published results in August 2010 from research focused on determining the relative contribution of natural sources and the oil sands industry of elements and polycyclic compounds into the Athabasca River. The conclusion was that, 'Contrary to claims made by industry and government in the popular press, the oil sands industry substantially increases loadings of toxic PPE [priority pollutants] to the AR [Athabasca River] and its tributaries via air and water pathways. This increase confirms the serious defects of the RAMP [the industry-led Regional Aquatics Monitoring Program] which has not detected such patterns in the AR

Environment in Focus

Box 12.3 Incomplete Understanding of the Causes of River Flow Variability

An interdisciplinary team of scientists has been examining the impacts since AD 1700 of climate and river flooding on the ecology of the Peace–Athabasca Delta. Part of the motivation for this research was the argument that regulation of the Peace River for hydroelectric power generation by the Bennett Dam since 1968 and the filling of the Williston Lake reservoir between 1968 and 1971 in BC had created major stress on the Peace–Athabasca Delta. Specifically, it has been suggested that the Bennett Dam has been largely responsible for reduced frequency of spring ice-jam and open-water flooding as well as an extended period of drying, both of which have had significant impacts on the ecology of the delta.

Paleo-environmental research was used to reconstruct regional climatic variability near the headwaters of the Athabasca River as well as the flood history of the Peace River in the northern section of the Peace–Athabasca Delta. Climate records were reconstructed from carbon and oxygen isotope analyses of a composite tree-ring chronology.

The climate reconstruction indicates the following climate situations: (1) cold and very dry conditions in the 1700s in association with the peak of the Little Ice Age; (2) subsequent warming and wetter conditions from about 1780 until 1940; and (3) progressively drier conditions.

In discussing the implications of the flow regulation of the Peace River for changes in the ecology of the delta, the scientists concluded that 'Although concerns have been focused on observed drying since the Peace River has been regulated since 1968, paleolimnological evidence from Spruce Island Lake indicates this to be part of an extended period of drying that was initiated in the early to mid-1990s. Resolving whether significant river regulation impacts are superimposed on natural climate-driven hydrological variability over this time interval, however, remains uncertain' (Wolfe et al., 2005: 160).

The research team also observed that 'Recent studies . . . suggest that the absence of high-magnitude floods between 1975 and 1995 may be a result of climatic variation that increased the temperatures during the ice-cover season, reduced the snow-pack depths, and altered the intensity and duration of the pre-melt period. Thus, reduced spring ice-jam flooding due to regulation of the Peace River and climate variability represent competing hypotheses to explain what are perceived to be unusually dry conditions in the PAD [Peace–Athabasca Delta] during much of the past 35 years' (Wolfe et al., 2006: 132).

The findings and conclusions of the interdisciplinary scientific team are a reminder that we often face considerable uncertainty and complexity regarding the behaviour of ecological systems. As a result, scientists are understandably cautious in drawing conclusions about cause-and-effect relationships. In this context, systematic monitoring of the impact of both climate variability and water withdrawals for oil sands production facilities will be needed to determine the long-term effect of the oil sands development on the Athabasca River system.

---

watershed. Detailed long-term monitoring is essential' (Kelly et al., 2010: 5).

The federal Minister of the Environment appointed an Oil Sands Advisory Panel on water monitoring for the lower Athabasca River and associated water systems on 30 September 2010, and directed it to report within 60 days on two aspects: (1) review and assess current scientific research and monitoring, and (2) identify strengths and weaknesses in the scientific monitoring, and reasons for them. The panel submitted its report in December 2010, observing that:

> Despite the myriad programs ongoing in the oil sands region . . . there was no evidence of science leadership to ensure that monitoring and research activities are planned and performed in a coordinated way, and no evidence that the vast quantities of data are analyzed in an integrated manner. Similarly there was a lack of leadership on reporting on oil sands environmental performance across media. (Oil Sands Advisory Panel, 2010: 34)

The panel concluded that there is not a first-class, state-of-the-art monitoring system in place for the oil sands, but emphasized that such a system could be created and offered recommendations.

The above conclusion was similar to the one reached by Gosselin et al. (2010: 7), who stated that the existing Regional Aquatics Monitoring Program (RAMP) 'needed ongoing external scientific oversight at a greater frequency than every five years, to demonstrate that it is using the best available monitoring methods with state-of-the-art detection methods.' A subsequent review by a panel of six North American scientists (Dillon et al., 2011), commissioned by the Alberta government, concluded that an enhanced monitoring program is needed.

The federal Minister of the Environment, John Baird, stated a few days after release of the report by the Oil Sands Advisory Panel that the federal government would begin to implement the panel's recommendations immediately, and he would place Environment Canada in overall charge of creating a new, first-class monitoring system. However, the

Ecologist David Schindler holds a deformed whitefish caught in Lake Athabasca, near Fort Chipewyan, during a press conference in September 2010.

Oil sands operation in Alberta. The Athabasca River is in the far background, adjacent to the end of the plant. In the foreground are large (soccer-field-sized) cakes of yellow sulphur, a by-product of the upgrading process.

Premier of Alberta, Ed Stelmach, stated that Alberta should have the lead role, and directed his Minister of Environment to redesign the monitoring system. Stelmach's view was that the oil sands are provincial resources, and while there is cross-jurisidictional responsibility, the province should lead. It appears, then, there could be challenges in having strong leadership to create a co-ordinated and integrated approach between the provincial and federal governments, an ongoing feature of resource and environmental management in Canada.

In October 2011, the Commissioner of the Environment and Sustainable Development (2011: 79) reported on an audit focused on cumulative environmental effects of oil sands projects, and observed: 'We have concluded that incomplete environmental baselines and environmental monitoring systems . . . have hindered the ability of Fisheries and Oceans Canada and Environment Canada to consider in a thorough and systematic manner the cumulative environmental effects of oil sands projects in the region.' Thus, work will need to continue to ensure capacity exists to monitor and assess environmental conditions related to oil sands development.

Air Quality

Development of the oil sands is a major contributor to air pollution emissions in Alberta. They represent 5 per cent of Canada's total GHG emissions and are the fastest-growing source. (Gosselin et al., 2010: 7).

Perspectives on the Environment
Greenhouse Gas Emissions from the Oil Sands

Greenhouse gas (GHG) emissions from the oil sands . . . are a major environmental issue. Although substantial progress has been made in reducing the quantity of GHG emitted per unit of production (emissions intensity) by the oil sands industry, and future reductions in emissions intensity will occur, the rapid pace of growth in bitumen production means direct oil sands GHG emissions have grown substantially. With current and projected developments, direct GHG emissions will continue to grow at a time when Canada has accepted targets for substantial overall reductions in response to the Copenhagen Accord. Technological solutions, such as carbon capture and storage (CCS), will not be sufficient to eliminate projected GHG emission increases from oil sands operations over the next decade.

—Gosselin et al. (2010: 4)

Particular attention has been directed to what are referred to as 'criteria air contaminants', or CACs. These contaminants are the ones most commonly emitted by heavy industry using fossil fuels, and they also negatively affect health. They include nitrogen oxides (NOx), sulphur dioxide (SO_2), volatile organic compounds (VOCs), and particulate matter ($PM_{2.5}$)—all released from oil sands operations.

Oils sands technology has improved, and this has led to a reduction in the volume of pollutants emitted per barrel of oil produced. Nevertheless, the emissions from producing synthetic oil from bitumen are higher than they are in conventional oil production processes. Furthermore, the rapid expansion of oil sands production has meant that overall emissions continue to increase even though emissions per barrel have decreased.

Modelling of air pollution based on approved future oil sands production expansion indicates that maximum emissions of NOx and SO_2 will exceed provincial, national, and international standards. VOCs are also of concern because in 2002 Alberta was one of the top four states or provinces in North America in terms of emissions, and these emissions are predicted to go up. Any additional development will make the situation worse. In contrast, forecast emissions related to $PM_{2.5}$ show that although they will increase, they will remain below accepted thresholds.

In early 2008, the federal Conservative government announced a 'green' plan that would allow GHG emissions from the oil sands to triple from 25 million to 75 million tonnes a year over the next decade, a period when the national goal is to reduce overall emissions by 150 million tonnes. The economic value of the oil sands development appears to override environmental concerns, even in regard to issues as serious as global climate change, as discussed in Chapter 7.

Wildlife

On 28 April 2008, ducks landed on a 12 km² tailings pond operated by Syncrude Canada in association with its oil sands operations, and 1,600 died. Syncrude was charged under provincial and federal regulations for failing to deter the ducks from landing in the tailings pond. The position of the governments was that it should have been apparent to Syncrude that deterrent systems (air cannons, scarecrows) to discourage landings by birds should have been in place in the spring as soon as reasonably possible. The lawyer for Syncrude argued that the company had followed all regulations, and finding the company guilty would have a serious negative impact on the entire oil sands industry.

Photos of the oil-covered ducks quickly appeared in the media around the world. Many ducks died because they could not get themselves out of the thick 'goop' on the surface of the tailings pond. Photos showed some being eaten alive by ravens while stuck on the surface, while it was reported others sank and drowned. Such images generated criticism from various groups, claiming the environmental costs of extraction from the oil sands were too high.

In June 2010, a provincial court judge found Syncrude guilty, and Syncrude was required to pay $3 million in penalties. The cost was broken down into a $500,000 provincial fine, a $300,000 federal fine, $1.3 million to support research on how to deter birds from oil sands operations, and $900,000 for habitat restoration. Some funds for habitat restoration will be used to purchase wetlands to the east of Edmonton, which will be managed by conservation groups. Also, one-half of the provincial fine will support an environmental diploma program at Keya College in Fort McMurray.

The above example illustrates challenges faced in extracting bitumen from the oil sands. It also highlights how public media and information can quickly become an issue.

Public Relations Skirmishes

In mid-July 2010, Corporate Ethics International, based in San Francisco, started an ad campaign urging possible visitors to Alberta to reconsider such a trip. The rationale? Executive Director Michael Max stated that the campaign had been triggered because, in the view of his organization, Alberta was the 'most environmentally unfriendly place in North America because of the tar sands' and 'the Alberta government had been pretty arrogant in ignoring the concerns of environmental groups' (Krugel, 2010). He hoped the campaign would encourage rethinking of the approach to the oil sands. The campaign used large billboard ads in Seattle, Portland, Denver, and Minneapolis, with follow-up billboards in Britain. Notably, opponents of what they call 'dirty oil' have used the original and more pejorative term, 'tar sands', to describe the bitumen deposits.

A second initiative to place pressure on the Alberta government came from a number of companies, such as Avon, Walgreen's, and Whole Food. Each stated it would purchase fuel for its trucks only from refineries not using feedstock from the oil sands. In addition, The Gap, Timberland, and Levi Strauss asked their transportation contractors to clarify what they were doing to eliminate use of high-carbon fuels.

In response, the Alberta government initiated its own ad campaign, including placing advertisements in major newspapers and providing information on its websites (e.g., www.oilsands.alberta.ca/tellitlikeitis.html). Alberta received support from the American Petroleum Institute, which announced a pro-oil sands campaign to emphasize economic benefits to the struggling US economy from using oil sands fuel.

Another initiative occurred in September 2011 when eight Nobel Peace Prize recipients wrote to Prime Minister Harper, asking him to take action to stop the growth of production from the oil sands. The Nobel laureates reminded the Prime Minister that on various occasions he had identified climate change as one of the biggest challenges facing humanity. They called for action, observing that 'It would be wrong for a rich minority of the world's inhabitants to create a problem like climate change and then refuse to do its fair share to fix it' (Weber, 2011).

The different campaigns highlight how resource management conflicts sometimes become a battle to influence values and behaviour of the general public, with contrary views being presented by competing stakeholder groups.

Uranium and Nuclear Power

Uranium

The main use of uranium, once it has been processed, is for fuel in nuclear reactors to generate electricity. When exploration for uranium began in the early 1940s, however, the demand for uranium was not for nuclear reactors, because none existed. Instead, uranium was required for the creation of atomic weapons then being developed. During World War II, the federal government established a Crown corporation that later became known as Eldorado Nuclear Ltd. For a short time, it was the only company approved to mine radioactive material. After the end of the war, other firms were given approval to mine and process uranium, and by the late 1950s more than 20 uranium mines were operating. The largest were near Elliot Lake, Ontario, and Uranium City, Saskatchewan. When commercial nuclear reactors became available in the 1960s, further exploration led to the discovery of major new deposits of uranium in northern Saskatchewan. One outcome was the closure of the lower-grade mines in Elliot Lake and Uranium City.

Uranium mining and production is controversial. The mining process results in tailings that need careful storage to prevent leakage into aquatic systems. After being used in power plants, the nuclear fuel wastes are highly radioactive, generating challenges in terms of long-term containment and storage. From an economic perspective, however, uranium mines have provided jobs for skilled workers in remote regions of the country where jobs have not been plentiful.

Canada mines about one-third of global uranium, making it the world leader in output. The next largest uranium producers are Australia, Kazakhstan, and Russia. Canada has some 524,000 tonnes of uranium oxide (U_3O_8), 9 per cent of the global total. Australia has two and a half times that amount.

> ### Perspectives on the Environment
> #### Uranium Mining in Saskatchewan
>
> A significant portion of the world's known uranium resources are located in Saskatchewan. Uranium deposits in Saskatchewan are large, contain high-grade ore and can be extracted at production costs below those in many other parts of the world. Saskatchewan's uranium resources are sufficient for more than 40 years at current rates of production.
>
> —Saskatchewan (2009: 1)

Most of the production in Canada is from two mines in northern Saskatchewan: the McArthur River underground mine, the largest uranium mine in the world, and the McClean Lake open-pit mine, to be followed by an underground mine. They began production in 2000 and 1999, respectively. In 2005, McArthur River contributed more than one-sixth of global uranium production. Another mine, at Rabbit Lake, is at the end of its productive life. A mine at Cigar Lake began production in 2011. Previous mines at Key Lake and Cluff Lake are no longer in production.

The provincial government in Saskatchewan supports uranium mining, subject to appropriate safeguards for the environment. This policy contrasts with that of the New Democratic government in the early 1990s, which aimed to end the mining of uranium. Four Saskatchewan uranium mines have been ISO 1401 certified: McClean Lake (2001), Key Lake (2003), McArthur River (2003), and Cluff Lake (2004), indicating that they meet international standards related to environmental considerations. AREVA (2007), operator of the McClean Lake mine, has stated that the 'industry's long-term goal is to return all operations, as close as possible, to a natural state suitable for future uses. All uranium mine site operators must post bonds with the federal government to ensure adequate funds are available for proper decommissioning of each site after the reserves have been mined out.'

Nuclear Power

A made-in-Canada experimental nuclear reactor was developed at Chalk River, Ontario, and began producing power in 1947. That reactor became the forerunner of the CANDU (Canada Deuterium Uranium) pressurized heavy-water reactors that are used around the world.

Canada produces between 14 to 15 per cent of its electricity from **nuclear power**, compared to about 20 per cent in the United States. Some 18 reactors produce more than 12,500 MW of power. Ontario, the province most dependent on nuclear power, has used it since the early 1970s. Quebec (Gentilly) and New Brunswick (Point Lepreau) each has a single-unit CANDU plant. As of 2010, Canada's nuclear power production represented 3.4 per cent of the world total, and Canada was ranked sixth globally, after the United States (30.7 per cent), France (16.1 per cent), Japan (9.4 per cent), Russian Federation (6.0 per cent), South Korea (5.5 per cent), and Germany (5.4 per cent) (International Energy Agency, 2010: 17).

Ontario has commercial nuclear reactors in operation in three multiple-unit locations (Pickering and Darlington, both on the shore of Lake Ontario, and Bruce on the shore of Lake Huron). During the early 1990s, nuclear reactors generated about two-thirds of Ontario's electricity. This had dropped to just over 50 per cent by 2011. Part of the reason is that several units at the Pickering and Bruce plants have been taken out of service for long periods for safety reasons. The refurbishing of nuclear units in these plants has been very expensive. For example, the cost to refurbish units at the Bruce plant to give them a further 25 years of life has been estimated at $4.25 billion. In 2005, the New Brunswick government decided to renew its Point Lepreau reactor at an estimated cost of $1.4 billion.

To place such costs in perspective, the Ontario Power Authority conducted a major review of energy requirements and in 2005 reported that the province would need to spend $83 billion to refurbish its electricity supply system over a 20-year period. The report concluded that nuclear power capacity would have to be increased to maintain its share of production at 50 per cent. Among the arguments in favour of building more nuclear power capacity are that nuclear power has less short-term environmental impact than coal- or petroleum-fuelled power plants and that it operates at a lower cost. As Ontario Power Generation (2010) has stated, nuclear power 'has two major benefits—low operating costs and virtually none of the emissions that lead to smog, acid rain or global warming.'

The Ontario government's position is that enhanced nuclear power capacity will be a key component of an overall plan to meet emerging shortages of electricity. About $40 billion is to be spent on enhancing nuclear capacity, including the addition of two new plants. It also decided to 'fast-track' an overall $83 billion plan, meaning that the plan was exempt from a full review under the Ontario Environmental Assessment Act. Proposals for individual nuclear plants would go through the federal government's assessment process, estimated as a two-year process.

Management of Used Nuclear Fuel

A major issue in the nuclear energy industry is the radioactive waste. Recognizing such concern, the federal government created the Nuclear Waste Management Organization (NWMO) to identify options for storage and disposal of **nuclear wastes**. The NWMO (2005: 2) has noted that Canada has used nuclear fuel to generate electricity for decades, and as a result Canada has about 2 million spent fuel bundles, which add up to 36,000 tonnes of uranium. This amount is estimated to increase by a factor of two in the future, assuming that all existing nuclear plants operate for their anticipated lifespan of 40 years.

The used fuel bundles from nuclear plants are stored in regulated facilities on the sites at which they are produced. This arrangement has always been viewed as a short-term one, with a need to determine what should be done in the long term. The NWMO explored such a long-term solution. In its view, both scientific and technical analyses are critical in finding a solution, but such analyses should not be the sole

Environment in Focus

Box 12.4 Implications of Nuclear Crisis in Japan, 2011

The 9.0 Richter scale earthquake, the largest to occur in Japan, and the associated tsunami that devastated the northeast coast of that country on 11 March 2011 had major impacts. Over 25,000 people lost their lives, more than 500,000 had to evacuate their homes, and an estimated 2 million households were left without electricity and 1.5 million households had no water. However, the impact on the Fukushima Daichi nuclear complex located on the coastline some 240 km north of Tokyo particularly drew attention since the combined natural disasters left the plant without the capacity to cool three nuclear reactors due to damage to the structure, loss of electricity, and swamping of back-up generators by the tsunami. In countries around the world, commentators suggested this event, considered to be a potential nuclear disaster, was a wake-up call and should create a pause to allow critical reassessment of the role and potential vulnerabilities of nuclear power plants.

The impact of the problems encountered at the Fukushima nuclear power plant is illustrated by a decision subsequently taken almost halfway around the world. In May 2011, Chancellor Angela Merkel announced that Germany intended to abandon its nuclear energy program over an 11-year period and would turn more to renewable energy sources, especially solar, wind, and hydroelectric. All 17 of Germany's nuclear power plants are to be shut down by 2022. This decision represented a remarkable shift in policy, as in 2010 Germany had announced a plan to extend the lifespan of its nuclear reactors, with the last one to go off-line about 2036. The rationale? The experience in Japan had led to a reassessment of the risks related to nuclear technology. In the light of this decision by the German government, what do you think should be done in Canada related to nuclear power?

Damage to the Unit 4 Nuclear Reactor Building at the Fukushima Daichi nuclear plant.

basis for choices (ibid., 3). Consequently, the NWMO spent considerable time listening to the views of Canadian citizens who are not technical specialists. The outcome was strong agreement between technical and non-technical commentators about two requirements: any approach must be (1) safe and secure for people, communities, and the environment and (2) fair to present and future generations.

The NWMO examined the benefits, costs, and risks of three approaches: (1) deep geological disposal in the Canadian Shield; (2) centralized storage above or below ground; and (3) storage at nuclear reactor sites. The following considerations guided the assessment: fairness, public health and safety, worker health and safety, community well-being, security, environmental integrity, economic viability, and adaptability. No option satisfied every consideration.

Deep disposal. Option 1, deep storage in the Canadian Shield, was judged to be better than Options 2 and 3 in the long term, because engineering design and natural barriers would isolate and confine the radioactive waste. The main disadvantage is lack of adaptability in terms of the capacity to respond to changing knowledge or evolving technology.

Centralized and on-site storage. Options 2 and 3 were judged to be appropriate in the short term. A concern, however, was that sites for nuclear power plants had never been selected with regard to their technical suitability as permanent storage sites. In addition, residents of the host communities had a reasonable expectation that the nuclear fuel waste would eventually be moved.

Hybrid. Given the limitations of the disposal and storage options, the NWMO identified a different option, labelled 'adaptive phased management', which includes a technical method and a management system.

Key features of the hybrid option are: (1) eventual centralized containment and isolation of used fuel in deep underground storage; (2) optional shallow storage at a central site as a contingency option; (3) continuous monitoring; (4) allowance for retrieval; (5) citizen engagement; and (6) phased and adaptive decision-making. It would be implemented in three phases: Phase 1 (30 years), preparation for centralized management of used nuclear fuel; Phase 2 (30 years), demonstration of central storage and technology systems; and finally, Phase 3 (beyond approximately 60 years), long-term containment and isolation of the used nuclear waste combined with monitoring.

The NWMO (ibid., 6) was explicit that a key component of the overall strategy would be 'to seek an informed, willing community to host the central facilities'. In the spirit of 'fairness', the NWMO concluded that it would focus on the provinces directly involved in nuclear fuel—Saskatchewan, Ontario, Quebec, and New Brunswick. And given that in 2007 Alberta announced it would build a nuclear power plant within eight to 10 years, Alberta should be added to the list. The Energy Alberta Corporation has chosen a site for two proposed twin-unit CANDU reactors, estimated to cost $6.2 billion, on private land about 30 kilometres west of Lac Cardinal some 480 kilometres northwest of Edmonton. Energy Alberta selected the Lac Cardinal site because of community support, as well as the presence of necessary infrastructure and support services.

To be inclusive, communities and regions in other provinces would be invited to indicate whether they were interested in hosting the central facilities. The NWMO emphasized that the siting process needed to be open, inclusive, and fair to all parties, meaning that all of those with an interest should have an opportunity to make their views known. This would include communities adjacent to routes used for transporting the used nuclear fuel wastes to the central location.

The work of the NWMO is a clear reminder that even if Canada stopped using nuclear plants to produce electricity tomorrow, a legacy of spent nuclear fuel needs to be stored and contained for a very long time.

Sustainable Energy Pathways

The Canadian Academy of Engineering (2007) examined various 'energy pathways' to explore how Canada's endowment of energy sources can be developed to meet end-use needs. It concluded that Canada, in the future, could become a **'sustainable energy superpower'**. The Academy noted that about 75 per cent of the energy demand within Canada was met by fossil fuels. This percentage is lower than that in many other industrialized countries because of the significant contribution in Canada from both hydro and nuclear power. Nevertheless, the Academy (ibid., 11) commented that 'there is clearly a global desire to shift away from fossil fuels.' In that regard, the Academy suggested that it would be reasonable for Canada to have a long-term target of satisfying no more than 33 per cent of its energy needs by fossil fuels. For such a target to be achieved, alternative energy pathways need to be considered.

To help identify what alternative pathways might be feasible, consider ideal outcomes identified by eight of the 15 criteria

---

Perspectives on the Environment

Adaptive Phased Management

Adaptive Phased Management tries to find an optimal balance of competing objectives. It embraces the precautionary principle and adaptive management. Societal goals and objectives and successful technology demonstration will determine the pace of implementation. We believe Adaptive Phased Management is the strongest possible foundation for managing the risks and uncertainties that are inherent in the very long time frames over which used nuclear fuel must be managed with care.

—NWMO (2005: 7)

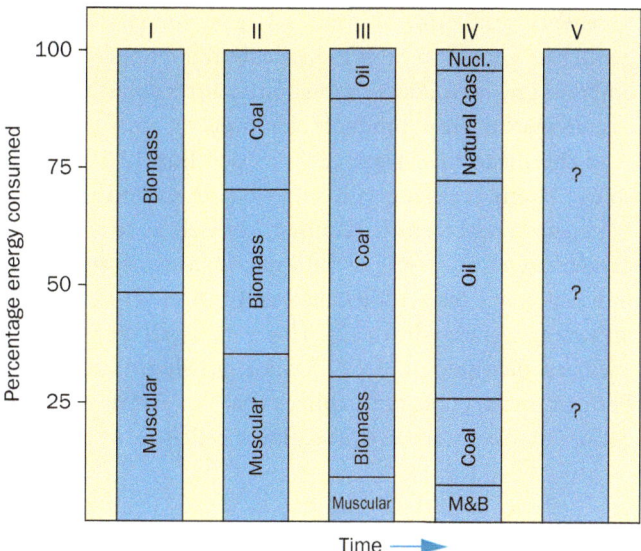

Figure 12.10 | The evolution of energy consumption mixes by source. Source: Chapman (1989: 6).

emissions will be moderately or significantly reduced; and (8) energy efficiency will be enhanced. We encourage you to consider whether these eight criteria represent all of the dimensions needing attention. For example, how would you incorporate the concept of 'resilience'? Could it be included into one of the above criteria, or should it appear as a separate criterion?

Figure 12.10 illustrates how the major sources of primary energy, over time and around the world, have evolved from muscular and biomass energy to coal, to oil and natural gas, and to nuclear fuels. Some countries, such as Japan, have moved rapidly from Phase I to Phase IV. In contrast, many developing countries struggle to move beyond Phase I except in their major cities. Canada is now well into Phase IV. What might a Phase V look like, especially in light of the experience in Japan related to the earthquake and tsunami in March 2011?

We encourage you to use the above criteria or some other *explicit* set of criteria and considerations to assess which options or pathways are most appropriate in the future. When using such criteria, you should decide whether all of them have equal value or whether to weight them in terms of relative importance. A systematic way of determining how to prioritize them is required. All of us, as citizens, should know the basis upon which assessments of energy options are made. This is important, because every option has a different

used by the Academy to judge energy pathways: (1) scientific principles and understanding underlie decisions; (2) relevant technology has been validated; (3) sustainable development is attainable; (4) social acceptability has been established; (5) positive economic impacts will be realized; (6) acceptable environmental impacts will occur; (7) greenhouse gas

Environment in Focus

Box 12.5 What You Can Do

1. Individuals can modify behaviour to reduce energy and materials use, waste production, and ecosystem degradation. Examples include:
   - Using a bucket and sponge and a trigger nozzle on a hose saves about 300 litres of water each time a car is washed, as well as saving some of the energy used to pump water through a municipal system.
   - Commuting to work by car pool, public transportation, or bicycle or on foot rather than by driving alone in a car reduces energy consumption and helps to reduce greenhouse emissions and other air pollutants.
2. Individuals can use more efficient technology or use products with lower environmental impact throughout their life cycle. Examples include:
   - Use smaller, more energy-efficient automobiles and major appliances with the lowest energy ratings. This will reduce greenhouse gas emissions and other air pollutants.
   - Install water-saving devices in the home, such as low-flow showerheads and toilet dams. Energy will be saved by heating less water and by moving less clean and grey water through the water supply system (discussed in Chapter 11).
   - Replace incandescent light bulbs with fluorescent bulbs, which use about 75 per cent less energy and last 10 times longer.
3. As part of a larger society, individuals can ask for appropriate information and insist that products, services, and planning explicitly address environmental implications. Examples include:
   - Individuals in their communities can promote better planning of urban transit and bicycle routes and reduced dependency on passenger vehicles (discussed in Chapter 13).
   - Individuals can lobby their local government and others to show leadership in educating the community about types of behaviour and products that are environmentally benign.
   - Individuals can support manufacturers committing to including environmental considerations into their production processes via life-cycle management and environmental management systems.

Source: Based on Environment Canada (2002: 68).

mix of strengths and weaknesses, and, ultimately, these factors will affect our lives now and for the generations to come.

Implications

If our ecological footprint is to become lighter in terms of our use of minerals and energy, individuals, institutions, and societies will have to change. Consumers complain that manufacturers do not build 'green cars', while manufacturers say that customer demand does not indicate that green cars are wanted in sufficient quantity to justify producing them. Thus, if change is to occur, adjustments are needed at all levels, with individuals taking initiative to reduce consumption of energy and mineral products, governments providing greater incentives to both individuals and manufacturers to embrace green products, and manufacturers showing leadership to market green products effectively.

In the meantime, what can you do? Box 12.5 identifies some actions or changes in behaviour to extend the life of non-renewable or stock resources, as well as to encourage greater use of renewable or flow resources. No simple recipe or formula will lead readily and easily to a society less materialistic and energy-intensive. However, small steps such as the ones identified in Box 12.5 can effect significant change. Perhaps most important, thinking about and taking such actions are first steps in shifting basic beliefs and values.

Summary

1. Non-renewable or stock resources take millions of years to form. Consequently, from a human viewpoint, they are for all practical purposes fixed in supply and therefore not renewable.

2. If done systematically and correctly, enhanced environmental management practices and extended social responsibility with regard to non-renewable resources almost always generate an economic return on investment for business, although usually over the long term.

3. The main environmental issues for the mining and energy sectors include: acid mine drainage; sulphur dioxide emissions; metal toxicity; disruption of remote ecosystems as a result of exploration, test drilling, and operation of oil fields or gas wells; disturbance to aquatic ecosystems from escape of waste heat produced by nuclear energy production; and threats to human and ecosystem health from radioactive waste associated with nuclear energy production over thousands of years.

4. 'Best practice' for environmental management in mining and fossil-fuel firms in Canada should include a combination of basic scientific research to ensure understanding of natural and social systems that can be affected by operations, design of appropriate mitigation measures, environmental impact assessments and reporting, environmental audits, corporate policies that explicitly include environmental aspects, environmental management systems, and life-cycle assessments.

5. Canada is one of the top five producers in the world of aluminum, diamonds, nickel, platinum, group metals, potash, uranium, and zinc.

6. In 1998, BHP's Ekati mine, located 200 kilometres south of the Arctic Circle in the Northwest Territories, became the first diamond operation in Canada. A second NWT diamond mine, Diavik, began producing in 2003. Jericho, in Nunavut, began production in July 2006 but then shut down in February 2008, and another mine, at Snap Lake, NWT, started in 2008. That same year the Victor diamond mine in northern Ontario began operations. It is anticipated that diamond mining will begin in Saskatchewan in 2016.

7. At the Ekati diamond mine, tailings are held in a lake. As the tailings settle, consolidate, and evolve to permafrost, rocks and soil will be spread over the surface. Revegetation will be started, with the goal of having the entire holding area become a wetland once the mining is completed.

8. Serious challenges were encountered by Ekati in incorporating traditional ecological knowledge into environmental research: (1) two Aboriginal groups were in the midst of land claim negotiations and as a result were reluctant to release traditional knowledge into the public domain because this knowledge was important for their negotiation strategy; (2) concern was expressed by Aboriginal people about using traditional knowledge outside of the context of their culture and broader system of knowledge that give it meaning and value.

9. The Bathurst caribou herd is the largest one in the NWT and includes about 350,000 animals. Since the caribou herd does not follow the same migration route each year and the areas affected by the Ekati mine represent less than 0.01 per cent of the range of the herd, it was believed that the mining activity would have a very small impact.

10. Beginning in 1992, BHP Billiton initiated scientific research to understand the impact of the proposed Ekati mining activity and to develop mitigation measures. In July 1994, the Minister of Indian Affairs and Northern Development referred the mining project for an environmental assessment, and in July 1995 the company submitted its environmental impact statement. From late January to late February 1996, an Environmental Assessment Panel held public meetings and in June submitted its report to the federal government. In February 1997, the federal government gave its formal approval, and construction started in May. In January 1999, the first diamonds from Ekati were sold in Antwerp.

11. Impact and benefit agreements were pioneered in Canada and are intended to ensure that Aboriginal communities benefit from mining projects and that if they contain compensation provisions, the communities are compensated for the negative impact on their communities, their land, and their traditional way of life. Targets for employment of Aboriginal people at the Ekati mine have been met.

12. The 'Ring of Fire' area in northwestern Ontario and the Horn Plateau in the Northwest Territories both highlight types of conflict that can emerge when exploration and/or extraction of minerals is pursued in remotes areas viewed to have high biodiversity value and that are traditional lands used by Aboriginal peoples.

13. Canada is ranked as the sixth largest user of primary energy in the world. Fossil fuels are the main type of energy consumed by Canadians, and one of the largest consumers of energy is the transportation sector.

14. The combustion of fossil fuels emits greenhouse gases, such as carbon dioxide and nitrous oxide, which accumulate in the atmosphere and contribute to climate change.

15. Alternative energy sources are solar, geothermal, hydro, tides, and wind.

16. Wind power is the fastest-growing sector in the world's energy market. Canada is ranked eleventh in the world in terms of installed wind power.

17. Wind turbines have minimal impact on flying birds, but various factors influence the seriousness of collisions with wind turbines. Fragmentation of habitats often is the most disruptive characteristic of wind farms, reflecting the cumulative effects of wind turbines along with related infrastructure (electrical transmission lines, roads).

18. There are growing complaints from individuals and communities about the noise and negative health impacts believed to be caused by wind farms.

19. Canada is ranked twenty-ninth in the world in total installed capacity of solar power.

20. Europe has the most diverse end uses of solar power. In Canada and the US, the main use of solar power is to heat water in swimming pools.

21. In Atlantic Canada, offshore oil production is based in the Hibernia, Terra Nova, and White Rose fields off the coast of Newfoundland, and natural gas extraction occurs near Sable Island, Nova Scotia.

22. Discoveries of petroleum and gas below the Beaufort Sea off the coast of the Northwest Territories and off the coast of British Columbia offer potential for development. However, there is a prohibition on transporting petroleum by ship along the coast of British Columbia because of the *Exxon Valdez* spill in 1989 and a moratorium in place since 1972 prevents exploration and production off the BC coast.

23. The oil spill from the *Deepwater Horizon* drilling rig explosion and wellhead blowout in the Gulf of Mexico in April 2010 highlighted the environmental risks associated with extraction of oil from significant depths on the ocean floor.

24. Various nations are challenging Canada's claims to Arctic sovereignty, not only for strategic military reasons but also to establish ownership over fossil fuels on the sea floor.

25. The Athabasca oil sands represent the second largest reserve of petroleum in the world, outranked only by the reserves in Saudi Arabia.

26. Extraction of the bitumen from the oil sands is having negative environmental impacts on the boreal forest and wetland systems, the Athabasca River, and air quality.

27. In 2008, 1,600 migratory birds died in a tailings pond operated by Syncrude Oil. Syncrude was fined $3 million for this incident.

28. Research findings from a team at the University Alberta led to calls for improved monitoring of the environmental impacts from the oil sands activities. The federal and provincial governments acknowledged that improved monitoring is needed, and committed to developing a better monitoring system.

29. In 2010, several public relations initiatives were taken in the United States and Britain to question the negative environmental impacts of the oils sands. The Alberta government responded with its own PR initiative.

30. Canada produces about one-third of the world's uranium; most of that production is from two mines in northern Saskatchewan whose reserves appear to be extractable for 40 years.

31. Canada is the sixth largest producer of nuclear energy in the world.

32. Twelve to 15 per cent of electricity in Canada is supplied from nuclear power plants in Ontario (Bruce Peninsula, Pickering, Darlington), Quebec (Gentilly), and New Brunswick (Point Lepreau).

33. The earthquake and tsunami that struck Japan in March 2011 and disabled nuclear power plants is pointed to as a reason why societies need to pause and reassess the vulnerability of nuclear power systems.

34. Most concern about nuclear energy focuses on how to dispose of used nuclear fuel.

35. Three strategies for disposing of used nuclear fuel have been identified: deep underground geological storage; centralized storage above or below ground; and storage at nuclear power plant sites. A hybrid approach, 'adaptive phased management', has been recommended.

36. The Canadian Academy of Engineering has examined 'sustainable energy pathways' for Canada and has concluded that Canada has the potential to become a 'sustainable energy superpower'.

Key Terms

acid mine drainage
bitumen
corporate social investment
crude bitumen
impact and benefit agreements (IBAs)
in situ recovery
kimberlite pipes
metal toxicity
non-renewable or stock resources
nuclear power
nuclear wastes
renewable or flow resources
solar power
steam-assisted gravity drainage
sulphur dioxide emissions
'sustainable energy superpower'
wind power

Questions for Review and Critical Thinking

1. What are the implications of non-renewable or stock resources for strategies related to 'sustainable development' or for 'resilience'?

2. How important are non-renewable resources for the Canadian economy?

3. What have been elements of 'best practice' related to the opening of diamond mines in the Canadian North?

4. What was learned from the environmental assessment for the Ekati mine regarding incorporation of local knowledge into scientific understanding of impacts?

5. Why is Canada so dependent on fossil fuels? What would have to change for there to be less dependence?

6. What are the main uses of primary energy in Canada?

7. What are the advantages and disadvantages of alternative energy sources?

8. What are the main objections from individuals and communities located adjacent to wind farms related to health? What scientific evidence exists related to health matters associated with wind turbines and wind farms?

9. What is the main use of solar power in Canada?

10. What are the greatest environmental risks associated with extracting fossil fuels from the seabeds of the Atlantic, Pacific, and Arctic oceans?

11. What is the significance of the *Exxon Valdez* incident off the coast of Alaska in 1989 and the *Deepwater Horizon* oil spill in the Gulf of Mexico during 2010 in terms of the development of offshore oil resources?

12. Why did the Russians plant a flag on the seabed at the North Pole in August 2007?

13. How much water and energy are required to produce one barrel of crude oil from the Athabasca oil sands?

14. Why is fragmentation of the boreal forests and wetlands in the area of the Athabasca oil sands of concern in terms of biodiversity?

15. Why does removal of water from the Athabasca River to support oil production pose a threat to the Peace–Athabasca Delta?

16. What are the strengths and weaknesses of the current monitoring systems for the lower Athabasca River system, and what key changes should be made to improve monitoring?

17. What are the best strategies for facilitating the economic benefits related to oil production from the Athabasca oil sands while minimizing negative environmental impacts?

18. What are the main controversies associated with mining of uranium?

19. Why did the earthquake and tsunami in March 2011 lead to calls for a review of the safety of nuclear power plants around the world?

20. Why are there concerns about storing used nuclear fuel on the sites of nuclear power plants?

21. What are the characteristics of 'adaptive phased management' in relation to spent nuclear fuel?

22. Do you agree or disagree that Canada has the potential to be a 'sustainable energy superpower'?

23. What changes should be made by individuals, organizations, businesses, and governments to reduce energy and mineral use?

Related Websites

Canadian Association for Renewable Energies
www.renewables.ca

Canadian Wind Energy Association
www.canwea.org

Citizens for Renewable Energy
www.cfre.ca

Danish Windpower Organization
www.windpower.org

Environment Canada, Wind Atlas
www.windatlas.ca/en/index.php

European Wind Energy Association
www.ewea.org

Global Wind Energy Council
www.gwec.net

Independent Environmental Monitoring Association, A Public Watchdog for Environmental Monitoring of the Ekati Diamond Mine
www.monitoringagency.net/Home/tabid/36/Default.aspx

Mining Association of Canada
www.mining.ca/www/index2.php

Mining Watch Canada
www.miningwatch.ca/index.php

National Commission on the BP Deepwater Horizon Oil Spill and Offshore Drilling
www.oilspillcommission.gov

Natural Resources Canada, Canadian Minerals Yearbook
www.nrcan.gc.ca/minerals-metals/business-market/canadian-minerals-yearbook/4070

Natural Resources Canada, Office of Energy Efficiency
oee.nrcan.gc.ca/english/index.cfm?attr=0

Natural Resources Canada Wind Energy Information Site
canmetenergy.nrcan.gc.ca/renewables/wind/2171

Ontario Wind Resistance
ontario-wind-resistance.org

Pollution Probe, Energy Program
www.pollutionprobe.org/whatwedo/Energy.html

Society for Wind Vigilance
www.windvigilance.com

Solar Energy Society of Canada
sesci.ca

Windpower Monthly
www.wpm.co.nz

Further Readings

Note: This list comprises works relevant to the subject of the chapter but not cited in the text. All cited works are listed in the Bibliography at the end of the book.

Adachi, C.W., and I.H. Rowlands. 2010. 'The effectiveness of policies in supporting the diffusion of solar photovoltaic systems: Experiences with Ontario, Canada's Renewable Energy Standard Offer Program', *Sustainability* 2, 1: 30–47.

Clarke, T. 2008. *Tar Sands Showdown: Canada and the New Politics of Oil in an Age of Climate Change.* Toronto: James Lorimer.

Das, S. 2009. *Green Oil: Clean Energy for the 21st Century?* Edmonton: Cambridge Strategies Inc.

Global Wind Energy Council. 2010. *Global Wind Energy Outlook.* Brussels: Global Wind Energy Council, with Greenpeace, Amsterdam.

Hanks, C., and S. Williams. 2002. 'Perceptions of reality: Cumulative effects and the Lac de Gras diamond field', in A.J. Kennedy, ed., *Cumulative Environmental Effects Management.* Edmonton: Alberta Society of Professional Biologists, 411–24.

International Energy Agency, Photovoltaic Power Systems Programme. 2006. *PVPS Annual Report 2006: Implementing Agreement on Photovoltaic Power Systems.* Paris: International Energy Agency. At: www.iea-pvps.org/products/download.rep_ar06.pdf.

Lavant, E. 2010. *Ethical Oil: The Case for Canada's Oil Sands.* Toronto: McClelland & Stewart.

Lorinc, J. 2009. 'On with the wind', *Canadian Geographic* 129, 3: 26–42.

McAllister, M.L., and P. Fitzpatrick. 2010. 'Canadian mineral resource development: A sustainable enterprise?', in B. Mitchell, ed., *Resource and Environmental Management in Canada: Addressing Conflict and Uncertainty.* Toronto: Oxford University Press, 356–81.

Natural Resources Canada. 2000. *Energy in Canada, 2000.* Ottawa: Canada Communication Group.

———. 2001. *Focus 2006: A Strategic Vision for 2001–2006.* Ottawa: Minister of Public Works and Government Services.

———, Minerals and Metals Sector. 1998. *From Mineral Resources to Manufactured Products.* Ottawa: Minister of Public Works and Government Services.

Nikiforuk, A. 2008. *Tar Sands: Dirty Oil and the Future of the Continent.* Vancouver: Greystone Books and David Suzuki Foundation.

Schindler, D.W. 2010. 'Tar sands need solid science', *Nature* 468, 7323: 499–501.

St Denis, G., and P. Parker. 2009. 'Community energy planning in Canada: The role of renewable energy', *Renewable and Sustainable Energy Reviews* 13: 2088–95.

Sweeny. A. 2010. *Black Bonanza: Canada's Oil Sands and the Race to Secure North America's Energy Future.* Toronto: John Wiley.

Tertzkaian, P., with K. Hollihan. 2009. *The End of Energy Obesity: Breaking Today's Energy Addiction for a Prosperous and Secure Tomorrow.* Hoboken, NJ: John Wiley.

Whiteman, G., and K. Maman. 2001. 'Community consultation in mining: A tool for community empowerment or for public relations', *Cultural Survival Quarterly* 25: 30–5.

Witteman, J., L.M. Davis, and C. Hanks. 1999. 'Regulatory approval process for BHP's Ekati Diamond Mine, Northwest Territories, Canada', in J.E. Udd and A. Keen, eds, *Proceedings of the International Symposium on Mining in the Arctic*, vol. 5. Rotterdam: Balkema, 7–11.

Wolfe, B.B., et al. 2007. 'From isotopes to TK interviews: Towards interdisciplinary research in Fort Resolution and the Slave River Delta, NWT', *Arctic* 60: 75–87.

Chapter 13
Urban Environmental Management

Learning Objectives

- To understand the nature and significance of urbanization.
- To understand the quality of environmental conditions in Canadian cities.
- To become aware of the impacts of urban areas on the environment.
- To understand the vulnerability of urban areas to natural and human-induced events.
- To become aware of 'best practices' related to urban environmental management.
- To identify strategies for governments, the private sector, civil societies, and individuals to reduce the impact of urban areas on the environment.

Introduction

According to the United Nations Population Fund (2007a: 1), the world passed a significant milestone during 2008, with more than half the global population, some 3.3 billion people, living in urban areas for the first time in the history of the planet. Looking forward to 2030, it is estimated that almost 5 billion people will be urban residents. The most dramatic changes will occur in Africa and Asia, where urban populations are expected to double between 2000 and 2030.

Regarding such dramatic change in the growth of urban areas, the United Nations Population Fund (ibid.) observed:

> Cities also embody the environmental damage done by modern civilization; yet experts and policymakers increasingly recognize the potential value of cities to long-term sustainability. If cities create environmental problems, they also contain the solutions. The potential benefits of urbanization far outweigh the disadvantages: The challenge is in learning how to exploit its possibilities.

Canada is a highly urbanized society, with four of five Canadians living in cities. As a result, both the environmental challenges and the opportunities identified by the UN Population Fund apply to Canada, making it important to be aware of the issues and alternative ways to address them.

William Rees, one of Canada's foremost ecologists, who first introduced the concept of ecological footprints, has written that 'the city might be described as a livestock feedlot' (Rees,

2010: 73). While in some regards cities do function as feedlots for humans, we differ from cattle and pigs in that we are not so willing to live in the muck we create, and humans have the capacity—and perhaps the willpower—to do something about it. In this chapter, we examine the nature of sustainable and resilient urban development at global and Canadian scales, determine what impacts cities have had on the environment, examine how urban areas become vulnerable to environmental variability, and consider strategies to enable cities to become part of the solution rather than the problem regarding environmental quality, sustainable development, and resilience. In other words, we want to learn not only 'what is the right thing to do' but also 'how to do the thing right' with regard to urban environmental management.

Sustainable Urban Development

The National Round Table on the Environment and the Economy (NRTEE, 2003c: 3) has defined urban sustainability as 'The enhanced well-being of cities or urban regions, including integrated economic, ecological, and social components, which will maintain the quality of life for future generations'. Achieving **sustainable urban development** requires attention to at least four key factors: *urban form*, *transportation*, *energy*, and *waste management*. Each is considered below.

Perspectives on the Environment
The State of the Canadian Urban Environment

The 2001 Canadian census revealed that 80 per cent of Canadians live in urban centres and that over half of them live in the four largest urban regions—the extended Golden Horseshoe [Greater Toronto Area plus], the Montréal region, the Lower Mainland of British Columbia, and the Calgary–Edmonton corridor. . . .

Yet the recent environmental performance of Canada's cities has been patchy at best. Despite improvements in areas such as the fuel efficiency of passenger vehicles (with the notable exception of SUVs and light trucks), most key indicators suggest negative trends: the use of cars is on the rise, urban transit ridership is down, and cities are using land less efficiently. Concentrations of ground-level ozone—which is linked to childhood asthma, respiratory illnesses, and a range of other health issues—are also increasing.

—NRTEE (2003c: xiii)

Urban Form

Urban form refers to the type and distribution of infrastructure (e.g., buildings, roads) and is a key factor influencing environmental quality in cities. For example, the configuration of roads and other transportation networks has a major impact on energy use for travel within cities. Furthermore, regulations related to buildings influence their energy efficiency. Energy use by both transportation and in buildings is a major contributor to greenhouse gas (GHG) emissions. Finally, **urban sprawl** contributes to loss and disruption to, or degradation of, adjacent agricultural land, environmentally sensitive areas, natural habitats, and water and air quality.

A compact urban form is more environmentally desirable than the 'sprawl' typical of many North American cities. But even though leaders in many Canadian cities have been working hard to revitalize their core areas, the trend continues to be towards sprawl. As the External Advisory Committee on Cities and Communities (2006) reported:

- The average home in Canada is farther away from a city centre than it was a decade ago.
- The proportion of low-rise, low-density homes, except in major cities, is expanding steadily and frequently accounts for more than two out of every three homes constructed.
- While house sizes have increased, the number of people in households has decreased, resulting in space and energy use per person going up significantly.
- Commuting times have increased, with traffic congestion costs estimated at $2.3 to $3.7 billion each year, with obvious negative consequences for productivity.
- Sprawl causes higher servicing and infrastructure costs and less effective public transit service, displaces large tracts of habitat and prime agricultural land, and contributes to water quality degradation.

The External Advisory Committee (ibid., 52) concluded that 'the principal land use challenge for the immediate present and the foreseeable future is to reduce sprawl in our growing places.' Urban areas with a high population density in their cores lead to more efficient and effective land use

Urban sprawl.

than in lower-density areas. They are also much more likely to be able to provide effective public transit. In addition, it is normally cheaper to provide services such as water supply and waste removal in higher-density areas. Later in this chapter, we examine alternative ways of reducing or minimizing urban sprawl and other important aspects of urban sprawl.

To illustrate the importance of urban form, the National Round Table (NRTEE, 2003c: 30) noted that part of the federal government's strategy to address climate change is based on calculations indicating that improved energy efficiency in federal government buildings would reduce GHG emissions by 0.2 megatons. It argued, however, that in the medium to long term, such a reduction in GHG emissions would be achieved or bettered by locating or relocating federal buildings to minimize travel time for workers and visitors, as well as by supplying the buildings with goods and services delivered by van or truck rather than by car. We examine considerations related to transportation in the next subsection.

Transportation

The negative consequences of low-density urban development are at the heart of many serious critiques of automobile-dependent cities and the adverse environmental impacts of cities. The strong relationship revealed by many studies between more compact, mixed-use urban form and reduced car use is reflected in efforts around the world to reduce urban sprawl and create more transit-oriented communities.

Several variables affect energy used for transportation in cities: distance travelled, vehicle loading, and vehicle mode. Each is significantly affected by urban form. Other influential variables are density of the urban area, urban structure, mixes of land use, and street patterns. All affect the number, length, and type of trips (that is, walking, cycling, using transit, or driving a car). The more spread out a city, the farther people have to travel between places. The lower the population density, the more challenging it is to provide high-quality public transit services. The usual outcome is higher reliance on automobiles as the preferred way of commuting between home and work, resulting in greater energy use.

Commuter traffic.

In contrast, for commercial and industrial buildings, the source of energy for heating, cooling, and lighting is often cleaner, such as natural gas or hydroelectricity. In Canada, almost 60 per cent of the energy used in transportation is for moving people, with automobiles accounting for the largest share. The National Round Table (ibid., 13) has stated that 'transit is a more environmentally sustainable form of urban transportation than the automobile.' However, the National Energy Board reports that between 1990 and 2006, 'passenger-kilometres', a measure of one passenger over a distance of one kilometre and therefore representing the total annual distance travelled by on-road passengers in Canada, increased by 1.8 per cent annually. Furthermore, the NEB (2009b: 5) stated that:

> Over time, Canadians have become more dependent on their automobiles. The number of Canadians aged 18 and over who travelled everywhere by car . . . rose from 68 per cent in 1992 to 74 per cent in 2005, and the number of people that made a trip under their own power by bicycle or on foot declined from 26 per cent in 1992 to 19 per cent in 2005. . . . Eleven per cent of Canadian workers used public transit to get to work in 2006, up from 10.1 per cent in 1996.

Various strategies can reduce energy use by transportation within cities. They include: (1) facilitating tele-working and tele-services to reduce travel time; (2) ensuring that parking arrangements encourage reduced car travel (providing ample parking adjacent to public transit departure nodes; setting appropriate [higher] charges for parking cars near workplaces); (3) encouraging development of ride-sharing programs; (4) initiating transit pass programs to provide a seamless public transit system, such as the systems in Hong Kong and various European cities that allow a single pass to be used on buses, trains, and ferries; and (5) facilitating the use of bicycles and other means with a small ecological footprint.

Perspectives on the Environment

Transportation and Energy Use in Toronto

In the City of Toronto, transportation is the largest source of emissions, responsible for an estimated 90 per cent of carbon monoxide emissions, 83 per cent of emissions of nitrogen oxides, and 60 per cent of sulphur dioxide emissions.

—NRTEE (2003c: 19)

Transportation is a major contributor to GHG emissions because most vehicles are powered by fossil fuels.

Perspectives on the Environment

Enhancing Use of Urban Transit

Shifting from automobile travel to public transit will likely have the single greatest impact on the environmental quality of Canadian cities and their effect on the global environment. Bringing about this shift involves making transit attractive and competitive relative to the automobile in terms of convenience, cost, and comfort.

—NRTEE (2003c: 33)

Energy Use

Promoting the green design, construction, renovation, and operation of buildings could cut North American greenhouse gas emissions more deeply, quickly, and cheaply than any other available measure (Commission for Environmental Cooperation, 2008). Buildings in North America release more than 2,200 megatons of CO_2 into the atmosphere annually, about 35 per cent of the continent's total. Rapid market uptake of currently available and emerging advanced energy-saving technologies could result in a reduction of more than 1,700 megatons by 2030 from the emissions projected for that year under a business-as-usual approach. A cut of that size would come close to equalling the CO_2 emitted by the entire US transportation sector in 2000.

In terms of end use, residential, commercial, and industrial buildings account for over 60 per cent of GHG emissions in Canada, and most of it occurs in urban areas. For the residential and commercial sectors, energy in buildings is used mostly for heating water and space and for cooling space. Indeed, for residential buildings, these three end uses are responsible for 80 per cent of energy use.

Energy use in residential buildings is influenced by the construction materials, shape, and orientation of the building, internal temperature settings, internal use activity, and climate conditions. Urban form also is an important influence. For example, townhouses and apartments are usually more energy-efficient than single detached houses. As the National Round Table (NRTEE, 2003c: 20) observed, 'evidence suggests that overall energy use is inversely related to the density of development: more compact, mixed-use cities, which support greater use of sustainable forms of transportation and less energy-intensive building types, tend to use less energy.' It is common now for more advanced green buildings to routinely reduce energy usage by 30, 40, or even 50 per cent over conventional buildings, with the most efficient buildings now performing more than 70 per cent better than conventional properties. Despite proven environmental, economic, and health benefits, however, green building today accounts for a only small fraction of new home and commercial building construction, just 2 per cent of the new non-residential building market, less than 0.5 per cent of the residential market in the US and Canada, and less than that in Mexico (Commission for Environmental Cooperation, 2008).

Perspectives on the Environment

Fossil Fuels and Sustainability

Canada's cities and communities are not environmentally sustainable on their current path. Use of fossil fuels is the main source of the problem. Canadian cities are also characterized by low densities of population, and Canada is still sprawling more than ever before.

—External Advisory Committee on Cities and Communities (2006: xiii)

Waste Management

Various factors affect total per capita amounts of waste generated. For residential areas, key factors include demographic characteristics such as household size, age structure, and annual income, as well as type of dwelling unit, geographical location, and time of year (Maclaren, 2010: 385). Some of these factors are affected by urban form. For example, presence or absence of yard wastes influences the amounts and composition of wastes produced by households. In that regard, apartment dwellers generate lower per capita wastes than single-family dwellings because the former do not have yards and instead often share a common area adjacent to the apartment building.

An integrated approach to waste management strives to divert as much waste as possible away from disposal through

This residential house in Waterloo, Ontario, features a BIPV (Building Integrated PV 7.6kW system) installed by the ARISE Technologies Corporation. The system uses roof vents and pipes to draw warm air down through the ground in order to cool it before it re-enters the house, thus avoiding the need for air conditioning. Also shown is a clothesline on the deck, another example of energy saving.

Chapter 13 | Urban Environmental Management

A garbage-filled moose on display at Parliament Hill as part of a protest against the dumping of Toronto's garbage in Adams Mine in northern Ontario.

Paper-recycling depot.

the 3Rs: source **reduction**, followed by **reuse**, **recycling** and biological treatment, thermal treatment (usually with energy **recovery**), and land treatment. Energy recovery is often challenging because it involves burning waste in incinerators. Establishing an incinerator facility is usually controversial, since most people are not enthusiastic about having one nearby. The other well-known option is disposal of waste in a landfill site. Finding such sites can also be controversial, since they, along with incinerators, are viewed as LULUs (locally unwanted land uses) that trigger **NIMBY** (not in my backyard) reactions. As Maclaren (ibid., 388) has observed, 'The precedence of a waste management option within the hierarchy is determined by its environmental performance.'

Urban Canadians are recycling and composting more than in the past but with significant variation across the country (Statistics Canada, 2007b). For example, in Montreal and Calgary, less than one-third of waste is diverted from landfill sites. Toronto diverts just over 40 per cent. In contrast, in Halifax about 55 per cent of waste is diverted and in Markham, Ontario, about 70 per cent. Markham officials attribute this success to a practice of collecting recyclable (**blue boxes**) and organic waste (**green bins**) twice as frequently as garbage.

Urban form, transportation, energy, and wastes are closely interrelated. As Kenworthy (2006: 67) has noted, 'Not only do urban form, transportation systems, and water, waste, and energy technologies have to change, but the value systems and underlying processes for urban governance and planning need to be reformed to reflect a sustainability agenda.'

Summary

The External Advisory Committee on Cities and Communities (2006: 46) concluded that:

- The way Canadians are living now in their cities and communities uses too much natural capital [the environment and natural resources] and is outstripping the capacity of the environment to recover from pressures placed on it.

Sign in Michigan protesting imports of Toronto garbage.

- Decisions and actions that enhance or degrade the environment also affect the Canadian economy, society, and culture.
- Investment in new environmental technologies could create major new opportunities for long-term prosperity and improved competitiveness.
- Cities and communities in Canada 'clearly lie at the heart of the problems and solutions for the sustainability of the Canadian environment'.

The ideas of the National Round Table (NRTEE, 2003c: 31) related to a sustainability checklist for location and site design of buildings offer useful ideas (see Box 13.1).

Guest Statement
Urban Waste Management
Virginia W. Maclaren

Urban waste management practices in Canada have come a long way since the 1980s when governments first started thinking about how to reduce, reuse, and recycle waste rather than just how to dispose of it. Residential curbside collection programs for recyclables are thriving across the country, and organic waste curbside collection programs are spreading quickly. Many provinces have implemented or are forging ahead with innovative extended producer responsibility (EPR) and product stewardship programs. EPR extends the producer's responsibility for a product to the post-consumer stage of the product's life cycle, while product stewardship involves shared responsibility among producers, retailers, users, and disposers for reducing a product's environmental impact across its life cycle. Both involve diversion of products from landfill by providing infrastructure for collection of the used product and incentives for its recycling or reuse. EPR or product stewardship programs now exist in several provinces for used pharmaceuticals, tires, paint, solvents, lead acid batteries, packaging, and e-waste (electronic waste). All provinces have some form of beverage container deposit system, which promotes reuse or recycling of the returned containers.

Even landfills are receiving attention. More and more municipalities are turning their old and existing landfills into energy sources by extracting methane, which is produced by the anaerobic decomposition of organic waste. Burning methane and turning it into electricity or heat not only provides energy but also reduces greenhouse gas emissions from landfills.

Where do we need to turn our attention to now? Many challenges exist, but I will address just three of them. The first and perhaps the most difficult is the issue of consumption. We live in a consumer society in which consumption is often associated with social status and personal happiness. For many people, the more they consume, the more status they attain and the happier they are. One consequence of excessive consumption is excessive waste production. Sustainable consumption is a relatively new concept that focuses on how to break the link between consumption and waste production. One way to do this is by simply purchasing less, an approach that is part of a lifestyle known as 'voluntary simplicity'. Another way to break the link is to purchase services rather than products. The rationale here is that if a company leases its products (i.e., provides product services) rather than selling them, it has an incentive to optimize the product's operation, increase its service life, and gain value from the old product once the lease is over by recycling or refurbishing it rather than sending it to landfill. Examples of leasing are most common in the commercial sector, such as for copy machines and flooring. A third way to break the link is to purchase products that produce less waste because they are more durable, they have less packaging, or they require fewer material inputs and are less pollution-intensive in their production. Education and promotion are important tools for encouraging all of these practices.

The second challenge is what to do about packaging waste. About one-third of the waste produced in Canada comes from packaging, yet there are no national targets for the recycling of packaging waste. This is in sharp contrast to efforts in Europe, where the European Union requires its member countries to meet national packaging waste reduction and recycling targets for many different types of packaging. For example, most member countries were expected to be recycling between 55 and 80 per cent of all packaging waste by 2008. Targets for specific materials varied, depending on the availability of recycling markets and technologies for recycling. By 2007, all member countries except Greece and 10 new members had exceeded the overall 2008 recycling targets (Quoden, 2010).

The third challenge is to improve residential waste diversion. Diversion of waste from landfills is working well in single-family households but not so well in apartments or multi-family dwellings (MFDs) (Toronto, 2011). For example, in Toronto in 2009, residents of single-family dwellings (SFDs) diverted 63 per cent of their waste, while residents of MFDs diverted only 18 per cent. Part of the reason for the wide difference in diversion rates was that SFDs had access to curbside collection of both recyclable and organic waste,

Toronto in-unit containers for recyclables in multi-family dwellings.

while most residents of MFDs only had recycling programs. Even accounting for the lack of an organics collection program in MFDs, waste diversion was still much lower than in SFDs. What makes waste diversion in MFDs so difficult? There are many reasons. It starts with the diversity of structures that comprise MFDs. Some have garbage chutes and recycling/organic waste rooms on individual floors, while others do not. These structural differences are important because they affect the convenience of participating in diversion programs. The less convenient a program is, the lower its participation rate. In MFDs with garbage chutes on each floor, garbage disposal is easy, but residents cannot use the same chute for recyclables or organics and have to take them down to bins on the ground floor or in the basement. This arrangement clearly makes it less convenient to divert waste than to throw it out. Another factor is the difficulty of finding storage space for recyclables and organic waste in apartment units. Some municipalities provide residents of MFDs with small containers, such as those shown in the photograph, to help with storage. A third factor affecting participation is that residents of MFDs do not face the same level of public scrutiny of their recycling habits by their neighbours as do residents of SFDs, who see their neighbours participating (or not) every time they set their recyclables and organics out at the curbside. Finally, many MFDs consist of rental units, and since renters tend to be more transient than owners, they may not receive the same amount of education about why, where, and how to divert waste locally or within their buildings.

Given the enormous progress that has been made in waste reduction and waste diversion since 1981, when the first blue box curbside recycling program was launched in Kitchener, Ontario, I am optimistic that municipalities will be successful in meeting the above challenges and in further transforming the face of modern urban waste management practices.

Virginia Maclaren is an associate professor and chair of the Department of Geography and Program in Planning at the University of Toronto. Her primary research interests are in the field of waste management, with a regional focus on Canada and Southeast Asia.

Environment in Focus

Box 13.1 Sustainability Checklist for Location and Site Design of Buildings

Location
- Use of existing buildings in already urbanized areas before new construction on **greenfields** (undeveloped land, as opposed to brownfields, or previously developed properties).
- Easy access to good transit service.
- Potential for walking and cycling access by employees and visitors.
- Proximity (walking distance) to amenities and services for workers (e.g., restaurants, personal services, and daycare).
- Potential to link to a community energy system.
- Potential to contribute to the regeneration of economically depressed urban areas.

Site Design
- Maximized building density.
- Integration with transit facilities (e.g., covered walkways connecting transit to the facility).
- Facilities for bicycle riders, such as racks and showers.
- Minimization and appropriate treatment of parking (e.g., creation of underground or structured parking lots; landscaping lots to maintain street frontages).
- Maximized site permeability.
- Easy pedestrian access to the facility.
- Integration of other uses into the facility (e.g., restaurants, services, amenities, and residences).

Environmental Issues in Cities

Air Pollutants

Concentrations of some common air pollutants in Canadian cities have been decreasing (Figure 13.1). The main reason for this drop is stricter regulation of emissions from automobiles combined with enhanced regulations regarding industrial emissions. One anomaly to this trend is ground-level ozone (also referred to as photochemical **smog** or summer smog) created when nitrogen oxides and volatile organic compounds combine in sunlight. There has been a striking increase in the number of 'smog advisory days' in many of Canada's major cities, especially in southern Ontario. The principal explanation is more ground-level ozone. Nevertheless, urban outdoor air quality is improving, as indicated in a 2011 report from the World Health Organization (2011) showing urban air quality in Canada as third best of 91 ranked countries in the world.

Environment Canada and other federal agencies monitor air quality through two measures: ground-level ozone and fine particulate matter ($PM_{2.5}$). Respiration and heart rates can be increased by ground-level ozone. Other health problems can be asthma attacks, bronchitis, and emphysema. Children are usually most vulnerable.

Environment Canada (Environment Canada et al., 2006: 5) reports that at a national level during the period 1990 to 2004, there was an average annual increase of 0.9 per cent for ozone as well as year-to-year variation. In urban areas, the triggers for ozone, such as nitric oxide and volatile organic compounds (VOCs), which are generated from local emissions, have dropped. The main explanation for such a decrease is improved quality of fuels and better emission control technology in vehicles. Nevertheless, over the same period there was an increasing trend in southern Ontario, which had 'the highest concentrations and fastest rise of all regions monitored' (ibid., 6) as well as about 30 per cent of Canada's population. Specifically, southern Ontario experienced an average annual increase of 1.3 per cent in ozone levels. Part of the explanation for this increase is proximity to the industrialized northeastern US, along with prevailing winds that result in long-range transport of ozone and its trigger chemicals.

Urban Heat Island Effect

The **urban heat island** effect occurs as a result of increased temperatures in core urban areas relative to surrounding areas. It is not uncommon for the temperature of city centres to be from 2° to 6°C higher than that of nearby rural areas. Such an effect can be reduced or countered by skilful use of green areas, given their cooling effect within urban areas. A secondary benefit of reducing the urban heat island effect is a decrease in the need for air conditioning buildings in city cores, thereby reducing use of electricity. Higher temperatures created by the urban heat island effect can also generate smog and enhance ground-level ozone, neither of which is confined to the urban area. The smog and ozone can drift to nearby rural areas and in some cases can reduce agricultural productivity, increase health risks, and even contribute to triggering tornadoes and thunderstorms (United Nations Population Fund, 2007a: 59).

The urban heat island effect has encouraged development of 'green roof technology', which involves creating a new roof or retrofitting an existing roof with a growing medium allowing plants, shrubs, or trees to grow on it. Leaders in this technology include countries in Europe and Japan. The Department of Geography at the University of Victoria persuaded the university to invest in a green roof for a new building on campus. It has been estimated that a one-degree reduction

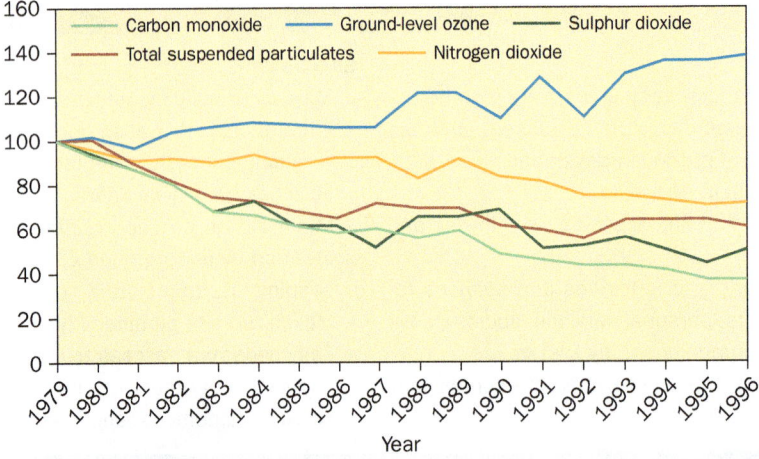

Figure 13.1 | Concentration of common air pollutants, Canada, 1979–96. Source: Centre for Sustainable Transportation. 2002. Sustainable Transportation Indicators, Report on Phase 3, 39. © Environment Canada 2002.

Environment in Focus
Box 13.2 Human Deaths from Air Pollution

Researchers at the University of British Columbia and the University of Alberta have concluded that up to 25,000 premature deaths annually in Canada are due to exposure to air pollution, hazardous chemicals, and pesticides. The associated costs to the health-care system are estimated at up to $9.1 billion related to illnesses such as cancers, respiratory diseases, heart problems, and congenital problems associated with pollutants, all of which have demonstrated connections to environmental contaminants.

They also concluded that such pollutants cause major disabilities and estimated that the types of pollution noted above cause between 1.1 million and 1.8 million 'restricted activity days' annually for people with asthma and result in Canadians every year spending between 600,000 and 1.5 million additional days in hospital.

David Boyd, one of the researchers, commented that 'In our cultural DNA, we think of Canada as a pristine nation, but this is at odds with our track record on the environment' (Mittelstaedt, 2007). Boyd argued that such deaths, impairments, and costs could be reduced if Canada introduced and enforced more stringent standards regarding air quality, drinking water, food, and consumer products. He suggested that Sweden represents best practice in this regard and would be a good model for Canada to strive to match.

---

in the urban heat island effect leads to significantly reduced demand for electricity for air conditioning and refrigeration, which in turn results in lower GHG releases.

Perspectives on the Environment
Urban Heat Island Effect in Canada

Regina and Saskatoon: The night temperatures in the city centres on average are three to four degrees warmer than those in the adjacent countryside. The heat island effect is greatest on calm and cloudless nights, when conditions are best for rural areas to cool more quickly than city surfaces.

—esask.uregina.ca/entry/urban_heat_islands.html

Winnipeg: The average low temperature at the Forks, in the centre of the city, is 2.73 degrees warmer than at the airport on the edge of the city. The comparable difference for the average warm high temperature is 1.57 degrees.

—www.fcpp.org/main/publication_detail.php?PubID=1884

Toronto: The average summer temperatures are four to 10 degrees higher in the city centre compared to temperatures in nearby rural communities.

www.team.gc.ca/english/dbProjects;viewProject.asp?id=5335&typ=comm

In addition to monitoring for energy savings, the proponents of green roofs have tracked their impact on stormwater retention. To deal with water runoff, Portland, Oregon, has used green roofs for some time. Most of the rain landing on a green roof is absorbed by plants and soil and eventually evaporates or transpires back into the atmosphere. Almost all the summer rain in Portland is held by green roofs, and in the fall and spring, retention is between 40 and 50 per cent. In the winter months, retention drops to between 10 and 20 per cent. This retention significantly reduces peak runoff flows following rain events, which reduces the possibility of localized flooding (Dawson, 2002).

Murphy and Martin (2001: 69–70) observed that 'solutions to heat island effects ultimately depend on reductions in energy consumption, pollution, and urban sprawl, but there are mitigation efforts that can work.' In that regard, they identify the following:

1. Design buildings and neighbourhoods to balance building structures with the geometric shapes and characters of the areas (or canyons) between buildings to reduce the amount of energy hitting the surface of buildings and roadways and thus the amount of energy to be re-radiated. North–south street orientations reduce the amount of energy that will reach roads.
2. Use light-coloured surfaces and less thermally absorptive exterior facing on buildings. However, this initiative reduces the potential for solar-based heating systems.
3. Provide vegetation surfaces in place of or to shade heat-absorbing surfaces. This can reduce the urban heat island effect by between 25 and 80 per cent. However, Murphy and Martin also note that there can be negative effects in terms of creating obstructions to walking and hiding places for assailants, injuries from falling branches, and an increase in pollen and mould that trigger allergies.

In assessing the third option, Murphy and Martin (ibid., 70) provide sound advice in observing that 'Ultimately, the question is about the relative short and long-term risks, costs, and benefits of an ecological mitigation of urban heat island effects versus doing nothing. We believe the cost of doing nothing is greater, especially since the benefits of ecological mitigation go beyond the urban heat island effect.'

The green roof of the Social Science and Mathematics Building at the University of Victoria is vegetated by native species such as sedums, yarrow, vine maples, sumac, grasses, wild flowers, wild strawberries, and wild rose. The roof results in a reduction of up to 95 per cent of the heat gain and 26 per cent of the heat loss, and adds 10 per cent R Value insulating properties. In terms of water, approximately 30 per cent is used by plants, 30 per cent percolates to aquifers, and 40 per cent is returned to the atmosphere with little to no surface runoff. For a typical urban non-green roof, 5 per cent goes to aquifers, 15 per cent to the atmosphere, and 75 per cent to surface runoff.

Hydrological Cycle

Urban areas affect the hydrological cycle in terms of both the quantity and quality of water.

Regarding quantity, one of the most obvious impacts is due to urban infrastructure creating an impervious rather than a pervious surface. The implications are twofold. First, expansion of roads and construction of parking lots and buildings results in precipitation running off the surface more quickly, since it is less likely to soak into the soil because of the human-made impervious surface. Second, the consequences are: (1) surface flooding and (2) reduced recharge of aquifers. Both outcomes have led to initiatives to build retention ponds in urban areas so that water can be collected during rainfalls and allowed to either percolate into the ground or be released more slowly than would otherwise be the case. These outcomes have also reminded urban residents that urban sprawl, especially when suburbs are built on the surface of aquifer recharge areas such as moraines, can significantly affect the amount of water available in aquifers for human use.

Water quality is negatively affected because pollutants such as oil and gas from vehicles and salt from winter applications (to make driving safer) get washed into surface streams and groundwater systems. The result is degraded water quality, with negative health consequences for humans and other species. Such negative outcomes have been summarized well by the United Nations Population Fund (2007a: 58): 'Urban areas can affect water resources and the hydrological cycle . . . through the expansion of roads, parking lots, and other impervious surfaces, which pollute runoff and reduce the absorption of rainwater and aquifer replenishment.'

The United Nations Population Fund (ibid.) also noted another negative impact 'through large-scale hydroelectric installations that help supply urban energy needs'. This perspective provides a counterview to those who argue that hydroelectric installations are 'clean' in terms of GHG emissions. Although this is true, it is important to recognize that hydroelectric dams and reservoirs can inundate significant amounts of habitat and can modify downstream flow regimes, which can affect both flora and fauna dependent on natural fluctuations in the hydrological cycle, fluctuations that are evened out when dams are introduced.

Perspectives on the Environment
Road Salt Challenges in Toronto

The GTA (Greater Toronto Area) is one of North America's fastest-growing regions, and serious questions are being raised regarding the environmental sustainability of the anticipated urban growth and the potential long-term impacts on the quality and quantity of ground and surface water resources. Degradation of groundwater quality by NaCl de-icing salt is the primary concern since there are no cost-effective alternatives to NaCl de-icing salt for large-scale use and there is little evidence that salt loadings to the subsurface can be significantly reduced. According to Environment Canada, almost 5 million tonnes of NaCl are released to the environment every year, with the City of Toronto alone receiving over 100,000 tonnes at a rate averaging 200 grams for every square metre of land surface.

—Howard and Maier (2007: 147)

Brownfield Sites

Many cities have a legacy of abandoned or active industrial sites. Surface or underground soils can be contaminated

Chapter 13 | Urban Environmental Management 469

Retention pond.

through disposal practices that were accepted in earlier times before people appreciated the long-term consequences. Such contaminated sites are often referred to as **brownfields**.

One specific problem results from LUST, or 'leaking underground storage tanks'. Industrial sites and garages usually have underground tanks in which materials, including gasoline and chemicals, have been or are stored. While in active use, storage tanks often develop leaks, resulting in some contents being gradually released into the surrounding soil. Alternatively, when an industrial plant or garage is closed, it was not uncommon in the past for the contents of storage tanks to be left in the tanks or for the tanks to be filled with unwanted liquids. Eventually, many of these tanks would develop leaks. Once out of the tanks, the liquids may move through the soil, often ending up in aquifers. When such aquifers are the source of well water, there is a risk to human health. Or if the contaminants are underneath fields on which crops are grown, the contaminants can be drawn into the plants by their root systems and eventually absorbed by the plant material, which is later consumed by humans or animals.

A similar contamination problem can occur through the burying of uncontained wastes under or on the property of a factory or production facility. The case study of the Sydney Tar Ponds in Chapter 1 highlights the huge cost of remediating large sites. As another example, the municipal government in Kitchener, Ontario, spent about $19 million to remove coal tar from beneath one square block in the centre of the city. The city had operated a coal gasification plant there from 1883 until 1958. Coal was heated in ovens, and the flammable gas created was used to provide heat and lighting for businesses and residences in the central part of the city. During the process of gasification, some coal became an oily tar, commonly called 'coal tar'. This coal tar, a waste byproduct, was placed in underground tanks or into open pits and then buried under the property.

After the gasification plant was closed, a Canada Post building was constructed over the coal tar site, and roads were built and paved on each side of the old site. Over decades, the buried coal tar gradually seeped laterally through the fine-grained silt beneath the property. Some was discovered under an adjacent property, leading to a $5 million lawsuit against the city. The suit was dropped when the city agreed to remove the coal tar from under the old gasification plant site and purchase the property.

Kitchener ended up paying almost $15 million more than was initially estimated for the remediation of the gasification site, an experience highlighting the uncertainty involved in such contaminants. The first estimate for the cleanup was $5.6 million. This amount was later increased to $9 million and then finally to $19.5 million. The reason for the extra cost was that there was much more coal tar and much more of it was hazardous waste than had originally been estimated.

Initial estimates of the amount and kind of soil contaminated by coal tar were based on test drilling of boreholes between 1986 and 2004. The conclusion was that the contaminated soil was not more than 5.5 metres below the surface. The test drilling also indicated that none of the contaminated soil was hazardous. In estimating remediation costs, the consultants included an allowance of 1,500 tonnes of hazardous waste to be removed in case some was present.

During May 2006, after work began on building a right-of-way to the site, it was discovered that the coal tar-contaminated soil extended at least seven metres below the surface. As a result, new boreholes were drilled between July and November 2006. The results verified that the contaminated soils extended to seven metres but also that there were 13 times more contaminated soils that were hazardous than estimated at the outset, as well as almost twice as much non-hazardous contaminated soils.

The cost of removing hazardous soils ($156 per tonne) was about three times as much as that of disposing of

Coal tar cleanup in Kitchener, Ontario.

non-hazardous soils ($48.50 per tonne). The hazardous material was trucked to a site in Quebec with the capacity to handle hazardous soil. The city of Kitchener subsequently reported that $9.37 million was required to deal with the higher than initially estimated remediation costs. The remediation was completed during the summer of 2007, and by the time it was finished about 2,500 truckloads of contaminated soil had been removed.

These examples illustrate that the presence of brownfield sites within urban areas can be a major challenge for city officials, private companies, residents, and provincial government regulatory agencies. They also highlight the considerable uncertainty and complexity involved in determining the nature of the problem and identifying solutions.

Vulnerability of Urban Areas to Natural and Human-Induced Hazards

Many cities are vulnerable to hazards, whether triggered by natural or human actions, because of their high concentration of people and, in some situations, the low-quality construction of buildings. The United Nations Population Fund (ibid., 59) states that natural disasters have become both more frequent and more severe since 1990.

Canadians experience many natural disasters. Geophysical hazards such as earthquakes have stayed constant over the past 50 years, but weather-associated hazards have increased dramatically. Environment Canada (2003c: 2) has noted that the following factors make us vulnerable: population growth, urbanization, environmental degradation (e.g., removing timber from hillsides, leading to landslides), urban sprawl in hazard-prone areas, loss of collective memory about hazardous events because of increased mobility, aging infrastructure, and historical overdependence on technological solutions. Specifically, Environment Canada (ibid.) observed that 'Higher concentrations of people living in urban areas mean that if disasters hit, they affect a larger number of individuals. Urban sprawl has led to more development in high-risk areas, such as flood plains.'

McBean and Henstra (2003) provide further information and insight on why urban Canada is susceptible to hazards. One of the core factors is that about 60 per cent of Canadians live in urban areas of 100,000 or more people and about 80 per cent in cities of 10,000 or more. Earthquakes are a significant hazard, with the most vulnerable areas in British Columbia and the St Lawrence Valley. However, in their view, 'about 80 per cent of the impacts are due to weather and weather-related hazards.' Weather hazards include tornadoes, hailstorms, winter storms, and heat waves, while what are termed 'weather-related hazards' are drought, storm surges, floods, and moving ice. In terms of specific weather and weather-related hazards, the probability of tornadoes and hailstorms is highest on the Prairies and in southern Ontario, storm surges are most frequent along the Atlantic coast, and winter storms are ubiquitous throughout the country.

McBean and Henstra (ibid., 2) have also observed that the viability of Canadian urban areas depends on the continuity of basic infrastructure such as sewer and water, transportation, and electrical power systems. In their words:

> Past experience in Canada has demonstrated that this urban machinery is highly susceptible to disaster damage. For example, the 1998 ice storm which struck Ontario, Quebec, and New Brunswick crippled power grids, caused structural damages to buildings, and brought transportation to a halt, becoming Canada's most expensive natural disaster, exceeding $5 billion in losses, and causing at least 28 deaths. Similarly, the British Columbia blizzard of 1996 shut down the cities of Vancouver and Victoria for days, with a major economic impact.

The following two subsections illustrate in more detail the variety of hazards that can be experienced by urban dwellers.

Hurricane Katrina and New Orleans

About 65 per cent of urban areas with populations of 5 million or more are located in low-elevation coastal areas throughout the world. A striking example of a low-elevation coastal area being damaged by a natural disaster occurred on 29 August 2005 when Hurricane Katrina hit the Gulf coast of the United States. More than 2,800 people died across the various states, thousands of homes were destroyed, and hundreds of thousands were left homeless. The city of New Orleans received the greatest damage, but almost 10 million people living in the Gulf coast states of Alabama, Louisiana, and Mississippi were affected by the winds from Katrina. In southeast Louisiana alone, 90 per cent of residents were evacuated.

In New Orleans, slightly less than 50 per cent of which is below sea level, about 80 per cent of the city had been flooded two days after the hurricane struck, with some areas under five metres of water. Much of the flooding occurred because the levees built over four decades to provide flood protection had been breached. Construction of the levees had started in 1965 and was scheduled to be completed in 2015. The Superdome stadium became a 'refuge of last resort', and by the evening of 28 August some 20,000 to 25,000 people were staying there.

As with many natural disasters, human behaviour contributed to the problems. Notwithstanding the mandatory evacuation order, many chose to stay in New Orleans, citing various reasons: belief that their homes were adequately protected, lack of access to transportation or sufficient money to cover the cost of leaving, and conviction that it was necessary to 'guard' their homes and possessions. Unfortunately, the third reason was valid. Within a day of Hurricane Katrina making landfall, violence and looting had spread throughout many areas of New Orleans. Those who remained in their homes often became stranded when Hurricane Katrina had passed by but flood waters had risen. Clean water was not available from the municipal water system, and power outages were common. Significant time and resources were needed to remove stranded people.

What has been learned from the Katrina experience? In November 2009, a federal judge stated that significant flooding had occurred in August 2005 because of negligence of the US Army Corps of Engineers, mainly responsible for the protective levees. Early in 2010, the federal government allocated billions of dollars for city hospitals and schools. A new city master plan was in place by the summer of 2010, and business people were starting new businesses at a rate above the national average.

Other indicators are worrisome. By the summer of 2010, the Greater New Orleans Community Data Center reported that 50,000, or 27 per cent, of the homes in New Orleans remained vacant. This proportion of vacant homes is the highest in the US. And it was estimated that about 100,000 residents of the city prior to Hurricane Katrina had not returned. As of July 2009, the population was estimated to be about 355,000, three-quarters of the pre-hurricane population. Scant data are available about those displaced.

Starting on 28 July and continuing throughout most of August 2010, the fifth anniversary of Katrina, we were further reminded of humans' vulnerability to natural hazards when the worst flooding since 1929 occurred in Pakistan, affecting both urban and rural communities from the northwest to the Punjab and to the southern Sind province. Triggered by heavy monsoon rains, the flooding destroyed an estimated 160,000 km² of crops, creating serious food security issues. The flooding also disrupted water supply systems, causing various water-based illnesses.

Canada agreed in early August to provide $2 million in immediate humanitarian assistance in response to the flooding, which the UN later stated affected over 20 million people (12 per cent of the population). The UN estimated an initial $460 million in foreign aid was needed, but donors were slow to provide assistance. One explanation was that the relatively low number of deaths (estimated to be 1,700) did not create a sense of urgency. Others suggested donors had concern about whether their assistance would reach those needing it, given the history of corruption in Pakistan. By late in August, the UN interim total had been reached, but in mid-September 2010 the UN indicated a total of $2 billion would be needed for relief and reconstruction. At that stage, leading donors were the United States ($345 million), the United Kingdom ($210 million), and Australia ($75 million). Canada was fourth, with an overall commitment of just under $40 million (US dollars).

While Hurricane Katrina and the flooding in Pakistan were extreme natural disasters, they emphasize that humans often are vulnerable because of where they live and work, sometimes by choice but more often because they have no alternatives or the means to relocate to a safer geographical area. Individuals living or working adjacent to rivers or shorelines are at risk of floods, just as those living in snow-prone areas are vulnerable to the risks of severe winter storms. The key message is that, as a species, humans do have choices and often have not been attentive enough to the risks posed by natural hazards. Individually, those who suffer the greatest harms are often people from the lower rungs in a society or in poorer countries—those living in substandard housing that cannot withstand an earthquake; those living in rural and

The 2010 floods in Pakistan, the result of heavy monsoon rains, left nearly one-fifth of Pakistan's total land area under water, killing approximately 2,000 and leaving 20 million bereft of their property, local infrastructure, and livelihoods. (a) A village destroyed by flooding. (b) A young boy struggles to clear away deep mud left by flood waters.

Haiti's 7.0-magnitude earthquake in January 2010 devastated this Caribbean country. (a) The destroyed capital city, Port-au-Prince. (b) Containing the crowds displaced by the earthquake.

low-lying areas with insufficient infrastructure to ameliorate flooding; those who have no way of escaping disaster before it strikes, as was the case for many in New Orleans; those living in a trailer park when a tornado sweeps through.

Earthquake and Haiti

Another dramatic example was the earthquake in Haiti on the late afternoon of 12 January 2010, with a magnitude of 7.0 on the Richter scale (Box 13.3). The epicentre was about 25 km west of Port-au-Prince. The impact was devastating, with estimates of more than 230,000 people dead, 300,000 injured, and 1 million homeless, and with damage to over 250,000 homes. However, disaster experts emphasize that such numbers could only be 'best guesses'. One remarked that it was unlikely the total numbers of dead and injured would ever be known. Domestic capacity to respond to an earthquake of this magnitude was modest, reflecting Haiti's being ranked 149th of the 182 countries recorded on the Human Development Index and being the poorest country in the western hemisphere. On 20 January the largest of many aftershocks occurred, recorded at 5.9 on the Richter scale.

The international community responded with humanitarian assistance. The federal Canadian government quickly

Environment in Focus

Box 13.3 Richter Scale

Earthquake severity is measured related to *magnitude* (amount of energy released at the hypocentre of an earthquake, measured through the amplitude of earthquake waves) and *intensity* (observed effect of movement of the ground on people, buildings, and natural landscape features). The **Richter scale**, developed in 1935, is a measure of magnitude.

The Richter scale is logarithmic, which means that a whole number increase (e.g., from 5 to 6) represents a tenfold increase in magnitude. Expressed in terms of energy released, the difference between two whole numbers is 31 times more energy. Earthquakes with a Richter scale number of up to 2.0 are not normally noticed by people and are only recorded on local seismographs. Quakes with a magnitude of 4.5 or higher are detected by seismographs around the world, and several thousand such events occur annually. Exceptionally high-magnitude earthquakes measure 8.0 or higher, and on average one such event happens each year. Examples are the 1906 earthquake in San Francisco, with a Richter number of 8.3 (calculated from historical data), the 'Good Friday' earthquake centred in Alaska in 1964, which measured 8.6 and affected the west coast of British Columbia and Vancouver Island, and the 7.9 quake in Mindanao, Philippines, in 1976. Damage from an 8.0 earthquake usually extends over at least 300 kilometres. Theoretically, there is no upper limit for the Richter scale, but the largest measured events have been between 8.8 and 9.0. The movement of the earth due to a high-magnitude earthquake can cause serious damage and disruption, but often more damage is caused by fires triggered from ruptured gasoline lines, etc. Most of the property damage in San Francisco after the 1906 earthquake was from fire.

Source: United States Geological Survey, 'The Severity of an Earthquake', at: pubs.usgs.gov/gip/earthq4/severitygip.html.

pledged $5 million, and then committed to match Canadian citizens' donations up to $50 million. However, getting support to those in need was challenging because infrastructure had been badly damaged. For example, air traffic control capacity was disrupted, as were ports and roads. Hospitals and communication systems were damaged. With regard to communication, social networking sites such as Facebook and Twitter became major conduits for many. Insufficient morgue facilities became a difficulty, leading to mass burials. Confusion also occurred related to co-ordination and leadership of the overall relief effort, given the large number of foreign governments and NGOs involved.

Implications

Regarding earthquakes in Canada, Environment Canada (2003c: 3–4) has observed that:

> Although the only significant earthquake in Canada occurred off the East Coast in 1929, triggering a tsunami that killed 28 people [on Newfoundland's Burin Peninsula], scientists predict that an earthquake in the Vancouver area is the most likely major disaster on our horizon. Since quakes occur where tectonic plates converge, only certain regions of the country are at risk: the West Coast, the St Lawrence and Ottawa valleys, off the coast of Nova Scotia and Newfoundland, and certain parts of the Arctic.

What is the nature of risk from earthquakes in your community? What about other natural hazards? In making your assessment, do you differentiate between high probability/risk and low probability/risk? To which type of potential disaster does our society devote the most resources? Is this the most appropriate strategy? To what extent is there awareness of natural hazards on the part of residents in your community? What are the main mitigation measures in place? What might 'next steps' to protect residents, buildings, and infrastructure be?

Rain, Wind, and Sedimentation and the Greater Vancouver Area

In mid-November 2006, the Lower Mainland and Vancouver Island in British Columbia experienced heavy rain from storms over one week. This led to boil-water advisories for some 2 million people in Greater Vancouver, Canada's third largest metropolitan area. The advisory applied to water for drinking and food preparation. It was issued because of mudslides on the mountains inland from North and West Vancouver that affected the reservoirs that provide much of the water to residents of the Greater Vancouver Area.

The mudslides increased turbidity levels in the reservoirs. At one time, turbidity in water was considered an aesthetic rather than a health issue. That view changed, however, once researchers learned that bacteria and parasites attach themselves to dirt suspended in water, with the result that, once attached, they are less likely to be killed by chlorine.

During the boil-water advisory period, restaurants, seniors' care homes, and daycare facilities used bottled water. Many homeowners also turned to bottled water. Rough jostling broke out in stores as supplies started to run out, indicating how quickly civil order can start to break down in the face of stress caused by hazards, as was also seen in New Orleans after Katrina, and in Haiti and Pakistan.

As turbidity levels dropped, the advisory was partially removed on 17 November for people receiving water from the Coquitlam reservoir, which had not been as severely affected. Nevertheless, at that stage the advisory continued for Vancouver, Burnaby, North Vancouver, and West Vancouver, affecting 900,000 people. The advisory remained for those cities because their water is drawn from the Seymour and Capilano reservoirs in the nearby Coast Mountains, which continued to have high turbidity levels.

This event highlights that basic infrastructure and related services in urban areas are vulnerable to natural events, even if they do not lead to the loss of life and destruction of buildings, as occurred with Hurricane Katrina along the Gulf coast, the monsoon-induced flooding in Pakistan, and the earthquake in Haiti. Such incidents are a reminder that environmental managers responsible for urban systems must understand and have the capacity to adapt to variations in natural systems. It is also a reminder that urban areas are often constructed with little attention to their vulnerability to hazards. What are the most significant lessons from the experiences outlined above?

Urban Sustainability

In this section, we turn our attention to cities or communities taking action to establish an urban sustainability trajectory.

Whistler, BC

The Resort Municipality of Whistler, popularly known as Whistler and one of Canada's best-known four-season destination resorts, is located 120 kilometres from Vancouver (Figure 13.2). With a permanent population of about 11,000 people, its economy is based primarily on tourism. Some 2 million people a year visit for both winter and summer activities.

Kelly and Williams (2007: 68) have commented that 'Energy use in tourism destinations is normally disproportionately greater than what is typically associated with other similar-sized communities. This is largely due to the extensive use of energy-intensive technologies that deliver tourism amenities.' The largest portion of energy use is related to travel from other regions to tourist destinations, an aspect over which destination communities have little influence. On the other hand, options exist to achieve energy efficiency within their communities. Kelly and Williams outline what Whistler has done in that regard by highlighting five strategies in Whistler's 2003 *Integrated Energy, Air Quality and Greenhouse Gas Management Plan*. These strategies have the potential for significantly reducing internal energy consumption and related GHG emissions: (1) comprehensive transportation strategy to create compact forms of development to reduce travel distances and encourage alternative forms of transportation; (2) municipal vehicle fleet efficiency to move towards hybrid vehicles; (3) new and redeveloped building energy efficiencies to promote LEED and CBIP (see Box 13.4); (4) natural gas as a primary energy supply to replace the piped propane system and thereby reduce GHG emissions; and (5) small-scale and localized renewable energy sources such as mini-hydro and geothermal to reduce GHG emissions.

Whistler has moved ahead with implementing actions consistent with the initiatives outlined above. For example, an inventory of GHG emissions was completed in the spring of 2006. While the inventory revealed that passenger vehicle transportation and space heating for residential and commercial buildings were the main contributors to GHG emissions, GHG emissions had decreased from the municipality's landfill site because of new processes to cap and capture gases. In the spring of 2007, an audit of appliances and heating systems was completed, and construction of a 50-kilometre natural gas pipeline from Squamish was completed along the Sea-to-Sky Highway. The pipeline provides the municipality with a reliable supply of natural gas. During the autumn of 2008, all existing propane appliances and heating systems were converted to natural gas. It has been estimated that the change from propane to natural gas will result in a 15 per cent decrease in GHG emissions from homes and businesses. A complementary step will be to introduce natural gas vehicle programs for the municipal fleet of vehicles.

To publicize sustainability issues, the municipality has introduced other activities. For example, Whistler has an annual 'dumpster dive' during which staff from the Public Works Department do an inventory of trash bins to determine which items could have been recycled. In addition, the municipality provides biodegradable 'doggy bags' and makes them available at many parks and areas designated for dog walking.

Whistler's actions have won several awards. At the end of 2005, Whistler was recognized as the first municipality in Canada to have completed all five milestones in the Partners for Climate Change Protection Program created by the Federation of Canadian Municipalities. Intended to help municipalities reduce GHG emissions through both municipal operations and community-wide activities, the five milestones include: (1) creating a GHG inventory and forecast; (2) setting an emissions reduction target; (3) developing a local action plan; (4) implementing the local action plan or a set of activities; and (5) monitoring progress and reporting results. Whistler has completed all of them. The community won a second award when it was one of seven receiving the inaugural Green City Award from the BC government in September 2007. Whistler was recognized for exemplary planning and development in balancing protection of the natural environment with promotion of its tourism economy.

Regarding the fifth step noted above, monitoring progress and reporting results, Whistler tracks and reports progress related to the Whistler 2020 Comprehensive Sustainability Plan, which identifies 16 strategic areas. Examples of indicators related to 'Protecting the Environment', one of the strategic areas, include: development footprint, energy use, greenhouse gas emissions, material (waste) use, water use, and sensitive habitat. Details can be found at the website for 'Whistler 2020 Monitoring Program'.

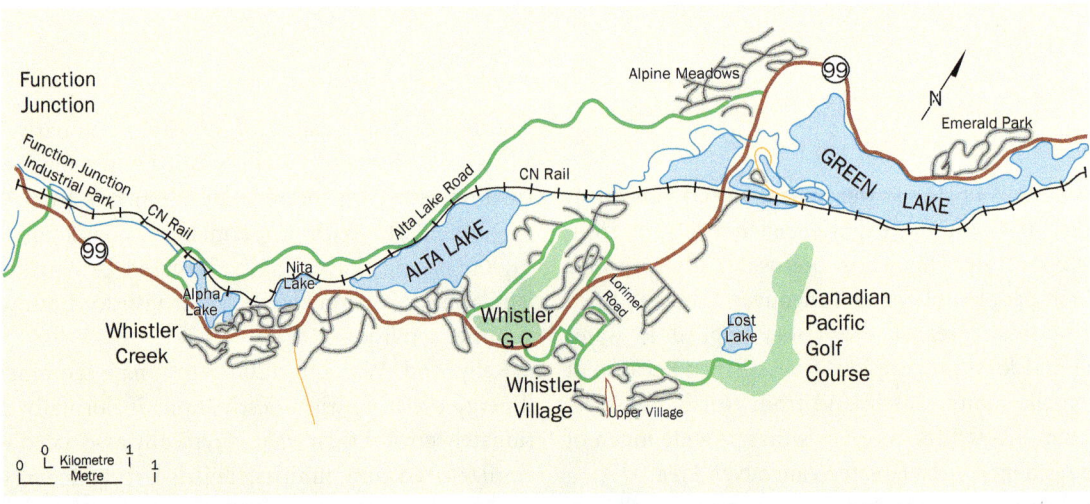

Figure 13.2 | Whistler, BC.

Environment in Focus
Box 13.4 LEED and CBIP

LEED stands for the Leadership in Energy and Environmental Design Green Building Rating System. It is a national standard, established in the US, related to the design, construction, and operation of 'high-performance green buildings'.

LEED provides benchmarks for performance related to five factors that impact human and environmental health: (1) sustainable site development; (2) water savings; (3) energy efficiency; (4) materials selection; and (5) indoor environmental quality.

CBIP, or the Commercial Building Initiative Program of Natural Resources Canada, is based on experience showing that new buildings can be designed to reduce overall energy consumption in a major way. Performance better than 25 per cent above the minimum requirements specified in the Model National Energy Code for Buildings (MNECB) is a common target in the marketplace. More than 1,000 building design models validated by Natural Resources Canada's Office of Energy Efficiency have a higher average design efficiency—35 per cent better than the MNECB.

Sudbury, Ontario: Remediating Mined Landscapes

Located about 400 kilometres north of Toronto and the twenty-fourth largest metropolitan area in Canada, Greater Sudbury at one time was 'notorious across the country for the air pollution and the barren, blackened landscape created by its smelters' (Richardson et al., 1989, 4). Starting in the 1970s, however, initiatives began to rehabilitate the landscape and restructure the economy of Sudbury.

Sudbury was established in the early 1880s as a construction camp for the Canadian Pacific Railway. During the building of the railway, copper and nickel deposits were discovered a few kilometres north of the construction camp. The camp soon evolved into a mining community, and Sudbury became the second largest producer of nickel in the world. In addition to Sudbury, other mining communities such as Falconbridge were established in the 30 x 60-kilometre Sudbury basin (Figures 13.3 and 13.4).

The strong economic growth, based on the mining industry, was offset by a dramatically degraded physical environment. Tens of thousands of hectares of the Sudbury area were devoid of significant vegetation because of air pollution and past mine practices. Many lakes in the area had become highly acidic and degraded by metal contaminants, partly as a result of acid deposition associated with the smelter, as described in Chapter 4. The city periodically experienced episodes of choking air pollution from the mining smelters.

On the environmental side, some initiatives had begun before the concerted effort in the late 1970s. During the 1970s, Inco and Falconbridge Ltd shut down part of their smelter capacity and significantly reduced emissions of sulphur dioxide. Inco also built its 381-metre 'Superstack'—the tallest smokestack in the world—to disperse the emissions farther afield. Other ore-processing measures were adopted in smelters, including 'removing sulphur from ore before smelting, increasing the efficiency of the roasting and smelting processes, and containing sulphur through the production of sulphuric acid, which is then sold' (Clean Air Sudbury, 2005: 8). The result was that the average level of sulphur dioxide in the air at Sudbury dropped from 54 to five parts per billion between 1971 and 2002, while total emissions fell by 88 per cent between 1960 and 2002. The outcomes were tangible. The city was no longer periodically subjected to air pollution fumigations. Vegetation could grow and began to return.

With local air quality significantly improved, opportunities for restoration presented themselves. Beginning in 1971, scientists established small plots to test various combinations of soils and plant species. Based on this experience, techniques were developed at Inco and in the Department of Biology at Laurentian University to grow grass, clover, and then tree seedlings on formerly degraded land. Applying crushed limestone resulted in neutralization of acidic soil, inhibition of the uptake of metals, and enhanced bacterial activity, all of which allowed some vegetation types to do well. As Lautenbach and his colleagues (1995: 112) later commented: 'There was little scientific information to guide the design and implementation of an effective reclamation program for this type of landscape. Therefore, testing and monitoring were essential for achieving the objectives of the reclamation program.' The objectives and outcomes of this work are highlighted in Box 13.5.

Perspectives on the Environment
Reclaiming Sudbury

Vegetation damage began with irresponsible logging, fire and roasting beds, but the decades of intense fumigation from smelters caused most of the damage. The poisoning of the soil by the addition of acids and toxic metals from smelter fumes created conditions that were unlikely to allow rapid natural recovery.

—Viewpoint: Perspectives on Modern Mining (2008: 1)

Inco's 'Superstack' in Sudbury, Ontario, built to address issues of air pollution. The world's tallest smokestack was also an engineering feat, as it was created in one continuous pour.

The regional municipality became involved in the reclamation program with the establishment of VETAC—a Vegetation Enhancement Technical Advisory Committee to Regional Council. This committee includes botanists, ecologists, landscape architects, horticulturalists, agriculturalists, planners, fisheries experts, gardeners, and interested citizens from the mining companies, university and college, provincial ministries, Hydro One, municipalities, and the general public. After surveying the degraded areas, they agreed on a target to rehabilitate about 30,000 hectares of barren and semi-barren land.

Perspectives on the Environment

Sudbury Barrens

Barrens: The areas within the City of Greater Sudbury that were impacted by past mining and smelting activities, resulting in virtually all of the vegetation cover being destroyed. Semi-barrens have slightly more vegetation cover than barrens, but are still considered heavily impacted.

—Greater Sudbury (2010: 1)

During the summer of 1978, 174 students worked to apply crushed limestone, fertilizer, and grass seed on an area adjacent to the highway close to the airport and on another area

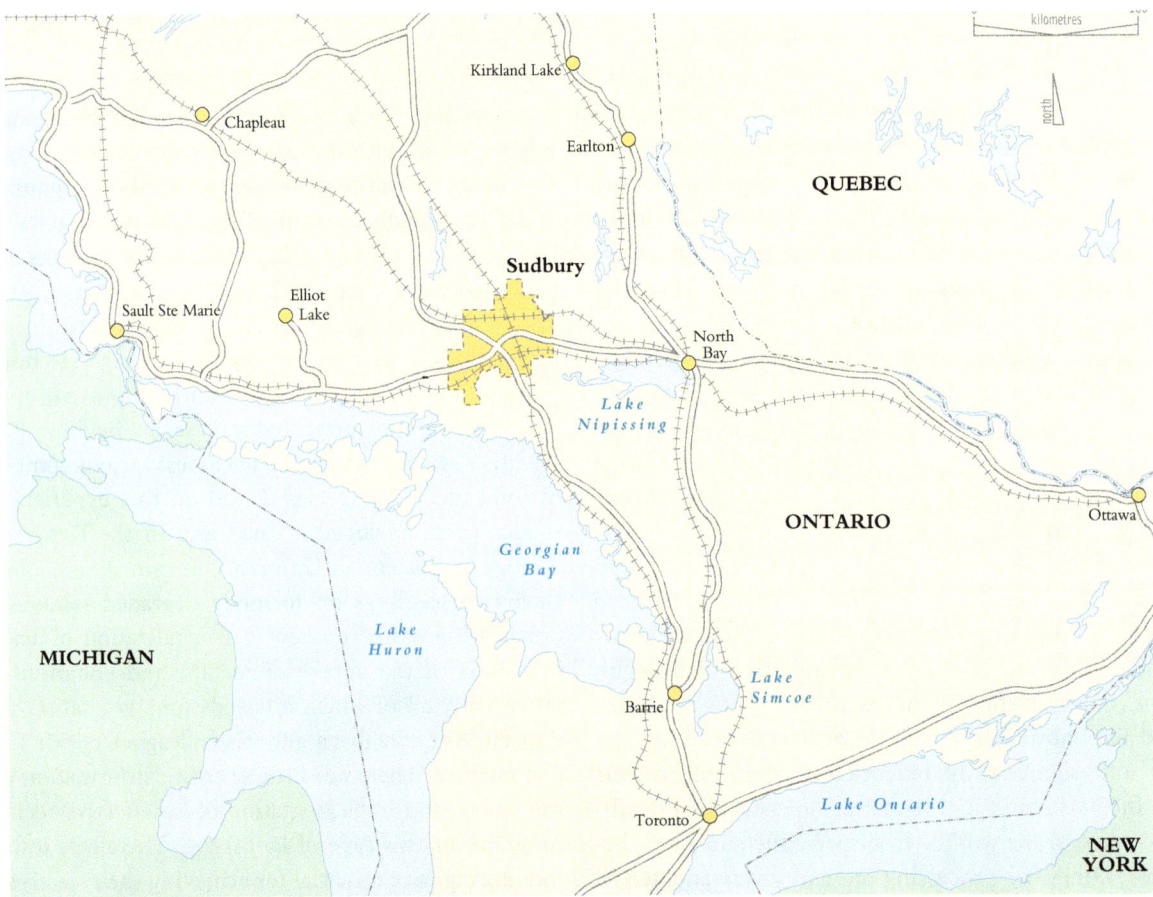

Figure 13.3 | Location of Sudbury.

Chapter 13 | Urban Environmental Management 477

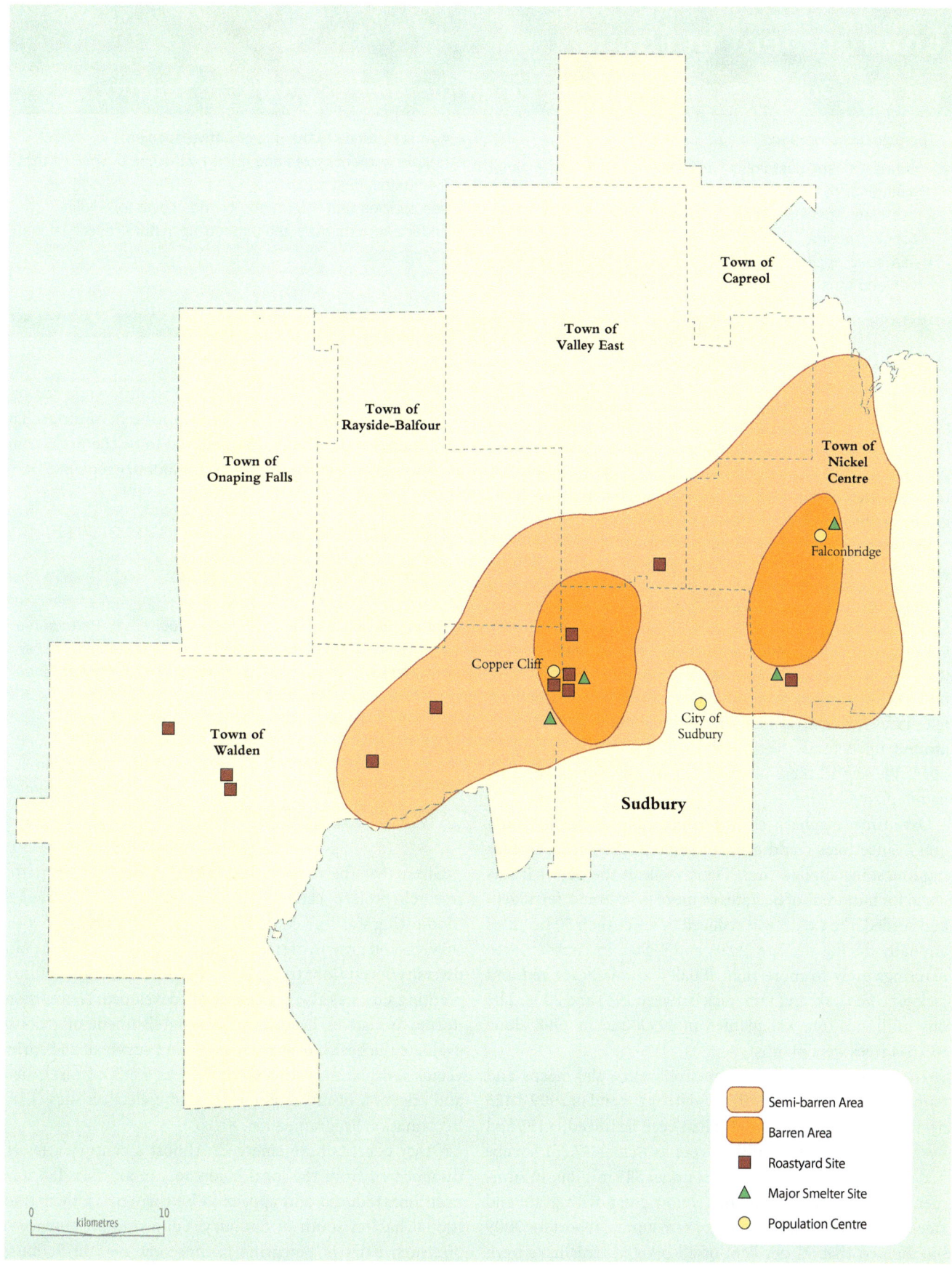

Figure 13.4 | Sudbury: Extent of barren and semi-barren landscape. Source: Lautenbach (1985: 4).

Environment in Focus

Box 13.5 Objectives of the Grassing and Reforestation Activities in Sudbury

The objectives were to:

- create a self-sustaining ecosystem requiring minimal maintenance;
- use plant species tolerant of acidic soils and low nutrient concentrations;
- use seed application rates that allow for natural colonization and thus increase species diversity;
- give preference to the use of native species;
- restore nutrient cycles and pools by the use of species that fix nitrogen (legumes);
- use species that attract and provide cover for wildlife;
- undertake initiatives that speed up natural successional changes.

---

beside the Trans-Canada Highway. Over the summer, they planted grass on 115 hectares of barren land, removed debris from a further 206 hectares, planted 6,000 trees, shrubs, and plants, collected some 30,000 samples for pH and nutrient testing, created 122 new test plots, and gathered 365 kilograms of native seeds.

In 1982, the regional municipality began to hire workers laid off from Inco and Falconbridge to plant trees in the spring before students could start to work. The work continued during the 1980s and 1990s, and in the 1990s, because of improved air quality, some tree species were colonizing previously barren areas. As Ross (2001: 60) reported, one VETAC member described how amazed team members felt about the natural regeneration: 'One day we wandered around up behind the smelter, and our mouths dropped open. We saw all these birch seedlings coming back, and we hadn't planted them.'

Over time, emphasis shifted to planting trees rather than grass, since trees could grow in some places without needing limestone applied first. Thus, while at the outset it was usual for hundreds of hectares or more to be limed, fertilized, and seeded, the extent was reduced to fewer than 50 hectares annually during the late 1980s and 1990s. In contrast, tree plantings grew to more than 100,000 annually for red and Jack pine, red oak, and tamarack between 1983 and 2000. The one-millionth tree was planted in 1990, and in 1998 alone 985,574 trees were planted.

The results have been impressive. While slag heaps and bare, blackened rock are still present, by the end of 2009 3,425 hectares had been limed, 3,212 had been fertilized, 3,193 had been seeded, and 9.11 million trees as well as 47,251 shrubs had been planted, at a cost of just under $25 million. In summer, this formerly barren land is now green with grass and trees. The Land Reclamation Program annual report for 2009 highlighted that 95 per cent of all planted trees have been coniferous, 4.5 per cent have been hardwood, and less than 0.5 per cent have been shrubs. Of the coniferous trees, some 75 per cent are pines (Jack, red, and white), while red oak trees represent just over 50 per cent of the hardwoods. The aim has been to plant species believed to be the main components of the original forests in the Sudbury region.

Perspectives on the Environment

Biodiversity and Ecological Recovery in Sudbury

During 2009, a Greater Sudbury Biodiversity Partnership was created by 26 organizations (www.greatersudbury.ca/biodiversity). Its purpose is to facilitate information sharing, broader participation through group activities, and public awareness and education on biodiversity issues.

In the same year, a Biodiversity Action Plan was developed related to ecological recovery in the Greater Sudbury area. The Plan (see above website) provides a vision and prioritizes goals for ecological recovery.

Given that the remediation initiative has been in part a research project, care has been taken to monitor results. Monitoring was started during the early years regarding the survival and growth of tree seedlings, as was tracking the biodiversity/forest floor transplant plots. Identification of future planting sites was also systematically developed. Newer monitoring initiatives have included establishment of plots to evaluate the health of shrubs, long-term survival, and spring versus fall planting success, as well as wetland assessment and selection of future potential seed collection sites (Land Reclamation Program, 2006: 8).

Other benefits have emerged. Almost a century after elk disappeared from the local landscape, more than 150 have been reintroduced and appear to be thriving in their traditional habitat south of Sudbury. Following an absence of 30 years, pairs of peregrine falcons now nest in Sudbury, and during the spring of 2000 a pair of trumpeter swans hatched cygnets.

Perspectives on the Environment
Importance of Science and Community Involvement

The science isn't really complicated. First lime is scattered onto the soil to help deal with the acidity. And then fertilizer is added to provide the nutrients that plants need. In the first four to five years of research, Keith Winterhalder [Laurentian University professor and former VETAC chairman] and his colleagues learned what were effective mixtures of lime and fertilizer, what grasses would provide cover, what trees would survive and how best to plant them.

Everything had to be done by hand, by armies of people carrying bags of lime. And after that was done, they would walk the land with bags of grass and fertilizer.

Getting the community involved is what has sustained the program. About 25 per cent of the trees have been planted by community groups—Scouts, schools, Lions and Rotary Clubs. Some groups volunteer over and over again.

The continuity of individuals, their determination, and the recognition that everyone would work has helped make the project succeed.

—David Pearson, founding director of Science North, quoted in *Viewpoint: Perspectives on Modern Mining* (2008: 3, 4)

The positive results outlined above have attracted recognition for Sudbury both within and outside Canada. Among the many awards Sudbury has received from within Canada are the Lieutenant-Governor's Conservation Award from the Council of Ontario and an award from Canadian Land Reclamation. Outside Canada, Sudbury has been recognized through a United Nations Local Government Honours Award and awards from the Society for Ecological Restoration and from Germany for Global Responsibilities through Local Action.

Significant progress has been made in remediating the landscape of Greater Sudbury. While much has been accomplished, the re-vegetation completed still leaves 60 to 70 per cent of the barren and semi-barren land to be re-vegetated. Another concern is the impact of climate change. The annual temperature has increased by 1 degree since 1970, which contributes to increased evaporation that may create drier soil conditions. The consequences could include damage to trees and shrubs in re-vegetated areas, because of the young age of the trees and shallow soil conditions. Drier summers also may lead to more planting in the autumn, and also changing the mix of trees to be planted.

Best Practice for Urban Environmental Management

In the following subsections, we highlight best practice related to urban environmental management. To begin, we present the ideas that the Canada Mortgage and Housing Corporation (CMHC, 2005: 1) developed regarding what it calls 'smart growth' for cities, or in its words, 'land use and development practices that limit costly urban sprawl, use tax dollars more efficiently, and create more livable communities'. In particular, it argues that 'in order to be sustainable, cities should alter their development patterns so as to be more compact, diverse in local/district land uses, with well-defined urban boundaries and clear internal structures.' CMHC (ibid., 2) uses the following 'indicators' to assess whether a municipality is using smart-growth principles:

- promote denser, mixed-use development in greenfield areas;
- intensify the existing fabric to moderate greenfield development;
- take advantage of specific intensification opportunities;
- increase transportation choice and reduce car usage;
- increase the supply of new affordable housing;
- improve the range of housing types;
- preserve agricultural lands;
- preserve lands essential to maintaining regional ecosystem functions;
- direct employment to strengthen the core and designated sub-centres;
- provide infrastructure to reduce ecological impacts of development.

In the following five subsections, we consider spatial scale, parks and pathways, air quality, waste management, and transportation. Earlier in the chapter, we examined urban form or structure in some detail. In the final subsection, we review approaches being taken in Victoria, Vancouver, Winnipeg, and Toronto.

Spatial Scale

The United Nations Population Fund (2007a: 53) has noted that:

> fragmentation of the urban territory brings both administrative inefficiency and environmental setbacks. The boundaries of the city's administration rarely coincide with its actual area of influence.... Without some sort of regional entity, the administration of key services, such as water and transport, that cut across different boundaries, is very difficult. By the same token, fragmentation breaks up the contiguity that natural processes require. Fragmentation also makes it difficult to protect ecologically fragile areas or regulate for environmental integrity.

The challenges outlined by the United Nations Population Fund have been recognized for many years but are difficult to overcome. The boundaries of urban municipalities usually reflect administrative or political, not ecological, considerations. For example, the cities of Ottawa in Ontario and Gatineau in Quebec are located across from each other on opposite sides of the Ottawa River. Their urban boundaries do not reflect the reality that both of them are within the Ottawa River catchment or basin. Furthermore, many cities are located in the upper portion of river basins and dispose of much of their waste in the river, which then carries the waste-laden water to downstream communities that have to bear the cost of treating the water if they use the river as their source of supply.

The most common way to address an appropriate scale for an ecosystem approach is to create a 'regional authority' based on landscape features in order to assist managers in considering the larger ecosystem within which a city is located. This is not a perfect solution, but it is a first step. For instance, many cities within a region are often reluctant to give up their authority or autonomy. This can subsequently lead to conflicts and requires a high level of capacity to facilitate co-operation and negotiation.

Parks and Pathways

Public parks and pathways in urban areas contribute to physical fitness and the well-being of residents. In addition, urban trees and other vegetation can filter air pollutants, moderate the urban heat island effect, and help to enhance water quality and reduce flood damage potential. Thus, parks or informal 'green spaces' offer considerable ecological value.

In addition to creating or enhancing parks, municipal governments can facilitate community gardens, ban the use of pesticides for 'cosmetic' reasons, and protect or restore wetlands and other natural areas that filter air pollutants. At the household level, residents can plant more trees and shrubs, which take up rainwater and help to keep buildings cooler. Using native species for home gardens instead of lawns and imported plants is beneficial because native species are adapted to local climate conditions and therefore need less watering or fertilizing. Another choice is to have a vegetable garden or grow vegetables in an allotment garden, since locally grown vegetables do not contribute as much to energy use and emissions. Finally, there are benign alternatives to pesticides, toxic cleaning products, and other chemicals.

Air Quality

The discussion earlier in the chapter about urban form, transportation, and energy use indicated that municipal governments have various options to enhance air quality. Local governments can use more energy-efficient vehicles in their municipal fleets, which will reduce fuel consumption and GHG emissions. Furthermore, in terms of energy for buildings or other purposes, options exist to develop renewable energy sources, such as wind or solar power. In terms of urban form, sprawl can be constrained through land-use regulations and by providing user-friendly and convenient public transit. Finally, building standards can be modified to improve the energy efficiency of new or renovated buildings.

Individual urban residents also have choices. They can walk, cycle, or use public transit, any of which contribute to reduced GHG emissions. Insulating or draft-proofing homes and switching to energy-efficient lighting and appliances will reduce the draw on energy sources, thereby contributing to lower GHG emissions. Setting the home furnace thermostat at a lower temperature and wearing a sweater in winter or setting the air conditioner to a higher temperature and wearing lighter clothing in summer reduces energy consumption and GHG emissions. Finally, purchasing food grown in the local area contributes to lower energy consumption than buying food transported from distant locations. These ideas highlight the importance of disseminating information to sensitize individuals to the impact of their behaviour, thus facilitating longer-term shifts in basic values, attitudes, and behaviour. Opportunities for individual actions are addressed in detail in the final chapter.

Waste Management

Municipal governments can enhance reduction of the waste stream by expanding the types of items that can be placed in blue boxes and by providing blue box services to apartments and office buildings. Reduction also can be improved by requiring manufacturers to reduce packaging for products and by establishing deposit refunds for glass bottles. Recycling can be enhanced by searching for and developing more markets for recycled materials. More aggressive options include the banning of certain products, as San Francisco did

Community garden.

International Guest Statement
Natural Landscapes in Urban Environments and the Role of *Feng Shui*
Lawal Marafa

Feng shui is an ancient art in Chinese culture that represents the balancing of natural, physical, and astronomical energies that can help to promote happiness and well-being. Traditionally, this concept has also been used in understanding that the entire environment is alive and full of energy and that balancing the elements in the environment can bring harmony and prosperity to the community. To this end, vegetation and natural ecosystems are protected and used in promoting the *feng shui* belief in societies and communities, especially in South China.

As urban areas continue to grow, the probability increases for a negative impact on ecosystem services that support and sustain urban populations. While ecosystem services can be diverse, a striking characteristic of *feng shui* forests is their rich biodiversity. Given that urban areas now are home to 50 per cent of the global population, this situation creates opportunity for inculcating environmental awareness and understanding of nature, in turn leading to more effective planning and management of urban landscapes.

When cities grow, some cause environmental degradation, reflecting lack of efficient urban planning. This outcome often results in problems that may include encroachment into agricultural and forest land, or simply onto natural ecosystems at the urban periphery. Originally, many cities sprang up out of or adjacent to natural areas. As they developed and expanded, they often left behind patches of natural habitat or some were even created. When such fragmentation happens, ecosystem services normally provided by forest landscapes and natural areas gradually erode. Although city life can be stressful, access and proximity to urban green areas have a beneficial influence on the health and well-being of the urban population, and thereby contribute to sustainable livelihood and quality of life. Nevertheless, in the past few decades, urban forests often have been given less priority and some have even become extinct. Consequently, there is a need to articulate a multi-managerial system that includes watersheds, wildlife habitats, landscape design, recycling of municipal wastes, tree care and tree management, etc.

The interaction of people with their local environment has normally been determined by the culture of the local people. According to the oriental culture from which *feng shui* emanates, humans are an integral part of nature. In Hong Kong and South China, the forested landscapes, often at the periphery of urban areas, have been modified as a result of adherence to the *feng shui*, or the science of spatial placement and traditional geomancy. In most areas where such urban forests exist, they provide opportunity for residents to develop a personal affinity as well as social identity and to foster environmentally sustainable attitudes. Indeed, the *feng shui* landscapes often become green and open spaces for recreation, social gathering, or other forms of amenities. In addition, the presence of such forests also brings ecological benefits.

Culture and nature are complex concepts, especially when related to one another. When nature and culture meet, landscapes emerge. Where culture is intertwined with nature, the natural elements, such as vegetation, water, and general

(a) The urban forest landscape in Tai Wai, Hong Kong, makes the area immune to development and attracts residential estates nearby in order to benefit from the *feng shui* status of the hill. (b) Luk Keng is a village in the North East New Territories of Hong Kong. The classical layout includes an agricultural field in the lower topography, a village at the foot of the hill, and an area of *feng shui* woodland on the elevated area adjacent to the village.

green cover, become part of urban landscapes. In cultural landscapes, human contributions to the land can be constructive and consistent with nature's own conditions and processes, but this outcome is not automatic. Intervention and management thus have to be reconsidered. Globally, the dramatic increase in urbanization in developed and developing countries has far-reaching impacts on the environment and ecosystem services, affecting biogeochemical cycles, hydrologic regimes, and climate at a range of different spatial scales. Urban areas are increasingly becoming mega-cities, usually creating intense pressure on local natural resources.

Natural landscapes within mega-cities provide services that may address various environmental and urban planning needs. As with *feng shui* forests, their spatial location and characteristics signify a belief in harmony between humans and nature, and their ecological structure contributes beneficial influences on the microclimate. These benefits include improvement of hydrological systems, mitigation of disaster risk, enhanced demand for local food and natural amenities, mitigation of climate change effects, amelioration of heat island effects, and carbon sequestration, among others. The relationship between the city and the natural environment is actually circular, with cities having massive effects on the natural environment, and with the natural environment, in turn, profoundly shaping urban configurations.

Cities have always placed demands on their sites and their hinterlands. In order to extend their usable territory, urban developers often reshaped natural landscapes by levelling hills, filling valleys and wetlands, and creating huge areas of human-made land. As population and resource demand increase, this trend will escalate. However, there is a need to intervene so that there can be harmony between the activities of humans in the urban areas and the existing urban forests and vegetation, as promoted by the *feng shui* philosophy.

Lawal Marafa was born and educated in Nigeria and completed graduate studies in Russia and Hong Kong. He is currently an associate professor in the Department of Geography and Resource Management at the Chinese University of Hong Kong, and his main research interests are ecotourism, tourism and environment, natural resource management, and sustainable development related to the Millennium Development Goals.

in banning plastic bags in larger supermarkets and pharmacies in 2006. Oakland, California, has banned polystyrene or Styrofoam containers, which has led to restaurants using paper, cardboard, or recyclable plastic containers.

Individuals can become more disciplined in their use of blue and green boxes and use reusable coffee containers, water bottles, and shopping bags. Other options include purchasing products with minimal packaging, buying reusable instead of disposable products, and, when appropriate, reusing paper, bottles, and other material.

Transportation

If people are to change their behaviour, incentives must be provided, and people need to be educated to understand why behavioural change is necessary. If it is clear that many people will continue to use automobiles, then they should be encouraged to use more fuel-efficient vehicles, which also would reduce GHG emissions.

What incentives might be provided to encourage people to purchase more fuel-efficient vehicles? Suggestions include charging a lower licence fee for such vehicles, providing free parking in municipal parking lots, and access to commuter lanes on highways normally reserved for buses or other vehicles with a minimum number of passengers. Another option is to encourage people to use other transportation means rather than cars. For example, it could be made more expensive to drive a car into the centre of a city, as was done in London, England, where drivers pay a fee for driving in the central area during the day. And first in Paris and now in other major cities, people can rent a bicycle for a small fee at one of the many docking stations throughout the city. When they are finished with the bicycle, they can leave it at any other docking station. Paris's Velib program was popular from its beginning in 2007, with one million customers during the first year. However, there also were challenges: 24,000 bicycles had been stolen by the summer of 2010.

By the end of 2011, Paris had 3,000 electric cars available for use in the same way as its bicycles. Called 'Autolib', after

BIXI Bikes at a docking station in downtown Toronto. Toronto and Montreal now participate in the BIXI program, which makes bicycles available throughout each city's core on a membership and per-use basis.

the Velib bicycle program, the program has the four-seater cars distributed among 1,200 stations in metro Paris, available 24 hours a day. Users must have a valid driver's licence and pay a subscription fee to borrow a car.

What information and education programs do you think should be developed to help individuals understand the impact of travel behaviour on the environment and why is it in everyone's interest to modify travel behaviour with the environment in mind?

Best Practice: Dockside Green (Victoria), Vancouver, Montreal, Winnipeg, and Toronto

Award-winning urban development in Victoria, Vancouver, Montreal, Winnipeg, and Toronto offer insight into best practice in environmental management.

Dockside Green, Victoria, BC

The Dockside Green development in Victoria received the Excellence in Urban Sustainability Globe Award in 2008, an international award presented to a local government, private-sector company, or consortium that has developed and applied outstanding urban sustainability principles. It is not the only award that the project has won; Dockside Green also received the international Brilliant Development award at the first Discover Brilliant Conference as well as several others.

Dockside Green is a multi-million dollar project designed for residential, home office, retail, office, and light industrial uses and includes extensive public amenities. With plans for 26 buildings (all aiming for LEED platinum certification) totalling 117,000 square metres, the development is being built over the next five years. The first two phases of the residential development both earned the highest LEED scores yet accorded any development in the world. Superior building practices and extensive community planning are transforming the seven-hectare waterfront site from a contaminated industrial wasteland into a healthy and lively community that 2,200 residents will call home. As the following list of attributes highlights, there are innovative features, such as training for Aboriginal people and a car-pooling program, rarely seen in conventional developments. Some of its notable attributes include:

- a wood waste gasification plant providing renewable heating to the development, making it 'greenhouse gas positive' and allowing sale of surplus heat;
- sewage treated on-site;
- heat recovered from a sewage treatment plant and a city sewage trunk line;
- treated sewage water used in toilets, irrigation, and water features;

Dockside Green, Victoria, British Columbia.

- projected potable water savings estimated at 70,318 million litres per year, equivalent to the entire region's use on the driest day of the year;
- green roofs on two residential buildings and one commercial building;
- construction training program for First Nations;
- housing affordability strategy offering qualified buyers one of 26 units at 25 per cent below market value;
- another 45 units promised as rental units;
- built on a brownfield—a former contaminated industrial site;
- no potable water used for landscaping;
- LED lights in corridors and compact fluorescents in suites and common areas;
- motion-sensor lights in closets, storage areas, and bathrooms;
- heat-recovery technology that captures heat from ventilated air to pre-warm incoming air;
- energy metering to individual suites, allowing individuals to control their energy use;

- every appliance Energy Star rated;
- solar trash compactor;
- 1,000 trees planted throughout the development;
- every resident a member of a car-share program with 10 cars;
- more than 93 per cent of construction waste diverted from the landfill.

Vancouver: Greenest City 2020

In early 2009, Mayor Gregor Robertson created a Greenest City Action Team, and directed it to determine necessary action to make Vancouver the greenest city in the world by 2020. In April 2009, the team identified 44 'quick start recommendations' focused on three aspects: jobs and the economy, greener communities, and human health (Vancouver, 2009a). Later the same year, the committee submitted a proposed action plan containing 10 long-term goals, as well as a set of targets for 2020 (Vancouver, 2009b). In February 2010, the City Council accepted the recommended long-term goals, which reflected the following vision:

> The greenest city in the world will be a vibrant place where residents live prosperous, healthy, happy lives with a one-planet footprint, so as not to compromise the quality of life of future generations or people living in other parts of the world. (Ibid., 11)

The team also outlined the rationale for the vision and its recommendations:

> Why green? Because in the highly competitive, highly mobile modern world, the elements that make a community healthy also make it wealthy. Functionally, a compact, efficient city with a well-organized transportation system and a light environmental footprint is cheaper to run and easier to maintain. The bright, creative people who are the key to conceiving and expanding a globally competitive economy also gravitate to the most desirable—livable—cities. (Ibid., 6)

The team recognized that the 10 goals were ambitious and could take a generation (20 to 30 years) to achieve. To ensure initiatives were kept on track, each goal was accompanied by a measurable 2020 target. The long-term goals were divided into three categories, shown below.

A. Green Economy, Green Jobs

Goal 1. Gain international recognition as a mecca of green enterprise
Target: Green economy capital: 20,000 new green jobs

Goal 2. Eliminate dependence on fossil fuels
Target: Climate leadership: Reduce GHG emissions 33 per cent from 2007 levels

Goal 3. Lead the world in green building design and construction
Target: Green buildings: All new construction carbon neutral: improve efficiency of existing buildings by 20 per cent

B. Greener Communities

Goal 4. Make walking, cycling, and public transit preferred transportation options
Target: Green mobility: Make the majority of trips (over 50 per cent) on foot, bicycle, and public transit

Goal 5. Create zero waste
Target: Zero waste: Reduce solid waste per capita going to landfill or incinerator by 40 per cent

Goal 6. Provide incomparable access to green spaces, including the world's most spectacular urban forest
Target: Easy access to nature: Every person lives within a five-minute walk of a park, beach, greenway, or other natural space; plant 150,000 additional trees in the city

Goal 7. Achieve a one-planet ecological footprint
Target: Lighter footprint: Reduce per capita ecological footprint by 33 per cent

C. Human Health

Goal 8. Enjoy the best drinking water of any major city in the world
Target: Clean water: Always meet or beat the strongest of BC, Canada, and World Health Organization drinking water standards; reduce per capita water consumption by 33 per cent

Goal 9. Breathe the cleanest air of any major city in the world
Target: Clean air: Always meet or beat World Health Organization air quality guidelines, which are stronger than Canadian guidelines

Goal 10. Become a global leader in urban food systems
Target: Local food: Reduce the carbon footprint of our food by 33 per cent per capita

The team concluded by observing that the recommended actions offered 'dividends in the form of better health, a more resilient economy, and a vibrant environment' (ibid., 63). We invite you to consider the recommendations for Vancouver and then to determine a comparable set of recommendations for your community or region.

Montreal

Sustainable development, or *développement durable*, has been a centrepiece for Montreal since its first Strategic Plan for Sustainable Development for the period from 2005 to 2009. Subsequently, through collaboration with more than 180 organizations in Montreal, its *Community Sustainable*

Development Plan, 2010–2015 was prepared. The second plan, shown below, is based on the same five orientations as in the first plan.

Perspectives on the Environment
Vision for Montreal's Community Sustainable Development Plan

Montreal is a city on a human scale, proud and respectful of its heritage, where everyone contributes to creating a vibrant, prosperous, united, viable, and democratic community. Montreal, its citizens and institutional leaders of the community are making sustainable development a priority.

—Montreal (2010)

Unlike the first sustainable development plan, the second one includes nine objectives, seven with specific targets. The city also recognized that the plan, by itself, was insufficient unless accompanied by systematic implementation initiatives. Below are the five orientations and nine specific objectives. For each objective, a set of initiatives also has been prepared, and can be viewed at the website for Montreal shown at the end of the chapter.

ORIENTATION 1: Improve Air Quality and Reduce Greenhouse Gas Emissions
   Objective 1: Reduce Montreal's greenhouse gas emissions by 30 per cent by 2020 compared with 1990.
   Objective 2: Achieve the Canadian standard for fine particle concentrations in the ambient air ($30\mu g/m^3$) by 2020.

ORIENTATION 2: Ensure Quality of Residential Living Environments
   Objective 3: Reduce the net migration between Montreal and the suburbs by 25 per cent by 2012, mainly by targeting Montrealers from 25 to 44 who each year leave the city.

ORIENTATION 3: Manage Resources Responsibility
   Objective 4: Reduce potable water production by 15 per cent by 2015 compared with 2000.
   Objective 5: Improve the quality of runoff water that flows into watercourses.
   Objective 6: Recover 80 per cent of recyclables and organic materials, household hazardous wastes, construction, renovation, and demolition waste and bulky refuse by 2019, as stipulated in Montreal's Municipal Waste Management Master Plan.

ORIENTATION 4: Adopt Good Practices for Sustainable Development in Companies, Institutions, and Businesses
   Objective 7: Make Montreal a North American leader in the environmental and clean-tech sector by 2020.
   Objective 8: Increase the number of environmental certifications and participation in voluntary environmental programs in Montreal by 30 per cent by 2030 compared with 2010.

ORIENTATION 5: Improve the Protection of Biodiversity, Natural Environments, and Green Spaces
   Objective 9: Improve Montreal's green infrastructure by increasing the canopy cover to 25 per cent from 20 per cent by 2025 compared with 2007.

In addition to the above five orientations, the plan includes what is termed a 'Social Component', explained as 'providing a place for family and the quality of the living environment'. Specifically, to further the social dimension, the city government and administration are committed to: (1) showing solidarity, especially through international co-operation and the social economy, (2) demonstrating equity, by efforts to reduce poverty, social marginalization, and inequality, and (3) addressing succession planning by engaging with young Montrealers.

The local government also acknowledges it cannot achieve sustainable development on its own, and therefore lobbies higher levels of government to develop complementary interventions. For example, Montreal calls on the province and federal governments to use fiscal or economic tools such as a regulated carbon market, to introduce measures to facilitate greenhouse gas reduction in the transportation and building sectors, to provide programs to encourage electric forms of transportation, to adopt new regulations regarding quality of the atmosphere, and to create awareness-building programs and incentives supported by regulatory and financial instruments for the sale and purchase of water-efficient equipment.

In terms of implementation, the plan identifies 37 specific actions to reach the objectives and specific targets. It also commits the local government to build awareness and provide information to residents, and to report on progress every two years.

The initiatives in Victoria, Vancouver, and Montreal provide a range of ideas for achieving sustainable development, whether for a specific area within a city or at a city-wide scale. Based on their ideas and experiences, we encourage you to consider what might be done in your community, whether for a designated area or for the entire community.

Winnipeg and Toronto

Innovative buildings in Winnipeg and Toronto provide examples of what can be expected of green cities in the future. Manitoba Hydro Place in downtown Winnipeg is viewed by many as the forerunner of the modern workplace. The 22-storey building includes various sustainability features, including:

- A site close to public transportation routes encourages employees living in the suburbs to use public transportation, thereby reducing GHG emissions and saving on travel time.

- Tubes sunk into the ground utilize the ground temperature to provide heating and cooling dependent on the season.
- A combination of energy efficiency features have led to a 65 per cent decrease in energy costs in comparison to a traditional building of the same size.
- A solar 'chimney' passively vents exhaust air and helps to draw in fresh air at other locations.
- Green roofs reduce air conditioner load, reduce storm water runoff, facilitate landscaping, and reduce the city heat island effect.
- A double skin building envelope involving a low-iron glass double facade creates a 'thermal buffer' and allows sunlight penetration within the building.
- Dimmable and programmable fluorescent bulbs save energy.
- Filtered water makes bottled water unnecessary.

Manitoba Hydro Place received the award for the best office tower in North America in 2009 from the Council on Tall Buildings and Urban Habitat and an award for one of the top 10 green projects in 2010 from the American Institute of Architects.

In Toronto, the City Council has introduced the 'Toronto Green Standard'. The Green Standard is a two-tiered set of performance measures to encourage improvements related to air and water quality, greenhouse gas emissions, energy efficiency, solid wastes, and the natural environment. The Green Standard became effective on 31 January 2010, and requires developers to meet minimum Tier 1 standards. If developers choose to meet the higher Tier 2 standards, they become eligible for a 20 per cent refund of development charges levied by the city.

Toronto became the first city in North America with a bylaw requiring inclusion of green roofs on new commercial, institutional, and residential developments with a minimum gross floor area of 2,000 m² effective 31 January 2010. Industrial developments became covered by the bylaw as of 31 January 2011. According to the city of Toronto, a green roof is:

> an extension of an above grade roof, built on top of a human-made structure, that allows vegetation to grow in a growing medium and which is designed, constructed and maintained in accordance with the Toronto Green Roof Construction Standard. A green roof assembly includes, as a minimum, a root repellent system, a drainage system, a filtering layer, a growing medium and plants, and shall be installed on a waterproof membrane of an applicable roof. (City of Toronto, at: www.toronto.ca/greenroofs/what.html)

Examples of measures in the Green Standard and Green Roof bylaw include using shade landscaping outside

(a) The Manitoba Hydro building; (b) schematic of the Manitoba Hydro building.

The 4,000-square-foot green roof on the fourteenth floor of the Fairmont Royal York Hotel in downtown Toronto boasts herbs, vegetables, edible flowers, and fruit trees—all are used in the hotel's kitchen. The rooftop garden also features an apiary that has produced award-winning honey. The project won a Green Toronto Award in 2008 in the Green Roof category.

buildings to reduce heat reflection, treating glass in windows to reduce injury to migratory birds, reducing parking places and increasing bicycle racks, enhancing water conservation, and ensuring energy efficiency is at least 25 per cent above building code minimum standards. These features increase the cost of the condominium units, but analysis indicates the energy savings will pay back such costs within 10 years.

Using the ideas for best practice from Whistler, Victoria, Vancouver, Montreal, Winnipeg, and Toronto, we challenge you to complete an inventory and assessment of sustainability initiatives in your community. How many of these best practices are in place in your community? Of those being used, how effectively have they been developed and implemented? If one new initiative might be undertaken, which one do you recommend? How can you become involved?

To motivate you to think imaginatively, we encourage you to check out the goals and accomplishments of Copenhagen (Copenhagen, 2007). Copenhagen, a city of more than 1.5 million people, is aiming to become the world's Eco-Metropole by 2015, or the capital city with the best urban environment in the world. Copenhagen's aspiration is a reminder that many cities, including Vancouver and Toronto, are seeking to be recognized as the leading green city in the world. Other cities aiming to become the most environmentally sound include Stockholm, Paris, Berlin, New York, Chicago, San Francisco, Portland, Seattle, and Sydney.

Thirty-six per cent of the residents of Copenhagen commute to work by bicycle along some 300 kilometres of dedicated bike lanes. The intent is to increase that to 50 per cent by 2015 and also to reduce the number of seriously injured cyclists by more than half. To achieve that ambition, about $40 million were to be spent between 2008 and 2011 on enhancing road arrangements for cyclists. And since 1995, Copenhagen has provided bicycles for use in the city. For a deposit of 20 Danish krone (a bit less than $5), people can use one of the more than 2,000 city-provided bicycles between April and November. The bicycles can be picked up and dropped off at one of more than 100 city bicycle stations.

When the 50 per cent target of Copenhagers bicycling to work or school each day is achieved, CO_2 emissions will have been reduced by 80,000 tonnes. The overall goal is to reduce CO_2 emissions by at least 20 per cent relative to 2005.

Copenhagen reuses about 90 per cent of all building waste and incinerates about 75 per cent of household waste. The energy from burning waste is used for both electricity generation and district heating. Another ambition is to clean up the harbour waters so that people can swim in it and catch fish.

What initiatives could or should be taken in your community?

Implications

If cities persist in the uncontrolled expansion of urban perimeters, indiscriminate use of resources, and unfettered consumption, without regard to ecological damage, the environmental problems associated with cities will continue to worsen. (UN Population Fund, 2007a: 67)

In this chapter, our goal has been to help you understand the implications of current values and behaviour with regard to urban environmental management. Understanding the implications of urbanization for the environment is important in Canada because four of five Canadians live in urban areas. We have choices regarding urban form and design (sprawl or compact), transportation (private vehicles or public transit), energy consumption (fossil fuel or hybrid engines for vehicles), waste generation, and green space. These choices have consequences for air and water quality, greenhouse gas emissions, and the health of humans and other species.

We invite you to develop a vision for urban sustainability and resilience for the community in which you live. If your community were on a trajectory towards sustainability and resilience, what would be different from what happens today? What changes would you have to make as an individual? What choices would have to be made by the entire community? What would be the costs of making such changes? What would be the costs of not making changes? In the short term, however, there are actions you can take as an individual. Examples are noted in Box 13.6.

Environment in Focus

Box 13.6 What You Can Do

1. Support elected officials in efforts to promote and introduce land-use planning practices that minimize urban sprawl.
2. Use public transit, car-sharing, bicycling, and/or walking as alternatives to travelling alone in your automobile.
3. Find opportunities to reuse, reduce, and recycle household and work-related waste.
4. Determine how green building technology could be incorporated in a new home or when renovating.
5. Purchase and use energy- and water-efficient units for home and in the workplace to conserve resources.
6. Replace low energy-efficient light bulbs with energy-efficient bulbs.
7. Set your furnace thermostat at a lower temperature during winter and your air conditioner at a higher setting during summer.
8. Use native trees and shrubs for your gardens and yards.
9. Volunteer to help create and maintain trails or walking paths in your community.

Summary

1. Fifty per cent of the world's population live in urban areas, and four out of five Canadians live in urban areas.
2. Urban sustainable development and resilience are based on four considerations: urban form, transportation, energy, and waste management.
3. In Canada, the proportion of low-rise, low-density homes is steadily increasing, and commuting times are increasing because of reliance on private automobiles. A major challenge is to reduce urban sprawl.
4. Urban sprawl can be reduced by compact, mixed-use urban form and reduced private car use.
5. Use of private automobiles is a major contributor to greenhouse gas emissions. Shifting from cars to other forms of transit is the one action likely to have the greatest single impact on improving environmental quality in Canadian cities.
6. Residential, commercial, and industrial buildings contribute more than 60 per cent of GHG emissions in Canada.
7. Advanced 'green' buildings can save up to 50 per cent in energy use relative to conventional buildings.
8. Key activities in waste management are reduction, reuse, and recycling, followed by energy recovery.
9. About one-third of waste in Canada is from packaging, but except for beverage containers, no national or provincial programs focus on reducing packaging waste. In contrast, the European Union has aggressive programs to reduce packaging waste.
10. Residential waste diversion is most challenging for apartments and multi-family dwellings.
11. To enhance sustainability of urban buildings, attention should be given to location and site design.
12. Major environmental issues in cities include air pollution, urban heat island effect, poor water quantity and quality, and brownfield sites.
13. Many cities are vulnerable to natural or human-induced hazards, and in Canada notable hazards include earthquakes, storm surges, floods, droughts, and snow or ice storms.
14. Numerous natural disasters in recent years have highlighted the vulnerability of urban infrastructure and residents.
15. Dockside Green in Victoria, Whistler, Vancouver, Montreal, Winnipeg, Toronto, Paris, and Copenhagen offer excellent examples of urban sustainability initiatives.

Key Terms

- blue boxes
- brownfields
- green bins
- greenfields
- LEED
- LUST
- NIMBY
- recovery
- recycling
- reduction
- reuse
- Richter scale
- smog
- sustainable urban development
- urban form
- urban heat island
- urban sprawl

Questions for Review and Critical Thinking

1. Why is Canada so urbanized? Why is urbanization a global phenomenon?
2. What are the principal attributes of urban sustainability or of urban resilience?
3. What is the significance of urban form, transportation, energy use, and waste management for urban sustainability and resilience, and what are the key connections among them?
4. What are the 3Rs?
5. How can the site selection of urban buildings help to achieve urban sustainability and resilience?
6. Why is air pollution generally increasing in highly urbanized areas?
7. How can the urban heat island effect be reduced?
8. What impact does urbanization have on the hydrological cycle?
9. What is the significance of LULUs, NIMBY, and LUST for urban sustainability?
10. What are the general lessons from Hurricane Katrina related to the vulnerability of urban areas to natural and human-induced hazards?
11. What are the lessons from Whistler, Sudbury, Victoria, Vancouver, Montreal, Winnipeg, and Toronto regarding how to evolve towards urban sustainability and resilience?

Related Websites

BCIT Centre for Architectural Ecology, Collaborations in Green Roofs and Living Walls
commons.bcit.ca/greenroof

Canada Green Building Council
www.cagbc.org

Canada Urban Institute
www.canurb.com

City of Montreal
ville.montreal.qc.ca/portal/page?_pageid=5977,87619593&_dad=portal&_schema=PORTAL

Environment Canada, Urban Environmental Management, Resource Library
www.ec.gc.ca/cppic/en'refView.cfm?refId=1186

International Green Roof Association, Greenroofs Project Database
www.greenroofs.com

National Round Table on Environment and Economy
nrtee-trnee.ca

Transport Canada, Environmental Affairs
www.tc.gc.ca/eng/environment-menu.htm

United Nations Population Fund
www.unfpa.org

World Health Organization, Health and Environment Linkages Initiative, The Urban Environment—A General Directory of Resources
www.who.int/heli/risks/urban/urbenvdirectory/en/index.html

Further Readings

Note: This list comprises works relevant to the subject of the chapter but not cited in the text. All cited works are listed in the Bibliography at the end of the book.

Alberti, M., and J.M. Marzluff. 2004. 'Ecological resilience in urban ecosystems: Linking urban patterns to human and ecological functions', *Urban Ecosystems* 7: 241–65.

Beckett, P.J. 2000. *A Reflection of Two Landscapes—The Copper and Nickel Mining Region of Sudbury before and after 25 Years of Land Reclamation Activity.* Sudbury, Ont.: Laurentian University Library.

Brown, R.D. 2010. *Design with Microclimate: The Secret to Comfortable Outdoor Space.* Washington: Island Press.

De Sherbinin, A., A. Schiller, and A. Pulsipher. 2007. 'The vulnerability of global cities to climate hazards', *Environment and Urbanization* 19, 1: 39–64.

Erell, E., D. Pearlmutter, and T. Williamson. 2010. *Urban Microclimate: Designing the Spaces between Buildings.* London: Earthscan.

Federation of Canadian Municipalities. 2002. *Municipal Governments and Sustainable Communities: A Best Practice Guide.* Ottawa: Federation of Canadian Municipalities and C2HMHill.

Gunn, J.M., ed. 1995. *Restoration and Recovery of an Industrial Region: Progress in Restoring the Smelter-Damaged Landscape near Sudbury, Ontario.* New York: Springer-Verlag.

Hollander, J., N. Kirkwood, and J. Gold. 2010. *Principles of Brownfield Regeneration: Cleanup, Design, and Reuse of Derelict Land.* Washington: Island Press.

Jabeen, H., C. Johnson, and A. Allen. 2010. 'Built-in resilience: Learning from grassroots coping strategies for climate variability', *Environment and Urbanization* 22: 415–31.

Kennedy, C., J. Cuddihy, and J. Engel-Yan. 2007. 'The changing metabolism of cities', *Journal of Industrial Ecology* 11, 2: 43–59.

Lautenbach, W.W. 1987. 'The greening of Sudbury', *Journal of Soil and Water Conservation* 42: 228–31.

Lehmann, S. 2010. *The Principles of Green Urbanism.* London: Earthscan.

Marafa, L.M. 2003. 'Identifying wilderness in the landscapes of Hong Kong urban periphery', *International Journal of Wilderness* 9, 3: 39–42.

———. 2003. 'Integrating natural and cultural heritage: The advantage of *feng shui* landscape resources', *International Journal of Heritage Studies* 9, 4: 307–23.

Muller, M. 2007. 'Adapting to climate change: Water management for urban resilience', *Environment and Urbanization* 19: 99–113.

Newman, P.W.G., and I. Jennings. 2006. *Cities as Sustainable Ecosystems.* Kobe, Japan: UNEP-IETC.

Richardson, N.H. 1991. 'Reshaping a mining town: Economic and community development in Sudbury, Ontario', in J. Fox-Przeworski, J. Goddard, and M. de Jong, eds, *Urban Regeneration in a Changing Economy: An International Perspective.* Oxford: Clarendon Press, 164–84.

Saarinen, O. 1992. 'Creating a sustainable community: The Sudbury case study'. In M. Bray and A. Thomson, Eds, *At the End of the Shift: Mines and Single Industry Towns in Northern Ontario*, 165–86. Toronto: Dundurn Press.

Satherthwaite, D. 2009. 'The implications of population growth and urbanization for climate change', *Environment and Urbanization* 21: 545–67.

Stewart, I. 2006. 'Influence of meteorological conditions on the intensity and form of the urban heat island effect in Regina', *Canadian Geographer* 44: 271–85.

Vegetation Enhancement Technical Advisory Committee (VETAC). 1999–2010. *Land Reclamation Program Annual Report.* Sudbury, Ont.: City of Greater Sudbury.

Wells, G. 2010. 'Calculating the green in green: What's an urban tree worth?', *Science Findings* 126 (Sept.). Portland: US Department of Agriculture, Pacific Northwest Research Station.

Winterhalder, K. 2002. 'The effects of the mining and smelting industry on Sudbury's landscape', in D.H. Rousell and K.J. Jansons, eds, *The Physical Environment of the City of Greater Sudbury.* Ontario Geological Survey Special Volume no. 6. Toronto: Ontario Ministry of Northern Development and Mines, Mines and Minerals Information Centre, 145–73.

Chapter 14
Endangered Species and Protected Areas

Learning Objectives

- To understand why endangered species are important and the factors leading to endangerment.
- To become aware of the extrinsic and intrinsic values of nature.
- To learn why some species are more vulnerable to extinction than others.
- To be able to discuss the main responses to endangerment at the international and national levels.
- To appreciate the many roles played by protected areas.
- To gain an international and a Canadian perspective on protected areas.
- To know some of the main management challenges faced by protected areas in Canada.

Introduction

Most people are aware that many more species are becoming **endangered** than is natural—that we have a 'biodiversity crisis'. What took hundreds of millions of years to evolve is disappearing in only generations. The United Nations declared 2010 as the International Year of Biodiversity to promote awareness and conservation of the biosphere, an essential component to ensuring functioning Earth-system processes. Biodiversity is the living underpinning of our lives: we depend on biodiversity for clear air, fresh water, medicine, and the various resources we consume every day. The United Nations recognized this significance and extended the Year of Biodiversity to the Decade of Biodiversity, running from 2011 to 2020. It signals the most ambitious international effort yet to end species loss and ecosystem destruction.

Around the world, extinction is occurring many times faster than natural rates. Rockström (2009) and his co-authors, in their global assessment of the resilience of key planetary systems, found biodiversity loss to be the most stressed process. A common perception, however, is that this is a problem more for the tropics than for countries such as Canada. Although it is true that threat levels and the numbers of endangered species are higher in the tropics, Canada also has plenty of challenges in this regard.

Canada has a long-established national parks system designed to protect species and their habitats, but parks do not necessarily afford adequate protection to all species needing it. The decline in turtle populations in Point Pelee National Park, Ontario, highlights some of the problems in our national parks and the challenges park managers face.

Turtles have evolved for hundreds of millions of years. Historically, their adaptations—terrestrial nesting, low adult mortality, late maturation, and longevity—have served them well. But these attributes are no longer adequate, and throughout the world many turtle species are experiencing dramatic declines. Historical records show that Point Pelee, the southern tip of Canada's mainland, at one time had seven different species of indigenous turtles. Research by Browne and Hecnar (2003) revealed that several species, including the stinkpot, map, and Blanding's turtles, now exist only in small populations, while other species may be headed towards extirpation—no spotted turtles were found in the park, and only one individual of the threatened spiny softshell turtle was recorded. Reasonably large populations exist for only one species, the painted turtle (Browne and Hecnar, 2007). The authors examined the age structures of the populations and found a preponderance of older animals for most species, indicating aging populations, especially for Blanding's and snapping turtles.

Although the populations are 'protected' within the park, at least three problems threaten the long-term viability of turtle populations. Roads are implicated in two of the three and illustrate the need to minimize development within parks (discussed later). Roads are a source of direct mortality. Some species are attracted by the soft shoulders of roads for nesting, and while migrating across roads in search of suitable sites, many turtles are killed. In addition, roadside nesting sites were found to be very vulnerable to nest predation (100 per cent loss), compared to the 62 to 64 per cent loss in more remote areas of the park. Roadsides are a favourite scavenging area for predators such as raccoons. These and other predators, such as striped skunks and opossums, are reportedly at higher levels in the park than previously, and raccoons may be the dominant limiting factor (Chapter 2) on population growth.

Contaminants are also implicated in turtle population declines. There are still elevated levels of DDT and DDE (see Chapter 10) in areas of the park from past agricultural practices, illustrating the vulnerability of parks to threats from surrounding land uses.

This example illustrates the plight of many species to which we give relatively little attention. Plant and animal populations are often assumed to be 'healthy', especially if they are within the boundaries of a national park. However, *no* national park—anywhere on the planet—is big or remote enough to exclude the impacts of modern society. This chapter focuses on endangered species and the factors behind endangerment. One of the main responses to endangerment is to protect habitats and species in park systems. The designation and management of park systems constitute the second main topic discussed.

Perspectives on the Environment
External Influences on Parks and Species

. . . it is a common fallacy that protected areas are unimpaired swaths of wilderness with pristine natural conditions. In reality, Ontario's provincial parks and conservation reserves are threatened by numerous ecological stresses and threats, some of which originate beyond their boundaries. Indeed, in many cases, the boundaries of protected areas are political constructs that do not reflect natural boundaries. As such, there may be an issue of concern outside of a protected area that affects its management.

—Wilkinson (2008: 185)

The Javan rhinoceros was found throughout Southeast Asia until recently. It has now been confirmed that the one shot in Vietnam in 2010 was the last surviving rhino in Vietnam and almost certainly in mainland Southeast Asia. One small population remains in Indonesia. Extinction is very real and happening all the time. (1874 artwork by British animal painter George Bouverie Goddard [1832–86].)

Extinction is a natural process that has been occurring since life first evolved on Earth (the Tyrannosaurus Rex, like this skeleton in the Royal Tyrell Dinosaur Museum near Drumheller, Alberta, disappeared during the Cretaceous-Tertiary extinction event approximately 65.5 million years ago). However, it is the speed of current extinction rate across many different forms of life that concerns scientists.

But why should we be concerned about endangerment? Extinction is a natural process that has been taking place since life first evolved on this planet more than four billion years ago (Chapter 3). Consequently, concern for the extinction of species does not focus on the process itself but on the increasing rates of extinction—i.e., what humans are doing to speed up the process. Before looking at some of the pressures responsible for this increase, we need to understand why high extinction rates are undesirable.

Valuing Biodiversity

Changes in biodiversity due to human activities were more rapid in the past 50 years than at any time in human history, and the drivers of change that cause biodiversity loss and lead to changes in ecosystem services are either steady, show no evidence of declining over time, or are increasing in intensity. Under the four plausible future scenarios developed by the Millennium Ecosystem Assessment, these rates of change in biodiversity are projected to continue or to accelerate (Millennium Ecosystem Assessment, 2005).

Extrinsic and Intrinsic Values

Humans derive **extrinsic values** from other species. These values can be *consumptive* (i.e., the organism is harvested) or *non-consumptive* (i.e., the organism is not harvested or the resource is not destroyed). There is no universally accepted framework for assigning value to biological diversity, but various approaches have been proposed. For example, the value of biodiversity can be calculated by examining import and export statistics for products bought and sold in markets. However, it is often difficult to assign an economic value to biodiversity. How do you put a price tag on environmental services provided by biological communities, such as photosynthesis, protection of watersheds, or climate regulation? These services, not directly consumed by humans, are vital for survival.

While important, the extrinsic reasons for species protection should not be allowed to dominate our thinking. Such thinking could lead to the protection of a selection of species believed to be of higher value, while species with less use value are afforded little or no protection. Thus, arguments for biological conservation focus on the **intrinsic value** of nature—nature has value in and of itself, apart from its value to humanity. The following discussion outlines some of the key ecological, economic, and ethical reasons for conservation.

Ecological Values

The elimination of species affects ecosystem functioning, which can lead to unfortunate consequences, such as the impact on coastal marine ecosystems on the Pacific coast when the sea otter was extirpated (discussed in Chapter 3). The important role species play in ecosystem functioning is

Perspectives on the Environment

Keeping Every Cog in the Wheel

If the land mechanism as a whole is good, then every part is good, whether we understand it or not. If the biota, in the course of aeons, has built something we like but do not understand, then who but a fool would discard seemingly useless parts? To keep every cog and wheel is the first precaution of intelligent tinkering.

—Aldo Leopold, *Round River* (1953)

another extrinsic value that humans derive from biodiversity. Species become **extirpated** when they have been eliminated from one part of their range but still exist somewhere else. A species is considered **ecologically extinct** in an area when it exists in such low numbers that it can no longer fulfill its ecological role in the ecosystem. For example, the eastern

The great hornbill is found throughout the tropical forests of Southeast Asia where it has been extirpated from many areas by hunting. Even where it exists in small numbers, it is often considered ecologically extinct as there are no longer sufficient numbers to crack and distribute the seeds of many of the tree species. Ultimately, this will also cause changes in the tree species composition of the forests.

mountain lion may still exist in the Maritime provinces and has not yet been declared extinct, but if this species does persist, it exists in such low numbers that it no longer acts as a significant control for species in the preceding trophic level. The eastern mountain lion may therefore be considered ecologically extinct in this region.

All species in a community combine to maintain the vital ecosystem processes that make human life possible on this planet—oxygen to breathe, water to drink, and food to eat. Humans are part of this web of life, but if we continue to eliminate components of the web, its strength will be compromised, with a significant impact on the ability of humans to survive. As we noted in Chapter 2, the elimination of species from an ecosystem is similar to the removal of rivets from an airplane—the system may continue to function after losing a few components, but sooner or later the system will crash. One role of science is to understand how these systems work, but it is difficult to achieve this understanding if components are missing as a result of extinction.

In addition to the value of species in ecosystem functioning, species should be protected for their evolutionary value, their value to future generations. Species evolve, as discussed in Chapter 3, and as more species become extinct, genetic variation in the ecosphere on which to base future adaptability is reduced. Fewer species means a more impoverished biosphere on which to base evolutionary adaptability for future generations. In short, the need to preserve species exemplifies the precautionary principle on a grand scale. In a fanciful manner, the movie *Star Trek: The Voyage Home* showed Captain Kirk, Spock, and their colleagues returning back in time to the present to rescue a humpback whale and transport it to the future, where it had become extinct, in order to save the planet. This is science fiction, to be sure, but it makes the point well—we cannot presume to know what importance or value a particular species might have for future generations.

Economic Values

Another extrinsic value of biodiversity is the economic benefit derived from preservation of ecosystem components and functions. Countless products used in agriculture and industry originate in the natural world. Naturally occurring plants in the tropics, for example, are the source of 90 per cent of the world's food supply. Corn, or maize, feeds millions of people and is now estimated to be worth at least $50 billion annually worldwide. Corn was first domesticated by indigenous people in Central America some 7,000 years ago.

More than 99.8 per cent of the world's plants have never been tested for human food potential. Some may become important food staples in the future, so preservation of their habitat is important. Furthermore, wild animals still provide an important source of food for millions of people worldwide, particularly indigenous peoples, including Canada's Aboriginal communities.

Other products besides food are also of economic importance. Many plant and animal products are used extensively in various industries. Rubber, for example, is an important commodity in the automotive sector. It is just one example of a chemical that tropical plants produce to prevent insect damage. Many other chemicals produced by plants are used in the pharmaceutical industry. Fifty-six per cent of the top 150 prescribed drugs in the United States contain ingredients from wild species, with an annual economic value of US$80 billion according to the UN. In 2010, half of all synthetic drugs had been traced to natural ingredients. One drug, a compound derived from a sea sponge to treat herpes, is valued at up to $100 million annually.

However, less than 1 per cent of the world's tropical plants have been screened for potential pharmaceutical application. Next time you take an aspirin tablet, thank the white willow, the species in which the active ingredient was first discovered. Taxol, found in the bark of the western yew—a small understorey tree in the forests of the Pacific Northwest—was recently discovered as a treatment for cancer. Prior to the discovery, the species was of little commercial value, and it was routinely cut down in clear-cuts and left to rot.

It is difficult to assess the economic value of many products we derive from nature. For example, natural gene pools provide a source of material to aid in the development of new genetic strains of crops needed to feed the world's burgeoning human population. Wheat, the mainstay of the western Canadian agricultural economy, originated in Mediterranean countries, where most of its wild forebears have disappeared. But preservation of such wild strains is necessary to allow selective breeding based on the widest range of genetic material to continue.

Ecosystems also provide humans with a wide array of economically important services. Natural pollinators, for example, provide an essential service to commercial crops. Pollination of flowers by diverse species of wild bees, wasps, butterflies, and other insects—not just managed honeybees—accounts for more than 30 per cent of all food production that humans depend upon. Environment Canada (2003b) has conservatively estimated the value of pollination services to crops in Canada at $1.2 billion annually. When New Brunswick switched from spraying its forests with DDT to fenitrothion in 1970 to control the spruce budworm (see Chapter 9), there was a devastating impact on pollination of the blueberry crop because fenitrothion is highly toxic to bees. The commercial crop fell by 665 tonnes per year, and growers successfully sued the government.

Natural predator–prey relationships also aid in food production. It has been shown, for example, that woodpeckers provide an economically important service in the control of pests such as coddling moths in the orchards of Nova Scotia. Such predator–prey relationships can significantly reduce the need to apply biocides to control pests. Similarly, the natural

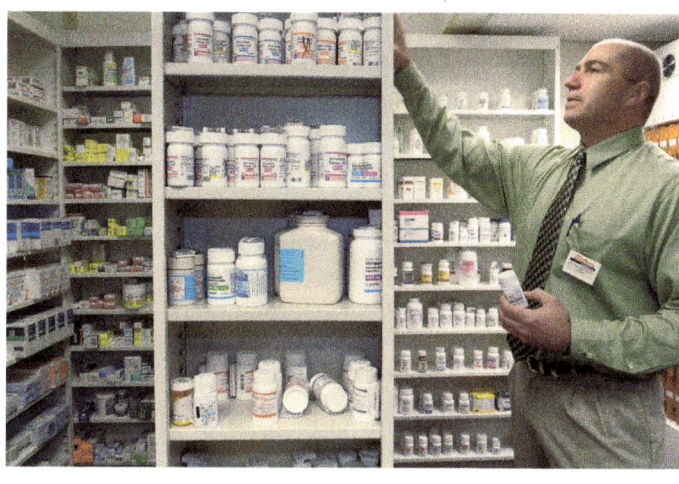 

Whether sold in a modern or a traditional pharmacy, many of our medicinal products are based on products found in nature.

toxicity found in some species can occasionally be refined into a natural biocide for use in agriculture. For example, a powerful insect repellent—trans-pulegol—was recently discovered in an endangered member of the mint family.

Other Extrinsic Values

The ecological and economic benefits mentioned above do not encompass all the values associated with protecting other species. How many of us are permanently enriched and emotionally uplifted by a wildlife encounter at some point in our lives? Wildlife contributes to the joy of life, but this joy often translates into a contribution to the economic values attached to wildlife. Viewing wildlife is a major reason for travel to some areas. The economic value attached to nature-based tourism can provide a significant impetus to conservation when managed so as to enhance biodiversity and educational values. This form of tourism can also provide a sustainable livelihood for local residents. The extrinsic ecological and economic values of biodiversity just discussed exemplify an anthropocentric view of life, a view that favours the protection of species providing a direct benefit to humans.

Ethical Values

Ethical arguments can be made for preserving all species, regardless of their use value to humans. Arguments based on the intrinsic value (value unrelated to human needs or desires) of nature suggest that humans have no right to destroy any species. In fact, humans have a moral responsibility to actively protect species from going extinct due to our activities. This philosophy reflects an ecocentric view—humans are part of the larger biotic community in which all species' rights to exist are respected. This view contrasts to the anthropocentric view of life presented in the previous section.

In the past, extinction has been viewed simply as a biological problem. The points raised above emphasize the need to make links among the biological process of extinction and the ethical and economic reasons why extinction is undesirable. However, decisions to protect species and communities

It is impossible to fully assess the value of natural processes such as pollination.

How much is it worth to see Orca whales surface near your boat?

often become arguments over money—how much will it cost, and how much is it worth? All too often, governments demonstrate a willingness to protect biodiversity only when its loss is perceived to cost money. Unfortunately, standard economic systems tend to undervalue natural resources, and as a consequence, the cost of environmental damage has been ignored and the depletion of natural resources has been disregarded. Ecosystems are being destroyed and species are now being driven to extinction at a rate greater than at any time in the past. An economic system that undervalues natural resources is a main underlying cause of extinction. The Millennium Ecosystem Assessment (2005) suggests that the amount of biodiversity remaining on the planet in another 100 years will reflect society's ability to understand and take into account the different values associated with biodiversity conservation (Figure 14.1). The next section reviews some of the main causes of biodiversity loss.

Main Pressures Causing Extinction

Chapter 3 described some human activities that have contributed to the increasing rates of extinction over the past 200 years. This section discusses some of the main pressures on biodiversity in greater detail. Rarely do these pressures act alone; they must therefore be seen as part of the overall stress that human demands are placing on the biosphere.

Humans expropriate more than 40 per cent of the net primary productivity (NPP) of the planet (Chapter 2). With global populations predicted to increase to 9.2 billion by 2050, this figure will only increase. As the amount of NPP increases to support one species—*Homo sapiens*—the amount available to support all other species decreases. The extinction vortex (Figure 14.2) is driven by these human pressures, manifested as habitat loss, overharvesting, pollution, and the introduction of exotic species (Chapter 3). These pressures result in small, isolated populations that become vulnerable to inbreeding and demographic instability. Unfortunately, there is a positive feedback loop (Chapter 3) between these factors and population decline. The more the population declines, the greater the impact of these factors, and ultimately this leads to extinction. The fragmented populations of the endangered mountain caribou in BC are a good example. As the populations have become progressively reduced and isolated, mainly because of logging, they become more vulnerable to extirpation through such factors as bad weather and increased predation, and fewer and fewer sites exist from which repopulation can take place (Apps and McClellan, 2006). The vortex eventually spins down to extinction.

The outer circle in the figure represents the present level of global biodiversity. Each inner circle represents the level of biodiversity under different value frameworks. Question marks indicate the uncertainties over where the boundaries exist and therefore the appropriate size of each circle under different value frameworks.

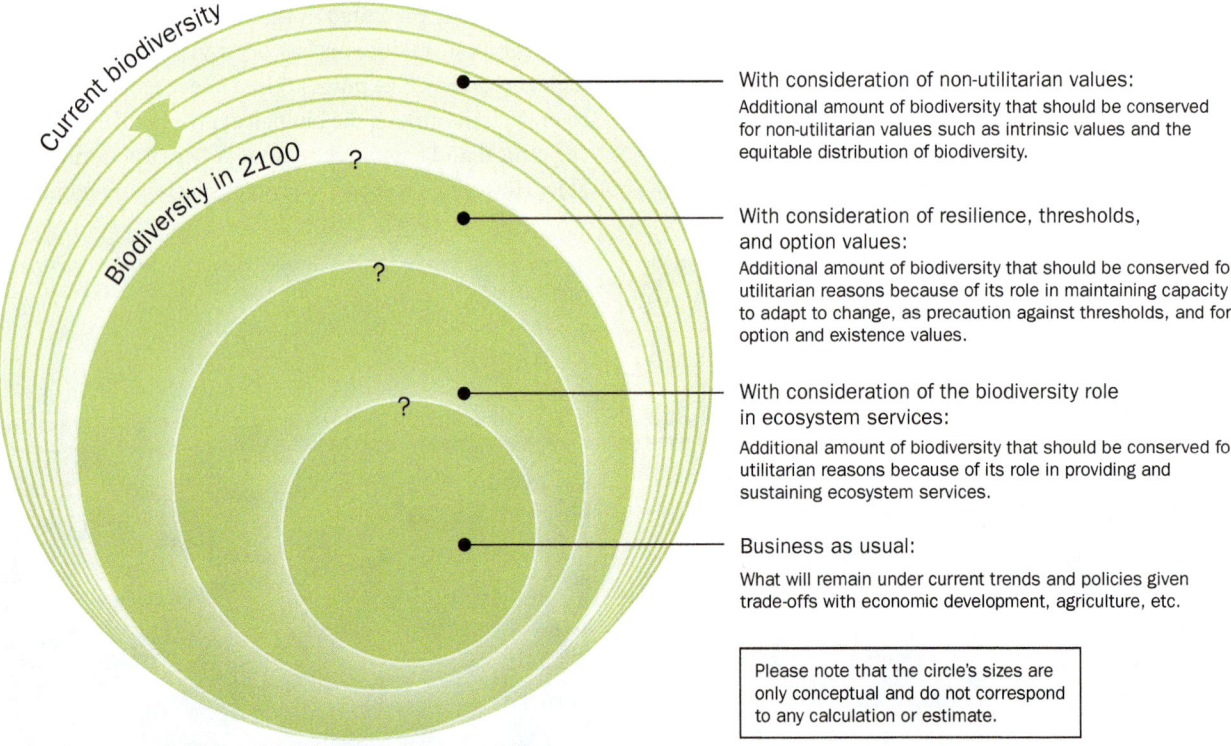

Figure 14.1 | How much biodiversity will remain a century from now under different value frameworks? *Source: Millennium Ecosystem Assessment (2005).*

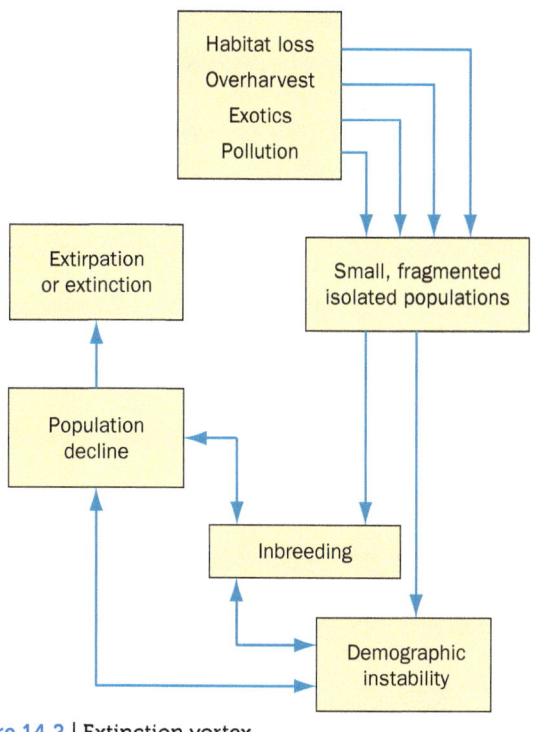

Figure 14.2 | Extinction vortex.

Most attention regarding extinction has been devoted to the tropical countries, mainly because they are 'hot spots' for biodiversity and because they are experiencing many pressures responsible for increasing rates of extinction. Of the 10 to 15 million terrestrial species thought to be on Earth, up to 90 per cent are estimated to exist in the tropics, particularly in the tropical rain forests. Tropical ecosystems are being degraded and/or destroyed at alarming rates (Box 14.1). Estimates suggest that perhaps 50 per cent of tropical rain forests have already disappeared, and at current rates of destruction, only a few forest fragments will remain in 30 years. Causes of tropical deforestation include:

- rapidly growing population levels—more people equals less biological diversity, since people use natural resources;
- overconsumption of resources—the rise of industrial capitalism and materialistic modern societies has greatly accelerated demands for natural resources, particularly in developing countries;
- inequality in the distribution of wealth—poor rural people with no land or resources of their own destroy biological communities and hunt endangered species just to stay alive.

Environment in Focus

Box 14.1 The Global Toll

Through its Species Survival Commission, the International Union for Conservation of Nature (IUCN) evaluates and categorizes species according to their relative level of extinction risk:

- Extinct
- Extinct in the wild
- Critically endangered
- Endangered
- Vulnerable
- Near threatened
- Least concern
- Data deficient
- Not evaluated

This so-called Red List is produced by thousands of scientific experts and is the best source of knowledge on the status of global biodiversity.

Some highlights from recent lists:

- The Red List includes 19,570 species threatened with extinction, falling into the critically endangered, endangered, or vulnerable categories.
- Close to 50 per cent of all primate species (204 out of 418) are listed as threatened.
- 9,156 plants were assessed as threatened in 2011, an increase of 3,384 since 2002. However, with only approximately 4.6 per cent of the world's described plants evaluated, the true percentage of threatened plant species is likely much higher. Most plant species listed are trees, since they have been relatively thoroughly assessed.
- There are now 801 plant and animal species recorded as extinct, and a total of 865 if species no longer existing in the wild are included. In 2002, the number of species assessed as extinct and extinct in the wild was 811.
- Countries with the most threatened species overall are Ecuador, the US, Malaysia, Indonesia, and Mexico.
- Birds and mammals are increasingly moving towards the higher-threat categories (i.e., more bird and mammal species are entering the critically endangered and endangered categories).
- It is more difficult to assess the trends for reptiles, amphibians, fish, and invertebrates because many information gaps still exist. Within these taxa, however, declines appear to be occurring. For example, a number of amphibian species have shown rapid and unexplained disappearances in Australia, Costa Rica, Panama, and Puerto Rico. There are similar indications that freshwater fish species are suffering serious deterioration, particularly river-dwelling species.
- Habitat loss and degradation affect 89 per cent of all threatened birds, 83 per cent of threatened mammals, and 91 per cent of threatened plants assessed. Habitats with the highest number of threatened mammals and birds are lowland and mountain tropical rain forests. Freshwater habitats are extremely vulnerable, with many threatened fish, reptile, amphibian, and invertebrate species.

Sources: International Union for Conservation of Nature (2003, 2007, 2011).

Recent endangered species statistics reveal another major area of concern—aquatic ecosystems (Box 14.2). In both the oceans and freshwater ecosystems, our knowledge of species abundance is very limited. As knowledge improves, many species have been found to be threatened, and these numbers will continue to increase. Freshwater fish and amphibians now hold the top two spots for animal species threatened worldwide (Table 14.1).

Extinctions also occur in developed and temperate countries. Since the European colonization of North America, more than 500 species and subspecies of native plants and animals have become extinct. Some comparative figures among Canada, North America, and the world are shown in Table 14.1. The Committee on the Status of Endangered Wildlife in Canada (COSEWIC) has identified 13 extinctions and 22 extirpations in Canada since the arrival of Europeans (Table 14.2). Most of the following examples have been chosen to illustrate extinction pressures in Canada.

Overharvesting

Many examples exist in Canada of species under pressure due to overharvesting. Several historical examples that took place at least partly in Canadian territory are well known.

The great auk. The great auk, a large flightless bird, inhabited the rocky islets of the North Atlantic. For many

Environment in Focus

Box 14.2 Aquatic Ecosystems in Trouble

Aquatic ecosystems are not only among the least known but also among those most inadequately protected in park systems. You will also see later that many of Canada's most recent extinctions have been from aquatic ecosystems. The following global examples highlight this plight.

- Fifty-six percent of the 252 endemic freshwater Mediterranean fish are threatened with extinction, the highest proportion in any regional freshwater fish assessment. Seven species, including carp relatives *Alburnus akili* in Turkey and *Telestes ukliva* in Croatia, are now **extinct**.
- Of the 564 dragonfly and damselfly species so far assessed, nearly one in three (174) are threatened, including nearly 40 per cent of endemic Sri Lankan dragonflies.
- In East Africa, human impacts on the freshwater environment threaten more than one in four (28 per cent) freshwater fish species. This could have major commercial and dietary consequences for the region. For example, in Malawi, 70 per cent of animal protein consumed comes from freshwater fish. The lake trout, or Mpasa (*Opsaridium microlepis*), from Lake Malawi is fished heavily during its spawning runs upriver but has suffered a 50 per cent decline in the past 10 years as a result of siltation of its spawning grounds and reduced flows because of water abstraction. It is now listed as **endangered**.
- Larger freshwater species, such as the common hippopotamus (*Hippopotamus amphibius*), are also in difficulty. One of Africa's best-known aquatic icons, it has been listed as threatened for the first time and is classified as a vulnerable species, primarily because of a catastrophic decline in the Democratic Republic of the Congo (DRC). In 1994, the DRC had the second largest population in Africa—30,000, after Zambia's 40,000—but numbers have plummeted by 95 per cent. The decline is due to unregulated hunting for meat and the ivory of their teeth.

Unregulated hunting has led to a catastrophic decline in the hippopotamus population in the Democratic Republic of Congo, putting this freshwater species on the Red List for the first time.

- As of 2010, 616 marine species have been listed as vulnerable, endangered, or critically endangered. Sharks and rays were among the first marine groups to be systematically assessed, and of the 547 species listed, 20 per cent are threatened with extinction. This confirms suspicions that these mainly slow-growing species are exceptionally susceptible to overfishing and are disappearing at an unprecedented rate across the globe. Curbing demand for shark products is a major step towards enhanced protection and in late 2011, for example, Toronto banned the sale of sharkfin soup.

Sources: International Union for Conservation of Nature, at: www.iucn.org/en/news/archive/2006/05/02_pr_red_list_en.htm; International Union for Conservation of Nature (2011).

Table 14.1 | Number of Threatened Species in Each Major Group of Organisms Worldwide

	2002	2011
Mammals	1,137	1,134
Birds	1,192	1,240
Reptiles	293	664
Amphibians	157	1,910
Fish	742	2,011
Molluscs	939	1,570
Other invertebrates	993	1,629
Plants	5,714	9,098
Total	11,167	19,265

Sources: World Conservation Union, 2002, 2011. © International Union for Conservation of Nature and Natural Resources. www.iucnredlist.org

Great auks, painted by John James Audubon.

years, fishers in these waters used the great auk as a source of meat, eggs, and oil. It was reasonably easy to catch and club these birds to death, and once the feathers became an important commodity for stuffing mattresses in the mid-1700s, extinction soon followed. On Funk Island off the east coast of Newfoundland, the species was extirpated by the early 1800s. The last two great auks were clubbed to death off the shores of Iceland in 1844.

The passenger pigeon. There are claims that the passenger pigeon was the most abundant land bird on Earth, totalling up to five billion birds. These great flocks used to migrate annually from their breeding grounds in southeastern Canada and the American northeast to their wintering grounds in the southeastern US. So great were their numbers that tree limbs would break under the pigeons' weight and trees would die as a result of the large amount of guano (bird excrement) deposited. Many eyewitness accounts tell of flocks so huge that they blotted out the sun for hours and sometimes days. Flocks of passenger pigeons were an easy target for hunters, and they were slaughtered in great numbers for food to feed growing urban populations. It is estimated that more than 1 billion birds were killed in Michigan alone in 1869. They were shot, netted, and clubbed into extinction. Hunting took place concurrently with a reduction in their breeding grounds as habitat was converted into agricultural lands. The last passenger pigeon sighted in Canada was at Penetanguishene, Ontario, in 1902; the last pigeon died in a zoo in Cincinnati

Table 14.2 | Summary of COSEWIC's Assessment Results for the Risk Categories

	Extinct	Extirpated	Endangered	Threatened	Special concern	Total
Mammals	2	3	20	16	29	70 (69)
Birds	3	2	29	26	20	80 (71)
Reptiles		4	17	11	9	41 (38)
Amphibians		2	9	5	6	22 (20)
Fish	7	3	48	37	49	144 (111)
Lepidopterans Arthropods		3	29	6	6	44 (13)
Molluscs	1	2	19	3	6	31 (26)
Plants		3	94	48	40	185 (168)
Mosses	1	1	8	3	4	17 (16)
Lichens			5	3	7	15 (9)
Total	14 (13)	23 (22)	278 (225)	158 (141)	176 (155)	649 (556)

Note: Figures shown are for 2011, with 2007 figures in parentheses.
Source: Adapted from Summary of COSEWIC's Assessment Results for the Risk Categories: www.cosewic.gc.ca/rpts/Full_List_Species.html

the Great Lakes (blue walleye, deepwater cisco, longjaw cisco) were at one time all very abundant, and millions of kilograms of the fish were harvested commercially. By 1950, they were fished into extinction. Similarly, the northern cod, once one of the most abundant fish on the planet, was fished into commercial extinction by the early 1990s. Two populations of Atlantic cod were designated as threatened and endangered by COSEWIC in 2003 (see Chapter 8). Overharvesting is the major cause of endangerment for marine species listed by COSEWIC and gives further support for Canada to fulfill its international commitments to establish a network of marine protected areas by 2012, as discussed in Chapter 8.

Perhaps extinction of the passenger pigeon and the great auk are sufficiently in the past that we can excuse their demise on the grounds of a lack of knowledge. However, such a case cannot be made for more recent extinctions. They stand as the ultimate symbol of the failure of resource managers and decision-makers to understand the natural dynamics of species supposedly being managed.

One aspect of overharvesting generally given little consideration is the demand for captive species. In the past, the zeal of zoo collectors to exhibit various species, particularly rare species such as pandas and orangutans that visitors would pay to see, was of serious concern. International regulations, such as the **Convention on International Trade in Endangered Species of Wild Fauna and Flora** (CITES), of which Canada is a signatory, make it difficult for this kind of trade to occur. However, despite international regulations, trade in rare and endangered species occurs because of demand from private collectors. Exotic species such as tigers, monkeys, parrots, and tropical fish belong in their native habitats, not in people's homes. Some of the most sought-after Canadian species are falcons, particularly the gyr and peregrine falcons; they can fetch thousands of dollars each on the international market. Although an allowable harvest of wild falcons exists in some provinces and territories, poaching is a problem because of the high price tags attached to these birds.

Hunting/fishing and the harvesting of live specimens for captivity can have a significant impact on populations. However, more subtle instances of 'non-consumptive' activities have detrimental impacts on species by causing displacement from valuable habitat or even death. For example, research is underway on both the Atlantic and Pacific coasts of Canada to assess the potential impact of whale-watching vessels on the well-being of whales. Such research requires detailed knowledge of a species' natural distribution and behaviour before it can be ascertained whether changes have occurred as a result of disturbance. Impacts associated with the non-consumptive use of natural resources are usually much more difficult to document than the more direct effects of consumptive use. Nonetheless, researchers have detected costs to killer whales in terms of energy use—for example, from boat traffic in BC (Williams et al., 2006).

A pair of passenger pigeons, painted by John James Audubon.

in 1914. The world will never again experience the sound and sight of millions of passenger pigeons darkening the heavens.

Perspectives on the Environment

The Passenger Pigeon

The noise they made, even though still distant, reminded me of a hard gale at sea, passing through the rigging of a close-reefed vessel. As the birds arrived and passed over me, I felt a current of air that surprised me. Thousands of the Pigeons were soon knocked down by the pole-men, while more continued to pour in. . . . The Pigeons, arriving by the thousands, alighted everywhere, one above another, until solid masses were formed on the branches all around. Here and there the perches gave way with a crash under the weight and fell to the ground, destroying hundreds of birds, beneath, and forcing down the loaded. The scene was one of uproar and confusion. I found it quite useless to speak, or even to shout, to those persons nearest me. Even the gun reports were seldom heard, and I was made aware of the firing only by seeing the shooters reloading.

—John James Audubon

This astonishing tale of a species going from such abundance to extinction is not that unusual. Three fish species of

Overhunting, especially for valuable products, is a main cause of endangerment for many species, such as this black rhinoceros in Kenya's Maasai Mara Conservation Area. Predator species such as the lion are often greatly reduced in numbers and distribution due to competition with humans.

Predator Control

Several species have been targeted for elimination by humans because they compete directly with humans for consumption of the same resource. One North American example is the Carolina parakeet, the only member of the parrot family native to North America. It was exterminated in the early part of the previous century because of its fondness for fruit crops. Another example is the prairie dog, which was extensively poisoned because of the mortality of horses and cattle after they stepped into prairie dog burrows and broke their legs. This extermination program has been highly successful—prairie dog populations have declined by 99 per cent.

Whales were hunted to the point of extinction around the world because of their commercial value. Baleen, or whalebone, pictured here, is a filter-feeding system inside of a whale's mouth. It was used for such 'indispensible' things as umbrellas, buggy whips, and ladies' corsets.

Once such a decline occurs, repercussions occur elsewhere in the food chain. In this case, the drastic decline in prairie dogs led to collapse of their main predator, the black-footed ferret, for which prairie dogs made up more than 90 per cent of their diet. Fortunately, the ferrets have been successfully bred in captivity, an example of **ex situ conservation**. The ferrets have been reintroduced to the wild in places where prairie dogs are protected. The swift fox is another example of predator control and ex situ conservation (Box 14.3).

Predator control is not a thing of the past. For example, deer are sometimes culled to prevent or stop over-browsing of vegetation. Across parts of their North American range, white-tailed deer populations are literally eating themselves out of house and home—deer populations are in excess of what can be supported by the natural resource base. This is due in large part to reductions in range and population declines of their predators, particularly wolves. In stark contrast, other parts of the country, such as Vancouver Island, do not have enough deer, and provincial governments have considered shooting wolves in order to increase the number of deer that humans can kill for sport (Box 14.4). A similar debate is also underway in Newfoundland, where coyotes, which first invaded the island in the 1980s, are now taking significant numbers of caribou.

Seal hunting off Canada's eastern seaboard is also regaining momentum, partly as a predator control program to protect cod in hopes of reviving the fishing industry (Chapter 8). Recently, there have been calls in Prince Edward Island and Ontario to kill double-crested cormorants, accused of taking too many fish and threatening endangered plant species. Point Pelee National Park has established a cormorant cull on select islands in Lake Erie as part of an **active management** program to protect endangered plants that exist in few other places. As the human population grows and as our

Environment in Focus

Box 14.3 Ex Situ Conservation at Work

Not all species subject to heavy pressures are pushed to extinction. Some, such as the beaver, may recover in numbers and start to repopulate their old range. The beaver was able to repopulate with relatively little help. The swift fox, on the other hand, was the target of a 20-year, $20 million reintroduction program, emphasizing the difficulties and costs associated with trying to reverse extinction trends.

The swift fox is so called because of its ability to run down rabbits and other prey in its home terrain, the dry, shortgrass prairie. The swift fox is small (about half the size of a red fox), and at one time roamed all the way from Central America to the southern prairies of Canada. Unlike most other members of the dog family, the swift fox uses dens throughout the year, preferably located on well-drained slopes close to a permanent water body. This may be for protection because of their small size. Natural enemies include coyotes and birds of prey such as eagles and red-tailed and rough-legged hawks.

The last swift fox in the wild was spotted in Alberta in 1938. A combination of factors led to its demise, including habitat degradation, overhunting, and predator control programs. The shortgrass prairie came under heavy pressure from cultivation, leading to a loss of habitat for the swift fox and many other species. In addition, the fox was heavily trapped in the mid- and late 1800s for its soft, attractive pelt. The Hudson's Bay Company sold an average of 4,681 pelts per year between 1853 and 1877; by the 1920s, the take had declined to just 500 pelts per year. However, predator control programs against the coyote and wolf finally removed the swift fox from the Canadian prairies. Predator control programs often relied on extermination methods not species-specific, such as poisoned bait and leg traps. As with many species, more than one factor typically drives a species to extinction, and these factors often interact synergistically.

Since 1978, efforts have been made to return the fox to the prairies. Foxes were bred in captivity, and wild populations from the United States were relocated to Alberta. The captive breeding program was initiated by two private citizens, illustrating the positive impact that individuals can have on environmental issues. More than 600 swift foxes are now living and breeding in the wild on the Canadian prairies.

The best strategy for the long-term protection of wildlife species is preservation of populations in the wild—only in natural communities are species able to continue their process of evolutionary adaptation to a changing environment. Conservation strategies focused on the organism within its natural habitat are referred to as in situ preservation. The Thelon Game Sanctuary in the Northwest Territories, for example, was established in 1927 to help protect the remaining population of muskoxen. Since that time, much of the mainland habitat of the animal has been recolonized by out-migration from this sanctuary. However, in situ preservation may not be a viable option for many rare species, including the swift fox. If remnant populations are too small, on-site preservation strategies will be ineffective. In such cases, the only way to prevent extirpation or extinction is to maintain individuals in artificial conditions under human supervision. This strategy is known as ex situ preservation. Zoos, game farms, aquariums, private breeders, and botanical gardens are all examples of ex situ facilities. The swift fox is an example of ex situ conservation, where the species is reintroduced to its natural habitat.

Given the scale of change evident in many species groups, some scientists believe that the only chance for survival for some is through the activities of zoos. One example is Amphibian Ark, which aims to prevent the world's more than 6,000 species of frogs, salamanders, and caecilians from disappearing. Scientists estimate that up to 170 species of frogs have become extinct over the past decade through fungal attack and other causes, and that an additional 1,900 species are threatened. Amphibian Ark wants zoos, botanical gardens, and aquariums in each country to take in at least 500 frogs from a threatened species to protect them from the killer fungus, thought to have originated in Africa. The fungus prevents amphibians from breathing through their pores and has wiped out frog populations from Australia to Costa Rica and the US. However, this is only a stopgap measure to buy time and prevent more species from going extinct while researchers figure out how to keep amphibians from dying off in the wild. Unfortunately, given the spread of alien species described in Chapter 3, combined with the effects of global climate change, such catastrophic measures are going to become much more common in the future.

Hunting of coyotes is allowed in most parts of Canada as part of predator control programs. However, numbers continue to increase in most areas.

Table 14.3 | Examples of Reintroductions of Endangered Species into Canadian National Parks

Species	Park
American beaver	Cape Breton Highlands National Park, Prince Edward Island National Park
American bison	Prince Albert National Park, Riding Mountain National Park
Plains bison	Elk Island National Park
Wood bison	Nahanni National Park Reserve, Jasper National Park, Waterton Lakes National Park, Elk Island National Park
Fisher	Georgian Bay Islands National Park, Riding Mountain National Park, Elk Island National Park
American marten	Fundy National Park, Kejimkujik National Park, Terra Nova National Park, Riding Mountain National Park
Moose	Cape Breton Highlands National Park
Muskox	Ivvavik National Park
Trumpeter swan	Elk Island National Park
Caribou	Cape Breton Highlands National Park
Swift fox	Grassland National Park

Source: Dearden (2001: 80).

demands increase, conflicts over who will consume another organism will escalate. So far, other species appear to be losing the battle.

Habitat Change

Habitat change is *the* most important factor causing biodiversity loss at national and international scales. For endangered species in Canada, habitat degradation is responsible for 100 per cent of the listed reptiles, amphibians, invertebrates, and lichens, 99 per cent of the listed plants, 90 per cent of the listed birds, 85 per cent of the listed fish, and 67 per cent of the listed mammals. Habitat loss is the most prevalent threat overall, accounting for 84 per cent of listings (Venter et al., 2006).

Human demands are causing both physical and chemical changes to the environment. Physical changes such as deforestation *remove* important habitat components, while chemical pollution may *degrade* habitats to the point that they are no longer able to support wildlife even if the physical structure of the habitat remains. Humans place further pressure on species by the introduction of alien species (Chapter 3).

Physical Changes

Some impacts arising from physical changes in the natural environment have already been discussed relative to forestry and agricultural practices (Chapters 9 and 10). It is difficult historically to separate these influences from the more general impact of colonization in North America. Large areas of forests in central and eastern Canada were cleared to make way for agriculture. Species dependent on these forests, such as the eastern cougar and wolverine, suffered accordingly. Forests are still being replaced by agriculture in some areas. Venter et al. (2006) found that agriculture, followed by urbanization, was the largest cause of endangerment in Canada, although 70 per cent of listed species are under pressure from more than one source.

Before the mid-1850s, some 101 million hectares of long-grass prairie existed in central North America; less than 1 per cent remains. Other prairie ecozones have not fared much better, with only 13 per cent of shortgrass prairie, 19 per cent of mixed-grass prairie, and 16 per cent of aspen parkland remaining. Millions of bison and antelope once grazed these regions, and the land trembled with their migrations. The bison and the antelope have been replaced by cattle. Not surprisingly, one-half of Canada's endangered and threatened mammal and bird species are prairie dwellers. In Canada, an overall loss of 44 per cent of the populations of grassland bird species has happened since the 1970s, with individual species showing declines of up to 87 per cent (Downes et al., 2010)

Accompanying the transformation of the natural prairie grassland for agricultural purposes, thousands of hectares of wetlands have been drained to create more agricultural land. Until the early 1990s, the Canadian Wheat Board Act made it financially attractive for farmers to expand cropland instead of managing their land more effectively, and as a result much marginal land was brought under the plough. It is now estimated that more than 70 per cent of prairie wetlands have been drained. Of the remaining wetlands, 60 to 80 per cent of the habitat surrounding the basins is affected by farming practices. Such changes have been a major factor behind declines in waterfowl breeding on the prairies, a trend only just being reversed by wildlife management practices (Box 14.5).

Draining of wetlands is not restricted to the prairies. Eighty per cent of the wetlands of the Fraser River Delta have been converted to other uses, as have 68 per cent of the wetlands in southern Ontario and 65 per cent of the Atlantic coastal marshes. Drainage not only has a negative impact on marsh-dwelling species but also serves to increase pollutant loads

Environment in Focus

Box 14.4 Predator Control

The wolf was vilified as a rapacious killer and enemy of humans for centuries. It was shot, poisoned, and extirpated throughout large areas of its range, particularly in parts of the United States, where it became an endangered species. Canada has some of the healthiest wolf populations in the world, numbering around 58,000, and they are being used for reintroductions, such as the transfer of wolves from Alberta to Yellowstone National Park. However, wolves are still being shot and poisoned in Canada for predator control programs. In 2010 for example, 95 wolves were killed in BC, partly to save the caribou and partly to safeguard the ranching industry. Trappers officially took a further 143 wolves in the winter of 2004–5.

Despite the estimated number of wolves in Canada, local populations are at risk. In Banff National Park, for example, the wolf population is estimated at about 60. Since 1981, 52 wolves have been killed on roads in Banff, Kootenay, and Yoho national parks, while many others have been shot or trapped legally outside park boundaries. At least 13 wolves were shot or trapped legally in close proximity to the boundary of Banff National Park in 1999–2000. A new provincial regulation now requires all hunters and trappers north of the Bow Valley to register their kills. A 10-kilometre buffer zone around the park where wolves cannot be killed has been suggested. This is also the approach taken in Ontario, where in the spring of 2004 the government announced a permanent moratorium on wolf hunting and trapping in the 39 townships surrounding Algonquin Provincial Park. A moratorium was first enacted in 2001, designed to protect the largest remaining population of the eastern wolf. However, a loophole allowed traps to be set if coyotes were the intended target species, which led to the death of several wolves. This protection issue is complicated in eastern Canada by the increasing populations of coywolves—hybrid pack animals resulting from wolf–coyote interbreeding that are considerably larger than the western coyote yet, like the coyote and unlike the wolf, do not shy away from heavily human-impacted areas. Also, genetic makeup or ancestry does not appear to distinguish definitively 'wolf' from 'coyote' or 'coywolf'; rather, like many early Aboriginal groups, it's who you run with that defines the individual. In 2010, British Columbia declared an open season on the wolf hunt in ranch country, allowing unlimited year-round trapping in some areas, while other areas are now allowed unlimited trapping on private land from 1 April to 14 October.

The grey wolf has now been removed from the endangered species list in the US, and an unlikely beneficiary will probably be the pronghorn antelope. In areas where wolves became re-established in Montana, Wyoming, and Idaho, pronghorn antelope numbers grew by more than 50 per cent. Research shows that the increase was due to increased calf survival, since wolves find a pronghorn calf too small to be worth the effort. The same is not true for coyotes, which take a high number of the calves. As the wolves move into an area, they displace the coyotes, and the pronghorns thrive. This relationship illustrates how complex predator control can be, as shown in the next example.

Another predator cull was carried out on Vancouver Island, the home of Canada's only endemic, endangered mammal, the Vancouver Island marmot. Numbers of the marmot were as low as 30 before ex situ breeding programs were established and began reintroducing animals into their mountain homelands. Of the 96 marmots released to the wild, 51 are still surviving. Released marmots have also successfully bred in the wild, resulting in five additional litters (14 pups) born in the wild to mixed (wild-born/captive-born) and captive-born parents. There are now an estimated 210 animals living in the wild with potential breeding pairs on 13 mountains—up from three mountains in 2006. However, this highly vulnerable population is threatened by predation. In a study conducted in 2002, six of 18 fitted with radio-transmitter collars were killed by predators—wolves killed four, an eagle killed one, and a cougar killed the other. A study between 2001 and 2005 showed that as marmot populations declined, their social structure began to disappear, compounding their risk for extinction. In 2003, despite government estimates that pointed to a decline in wolf and cougar populations on the island, a cull of up to 30 wolves and 20 cougars was approved. Some environmentalists suggest that if the government was truly concerned about recovering marmot populations, it would advocate an end to clear-cut logging at higher elevations. Non-lethal predator management techniques are now being tested, such as the use of human shepherds, which has shown some success in deterring cougars and wolves but not golden eagles. New release sites are also being tested to see if predation rates vary from site to site. Similarly, new breeding programs are breeding in natural conditions and within colonies in an attempt to help the marmots regain their sociality.

Do you think that it is acceptable to cull one endangered species if it is threatening the survival of an even more endangered species? The marmot case is not unique; the same problem arises on the west coast with the reintroduced sea otter that is feasting on another endangered species, abalone. What would you do?

and sediment inputs accumulated in drainage water. High pesticide, fertilizer, sediment, and salt levels may negatively affect organisms further downstream. Wetland drainage is recognized as being one of the main pressures on biodiversity declines both nationally and internationally. Canada has no national inventory or monitoring program, but is thought to harbour one-quarter of the world's remaining wetlands (Federal, Provincial, and Territorial Governments of Canada, 2010).

Another ecozone particularly hard-hit by habitat destruction is the Carolinian forests of southwestern Ontario. These southern deciduous forests support a greater variety of wildlife than any other ecosystem in the country, including 40 per cent of the breeding birds. More than 90 per cent of this habitat has now been transformed by forestry, agriculture, and urbanization; less than 5 per cent of the original woodland remains. An estimated 40 per cent of Canada's species at risk are in this zone. Most of the remaining forests are in tracts belonging to regional conservation authorities or in privately owned woodlots potentially open to logging. Landscapes such as these are particularly suited to stewardship initiatives, discussed in greater detail later in the chapter.

The Canadian prairies have been virtually completely transformed into an agricultural landscape, with the result that many species native to this habitat are now endangered.

Environment in Focus

Box 14.5 Is Canada's 'Duck Factory' Disappearing?

Millions of ducks, geese, and swans darken the skies every year as they migrate across the length of the continent and back again. This annual migration evokes a sense of wonder and mystery in the more than 60 million North Americans who watch migratory birds each year. But for some, wonder and mystery is accompanied by anxiety over the future status of the 35 species of waterfowl that spend part of each year in Canada. Waterfowl depend on a complex and increasingly vulnerable chain of habitats extending across international borders. Accelerated conversion and degradation of habitat caused by human activities have led to record low populations of most duck species, as can be seen in Figure 14.3.

Many of the most productive wetlands in Canada have been drained to bring more land under cultivation. Wetlands in the Prairie provinces are particularly productive. The retreat of glaciers that at one time covered all of Canada left behind significant nutrient deposits, which have formed the basis of richly productive ecosystems. Waterfowl such as mallards and pintails feed on the plants and invertebrates that feed on the nutrients. But as farm intensification has increased, prairie wetlands have diminished in number and extent, making it difficult for ducks to secure adequate food supplies and nesting sites along their long migratory routes.

Unfortunately, habitat loss and degradation are not the only pressures on migrating waterfowl. Duck mortality rates also vary in response to weather, climate, competition for resources, environmental contamination, and hunting. In the Canadian and US prairies, weather has a particularly strong influence on the habitat conditions for waterfowl breeding and consequently on the abundance of waterfowl populations. Drought in the late 1980s and early 1990s created difficult breeding conditions for ducks. Spring habitat conditions improved into the late 1990s from the low levels during the drought of the 1980s but declined again in the early years of this century. In 2002, pond numbers were 58 per cent below the 10-year and the long-term (1961–2002) averages and total duck populations declined by 33 per cent (to 7.2 million ducks), illustrating the dramatic impact that weather can have on the reproductive potential of waterfowl. However, approximately 13 per cent more ponds were recorded in 2006 compared to 2005, 35 per cent above the long-term (1955–2005) average. By 2009 and 2010, numbers were at all-time lows with more than 34 per cent of the total ducks that usually settle in the Canadian prairies staying in the US. Major flooding events in the spring of 2011 saw a record-breaking 45.6 million ducks in the Canadian and US prairies, with 43 per cent more ducks in Canada than the long-term average. This does not indicate that all is well—if wetlands continue to be lost at the current rate, populations could plummet faster than they have increased.

With more than 3 million Canadians and Americans shooting migratory waterfowl each year, hunting also has a significant impact on waterfowl populations. For example, in 2010 more than 440,000 mallard ducks were killed by hunters.

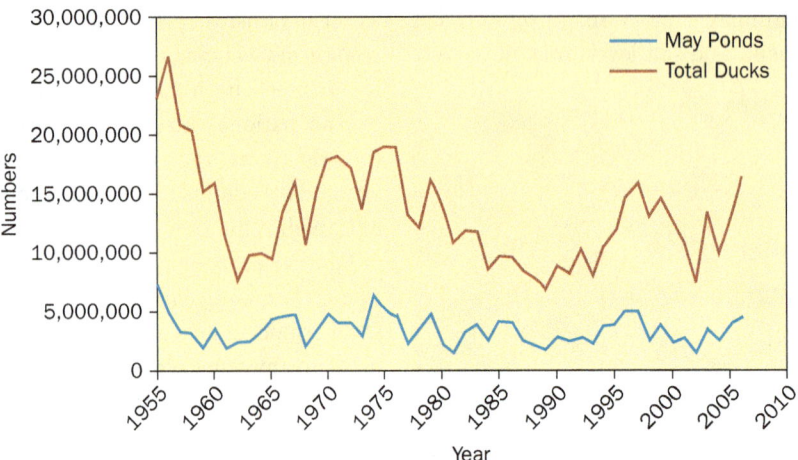

Figure 14.3 | Estimates of total ducks and May ponds in the southern prairie region of Canada. *Source:* Environment Canada (2007a).

Hunting regulations were introduced decades ago by the Canadian and American governments to protect waterfowl populations, but governments were slow to recognize the impact of land-use practices on waterfowl habitat and therefore abundance. The issue was not formally addressed until 1986, when Canada and the US signed the North American Waterfowl Management Plan (NAWMP) (Mexico joined in 1994). A distinctive feature was the focus on public–private stewardship initiatives, and it remains one of the most successful examples of this kind of stewardship approach to conservation. The priority goals of the plan were to:

- sustain average waterfowl populations of the 1970s;
- stop further wetland loss;
- stop further loss of native lands, especially native grasslands;
- restore lost wetlands, especially small basins;
- restore the function of upland habitats in landscapes conducive for maintenance of bird populations.

Conservation efforts under the NAWMP include involvement from various stakeholder groups—government agencies at all levels, industry, conservation groups, hunters, farmers, and other landowners. Duck conservation practices include maintaining nesting areas on land close to shallow water for land breeders such as mallards, pintails, teal, gadwalls, wigeons, and shovellors and ensuring that water levels are managed for diving ducks such as redheads and canvasbacks. Encouraging better cropping practices is also important. More than 17,000 landowners participate in habitat conservation programs on their lands.

Concern about the loss of habitat for Canadian species extends beyond the Canadian border. The harsh winters and productive summers that characterize much of Canada mean that many species in Canada, especially birds, are migratory. Over the past few decades, significant population declines have occurred for species that spend most of the year in tropical habitats but migrate to Canada to breed. In BC, for example, significant declines have occurred among northern flickers, Swainson's thrushes, chipping sparrows, yellow warblers, and dark-eyed juncos. These declines probably involve several factors, including loss of winter range through tropical deforestation and increased **fragmentation** within their northern breeding habitat. More long-term data and detailed studies are required to sort out the complexities of these changes.

Sometimes the impact of physical habitat change can be indirect. A good example is the parasitic habit of the brown-headed cowbird. This species lays its eggs in the nests of other species. The unsuspecting parents often lavish more attention on this large interloper and neglect their own young, leading to their death. The cowbird is an indigenous grassland species, but its distribution has expanded dramatically with human disturbance, since it prefers fragmented habitats and is adept at interloping on the forest-edge nesting sites of other birds. Some of its favourite targets are endangered species, such as the Kirtland's warbler, for which rates of up to 70 per cent parasitism have been recorded.

Physical habitat change has usually focused on aspects of the land or water that a species inhabits. An integral component of this habitat is the climate of the area. Ecological theory, as outlined in Chapter 2, tells us that vegetation growth is primarily controlled by climate in most areas and that animal communities depend on the vegetation for their sustenance. Global climate change will have a dramatic effect on these relationships.

The climate change scenarios for Canada show that we will be one of the most affected countries in the world because of our high latitudes, as discussed in Chapter 7. The northern regions are already showing large-scale reductions in snow and ice cover, reductions in permafrost, coastal inundation, and stressed populations of northern species such as polar bears. The US government has listed the polar bear under its Endangered Species Act as a result of thinning ice sheets due to global warming.

Scientists predict that each 1°C rise in temperature will cause biomes to migrate northward some 300 kilometres. Given the predicted minimum increase of 2 to 5°C in 70 to 100 years, this will translate into 600 to 1,500 metres in elevation and 300 to 750 kilometres in distance. Species will either be able to migrate fast enough to keep up with these changes, evolve to deal with them, or go extinct. Certain biomes, such as Arctic-alpine and the boreal forest, will be very vulnerable to these changes. Parmesan (2006) reviewed 866 scientific studies on the effects of climate change on biodiversity and found that at least 70 species of frogs, mostly mountain dwellers that had nowhere to escape, have become extinct. A further 100 to 200 cold-dependent species, such as the polar bear, are showing signs of increasing stress. Emperor penguin numbers, for example, dropped from 300 breeding pairs to the current nine pairs in the western Antarctic Peninsula.

In Canada, many of the great caribou herds have plummeted in numbers and there is concern that they may be in danger of extirpation over a large part of their range (Festa-Bianchet et al., 2011). In the High Arctic, the most northerly caribou, Peary's, numbered more than 50,000 in the 1970s and are now down to about 8,000. Unusually warm weather has caused an increase in freezing rains, creating a surface that the caribou cannot penetrate to access the tundra vegetation beneath it, and they starve to death. Although scientists had expected these impacts, they had not anticipated that they would occur so rapidly or to such an extent. Global climate change is not the only problem, but when linked with other causes, such as logging, increased predation due to habitat changes, disturbance by mining and oil exploration, and overhunting, the pressures may well reach a critical threshold for many populations.

Some effects of climate change are subtle. Mismatch between food supply and brood arrival in birds is one example. Many bird species are returning from southern migrations earlier and also producing earlier offspring. These changes may not be reflected in the abundance of the food supply, which may not be so tightly keyed to climate change. For example, on the west coast several fish-eating birds have declined rapidly in numbers over the last few decades. Their hatching dates have moved earlier by over a month in some species and these dates no longer match peak food supply in their ocean habitat (Gaston et al., 2009).

Hybrids occur in nature, but what happens when hybrids occur much more frequently because of human interference, such as through climate change? Are they to be celebrated as a natural evolutionary adjustment to climate change or despised as the ultimate symbol of human interference with the greatest biological process on Earth, evolution? Animal hybrids are often infertile, as is most often the case, for example, of the mule, the offspring of a donkey and a horse. Even if not, they might have trouble finding a mate. Species with small populations can rarely afford the luxury of wasted reproductive effort, so this is particularly troubling for endangered species. Furthermore, interbreeding may lead to genetic swamping of a rare species by a more common one. Thus, fears exist that the endangered red wolf of eastern Canada may become genetically swamped by interbreeding with the more common coyote.

Interaction among many different changes caused by global climate change is also important. For example, the whitebark pine of the Pacific coast and Rocky Mountains may run out of habitat as temperatures rise. Its population has already fallen by 70 per cent as a result of an infectious fungus called blister rust, mountain beetle infestation, rising temperatures, and fire suppression.

Chemical Changes

As the number of chemicals introduced into the environment continues to increase, concern over chemical degradation of habitats intensifies. Pollution is the second most important cause of endangerment for freshwater species in Canada. More than 20,000 chemicals are in use in Canada, and more than 1,000 new chemicals are introduced every year. The effects of chemical pollution are often more difficult to assess than those of physical destruction. Unless there is a catastrophic chemical spill, the signs of declining populations often go unnoticed for several years, even decades. Even after population declines have been documented, it may take many years of careful analysis before a conclusive link to chemical pollution can be established. This was the case with the decline in the numbers of birds at the top trophic level (Box 14.6) after pesticide biomagnification (Chapter 10) led to thinner eggshells and ultimately lower breeding success. Bald eagles, for example, had been killed for a long time around the Great Lakes, but it was the total breeding failure due to high chemical levels that led to their extirpation from the Ontario side of Lake Erie by 1980. In 1980, only seven nests existed along the entire Canadian shoreline of the Great Lakes, including Lake Superior, and not one healthy chick was produced.

Bald eagles have now recolonized many areas in the Great Lakes region where they were extirpated as a result of chemical use. There are now at least 136 nests, and the birds are producing strong, healthy chicks. But the birds are often dying young—at 13 to 15 years, less than half their normal natural lifespan. Autopsies completed on dead birds show high levels of lead and mercury contamination. The former is likely persisting in the environment from the time when lead was used in the manufacture of bullets and fishing lures; it should decrease over time. The source of mercury, which is highly toxic

Bald eagles in the Great Lakes region are dying at half their natural lifespan, often with high levels of lead and mercury in their systems as result of the persistence of these pollutants in the environment.

and accumulates through the food chain, is undetermined although mercury is a naturally occurring element. It has been eliminated from most products in which it was once used, and discharges from human sources are down about 80 per cent. However, the metal is emitted as a by-product of burning coal to produce electricity. Mercury also tends to build up in fish that live in the reservoirs behind power dams, as discussed in Chapter 11 regarding the James Bay Project. Scientists are now investigating these sources and their links to bald eagles.

There are also interesting and disturbing links between contaminant pollution and climate change. For example, in western Hudson Bay, due to declining sea ice, polar bears are feeding more on fish-eating open-water seals rather than on invertebrate-eating ice seals. Fish-eating seals have larger concentrations of contaminants, resulting in larger concentrations in polar bears. One contaminant flame retardant chemical increased in concentration among polar bears sampled by 28 per cent between 1991 and 2007 (Mckinney et al., 2009).

Alien Species

Invasive alien species are responsible for about 40 per cent of animal extinctions for which the cause is known, and globally these species are second only to habitat destruction as a main cause of endangerment. Introduced species have a significant impact by out-competing native species for necessary resources or by direct predation on native species (Chapter 3). The introduction of new species to insular habitats provides graphic examples of the destruction that can be wrought. On Haida Gwaii, for example, the introduction of both raccoons and Norway rats is having a catastrophic impact on ground-nesting seabirds. The breeding population of ancient murrelets on Langara Island off the north coast of Haida Gwaii declined by approximately 40 per cent between 1988 and 1993, leaving the population at less than 10 per cent of its original size. Further south in the new national park reserve of Gwaii Hanaas, the main ancient murrelet colony on Kunghit Island decreased in size by approximately one-third between 1986 and 1993. In both cases, predation by Norway rats appears to be mainly responsible for the declines.

Aside from direct predation, alien species can affect native species in other ways. For example, populations of the Newfoundland crossbill have declined significantly; competition for food (pine cones) with the introduced red squirrel is believed responsible. However, in contrast to the US, Venter et al. (2006) found that, currently, alien species are the least influential causes of endangerment in Canada. This is likely to change as warming climates make for a more hospitable environment for many more potential invaders.

Vulnerability to Extinction

The effects of overhunting and habitat degradation differ among species, since not all species are equally vulnerable to extinction. Ecologists have identified a set of extinction-prone characteristics to identify species most vulnerable to extinction. Using such characteristics, conservationists are better able to anticipate the need for protection. Species with one or more of the following characteristics are more vulnerable to extinction.

- *Specialized habitats for feeding or breeding.* Once a habitat is altered, the environment may no longer be suitable for specialized species. A good example is the northern spotted owl, discussed in Chapter 9. Northern spotted owls require old-growth habitat for survival and reproduction.
- *Migratory.* Many songbirds are experiencing population declines in Canada as a result of their long and hazardous migrations to South and Central America. Species that migrate seasonally depend on two or more distinct habitat types, and if either one of these habitats is damaged, the species may be unable to persist.
- *Insular and local distributions.* Dawson's caribou, endemic to the Haida Gwaii archipelago, became extinct because of the ease with which it could be hunted in such a restricted habitat, with no hope of an emigrating population for replacement.
- *High economic value.* Many organisms are overharvested to the point of extinction because of their high economic value. The American ginseng was once abundant in the forests of eastern North America but is now rare because of demand in Asian countries for dried ginseng roots for medicinal purposes. Similarly, populations of Asian bears have been all but eliminated across their range. Their gall bladders are highly valued in Asian markets, and North American bears are now coming under pressure from the same markets. A single gall bladder can be worth more

than $5,000. One illegal dealer in BC was found with 1,125 gall bladders in his possession. BC has passed a law making possession of endangered animal contraband an offence.
- *Animals with large body size.* Large animals tend to have large home ranges, require more food, and are more easily hunted by humans. Top carnivores, for example, depend on abundance of many different species lower in the food chain. If the numbers of prey species are disrupted, the impacts are felt at the top of the food chain. Furthermore, animals higher up the food chain are more vulnerable to the concentration of toxic materials (Chapter 10). Killer whales are a good example of this vulnerability.

Environment in Focus

Box 14.6 Raptors as Indicators of Chemical Degradation of Habitat

Sitting as they do at the top of the food chain, birds of prey or raptors are powerful indicator species of ecosystem health. Raptors were discovered to be useful indicators of environmental health during the 1960s when research into drastic population declines in bird- and fish-eating species revealed that eggshell-thinning and reproductive failure were caused by organochlorine pesticides. The decline of peregrine falcon populations is particularly well documented.

Canada has three subspecies of peregrine falcon: the continental *anatum* subspecies breeding south of the treeline from the Atlantic to the Pacific Ocean; the northern *tundrius* subspecies nesting along Arctic rivers, lakes, coastline, and inland escarpments; and the western *pealei* subspecies occupying coastal islands and areas of adjacent mainland BC. Peregrine falcons are extremely powerful birds of prey, catching other birds in flight while attaining speeds as great as 300 kilometres per hour. Prey is killed by a direct blow of the closed fist delivered at great speed. Favourite prey include songbirds, waterfowl, pigeons, shorebirds, and seabirds, and, especially among the Arctic peregrines, small mammals such as lemmings. Falcons nest on cliffs or in trees where they can look down over water bodies. Tall buildings in cities may serve as a substitute, in which case urban pigeons are the main prey. Falcons are quite territorial during the breeding season, with nests seldom closer than one kilometre apart.

The peregrine falcon once bred all across Canada. Populations appeared remarkably stable until the 1940s when they started to crash, linked to the bioaccumulation of pesticides (Chapter 10) such as DDT, DHC, dieldrin, and heptachlor epoxide (Rowell et al., 2003). Surveys in the 1970s documented the continuing downfall of the peregrine, and by that time the species had been extirpated from large areas of its previous range. In 1978, COSEWIC classified *anatum* peregrines as endangered, *tundrius* as threatened, and *pealei* as rare.

In the late 1980s, urban populations were established in southern Canada through the reintroduction of captive-raised young (an example of ex situ conservation). The program was expanded, and to date more than 700 birds have been released to the wild at more than 20 sites from the Bay of Fundy to southern Alberta and the Okanagan Valley in BC.

Peregrine falcons appear to be recovering. Nevertheless, chemical habitat degradation is not the only stress threatening the long-term viability of peregrine falcons across Canada. In BC, for example, falcon populations are threatened by declines in the raptor's supply of colonizing seabirds due to habitat loss and competition from alien predators such as rats. Alien species are discussed in more detail in a later subsection.

In 1996, the captive breeding station at Wainwright, Alberta, was closed in the belief that it was no longer needed, and in 1999 COSEWIC downlisted *anatum* peregrines because extinction was no longer a threat. Following a 2005 country-wide survey, none of the three subspecies is now listed by COSEWIC.

The peregrine falcon suffered badly as a result of the biomagnification of agricultural chemicals but is now recovering in numbers in many areas.

Species that have high economic values are targeted by hunters and poachers. Here a leopard has fallen victim to a poacher's snare in a park in Sri Lanka, and these cobras in Laos are now part of a local drink.

- *Need for a large home range.* Species that need to forage over a wide area are prone to extinction when part of their range is damaged or fragmented. Grizzly bears, for example, are very sensitive to fragmentation caused by logging and agricultural clearance.
- *Only one or a few populations and/or small population size.* Any one population may 'blink out' as a result of chance factors (e.g., earthquakes, fire, disease), increasing the species' vulnerability to extinction. Small populations are also more likely to become extinct locally because of their greater vulnerability to demographic and environmental variation. This is why one of the main goals of the Vancouver Island marmot reintroduction has been to start a number of geographically dispersed different colonies.
- *Not effective dispersers.* Species unable to adapt to changing environments must migrate to a more suitable habitat or face extinction. Species that cannot migrate quickly have a greater chance of extinction. This factor will become much more important as the impact of climate change increases.

- *Behavioural traits causing susceptibility.* Some species have behavioural traits that make them particularly vulnerable to clashes with human activities. For example, the red-headed woodpecker flies in front of cars, and the Florida manatee appears to be attracted by motorboats, a main cause of death for the animal. Other species, although not attracted by human activities, may be too slow to get out of the way. Tragically, this is the case with the few remaining right whales off the east coast (Box 14.7).

Although some species are more vulnerable to extinction than others, some species also naturally occur at lower population densities than others. The cougar, for example, was once found all across North America but at very low densities because of its need for an area large enough for each individual to secure sufficient food. How do scientists determine whether a species is just naturally rare or declining and in danger of extinction? How small does a population have to be before it is considered endangered? These and similar questions are answered by scientists on the basis of standardized, quantitative criteria. In Canada, COSEWIC uses criteria based on those suggested by IUCN and used in the global Red List.

The extinction of species may now be occurring roughly 100 to 1,000 times faster than the natural rate of extinction. This rate of extinction is much faster than the evolution of new species, so we are in a period in which the world's biological diversity is in decline. Even high-profile species, such as the tiger, are declining rapidly, despite millions of dollars devoted to their protection, as discussed in the accompanying international guest statement. What are nations doing to arrest this decline in biodiversity? What are the best strategies for the long-term preservation of biological diversity? In the sections that follow, we will discuss the international and Canadian responses to our biodiversity crisis.

Alive or dead, exotic species, such as these porcupine fish made into ornaments in Thailand, do not belong in your home.

Environment in Focus

Box 14.7 North Atlantic Right Whales

At one time the right whale ranged throughout the northern hemisphere, but it was extirpated from most areas as a result of hunting it for oil. The largest remaining population summers off the east coast of Canada and winters along the US coast and into the eastern Caribbean. The Roseway Basin, located approximately 30 kilometres south of Cape Sable Island, Nova Scotia, is one of only two known areas where large numbers of North Atlantic right whales gather on a seasonal basis in Canadian waters. The lower Bay of Fundy and three critical habitat areas in US waters are the only areas in the western North Atlantic where right whales are known to repeatedly gather on a seasonal basis for several months at a time. The Roseway Basin is an important feeding and socializing habitat for this critically endangered species, which scientists estimate number only about 400 individuals today. Unfortunately, each of these important habitats is intersected by or located near major shipping routes.

New shipping lanes in the Bay of Fundy, designed to protect the endangered North Atlantic right whale population from ship strikes, were put into operation in 2003. Right whales, however, were still being struck, leading Canada to propose the establishment of a recommended seasonal Area to be Avoided (ATBA) for ships of 300 gross tonnage and more during the seven-month period from 1 June to 31 December when the largest percentage of right whales is known to be in the area and when the risk of ship strikes is greatest. The proposal to the International Maritime Organization was adopted and came into effect in the spring of 2008.

International Guest Statement

Tiger Conservation in Thailand

Anak Pattanavibool

The tiger is the pride of Asia's natural heritage. They used to roam across Asia from the Middle East to Southeast Asia and from the Russian Far East down to Indonesia. They have become endangered mainly because of the clash with human civilization and exploitation. Their survival now depends on us. At the beginning of the twentieth century approximately 100,000 tigers existed across Asia. The global population has now declined to about 3,500 individuals. The habitat remaining is only 7 per cent of its historical range. The majority of breeding populations are in the Indian subcontinent, including India, Nepal, and Bangladesh. In Southeast Asia breeding populations are restricted to either very large or well-protected landscapes (Walston et al., 2010).

In Thailand, an estimated 200–250 wild tigers are scattered in protected areas, which cover about 25 per cent of the country's land area. Most of the remaining tigers exist in small numbers in heavily fragmented landscapes. Only one place in Thailand, as detailed below, now contains a breeding population of over 100 tigers.

Tigers face three major threats: (1) direct poaching for tiger parts and traditional Chinese medicines; (2) poaching of main prey species, particularly sambar, gaur, and banteng; and (3) habitat alteration from forest to agricultural landscapes. In Thailand, most primary forests have been fragmented. Poaching still penetrates deep inside many protected areas. Tigers and other large animals have been either wiped out or

Tigers at Huai Kha Khaeng Wildlife Sanctuary.

severely depleted from many protected areas due to poaching and fragmentation (Lynam, 2010). Tigers will survive in ecologically functioning numbers only in areas with strong law enforcement. In India, tigers remain only in well-guarded national parks. The recent extinctions of tigers from famous Indian tiger reserves (e.g., Sariska and Panha national parks) happened mainly because of inadequate protection.

In Thailand, the situation is desperate. Half of the country's tiger population exists in a large (18,000 km^2) and well-guarded forest landscape named the Western Forest Complex (WEFCOM). More than 2,000 guards are employed and stationed in 17 locations to protect WEFCOM. However, the large number of guards will not guarantee saving tigers and other wildlife. Many times they lack capacity and law enforcement training to cope with the poaching pressure. Therefore, a program has been created to monitor the performance of park guards. Currently, park guards in WEFCOM use a 'Smart' patrol system, which involves a suite of implementation components necessary for effective law enforcement, including strategic planning, adequate training, staff levels, equipment and other resource needs, standardized law enforcement (LEM) protocols, and full integration of LEM data into an adaptive management cycle. An effective Smart patrol promotes 'good governance' and 'best practice' by empowering park guards to engage fully in decision-making processes with park managers (Department of National Parks, Wildlife and Plant Conservation, 2010).

Wildlife scientists have tried to develop effective techniques to monitor tiger populations for many years. In India, the pugmark census that has been used to estimate numbers of tigers from tigers' footprints has proved inadequate to give reliable tiger numbers. In recent decades, scientists have adopted a technique using infrared-triggered camera traps with capture analysis as a reliable method to count tigers. Individual tigers have a unique stripe pattern the same as human fingerprints. With this technology scientists can estimate the number of tigers more precisely and reliably than the old pugmark technique. In WEFCOM, the tigers have been monitored annually with camera trapping and capture analysis, a methodology that has revealed around 100 tigers.

Since 2010 (the Year of the Tiger), leaders from governments, NGOs, and various other organizations have met several times to discuss how to save and assist the recovery of wild tigers in important tiger source sites and landscapes. Although the meetings have led to various tiger action plans, the implementation to save wild tigers on the ground remains to be seen.

An additional challenge for WEFCOM is that it is on the border with Myanmar. Both countries need to work together to reduce impacts from future development projects (e.g., roads, dams) that will fragment the tiger habitat. People have to understand the long-term benefits of conserving tigers as a key part of the ecosystem for this and future generations.

Dr Anak Pattanavibool obtained his Ph.D. in Canada at the University of Victoria, and is now Director of the Wildlife Conservation Society in Thailand.

Responses to the Loss of Biodiversity

The International Response

Awareness that we are living in a period of mass extinction unprecedented in human history has led to several international conventions and programs. Some programs have a regional orientation, such as the North American Waterfowl Management Plan with the United States and Mexico (Box 14.5), while others include many different nations. Some conventions have a long history. Established in 1973, the Convention on International Trade in Endangered Species of Wild Fauna and Flora (CITES) is one of the longest-standing treaties. Ratified by more than 120 countries (including Canada), this treaty establishes lists of species for which international trade is to be controlled or monitored (e.g., orchids, cacti, parrots, large cat species, sea turtles, rhinos, primates). International treaties such as CITES are implemented when a country passes laws to enforce them. Once CITES laws are passed within a country, police, customs inspectors, wildlife officers, and other government agents can arrest and prosecute individuals possessing or trading in CITES-listed species and seize the products or organisms involved. Countries that ratify CITES also vote on the species that should be protected by the treaty.

Several well-known Canadian species are listed under Appendix II of the Convention, such as the lynx, bobcat, cougar, polar bear, river otter, and burrowing owl. These species may only be traded with a valid permit from the Canadian Wildlife Service. When travelling through international airports in Canada, you will commonly see information on CITES to warn travellers about trying to import listed plants or animals.

CITES has been instrumental in restricting trade in certain endangered wildlife species. Its most notable success has been a global ban on the ivory trade, instituted in 1989. Without this ban, it is unlikely that any elephants would be left in East Africa, although large-scale smuggling is still evident. Despite the treaty's success in protecting some species, Canada's support has been disappointing. According to Le Prestre and Stoett (2001: 197), the Canadian voting record on whether species should be listed has reflected 'domestic economic and/or cultural interests at every turn', and Canada has consistently failed to pay its dues to support the Convention.

Other international treaties focused on conserving biodiversity include:

- Convention on Conservation of Migratory Species of Wild Animals (the Bonn Convention, 1979);
- Convention on Conservation of Antarctic Marine Living Resources (1982);
- International Convention for the Regulation of Whaling, which established the International Whaling Commission (1946);
- International Convention for the Protection of Birds (1950);
- Benelux Convention on the Hunting and Protection of Birds (1970).

Unfortunately, participation in these treaties is voluntary, and countries can withdraw at any time to pursue their own interests when they find the conditions of compliance too arduous. Canada, for example, withdrew from the Whaling Convention in order to unilaterally permit indigenous whaling in the Arctic. Although Japan still participates, it acts against the spirit of the Convention by continuing to hunt whales under scientific pretenses. The excess whale meat from these putative scientific expeditions is sold commercially.

Perspectives on the Environment
Global Imperative for Biodiversity Conservation

The action taken over the next decade or two, and the direction charted under the Convention on Biological Diversity, will determine whether the relatively stable environmental conditions on which human civilization has depended for the past 10,000 years will continue beyond this century. If we fail to use this opportunity, many ecosystems on the planet will move into new, unprecedented states in which the capacity to provide for the needs of present and future generations is highly uncertain.

—Secretariat of the Convention on Biological Diversity (2010)

One of the most important international agreements to protect biodiversity—the **Convention on Biological Diversity (CBD)**—emerged from the World Summit on Sustainable Development held in Rio de Janeiro in 1992 (Chapter 1). Canada was the first industrialized nation to sign the Convention and provided extra funding to house the secretariat in Montreal. The CBD is legally binding and requires signatories to develop biodiversity strategies, identify and monitor important components of biodiversity, develop endangered species legislation and **protected areas** systems, and promote environmentally sound and sustainable development in areas adjacent to protected areas. The Conference of the Parties adopted a motion to achieve by 2010 a significant reduction of the current rate of biodiversity loss at the global, regional, and national levels and met in Nagoya, Japan, in late 2010 to assess progress. At that time, it became clear that the countries of the world not only had failed to halt biodiversity erosion but had even failed to slow down the rate of biodiversity erosion (Secretariat of the Convention on Biological Diversity, 2010).

The question is why, and what can be done about it? There are many answers to this question, but a main one is shown in Figure 14.4. Tremendous effort has been invested over the last decade in trying to assess the state of biodiversity and in dealing with direct causes of biodiversity loss, such as habitat loss, by setting aside protected areas, as discussed in the next section. However, relatively little attention has been devoted to addressing the main drivers of biodiversity loss, such as poverty, increased spread of invasive species, and climate change. Future strategies must address these challenges as well as documenting and increasing the benefits that humans derive from biodiversity if the required progress is to occur in terms of preventing further biodiversity erosion. New targets for 2020 for the CBD were adopted at Nagoya (see box).

Perspectives on the Environment
CBD Targets for 2020

The new target is to 'take effective and urgent action to halt the loss of biodiversity in order to ensure that by 2020 ecosystems are resilient and continue to provide essential services, thereby securing the planet's variety of life, and contributing to human wellbeing and poverty eradication'. The subtargets include, among others, that by 2020 'the rate of loss of all natural habitats . . . is at least halved and, where feasible, brought close to zero and degradation and fragmentation is significantly reduced'; 'at least 17 per cent of terrestrial and inland water, and 10 per cent of coastal and marine areas . . . are conserved through effectively and equitably managed, ecologically representative and well connected protected area systems and other effective area-based conservation measures, and integrated into the wider landscape and seascapes'; 'the extinction of known threatened species has been prevented and their conservation status . . . has been improved and sustained'; and 'ecosystem resilience and the contribution of biodiversity to carbon stocks has been enhanced, through conservation and restoration, including restoration of at least 15 per cent of degraded ecosystems'.

—Herkenrath and Harrison (2011)

Canada has been very slow in implementing many strategies required by the Convention. The *Canadian Biodiversity Strategy* (Canada, 1995) was developed as a response to the requirements of the CBD, and other developments, such as the new Species at Risk Act (described below), are also consistent with these requirements. However, not until 2006 was Canada's Biodiversity Outcomes Framework agreed upon

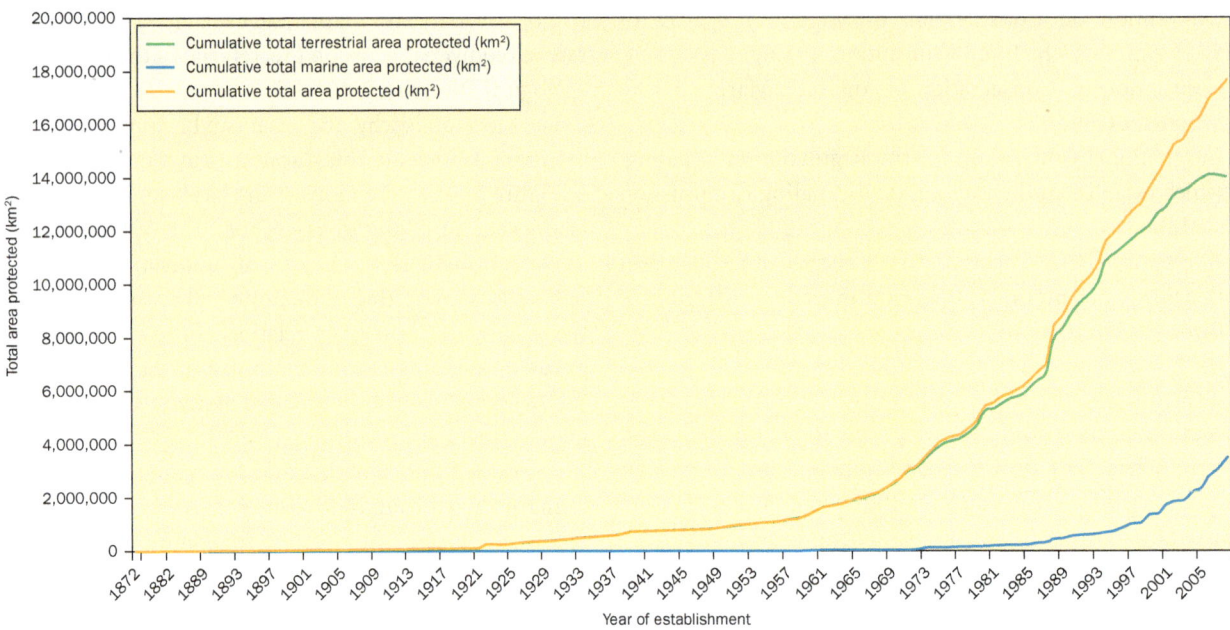

Figure 14.4 | Growth in nationally designated protected areas, 1872–2008. Source: *IUCN* and *UNEP-WCMC (2009) The World Database on Protected Areas (WDPA): January 2009.* Cambridge, UK: *UNEP-WCMC*.

Sariska Tiger Reserve (left) in India is protected by Project Tiger and yet had all 26 tigers poached out of it in the early years of this century. Authorities have now reintroduced tigers to the reserve, but they have yet to successfully raise young. Tiger reintroductions are very difficult and it is much better to invest in saving tiger populations than trying to reintroduce them after extirpation. The Chinese government is now planning to reintroduce tigers into southern China, as in this reserve in Hunan (right), but first the prey base must be re-established to ensure they have food to eat.

and begun to be implemented, and only in 2010 was the first report released on the status of Canada's ecosystems (Federal, Provincial, and Territorial Governments of Canada, 2010). The report is useful, but one of the main lessons was the lack of adequate data on many aspects of biodiversity in Canada to be able to understand trends, let alone design strategies to address the needs. These responses are discussed more in the next section.

The Canadian Response

In 1973, the United States became the first country to pass endangered species legislation. Australia followed some 20 years later, and a host of other countries, including the European Union and Japan, have developed similar legislation. In Canada, six provinces have endangered species legislation (New Brunswick, 1974; Quebec, 1989; Manitoba, 1990; Nova Scotia, 1998; Newfoundland and Labrador, 2001; Ontario, 2007). However, this provincial legislation is fairly weak in that many of the provisions to protect endangered species are discretionary. It has taken many years for the federal government to respond to the legislative challenge of protecting endangered species. The **Species at Risk Act (SARA)** was finally passed in 2002. The federal government had little choice; as a signatory to the CBD, it was required to enact such legislation.

One reason the government was so reluctant to introduce and pass federal endangered species legislation related to the Canadian Constitution. Unlike the situation in many other countries, most of the land in Canada is publicly owned, with 71 per cent held by the provinces and 23 per cent held by the federal government (see Chapter 1). Most responsibilities are shared by these two levels of government, with the federal government responsible for oceans and freshwater ecosystems, migratory birds, and the management of federal lands, including the Northwest Territories and Nunavut. Yukon now has responsibilities for its own land base, and a similar devolution is underway for the NWT and Nunavut. The federal government also has responsibility for Aboriginal lands in the provinces, although this continues to change as land claim negotiations are settled. Therefore, much of Canada's public lands, and their resources, are under provincial jurisdiction. Federal legislation may not be easily enforced in Canada.

The provinces have numerous individual programs related to nature conservation, and they also participate in joint programs with the federal government. Ontario has passed a comprehensive package that it hopes will help protect endangered species:

- The Endangered Species Act, 16 May 2007. Compared to Ontario's previous legislation, the new Act provides broader protection provisions for species at risk and their habitats, greater support for volunteer stewardship from private landowners and partners, a stronger commitment to recovery of species, and more effective enforcement provisions. Significantly, the Ontario approach avoids many of the criticisms levelled at the new federal legislation by allowing science-based assessments and listings of endangered species rather than allowing politicians to make the final decisions.
- Programs and policies to implement the new legislation.
- Enhanced stewardship programs that support volunteer efforts to protect and restore species at risk and their habitats. The $18 million, four-year Species at Risk Stewardship Fund is part of Ontario's stewardship-first approach to species protection. The Fund is open to landowners, farmers, **Aboriginal peoples**, academic institutions, industries, municipalities, conservation organizations, stewardship councils, and others across the province for eligible protection and recovery activities. To be eligible for funding, a proposal must aim to do one or more of the following:
- improve the status of species at risk and their habitats through stewardship and recovery activities;
- encourage involvement in stewardship activities through outreach, education, or youth employment;
- increase stewardship-related knowledge and skills of interested landowners or groups.

The Canadian Endangered Species Conservation Council (CESCC) comprises the three federal ministers responsible for environment, Canadian heritage, and fisheries and oceans, as well as provincial and territorial ministers responsible for the conservation and management of wildlife. The CESCC co-ordinates federal, provincial, and territorial government activities related to the protection of species at risk and provides general direction on the activities of COSEWIC and the preparation of recovery strategies and action plans.

Since 1976, COSEWIC has been responsible for determining the status of endangered species. The Committee—which includes representatives from relevant federal agencies, provincial and territorial wildlife agencies, and the Aboriginal community, as well eight scientific subcommittees that are species specialist groups—meets annually to consider status reports on candidate species and to assign them to various categories. In 2007, 552 species were designated in five risk categories and by 2011 this had risen to 649 species (Table 14.2). Ideally, the numbers of threatened species would be getting fewer over time and their category of endangerment becoming less severe. Of the species reassessed in 2010, 66 per cent were in the same category as the previous assessment, 12 per cent had become more secure, but 22 per cent had become more endangered.

The Committee's assessment is the first step in the process for protecting a proposed species at risk under SARA (Box 14.8). However, even if COSEWIC lists a species, this does not guarantee that the species will receive protection. The ultimate decision is in the hands of the politicians who make up the CESCC, and for this reason SARA has been strongly criticized by many who feel that the process should be scientific, not political.

The influence of politics was clear even when the first new species listings under SARA were made in April 2004, when the federal Fisheries Minister decided to delay by nine months a decision on whether to list 12 aquatic species recommended by COSEWIC. The Atlantic cod, for example, is estimated to have a population of less than 1 per cent of historic levels and shows no signs of recovery (Chapter 8), but its listing has not been confirmed. By 2009, while 77 per cent of the species suggested for listing by COSEWIC had been listed under SARA, only 35 per cent of marine fish species recommended for protection on biological grounds were protected.

An analysis of the kinds of species assessed for inclusion on the list found that 93 per cent of non-harvested species recommended by COSEWIC have been listed by SARA, while only 17 per cent of harvested species are listed (Findlay et al., 2009). Furthermore, species in the North were significantly less likely to be listed than species in southern Canada. Peary's caribou, discussed earlier, were put forward by COSEWIC in 2004, but have not yet been ascribed to SARA. Mooers et al. (2007) suggest that these patterns result from the lack of capacity or willingness on the part of certain agencies, particularly the Department of Fisheries and Oceans (DFO) and wildlife management boards in the North, to accept the additional stewardship responsibilities required by SARA. Delays for further consultation with management boards may lead to the extinction of some species, such as the eastern beluga. A second reason is that each suggested species undergoes a cost-benefit analysis in which very little attention is given to the benefits, both tangible and intangible, whereas the costs are studied in detail. Furthermore, these analyses are not open to peer review.

In 2005, the Commissioner of the Environment and Sustainable Development undertook an audit and recommended that Environment Canada develop a decision-making process by 2006 on the listing of species at risk that would be consistent, systematic, and transparent. A subsequent audit in 2008 found that such a listing process had yet to be established (Auditor General of Canada, 2008). A five-year report card on SARA released by four prominent conservation groups gave the federal government an 'F' in two

Environment in Focus

Box 14.8 The Process for Protecting a Species at Risk

1. COSEWIC assesses and classifies a wildlife species as extinct, extirpated, endangered, threatened, of special concern, data-deficient, or not at risk. COSEWIC provides its report to the Minister of the Environment and the Canadian Endangered Species Conservation Council, and a copy is deposited in the Public Registry.
2. Within 90 days, the minister indicates how he or she intends to respond to a COSEWIC assessment. Within nine months, the government makes a decision about whether or not to add the species to the List of Wildlife Species at Risk. If no government action is taken, the species is automatically added.
3. When a species is on or added to the List of Wildlife Species at Risk, then extirpated, endangered, or threatened species and their habitats have:
   - immediate protection on federal lands (except for those species in the territories that go through the safety net process described below);
   - immediate protection if it is an aquatic species;
   - immediate protection if it is a migratory bird;
   - protection through a safety net process if it is any other species in a province or territory.
4. For all species included on the List of Wildlife Species at Risk on 5 June 2003:
   - a recovery strategy must be prepared within three years for endangered species and within four years for threatened species or extirpated species (progress regarding such strategies is discussed below);
   - a management plan must be prepared within five years for a special concern species.

   For all species added to the List of Wildlife Species at Risk after 5 June 2003:
   - a recovery strategy must be prepared within one year for endangered species and within two years for threatened or extirpated species;
   - a management plan must be prepared within three years for a special-concern species.
5. Recovery strategies and action plans, which must include the identification of critical habitat for the species if possible, and management plans are published in the Public Registry. The public has 60 days to comment on these documents.

Five years after a recovery strategy, action plan, or management plan comes into effect, the minister must report on the implementation and the progress towards meeting objectives.

Source: Canada, 'Species at Risk Act' (2004), at: www.sararegistry.gc.ca/background/process_e.cfm.

categories: first, for failing to protect the habitat of species-at-risk, and second, for refusing to employ the federal safety net meant to protect SARA-listed species under provincial jurisdiction when provincial governments fail to protect them. A listing of species submitted on ecological grounds but yet to be ascribed to SARA can be found at: www.sararegistry.gc.ca/sar/assessment/batchreportHTML_0510_e.cfm.

For species listed under SARA, recovery and management plans must be developed and implemented, unless the minister responsible feels that recovery is not 'feasible', a caveat that provides another political opportunity to block action. Under the Act, proposed recovery strategies allow for a 60-day comment period during which any person may file written comments with the minister responsible. Within 30 days of the closing of the public comment period, the proposed recovery strategy must be finalized. Recovery strategies are evaluated every five years and updated as necessary.

In 2011 COSEWIC had 436 species listed as endangered or threatened. 'Recovery strategies' are to be devised for these species, complete with 'action plans' that will be subject to a cost-benefit analysis. Recovery strategies are due for 282 species, yet only 99 have been completed despite the legal time periods noted above. These delays appear to be a direct consequence of a budget freeze on the Canadian Wildlife Service applied by the Conservative government in 2007.

The Act has been further criticized because even when a species is listed, it receives automatic protection only on federal lands. In southern Canada, where many endangered species live, a significant proportion of federal lands are national parks in which the species are already protected. Therefore, no incremental gain in protection occurs unless provincial jurisdictions agree to provide it. There is a so-called 'safety net' whereby the federal government can invoke powers to act if a provincial government refuses to do so and if the case is seen as critical. It has never been used.

Recovery plans that determine critical habitat must be established for listed species. If the habitat is not already protected, the minister must order its protection if the habitat is on federal land, and if not, must report on steps taken to protect habitat. In only two cases has SARA been used to protect habitat. One was the protection of killer whale habitat on the west coast and only came about after the agency responsible, DFO, had been taken to court by a consortium of environmental groups. The judge in the case said 'DFO behaved in an evasive and obstructionist way and unnecessarily provoked and prolonged the litigation in this case . . . for no other purpose than to thwart attempts to bring important public issues before the court.' It would appear that the government agencies charged with protecting endangered species in Canada are unwilling to comply with their own legislation.

The Species at Risk Act is different from the American approach to endangered species protection in that it lays out a framework for co-operation on the protection of endangered species and relies primarily on volunteerism. The federal government has adopted the same approach for habitat protection, establishing the Habitat Stewardship Program to provide information to landowners on how best to manage their lands to protect endangered species. Landowners may receive compensation for any economic losses incurred. Warnock and Skeel (2004), in their evaluation of Operation Burrowing Owl in Saskatchewan, present evidence that some voluntary habitat stewardship programs can be effective.

Only time will tell whether the Canadian approach will be successful in protecting endangered species. If there is political interference at any of the numerous opportunities available, if the provinces disagree with the federal government, or if cost-benefit analysis indicates that it is less expensive to let a species become extinct, we will have failed in this experiment. The evidence suggests all three are occurring. The costs of such failure will be extremely high. Fortunately, endangered

Perspectives on the Environment

Progress on the Species at Risk Act

Environment Canada and Fisheries and Oceans Canada have made unsatisfactory progress in responding to our 2001 recommendation relating to the development of a comprehensive inventory of species at risk, while Parks Canada has made satisfactory progress on this recommendation.

- The three organizations have made unsatisfactory progress in responding to our 2001 recommendation relating to the development of recovery strategies and have not complied with specific deadline requirements established by the Species at Risk Act. As of June 2007, recovery strategies should have been completed for 228 species at risk, but recovery strategies completed at that date address only 55 of those species.
- Departments and organizations are also required under the Act to identify, to the extent possible, critical habitat necessary for the survival or recovery of species at risk. As of June 2007, critical habitat had been identified for 16 of the 228 species at risk for which recovery strategies were due.
- Despite the progress noted at Parks Canada, the federal government as a whole has made unsatisfactory progress in responding to our 2001 recommendations relating to the development of a comprehensive inventory of species at risk and of recovery strategies. While work is underway to develop appropriate data-sharing agreements with third parties, such as provincial and territorial governments, and non-governmental organizations such as Nature Serve, inventory data collections vary across Canada. Ongoing improvements to data quality and data consistency are needed.

—Auditor General of Canada (2008)

species legislation is not the only means of protecting biodiversity (Box 14.9). However, in its review of progress by the federal government on protecting endangered species, the status report of the Commissioner of the Environment and Sustainable Development, released in 2008, concluded that 'the federal government as a whole has made unsatisfactory progress in responding to our 2001 recommendations relating to the development of a comprehensive inventory of species at risk and of recovery strategies.' This further underlines a key message in the Millennium Ecosystem Assessment (2005) that 'Science can help ensure that decisions are made with the best available information, but ultimately the future of biodiversity will be determined by society.'

Protected Areas

Protected areas have emerged as one of the key strategies to combat the erosion of biodiversity both internationally and in Canada (Box 14.11). Protected areas play different roles in society (Box 14.10), and in many cases their conservation role has been recognized only recently as the dominant one. This is especially true in Canada, where Banff, our first national park (1885), was set aside mainly to promote tourism and generate income rather than to protect species and ecosystems. However, since that time, the crucial role in species and ecosystem protection played by protected areas in Canada has led to both legislation and policy directives that make it clear that biodiversity protection is the prime mandate for the national park system.

Canada's national park system is central to the protection of rare and endangered species. Although the 47 parks cover only about 4 per cent of the land base, they contain more than 70 per cent of the native terrestrial and freshwater vascular plants and 80 per cent of the native vertebrate species. More than 50 per cent of the endangered vascular plant species and almost 50 per cent of the endangered vertebrate species are found in national parks. Canada's national parks have also played a critical and increasingly important role as sites for reintroduction of endangered species (Table 14.3).

Protected Areas: A Global Perspective

The International Union for Conservation of Nature (IUCN) is an international body that draws together governments, non-governmental organizations, and scientists concerned with nature conservation. The IUCN helps to provide leadership and set global standards for conservation. A protected area, as defined by the IUCN, is 'a clearly defined geographical space, recognized, dedicated, and managed, through legal or other effective means, to achieve the long-term conservation of nature with associated ecosystem services and cultural values'. There are many different kinds of protected areas, such as national and provincial parks, wilderness areas, and biosphere reserves. They all offer some form of protection but with differing degrees of stringency. To help bring some order and understanding to the different types of protected areas, the IUCN has developed a system of classification that ranges from minimal to more intensive use of the habitat by humans (Table 14.4).

The growth of protected areas has been strong, especially over the past decade (Figure 14.5). In 1994, approximately 8,600 protected areas had been designated worldwide, covering some 8 million km^2; by 2010, 107,034 protected areas had been designated, covering more than 19.5 million km^2. This is equivalent to the combined area of Southeast and South Asia and China and represents almost 13 per cent of the Earth's land surface. Several factors explain the rapid growth in protected area establishment:

The 'playground' role is one of the most controversial issues for park managers. This Parks Canada golf course sign in Pacific Rim National Park, for example, is advertising a facility that is not actually located on park lands. Nonetheless, by advertising it, a link is created in the public mind between national parks and certain recreational activities. What about traditional activities in some parks, such as horseback riding, which can have a significant impact on the environment, or are national parks places where we should rely on our own feet to explore the beauty of the landscape?

Environment in Focus

Box 14.9 What You Can Do for Endangered Species

Although the challenges created by endangered species can seem daunting to the individual, you can do several things.

1. If you own land, even your own backyard, try to encourage the growth of native species and promote high diversity among these species. Provide the three staples—food, water, and shelter. Plant perennials such as fruit and nut trees, nectar-producing flowers, and berry bushes. Don't use chemicals!
2. Write letters to politicians at all levels encouraging them to adopt specific measures. For example, write to local politicians urging protection for a natural habitat in your area.
3. Join and support an environmental group with a special interest in endangered species.
4. Take part in an active biodiversity monitoring project such as the Christmas Bird Count, the Canadian Lakes Loon Survey, Frogwatch, Project FeederWatch, or one of the other many organized activities that take place across the country. Details on these projects are available from local NGOs and university and college departments.
5. Don't keep exotic pets.
6. If you have a pet, try to make sure that it does not injure or harass wildlife. Put a bell on your cat. Domestic cats kill large numbers of songbirds every year.
7. Don't buy products made of endangered animals or plants.
8. Vote for political candidates who share your views on conservation matters.
9. Keep informed of biodiversity issues by watching nature programs on television, reading books, attending public lectures, and having discussions with local conservationists.
10. Actively learn more about wildlife, not just by reading and watching television but also by becoming more aware of the wildlife in your region through field observation. Encourage others, especially children, to do likewise.

Environment in Focus

Box 14.10 The Many Roles of Protected Areas

Art gallery: Many parks were designated for their scenic beauty, still a major reason why people visit parks.

Zoo: As one component of the art gallery, parks are usually places to view wildlife easily in relatively natural surroundings. Because it is protected from hunting in most parks, the wildlife is not as shy of humans as wildlife outside parks.

Playground: Parks provide excellent recreational settings for many outdoor pursuits.

Movie theatre: Just like a movie, parks can lift us into a setting different from that of our everyday life.

Cathedral: Many people derive spiritual fulfillment from communing with nature, just as others go to human-built places of worship.

Factory: The first national parks in Canada were designated with the idea of generating income through tourism. Since these early beginnings, the economic role of parks has been recognized, although it is a controversial one because of potential conflict with most other roles.

Museum: In the absence of development, parks serve as museums, reminding us of how landscapes might have looked to early settlers. These museums also perform a valuable ecological function, since they encompass important areas against which ecological change in the rest of the landscape can be measured.

Bank: Parks are places in which we store and protect our ecological capital, including threatened and endangered species. We can use the 'interest' from these 'accounts' to repopulate areas with species that have disappeared.

Hospital: Ecosystems are not static and isolated phenomena but are linked to support processes all over the planet. Protected areas constitute one of the few places where such processes still operate in a relatively natural manner. As such, they may be considered ecosystem 'hospitals' where air is purified, carbon stored, oxygen produced, and ecosystems 'recreated'. About 15 per cent of the carbon sequestered in North America is in protected areas (www.cbd.int/lifeweb/carbon/).

Laboratory: As relatively natural landscapes, parks represent outdoor laboratories for scientists to use in unravelling the mysteries of nature. Killarney Provincial Park in Ontario, for example, was an important laboratory for early research on acidic precipitation in Canada.

Schoolroom: Parks can play a major role in education as outdoor classrooms.

Source: Dearden (1995).

Environment in Focus

Box 14.11 People and Protected Areas: A Global Perspective

From the origins of the conservation movement in the US, the idea of having areas protected by government for conservation and public benefit, education, and enjoyment has spread throughout the world. Implementation has differed to reflect local conditions, but one ubiquitous concern for managers is the relationship between protected areas and local populations. As human populations grow, so do pressures for increased use of protected areas. In the UK, this overuse might be mainly recreational, and significant biophysical impacts may result just from the sheer numbers of people enjoying the parks. In many tropical areas, conflicts arise as local people, often driven by poverty and land-use pressures, encroach on the parks in large numbers, hunting wild animals and cutting down trees to make way for agriculture, obtain firewood, and/or sell on international markets. Estimates suggest that more than half a million people are illegally occupying national parks and wildlife sanctuaries in Thailand, with almost half the area of some parks suffering environmental degradation.

Such management problems are challenging. There is little point in trying to manage the area of land officially designated as a protected area if, in fact, it is not protected from resource use. In the past in Thailand, management activities focused on a preventive approach, with armed guards patrolling boundaries. Since most large remaining areas of forest and most wild animal populations are within the protected area system, there have been some benefits to this approach. Nevertheless, large-scale poaching continues in many areas, and shootouts between poachers and park guards are not an ideal management tool. Attention, therefore, has also spread to trying to address underlying motives behind poaching, such as poverty, although here, too, there are substantial challenges. Economic development programs initiated in some villages have triggered an increase in land prices, leading some villagers to sell their lands and encroach further into park lands. Unscrupulous local leaders may also encourage villagers to sell so that they can gain control over more land.

As with many environmental management problems, the answer does not lie in one single solution. Each case is different, and an adaptive management approach (see Chapter 6) to the protected area ecosystem is essential. In the long term, education must play a lead role. Many people are unaware of the vital functions played by protected areas. It is better to achieve voluntary compliance with more flexible management regimes than to have armed standoffs and mass non-compliance, as has often occurred in the past.

Park wardens in Thailand receive little pay and risk their lives to protect what remains of the wildlife. Every year lives are lost in battles with poachers.

1. *Increased realization of the rate of biodiversity loss and the severity of the issue.* In 1990, for example, statistics on endangered wildlife in Canada listed 194 species at risk, compared to over 600 by 2011.
2. *Growing awareness at the political level of the links between environmental and societal health.* When ecosystems collapse, livelihoods and economies collapse too. This interdependency between ecosystem protection and poverty is now recognized by many international development agencies and has helped to spur the worldwide interest in protected area establishment. For example, the Global Environmental Facility (GEF), an international agency formed in 1991 to assist developing countries in undertaking activities that benefit the global environment, provided approximately $9.5 billion in grants and leveraged about $42 billion in co-financing in support of more than 2,700 biodiversity projects in 165 countries between 1991 and 2011. In addition, a small grants program has started more than 6,500 biodiversity projects at the community level in 120 countries since 1992.
3. *Realization of the value of ecosystem services.* Increased methodological sophistication has allowed monetary values to be placed on ecosystem values. One team of researchers put an average price tag of US$33 trillion a year on fundamental ecosystem services such as nutrient cycling, soil formation, and climate regulation. This figure is nearly twice the annual global GDP of US$18 trillion and demonstrates, in economically understandable terms, the value of the so-called 'free' services of functioning environments (Costanza et al., 1997).

4. *Growing evidence of the effectiveness of protected areas in helping to combat environmental degradation.* Many studies demonstrate the effectiveness of protected areas for biodiversity protection. Protected areas are a main cornerstone for biodiversity protection under international treaties, such as the Convention on Biological Diversity, and a main recommendation of the Millennium Ecosystem Assessment (2005).

Although major gains have been seen in terms of terrestrial park systems, greater progress is required in the marine realm. Less than 1 per cent of freshwater and oceanic ecosystems enjoy any effective protection (Chapter 8).

Protected Areas: A Canadian Perspective

Canada has a large variety of protected areas, ranging from small ecological reserves to vast multiple-use areas, and the growth in protected areas has been strong (Figure 14.4). These areas are protected by a wide range of authorities, from municipal to federal and even international agencies. The most important protected areas are in our national and provincial park systems, and the total area of land protected overall by these two levels of government is about equal. Parks Canada is the main federal agency in charge of federal protected areas, such as national parks, although Environment Canada has jurisdiction over large areas, such as migratory bird sanctuaries and national wildlife areas, set aside primarily for wildlife. The location of national parks is strongly influenced by Parks Canada's national park **system plan**, which divides the country into 39 physiographic regions representative of Canada's natural heritage. The goal is to have at least one national park in each of these regions (Figure 14.6). All provinces have similar system plans, and the overall total amounts to 486 natural regions across the country. Malcolm (2009) discusses the various provincial park systems in more detail.

In 1969, the federal minister in charge of parks, Jean Chrétien, announced a goal of achieving system completion by 1985. However, by 1985 the system was less than half complete. In 1992, Canada's federal, provincial, and territorial ministers of environment, parks, and wildlife signed a Statement of Commitment to Complete Canada's Network of Protected Areas. Terrestrial systems were to be completed by 2000, whereas marine designation was to be 'accelerated'. Again, the goal was not met, and in 2002 Prime Minister Jean Chrétien announced a five-year plan to establish 10 new terrestrial parks, increasing representation to 35 of the 39 regions, and five marine protected areas under the 29-region marine system plan.

Table 14.4 | IUCN Classification of Protected Areas

Category	Name	Description
Ia	Strict nature reserve	Strictly protected areas set aside to protect biodiversity and also possibly geological/geomorphological features, where human visitation, use, and impacts are strictly controlled and limited to ensure protection of the conservation values.
Ib	Wilderness area	Usually large unmodified or slightly modified areas, retaining their natural character and influence, without permanent or significant human habitation, which are protected and managed so as to preserve their natural condition.
II	National park	Large natural or near-natural areas set aside to protect large-scale ecological processes, along with the complement of species and ecosystems characteristic of the area, which also provide a foundation for environmentally and culturally compatible spiritual, scientific, educational, recreational, and visitor opportunities.
III	Natural monument or feature	Areas set aside to protect a specific natural monument, which can be a landform, sea mount, submarine cavern, geological feature such as a cave, or even a living feature such as an ancient grove.
IV	Habitat/species management area	Areas that aim to protect particular species or habitats and where management reflects this priority. Many category IV protected areas will need regular, active interventions to address the requirements of particular species or to maintain habitats, but this is not a requirement of the category.
V	Protected landscape or seascape	An area where the interaction of people and nature over time has produced a distinct character with significant ecological, biological, cultural, and scenic value, and where safeguarding the integrity of this interaction is vital to protecting and sustaining the area and its associated nature conservation and other values.
VI	Protected areas with sustainable use of natural resources	Areas conserving ecosystems and habitats, together with associated cultural values and traditional natural resource management systems. They are generally large, with most of the area in a natural condition, where a proportion is under sustainable natural resource management and where low-level, non-industrial use of natural resources compatible with nature conservation is seen as one of the main aims of the area.

Source: Dudley et al. (2010: 34).

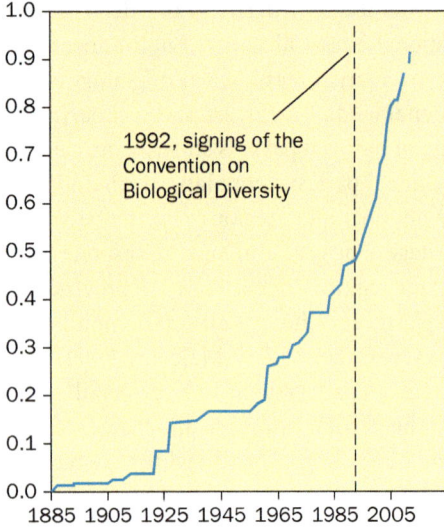

Figure 14.5 | Terrestrial protected areas in Canada. *Source: Federal, Provincial, and Territorial Governments of Canada (2012:48).*

Canada is still far from meeting these commitments. Parks Canada has fallen behind in reaching the goal, with about 60 per cent of the terrestrial park system complete (Figure 14.6) and 15 per cent of the marine system (Figure 14.7). The most recent Corporate Plan targets the number of represented terrestrial regions to increase by one, to 29, and the number of marine regions to increase from three to five between 2007 and 2013, both very modest targets in a time of increasing urgency. However, there has been major progress at the provincial level. In 1968, Ontario had 90 regulated protected areas totalling 1.6 per cent of the province by area. Currently, Ontario has 631 protected areas totalling more than 9.4 million hectares, or 8.7 per cent of the province. In British Columbia, the area of parkland doubled between 1977 and 2005 and now totals more than 12 million hectares (Figure 14.8). BC is the only jurisdiction to accomplish the 12 per cent target set by the World Commission on Environment and Development (WCED, 1987). Yet, in terms of overall protection of Canada's 177 ecoregions, only 29 per cent have a high level of protection

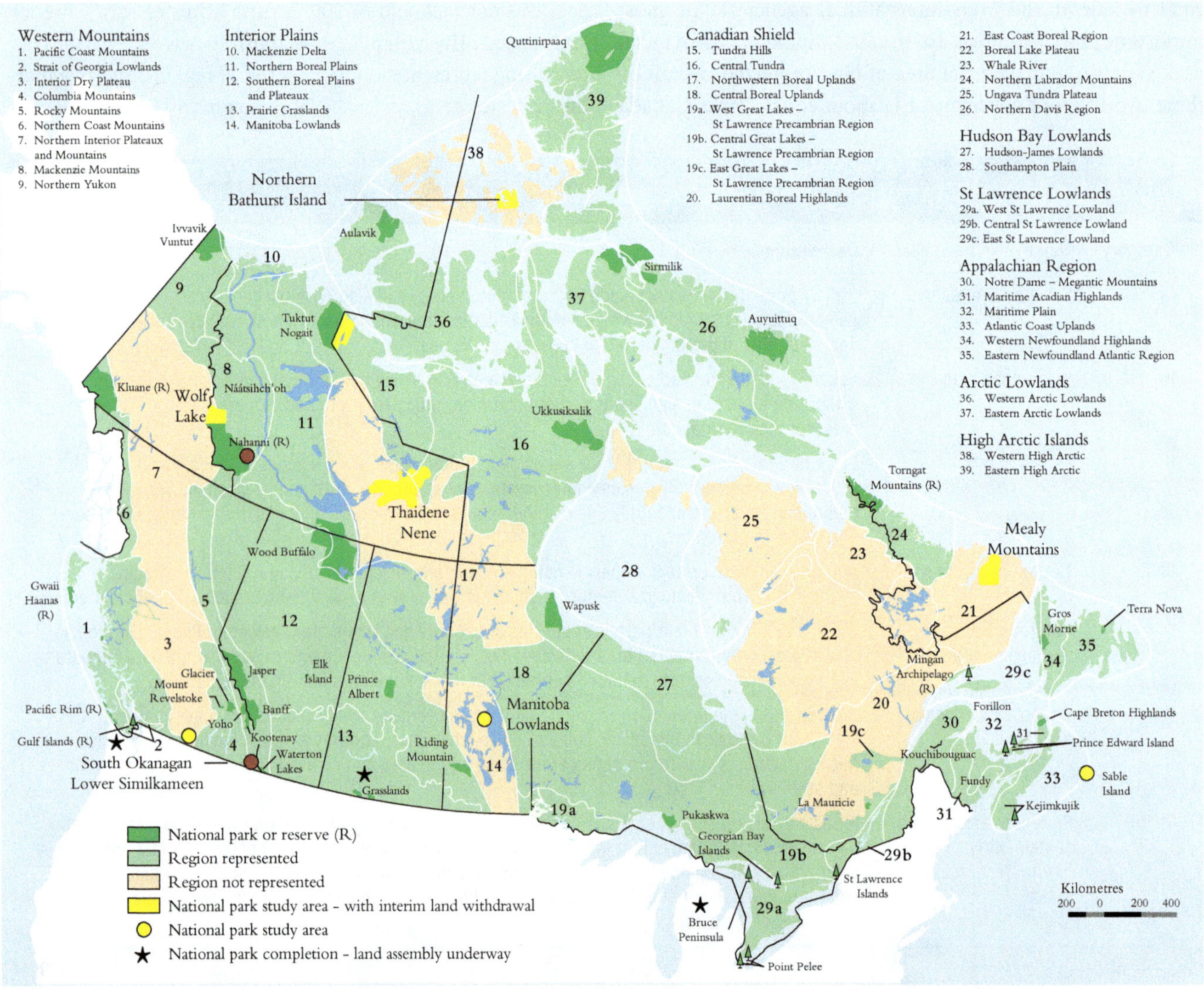

Figure 14.6 | Canada's national park system and state of completion as of 2010. *Source: Parks Canada Agency (2011: 7).*

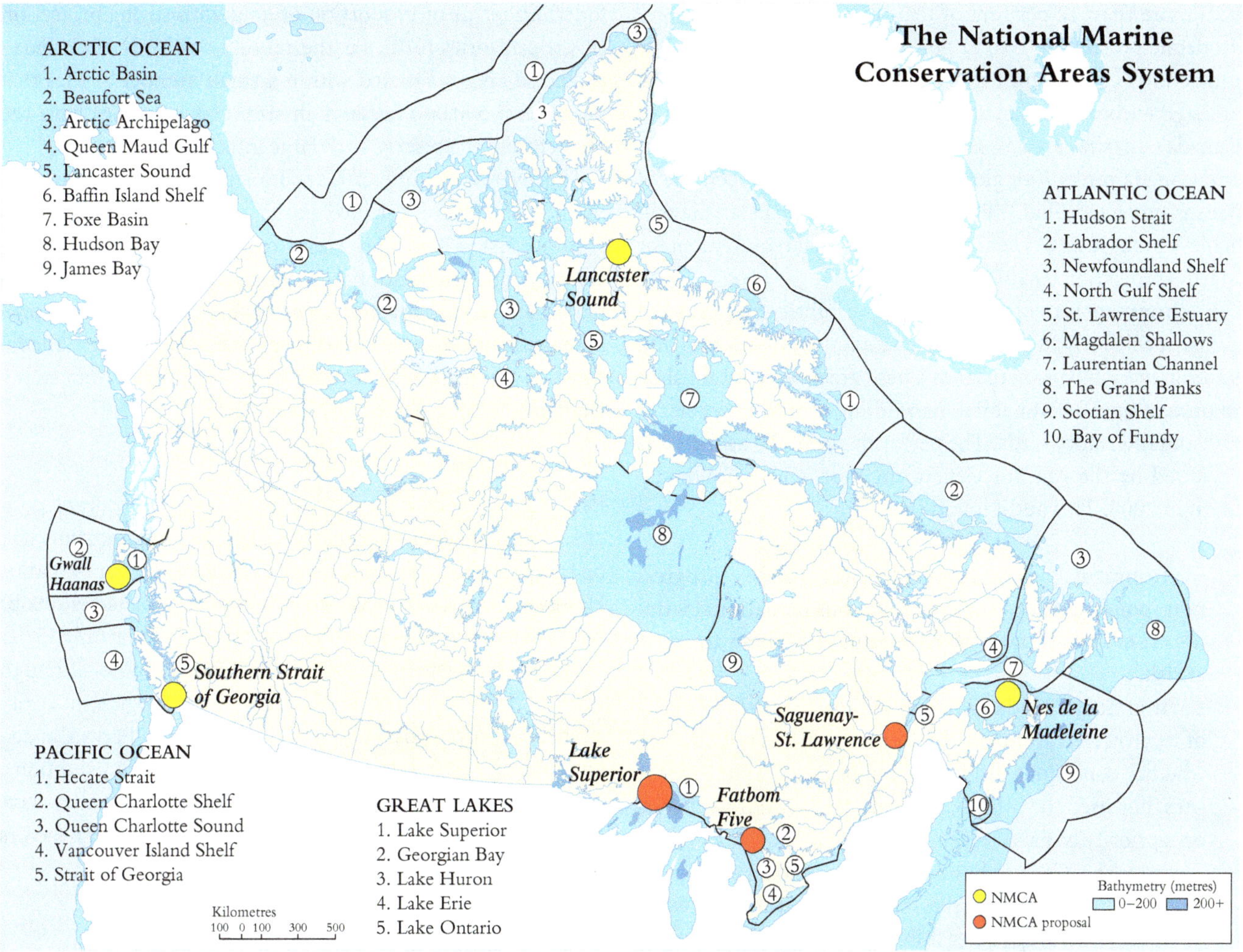

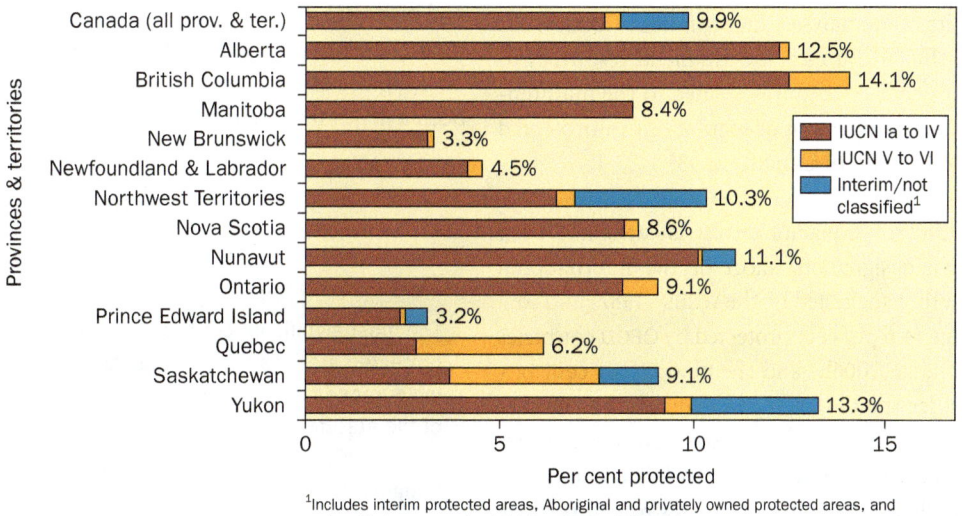

The National Marine Conservation Areas System
ARCTIC OCEAN
1. Arctic Basin
2. Beaufort Sea
3. Arctic Archipelago
4. Queen Maud Gulf
5. Lancaster Sound
6. Baffin Island Shelf
7. Foxe Basin
8. Hudson Bay
9. James Bay
ATLANTIC OCEAN
1. Hudson Strait
2. Labrador Shelf
3. Newfoundland Shelf
4. North Gulf Shelf
5. St. Lawrence Estuary
6. Magdalen Shallows
7. Laurentian Channel
8. The Grand Banks
9. Scotian Shelf
10. Bay of Fundy
PACIFIC OCEAN
1. Hecate Strait
2. Queen Charlotte Shelf
3. Queen Charlotte Sound
4. Vancouver Island Shelf
5. Strait of Georgia
GREAT LAKES
1. Lake Superior
2. Georgian Bay
3. Lake Huron
4. Lake Erie
5. Lake Ontario
Lancaster Sound
Gwaii Haanas
Southern Strait of Georgia
Lake Superior
Fatbom Five
Saguenay-St. Lawrence
Nes de la Madeleine
Kilometres
100 0 100 300 500
NMCA
NMCA proposal
Bathymetry (metres)
0–200
200+

Figure 14.7 | The system of national marine conservation areas of Canada. *Source: Parks Canada Agency (2011: 10).*

Canada (all prov. & ter.)
Alberta
British Columbia
Manitoba
New Brunswick
Newfoundland & Labrador
Northwest Territories
Nova Scotia
Nunavut
Ontario
Prince Edward Island
Quebec
Saskatchewan
Yukon
9.9%
12.5%
14.1%
8.4%
3.3%
4.5%
10.3%
8.6%
11.1%
9.1%
3.2%
6.2%
9.1%
13.3%
IUCN Ia to IV
IUCN V to VI
Interim/not classified[1]
Provinces & territories
0
5
10
15
Per cent protected

[1]Includes interim protected areas, Aboriginal and privately owned protected areas, and unclassified government administered protected areas.

Figure 14.8 | Percentage of land in protected areas in each province and territory. *Source: Environment Canada (2006b).*

(i.e., more than 12 per cent of their area), 12.4 per cent have moderate protection (6 to 12 per cent), 41.9 per cent have low protection (<6 per cent), and 16.6 per cent have no protected areas (Environment Canada, 2006b). Less than 1 per cent of Canada's marine area is set aside in protective designation, and Canada ranks 70th globally in terms of the percentage of oceans protected (ibid.). Projections suggest that Canada will achieve perhaps 33 per cent of its international requirement for marine protection by 2012.

Environment Canada also has responsibilities for protected areas; specifically, it manages 51 national wildlife areas and 92 migratory bird sanctuaries. These areas were established to protect significant habitat for wildlife, including species at risk and migratory birds. The performance of the agency was reviewed by the Auditor General in 2001, with a follow-up audit in 2008. This audit found that:

- Environment Canada has made unsatisfactory progress in responding to our recommendations on national wildlife areas and migratory bird sanctuaries. These areas are at risk.
- Environment Canada has identified specific threats to each of its protected areas, but the Department has not assessed whether conditions are improving or deteriorating at the sites, nor used the information collected to address threats on a priority basis.
- Environment Canada has developed a national strategy to guide the management of sites in its protected areas network, but the strategy is not being fully implemented. For example, most protected areas still lack up-to-date management plans.
- Environment Canada has not established explicit performance expectations against which progress can be assessed and does not comprehensively monitor or regularly report on the condition and management of its network of protected areas.
- According to its own analyses, Environment Canada has allocated insufficient human and financial resources to address urgent needs or activities related to the maintenance of sites and enforcement of regulations in protected areas. (Auditor General of Canada, 2008)

About 10 per cent of Canada's terrestrial area has been awarded protective designation, short of the international goal of 12 per cent first suggested by the WCED (1987) and well short of the average 14.6 per cent protected by OECD countries (Environment Canada, 2006b) and the new 17 per cent target established under the CBD. Canada manages 5.1 per cent of the world's terrestrial protected area estate. However, 95 per cent of Canada's terrestrial protected areas fall within IUCN categories I–IV (Table 14.4) and hence have a strong protective mandate. Among OECD countries, Canada ranks sixteenth out of 30 in terms of the proportion of land protected. The US protects almost 25 per cent compared to our 10 per cent, yet ranks fourth in terms of proportion of land with strong protection (IUCN categories I–IV). Furthermore, two-thirds of Canada's protected area is situated within a small number of sites that are at least 300,000 hectares in size. Few countries have the opportunity to preserve such large intact landscapes.

The overall quality of Canadian protected areas is very high, with several national parks appearing on the World Heritage list. These sites are of national and global significance. High international accolades depend on first-rate park management practices. The next section identifies some management challenges facing protected areas in Canada. Most of the section is devoted to national parks, but many challenges also apply to provincial parks.

Park Management Challenges

Park management is guided by legislation and relevant policies. Park management plans articulate how requirements will be translated into on-the-ground activities in different parks. However, a national survey by Environment Canada (2006b) on the status of Canada's protected areas found that only 25 per cent had management plans in place. Furthermore, although most park systems in Canada recognize ecological integrity as their main purpose, only two jurisdictions (Parks Canada and Ontario) have measures to monitor changing conditions. Ecosystems are considered to have integrity when they have their native components and processes in place. Several factors make park management a challenging process.

Development within the Parks

Management in national parks is determined by the National Parks Act. Although the first national park was created in Banff, Alberta, in 1885, the first National Parks Act was not passed until 1930. Both this Act and earlier legislation dedicated the parks to 'the people of Canada for their benefit, education and enjoyment . . . such Parks shall be maintained and made use of so as to leave them unimpaired for the enjoyment of future generations.' Reconciling the balance between 'making use of' and maintaining the parks 'unimpaired' has been a major topic of debate ever since.

Perspectives on the Environment

Significance of Ecological Integrity in Protected Areas

The most significant change to the government of Ontario's protected areas is that ecological integrity is now the guiding purpose for planning and management. Subsection 3(1) of the statute [Provincial Parks and Conservation Reserves Act, 2006] states that the 'maintenance of ecological integrity shall be the first priority and the restoration of ecological integrity shall be considered' for all provincial parks and conservation reserves.

—Wilkinson (2008: 182)

Tourism and income generation were the main reasons behind the establishment of many parks, including Banff. Thus, catering to the demands of tourists was the most important management priority.

By the 1960s, the visitors to Canada's national parks had increased tremendously, as had developments to serve them, including ski hills, golf courses, roads, and hotels. A massive proposal to expand the Lake Louise ski area was rejected in the early 1970s, signifying that the environmental movement was at last starting to be heard in the parks. The next 40 years witnessed many debates about controlling development in protected areas. Although the Lake Louise expansion was thwarted, smaller developments permeated the parks. In 1988, the balance between development and protection was clarified in amendments to the National Parks Act. Protection of ecological integrity became the primary mandate. Despite this legislative mandate, development pressures continued.

Perspectives on the Environment
On Banff

I do not suppose in any portion of the world there can be found a spot, taken all together, which combines so many attractions and which promises in as great a degree not only large pecuniary advantage to the Dominion, but much prestige to the whole country by attracting the population, not only on this continent, but of Europe to this place. It has all the qualifications necessary to make it a place of great resort. . . . There is beautiful scenery, there are curative properties of the water, there is a genial climate, there is prairie sport, and there is mountain sport; and I have no doubt that it will become a great watering-place.

—Sir John A. Macdonald on Banff, 1887

Although the environmental lobby was successful in its fight against the construction of an upper and lower village at Lake Louise founded by Imperial Oil and supported by Parks Canada in the early 1970s, many would argue that the subsequent incremental developments have achieved almost the same result. This photograph shows the enlarged Château Lake Louise in front of what is advertised as the largest ski hill in Canada. Is this a national park landscape?

Gros Morne National Park in western Newfoundland is a World Heritage Site.

Wildlife overpass built over the Trans-Canada Highway in Banff National Park.

In response, in 1998 the federal minister created an Ecological Integrity Panel—to look at development pressures in all of Canada's national parks. The Ecological Integrity Panel's report concurred with an earlier study on Banff, and strongly recommended a more adaptive approach to park management, with greater attention to ecosystem-based management and greater consultation with stakeholders (Chapter 6). The panel's 127 recommendations delivered one central message: *ecological integrity in all the national parks is in peril*. The minister accepted the panel's findings and began implementing the recommendations. The proclamation of a new National Parks Act in 2000, which further strengthens the ecological mandate of the parks, was one response.

The findings of the panel and subsequent recommendations emphasized what had already been well known to biogeographers. The theory of **island biogeography** suggests that small islands are unable to support as many species as large islands of similar habitat. Given that many terrestrial parks are separated from colonizing sources from outside the park, they are analogous to islands. Development inside the parks essentially makes them smaller, and smaller parks are more likely to experience extinctions.

Research has shown that development within parks is detrimental to many species, and over the past 15 years the government has revised the National Parks Act twice in favour of a mandate that supports wildlife protection over recreational opportunities. But while these changes are positive, there has been a growing realization that development is not the only threat to ecological integrity within our parks. Many management challenges arise from threats originating beyond park boundaries.

External Threats

Parks do not exist in isolation—they are intimately linked to surrounding and global ecosystems. It is therefore necessary to be aware of any influences from outside park boundaries that may have a detrimental impact on wildlife resources within the park (Box 14.12). This awareness is relatively recent (Figure 14.9). In the early days, boundaries were easily penetrated as society adjusted to the idea of preserving nature. But as time passed, park boundaries became less permeable, and protection of wildlife resources was more assured. However, as development surrounding parks intensified, park managers began to realize that development pressures outside park boundaries were affecting resources within parks. In response, they began to develop integrated management plans that took external threats into account.

Although the creation of integrated management plans is a step in the right direction, external threats to parks are often difficult to eliminate or even control. Invasions by exotic species constitute one of the most challenging problems, and alien species make up to 50 per cent of the flora in some national parks. External threats may originate from private landowners surrounding park boundaries, and it may be difficult or impossible to restrict activities on private land. External threats also come in all shapes and sizes—to tackle them all would require significant human and economic resources. Examples of external threats include mining, logging, agriculture, urbanization, water projects, hunting, exotic species, tourism, acid precipitation, and chemical pollution.

Some external threats can be readily identified and even managed, such as forestry activities along a park boundary. However, in other instances the influences of external factors are too distant and diffuse for park managers to control. Global climate change (see Chapter 7) is a good example, and it will obviously have serious implications for protected areas. On the one hand, protected areas will have a huge role to play in terms of their *hospital role* (Box 14.10) in helping sequester carbon from the atmosphere. On the other hand, the *bank role*,

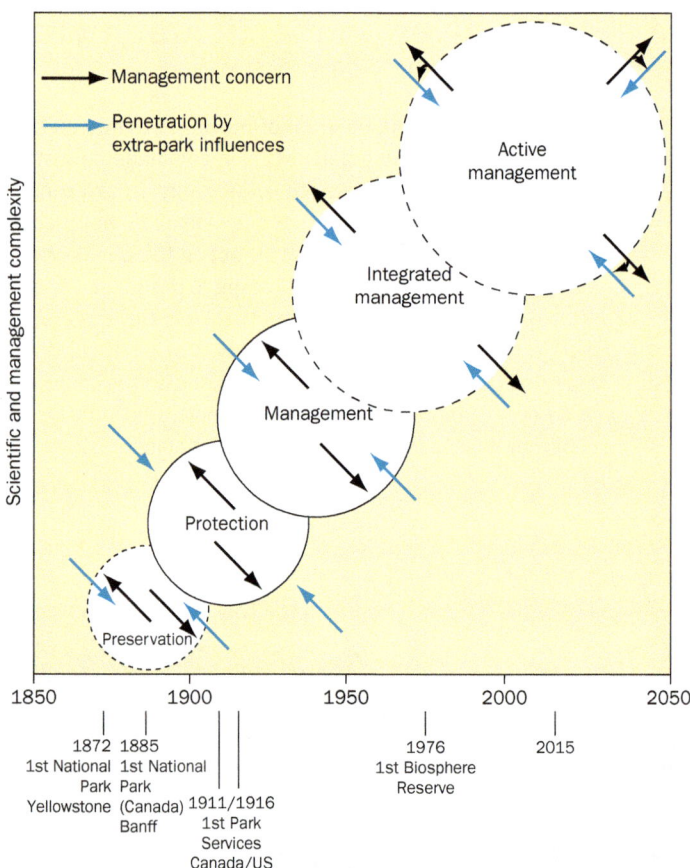

Figure 14.9 | Protected area: evolving relationships from isolation to active management. The circles represent the growing size of the protected system over time. Boundaries (circle circumferences) were initially of little importance but assumed greater significance in the protection and management phases. It is now realized that for park management (arrows) to be effective, it must pay equal attention to environmental changes outside park boundaries and take an active role in managing park ecosystems within boundaries. *Source: Augustine and Dearden (2012).*

Environment in Focus

Box 14.12 The Role of Parks in Endangered Species Protection: Wood Buffalo National Park

Straddling the Alberta–Northwest Territories boundary, Wood Buffalo covers 44,807 square kilometres. The park is both a World Heritage Site and a Ramsar Site. Ramsar Sites are wetlands of global significance. Wood Buffalo contains critical habitat for two endangered species: North America's largest terrestrial mammal, the bison, and the tallest bird, the whooping crane.

> We came to places where, as far as the eye could see, untold thousands [of buffalo] were in sight; the country being fairly black with them . . . these immense herds were moving north and there seemed no end to them.
>
> —Cecil Denny in Saskatchewan, 1874

Bison: The image of vast herds of bison ranging back and forth along the Great Plains of North America is one never to be seen again. With up to 60 million animals, bison herds probably constituted the greatest large-mammal congregations that ever existed on Earth and were important in the subsistence lifestyles of many Aboriginal peoples in western Canada. But by the 1860s, the bison had been extirpated from the plains of Manitoba. As American Indians flooded into Canada to seek the protection of the Great White Mother (Queen Victoria), the pressure on the remaining herds increased dramatically, and the wild bison herds were extirpated from the Canadian Prairies.

> All through today's journey, piled up at the leading stations along the road, were vast heaps of bones of the earliest owners of the prairie—the buffalo. Giant heads and ribs and thigh bones, without one pick of meat on them, clean as a well-washed plate, white as driven snow, there they lay, a giant sacrifice on the altar of trade and civilization.
>
> —traveller on the Canadian Pacific Railway, 1888

Several remnants remained, however. A small number had been protected by the earlier establishment of Yellowstone National Park. Yellowstone is the only place where wild, free-ranging plains bison have survived since colonial times. Banff also had a growing population kept as a tourist attraction in an animal compound. In addition, two remnants had been brought together by an American rancher. The herd was bought by the Canadian government, and the 703 animals were transported to a national park (created for that purpose) adjacent to the railway near Wainwright, Alberta. In the mid-1920s, the herd, then numbering 6,673, was relocated to Wood Buffalo National Park.

These were plains bison (*Bison bison*). Less well known are their non-migratory, taller, and darker cousins, the wood bison (*Bison bison athabascae*). These bison were once widely distributed from the aspen parklands of Saskatchewan and Alberta to the eastern slopes of the Rockies and British Columbia and north to the coniferous forests of the Mackenzie Valley. They are endemic to Canada. Estimates suggest that more than 168,000 wood bison were once in Canada. The wood bison was never as abundant as the plains bison, and by 1891 fewer than 300 of them remained. Wood Buffalo National Park was established at least partly to protect this remnant, and by 1922 the herd had grown from 1,500 to 2,000 animals. Shortly thereafter, the herd of plains bison was imported from Wainwright, and interbreeding led to the disappearance of the distinctive wood bison characteristics. Wood bison were believed to have become extinct.

In 1957, however, an isolated group of wood bison was located in a remote area of the vast park. This herd was relocated to guard against further interbreeding within Wood Buffalo. Some animals were removed to the Mackenzie Bison Sanctuary in the North; the herd now numbers more than 200, and individuals have expanded their range outside the sanctuary. Other animals were removed to Elk Island National Park near Edmonton, where their numbers have to be closely controlled because of the small area available. This herd has provided animals for satellite herds in the Yukon, the Northwest Territories, northwestern Alberta, and Manitoba. Some have even been sent to repopulate parks in Russia. As of 2006 an estimated 4,188 wood bison lived in seven free-ranging, disease-free herds; 6,216 animals in four diseased, free-ranging herds; and 1,029 animals in captive conservation (public and private) and research herds. Two wild herds exceed the minimum viable population of 400 individuals. In 1988, COSEWIC downlisted the wood bison from endangered to threatened. In May 2000, the species was reassessed; its status did not change.

The dangers for the bison are not yet over, however. When the plains bison were imported from Wainwright, they brought with them bovine diseases such as brucellosis and tuberculosis. These diseases have already taken a toll on bison populations—from highs of more than 12,000 animals, they dropped to a quarter of this number by the early 1990s. Diseases afflicting the bison have raised concerns from the agricultural sector. Bison represent the last focus for both diseases in Canada, and as agriculture has impinged on the western boundary of Wood Buffalo, farmers are concerned that domestic stock will become infected. This has led to calls from the agricultural lobby for elimination of the herd.

Several factors besides disease also threaten bison populations. The Peace–Athabasca Delta, for example, supported the highest concentrations of bison during the twentieth

The vast herds of plains bison had been extirpated from Canada until efforts were made to reintroduce them from the US and eventually transport them to Wood Buffalo National Park, where they mixed with the wood bison population.

century. However, since the completion in 1968 of the Bennett Dam upstream in British Columbia, water levels on the Delta have fallen considerably, causing habitat changes that have negatively affected many animal species, including the bison. A more recent impact on the Delta has been changes in the Athabasca River system as a result of extractions of water for oil production at the oil sands in Alberta, discussed in Chapter 12. This impact from outside Wood Buffalo National Park again emphasizes the need for an ecosystem-based perspective on park management (Chapter 6).

Whooping crane: Unlike the bison, whooping cranes (*Grus americana*) were never numerous. Historical accounts suggest a population of 1,500. What they lacked in numbers, they made up for in presence. More than 1.5 metres high and pure white except for black wing tips, black legs, and a red crown, with wing spans in excess of two metres, these majestic birds migrate annually from wintering grounds on the Gulf of Mexico coast of Texas to the Northwest Territories. These wintering grounds are all that remain of a winter range that included marshes from southern Louisiana into central Mexico, and at one time these birds had a breeding summer range that stretched from New Jersey in the east to Salt Lake City in the west and as far north as the Mackenzie Delta and (see Figure 14.10). Requiring undisturbed breeding habitat, the cranes soon declined under the expansion of agriculture. Unrestricted hunting along their long migration routes also contributed to the decline. By 1941, there were only 22 whooping cranes left.

The governments of the US and Canada agreed to a joint program to try to save the species from extinction. They used the 1916 Migratory Bird Treaty between the US and Canada to stop legal hunting. In 1937, the US government bought the Aransas National Wildlife Refuge to protect the wintering habitat on the Gulf coast. In 1954, the only known nesting area was discovered in the northern part of another protected area, Wood Buffalo. Finding the breeding grounds enabled direct human interventions, such as artificial incubation of eggs. Whooping cranes generally lay two eggs, but usually only one chick survives. A captive propagation program in the 1960s and 1970s moved one of the eggs for incubation. By 2008, the North American population of wild whooping cranes had risen

Figure 14.10 | The original range of the whooping crane in recent times.

The whooping crane, the tallest North American bird and one of the rarest, makes its habitat in muskeg, prairie pools, and marshes.

to 388 including introduced populations, with 270 in Canada. The species is currently listed as endangered by COSEWIC.

If you turn back to the discussion on factors influencing the vulnerability of species to extinction earlier in the chapter, you will see that the whooping crane possesses many of the characteristics that make species vulnerable. They are especially at risk because of their long migration and the vulnerability of their wintering grounds to both natural (e.g., hurricanes) and human-caused destruction.

An interesting dilemma arose with their recovery program. As mentioned above, the cranes lay two eggs, and usually only one chick survives. As part of the recovery program, second eggs were removed from the nests and hatched separately. However, Parks Canada, in an effort to maintain natural processes in Wood Buffalo, disallowed the removal of the second egg. This has slowed down the population growth rate (Boyce et al., 2005). What would you do—maintain natural processes or speed up recovery?

providing refuge for natural populations, will be highly vulnerable to the changes. Protected area networks must be made as resilient as possible against these changes. One main mechanism for doing this is through large-scale bioregional planning illustrated by the Yellowstone-to-Yukon initiative discussed below, which emphasizes connectivity, especially north–south connectivity, among protected areas. We will also require new protected areas that help to facilitate migration, provide source populations, and offer suitable habitat for incoming populations. Including private lands in planning will also be important. The biodiversity implications of climate change are likely to be especially severe in the oceans, yet Canada has created very few marine protected areas, let alone functioning networks of marine protected areas, as discussed in Chapter 8.

Suffling and Scott (2002) used climate change models to indicate some possible effects on the national parks system over the next century. They conclude that all regions and parks will be dramatically affected and suggest that Parks Canada will need to intervene in park ecosystems to a greater degree in order to maximize the capacities of ecosystems and species to adapt to climate change. A survey of protected area jurisdictions in Canada found that three-quarters of the agencies already reported climate change impacts and that 94 per cent felt that climate change will significantly alter protected areas planning and policy over the next 25 years. Furthermore, 91 per cent felt that they did not have the capacity to deal with climate change issues (Lemieux et al., 2011). Hannah (2003) suggests that one of the most important steps in dealing with global change in protected areas is to close this management gap before it becomes too wide to bridge. Unfortunately, in almost all provincial jurisdictions in Canada, politicians have been driving things the other way by consistently cutting the funding available to park agencies. One often overlooked role for protected areas in addressing climate change is their *schoolroom role* (Box 14.10), making visitors more aware of the challenges of global change and of the things they can do to help. One of these things is to help persuade politicians to restore funding to park programs.

The idea of park managers actively intervening in park ecosystems rather than leaving change to the vagaries of nature is known as active management. It recognizes that the human forces of change are so prevalent throughout the landscape that even parks are affected. Active management activities include habitat restoration, creation of wildlife corridors, reintroduction of extirpated species (Table 14.3), prescribed burning, and management of hyper-abundant species, such as culling white-tailed deer populations in Point Pelee National Park in Ontario. However, because of limited knowledge of ecosystem processes, considerable debate often occurs among scientists regarding how such programs should be implemented.

Effective management of threats originating from outside parks requires an ecosystem-based approach (Chapter 6), combined with methods that protect wildlife resources along ecosystem rather than legal/political boundaries. Such approaches attempt to mitigate external threats while also counteracting the forces of fragmentation.

Fragmentation

Parks are increasingly becoming islands of natural vegetation totally surrounded by human-modified landscapes, as

Building greater awareness among visitors of the role of national parks in society is a central facet of sound management. Parks Canada has developed some of the best interpretive facilities in the world. This is the visitor centre at Greenwich in PEI National Park.

Logging on the border of Pacific Rim National Park Reserve shows both the impacts of activities outside park boundaries as well as the fragmentation of habitat that follows. Many parks are now islands of natural habitat with no connection to natural habitat elsewhere.

illustrated by the Riding Mountain case discussed in Box 14.13. Studies of Fundy National Park in New Brunswick showed that only 20 per cent of the surrounding area remained in forest patches large enough to be 500 metres from disturbed areas. This situation creates several problems, since many animal species and some bird species cannot cross modified landscapes. As a result, they become an isolated breeding population, leading to genetic inbreeding and a higher susceptibility to extinction. This raises two questions: (1) How many individuals are necessary to ensure the long-term survival of a species? (2) How large an area of habitat is required to sustain the population?

The first question, related to the **minimum viable population** (MVP) of a species (i.e., the smallest population size predicted to have a very high chance of persisting for the foreseeable future), can be estimated using genetic and demographic models. Estimates of MVP are then multiplied by the area required to support each animal. In western Canada, for example, calculations suggest that 15,000 km^2 would be required to support a viable wolf population. A study by Landry et al. (2001: 19) concludes that 'the small size, high visitation rates, and ecological isolation of southern parks mitigate against minimum viable populations of wolves, black bears, and grizzly bears being maintained there . . . most of Canada's national parks cannot indefinitely sustain [MVPs] of these large carnivorous mammals.' Similar conclusions have also been reached regarding large herbivores in many parks. Flanagan and Rasheed (2002), for example, used two different models, and both models indicated that caribou in Jasper National Park will become extinct within the next 40 years if conditions do not change.

MVP analysis has now been broadened to include a wider range of factors and is being replaced by **population viability analysis** (PVA), a means of quantifying risk of extinction, elucidating factors contributing to numerical decline, and helping in prioritizing conservation actions among endangered species and populations. These methods enable assessment of extinction risk relative to uncertainty in source data and under a variety of environmental or management scenarios.

Island biogeography theory also suggests that the number of species surviving on an island represents an equilibrium between species immigration and extinction, and this depends on its distance from a colonizing source. In theory, the number of species on an island will be greater if the island is large and sources of immigration are close. There is some debate as to the veracity of this assertion, but in association with other research it has given rise to principles regarding reserve design. In general, (1) blocks of habitat close together are better than blocks far apart; (2) habitat in contiguous blocks is better than fragmented habitat; and (3) interconnected blocks of habitat are better than isolated blocks.

The importance of connectivity can be seen in the efforts of Parks Canada to mitigate the impacts of the Trans-Canada Highway cutting through Banff National Park. The highway was a major sink for wildlife populations, with more than 800 collisions per year. Following the installation of 22 underpasses and two overpasses in the late 1990s, wildlife mortality overall was reduced by 80 per cent. More than 90,000 uses of the underpasses and overpasses by 10 species of large mammals have been recorded in the past 10 years. The most significant factor affecting the amount of wildlife use has nothing to do with the crossing design but rather whether humans also use the crossing. As a result, Parks Canada is now concentrating on managing human use of the crossings. Given the success of the Banff experiment, Waterton Lakes National Park in the southwest corner of Alberta is now planning to install four tunnels to help in the survival of long-toed salamanders, often squished by the hundreds as the 13-centimetre-long amphibians try to cross the road to and from their breeding ground.

These examples show how connectivity can be improved with enough scientific information and resources to build mitigating structures. However, overall our parks are too small, too few, and too far apart to sustain populations of many species throughout the next century. Attention is being directed towards ways of linking the parks through corridors of natural habitat. One such scheme would extend American parks such as Yellowstone north through the Canadian Rockies and into Yukon and Alaska. There are 11 national parks and dozens of state, provincial, and territorial parks in the Yellowstone-to-Yukon corridor. One wolf marked for tracking in Montana was actually shot on the Alaska Highway along the corridor. Many other threats exist besides hunting along such corridors, including mining, industrial development, and resort and housing developments. It is much easier to maintain connectivity before development starts than to restore it afterward.

These kinds of bioregional schemes explicitly acknowledge the limitations of park systems and encourage a more integrated perspective towards resource management on lands outside the parks that involves other actors, such as landowners and private foundations. This approach is often called stewardship and refers, in general, to many different activities that can be undertaken to care for the Earth. In the context of protected areas, it generally means encouraging landowners to modify their activities to help protect ecosystems. In practice, stewardship takes many forms, including:

- landowners voluntarily restricting damaging use of land, planting native species rather than exotic ones, and placing protective covenants on their land;
- community members contributing to wildlife monitoring programs, providing passive education for tourists and visitors, and participating in collective restoration;
- park visitors voluntarily choosing to avoid hikes along sensitive trails or participating in park host programs;
- corporations introducing sustainable land practices that reduce damage to wildlife habitat.

A diversity of land trust and other conservation organizations is emerging across Canada, using a variety of tools to conserve lands under private ownership. Over a thousand stewardship groups and over one million people in Canada participate in thousands of initiatives on private and public lands. Stewardship Canada is an on-line portal that provides access to many resources related to stewardship programs in Canada (www.stewardshipcanada.ca/). The Nature Conservancy of Canada—one of many conservation organizations—has protected almost 1 million hectares since 1962. Ducks Unlimited has been responsible for the protection of more than 2.5 million hectares of Canadian wetlands since 1938. Smaller, provincially based and local land trusts are also increasing. In the southern Vancouver Island–Gulf Islands area, where natural heritage is fading quickly and land prices are increasing, at least eight land trusts are operating to protect endangered spaces, most of which have been established since 1998. Government agencies such as Environment Canada and various provincial ministries are also embracing stewardship. For example, numerous funding programs exist for community stewardship, and some ministries publish guides or maintain websites to educate and to support local initiatives. Legislative changes, particularly tax deductions, have also been implemented, resulting in incentives and encouragement for ecological gifts and donations.

These initiatives are critically important for the future of conservation in Canada and will play a significant role in restoring connectivity among other protected areas. Whitelaw and Eagles (2007) explain how this might work in southern Ontario. Private lands will never replace the role played by strictly protected areas, but they do play an essential role in 'gluing together' the larger wilderness areas set aside in government parks. The international Biosphere Reserve Program is one of the best-known initiatives promoting greater stewardship surrounding protected areas (Box 14.13).

Stakeholder Interests

Balancing the interests of the range of stakeholders that may be affected by the establishment and/or management of parks is a formidable challenge. Private landowners, local communities, Aboriginal peoples, industry, tourists, conservation organizations, and government agencies all have an influence over how parks are managed. Environment Canada (2006b) reports that 13 out of 15 jurisdictions (which includes two federal agencies) provide opportunities for community involvement in most or all of their protected areas. Community participation is enshrined in the legislation governing eight protected areas agencies.

The totem poles of Ninstints, an abandoned Haida village on Gwaii Hanaas (Moresby Island in the Queen Charlotte Islands), give some impression of the Haida's spiritual connection with the environment. The village is now part of Gwaii Hanaas National Park Reserve and is co-managed by Parks Canada and the Haida.

Environment in Focus

Box 14.13 Riding Mountain National Park and Biosphere Reserve

Riding Mountain National Park illustrates the kinds of external pressures that threaten the ecological integrity of many of our national parks. The park is located on the Manitoba Escarpment and is an isolated boreal forest area completely surrounded by agricultural land. Nonetheless, the 3,000-square-kilometre park provides habitat for some 5,000 elk, 4,000 moose, more than 1,000 black bears, and populations of cougars and wolves.

Large mammal populations have become increasingly threatened by the intensification of agricultural activities surrounding the park. Between 1971 and 1986, the amount of land under agriculture within 10 kilometres of the boundary increased from 77 to 93 per cent. Not only did the amount of farmland increase, the intensity of use did as well, with a 42 per cent increase in cropland area in the same zone over the same period. During the same 15-year period, the area of woodland declined by 63 per cent within a 70-kilometre radius of the boundary, the volume of agricultural fertilizers used quintupled, and pesticide expenditures indicate an increase of 744 per cent in pesticide applications.

Agricultural expansion is not the only challenge confronting park wildlife. Until recently, bear-baiting was permitted directly on the park boundary. Farmers conditioned bears to feed from barrels full of meat. In the hunting season, the bears formed easy targets for 'sportsmen'. Some 70 'bear-feeding' stations existed around the park, causing unnatural bear distributions, very large bears, irregularities in breeding behaviour, and death. On average, 122 bears are killed each year in this way. Scientists suggest that these mortality levels cannot be maintained if the bear population of the park is to survive. Farmers are now required to move bait barrels back from the park boundary itself. However, bears are highly mobile animals and will have little difficulty in locating the barrels.

These types of external pressures can have significant impacts on biodiversity within park boundaries. The Biosphere Reserve Program of the United Nations Educational, Scientific and Cultural Organization (UNESCO) is one of the most highly touted means of dealing with such external threats to protected areas. The concept behind the program is sound. Biosphere reserves exist to represent global natural regions and should consist of a protected core area, such as a national park, surrounded by a zone of co-operation where socio-economic activities may take place but are modified to help protect the integrity of the core area. The reserves also have important educational and scientific roles. Unfortunately, no legislation ensures co-operation on the privately owned lands in the zone of co-operation. Continued hunting around the park boundary at Riding Mountain National Park, designated as a biosphere reserve in 1986, graphically illustrates the need for landowners to co-operate. Similar challenges face other biosphere reserves, such as Georgian Bay Islands National Park in Ontario, the Niagara Escarpment Biosphere Reserve in Ontario, the Greater Fundy Ecosystem in Nova Scotia, and Waterton Lakes National Park in southern Alberta, Canada's oldest biosphere reserve.

Aboriginal peoples have a particularly powerful role in the designation and management of many parks and are considered another level of government rather than a stakeholder. An amendment to the National Parks Act in 1972 created a special category of park, the *national park reserve*, which does not prejudice future land-claim negotiations. Several large parks have been created in the Arctic as a result. In the south, on the other hand, where the provincial governments have jurisdiction over the land base, progress has been slow. It was not until after protests and several court cases that the legitimacy of Aboriginal claims over land and resources in southern Canada was taken seriously. More than 50 per cent of the land area in Canada's national park system has now been protected as a result of Aboriginal peoples' support for conservation of their lands, and this proportion will only increase in the future as the remaining Native land claims are settled.

There are also significant additions to protection at the provincial level through Aboriginal commitments. In Manitoba, the government announced in 2011 that it would support the efforts of one Aboriginal group, the Poplar River First Nation, which developed a land-use plan for its territory on the east side of Lake Winnipeg that resulted in the protection of 807,650 hectares of boreal forest and wetlands.

Caribou from the Porcupine caribou herd at Vuntut National Park in Yukon.

This brings the total of land protected in that province to over 10 per cent of its land area.

The goals of Aboriginal peoples and conservationists, however, are not always identical. Aboriginal peoples, for example, retain the right to hunt in many parks. At the moment, there are few restrictions on the size or means of harvest, a source of concern for some conservation scientists. Several parks in both northern and southern Canada have some form of co-management arrangement between Parks Canada and Aboriginal peoples. In general, they have worked to the satisfaction of both groups, but the issue of shared responsibility remains an ongoing challenge.

Guest Statement
Canada's Great Bear Rainforest
Ian McAllister

Environmental campaigns do not originate in boardrooms or around coffee tables. They begin when one person or a small group of people fall head-over-heels in love with a mountain, a river, an animal, or some combination of all three. You can trace the roots of any successful environmental campaign back to the simple emotion of love, and that is why there is such passion in the environmental movement today.

I fell in love with the central coast of British Columbia about 18 years ago. My father had chartered a sailboat to visit the Koeye River Valley, a mysterious estuary halfway up the BC coast where grizzly bears were rumoured to walk white-sand beaches and to fish in waters teeming with salmon. I had managed to squeeze aboard that boat, and by the end of the trip I was hooked. It turned out that the rumours had been true.

At that time, none of the river valleys on the coast had been extensively explored or protected by legislation, yet the majority had already been licensed out to logging companies. In other words, British Columbia didn't know what it was about to lose! Six months after that fateful voyage, along with my wife Karen, my father Peter, and some friends, we launched the campaign to protect the Great Bear Rainforest.

Now, over 20 years later, the campaign to save the Great Bear Rainforest (GBR) is one of the largest environmental movements in Canadian history. It has had a huge impact on the international marketplace for timber and has become a keystone example of how a group of normal individuals can combine their strengths to bring about change. From scientists to celebrities, from letter-writers to lawyers, people from all walks of life joined together in a mutual love of nature and the commitment to help preserve it. I guess it makes sense that we are such a diverse group. After all, diversity is what we strive to protect.

In 2009, approximately 30 per cent of the GBR was formally protected from logging and mining, and while this is considered a significant step forward it still leaves unfinished business in the rain forest. Trophy hunting of large carnivores like grizzly bears and wolves is still commonplace, even within some of the new protected areas, and marine protected areas are only now being considered.

The greatest threat now, however, is coming from the energy sector. Canada has the second largest known oil reserves in the world and China and India want access to it. In order for Canada to meet this demand it would involve expanding the Alberta tar sands, considered the 'dirtiest' oil production process in the world, and building a pipeline to the Great Bear Rainforest where oil tankers would wait for shipping to Asia. This new threat facing the Great Bear Rainforest could bring massive oil spills that would destroy coastal ecosystems and Aboriginal culture.

There's no question that progress has been made, but with threats like oil spills we have a ways to go before success can be achieved. Over the years, we have been called heroes, and we have been called traitors (we've even been called colonialists for renaming the coast the Great Bear Rainforest!). But now, governments and businesses are beginning to listen to our voice and recognize the strength in numbers behind it. And these numbers continue to grow.

Environmental work is one of the most exciting and creative occupations I can think of. It is a constant challenge filled with unrivalled spiritual and physical rewards. Perseverance is a prerequisite, as is a sensitivity to other people's points of view, and the positive result of this is that environmentalism attracts fun-loving, hard-working, passionate individuals who believe very strongly in the value of teamwork.

I didn't know what I was getting into when I hopped aboard that boat over 20 years ago. But that's the best thing about environmental work: you can learn everything you need to know along the way, and I wouldn't trade it for anything.

Ian McAllister is a conservationist and founder of PacificWild.org.

A Biosphere Reserve surrounding Waterton Lakes National Park has been set aside to allow for joint decision-making between park authorities and local landowners. The agreement has been particularly useful in reducing conflicts between grizzly bears and local ranchers.

Implications

There is no debate about whether rates of extinction have increased as a result of human activities over recent times. Some uncertainty still exists, however, regarding the impact of the extinctions. Is there ecological redundancy such that the Earth can afford to lose some species without major impacts? Is the loss of any species as a result of human activities ethically and morally acceptable? There are many unanswered questions regarding the implications of reduced biodiversity.

Extinction is commonly viewed as simply a biological problem. Yet there is a need to link the biological process of extinction with the economic and ethical reasons why extinction is undesirable, the reasons why human-caused extinctions are increasing, and the kinds of measures needed to prevent this from happening. In other words, extinction is not just the domain of biologists but involves consideration from a broad range of perspectives, including all the social sciences, geography, law, and ethics.

Environment in Focus

Box 14.14 What You Can Do for Protected Areas

1. Visit parks and other protected areas often throughout the year. Enjoy yourself. Tell others that you have enjoyed yourself, and encourage them to visit.
2. Always follow park regulations regarding use. Feeding wildlife, for example, may seem kind or harmless, but it can lead to death of the animal.
3. If you have questions regarding the park's management or features, do not be afraid to ask. A questioning public is a concerned public.
4. Many park agencies have public consultation strategies relating to topics ranging from park policy to the management of individual parks. Let them know your interests so that you can be placed on the mailing list to receive more information.
5. Join a non-governmental organization, such as the Canadian Parks and Wilderness Society or Nature Canada, with a strong interest in parks issues.
6. Many parks now have co-operating associations in which volunteers can help with various tasks. Find out whether a park near you has such an organization.
7. Write to politicians to let them know of your park-related concerns.

Summary

1. Extinction levels have reached unprecedented levels. There are several reasons why we should be concerned. Life-supporting ecosystem processes depend on ecosystem components. As we lose components through extinction, these processes become more impaired. We also derive many useful and valuable products from natural biota, including medicines. In addition to these utilitarian reasons, there are ethical and moral reasons why we should be concerned about species extinction.
2. Many factors are behind current declines. The underlying factor is human demand as population and consumption levels grow. Much attention has concentrated on the tropics because of the high biodiversity levels and high rates of destruction there. However, Canada has experienced 13 extinctions and 22 extirpations since European colonization.
3. Main pressures causing extinction include overharvesting, predator control, and habitat change. Habitat change

includes not only physical changes (e.g., conversion of habitat into agricultural land) but also those caused by chemicals and the introduction of alien species.

4. Not all species are equally vulnerable to extinction. Species with specialized habitat requirements, migratory species, species with insular and local distributions, species valued by humans for commercial reasons, animal species with a large body size, species needing a large home range, species not effective as dispersers, and species with low reproductive potential tend to be the most vulnerable.

5. Canada is party to several international treaties for the protection of biodiversity, including the legally binding Convention on Biological Diversity. None of the CBD goals set for 2010 were met, including the overriding mission of slowing down the rate of biodiversity loss.

6. As a signatory to the CBD, Canada was required to introduce legislation to protect endangered species. In 2002, the federal government passed the Species at Risk Act (SARA).

7. The Committee on the Status of Endangered Wildlife in Canada (COSEWIC) is responsible for determining the status of rare species and categorizing them as extinct, extirpated, endangered, threatened, or vulnerable. As of 2011, over 600 species had been classified as at risk. The Committee's assessment is the first step in the process for protecting a proposed species at risk under SARA.

8. For species listed under SARA, recovery and management plans must be developed and implemented, unless the minister responsible feels that recovery is not 'feasible'. By 2011, only 99 out of 282 plans had been completed.

9. Protected areas are one of the key strategies to combat the erosion of biodiversity, both internationally and in Canada. Protected areas fulfill many roles in society, including species and ecosystem protection, maintenance of ecological processes, and as places for recreation and spiritual renewal, aesthetic appreciation, tourism, and science and education in natural outdoor settings.

10. There are many different kinds of protected areas in Canada, including national and provincial parks, wilderness areas, tribal parks, wildlife refuges, ecological reserves, and regional and municipal parks. The amount of protection given to ecosystem components varies among these different types.

11. National parks are outstanding natural areas protected by the federal government because of their ecological importance and aesthetic significance. There are 42 national parks in Canada. The goal is to have at least one national park in each of the 39 regions of the national system plan. The CBD required the establishment of completed networks of protected areas by 2010. At present, Canada's terrestrial system is about 60 percent complete and the marine system about 15 per cent complete. The CBD target is to establish 17 per cent of the land base and 10 per cent of the marine area of each country in protected areas by 2020. Canada currently has about 10 per cent and 1 per cent in these categories.

12. Banff, the first national park in Canada, was established in 1885. Since that time, the national parks fulfilled a dual mandate that required protection of park resources in an unimpaired state but also permitted their use. This conflicting mandate was clarified in an amendment to the National Parks Act in 1988 and further clarified in the National Parks Act of 2000, giving first priority to protecting the ecological integrity of the parks.

13. Management challenges to the national parks system include external threats and fragmentation. An ecosystem approach to management is required to address these challenges by embracing stewardship of park lands.

14. Aboriginal peoples have been integral to the development of many protected areas in Canada and constitute a third level of government that needs to be involved in the designation and management of protected areas in most areas of Canada.

Key Terms

Aboriginal peoples
active management
Convention on Biological Diversity (CBD)
Convention on International Trade in Endangered Species of Wild Fauna and Flora (CITES)
ecologically extinct
endangered
ex situ conservation
ex situ preservation
extirpated
extrinsic values
fragmentation
in situ preservation
intrinsic value
island biogeography
minimum viable population (MVP)
population viability analysis (PVA)
protected areas
Red List
Species at Risk Act (SARA)
stewardship
system plan
threatened
vulnerable

Questions for Review and Critical Thinking

1. Why do you think the signatories to the CBD failed to meet all the 2010 targets? What changes do you think need to be made to meet the targets set for 2020?
2. What are the main reasons why we should be concerned about species extinctions?
3. Why are some species more vulnerable to extinction than others?
4. What is being done to protect endangered species in your province?
5. What are some of the strengths and weaknesses of Canada's Species at Risk Act?
6. What do you think should be the relative importance of the various roles played by protected areas?
7. What different classifications of protected areas exist in your province, and what kinds of protection are offered by these different systems?
8. What is your province doing to achieve the 12 per cent protected area that all jurisdictions in Canada have committed to establishing?

Related Websites

Canadian Council on Ecological Areas
ccea.org

Canadian Environmental Assessment Agency
www.ceaa.gc.ca

Canadian Parks and Wilderness Society
www.cpaws.ca

Committee on the Status of Endangered Wildlife in Canada
www.cosewic.gc.ca

Ducks Unlimited Canada
www.ducks.ca

Environment Canada
www.ec.gc.ca

Global Environment Facility
www.gefweb.org

International Union for Conservation of Nature, Red List
www.iucnredlist.org

Nature Canada
www.naturecanada.ca

North American Waterfowl Management Plan
www.nawmp.ca

Northwest Territories Wildlife Division
www.nwtwildlife.com

Ontario Ministry of Natural Resources
www.mnr.gov.on.ca

Parks Canada
www.pc.gc.ca

Sierra Youth Coalition
www.syc-cjs.org

Species at Risk Act Public Registry
www.sararegistry.gc.ca

Species at Risk, Environment Canada
www.ec.gc.ca/nature/default.asp?lang=En&n=FB5A4CA8-1

United Nations Environment Programme, World Conservation Monitoring Centre
www.unep-wcmc.org

Vancouver Island Marmot Recovery Foundation
www.marmots.org

Western Canada Wilderness Committee
www.wildernesscommittee.org

World Commission on Protected Areas
www.iucn.org/about/union/commissions/wcpa

World Wildlife Fund Canada
www.wwf.ca

Further Readings

Note: This list comprises works relevant to the subject of the chapter but not cited in the text. All cited works are listed in the Bibliography at the end of the book.

Abbey, E. 1968. *Desert Solitaire: A Season in the Wilderness*. New York: Simon and Schuster.

Dearden, P., and S. Langdon. 2009. 'Aboriginal people and national parks', in P. Dearden and R. Rollins, eds, *Parks and Protected Areas in Canada: Planning and Management*, 3rd edn. Toronto: Oxford University Press, 373–402.

Dempsey, J., and P. Dearden. 2009. 'Stewardship: Expanding ecosystem protection', in P. Dearden and R. Rollins, eds, *Parks and Protected Areas in Canada: Planning and Management*, 3rd edn. Toronto: Oxford University Press, 432–54.

Dudley, N. 2008. *Guidelines for Applying Protected Area Management Categories*. Gland, Switzerland: IUCN.

McNamee, K. 2009. 'From wild places to endangered spaces', in P. Dearden and R. Rollins, eds, *Parks and Protected Areas in Canada: Planning and Management*, 3rd edn. Toronto: Oxford University Press, 24–55.

Parks Canada Agency. 2000. *Unimpaired for Future Generations? Protecting Ecological Integrity with Canada's National Parks*, vol. 2, *Setting a New Direction for Canada's National Parks*. Report of the Panel on the Ecological Integrity of Canada's National Parks. Ottawa: Minister of Public Works and Government Services.

———. n.d. *National Park System Plan*. Ottawa: Minister of Supply and Services.

Rivard, D.H., et al. 2000. 'Changing species richness and composition in Canadian national parks', *Conservation Biology* 14: 1099–109.

Willcox, L., and P. Aengst. 1999. 'Yellowstone to Yukon: Romantic dream or realistic vision of the future?', *Parks* 9: 17–24.

Be the change you want to see in the world.
—Mahatma Gandhi

We must always change, renew, rejuvenate ourselves; otherwise we harden.
—Johann Wolfgang von Goethe

Part E
Environmental Change and Challenge Revisited

Change is ubiquitous. This book provides an overview of environmental change and challenge in Canada. Change occurs as a result of both natural and human-induced pressures, and it is often difficult to determine the balance between them. However, it seems that human-induced changes are becoming the dominant driving factor for many aspects of environmental change. In many cases, these changes, such as extinction and climate change, are irreversible. Current activities are serving to impoverish the planet for future generations.

The preceding chapters have emphasized the need to understand the ecological aspects of environmental change, along with the various management approaches that may be useful. Reading this book should enable you to understand the background to many of the environmental problems we face and also to appreciate the different management approaches to their resolution. In each chapter, we have attempted to make you aware not only of the nature of the challenges being faced but also of some of the solutions being tried.

This final part contains one chapter, and its main focus is on solutions. We provide an assessment of progress at the global and national levels in coming to terms with environmental change. We end with some suggestions on the kinds of actions that you can take. We hope that you will not only read the chapter carefully but also take action to try to improve your balance sheet with or your footprint on the environment! Your efforts, combined with those of thousands of others acting individually, can make substantial changes in the environment of tomorrow.

Chapter 15
Making it Happen

Learning Objectives

- To identify some of the main global responses to environmental degradation.
- To understand some of the main Canadian responses to environmental degradation.
- To place Canada within the global context for environmental response.
- To assess how important environment is to the administration of your university.
- To make better decisions to minimize your impact on the environment.
- To use your influence more effectively to benefit the environment.
- To clarify what 'the good life' means for you.

When I call to mind my earliest impressions, I wonder whether the process ordinarily referred to as growing up is not actually a process of growing down; whether experience, so much touted among adults as the thing children lack, is not actually a progressive dilution of the essentials by the trivialities of life.

—Aldo Leopold, *A Sand County Almanac* (1949)

Introduction

Aldo Leopold, one of the greatest conservation thinkers and writers, points out that as we get older and our lives get busier, we often get distracted from the important things in life, like protecting the environment. Nearly everyone says that environmental protection is important, but most devote minimal effort to actually doing anything about it. We are all members of NATO: No Action, Talk Only. And the same is true of our country. On paper, Canada has impressive legislation, policies, strategies, and action plans regarding the environment. Sadly, the translation of these into 'on-the-ground' improvements is often chronically under-resourced. Many examples have been cited in this book, ranging from lack of resources to implement Canada's Oceans Strategy, as mandated under the Oceans Act (Chapter 8), to the failure to follow through on how to meet our obligations under the Kyoto Protocol (Chapter 7).

In this chapter, we provide a brief overview of global and Canadian responses to environmental change. But governments are only part of the answer. This final chapter rests on the firm conviction that individuals can make a *significant difference* in how the environmental challenges presented in this book will develop over the next decade, if we are aware of the problems and are willing to do something about them. This chapter provides some ideas about how *you* can become involved in creating change.

Perspectives on the Environment
The Pattern of Change

We can say this much with confidence: When change works, it tends to follow a pattern. The people who change have a clear direction, ample motivation, and a supportive environment.

—Heath and Heath (2010: 255)

Global Perspectives

The previous century witnessed many changes. It may be characterized as an age of diminishing imperial powers, ongoing wars, atomic bombs, the harnessing of the entire globe into an interconnected economic system, rising consumer demands, and an exploding human population. This century will witness the continuation of some of these trends, but many scientists seem convinced that global climatic change, water shortages, biological impoverishment, declining food yields per capita, desertification, pollution, and overpopulation will constitute the backdrop for events.

Many of these trends are driven by consumption of material goods, which emerged over the last couple of decades of the past century as the dominant international ideology. From its heartland in Europe, North America, and Japan, the globalization of consumption will be one of the main developments of the next couple of decades, if not the entire century. Although population growth is still a concern, convincing signs point to falling rates of increase and the stabilization of populations, probably within the next 50 years. In contrast, consumption knows no bounds. Indeed, our whole global economic system is focused on increasing consumption levels. At the individual level, our psyches are dominated by images of the consumer goods we hanker for. The shopping mall has become the new place of worship.

Roughly one-quarter of humanity is now within this consumer class, and this number is divided more or less equally between those in developed countries and the rapidly increasing numbers of consumers in developing countries such as China and India.

The impacts of growing consumer demands are far-reaching. The Millennium Ecosystem Assessment (2005), introduced in Chapter 1, provided a detailed review of the state of planetary ecosystems, calculating that 15 of the 24 major ecosystem services that support humanity, such as climate regulation, water provision, and soil production, have been pushed beyond their limits and are in a degraded state. The UNEP (2010) estimates that the services provided by ecosystems are worth between $21 and $72 trillion per year

Until we find an effective way to incorporate these ecosystem values into the decision-making process, they will continue to be eroded. There is clear scientific evidence of the deterioration of global ecosystems, but degradation cannot be seen outside the human context that drives the activities causing degradation. The Millennium Ecosystem Assessment's framework shown in Figure 15.1 illustrates these connections. The box in the bottom left, the ecosystem services provided by functioning ecosystems, is influenced by the direct drivers of change promoting indirect drivers that affect ecosystems and reduce human well-being. In turn, the desire for human well-being is the force behind the indirect drivers. The pincers on the arrows show links that are amenable to strategic interventions.

The **Millennium Development Goals (MDGs)** introduced in Chapter 1 are globally accepted goals for development. The target date for meeting the goals is 2015, and every few years the UN issues progress reports outlining some of the positive developments:

- The proportion of people living in extreme poverty fell from nearly a third to less than one-fifth between 1990 and 2004. If the trend is sustained, the MDG poverty reduction target will be met for the world as a whole and for most regions.
- The number of extremely poor people in sub-Saharan Africa has levelled off, and the poverty rate has declined by nearly six percentage points since 2000. Nevertheless, the region is not on track to reach the goal of reducing poverty by half by 2015.
- Progress has been made in getting more children into school in the developing world. Enrolment in primary education grew from 80 per cent in 1991 to 89 per cent in 2008. Most of this progress has taken place since 1999.

Resist the temptation to give in to the consumer binge.

- Women's political participation has been growing. Between 1995 and 2010 there was a 73 per cent increase in the share of women filling political office, from 11 per cent to 19 per cent worldwide (UNDP, 2010: 70).
- Child mortality has declined globally, and life-saving interventions are proving effective in reducing the number of deaths caused by the main child killers, such as measles.

The United Nations Development Programme's *Human Development Report* for 2011 points out that tremendous improvements have been made in the living conditions in many poor nations over the last few decades, but that a major gap remains between these achievements and meeting the Millennium Development Goals. Furthermore, there are close links between the MDGs and the environment (Table 15.1),

Changes in drivers that indirectly affect biodiversity, such as population, technology, and lifestyle (upper right corner of figure), can lead to changes in drivers directly affecting biodiversity, such as the catch of fish or the application of fertilizers (lower right corner). These result in changes to ecosystems and the services they provide (lower left corner), thereby affecting human well-being. These interactions can take place at more than one scale and can cross scales. For example, an international demand for timber may lead to a regional loss of forest cover, which increases flood magnitude along a local stretch of a river. Similarly, the interactions can take place across different time scales. Different strategies and interventions can be applied at many points in this framework to enhance human well-being and conserve ecosystems.

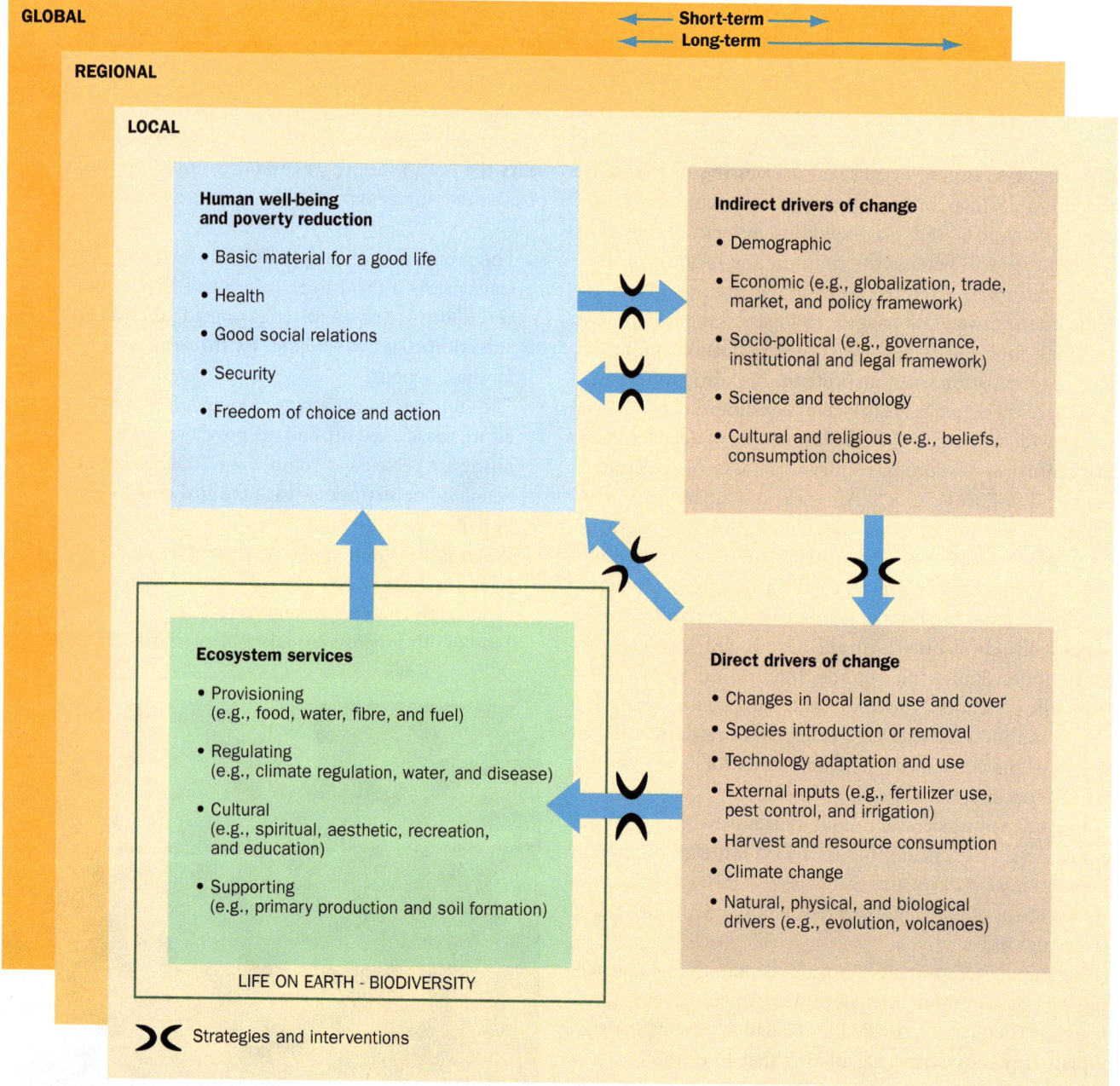

Figure 15.1 | Conceptual framework of the Millennium Ecosystem Assessment. *Source: Millennium Ecosystem Assessment (2005: vii).*

Table 15.1 | Links between the MDGs and the Environment

1.	Eradicate extreme poverty and hunger	Livelihood strategies and food security of the poor often depend directly on healthy ecosystems and the diversity of goods and ecological services they provide. Natural capital accounts for 26 per cent of the wealth of low-income countries. Climate change affects agricultural productivity. Ground-level ozone damages crops.
2.	Achieve universal primary education	Cleaner air will decrease the illnesses of children due to exposure to harmful air pollutants. As a result, they will miss fewer days of school. Water-related diseases such as diarrheal infections cost about 443 million school days each year and diminish learning potential.
3.	Promote gender equality and empower women	Indoor and outdoor air pollution is responsible for more than 2 million premature deaths annually. Poor women are particularly vulnerable to respiratory infections, since they have high levels of exposure to indoor air pollution. Women and girls bear the brunt of collecting water and fuelwood, tasks made harder by environmental degradation, such as water contamination and deforestation.
4.	Reduce child mortality	Acute respiratory infections are the leading cause of death in children. Pneumonia kills more children under the age of five than any other illness. Environmental factors such as indoor air pollution may increase children's susceptibility to pneumonia. Water-related diseases such as diarrhea and cholera kill an estimated 3 million people per year in developing countries, the majority being children under the age of five. Diarrhea has become the second biggest killer of children, with 1.8 million children dying every year (almost 5,000 per day).
5.	Improve maternal health	Indoor air pollution and carrying heavy loads of water and fuelwood adversely affect women's health and can make women less fit for childbirth and at greater risk of pregnancy complications. Provision of clean water reduces the incidence of diseases that undermine maternal health and contribute to maternal mortality.
6.	Combat major diseases	Up to 20 per cent of the total burden of disease in developing countries may be associated with environmental risk factors. Preventive environmental health measures are as important and at times more cost-effective than health treatment. New biodiversity-derived medicines hold promises for fighting major diseases.
7.	Ensure environmental sustainability	Trends in environmental degradation must be reversed in order to sustain the health and productivity of the world's ecosystems.
8.	Develop a global partnership for development	Poor countries and regions exploit their natural resources to generate revenue and make huge debt repayments. Unfair globalization practices export their harmful side effects to countries that often do not have effective governance.

Source: Adapted from UNDP (2007).

and achieving the goals will not be possible without giving greater attention to environmental protection. However, a team of Canadian researchers recently looked at this link between environment and human welfare and questioned the relationship (Raudsepp-Hearne et al., 2010). As shown above, several indicators suggest that overall human welfare is increasing on a global scale. If this is so and yet global ecosystems continue to be degraded, is there really any relationship between livelihoods and environmental integrity?

To examine the relationship in more detail, the researchers advanced four explanations. First, humans are really worse off than we realize; second, well-being mostly depends on food, and food production has been increasing; third, humans have averted the worst consequences of environmental degradation through technology; and, finally, due to a time lag, the worst is yet to come in terms of experiencing the full impacts of global degradation. They found little support for the first explanation, although other scientists disagree, pointing out that the data were based almost entirely on the Human Development Index, which might not be the most effective comparison. The other three explanations all had some support, but the main finding was in the last one—the worst is yet to come. In other words, humans have not yet felt the true impacts of environmental degradation, but have continued to enjoy the benefits of environmental over-exploitation. Turn back to Figure 1.18. It indicates we are in the space of the 'accumulated ecological debt' and waiting for the full repercussions of the debt to be realized.

Some environmentalists have pointed out that, while having Millennium Development Goals is a good thing, these goals only address the challenge from one perspective. What we really need are Millennium Consumption Goals that identify the most important cuts in consumption that are needed. Infinite development cannot occur in a finite world without cuts in consumption. Suggested goals include:

- Halve total energy use by 2025.
- Halve the fossil fuels used by 2020.
- Halve household energy use by 2020.
- Cut military spending by 75 per cent by 2025.
- Replace GNP with a genuine progress indicator or well-being index by 2015.
- Halve obesity and overweight rates by 2020.
- Produce half of food organically by 2020.

- Reduce consumption of animal products by 50 per cent by 2020.
- Increase local resilience of food supply, producing more crops locally where appropriate.
- Halve the work week from the current 40+ hours per week to 20 hours per week.
- Better distribute wealth by raising taxes on the wealthiest members of society.
- Double the rate of use of non-motorized transport (bikes, walking, etc.).
- Increase density of suburban housing by 50 per cent by 2020.
- Guarantee access to health care for all.

Although some landmark international agreements have been signed in the past 20 years, such as the Kyoto Protocol, the Convention on Biological Diversity, and the Stockholm Convention on Persistent Organic Pesticides, progress on many of them has been slow. For example, the Convention on Biodiversity requires signatory nations to establish networks of marine protected areas by 2012. Yet, at the current rate this will not be achieved until 2085, or three decades *after* the predicted collapse of the global fisheries that marine protected areas are intended to forestall (UNEP, 2007). In contrast, global change is continuing apace. In 2007, the Arctic ice retreated to a record-setting minimum area of about 4.3 million km^2. This is a loss of 10.4 million km^2 from the mid-March maximum. Another record loss of ice occurred in 2011 when the total coverage in late summer once again went below 5 million km^2. The five largest thaws recorded since 1970 have occurred in the last five years and scientists are predicting that for the first time in history, the North Pole could be ice-free by as early as 2015.

However, progress is occurring in some areas, such as product certification as discussed in Chapters 8 and 9. By 2011, the Forest Stewardship Council (FSC) had certified more than 140 million hectares of commercial forest in 80 countries and the Marine Stewardship Council had certified 131 fisheries offering 500 certified seafood products. Furthermore, significant barriers to international and national efforts to enact environmentally sound practices are increasingly being challenged as 'protectionist' (Box 15.1).

Obviously, given the very different conditions prevailing in different parts of the world, different solutions are needed to find the most effective way to improve resilience and achieve sustainability. For some countries, increased consumption is required if people are to meet their basic needs. For others, drastic reductions in consumption are necessary. Such reduction, however, does not necessarily imply a reduction in quality of life. The Human Development Index (HDI) of the United Nations, for example, discussed in Chapter 1, takes into consideration education, longevity, and living standards rather than just GNP as measures of development. For very poor people, even a small increase in energy consumption can make a major difference to their living standard. It is difficult to spend long hours studying at night, for example, if no electricity is available. This retards educational levels, which in turn hold back economic development. The benefits of additional energy availability and income increase up to a certain point. After that point, no relationship exists between consumption and the HDI.

Perpsectives on the Environment
Degradation of Ecosystem Services

The degradation of ecosystem services is already a significant barrier to achieving the Millennium Development Goals agreed to by the international community in September 2000, and the harmful consequences of this degradation could grow significantly worse in the next 50 years. The consumption of ecosystem services, which is unsustainable in many cases, will continue to grow as a consequence of a likely three to sixfold increase in global GDP by 2050 even while global population growth is expected to slow and level off in mid-century. Most of the important direct drivers of ecosystem change are unlikely to diminish in the first half of the century, and two drivers—climate change and excessive nutrient loading—will become more severe.

—Millennium Ecosystem Assessment (2005)

Perspectives on the Environment
Treaties, Agreements, and Commitment

The number of international environmental accords has exploded as countries awaken to the seriousness of transboundary and global ecological threats. The UN Environment Programme (UNEP) estimates that there are now more than 500 international treaties and other agreements related to the environment, more than 300 of them negotiated in the last 30 years.

But reaching agreements is only the first step. The larger challenge is seeing that the ideals expressed in them become reality. What is needed is not more agreements but a commitment to breathe life into the hundreds of existing accords by implementing and enforcing them.

—Mastny and French (2002: 13)

Other research points to similar conclusions regarding the alleged need for energy consumption. Robert Prescott-Allen (2001) measured 87 indicators ranging from life expectancy and school enrolment to deforestation in 180 countries to develop his Wellbeing Index. He found no relationship

Environment in Focus

Box 15.1 Trade and the Environment in Conflict?

Meetings of the World Trade Organization (WTO) have attracted increasing attention since the organization was first formed in 1995. Environment has become a central concern. One goal of the WTO is to break down unfair trade barriers that prevent countries from gaining access to the markets of others. This is particularly important to many developing countries that want to market and sell their goods to the rich consumers of the developed world. Over the years, many developed nations have erected trade barriers to protect their own industries from such overseas competition.

However, many countries have also established laws and policies that relate to minimum standards, including environmental standards, that must be met by locally produced and imported goods sold within the country. Less developed countries argue that such laws and policies discriminate against their products, since it is too costly for them to meet these standards. Therefore, setting such standards is protectionist and violates international trade agreements.

This difference of perspective has set the stage for some high-profile international disputes relating to the conflict between environmental protection and trade barriers. The US, for example, placed an embargo on tuna from Mexico following declaration of the domestic Marine Mammal Protection Act. Mexican tuna fishing led to the deaths of thousands of dolphins through the practice of setting tuna nets around the highly visible dolphins that school with the tuna. Mexico claimed, successfully, that the embargo discriminated against their fishers. Several Asian countries recently won a similar case involving the shrimp fishery. US law requires shrimp fishers to have turtle-excluding devices on their nets to prevent the drowning of turtles. All species of marine turtle are listed as endangered.

International trade and the rulings of the WTO obviously have major implications for future environmental goals. How they will be played out is far from certain. On the one hand, there is enormous potential for international trade law to actively encourage a transition to a more sustainable society. Some of the most damaging environmental activities, for example, are the perverse subsidies that many governments use to support unsustainable fishing, agricultural, and forestry activities. On the other hand, past rulings do not cause much optimism that this proactive environmental role will be played in the near future.

---

between energy consumption and the well-being of a country. In fact, the United Arab Emirates, with the world's second highest per capita energy consumption, was ranked 173rd in terms of well-being. Austria was ranked fifth in well-being and 26th in energy consumption. Sweden and the Netherlands are ranked very closely in terms of human well-being, yet the Netherlands has a much lower environmental rating. This suggests that a high degree of human well-being can be achieved independent of environmental costs as illustrated by energy consumption.

Imagine if a direct correlation existed between energy consumption and societal well-being. This would trigger an

Extreme poverty forces people in some parts of the world to consume the resources today that they and later generations will need to survive in the future. Here a poor fisher from Sri Lanka's east coast wheels dynamited coral reef fragments to the brick factory for processing. He knows he is destroying the reef that should feed his children in the future, but the reef has been so overfished that it no longer provides for the needs of today. Intact, the reef could have been a protective barrier against the Asian tsunamis of 26 December 2004.

imperative to raise the level of energy consumption throughout the world at least to North American levels, which would entail a fivefold increase in energy use and a probable fivefold increase in all the attendant problems, including acid rain generation (Chapter 4) and global climatic change (Chapter 7). The scale of disruption to planetary biogeochemical cycles (Chapter 4) would be catastrophic.

Thankfully, this is not the case. The studies cited above confirm what many people already know intuitively, even though the truism is something we struggle to address in terms of national policies or individual choices: *The best things in life aren't things.* Consuming more will not, after a certain threshold, improve the overall well-being of society or individuals. The challenge at the international level is to enact policies and programs that will see consumption levels raised in needy countries but reduced in over-consuming countries, such as Canada. Reduction in material consumption in wealthier nations will have to be in the order of 90 per cent for some semblance of sustainability and equity to emerge. In short, people in developed countries need to rediscover the importance of the *quality* of life as opposed to the current emphasis on the *quantity* of goods that can be acquired.

Global challenges require different solutions in different parts of the world. While most of us need to cut our consumption, this peasant family in Myanmar has nothing to cut. They need to increase their levels of consumption just to meet what many of us would classify as basic needs.

Perspectives on the Environment
The Good Life

Rethinking what 'constitutes the good life' is overdue in a world on a fast track to self-inflicted ill-health and planet-wide damage to forests, oceans, biodiversity, and other natural resources. By redefining prosperity to emphasize a higher quality of life rather than the mere accumulation of goods, individuals, communities, and governments can focus on what people most desire. Indeed, a new understanding of the good life can be built not around wealth but around well-being: having basic survival needs met, along with freedom, health, security, and satisfying social relations. Consumption would still be important, to be sure, but only to the extent that it boosts quality of life. Indeed, a well-being society might strive to minimize the consumption required to support a dignified and satisfying life.

—Gardner and Assadourian (2004: 165)

Perspectives on the Environment
Imagine

Imagine no possessions
I wonder if you can
No need for greed or hunger
A brotherhood of man.
Imagine all the people
Sharing all the world . . .

—John Lennon

The **Happy Planet Index** (HPI) is one index that attempts to provide a different perspective on human well-being and environmental impact and focus on achieving sustainability. The HPI is based on the idea that most people want to live long and fulfilling lives and that the country doing the best is one that allows its citizens to do so while avoiding infringing on the ability of people in the future and in other countries to do the same. Human well-being is assessed as Happy Life Years and impacts by measuring ecological footprints per capita, as discussed in Chapter 1. The HPI is not a measure of the 'happiest' countries in the world. Countries with relatively high levels of life satisfaction are found throughout the index, but the index provides an assessment of the

'The best things in life aren't things.' This Bali rice farmer has none of the mechanical aids of our modern farmers, none of the household appliances that you and I have, but he is a happy man.

environmental efficiency of supporting well-being in a given country. Such efficiency could emerge in a country with a medium environmental impact and very high well-being, but it could also emerge in a country with only mediocre well-being but very low environmental impact (such as Vietnam). Each country's HPI value is a function of its average subjective life satisfaction, life expectancy at birth, and ecological footprint per capita. The exact function is a little more complex, but conceptually it approximates multiplying life satisfaction and life expectancy and dividing that by the ecological footprint. In 2009, Costa Rica scored the highest, followed by Dominican Republic and Jamaica, while Botswana, Tanzania, and Zimbabwe ranked at the bottom of the list. Canada ranked eighty-ninth.

These kinds of indices are useful, if controversial, because they make us think of different ways of looking at progress and re-evaluate what we are trying to achieve. The traditional way of looking at global development along the consumer-based lines that Western societies have 'progressed' is simply untenable for the rest of the world. All evidence shows that planetary ecosystems are collapsing under current burdens and can no longer provide the services required for future generations. Adding further to these stresses is an option only for the most short-sighted and self-centred of societies. Unfortunately, as Jared Diamond points out in his book *Collapse* (2005), some societies took exactly that route in the past. It usually occurred where a ruling class benefited greatly from the existing state of affairs and preferred to see it continue even over the short term rather than risk their extravagant lifestyle.

The challenge, then, is to reduce the level of stress below current levels while raising the standard of living of the world's poor to acceptable levels. Although international agencies such as the World Bank do define a 'poverty level', analysis shows that this level varies remarkably among societies and that a broader range of values has to be taken into account. The Happy Planet Index is one example of such an approach, and it applies not only to the so-called underdeveloped nations but equally to developed nations such as Canada. The end goal is to produce as much human well-being as possible for society with minimal environmental impact. The next section discusses some national perspectives on this challenge.

National Perspectives

Canada is a remarkable country. It is the world's second largest country by area, with some of the largest remaining wilderness areas. Canada is ranked the highest in the world in terms of estimated value per capita of natural capital (Statistics Canada, 2011a). Canadians enjoy one of the highest standards of living in the world. No country has topped the global Human Development Index rankings more often than Canada, although it has not done so since 2000 and our ranking has consistently declined since then (UNDP, 2011) and now is eighth. Norway, Australia, New Zealand, and the US top the list. However, we seem to be doing fine, and this is exactly the main challenge we face. We are few in number with a big country and a high standard of living. Most Canadians have no idea of the scope of the various global challenges we face because of our isolation from these global pressures. As a result, our performance in most areas of environmental management, from failing to comply with global treaties or to pass and enforce effective legislation to the everyday statistics that document our excessive per capita global impact, is largely overlooked. Throughout this book, we have cited many examples. The audits undertaken on environmental and sustainability issues by the government's own Office of the Auditor General also provide many examples. The audits repeatedly show failures by government departments to comply with their own policies and legislation. We make commitments at international and national levels that we consistently fail to fulfill. To emphasize the seriousness of these shortcomings, the main findings of the Commissioner of the Environment and Sustainable Development for 2010 are shown in Box 15.2. In 2010 the report focused on climate change, oil spill cleanup capability, and drinking water provision, which is why the Commissioner particularly identifies these topics as long-standing problems.

Violations of our commitments under the Kyoto Protocol, discussed in Chapter 7, are well known, but there are many others. In terms of marine conservation, projections suggest that we will manage less than a third of our commitments under the Convention on Biological Diversity to establish a network of marine protected areas. Canada would not support the global moratorium on bottom trawling promoted by George W. Bush, not a conservationist US President, and voted against declaring bluefin tuna endangered under the Convention on International Trade in Endangered Species. Canada also refused to support the US proposal to ban trade in polar bears under the same treaty.

On global change, we are charting a course diametrically opposed to that of much of the rest of the world, as discussed in Chapter 7, and have been recognized as a major impediment to progress on climate change regulations through the numerous mock 'Fossil of the Day' awards given by a coalition of hundreds of international and non-governmental organizations at climate talks ranging from Poland in 2008 through Copenhagen and Bangkok in 2009 to Cancun in 2010 and Durban at the end of 2011, where Canada was expected to hold 'a virtual mortgage' on this award for obfuscation and delay (Parkinson, 2011). This poor performance was emphasized at Copenhagen, the most important talks aiming to find a successor to Kyoto, when Canada received the 'Fossil of the Year' award as the country that had done the most to block progress on climate action. Canada has reduced its emissions target from 282 million tonnes in the

Environment in Focus

Box 15.2 Commissioner of the Environment and Sustainable Development: Main Findings for 2010

The chapters in this report point to some common weaknesses in how the federal government is managing environmental and sustainability issues. Specifically, this report identifies a pattern of unclear and uncoordinated actions. This has been aggravated by the overriding problem of a lack of sustained leadership.

The concerns we have raised in this report are hardly new. About 20 years ago, the federal government acknowledged that the impacts of climate change would pose significant, long-term challenges throughout Canada, from more frequent and severe storms in Atlantic Canada to changes in the amount of rain available to farmers. And today, the federal government still lacks an overarching federal strategy that identifies clear, concrete actions supported by coordination among federal departments.

Also 20 years ago, the federal government recognized the need for a national strategy to respond to the risks of spills from vessels transporting all kinds of hazardous and noxious substances. The volume of such substances—from industrial chemicals to solvents and pesticides—transported in Canadian waters continues to increase. Yet Canada still does not have a national plan to ensure the federal government is ready to respond to major incidents.

Environment Canada has been running the federal water quantity and water quality monitoring programs for about 40 years without knowing who—if anyone—is monitoring the quality of fresh water on federal lands. As a consequence, there are unacceptable gaps in the federal monitoring of fresh water—notably, that Environment Canada has water quality monitoring stations on only 12 of some 3,000 First Nation reserves.

Federal leadership for water monitoring needs to be revisited, and Environment Canada needs to set out clearly how it will meet its responsibilities. In my view, this is long overdue.

Sustained leadership begins by knowing what the major environmental problems are, setting out a concrete plan with sufficient resources to tackle them consistently over time, and having the management systems needed to direct the work and monitor the achievement of those goals. Acquiring reliable environmental data and information is the first step in addressing the most pressing environmental priorities.

Solid, objective, and accessible information is essential to identify and respond to the quickening pace and complexity of environmental change, in Canada and globally. Managing Canada's environment without scientifically sound environmental information is akin to trying to steer the country's economy without using indicators such as the gross domestic product, unemployment rates, and trade balances. As noted in previous reports to Parliament, critical gaps in the federal government's environmental information hinder both its capacity to inform Canadians about key environmental conditions, and its ability to know if the billions of dollars it spends each year on environmental protection are making a difference. . . .

Conclusions

. . . There is little in our findings to offset a discouraging picture, as most suggest underlying problems in how these federal programs are being managed. In short, the two fundamental problems we identified are a lack of effective and sustained leadership, especially when responsibilities are shared, and inadequate information.

Source: Auditor General of Canada (2010: 2–3, 6).

---

government's first plan in 2007 to 28 million tonnes in 2010, a drop of approximately 90 per cent. It has also made new commitments set out under the Copenhagen Accord, the 2010 Federal Sustainable Strategy, and the Cancun action plan to reduce its greenhouse gas emissions by 17 per cent, from 2005 levels, by 2020. A recent audit by the Auditor General (2011) concludes that Canada will not be able to make even these very reduced commitments until there is a clear strategic plan including clear objectives, timelines, interim targets, and expectations with key partners.

Pollution does not come without costs, not only to the environment but also to people. Boyd and Genuis (2008) suggest that pollution could be causing up to 25,000 premature deaths in Canada each year and burdening the health-care system with up to $9.1 billion per year in extra costs. Pollution is responsible for 8,000 to 24,000 new cases of cancer and 500

Perspectives on the Environment
Canada's International Performance

Canada's image now lies in tatters. It is now to climate what Japan is to whaling. The tar barons have held the nation to ransom. This thuggish petro-state is today the greatest obstacle to a deal in Copenhagen. In Copenhagen next week, this country will do everything in its power to wreck the talks. The rest of the world must do everything in its power to stop it. But such is the fragile nature of the climate agreements that one rich nation, especially a member of the G8, the Commonwealth, and the Kyoto group of industrialized countries, could scupper the treaty. Canada now threatens the well-being of the world.

—*The Guardian*, 30 November 2009

Perspectives on the Environment
International Attention on Canada and the Need for Activist Scientists

Canada's international reputation as a green and gentle nation has long been a matter of national pride. But is that reputation deserved? Canada's actions on environmental issues—from ignoring Kyoto Protocol targets to obstructing progress at United Nations climate-change talks—are increasingly raising eyebrows, both at home and abroad. Perhaps nothing is more emblematic of this reality gap than Canada's determination to mine its tar sands at a frantic rate. The sands are a dirty source of oil. They require more energy for oil extraction than do conventional reserves, producing extra greenhouse-gas emissions. The industry has torn up vast swathes of landscape, created toxic ponds of waste, and released pollutants into waterways. Where such issues justify pressure for action, it is crucial that scientists such as David Schindler . . . highlight them.

—Editorial in *Nature* (2010)

to 2,500 low-birth-weight babies annually. The study estimated that pollution causes about 1.1 million to 1.8 million restricted activity days for asthma sufferers each year and that Canadians as a group spend from 600,000 to 1.5 million extra days in hospital annually.

The availability of information to interested citizens also seems increasingly restricted in Canada, in contrast to many other countries where growing Internet use has promoted greater citizen access to government data. In fact, an international study of the effectiveness of Freedom of Information Acts in five parliamentary democracies placed Canada last (Hazell and Worthy, 2010). Canada was criticized for its low political support, weak Information Commissioner, and an antiquated and expensive system for request that deters many from seeking information. Canada was among the first countries in the world to introduce freedom of information legislation, in 1983, but several studies have shown that we are far from a leader in this field. In 2010 only 16 per cent of the 35,000 requests made by Canadian citizens resulted in full disclosure of the information, compared with 40 per cent in the year 2000 (*Globe and Mail*, 10 January 2011, A7).

The lack of good information was highlighted by the Auditor General (Box 15.2), and Canada has been the subject of international condemnation in premier international journals such as *Nature* for the barriers it has put in place against federal government scientists being able to talk openly about their work. These restrictions apply not only to research that might be considered 'sensitive', such as on climate change, but also to pure research such as the report in *Nature* by a natural resources scientist on a flood that occurred in northern Canada 13,000 years ago (*Nature* 464, 2010: 740–3).

Fisheries scientists have also been particularly constrained in their ability to discuss their findings in public. A federal fisheries biologist published a paper in *Science* (Miller, 2011), linking the crash in fisheries stocks in the Fraser River to a virus infection. *Science* highlighted the study and notified over 7,400 journalists worldwide about it, telling journalists to contact a media officer in Fisheries and Oceans Canada. Major media outlets throughout the world requested interviews with Miller but were denied access by the Privy Council Office.

Perspectives on the Environment
Federal Scientists Muzzled

All federal scientists must get pre-approval from their minister's office before speaking to journalists who represent national or international media. The pre-approval process requires time-consuming drafting of questions and answers, scrutinized by as many as seven people, before a scientist can be given the go-ahead by the minister's staff. This is to spare the minister 'any surprises'. What kind of politician needs that sort of pampering? And what kind of journalist submits questions for a scientist to a ministerial clearing house? This message manipulation shows a disregard for the values and virtues of both journalism and science, and subverts timely disclosure and access to scientific data.

—O'Hara (2010: 501)

However, some progress is being made. British Columbia initiated a carbon tax (Box 15.3) on the rationale that people should be rewarded for adopting behaviour less likely to exacerbate the effects of global warming. This means taxing carbon-based activities, such as those using fossil fuels. In BC, for example, people pay such a tax at the gas pump every time they fill up. Details of the carbon tax in BC are provided in Chapter 7, and recent evaluations indicate that the tax is effective and that no noticeable loss of quality of life has occurred in the province. Yet there was major opposition to the tax when it was first introduced. Despite such opposition, BC politicians moved ahead with the tax in the belief that it was the right thing to do.

Other provinces are moving forward with innovative ideas. Nova Scotia, for example, is hoping to harness the power of the world's largest tides at the Bay of Fundy to develop an electricity supply that could meet up to 15 per cent of the province's needs. The idea is not new. In the 1970s, a facility producing up to 20 megawatts daily was built, which acted much like a giant hydroelectric plant. In contrast, the new projects involve using fixed or floating turbines within the tides. The province is also exploring an equally innovative project to see whether the old coal mines in Cape Breton, some 3,200 kilometres in length, can be used to produce geothermal energy.

Environment in Focus
Box 15.3 Addressing Climate Change in BC

British Columbia has set a target to reduce greenhouse gases by at least 33 per cent below current levels by 2020. This is 10 per cent less than the 1990 level, bettering California's goal of returning to 1990 levels by 2020. On 30 June 2011, British Columbia became the first province/state in North America to be carbon-neutral in the public sector. Some initiatives to continue these efforts and meet their goal include:

- Interim targets were set for 2012 and 2016 through a Climate Action Team that determined the most credible, aggressive, and economically viable targets.
- A longer-term emissions reduction target will be set for 2050.
- All electricity produced in BC will be required to have net zero greenhouse gas emissions by 2016.
- Greenhouse gas emissions from the oil and gas industry will be reduced to 2000 levels by 2016, including a zero-flaring requirement at producing wells and production facilities.
- A new $25 million Innovative Clean Energy Fund will be established to encourage the commercialization of alternative energy solutions such as bioenergy, geothermal energy, tidal, run-of-the-river, solar, and wind power.
- Tailpipe emission standards for all new vehicles sold in BC will be phased in by 2016, reducing carbon dioxide emissions from autos by 30 per cent.
- A low-carbon fuel standard will be established that will reduce carbon intensity of all passenger vehicles by at least 10 per cent by 2020.
- The energy plan will require that 90 per cent of BC's electricity needs come from clean, renewable sources.
- BC will become the first jurisdiction in North America, if not the world, to require 100 per cent carbon sequestration for any coal-fired electricity project.
- The $2,000 sales tax exemption on new hybrid vehicles will be extended.
- All new cars leased or purchased by the province will be hybrids.
- A new unified BC Green Building Code will be developed with industry and communities.
- The Emissions Standards Statutes Amendment Act was passed to phase in requirements for methane capture at landfills, the source of about 9 per cent of BC's greenhouse gas emissions.
- New incentives were introduced to retrofit existing homes and buildings to make them energy-efficient.
- New measures were introduced to help homeowners undertake 'energy audits' to identify possible energy savings.
- Real-time, in-home smart metering is being installed to help homeowners measure and reduce energy consumption.
- New strategies will be launched to promote Pacific Green universities, colleges, hospitals, schools, prisons, ferries, and airports. The province will substantially increase its tree-planting efforts and passed new legislation in 2010 to achieve zero net deforestation The province will ensure that school curricula inform students about how they can reduce individual impacts on the environment at home and at work.
- Beehive burners for burning wood waste have been eliminated.
- Trees infested by the mountain pine beetle have been used to create new, clean energy.
- A federal–provincial partnership invested $89 million in hydrogen fuelling stations and the world's first fleet of 20 fuel cell buses, the world's largest. The new fuelling stations are part of the initial phase of the hydrogen highway from Whistler to Vancouver, Surrey, and Victoria.
- The province will work with Pacific states to encourage a hydrogen highway from Whistler to San Diego by 2020. It would be the longest hydrogen highway in the world.
- BC is working with its neighbours to create electrified truck stops to reduce idling.
- The Premier has met with governors to assess and address the impact of climate change on our oceans and establish common standards for Pacific ports.
- The province is seeking federal co-operation to electrify ports and reduce container ship carbon emissions in all Canadian ports.
- The province is working with Pacific partners to develop a sensible, efficient system to register, trade, and purchase carbon offsets and credits and was the first full Canadian partner in the Western Climate Initiative.

Source: British Columbia (2008) and subsequent updates.

The mines are flooded with water at a temperature of 9 to 15°C, and the idea is to turn the mines that once supplied half of Canada's coal into a new energy resource. A related example in eastern Ontario is a $660 million plan proposed for Marmora called 'pumped storage'—the large reservoir in an abandoned open-pit iron mine would be pumped to another reservoir at the top during off-peak energy hours and then sent back down through turbines.

These examples illustrate two key principles for moving forward. First, enlightened political leadership can play a major role in visioning and implementing change at the societal level, and second, human ingenuity is vast but needs to be applied to the key challenges facing society.

The bottom line is that Canadians are too often not rising to meet environmental challenges and, in many cases, are falling behind. We are followers and laggards rather than leaders.

An example is the Canadian response to the International Chemical Management Agreement signed in Dubai in 2006. The agreement prompted many governments to institute new measures to control chemical use. Leading the way has been the European Union, which has a program requiring industries to prove the safety of their products before they reach the market. Companies must pay for studies, reveal the chemical composition and toxicity of their consumer products, and enter the data in a public registry. In Canada, under what was boasted as flagship legislation, the 1999 Canadian Environmental Protection Act, the taxpayer foots the bill for the assessment of new chemicals in a restricted number of tests (which were highly criticized by the Office of the Auditor General in 2008), companies can sell their products before tests are complete, and there is no obligation for the government to ban chemicals found dangerous. In the EU, if a company cannot prove the safety of a product, it cannot be sold. In Canada, the product can be sold before testing, while the taxpayer must pay for the tests and can only hope that the product is withdrawn if it is found to be unsafe.

A larger-scale global comparison is provided by the Environmental Performance Index (epi.yale.edu/), which ranks 163 countries in terms of overall environmental performance. Canada's overall rank in 2011 is 46th, a very modest position for a rich, large country. Canada scores well on some attributes, such as agriculture, forestry, and water management. Unfortunately, for greenhouse gas emissions we ranked 151st, 146th for sulphur dioxide emissions, and 141st for nitrous oxides. We did slightly better in biodiversity, ranked 80th, but for fisheries only two countries were below us. Canada's performance in the environmental field is also reflected in the newly released Canadian Index of Wellbeing discussed in Chapter 1. This index tracks eight factors. From 1994 to 2008, it shows an increase of 11 per cent in the CIW, with 'environment' as one of three factors that have dropped over this time period. Details are available at: www.ciw.ca.

The image we have of ourselves—that this country is a leader in global environmental management—is strongly supported by governments, but it is inconsistent with the facts. Macdonald (2009) attributes this to the governmental 'search for environmental legitimacy', or what environmentalists might term 'greenwashing', that is, controlling the message rather than addressing the problem. Some suggest that the appointment in 2011 of an ex-TV broadcaster as the Minister of Environment, with his first public speech defending the 'ethical oil' produced by the oil sands, epitomizes the concern that Canadian governments have with image rather than substance regarding environmental management.

Unfortunately, the longer we delay action to address environmental degradation, the more dramatic (and often more costly) the remedial actions will be when they do occur. In some cases, remedial action is delayed so long that the desired environmental conditions cannot be restored. Extinction is the ultimate example, but there are many others. Scientists have found that although politicians have met agreed-upon targets for reduced sulphate depositions, large areas are still receiving excessive sulphates, pH levels are not improving in many lakes, and biological indicators of recovery are even further behind. Reductions were made, but they were too little and too late to halt the acidification of many lakes in eastern Canada. The real challenge is to identify and mitigate these problems before they develop so far that they become either irreversible or so costly to reverse that they cannot be resolved. A 2011 report on the impacts of global climate change on the Canadian economy, for example, suggests that by 2020 it could cost the economy $5 billion per year and that by 2050, annual costs could range from $21 billion to $43 billion (NRTEE, 2011b).

Despite our poor response as a nation in many areas of environmental management, individual Canadians can take many actions that will have a significant impact not only on the Canadian environment but also globally. If an average Cambodian were to halve his or her consumption of global resources, the incremental gains would be small. But because of the scale of our personal consumption, even individual Canadians can make significant changes. If 34 million Canadians made changes, the cumulative impact would be staggering. The next section discusses some perspectives on environmental change at the personal level, to counter Leo Tolstoy's observation that 'Everyone thinks of changing the world, but no one thinks of changing himself', and instead to endorse the view of Norman Vincent Peale that 'You change your thoughts and you change your world.'

Personal Perspectives

Pick Up That Degree

> The future is increasingly a race between education and catastrophe.
>
> —H.G. Wells

The difficulty with many environmental challenges is the vastness of their scale. They are so widespread that most Canadians do not realize they are out there, especially because we are more sheltered from their effects than people in smaller, more densely populated countries. Most of these problems also have long lag times (the period between the time when the processes are set in motion and when the effects are felt), especially when the effects may have different impacts in different parts of the globe. Global warming, for example, may lead to cooling and increased precipitation in some areas, while the effect may be just the opposite in other areas. Changes of this complexity and magnitude require long-term study before they can be understood. The scale of change also suggests that within the human lifespan, many of these trends may be irreversible.

Guest Statement
A Generation of Possibility
Skye Augustine

There has never been a better time to be a young person in Canada!

Unlike previous generations, we young people have the fortune of being limited *only by our imagination*. Today over 3 billion people are under the age of 25, comprising almost half the world population. In Canada, we are among the world's wealthiest people. Most young Canadians are no longer limited by gender oppression, by child labour, or by strict family social roles. In addition, we are fortunate to have the most global world view of any generation before us.

Today, youth are looked to as catalysts of innovation and creativity. We have the opportunity to think of solutions that have never been tried. There are more young people in politics, more young CEOs, and more socially engaged young people than ever before. This is a generation of possibility!

Many are using their growing voice to make substantial change in the world. Young Canadians serve on youth councils and groups for environmental professionals, and attend international negotiations for current world issues. Ontario youth Craig Kielburger was 12 years old in 1995 when he began what is now Free The Children, an organization that has built over 650 schools for children in developing countries and works globally to end exploitative child labour practices. Raised in British Columbia, Severn Cullis-Suzuki stole the stage at the United Nations Earth Summit in Rio de Janeiro in 1992 at the age of 12 when she spoke about the importance of the environment to young citizens of the world.

> Our deepest fear is not that we are inadequate, it is that we are powerful beyond measure.
>
> —Marianne Williamson

As Canadians of the generation of possibility, we also have a level of responsibility to both our country and the planet. Youth are often marginalized in decision-making processes and in many cases our future is being decided for us. Changes are occurring more rapidly than ever before, and many of these changes could have dire long-term consequences. As young people, it is paramount that we engage and represent our ideas to ensure that we receive an adequate future, and pass on a better future to our children.

There are many instances where young Canadians have banded together to shape policy, stop a company, protect a significant natural space, and raise awareness about important topics. In 2009, and again in 2012, over a thousand youth from across the country gather in Ottawa for a conference called PowerShift. Together they learn about climate change, inspire one another and lobby politicians to take climate change, and the future of the planet, seriously.

Universities, such as yours, are often hubs of change. They present opportunities to create networks with like-minded people, and develop the capacity to influence change in the surrounding communities.

Now is the time to get involved! What can you do? (1) Get informed. (2) Join a group that represents an issue you care about. (3) Ask your professors difficult questions. (4) Vote. (5) Get your friends to vote. (6) Stay inspired!

As the environmentalist and entrepreneur Paul Hawken put it, 'If you look at the science about what is happening on earth and aren't pessimistic, you don't understand the data. But if you meet the people who are working to restore this earth and the lives of the poor, and you aren't optimistic, you haven't got a pulse.'

Skye Augustine is a Master's student at the University of Victoria studying climate change impacts on marine conservation in Thailand. She was a regional co-ordinator for PowerShift 2009.

Urgent actions are required if we are to help defuse these trends, but actions do not occur in a vacuum. They require understanding of the road we are on, where it goes, and how we can get on other, more desirable roads. Educational systems should help to bring about this understanding. Many schools, colleges, and universities graduate students who have little or no idea about how the ecosphere functions and how human activities impair those functions. They shop, travel, eat, drink, work, and play in blissful ignorance of the impact they may be having on life support systems.

Colleges and universities are frequently accused of not being part of the 'real' world. By 'real' world, people usually mean the economic realities of today's society. However, that is not the real 'real' world; it is essentially a game that humans invented to facilitate barter and exchange. Important? Yes! But is it the real world? Only partly! The real world includes the air we breathe, the water we drink, the organisms that keep the life support systems going, and the ground we stand on. Without these things, there can be no invented world, however 'real' it might seem. We concentrate on balancing these 'play' budgets when, in reality, the more significant budgets of energy throughflow and material balance determine the future of society. The deficit? Yes, Canadians have a huge deficit, but it relates to the 60 million bison no longer roaming the prairies, the skies cleared of all but a fraction of the birds that used to flock in such large numbers that day turned to night, and the seas once home to the largest animals ever to evolve on the planet but where not even the smallest fish are safe from sonar detection, vacuum trawling, and human consumption. Yes, we have a deficit, most of which remains unrecognized in the routine financial and economic accounting procedures that help to shape national policies.

One major challenge contributing to lack of understanding and action on environmental matters is **nature deficit disorder**. The term was coined by Richard Louv (2006) to describe an increasing gap in understanding of the *real* world on the part of the younger generation. Instead of playing outdoors in fields, woods, streams, lakes, or the ocean, an increasing proportion of the youth of today are glued to their computer or TV screens. They seldom visit the outside world, especially areas dominated by nature rather than human activities. The decline in the number of visitors to national parks and similar areas across North America bears testimony to this trend. A feedback loop develops. The less exposure that young people have to the natural world, the less they understand it and the less comfortable they feel outdoors. As a result, they tend to avoid encounters with nature and become even further estranged.

At a time when it is critical that more people become involved in significant environmental action, we seem to be producing a new generation that is further than ever from developing any attachment to the environment. This does not bode well for society in the future as this generation matures and becomes the main economic drivers and decision-makers. Concern over this situation has prompted movements across North America to provide opportunities and facilities to encourage outdoor re-engagement by younger people (e.g., see The Kesho Trust, www.naturechildreunion.ca). British Columbia, for example, is looking at a program (Kesho Trust, 2007) to reconnect children and families with nature and the outdoors, with these four objectives:

1. Children have outdoor playtime included in their school and out-of-school lives and have opportunities to freely access the outdoors and/or wild nature with friends and significant adults.
2. We design and build neighbourhoods and environments that allow and recognize nature to be nearby, accessible, and attractive to children and their families and seen as friendly, safe, joyful, and beneficial.
3. Children are seen as independent and competent and able to handle and benefit from the challenges of being outdoors and/or in wild nature.
4. We are seen as part of nature and are comfortable with the environment in which we live.

A primary function of our educational system should be to give students a general level of understanding about the nature of our environment and resources. We have requirements for general levels of language and mathematical competence but require nothing from our students in terms of this most fundamental challenge of the future. Indeed, pressures grow, especially on universities, to put greater and greater emphasis on meeting the short-term economic demands of society. Business schools and faculties of commerce flourish, yet few additional resources are allocated for programs dealing with the environment.

Even in such programs, colleges and universities have seldom done a good job of instilling in students an appreciation for and love of the planet as discussed above. Increasingly, science programs have become a process of learning more and more about less and less. They have produced technically competent scientists, but these programs often miss the mark considerably in terms of maintaining students' wonder about the natural world and combining the rigour of scientific inquiry with deep moral questioning. Science programs often mistake the laboratory for the 'real' world and cut students off from a more comprehensive understanding of and passion for their environment.

Perspectives on the Environment
On Education

. . . without significant precautions, education can equip people merely to be more effective vandals of the Earth. If one listens closely, it may even be possible to hear the Creation groan every year in May when another batch of smart, degree-holding, but ecologically illiterate *Homo sapiens* who are eager to succeed are launched into the biosphere.

—Orr (1994)

You can create change on your campus. Are there sufficient courses on the environment? Do these courses cover a wide spectrum from the technical to the philosophical, and more important, are students encouraged or even required to select from courses all along this spectrum? You should also remember that campuses are large consumers and processors of matter and energy. How efficient are they? Has anyone undertaken an environmental audit of your campus? How are wastes disposed of? How much recycling occurs? Are chemicals used for landscaping? Does the faculty pension fund invest in businesses with unsound environmental practices? You can investigate many questions through course work, in environmental clubs, or as an individual. If you are interested in pursuing these ideas, you may wish to draw on such experiences elsewhere (e.g., see the guest statement in this chapter by Darren Bardati, 1995, 2006) and start some activities on your own campus.

So pick up that degree and encourage others to increase their understanding of environmental challenges. Do not be intimidated by people who think that interest in the environment and higher learning is not the 'real' world. Challenge your teachers to inspire you. Be interested. Apply what you learn to your life. Do not be misled, however, into thinking that the formal education system is the only source of learning. Keep on reading. Many of the most inspiring works on the environment do not make their way onto college or university reading lists, and most of the rewarding environmental experiences are certainly not part of the curriculum. Challenge society to change, and seek a new kind of relationship between humanity and our home, planet Earth.

One of the main ways to fight environmental degradation is to become aware of the implications of your own actions.

Guest Statement

Sustainability in Higher Education: Learning It, Teaching It, and Doing It

Darren Bardati

> Our greatest challenge lies in rethinking what kind of education is appropriate for a species whose standards of success threaten its ecological foundations.
>
> —Orr (1992)

I first read the above quotation in the summer of 1993. I was trying to come up with a topic for a senior undergraduate independent study project and had stumbled across a special issue of the journal *New Directions for Higher Education* called 'The campus and environmental responsibility'. Dr Orr's opening article, in particular, impressed upon me a new perspective on my university education. He wrote of the irony of so-called 'higher education' graduates who were ignorant about their individual impacts on the earth's ecosystems. I asked myself: Would I also become an ignorant graduate?

The special issue contained 11 papers on how university and college campus communities had begun to respond to the environmental challenges of the times. The authors observed that for decades, academics had been 'pointing the finger' at corporations and industrial managers for poor environmental practices that damage the environment, while turning a blind eye to the impacts of their own campus activities. Noting that the campus is a major natural resource user and a microcosm of environmental problems linked to the larger community in many ways, these papers discussed practical experiences towards 'greening' campuses through improvements such as energy efficiency, waste reduction, purchasing and investment decisions, and raising environmental awareness. I asked myself: What was the state of the environment on my university campus?

Thus began my journey into the fascinating world of sustainability in higher education. That summer's independent study turned into an undergraduate thesis the following year, where I performed the first sustainability audit at Bishop's University and crafted an environmental action plan with specific recommendations based on my findings.

When I began teaching at Bishop's in 1996, I incorporated the periodic campus sustainability audit into a senior undergraduate environmental planning course. It was a natural fit to involve students in a 'hands-on' project that would apply the theory from the class lectures. Thanks to the collaboration of every sector of the campus community—including other students, faculty, staff, administrators, and members of the Board of Governors, a slow and gradual but systemic transformation took place towards a more sustainable campus. Major barriers to progress were identified and resources mobilized to overcome them. Students played the leading role in this transformation by bringing the environmental awareness issues to the forefront of every campus user's mind and by voluntarily doing a lot of the groundwork that university employees did not have time to do. The sustainability audit was performed every three years, and the number of people involved grew from a handful to more than 100. Several actions were taken every semester to raise environmental awareness and to reduce the university's ecological footprint.

By 2006, when the fifth edition of the student-led sustainability audit was published, it had the full support of the university's Board of Governors and its administration. Now sustainability assessments, policy statements, measurable targets, and much improved practices are firmly established in strategic planning documents, in daily operations, and in the overall culture of the university.

In 2008, Bishop's University became the first university in Quebec to ban bottled water. It also instituted a 'green levy' as part of the tuition in order to fund numerous initiatives. Two 'sustainability internships' are now available for students to lead various campus sustainability initiatives and conduct sustainability audits. In 2011–12, the university invested heavily in converting entirely from oil and gas to geothermal energy, thereby greatly reducing its carbon footprint.

The experience at Bishop's University was not unique, as many universities are now taking up these responsibilities. While my students and I struggled to integrate sustainability theory with practical experience on our own campus, the scope of the sustainability assessments grew, drawing from experiences in other parts of North America and Europe, where a 'greening the campus' movement had taken hold. In the past decade or so, dozens of books have been published on the topic of campus sustainability, ranging from the fundamental reasons for such a transformation to practical 'how to' manuals. A refereed journal—the International Journal of Sustainability in Higher Education—was launched, and numerous associations sprang to life or grew in stature, such as the Association of University Leaders for a Sustainable Future, the Sierra Youth Coalition, and the Association for the Advancement of Sustainability in Higher Education. Their websites will convince anyone that the expectation is fully established for institutions of higher education to lead by example.

One lesson is that progress towards a more sustainable campus happens best when enthusiasm and patience walk hand in hand. The ability to create, maintain, and harness the enthusiasm required to make positive changes must be matched with sufficient patience, communicative skill, and logic to enable these desired changes to happen with long-term significance, or 'staying power'. No one group, whether students, faculty, or administrators, can do it on its own or should be held responsible for doing it all. Although 'paper changes' such as adopting internal policy statements and signing international declarations like the Talloires and Halifax Declarations are useful tools in the transformation, as are standardized criteria to compare sustainability progress among universities, each campus community must create its own individual path towards sustainability.

Thankfully, nowadays, sustainability assessments are well entrenched on most university campuses, and students have many opportunities to get involved in practical sustainability work on their campuses. It seems obvious to repeat it: Everyone has a part to play! So join the local environmental club and get involved! If none exists, start one! And please, please do not be afraid to contact your university's administration and ask them about their environmental commitments and targets and how you can help the university to achieve them.

Despite so much progress in the past two decades regarding campus sustainability, I still wonder about Dr Orr's challenge of rethinking our university education. Are university graduates more environmentally literate than in the early 1990s? Are we moving from promoting campus sustainability to achieving more in-depth knowledge and concern for sustainability in higher education? Regardless of what degree our students earn, are they more personally responsible for reducing their own ecological footprint? I hope so.

Darren Bardati is Chair of the Department of Environmental Studies and Geography at Bishop's University.

Light Living

Light living expresses the need to tread as lightly as possible, to minimize our own ecological footprints (Chapter 1). A brief selection of ideas follows that you may wish to try. Given the statistics elsewhere in the text regarding per capita energy consumption, waste production, and water consumption levels in Canada, you can be assured that Canadians contribute greatly to overconsumption. Light living is often characterized by the four R's: refuse, reduce, reuse, and recycle.

Refuse

Our society is geared towards making consumption easy. Our newspapers are full of advertisements regarding the best buys. Turn on the radio, and you hear from the sponsor, or you are bombarded with commercials on TV. Estimates suggest that the average American will see 35,000 television ads every year. Most of us are surrounded by shopping opportunities on a daily basis. We can drive to one of several mega-malls in most large Canadian cities, park with ease and at little expense, and consume from a wide variety of stores, paying by credit

Perspectives on the Environment

Master of Change

Resolve to be a master of change rather than a victim of change.

—Brian Tracey

Table 15.2 | Annual Expenditure on Luxury Items Compared with Funding Needed to Meet Selected Basic Needs

Product	Annual Expenditure	Social or Economic Goal	Additional Annual Investment Needed to Achieve Goal
Makeup	$18 billion	Reproductive health care for all women	$12 billion
Pet food in Europe and the US	$17 billion	Elimination of hunger and malnutrition	$19 billion
Perfumes	$15 billion	Universal literacy	$5 billion
Ocean cruises	$14 billion	Clean drinking water for all	$10 billion
Ice cream in Europe	$11 billion	Immunizing every child	$1.3 billion

Source: Worldwatch Institute (2004a: 10).

card, ATM, cheque, or even cash. We frequently shop not to fulfill basic needs but to indulge frivolous and petty whims (Table 15.2). Clothes are discarded when no longer fashionable rather than when they lose their durability. Gadgets are discarded in favour of newer and shinier models. Christmas as a religious holiday is now the time when we pay homage to our greatest god, **consumerism** (Box 15.4).

Resist and *refuse* to buy anything that you do not really, really *need*. If the purchase is necessary, shop carefully. Buy items that are less harmful to the environment during all stages of the product life cycle, from manufacture to consumption and disposal. There are products certified by the government for their low impacts, as discussed later. Buy one quality item rather than a succession of several shoddy ones to fulfill your needs. Buy organic produce wherever available, or better still and if possible, refuse to buy any and grow your own. It is difficult, because consuming is easy. All the messages that we receive from society extol the virtues of buying things.

Reduce

Can you reduce your consumption of certain items? Energy is a good place to start and a major contributor to greenhouse gas emissions. More than a quarter of Canada's GHG emissions occur as a result of the everyday activities of Canadians. A lot of energy is used in space heating in Canada, but must you set that thermostat so high? Canadians tend to keep their houses much warmer inside than Northern Europeans do, for example. These cultures (New Zealanders, too) are accustomed to setting a low thermostat and wearing warmer clothing in the house in winter. They would not expect to be comfortable wearing just a T-shirt. Turn down the thermostat, wear warmer clothes, and turn the thermostat down further when you go out or go to bed. Statistics Canada (2011b) reports that in 2009, 61 per cent of Canadian households turned down the thermostat at night while sleeping in winter, compared with 55 per cent in 2007.

Reduce lighting costs by replacing burnt-out bulbs with long-life bulbs. They are more expensive to buy, but they last 10 years or longer and use 75 per cent less electricity. An incandescent light bulb converts only about 5 per cent of the energy it produces into light; the rest is wasted heat energy, as you will recall from the discussion of the laws of thermodynamics in Chapter 2. In 2007, Australia became the first country in the world to ban incandescent light bulbs to prevent wasted energy and cut greenhouse gas emissions. Calculations suggest that phasing out incandescent bulbs over the next three years could save about 800,000 tonnes of carbon dioxide emissions a year by 2012 and as much as 4 million by 2015. If you want a simple and easy way to make your own contribution, changing your light bulbs is a good place to start. Several provincial governments have already phased out incandescent light bulbs and the federal government will move to a national ban in 2012. This may well be necessary. A Statistics Canada report comparing environmental behaviours between years in Canada found that having at least one halogen bulb in the house was one of the very few indices that showed no change between 2007 and 2009 (Statistics Canada, 2011b).

When replacing electrical appliances, energy efficiency should be a major factor in your choice. Before you turn on a light or an appliance, think about whether you really need it. Every time you flick the switch or plug in an appliance, you are sending out the message of demand. Electricity-producing utilities and governments will react to your message by building new production facilities with all their attendant environmental costs. If you do not want those costs, try not to send along as many messages signalling your demand. Of households buying major appliances in Canada in the past five years, almost two-thirds said that water and energy consumption was the biggest factor in their decision-making (Statistics Canada, 2011b)

Transportation is also a big energy consumer, accounting for one-quarter of all energy used in Canada. Road vehicles are responsible for 83 per cent of that share. Canadians' average per capita gasoline consumption is 1,100 litres per year, compared with 350 to 500 litres in European countries. It is

Environment in Focus

Box 15.4 Dreaming of a 'Green' Christmas?

Christmas heralds the biggest consumer bash of the year, although merchants are also trying to persuade us to be equally excessive at other times. Take control of your consumer lifestyle at Christmas. Consider the following gifts:

1. Arrange an event, an outing, or a personal service rather than giving a material item.
2. Increase 'green' education by giving a book or subscription to a magazine. You could also buy someone a membership in a 'green' organization, such as Pollution Probe, the Canadian Parks and Wilderness Society, or Greenpeace.
3. Give a houseplant, a backyard composting kit, or an unbreakable coffee mug to replace the use of disposable ones. The World Wide Fund for Nature also enables you to protect an acre of rain forest by making a donation to fund a project that was started by unemployed students.
4. Give, in your gift recipient's name, a goat, a llama, a water buffalo, or any number of other animals through Heifer International, a 65-year-old NGO with a mission of alleviating hunger, poverty, and environmental degradation by providing food-producing animals and sustainable agriculture education to families in need.
5. Give something that conserves energy, such as a bus pass or an energy-saving showerhead.
6. Give second-hand items.
7. Make your own gifts, such as a sweater, dried flowers, or jam.
8. Give items that display the EcoLogo of three doves.
9. Choose gifts that require little wrapping. Reuse old wrapping paper or use reusable fabric gift bags instead.
10. Put 'Planet Earth' at the top of your list. If we all gave the planet an offering for Christmas that would make it feel better—the Earth would be a little more loved and a little less stressed.

also estimated that each kilometre of road or highway takes up about 6.5 hectares of land. In Ontario, which has 160,000 kilometres of highways, roads, and streets, this would add up to one million hectares for motorized vehicles. Whenever feasible, walk or ride a bicycle. If you have to use motorized transport, use public transport such as buses and trains. If you have a car, get a small economical one with a standard transmission, use it sparingly, and try to car-pool. Most people are aware that larger cars (and SUVs) consume more gasoline. However, since it takes 18 litres of water to produce one litre of gasoline, they also contribute to water deficiencies.

Food choices are second only to transportation in terms of their environmental impacts. Agriculture covers more than 25 per cent of the world's surface and profoundly affects the health of natural ecosystems. A meat-rich meal with ingredients imported from afar generates as much as nine times the carbon emissions of a vegetarian meal made from local produce and requires two to four times the land to produce. It is not only what we choose to eat but how much we eat as well. In Canada, as in other Western countries and in the consumer classes of developing countries, growing prevalence of obesity is one of the main challenges not only to personal

In Europe, few students have cars. Cities are designed for people, and it's easy to get to campus by bicycle or mass transit. Even when people buy cars, they prefer small ones.

health but also to global health. In fact, the number of clinically obese people now exceeds the number of chronically hungry people.

In 2008, researchers calculated that obese people consume 18 per cent more calories than the average, which overall results in more fuel being consumed to produce and transport food. Overeating drives up the cost of food by increasing demand. It also contributes to global warming by boosting food production. Edwards and Roberts (2008) call for policies to promote walking and cycling that would help to reduce obesity in the population as well as reducing fuel consumption by the transportation sector. Reducing the prevalence of obesity would, they say, reduce the global demand for both fuel and food. They conclude that decreased car use would reduce greenhouse gas emissions and thus the need for biofuels produced from vegetative matter. It would also mean more physical activity and reduce the risk of injury from traffic accidents as well as air pollution, thereby improving population health.

Perspectives on the Enviroment
Excess Consumption

The excess consumption by the world's obese costs $20 billion annually to which must be added indirect costs of $100 billion resulting from premature death and related diseases.

—Jacques Diouf, director general, FAO, World Food Summit, 2008

Eat only as much as your body needs to function in an effective and healthy manner. Become part of the food democracy movement. This means using your purchasing power to help establish connections between consumers and producers in the local area and making sure that you eat the kinds of food you wish to eat, not the kinds promoted by government subsidies and industrial agriculture. See how close you can come to the **100-mile diet**. The term was introduced in Vancouver in 2005 and refers to buying and eating food grown, manufactured, or produced entirely within a 100-mile radius of one's residence (Smith and MacKinnon, 2007) The average distance in the modern North American industrial food system is 1,500 miles from farm to plate (Halweil, 2002), which obviously involves significant environmental costs.

Water use is also where individuals and households can make important contributions to conservation, and appear to be doing so in Canada. Statistics Canada (2011b) reports that between 1991 and 2009 use of low-flow showers in Canada increased from 28 to 63 per cent and low-flow toilets from 9 to 42 per cent. Furthermore, only 24 per cent of households reported primarily drinking bottled water compared with 30 per cent in 2007. All these statistics show changes in the right direction. However, there is a long way to go. A recent study of consumer behaviour in 10 OECD countries showed that Canada's household water use was still almost double that of the OECD average (OECD, 2011).

You can also think about reducing the waste associated with the things you buy. Many products are overpackaged. They may look good on the store shelf but will only add to the amount of waste sent to the landfill. When you can, buy groceries in bulk to help reduce packaging. Reduce your waste by starting a compost heap for kitchen wastes. Reduce and if possible eliminate your use of toxic materials. Products that may seem innocuous (paint, solvents, and cleaning agents, for example) become hazardous wastes when disposed. Try to find alternatives. Do not buy more products than you need, and dispose of them in full accordance with the instructions from your local municipality.

Some countries, such as Bangladesh and China, have banned plastic bags while others, such as Ireland, South Africa, and Taiwan, have legislation to discourage plastic bag use, with authorities either taxing shoppers who use them or imposing fees on companies that distribute them. Canada has no such national law, but Leaf Rapids, Manitoba, was the first municipality in North America to propose, pass, and adopt a law forbidding shops to use plastic bags. This was followed by bans in Huntington and Amqui, Quebec, in 2008 and Wood Buffalo, Alberta, in 2010. Many other cities have since followed, or else stores now charge for plastic bags as a disincentive for their use. Statistics Canada (2011b) reports that 49 per cent of Canadian households used recycled or reusable bags for shopping in 2009 compared with 30 per cent in 2007.

Reuse

Buy products that can be reused, such as rechargeable batteries. Try to find another use for something no longer useful in its original state. Use plastic food containers to store items in your fridge or workshop. Return to the store with the same plastic bags that you used the previous time and use them for your next load of groceries, or better still, use cotton bags. When you are finished with something, it may still be useful to someone else. Organize a garage sale or donate the items to charity rather than throwing them out.

Recycle

Recycling facilities have sprung up across the country over the past decade and a half. Recyclable materials include newspaper, cardboard, mixed paper, glass, various metals, some plastics, car batteries, tires, and oil. These materials can be reprocessed into new goods. It takes 30 to 55 per cent less energy, for example, to make new paper from old paper than to start fresh from a new tree. Estimates suggest that if we recycled all the paper we use in Canada, we would save 80 million trees annually. Canadians consume more than 50 kilograms of newsprint per person per year—enough to account for one whole mature tree. Similar efficiencies can be obtained by recycling other materials that require less energy for remanufacture. Oil, for example, fuels many industrial processes. Estimates suggest that if 1 per cent of the Canadian population recycled instead of trashing one aluminum can a day, the oil saved by remanufacturing would produce 21 million litres of gasoline. Recycling one aluminum can saves enough energy to run your TV for three hours. Making steel from recycled material uses only one-quarter of the energy it takes to make steel from virgin ore, while recycling copper results in energy savings of up to 85 per cent.

The vast majority of Canadian households with access to recycling programs use them, regardless of household income, the occupants' education levels, or the type of dwelling (Statistics Canada, 2007c). Canada has one of the highest levels of recycling on the planet. This is one of the ratings you can find on the 'Greendex' developed by the National Geographic Society (2010) to measure and monitor consumer progress towards sustainability. Based on surveys in 17 countries, it focuses on actual behaviour and material lifestyles, including measures such as the relative penetration of green products versus traditional products, household footprint, energy use, transportation habits, and food consumption. Canadians led in recycling and in their ability to answer a knowledge question regarding the water demands of different types of food. Unfortunately, they fared much worse in the other categories, and Canada ended up ranking second from the bottom, surpassing only the US, in the previous survey in 2008. Brazil, India, and China topped the list. Among developed countries, average consumers in the UK, Germany, and Australia received the highest scores. Overall, though, the survey found that all 14 countries surveyed in both 2008 and 2010, except one, had more environmentally friendly consumer behaviours in 2010 than in the previous survey.

Consumers can adopt more sustainable consumption habits and help to raise Canada's Greendex score by doing any of the following:

- eating less meat, more locally produced foods, and more fruits and vegetables and drinking less bottled water;
- improving the energy efficiency of their homes by sealing drafts, upgrading windows, and installing more efficient water heaters and other appliances;
- keeping air heating at lower and cooling at higher settings;
- using only cold water to wash laundry and minimizing water use overall;
- driving alone less often (e.g., car pooling);
- driving less overall;
- driving smaller and/or more fuel-efficient vehicles;
- having fewer vehicles for their household;
- walking or riding a bicycle when distance allows;
- maximizing the lifespan of household items and minimizing disposal;
- avoiding environmentally harmful products and packaging and seeking out environment-friendly alternatives;
- recycling whenever possible;
- using reusable shopping bags rather than accepting new disposable ones;
- having fewer televisions sets and personal computers in their household.

Nonetheless, the bottom line remains that while it is better to recycle than not to, reducing consumption levels in the first place is still the preferred option, and significant progress in this area has yet to be achieved.

The Law of Everybody

You can reduce the pressure on the environment in many other ways. We have offered just a few suggestions to get you started. The key point to remember is that the accumulated actions of many concerned individuals will make a difference. In this book, we have introduced you to many scientific laws and principles related to understanding and managing the environment. However, if everyone knew and enacted the following law, this would go a long way towards ameliorating our environmental challenges: it is, simply, the **law of everybody**.

'Everybody's Got to Do Something'

If we all did many small things that could be done easily, without having much negative or even noticeable influence on our lifestyle (e.g., drive 10 per cent less a year, buy fewer toys, shower with a friend), we would find that many of our environmental challenges would be greatly reduced in scale. It is not important that all people do everything, but everyone has to do something!

Perspectives on the Environment
The Paradox of Change

Only the wisest and stupidest of men [and women] never change.

—Confucius

Several websites can help you to find out more about your own environmental impact and suggest concrete ways in which your impact can be reduced. For example, try the *personal challenges* on the website of the Canadian environmentalist, David Suzuki (www.davidsuzuki.org), and explore some of the websites at the end of this and other chapters. One site you will enjoy is 3rd Whale. This 'green facebook' run out of Simon Fraser University in Vancouver lets you explore everything from finding a green partner to finding a green job. Every year, this website runs a 'greenest person on the planet' competition. Nearly every contestant eats locally grown organic foods, recycles, and walks, cycles, or takes transit everywhere. A few go out of their way to power-down devices that sip away at electricity even when they are turned off, like TVs, radios, DVD players, and computers.

Consider these contestants the triathletes of the green world. We applaud their inventiveness and dedication, but unlike 3rd Whale, we are not sure that everyone is ready to compete in the triathlon. But people who, comparatively speaking, are just out for a quiet stroll can still help in achieving sustainability and resilience. We feel that engaging as broad a segment of society as possible in as great a number as possible, with each making a contribution, is a better goal than either doing the triathlon—or doing nothing.

Influence

One of the best ways to influence business is through your purchasing power as a consumer. If consumers refuse to buy certain products because of their impact on the environment, the manufacturer will either have to respond to these concerns or go out of business. Many successful examples exist of these kinds of actions. During the 1970s and 1980s, for example, conservationists were able to exert pressure on hamburger chains to change their source of beef supply to ensure that tropical rain forests were not being cut down and replaced by grass to feed cattle for hamburgers. Following boycotts, all the major chains were persuaded to ensure that their sources did not contribute to the destruction of the rain forests. Similar campaigns have been directed at tuna canners to ensure that tuna is not caught by methods that kill dolphins, which often swim with schools of tuna.

Perspectives on the Environment
Managing Planet Earth

A second myth is that with enough knowledge and technology, we can, in the words of *Scientific American* (1989), 'manage planet Earth'. Higher education has largely been shaped by the drive to extend human domination to its fullest. In this mission, human intelligence may have taken the wrong road. Nonetheless, managing the planet has a nice ring to it. It appeals to our fascination with digital readouts, computers, buttons, and dials. But the complexity of Earth and its life systems can never be safely managed. The ecology of the top inch of topsoil is still largely unknown, as is its relationship to the larger systems of the biosphere. What might be managed, however, is us: human desires, economies, politics, and communities. But our attention is caught by those things that avoid the hard choices implied by politics, morality, ethics, and common sense. It makes far better sense to reshape ourselves to fit a finite planet than to attempt to reshape the planet to fit our infinite wants.

—Orr (1994: 9)

Consumer boycotts can be an effective way of influencing business practices and reducing environmental impacts. When choosing fish to eat, for example, check out the websites at the end of Chapter 8. Download Canada's Seafood Guide (www.Seachoice.org). This guide lists three categories of seafood—best choices, concerns, and avoid—based on independent assessments of the sustainability of that particular fishery. If consumers followed these guidelines, it would alleviate much of the pressure on fishery resources.

Not only should we be sending messages to producers about our environmental standards through purchasing power, we should also let non-conforming producers know why we are not buying their products. If you believe in eating wild rather than farmed salmon, always ask the origin of the salmon before you buy it. If you believe in fair-trade coffee or certified wood products, always let retailers know that is what you are shopping for. Home Depot, for example, is the largest retailer of wood products in the world. Following a consumer boycott organized by a non-governmental organization, Home Depot committed to phase in sale of only certified wood products. This decision, rapidly echoed by two other major retailers, probably did more to protect old-growth forests in BC than all the protests of the past 25 years. However, the change was slow in coming, and opportunity still exists for improvement in Home Depot's purchasing practices. Home Depot argues that one of the main reasons behind its tardy behaviour is simply that customers do not ask for certified wood products and often do not buy them preferentially.

While boycotts can send a strong message and can influence corporations to change their practices, **Carrot mobs** developed as a way to support businesses doing a great job. Carrot mobs, like boycotts, are a form of organized consumer purchasing. Members of the public gather to encourage businesses demonstrating environmental and social leadership. By compiling the dollars of many consumers, organized purchasing can provide an economic incentive for businesses to provide energy-efficient, socially just, and long-lasting alternatives to mainstream corporate stores.

As a responsible consumer, you should let your preferences be known. This also raises the question of **corporate responsibility** (Box 15.5). Is it the responsibility of the retailer to take part in, or even lead, the switch to a more sustainable and resilient way of doing things by informing its customers about the environmental implications of different products? Or is it sufficient merely to react to customer demands once they are manifest in the marketplace?

Many countries have enacted what are known as **extended producer responsibility** laws that require manufacturers and importers to accept responsibility for their products at the end of their useful lifespan. These laws provide an incentive for companies to design their products so that they can be recycled or reused and to eliminate toxic materials, since they would have to dispose of them. Ultimately, products should be designed to be either biodegradable or disassembled into their components for reuse in the future. Canada has no such laws, but some companies have taken the initiative, as discussed in Box 15.5.

Life-cycle assessments (LCAs) are now gaining greater support from both government and industry. LCAs identify inputs, outputs, and potential environmental impacts of a product or service throughout its lifetime. The 2002

Environment in Focus
Box 15.5 Corporate Contributions

Large corporations and retailers are highly visible targets for environmental action and have attracted a lot of attention. However, many corporations do much better than we do personally, or than our governments do, in systematically looking at their environmental impacts and moving to address them. For example, the global fast-food corporation McDonald's has the Earth Effort program, which embraces the reduce, reuse, and recycle philosophy. Every year, McDonald's is committed to buying at least $100 million worth of recycled products for building, operating, and equipping its facilities. Carry-out bags are made from recycled corrugated boxes and newsprint; take-out drink trays are made from recycled newspapers. New restaurants have been constructed with concrete blocks made from recycled photographic film and roofs made from computer casings. McDonald's has also cut down on the amount of waste produced; for example, sandwich packaging has been reduced by more than 90 per cent by switching from foam packaging to paper wraps. It also uses compostable food packaging made from reclaimed potato starch and other materials.

McDonald's also has a strict policy to not buy the beef of cattle raised on land converted from rain forests for that purpose. In addition, it has programs to reduce energy consumption and to take part in a wide range of local initiatives ranging from tree planting to local litter drives. For its efforts, McDonald's in the US has won White House awards and the National Recycling Coalition's Award for Outstanding Corporate Leadership.

Another company that environmentalists love to hate, Wal-Mart, also has tremendous potential to influence sustainability. The chain has 4,100 stores in the US and more than 300 in Canada. It operates the continent's largest trucking company, is the largest consumer of electricity, and with 1.3 million employees in the US and 75,000 in Canada stands as the biggest employer. More than 100 million customers per week visit Wal-Mart stores. Wal-Mart has introduced some of the most far-reaching programs and demands that its suppliers also achieve sustainability thresholds. It has launched a $30 million Personal Sustainability Program that raises awareness among employees of individual measures to live more sustainable lifestyles. The combined impact of these kinds of programs can be many times that of government agencies.

Many other well-known corporations are making contributions. Eddie Bauer, for example, has joined forces with an NGO, American Forests, in a tree-planting effort on damaged forest ecosystems in eight reforestation sites in the US and Canada. Consumers can opt to add a dollar to the price of their purchase at the store, and in turn the store will plant a tree for each dollar and donate another tree. Nike, the shoe, clothing, and sports equipment giant, has not only boosted the proportion of organic cotton in its own products but also helped to launch Organic Exchange, a network of 55 businesses committed to significantly expanding the use of organic cotton. Nike has programs to achieve zero toxics, zero waste, and 100 per cent recovery, recycling, and reuse of the products it sells. The company is designing a running shoe with a biodegradable sole and an upper portion that can be endlessly remanufactured into a new shoe. Other companies, such as Texas Instruments, Levi Strauss, and the Ford Motor Company, have banded their purchasing power together to buy recycled paper.

Also consider Valhalla Pure Outfitters of Vernon, BC, makers of high-quality outdoor clothing. Valhalla makes all of its polyester fleece jackets from recycled materials that look, feel, wear, and last the same as new fabric, even though they are made from recycled pop bottles. Each pop bottle diverted from a landfill site for recycling is sent to a factory in South Carolina, where they are separated by colour and then reprocessed before being sent to finishing plants in the US and Quebec. The final cost is similar to that of making new material but has added environmental advantages. Each jacket is equivalent to about 25 pop bottles.

Some in the corporate and business world also have created substantial foundations to which environmental groups can apply for funding for specific problems. As government coffers are drained, more pressure is put on these sources. The Laidlaw Foundation, for example, used to give $50,000 a year for environmental projects, and requests rarely exceeded this amount. The Foundation now disburses $350,000 but has requests exceeding $1 million. The Richard Ivey Foundation in London, Ontario, provides up to 80 per cent of its $2 million budget for environmental projects each year but still manages to meet only 5 per cent of the amount requested. With such pressure on resources, it is essential to ensure that the money is used effectively to address the most pressing problems.

If there are environmental problems in your area that you think need to be addressed, do not be afraid to approach local businesses for support. They may well agree with you and be happy to make a contribution to a carefully crafted solution.

World Summit on Sustainable Development (WSSD) called for greater investment in LCAs, and the United Nations Environment Programme has developed a program to assist in the development and dissemination of practical tools to help in life-cycle assessment. One industrial example is Volvo, which provides LCAs for the various components involved in vehicle manufacture. Some NGOs, such as Green Seal, provide standards for various products that manufacturers must meet to gain their endorsement. Others have excellent programs to assist consumers, industry, and government in making wise procurement decisions, such as the US-based Center for a New American Dream. Among useful websites are Fair Trade, Ethiquette, and Climate Counts, all listed at the end of the chapter. Climate Counts produces a scorecard rating various companies on their performance in measuring, assessing, and reducing the carbon print of their products and on how supportive of climate change initiatives and policies they are. In 2010, scores were up 14 per cent over the previous year. In Canada, the product ecolabelling program EcoLogo aims to influence business practices by shifting consumer and institutional purchasing to products and services conforming to stringent environmental standards. A diverse stakeholder committee establishes the criteria that each category of product must at minimum meet to obtain certification, signified by the three-doves logo. EcoLogo standards can contain, for example, criteria for recyclability, design of packaging material, product emissions, use of recyclate in manufacturing, effluent toxicity, and energy consumption. More than 70 categories are available (www.ecologo.org).

Governments make many decisions that can promote or retard sustainability and resilience. Public transport can be subsidized instead of private cars, land-use planning and building codes can be designed to minimize energy and material use, and governments can facilitate shared public consumption, such as provision of libraries and swimming pools, rather than encouraging private acquisition. Many municipalities have inherited what has been called the 'infrastructure of consumption' from previous generations. However, it is necessary now to redesign the way we live in accordance with our increased understanding of planetary limits. Box 15.5 offers several examples of this kind of approach.

What can you do to encourage governments to act? Election times offer a major opportunity. Make a point of asking your local candidates at all levels of government for their views on sustainable development and resilience issues. Help to publicize these views. It is much more difficult for politicians to change their positions if their views are well known. Make sure that politicians follow through on their pre-election promises on environmental matters. Make sure that you express your views on the environment to politicians between elections. Write them letters, call them, go and see them, and

The symbol of three doves in the shape of a maple leaf identifies products certified under Canada's EcoLogo Program. The three doves represent Canada's consumers, industry, and government working together to improve the environment. (*Printed with permission from TerraChoice, manager of the EcoLogo Program.*)

organize demonstrations. The position of the Commissioner of the Environment and Sustainable Development was specifically created by the federal government within the Auditor General's office to provide a place where *any Canadian* can go with a question about the environmental performance of a particular federal ministry or program and ask to have it audited. If you have such a question, go to the Commissioner's website listed at the end of this chapter and review the simple process on how to proceed.

A wind farm in Pincher Creek, Alberta.

Governments are often more than willing to adopt environment-friendly, voter-popular measures—as long as they do not involve any significant economic cost. Issues are often very complex, too complex for the understanding of many individuals who nonetheless care about acting in an environmentally responsible fashion. For this reason, concerned people may band together in **non-governmental organizations** (NGOs). Such organizations represent the collective concern and resources of many people and consequently are in a much better position to attack a problem than an individual. One of the best ways to spend your conservation dollar is to support such a group. They have had significant impacts on government policies in Canada. The appendix lists of some of these organizations.

Implications

Perspectives on the Environment
Ongoing Change

When you're finished changing, you're finished.

—Benjamin Franklin

The overall implications raised in this book should be brutally clear by now. Humanity is facing major challenges over the next decade regarding our relationship with the Earth and its life support systems. So great are our capabilities to affect these systems that human-controlled influences now dominate many natural processes, resulting in the many critical environmental problems described throughout this book. Government programs at the international and national levels continue to emerge, but actions that make a difference on the ground are few and far between. Still, we are not powerless to change the direction of society. As individuals and as concerned individuals banding together, we can affect many of the changes that need to be made, and we can strive for greater well-being, not wealth, in the future (see Guest Statement by Skye Augustine earlier in the chapter). So, despite the severity of the challenges, it is important to remain optimistic.

Perspectives on the Environment
On Well-Being and Accumulation

Societies focused on well-being involve more interaction with family, friends, and neighbours, a more direct experience of nature, and more attention to finding fulfillment and creative expression than in accumulating goods. They emphasize lifestyles that avoid abusing your own health, other people, or the natural world. In short, they yield a deeper sense of satisfaction with life than many people report experiencing today.

—Gardner and Assadourian (2003: 16)

Above all, stay cheerful, stay active, look after yourself, look after others, love this planet, and do not give up! We are on the most beautiful planet we know about. Canada has some of the most breathtaking wonders in the universe. We have a responsibility. We can think of no better advice than that offered by Edward Abbey (1977):

> One final paragraph of advice: *Do not burn yourselves out. Be as I am—a reluctant enthusiast . . . a part-time crusader, a half-hearted fanatic. Save the other half of yourselves and your lives for pleasure and adventure. It is not enough to fight for the land; it is even more important to enjoy it. While you can. While it's still here. So get out there and . . . ramble out yonder and explore the forests, encounter the grizz, climb the mountains, bag the peaks, run the rivers, breathe deep of that yet sweet and lucid air, sit quietly for a while and contemplate the precious stillness, that lovely, mysterious, and awesome space. Enjoy yourselves, keep your brain in your head and your head firmly attached to the body, the body active and alive, and I promise you this much: I promise you this one sweet victory over our enemies, over those desk-bound people with their ears in a safe deposit box and their eyes hypnotized by desk calculators. I promise you this: you will outlive the bastards.*

Summary

1. The previous century witnessed the start of a consumer-dominated society in several parts of the world that is rapidly spreading and threatens to engulf most of the world in the early part of this century.

2. Population growth rates are showing declines, but consumerism continues to expand. Environmental conditions continue to deteriorate and are likely to do so even more as consumerism spreads.

3. Today, most emphasis is on meeting the Millennium Development Goals, but overall progress is slow. There is potential for significant conflict between the movements to improve global environmental conditions and those seeking reductions in international trade barriers.

4. Important global agreements have been forged, including the Kyoto Protocol and the Convention on Biological Diversity, that set frameworks for global actions, but progress on their implementation, by and large, is very slow.

5. Progress has been made in some areas. The Forest Stewardship Council (FSC) has now certified more than 140 million hectares of commercial forest in 80 countries. A Marine Stewardship Council, modelled on the FSC, has been initiated and has certified 131 fisheries offering 500 certified seafood products covering 7 per cent of the world's marine catch.

6. Research examining the relationship between consumerism and well-being shows that a positive relationship does exist when incomes are small. However, after a certain point is reached, there is no relationship. Most of the developed world has already attained triple this income level.

7. Only one of the eight global regions is on track to achieve all of the Millennium Development Goals by 2015, with severe shortfalls in many areas such as sub-Saharan Africa.

8. We need Millennium Consumption Goals as well as Millennium Development Goals that indicate where consumption needs to fall and by when.

9. The Happy Planet Index is a measure of human well-being in various countries against the impacts caused. In 2009, Canada ranked 89th on the scale.

10. Canada is ranked the highest in the world in terms of estimated value per capita of natural capital.

11. Canada suffers from a lack of sustained leadership for environmental protection and our response to environmental degradation at both international and national scales continues to be very slow despite clear signals regarding the state of the environment. We are in violation of many international environmental agreements.

12. Canada ranks 46th on the international Environmental Performance Index

13. Poor quality of information and the difficulty of accessing government information are serious challenges to improved environmental management in Canada. Canada ranked at the bottom of an international comparison of freedom-of-information legislation. There is increased control by the federal government on the ability of government scientists to discuss their research.

14. BC has introduced the most ambitious plan to address global climate change in North America, including carbon taxes. BC has set a target of reducing greenhouse gases by at least 33 per cent below current levels by 2020. This would be 10 per cent lower than the 1990 level, exceeding California's goal of returning to 1990 levels by 2020.

15. Ordinary citizens can have a positive impact on the environmental challenges facing society. A first step is building awareness of these problems. Universities and colleges should be intimately involved in this process by ensuring that all students have a measure of environmental literacy before they graduate. Universities and colleges are also large consumers of matter and energy and should lead by example in reducing their impact on the environment.

16. There is growing evidence of 'nature deficit disorder' among younger people as they spend more time on their computers and in front of TV screens and less time interacting with nature. This makes building awareness of environmental change and getting people emotionally engaged with change even more challenging.

17. Individuals can help by living in accordance with the four R's: *refuse* to be goaded into overconsumption; *reduce* consumption of matter and energy whenever possible; *reuse* materials whenever possible; and *recycle* those that you cannot reuse.

18. Individuals can also wield influence by banding together and taking collective action—for example, through consumer boycotts of products and companies that engage in environmentally destructive practices. One of the most effective ways of doing this is to join a non-governmental organization (NGO) composed of and supported by like-minded individuals. As an individual, you will probably not be able to afford to support an environmental lobbyist in Ottawa. However, if thousands of people contribute, it can become a reality.

19. Big corporations have a lot of opportunity to influence sustainability and resilience, and many of them have accepted that responsibility.

20. In spite of the serious nature of many of the environmental problems described in this book, it is important to remain optimistic that they can be solved and to take time to get out and enjoy the beauty and challenge of one of the most splendid parts of the planet, Canada, our home.

Key Terms

carrot mobs
consumerism
corporate responsibility
extended producer responsibility
Happy Planet Index (HPI)
law of everybody

life-cycle assessments (LCAs)
light living
Millennium Development Goals (MDGs)
nature deficit disorder

non-governmental organizations (NGOs)
100-mile diet
World Summit on Sustainable Development (WSSD)

Questions for Review and Critical Thinking

1. What do you think are the main factors driving global environmental degradation?

2. Which factors do you think are the most serious, and why?

3. Discuss what you think should be the top priorities for global action.

4. Outline some of the main responses to environmental degradation at both the international and national levels.

5. Discuss Canada's current and potential role at the international level.

6. What do you think are the main barriers to more effective government response in Canada?

7. Discuss your institution's present and potential role in raising environmental awareness.

8. What are three concrete steps towards 'light living' that you are willing to take over the next month?

9. What are some initiatives that you could take as an individual or as part of your community to reduce negative environmental consequences from human activities?

10. Find out the names and mandates of the environmental NGOs in your area. Is there anything you can do to help them?

11. Who is your local MP? What is his or her view on environmental issues? Exercise your democratic right—phone and find out.

Related Websites

Auditor General of Canada
www.oag-bvg.gc.ca

BC Sustainable Energy Association
www.bcsea.org

Center for a New American Dream
www.newdream.org

City Green
www.citygreen.ca

Climate Counts
www.ClimateCounts.org

Commissioner of the Environment and Sustainable Development
www.oag-bvg.gc.ca

David Suzuki Foundation, what you can do
www.davidsuzuki.org/what-you-can-do

EcoLogo
www.EcoLogo.org

Environment Canada, What You Can Do
www.ec.gc.ca/eco/main_e.htm

Environmentally Sound Transportation
www.best.bc.ca

Fair Trade Certified
www.transfair.ca

Happy Planet Index
www.happyplanetindex.org

Kesho Trust
www.naturechildreunion.ca

Marine Stewardship Council
www.msc.org

Natural Resources Canada, Office of Energy Efficiency
oee.nrcan.gc.ca/home

Nature Serve
www.natureserve.ca

Nature Watch
www.naturewatch.ca/english

Oceanwise
www.oceanwise.ca/iphone-app

Sierra Youth Coalition
www.syc-cjs.org

United Nations Conference on Sustainable Development
www.uncsd2012.org

United Nations Environment Programme
www.unep.org

United Nations Human Development Index
hdr.undp.org/en/statistics/hdi

World Summit on Sustainable Development
www.un.org/events/wssd

Further Readings

Note: This list comprises works relevant to the subject of the chapter but not cited in the text. All cited works are listed in the Bibliography at the end of the book.

Barber, J. 2003. *Production, Consumption and the World Summit for Sustainable Development*. Rockville, Md: Integrative Strategies Forum.

Cross, G. 2002. *An All-Consuming Century: Why Commercialism Won in Modern America*. New York: Columbia University Press.

French, H. 2004. 'Linking globalization, consumption and governance', in Worldwatch Institute, *State of the World 2004*. New York: W.W. Norton, 144–61.

United Nations. 2003. *Plan of Implementation of the World Summit on Sustainable Development*. New York: UN.

———. 2007. *The Millennium Development Goals Report 2007*. New York: UN.

Appendix: Conservation Organizations

International Organizations

Alliance for the Wild Rockies
P.O. Box 505
Helena, MT 59624
Phone: 406 459-5936
E-mail: awr@wildrockiesalliance.org
Website: www.wildrockiesalliance.org

Antarctic and Southern Ocean Coalition
1630 Connecticut Avenue NW, 3rd Floor
Washington, DC 20009
Phone: 202 234-2480
Fax: 202 387-4823
E-mail: secretariat@asoc.org
Website: www.asoc.org

Center for Plant Conservation
P.O. Box 299
St Louis, MO 63166-0299
Phone: 314 577-9450
Fax: 314 577-9465
E-mail: cpc@mobot.org
Website: www.centerforplantconservation.org

Clean Water Action Project
1010 Vermont Avenue, Suite 400
Washington, DC 20005
Phone: 202 895-0420
Fax: 202 895-0438
Website: www.cleanwateraction.org

Conservation International
2011 Crystal Drive, Suite 500
Arlington, VA 22202
Phone: 703 341-2400
Website: www.conservation.org

Cousteau Society
732 Eden Way North, Suite E, #707
Chesapeake, VA 23320
Phone: 757 523-9335
Fax: 757 523-8785
E-mail: cousteau@cousteausociety.org
Website: www.cousteau.org

Durrell Wildlife Conservation Trust
Les Augrès Manor, La Profonde Rue
Trinity, Jersey
Channel Islands JE3 5BP
Phone: 44 1534 860 000
Fax: 44 1534 860 001
Website: www.durrell.org

Earth Island Institute
2150 Allston Way, Suite 460
Berkeley, CA 94704-1375
Phone: 510 859-9100
Fax: 510 859-9091
Website: www.earthisland.org

Earthwatch
114 Western Avenue
Boston, MA 02134
Phone: 978 461-0081
Toll-free: 800 776-0188
Fax: 978 461-2332
E-mail: info@earthwatch.org
Website: www.earthwatch.org

Ecological Agricultural Projects
Macdonald Campus
McGill University
Ste-Anne-de-Bellevue, QC H9X 3V9
Phone: 514 398-7771
Fax: 514 398-7621
E-mail: ecological.agriculture@mcgill.ca
Website: eap.mcgill.ca

Friends of Animals
777 Post Road, Suite 205
Darien, CT 06820
Phone: 203 656-1522
Fax: 203 656-0267
E-mail: info@friendsofanimals.org
Website: www.friendsofanimals.org

Friends of the Earth International
International Secretariat
P.O. Box 19199 1000 GD
Amsterdam
The Netherlands
Phone: 31-20-622-1369
Fax: 31-20-639-2181
Website: www.foei.org

Greenpeace International
Ottho Heldringstraat 5 1066 AZ
Amsterdam
The Netherlands
Phone: 31-20-718-2000
Fax: 31-20-718-2002
E-mail: supporter.services.int@greenpeace.org
Website: www.greenpeace.org

International Fund for Animal Welfare
Canadian Office
301 1/2 Bank Street, Unit 2
Ottawa, ON K2P 1X7
Toll-free: 888 500-4329
Fax: 613 241-0641
E-mail: info-ca@ifaw.org
Website: www.ifaw.org

International Water Resources Association
Domaine de Lavalette
859 rue Jean-François Breton
34093 Montpellier Cedex 5
France
Phone: 33-4-676-12945
Fax: 33-4-675-22829
E-mail: office@iwra.org
Website: www.iwra.org

IUCN (World Conservation Union)
Rue Mauverney 28 CH 1196
Gland, Switzerland
Phone: 41-22-999-0155
Fax: 41-22-999-0015
Website: www.iucn.org

Jane Goodall Institute
4245 North Fairfax Drive, Suite 600
Arlington, VA 22203
Phone: 703 682-9220
Fax: 703 682-9312
Website: www.janegoodall.org

National Audubon Society
225 Varick Street
New York, NY 10014
Phone: 212 979-3000
Website: www.audubon.org

Ocean Conservancy
1300 19th St NW, 8th Floor
Washington, DC 20036
Phone: 202 429-5609
Toll-free: 800 519-1541
E-mail: membership@oceanconservancy.org
Website: www.oceanconservancy.org

Ocean Alliance
191 Weston Road
Lincoln, MA 01773
Phone: 781 259-0423
Toll-free: 800 969-4253
Fax: 781 259-0288
E-mail: question@oceanalliance.org
Website: www.oceanalliance.org

Rainforest Action Network
221 Pine Street, 5th Floor
San Francisco, CA 94104
Phone: 415 398-4404
Fax: 415 398-2732
E-mail: answers@ran.org
Website: www.ran.org

Sierra Club
85 Second Street, 2nd Floor
San Francisco, CA 94105
Phone: 415 977-5500
Fax: 415 977-5797
E-mail: information@sierraclub.org
Website: www.sierraclub.org

Soil and Water Conservation Society
945 SW Ankeny Road
Ankeny, IA 50023-9723
Phone: 515 289-2331
Toll-free: 800 843-7645
Fax: 515 289-1227
Website: www.swcs.org

United Nations Environment Programme
Regional Office for North America
900 17th Street, NW, Suite 506
Washington, D.C. 20006
Phone: 202 785-0465
Fax: 202 785-2096
Website: www.unep.ch
Convention on Biological Diversity: www.cbd.int

World Commission on Forests and Sustainable Development
International Institute for Sustainable Development
161 Portage Avenue East, 6th Floor
Winnipeg, MB R3B 0Y4
Phone: 204 958-7700
Fax: 204 958-7710
Website: www.iisd.org/wcfsd

World Resources Institute
10 G Street NE, Suite 800
Washington, DC 20002
Phone: 202 729-7600
Fax: 202 729-7610
Website: www.wri.org

World Society for the Protection of Animals
5th Floor 222 Grays Inn Road
London WC1X 8HB
Phone: 44 20 7239 0500
Fax: 44 20 7239 0653
E-mail: wspa@wspa-international.org
Website: www.wspa-international.org

Worldwatch Institute
1776 Massachusetts Avenue NW, Suite 800
Washington, DC 20036-1904
Phone: 202 452-1999
Fax: 202 296-7365
E-mail: worldwatch@worldwatch.org
Website: www.worldwatch.org

Canadian National Organizations

Animal Alliance of Canada
221 Broadview Avenue, Suite 101
Toronto, ON M4M 2G3
Phone: 416 462-9541
Fax: 416 462-9647
E-mail: contact@animalalliance.ca
Website: www.animalalliance.ca

Assembly of First Nations
Trebla Building
473 Albert Street, Suite 900
Ottawa, ON K1R 5B4
Phone: 613 241-6789
Toll-free: 866 869-6789
Fax: 613 241-5808
Website: www.afn.ca

Association for the Protection of Fur-Bearing Animals
215- 3989 Henning Drive
Burnaby, BC V5C 6P8
Phone: 604 435-1850
Fax: 604 435-1840
E-mail: fbd@furbearerdefenders.com
Website: www.furbearerdefenders.com

Canadian Arctic Resources Committee
488 Gladstone Ave
Ottawa, Ontario K1N 8V4
Phone: 613 759-4284
Toll-free: 866 949-9006
Fax: 613 237-3845
Website: www.carc.org

Canadian Council on Ecological Areas
c/o Vice Chairperson Robert Hélie
Wildlife Conservation Branch, Canadian Wildlife Service
Environment Canada
3-351 St. Joseph Boulevard
Gatineau, QC K1A 0H3
Phone: 819 953-7935
E-mail: robert.helie@ec.gc.ca
Website: www.ccea.org

Canadian Environmental Law Association
130 Spadina Avenue, Suite 301
Toronto, ON M5V 2L4
Phone: 416 960-2284
Fax: 416 960-9392
Website: www.cela.ca

Canadian Global Change Program
University of Victoria
P.O. Box 1700 STN CSC
Victoria, BC V8W 2Y2
Phone: 250 472-4337
Fax: 250 472-4830
E-mail: cgcp@uvic.ca
Website: www.globalcentres.org/cgcp

Canadian Parks and Wilderness Society (CPAWS)
506-250 City Centre Ave
Ottawa, ON K1R 6K7
Phone: 613 569-7226
Toll-free: 800 333-WILD (9453)
Fax: 613 569-7098
Website: www.cpaws.org

Canadian Water Resources Association
9 Corvus Court
Ottawa, ON K2E 7Z4
Phone: 613 237-9363
Fax: 613 594-5190
Website: www.cwra.org

Canadian Youth Climate Coalition
Phone: 514 467-6413
E-mail: cycc.director@gmail.com
Website: www.ourclimate.ca/wordpress/

City Farmer—Canada's Office of Urban Agriculture
Box 74567, Kitsilano RPO
Vancouver, BC V6K 4P4
Phone: 604 685-5832
E-mail: cityfarm@interchange.ubc.ca
Website: www.cityfarmer.info

Commission for Environmental Co-operation
393, rue Saint-Jacques ouest, bureau 200
Montreal, QC H2Y 1N9
Phone: 514 350-4300
Fax: 514 350-4314
E-mail: info@cec.org
Website: www.cec.org

Earth Day Canada
111 Peter Street, Suite 503
Toronto, ON M5V 2H1
Phone: 416 599-1991
Toll-free: 888 283-2784
Fax: 416 599-3100
E-mail: info@earthday.ca
Website: www.earthday.ca

Energy Probe Research Foundation
225 Brunswick Avenue
Toronto, ON M5S 2M6
Phone: 416 964-9223
Fax: 416 964-8239
Website: www.epresearchfoundation.wordpress.com

Environment Canada
Inquiry Centre
10 Wellington, 23rd Floor
Gatineau, QC K1A 0H3
Phone: 819 997-2800
Fax: 819 994-1412
TTY: 819 994-0736 (Teletype for the hearing impaired)
E-mail: enviroinfo@ec.gc.ca
Website: ec.gc.ca
Environmental Indicators: www.ec.gc.ca/indicateurs-indicators/
Sustainable Development: www.ec.gc.ca/dd-sd/

Fisheries and Oceans Canada
Communications Branch
200 Kent Street, 13th Floor, Station 13E228
Ottawa, ON K1A 0E6
Phone: 613 993-0999
Fax: 613 990-1866
TTY: 800 465-7735
E-mail: info@dfo-mpo.gc.ca
Website: www.dfo-mpo.gc.ca/index-eng.htm

Friends of the Earth Canada
300-260 St Patrick Street
Ottawa, ON K1N 5K5
Phone: 613 241-0085
Fax: 613 241-7998
E-mail: foe@foecanada.org
Website: www.foecanada.org

Greenpeace Canada
33 Cecil Street
Toronto, ON M5T 1N1
Phone: 416 597-8408
Toll-free: 800 320-7183
Fax: 416 597-8422
Website: www.greenpeace.org/canada/en/

International Institute for Sustainable Development
161 Portage Avenue East, 6th Floor
Winnipeg, MB R3B 0Y4
Phone: 204 958-7700
Fax: 204 958-7710
E-mail: info@iisd.ca
Website: www.iisd.org

Nature Canada
75 Albert Street, Suite 300
Ottawa, ON K1P 5E7
Phone: 613 562-3447
Toll-free: 800 267-4088
Fax: 613 562-3371
E-mail: info@naturecanada.ca
Website: www.naturecanada.ca

Nature Conservancy of Canada
36 Eglinton Avenue West, Suite 400
Toronto, ON M4R 1A3
Phone: 416 932-3202
Toll-free: 800 465-0029
Fax: 416 932-3208
E-mail: nature@natureconservancy.ca
Website: www.natureconservancy.ca

Parks Canada
25-8-N Eddy Street
Gatineau, QC K1A 0M5
Phone: 888 773-8888
TTY: 866 787-6221
E-mail: information@pc.gc.ca
Website: www.pc.gc.ca

Sea Shepherd Conservation Society
Canadian Office
P.O. Box 48446
Vancouver, BC V7X 1A2
Phone: 604 688-7325
E-mail: canada@seashepherd.org
Website: www.seashepherd.org

Sierra Club of Canada
1 Nicholas Street, Suite 412
Ottawa, ON K1N 7B7
Phone: 613 241-4611
Toll-free: 888 810-4204
Website: www.sierraclub.ca

Wildlife Habitat Canada
120 Iber Road, Suite 207
Ottawa, ON K2S 1E9
Phone: 613 722-2090
Toll-free: 800 669-7919
Fax: 613 722-3318
Website: www.whc.org

Wildlife Preservation Canada
RR 5, 5420 Highway 6 North
Guelph, ON N1H 6J2
Phone: 519 836-9314
Toll-free: 800 956-6608
Fax: 519 836-8840
E-mail: admin@wildlifepreservation.ca
Website: www.wildlifepreservation.ca

World Wildlife Fund Canada
245 Eglinton Avenue East, Suite 410
Toronto, ON M4P 3J1
Phone: 416 489-8800
Toll-free: 800 267-2632
Fax: 416 489-3611
E-mail: ca-panda@wwfcanada.org
Website: www.wwf.ca

Zoocheck Canada Inc.
788 1/2 O'Connor Drive
Toronto, ON M4B 2S6
Phone: 416 285-1744
E-mail: zoocheck@zoocheck.com
Website: www.zoocheck.com

British Columbia

BC Spaces for Nature
Box 673
Gibsons, BC V0N 1V0
E-mail: info@spacesfornature.org
Website: www.spacesfornature.org

CPAWS, BC Chapter
410-698 Seymour Street
Vancouver, BC V6B 3K6
Phone: 604 685-7445
Fax: 604 629-8532
E-mail: info@cpawsbc.org
Website: www.cpawsbc.org

Friends of Clayoquot Sound
P.O. Box 489, 331 Neill Street
Tofino, BC V0R 2Z0
Phone: 250 725-4218
E-mail: info@focs.ca
Website: www.focs.ca

Ministry of Environment
P.O. Box 9339, Stn Prov Govt
Victoria, BC V8W 9M1
Phone: 250 387-1161
Fax: 250 387-5669
E-mail: www.envmail@gov.bc.ca
Website: www.gov.bc.ca/env/

Nature Trust of British Columbia
260-1000 Roosevelt Crescent
North Vancouver, BC V7P 1M3
Phone: 604 924-9771
Toll-free: 866 288-7878
Fax: 604 924-9772
E-mail: info@naturetrust.bc.ca
Website: www.naturetrust.bc.ca

Sierra Club of Canada, BC Chapter
302-733 Johnson Street
Victoria, BC V8W 3C7
Phone: 250 386-5255
E-mail: info@sierraclub.bc.ca
Website: www.sierraclub.bc.ca

Western Canada Wilderness Committee
P.O. Box 2205, Station Terminal
Vancouver, BC V6B 3W2
Phone: 604 683-8220
Toll-free: 800 661-9453
Fax: 604 683-8229
E-mail: info@wildernesscommittee.org
Website: www.wildernesscommittee.org

Wildlife Rescue Association of British Columbia
5216 Glencairn Drive
Burnaby, BC V5B 3C1
Phone: 604 526-2747
Fax: 604 524-2890
E-mail: info@wildliferescue.ca
Website: www.wildliferescue.ca

Alberta

Alberta Environment and Water
10th Floor, Petroleum Plaza South Tower
9915-108 Street
Edmonton, AB T5K 2G8
Phone: 780 427-2700 (Toll-free by first dialing 310-0000)
Fax: 780 422-4086
Website: environment.alberta.ca

Alberta Sport, Recreation, Parks, and Wildlife Foundation
#903 Standard Life Centre
10405 Jasper Avenue
Edmonton, AB T5J 4R7
Phone: 780 415-1167
Fax: 780 415-0308
Website: www.asrpwf.ca

Alberta Wilderness Association
455-12 Street, NW
Calgary, AB T2N 1Y9
Phone: 403 283-2025
Toll-free: 866 313-0713
Fax: 403 270-2743
E-mail: awa@shaw.ca
Website: www.albertawilderness.ca

Bow Valley Naturalists
Box 1693
Banff, AB T1L 1B6
Phone/Fax: 403 762-4160
E-mail: info@bowvalleynaturalists.org
Website: www.bowvalleynaturalists.org

CPAWS, Northern Alberta Chapter
P.O. Box 52031
Edmonton, AB T6G 2T5
Phone: 780 432-0967
Fax: 780 439-4913
E-mail: infonab@cpaws.org
Website: www.cpawsnab.org

CPAWS, Southern Alberta Chapter
425–78th Avenue SW
Calgary, AB T2V 5K5
Phone: 403 232-6686
Fax: 403 232-6988
E-mail: info@cpawscalgary.org
Website: www.cpaws-southernalberta.org

Nature Alberta
11759 Groat Road
Edmonton, AB T5M 3K6
Phone: 780 427-8124
Fax: 780 422-2663
E-mail: info@naturealberta.ca
Website: www.naturealberta.ca

Saskatchewan

CPAWS, Saskatchewan Chapter
P.O. Box 25106, River Heights RPO
Saskatoon, SK S7K 8B7
Phone: 306 975-7005
E-mail: info@cpaws-sask.org
Website: www.cpaws-sask.org

Ministry of Environment
3211 Albert Street
Regina, SK S4S 5W6
Toll-free: 800 567-4224
E-mail: Centre.Inquiry@gov.sk.ca
Website: www.environment.gov.sk.ca

Nature Saskatchewan
1860 Lorne Street, Suite 206
Regina, SK S4P 3W6
Phone: 306 780-9273
Toll-free: 800 667-4668
Fax: 306 780-9263
E-mail: info@naturesask.ca
Website: www.naturesask.ca

Saskatchewan Environmental Society
P.O. Box 1372
Saskatoon, SK S7K 3N9
Phone: 306 665-1915
E-mail: info@environmentalsociety.ca
Website: www.environmentalsociety.ca

Manitoba

CPAWS, Manitoba Chapter
3-303 Portage Avenue
Winnipeg, MB R3B 2B4
Phone: 204 949-0782
E-mail: info@cpawsmb.org
Website: www.cpawsmb.org

Manitoba Conservation
Phone: 204 945-3744
Toll-free: 866 626-4862 (MANITOBA)
TTY: 204 945-4796
E-mail: mgi@gov.mb.ca
Website: www.gov.mb.ca/conservation/

Nature Manitoba
63 Albert Street, Suite 401
Winnipeg, MB R3B 1G4
Phone: 204 943-9029
E-mail: info@naturemanitoba.ca
Website: www.naturemanitoba.ca

Ontario

CPAWS, Wildlands League (Ontario Chapter)
401 Richmond Street West, Suite 380
Toronto, ON M5V 3A8
Phone: 416 971-9453 (WILD)
Toll-free: 866 510-9453 (WILD)
Fax: 416 979-3155
E-mail: info@wildlandsleague.org
Website: www.wildlandsleague.org

CPAWS, Ottawa Valley Chapter
190 Bronson Avenue
Ottawa, ON K1R 6H4
Phone: 613 232-7297
Fax: 613 569-7098
E-mail: jmcdonnell@cpaws.org
Website: www.cpaws-ov-vo.org

Earthroots
401 Richmond Street West, Suite 410
Toronto, ON M5V 3A8
Phone: 416 599-0152
Fax: 416 340-2429
E-mail: info@earthroots.org
Website: www.earthroots.org

Elora Environment Centre
75 Melville Street, 2nd floor
Elora, ON N0B 1S0
Phone: 519 846-0841
Fax: 519 846-2642
Toll-free: 866 865-7337
E-mail: info@ecee.on.ca
Website: www.ecee.on.ca

Environment North
P.O. Box 10307
Thunder Bay, ON P7B 6T8
E-mail: environmentnorth@gmail.com
Website: www.environmentnorth.ca

Ministry of the Environment
Public Information Centre
135 St Clair Avenue West, 1st Floor
Toronto, ON M4V 1P5
Phone: 416 325-4000
Toll-free: 800 565-4923
Fax: 416 325-3159
Website: www.ene.gov.on.ca

Ontario Nature
214 King Street West, Suite 612
Toronto, ON M5H 3S6
Phone: 416 444-8419
Fax: 416 444-9866
E-mail: info@ontarionature.org
Website: www.ontarionature.org

Water Environment Association of Ontario
P.O. Box 176
Milton, ON L9T 4N9
Phone: 416 410-6933
Fax: 416 410-1626
E-mail: julie.vincent@weao.org
Website: www.weao.org

Quebec

CPAWS (SNAP), Quebec chapter
7275, St-Urbain suite 303
Montréal, QC H2R 2Y5
Phone: 514 278-7627
Website: www.snapqc.org

La Fondation québécoise en environnement
1255 carré Phillips, bureau 706
Montréal, QC H3B 3G1
Phone: 514 849-3323
Toll-free: 800 361-2503
Fax: 514 849-0028
E-mail: info@fqe.qc.ca
Website: www.fqe.qc.ca

Nature Québec/UQCN
870 avenue de Salaberry, bureau 207
Québec, QC G1R 2T9
Phone: 418 648-2104
Fax: 418 648-0991
E-mail: conservons@naturequebec.org
Website: www.naturequebec.org

Ministry of Natural Resources and Wildlife
880, chemin Sainte-Foy, RC 120-C
Québec, QC G1S 4X4
Phone: 418 627-8600
Toll-free: 866 248-6936
Fax: 418 644-6513
E-mail: services.clientele@mrnf.gouv.qc.ca
Website: www.mrn.gouv.qc.ca

Ministry of Sustainable Development, Environment and Parks
Édifice Marie-Guyart, 29th Floor
675, boulevard René-Lévesque Est
Québec, QC G1R 5V7
Phone: 418 521-3830
Toll-free: 800 561-1616
Fax: 418 646-5974
E-mail: info@mddep.gouv.qc.ca
Website: www.mddep.gouv.qc.ca

New Brunswick

Conservation Council of New Brunswick
180 St John Street
Fredericton, NB E3B 4A9
Phone: 506 458-8747
Fax: 506 458-1047
E-mail: info@ccnbaction.ca
Website: www.conservationcouncil.ca

Department of Environment
Marysville Place, P.O. Box 6000
Fredericton, NB E3B 5H1
Phone: 506 453-2690
Fax: 506 457-4994
E-mail: env-info@gnb.ca
Website: www.gnb.ca/0009/index-e.asp

Nature Trust of New Brunswick
P.O. Box 603, Station A
Fredericton, NB E3B 5A6
Phone: 506 457-2398
Fax: 506 450-2137
E-mail: naturetrust@ntnb.org
Website: www.naturetrust.nb.ca

New Brunswick Federation of Naturalists
924 Prospect Street, Suite 110
Fredericton, NB E3B 2T9
Phone: 506 459-4209
E-mail: nbfn@nb.aibn.com
Website: www.naturenb.ca

Prince Edward Island

Environment, Energy, and Forestry
Jones Building, 4th Floor
11 Kent Street, P.O. Box 2000
Charlottetown, PEI C1A 7N8
Phone: 902 368-5000
Fax: 902 368-5830
Website: www.gov.pe.ca/eef/

Island Nature Trust
P.O. Box 265
Charlottetown, PEI C1A 7K4
Phone: 902 566-9150
Fax: 902 628-6331
E-mail: admin@islandnaturetrust.ca
Website: www.islandnaturetrust.ca

Nova Scotia

CPAWS, Nova Scotia Chapter
5435 Portland Place, Suite 101
Halifax, NS B3K 6R7
Phone: 902 446-4155
Fax: 902 446-4156
Website: www.cpawsns.org

Ecology Action Centre
2705 Fern Lane
Halifax, NS B3K 4L3
Phone: 902 429-2202
Fax: 902 405-3716
E-mail: info@ecologyaction.ca
Website: www.ecologyaction.ca

Nature Nova Scotia
c/o Nova Scotia Museum of Natural History,
1747 Summer Street
Halifax, NS B3H 3A6
Phone: 902 582-7176
Website: www.naturens.ca

Nova Scotia Environment
PO Box 442, 5151 Terminal Road
Halifax, NS B3J 2P8
Phone: 902 424-3600
Fax: 902 424-0503
Website: www.gov.ns.ca/nse/

Newfoundland and Labrador

Department of Environment and Conservation
4th Floor, West Block Confederation Building
P.O. Box 8700
St John's, NL A1B 4J6
Phone: 709 729-2664
Toll-free: 800 563-6181
Fax: 709 729-6639
Website: www.env.gov.nl.ca

Protected Areas Association of Newfoundland and Labrador
Box 1027, Station C
St John's, NL A1C 5M5
Phone: 709 726-2603
Fax: 709 726-2764
E-mail: paa@nf.aibn.com
Website: www.paanl.org

Wilderness and Ecological Reserves Advisory Council
c/o Parks and Natural Areas Division,
Department of Environment and Conservation
33 Reid's Lane
Deer Lake, NL A8A 2A3
Phone: 709 635-3854
Fax: 709 635-4541
E-mail: werac@gov.nl.ca
Website: www.env.gov.nl.ca/env/parks/wer/adc/index.html

Yukon

CPAWS, Yukon Chapter
P.O. Box 31095
211 Main Street
Whitehorse, YT Y1A 5P7
Phone: 867 393-8080
Fax: 867 393-8081
Website: www.cpawsyukon.org

Department of Environment
Box 2703 (V-3A)
Whitehorse, YT Y1A 2C6
Phone: 867 667-5652
Toll-free (in Yukon): 800 661-0408, local 5652
Fax: 867 393-7197
E-mail: environment.yukon@gov.yk.ca
Website: www.env.gov.yk.ca

Yukon Conservation Society
302 Hawkins Street
Whitehorse, YT Y1A 1X6
Phone: 867 668-5678
Fax: 867 668-6637
E-mail: ycs@ycs.yk.ca
Website: www.yukonconservation.org

Northwest Territories

Department of Environment and Natural Resources
P.O. Box 1320
Yellowknife, NT X1A 2L9
Phone: 403 669-2302
Website: www.enr.gov.nt.ca

Ecology North
5013 51st Street
Yellowknife, NT X1A 1S5
Phone: 867 873-6019
E-mail: admin@ecologynorth.ca
Website: www.ecologynorth.ca

Nunuvut

Department of Environment
P.O. Box 1000, Stn. 1300
Iqaluit, NU X0A 0H0
Phone: 867 975-7700
Fax: 867 975-7742
E-mail: environment@gov.nu.ca
Website: env.gov.nu.ca

Nunavut Parks
P.O. Box 1000, Stn. 1340
Iqaluit, NU X0A 0H0
Phone: 867 975-7700
Fax: 867 975-7747
E-mail: parks@gov.nu.ca
Website: www.nunavutparks.com

Glossary

abiotic components Non-living parts of the ecosystem, including chemical and physical factors, such as light, temperature, wind, water, and soil characteristics.

Aboriginal peoples The Indian (First Nations), Inuit, and Métis peoples of Canada.

acid deposition Rain or snow that has a lower pH than precipitation from unpolluted skies; also includes dry forms of deposition, such as nitrate and sulphate particles.

acidification The increased acidic content of waters, notably the world's oceans, so that the concentration of available carbonate ions will be too low for marine calcifiers, such as coral reefs, molluscs, crustaceans, and some algae, to build their shells and skeletons.

acid mine drainage Acidic drainage from waste rock and mine tailings caused by the oxidization of iron sulphides to create sulphuric acid, which in turn dissolves residual metals.

acid shock The buildup of acids in water bodies and standing water over the winter, resulting in higher acidity than experienced through the rest of the year.

active management Purposeful interference by resource and environmental managers in ecosystems, which recognizes that the human forces of change are now so ubiquitous that even protected areas are affected; includes habitat restoration, creation of wildlife corridors, reintroduction of extirpated species, prescribed burning, and management of hyper-abundant species.

adaptation Adjustment to different or changing circumstances, such as when insurance companies modify their claims forecasting and setting of premiums with regard to future climate change conditions. The largest challenge for adaptation strategies will occur in the future when the most significant consequences from climate change will appear.

adaptive co-management Management concept including such key attributes as learning-by-doing, integrating different knowledge systems, collaborating and power-sharing among community, regional, and national levels, and managing for flexibility.

adaptive environmental management An approach that develops policies and practices to deal with the uncertain, the unexpected, and the unknown; approaches management as an experiment from which we learn by trial and error.

aerobic Requiring oxygen.

albedo The extent to which the surface of the Earth reflects rather than absorbs incoming radiation from the sun. Snow has a high albedo, but as temperatures rise, the area covered in snow will be replaced by areas free of snow, uncovering rocks and vegetation with lower albedo values that absorb radiation and thus add to warming.

alien species Any organism, such as zebra mussels, purple loosestrife, and Eurasian water milfoil in Canada, that enters an ecosystem beyond its normal range through deliberate or inadvertent introduction by humans; also known as exotic, introduced, invader, or non-native species.

allelopathic A plant that directly inhibits the growth of surrounding species through production of chemicals in the soil.

alternative dispute resolution (ADR) A non-judicial approach to resolving disputes that uses negotiation, mediation, or arbitration, with a focus on reparation for harm done and on improving future conduct.

anadromous Aquatic life, such as salmon, that spend part of their lives in salt water and part in fresh water.

anaerobic Lacking oxygen.

annual allowable cut (AAC) The amount of timber that is allowed to be cut annually from a specified area.

anthropocentric view Human-centred, in which values are defined relative to human interests, wants, and needs.

aquaculture Seafood farming, the fastest-growing food production sector in the world.

aquifer A formation of permeable rocks or loose materials that contains usable sources of groundwater and may extend from a few square kilometres to several thousand square kilometres.

arbitration A procedure for dispute resolution in which a third party is selected to listen to the views and interests of the parties in dispute and develop a solution to be accepted by the participants.

artisanal Small-scale fisheries.

aspirational approach With reference to climate change, an approach emphasizing long-term but unspecific and non-binding targets for reducing greenhouse gas emissions. Advocated by developed countries.

assimilated food energy The proportion of ingested energy actually absorbed by an organism.

Atlantic Maritime Ecozone stretching from the mouth of the St Lawrence River across New Brunswick, Nova Scotia, and Prince Edward Island that is heavily influenced by the Atlantic Ocean, which creates a cool, moist maritime climate but with quite variable conditions between the upland masses, such as the Cape Breton and New Brunswick highlands, and the coastal lowlands that support most of the population.

atmosphere Layer of air surrounding the Earth.

autotrophs Organisms, such as plants, that produce their own food, generally via photosynthesis.

Bali Conference A UN-sponsored climate change conference held in the first two weeks of December 2007 in Bali, Indonesia, to start a process to create a new framework to replace the Kyoto Protocol, which ends in 2012. There was agreement that both developed and developing countries must participate in reducing greenhouse gas emissions but reluctance from key developed countries to commit to binding targets.

benthic Of or living on or at the bottom of a water body.

bioaccumulation The storage of chemicals in an organism in higher concentrations than are normally found in the environment.

biocapacity The amount of biologically productive area—cropland, pasture, forest, and fisheries—available to meet humanity's needs.

biocentric perspective A view that values aspects of the environment simply because they exist and accepts that they have the right to exist.

biocides Chemicals that kill many different kinds of living things; also called pesticides.

bioconcentration The combined effect of bioaccumulation and biomagnification.

biodiversity The variety of life forms that inhabit the Earth. Biodiversity includes the genetic diversity among members of a population or species as well as the diversity of species and ecosystems.

biodiversity hot spots Areas with high numbers of endemic species, as in tropical forests.

biofuels Solid, liquid, or gas fuel derived from relatively recently dead biological material and distinguished from fossil fuels, which are derived from long-dead biological material.

biogeochemical cycles Series of biological, chemical, and geological processes by which materials cycle through ecosystems.

biological oxygen demand (BOD) The amount of dissolved oxygen required for the bacterial decomposition of organic waste in water.

biomagnification Buildup of chemical elements or substances in organisms in successively higher trophic levels.

biomass The sum of all living material, or of all living material of particular species, in a given environment.

biomass pyramid Related to the fact that in terrestrial ecosystems, greater biomass generally exists at the level of primary consumers, with the least total biomass at the highest trophic levels; in marine ecosystems, the reverse is true, and the pyramid is inverted—greater biomass is at the highest trophic level, while the primary consumers, phytoplankton, at any given time comprise much less biomass but reproduce rapidly.

biomes Major ecological communities of organisms, both plant and animal, that are usually characterized by the dominant vegetation type; for example, a tundra biome and a tropical rain forest biome.

biosphere The zone of all living matter on Earth, including animals, vegetation, and the soil layer.

biotic components Those parts of ecosystems that are living; organisms.

biotic potential The ability of species to reproduce regardless of the level that an environment can support, i.e., regardless of the carrying capacity of the environment.

bitumen Any of various minerals that will burn, such as asphalt or petroleum.

blue boxes Blue plastic boxes or bins introduced for curbside collection programs in urban neighbourhoods to divert selected household waste material (e.g., bottles, cans, paper and plastic products) from landfill sites to be recycled.

Boreal Cordillera Ecozone, to the south and west of the Tundra Cordillera, in northern BC and southern Yukon made up of mountains in the west and east, separated by intermontane plains, with a wet climate that gives rise to tree growth.

boreal forest One of the largest forest belts in the world, extending all across North America and Eurasia, encompassing roughly a third of the Earth's forest land and 14 per cent of the world's forest biomass and separating the treeless tundra regions to the north from the temperate deciduous forests or grasslands to the south.

Boreal Plains Ecozone extending from the southern part of the Yukon in a wide sweeping band down into southeastern Manitoba, consisting of a generally flat to undulating surface similar to the Prairie zone to the south.

Boreal Shield The largest ecozone in Canada, stretching along the Canadian Shield from Saskatchewan to Newfoundland.

bottom trawling One of the most destructive means of fishing in which heavy nets are dragged along the sea floor scooping up everything in their path.

brownfields Abandoned or active industrial sites. On the surface or underground are soils contaminated through disposal practices accepted in earlier times before people appreciated the long-term consequences.

buffering capacity The factors in an environment, such as carbonate-rich rocks and deep soils, that ameliorate the harms caused by acid deposition,

butterfly effect A central example in chaos theory that postulates the effect of a butterfly flapping its wings in South America might affect weather systems in North America.

bycatch Non-target organisms caught or captured in the course of catching a target species, as in the fisheries, where estimates suggest that 25 per cent of the world's catch is dumped because it is not the right species or size.

calorie A unit of heat energy; the amount of heat required to raise the temperature of one gram of water by 1°C.

Cancún Summit Meeting of representatives from 193 countries and other interested parties held in Mexico in December 2010 to seek to advance mitigative action on climate change. Canada continued to be a laggard, and only incremental progress was made.

carbon balance A balance between the amount of CO_2 in the atmosphere and bicarbonate in the water.

carbon offsets A measure of greenhouse gas reduction, measured in metric tons of carbon dioxide equivalent. One carbon offset equals a reduction in one metric ton of carbon dioxide equivalent of greenhouse gas emission. Companies or governments can arrange to receive credit against emission caps from carbon offsets. At a personal level, some travellers pay for a carbon offset related to greenhouse gas emissions produced when they travel by air. The amount paid is used by airlines for planting trees or other initiatives that reduce greenhouse gas emissions.

carbon sequestration Reforestation and afforestation to ameliorate carbon dioxide loadings in the atmosphere because trees and shrubs use the excess CO_2.

carbon tax An approach in which greenhouse gas emissions by individuals or companies are taxed. The purpose is to change human behaviour towards activities that produce fewer greenhouse gas emissions.

carnivore An organism that consumes only animals.

carrot mobs Organized consumer purchasing in support of businesses demonstrating environmental and social leadership.

carrying capacity Maximum population size that a given ecosystem can support for an indefinite period or on a sustainable basis.

certification The confirmation of certain characteristics of an object, person, or organization, as with various forestry programs certifying that wood products have come from sustainably managed forests.

chain-of-custody Procedures for verification of compliance with sustainable practices from product origin through to the final product, as with wood products from the forest to Home Depot.

chemoautotroph A producer organism that converts inorganic chemical compounds into energy.

chlorophylls Pigments of plant cells that absorb sunlight, thus enabling plants to capture solar energy.

clear-cutting A forest harvesting technique in which an entire stand of trees is felled and removed.

climate The long-term weather pattern of a particular region.

climate change A long-term alteration in the climate of a particular location or region or for the entire planet.

climate change deniers Those who, for ideological and economic reasons, use communication tactics to question the science underlying climate change and therefore delay action to mitigate this change.

'Climategate' The controversy surrounding leaked e-mails from a climate research centre at the University of East Anglia, just weeks prior to the Copenhagen Summit, which appeared, incorrectly, to suggest that researchers had manipulated their data to make climate change appear more severe.

climate modelling Various mathematical and computerized approaches for determining past climate trends in an effort to build scenarios predicting future climate, which use any or all of the following factors in measurement: incoming and outgoing radiation; energy dynamics or flows around the globe; surface processes affecting climate, such as snow cover and vegetation; chemical composition of the atmosphere; and time step or resolution (time over which the model runs and the spatial scale to which it applies).

climatic climax The situation in a mature or climax community where the vegetative growth is largely influenced by climate.

climax community Last stage of succession; a relatively stable, long-lasting, complex, and interrelated community of organisms.

co-evolution Process whereby two species evolve adaptations as a result of extensive interactions with each other.

collaboration The art of working together.

co-management An arrangement in which a government agency shares or delegates some of its legal authority regarding a resource or environmental management issue with local inhabitants of an area.

commensalism An interaction between two species that benefits one species and neither harms nor benefits the other.

community Populations in a particular environment.

competitive exclusion principle The principle that competition between two species with similar requirements will result in the exclusion of one of the species.

complete-tree harvesting The harvesting of all of the above- and below-ground biomass of a tree.

compound The coming together of two different atoms to form a different substance, such as water (H_2O), a compound made up of two hydrogen atoms (H) and one oxygen atom (O).

condensation nuclei Particles in the atmosphere that provide a starting point for water moving from the gaseous to liquid phase.

consumerism Wasteful consumption of resources to satisfy wants rather than needs.

consumers Organisms that cannot produce their own food and must get it by eating or decomposing other organisms; in economics, those who use goods and services.

contemporary evolution Evolution that occurs on short time scales.

context Specific characteristics of a time and place.

contour cultivation The cultivation and seeding of fields parallel to the contour of the slope, which serves to reduce the speed of runoff by catching soil particles in the plough furrows.

Convention on Biological Diversity (CBD) International treaty that emerged from the World Summit on Sustainable Development in Rio de Janeiro in 1992 that requires signatories, including Canada, to develop biodiversity strategies, identify and monitor important components of biodiversity, develop endangered species legislation/protected areas systems, and promote environmentally sound and sustainable development in areas adjacent to protected areas.

Convention on International Trade in Endangered Species of Wild Fauna and Flora (CITES) A 1973 treaty currently ratified by more than 120 countries (including Canada) that establishes lists of species for which international trade is to be controlled or monitored (e.g., orchids, cacti, parrots, large cat species, sea turtles, rhinos, primates).

co-ordination The effective or harmonious working together of different departments, groups, and individuals.

Copenhagen Summit Two-week meeting of world leaders, environment ministers, and other interested parties held in Copenhagen, Denmark, in late 2009, which sought unsuccessfully to advance the agenda for action on climate change. Canada showed itself at this conference to be among the greatest laggards in seeking action for improved GHG emissions standards.

coral bleaching Death of corals caused by water temperatures becoming too warm.

coral polyps Individual biotic members of a coral reef.

corporate responsibility Occurs when corporations systematically examine the environmental impact of their activities, then take action to reduce the negative impacts. Initiatives include reducing, reusing, and recycling, specifying environment-friendly production practices for suppliers, and providing funds to environmental groups.

corporate social investment Contributions or actions of a company to assist the affected community, such as a mining community, in achieving its own development priorities in ways sustainable and supportive of both parties' objectives.

critical load The maximum level of acid deposition that can be sustained in an area without compromising ecological integrity.

crop rotation Alternating crops in fields to help restore soil fertility and also control pests.

crude birth rate (CBR) Number of births in a population per 1,000 individuals per year.

crude bitumen A thick and heavy oil.

crude death rate (CDR) Number of deaths in a population per 1,000 individuals per year.

crude growth rate (CGR) Produced by subtracting the crude death rate (CDR) from the crude birth rate (CBR).

cryosphere Based on the Greek word *kryos* meaning 'cold', those parts of the Earth's surface where water is in solid form as ice or snow. The cryosphere includes sea, lake, and river ice, snow cover, glaciers, ice caps and ice sheets, as well as frozen ground (permafrost).

culmination age The age of economic maturity of a tree crop, which varies widely but usually falls within the 60- to 120-year range in Canada.

cumulative environmental effects The combined effects of an action with the effects of other past actions and that have implications for the present and future.

custom-designed solutions Management approach in which the specific conditions of a place and time are recognized and the attempt to ameliorate or resolve a problem takes these specifics into account.

cyclic succession Where a community progresses through several seral stages but is then returned to earlier stages by natural phenomena such as fire.

DDT (dichlorodiphenyltrichloroethane) An organochlorine insecticide used first to control malaria-carrying mosquitoes and lice and later to control a variety of insect pests but now banned in Canada because of its persistence in the environment and ability to bioaccumulate.

decomposer food chain A specific nutrient and energy pathway in an ecosystem in which decomposer organisms (bacteria and fungi) consume dead plants and animals as well as animal wastes; essential for the return of nutrients to soil and carbon dioxide to the atmosphere; also called detritus food chain.

demand management Emphasizes influencing human behaviour so that less water or energy is used.

demographic transition Transition of a human population from high birth rate and high death rate to low birth rate and low death rate.

denitrification The conversion of nitrate to molecular nitrogen by bacteria in the nitrogen cycle.

detritus Organic waste, such as fallen leaves.

disturbances Natural or human-induced events or processes that interrupt ecological succession.

diversions Movements of water from one water system to another in order to enhance water security, reduce flood vulnerability, or generate hydroelectricity.

dominant limiting factor The weakest link in the chain of various factors necessary for an organism's survival.

double-loop learning Situations for which there is a mismatch between intention and outcome and when such a mismatch is addressed by challenging underlying values and behaviour rather than assuming that the prevailing values and behaviour are appropriate.

drought Condition in which a combination of lack of precipitation, temperature, evaporation, evapotranspiration, and the inability of soil to retain moisture leads to a loss of resilience among flora and fauna in dry conditions.

dyke A wall or earth embankment along a watercourse to control flooding (running dyke), or encircling a town or a property to protect it from flooding (ring dyke), or across a stream so that the flow of water is stopped from going upstream by a sluice gate (cross dyke).

dynamic equilibrium Occurs when two opposing processes proceed at the same rate.

E. coli *Escherichia coli*, a bacterium present in fecal matter that can get into a water supply and pollute it, as happened in Walkerton, Ontario, in May 2000.

ecocentric (biocentric) values The view that a natural order governs relationships between living things and that a harmony and balance reflect this natural order, which humankind tends to disrupt.

ecological footprint The land area a community needs to provide its consumptive requirements for food, water, and other products and to dispose of the wastes from this consumption.

ecologically extinct A species that exists in such low numbers that it can no longer fulfill its ecological role in the ecosystem.

ecological redundancy The situation, as in a tropical rain forest, where there are many times more species than in more northerly ecosystems and the chance of other species combining to fulfill the ecological role of a depleted one is much higher.

ecological restoration Renewing a degraded, damaged, or destroyed ecosystem through active human intervention.

ecological succession The gradual replacement of one assemblage of species by another as conditions change over time.

ecosphere Refers to the entire global ecosystem, which comprises atmosphere, lithosphere, hydrosphere, and biosphere as inseparable components.

ecosystem Short for ecological system; a community of organisms occupying a given region within a biome, including the physical and chemical environment of that community and all the interactions among and between organisms and their environment.

ecosystem-based management Holistic management that takes into account the entire ecosystem and emphasizes biodiversity and ecosystem integrity, as opposed to focusing primarily or solely on a resource or resources, such as water or timber, within an ecosystem.

ecosystem diversity The variety of ecosystems in an area.

ecotone The transitional zone of intense competition for resources and space between two communities.

edaphic climaxes The situation in mature or climax communities where the vegetative growth is principally influenced by underlying geologic features, such as soils.

El Niño A marked warming of the waters in the eastern and central portions of the tropical Pacific that triggers weather changes and events in two-thirds of the world.

emission credits Can be earned by a nation based on land-use or forestry (afforestation, reforestation) initiatives that reduce measurable greenhouse gas emissions.

emissions trading Under the Kyoto Protocol, a system whereby one country that will exceed its allotted limit of greenhouse gas emissions can buy an amount of greenhouse gas emissions from another country that will not reach its own established emissions limit.

endangered An official designation assigned by the Committee on the Status of Endangered Wildlife in Canada to any indigenous species or subspecies or geographically separate population of fauna or flora that is threatened with imminent extinction or extirpation throughout all or a significant portion of its Canadian range.

endemic species A plant or animal species confined to or exclusive to a specific area.

endocrine disruption The interference of normal bodily processes such as sex, metabolism, and growth by chemicals in such products as soaps and detergents that are released into an ecosystem, as happens among aquatic species, often causing feminization.

energy The capacity to do work; found in many forms, including heat, light, sound, electricity, coal, oil, and gasoline.

energy efficiency Amount of total energy input of a system that is transformed into work or some other usable form of energy.

entropy A measure of disorder. The second law of thermodynamics applied to matter says that all systems proceed to maximum disorder (maximum entropy).

environment The combination of the atmosphere, hydrosphere, cryosphere, lithosphere, and biosphere in which humans, other living species, and non-animate phenomena exist.

environmental impact assessment Part of impact assessment that identifies and predicts the impacts from development proposals on both the biophysical environment and on human health and well-being.

epidemiological transition Change in mortality rates from high to low in a human population.

epiphytes Plants that use others for support but not nourishment.

estuary Coastal regions, such as inlets or mouths of rivers, where salt water and fresh water mix.

euphotic zone Zone of the ocean to which light from the sun reaches.

eutrophic Pertaining to a body of water rich in nutrients.

eutrophication The over-fertilization of a body of water by nutrients that produce more organic matter than the water body's self-purification processes can overcome; also called nutrient enrichment.

evapotranspiration Evaporation of water from soil and transpiration of water from plants.

evolution A long-term process of change in organisms caused by random genetic changes that favour the survival and reproduction of those organisms possessing the genetic change; organisms become better adapted to their environment through evolution.

exclusive economic zones (EEZs) Areas off the coasts of a nation that are claimed by that nation for its sole responsibility and exploitation, as permitted by the UN Convention on the Law of the Sea.

exponential growth The growth when a population increases by a certain percentage rather than an absolute amount, producing a J-shaped curve.

ex situ conservation The conservation of species outside their natural habitat, including breeding in captivity, so that they can be reintroduced to their natural habitat, as has been done, for example, with the black-footed ferret and swift fox.

ex situ preservation The preservation of representatives of a species, often endangered, outside their natural habitat, as in a zoo, aquarium, or game farm.

extended producer responsibility The concept underlying laws or regulations that require manufacturers and importers to accept responsibility for their products at the end of their useful lifespan. They provide an incentive for companies to design their products so that they can be recycled or reused and to eliminate toxic materials, since they would have to dispose of them.

extinction The elimination of all the individuals of a species.

extirpated An official designation assigned by the Committee on the Status of Endangered Wildlife in Canada to any indigenous species or subspecies or geographically separate population of fauna or flora no longer known to exist in the wild in Canada but occurring elsewhere.

extrinsic values Values that humans derive from other species, including consumptive and non-consumptive values.

falldown effect The lower volume of harvestable timber at the culmination age for second growth on sites where old-growth forest was previously harvested.

feng shui landscapes Landscapes reflecting the East Asian belief that humans should live in harmony with nature and that trees bring good fortune. Thus, *feng shui* landscapes involve the inclusion of trees and other vegetation within urbanized or settled areas.

fishing down the food chain Harvesting at progressively lower trophic levels as higher trophic levels become depleted.

flood plain Low-lying land along a river, stream, or creek or around a lake that under normal conditions is flooded from time to time.

floodway (diversion) An excavated channel to divert flood waters away from a population centre.

food chain A specific nutrient and energy pathway in an ecosystem proceeding from producer to consumer; along the pathway, organisms in higher trophic levels gain energy and nutrients by consuming organisms at lower trophic levels.

food webs Complex intermeshing of individual food chains in an ecosystem.

forest tenure The conditions that govern forest ownership and use.

fossil fuels Organic fuels (coal, natural gas, oil, tar sands, and oil shale) derived from once-living plants or animals.

fragmentation The division of an ecosystem or species habitat into small parcels as a result of human activity, such as agriculture, highways, pipelines, and population settlements.

frozen core dam Type of dam used in mining in the North, as at the Ekati diamond mine in the Northwest Territories, with a central core of frozen soil saturated with ice and bonded to the natural permafrost and surrounded by granular fill to ensure both stability and thermal protection.

full-tree harvesting Timber-cutting where trees are felled and transported to roadside with branches and top intact.

functional compensation The situation where a given role in an ecosystem, e.g., as decomposer or as prey, can be fulfilled by more than one species within that system.

Gaia hypothesis View that the ecosphere itself is a self-regulating homeostatic system in which the biotic and abiotic components interact to produce a balanced state; the ecosphere as Mother Nature.

gaseous cycles Cycles of elements that have most of their matter in the atmosphere.

general circulation models (GCMs) The most prominent and most complex type of climate modelling, which takes into account the three-dimensional nature of the Earth's atmosphere and oceans or both.

generalist species Species, like the black bear and coyote, with a very broad niche where few things organic are not considered a potential food item.

genetically modified organisms (GMOs) Organisms created by humans through genetic manipulation combining genes from different and often totally unrelated species to create a different organism that is economically more productive and/or has greater resistance to pathogens.

genetic diversity The variability in genetic makeup among individuals of the same species.

geo-engineering Various technologies, from as simple as tree planting to as complex as stratospheric aerosols and space mirrors, that are used or have been proposed to mitigate the effects of climate change.

glaciation Period of global cooling when alpine glaciers increase and continental ice sheets cover and scour vast land masses.

global climate change Impacts of accumulation of greenhouse gases on the Earth's climate.

global warming Changes in average temperatures of the Earth's surface, although these changes are not uniform (i.e., some regions experience significantly higher temperatures, others only slight changes upward, and still others might experience somewhat cooler temperatures).

governance The processes used to determine how policy decisions are taken and by whom. Governance arrangements identify how disputes will be resolved and require capacity to identify trade-offs and compromises.

government The formal rules or authority over a country, state, or other jurisdiction, facilitated by people, structures, and processes and designed to provide transparency and accountability for decisions taken.

grasshopper effect Atmospheric transport and deposition of persistent and volatile chemical pollutants whereby the pollutants evaporate into the air in warmer climates and travel in the atmosphere towards cooler areas, condensing out again when the temperature drops. The cycle then repeats itself in a series of 'hops' until the pollutants reach climates where they can no longer evaporate.

grazing food chains Energy transfer among organisms that is directly dependent on solar radiation as the primary source of energy and the producers (green plants) are eaten by organisms that are subsequently eaten by other organisms.

green bins The second stage of urban curbside collection programs involving residents placing organic and similar wastes into green plastic boxes or bins with the contents then composted and sold or made available at nominal cost for residents to use as soil.

greenfields Areas not yet developed and known not to have surface or underground soil contamination.

greenhouse effect A warming of the Earth's atmosphere caused by the presence of certain gases (e.g., water vapour, carbon dioxide, methane) that absorb radiation emitted by the Earth, thereby retarding the loss of energy to space.

greenhouse gas (GHG) A gas that contributes to the greenhouse effect, such as carbon dioxide.

green manure Growing plants that are ploughed into the soil as fertilizer.

Green Revolution Development in plant genetics (hybridization) in the late 1950s and early 1960s resulting in high-yield varieties producing three to five times more grain than previous plants but requiring intensive irrigation and fertilizer use.

gross national product (GNP) The total value of all goods and services produced for final consumption in an economy, used by economists as an index or indicator to compare national economies or periods of time within a single national economy.

gross primary productivity (GPP) The total amount of energy produced by autotrophs over a given period of time.

groundwater Water below the Earth's surface in the saturated zone.

guano The phosphorus-rich droppings of seabirds that, in quantity, as from offshore islands of Peru, is mined for fertilizer.

habitat The environment in which a population or individual lives.

Happy Planet Index (HPI) An index that attempts to provide a perspective on human well-being and environmental impact and to focus on achieving sustainability. Each country's HPI value is a function of its average subjective life satisfaction, life expectancy at birth, and ecological footprint per capita—it approximates multiplying life satisfaction and life expectancy and dividing that by the ecological footprint.

heat The total energy of all moving atoms.

herbivores Animals that eat plants—that is, primary consumers.

heritage rivers Rivers designated for special protection by the Canadian Heritage Rivers Board because of their historical, cultural, ecological, and recreational significance.

heterotroph An organism that feeds on other organisms.

high-quality energy Energy that is easy to use, such as a hot fire or coal or gasoline, but that disperses quickly.

humus Decomposed organic material found in some soils.

hybridization The crossbreeding of two varieties or species of plants or animals.

hydrochlorofluorocarbons (HCFCs) Compound containing hydrogen, chlorine, fluorine, and carbon. HCFC is one type of chemical being used to replace chlorofluorocarbons (CFCs) because HCFCs have less impact on reducing ozone in the stratosphere. CFCs are formed by chlorine, fluorine, and carbon and are broken down by ultraviolet light in the stratosphere. When broken down, CFCs release chlorine atoms, which deplete the ozone layer. CFCs are used as refrigerants, solvents, and foam-blowing agents.

hydrological cycle The circulation of water through bodies of water, the atmosphere, and land.

hydrosolidarity An approach that recognizes the interconnections among aquatic, terrestrial, and other resource systems, leading to management that is integrated, participative, collaborative, co-ordinated, and shared, whether at local, provincial, national, or international levels.

hydrosphere One of three main aspects of the ecosphere, containing all the water on Earth.

hypoxic Oxygen-deficient.

impact and benefit agreements (IBAs) Voluntary agreements between extractive industries and communities that go beyond formal impact assessment requirements and are intended to facilitate extraction of resources in a way that contributes to the economic and social well-being of local people and communities.

impact assessment Thorough consideration of the effects of a project that takes into account its potential and probable impacts on the environment and on society or a community and that assesses the technology proposed for the project as well as the technology available for dealing with any negative impacts.

'implementation gap' The situation that occurs when it becomes difficult or impossible to implement the ideas contained in a strategy or a plan, resulting in a 'gap' between intention and action.

indicators Specific facets of a particular system, such as the population of a key species within an ecosystem, that tell us something about the current state of the system but do not help us understand why the system is in that state.

indigenous knowledge Understanding of climate, animals and animal behaviour, soil, waters, and/or plants within an ecosystem based on experiential knowledge of a people who have lived or worked in a particular area for a long period of time; also referred to as traditional ecological knowledge (TEK) or local knowledge.

inertia The tendency of a natural system to resist change.

ingenuity gap Refers to the critical gap between the need for ideas to fix complex problems resulting from resource scarcity and the actual supply of such ideas. This concept highlights the importance of being able to generate and disseminate ingenuity to ensure that scarcity of resources does not negatively affect well-being for humans and other living organisms.

in situ preservation Conservation strategies that focus on a species within its natural habitat.

in situ recovery Refers to the general practice used at depths greater than 100 metres to remove crude bitumen from oil sands by the specific technique of steam-assisted gravity drainage.

integrated pest management (IPM) The avoidance or reduction of yield losses caused by diseases, weeds, insects, etc., while minimizing the negative impacts of chemical pest control.

integrated plant nutrient systems (IPNSs) Maximization of the efficiency of nutrient use by recycling all plant nutrient sources within the farm and by using nitrogen fixation by legumes.

integrated water resource management (IWRM) An approach that promotes the co-ordinated development and management of water, land, and related resources in order to maximize the resultant economic and social welfare in an equitable manner without compromising the sustainability of vital ecosystems.

intensive livestock operations (ILOs) Factory farms, feedlots, etc. where large quantities of external energy inputs are required to raise for market larger numbers of animals than the area in which they are raised can support, which can result in problems of disease and dealing with animal waste.

intermediate disturbance hypothesis Hypothesis suggesting that ecosystems subject to moderate disturbance generally maintain high levels of diversity compared to ecosystems with low levels of disturbance or those with high levels of disturbance.

International Joint Commission A bilateral institution, consisting of three Canadian commissioners and three American commissioners, established by the 1909 Boundary Waters Treaty to manage interjurisdictional resource issues between Canada and the United States.

interspecific competition Competition between members of different species for limited resources such as food, water, or space.

intraspecific competition Competition between members of the same species for limited resources such as food, water, or space.

intrinsic value A belief that nature has value in and of itself apart from its value to humanity; a central focus for the preservation of species.

invasive An introduced species that spreads out and causes harmful effects on other species and ecosystems.

island biogeography A field within biogeography that attempts to establish and explain the factors that affect the species richness of natural communities.

James Bay and Northern Quebec Agreement Treaty signed in 1975 by the James Bay Cree and Quebec Inuit with the Quebec government permitting the continuance of the James Bay Project and granting to the Natives, among other things, $232.5 million in compensation and outright ownership of 5,543 km^2, as well as exclusive hunting, fishing, and trapping rights to an additional 62,160 km^2; often considered the first modern Aboriginal land claim settlement in Canada.

James Bay Project A hydroelectric megaproject in northern Quebec, begun in the 1970s, that has involved extensive dams on the La Grande and other rivers and has flooded thousands of square kilometres of the James Bay Cree homeland.

keystone species Critical species in an ecosystem whose loss profoundly affects several or many others.

kimberlite pipes Rare, carrot-shaped igneous rock formations sometimes containing diamonds and found in parts of northern Canada.

kinetic energy The energy of objects in motion.

K-strategists Species that produce few offspring but make considerable effort to ensure that the offspring reach maturity.

Kyoto Protocol An international agreement reached in Kyoto, Japan, in 1997 that targets 38 developed nations as well as the European Community to ensure that 'their aggregate anthropocentric carbon dioxide equivalent emissions of the greenhouse gases [e.g., carbon dioxide, methane, nitrous oxide, hydrofluorocarbons, perfluorocarbons, sulphur hexafluoride] . . . do not exceed their assigned amounts'. The Protocol came into effect in 2004 when 55 countries accounting for 55 per cent of 1990 global carbon dioxide emissions had ratified it.

landscape connectivity The degree to which the landscape facilitates or restricts movement between and among habitat patches.

landscape ecology The science of studying and attempting to improve the relationships between spatial patterns and ecological processes on a multitude of spatial scales and organizational levels.

law of conservation of energy Law stating that energy cannot be created or destroyed; it is merely changed from one form to another; also known as the first law of thermodynamics.

law of conservation of matter Law that tells us that matter cannot be created or destroyed, but merely transformed from one form into another.

law of everybody The understanding that if everyone did many small things of a conserving and environmentally aware nature, major environmental problems, threats, and dangers would be ameliorated or alleviated.

leaching The downward movement of dissolved nutrients to the hydrological system.

LEED Acronym for the Leadership in Energy and Environmental Design Green Building Rating System. It is a national standard, established in the US, related to the design, construction, and operation of 'high-performance green buildings'. LEED provides benchmarks for performance related to five issues regarding human and environmental health: sustainable site development, water savings, energy efficiency, materials selection, and indoor environmental quality.

life-cycle assessments (LCAs) Identification of inputs, outputs, and potential environmental impacts of a product or service throughout its lifetime, from manufacture to use and ultimate disposal.

light living Treading as lightly as possible, to minimize our ecological footprints, often characterized by the four R's: refuse, reduce, reuse, and recycle.

limiting factor A chemical or physical factor that determines whether an organism can survive in a given ecosystem. In most ecosystems, rainfall is the limiting factor.

limiting factor principle Stipulates that all factors necessary for growth must be available in certain quantities if an organism is to survive.

lithosphere The Earth's crust.

Livestock Revolution The shift in production units from family farms to factory farms and feedlots that depend on outside supplies of feed, energy, and other inputs to produce vastly more livestock, a shift that has fuelled the growth in meat consumption worldwide, which has doubled since 1977.

Living Planet Index An index that quantifies the overall state of planetary ecosystems.

loams Soils that contain a mixture of materials of different sizes, including humus.

longline Type of commercial fishing using lines with many baited hooks.

long-run sustained yield The yield for an area that is equal to the culmination of mean annual increment weighted by area for all productive and utilizable forest land types in that area; what a given unit of land, such as a forest, should yield in perpetuity.

low-quality energy Energy that is diffuse, dispersed, at low temperatures, and difficult to gather; most of the energy available to us.

LULUs Acronym for 'locally unwanted land uses', often the source of a NIMBY ('not in my backyard') reaction.

LUST Acronym for 'leaking underground storage tanks', which results in contaminated aquifers.

macronutrient A chemical substance needed by living organisms in large quantities (for example, carbon, oxygen, hydrogen, and nitrogen).

marine protected areas (MPAs) Underwater reserves set aside and protected from normal human exploitation because of the fragility, rarity, or valued biodiversity of their ecosystems.

matter What things are made of—92 natural and 17 synthesized chemical elements such as carbon, oxygen, hydrogen, and calcium.

mature community A collection of plants and associated animal species that, over time, see replacement of individuals by similar species.

mediation A negotiation process guided by a facilitator (mediator).

megaprojects Large-scale engineering or resource development projects that cost at least $1 billion and take several years to complete.

mesosphere Layer of the atmosphere extending from the stratosphere, from about 50 to about 80 kilometres above Earth.

mesotrophic Water bodies with nutrient levels between oligotrophic (low levels) and eutrophic (high levels).

metal toxicity The poisonous or harmful nature of metals and minerals, such as asbestos and lead, both to humans and to ecosystems.

micronutrient An element needed by organisms but only in small quantities, such as copper, iron, and zinc.

Millennium Development Goals (MDGs) Globally accepted goals for development agreed to by member states of the United Nations. The target date for meeting the goals is 2015. One of the goals is to 'ensure environmental sustainability'.

Millennium Ecosystem Assessment Assessment carried out by the UN to assess the consequences of ecosystem change for human well-being and to establish the scientific basis for actions needed to enhance the conservation and sustainable use of ecosystems and their contributions to human well-being.

mineralization The process by which biomass is converted back to ammonia (NH_3) and ammonium salts (NH_4) by bacterial action and returned to the soil when plants die.

minimum viable population (MVP) The smallest population size of a species that can be predicted to have a very high chance of persisting for the foreseeable future.

mitigation Strategies to reduce or minimize the negative consequences from a hazard such as climate change. Mitigation requires action today in order for initiatives to be able to reduce the most serious negative impacts in the future.

Mixed Wood Plains The most urbanized ecozone in Canada, spreading from the lower Great Lakes north and east through the St Lawrence Valley, with gently rolling topography and a continental climate characterized by warm, humid summers and cool winters.

monitoring Explicit and systematic checking of outputs and outcomes related to a management initiative in order to understand what works and what does not work and to determine needed modifications to enhance effectiveness.

monoculture cropping Cultivation of one plant species (such as corn) over a large area, which leaves the crop highly susceptible to disease and insects, especially when all of the individual plants are genetically identical.

Montane Cordillera Ecozone in the BC Interior with considerable contrast between the summits of the snowbound peaks and high montane valleys, rolling plateaus, and deeply entrenched desert-like conditions and a climate generally characterized by long, cold winters and short, warm summers.

Montreal Protocol Signed in 1987 by 32 nations, established a schedule for reducing use of chlorofluorocarbons and halons to reduce the rate of depletion of the ozone layer.

multi-barrier approach A method of ensuring the quality of a water supply by using a series of measures (e.g., system security, source protection through pollution regulations within a watershed, water treatment and filtration, testing), each independently acting as a barrier to water-borne contaminants through the system.

mutualism Relationship between two organisms having to do with food supplies, protection, or transport that is beneficial to both.

natural selection The selection by nature of that segment of a population whose genetic attributes favour its success in a changing or changed environment.

nature deficit disorder The increasing gap in understanding of the 'real world' on the part of the younger generation. Instead of playing outdoors in fields, woods, streams, lakes, or the ocean, an increasing proportion of the children of today are glued to their computer or TV screens. They seldom visit the outside world, especially areas dominated by nature rather than human activities.

negative feedback Control mechanism present in the ecosystem and in all organisms—information in the form of chemical, physical, and biological agents influences processes, causing them to shut down or reduce their activity.

negotiation One of the two main types of alternative dispute resolution when two or more parties involved in a dispute join in a voluntary, joint exploration of issues with the goal of reaching a mutually acceptable agreement.

neo-liberalism A political or policy perspective that places high value on the role of free markets to allocate resources efficiently, leading to a belief that it is best to allow markets to function with minimum intervention by government regulations.

net community productivity (NCP) The rate of accumulation of organic material, allowing for both plant respiration and heterotrophic predation during the measurement period.

net primary productivity (NPP) Gross primary productivity (the total amount of energy that plants produce) minus the energy plants use during cellular respiration.

new forestry A silvicultural approach that mimics natural processes more closely through emphasizing long-term site productivity by maintaining ecological diversity.

niche An organism's place in the ecosystem: where it lives, what it consumes, and how it interacts with all biotic and abiotic factors.

NIMBY 'Not in my backyard', a phrase used to describe local people's reactions when a noxious or undesired facility—for example, a landfill site, a sand and gravel pit, or an expressway—is proposed in an area adjacent to or near their property.

nitrogen fixation Conversion of gaseous (atmospheric) nitrogen (N_2) into ammonia (NH_3) by bacteria, such as those that grow on the root nodules of legumes.

non-governmental organizations (NGOs) Organizations outside of the government and private sectors, usually established to address a specific societal issue or need; also referred to as 'not-for-profit' or 'social profit' organizations. They are one element of 'civil society' and when focused on environmental matters can be referred to as ENGOs (environmental non-governmental organizations).

non-point sources Sources of pollution from which pollutants are discharged over a widespread area or from a number of small inputs rather than from distinct, identifiable sources.

non-renewable or stock resources Resources, such as oil, coal, and minerals, that take millions of years to form and thus, for practical purposes, are fixed in supply and therefore not renewable.

non-timber forest products (NTFPs) Forest resources of economic value but not related to the lumber and pulp and paper industries, such as wild rice, mushrooms and berries, maple syrup, edible nuts, furs and hides, medicines, and ornamental cuttings.

no-till/conservation agriculture (NT/CA) Zero, minimum, or low tillage to protect and stimulate the biological functioning of the soil while maintaining and improving crop yields, which includes direct sowing or drilling of seeds instead of ploughing, maintenance of permanent cover of plant material on the soil, and crop rotation.

nuclear power Power, usually electric power, produced by atomic energy. Atoms contain atomic energy, which can be released slowly in reactors or rapidly in bombs through alteration of the nuclei of atoms.

nuclear wastes The radioactive wastes remaining from the uranium used to fuel nuclear fuel reactors. Such wastes have an extremely long life and are life-threatening, thereby creating significant storage and containment challenges.

nutrients Elements or compounds that an organism must take in from its environment because it cannot produce them or cannot produce them as quickly as needed.

old-growth forests Forests that generally have a significant number of huge, long-lived trees; many large standing dead trees; numerous logs lying about the forest floor; and multiple layers of canopy created by the crowns of trees of various ages and species.

oligotrophic Nutrient poor.

omnivores Organisms that eat both plants and animals.

100-mile diet A term introduced in 2005 referring to buying and eating food grown, manufactured, or produced entirely within a 100-mile radius of one's residence.

optimal foraging theory The relationship between the benefit of making a kill and feeding and the cost of the energy expended to make the kill.

optimum range The ideal conditions for the survival of a species.

organic farming An agriculture production management system that focuses on food web relations and element cycling to maximize agro-ecosystem stability and to promote and enhance ecosystem health. It is based on minimizing the use of external inputs.

organism A living entity; one of a population.

oxygen sag curve The drop in oxygen levels in a body of water when organic wastes are added and the number of bacteria rises to help break down the waste.

ozone An atmospheric gas (O_3) that when present in the stratosphere helps to protect the Earth from ultraviolet rays. However, when present near the Earth's surface, it is a primary component of urban smog and has detrimental effects on both vegetation and human respiratory systems.

ozone layer Thin layer of ozone molecules in the stratosphere that absorbs ultraviolet light and converts it into infrared radiation, effectively screening out 99 per cent of the ultraviolet light.

Pacific Maritime Ecozone characterized by the influence of the Pacific Ocean, with the highest rainfall figures in Canada (up to 3,000 mm) and the warmest average temperatures.

Palliser's Triangle Roughly triangular-shaped semi-arid area of southeast Alberta and southwest Saskatchewan, south of the Saskatchewan River, first identified by Captain John Palliser during an expedition to the Canadian West in 1857–60 sponsored by the Royal Geographical Society and the British Colonial Office.

parasitism Relationship in which one species lives in or on another that acts as its host.

parent material The material from which soil forms, such as sediment or weathered bedrock.

partnerships A sharing of responsibility and power between two or more groups, especially a government agency and a second party, regarding a resource or environmental issue; co-management is an example of a partnership.

pelagic Marine life, such as cod and whales, that live in the upper layers of the open sea.

permaculture Agricultural designs, such as urban farming and organic farming, based on ecological relationships with the fundamental principle of minimizing wasted energy and with the wastes of one component becoming the inputs for another.

pheromones Volatile compounds, or 'scents', used by insects of a given species to communicate with each other.

photosynthesis A two-part process in plants and algae involving (1) the capture of sunlight and its conversion into cellular energy and (2) the production of organic molecules, such as glucose and amino acids from carbon dioxide, water, and energy from the sun.

phototrophs Organisms that produce complex chemicals through photosynthesis.

phytoplankton Single-celled algae and other free-floating photosynthetic organisms.

planetary carrying capacity The ability of Earth and its various systems to sustain the number of people and other organisms on the planet and their effects on these systems.

point sources Easily discernible 'end-of-pipe' sources of pollution, such as a factory or a town sewage system.

polar amplification The effect of a positive feedback loop in the North whereby increased temperatures lead to a greater area of snow-free land in summer, which in turn leads to increased temperatures because of the lower albedo.

policy target value A target set as a result of compromise among scientific, social, economic, cultural, and political objectives.

polynyas Ice-free areas of permanent open water in the Arctic surrounded by ice, created by tides, currents, ocean-bottom upwellings, and winds, that are biologically productive and vary greatly in size from 60 to 90 metres in diameter to areas as large as the North Water polynya between Ellesmere Island and Greenland that may cover as much as 130,000 km^2.

population A group of organisms of the same species living within a specified region.

population age structure The relative distribution of age cohorts in the population.

population density The number of individuals of a population within a certain defined area, such as sea otters per hectare or humans per square kilometre.

population viability analysis (PVA) A process that determines the probability that a population will go extinct within a given number of years.

positive feedback loop A situation in which a change in a system in one direction provides the conditions to cause the system to change further in the same direction.

potential energy Stored energy that is available for later use.

precautionary principle A guideline stating that when there is a possibility of serious or irreversible environmental damage resulting from a course of action, such as a development project, lack of scientific certainty is not an acceptable reason for postponing a measure to prevent environmental degradation or for assuming that damage in the future can be rectified by some kind of technological fix.

predator An organism that actively hunts its prey.

prescribed burning Burning purposely initiated, usually to achieve ecological goals of restoring natural fire regimes.

prey An organism (e.g., deer) that is attacked and killed by a predator.

prey switching A familiar foraging behaviour whereby a predator shifts from its target species after it is depleted or not available in an area to the next most preferred or profitable species until that, too, is depleted and then continuing to move down the food chain, as wolves do in moving from caribou to Arctic hare to small rodents or as humans have done in fishing down the food chain in commercial fisheries.

primary consumers The first consuming organisms in a given food chain, such as a grazer in grazer food chains or a decomposer organism or insect in decomposer food chains; primary consumers belong to the second trophic level.

primary succession The development of a biotic community in an area previously devoid of organisms.

producers Autotrophs capable of synthesizing organic material, thus forming the basis of the food web.

protected areas Areas such as national and provincial parks, wildlife sanctuaries, and game preserves established to protect species and ecosystems.

radiant energy Energy from the sun.

rainshadow effect The decrease in precipitation levels as the air warms up in its descent from the mountains and can hold more moisture (i.e., there is considerably less precipitation on the leeward side of a mountain or a mountain range than on the windward side).

range of tolerance Range of abiotic factors within which an organism can survive, from the minimum amount of a limiting factor that the organism requires to the maximum amount that it can withstand.

reclamation The process of bringing an area back to a useful, good condition—similar to rehabilitation.

recovery Involves burning waste in incinerators and then recovering the generated energy to provide heating for homes, offices, and other buildings.

recycling The third stage in waste management, involving the return of used products (glass, plastic, or metal containers; newspapers) to be processed so that the glass, plastic, metal, or newsprint can be used for other products; movement of elements in characteristic repetitive paths through ecosystems.

REDD A mechanism for compensating countries for reducing emissions from deforestation and forest degradation.

Red List An annual listing of species at risk prepared by the IUCN and produced by thousands of scientific experts; the best source of knowledge on the status of global biodiversity.

reduction The first stage in waste management, involving using less material or products to meet needs.

relative humidity The amount of moisture held in the air compared to how much could be held if fully saturated at a particular temperature.

renewable or flow resources Resources that are renewed naturally within a relatively short period of time, such as water, air, animals, and plants, as well as solar radiation, wind power, and tidal energy.

renewable water supply Supply based on precipitation that falls, then runs off into rivers, often being held in lakes before draining to the ocean or moving downward into aquifers. The flows associated with precipitation or snowmelt should be identified as the renewable supply.

replacement-level fertility Fertility rate that will sustain a population.

resilience Ability of an ecosystem to return to normal after a disturbance.

resource partitioning A situation in which resources are used at different times or in different ways by species with an overlap of fundamental niches, such as owls and hawks, which seek the same prey but at different times during the day.

resources Such things as forests, wildlife, oceans, rivers and lakes, minerals, and petroleum.

reuse The second stage in waste management, involving the reuse of a product rather than discarding and replacing it.

Richter scale A scale developed in 1935 by C.F. Richter for measuring the magnitude of earthquake severity on a logarithmic scale. Although theoretically open-ended, beginning at zero, no earthquakes have registered higher than 9.0 on the scale.

risk assessment Determining the probability or likelihood of an environmentally or socially negative event of some specified magnitude.

rock cycle The relationship among three rock-forming processes and how each rock type can be transformed into another type.

r-strategists Species that produce large numbers of young early in life in a short time but invest little energy in their upbringing.

salinization Deposition of salts in irrigated soils, making soil unfit for most crops; caused by a rising water table due to inadequate drainage of irrigated soils.

scientific target value A target set on the basis of scientific information.

secondary consumers Second consuming organisms in a food chain and belonging to the third trophic level.

secondary succession The sequential development of biotic communities after the complete or partial destruction of an existing community by natural or anthropogenic forces.

second growth A second forest that develops after harvest of the original forest.

second law of thermodynamics Law stating that when energy is converted from one form to another, it is degraded—that is, it is converted from a concentrated to a less concentrated form. The amount of useful energy decreases during such conversions.

sedimentary cycles Those cycles of elements, such as the phosphorus and sulphur cycles, that hold most of their matter in the lithosphere.

seed bank Locations where plant seeds accumulate.

seral Each stage in a successional process.

serial depletion When one stock after another becomes progressively depleted as a result of prey switching, even if the total catch remains the same.

serotiny Behaviour of some plant species that retain their non-dormant seeds in a cone or woody fruit for up to several years but release them after exposure to fire.

shifting baseline When scientists have no other option than to take the current or recent degraded state as the baseline for stock biomass rather than the historical ecological abundance.

Silent Spring A book written by Rachel Carson and published in 1962 that detailed the disastrous effects of biocides on the environment.

silviculture The practice of directing the establishment, composition, growth, and quality of forest stands through a variety of activities, including harvesting, reforestation, and site preparation.

single-loop learning Learning that emphasizes ensuring a match between intent and outcome.

smog Originally, a mixture of smoke and fog in urban areas, mainly due to burning of coal to heat homes and power factories and the subsequent mixing of smoke with the humid air. Now more usually due to photochemical reactions of sunlight with hydrocarbons and nitrogen oxide emitted into the atmosphere from vehicles and industries.

social learning Learning applied not only to individuals but also to social collectives, such as organizations and communities. The implication is that resource and environmental management processes should be designed so that both individuals and organizations are able to learn from their experience and thereby become more knowledgeable and effective in the future.

soft path A management approach to improving water use efficiency by challenging basic patterns of consumption. While demand management emphasizes the question of 'how', or how to do the same with less water, the soft path asks why water is even used for a function. The 'why' question normally leads to consideration of a broader range of methods.

soil compaction The compression of soil as a result of frequent heavy machinery use on wet soils or the overstocking of cattle on the land.

soil erosion A natural process whereby soil is removed from its place of formation by gravitational, water, and wind processes.

soil horizons Layers found in most soils.

soil permeability The rate at which water can move through a soil, largely determined by soil texture, i.e., the size of the materials that make up the soil.

soil profile A view across soil horizons.

specialist Organism that has a narrow niche, usually feeding on one or a few food materials and adapted to a particular habitat.

speciation Phyletic evolution, i.e., formation of new species when evolution within a population is so great that interbreeding with the original population is no longer possible.

species A group of individuals that share certain identical physical characteristics and are capable of producing fertile offspring.

species-area curves A graph showing the numbers of species found in areas of different size.

Species at Risk Act (SARA) Canadian legislation passed in 2002 that mandates the Committee on the Status of Endangered Wildlife in Canada to maintain lists of species at risk and to recommend to the minister responsible that particular species be given special protection in their environment.

species diversity The total number of different species in an area.

stakeholders Persons or groups with a legal responsibility relative to a problem or issue, or likely to be affected by decisions or actions regarding the problem or issue, or able to pose an obstacle to a solution of the problem or issue.

steam-assisted gravity drainage Injection of high-pressure steam into tar sands at depths greater than 100 metres to separate the bitumen from the sand, silt, and clay. Exposed to the steam, bitumen becomes liquefied and can flow to a well from which it can be pumped to the surface.

stewardship Activities undertaken by humans towards caring for the Earth.

strategic environmental assessment Focuses on policies, plans, and programs (PPPs) in order to integrate environmental considerations at the earliest possible stage of decision-making. Such assessment occurs before development decisions are made and when alternative futures and options for development are still open. Emphasis is on opportunities, regions, and sectors as opposed to projects. The objective is to integrate environmental considerations in the development of PPPs and to identify preferred futures and the means to achieve them, rather than focusing on mitigating the most likely outcomes of an already taken development decision.

stratosphere The layer of the atmosphere (about 10 to 50 kilometres above the Earth's surface) in which temperatures rise with increasing altitude.

strip cropping A technique similar to contour cultivation in which different crops are planted in strips parallel to the slope.

sublimation The process for direct transfer between the solid and vapour phases of matter, regardless of direction.

subsidiarity A policy and management approach stipulating that decisions should be taken at the level closest to where consequences are most noticeable or have the most direct impact.

subsistence farming The production of food and other necessities to satisfy the needs of the farm household.

sulphur dioxide emissions Release into the atmosphere of huge quantities of sulphur, mainly in the form of dioxides, as a result of smelting sulphide ores and burning fossil fuels, which causes air pollution and climate change.

summer fallow A practice common on the prairies in which land is ploughed and kept bare to minimize moisture losses through evapotranspiration but which leads to increased salinization.

supply management Approach based on manipulating the natural system to create new sources of supply, normally through either augmenting an existing supply or developing a new supply.

sustainability assessment Analysis that seeks to determine the environmental sustainability of a proposed course of action.

sustainable development Economic development that meets current needs without compromising the ability of future generations to meet their needs, a concept popularized by the 1987 World Commission on Environment and Development headed by Norwegian Prime Minister Gro Harlem Brundtland.

'sustainable energy superpower' A view that Canada, with its huge energy resources, should be able to produce upgraded energy products at reasonable prices with acceptable environmental impacts, based on new technology, effective public policy, new concepts of risk-sharing, and individual companies and governments working together.

sustainable livelihoods A human-centred approach to broad environmental management directed towards ways for local people to meet basic needs (food, housing), as well as other needs related to security and dignity, through meaningful work, at the same time minimizing environmental degradation, rehabilitating damaged environments, and addressing concerns about social justice.

sustainable urban development The enhanced well-being of cities or urban regions, including integrated economic, ecological, and social components, which will maintain the quality of life for future generations.

sustained yield The amount of harvestable material that can be removed from an ecosystem over a long period of time with no apparent deleterious effects on the system.

synergism An interaction between two substances that produces a greater effect than the effect of either one alone; an interaction between two relatively harmless components in the environment.

system plan An idealized blueprint of the distribution of protected areas within a given jurisdiction.

taiga The portion of the boreal forest lying between the southern boundary of the tundra and the closed-crown coniferous forest to the south, characterized by coniferous forests, soil that thaws during the summer months, abundant precipitation, and high species diversity.

Taiga Plains Gently rolling northern ecozone with a high proportion of surface water storage, wetlands, and organic soils, a cold and relatively dry climate, and on better-drained localities and uplands, mixed coniferous-deciduous forests.

technocentric perspective The assumption that humankind is able to understand, control, and manipulate nature to suit its purposes and that nature and other living and non-living things exist to meet human needs and wants.

territory A specific area dominated by a specific individual of a species.

tertiary consumers In a food chain, organisms at the top that consume other organisms.

thermocline Sharp transition in temperature between the warmer surface waters of the ocean and the cooler waters underneath, generally occurring at a depth of 120 to 240 metres.

thermohaline circulation The movement of carbon-saturated water around the globe, mainly as a result of differing water densities.

thermosphere Uppermost layer of the atmosphere, beyond the mesosphere.

threatened species A species designated by the Committee on the Status of Endangered Wildlife in Canada is likely to become endangered in Canada if factors threatening its vulnerability are not reversed.

threshold A point or limit beyond which something is unsatisfactory relative to a consideration, such as health, welfare, or ecological integrity.

total allowable catch (TAC) The amount, in tonnage, of a particular aquatic species that the federal Department of Fisheries and Oceans, for example, determines can be landed within a particular fishery in a given year.

total fertility rates Average number of children each woman has over her lifetime.

traditional ecological knowledge (TEK) Belief, knowledge, and practice gained through experience, normally shared and transmitted verbally; often referred to as 'indigenous knowledge'.

transpiration The loss of water vapour through the pores of a plant.

tree-length harvesting Felling, delimbing, and topping the trees in the cut-over area.

trophic level Functional classification of organisms in a community according to feeding relationships: the first trophic level includes green plants, the second level includes herbivores, and so on.

troposphere Innermost layer of the atmosphere that contains 99 per cent of the water vapour and up to 90 per cent of the Earth's air and is responsible for our weather, extending about 6 to 17 kilometres up from the Earth, depending on latitude and season.

uncertainty A situation in which the probability or odds of a future event are not known and therefore that indicates the presence of doubt.

urban form The type and distribution of infrastructure (e.g., buildings, roads) in communities and a key factor influencing environmental quality in cities.

urban heat island Increased temperatures in core urban areas relative to surrounding areas resulting from heat absorbed and radiated from the built environment (e.g., buildings, roads). It is not uncommon for the temperature of city centres to be from 2°C to 6°C higher than that of nearby rural areas.

urban sprawl Urban areas characterized by low population densities and significant travel costs because of the high priority the majority of residents place on living in single-family, detached homes with large properties. Sprawl contributes to loss of, disruption to, or degradation of adjacent agricultural land, environmentally sensitive areas, natural habitats, and water and air quality.

virtual water A concept recognizing that because significant amounts of water are required to grow some foodstuffs, nations can reduce pressure on their water resources by importing such products, allowing use of water for other, higher-value products.

vision A view for the future for a region, community, or group that is realistic, credible, attractive, and attainable.

vulnerable species An official designation assigned by the Committee on the Status of Endangered Wildlife in Canada to any indigenous species or subspecies or geographically separate population of fauna or flora that is particularly at risk, though not at present 'threatened', because of low or declining numbers, because it occurs at the fringe of its range or in restricted areas, or for some other reason.

water ethics A statement of principles or values to guide behaviour by individuals or groups with regard to water.

water footprint An indicator of water consumption that tracks and totals both direct and indirect yearly water use by a consumer or a nation, or of a product over the course of its life cycle.

water rights The view that water is a basic human right, as put forth several times in recent years by the UN Human Rights Council, a view that the Canadian government has strongly opposed, presumably for geopolitical and internal political and administrative reasons.

water table The top of the zone of saturation.

water terrorism Deliberate disruption of water infrastructure or related services by directly attacking the infrastructure or by contaminating water with poison or disease-bearing agents.

weather The sum total of atmospheric conditions (temperature, pressure, winds, moisture, and precipitation) in a particular place for a short period of time.

wetlands Areas that are hybrid aquatic and terrestrial systems, such as swamps and marshes, where the ground is saturated with water much or all of the time.

wind power The fastest-growing sector in the world's energy market, which uses wind turbines to generate electricity.

windthrow Uprooting and blowing down of trees by wind.

World Summit on Sustainable Development (WSSD) Conference held in Johannesburg in 2002, 10 years after the Earth Summit in Rio, that made various commitments, such as halving the proportion of people without access to adequate sanitation by the year 2015.

zone of physiological stress Upper and lower limits of the range of tolerance in which organisms have difficulty surviving.

zooplankton Non-photosynthetic, single-celled aquatic organisms.

zooxanthellae Unicellular algae.

References

Abbey, E. 1977. *The Journey Home: Some Words in Defense of the American West.* © 1977 by Edward Abbey. Used by permission of Dutton, a division of Penguin Group (USA) Inc. New York: Dutton.

Agreement Review Committee. 2007. *Review of the Great Lakes Water Quality Agreement, Volume 1, Final Report to the Great Lakes Binational Executive Committee.* Windsor, Ont.: Environment Canada–Ontario, Great Lakes Office.

Agriculture and Agri-Food Canada. 2003. 'Canada's Agriculture, Food and Beverage Industry: Organic Industry'. At: www.ats.agr.gc.ca/pro/fhs-eng.htm.

———. 2008. 'An Overview of the Canadian Agriculture and Agri-Food System 2007'. At: www4.agr.gc.ca/AAFC-AAC/display-afficher.do?id=1201291159395&lang=eng.

———. 2010a. 'Producers'. At: www4.agr.gc.ca/AAFC-AAC/display-afficher.do?id=1165871799386&lang=eng.

———. 2010b. 'Government of Canada invests in organic sector', news release, 6 Mar. At: www.agr.gc.ca/cb/index_e.php?s1=n&s2=2010&page=n100306.

Alberta Energy. 2008. 'Electricity frequently asked questions'. At: www.energy.gov.ab.ca/Electricity/683.asp#where.

Alberta Environment. 2005. *Report on the Implementation Progress of Water for Life: Alberta's Strategy for Sustainability.* Edmonton: Alberta Environment.

Alberta Food and Rural Development. 2001. 'Woodlot harvesting'. At: www1.agric.gov.ab.ca/$department/deptdocs.nsf/all/apa3316.

Aldaya, M.M., P. Martinez-Santos, and M.R. Llamas. 2010. 'Incorporating the water footprint and virtual water into policy: Reflections from the Mancha Occidental Region, Spain', *Water Resources Management* 24: 941–58.

Allan, T. 2011. *Virtual Water.* New York: I.B. Tauris.

Anderegg, W.R.I. 2010. 'The ivory lighthouse: Communicating climate change more effectively', *Climatic Change* 101, 3 and 4: 655–62.

Andrey, J., and L. Mortsch. 2000. 'Communicating about climate change: Challenges and opportunities', in D. Scott et al., *Climate Change Communication: Proceedings of an International Conference.* Waterloo, Ont.: University of Waterloo and Environment Canada, Adaptation and Impacts Research Group, wp1–wp11.

Anielski, M., and S. Wilson. 2005. *Assessing the Real Value of Canada's Boreal Ecosystem.* Edmonton: Pembina Institute.

——— and ———. 2009. *Counting Canada's Natural Capital: Assessing the Real Value of Canada's Boreal Ecosystems.* Ottawa: Pembina Institute and Canadian Boreal Initiative.

Apps, C.D., and B.N. McClellan. 2006. 'Factors influencing the dispersion and fragmentation of endangered mountain caribou populations', *Biological Conservation* 130: 84–97.

AREVA. 2007. 'Uranium in Saskatchewan'. At: www.arevaresources.com/publications/uranium_in_sask_01/environment and safety.html.

Armitage, D., F. Berkes, and N. Doubleday, eds. 2007. *Adaptive Co-management: Collaboration, Learning and Multi-Level Governance.* Vancouver: University of British Columbia Press.

———, ———, and ———. 2007. 'Introduction: Moving beyond co-management', in Armitage et al. (2007: 1–15).

Armstrong, A. 2009. 'Further ideas towards a water ethic', *Water Alternatives* 2: 138–47.

Arndt, D.S., M.O. Baringer, and M.R. Johnson. 2010. 'State of the climate in 2009', *Bulletin of the American Meteorological Society* 91, 7: S1–S224.

Arnstein, S. 1969. 'A ladder of citizen participation', *Journal of the American Institute of Planners* 35, 4: 216–24.

Auditor General of Canada. 2002. *2002 Report of the Commissioner of the Environment and Sustainable Development.* Ottawa: Minister of Supply and Services.

———. 2005a. *2005 Report of the Commissioner of the Environment and Sustainable Development,* Chapter 5: 'Drinking water in First Nations communities'. At: www.oag-bvg.gc.ca/internet/English/parl_cesd_200509_05_e_14952.html.

———. 2005b. *2005 September Report of the Commissioner of the Environment and Sustainable Development.* Ottawa: Minister of Supply and Services.

———. 2006. *2006 Report of the Commissioner of the Environment and Sustainable Development.* www.oag-bvg.gc.ca/internet/English/parl_cesd_200609_e_936.html.

———. 2008. *Status Report of the Commissioner of the Environment and Sustainable Development to the House of Commons.* Ottawa: Minister of Supply and Services. Reproduced with the permission of the Minister of Public Works and Government Services, 2011.

———. 2010. *Report of the Commissioner of the Environment and Sustainable Development.* Fall 2010. Office of the Auditor General of Canada. Reproduced with the permission of the Minister of Public Works and Government Services, 2011.

———. 2011. *Report of the Commissioner of the Environment and Sustainable Development.* Fall 2010. Office of the Auditor General of Canada. Reproduced with the permission of the Minister of Public Works and Government Services, 2011.

Augustine, S., and P. Dearden. 2012. 'A new management paradigm in marine and coastal conservation: A case study of clam gardens in the southern Gulf Islands', *Canadian Geographer.*

Babbage, M. 2010. 'Provinces to take action on climate', *The Record,* 9 Nov., D4.

Bakker, K. ed. 2007. *Eau Canada: The Future of Canada's Water.* Vancouver: University of British Columbia Press.

Banff–Bow Valley Study. 1996. *Banff–Bow Valley: At the Crossroads. Summary Report of the Banff–Bow Valley Task Force,* R. Page et al., eds. Ottawa: Ministry of Canadian Heritage.

Bardati, D.R. 1995. 'An environmental action plan for Bishop's University', *Journal of Eastern Townships Studies* 6: 19–38.

———. 2006. 'The integrative role of the campus environmental audit: Experiences at Bishop's University, Canada', *International Journal of Sustainability in Higher Education* 7, 1: 57–68.

Barlow, M., and E. May. 2000. *Frederick Street: Life and Death on Canada's Love Canal.* Toronto: HarperCollins.

Barnosky, A.D., et al. 2004. 'Assessing the causes of late Pleistocene extinctions on the continents', *Science* 306: 70–5.

——— et al. 2011. 'Has the Earth's sixth mass extinction already arrived?', *Nature* 471: 51–7.

Beamish, R., and H. Harvey. 1972. 'Acidification of the La Cloche Mountain lakes, Ontario, and resulting fish mortalities', *Journal of the Fisheries Research Board of Canada* 29, 8: 1135.

Beanlands, G.E., and P.N. Duinker. 1983. *An Ecological Framework for Environmental Impact Assessment in Canada.* Halifax: Institute for Resource and Environmental Studies, Dalhousie University.

Berch, S.M., D. Morris, and J. Malcolm. 2011. 'Intensive biomass harvesting and biodiversity in Canada: A summary of relevant issues', *Forestry Chronicle* 87: 479–87.

Berger, T.R. 1977. *Northern Frontier, Northern Homeland: The Report of the Mackenzie Valley Pipeline Inquiry,* 2 vols. Ottawa: Minister of Supply and Services.

Bergeron, Y., et al. 2001. 'Natural fire frequency for the eastern Canadian boreal forest: Consequences for sustainable forestry', *Canadian Journal of Forest Research* 31: 384–91.

Berkes, F. 1988. 'The intrinsic difficulty of predicting impacts: Lessons from the James Bay hydro project', *Environmental Impact Assessment Review* 8: 201–20.

———, D. Armitage, and N. Doubleday. 2007. 'Synthesis: Adapting, innovating, evolving', in Armitage et al. (2007: 308–27).

BHP Billiton. 2001. *Environmental Agreement Annual Report 2001*. At: www.corporateregister.com/a10723/ekati01-env-ca.pdf.

Birkedal, T. 1993. 'Ancient hunters in the Alaskan wilderness: Human predators and their role and effect on wildlife populations and the implications for resource management'. In W.E. Brown and S.D. Veirs Jr, Eds, *Partners in Stewardship: Proceedings of the 7th Conference on Research and Resource Management in Parks and on Public Lands*, 228–34. Hancock, MI: The George Wright Society.

Biswas, A.K. 2004. 'Integrated water resource management: A re-assessment', *Water International* 29: 248–56.

———. 2008. 'Integrated water resources management: Is it working?', *Water Resources Development* 24: 5–22.

Blatchford, C. 2010. *Helpless: Caledonia's Nightmare of Fear and Anarchy, and How the Law Failed All of Us*. Toronto: Doubleday Canada.

Bone, R.M. 2005. *The Regional Geography of Canada*, 3rd edn. Toronto: Oxford University Press.

Boyce, M.S., S.R. Lele, and B.W. Johns. 2005. 'Whooping crane recruitment enhanced by egg removal', *Biological Conservation* 126: 395–401.

Boychuk, Rick. 2008. 'Seeing the light', *Canadian Geographic*. At: www.canadiangeographic.ca/Magazine/jun08/ednotebook.asp.

Boyd, D.R., and S.J. Genuis. 2008. 'The environmental burden of disease in Canada: Respiratory disease, cardiovascular disease, cancer, and congenital affliction', *Environmental Research* 106: 240–9

Boyer, M. 2008. *Freshwater Exports for the Development of Quebec's Blue Gold*. Montreal: Montreal Economic Institute.

Branch, T.A. 2008. 'Not all fisheries will be collapsed in 2048', *Marine Policy* 32, 1: 38–9.

Brandes, O.M., and D.B. Brooks. 2006. *The Soft Path for Water in a Nutshell*. Ottawa: Friends of the Earth; Victoria: University of Victoria, POLIS Project.

———, R. Maas, and E. Reynolds. 2006. *Thinking beyond Pipes and Pumps: Top 10 Ways Communities Can Save Water and Money*. Victoria: University of Victoria, POLIS Project.

Bregha, F., et al. 1990. *The Integration of Environmental Considerations into Government Policy*. Prepared for the Canadian Environmental Assessment Research Council. Ottawa: Minister of Supply and Services.

Briggs, D., et al. 1993. *Fundamentals of Physical Geography*, 2nd Canadian edn. Toronto: Copp Clark Pitman.

British Columbia. 2008. 'Budget Speech: Turning to the Future, Meeting the Challenge', 19 Feb. Victoria: Ministry of Finance.

British Columbia Ministry of Environment. 2006. *BC Seafood Industry Year in Review*. At: www.env.gov.bc.ca/omfd/reports/YIR-2006.pdf.

British Columbia Ministry of Finance. 2011a. 'How the carbon tax works'. At: www.fin.gov.bc.ca/tbs/tp/climate/A4.html.

———. 2011b. 'Myths and facts about the carbon tax'. At: www.fin.gov.bc.ca/tbs/tp/climate/A6.htm.

Brook, B.W., and D.M.J.S. Bowman. 2004. 'The uncertain blitzkrieg of Pleistocene megafauna', *Journal of Biogeography* 31: 517–23.

Brooks, D.B. 2005. 'Beyond greater efficiency: The concept of water soft paths', *Canadian Water Resources Journal* 30: 83–92.

———, O.M. Brandes, and S. Gurman. 2009. *Making the Most of the Water We Have: The Soft Path Approach to Water Management*. London: Earthscan.

——— and S. Holtz. 2009. 'Water soft path analysis: From principles to practice', *Water International* 34: 158–69.

Brooks, G.R., and E. Nielson. 2000. 'Red River, Red River Valley, Manitoba', *Canadian Geographer* 44: 306–11.

Brown, S., H. Schreier, and L.M. Lavkulich. 2009. 'Incorporating virtual water into water management: A British Columbia example', *Water Resources Management* 23: 2681–92.

Brown, V.A., J.A. Harris, and J.Y. Russell, eds. 2010. *Tackling Wicked Problems: Through the Transdisciplinary Imagination*. London: Earthscan.

Browne, C.L., and S.J. Hecnar. 2003. 'Dwindling turtle populations', *National Park International Bulletin* no. 9 (May).

———. 2007. 'Species loss and shifting population structure in freshwater turtles despite habitat protection', *Biological Conservation* 138: 421–9.

Bruno, J.F., and E.R. Selig. 2007. 'Regional decline of coral cover in the Indo-Pacific: Timing, extent, and subregional comparisons', *PLoS ONE* 2, 8: e711. At: doi:10.1371/journal.pone.0000711.

Bryant, R. 2007. 'International chronology of environmental justice'. At: www-personal.umich.edu/~bbryant/iejtimeline.html.

Bullock, R., and K. Hanna. 2008. 'Community forestry in British Columbia: Mitigating or creating conflict?', *Society and Natural Resources* 2: 77–85.

——— and A. Watelet. 2006. 'Exploring conservation authority operations in Sudbury, Northern Ontario: Constraints and opportunities', *Environments* 34, 2: 29–50.

Burn, D.H., and N.K. Goel. 2001. 'Flood frequency analysis for the Red River at Winnipeg', *Canadian Journal of Civil Engineering* 28: 355–62.

Burt, B. 2007a. 'Land claim stalls wind project', *The Record*, 12 Aug., B1–B2.

———. 2007b. 'Wind farm claim riles Tory', *The Record*, 16 Aug., A1–A2.

Butchart, S.H.M., et al. 2010. 'Global biodiversity: Indicators of recent declines', *Science* 328: 1164–8.

Butterworth, J., J. Warner, P. Moriarty, S. Smits, and C. Batchelor. 2010. 'Finding practical approaches to integrated water resources management', *Water Alternatives* 3: 68–81.

Cameron, R.P. 2006. 'Protected area–working forest interface: Ecological concerns for protected area management in Canada', *Natural Areas Journal* 26: 403–7.

Cameron, S.D. 1990. 'Net losses: The sorry state of our Atlantic fishery', *Canadian Geographic* 110, 2: 28–37.

Canada. 1995. *Canadian Biodiversity Strategy: Canada's Response to the Convention on Biological Diversity*. Ottawa: Minister of Supply and Services.

Canada and Ontario. 1988. *First Report of Canada under the 1987 Protocol to the 1978 Great Lakes Water Quality Agreement*. Toronto: Environment Canada, Communications Directorate.

Canada Mortgage and Housing Corporation. 2005. *Smart Growth in Canada: A Report Card*. Research Highlight Socio-economic Series 05-036. Ottawa: Canada Mortgage and Housing Corporation.

Canada–Newfoundland Offshore Petroleum Board, Terra Nova Development Project Environmental Assessment Panel. 1997. *Terra Nova Development: An Offshore Petroleum Project*. Ottawa: Minister of Public Works and Government Services.

Canadian Academy of Engineering. 2007. *Energy Pathways Task Force Phase 1: Final Report*. Ottawa: Canadian Academy of Engineering.

Canadian Biodiversity Strategy. 2010. 'Ecosystem status and trends'. At: www.biodivcanada.ca/default.asp?lang=En&n=560ED58E-1&offset=3&toc=show.

Canadian Council of Forest Ministers (CCFM). 2006. *Facing the Challenge of Forest Industry Restructuring*. Ottawa: Canadian Council of Forest Ministers.

———. 2007. 'Compendium of Canadian Forestry Statistics'. At: www.nfdp.ccfm.org.

———. 2008. *A Vision for Canada's Forests: 2008 and Beyond*. Ottawa: Canadian Council of Forest Ministers. At: www.ccmf.org/pdf/Vision_EN.pdf. Reproduced with permission, 2012.

Canadian Council of Ministers of the Environment. 2003. *Climate, Nature, People: Indicators of Canada's Changing Climate*. Winnipeg: Canadian Council of Ministers of the Environment.

———. 2005. *Five-Year Review of the Canada-Wide Acid Rain Strategy for Post-2000*. At: www.ccme.ca/assets/pdf/5_year_review_acid_rain_strategy_e1.0_web.pdf.

———. 2006. *2004–2005 Progress Report on the Canada-Wide Acid Rain Strategy for Post-2000*. At: www.ccme.ca/assets/pdf/2004_2005_ar_progrrpt_1.0_e_web.pdf.

———. 2008. *2006–2007 Progress Report on the Canada-Wide Acid Rain Strategy*. Ottawa: Canadian Council of Ministers of Environment.

Canadian Environmental Assessment Agency. 1996. *NWT Diamonds Project: Report of the Environmental Assessment Panel*. Cat. En105-53 1996.

———. 2003. 'Federal environmental assessment making a difference: Sable Island offshore gas project'. At: http://www.ceaa.gc.ca/default.asp?lang=En&n=B79A2A7A-1&offset=3&toc=show.

Canadian Food Inspection Agency (CFIA). 2004. *Action Plan for Invasive Alien Terrestrial Plants and Plant Pests*. Ottawa: Canadian Food Inspection Agency.

———. 2010. 'Imported food sector regulatory proposal'. At: www.inspection.gc.ca/english/fssa/imp/lic/queste.shtml.

Canadian General Standards Board. 2003. *Draft National Standard: Organic Agriculture*. Ottawa: Canadian General Standards Board, Committee on Organic Agriculture.

Canadian Heritage Rivers Board. 2002. *The Canadian Heritage Rivers System: Annual Report 2001–2002*. Ottawa: Minister of Public Works and Government Services.

———. 2010. *Canadian Heritage Rivers System, Annual Report 2009–2010*. Ottawa: Minister of Public Works and Government Services Canada, Apr.

Canadian Institute of Forestry. 2003. 'Silviculture systems' At: web.archive.org/web/20030428183657/http://cif-ifc.org/practices/silviculture.htm.

Canadian Parks and Wilderness Society (CPAWS). 2003. 'Boreal forests: For the birds'. Boreal Forest Factsheet Series. At: www.borealbirds.org/resources/factsheet-bsi-overview.pdf.

Canadian Press. 2007. 'No end in sight to stand-off', *The Record*, 27 Feb., C11.

———. 2008. 'Walkerton victims waiting for help', *The Record*, 10 Mar., A1, A3.

Canadian Wildlife Service. 'Contaminants levels in double-crested cormorant eggs, 1970–2000'. At: www.ecoinfo.org/env_ind/region/cormorant/pcbs_e.cfm.

Canadian Wind Energy Association. 2008. 'Wind energy sets global growth record in 2007'. At: www.canwea.ca/news_releases.cfm?ID=58.

Carlson, M., et al. 2010. 'Maintaining the role of Canada's forests and peatlands in climate regulation', *Forestry Chronicle* 86, 4: 1–10.

Carpenter, R.A. 1995. 'Communicating environmental science uncertainties', *Environmental Professional* 17: 127–36.

Carson, R. 1962. *Silent Spring*. Boston: Houghton Mifflin.

Casselman, A. 2011. 'The source of life: Canada's watershed protection action guide', *Canadian Geographic* (June): 69–76.

CBC News. 2011. 'Labrador Innu vote on contentious land claim deal', 30 June. At: www.cbc.ca/news/canada/newfoundland-labrador/story/2011/06/30/nl-innu-labrador-claims-churchill-vote-630.html.

Chalecki, E.L. 2000. 'Same planet, different worlds: The climate change information gap', in D.N. Scott et al., *Climate Change Communication: Proceedings of an International Conference*. Waterloo, Ont.: University of Waterloo and Environment Canada, Adaptation and Impacts Research Group, A2, 15–22.

Chambers, P., et al. 2001. 'Agricultural and forestry land use impacts', in Environment Canada, *Threats to Sources of Drinking Water and Aquatic Ecosystem Health in Canada*. Ottawa: Environment Canada, 57–62.

Chapman, J.D. 1989. *Geography and Energy: Commercial Energy Systems and National Policies*. Harlow, UK: Longman Scientific and Technical. Reprinted with permission of the author.

Chestnut, L.G., and D.M. Mills. 2005. 'A fresh look at the benefits and costs of the US acid rain program', *Journal of Environmental Management* 77: 252–66.

Chevalier, G., et al. 1997. 'Mercury in northern Québec: Role of the Mercury Agreement and the status of research and monitoring', *Water, Air and Soil Pollution* 97: 75–84.

Chomitz, K.M., et al. 2007. *At Loggerheads? Agricultural Expansion, Poverty Reduction, and Environment in the Tropical Forests*. Jakarta, Indonesia: World Bank.

Clark, W.F., J.M. Sontrop, J.M. Macnab, M. Salvador, L. Moist, R. Suri, and A.X. Garg. 2010. 'Long-term risk for hypertension, renal impairment, and cardiovascular disease after gastroenteritis from drinking water contaminated with *Escherichia coli* 0157:H7: A prospective cohort study', *British Medical Journal* 341. At: www.bmj.com/content/341/bmj.c6020.full.

Clarke, A., and C.M. Harris. 2003. 'Polar marine ecosystems: Major threats and future change', *Environmental Conservation* 30: 1–25.

Clean Air Sudbury. 2005. *Clearing the Air: Air Quality Trends in Sudbury*. Sudbury, Ont.: Clean Air Sudbury.

Cline, W.R. 2007. *Global Warming and Agriculture: Impact Estimates by Country*. Washington: Center for Global Development.

Collinge, S. 2010. 'Spatial ecology and conservation', *Nature Education Knowledge* 1, 8: 69.

Commissioner of the Environment and Sustainable Development. 2003. 'Managing the safety and accessibility of pesticides'. *Report of the Commissioner of the Environment and Sustainable Development*, Office of the Auditor General of Canada. Reproduced with the permission of the Minister of Public Works and Government Services, 2011.

———. 2008. 'Selected aspects of managing the safety and accessibility of pesticides'. *Status Report of the Commissioner of the Environment and Sustainable Development*, Office of the Auditor General of Canada. Reproduced with the permission of the Minister of Public Works and Government Services, 2011.

———. 2010. *2010 Fall Report of the Commissioner of the Environment and Sustainable Development*. Ottawa: Office of the Auditor General, 7 Dec. At: www.oag-bvg.ca/internet/English/parl_cesd_201012_00_e_34423.html.

———. 2011. *2011 Report of the Commissioner of the Environment and Sustainable Development*. Ottawa: Office of the Auditor General, 4 Oct.

Commission for Environmental Cooperation. 2008. *Green Building in North America: Opportunities and Challenges*. Montreal: Commission for Environmental Cooperation.

———. 2011. *North American Environmental Outlook to 2030*. Montreal: CEC.

Committee on the Status of Endangered Wildlife in Canada (COSEWIC). 2003. 'Canadian species at risk'. At: www.cosewic.gc.ca.

———. 2007. 'Canadian species at risk'. At: www.cosewic.gc.ca.

———. 2011. 'Canadian species at risk'. At: www.cosewic.gc.ca/rpts/Full_List_Species.html.

Conference Board of Canada. 2007. *Mission Possible*, vol. 2. Ottawa: Conference Board of Canada.

Conservation Authorities of Ontario. 1993. *Restructuring Resource Management in Ontario: A Blueprint for Success*. Mississauga, Ont.: Credit Valley Conservation Authority.

Copenhagen. 2007. *Eco-Metropole: Our Vision for Copenhagen 2015*. Municipality of Copenhagen: Technical and Environmental Centre.

Cornia, G.A. 1985. 'Farm size, land yields and the agricultural production function: An analysis for fifteen developing countries', *World Development* 13, 4: 513–34.

Costanza, R., et al. 1997. 'The value of the world's ecosystem services and natural capital', *Nature* 387: 253–60.

Couch, W.J. 2002. 'Strategic resolution of policy, environmental and socio-economic impacts in Canadian Arctic diamond mining: BHP's NWT diamond project', *Impact Assessment and Project Appraisal* 20: 265–78.

Cressman, D.R. 1994. 'Remedial action programs for soil and water degradation problems in Canada', in T.L. Napier, S.M. Camboni, and S.A. El-Swaify, eds, *Adopting Conservation on the Farm*. Ankeny, Iowa: Soil and Water Conservation Society, 413–34.

Crutzen, P., A. Mosier, K. Smith, and W. Winiwarter. 2007. 'N_2O release from agro-biofuel production negates global warming reduction by replacing fossil fuels', *Atmospheric Chemistry and Physics Discussions* 7: 11191–205.

Dale, A., and L. Newman. 2007. 'Governance for integrated resource management', in K.S. Hanna and D.S. Slocombe, eds, *Integrated Resource and Environmental Management: Concepts and Practice*. Toronto: Oxford University Press, 56–71.

Darwin, C.R. 1859. *On the Origin of Species*. London: John Murray.

David Suzuki Foundation. 2007. 'What is carbon offset?' At: www.davidsuzuki.org/Climate_Change/What_You_Can_Do/carbon_offsets.asp.

Davis, A., and J. Wagner. 2006. 'A right to fish for a living? The case for coastal peoples' determination of access and participation', *Ocean and Coastal Management* 49: 476–97.

Dawson, D. 2002. 'Plant-covered roofs ease urban heat', *National Geographic News*, 15 Nov. At: news.nationalgeographic.com/news/2002/11/1115_021115_GreenRoofs.html.

Day, J.C., and F. Quinn. 1992. *Water Diversion and Export: Learning from Canadian Experience*. Department of Geography Publication Series no. 36. Waterloo, Ont.: University of Waterloo.

Dearden, P. 1995. 'Park literacy and conservation', *Conservation Biology* 9: 1654–6.

———. 2001. 'Endangered species and terrestrial national parks', in K. Beazley and R. Boardman, eds, *Politics of the Wild: Canada and Endangered Species*. Toronto: Oxford University Press Canada, 75–93. Reprinted by permission of the publisher.

——— and R. Rollins, eds. 2009. *Parks and Protected Areas in Canada: Planning and Management*, 3rd edn. Toronto: Oxford University Press Canada. Reprinted by permission of the publisher.

de Loë, R.C., S. Di Giantomasso, and R.D. Kreutzwiser. 2002. 'Local capacity for groundwater protection in Ontario', *Environmental Management* 29, 2: 217–33.

Department of Fisheries and Oceans Canada (DFO). 2008. 'Canadian fisheries statistics 2005'. At: www.dfo-mpo.gc.ca/stats/commercial/cfs/2005/cfs05-eng.htm.

———. 2010. *Canadian Marine Ecosystem Status and Trends Report*. Ottawa: Her Majesty the Queen in Right of Canada.

———. 2011. *2010 Canadian Marine Ecosystem Status and Trends Report*. Ottawa: DFO.

Department of National Parks, Wildlife and Plant Conservation (DNP). 2010. *Thailand Tiger Action Plan: 2010–2022*. Bangkok: DNP.

Desjardins, R.L., et al. 2008. 'Moving Canadian agricultural landscapes from GHG source to sink', in H. Hengeveld et al., *Enhancement of Greenhouse Gas Sinks: A Canadian Science Assessment*. Ottawa: Ministry of Environment, 19–37.

de Villiers, M. 1999. *Water Wars: Is the World Running out of Water?* London: Weidenfeld and Nicolson.

Diamond, B. 1990. 'Villages of the dammed', *Arctic Circle* (Nov.–Dec.): 24–34.

Diamond, J. 2005. *Collapse: How Societies Choose to Fail or Succeed*. New York: Viking.

Diduck, A. 2010. 'Incorporating participatory approaches and social learning', in Mitchell (2010: 495–525).

——— and B. Mitchell. 2003. 'Learning, public involvement and environmental assessment: A Canadian case study', *Journal of Environmental Assessment Policy and Management* 5: 339–64.

Dillon Consulting Ltd. 2000. *Wind Turbine Environmental Assessment: Draft Screening Document*. Toronto: Dillon Consulting Ltd.

Dillon, P., G. Dixon, C. Driscoll, J. Giesy, S. Hurlbert, and J. Nriagu. 2011. *Evaluation of Four Reports on Contamination of the Athabasca River System by Oil Sands Operations*. Edmonton: Government of Alberta.

Doubleday, W.G. 2000. 'Seals & cod', *Isuma* 1, 1. At: web.archive.org/web/20070422225604/http://www.isuma.net/v01n01/doubleda/doubleda_e.shtml.

Downes, C., P. Blancher, and B. Collins. 2010. *Landbird Trends in Canada, 1968–2006*. Ottawa: Environment Canada.

Dubreuil, C. 2006. *The Right to Water: From Concept to Implementation*. Marseille: World Water Council.

Dudley, N., ed. 2008. *Guidelines for Applying Protected Area Management Categories*. Gland, Switzerland: IUCN.

———, J.D. Parrish, K.H. Redford, and S. Stolton. 2010. 'The revised IUCN protected area management categories: The debate and ways forward', *Oryx* 44: 485–90.

Duinker, P. 2011. 'Advancing the cause? Contributions of criteria and indicators to sustainable forest management in Canada', *Forestry Chronicle* 87, 4: 488–93.

——— et al. 1991. 'Community forestry and its implications for northern Ontario', *Forestry Chronicle* 67: 131–5.

——— et al. 1994. 'Community forests in Canada: An overview', *Forestry Chronicle* 70: 711–20.

Dumont, C., et al. 1998. 'Mercury levels in the Cree population of James Bay, Quebec, from 1988 to 1993/94', *Canadian Medical Association Journal* 158: 1439–45.

Dunbar, D., and I. Blackburn. 1994. *Management Options for the Northern Spotted Owl in British Columbia*. Victoria: Ministry of Environment.

Durner, G., J. Whiteman, H. Harlow, S. Amstrup, E. Regehr, and M. Ben-David. 2011. 'Consequences of long-distance swimming and travel over deep-water pack ice for a female polar bear during a year of extreme sea ice retreat', *Polar Bear* (US government) 34: 975–84.

Dwivedi, O.P., and R. Khator. 2006. 'Sustaining development: The road from Stockholm to Johannesburg', in G.M. Mudacumura, D. Mebratu, and M.S. Haque, eds, *Sustainable Development Policy and Administration*. Boca Raton, Fla: Taylor and Francis, 114–33.

Dyer, G. 2007. 'Arctic scramble is about seabed resources', *The Record*, 14 Aug., A9.

———. 2009. 'A missed opportunity', *The Record*, 22 Dec., A11.

Earth Policy Institute. 2007. *Earthtrends*. At: www.earth-policy.org/Updates/2006/Update55_data.htm.

Easterling, W.E., et al. 2007. 'Food, fibre and forest', in Intergovernmental Panel on Climate Change, *Climate Change 2007: Impacts, Adaptations and Vulnerability*. Cambridge, UK: IPCC.

Ebeling, J. 2006. *Tropical Deforestation and Climate Change: Towards an International Mitigation Strategy*. Oxford: Oxford University Press.

Edwards, P., and I. Roberts. 2008. 'Transport policy is food policy', *The Lancet* 371: 1661.

Edwards, P., and T. Talaga. 2011. '$20M settlement in Caledonia suit', *The Record*, 9 July, A3.

ÉEM Inc. 2007. 'Environmental paper procurement: A review of forest certification schemes in Canada'. At: marketsinitiative.org/uploads/MI-EEMcert-2.pdf.

Eggertson, L. 2008. 'Despite federal promises, First Nations' water problems persist', *Canadian Medical Association Journal* 178: 985.

Ehrlich, P., and A. Ehrlich. 1996. *Betrayal of Science and Reason: How Anti-Environmental Rhetoric Threatens Our Future*. Washington: Island Press.

El Ayoubi, F., and J. McNiven. 2006. 'Political, environmental and business aspects of bulk water exports: A Canadian perspective', *Canadian Journal of Administrative Sciences* 23: 1–16.

El Lakany, H., M. Jenkins, and M. Richards. 2007. 'Background paper on means of implementation', contribution by PROFOR to discussions at UNFF-7, Apr. Program on Forests (PROFOR).

El-Sadek, A. 2010. 'Virtual water trade as a solution for water scarcity in Egypt', *Water Resources Management* 24: 2437–88.

Environmental Assessment Panel. 1991. *Rafferty-Alameda Project: Report of the Environmental Assessment Panel*. Ottawa: Federal Environmental Assessment Review Office.

Environmental Commissioner of Ontario. 2003. *2002–2003 Annual Report*. Toronto: Environmental Commissioner of Ontario.

Environmental Protection Agency. 1997. 'About environmental justice'. At: epa.gov/swerops/ej/aboutej/html.

Environment Canada. 1985. *Currents of Change: Final Report, Inquiry on Federal Water Policy*. Ottawa: Environment Canada.

———. 2002a. *Population Status of Migratory Game Birds in Canada*. CWS Migratory Birds Regulatory Report no. 7. Ottawa: Canadian Wildlife Service Waterfowl Committee.

———. 2002b. 'Trichloroethylene and tetrachloroethylene in solvent degreasing'. At: www.ec.gc.ca/Publications/C835ECFF-D5EA-40A3-9E09-2784A982E525/StrategicOptionsForTheManagementOfToxicSubstancesTrichloroethyleneAndTetrachloroethyleneInSolventDegreasing.pdf.

———. 2003a. 'Atmospheric science: Acid rain'. At: www.ec.gc.ca/air/default.asp?lang=En&n=7E5E9F00-1.

———. 2003b. 'Environmental signals: Canada's National Environmental Indicator Series 2003'. Ottawa: Environment Canada. At: publications.gc.ca/collections/Collection/En40-775-2002E.pdf.

———. 2003c. 'Natural disasters on the rise', *Science and Environment Bulletin* (Mar.–Apr.).

———. 2003d. 'NPRI: Substance information, dioxins and furans'. At: www.ec.gc.ca/pdb/npri/npri_dioxins_ecfm.

———. 2004. 'Canadian acid deposition science assessment'. At: publications.gc.ca/collections/Collection/En4-46-2004E.pdf.

———. 2006a. Canada–US Air Quality Agreement Progress Report 2006, p. 40, at: www.ec.gc.ca/Publications/default.asp?lang=En&xml=4B98B185-7523-4CFF-90F2-5688EBA89E4A, 2006. Reproduced with the permission of the Minister of Public Works and Government Services Canada, 2011.

———. 2006b. *Canadian Protected Areas Status Report 2000–2005*. Environment Canada, Cat. No.: En81-9/2005E, ISBN: 0-662-44299-7, 81 pages, 2006. Reproduced with the permission of the Minister of Public Works and Government Services Canada, 2011.

———. 2006c. *National Inventory Report*. Ottawa: Environment Canada.

———. 2006d. *Impacts of Sea-Level Rise and Climate Change on the Coastal Zone of New Brunswick*. Ottawa: Minister of the Environment.

———. 2007a. *2006 Prairie Waterfowl Status Report*. Ottawa: Environment Canada.

———. 2007b. *Canadian Environmental Sustainability Indicators*. Ottawa: Environment Canada.

———. 2009a. 'Water conservation—every drop counts', 26 Nov. At: www.ec.gc.ca/eau-water/default.asp?lang=Eng&n=3377BC74-1.

———. 2009b. *Canadian Environmental Sustainability Indicators, 2008: Freshwater Quality Indicator, Data Sources and Methods*. Ottawa: Environment Canada.

———. 2009c. *Canada's 4th National Report to the UN Convention on Biological Diversity*. Ottawa: Environment Canada.

———. 2006a. 'Canada-U.S. Air Quality Agreement Progress Report 2010', p. 45, www.ec.gc.ca/Publications/default.asp?lang=En&xml=4B98B185-7523-4CFF-90F2-5688EBA89E4A. Reproduced with the permission of the Minister of Public Works and Government Services Canada, 2012

———. 2010b. *Canada Water Act: Annual Report for April 2008 to March 2009*. Ottawa: Environment Canada.

———. 2010c. 'Dams and diversions', 30 July. At: www.ec.gc.ca/eau-water/default.asp?lang-Eng&n=9D404A01-1.

———. 2011. 'Canada's protected areas'. At: www.ec.gc.ca/indicateurs-indicators/default.asp?lang=en&n=478A1D3D-1.

———, Statistics Canada, and Health Canada. 2006. *Canadian Environmental Sustainability Indicators*. Ottawa: Statistics Canada, Environment Accounts and Statistics Division.

——— and US Environmental Protection Agency. 2003. *State of the Great Lakes, 2003*. Ottawa: Environment Canada; Washington: US Environmental Protection Agency.

——— and ———. 2007. *State of the Great Lakes 2007*. Ottawa: Environment Canada.

Estes, J.A., D.O. Duggins, and G.B. Rathbun. 1989. 'The ecology of extinctions in kelp forest communities', *Conservation Biology* 3: 252–64.

External Advisory Committee on Cities and Communities. 2006. *From Restless Communities to Resilient Places: Building a Stronger Future for All Canadians*. Ottawa: Infrastructure Canada.

Fargione, J., J. Hill, D. Tilman, S. Polasky, and P. Hawthorne. 2008. 'Land clearing and the biofuel carbon debt', *Science* 319: 1235–8.

Farrell, A.E., et al. 2006. 'Ethanol can contribute to energy and environmental goals', *Science* 311: 506–8.

Federal, Provincial, and Territorial Governments of Canada. 2010. *Canadian Biodiversity: Ecosystem Status and Trends 2010*. Ottawa: Canadian Council of Resource Ministers. Reproduced with the permission of the Minister of Public Works and Government Services Canada, 2012.

———. 2012. *Canadian Biodiversity: Ecosystem Status and Trends, 2010* [pg. 48]. Ottawa: Canadian Councils of Resource Ministers. Reproduced with the permission of the Minister of Public Works and Government Services Canada, 2012. Source data for this graph was obtained from: *Canadian Council on Ecological Areas*. 2009. Conservation Areas Reporting and Tracking System (CARTS). V.2009.05. http://ccea.org

Festa-Bianchet, M. 2003. 'Exploitative wildlife management as a selective pressure for the life-history evolution of large mammals', in M. Festa-Bianchet and M. Apollonio, eds, *Animal Behavior and Wildlife Conservation*. Washington: Island Press, 191–207.

———, J.C. Ray, S. Boutin, S.D. Côté, and A. Gunn. 2011. 'Conservation of caribou (*Rangifer tarandus*) in Canada: An uncertain future', *Canadian Journal of Zoology* 89: 419–34.

Field, J.G., G. Hempel, and C.P. Summerhayes. 2002. *Oceans 2020*. Washington: Island Press. Reproduced with permission.

Findlay, C.S., E. Elgie, B. Giles, and L. Burr. 2010. 'Species listing under Canada's Species at Risk Act', *Conservation Biology* 23: 1609–17.

Fischer, G., M. Shah, F. Tulbello, and H. van Velhulzen. 2005. 'Socio-economic and climate change impacts on agriculture: An integrated assessment', *Philosophical Transactions of the Royal Society B* 360: 2067–83.

Fisheries and Oceans Canada, *see* Department of Fisheries and Oceans (DFO).

Flanagan, K., and S. Rasheed. 2002. 'Population viability analysis applied to woodland caribou in Jasper National Park', *Research Links* 10: 16–18.

Fleischhauer, M. 2008. 'The role of spatial planning in strengthening urban resilience', in H.J. Pasman and L.A. Kirillov, eds, *Resilience of Cities to Terrorist and Other Threats*. Dordrecht, The Netherlands: Springer, 273–98.

Food and Agriculture Organization (FAO). 2001. 'Conservation agriculture', *FAO Magazine Spotlight*. At: www.fao.org/ag/magazine.

———. 2003a. 'FAOSTAT statistical database'. At: faostat.fao.org.

———. 2003b. 'Water management: Towards 2030'. At: www.fao.org/ag/magazine.

———. 2006. *The State of World Fisheries and Aquaculture*. Rome: FAO.

———. 2007a. *The State of the World's Animal Genetic Resources for Food and Agriculture*. Rome: FAO.

———. 2007b. *World Food Outlook*. Rome: FAO.

———. 2010. *Global Forest Resources Assessment 2010*. Rome: FAO. At: www.fao.org/forestry/fra/fra2010/en/.

———. 2011a. 'Global wheat production to increase in 2011', press release, 23 Mar. At: www.fao.org/news/story/en/item/53813/icode/.

———. 2011b. *FAO Cereal Supply and Demand Brief*. World Food Situation. Rome: FAO. At: www.fao.org/worldfoodsituation/en/.

Forest Products Industry Competitiveness Task Force. 2007. *Industry at a Crossroads: Choosing the Path to Renewal*. Ottawa: Forest Products Association of Canada.

Foster, H.D., and W.R.D. Sewell. 1981. *Water: The Emerging Crisis in Canada*. Ottawa: Canadian Institute for Economic Policy.

Fraser Basin Council. 1997. *Charter for Sustainability*. Vancouver: Fraser River Basin Council.

Fraser Institute. 1999. *Environmental Indicators for Canada and the United States*. Vancouver: Fraser Institute.

Freedman, B. 1981. *Intensive Forest Harvest: A Review of Nutrient Budget Considerations*. Information Report M–X–121. Fredericton: Maritimes Forest Research Centre.

———, P.N. Duinker, and R. Morash. 1986. 'Biomass and nutrients in Nova Scotia forests, and implications of intensive harvesting for future site productivity', *Forest Ecology and Management* 15: 103–27.

Freiwald, A., et al. 2004. *Cold-water Coral Reefs*. Cambridge, UK: UNEP-WCMC.

Fritz, S.M., J. Povlacs Lunde, W. Brown, and E.A. Banset. 2005. *Interpersonal Skills for Leadership*, 2nd edn. Englewood Cliffs, NJ: Pearson Prentice-Hall.

Gabriel, A.O., and R.D. Kreutzwiser. 1993. 'Drought hazard in Ontario: A review of impacts, 1960–1989, and management implications', *Canadian Water Resources Journal* 18: 117–32.

Gardner. G. 2011. 'Roundwood production plummets', in Worldwatch Institute (2011: 84–8).

——— and A. Assadourian. 2004. 'Rethinking the good life', in Worldwatch Institute (2004a: 164–80).

Garshelis, D.L. 1997. 'Sea otter mortality estimated from carcasses collected after the *Exxon Valdez* oil spill', *Conservation Biology* 11, 4: 905–16.

Gaston, A.J., et al. 2009. 'Changes in Canadian seabird populations and ecology since 1970 in relation to changes in oceanography and food webs', *Environmental Reviews* 17: 267–86.

Gauthier, G., et al. 2001. 'Seasonal survival of greater snow geese and effect of hunting under dependence in sighting probability', *Ecology* 82: 3105–19.

Gedalof, Z., and A.A. Berg. 2010. 'Tree ring evidence for a limited direct CO_2 fertilization effect', *Global Biogeochemical Cycles* 24: GB3027, doi:10.1029/2009GB003699.

Gibson, R.B. 2007. 'Integration through sustainability assessment: Emerging possibilities at the leading of edge of environmental assessment', in K.S. Hanna and D.S. Slocombe, eds, *Integrated Resource and Environmental Management: Concepts and Practice*. Toronto: Oxford University Press, 72–96. Copyright © Oxford University Press Canada 2007. Reprinted by permission of the publisher.

Gleick, P.H. 2003. 'Global freshwater resources: Soft-path solutions for the 21st century', *Science* 302: 1524–8.

———. 2006. 'Water and terrorism', *Water Policy* 8: 481–503.

———. 2010. *Bottled and Sold: The Story behind Our Obsession with Bottled Water*. Washington: Island Press.

Global Water Partnership. 2000. *Integrated Water Resources Management*. Technical Advisory Committee Background Paper no. 4. Stockholm: Global Water Partnership.

Global Wind Energy Council. 2006. *Global Wind 2006 Report*. Brussels: Global Wind Energy Council.

Goodarzi, F., and M. Mukhopadhyay. 2000. 'Metals and polyaromatic hydrocarbons in the drinking water of the Sydney Basin, Nova Scotia, Canada: A preliminary assessment of their source', *International Journal of Coal Geology* 43: 357–72.

Gorrie, P. 1990. 'The James Bay power project: The environmental cost of reshaping the geography of northern Quebec', *Canadian Geographic* 110, 1: 21–31.

———. 2008. 'Green giant', *Toronto Star*, 8 Mar., ID1–ID2.

Gosselin, P., S.E. Hrudey, M.A. Naeth, A. Plourde, R. Therrien, G. Van der Kraak, and Z. Xu. 2010. *Environmental and Health Impacts of Canada's Oil Sands Industry*. Ottawa: Royal Society of Canada.

Gould, S.J. 1989. *Wonderful Life: The Burgess Shale and the Nature of History*. New York: W.W. Norton.

———. 1994. 'The evolution of life on the earth', *Scientific American* 271: 84–91.

Graenfeldt, D. 2010. 'The next nexus? Environmental ethics, water policies, and climate change', *Water Alternatives* 3: 575–86.

Gray, B. 1989. *Collaborating: Finding Common Ground for Multiparty Problems*. San Francisco: Jossey-Bass.

Greater Sudbury. 2010. *Living Landscape: A Biodiversity Action Plan for Greater Sudbury*. Sudbury, Ont.: Vegetation Enhancement Technical Advisory Committee.

Greer, A., V. Ng, and D. Fisman. 2008. 'Climate change and infectious diseases in North America: The road ahead', *Canadian Medical Association Journal* 178: 715–22.

Gruber, N., and J.N. Galloway. 2008. 'An Earth-system perspective of the global nitrogen cycle', *Nature* 451: 293–6.

Gutberlet, J. 2008. 'Empowering collective recycling initiatives: Video documentation and action research with a recycling co-op in Brazil', *Resources, Conservation and Recycling* 52: 659–70.

———. 2009. 'The solidarity economy of recycling co-ops: Micro-credit to alleviate poverty', *Development in Practice* 19, 6: 737–51.

———. 2010. 'Waste, poverty and recycling', *Waste Management* 30, 2: 171–3.

——— and A. Baeder. 2008. 'Informal recycling and occupational health in Santo André, Brazil', *International Journal of Environmental Health Research* 18, 1: 1–15

Hall, J. 2004. 'Hazel's gift wrapped in green', *Toronto Star*, 15 Oct., B1–B2.

Halweil, B. 2002. *Home Grown: The Case for Local Food in a Global Market*. Worldwatch Paper 163. Washington: Worldwatch Institute.

Hambler, C., M.A. Speight, and P.A. Henderson. 2011. 'Extinction rates, extinction-prone habitats and indicator groups in Britain and at larger scales', *Biological Conservation* 144, 2: 713–21.

Hangs, R.D., J.D. Knight, and K.C.J. Van Rees. 2003. 'Nitrogen uptake characteristics for roots of conifer seedlings and common boreal forest competitor species', *Canadian Journal of Forest Research* 33: 156–63.

Hannah, L. 2003. 'Protected areas management in a changing climate', in N.W.P. Munro et al., eds, *Making Ecosystem-Based Management Work*. Proceedings of the Fifth International Conference on Science and Management of Protected Areas, Victoria, BC, May 2003. Wolfville, NS: Science and Management of Protected Areas Association. At: www.sampaa.org/publications.

Harden, A., and H. Levalliant. 2008. *Boiling Point: Six Community Profiles of the Water Crisis Facing First Nations within Canada*. Ottawa: Polaris Institute.

Harris, Michael. 1998. *Lament for an Ocean: The Collapse of the Atlantic Cod Fishery, A True Crime Story*. Toronto: McClelland & Stewart.

Hazell, R., and B. Worthy. 2010. 'Assessing the performance of freedom of information', *Government Information Quarterly* 27: 352–9.

Health Canada. 2007. *Canada's Health Concerns from Climate Change and Variability*. Ottawa: Health Canada. At: www.taiga.net/nce/resources/newsletters/NCE_Newsletter_Winter2003.pdf.

———. 2008. *Canada's Food Guide: Food and Nutrition*. Ottawa: Government of Canada.

———. 2010. *First Nations, Inuit and Aboriginal Health: Drinking Water and Wastewater*. At: www.hc-sc.gc.ca/fniah-spnia/promotion/public-publique/water-eau-eng.php.

Hearnden, K.W., et al. 1992. *A Report on the Status of Forest Regeneration*. Toronto: Ontario Independent Forest Audit Committee.

Heath, C., and D. Heath. 2010. *Switch: How to Change Things When Change Is Hard*. Toronto: Random House Canada.

Heimann, M., and M. Reichstein. 2008. 'Terrestrial carbon ecosystem dynamics and climate feedbacks', *Nature* 451: 289–92.

Hengeveld, H. 1991. *Understanding Atmospheric Change: A Survey of the Background Science and Implications of Climate Change and Ozone Depletion*. SOE Report no. 91–2. Ottawa: Minister of Supply and Services.

———. 2006. CO_2/*Climate Report*. Toronto: Environment Canada, Atmospheric Science Assessment and Integration Branch.

———, E. Bush, and P. Edwards. 2002. *Frequently Asked Questions about Climate Change Science*. Ottawa: Minister of Supply and Services.

———, B. Whitewood, and A. Fergusson. 2005. *An Introduction to Climate Change: A Canadian Perspective*. Toronto: Environment Canada, Atmospheric Science Assessment and Integration Branch.

——— et al. 2008. *Enhancement of Greenhouse Gas Sinks: A Canadian Science Assessment*. Ottawa: Ministry of Environment.

Herkenrath, P., and J. Harrison. 2011. 'The 10th meeting of the Conference of the Parties to the Convention on Biological Diversity—a breakthrough for biodiversity?', *Oryx* 45: 1–2.

Hilson, G. 2000. 'Sustainable development policies in Canada's mining sector: An overview of government and industry efforts', *Environmental Science and Policy* 3: 201–11.

Hilson, G., and V. Nayee. 2002. 'Environmental management system implementation in the mining industry: A key to achieving cleaner production', *International Journal of Minerals and Processes* 64: 19–41.

Himmelman, A.T. 1996. 'On the theory and practice of transformational collaboration: From social service to social justice', in C. Huxham, ed., *Creating Collaborative Advantage*. Thousand Oaks, Calif.: Sage, 19–43.

Hites, R.A., et al. 2004. 'Global assessment of organic contaminants in farmed salmon', *Science* 303: 226–9.

Hobson, K.A., E.M. Bayne, and S.L. Van Wilgenburg. 2002. 'Large-scale conversion of forest to agriculture in the boreal plains of Saskatchewan', *Conservation Biology* 16, 6: 1530–41.

Hockstra, A.Y., and A.K. Chapagain. 2007. 'Water footprints of nations: Water use by people as a function of their consumption pattern', *Water Resources Management* 21: 35–48.

Hofmann, N., and M.S. Beaulieu. 2006. *A Geographical Profile of Manure Production in Canada, 2001*. Report 21–601–MIE–N.07. Ottawa: Statistics Canada.

———, G. Filoso, and M. Schofield. 2005. *The Loss of Dependable Agricultural Land in Canada*. Rural and Small Town Canada Analysis Bulletin 6. Catalogue no 21–006–XIE. Ottawa: Statistics Canada.

Hoggan, J., with R. Littlemore. 2009. *Climate Cover-up: The Crusade to Deny Global Warming*. Vancouver: Greystone Books.

Holling, C.S., ed. 1978. *Adaptive Environmental Assessment and Management*. Chichester: John Wiley.

———. 1986. 'The resilience of terrestrial ecosystems: Local surprise and global change', in W.C. Clark and R.E. Munn, eds, *Sustainable Development in the Biosphere*. Cambridge: Cambridge University Press, 292–317.

Homer-Dixon, T. 2000. *The Ingenuity Gap*. New York: Alfred A. Knopf.

———. 2007. *The Upside of Down: Catastrophe, Creativity and the Renewal of Civilization*. Toronto: Vintage Canada.

Howard, K.W.F., and H. Maier. 2007. 'Road de-icing salt as a potential constraint on urban growth in the Greater Toronto Area, Canada', *Journal of Contaminant Hydrology* 91, 1 and 2: 146–70.

Human Development Resource Office. 2007. *Human Development Report 2007/2008: Fighting Climate Change*. Houndmills, UK: Palgrave Macmillan.

Huntington, H.P., et al. 2005. 'The changing Arctic: Indigenous perspectives', in L. Arris, ed., *Arctic Climate Impact Assessment*. Cambridge: Cambridge University Press, 61–98.

Hutchings, J.A., and R.A. Myers. 1994. 'What can be learned from the collapse of a renewable resource? Atlantic cod, *Gadus morhua*, of Newfoundland and Labrador', *Canadian Journal of Fisheries and Aquatic Sciences* 51: 2126–46.

Hyde, D., H. Hermann, and R.A. Lautenschlager. 2010. *The State of Biodiversity in Canada*. Ottawa: Natureserve Canada.

Independent Environmental Monitoring Agency (IEMA). 2007. *Plain Language Annual Report 2006–07*. At: www.monitoringagency.net/AgencyDocumentsPresentations/AnnualReports/tabid/64/Default.aspx.

———. 2010a. 'Independent Environmental Monitoring Agency: Why was it created?' At: www.monitoringagency.net/AboutUs/WhytheAgencywascreated/tabid/61/Default.aspx.

———. 2010b. *Plain Language Annual Report 2009–10*. Yellowknife, NWT: Independent Environmental Monitoring Agency.

Indian and Northern Affairs Canada. 1997a. *Highlights of the Canadian Arctic Contaminants Assessment Report: A Community Reference Manual*. Ottawa: Minister of Public Works and Government Services.

———. 1997b. 'Pathways of transportation of persistent organics and metals to Arctic freshwater and marine ecosystems', in Indian and Northern Affairs Canada (1997a).

Intergovernmental Panel on Climate Change (IPCC). 2001a. *Climate Change 2001: Impacts, Adaptation, and Vulnerability. Contribution of Working Group II to the Third Assessment Report of the Intergovernmental Panel on Climate Change*, J.J. McCarthy et al., eds. Cambridge: Cambridge University Press.

———. 2001b. *Climate Change 2001: The Scientific Basis. Contributions of Working Group I to the Third Assessment Report of the Intergovernmental Panel on Climate Change*, J.T. Houghton et al., eds. Cambridge: Cambridge University Press.

———. 2001c. *Climate Change 2001: Synthesis Report. A Contribution of Working Groups I, II and III to the Third Assessment Report of the Intergovernmental Panel on Climate Change*, R.T. Watson and the Core Writing Team, eds. Cambridge: Cambridge University Press.

———. 2007a. *Climate Change 2007: Impacts, Adaptation and Vulnerability*. Paris and Geneva: IPCC Secretariat.

———. 2007b. *Climate Change 2007: The Physical Science Basis. Working Group I Contribution to the Fourth Assessment Report of the Intergovernmental Panel on Climate Change*, Figure SPM.2. Cambridge: Cambridge University Press.

———. 2007c. 'Summary for policy makers of the synthesis report of the IPCC fourth assessment report'. At: www.ipcc.ch.

International Energy Agency. 2010. *Key World Energy Statistics*. Paris: OECD/IEA.

International Joint Commission (IJC). 2000. *Living with the Red*. Ottawa and Washington: IJC.

———. 2004. *Twelfth Biennial Report on Great Lakes Water Quality*. Windsor, Ont.: IJC.

International Organization for Migration. 2010. 'Facts and figures: Global estimates and trends'. At: www.iom.int/jahia/Jahia/about-migration/facts-and-figures/lang/en.

International Reference Group on Great Lakes Pollution from Land Use Activities. 1978. *Environmental Management Strategy for the Great Lakes System: Final Report to the International Joint Commission*. Windsor, Ont.: IJC.

International Strategy for Disaster Reduction. 2005. *Hyogo Framework for Disaster Reduction 2005–2015: Building Resilience of Nations and Communities to Disaster*. Report of the World Conference on Disaster Reduction, 18–21 Jan. Kobe, Hyogo, Japan.

International Water Resources Association. 2000. 'Towards hydrosolidarity', *Water International* 25, 2 (special issue).

IUCN, *see* World Conservation Union.

Jachmann, H., P.S.M. Berry, and H. Imae. 1995. 'Tusklessness in African elephants: A future trend', *African Journal of Ecology* 33: 230–5.

Jackson, J.B.C., et al. 2001. 'Historical overfishing and the recent collapse of coastal ecosystems', *Science* 293: 629–38.

Jacques Whitford Engineering, Scientific, Planning and Management Consultants. 2006. *Final Report: 100% Design Report, Demolition and Disposal of the Old SYSCO Cooling Pond*. Prepared for the Sydney Tar Ponds Agency. Dartmouth, NS: Jacques Whitford.

James, C. 2006. *Global Status of Commercialized Biotech/GM Crops: 2006*. ISAAA Brief no. 35. Ithaca, NY: International Service for the Acquisition of Agri-biotech Applications.

———. 2010. *Global Status of Commercialized Biotech/GM Crops: 2010*. ISAAA Brief no. 42. Ithaca, NY: ISAAA. At: www.isaaa.org/resources/publications/pocketk/16/default.asp.

Jayyousi, O.A. 2007. 'Water as a human right: Towards civil society globalization', *Water Resources Development* 23: 329–39.

Jeffries, D.S., et al. 2003. 'Assessing the recovery of lakes in southeastern Canada from the effects of acidic deposition', *Ambio* 32: 176–83.

———, I. Wong, I. Dennis, and M. Sloboda. 2010. 'Terrestrial and aquatic critical loads map'. Ottawa: Environment Canada, Water Science and Technology Branch, unpublished.

———, ———, and M. Sloboda. 2010. 'Boreal Shield steady-state exceedances for forest soils or lakes map', prepared for Boreal Shield Ecozone status and trends report. Ottawa: Environment Canada, Water Science and Technology Branch, unpublished.

Jeziorski, A., et al. 2008. 'The widespread threat of calcium decline in fresh water', *Science* 322: 1374–7.

Jones, C.G., R.S. Ostfield, M.P. Richard, E.M. Schauber, and J.O. Wolff. 1998. 'Chain reactions linking acorns to gypsy moth outbreaks and Lyme disease risk', *Science* 279: 1023–6.

Jørgensen, C., et al. 2007. 'Managing evolving fish stocks', *Science* 318: 1247–8.

Joseph, C., T.I. Gunton, and J.C. Day. 2006. 'Implementation of resource management plans: Identifying keys to success', *Journal of Environmental Management* 88: 594–606.

Kanninen, M., et al. 2007. *Do Trees Grow on Money? The Implications of Deforestation Research for Policies to Promote REDD*. Bogor, Indonesia: Center for International Forestry Research.

Katz, D. 2010. *Making Waves: Examining the Case for Sustainable Water Exports from Canada*. Vancouver: Fraser Institute.

Kay, C.E. 1994. 'Aboriginal overkill: The role of Native Americans in structuring western ecosystems', *Human Nature* 5: 359–98.

KBM Forestry Consultants. 2007. *National Forest Strategy (2003-2008) Evaluation: Final Report*. Ottawa: Canadian Council of Forest Ministers.

Keith, L.B., et al. 1984. 'Demography and ecology of a declining snowshoe hare population', *Wildlife Monographs* 90: 1–43.

Kelly, E.N., D.W. Schindler, P.V. Hudson, J.W. Short, R. Radmanovich, and C.C. Nielsen. 2010. 'Oil sands development contributes elements toxic at low concentrations to the Athabasca River and its tributaries', *Proceedings of the National Academy of Sciences* 107, 37: 16178–83.

Kelly, J., and P.W. Williams. 2007. 'Modelling tourism destination energy consumption and greenhouse gas emissions: Whistler, British Columbia, Canada', *Journal of Sustainable Tourism* 15, 1: 67–90.

Kenworthy, J.R. 2006. 'The eco-city: Ten key transport and planning dimensions for sustainable city development', *Environment and Urbanization* 18, 1: 67–85.

Kesho Trust, The. 2007. 'Reconnecting children and families with nature'. Report of workshops sponsored by ActNow BC, 28 June and 17 July, Royal Roads University, Victoria, BC.

Khadka, A.K. 2010. 'The emergence of water as a "human right" on the world stage: Challenges and opportunities', *Water Resources Development* 26: 37–49.

Kimmins, J.P. 1977. 'Evaluation of the consequences for the future tree productivity of the loss of nutrients in whole-tree harvesting', *Forest Ecology and Management* 1: 169–83.

Kitasei, S. 2011. 'Wind power growth continues to break records despite recession', in Worldwatch Institute (2011: 26–8).

Klironomos, J.N., M.F. Allen, M.C. Rillig, J. Piotrowski, S. Makvandi-Nejad, B.E. Wolfe, and J.R. Powell. 2005. 'Abrupt rise in atmospheric CO_2 overestimates community response in a model plant–soil system', *Nature* 433: 621–4.

Knight, W.A., and T. Keating. 2010. *Global Politics: Emerging Networks, Trends, and Challenges*. Toronto: Oxford University Press.

Krantzberg, G., and J.P. Manno. 2010. 'Renovation and innovation: It's time for the Great Lakes Regime to respond', *Water Resources Management* 24: 4273–85.

Krebs, C.J., et al. 2001. 'What drives the 10-year cycle of snowshoe hares?', *Bioscience* 51: 25–35.

Kreutzwiser, R., and R. de Loe. 2004. 'Water security: From exports to contamination of local water supplies', in B. Mitchell, ed., *Resource and Environmental Management in Canada: Addressing Conflict and Uncertainty*. Toronto: Oxford University Press, 166–94.

Krkošek, M., et al. 2007. 'Declining wild salmon populations in relation to parasites from farm salmon', *Science* 318: 1772–5.

Krugel, L. 2010. 'U.S. group hopes to tar Alberta with image of oily ducks', *The Record*, 15 July, B3.

Kupchella, C.E., and M.C. Hyland. 1989. *Environmental Science: Living within the System of Nature*. 2nd ed. Reprinted by permission of Pearson Education Inc., Upper Saddle River, NJ.

Kurz, W.A., et al. 2008. 'Mountain pine beetle and forest carbon feedback to climate change', *Nature* 452: 987–90.

Kuvlesky, W.P., L.A. Brennan, M.L. Morrison, K.K. Boydson, B.M. Ballard, and R.C. Bryant. 2007. 'Wind energy development and wildlife conservation: Challenges and opportunities', *Journal of Wildlife Management* 71: 2487–98.

Kyoto Protocol to the United Nations Framework Convention on Climate Change. 1997. At: unfccc.int/resource/docs/convkp/kpeng.pdf.

Lahey, A. 1998. 'Black lagoons: They're ugly. They stink. But are the tar ponds really killing the people of Sydney, Nova Scotia?', *Saturday Night* 113, 8: 37–40.

Land Reclamation Program. 2007. *Land Reclamation Program Annual Report, 2007*. Sudbury, Ont.: Environmental Planning Initiatives.

Landry, M., V.G. Thomas, and T.D. Nudds. 2001. 'Sizes of Canadian national parks and the viability of large mammal populations: Policy implications', *The George Wright Forum* 18, 1: 13–23.

Lasserre, F. 2007. 'Drawers of water: Water diversions in Canada and beyond', in Bakker (2007: 143–62).

———. 2009. 'Transfers massifs d'eau au Canada: Entre mythe et réalité', *Policy Options* (July–Aug.): 53–9.

Lautenbach, W.W. 1985. *Land Reclamation Program 1978–1984*. Sudbury, Ont.: Vegetation Enhancement Technical Advisory Committee.

——— et al. 1995. 'Municipal land restoration program: The regreening process', in J.M. Gunn, ed., *Restoration and Recovery of an Industrial Region: Progress in Restoring the Smelter-Damaged Landscape near Sudbury, Ontario*. New York: Springer-Verlag, 109–22. Reprinted with kind permission from Springer Science + Business Media B.V.

Lautenschlager, R.A., and T.P. Sullivan. 2002. 'Effects of herbicide treatments on biotic components in regenerating northern forests', *Forestry Chronicle* 78: 695–731.

Lawrence, D., ed. 2003. *Environmental Impact Assessment: Practical Solutions to Recurrent Problems*. Toronto: John Wiley.

Lebeuf, M., et al. 2007. 'Temporal trends (1987–2002) of persistent, bioaccumulative and toxic (PBT) chemicals in beluga whales (*Delphinapterus leucas*) from the St Lawrence estuary, Canada', *The Science of the Total Environment* 383: 216–31.

Lee, K.N. 1993. *Compass and Gyroscope: Integrating Science and Politics for the Environment*. Washington: Island Press.

Leeder, J. 2010. 'As nations squabble, bluefin is fished closer to extinction', *Globe and Mail*, 26 Nov., A8.

Leggett, W.C., and K.T. Frank. 2008. 'Paradigms in fisheries oceanography', *Oceanography and Marine Biology: An Annual Review* 46: 331–63.

Lemieux, A. 2005. 'Canada's global mining presence', in *Canadian Minerals Yearbook, 2005*. Ottawa: Minister of Public Works and Government Services.

Lemieux, C., T. Beechey, and D. Scott. 2007. 'A survey on protected areas and climate change (PACC) in Canada: Survey update', *ECO* 16: 2–3.

———, ———, ———, and P. Gray. 2011. 'The state of climate change adaptation in Canada's protected area sector', *Canadian Geographer* 55: 301–17.

Lennon, John. 1971. 'Imagine'. Words and music by John Lennon © 1971 (renewed 1999) LENONO MUSIC. All rights controlled and administered by EMI BLACKWOOD MUSIC INC. All rights reserved. International Copyright Secured. Reprinted by Permissions of Hal Leonard Corporation.

Lenton, R., and M. Muller, eds. 2009. *Integrated Water Resources Management in Practice*. London: Earthscan.

Leopold, A. 1949. *A Sand County Almanac*. Oxford: Oxford University Press.

Le Prestre, P.G., and P. Stoett. 2001. 'International initiatives, commitments, and disappointments: Canada, CITES, and the CBD', in K. Beazley and R. Boardman, eds, *Politics of the Wild: Canada and Endangered Species*. Toronto: Oxford University Press, 190–216.

Lesieur, D., S. Gauthier, and Y. Bergeron. 2002. 'Fire frequency and vegetation dynamics for the south-central boreal forest of Quebec, Canada', *Canadian Journal of Forest Research* 32: 1996–2002.

Lin, B. 2011. 'Resilience in agriculture through crop diversification: Adaptive management for environmental change', *Bioscience* 61, 3: 183–93.

Liu, J., A.J.B. Zehnder, and H. Yang. 2007. 'Historical trends in China's virtual water trade', *Water International* 32: 78–90.

Longhurst, A. 2002. 'Murphy's law revisited: Longevity as a factor in recruitment to fish populations', *Fisheries Research* 56: 125–31

Louv, R. 2006. *Last Child in the Woods*. Toronto: Algonquin Books.

Lovelock, J.E. 1988. *The Ages of Gaia*. New York: W.W. Norton.

Lynam, A.J. 2010. 'Securing a future for wild Indochinese tigers: Transforming tiger vacuums into tiger source sites', *Integrative Zoology* 5: 324–34.

McBean, G., and D. Henstra. 2003. *Climate Change, Natural Hazards, and Cities*. Paper Series no. 31. Toronto: Institute for Catastrophic Loss Reduction.

Macdonald, D. 2009. 'The Government of Canada's search for environmental legitimacy: 1971–2008', *International Journal of Canadian Studies* 39 and 40: 191–210.

McGinnity, P., et al. 2003. 'Fitness reduction and potential extinction of wild populations of Atlantic salmon (*Salmo salar*) as a result of interactions with escaped farm salmon', *Proceedings of the Royal Society: Biological Sciences* 270: 2443–50.

McIntosh, R.P. 1980. 'The relationship between succession and the recovery process in ecosystems', in J. Cairns, ed., *The Recovery Process in Damaged Ecosystems*. Ann Arbor, Mch.: Ann Arbor Science Publishers, 11–62.

McIver, J.D., A.R. Moldenke, and G.L. Parsons. 1990. 'Litter spiders as bio-indicators of recovery after clear-cutting in a western coniferous forest', *Northwest Environmental Journal* 6: 410–12.

McKibben, B. 2010. 'Science under siege: Climate change skeptics have presented their case too well', *The Record*, 3 Mar., A9.

Mckinney, M.A., E. Peacock, and R.J. Letcher. 2009. 'Sea ice-associated diet change increases the levels of chlorinated and brominated contaminants in polar bears', *Environmental Science and Technology* 43: 4334–9.

Mackintosh, W.A. 1934. *Prairie Settlement: The Geographic Background*. Toronto: Macmillan.

Maclaren, V.W. 2010. 'Waste management: Moving up the hierarchy', in Mitchell (2010: 382–406).

McLeman, R., and B. Smit. 2003. *Climate Change, Migration and Security*. Commentary no. 86. Ottawa: Canadian Security Intelligence Service.

Malcolm, C. 2003. 'The current state and future prospects of whale-watching management with special emphasis on whale-watching in BC, Canada', PhD dissertation, University of Victoria.

———. 2009. 'Provincial parks', in Dearden and Rollins (2009: 56–82).

Malcolm, J.R., et al. 2006. 'Global warming and extinctions of endemic species from biodiversity hotspots', *Conservation Biology* 20: 538–48.

Manitoba Round Table on Environment and Economy. 1992. *Sustainable Development: Towards Institutional Change in the Manitoba Public Sector*. Winnipeg: Manitoba Round Table on Environment and Economy.

Manitoba Water Stewardship. 2006. 'About us, and minister's message'. At: www.gov.mb.ca/waterstewardship/misc/about.htm.

———. 2007. 'Potential transboundary water projects'. At: www.gov.mb.ca/waterstewardship/water_info/transboundary/potential.html.

Martin, P.S. 1967. 'Prehistoric overkill', in P.S. Martin and H.E. Wright Jr, eds, *Pleistocene Extinctions: The Search for a Cause*. New Haven: Yale University Press, 75–120.

Mascarenhas, M. 2007. 'Where the waters divide: First Nations, tainted water and environmental justice in Canada', *Local Environment* 12: 565–77.

Masters, M., A. Tikina, and B. Larson. 2010. 'Forest certification audit results as potential changes in forest management in Canada', *Forestry Chronicle* 86: 455–60.

Mastny, L. 2004. 'Purchasing for people and the planet', in Worldwatch Institute (2004a: 122–42).

——— and H. French. 2002. 'Crimes of (a) global nature', *Worldwatch* (Sept.–Oct.): 12–23.

Matthews, C., R.B. Gibson, and B. Mitchell. 2007. 'Rising waves, old charts, nervous passengers: Navigating toward a new water ethic', in Bakker (2007: 335–58).

May, E., and M. Barlow. 2001. 'The tar ponds', *Alternatives Journal* 27, 1: 7–11.

Mayhew, P.J., G.B. Jenkins, and T.G. Benton. 2008. 'A long-term association between global temperature and biodiversity, origination and extinction in the fossil record', *Proceedings of the Royal Society of London B* 275: 47–53.

Meadows, D.H., et al. 1972. *The Limits to Growth*. London. Earth Island.

Ménard, M. 2009. 'Canada, a big energy consumer: A regional perspective', Statistics Canada, 12 Nov. At: www.statcan.gc.ca/pub/11-621-m/11-621-m2005023-eng.htm.

Mercredi, O., and M.E. Turpel. 1993. *In the Rapids: Navigating the Future of First Nations*. Toronto: Viking. Copyright © Ovide Mercredi and Mary Ellen Turpel, 1993. Reprinted by permission of Penguin Group (Canada), a Division of Pearson Canada Inc.

Millennium Ecosystem Assessment. 2003. *Ecosystems and Human Well-being: A Framework for Assessment*. Washington: Island Press.

———. 2005. *Ecosystems and Human Well-being: Synthesis*. Washington: Island Press.

Miller, G. 2002. *Climate Change: Is the Science Sound?* Special Report to the Legislative Assembly of Ontario. Toronto: Environmental Commissioner of Ontario.

———. 2003. *2002–03 Annual Report of the Environmental Commissioner of Ontario*. At: www.eco.on.ca/uploads/Reports%20-%20Annual/2002_03/03ar.pdf.

Miller, K., et al. 2011. 'Genomic signatures predict migration and spawning failure in wild Canadian salmon', *Science*, 14 Jan., 214–17.

Mills, T.J., T.M. Quigley, and F.J. Everest. 2001. 'Science-based natural resource management decisions: What are they?', *Renewable Resources Journal* 19, 2: 10–15.

Mitchell, B. 2009. 'Implementation gap', *IWRA Update* 22, 3: 7–12.

———, ed. 2010. *Resource and Environmental Management in Canada: Addressing Conflict and Uncertainty*, 4th edn. Toronto: Oxford University Press.

Mittelstaedt, M. 2007. 'Pollution causing premature deaths', *Globe and Mail*, 3 Oct., A1. At: www.theglobeandmail.com/servlet/story/LAC.20071003.TOXIC03/TPStory/TPNational/BritishColum.

Molle, F. 2007. 'Scales and power in river basin management: The Chao Phraya River in Thailand', *Geographical Journal* 173: 358–73.

Montreal. 2010. *Montreal: Community Sustainable Development Plan, 2010–2015*. Montreal: Environment and Sustainable Development.

Mooers, A.O., et al. 2007. 'Biases in legal listing under Canadian endangered species legislation', *Conservation Biology* 21: 572–5.

Moreno-Sánchez, R.D.P., and J.H. Maldonado. 2006. 'Surviving from garbage: The role of informal waste-pickers in a dynamic model of solid-waste management in developing countries', *Environment and Development Economics* 11: 371–91.

Morris, T.J., et al. 2007. *Changing the Flow: A Blueprint for Federal Action on Freshwater*. Toronto: The Gordon Water Group of Concerned Scientists and Friends.

Morrison, K., T. Nelson, and A. Ostry. 2011. 'Methods for mapping local food production capacity from agricultural statistics', *Agricultural Systems* 104: 491–9.

Mosquin, T., and P.G. Whiting. 1992. *Canada Country Study of Biodiversity*. Ottawa: Canadian Museum of Nature.

———, ———, and D.E. McAllister. 1995. *Canada's Biodiversity: The Variety of Life, Its Status, Economic Benefits, Conservation Costs and Unmet Needs*. Ottawa: Canadian Museum of Nature.

Mulkins, L.M., et al. 2002. 'Carbon isotope composition of mysids at a terrestrial-marine ecotone, Clayoquot Sound, British Columbia, Canada', *Estuarine, Coastal and Shelf Science* 54: 669–75.

Munro, M. 2008. 'Environment Canada "muzzles" scientists' dealings with media', *Ottawa Citizen*, 1 Feb.

Murphy, S.D., and L.R.G. Martin. 2001. 'Urban ecology in Ontario, Canada: Moving beyond the limits of city and ideology', *Environments* 29, 1: 67–83.

Myers, R.A., and B. Worm. 2003. 'Rapid worldwide depletion of predatory fish communities', *Nature* 423: 280–3.

Nakashima, D.J. 1990. *Application of Native Knowledge in EIA: Inuit, Eiders and Hudson Bay Oil*. Prepared for the Canadian Environmental Assessment Research Council. Ottawa: Minister of Supply and Services.

Nanus, B. 1992. *Visionary Leadership*. San Francisco: Jossey-Bass.

National Commission on the BP Deepwater Horizon Oil Spill and Offshore Drilling. 2011. *Deep Water: The Gulf Oil Disaster and the Future of Offshore Drilling: Report to the President*. Washington: National Commission on the BP Deepwater Horizon Oil Spill and Offshore Drilling, Jan.

National Energy Board (NEB). 2009a. 'Global and Canadian context for energy demand analysis—energy brief'. At: www.neb-one.gc.ca/clf-nsi/rnrgynfmtn/nrgyrprt/nrgdmnd/glblcndncntxt2008/glblcndncntxtnrgbrf-eng.html.

———. 2009b. *Canadian Energy Demand: Passenger Transportation*. Energy Briefing Note. Calgary: NEB, Publications Office.

National Forest Strategy Coalition (NFSC). 2005. *Highlights of Accomplishments: Advancing the National Forest Strategy*. Ottawa: Canadian Council of Forest Ministers.

National Geographic Society. 2008. *Greendex 2008: Consumer Choice and the Environment: A Worldwide Tracking Survey*. Washington: National Geographic Society.

National Oceanic and Atmospheric Administration, National Climate Data Center. 2010. *State of the Climate Global Analysis, Annual 2010*. At: www.ncdc.noaa.gov/sotc/global/2010/13.

National Research Council, Division of Earth and Life Studies, Committee on the Science of Climate Change. 2001. *Climate Change Science: An Analysis of Some Key Questions*. Washington: National Academy Press.

National Round Table on the Environment and the Economy (NRTEE). 2003a. *2003 Environment and Sustainable Development Indicators for Canada*. Ottawa: NRTEE.

———. 2003b. *Securing Canada's Natural Capital: A Vision for Nature Conservation in the 21st Century*. Ottawa: NRTEE.

———. 2003c. *State of the Debate: Environmental Quality in Canadian Cities: The Federal Role*. Ottawa: NRTEE.

———. 2006. *The State of the Debate on the Environment and the Economy: Environmental Quality in Canadian Cities: The Federal Role*. Ottawa: NRTEE.

———. 2010. *Degrees of Change: Climate Warming and the Stakes for Canada*. Ottawa: NRTEE.

———. 2011a. *Parallel Paths: Canada–U.S. Climate Policy Choices*. Ottawa: NRTEE.

———. 2011b. *Paying the Price: The Economic Impacts of Climate Change for Canada*. Ottawa: NRTEE.

Natural Resources Canada. 1995. 'Silviculture terms in Canada'. At: cfs.nrcan.gc.ca/pubwarehouse/pdfs/24216_e.pdf.

———. 2005a. 'Canadian vehicle survey 2005'. At: oee.nrcan.gc.ca/Publications/statistics/cvs05/index.cfm?attr=0.

———. 2005b. *The State of Canada's Forests: Annual Report 2004–5: The Boreal Forest*. Ottawa: Canadian Forest Service, Natural Resources Canada.

———. 2005c. 'Technologies and applications: About solar energy'. At: canmetenergy.nrcan.gc.ca/renewables/solar-photovoltaic/2438.

———. 2006a. 'Canadian minerals yearbook'. Minerals and Mining Statistics On-line. At: web.archive.org/web/20071026150232/http://mmsd1.mms.nrcan.gc.ca/mmsd/intro_e.asp.

———. 2006b. 'Energy efficiency trends in Canada 1990–2004'. At: oee.nrcan.gc.ca/publications/statistics/trends06/chapter6.cfm?attr=0.

———. 2006c. 'Technologies and applications: About wind energy'. At: canmetenergy.nrcan.gc.ca/renewables/wind/2171.

———. 2007a. 'Employment in Canada's mining and mineral processing industries remains robust in 2006', Information Bulletin. At: publications.gc.ca/collections/collection_2009/nrcan/M31-6-2007E.pdf.

———. 2007b. *The State of Canada's Forests: Annual Report 2007*. Ottawa: Canadian Forest Service, Natural Resources Canada.

———. 2007c. 'Strong growth expected for Canada's mining sector', Natural Resources Canada's News Room. At: web.archive.org/web/20080221193508/http://www.nrcan-rncan.gc.ca/media/newsreleases/2007/200789a_e.htm.

———. 2009a. 'Canadian mining industry: 2009 general review', *Canadian Minerals Yearbook—2009*. At: www.nrcan.gc.ca/minerals-metals/business-market/canadian-minerals-yearbook/4033.

———. 2009b. 'Partnership agreements', *Ekati Diamond Mine. Partnership Agreements*. At: www.nrcan.gc.ca/minerals-metals/aboriginal/bulletin/3719.

———. 2010a. 'Non-timber forest products'. At: canadaforests.nrcan.gc.ca/article/borealntfp.

———. 2010b. *State of Canada's Forests: Annual Report 2010*. Ottawa: Natural Resources Canada.

———. 2010c. 'Energy efficiency: Energy use'. At: web.archive.org/web/20101227104424/http://www.nrcan.gc.ca/eneene/effeff/transuse-eng.php.

———. 2011. *The State of Canada's Forests: Annual Report 2011*. Ottawa: Natural Resources Canada. Reproduced with the permission of the Minister of Public Works and Government Services Canada, 2011, and the courtesy of the Canadian Forest Service.

Nature. 2010. 'Citizen scientists' (editorial), 468: 476.

Neave, D., et al. 2002. 'Forest biodiversity in Canada', *Forestry Chronicle* 78, 6: 779–83.

Neilson, E.T., et al. 2007. 'Spatial distribution of carbon in natural and managed stands in an industrial forest in New Brunswick, Canada', *Forest Ecology and Management* 253: 148–60.

Nellemann, C., S. Hain, and J. Alder, eds. 2008. *In Dead Water: Merging of Climate Change with Pollution, Over-harvest, and Infestations in the World's Fishing Grounds*. Arendal, Norway: UNEP/GRID-Arendal. At: www.grida.no.

Newman, P.W.G. 2006. 'The environmental impact of cities', *Environment and Urbanization* 18, 2: 275–95.

Newmark, W.D. 1995. 'Extinction of mammal

populations in western North American national parks', *Conservation Biology* 9: 512–26.

Nikiforuk, A. 2007. *On the Table: Water, Energy and North American Integration*. Toronto: University of Toronto, Munk Centre for International Studies, Program on Water Issues, Sept.

Nobel, P.G. 1990. *Contaminants in Canadian Seabirds*. SOE Report no. 90–2. Ottawa: Environment Canada.

Nova Scotia Department of Environment and Labour. 2002. *A Drinking Water Strategy for Nova Scotia: A Comprehensive Approach to the Management of Drinking Water*. Halifax: Nova Scotia Environment and Labour.

Nova Scotia Tourism Partnership and Nova Scotia Tourism, Culture and Heritage. 2007. *Nova Scotia Strategy for Sustainable Coastal Tourism Development*. Halifax: Nova Scotia Tourism, Culture and Heritage.

Nuclear Waste Management Organization (NWMO). 2005. *Choosing a Way Forward: The Future Management of Canada's Used Nuclear Fuel: A Summary*. Toronto: NWMO.

O'Connor, D.R. 2002a. *Report of the Walkerton Inquiry: Part One, The Events of May 2000 and Related Issues*. Toronto: Ontario Ministry of the Attorney General.

———. 2002b. *Report of the Walkerton Inquiry: Part Two, A Strategy for Safe Drinking Water*. Toronto: Ontario Ministry of the Attorney General.

Odum, E.P. 1969. 'The strategy of ecosystem development', *Science* 164: 262–70.

———. 1971. *Fundamentals of Ecology*, 3rd edn. Toronto: Holt, Rinehart and Winston, CBS College Publishing.

O'Hara, K. 2010. 'Canada must free scientists to talk to journalists', *Nature* 467: 501.

Oil Sands Advisory Panel. 2010. *A Foundation for the Future: Building an Environmental Monitoring System for the Oil Sands*. A report submitted to the Minister of Environment, Dec.

Olsson, P. 2007. 'The role of vision in framing adaptive co-management processes: Lessons from Kristianstads Vattenrike, southern Sweden', in Armitage et al. (2007: 168–85).

O'Neill, D. 2004. 'Threats to water availability in Canada: A perspective', in *Threats to Water Availability in Canada*. NWRI Scientific Assessment Report Series no. 3 and ACSD Science Assessment Series no. 1. Ottawa and Burlington, Ont.: Environment Canada and National Water Research Institute, xi–xvi.

Ontario Ministry of Energy. 2006. 'Ontario leads Canada in wind power generation', press release. At: news.ontario.ca/archive/en/2006/11/22/Ontario-Leads-Canada-In-Windpower-Generation.html.

Ontario Power Generation. 2010. 'Nuclear power'. At: www.opg.com/power/nuclear/.

Organization for Economic Co-operation and Development (OECD). 2002. *Towards Sustainable Consumption: An Economic Conceptual Framework*. Paris: OECD Environment Directorate.

———. 2011. *Greening Household Behaviour*. Paris: OECD.

O'Riordan, T. 1976. *Environmentalism*. London: Pion.

Orr, D.W. 1992. 'The problem of education', *New Directions for Higher Education* 77 (Spring): 3–8.

———. 1994. *Earth in Mind: On Education, Environment and the Human Prospect*. Covelo, Calif.: Island Press.

Pappas, S. 2011. 'Global chronic hunger rises above 1 billion', in Worldwatch Institute (2011: 92–5).

Paré, D., P. Bernier, E. Thiffault, and B.D. Titus. 2011. 'The potential of forest biomass as an energy supply for Canada', *Forestry Chronicle* 87, 1: 71–6.

Parkinson, G. 2011. 'Durban talks off to a bad start', *Climate Spectator*, 28 Nov. At: www.climatespectator.com.au/commentary/durban-talks-bad-start.

Parks Canada Agency. 2011. *Corporate Plan 2011/2012 to 2015/2016*. Ottawa: Her Majesty the Queen in Right of Canada.

Parmesan, C. 2006. 'Ecological and evolutionary responses to recent climate change', *Annual Review of Ecology and Systematics* 37: 637–69.

Parson, E.A. 2000. 'Environmental trends and environmental governance in Canada', *Canadian Public Policy* 26 (supplement): S123–S143.

Partidario, M.F., and R. Clark, eds. 2000. *Perspectives on Strategic Environmental Assessment*. Boca Raton, Fla: Lewis Publishers.

Pauly, D. 2006. 'Major trends in small-scale fisheries, with emphasis on developing countries, and some implications for the social sciences', *Maritime Studies* 4, 2: 7–22.

Perkel, C. 2011. 'War over fresh water possible, meeting told', *The Record*, 23 Mar., A8.

Peterson, C.H., et al. 2003. 'Long-term ecosystem response to the *Exxon Valdez* oil spill', *Science* 302: 2082–6.

Phillips, D. 1990. *The Climate of Canada*. Ottawa: Minister of Supply and Services Canada.

Ping, T.S.T. 2010. 'Terrorism—A new perspective in the water management landscape', *Water Resources Development* 26: 51–63.

Plummer, R., and J. FitzGibbon. 2007. 'Connecting adaptive co-management, social learning, and social capital through theory and practice', in Armitage et al. (2007: 38–61).

Pollution Probe. 2007. *Towards a Vision and Strategy for Water Management in Canada*. Toronto: Pollution Probe.

Population Reference Bureau. 2007. 'World population highlights', *Population Bulletin* 62: 2.

Prescott-Allen, R. 2001. *The Well-being of Nations: A Country by Country Index of Quality of Life and the Environment*. Washington and Ottawa: Island Press and IDRC.

Priscoli, J.D., J. Dooge, and M.R. Llamas. 2004. *Water and Ethics: An Overview*. Series on Water and Ethics, Essay 1. Paris: UNESCO.

Quinn, F. 2007. *Water Diversion, Export and Canada–US Relations: A Brief History*. Toronto: University of Toronto, Munk Centre for International Studies, Program on Water Issues, Aug.

——— et al. 2004. 'Water allocation, diversion and export', in *Threats to Water Availability in Canada*. NWRI Scientific Assessment Report Series no. 3 and ACSD Science Assessment Series no. 1. Ottawa and Burlington, Ont.: Environment Canada and National Water Research Institute, 1–8.

Quoden, J. 2010. 'Multi-instrumental EU environmental policy', presentation at conference on Europe Environment Policy: What's Next? . . . Towards a 7th Environment Action Program, Brussels, 25–6 Nov. At: www.eapdebate.org/files/files/Session4-Joachim-Quoden.pdf.

Rahman, M.M., and O. Varis. 2005. 'Integrated water resources management: Evolution, prospects and future challenges', *Sustainability: Science, Practice and Policy* 1: 15–21.

Rainham, D. 2002. 'Risk communication and public response to industrial chemical contamination in Sydney, Nova Scotia: A case study', *Journal of Environmental Health* 65, 5: 26–32.

Raudsepp-Hearne, C., et al. 2010. 'Untangling the environmentalists' paradox: Why is human well-being increasing as ecosystem services are degrading?', *BioScience* 60, 8: 576–89.

RBC. 2011. '2011 Canadian Water Attitudes Study: Three quarters of Canadians using toilet as garbage can'. At: smr.newswire.ca/en/rbc-financial-group/canadian-water-attitudes-study.

Rees, J. 1985. *Natural Resources: Allocation, Economics and Policy*. London: Methuen.

Rees, W.E. 2010. 'Getting serious about urban sustainability: Eco-footprints and the vulnerability of twenty-first-century cities', in T. Bunting, P. Filion, and R. Walker, eds, *Canadian Cities in Transition: New Directions in the Twenty-First Century*, 4th edn. Toronto: Oxford University Press, 70–86.

Reeves, H.W. 2010. *Water Availability and Use Pilot: A Multiscale Assessment in the U.S. Great Lakes Basin*. Professional Paper 1778. Reston, Virginia: US Geological Survey.

Regehr, E.V., N.J. Lunn, S.C. Amstup, and I. Stirling. 2007. 'Effects of earlier sea ice breakup on survival and population size of polar bears in Western Hudson Bay', *Journal of Wildlife Management* 71, 8: 2673–83.

Reguly, E. 2009. 'We're in the doghouse—but we can buy our way out', *Globe and Mail*, 16 Dec., A16.

Reichel, J. 2011. 'New oil sands extraction method cheaper, more sustainable', *Epoch Times*, 22–8 Sept., A5.

Reynolds, M. 2010. 'Who sets climate policy? The U.S.?', *The Record*, 9 Feb., A11.

Richardson, N.H., B.I. Savan, and L. Bodnar. 1989. *Economic Benefits of a Clean Environment: Sudbury Case Study*. Prepared for the Department of Environment, Canada. Toronto: N.H. Richardson Consulting.

Rivers, N., and D. Sawyer. 2008. *Pricing Carbon: Saving Green—A Carbon Price to Lower Emissions, Taxes and Barriers to Green Technology*. Vancouver: David Suzuki Foundation.

Roberts, C.M. 2003. 'Our shifting perspectives on the oceans', *Oryx* 37: 166–77.

Rochette, P., et al. 2008. 'Estimation of N_2O emissions from agricultural soils in Canada, II: 1990–2005 inventory', *Canadian Journal of Soil Science* 88, 5: 655–69.

Rockström, J., et al. 2009. 'A safe operating space for humanity', *Nature* 461: 472–5. Reproduced with permission of Nature Publishing Group in the format Journal via Copyright Clearance Center.

Rode, K., S. Amstrup, and E. Regehr. 2010. 'Reduced body size and cub recruitment in polar bears associated with sea ice decline', *Ecological Applications* 20, 3: 768–82.

Rohde, K. 1992. 'Latitudinal gradients in species diversity: The search for the primary cause', *Oikos* 65: 514–27.

Ross, N. 2001. *Healing the Landscape: Celebrating Sudbury's Reclamation Success*. Sudbury, Ont.: Vegetation Enhancement Technical Advisory Committee.

Ross, P.S., et al. 2000. 'High PCB concentrations in free-ranging Pacific killer whales, *Orcinus orca*: Effects of age, sex and dietary preference', *Marine Pollution Bulletin* 40: 504–15.

Rowe, S. 1993. 'In search of the holy grass: How to bond with the wilderness in nature and ourselves', *Environment Views* (Winter): 7–11.

Rowell, P., G.L. Holroyd, and U. Banasch. 2003. 'Summary of the 2000 Canadian peregrine falcon survey', in J. Kennedy, Ed., *Bird Trends: A Report on Results of National Ornithological Surveys in Canada*. Ottawa: Canadian Wildlife Service, 52–6.

Royal Commission on the Future of the Toronto Waterfront. 1992. *Regeneration: Toronto's Waterfront and the Sustainable City, Final Report*. Ottawa and Toronto: Minister of Supply and Services Canada and Queen's Printer of Ontario.

Ruddiman, W.F. 2003. 'The anthropogenic greenhouse era began thousands of years ago', *Climatic Change* 61: 261–93.

Rylatt, M.G. 1999. 'Ekati diamond mine background and development', *Mining Engineer* 51: 37–43.

Sanchez, L.E. 1998. 'Industry response to the challenge of sustainability: The case of the Canadian nonferrous mining sector', *Environmental Management* 22: 521–31.

———. 2001. 'The climate change: Soil fertility–food security nexus', Address at the Sustainable Food Security for All by 2020 Conference, Bonn, Sept.

Sandford, R., and M.-A. Phare. 2011. 'A new water ethic', *Alternatives Journal* 37, 1: 12–14.

Saskatchewan. 2008. 'Mineral resources of Saskatchewan'. At: www.er.gov.sk.ca/mineralsandenergy.

Saskatchewan Energy and Resources. 2009. 'Mineral resources of Saskatchewan, uranium'. At: www.er.gov.sk.ca.

Saskatchewan Round Table on Environment and Economy. 1992. *Conservation Strategy for Sustainable Development in Saskatchewan*. Regina: Saskatchewan Round Table on Environment and Economy.

Saumure, R.A., T.B. Herman, and R.D. Titman. 2007. 'Effects of haying and agricultural practices on a declining species: The North American wood turtle (*Glyptemis insculpta*)', *Biological Conservation* 135: 565–75.

Sawin, J.L. 2004. 'Making better energy choices', in Worldwatch Institute (2004a: 24–43).

Schindler, D., and P. Lee. 2010. 'Comprehensive conservation planning to protect biodiversity and ecosystem services in Canadian boreal regions under a warming climate and increasing exploitation', *Biological Conservation* 143: 1571–86.

Schueler, F.W., and D.E. McAllister. 1991. 'Maps of the number of tree species in Canada: A pilot project GIS study of tree diversity', *Canadian Biodiversity* 1: 22–9.

Scialabba, N.E. 2003. *Organic Agriculture: The Challenge of Sustaining Food Production While Enhancing Biodiversity*. Rome: Food and Agriculture Organization.

——— and C. Hattam, eds. 2002. *Organic Agriculture, Environment and Food Security*. Rome: Food and Agriculture Organization.

Science Daily. 2011. 'World population to surpass 7 billion in 2011', 28 July. At: www.sciencedaily.com/releases/2011/07/110728144933.htm.

Scientific Review Panel. 2002. *British Columbia Offshore Hydrocarbon Development: Report of the Scientific Review Panel*. Submitted to the BC Minister of Energy and Mines, 15 Jan.

Scott, D., B. Jones, J. Andrey, L. Mortsch and K. Warriner, eds. 2000. *Climate Change Communication: Proceedings of an International Conference*, Waterloo, ON: University of Waterloo and Environment Canada, Adaptation and Impacts Research Group.

Seaweb. 2003. 'Danger at sea: Our changing ocean'. At: www.seaweb.org.

Secretariat of the Convention on Biological Diversity. 2010. *Global Biodiversity Outlook 3*. Montreal: Convention on Biological Diversity.

Selbourne, Earl of. 2000. *The Ethics of Freshwater Use: A Survey*. Paris: UNESCO.

Selin, S., and D. Chavez. 1995. 'Developing a collaborative model for environmental planning and management', *Environmental Management* 19: 189–95.

Shaftoe, D., ed. 1993. *Responding to Changing Times: Environmental Mediation in Canada*. Waterloo, Ont.: Conrad Grebel College.

Shang, E.H., R.M.K. Yu, and R.S. Wu. 2006. 'Hypoxia affects sex differentiation and development, leading to a male-dominated population in zebrafish', *Environmental Science and Technology* 40: 3118–22.

Shaw, R.W., and CCAF A041 Project Team. 2001. *Coastal Impacts of Climate Change and Sea-Level Rise on Prince Edward Island: Synthesis Report*. Climate Change Action Fund Project CCAF A041. Ottawa: Ministry of Supply and Services.

Shiklomanov, I.A. 2000. 'Appraisal and assessment of world water resources', *Water International* 25: 11–32.

Shrubsole, D. 2001. 'The cultures of flood management in Canada: Insights from the 1997 Red River experience', *Canadian Water Resources Journal* 26: 461–79.

Sierra Club of Canada. 2006. *National Forest Strategy 2003–2008: An Assessment in 2006—Is It Making a Difference?* Toronto: Sierra Club of Canada.

Simpson, C.G. 1964. 'Species density of North American recent mammals', *Systematic Zoology* 13: 15–73.

Sinclair, A.J., and M. Doelle. 2010. 'Environmental assessment in Canada: Encouraging decisions for sustainability', in Mitchell (2010: 462–94).

Slocombe, D.S. 1993. 'Implementing ecosystem-based management: Development of theory, practice and research for planning and managing a region', *BioScience* 43: 612–22.

———. 2010. 'Applying an ecosystem approach', in Mitchell (2010: 409–33).

Smith, A., and J.B. MacKinnon. 2007. *The 100-Mile Diet: A Year of Local Eating*. Toronto: Random House Canada.

Smith, J.N.M., et al. 1998. 'Population biology of snowshoe hares II: Interactions with winter food plants', *Journal of Animal Ecology* 57: 269–86.

Smith, L.C. 2010. *The World in 2050: Four Forces Shaping Civilization's Northern Future*. New York: Dutton Books/Penguin.

Sprague, J.B. 2007. 'Great Wet North: Canada's myth of water abundance', in Bakker (2007: 23–35).

Standing Committee on Environment and Sustainable Development. 2000. *Pesticides:*

Making the Right Choice for the Protection of Health and the Environment. Ottawa: House of Commons.

Statistics Canada. 2001. '2001 Census of agriculture'. At: www.statcan.gc.ca/ca-ra2001/index-eng.htm.

———. 2003a. *Farming Facts 2002*. Ottawa: Ministry of Industry.

———. 2003b. *Human Activity and the Environment: A Statistical Compendium*. Catalogue no. 16-201-XIE. Ottawa: Minister of Supply and Services.

———. 2007a. *2006 Census of Agriculture*. Ottawa: Statistics Canada.

———. 2007b. 'Households and the environment survey 2006'. At: www.statcan.gc.ca/daily-quotidien/070711/dq070711b-eng.htm.

———. 2007c. 'Recycling in Canada', *EnviroStats* 1, 1: 3–7.

———. 2009. *Food Statistics*. Catalogue no. 21-020-X. Ottawa: Statistics Canada.

———. 2010. 'Agriculture', in *2010 Annual Report*. Ottawa: Statistics Canada.

———. 2011a. *Human Activity and the Environment 2010*. Ottawa: Statistics Canada.

———. 2011b. *Households and the Environment 2009*. Ottawa: Statistics Canada.

———, Agriculture Division. 2003. 'Food statistics, 2003'. At: publications.gc.ca/Collection/Statcan/21-020-X/21-020-XIE2003002.pdf.

Steele, D.H., R. Andersen, and J.M. Green. 1992. 'The managed commercial annihilation of northern cod', *Newfoundland Studies* 8, 1: 34–68.

Stewart, E.J., et al. 2007. 'Sea ice in Canada's Arctic: Implications for cruise tourism', *Arctic* 60: 370–80.

Stinchcombe, K., and R.B. Gibson. 2001. 'Strategic environmental assessment as a means of pursuing sustainability: Ten advantages and ten challenges', *Journal of Environmental Assessment Policy and Management* 3: 343–72.

Stirling, A., et al. 2008 'Unusual predation attempts of polar bears on ringed seals in the southern Beaufort Sea: Possible significance of changing spring ice conditions', *Arctic* 61: 14–22.

Stockton, A.S., et al. 2005. 'A natural experiment on the impacts of high deer densities on the native flora of coastal temperate rain forests', *Biological Conservation* 126: 118–28.

Stockwell, C.A., P. Hendry, and M.T. Kinnison. 2003. 'Contemporary evolution meets conservation biology', *Trends in Ecology and Evolution* 18: 94–101.

Stokes, K., and R. Law. 2000. 'Fishing as an evolutionary force', *Marine Ecology Progress Series* 208: 307–9.

Stuart-Smith, J., et al. 2002. 'Conserving whitebark pine in the Canadian Rockies', *Research Links* 10: 11–14.

Suffling, R., and D. Scott. 2002. 'Assessment of climate change effects on Canada's national park system', *Environmental Monitoring and Assessment* 74: 117–39.

Sumaila, U.R., et al. 2007. 'Potential costs and benefits of marine reserves on the high seas', *Marine Ecology Progress Series* 345: 305–10.

——— and D. Pauly, eds. 2007. *Catching More Bait: A Bottom-up Re-estimation of Global Fisheries Subsidies (2nd version)*. Fisheries Centre Research Reports 14(6). Vancouver: Fisheries Centre, University of British Columbia.

Tahera Diamond Corporation. 2008. 'Mining—Jericho Diamond Mine'. At: web.archive.org/web/20080915022408/http://www.tahera.com/Operations/Mining/JerichoDiamondMine/default.aspx.

Taylor, L. 2009. 'Water challenges in oil sands country: Alberta's Water for Life strategy', *Policy Options* (July-Aug.): 44–7.

Taylor, P.D., L. Fahrig, K. Henein, and G. Merriam. 1993. 'Connectivity is a vital element of landscape structure', *Oikos* 68, 3: 571–2.

Theobold, M. 2009. 'Fish production reaches a record', *Vital Signs*, 3 Dec. At: vitalsigns.worldwatch.org/vs-trend/fish-production-reaches-record.

The Record (Kitchener–Waterloo). 2004. 'Harvesting the wind: Local farmers say they can provide power—if regulations let them', 9 Aug., A1–A2.

———. 2007. 'Six Nations can't dictate new laws', 15 Sept., A16.

———. 2010. 'Hopeful progress out of Cancun', 14 Dec., A10.

Thérival, R. 1993. 'Systems of strategic environmental assessment', *Environmental Impact Assessment Review* 13, 3: 145–68.

Thompson, I.D., J.A. Baker, and M. Ter-Mikaelian. 2003. 'A review of the long-term effects of post-harvest silviculture on vertebrate wildlife, and predictive models, with an emphasis on boreal forests in Ontario, Canada', *Forest Ecology and Management* 177: 441–69.

Tinker, J. 1996. 'Introduction', in J.B. Robinson et al., eds, *Life in 2030: Exploring a Sustainable Future for Canada*. Vancouver: University of British Columbia Press, ix–xv.

Todhunter, P.E. 2001. 'A hydroclimatological analysis of the Red River of the North snowmelt flood catastrophe of 1997', *Journal of the American Water Resources Association* 37: 1263–78.

Toronto, City of. 2011. '2010 residential waste diversion rates'. At: www.toronto.ca/garbage/residential-diversion.htm.

Toronto and Region Conservation Authority. 2006. *Moving toward the Living City*. Toronto: Toronto and Region Conservation Authority.

Toronto Public Health. 2007. *The State of Toronto's Food: Discussion paper for Toronto Food Strategy*. Toronto: Toronto Public Health.

Trainor, S.F., et al. 2007. 'Arctic climate impacts: Environmental injustice in Canada and the United States', *Local Environment* 12: 627–43.

Tremblay, C., J. Gutberlet, and A.M. Peredo. 2010. 'United we can: Resource recovery, place and social enterprise', *Resources, Conservation & Recycling* 54, 7: 422–8.

Trewartha, G.T. 1954. *An Introduction to Climate*. New York: McGraw-Hill.

Trites, A.W., et al. 2007. 'Killer whales, whaling, and sequential megafaunal collapse in the North Pacific: A comparative analysis of the dynamics of marine mammals in Alaska and British Columbia following commercial whaling', *Marine Mammal Science* 23, 4: 751–65.

Trostel, K., et al. 'Can predation cause the 10-year hare cycle?', *Oecologia* 74: 185–92.

Tynan, C.T., and J.L. Russell. 2008. 'Assessing the impacts of future 2°C global warming on Southern Ocean cetaceans', paper presented to the International Whaling Commission Scientific Committee, Chile, June.

UNAIDS. 2007. *AIDS Epidemic Update 2007*. Geneva: UNAIDS.

UNAIDS. 2010. *Report on the Global AIDS Epidemic*. At: www.unaids.org/globalreport/Global_report.htm.

United Nations. 2010. 'Millennium Goal 1 Update'. At: www.undp.org/mdg/goal1.shtml.

United Nations Development Programme (UNDP). 2005. *Human Development Report 2005: International Cooperation at a Crossroads: Aid, Trade and Security in an Unequal World*. New York: UNDP.

———. 2007. *Human Development Report 2007/8. Fighting Climate Change: Human Solidarity in a Divided World*. New York: UNDP. p. 5. Reproduced with permission of Palgrave Macmillan.

———. 2010. *Human Development Report 2010. The Real Wealth of Nations: Pathways to Human Development*. New York: UNDP.

———. 2011. *Human Development Report 2011. Sustainability and Equity: A Better Future for All*. New York: UNDP.

United Nations Environment Programme (UNEP). 2002. 'Global environmental outlook 3 (GEO 3)'. At: www.unep.org.geo/geo3.asp.

———. 2007. *Global Environment Outlook: Environment for Development (GEO-4)*. Nairobi: UNEP.

———. 2008. 'Meltdown in the mountains'. At: www.unep.org/Documents.Multilingual/Default.asp?DocumentID=530&ArticleID.

———. 2010. *Annual Report 2009: Seizing the Green Opportunity*. Nairobi: UNEP.

———. 2011. *Annual Report 2010*. Nairobi: UNEP.

United Nations General Assembly. 2010. 'General Assembly adopts resolution recognizing access to clean water, sanitation as a human right, by recorded vote of 122 in favor, none against, 41 abstentions'. At: www.un.org/News/Press/docs/2010/ga10967.doc.htm.

United Nations Population Division. 2011. 'World population to reach 10 billion by 2100 if fertility in all countries converges to replacement level'. At: esa.un.org/unpd/wpp/Other-Information/Press_Release_WPP2010.pdf.

United Nations Population Fund. 2007a. *State of the World Population 2007*. New York: United Nations Population Fund.

———. 2007b. *World Population Prospects: The 2006 Revision*. Geneva: UN.

United Nations Statistics Division. 2009. 'Agricultural support estimate for OECD countries from OECD database 2009'. Millennium Development Goal Indicators. At: unstats.un.org/unsd/mdg/SeriesDetail.aspx?srid=601.

United States Agency of International Development (USAID). 2011. 'USAID Fact Sheet 2011. Food for Peace 2010 Programs', news release. Washington: USAID.

United States Department of Energy. 2011. *Biomass Energy Data Book: Energy Efficiency and Renewable Energy*. Washington: US Government.

Vancouver, City of. 2009a. *Greenest City: Quick Start Recommendations*. Vancouver: City of Vancouver.

——— 2009b. *Vancouver 2020: A Bright Green Future—An Action Plan for Becoming the World's Greenest City by 2020*. Vancouver: City of Vancouver.

Vancouver Sun. 2003. 'Spotted owl survival threatened', 11 July.

Vanderploeg, H.A., T.H. Johengen, and J.R. Liebig. 2009. 'Feedback between zebra mussel selective feeding and algal composition affects mussel condition: Did the regime changer pay a price for its success?', *Freshwater Biology* 54: 47–63.

Vegetation Enhancement Technical Advisory Committee (VETAC). 2005. *Annual Report*. Sudbury, Ont.: Greater Sudbury Land Reclamation Program.

———. 2007. *Annual Report*. Sudbury, Ont.: Greater Sudbury Land Reclamation Program.

———. 2009. *Land Reclamation Program Annual Report 2009*. Sudbury, Ont.: City of Greater Sudbury.

Venter, O., et al. 2006. 'Threats to endangered species in Canada', *BioScience* 56, 11: 903–10.

Viewpoint: Perspectives on Modern Mining. 2008, issue 4. 'The reclamation of Sudbury: The greening of a moonscape', Caterpillar Global Mining. At: https://mining.cat.com/cda/files/2785515/7/Sudbury_Eng.pdf.

Volpe, J.P., et al. 2000. 'Evidence of natural reproduction of aquaculture escaped Atlantic salmon (*Salmo salar*) in a coastal British Columbia river', *Conservation Biology* 14, 3: 899–903.

von Braun, J. 2007. *The World Food Situation: New Driving Forces and Required Actions*. Washington: International Food Policy Research Institute.

Voynick, S. 1999. 'Diamonds on ice', *Compressed Air* 104: 60–8.

Wada, Y., et al. 2010. 'A worldwide view of groundwater depletion', *Geophysical Research Letters* 37, L20402, doi:10.1029/2010GL044571.

Walker, B., and D. Salt. 2006. *Resilience Thinking: Sustaining Ecosystems and People in a Changing World*. Washington: Island Press.

Walker, I.J., and R. Sydneysmith. 2008. 'British Columbia', in D.S. Lemmen et al., eds, *From Impacts to Adaptation: Canada in a Changing Climate 2007*. Ottawa: Government of Canada, 329–86.

Walsh, J.E., et al. 2004. 'Cryosphere and hydrology', in *Arctic Climate Impact Assessment: Impacts of a Warming Climate*. New York: Cambridge University Press, 183–242.

Walston, J., et al. 2010. 'Bringing the tiger back from the brink—The six percent solution', *PLoS Biology* 8, 9: e1000485.

Walters, C.J. 2007. 'Is adaptive management helping to solve fisheries problems?', *Ambio* 36, 4: 304–7.

Warner, J.F., and C.L. Johnston. 2007. 'Virtual water—real people: Useful concept or prescriptive tool?', *Water International* 32: 63–77.

Warnock, R.G., and M.A. Skeel. 2004. 'Effectiveness of voluntary habitat stewardship in conserving grassland: Case of Operation Burrowing Owl in Saskatchewan', *Environmental Management* (25 Mar.). At: wwwspringerlink.com.

Water Footprint Network. 2011. At: www.waterfootprint.org.

Watson, R., and D. Pauly. 2001. 'Systematic distortion in world fisheries catch trends', *Nature* 414: 534–6.

Watt-Cloutier, S. 2000. 'Wake-up call'. At: www.ourplanet.com/imgversn/124/watt.html.

Weber, B. 2011. 'Nobel Peace Prize winners urge less oilsands growth', *The Record*, 20 Sept., A8.

Weiss, W., I. Bergmann, and G. Faninger. 2007. *Solar Heat Worldwide: Markets and Contribution to the Energy Supply 2005*. Paris: International Energy Agency.

Wells, J., D. Roberts, P. Lee, R. Cheng, and M. Darveau. 2010. *A Forest of Blue—Canada's Boreal Forest: The World's Waterkeeper*. Seattle: International Boreal Conservation Campaign. Courtesy of Pew Environment Group, 2011.

Werniuk, J. 1998a. 'Great Canadian diamonds', *Canadian Mining Journal* (Oct.): 8–22.

———. 1998b. 'Where the smart mining is going', *Canadian Mining Journal* (Dec.): 14–18.

Whitelaw, G.S., and P.E.J. Eagles. 2007. 'Planning for long, wide conservation corridors on private lands in the Oak Ridges Moraine, Ontario, Canada', *Conservation Biology* 21: 675–83.

Wichelns, D. 2010. 'Virtual waters: A helpful perspective, but not a sufficient policy criterion', *Water Resources Management* 24: 2203–19.

Wielgus, R.B., and P.R. Vernier. 2003. 'Grizzly bear selection of managed and unmanaged forests in the Selkirk Mountains', *Canadian Journal of Forest Research* 33: 822–9.

Wiken, E. 1986. *Terrestrial Ecozones of Canada*. Ottawa: Environment Canada. http://sis.agr.gc.ca/cansis/nsdb/ecostrat/zones.gif, Agriculture and Agri-Food Canada © 1995. Reproduced with the permission of the Minister of Public Works and Government Services, 2011.

Wiley-Blackwell. 2008. 'Birds migrate earlier, but some may be left behind as climate warms rapidly', *Science Daily*, 22 June. At: www.sciencedaily.com/releases/2008/06/080620115925.htm.

Wilkinson, C.J.A. 2008. 'Protected areas law in Ontario, Canada: Maintenance of ecological integrity as the management priority', *Natural Areas Journal* 28: 180–6.

Williams, R., D. Lusseau, and P.S. Hammond. 2006. 'Estimating relative energetic costs of human disturbance to killer whales (*Orcinus orca*)', *Biological Conservation* 133: 301–11.

Winfield, M.S., and G. Jenish. 1998. 'Ontario's environment and the "common sense revolution"', *Studies in Political Economy* 57 (Autumn): 129–45.

Wolfe, B.B., et al. 2005. 'Impacts of climate and river flooding on the hydro-ecology of a floodplain basin, Peace–Athabasca Delta, Canada, since A.D. 1700', *Quaternary Research* 64: 147–62.

——— et al. 2006. 'Reconstruction of multi-century flood histories from oxbow lake sediments, Peace–Athabasca Delta, Canada', *Hydrological Processes* 20, 4: 131–53.

———, T.W.D. Edwards, R.I. Hall, and J.W. Johnston. 2011. 'A 5200-year record of freshwater availability for regions in western North America fed by high-elevation runoff', *Geophysical Research Letters* 38: L11404. At: doi:10.1029/2011GL047599,2011.

Wong, P.Y., and I. Brodo. 1992. 'The lichens of southern Ontario', *Syllogeus* 69: 1–79.

Woods, A.J., D. Heppner, H. Kope, J. Burleigh, and L. Maclauchlan. 2010. 'Forest health and climate change: A British Columbia perspective', *Forestry Chronicle* 86, 4: 412–22.

World Bank. 2008. *World Development Indicators 2008*. Washington: World Bank.

World Commission on Environment and Development (WCED). 1987. *Our Common Future*. Oxford: Oxford University Press.

World Conservation Union (IUCN—International Union for the Conservation of Nature and Natural Resources). 2002. '2002 IUCN Red List of Threatened Species'. At: www.redlist.org.

World Conservation Union (IUCN—International Union for the Conservation of Nature and Natural Resources). 2003. 'Release of the 2003 IUCN Red List of Threatened Species', press release. At: www.iucn.org.

———. 2007. '2007 IUCN Red List of Threatened Species'. At: www.redlist.org.

———. 2011. 'The IUCN Red List of Threatened Species'. At: www.iucnredlist.org.

——— and UNEP-WCMC. 2009. *The World Database on Protected Areas (WDPA)*. Cambridge: UNEP-WCMC, Jan.

World Health Organization, Public Health and Environment. 2011. 'Database: Outdoor air pollution cities'. At: www.who.in/phc/health_topics/outdoorair/databases/en.

Worldwatch Institute. 2003a. 'Earth trends: Agricultural inputs 2003'. At: earthtrends.wri.org/datatables/index.cfm?theme=8&CFID=504446&CFTOKEN=74873335.

———. 2003b. *Vital Signs*. New York: W.W. Norton.

———. 2004a. *State of the World 2004*. New York: W.W. Norton.

———. 2004b. *Vital Signs: The Trends That Are Shaping Our World*. New York: W.W. Norton.

———. 2007a. *Biofuels for Transport: Global Potential and Implications for Energy and Agriculture*. London: Earthscan.

———. 2007b. *Vital Signs: The Trends That Are Shaping Our Future, 2007*. New York: W.W. Norton.

———. 2011. *Vital Signs 2011: The Trends That Are Shaping Our Future*. Washington: Worldwatch Institute.

World Wildlife Fund. 2010. *Living Planet Report 2010*. At: wwf.panda.org/about_our_earth/all_publications/living_planet_report/2010_lpr/. Some rights reserved.

———, Zoological Society of London, and Global Footprint Network. 2006. *Living Planet Report 2006*. Gland, Switzerland: World Wildlife Fund.

Worm, B., et al. 2006. 'Impacts of biodiversity loss on ocean ecosystem services', *Science* 314, 7: 87–90.

Woynillowicz, D., and C. Severson-Baker. 2006. *Down to the Last Drop: The Athabasca River and Oil Sands*. Drayton Valley, Alta: Pembina Institute.

———, ———, and M. Raynalds. 2005. *Oil Sands Fever: The Environmental Implications of Canada's Oil Sands Rush*. Drayton Valley, Alta: Pembina Institute. Courtesy of the Pembina Institute.

Wu, J., and R. Hobbs, eds. 2007. *Key Topics in Landscape Ecology*. Cambridge: Cambridge University Press.

Yates, J.S., and J. Gutberlet. 2011. 'Enhancing livelihoods and the urban environment: The local political framework for integrated organic waste management in Diadema, Brazil', *Journal of Development Studies* 47, 4: 1–18.

Zhang, X., and X. Cai. 2011. 'Climate change impacts on global agricultural land availability', *Environmental Research Letters* 6: 1–8.

Zhu, Y., et al. 2000. 'Genetic diversity and disease control in rice', *Nature* 406: 718–22.

Index

Abbey, E., 39, 563
abiotic, 60, 66–71, 73, 80, 92, 101, 104, 117, 119, 169, 341
Aboriginal peoples, 13, 15, 39, 46, 108, 154, 159, 174, 183, 270, 400, 455, 483, 494, 515; and conflict, 186–8, 190, 238, 266–7, 431, 433–4; and forests, 282, 284, 294, 320, 327, 328; hiring of, 425, 431; and hydroelectric development, 377–81; and marine resources, 238, 259, 266–7, 279; and protected areas, 531–3, 535; and water security, 374, 390–1; *see also* indigenous knowledge, James Bay Cree
abundance, 49, 69, 79, 83, 111, 112, 143, 248, 273, 305, 309, 328, 343, 388, 413, 498, 505–6, 508; hyper-, 98, 529; super-, 374–5, 388, 416; of water, 374–6
acid deposition, *or* acid rain, 52, 100, 114, 115, 124, 139–49, 165, 242, 282, 420–1, 475, 546, 551; aquatic effects of, 141–2, 149; and ecosystem sensitivity, 144; socio-economic effects of, 144–5, 149; solution for, 145–8, 149; terrestrial effects of, 142–4, 149
acid mine drainage, 420, 454
acid shock, 142
acidification, 242, 350, 371
active management, 283, 501, 526, 529
adaptation, 178, 191, 201, 229, 231, 232–3, 234; to climate change, 218
adaptive co-management, 172, 177–8, 194
adaptive management approach, 39, 77, 79, 92, 111, 153–71, 165, 172–97, 266, 328, 379, 411, 440; learning dimension of, 177, 183; phased, 452, 456
Adeniyi, P.O., 23–4
aerobic, 54, 122
agriculture, 90, 158, 165, 270, 332–73, 377, 386, 387, 408, 410, 492, 503, 505, 509, 528, 532, 557–8; and climate change, 201, 212, 214; and compost, 125; as ecosystem, 334, 337, 338–9, 340, 365, 367–70; environmental implications of 332, 337; global, 347; industrialized, 334, 340, 342, 343, 346, 347, 362–4, 368, 371, 385–6; local, 367, 368–9, 370; and soil degradation, 349–52, 371; sustainable, 332, 365–70; trends in, 347–8; and water, 340, 342, 343
albedo, 8, 99, 207
Aldaya, M.M., 394
alien species, 93–8, 112, 289, 502, 503, 508, 513, 526; aquatic, 97, 98, 410; *see also* invasive
Allan, T., 395
alleopathic, 94, 100
alternative, approaches, 154, 172, 194; energy, 419–21, 455; possible futures, 176
alternative dispute resolution (ADR), 189–90, 191;
anadromous, 121
anaerobic, 54, 122, 125, 127
Anderegg, W.R.I., 220

Anderson, D., 224
Andrey, J., 175, 218–19, 220
Anielski, M., 283, 293–4
annual allowable cut (AAC), 145, 295, 305, 327
Anthropocene age, 24
anthropocentric view, 5, 39, 154, 163, 165, 223, 240, 242, 495
Apps, C.D., 496
aquaculture, 238, 259, 274, 275–7, 279
aquatic ecosystems, 137, 342, 343, 374–418; disturbances to, 421, 435, 446, 454; and extinction, 498; *see also* freshwater ecosystems, oceans
aquifers, 131, 132, 133, 388, 403, 445, 468, 469
arbitration, 190
Arctic, xi, 46, 57, 61, 62, 79, 93, 102, 103, 108, 122, 125, 132, 165, 175, 176, 240, 245, 255, 257, 269, 376, 425, 507; and chemicals, 355–8; and climate change, 204–6, 210, 214–15, 216, 217–18; sovereignty, 443, 455
Armitage, D., 177
Armstrong, A., 412
Arnstein, S., 174, 175
artisanal fisheries, 357–8
aspirational approach, 224
Assadourian, A., 546, 563
assimilated food energy, 61
Athabasca River, 443–8, 455, 528
Atlantic Maritime ecozone, 289–90, 327
atmosphere, definition of, 48
Audubon, J.J., 500
Augustine, S., 526, 552, 563
autotrophs, 54, 58, 64, 303
auxiliary energy flows, 64, 68, 90, 99, 120, 339, 340, 346, 347, 369, 371

Babbage, M., 229
Babcock, T., 12–13
backcasting, 162, 394
Baeder, A., 116
Baird, J., 158, 447
Bakker, K., 375, 412
Bali Conference, 225, 234
ballast water, 97–8
Bardati, D., 554–5
Barlow, M., 8, 13
Barnosky, A.D., 106, 239
Bay of Fundy ecosystem, 161–2
Beanlands, G.E., 178
benthic, 137, 343
Berch, S.M., 311
Berg, A., 88–9
Berger, T., 178
Bergeron, Y., 297
Berkes, F., 157, 380, 381
best practice, xii, 12–13, 153–5, 156, 168, 179, 415, 421, 454, 459, 467, 479–80, 483–7
Bierkens, M., 132

Big Old Fat Fecund Female Fish (BOFFFF), 104, 248
bioaccumulation, 352, 357, 384, 385, 509
biocapacity, 24, 33–7, 39, 41
biocentric perspective, 5, 39, 154, 163, 165; *see also* ecocentric view, value
biocides, 297, 299–300, 327, 339 342–3, 347, 349, 352–62, 379, 371; regulation of, 358–62, 364, 371
bioconcentration, 254, 358, 365
biodiversity, 49–82, 304; in forests, 303–9, 328; hotspots, 76, 108, 242, 497; loss of, 25, 49–82, 91, 94, 98, 321, 343, 345, 364, 491–537; oceanic, 55; protection of, 491–537
biofuels, 332, 343–6, 379, 371, 550, 558
biogeochemical cycles, xii, 38, 52, 54, 114–52, 302, 328, 337, 347, 546
biogeography, 248, 258
biological oxygen demand (BOD), 137, 139
biomagnification, 255, 270, 343, 352, 357–8, 359, 371, 384, 385, 507, 509
biomass, cycles, 56, 60–1; definition of 54, 61; pyramid, 61
biomes, 65, 66, 80, 88
biosphere, definition of, 5; reserves, 532, 534
biotechnology, 341–2
biotic, community, 494; components, 66–71, 80, 92, 101, 117, 119, 122, 124, 148, 169; potential, 102; pyramids, 60–1, 240; relationships, 70–71, 73; responses, xii
Birkedal, T., 106
bison, 503, 527–9
Biswas, A.K., 410
bitumen, 444–5, 448, 455; crude, 444
Blackburn, I., 308
Blatchford, C., 188
Blusson, S., 423
Bone, R.M., 402
Boreal Cordillera ecozone, 286, 327
boreal forest, 88, 91, 111, 119, 210, 287–331, 444, 445–6, 532; Agreement, 283–4, 309, 328; Conservation Framework, 284
Boreal Plains ecozone, 287, 288, 327
Boreal Shield, 145, 147, 282–4, 288–9, 327
bottom trawling, 242, 249, 252–3, 274, 547
Bourassa, Premier R., 377, 379
Bowman, D.M.J.S., 106
Boyce, M.S., 529
Boyd, D., 467, 548
Boyer, M., 381
Branch, T.A., 246
Brandes, O.M., 392, 393, 414
Briggs, D., 134
Brodo, I., 85
Brook, B.W., 106
Brooks, D.B., 392, 393, 394
Brooks, G.R., 398
Brown, S., 395

Brown, V.A., 6
Browne, C.L., 492
brownfields, 468–70, 483, 488
Bruno, J.F., 241
buffering capacity, 144, 290
Bullock, R., 317
Burgess Shales, 107
Burn, D.H., 399
burrowing owl, 354, 512, 517
Burt, B., 188
Bush, President G.W., 161, 222, 234, 238, 547
butterfly effect, 100
Butterworth, J., 410
bycatch, 246, 249, 274

Cai, X., 334
calorie, definition of, 50
Cameron, R.P., 309
Cameron, S.D., 262, 264
Canadian Heritage Rivers Program (CHRP), 403–7, 415
Cancún Summit, 225, 227, 234, 547
carbofuran, 354
carbon, balance, 240, 279; bank account, 283; cycle, 114, 117, 120, 127–9, 135, 207, 240, 242–3, 314; -neutral, 550; offsets, 233; sequestration, 230–1, 290, 315, 325, 327, 328, 339, 482, 550; tax, 229–30, 549, 564
carbon dioxide, levels of, 88–9, 129, 148 , 314, 336, 338
Cardow, C., 360
caribou, 88, 102, 283, 307, 308, 328, 425, 427–8, 433, 437–8, 454, 496, 507, 508, 516, 532
Carlson, M., 313
carnivores, 54, 56, 57, 60–1
Carpenter, R.A., 175, 176
carrot mobs, 560
carrying capacity, ecological, 101; of planet, 24, 64; social, 101; of species, 37
Carson, R., 26, 352, 353
cellular respiration, 54, 61, 62, 127, 148
certification of forest products, 258, 321–4, 329, 545, 560
chain-of-custody procedures, 323
Chalecki, E.L., 219
Chambers, P., 343
Chapagain, A.K., 395, 396
Chapman, D., 434, 449, 453
Charest, Premier J., 228, 284
Chavez, D., 173
chemoautotrophs, 54, 122, 240
Chevalier, G., 380, 381
chlorophylls, 53
Chomitz, K.M., 325
Chrétien, Prime Minister J., 275, 521
Cinq-Mars, J., 319
Clark, R., 179
Clark, W.F., 388
clear-cutting, 296–7, 298, 305–8, 311, 315, 327; *see also* forestry
Clement, T., 422
climate, definition of, 202, 234

climate change, xii, 4, 24–8, 52, 53, 55, 62, 83, 99–100, 102, 106, 107–12, 124, 127, 134, 135, 148, 158, 165, 168, 175, 177, 199–237, 239, 240, 244, 245, 256–7, 265, 279, 282, 313–5, 318, 325, 332, 334, 336–7, 370, 378, 407, 436, 449, 461, 479, 482, 493, 502, 506–8, 513, 526, 529, 544, 546, 548, 551, 564; in atmospheric system, 199–237; and China, 225; definition of, 202; deniers, 201, 220–1, 234; implications of, 209–18; and infectious diseases, 216–17
climate hypocrite, 225, 234
climate modelling, *or* measuring, 207–9, 234, 266, 529
Climategate, 221
climatic climax, 86
climax community, 86, 182
Cline, W.R., 336
coastal zones, 239–40, 246, 252, 256–9, 279
cod, 246–7, 253, 259, 260–6, 279, 419, 500, 501, 516; *see also* fisheries, problems in
collaboration, 39, 46, 154, 172–97, 220, 227, 229, 441; cross-disciplinary, 6, 39; disciplinary, 6, 39; interdisciplinary, 6, 39, 173, 447; multidisciplinary, 6, 39, 173; transdisciplinary, 6, 39, 173
collaborative management, 1–2, 5–6, 159, 172–97, 406, 407–11; linkage dimension of, 177
Collinge, S., 72
Collins, H., 404
co-management, 46, 175, 177–8, 194, 533
commensalism, 73, 81
Commissioner of Environment and Sustainable Development, 29–30, 98, 227, 228, 353, 360, 361, 384, 390, 392, 448, 516, 518, 547, 548, 562
Committee on the Status of Endangered Wildlife in Canada (COSEWIC), 55, 261, 308–9, 498, 499, 500, 515–17, 527, 535
communication, 154, 172, 173, 175–6, 191, 194, 200, 218–21, 234, 319
community, 65, 80, 90, 169, 182
community forest, 316–17
competition, 70, 278; interspecific, 70; intraspecific, 70; for resources, 292, 294, 343, 411–12, 508
competitive exclusion principle, 69–70, 81, 90
complete-tree harvesting, 310–11
complexity, 2, 3, 4, 7, 13, 37, 39, 40, 52, 79, 80, 86, 95, 99, 107, 111, 117, 122, 148, 157, 169, 200, 207, 218, 265, 379, 407, 433, 470, 551, 560
compound, 114
condensation nuclei, 131
conflict, 3, 5, 13, 23, 28, 37, 39, 92, 93, 109, 153–97 (*passim*), 200, 212, 265–6, 293, 303, 375, 415, 431, 433–4, 437, 449, 503, 534, 545, 564; in Caledonia, 186–8, 190; in Oka, 188
conservation authorities, 29, 405, 406–7, 505, 505
conservation of energy, law of, 51
conservation of matter, law of, 115, 137, 299, 302
consumerism, *or* consumption, 2, 18, 21–2, 29, 37, 51, 332, 497, 560–66
consumers, 53–61, 347, 370; primary, 56, 60–1; responsibility of, 560–1, 565; secondary, 56, 60–1; tertiary, 56, 60–1

context, 156, 157–61, 163, 169, 191
contour cultivation, 366
Convention on Biological Diversity (CBD), 77, 80, 93, 97, 275, 321, 328, 513, 515, 521, 524, 535, 544, 547, 564
Convention on International Trade in Endangered Species of Wild Fauna and Flora (CITES), 108, 110, 238, 500, 512, 547
Cook, J., 99
co-operative approach, 138, 320, 410, 411, 480, 532, 550
co-ordinated approach, 154, 172–97, 375, 448, 515
Copenhagen Summit, 202, 221, 225–7, 228, 234, 256, 547
coral, bleaching, 241; polyps, 241; reefs, 17, 55, 128, 203, 204, 207, 239, 240, 241–2, 256, 278, 545
Cornia, G.A., 368
corporate social investment (CSI), 432–3; *see also* responsibility
Costanza, R., 239, 241, 520
Couch, W.J., 423, 428
Cressman, D.R., 387
critical load, 144, 145–6
crop rotation, 365
crude birth rate (CBR), 18
crude death rate (CDR), 18
crude growth rate (CGR), 18
Crutzen, P., 345
cryosphere, 48, 214–5, 234
culmination age, 295
cumulative effects, *or* impacts, 180–1, 381, 440, 445, 448, 551
custom-designed solutions, 157, 158
cyclic succession, 86

Dale, A., 410
Darwin, C., 47, 78, 104, 107
Davis, A., 266
Dawson, D., 467
Day, J., 377, 381
DDT (dichlorodiphenyltrichloroethane), 255, 270, 353, 356–8, 386, 492, 494
dead zones, 138, 149, 255
Dearden, P., 72, 503, 519, 526
de Boer, Y., 225
decomposer, 56–7, 90, 119, 148; food chain, 56–7, 58, 119
Deepwater Horizon, 182, 254, 256, 442, 455
deforestation, 135, 285, 315, 325–6, 497, 503, 544, 550
demand management approach, 374, 392–4, 415
demographic transition, 19, 20
denitrification, 124, 125, 127, 311
Denny, C., 527
density, dependent, 101; independent, 101
Desjardins, R., 339
detrital food chain, *see* decomposer
detritus, 56–7, 58
Diamond, B., 380
Diamond, J., 25, 547
Diduck, A., 167
Dillon, P., 447

Diouf, J., 558
direction, *see* vision
disaster risk reduction (DRR), 193
dispute resolution, 39, 154, 184, 186–90, 195; *see also* alternative dispute resolution, conflict
disturbances, 55, 76, 83, 86, 87–8, 89, 90–2, 93, 98, 100, 111, 114, 125, 131, 148, 149, 178, 181, 182, 241, 273, 298–314, 438, 446, 500, 506; human, 302, 304, 308, 310, 311, 347, 365–6, 507
diversions, water, 374, 376–83, 399–401, 404, 428, 468, 508
diversity, level of, 90, 163, 305, 347, 348, 478
Dockside Green, 483–4, 488
Doelle, M., 179
Doubleday, W.G., 265
double-loop learning, 167, 169
Downes, C., 503
drivers of environmental change, 32, 33, 325, 410, 492, 513, 540, 541, 542, 544; direct, 542–3, 544; indirect, 542–3
Drivers-Pressures-State-Impact-Response framework (DPSIR), 32, 33
drought, 80, 88, 102, 108–9, 203, 212, 220, 232, 290, 314, 332–4, 339–40, 367, 374, 401–3, 415, 470, 488, 505
Dubreiul, C., 413
Dudley, N., 521
Duinker, P.N., 178, 317
Dumont, C., 380, 381
Dunbar, D., 308
Durner, G., 257
Dyer, G., 226, 443
dynamic ecosystem, 83–113, 165, 191, 207
dynamic equilibrium, 83–113 (*passim*)

Eagles, P.E.J., 531
Earth Summit (Rio de Janeiro), 26, 158, 179, 208, 222, 341, 552
earthquakes, 470–3, 488; in Haiti, 472–3; in Japan, 451, 453, 456; Richter scale of, 193, 472
Easterling, W.E., 336
Ebeling, J., 325
E.coli, 363, 388–92
ecological, integrity, 72, 524–6, 532; *see also* health; redundancy, 57; restoration, 100; succession, *see* succession
ecologically extinct, 493–4
EcoLogo, 557, 562
economic considerations, *or* systems, 108, 165, 182, 386, 420, 428, 431, 454
ecosphere, 45–151; components of, 115, 148; layers of, 47–8, 80; processes of, 37
ecosystem, definition of, 65, 80; disruption of, 421; diversity in, 74–5
ecosystem-based management, 39, 77, 79, 92, 138, 154, 156, 163–6, 172, 191, 199, 266, 284, 291, 310, 315, 320, 325, 328, 400, 410, 415, 480, 526, 528
ecotones, 86
ecozones, definition of, 65; terrestrial, 65, 67, 106–7, 283, 286–90, 327
edaphic climax, 91, 91
Edwards, P., 558

Ehrlich, A., 219
Ehrlich, P., 219
Ekati diamond mine, 419, 423–9, 430, 433–4, 454–5; environmental impacts of, 425, 427–8, 454–5; and frozen core dams, 425, 427
El Ayoubi, F., 382
El Lakany, H., 325
El Niño, 203, 207
El-Sadek, A., 395
emissions, assigned amounts of, 222–3; carbon, 314, 325, 550; control of, *or* reduction of, 148–9, 149, 165, 166, 221–5, 227–30, 234, 270, 474, 480, 485, 550, 556, 558; credits, 223; nitrogen oxide, 141, 142; sulphur dioxide, 141, 142, 420–1, 454, 475, 551; trading, 223
endangered species, 94, 100, 102, 108, 145, 199, 253, 261, 269, 283, 307, 309, 354, 433, 491–53, 545; and government, 515–18; recovery of, 358, 503, 504, 509, 517, 529, 535; reintroduction of, 509, 509
endemic species, 55, 76–7, 79, 283, 307, 309
endocrine disruption, 254
energy, 419–21, 434–58; conservation of, 51; consumption of, 22, 28, 29, 39, 49, 51–3, 255–6, 420, 435–6, 453–4, 455, 467, 480, 487, 488, 544–6; definition of, 50, 80; demand for, 377; efficiency, 60–1; flows, xii, 38, 49–82, 108, 114, 115, 143, 164, 239, 302, 303, 333, 338, 343, 347; high-quality, 50; kinetic, 50; low-quality, 50; non-renewable, 53, 419, 454; pathways, 452–3, 456; potential, 50; pyramid, 60–1, 63; radiant, 50, 53; renewable, *or* recovery of, 51–3, 230, 233, 419–20, 421, 455, 463, 480, 488; sustainable, 420–1, 437, 452–4; transformation of, 50–82, 345; types of, 50, 54, 80
energy balance models (EBMs), 207–9
entropy, 51, 54
environment, definition of, 5, 39
environmental impact assessment (EIA), 178–80, 181, 194, 379, 386, 410, 421, 429, 441, 454, 455
environmental justice, 156, 167–8, 169
environmental refugees, 27, 212, 234
epidemiological transition, 20–1
epiphytes, 73
erosion, 118, 206, 219, 311, 313, 328, 338, 343, 349, 352, 365–6, 371, 386, 387
Estes, J.A., 99
estuaries, 2, 8–14, 62, 64, 121, 379, 381
ethics, 163, 534; of water, 374, 411–3, 415; *see also* value, vision
euphotic zone, 58, 121, 240
Eurasian water milfoil, 92, 93
eutrophic, 135, 137, 254, 386
eutrophication, 90, 114, 120, 134, 135, 137–40, 146, 148, 149, 165, 341, 342, 343, 364, 386, 387; cultural, 135, 138
evapotranspiration, 130, 135, 214, 313, 352, 396, 401, 415; *see also* transpiration
evolution, 53, 74, 76, 78, 83, 84, 104–8, 112, 241, 248, 253–4, 445, 494, 502, 507, 493–4, 507; co-, 78, 104, 105, 112, 124; contemporary, 104; and natural selection, 47, 104

exclusive economic zones (EEZs), 97, 257, 258
exotic species, 496, 510, 526, 531
Experimental Lakes Research Area (ELRA), 184
exponential population growth, 18, 101; *see also* population
ex situ conservation, 501–3, 504, 509; and swift fox, swift fox, 501–3
ex situ preservation, 501–3
extinction, 17, 57, 76, 83, 84, 94, 85, 104–8, 112, 217, 239, 308, 491–537, 551; vortex, 278, 496–7
extirpation, 71, 73, 99, 106, 491–537
Exxon Valdez, 255–6, 441–2, 455

Fa. J.E., 104
falldown effect, 296
Fargione, J., 345
Farrell, A.E., 345
feedback loop, xi, 87, 90, 99–100, 101, 111, 207–8, 553; negative, 99–100, 111, 127, 243, 245; positive, 87, 99–100, 101, 111, 129, 134–5, 208, 243, 245, 496
fertilizer, 64, 68, 120, 124, 125, 137, 148, 165, 270, 332–73 (*passim*), 385
Festa-Bianchet, M., 104, 507
Field, J.G., 243
Findlay, C.S., 516
Fipke, C., 423
fire, 86, 87, 99, 102, 110–11, 300, 302, 314; *see also* prescribed burn
fire suppression, 300, 302, 328, 507
Fischer, G., 336
fish aggregating devices (FADs), 247
Fish Lake, 184, 185
fish size, decline in, 248–9, 251
fisheries, 238–81; and climate change, 214; IUU (illegal, unregulated, and unreported), 246; problems in, 39, 46, 176, 200, 238, 246–7, 260–6, 279, 419, 544, 549; *see also* cod, salmon
fishing down the food chain, 249, 252, 253, 265
FitzGibbon, J., 177
Fitzpatrick, P., 430–1
Flanagan, K., 530
Fleischauer, M., 193
flooding, 86, 109, 135, 200, 206, 207, 214, 232, 256, 290, 298, 305, 311, 313, 321, 349, 374, 376, 391, 302, 307–401, 410, 415, 467, 470–3, 480, 488
and dykes, 382, 392, 392, 397–400, 401
floodway, 399–401
flow resources, 419–20, 454; *see also* renewable resources
food, costs, 334, 336–7, 370; security, 332, 369, 394; supply, 18, 102, 108, 124, 141, 332–7, 370, 375, 543
food chains, 54–61, 70, 80, 90, 96, 121, 122, 141, 148, 240, 249, 253, 270, 338, 347, 359, 370, 371, 384, 508, 509
food webs, 57, 60, 83, 99, 100, 243, 367
footprint, carbon, 26–7, 484, 555; ecological, 30–4, 35, 36, 39, 40, 41, 395, 419, 459, 484, 546, 555; water, 374, 385–6, 415
forecasting, 162

forest tenure, 294–5, 317
forestry, 205, 302–28, 380; and biodiversity, 303–9; and climate change, 313–15; and hydrological change, 305, 311, 313, 328; impacts of, 302–14, 327–8, 503, 505, 526; intensive, 300, 327; and logging, 117, 121, 125, 223, 287, 292, 302, 309, 311, 313, 496; new, 282, 303, 315–17, 328; and soil erosion, 311, 313, 328; and soil fertility, 309–11, 312, 328; sustainable, 318–20, 321, 328; *see also* clear-cutting
forests, 282–331; and carbon storage, 283, 327; and conflict, 293, 294, 296–7, 317; and global objectives, 320–3, 328, 329; and government, 294, 327; harvesting of, 29, 282, 285, 292–300, 302, 327; products of 283, 285, 290–4, 327; reclamation of, 294; and recreation, 283, 288–92, 304, 327; regeneration of, 295, 297, 300, 305–6, 308, 445; sustainable, 282–331 (*passim*); and urbanization, 287; *see also* reforestation
Fossil of the Day, 225, 234, 547
fossil fuels, 140, 148, 149, 168, 220, 230, 233, 270, 314, 340, 345, 347, 371, 420, 421, 435–6, 438, 452, 455, 462, 487, 508, 549; alternatives to, 422, 435, *see also* solar, wind
Foster, H.D., 374
fragmentation, 72–4, 79, 104, 109, 164, 181, 296, 298, 308, 438, 445, 455, 479, 481, 496, 497, 506, 510, 511–12, 513, 529–31, 535; of management, 256, 260
Francis, D., 76
Frank, K.T., 104
free trade, 32, 382–3, 412, 415
Freedman, B., 311
Freiwald, A., 242
French, H., 544
freshwater ecosystems, 17, 24, 57, 65, 120, 121, 135, 374–418, 521; and climate change, 214; *see also* aquatic ecosystems
Fritz, S., 160
full-tree harvesting, 310–11
functional compensation, 57
fungi, 119, 120; and button mushroom, 291; destructive, 94–5, 289, 507

Gabriel, A.O., 401
Gaia hypothesis, 92, 93
Galloway, J.N., 120, 125, 127, 140
Gardner, G., 546, 563
gaseous cycles, 117, 124–9, 148
Gaston, A.J., 507
Gauthier, G., 104
Gedalof, Z., 88–9
general circulation models (GCMs), 207–9, 234
generalist species, 70
genetic diversity, 74, 81, 104; loss of, 248, 308, 328, 348
genetic engineering, 339, 341–2
genetically modified organisms (GMOs), 340, 341–2, 363, 368
Genuis, S.J., 548
geo-engineering, 201, 227, 231–2, 234

George, D., 188
geothermal energy, 233, 434–5, 455, 474, 549, 550, 555
Gibson, R.B., 163, 184, 185
glaciation, 78, 205, 206
Gleick, P., 388, 392, 396
global warming, 87, 99–100, 106, 110, 115, 120, 148, 149, 161, 168, 199–237, 242, 243, 245, 254–7, 305, 325, 336, 339, 345, 346, 549, 551, 558
globalism and environment, xi, 2, 4, 14–28, 40, 62, 88–9; *see also* climate change, global warming
Goel, N.K., 399
Goodarzi, F., 10
Gore, Vice President A., 225
Gorrie, P., 225, 380
Gosselin, P., 447, 448–9
Gould, J.S., 105, 107
governance, 2, 24, 28–30, 39, 77, 156, 157, 159, 166; of energy, 432–3; of forests, 320; urban, 463, 483–7, 488; of water, 159, 374–418
government, 156, 157, 159, 161, 174, 430–1, 448, 485, 496, 521, 547–51; and efficiency concerns, 159, 161; and water, 374–418 (*passim*)
Graenfeldt, D., 412
Grand Canal, China, 408–9
Grand River, 404, 405–6
grasshopper effect, 255, 355–6
grazing food chains, 56–7
great auk, 498–9
Great Bear Rainforest (GBR), 268, 533
Great Lakes, 60, 95–8, 102, 138–9, 166, 214, 375, 382, 386–7, 402–3, 410–11, 415, 507
Great Lakes Charter, 138, 382–3
Great Lakes Water Quality Agreement, 138, 184, 387, 410–11, 415
green, bins, 463; buildings, 462, 475, 488, 550; energy, 230; manure, 367; policies, 219; urban policies, 473–87
Green Revolution, 332, 334, 339–40, 341, 342–3, 363, 371
green roof technology, 466–7, 468, 483, 486–7
greenfields, 465, 479
greenhouse effect, 58, 203–4
greenhouse gases (GHGs), xi, 26–8, 39, 83, 88–9, 100, 108, 125, 166, 203–37, 240, 256, 283, 314, 315, 338, 339, 344, 345, 371, 437, 444, 448, 453, 455, 460, 462, 474, 480, 482, 485, 488, 550–1, 556, 558; *see also* global warming
Greer, A., 216
grizzly bears, 268, 310
Groombridge, B., 75
gross primary productivity (GPP), 62–5
groundwater, 129, 131, 133, 135, 340, 342, 361, 375, 376, 388, 394, 403, 405, 410, 411, 415, 425
Gruber, N., 120, 125, 127, 140
guano, 98, 122
Gutberlet, J., 116

habitat, definition of, 69; destruction of, 93–4, 99, 107, 239, 302, 334, 343, 345, 364, 381, 384, 428, 438, 445, 491–537; protection of, 491–537
Halweil, B., 558

Hambler, C., 106
Hanna, K., 294–5, 317
Hannah, L., 529
Happy Planet Index, xii, 546–7, 564
Harden, A., 390
Hardin, G., 246
Harper, Prime Minister S., 222, 224–5, 226, 267, 445, 449
Harris, L., 264
Harris, Premier M., 383
Harrison, J., 513
Harvey, H., 139–40
hazardous materials, *see* toxic
Hazell, R., 549
health, aquatic ecosystem, 181, 270–1, 274, 278, 385, 406, 413, 446, 475; ecosystem, 2, 32, 40, 73, 74, 77, 79, 83, 86, 143, 178, 194, 274, 307, 308, 349, 363, 364, 367, 369, 371, 385, 386, 394, 405, 406–7, 421, 454, 475; human, 9–10, 30, 97, 125, 144–5, 149, 168, 178, 189, 194, 216, 271, 277, 301, 336, 344, 352, 360–4, 367, 369, 381, 385, 388–91, 410–11, 421, 435, 436–8, 448–9, 452, 455, 460, 466, 467–8, 469, 475, 481; *see also* ecosystem integrity, wellbeing
Hearnden, K.W., 304
heat, 60, 345
Heath, C., 541
Heath, D., 541
Hecnar, S.J., 492
Heimann, M., 129
Hengeveld, H., 202, 203, 205, 212, 213, 214, 215, 219, 232, 339
Henstra, D., 470
herbicide, 15, 90, 165, 299–300, 305, 342, 353; *see also* biocides
herbivores, 54, 61–2
heritage rivers, 403–7, 415; *see also* Grand River
Herkenrath, P., 513
heterotroph, 54, 64, 83, 119, 143–4, 340
Hibernia, 255–6, 439–43, 455
Hilson, G., 420
Himmelman, A.T., 173
Hitch, M., 430, 431, 432–3
Hites, R.A., 277
Hobbs, R., 72
Hocking, M.D., 56
Hockstra, A.Y., 395, 396
Hofmann, N., 364
Hoggan, J., 220–1
holistic perspective, 6, 72, 164–5, 169, 172–3, 310, 363
Holling, C.S., 176–7
Holtz, S., 393, 394
Homer-Dixon, T., 157, 434
Horn Plateau, 433–4
Howard, K.W.F., 468
Human Development Index, 23, 34–5, 472, 543, 544, 547
humus, 68
hunger, 336–7, 369–70; *see also* food supply, poverty
Hurricane Katrina, 470–3

Hutchings, J.A., 265
Huxley, T., 246
hybridization, 278, 339, 452, 456, 507
Hyde, D., 79
hydroelectric power, 377–81, 414–15, 435, 468, 508, 549
hydrogen economy, 230
hydrological cycle, 114, 118, 129–35, 148–9, 305, 311, 313, 328, 302, 353, 374, 375–81; and urbanization, 468
hydrosolidarity, 374, 407–11, 415
hydrosphere, definition of, 48
hydrothermal vents, 55; at Endeavour, 240, 273
Hyland, M.C., 118
hypoxic areas, 255

ice, melting of, 108, 110, 111, 131–2, 149, 176, 204–6, 214–15, 216, 231, 245, 255, 256–7, 380–1, 507, 508, 544
immediate disturbance hypothesis, 90
impact assessment, 38–9, 154, 172, 178–80, 182–4, 193, 194, 381
impact and benefit agreements (IBAs), 431, 432–3, 455
implementation gap, 155, 191, 193, 195
independent oversight, 528, 429, 430–1
independent resource use, 292
indicators, 2, 4, 26, 30–6, 39, 479, 544, 551; aggregate indexing of, 32–6, 39; composite, 35–6, 39; of environmental sustainability, 31–2, 138, 383–4; and users, 34; of wellbeing, 36, 551
indigenous, *see* Aboriginal
indigenous knowledge, 183, *see also* traditional ecological knowledge (TEK)
inertia, 92–3, 111
ingenuity gap, 157
insecticide use, 300, 301, 302
in situ, preservation, 502; recovery, 444–5, 446
integrated management approach, 94, 272, 448, 526
integrated pest management (IPM), 365, 367, 371
integrated plant nutrient systems (IPNSs), 365, 367
integrated water resource management (IWRM), 374, 394, 400, 407–11, 415
intensive livestock operations (ILOs), 334, 346, 347, 362–4; *see also* agriculture
Intergovernmental Panel on Climate Change (IPCC), 205, 208, 209, 210, 216, 221, 223, 225, 234, 243, 256
intermediate disturbance hypothesis, 90, 92
International Joint Commission (IJC), 174, 383, 386, 387, 400, 410, 411
invasive species, 83, 93–8, 276, 410, 411, 508, 513; *see also* alien
irrigation, 342, 343, 347, 350–1, 376
island biogeography theory, 530

Jachmann, H., 105
Jackson, J.B.C., 247
James, C., 341
James Bay Cree, 377–81, 390–1

James Bay hydroelectric project, 200, 377–81, 415, 508
James Bay and Northern Quebec Agreement, 379–81
Jamieson, J., 186–7
Jayyousi, O.A., 413
Jenish, G., 161
Jeriorski, A., 148
Jessen, S., 274
Jones, C.G., 72
Jorgensen, C., 248
Joseph, C., 191
judicial approach, 189–90
jurisdictional environmental management, 2, 28–30, 39; cross-, 448; *see also* government

Kanninen, M., 325
Kasechewan, 390–1, 392, 415
Katz, D., 381
Kay, C., 106
Keating, T., 321
Keith, I.B., 62
Kelly, E.N., 447
Kelly, J., 473–4
Kennedy, President J.F., 4
Kent, P., 228
Kenworthy, J.R., 463
keystone species, 73, 81, 95, 99, 111
Khadka, A.K., 413
kimberlite pipes, 423, 428
Kimmins, J.P., 312
Ki-moon, B., 225
King, M.L., Jr., 160
Kitasei, S., 436
Klein, Premier R., 166
Klironomos, J.N., 88
knapweed, 94, 100
Knight, W.A., 321
Krantzberg, G., 411
Kreutzwiser, R.O., 401
Krkošek, M., 276
Krugel, L., 449
K-strategists, 102, 103, 105, 112, 251
Kublesky, W.P., 437, 438
Kupchella, C.E., 118
Kurz, W.A., 305
Kuvlesky, W.P., 437
Kyoto Protocol, 26, 39, 161, 166, 201, 208, 221–5, 227, 228, 229, 230, 231, 234, 325, 540, 544, 547, 549, 564

La Grande River, 377–81
Lahey, A., 8
Lake Erie, 138–9, 140, 149, 386, 387, 407, 507
Landry, M., 530
landscape connectivity, 72–3; functional, 72; structural, 72
landscape ecology, 72–3, 81
land-use planning, 172, 190, 284; *see also* regional
Lasserre, F., 381, 383
Lautenbach, W.W., 475, 477
Lautenschlager, R.A., 299–300

Law, R., 104
law of everybody, xii, 559–63
Lawrence, D., 178
leaching, 311, 339
Lebeuf, M., 270
LeBreton, M., 227
Lee, K.N., 177
Lee, P., 283
LEED (Leadership in Energy and Environmental Design), 474, 475, 483
Leggett, W.C., 104
Lemieux, C., 529
Lennon, J., 546
Lenton, R., 410
Leopold, A., 540
Le Prestre, P.G., 529
Lesieur, D., 297
Levalliant, A., 390
Li, S., 408–9
lichens, 38, 85, 88, 311
life cycle assessments (LCAs), 421, 454, 561–2
life support system, planetary, xi, xii, 28, 37, 47, 53, 76, 80, 369, 534, 552
light living, 555–9
limiting factor, 68, 80, 268, 381; dominant, 68, 120, 311; major, 127
lithosphere, definition of, 48
Liu, J., 395
Livestock Revolution, 346, 371
Living Planet Index, 2, 32, 34, 37, 40
locally unwanted land uses (LULUs), 167, 463
Longhurst, A., 248
longline fisheries, 251, 257
long-term view, 162, 166–7, 169, 173, 191, 192–3, 238, 295, 310, 315, 405, 484
Lophelia pertusa, 242
Louv, R., 553
Lovelock, J., 92, 93
Lumsden Lake, 139–40, 142
lurching change, *see* non-linear
LUST (leaking underground storage tanks), 469
Lynam, A.J., 511

McAllister, I., 533
McBean, G., 470
McClellan, B.N., 496
Macdonald, D., 551
Macdonald, Sir J.A., 525
McGinnity, P., 276, 278
McGuinty, Premier D., 228
McIntosh, R.P., 182
McKibben, B., 221
Mckinney, M.A., 508
MacKinnon, J.B., 558
Mackintosh, R.P., 401
Mackintosh, W.A., 401
Maclaren, V.W., 462, 463, 464–5
McLeman, R., 212
McLuhan, M., 45
McNiven, J., 380
macronutrients, 115
Maier, H., 468

Malcolm, C., 72, 58, 59, 61, 61, 72–3, 521
Malcolm, J.R., 108
Maldonado, J.H., 116
Malthus, T., 18
management of environment, xi, 2, 3–42; philosophy of, 153–71; processes of, 153–5, 156, 172–97; products of, 153–5, 156, 166, 172–97; systems of, 46
Manno, J.P., 411
Marafa, L., 481–2
marine ecosystems, *see* oceans
Marshall, D. Jr., 266
Martin, L.R.G., 467
Martin, P., 106
Mascarenhas, M., 161
Masters, M., 323
Mastny, L., 544
matter, 114–51; cycles of 53, 146, 239
Matthews, C., 163, 412
mature community, 86; ecosystem, 90, 92
Max, M., 449
May, E., 8, 13
Mayhew, P.J., 108
Meadows, D.H., 26
meat consumption, 332, 338, 557, 559
mediation, 190, 192–3
megaprojects, 377–81, 415
Memon, A., 159
Ménard, M., 435
Mercredi, O., 14, 15, 190
mercury, 379–81, 385, 435, 507–8
mesosphere, definition of, 48
mesotrophic, 135
micronutrients, 115
Millennium Development Goals (MDGs), 26, 27, 34, 39, 336, 337, 340, 344, 541, 543, 544, 564
Millennium Ecosystem Assessment, 16, 17, 76, 127, 252, 263, 493, 496, 518, 521, 541, 542, 544
Miller, G., 161, 162, 205
Miller, K., 549
Mills, T.J., 7
mineralization, 125
minerals, 419–34, 454–5; export of, 421, 422
minimum viable population (MVP), 530
mining, 419–34, 454–5; conflicts in, 431–4; and environmental concerns, 425, 428–9, 431, 433–4; and tailings management, 425, 428, 429, 449, 450, 454, 455; and water, 428
Mitchell, B., 163, 191
mitigation, 180, 183, 223, 229–32, 240, 284, 313, 421, 422, 431, 441, 454, 467
Mittelstaedt, M., 467
Mixed Wood Plains ecozone, 289, 327
Model Forest Program, 320–1, 328
Molle, F., 392
monitoring, 29–30, 157, 183–4, 191, 208, 245, 257, 258, 266, 380, 383–92, 428–31, 441, 447–8, 455, 467, 474, 478, 505, 531, 548
monoculture cropping, 347
Montane Cordillera ecozone, 287–8, 327
Montreal Protocol, 212
Mooers, A.O., 516

Moreno-Sánchez, R.D.P., 116
Morrison, K., 369
Mortsch, L., 175, 218–19, 220
Mosquin, T., 76, 77
mountain pine beetle, 86, 185, 305, 324, 507, 550
Mukhopadhyay, M., 10
Muller, M., 410
multi-barrier approach, 374, 389, 415
Murphy, S.D., 467
muskox, 102, 103
mutualism, 57, 71, 78, 81, 95, 125
Myers, R., 248, 265

Nakashima, D.J., 183
Nanus, B., 161
National Forest Strategy (NFS), 316, 317–21, 328
national marine conservation areas (NMCAs), 273–5
national parks, 100, 210, 273–5, 308, 491–537; Banff, 86, 87, 287, 357, 504, 518, 525–6, 530, 535; Point Pelee, 74, 98, 491–3, 501; Riding Mountain, 532 ; Wood Buffalo, 527–9; Yoho, 107
nature deficit disorder, 553, 564
Neave, D., 302
negotiation, 175, 177, 187, 190, 425, 432, 454, 480
Neilson, E.T., 315
Nellemann, C., 248
neo-liberalism, 161, 169
net community productivity (NCP), 64
net primary productivity (NPP), 62–5, 90, 496
Newman, L., 410
niche, 69–70, 80, 81
Nielsen, E., 398
Nikiforuk, A., 381
NIMBY (not in my back yard), 463
nitrification, 125, 127
nitrogen cycle, 25, 38, 71, 114, 120, 124–7, 135, 148, 149, 311
nitrogen fixation, 124–5, 143, 299, 311, 348, 352, 365
Noble, B., 181
Noble, P.G., 359
Nollen, G., 325–6
non-governmental organizations (NGOs), 168, 175, 225, 227, 234, 251, 278, 320, 326, 370, 473, 512, 519, 561, 562–3
non-linear change, 15, 24
non-point sources, 137–8, 158, 254, 374, 384, 385–7, 405, 410, 415; agricultural, 387
non-renewable resources, 199, 367, 419–58
non-structured management approach, 374, 398, 400, 415
non-target organisms, 249, 253, 300, 353, 365, 371, 353
non-timber forest products (NTFP), 291–2, 327
North American Water and Power Alliance (NAWAPA), 381–2, 415
no-till/conservation agriculture (NT/CA), 339, 365–6, 367, 387
nuclear power, 419–20, 421, 450–2, 456
nuclear wastes, 451–2, 454, 456

nutrients, 115, 148; cycling of, 90–1, 290, 397, 478; loss of, 343, 351–2, 384

Obama, President B., 228, 445
oceans, Canadian, 258–79, 540; and climate change, 216; currents in, 203, 269; ecosystems of, 55, 58, 59, 60, 61, 70, 110, 121, 127, 149, 238–81, 521; health of, 278; management of, 245–81; pollution in, 254–7, 270–1; warming of, 253–4, 262, 265, 269; *see also* aquatic ecosystems
Oceans Act, 62, 271–2, 279, 540
O'Conner, Justice D., 388–90
Odum, E.P., 64, 92
offshore oil, 419–20, 439–43
O'Hara, K., 549
oil prices, 176, 343, 444
oil sands (Alberta), 147, 200, 377, 419–20, 443–9, 455, 528, 551; impacts of, 445–9; investment in 444
old-growth forest, 282–331 (*passim*), 560
oligotrophic, 135, 137
Olsson, P., 157, 177–8
omnivores, 56
100-mile diet, 558
O'Neill, D., 375
optimal foraging theory, 70
optimum range, 68
organic farming, 342, 363, 367–8, 371
organism, definition of, 65
O'Riordan, T., 163
Orr, D.W., 553, 554, 560
Osano, P., 109–10
Our Common Future, 2
overharvesting, 39, 264, 289, 419, 496, 498, 496, 498–500, 508–9, 534
oxygen depletion, 137, 255, 384
oxygen sag curve 137, 139
ozone, levels of, 124, 125, 143, 148, 204, 212, 282; ground level, 88, 100, 460, 466
ozone layer, 125, 148, 209, 212

Pachauri, R., 225
Pacific Maritime ecozone, 286–7
PAHs (polycyclic aromatic hydrocarbons), 8–13
Palliser's Triangle, 401–2
Pappas, S., 337
parasitism, 71, 81, 506
Paré, D., 311
parent material, 66
Parkinson, G., 547
Parmesan, C., 107, 507
Parson, E.A., 174
participation, 159, 172–97, 328, 401, 406, 407–11, 431, 513, 531; and empowerment, 173; and power, 173–7, 191
participatory approach, 10–13, 14, 39, 154, 159, 160, 164, 172–97
Partidario, M.F., 179
partnerships, 173, 191, 406, 515, 543, 548, 550
passenger pigeon, 499–500
Pasteur, L., 21

pathogens, 343, 362, 363, 365, 396, 411
Pattanavibool, A., 511–12
Pauly, D., 239, 246–7, 248, 257–8, 265
PBT (persistent, bioaccumulative, and toxic), 270
PCBs (polychlorinated biphenyls), 8–13, 167–8, 269, 270, 356, 386
Pearson, D., 479
pelagic marine organisms, 357
Penashue, P., 377
peregrine falcon, reintroduction of, 509
permaculture, 369
Pest Management Regulatory Agency (PMRA), 359–61
pesticides, 15, 165, 270, 311, 332–73 (*passim*), 509
Peterson, C.H., 255–6
Phare, M.A., 412
pheromones, 300
Phillips, D., 134
phosphorous, cycle, 114, 119–22, 137, 138, 148, 149; levels of, 384, 386, 387, 410
photosynthesis, 16, 53–4, 57, 58, 80, 88, 119, 148, 240, 243, 338, 339, 493
phototrophs, 54
phytoplankton, 57, 58, 61, 96, 100, 124, 137, 240, 253–4
Ping, T.S.T., 397
pioneer species, 92
Plummer, R., 177
point sources, 137–8, 141, 254, 374, 384–5, 415
polar amplification, 99
polar bear, 46, 108, 210, 214, 215, 508, 547
POPs (persistent organic pollutants), 270–1, 355–6, 357–8; Stockholm convention on, 544
population, age structure, 17–19, definition of, 65, 100; density, 100–1; duck, 505–6; human, growth of, 2, 16–21, 27, 39, 63–4, 333, 335, 339, 375, 459, 497; pyramid, 19–20; species, 83, 84, 100–3, 371, 496, 503, 505–8, 509, 534
population viability analysis (PVA), 530
potash, 422
poverty, 14–15, 18, 22–4, 39, 109–10, 337, 345, 375, 497, 513, 541, 545, 547
precautionary principle, 39, 179, 269, 272, 440
precipitation, 106, 131, 132, 149, 202, 214
predation, *or* predator–prey relationship, 57, 60, 62, 70–1, 78, 81, 83, 98–9, 102, 104–6, 110, 247, 248, 249, 265, 277, 279, 269, 347, 357, 492, 494–5, 496, 504, 508, 509
predator, *see* predation
predator control, 501–3, 504, 534
Prescott-Allen, R., 544
prescribed burning, 87–8, 87, 111, 300, 327, 529
prey, *see* predation
prey switching, 249
primary colonizers, 84–6, 90, 125
Priscoli, J.D., 412
producers, 53–61; primary, 347
productivity, 39, 61–5, 80, 90, 99, 108, 110, 134, 240, 248, 339, 366, 369; primary global, 239, 240, 245–6
progress, 539–66; in British Columbia, 549, 500; Canadian, 201, 223–9, 540–1, 547–51, 564;

corporate contributions to, 561; and education, 551–5, 564; global, 540–7, 564
pronghorn antelope, 71
protected areas, 77, 199, 377, 433, 491–537; agricultural, 335; forest, 321; marine (MPAs), 55, 62, 248–9, 258, 272–5, 279, 500, 523, 529, 547; role of, 526, 529
protected species, 491–537
provincial parks, rationale for, 210, 521
puffins, 49–50, 61, 101, 114
purple loosestrife, 94

Quinn, F., 376, 377, 381
Quoden, J., 464

radiative forcings, 209
Raeside, A., 251
Rahman, M.M., 410
rain forests, 75, 76, 78, 106, 117, 118, 186, 287, 302, 306, 497, 560
rainshadow effect, 132
Ramsay, D., 391
range of tolerance, 68–9, 80, 83, 90, 92, 107, 140, 269, 310, 358
Rasheed, S., 530
Raudsepp-Hearne, C., 543
recovery, of resources, 116–17
recycling, 116–17; and blue boxes, 463; of metals, 423; of nutrients, 119; of resources, 420, 421; of waste, 463, 464–5, 482, 488; *see also* reduce, reuse, recycle
Red List of World Conservation Union (IUCN), 94, 251, 253, 497, 499, 514, 518
Red River flooding, 200, 398–401
REDD (Reduced Emissions from Deforestation and Forest Degradation), 325–6
reduce, reuse, recycle, 463, 464–5, 482, 488, 555–9, 564; and refuse, 555–6, 564
Rees, J., 419, 420
Reeves, H.W., 383
reforestation, 223, 287, 294, 295, 297; *see also* forests, regeneration of
Regehr, E.V., 215
regional land-use planning, 155, 172, 190–3, 193, 195, 410
Regional Scale Nodes (RSN), 244–5
Reguly, F., 227
Reichel, J., 445
Reichstein, M., 129
relative humidity, 131
renewable resources, 51–3, 199, 294, 295, 367, 376, 419–58
replacement-level fertility, 19
resilience, 2, 13, 15, 39, 92–3, 154, 161, 173, 178, 183, 248, 337, 369, 453, 487, 488, 513, 544, 560, 565
resource partitioning, 70
resources, definition of, 5
responsibility, 27; and consumer boycotts, 560, 565; corporate, 560, 561; extended producer, 560–1; governmental, 28–9, 40; individual, *see* What You Can Do; shared, 173

Reynolds, J.D., 56
Reynolds, M., 228
Rhode, K., 76
Richardson, N.H., 475
Ring of Fire, 431, 433, 455
risk, 179, 194, 202, 392, 455, 482; assessment, 154, 172, 178–9, 193, 194, 219; management, 397
Rivers, N., 230
Roberts, C.M., 247
Roberts, I., 558
Robertson, G., 484
Rochette, P., 339
rock cycle, 122
Rockstrom, J., 24–5, 76, 135, 491
Rode, K., 257
Rohde, K., 75
Ross, N., 478
Ross, P.S., 270
Rowe, S., 115
Rowell, P., 509
r-strategists, 102, 103, 105, 112
Ruddiman, W.F., 338
runoff, agricultural, 125, 137, 138, 181, 270, 339, 343, 364, 384–5; urban, 138, 386, 468
Russell, J.L., 110
Russell, Sir M., 221
Rylatt, M.G., 423

Sable Gulley, 55
Sable Island, 439–43, 455
salination, 350–1, 371
salmon, 268–9, 275–7, 279; *see also* fisheries
Salt, D., 15
Sanchez, I.E., 336
Sandford, R., 375, 412
Sawyer, D., 230
Schindler, D.W., 282, 283, 446, 448
Schroeder, P., 362
Scialabba, N.E., 365
science, credible, 386–7, 415; use of, 1–2, 3–42, 45, 46, 47, 79–80, 156, 169, 174, 182–3, 200, 201–37, 264, 265, 269, 274, 313, 328, 374, 379, 386–7, 413, 415, 419, 421, 422, 453, 454, 518
scientific target value (STV), 145, 212
Scott, D., 175
sea level, rising, 205, 206, 234, 256, 279
sea otters, 99, 100, 253, 255, 493
second growth timber, 295, 296
sedimentary cycles, 117, 119–24, 148
sedimentation, 121, 181, 313, 343, 384, 387, 473
seed bank, 85
Selbourne, Earl of, 412
Selig, E.R., 241
Selin, S., 173
seral stage, 86, 102, 347
serial depletion, 249
serotiny, 87
Setiawan, B., 192–3
Severson-Baker, C., 446
sewage, 137, 165, 225, 270–1, 343, 384–5, 391, 483; *see also* waste
Sewell, W.R.D., 374

Shaftoe, D., 189
Shang, E.H., 255
Shaw, G.B., 161
shifting baseline, 248
Shiklomanov, I.A., 375
short-term perspective, 162, 166–7, 169, 192–3
Shrubsole, D., 160, 163, 399, 401
Silent Spring, 26, 352, 353
silo effect, 154, 410
silviculture, 295–300, 302, 308, 327; see also forestry
Sinclair, A.J., 179
single-loop learning, 167, 169
sinks, carbon, 127, 223, 230, 231, 290, 314–15, 328; forest, 223, 290, 314–15, 328; ocean, 254, 279; soil, 254, 279; wildlife, 530
Skeel, M.A., 354, 517
Slocombe, D.S., 164, 165
smelting, 124, 140–1, 149, 385, 420, 475
Smit, B., 212, 230
Smith, A., 558
Smith, J.N.M., 62
Smith, L.C., 217–18
smog, 466, 480
Snuffling, R., 529
social learning, 156, 167, 169, 177
socio-economic systems, 1, 3, 5–6, 92, 144–5, 164, 179, 208, 218, 429, 431, 485, 532
soft path management approach, 374, 392–4, 415
soil, 66–9; chemistry, 91; compaction, 349, 365–6; erosion, *see* erosion; fertility, 309–11, 312, 328, 338–9, 347, 349–52, 371; horizons, 67; loam, 68; permeability, 68; profile, 67, 68
solar power, 419–20, 438–9, 455, 467, 480, 550
specialist species, 69–70
specialized species, 92, 104–5
speciation, 104–5
species, diversity, 74, 81; dominant, 90; removal, 98–9; see also extirpation
species-area curves, 75
species at risk, 258, 269, 283, 307
Species at Risk Act (SARA), 46, 108, 269, 513, 515–18, 535
spotted owl, 307, 308–9, 328, 508
Sprague, J.B., 376
spruce budworm, 290, 300, 301, 327
stakeholders, 266, 392, 420, 434; competing, 449; and conflict, 92, 93; interests of, 39, 531–3; in management, 154, 161, 172–97; multi-, 191, 323; participation of, 10–13, 39, 156–69, 172–97, 316, 320, 323, 407–11, 506, 526, 531–3
steam-assisted gravity drainage (SAGD), 444–5
Steele, D.H., 262
Steiner, A., 215
Stelmach, Premier E., 448
Stern, Sir N., 231
Stevenson-Baker, C., 446
stewardship, 158, 313, 505; agricultural, 387; of ecosystem, 531; forest, 294–5, 321–2, 323, 544, 564; habitat, 354, 517; marine, 258, 544, 564; of nature, 163; product, 464; of protected areas, 531; of species at risk, 515–17

Stewart, E.J., 216
Stirling, A., 108
stock resources, 419, 454; *see also* renewable resources
Stockton, A.S., 95
Stockwell, C.A., 104
Stoett, P., 512
Stokes, K., 104
Stonehouse, D.P., 158
strategic environmental assessment (SEA), 142, 179–80
stratosphere, definition of, 48
strip cropping, 366
structured approach, 374, 397, 400, 415
Stuart-Smith, J., 95
sublimation, 131
subsidiarity, 28, 158–9, 161, 169
subsistence farming, 340
succession, 83–111, 135, 182, 478; cyclic, 86; primary, 83, 84–6, 90, 111, 299, 304–5; secondary, 83, 85, 90, 111, 297, 300, 302, 310
successional processes, 62, 297, 304–5, 306–8, 347
Sudbury, 100, 115, 139, 141, 142, 147; regeneration of, 475–9
Sullivan, T.P., 299–300
sulphur cycle, 114, 122–4, 148
sulphur dioxide, 123, 124, 141–2, 146–7, 149, 204, 420–1, 448, 454, 461, 475, 551; *see also* climate change, emissions
Sumaila, U.R., 258
summer fallow, 339, 351
superstack, 115, 141, 475, 476
supply management approach, 374, 381, 392–4, 415
sustainability, 316, 326, 294–5, 363, 366, 394, 404, 410, 431, 432, 460–5, 483, 487, 488, 531, 546, 554–5, 560, 565; assessment, 172, 184, 195
sustainable development, 2, 13–14, 15, 39, 154, 161, 173, 272, 284, 320, 325–6, 453, 513; in Montreal, 484–5; progress towards, 26, 29–36
sustainable energy superpower, 452
sustainable environment, 26, 348, 369
sustainable forest management (SFM), 283, 320, 321
sustainable livelihoods, 2, 13–15, 39, 117, 154, 161, 173
sustained yield, 316; long-run, 295
Sydney Tar Ponds, xi, 2, 4, 7, 8–13, 23–4, 37, 39, 100, 172, 176, 184, 190, 469
Sydneysmith, R., 232
synergistic, 100, 120, 143, 358, 359, 371
systems, approach, 154, 156, 169, 191, 194, 199; definition of, 37; models of, 45–151, 266; relationships between, 37–9
Suzuki, D., 44, 233, 559

Table Mountain, 91
Taiga Plains ecozone, 277, 327
target organisms, *or* species, 104, 246, 248, 249, 262, 342, 353, 365, 371, 504, 510
Taylor, L., 413
Taylor, P.D., 72

temperatures, 202, 203, 206, 208, 214
Terra Nova, 439–43, 455
territory, definition of, 70
Theobold, M., 249
Thérival, R., 180
thermocline, 240
thermodynamics, laws of, 50–3, 60–1, 80, 115, 203, 249, 277, 338, 345, 346, 370, 556
thermohaline circulation, 243, 256–7
thermosphere, definition of, 48
Thibault, R., 261
Thompson, I.D., 300
threatened species, 269, 307, 309, 491–537
thresholds, 79–80, 145, 165, 446; critical, 79–80, 165, 507
Tinker, J., 162
Todhunter, P.E., 399
Toronto, Green Standard, 485–7, 488; waterfront, 154, 164, 166
total allowable catch (TAC), 145, 260, 261
total fertility rate, 19
toxic, *or* toxicity, 384; burden, 358; metal, 420–1, 454; natural, 494–5; substances, 167–8, 270–1, 386–7, 400, 410, 420–1, 428, 437, 446, 507–9, 558, 560
traditional ecological knowledge (TEK), 46, 183, 206, 265, 269, 425, 454
Trainor, S.F., 175
transpiration, 129–30, 313, 375; *see also* evapotranspiration
tree-length harvesting, 310–11
Tremblay, C., 116
Trewartha, G.T., 202
Trist, E., 176
Trites, A.W., 253
trophic level, 54, 55, 60–2, 70, 80, 92, 240, 249, 270, 343, 347, 371, 507
troposphere, definition of, 48
Trostel, K., 62
tuna, 238–9, 249, 545, 547, 560
turbulence, 100, 176, 193
Turpel, M.E., 14, 15
Tynan, C.T., 110

uncertainty, 2, 3, 4, 7, 12–14 (*passim*), 37–40 (*passim*), 62, 76, 80, 88, 146, 154, 157, 164, 172–8 (*passim*), 181, 183, 185, 193, 194, 200, 202, 219, 245, 265, 266, 342, 379, 381, 392, 434, 470
uranium, 450, 456
urban development, 13, 22, 199, 287, 339, 345, 459–90, 505, 505; and air pollution, 459–90 (*passim*); and energy, 460, 462, 463–4, 468, 487, 488; environmental issues in, 466–70, 479–80, 488; and feng shui landscapes, 481–2; form of, *or* scale, 460–1, 463–4, 479–80, 487, 488; governance of, 463, 483–7, 488; hazards of, 470–3, 488; sustainable, 460, 463–5, 473–9, 487, 488; transportation in, 460–4, 466, 479, 480, 482–3, 485, 487, 488; waste management of, 460, 462–4, 480, 482, 487, 488
urban heat island, 466–7, 480, 482, 488

urban sprawl, 460, 461, 467, 468, 470, 479, 487, 488
urbanization, *see* urban development

value, 2, 3–4, 5, 7, 14, 39, 46, 90; consumptive, 493; ecocentric, 5, 39, 156, 163, 165, 169; ecological, 493–4; economic, 22, 33, 36, 94, 144, 153–4, 157, 160, 161, 163, 165–8, 169, 173–4, 178, 180, 182, 184, 189, 191, 194, 200, 220, 234, 493–5, 508–9; ethical, 495–6; extrinsic, 491, 493–5; intrinsic, 5, 491, 493–5; non-consumptive, 493, 500; policy target (PTV), 145; technocentric, 156, 163, 165, 169; *see also* ethics, vision
Vancouver, green city, 484, 488
Vanderploeg, H.A., 139
Varid, O., 410
Veale, B., 404, 405–6
Venter, O., 94, 503, 508
Vernier, P.R., 308
virtual water, 374, 394–5, 415
vision, 13, 154, 16, 160, 161–3, 169, 177, 178, 191, 318–19, 484, 485, 487; and ethics, 163; shared, 154, 161, 169, 173, 177, 284, 432–3; statement, 160, 318–9; *see also* ethics, value
Volpe, J.P., 275
von Braun, J., 334, 337, 345
Voynick, S., 423
vulnerability, 425, 491–2, 496, 504, 491–537

Wagner, J., 266
Walker, B., 15
Walker, I.J., 232
Walkerton, 200, 363, 388–90, 391, 415; *see also* E.coli
Wallace, A.R., 47
Walsh, J.E., 215
Walston, J., 511
Walters, C., 266
Warnock, R.G., 354, 517

waste, agricultural, 385, 387; industrialized, 385; management of, 116–17; radioactive, 421, 454; urban, 384, 385, 462–5, 479, 480, 482, 487, 488; and water, 387–92; *see also* run-off, toxic
water, 374–418; consumption, 374–418, 558; cycling, 290; and decision-making, 159, 163; ethics, 374, 411–13, 415; export of, 374, 382–3, 415; governance of, 374–418; as hazard, 374, 397–403; as human right, 374, 412–13, 415; management, 375–418 (*passim*); quality, 383–92, 394, 415, 428, 460, 468, 480, 488, 548; quantity, 135, 375, 383, 387, 392–5; as renewable, 376; as resource, 374–97; rights, 412; security, 200, 374, 387–97, 415; table, 401, 403; terrorism, 176, 374, 396–7, 415
watershed, 143, 290, 407, 410, 411, 387; management, 389; protection, 493
Watson, R., 248
Watt-Cloutier, S., 355
weather, 201–6, 217, 496, 505; definition of, 202, 234; elements of, 202
weathering, 118, 122, 311
Weber, B., 449
wellbeing, human, 540–66; *see also* health, human
Wells, J., 282, 284
wetlands, 135, 178, 206, 216, 343, 376, 408, 445–6, 454, 503–6; creation of, 425, 454
Weynillowicz, D., 444, 446
whales, 16, 46, 55, 61, 73, 102, 110, 102, 110, 240, 250–1, 253, 269, 270, 279, 358, 494, 495, 500, 510, 511, 517
What You Can Do, 111, 146, 160, 168, 194, 221, 232–3, 278, 326, 359, 370, 393, 413, 453, 454, 488, 519, 534, 540–1, 551–63; *see also* responsibility
Whistler, 573–4, 488
White Rose, 440, 455
Whitelaw, G.S., 531
Whitford, J., 10

whooping crane, 528–9
Wichelns, D., 395
Wielgus R.B., 308
Wiken, E., 286
Wiley-Blackwell, 79
Wilkinson, C.J.A., 492, 524
Wilkinson, J., 229
Wilkinson, M., 316
Williams, Premier D., 377
Williams, P.W., 473–4
Williams, R., 500
Wilson, S., 283, 293–4
wind energy, 188–9, 200, 217, 230, 419–20, 436–8, 455, 480, 550, 562
windthrow, 315
Winfield, M.S., 161
Winterhalder, K., 479
Wipond, K., 72
Wolfe, B.B., 402
wolves, reintroduction of, 504
Wong, P.Y., 85
Woods, A.J., 306
World Summit on Sustainable Development (WSSD, Johannesburg), 26, 257, 258, 275, 513, 562
Worm, B., 248
Worthy, B., 549
Woynillowicz, D., 443, 444, 446
Wu, J., 72

Yates, J.S., 116

Zhang, X., 334
zebra mussels, 957, 102, 139
zone of physiological stress, 68
zooplankton, 57, 61, 137, 242, 253
zooxanthellae, 241